A&P TECHNICIAN
AIRFRAME
TEXTBOOK

JEPPESEN

Westwind cover photo taken in cooperation with Straight Flight, Inc.
Centennial, Colorado

Jeppesen Sanderson, Inc.

Published in the United States of America
Jeppesen Sanderson, Inc.
55 Inverness Drive East, Englewood, CO 80112-5498
www.jeppesen.com

All rights reserved. No part of this publication may be reproduced, stored in a retrieval system, or transmitted in any form or by any means, electronic, mechanical, photocopying, recording or otherwise, without the prior permission of the publisher.

ISBN-13: 978-0-88487-331-0
ISBN-10: 0-88487-331-5

Jeppesen Sanderson, Inc.
55 Inverness Dr. East
Englewood, CO 80112-5498
Web Site: www.jeppesen.com
Email: Captain@jeppesen.com
© Jeppesen Sanderson Inc.
All Rights Reserved. Published 2002, 2003
Printed in the United States of America

PREFACE

Congratulations on taking the first step toward learning to becoming an Aviation Maintenance Technician. The *A&P Technician Airframe Textbook* contains the answers to many of the questions you may have as you begin your training program. It is based on the "study/review" concept of learning. This means detailed material is presented in an uncomplicated way, then important points are summarized through the use of bold type and illustrations. The textbook incorporates many design features that will help you get the most out of your study and review efforts. These include:

Illustrations — Illustrations are carefully planned to complement and expand upon concepts introduced in the text. The use of bold in the accompanying caption flag them as items that warrant your attention during both initial study and review.

Bold Type — Important new terms in the text are printed in bold type, then defined.

Federal Aviation Regulations — Appropriate FARs are presented in the textbook. Furthermore, the workbook offers several exercises designed to test your understanding of pertinent regulations.

This textbook is the key element in the training materials. Although it can be studied alone, there are several other components that we recommend to make your training as complete as possible. These include the *A&P Technician Airframe Test Guide with Oral and Practical Study Guide*, the *A&P Technician Airframe Workbook, AC 43.13-1A/2A,* and the *FAR Handbook for Aviation Maintenance Technicians*. When used together, these various elements provide an ideal framework for you and your instructor as you prepare for the FAA computerized and practical tests.

The A&P Technician Airframe course is one of three segments of your training as an aviation maintenance technician. The airframe section introduces you to the concepts, terms, and common procedures used in the inspection and maintenance of airframe structures and systems.

TABLE OF CONTENTS

PREFACE iii

CHAPTER 1   **Aircraft Structural Assembly and Rigging** 1-1
    Section A   Aircraft Design and Construction 1-2
    Section B   Airplane Assembly and Rigging 1-20
    Section C   Fundamentals of Rotary-Wing Aircraft 1-52

CHAPTER 2   **Sheet Metal Structures** 2-1
    Section A   Metallic Aircraft Construction 2-2
    Section B   Sheet metal Tools and Fasteners 2-16
    Section C   Sheet Metal Fabrication 2-51
    Section D   Instection and Repair of Metallic Aircraft Structures 2-83

CHAPTER 3   **Wood, Composite, and Transparent Plastic Structures** .. 3-1
    Section A   Aircraft Wood Structures 3-2
    Section B   Composite Structures 3-22
    Section C   Transparent Plastic Materials 3-63

CHAPTER 4   **Aircraft Welding** 4-1
    Section A   Welding Processes 4-2
    Section B   Advanced Welding and Repairs 4-15
    Section C   Basic Gas Welding 4-22

CHAPTER 5   **Aircraft Fabric Covering** 5-1
    Section A   Fabric Covering Process 5-2
    Section B   Covering Procedures 5-12
    Section C   Inspection and Repair of Fabric Covering 5-26

CHAPTER 6   **Aircraft Painting and Finishing** 6-1
    Section A   Fabric Finishing Processes 6-2
    Section B   Aircraft Painting Processes 6-5
    Section C   Finishing Equipment and Safety 6-15

CHAPTER 7   **Airframe Electrical Systems** 7-1
    Section A   Airborne Sources of Electrical Power 7-2
    Section B   Aircraft Electrical Circuits 7-35
    Section C   Wiring Installation 7-53
    Section D   Electrical System Components 7-66

CHAPTER 8   **Hydraulic and Pneumatic Power Systems** 8-1
    Section A   Principles of Hydraulic Power 8-2
    Section B   Hydraulic System Components and Design 8-6
    Section C   Hydraulic Power Systems 8-11
    Section D   Aircraft Pneumatic Systems 8-49

CHAPTER 9	**Aircraft Landing Gear Systems**	**9-1**
Section A	Landing Gear Systems and Maintenance	9-2
Section B	Aircraft Brakes	9-18
Section C	Aircraft Tires and Tubes	9-39

CHAPTER 10	**Position and Warning Systems**	**10-1**
Section A	Antiskid Brake Control Systems	10-2
Section B	Indicating and Warning Systems	10-9

CHAPTER 11	**Aircraft Instrument Systems**	**11-1**
Section A	Principles of Instrument Systems	11-2
Section B	Instrument System Installation and Maintenance	11-43

CHAPTER 12	**Aircraft Avionics Systems**	**12-1**
Section A	Avionics Fundamentals	12-2
Section B	Autopilots and Flight Directors	12-34
Section C	Installation and Maintenance of Avionics	12-48

CHAPTER 13	**Airframe Ice And Rain Control**	**13-1**
Section A	Airframe Ice Control Systems	13-2
Section B	Rain Control Systems	13-16

CHAPTER 14	**Cabin Atmosphere Control Systems**	**14-1**
Section A	Flight Physiology	14-2
Section B	Oxygen and Pressurization Systems	14-6
Section C	Cabin Climate Control Systems	14-27

CHAPTER 15	**Aircraft Fuel Systems**	**15-1**
Section A	Aviation Fuels and Fuel Systems requirements	15-2
Section B	Fuel System Operation	15-10
Section C	Fuel System Repair, Testing, and Servicing	15-33

CHAPTER 16	**Fire Protection Systems**	**16-1**
Section A	Fire Detection	16-2
Section B	Fire-Extinguishing Systems	16-16

CHAPTER 17	**Aircraft Airworthiness Inspection**	**17-1**
Section A	Required Airworthiness Inspections	17-2
Section B	Inspection Guidelines and Procedures	17-20
Section C	Aircraft Maintenance Records	17-43

INDEX **I-1**

CHAPTER 1

AIRCRAFT STRUCTURAL ASSEMBLY AND RIGGING

INTRODUCTION

Aircraft assembly involves the joining of various components and structures that form an entire aircraft, while rigging generally refers to the positioning and alignment of an aircraft's major sub-assemblies to produce a synergistic design. For example, airplanes are typically fabricated in a number of major sub-assemblies, such as the fuselage or main body, an empennage or tail section, wings, landing gear, and an engine or powerplant section. These components provide stability and maneuverability when assembled and rigged in accordance with the manufacturer's specifications. As an aviation maintenance technician, you will likely be involved daily in the dismantling and re-assembly of aircraft. In addition, you may also be required to perform rigging procedures on a variety of aircraft types during your career.

SECTION A

AIRCRAFT DESIGN AND CONSTRUCTION

Throughout the 20th century, manufacturers improved aircraft durability and safety by implementing advancements in construction techniques and materials technology into their products. As a result, a large number of older aircraft are still in service, even after decades of use. During the 21st century, it will be the aircraft technician's responsibility to maintain the structural integrity of these aging aircraft as well as the structures of currently manufactured and future aircraft designs.

STRUCTURAL DESIGNS

The early dreamers of flight had little concept of a practical structure for their machines. The Greeks had Daedalus and his son Icarus flying with wings made of feathers and wax, while other dreamers conjured up machines resembling birds. Even the genius, Leonardo da Vinci, conceived of a flying machine, which had flapping wings attached to a body that was modeled after a bird.

It was only with the discovery that lift could be produced by causing air to flow over a cambered surface, that aerodynamics took a practical turn. The gliders of Lilienthal and Chanute proved that manned flight was possible. By using the results of their experiments, the Wright brothers developed a biplane glider with which they solved one of the biggest problems of the time — the problem of control.

The early flying machines produced by the Wrights, Glenn Curtiss, Henri Farman, Alberto Santos-Dumont, and the Voisin brothers all had a common type of structure. Each aircraft had wings composed of ribs made from wood, which were covered with organic fabrics such as cotton or linen to form the lifting surfaces. The bodies of these machines were little more than open girder frameworks made from bamboo or strips of wood and held together with piano wire. Auxiliary structures, such as stabilizers and flight controls, were similarly constructed and attached to the body either in front of or behind the wings. [Figure 1-1]

Once the basic problems of flight and control were refined, airplanes evolved into a somewhat stan-

Figure 1-1. The Wright Flyer used an open truss frame to hold the occupants and to provide an attachment point for the engine and lift producing surfaces. This was a common structural design for many early aircraft.

dardized configuration. Up through World War I, most airplanes were built with a **truss structure** that used struts and wire-braced wings. The occupants sat in open cockpits within a fabric-covered hull, or **fuselage**. Almost all of these airplanes had the engine installed up front and auxiliary surfaces mounted aft of the wings to form the tail, or **empennage**, of the airplane.

Increased knowledge of flight and the experience gained in building strong, lightweight structures allowed builders to turn their attention to the problem of decreasing the air resistance of their machines. This air resistance robbed much of the potential speed of the early airplanes. To minimize wind resistance and yet retain the strength provided by a truss structure, designers constructed a superstructure of wooden **formers** and **stringers** over the framework to produce a more streamlined shape. Formers provide the contoured cross-sectional shape to a structure while stringers run the length between the formers to fill in the shape.

One of the major breakthroughs in structural designs was made in the latter years of World War I, when thin-walled steel tubing was welded together to form the fuselage truss. When fabricated in this fashion, the structure reduced the overall weight of the aircraft while increasing the structural strength. Ultimately, the combination resulted in significant improvements in the strength-to-weight ratio of subsequent aircraft designs.

The next advance in structural designs came with the development of a construction technique that allowed the aircraft to be formed without a truss frame. This design, generally known as a **stressed-skin structure**, allowed the aircraft to be built with a more streamlined shape and provided further reductions in weight because the skin itself carried the structural loads. When constructed in this fashion, the aircraft was referred to as having a **monocoque** design. The term monocoque is derived from the French meaning "single-shell."

The Lockheed Aircraft Company pioneered stressed-skin construction with the popular Vega series airplanes during the 1920s and 1930s. In the Vega's construction, strips of spruce wood were glued together then cured under heat and pressure in a large concrete mold. Once fabricated, the wood strips formed eggshell-like plywood structures. To provide additional support in engine mounts, wings, and the landing gear attachment areas, laminated wood rings were added to the interior of the plywood shell. [Figure 1-2]

Figure 1-2. Concrete molds, similar to those shown here, were used to form strips of wood to a contoured shape. Once glue was applied, the wood strips were positioned in the mold in an over-lapping fashion. The wood was held under pressure until the glue cured by placing a rubber bag or bladder inside the mold and inflating it. Once cured, the plywood formed a durable laminated shell that retained its shape after being removed from the mold.

Thin aluminum-alloy sheets were next used for the exterior of monocoque stressed-skin structures. These sheets had compound curves formed in them by using hydropresses or drop hammers to forge complex shapes. The formed skins were then riveted onto thin sheet metal formers. The designs provided a lightweight and reasonably durable structure that manufacturers used for many years. In fact, many aircraft constructed in this manner remain in service today. [Figure 1-3]

Figure 1-3. The empennage structure of this airplane is one example of an aluminum monocoque design. In true monocoque designs, formers and other interior structures give the skin of the aircraft its shape but do not carry the structural loads. Instead, the skin carries the load.

A disadvantage of monocoque designs is that they can fail once subjected to relatively minor dents or creases. To further increase the strength of the structure, manufacturers improved their designs by developing **semi-monocoque** construction techniques. In these aircraft, the skin is fastened to a sub-structure or skeletal framework, which allows the loads to be distributed between the structural components and the skin of the aircraft. These designs proved to be so successful that they continue to be the primary method of modern aircraft construction. [Figure 1-4]

In combination with improvements in airframe structural designs, aircraft powerplant performance and dependability also increased. As aircraft became capable of flying at high altitude, a means of pressurizing the cabin and cockpit area became necessary to increase the safety and comfort of the passengers and crew. By making further modifications, aircraft structures have been designed to be capable of sustaining high internal air pressures, which are produced to obtain a lower altitude environment in the cabin.

However, soon after the first pressurized jet transport aircraft started flying in the early 1950s, three of them broke apart in flight under mysterious circumstances; two of them in relatively non-turbulent air. An extremely thorough investigation disclosed

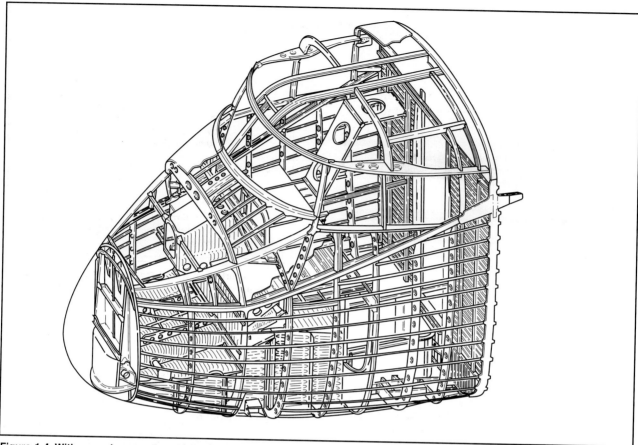

Figure 1-4. With a semi-monocoque construction technique, the skin is reinforced by the use of a sub-frame of internal components consisting of bulkheads, formers, stringers and longerons. With the use of these components, the loads imposed on the aircraft are carried from the skin into the supporting structure.

that the cause of the breakups was metal fatigue brought about by the flexing of the structure during the cabin pressurization and depressurization cycles. As a result of this investigation, **rip-stop doublers** were installed at strategic locations throughout the airplane structure, especially around windows and doors. Because of this design improvement, if a crack begins to develop in the structure, it will stop at a doubler, which will carry the load to help produce a fail-safe design.

TYPES OF AIRCRAFT STRUCTURES

An aircraft technician holding an airframe rating is authorized to work on all types of aircraft ranging from lighter-than-air equipment such as balloons and dirigibles, to rotorcraft and fixed wing airplanes. While gliders and airplanes have reasonably similar designs, rotorcraft and lighter-than-air aircraft are significantly different. Although it is beyond the scope of this text to cover all design types, technicians must be thoroughly familiar with each aircraft they maintain. In this section, a general overview of aircraft structures is presented to help you become familiar with factors regarding how various structures provide lift, stability or control. Section B then provides general information with regard to properly assembling and rigging airplane structures, while section C provides assembly and rigging information with regard to helicopter designs.

AIRFOIL SECTIONS

The lift producing surfaces of an aircraft, such as the wings of an airplane or the rotor of a helicopter, have an aerodynamically efficient shape called an airfoil. An airfoil provides the lifting force when it interacts with a moving stream of air. Some of the terms used to describe an airfoil, and the interaction of the airflow about it, are defined in figure 1-5.

The airfoils of some airplanes have more curvature on the top than on the bottom, but most helicopter rotors and many high-speed airplanes use airfoil sections that are **symmetrical**; that is, the curvature on the top of the airfoil is the same as that on the bottom. The movement of the air stream around the airfoil causes changes in the surrounding air pressure distribution to create lift. [Figure 1-6]

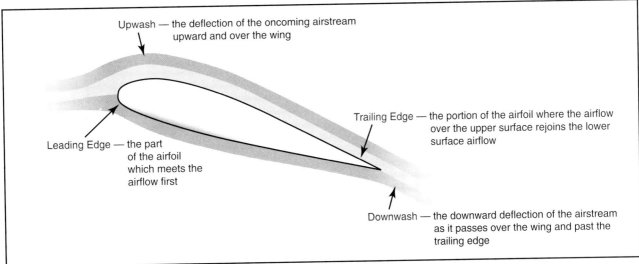

Figure 1-5. As an airfoil moves through the air, it alters the air pressure around its surface. A typical subsonic airfoil has a rounded nose, or leading edge, a maximum thickness about one-third of the way back, and a smooth taper into a relatively sharp point at the rear or trailing edge.

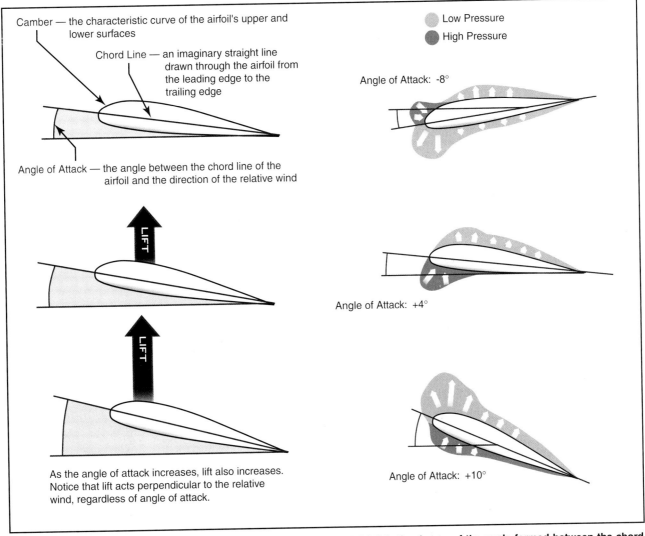

Figure 1-6. One factor that affects the amount of lift produced by an airfoil is the degree of the angle formed between the chord line and the direction of the relative wind. This angle is referred to as the angle of attack.

As air passes below the airfoil, it is deflected downward, and its velocity is slightly decreased. By slowing, the energy in the air converts from velocity energy (kinetic energy) into pressure energy (potential energy), and so there is a slight increase in pressure below the surface, with the maximum increase just behind the leading edge.

The air that passes over the top of the airfoil adheres to the upper camber, which causes it to accelerate because of the camber's curvature. In speeding up, some of the air's energy is converted from pressure energy into velocity energy, which causes a decrease in the static air pressure over the upper camber of the airfoil. This causes an area of low pressure to form over the airfoil surface, with the lowest pressure near the thickest section.

The air above the surface is pulled down into this low-pressure area and, as a result, is forced down as the airfoil moves through the atmosphere. This deflection of air, called the **downwash**, produces a large portion of lift. In order to support an aircraft, the total pressure of the air forced downward must be sufficient to support the weight of the aircraft. One way to increase the production of lift is to increase the angle of attack, and thus the downwash angle. However, this is no longer true beyond a maximum angle of attack. Above this critical angle, the air no longer flows smoothly over the upper camber, and the airfoil reaches a stalled condition.

The amount of lift produced is also affected by the velocity of the airfoil traveling through the atmosphere and the airfoil's surface area. In fact, the greatest change in lift is affected by the speed of the relative airflow over the airfoil. For example, with all other factors such as air density and angle of attack remaining constant, if the airspeed is doubled, the amount of lift produced will increase four times. On the other hand, if the area of the airfoil is doubled, the amount of lift will also double. Since at slow speeds the amount of lift may not be sufficient to support the aircraft, airplanes usually have a method of changing the shape of the airfoil to increase the camber shape and/or wing area. This is done with leading or trailing edge devices such as flaps or slats, described further in section B of this chapter.

To obtain maximum performance, the airflow over the leading edge of an airfoil is critical. A truss-type wing used on many slow-speed airplanes has its leading edge covered with thin sheet metal. **Ribs** are spaced at intervals throughout the wing structure and form the shape of the leading edge, camber, and trailing edge. Each rib is also attached to the wing **spar**, which runs the length of the wing from the root to the tip. The spar is the main spanwise member of the wing structure and carries the aerodynamic loads to the fuselage structure. [Figure 1-7]

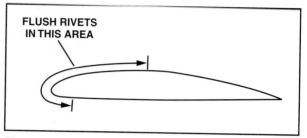

Figure 1-7. Some all-metal general aviation airplanes, which have a relatively high speed, have the wing skin attached to the ribs and spars with flush rivets along the leading edge and back to about one-third of the upper camber. Behind this, and on the bottom of the wing where the airflow is not as critical to produce a smooth or laminar flow, protruding head rivets are used for economy of construction.

TRANSMITTING LIFT INTO THE STRUCTURE

Air deflected by the wing produces the lift that supports an airplane, but lift must be transmitted into the structure in such a manner that the airplane can be balanced in every condition of flight. In addition, the structure must be built to support all of the loads without any damaging distortion. To do this, wings are mounted on an airplane in a location that places its **center of lift** just slightly behind the center of gravity. The center of lift is the point at which the air pressures produced by the wing can be considered concentrated.

As an airplane is maneuvered and the angle of attack changes, the center of lift also changes and produces some rather large torsional loads on the wing structure. This is especially true at the point where the wing attaches to the fuselage.

In addition to the twisting loads imposed on the structure, the wing is also subjected to bending loads. While weight is essentially concentrated at the fuselage, lift is produced along the full length of the wing. With the generation of lift, the wing tends to bend upward from the root toward the tip. The wing spars are designed to flex to carry these bending loads.

TRUSS-TYPE WING CONSTRUCTION

Fabric-covered airplane wings utilize truss-type structures that have changed very little throughout the design development of aircraft. As with other wing designs, spars are the main load-carrying

Aircraft Structural Assembly and Rigging

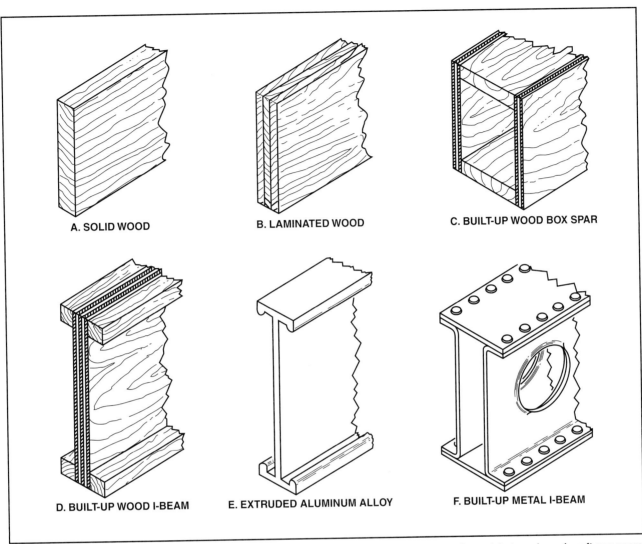

Figure 1-8. Spars for fabric covered truss-type wings. In the past, spars were mainly made of wood, but modern aircraft use spars constructed with aluminum alloy.

members in a wing truss. In the past, spars were mainly manufactured of wood, but the majority of modern aircraft incorporate spars fabricated from extruded aluminum alloy. [Figure 1-8]

Wood spars are usually made of Sitka spruce and may be either solid or laminated. Because of the difficulty in finding a single piece of near-perfect wood in the size needed for wing spars, many manufacturers produce laminated spars. A laminated spar is constructed of strips of wood that are glued together with the grain running in a parallel direction. A laminated spar is just as strong as a solid spar as long as it is manufactured from the same quality wood as a solid spar and manufactured to aviation standards.

The spars are separated by compression members, or **compression struts**, that may be either steel tubing or heavy-wall aluminum alloy tubing. **Compression ribs** are sometimes used, which have been specially strengthened to take compressive loads.

The truss is held together with high-strength solid steel wires that cross the bays formed by the compression struts. The wires that extend from the front spar to the rear spar and that are running diagonally from inboard to outboard oppose the forces that tend to drag against the wing and pull it backward. These wires are typically called **drag wires**. Conversely, wires that run between the front and rear spar and run diagonally from outboard to inboard are called **anti-drag wires**, since they oppose any force that tends to move the tip of the wing forward. A wing truss consisting of spars, compression members, and drag and anti-drag wires, when properly assembled and rigged, pro-

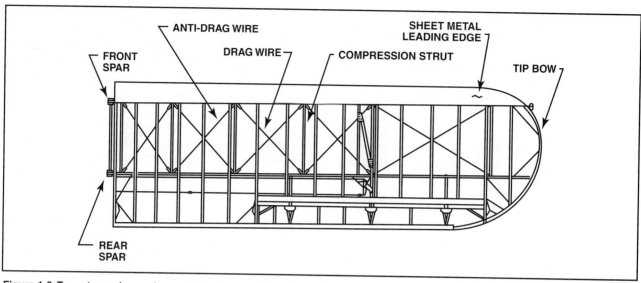

Figure 1-9. Truss-type wings, when properly assembled and rigged, provide the strong structure needed for fabric-covered wings.

vides the lightweight and strong foundation needed for fabric-covered wings. [Figure 1-9]

A variation of the truss-type wing is one using a box spar. The use of the box spar was pioneered in World War I on some of the all-metal Junkers airplanes in addition to the wood and fabric Ford triplane. A box structure built between the spars stiffens the spars so they can carry all of the bending and torsional loads imposed on the wing during flight. The ribs in the wing attach to the spars to give it the aerodynamic shape needed to produce lift when air flows over the fabric covering.

Before the cost of labor became so high, some wing ribs were built of Sitka spruce strips. **Cap strips** form the top and bottom of the rib and cross members form the connection between the top and bottom cap strips. Because wooden end-grain joints produce weak glue joints, **gussets** are attached to each intersection of a cap strip and a cross member. A gusset is a thin mahogany plywood-plate attached to two or more members to carry the stresses from one member to the other.

Metal wing ribs may be either built up by riveting together cap strips and cross members made of formed, thin sheets of aluminum alloy, or pressed from aluminum alloy sheets in a hydropress. Again, the most critical part of a wing, as far as the production of lift is concerned, is the front end, or the leading edge. To prevent air loads from distorting the leading edge, most wings have **nose ribs**. Nose ribs, sometimes called **false ribs**, extend from the front spar forward and are placed between each of the full-length former ribs. A sheet of thin aluminum alloy is wrapped around the leading edge so the fabric will conform to the desired shape between the ribs. [Figure 1-10]

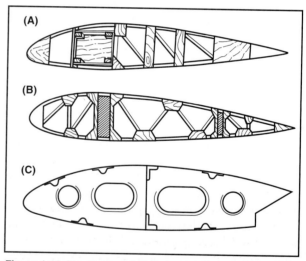

Figure 1-10. Illustrations A & B are wood wing ribs while C is a wing rib made of pressed sheet metal. Note the built-up wood box spar in figure A.

The trailing edge is normally formed of aluminum alloy and ties the back end of the ribs together to give the wing its finished shape. Cloth reinforcing tape is laced diagonally between the ribs, from the top of one rib to the bottom of the adjacent rib. Attachments are made near the point of their greatest thickness to hold the ribs upright until the fabric is stitched to them. The fabric covering is placed over the wing and is laced to each of the ribs with strong rib-lacing cord to hold them in place.

STRESSED-SKIN WING CONSTRUCTION

In the same manner as the fuselage, wings generally evolved from the truss form of construction to one in which the outer skin carries the greatest amount of the stresses. Semi-monocoque construction is generally used for the main portion of the wing, while the simple monocoque form is often used for control surfaces.

Wing ribs may be pressed from sheet aluminum alloy in a hydropress, or built up of sheet metal channels and hat sections riveted to the skin to give it both the shape and rigidity it needs. One of the advantages of an all-metal wing is that it is designed to carry all of the flight loads within the structure, so it does not need any external struts or braces. Internally braced wings not requiring external support are called **cantilever** wings. [Figure 1-11]

Figure 1-11. This modern airplane uses a cantilever wing construction, which eliminates the need for struts to support the main wing.

The Douglas DC-2 was one of the first highly successful airplanes to use the configuration that has become standard for modern transport category aircraft — cantilever low-wing construction, with retractable landing gear. The airfoil section of a cantilever wing is normally quite thick, and has a strong center section built into the fuselage. The engines and landing gear attach to this center section. Rather than using a common two-spar construction, most of these types of wings use multi-spar construction. Several spars carry the flight loads, and span wise stiffeners run between the spars to provide even greater strength.

As airplane airspeeds increased with their resulting higher flight loads, it became apparent that not only was more strength needed for the skins of all-metal wings, but more stiffness was also necessary. To gain the strength and stiffness needed, and yet keep the weight down, manufacturers of some of the high-speed military aircraft began the construction of wing skins with thick slabs of aluminum alloy. With these designs, the slab of aluminum is machined away, but enough material is left in the proper places to provide the desired strength and stiffness.

There have been two major improvements made over the conventional method of machining wing skins. The first improvement is termed **chemical milling**. In chemical milling, a slab of aluminum alloy is treated with an acid-resisting coating where the full thickness of the material is needed. The slab is then immersed in a vat of acid and unnecessary aluminum is chemically eaten away. Chemical milling is good for quickly removing large amounts of material, but when complex shapes or deep grooves must be cut, a second process, called **electrochemical machining**, may be used. With electrochemical machining, after the skin is immersed in a salty electrolyte, an electrode-cutting tool made from soft copper and carrying a large amount of electrical current, is passed near the surface of the skin. This electrolytic process eats away the metal at a rapid rate without actually touching the metal, leaving no tooling marks that could cause stress concentration points where cracks could form. [Figure 1-12]

The chemical milling process provided a much stronger wing skin that had a reduced tendency to crack. Not only were the stringers built-in by the milling process, but there were also no gaps or tool marks between rivet locations to produce stress points. When the wing skins were fabricated into multiple-spar box assemblies, the result was a reasonably flexible, very strong, main wing. These sections of the wing could also be sealed off to create integral fuel cells, where fuel could be carried without the added weight of a fuel tank.

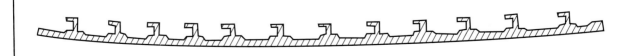

Figure 1-12. Milled wing skins give maximum strength and rigidity with minimum weight.

To gain the maximum amount of stiffness for the weight, some aircraft have wing skins made of laminated structure in which thin sheets of metal are bonded to a core of fiberglass, paper, or metal honeycomb material. For example, some airplanes that travel at supersonic speeds have outer skins made of stainless steel, brazed to cores of stainless steel honeycomb. With the laminated structure wings, the leading edges and even box spar sections may be made of bonded honeycomb-type material and the inside of the structures can be sealed to carry fuel. The greatest advantage of this type of construction for integral fuel tanks is that there is no need for sealing around thousands of rivets, as must be done with tanks made from conventional riveted sheet metal construction.

Some of the extremely light wing structures, such as those used for high-performance gliders and for some experimental airplanes, are built using a composite structure. A polystyrene foam core is covered with layers of reinforcing material and bonded to the foam with a matrix of epoxy or polyester resin.

CONTROL SURFACE CONSTRUCTION

This section describes basic design and construction of flight controls. These control surfaces produce aerodynamic forces to redirect an aircraft's flight path. The aerodynamics and operation of various types of airplane controls is covered in Section B of this chapter. Helicopter controls are covered in Section C.

Several of the higher-speed airplanes of World War II vintage were of all-metal construction except for the control surfaces. To keep the control surface weight to a minimum, they were covered with cotton or linen fabric. Today, almost all new metal airplanes have their control surfaces covered with either thin aluminum alloy, magnesium alloy sheets, or in some cases, advanced composite materials.

Flutter is a primary design consideration for any control surface. Flutter occurs when an out-of-balance condition causes a control surface to oscillate in the air stream, typically increasing in frequency and amplitude until the control surface fails catastrophically. To eliminate flutter, it is extremely important that control surfaces be balanced so that their center of gravity does not fall behind the hinge line. For this reason, many surfaces have extensions ahead of the hinge line on which lead weights are installed. To retain the flutter resistance, most control surfaces must be statically balanced anytime repairs or modifications are made, including painting.

FABRIC-COVERED CONTROL SURFACES

Most of the simpler truss-type fabric-covered airplanes have all of their tail surface internal structures made of welded thin-wall steel tubing. The vertical fin of this type airplane is built as an integral part of the fuselage, and the rudder attaches to the fin with hinge pins through steel tubes welded to both the fin and the rudder. [Figure 1-13]

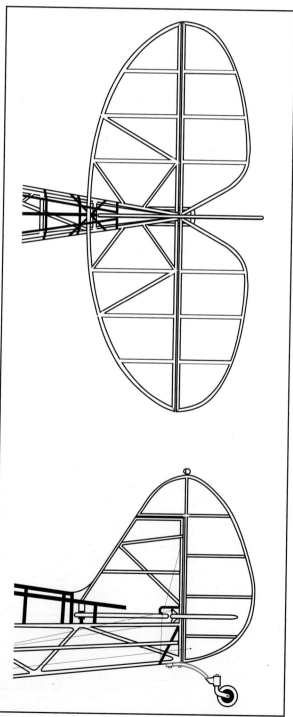

Figure 1-13. Vertical and horizontal surfaces made of welded thin-walled tubing are covered with cloth or synthetic fabrics.

The horizontal stabilizer bolts to the fuselage and is held rigid with high-strength steel wires. Elevators hinge to the stabilizer's trailing edge in the same way the rudder hinges to the vertical fin, while the ailerons are built up in much the same way as the wings. Aileron ribs conform to the shape of the rear end of the wing former ribs, and the aileron trailing edge is made of the same material as the trailing edge of the wing. The aileron leading edge is normally covered with thin sheet aluminum alloy to retain its shape under all flight loads. The hinge line of the aileron is usually well behind its leading edge.

METAL-COVERED CONTROL SURFACES

Most small modern airplanes use thin sheet metal for the control surfaces. To gain rigidity from the thin metal covering, many manufacturers corrugate the external skins. The stiffness provided by the corrugation minimizes the amount of substructure needed, thereby reducing the weight of the control. Where a substructure is required, the control surfaces are constructed with stamped or forged ribs and spars to form a monocoque or semi-monocoque frame. The hinges typically use thin wire to hold the hinge halves together, which require periodic checks to verify their condition, security of attachment, and wear. Excessively worn hinges often cause flutter to be induced into the control, in a similar manner to an out-of-balance condition. [Figure 1-14]

Always lock the control surfaces into a fixed position when parking the aircraft. **Control locks** can be installed inside the cockpit, or with external locks directly on the controls themselves. This prevents damage to the control surfaces by preventing the wind from blowing the controls against the stops. Using control locks also reduces wear to the control hinges. All control locks should be marked in a distinctive fashion to preclude being inadvertently left in place during preflight inspection.

AIRFOIL CONTROL AND AERODYNAMIC CONFIGURATIONS

The largest percentage of airplanes use standard primary control surface configurations consisting of the ailerons, rudder, and elevator. However, a number of aircraft use other control system designs. For example, large, transport category aircraft use additional control surfaces to provide different amounts of control authority during high- and low-speed flight.

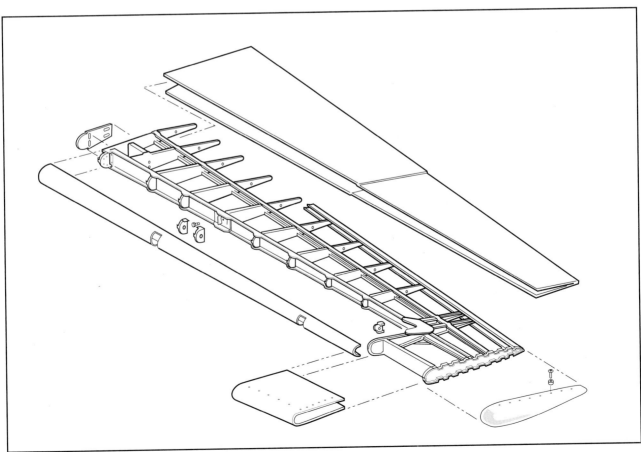

Figure 1-14. Thin sheet metal covers a lightweight substructure of stamped or forged wing ribs and spars.

AILERONS

Ailerons on almost all airplanes are located near the wing tips and hinge to the aileron spar to become part of the trailing edge of the wing. However, many large jet transport aircraft have two sets of ailerons; one in the conventional outboard location, and one inboard. For slow-speed flight, both sets of ailerons operate to provide the needed lateral control, but for high-speed flight, only the inboard, or high-speed, ailerons are active. If both sets were active at high speed, the aerodynamic effectiveness of the outboard ailerons would be too great, possibly causing too rapid movement, thereby inducing over-control.

SPOILERS

Spoilers are control devices that destroy lift by disrupting the airflow over a portion of the wing. They are simply structural slabs that are stowed flush with the airfoil surface that can be deployed by the pilot to swing upward into the air stream. Common types of aircraft that use spoilers include gliders. For these aircraft, the spoilers can be extended into the air stream by the pilot to reduce lift on a portion of the wing, thereby allowing a rapid rate of descent, while still providing full speed- and directional-control. When the spoilers are retracted, they fold down to eliminate the disrupted airflow and drag.

Transport category aircraft use spoilers as a part of the secondary flight control system. They can be used as an aid for the ailerons, to relieve control pressures, and to increase and decrease lift. Operating together, they can be used as **speed brakes**, which allow the pilot to slow the airplane by increasing parasite drag. On the ground, spoilers can be raised to help increase braking efficiency. [Figure 1-15]

FLAPERONS AND ELEVONS

Flaperons utilize a linkage that combines the trailing edge flaps with the ailerons. Generally, the entire trailing edge of the wing is lowered to increase lift. The outer sections of the flap are deflected in opposite directions to act as ailerons, except that they start from an extended flap angle instead of a streamlined position.

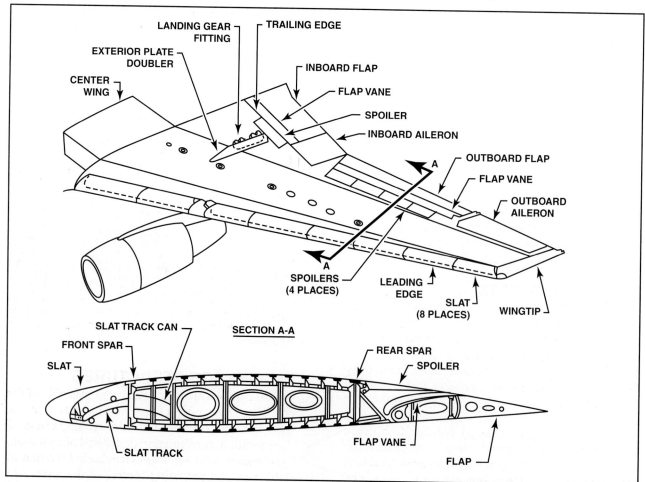

Figure 1-15. Spoilers are a part of the secondary flight control system as shown in the wing construction of a DC-10. They work in combination with flaps, ailerons, leading edge slats and other components of the wing.

Although **elevons** are not commonly found on civilian aircraft, they appear as primary control systems on flying-wing aircraft designs. A linkage similar to the ruddervator system combines the actions of the ailerons and elevator. Several designs of variable-geometry military aircraft use elevons when the wings are swept back for high-speed flight.

WINGLETS

Because wingtip vortices cost airplanes so much efficiency and performance, there has been much research to diffuse them. One of the most effective devices is the winglet, developed by Richard Whitcomb of NASA. These nearly vertical extensions on the wingtips are actually carefully designed, proportioned, and positioned airfoils with their camber toward the fuselage. The span, taper, and aspect ratio of the winglets are optimized to provide maximum benefit at a specific speed and angle of attack. On most jets, this is cruise speed, but turboprop airplanes use winglets to improve lift and reduce drag at low speeds.

Figure 1-16. Winglets are used to increase performance by improving lift and reducing drag at low speeds.

The winglet combines many small factors to increase performance. Downwash from the trailing edge of the winglet blocks the vortices. Even the winglet vortex is positioned to counteract a portion of the main wingtip vortex. The leading edges of many winglets are actually canted outward about 4 degrees, but because of the relative wind induced by the wingtip vortex, the winglet is actually at a positive angle of attack. Part of the lift generated by the winglet acts in a forward direction, adding to the thrust of the airplane. Other winglets may be canted outward around 15 degrees, which adds to vertical lift and increases their aerodynamic efficiency, while also contributing to dihedral effect. Depending on the application, performance improvements due to winglets can increase fuel efficiency at high speeds and altitudes as much as 26°. [Figure 1-16]

VORTEX GENERATORS

Stalls are usually associated with high angle of attack flight conditions, but a special type of stall, called **shock-induced separation**, can occur on the wing of a high-speed airplane when it approaches its critical Mach number. Critical Mach number is the speed at which the airflow over any portion of an airfoil surface reaches the speed of sound.

When an airfoil approaches its critical Mach number, a shock wave begins to form just behind the point at which the air is moving the fastest. A shock wave typically first forms somewhere on the upper camber of the wing and tends to oscillate back and forth. The oscillating wave causes the airflow over the airfoil to induce buffeting to control surfaces while also reducing control surface effectiveness. To reinvigorate the airflow toward the surface of the airfoil, vortex generators may be installed on the airfoil at the point where this separation is most likely to occur. Vortex generators are short, low-aspect-ratio airfoils arranged in pairs. The tip vortices of these airfoils pull high-energy air down into the **boundary layer** helping prevent the separation. The boundary layer is the region of air that flows immediately adjacent to the surface of the airfoil. [Figure 1-17]

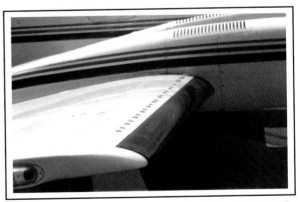

Figure 1-17. Vortex generators are short, low-aspect-ratio airfoils installed on the wing to help prevent boundary layer separation.

EMPENNAGE STRUCTURES

The empennage of an airplane is the assembly of the tail structures and includes components that are used both for control and stability. Regardless of their location, configuration or method of operation, these components serve the same basic functions of stabilizing and providing control of the aircraft in flight, both longitudinally and vertically. Longitudinal stability and control are provided by the horizontal surfaces, while directional stability and control are provided by vertical surfaces.

For propeller-driven airplanes, the location of the horizontal control surfaces must take into consideration both the effect of the propeller slipstream and the turbulence produced by the airflow over the wings. Some airplanes have these surfaces located quite low on the fuselage. For example, the vertical fin on some airplanes is quite large and is swept back to increase its effective aerodynamic arm. This is accomplished by moving the center of its area back as far from the vertical axis of the airplane as practical. The extension of the vertical fin forward, nearly to the back window, is called a dorsal fin. Large vertical fins are often needed to counteract the surface area ahead of the vertical axis, sometimes even to offset the forces caused by a nose wheel fairing.

Turbine-powered aircraft often have a conventional vertical fin and rudder, but the horizontal surfaces may be moved up on the fin to remove them from the exhaust area when the engines are mounted on the sides of the tail. [Figure 1-18]

Figure 1-18. Some turbine-powered aircraft have the engines mounted on each side of the empennage. One benefit of this design is that, if an engine fails, the yawing tendency from adverse thrust is reduced compared to an aircraft with the engines mounted further outboard on the wings. To accommodate the engine location, manufacturers sometimes use a mid-tail structure.

The T-tail configuration is a popular design on both turbine and propeller-driven airplanes. The horizontal tail surfaces are mounted on top of the vertical fin, keeping them out of the turbulence caused by the wing. In the case of single-engine propeller-driven aircraft, an additional benefit is that the horizontal surface is out of the propeller slipstream, which reduces vibration and noise inside the aircraft. [Figure 1-19]

The stabilator, or all-movable tail, is used extensively on turbine-powered airplanes, and also on some propeller-driven aircraft. This type of horizontal tail surface has no fixed stabilizer, but rather, there is an almost full-length **anti-servo tab** on its trailing edge. The tab can be adjusted from the cockpit to change the longitudinal trim for hands-off flight at various airspeeds and flight attitudes.

Figure 1-19. The horizontal tail surfaces are removed from wing turbulence and propeller slipstream by locating them at the top of the vertical fin.

Another type of tail is the V-tail that appears on older Beechcraft Bonanza airplanes. It uses two slanted tail surfaces to perform the same functions as the surfaces of a conventional elevator and rudder configuration. The fixed surfaces act as both horizontal and vertical stabilizers. The movable surfaces, commonly called **ruddervators**, are connected through a special linkage that allows in and out movement of the control wheel to move both surfaces simultaneously. On the other hand, movement of the rudder pedals moves the surfaces differentially, thereby providing directional control.

Another interesting configuration of empennage control surfaces is seen in a center-line-thrust twin-engine airplane. The empennage of this airplane is two booms extending back from the wing. The controls are conventional, with fixed vertical fins, movable rudders, a fixed horizontal stabilizer, and a movable elevator. [Figure 1-20]

Figure 1-20. A Cessna, model 337, uses a twin tail-boom configuration to allow one forward and one aft mounted engine. When mounted in this configuration, the engines provide thrust in-line with the center of the aircraft.

FUSELAGE STRUCTURES

The fuselage is the body of the aircraft, to which the wings, tail, engine and landing gear are attached. Since tremendous loads are imposed upon the fuselage structure, it must have maximum strength and, as with all of the parts of an aircraft, it must be lightweight. As previously mentioned, there are two types of construction used in modern aircraft fuselages: the truss- and stressed-skin-type.

TRUSS-TYPE FUSELAGE

By definition, a truss is a form of construction in which a number of members are joined to form a rigid structure. Many early aircraft used the Pratt truss, in which wooden longerons served as the main longitudinal structural members. Wood struts supported and held the longerons apart. Two pianowire stays crossed each bay, or space between the struts. The tension of the piano wire was adjusted using brass turnbuckles. A defining characteristic of the Pratt truss is that struts only carry compressive loads, while stays only carry tension loads.

When technology progressed to the extent that fuselage structures could be built of welded steel tubing, the Warren truss became popular. In this type of truss, longerons are separated by diagonal members that can carry both compressive and tensile loads. [Figure 1-21]

Figure 1-21. The Warren truss features longerons separated by diagonal members that carry both compressive and tensile loads.

The smooth aerodynamic shape required by an airplane fuselage is provided using both Pratt and Warren trusses by the addition of a non-load-carrying superstructure, and the entire fuselage covered with cloth fabric.

STRESSED-SKIN FUSELAGE

The necessity for having to build a non-load-carrying superstructure over the structural truss led designers to develop the stressed-skin form of construction, in which all of the loads are carried in the exterior skin. Stressed skin does not require the angular shape that is necessary for a truss, but can be built with a very clean, smooth, and aerodynamically efficient shape.

One of the best examples of a natural stressed-skin structure is the common hen egg. The fragile shell of an egg can support an almost unbelievable load, when applied in the proper direction, as long as the shell is not cracked. A significant limitation of a stressed-skin structure is that it cannot tolerate any dents or deformation in its surface. A thin aluminum can that is used for beverages may be used to demonstrate this concept. When a can is free of dents, it withstands a great amount of force applied to its ends. However, with only a slight dent in its side, it can be crushed very easily from top or bottom.

MONOCOQUE FUSELAGE

A full monocoque structure is one in which the fuselage skin carries all of the structural stresses. The portion of the fuselage behind the cabin of some of the smallest training airplanes is built with a monocoque-type construction. The upper and lower skins are made of thin sheet aluminum alloy that have been formed into compound curved shapes with a drop hammer or a hydropress. The edges of both of these skins are bent to form a lip that gives the skin rigidity and then riveted to former rings that have been pressed from thin sheet aluminum in a hydropress. The sides of the fuselage between the top and bottom skins are made of flat sheet aluminum, riveted to the skins and to the former rings.

Monocoque construction is economical and has sufficient strength for relatively low-stress areas. It is extremely important that all repairs to monocoque structures restore the original shape, rigidity and strength to any area that has been damaged.

SEMI-MONOCOQUE FUSELAGE

Most aircraft structures require more strength than can be provided by pure monocoque construction. For enhanced strength, a substructure of formers and stringers is built and the skin is riveted to it. Former rings and bulkheads, which are formers that also serve as compartment walls, are made of relatively thin sheet metal that have been formed in hydropresses, and stringers are made of extruded aluminum alloy. Stringers usually have a bulb on one of their sides to provide added strength to oppose bending loads. Longerons are also made of extruded aluminum alloy, but are heavier than the stringers in order to carry a large amount of the structural loads in the fuselage.

PRESSURIZED FUSELAGE

High-altitude flight places the occupants in a hostile environment in which life cannot be sustained unless supplemental oxygen is supplied. Since wearing an oxygen mask is both uncomfortable and inefficient, improved methods were developed for increasing passenger and crew comfort. One of the most significant improvements was achieved by pressurizing the interior of the fuselage. With increased cabin and cockpit air pressure, the occupants could be assured of receiving enough oxygen so that supplemental breathing equipment would not be required under normal conditions.

The first airliners to be pressurized were powered by piston engines and were unable to cruise at the extremely high altitudes that are common for modern jet transports. Cabins were pressurized to a pressure differential of only about two psi. Low pressurization created no major structural problems, but when the first jet transports, the British Comets, were put into service with pressurization of 8 psi, significant problems did arise. Continual flexing of the structure caused by the pressurization and depressurization cycles fatigued the metal. For a number of these aircraft, a crack developed at a square corner of a cutout in the structure, and the large amount of pressure differential caused the structure to virtually rip apart and explode. When the cause of the structural failure was determined, new emphasis was placed on fail-safe design of aircraft structures. Stress risers or portions of the structure where the cross section changes abruptly were eliminated. Joints and connections were carefully pre-stressed to minimize the cyclic stresses from the cyclic pressurization loads and, most important, the structure was designed with more than one load path for the stresses. If a crack did develop and weaken the structure in one place, another path existed through which the stresses could be supported. The improvements were so successful that each of them continues to be utilized in today's modern airplane designs. [Figure 1-22]

LANDING GEAR

Although an airplane is designed primarily to be operated in the air, it also must be manageable for ground operations. The 1903 Wright Flyer had extremely simple landing gear. It was launched from a rail and landed in the soft sand on wooden skids. Inventors later turned to bicycle wheels to support the airframe on the ground. To land on water, some designers equipped their flying machines with floats.

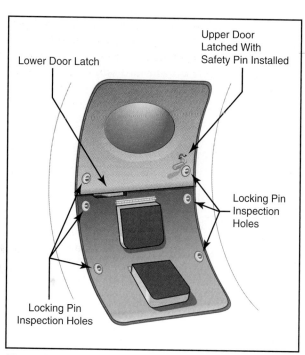

Figure 1-22. Pressurized aircraft require greater structural strength especially in windows and doors to accommodate the cyclic pressurization stresses.

The majority of airplanes through World War II used the tail wheel landing gear configuration, also called a conventional gear arrangement. Two main wheels are attached to the airframe, ahead of the center of gravity, to support most of the aircraft weight, and a small tail-skid or wheel at the very back of the fuselage provides a third point of support. This arrangement allows adequate ground clearance for a long propeller and provides the lightest-weight landing gear available. Before hard-surfaced runways became commonplace, a steel-shoed tail-skid provided adequate braking action for airplanes that had no regular wheel brakes. Taxiing these airplanes required a high degree of skill, and a two-wheel dolly was placed under the tail-skid to maneuver the airplane by hand for ground maneuvering. The main drawback of the conventional landing gear is that the airplane's center of gravity is behind the point of contact of the main wheels. This makes it easy for the airplane to ground loop if the pilot allows the airplane to swerve slightly while rolling on the ground. If the speed is below that which the rudder has sufficient control to counteract the motion, the center of gravity attempts to move ahead of the point of contact. The tail moves forward, causing the airplane to spin around.

The demand for airplanes that were easy to handle on the ground, the availability of paved runways on most airports and engines that turn with a high enough rpm to allow the use of short propellers, have made the tricycle landing gear popular. With this configuration, a nose wheel is installed in the front of the airplane, and the two main wheels are moved behind the center of gravity. The natural tendency for a nosewheel airplane is to move straight down the runway, rather than attempting to spin around.

With current production airplanes, the tailwheel landing gear is found only on those airplanes used for special purposes such as agricultural operations. A conventional landing gear is preferred for rough field applications where rugged or muddy sod surfaces make taking off and landing difficult. Without the heavy structure required for a nose wheel and the reduced wheel contact, tailwheel airplanes negotiate rough ground conditions much more easily.

Parasite drag is most felt on an airplane at high speed, and for low-speed airplanes, the simplicity and light weight of a fixed landing gear make the fixed-tricycle gear configuration a logical choice. [Figure 1-23]

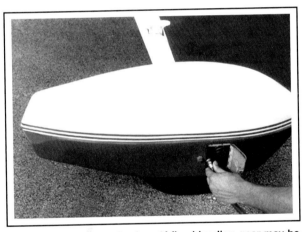

Figure 1-23. The parasite drag of fixed landing gear may be decreased by the use of streamlined speed fairings, or wheel pants, over the wheels. The decrease in drag provided by these streamlined fairings more than compensates for the additional weight.

At airspeeds where the drag of an exposed landing gear becomes appreciable, performance can be significantly increased by retracting the wheels into the structure. Generally, retractable landing gear fold up into the wing or fuselage. Retraction systems may be actuated with hydraulic cylinders or electric motors, but some lighter airplanes employ mechanical linkage to pull the wheels up. [Figure 1-24]

One of the biggest problems with retractable landing gear is the human factor; the failure of the pilot to lower the wheels before landing. To overcome this problem, some manufacturers have built into the

Figure 1-24. Electrically or hydraulically operated landing gear is commonly found on modern airplanes. However, the complexity and weight of a retractable landing gear system makes them impractical for smaller, slower aircraft.

landing gear system an airspeed sensor that automatically lowers the landing gear when the airspeed drops below a preset value.

WATER OPERATIONS

Before thousands of hard surfaced airports were built throughout the world, the airlines flying across both the Atlantic and Pacific Oceans used large flying boats such as the Boeing and Martin Clippers. However, the excessive size of the support structure required to hold these airplanes up in the water produced so much drag that the aircraft were unable to carry a profitable payload with the amount of engine power then available.

The availability of hard-surfaced runways and the progress made with long-range land-planes caused the flying boat to pass from the scene of practical transportation. Most land-planes today can be fitted with twin floats that support them on the water. Due to the compromises required to make an airplane suitable for water operations, an efficient amphibian airplane is a challenge to designers. The airplane must meet both structural and aerodynamic requirements in order to fly. When water and land handling requirements are added, the resulting machine becomes less efficient. [Figure 1-25]

Figure 1-25. Amphibious planes must be designed with both land and water operations in mind.

One successful approach to land and water operations is to use amphibious floats. Amphibious floats are installed in the same manner as normal floats, but they have built-in retractable wheels that may be extended for operations on land or retracted for water operations. True amphibian airplanes have a hull much like that of a flying boat, but they also have wheels that may be extended for hard surface landing.

SNOW OPERATIONS

Skis may be fitted to an airplane to further extend its utility. The simplest type of airplane ski is the wheel-replacement type, in which the wheel is removed and the ski is installed on the normal landing gear axle. Retractable skis are far more useful than the wheel-replacement type, since they allow the airplane to land on either a hard-surface runway or on snow. This type of ski is installed on the landing gear with the wheel in place. For landing on a hard-surfaced runway, the ski is pulled up so that the wheel sticks out below the ski where the weight of the airplane can be supported by the tire rather than the ski. For landing on snow, the ski is lowered, making contact with the ground first and supporting the airplane. [Figure 1-26]

Figure 1-26. Attaching skis to an airplane further extends its capability by allowing the pilot to take-off and land on snow and ice.

POWERPLANT SUPPORT STRUCTURES

A number of accessories and systems are connected to the engine(s) to make them work. Engines must be started, cooled, controlled, and mounted in places where they can efficiently provide thrust.

PISTON ENGINES

After basic problems of control were solved with experiments on gliders, attention was turned to providing adequate power for newer airplanes. Because there was so little knowledge of the requirements for flight in the early development of aviation, airplanes often went through many evolutionary steps that brought poor results. Most of the earliest airplanes had the engine mounted behind the pilot or on the wing. In the case of the Wright Flyer, the wing-mounted engines drove the propellers by linked chains. As airplane development progressed, the engine was moved up front where the propeller could operate in undisturbed air.

The early air-cooled engines merely had finned cylinders sticking out into the airstream to remove excess heat. However, as the power developed by these engines increased, efficient cowling enclosures had to be designed to increase the airflow for cooling. Radial engines were enclosed in a Townend ring and later in a full NACA cowling that directed the maximum amount of air through the finned cylinders. While increasing airflow, cowlings also minimized drag for engines that have large frontal areas. [Figure 1-27]

Figure 1-27. Cowlings on radial engines are specifically designed to increase airflow around the cylinders and to reduce drag.

Today, almost all piston-powered airplanes use horizontally-opposed engines enclosed inside a pressure cowling. The cooling air for these engines enters the cowling from the front, above the engine, and then passes through baffles and fins to remove the heat. A low-pressure area below the engine created by air flowing over the bottom of the cowling draws air through the engine to increase the amount of cooling.

The amount of airflow through a high-powered engine is controlled by cowl flaps at the air exit. Cowl flaps may be actuated either by the pilot or, in some installations, automatically, by actuators that sense the engine temperature. Cowl flaps are normally left wide open for ground operations, but closed in flight to keep the engine temperatures within the proper operating range.

TURBINE ENGINES

Turbine engines are truly revolutionary powerplants for aircraft. They are far smaller and lighter than piston engines of the same power, while also producing far less vibration.

The first jet engines were mounted inside the fuselage consistent with the principles learned in reciprocating-engine installations. As experience increased with these engines, designers realized that space and weight could be saved by mounting the engines outside the fuselage in self-contained pods. The two most common locations for turbojet installations are beneath the wing and at the rear end of the fuselage. The pods are suspended by a pylon mounted to the wing's basic structure. Engine pylons are critical to the safe operation of the aircraft, and are subject to the same inspection and maintenance techniques as other airframe structures. [Figure 1-28]

Figure 1-28. Engines on modern turbine-engine aircraft are attached to the aircraft main structure through pylons under the wings or on the rear section of the fuselage.

ENGINE MOUNTS

Engine mounts consist of the structure that transmits the thrust provided by either the propeller or turbojet, to the airframe. The mounts can be constructed from welded alloy steel tubing, formed sheet metal, forged alloy fittings, or a combination of all three. Engine mounts are required to absorb not only the thrust, but also the vibrations produced by the particular engine, or engine-propeller combination. There are several types of vibration isolators used and their inspection, repair and replacement is a regular part of any structural maintenance program.

ACCESS AND INSPECTION

Openings in structures are necessary for entrance and egress, servicing, inspection, and repair. Electrical wiring, fuel and oil lines, air ducting, and many other items also require openings for safe routing. The complexity of the structure and its strength requirements determine how an opening is fabricated. For instance, a passenger door opening reinforcement on a training airplane has the same parts as one on a transport category airplane, but the strength differs considerably. Differences are also apparent in other parts of the airplane. As an example, the rear bulkhead access panel on a pressurized airplane is not the same as on a non-pressurized one, though both serve the same function. Installation, or reinstallation, of any access panel or inspection cover should be done carefully in accordance with the manufacturer's maintenance manual.

SECTION B

AIRPLANE ASSEMBLY AND RIGGING

The stability and controllability of an aircraft, and ultimately, the safety of flight, depends on the proper assembly and rigging of the flight controls. A pilot expects any aircraft to respond to control inputs in a consistent and predictable manner. An aircraft that is out-of-rig will be difficult or impossible to trim for stable flight and will require constant attention by the pilot. Improper rigging can also create problems for the pilot during specific flight operations. For example, if the elevator has insufficient upward travel, the pilot may not be able to raise the nose sufficiently during the landing flare to make a safe landing.

If the flight control system is improperly assembled, the potential for disaster is even greater. Imagine the confusion that would result if the pilot commands a left turn and the airplane responds by turning right!

AIRPLANE AXES

An aircraft in flight is free to rotate about three axes, and its flight controls are designed to allow the pilot to control its rotation about each axis. The three axes pass through a common reference point called the center of gravity (CG), which is the theoretical point where the entire weight of the aircraft is considered to be concentrated. Since all three axes pass through the CG, the airplane always moves about its CG, regardless of which axis is involved.

LONGITUDINAL AXIS

The longitudinal axis is a straight line passing through the fuselage from nose to tail. Motion about the longitudinal axis is called roll, and this axis is often referred to as the roll axis. [Figure 1-29]

LATERAL AXIS

The lateral axis is a straight line extending parallel to the wing span at right angles to the longitudinal axis. You can think of it as extending from wingtip to wingtip. Motion about the lateral axis is called pitch and this axis is referred to as the pitch axis. [Figure 1-30]

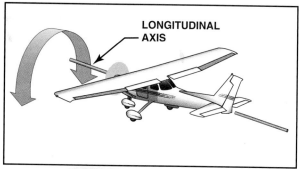

Figure 1-29. The ailerons cause an airplane to roll about the longitudinal axis. The primary purpose of the ailerons is to bank the wing, causing the airplane to turn.

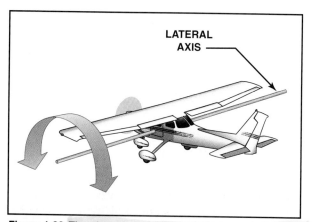

Figure 1-30. The elevators cause an airplane to pitch about the lateral axis. The primary purpose of the elevators is to change the angle of attack, and thereby control airspeed.

VERTICAL AXIS

The vertical axis is the third axis, and is perpendicular to the other two. It can be envisioned as a straight line from the top to the bottom of the airplane. Motion about the vertical axis is called yaw, and so this axis is sometimes called the yaw axis. [Figure 1-31]

STABILITY AND CONTROL

Stability, maneuverability and controllability all refer to movement of the aircraft about one or more of the three axes of rotation.

Aircraft Structural Assembly and Rigging

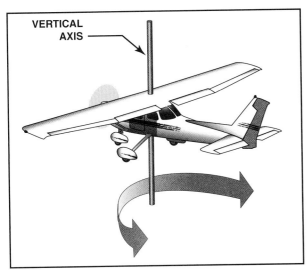

Figure 1-31. The rudder causes an airplane to yaw about the vertical axis. The primary purpose of the rudder is to counteract aileron drag and keep the fuselage streamlined with the relative wind. This improves the quality of turns and reduces drag.

Stability is the characteristic of an airplane in flight that causes it to return to a condition of equilibrium, or steady flight, after it is disturbed. Maneuverability is the characteristic of an airplane that permits the pilot to easily move the airplane about its axes and to withstand the stress resulting from these maneuvers. Controllability is the capability of an airplane to respond to the pilot's control inputs.

Aircraft are not designed to be stable in their attitude with respect to the earth, but they are stable with respect to the relative wind. A aircraft that is stable in pitch will return to the angle of attack for which it is trimmed any time it is disturbed from this angle. An aircraft that is stable in roll will tend to return to a wings-level attitude. An aircraft that is directionally stable will tend to weathervane, so as to align its fuselage with the relative wind.

TYPES OF STABILITY

STATIC STABILITY
An airplane is in equilibrium when there are no forces trying to disturb its condition of steady flight. If the plane is disturbed from steady flight, static stability will try to return it to its original attitude.

DYNAMIC STABILITY
While static stability creates a force that tends to return the aircraft to its original attitude, dynamic stability determines how it will return. Dynamic stability is concerned with the way the restorative forces act with regard to time.

CONDITIONS OF STABILITY

Both static and dynamic stability can be designed in one of three states; positive, negative or neutral. At times, turbulence or erratic movement causes buffeting in an airplane. If an aircraft is designed with positive stability, it will return to its original flight condition when turbulence ceases. Positive stability is desirable for most aircraft but advanced fighter aircraft with computer augmented flight controls may employ neutral or negative stability to enhance combat maneuverability.

POSITIVE STABILITY

Positive stability can be illustrated by considering the action of a ball in a U-shaped trough as shown in figure 1-32. If the ball is rolled up to one edge of the trough and released, positive static stability will cause it to roll back down towards its original position.

When the ball returns to its original position at the bottom of the trough, it will probably overshoot its position of equilibrium and start up the opposite side. As soon as it starts up the opposite slope, positive static stability will tend to return it to the bottom. The ball will rock back and forth, each time moving a shorter distance up the slope, until it finally stops at the bottom — a demonstration of positive dynamic stability. [Figure 1-32]

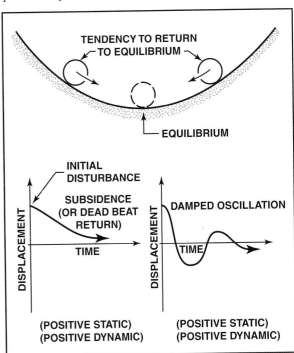

Figure 1-32. Positive static and dynamic stability, as illustrated by the ball in a trough, is a desirable characteristic for most airplanes. Most airplanes are designed to exhibit the damped oscillation form of stability when disturbed from pitch equilibrium.

NEGATIVE STABILITY

If the ball is released from the top of a hill, it will not tend to return to its original position. In this condition, the ball is said to have negative static stability, or to be statically unstable.

If the corrective forces increase with time, the body has negative dynamic stability. Depending on the static stability of the body, the corrective forces can either diverge or set up an oscillation that grows larger over time. Neither condition is desirable in most airplanes. [Figure 1-33]

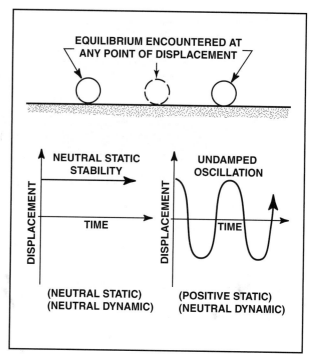

Figure 1-34. An object that has neutral stability remains displaced from its original state whenever a force is applied. A neutrally stable airplane would be difficult to control and would probably require computer-augmented flight controls.

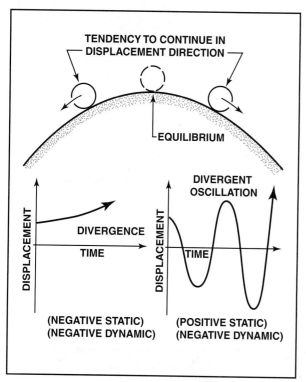

Figure 1-33. Negative stability, as illustrated by a ball rolling off the crest of a hill, is an undesirable characteristic in airplanes. A pilot would be very likely to lose control of an airplane with negative stability.

NEUTRAL STABILITY

If the ball is moved across a perfectly level surface, it will not tend to return to its original position, nor will it tend to move further away. This illustrates neutral static stability.

The corrective forces of a body with neutral dynamic stability neither increase or decrease with time. Energy is not added to the body or taken away by any form of damping. [Figure 1-34]

STABILITY ABOUT THE AXES

The longitudinal, or pitch, stability of an airplane determines its ability to be trimmed to fly hands-off at any airspeed. Because the wing's center of lift is behind the center of gravity, the wing produces a nose-down pitching moment. This pitching moment is counter-acted by a down load produced by the horizontal tail surfaces. Elevator trim can be adjusted by the pilot to produce the required down load at any speed, thereby balancing the plane so that it will maintain level flight with little or no control input. [Figure 1-35]

Lateral, or roll, stability is provided primarily by dihedral in the wings. Dihedral is the upward angle between the wing and the lateral axis of the airplane. The dihedral angle of most airplanes is usually just a few degrees. When an airplane enters a downward sideslip toward the low wing, the direction of relative wind changes. The low wing experiences a higher angle of attack while the angle of attack on the high wing is reduced. The changes in angle of attack cause the low wing to generate more lift, at the same time the high wing generates less lift, and the combined forces tend to roll the aircraft back to a wings-level attitude. [Figures 1-36, 1-37]

Directional stability is stability about the vertical axis and is provided primarily by the vertical tail, which causes the airplane to act much like a weather vane. Most airplanes are designed with

Aircraft Structural Assembly and Rigging

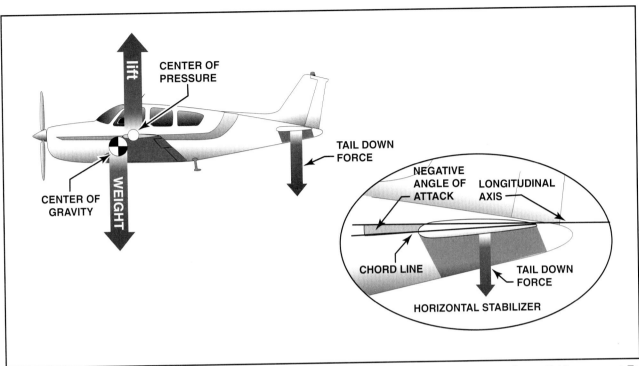

Figure 1-35. With the center of pressure aft of the center of gravity, an airplane wing produces a nose-down pitching moment. To counter-balance this moment, the tail down load can be varied to make the airplane longitudinally stable over a wide range of airspeeds.

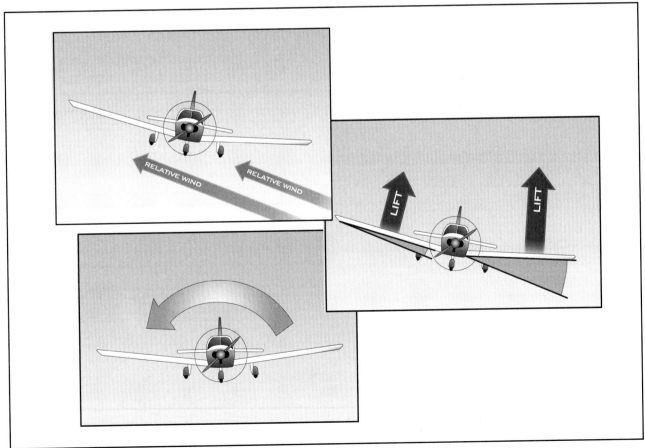

Figure 1-36. Wing dihedral is a major contributor to lateral stability. Airplanes generally have less roll stability than pitch stability.

Figure 1-37. Because they are inherently more stable laterally, high-wing aircraft such as the Cessna 172 on the left are designed with less dihedral than the typical low-wing aircraft, as exhibited by the Beechcraft Bonanza on the right.

more fuselage surface area behind the CG than forward of it.

When the airplane enters a sideslip, the relative wind strikes the side of the fuselage and the vertical tail. Since the force exerted on the airplane aft of the CG tends to cause the nose to turn towards the sideslip, this aligns the fuselage with the relative wind. [Figure 1-38]

CONTROL SYSTEMS

The primary flight controls of an aircraft do no more than modify the camber, or aerodynamic shape, of the surface to which they are attached. This change in camber creates a change in the lift and drag produced by the surface, with the immediate result of rotating the airplane about one of its axes. This rotation produces the desired changes in the flight path of the aircraft.

LONGITUDINAL CONTROLS

The most conventional longitudinal control system consists of a fixed horizontal stabilizer on the rear end of the fuselage, with movable elevators hinged to its trailing edge. The trailing edge of the elevator may have a trim tab to adjust the down load of the tail for hands-off flying at any desired airspeed. Another means of providing the necessary trimming force is to adjust the entire horizontal stabilizer by rotating it about a pivot point.

The elevator is connected to the control wheel or yoke in the cockpit with steel control cables, and moves up or down as the wheel is moved forwards or backwards. Pulling back on the wheel pulls the elevator-up cable and rotates the top of the elevator bell crank forward. The control horn on the bottom of the elevator torque tube is attached to the bell crank with a push-pull rod, and as the bottom of the bell crank

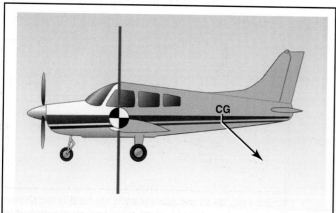 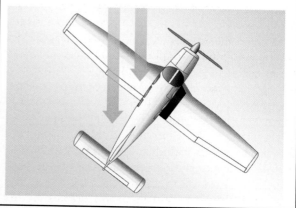

Figure 1-38. To be directionally stable, an airplane must have more surface area behind the CG than in front of it. When an airplane enters a sideslip, the greater surface area behind the CG helps keep the airplane aligned with the relative wind.

Aircraft Structural Assembly and Rigging

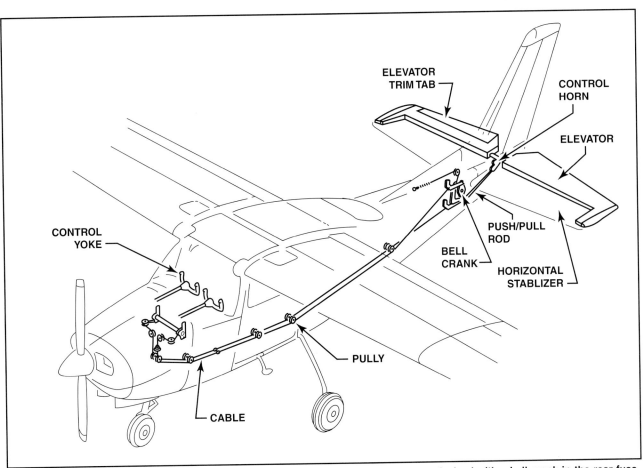

Figure 1-39. A typical elevator-control system consists of cables connecting the control wheel with a bell crank in the rear fuselage. The bell crank is connected to the elevator control horn by a push rod.

moves back, it pushes the elevator up. Pushing in on the control wheel has the opposite result. The elevator-down cable is pulled and the bottom of the bell crank moves forward, causing the push-pull rod to pull the elevator down. [Figure 1-39]

Many modern airplanes have several rows of seats or otherwise have a possibility of inadvertently being loaded with their center of gravity too far back. For such airplanes, an elevator down spring automatically lowers the nose to prevent an approach-to-landing stall caused by the excessively far-aft center of gravity. If an airplane with the center of gravity too far aft is slowed down for landing, the trim tab will be working hard to hold the nose down, and the elevators will actually be in a slightly down position. If the airplane in this unstable condition encounters turbulence and slows down further, the elevator will streamline, and at this slow speed the trim tab cannot force it back down. The nose of the airplane will pitch up, aggravating the situation and possibly causing a stall at this critical altitude.

The elevator down spring holds a mechanical load on the elevator, forcing it down. This mechanical force is balanced by the aerodynamic force of the trim tab, and the airplane may be trimmed for its approach speed in the normal way. If the airplane encounters turbulence that slows it down, the trim tab will lose its effectiveness. The down spring will pull the elevator down, lowering the nose so the airspeed will build up and prevent a stall. [Figure 1-40]

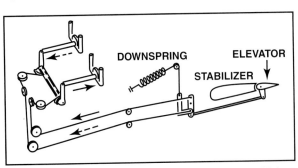

Figure 1-40. When the aerodynamic tail load is insufficient due to an aft center of gravity condition, an elevator down spring may be used to supply a mechanical load to lower the nose.

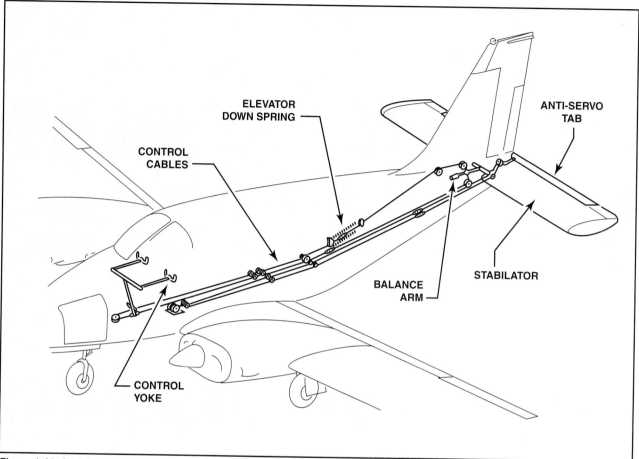

Figure 1-41. A typical stabilator control system is little different from an elevator control system. The main difference is that the entire horizontal surface pivots when control forces are applied.

Airplanes with **stabilators** have essentially the same type of control system as those with elevators. When the control yoke is pulled back, it pivots and pulls on the stabilator-up cable. This pulls down on the balance arm of the stabilator and raises its trailing edge, rotating the airplane's nose up. Pushing the yoke in lifts the stabilator balance arm and lowers the trailing edge of the stabilator. This system uses stabilator down springs in the same way as those just described for the elevator, to improve the longitudinal stability of the airplane during conditions of low airspeed with a far-aft center of gravity. [Figure 1-41]

LATERAL AND DIRECTIONAL CONTROLS

Turning an airplane requires rotation about both the longitudinal and vertical axes, and so both the rudder and ailerons must be used. In many modern airplanes there is some form of mechanical interconnection between these two systems, usually not a positive one, but one that can be overridden if it is necessary to slip the airplane.

Rotation of the control wheel turns the drum to which the aileron control cables are attached. If the wheel is rotated to the right, the right cable is pulled and the left one is relaxed. The cable rotates the right aileron bell crank, and the push-pull tube connected to it raises the right aileron. A balance cable connects both aileron bell cranks, and as the right aileron is raised, the balance cable pulls the left bell crank and its push-pull tube lowers the left aileron. [Figure 1-42]

Aileron drag is a big problem caused by the displacement of the ailerons The aileron that moves downward is the one that causes the problem, as it creates both more lift and drag, and this drag way out near the wing tip pulls the nose of the airplane around in the direction opposite to the way the airplane should turn. The geometry of the bell cranks is such that the aileron moving upward travels a greater distance than the one moving down, and it produces enough parasite drag to counteract some or all of the induced drag on the opposite wing.

Aircraft Structural Assembly and Rigging

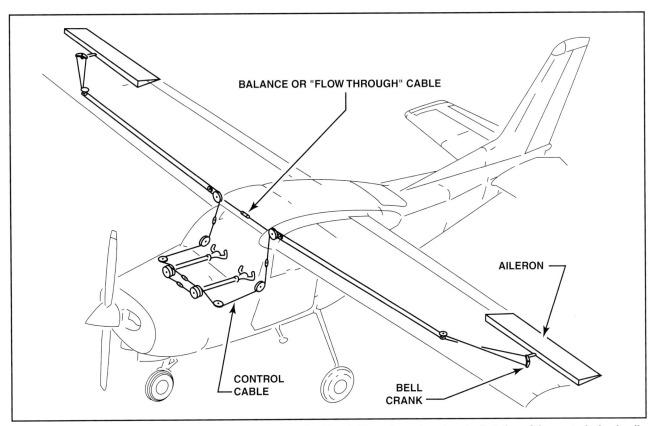

Figure 1-42. A typical aileron control system also consists of cables, bell cranks and push rods. Rotation of the control wheel pulls causes one aileron to rise and the other to lower.

There are two expedients in use to decrease aileron drag: the **Frise aileron** and **rudder-aileron interconnect springs**. The Frise aileron is the type most commonly used today, and it minimizes aileron drag because of the location of its hinge point. These ailerons have their hinge point some distance back from the leading edge. When the aileron is raised, its nose sticks out below the lower surface of the wing and produces enough parasite drag to counter the induced drag from the down aileron. [Figure 1-43]

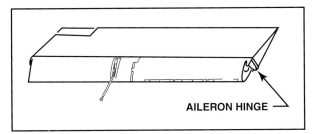

Figure 1-43. The hinge line of a Frise aileron is aft of the leading edge, allowing its nose to protrude slightly from the lower wing surface to counteract aileron drag.

Since aileron drag is produced each time the control wheel deflects the ailerons, many manufacturers connect the control wheel to the rudder control system through an interconnecting spring. When the wheel is moved to produce a right roll, the interconnect cable and spring pulls forward on the right rudder pedal just enough to prevent the nose of the airplane yawing to the left. [Figure 1-44]

Figure 1-44. Rudder-aileron interconnect springs may also be used to help correct for aileron drag by automatically deflecting the rudder at the same time as the ailerons are deflected.

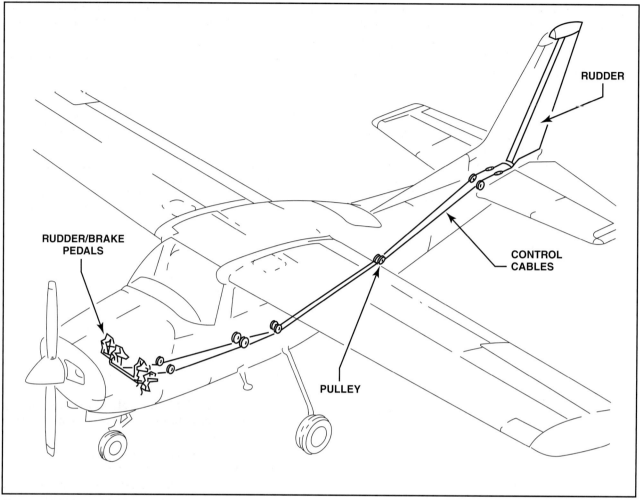

Figure 1-45. In a typical rudder-control system, cables transmit control inputs from the pilot's rudder pedals to the rudder-control horn.

Airplanes whose rudder pedals are connected rigidly to the nosewheel for steering have the interconnect cables attached to the rudder cables with connector clamps in the aft end of the fuselage. The effect is the same for connection at either location. A small amount of rudder force is applied when the ailerons are deflected, but this force can be overridden because it is applied through a spring.

The rudder pedals are connected to the rudder horn with steel control cables and, on an airplane with a nosewheel, also to the nosewheel steering mechanism. Forward movement of the right rudder pedal will deflect the rudder to the right. [Figure 1-45]

AUXILIARY OR TRIM CONTROLS

TRIM TABS

Trim tabs are small movable portions of the trailing edge of a control surface. These tabs are controlled from the cockpit to alter the camber of the surface and create an aerodynamic force that will hold the control surface deflected. [Figure 1-46]

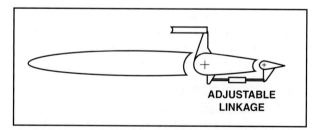

Figure 1-46. Trim tabs on the trailing edge of control surfaces can be adjusted to provide an aerodynamic force to hold the surface in a desired position.

Trim tabs may be installed on any of the primary control surfaces. If only one tab is used, it is normally on the elevator, to permit adjustment of the tail load so the airplane can be flown hands-off at any given airspeed. The airplane speed is set with the control wheel, and then the trim tab is adjusted until the airspeed can be maintained without exerting force on the wheel. FAR 23.677 states that a pilot must be able to determine the current position of the trim tabs as well as the neutral position of trim controls for any cockpit adjustable lateral and directional trim.

BALANCE TABS

The control forces may be excessively high in some airplanes, and in order to decrease them, the manufacturer may use a balance tab. This tab is located in the same place as a trim tab. In many installations, one tab serves both functions. The basic difference is that the control rod for the balance tab is connected to the fixed surface on the same side as the horn on the tab. If the control surface is deflected upward, the connecting linkage will pull the tab down. When the tab moves in the direction opposite that of the control surface, it will create an aerodynamic force that aids the movement of the surface. [Figure 1-47]

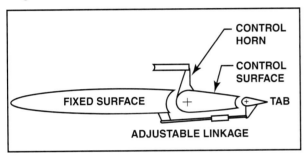

Figure 1-47. Balance tabs provide a means for the airflow across the surface to provide a "power-assist" to reduce high control forces.

If the linkage between the tab and the fixed surface is adjustable from the cockpit, the tab will act as a combination trim and balance tab. It can be adjusted to any desired deflection to trim the airplane for a steady flight condition. Any time the control surface is deflected, the tab will move in the opposite direction and ease the load on the pilot.

ANTI-SERVO TABS

All-movable horizontal tail surfaces do not have a fixed stabilizer in front of them, and the location of their pivot point makes them extremely sensitive. To decrease this sensitivity, an anti-servo tab may be installed on the trailing edge. This tab works in the same manner as the balance tab except that it moves in the opposite direction. The fixed end of the linkage is on the opposite side of the surface from the horn on the tab, and when the trailing edge of the stabilator moves up, the linkage forces the trailing edge of the tab up. When the stabilator moves down, the tab also moves down. The fixed end of the linkage may be attached to a jackscrew so the tab may be used as a trim tab as well as an anti-servo tab. [Figure 1-48]

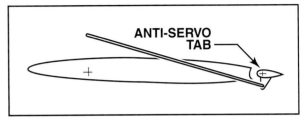

Figure 1-48. An anti-servo tab attempts to streamline the control surface and is used to make an all-moving horizontal tail surface less sensitive by opposing the force exerted by the pilot.

SERVO TABS

Large aircraft are usually equipped with a power-operated irreversible flight control system. In these systems, the control surfaces are operated by hydraulic actuators controlled by valves moved by the control yoke and rudder pedals. An artificial feel system gives the pilot resistance that is proportional to the flight loads on the surfaces.

Control forces are too great for the pilot to manually move the surfaces. In the event of a hydraulic system failure, they are controlled with servo tabs, in a process known as manual reversion. In the manual mode of operation, the flight control column moves the tab on the control surface, and aerodynamic forces caused by the deflected tab move the main control surface. [Figure 1-49]

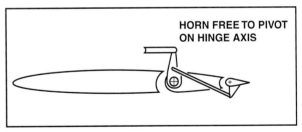

Figure 1-49. Servo tabs provide a force to assist the pilot in moving a primary control surface of a large aircraft in the event of a hydraulic system failure.

SPRING TABS

Another device for aiding the pilot of high-speed aircraft is the spring tab. The control horn is free to pivot on the hinge axis of the surface, but it is restrained by a spring. For normal operation when control forces are light, the spring is not compressed. The horn acts as though it were rigidly attached to the surface. At high airspeeds when the control forces are too high for the pilot to operate properly, the spring collapses and the control horn deflects the tab in the direction to produce an aerodynamic force that aids the pilot in moving the surface. [Figure 1-50]

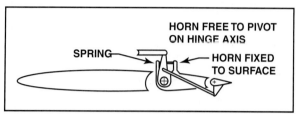

Figure 1-50. A spring tab is another means for assisting the pilot of a high-speed airplane to overcome high control forces.

GROUND-ADJUSTABLE TABS

Many small airplanes have a non-moveable metal trim tab on the rudder. This tab is bent in one direction or the other on the ground to apply a trim force to the rudder. The correct displacement is determined by trial-and-error until the pilot reports that the airplane is no longer skidding left or right during normal cruising flight. [Figure 1-51]

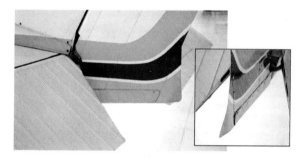

Figure 1-51. A ground-adjustable tab is used on the rudder of many small airplanes to correct for a tendency to fly with the fuselage slightly misaligned with the relative wind.

ADJUSTABLE STABILIZER

Rather than using a movable tab on the trailing edge of the elevator, many airplanes pivot the horizontal stabilizer about its rear spar, and mount its leading edge on a jackscrew that is controllable from the cockpit. On smaller airplanes, the jackscrew is cable-operated from a trim crank, and on larger airplanes it is motor driven. The trimming effect of the adjustable stabilizer is the same as that obtained from a trim tab. [Figure 1-52]

HIGH LIFT DEVICES

An airplane is a series of engineering compromises. Designers must choose between stability and maneuverability, and between high cruising speed and low landing speed, as well as between high utility and low cost. Lift-modifying devices give us some good compromises between high cruising

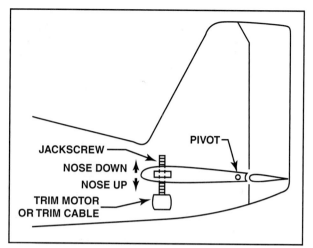

Figure 1-52. Many airplanes, including most jet transports, use an adjustable stabilizer to provide the required pitch trim forces.

speed and low landing speed, because they may be extended only when needed, then tucked away into the structure when not needed.

FLAPS

Perhaps the most universal lift-modifying devices used on modern airplanes are flaps on the trailing edge of the wing. These surfaces change the camber of the wing, increasing both lift and drag for any given angle of attack. The basic airfoil section, at 15 degrees angle of attack, has a lift coefficient of 1.5 and a drag coefficient of about 0.05. If plain flaps are hinged to the trailing edge of this airfoil, we get a maximum lift coefficient of 2.0 at 14 degrees angle of attack and a drag coefficient of 0.08. Slotted flaps are even better, giving a maximum lift coefficient of 2.6 at the same angle of attack and a drag coefficient of 0.10. The total effect of Fowler flaps is not seen in just the lift and drag coefficients, because they not only provide an excellent increase in the lift coefficient, but they also increase the wing area. Increased wing area has an important effect on both lift and drag. [Figure 1-53]

PLAIN FLAPS

These simple devices are merely sections of the trailing edge of the wing, inboard of the ailerons. They are about the same size as the aileron and are hinged so they can be deflected, usually in increments of 10, 25, and 40 degrees. Generally speaking, the effect of these flaps is minimal, and they are seldom found on modern airplanes.

SPLIT FLAPS

This is another design of flap that was used with a great deal of success in the past, but is seldom used today. On the extremely popular Douglas DC-3, a

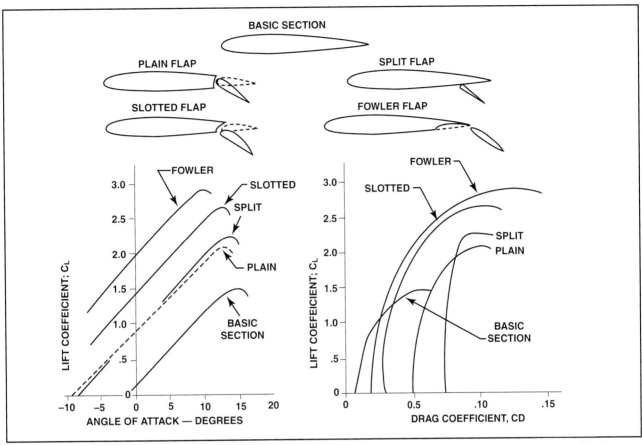

Figure 1-53. Different types of flaps provide differing amounts of lift and drag. Slotted and Fowler flaps are the most effective type and are used almost exclusively on modern aircraft.

portion of the lower surface of the trailing edge of the wing from one aileron to the other, across the bottom of the fuselage, could be hinged down into the airstream. The lift change was similar to that produced by a plain flap, but it produced much more drag at low lift coefficients. This drag coefficient changed very little with the angle of attack.

SLOTTED FLAPS

The most popular flap on airplanes today is the slotted flap. Variations of this design are used for small airplanes as well as for large ones. Slotted flaps increase the lift coefficient a good deal more than the simple flap. On small airplanes, the hinge is located below the lower surface of the flap, and when it is lowered, it forms a duct between the flap well in the wing and the leading edge of the flap.

When the flap is lowered all of the way and there is a tendency for the airflow to break away from its surface, air from the high-pressure area below the wing flows up through the slot and blows back over the top of the flap. This high energy flow on the surface pulls air down and prevents the flap stalling. It is not uncommon on large airplanes to have double- and even triple-slotted flaps to allow the maximum increase in drag without the airflow over the flaps separating and destroying the lift they produce. [Figure 1-54]

FOWLER FLAPS

Fowler flaps are a type of slotted flap. The design of this wing flap not only changes the camber of the wing, it also increases the wing area. Instead of rotat-

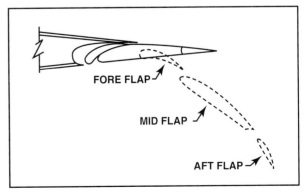

Figure 1-54. Triple-slotted flaps are used on many jet transports to balance the lift and drag necessary for reasonable takeoff and landing speeds with the requirements for high-speed cruising flight.

ing down on a hinge, it slides backwards on tracks. In the first portion of its extension, it increases the drag very little, but increases the lift a great deal as it increases both the area and camber. As the extension continues, the flap deflects downward, and during the last portion of its travel, it increases the drag with little additional increase in lift.

LEADING EDGE DEVICES

SLOTS

Stalls occur when the angle of attack becomes so great that the energy in the air flowing over the wing can no longer pull air down to the surface. The boundary layer thickens and becomes turbulent and the airflow separates from the surface.

This separation can be delayed to a higher angle of attack by any means that increases the energy of the air flowing over the surface. One method used is a slot in the leading edge of the wing. This slot is simply a duct for air to flow from below the wing to the top where it is directed over the surface in a high-velocity stream. Slots are usually placed ahead of the aileron to keep the outer portion of the wing flying after the root has stalled. This keeps the aileron effective and provides lateral control during most of the stall. [Figure 1-55]

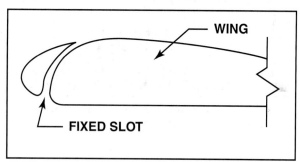

Figure 1-55. A fixed slot ducts air from the lower surface to the upper surface of a wing at high angles of attack

SLATS

Many high-performance airplanes have a portion of the wing leading edge mounted on tracks so it can extend outward and create a duct to direct high-energy air down over the surface and delay separation to a very high angle of attack.

In many airplanes these slats are actuated by aerodynamic forces and are entirely automatic in their operation. As the angle of attack increases, the low pressure just behind the leading edge on top of the wing increases and pulls the slat out of the wing. When the slat moves out, it ducts the air from the high-pressure area below the wing to the upper surface and increases the velocity of the air in the boundary layer. When the angle of attack is lowered, air pressure on the slat moves it back into the wing where it has no effect on the airflow.

Some airplanes have slats operated by either hydraulic or electric actuators, and the slats are extended automatically when the trailing edge flaps are lowered. These slats prevent the airflow breaking away from the upper surface when the flaps increase the camber of the wing. Flaps that are used with slats are usually slotted. They duct high-energy air over the deflected flap sections so the airflow will not break away from their surface. [Figure 1-56]

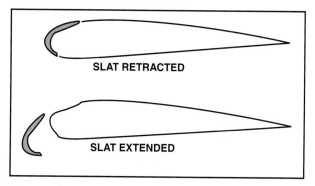

Figure 1-56. Slats extend out of the leading edge of the wing at high angles of attack and serve the same function as a slot. They may be actuated automatically by aerodynamic forces or mechanically operated, usually in conjunction with the trailing edge flaps.

LEADING EDGE FLAPS

The leading edge of some wings may be deflected downward to increase their camber. These leading edge flaps are usually electrically or hydraulically actuated and are used in conjunction with the trailing edge flaps.

STALL STRIPS

It is important that a wing stall at the root first so the ailerons will still be able to provide lateral control throughout the stall. If the wing does not have this characteristic naturally, it can be given it by installing small triangular stall strips on the leading edge of the wing in the root area. When the angle of attack is increased enough for a stall to occur, the strips provide enough air disturbance to hasten the stall on the section of wing behind them. This loss of lift will usually cause the nose of the airplane to drop while the outer portion of the wing is still flying and the ailerons are still effective. It also causes the disturbed air to buffet the horizontal tail surfaces, thus providing the pilot with a feeling in the controls of the impending stall. [Figure 1-57]

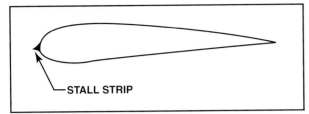

Figure 1-57. A small triangular stall strip may be installed on the leading edge of the wing near the wing root to cause area behind it to stall first.

SPECIAL WING TIPS

Air flowing over the top of a wing creates a low pressure, while the air passing below the wing has been slowed down somewhat and its pressure has increased. This difference in pressure causes air to spill over the wing tip and create vortices that effectively kill some of the lift and create drag, especially at high angles of attack and low airspeed. The wing tip vortex will continue to expand in the form of a giant cone and trails far behind the airplane. At this point it ceases to be a wing tip vortex and becomes wake turbulence, which can be dangerous to any aircraft. [Figure 1-58]

Figure 1-58. A wing tip vortex develops as a result of air flowing around the tip due to pressure differences. All airplanes generate them. These vortices are strongest when the airplane is flying slowly at high angles of attack and can be very dangerous to other airplanes.

There are a number of methods that have been used to reduce the effects of wing-tip vortices. Some manufacturers install fuel tanks on the wing tips that serve the triple function of increasing the range of the airplane, distributing the weight over a greater portion of the wing and preventing the air spilling over the wing tip. Smaller airplanes that do not use tip tanks may have tip plates installed on the tip. These plates have the same shape as the airfoil but are larger and prevent the air spilling over the tips. [Figure 1-59]

Figure 1-59. Tip tanks minimize the amount of high pressure air spilling over from the high-pressure area beneath the wing to the low-pressure area above it.

Far less drastic than tip tanks or tip plates are specially shaped wing tips. Some wing tips have a special droop and a square trailing edge to tighten the vortices and spin them away so they will not contaminate the upper surface so much. [Figure 1-60]

Figure 1-60. Drooped wing tips are a simpler method of reducing losses from wing tip vortices.

WINGLETS

Another popular method of controlling, or reducing, wing tip vortices is by the use of winglets. Used principally on high-speed airplanes, they also allow for drag reduction and better airflow control. [Figure 1-61]

WING FENCE

For the most efficient lift production, the airflow should always be chordwise across the wing. This is not always possible and some planforms will allow air to travel spanwise at the same time. The wing fence is a simple method to stop, or reduce, the spanwise flow.

CANARD SURFACES

Any aircraft that has the equivalent of two lifting surfaces, instead of the conventional horizontal

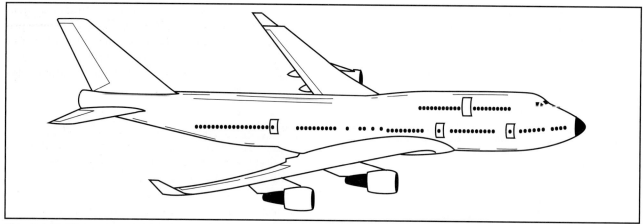

Figure 1-61. Winglets actually recover some of the energy that would be lost to wing tip vortices and not only reduce the vortex strength but also reduce the total drag on the airplane.

stabilizer that provides a down load, can be classified as a canard. The canard is the forward surface, and frequently is also a control surface.

In a conventional airplane, the wing stalls, aileron control is lost, the CG shifts forward, then speed builds up and control is regained. During this sequence of events there is always the chance that lateral control will also be lost, particularly when the airplane is in a turn, causing an accidental spin. With a canard configuration, the sequence changes somewhat. The canard stalls first, the nose drops and speed builds back up. The canard regains full lift and the nose comes back up. The CG never changed, and full aileron control is available at all times. This virtually eliminates the chance of an inadvertent stall/spin accident. [Figure 1-62]

Figure 1-62. The canard, seen at the nose of this state-of-the-art Beech Starship, was used by the Wright Brothers on their first airplanes.

T-TAILS

Many aircraft today use the T-tail configuration. Although somewhat heavier, this arrangement has several desirable characteristics. The stabilizer is moved away from the disturbed airflow of the wing, rudder effectiveness is improved because of the cap on its end. Many jet aircraft use this configuration because it allows the engines to be mounted on the aft fuselage. Spin recovery may be improved because of better airflow with the stabilizer moved higher and out of the turbulence of the wing and fuselage.

However, T-tail aircraft may also experience a phenomena known as deep-stall. In a deep stall, the airflow over the horizontal tail is blanketed by, or in the shadow of, the disturbed airflow from the wing. Elevator control is ineffective and it may be impossible to recover from the stall.

CONTROL SYSTEMS FOR LARGE AIRCRAFT

As aircraft increase in size and weight, their controls become more difficult to operate and systems must be used to aid the pilot. The power-boosted control system is similar in principle to power steering in an automobile. A hydraulic actuator is in parallel with the mechanical operation of the controls. In addition to moving the control surface, the normal control movement by the pilot also moves a control valve that directs hydraulic fluid to the actuator to help move the surface. A typical boost ratio is about 14, meaning that a stick force of one pound will apply a force of 14 pounds to the control surface.

TYPES OF CONTROL SYSTEMS

The problem with a power-boosted control system is that during transonic flight shock waves form on the control surfaces and cause control surface buffeting. This force is fed back into the control system. To prevent these forces reaching the pilot,

Aircraft Structural Assembly and Rigging

many airplanes that fly in this airspeed region use a power-operated irreversible control system. The flight controls in the cockpit actuate control valves which direct hydraulic fluid to control surface actuators. Since the pilot has no actual feel of the flight loads, some form of artificial feel must be built into the system that will make the control stick force proportional to the flight loads on the control surfaces.

As mentioned earlier, if a mechanical system exists that links the pilot's controls with tabs on the control surfaces so that the pilot can control the airplane in the event of a complete hydraulic failure, the airplane is said to have "manual reversion" capability. If sufficient redundancy is built into the system so that loss of one or more hydraulic systems will not seriously affect the pilot's ability to control the airplane, manual reversion is not necessary.

BOEING 747 CONTROL SYSTEM

The flight control system for the Boeing 747 jet transport aircraft is typical for large aircraft. Rotation about the three axes is controlled by conventional ailerons, rudders and elevators. These primary controls are assisted by spoilers which also double as speed brakes, by an adjustable horizontal stabilizer and by both leading and trailing edge flaps. [Figure 1-63]

The primary control surfaces are moved hydraulically by dual-tandem irreversible actuators. Hydraulic power for each primary flight control is supplied by the four independent hydraulic systems. Manual or electrical inputs from the cockpit direct hydraulic pressure to the control surface actuators. There are no trim tabs installed on any control surface and there is no manual reversion capability such as found in other aircraft like the Boeing 727. Manual reversion is not necessary due to hydraulic redundancies and duplication of control surfaces. Control system feel forces are computed and generated artificially. The flight controls are programmed so that the airplane response to control inputs is the same regardless of speed, center of gravity or gross weight.

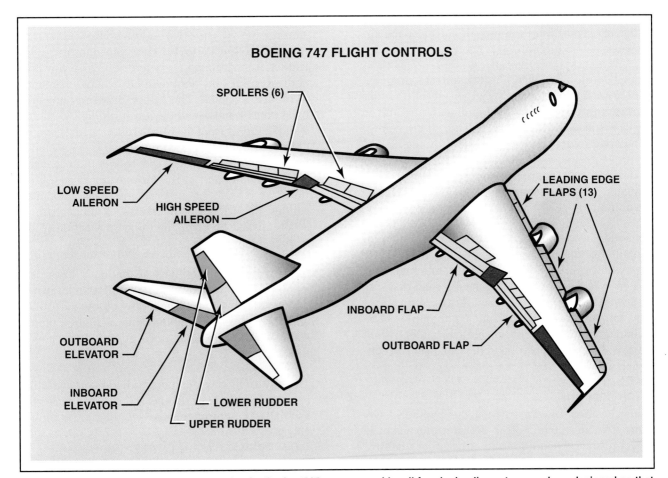

Figure 1-63. The flight controls for each axis of a Boeing 747 are powered by all four hydraulic systems and are designed so that the airplane remains controllable even after failure of any two hydraulic systems.

ROLL CONTROL

The 747 has two sets of ailerons and six spoilers on each wing. During high-speed flight, the inboard ailerons and flight spoilers provide roll control and an outboard aileron lockout system isolates the outboard ailerons from the lateral control system. When the trailing-edge flaps are extended out of the full up (retracted) position, the outboard ailerons are unlocked and both sets of ailerons and the flight spoilers operate together.

The five outboard spoilers on each wing operate in flight. Their movement is proportional to aileron displacement and they decrease lift on the wing the pilot wants to move down. When the speed brake control is actuated in flight, the four inboard sections on each wing are raised, and they produce a great deal of drag. The innermost spoiler on each wing rises only when the speed brake control is at its maximum flight setting. If the control wheel is rotated while the spoilers are deployed as speed brakes, they will move differentially, decreasing the lift on the wing that should go down, and decreasing some of the drag on the wing that should rise. When the airplane lands, the ten flight spoilers and the two ground spoilers automatically extend to the fully raised position to dump lift and slow the airplane in its landing roll.

Both control wheels in the cockpit are interconnected so that movement of either wheel provides an input through a trim, feel and centering unit to two central control actuators. The pilot's control inputs are hydraulically augmented by the central control actuators and transmitted to the aileron programmers and a spoiler mixer. The central control actuators also receive and augment inputs from the autopilot.

The aileron programmers control the movement of the ailerons through individual dual-tandem actuators which are powered by two separate hydraulic systems. The design of the system is such that loss of any two hydraulic systems will cause only one of the four ailerons to become inoperative. The ailerons are trimmed by an electric motor which adjusts the neutral position of the trim, feel and centering unit.

PITCH CONTROL

Four elevators, powered by all four hydraulic systems, provide pitch control. Elevator movement is accomplished by inputs from either pilot's control column through a system of cables and pulleys to the inboard elevator control valves. When the inboard elevators move, they mechanically position the opposite outboard elevator control valves to move the outboard elevator. The inboard elevators are powered by two hydraulic systems and the outboard elevators by a single system. The autopilot controls the elevators through transfer valves which operate a common linkage to both inboard elevator control valves.

Elevator feel is provided by a computer which receives airspeed, hydraulic and stabilizer trim inputs. The computer varies control column pressure relative to airspeed and stabilizer position.

Pitch trim is provided by hydraulically positioning the horizontal stabilizer. Pilot or autopilot inputs move the stabilizer through two independent hydraulic motors and a gearbox which rotates a trim jackscrew to move the leading edge of the stabilizer up or down. Arming and directional valves control the hydraulic motors and these valves are actuated electrically by trim switches on the pilot's control wheels or by the autopilot. Cables from a trim lever in the cockpit provide backup manual control of these valves.

YAW CONTROL

The Boeing 747 has two independent rudders, an upper and lower, to provide yaw control. Each rudder is positioned by a hydraulically powered control unit and is powered by two separate hydraulic systems. The upper rudder is powered by systems 1 and 3, and the lower rudder by systems 2 and 4. Inputs from the pilot's rudder pedals control a trim, feel and centering unit which is connected to two rudder ratio changers. The rudder ratio changers control the amount of rudder movement as a function of airspeed. As airspeed increases, maximum rudder displacement decreases. Rudder trim is accomplished by changing the zero position of the trim, feel and centering unit.

Swept-wing airplanes have a tendency towards a combined rolling and pitching oscillation known as dutch roll. Dutch roll is generally of low magnitude and, while objectionable from a flight comfort standpoint, it is usually not a serious flight condition.

Two full-time yaw dampers provide compensation for dutch roll. Yaw damper computers sense the oscillation and provide signals to the rudder control valves to cancel it.

FLAPS

Each wing of the Boeing 747 is equipped with two triple-slotted Fowler flaps on the trailing edge. The leading edge is equipped with three Krueger flaps on the section between the fuselage and the inboard engines and unique, variable-camber leading edge flaps from the inboard engines out to the wingtips. The trailing edge flaps are actuated by hydraulic motors and have an emergency backup system of electric motors that will raise or lower them.

There are detents in the cockpit flap selector control at 0, 1, 5, 10, 20, 25 and 30. These numbers are references only and do not represent the flap extension in degrees. Gates are located at the Flaps 1 and Flaps 20 positions to remind the pilot to check airspeed before extending or retracting the flaps further. Flaps 10 and 20 are used for takeoff and Flaps 25 and 30 are used for landing.

The Fowler flaps extend out of the trailing edge of the wing to nearly their full chord before they deflect downward an appreciable amount. This provides a large increase in wing area and lift, with a minimum amount of drag, and this position is used for takeoff. When the flaps are extended to the landing position, they deflect fully downward to provide the increased lift and drag needed to slow the airplane down for landing.

Both the Krueger flaps and the variable-camber leading edge flaps alter the shape of the wing's leading edge to provide the added camber needed for high lift at slow speed. Operation of the leading edge flaps is controlled automatically by the position of the trailing edge flaps. The leading edge flaps are normally powered by pneumatic motors with electric motors providing a backup means to extend or retract the flaps. [Figures 1-64, 1-65]

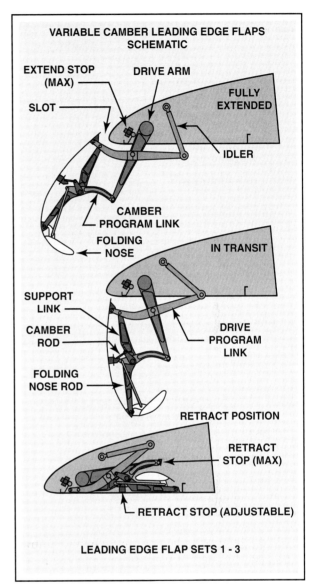

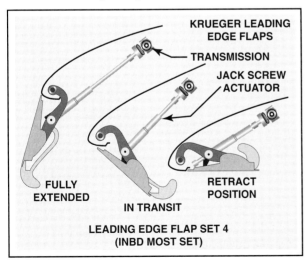

Figure 1-64. A set of Krueger flaps, consisting of a flat center section and a rounded "bull nose" with a pivot point inside the leading edge of the wing allows them to simply and effectively alter the camber of the inboard wing leading edge.

Figure 1-65. The flexible, composite construction of the center section of the variable camber leading edge flaps allows them to retract into the relatively thin wing leading edge, yet assume the most effective aerodynamic shape when extended. These flaps are used from the inboard engines out to the wingtip.

AIRPLANE ASSEMBLY AND RIGGING SPECIFICATIONS

The most important source of rigging specifications is the Type Certificate Data Sheets published by the Federal Aviation Administration for every aircraft that has been certificated. [Figure 1-66]

The data that is approved by the FAA was originated by the manufacturer of the aircraft, and these specifications are included in an expanded form in the manufacturer's service manual. The service manual includes not only this same information, but also details the way to make all of the

310
Revision 61
November 15, 1997

DEPARTMENT OF TRANSPORTATION FEDERAL AVIATION ADMINISTRATION
TYPE CERTIFICATE DATA SHEET NO. 3A10
This data sheet which is part of Type Certificate No. 3A10 prescribes conditions and limitations under which the product for which the type certificate was issued meets the airworthiness requirements of the Federal Aviation Regulations.

Cessna Aircraft Company
P. O. Box 7704
Wichita, Kansas 67277

I - Model 310 (Normal Category), Approved March 22, 1954

Engines
2 Continental O-470-B or O-470-M (installed per Cessna Dwg. 0850000, 0951560, 0851000 and 0851755)

*Fuel
Grade 100 or 100LL aviaiton gasoline

*Engine Limits
For all operations, 2600 r.p.m. (240 hp.)

Propeller and Propeller Limits

```
(a) Hartzell hub HC82XF or HC-A2XF-2                              68 lb. ea. (-25)
    with 8433 blades
    Diameter: not over 84 in., not under 78 in.
    Pitch settings at 30 in. sta.:
    low 12.5°, high 22.0°, feathered 82.0°
(b) Hydraulic governor, Woodward 210105, 210155,                   4 lb. ea. (-17)
    210280, 210444, A210438, 210290 or C210355;
    McCauley DCFU290D1/T2, DCFU290D2/T2
(c) Propeller spinner, Hartzell C-888 dome with                    4 lb. ea. (-23)
    C-807-1 bulkhead or
    Cessna 0752006 dome with 0850300 bulkhead or
    Cessna 0850311 dome with 0850300 bulkhead or
    Cessna 0850313 dome with 0850300 bulkhead
```

*Airspeed Limits (TIAS)

Maneuvering

qt. (12 qt. in each engine at (0), 6 qt. unusable per engine)
See <u>NOTE 1</u> for data on system oil.

Control Surface Movements

```
                                                  Down    45°
Wing flaps
Main surfaces
  Aileron                         Up     20°      Down    20°
  Elevator                        Up     25°      Down    15°
  Rudder                          Right  25°      Left    25°
Tabs (main surface in neutral)
  Aileron                         Up     20°      Down    20°
  Elevator                        Up     20°      Down    28°
  Rudder                          Right  20°      Left    26°
```

Serial Nos. Eligible
35000 through 35546. Delegation Option Manufacturer No. CE-1 authorized to issue airworthiness certificates for S/N 35216 through 35546 and approve repairs and alterations of airplanes S/N 35000 through 35546 under delegation option provisions of <u>Part 21</u> of the Federal Aviation Regulations.

<u>Data Pertinent to All Models</u>

Figure 1-66. This partial example of an Airplane Type Certificate Data Sheet illustrates typical rigging information such as the control surface movement and the leveling means and all of the needed weight and balance information.

adjustments to get the required control surface movement, the correct way to install all of the components, and the proper way to connect and adjust the controls. In short, all of the needed information for assembling and rigging an aircraft is prepared by the manufacturer and is available to the technician.

Aircraft Structural Assembly and Rigging

Normally, rigging and alignment checks should be performed inside a closed hangar. If this cannot be avoided, the aircraft should be positioned with the nose into the wind.

Each of the required inspections for an aircraft — the 100-hour, the annual, the progressive, and the continuous airworthiness inspection program — must include all checks that are needed to determine whether or not all of the rigging specifications published in the Type Certificate Data Sheets are actually met on the aircraft.

AIRPLANE ASSEMBLY

Modern airplanes are almost all of the **full cantilever** type. This type has no external struts or bracing, or are braced with a minimum of struts, generally of a fixed length. Earlier airplanes, especially biplanes, were braced by a maze of wires and struts, and their assembly and rigging was a time-consuming process and required much skill on the part of the technician.

The assembly of a cantilever airplane consists of following the instructions in the manufacturer's maintenance manual in detail. No attempt should ever be made to assemble an aircraft without this vital information. In general, the fuselage is leveled in the manner specified in the maintenance manual, and the wings are installed and the attachment bolts are torqued to specification. The fixed horizontal and vertical tail surfaces are installed and all of their attachment bolts are torqued, again to the specified values. After all of the fixed surfaces are installed, the movable surfaces are installed and the control actuating mechanism is attached and adjusted.

WING ALIGNMENT

Cantilever wings have very little adjustment potential, as this is all taken care of when the airplane is built. Some airplanes have either a cam arrangement or a serrated washer at the rear spar attachment bolt, and a few degrees of **wash-in** may be set in the wing to correct for a wing-heavy flight condition.

Strut-braced wings using **V-struts** normally have provisions for adjusting both the **dihedral angle** and the **incidence angle** of the wings. Install the wing and check the fuselage to be sure that it is level both longitudinally and laterally, and adjust the fittings in the end of the front struts to get the correct dihedral. This is determined by using a dihedral board that has a specific taper. It is held against the main spar on the bottom of the wing at the location specified by the manufacturer, and the fitting in the end of the strut is screwed either in or out until the bottom of the dihedral board is level. Some aircraft,

rather than measuring the dihedral with a dihedral board, use a string stretched between the wing tips at the front spar. When the dihedral is correctly adjusted, there will be a specific distance between the wing root fitting and the string. [Figure 1-67]

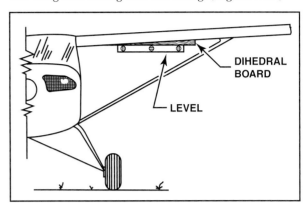

Figure 1-67. When the bottom of the board is level, the wing has the proper dihedral. Before this check can be made, the aircraft must be leveled according to specifications.

When the dihedral is correctly adjusted, the wash-in or wash-out may be set. This is normally done by adjusting the length of the rear strut. An incidence board similar to a dihedral board is held under a specified wing rib, and the strut length is adjusted until the bottom of the board is level. On airplanes having this adjustment, the initial setting will likely have to be changed after the first flight to trim the airplane for straight and level hands-off flight. Increasing the **angle of incidence**, that angle between the chord line of the wing and the longitudinal axis of the airplane, is called **"washing the wing in"** and it increases the lift. **Washing out** a wing is done by rigging it with a lower angle of incidence to decrease its lift. [Figure 1-68]

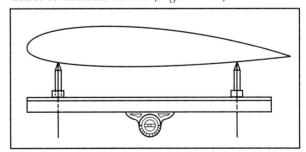

Figure 1-68. An incidence board is used to check for wing warpage and for the proper wash-in or wash-out.

AILERON INSTALLATION

After the wing is installed and aligned, and all of the attachment bolts torqued and safetied, the ailerons may be installed and rigged. It is important that the hinges be shimmed in exact accordance with the manufacturer's specifications, and that the control rods or cable be attached to the horns with the proper type of bolt. The attachment must be free to pivot, yet have no excessive looseness. The cable tension must be adjusted, and the stops checked to be sure that the travel is that specified in the Type Certificate Data Sheet. Level the aileron and install a protractor such as a universal propeller protractor or a special control surface protractor. Zero the protractor when the aileron is in the exact trail position, and then deflect it upward until it contacts its stop and measure its travel. Next, deflect it downward until it contacts the stop, and measure its travel. The travel must be within the specified range. If it is not, the stops must be adjusted. When the stops have been reset, move the ailerons through their full travel from the control wheel to be sure they travel through the entire range and that they are stopped by the stops in the wing and not by those at the control wheel. Check to be extremely sure that the control wheel moves in the proper direction for the aileron action. Serious accidents have been caused by ailerons being rigged backward. [Figure 1-69]

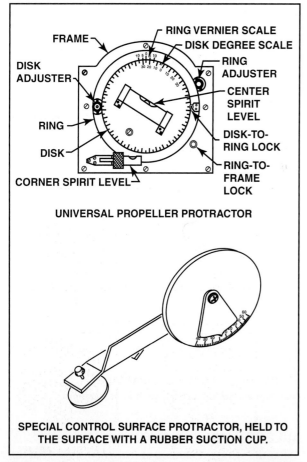

Figure 1-69. A universal propeller protractor or a special control surface protractor are common tools used to set the correct control surface travel. In all cases, follow the manufacturer's recommendations for both the tools and the procedures to be used for any rigging operation.

Some ailerons are required to have a few degrees of droop. This means that when there is no airload they should both be a few degrees below the trailing edge of the wing. If the ailerons you are rigging have this requirement, be sure that they are properly drooped. The final check in rigging the ailerons is to be sure that all of the turnbuckles and any bolt or connector in the entire system are properly adjusted and safetied. Leave nothing to chance. Start at the control wheel, and systematically check every connection in the system to the aileron, through the balance cable to the other aileron and back to the control wheel.

FLAP INSTALLATION

The flaps are connected to their hinges and actuator rods in a manner similar to that of the ailerons. However, Fowler flaps are normally mounted on rollers that ride in tracks, and these must be adjusted so they ride up and down smoothly with no binding or interference.

There are a number of actuation methods for wing flaps. The simplest flaps are actuated by either cables or a torque tube directly from a hand lever in the cockpit. Other airplanes use electric motors to drive jackscrews that move the flaps up or down, and many of the larger aircraft use hydraulic actuators to provide the muscle to move the flaps against the air loads.

In a single-engine airplane flap system, the flaps are moved by cables from an electric motor-driven jackscrew. Limit switches shut the motor off at the full up and down position, and a cam-operated follow-up system allows the pilot to select various intermediate flap positions. When the flaps reach the selected deflection, the motor will stop. [Figure 1-70]

EMPENNAGE INSTALLATION

The tail surfaces on almost all modern airplanes are of the cantilever type and are bolted to fittings in the fuselage. Special care must be exercised that all the bolts used have the proper part number and that all of them are tightened in the proper sequence and to the correct torque.

Some lighter aircraft have wire-braced tail surfaces. When securing the horizontal stabilizer with streamlined wires, those having a flattened oval cross section, it is important that the wires be adjusted to the proper tension and that they be streamlined in the line of flight to minimize vibration.

Aircraft Structural Assembly and Rigging

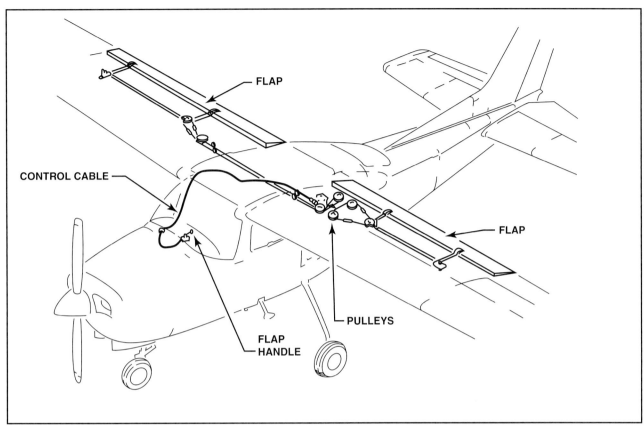

Figure 1-70. A typical small aircraft flap actuation system uses a series of cables, bell cranks and push rods and may be actuated manually by a hand lever in the cockpit or by an electric motor.

CABLE SYSTEM

Aircraft control cable is available in both corrosion-resistant steel and carbon steel. The corrosion-resistant steel is somewhat more expensive and has a slightly lower strength, but its longer life makes it the better of the two cables for use where corrosion may be a problem, such as in agricultural aircraft and seaplanes.

CABLE CONSTRUCTION

There are three types of steel cable used for aircraft control systems: nonflexible, flexible, and extra-flexible. **Nonflexible cable** may be of either the 1 × 7 or 1 × 19 type. This designation means that the 1 × 7 cable is made up of seven strands, each having only one wire. The 1 × 19 cable is made of 19 strands of one wire each. Nonflexible cable may be used only for straight runs where the cable does not pass over any pulleys. **Flexible cable** is made up of seven strands, each of which has seven wires. Flexible cable may be used only for straight runs or where the pulleys are large. When cables must change direction over relatively small diameter pulleys, **extra-flexible cable** must be used. This type of cable is made up of seven strands, each having 19 separate wires. All aircraft control cable is pre-formed, which means that the wires were shaped in their spiral form before the cable was wound, and they will not spring out when the cable is cut. [Figures 1-71, 1-72]

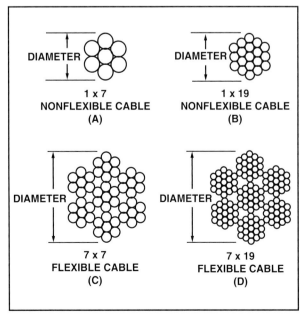

Figure 1-71. This cross-sectional drawing illustrates the construction of various types of aircraft control cable.

	1 x 7 and 1 x 19				7 x 7, 7 x 19, and 6 x 19 (1 WRC)			
Diameter (inch)	Nonflexible, corbon		Corrosion, resisting		Flexible, corbon		Flexible, corrosion resisting	
	MIl -W-6940		MIL-C-5693		MIL-W-1511		MIL-C-5424	
	Weight, pounds per 100 feet	Breaking strength, pounds	Weight, pounds per 100 feet	Breaking strength, pounds	Weight, pounds per 100 feet	Breaking strength, pounds	Weight, pounds per 100 feet	Breaking strength, pounds
1/32	0.25	185	0.25	150	....	....	....	....
3/64	.55	375	.55	375	.....	.....	.....	.....
1/16	.85	500	.85	500	0.75	480	0.75	480
5/64	1.40	800	1.40	800	....	....	....	....
3/32	2.00	1,200	2.00	1,200	1.60	920	1.60	920
7/64	2.70	1,600	2.70	1,600	....	....	....	....
1/8	3.50	2,100	3.50	2,100	2.90	2,000	2.90	1,760
5/32	5.50	3,300	5.50	3,300	4.50	2,800	4.50	2,400
3/16	7.70	4,700	7.70	4,700	6.50	4.200	6.50	3,700
7/32	10.20	6,300	10.20	6,300	8.60	5,600	8.60	5,000
1/4	13.50	8,200	13.50	8,200	11.00	7,000	11.00	6,400
9/32	.....	....	.....	....	13.90	8,000	13.90	7,800
5/16	21.00	12,500	21.00	12,500	17.30	9,800	17.30	9,000
11/32	....	....	....	....	20.70	12,500	....	....
3/8	....	....	....	....	24.30	14,400	24.30	12,000
7/16	....	....	....	....	35.60	17,600	35.60	16,300
1.2	....	....	....	....	45.80	22,800	45.80	22,800

* The strength values listed were obtained from straight tension tests and do not include the effects of wrapped ends.

Figure 1-72. The breaking strength and weight per 100 feet of various sizes and types of steel aircraft control cables can be determined from this table.

TERMINATION

Woven Splice

Control cables were originally terminated with a hand-woven splice, using either the Army-Navy five-tuck splice or the Roebling roll. Both of these systems were time consuming and produced a termination that was certificated for only 75% of the cable strength. [Figure 1-73]

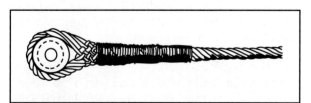

Figure 1-73. Typical woven splice is seldom used because of its low strength and high cost of production.

Nicopress Process

Copper Nicopress sleeves may be compressed onto the cable with a special tool. The cable is put through the sleeve and around the thimble, and the end is passed back through the sleeve and it is compressed. The center compression is made first, then the compression nearest the thimble, and finally the compression at the end of the sleeve. After all three compressions are made they are checked with a special gauge. If they are properly made, the sleeve will exactly fit the slot. [Figures 1-74, 1-75]

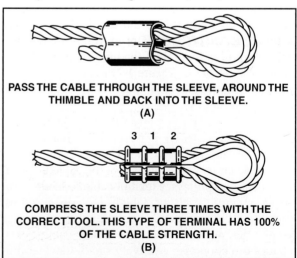

Figure 1-74. Nicopress sleeves provide a simple and strong method of terminating cables, provided the proper tools and procedures are used.

Aircraft Structural Assembly and Rigging

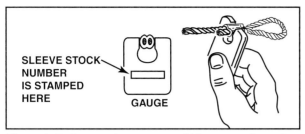

Figure 1-75. A properly compressed Nicopress sleeve has the dimension of the slot in the proper gauge.

Running splices may be made to join two lengths of cable using two Nicopress sleeves. When two sleeves are used, the splice is approved as having the full cable strength. [Figure 1-76]

Figure 1-76. A splice using two Nicopress sleeves has the full strength of the cable.

A safety feature to aid during an inspection for Nicopress sleeves as well as for swaged terminals is a spot of paint across the cable and the end of the sleeve. If this paint is ever broken, there is reason to suspect that the cable has slipped.

Swaged Terminals

The vast majority of aircraft control cables are terminated with swaged terminals. To install one of these terminals, cut the cable to the proper length using a cable cutter or a chisel rather than cutting it with a torch. Bend the end of the cable so it will not slide out of the terminal during the swaging operation, and then slip it into the terminal until it bottoms at the end of the hole, or until it is almost even with the end of the hole if the terminal is drilled all of the way through. [Figure 1-77]

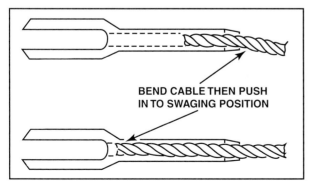

Figure 1-77. When installing a swaged-on terminal, first bend the cable to prevent it slipping out, and then push it in, almost to the end of the hole.

Use the proper swaging tool to compress the sleeve into the cable. A special "before and after" gauge is used to determine that the terminal is properly compressed. The sleeve or ball should fit into the "before" portion of the gauge before it is swaged, and after it is swaged, it must slip through the "after" slot. When the swaging is complete, mark the end of the terminal and the cable with a dot of paint to show if the cable ever slips. [Figure 1-78]

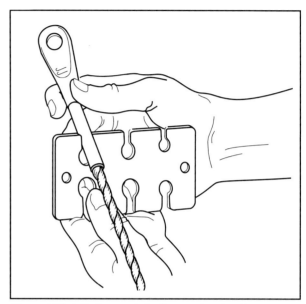

Figure 1-78. A "before and after" gauge for checking the correctness of the swaging of a swaged-on terminal.

There are a number of different swage-type terminals available, the most popular being the fork end, the clevis end, the turnbuckle ends, and balls with either a single or a double shank. [Figure 1-79]

Proof Load Test

After the cable is completed with the terminals properly installed, checked, and marked with paint, it should be tested by loading it to 60% of its breaking strength, with the load applied gradually and held for at least three minutes.

CABLE INSPECTION

Since the cables are such a vital part of an aircraft control system, they should be carefully examined at each required inspection. Cables usually wear or break where they pass over pulleys or through fairleads. Wear and corrosion are the two most common problems with of control cables. To inspect for cable breaks, hold a rag in your hand and wipe every portion of the cable within reach. The rag snags on any broken strands that stick out of the cable. Move the controls and check the cable as near the pulleys as you can reach. If there is any possibility of corro-

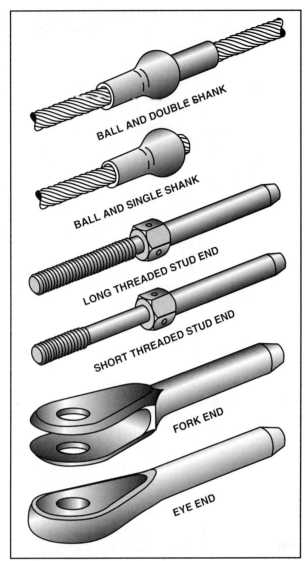

Figure 1-79. Various types of swaged-on control cable terminals enable the designer to use the termination method most suitable for a specific application

sion, disconnect the cable and bend it into a loop to make any broken strands pop out. [Figure 1-80]

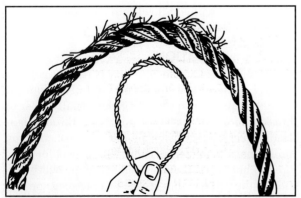

Figure 1-80. When inspecting a steel cable for broken strands, bend it into a loop and check for wire ends that pop out of the cable.

Both seaplanes and agricultural aircraft operate in an environment that is highly conducive to corrosion in the control cables, and both of these types of aircraft normally use corrosion-resisting steel cable. However, it is still a good idea to wipe these cables with Par-al-ketone, a waxy grease to protect the cable from moisture and corrosion.

INSTALLATION

PULLEYS AND FAIRLEADS

Aircraft manufacturers route the control cables in the most direct manner possible, and at each point where a change in direction is required, they use a pulley. At any point within a run of cable where there is the likelihood of the cable contacting the structure, it is run through a fairlead made of some form of relatively soft plastic or fiber. The cable may touch or rub on a fairlead, but fairleads must never be used to change the direction of the cable. [Figure 1-81]

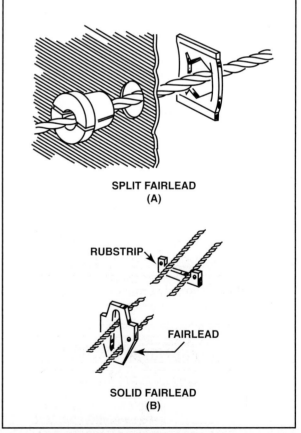

Figure 1-81. A fairlead prevents contact between a control cable and the aircraft structure to prevent wear on the cable and damage to the structure.

Pressure seals are used at each location at which a control cable penetrates the structure of a pressure vessel. These seals are filled with grease, which allows the cable to pass freely, yet prevent the leakage of the pressurization air. [Figure 1-82]

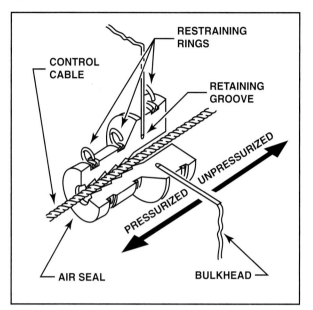

Figure 1-82. A pressure seal is used whenever a moving control cable must pass through the aircraft structure between a pressurized and an unpressurized area.

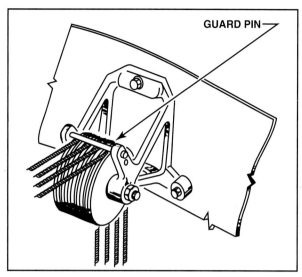

Figure 1-83. In a typical control cable installation, guard pins are used to prevent the cable from jumping off the pulley and becoming jammed.

Control pulleys must be carefully aligned with the cable so it rides squarely in the center of the groove, and all of the pulleys must be free to rotate through their full travel. All pulleys must be equipped with guards to prevent the cable from jumping out of the groove when the cable is slack. It is good practice at each inspection to rotate any pulley that does not turn through its full rotation with normal cable travel so the cable will contact a different portion of the pulley. This will even out the wear. [Figures 1-83, 1-84]

TRAVEL ADJUSTMENT AND CABLE TENSION

The Type Certificate Data (TCD) sheets specify the required control surface deflection, while the aircraft maintenance manual describes the procedure to set this required deflection. When rigging the controls, be sure to follow the manufacturer's instructions explicitly. When cables are rigged excessively tight, it creates a great deal of strain on the system by prematurely wearing pulleys and making it difficult to move the flight controls.

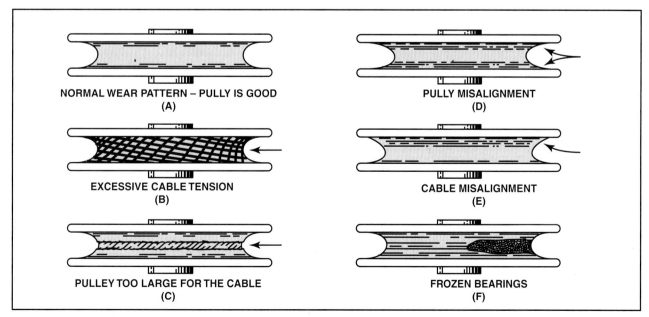

Figure 1-84. Maintenance technicians can detect a variety of control cable adjustments problems by observing control cable wear patterns on the pulleys in the system.

Most of the larger aircraft use rigging pins to lock the control system in place when rigging the controls. The control actuation system is locked in place with the pins, and the cables are adjusted to get the correct position of the surfaces. The cable tension is then adjusted to the manufacturer's specification and checked with a tensiometer. These instruments have the riser installed that is specified for the cable size being checked, and the cable is slipped between the riser and the two anvils, and the trigger is clamped against the housing. The pointer indicates the cable tension in pounds. [Figure 1-85]

Large aircraft whose dimensions change appreciably with temperature must have the tension adjusted for a given ambient temperature. To adjust the tension on a 1/8 inch extra-flexible (7 × 19) cable at an aircraft temperature of 85 degrees Fahrenheit, adjust the tension to 70 pounds. [Figure 1-86]

The maintenance manuals for many smaller aircraft specify that the cable tension should be adjusted at the "average temperature at which the aircraft is operated." Some of the larger aircraft whose dimensions change rather drastically as they fly at very high altitudes and low temperatures have automatic tension adjusters to maintain a relatively constant cable tension as the temperature of the structure changes.

SPRINGBACK

All of the controls have a stop located at the surface itself, and many systems have a secondary stop on

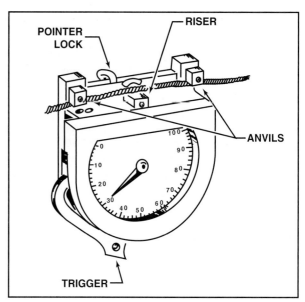

Figure 1-85. Verify that the correct riser is installed before checking cable tension with a control cable tensiometer. Failure to do so may result in an incorrectly tensioned cable.

the actuating control. It is important that the stop at the surface be contacted first and there be some springback in the system before the actuating control reaches its stop. When you feel this slight springback, you can be sure that the surface has reached its full travel.

TURNBUCKLE SAFETYING

After the cable tension has been adjusted with the turnbuckles, they must all be checked for the

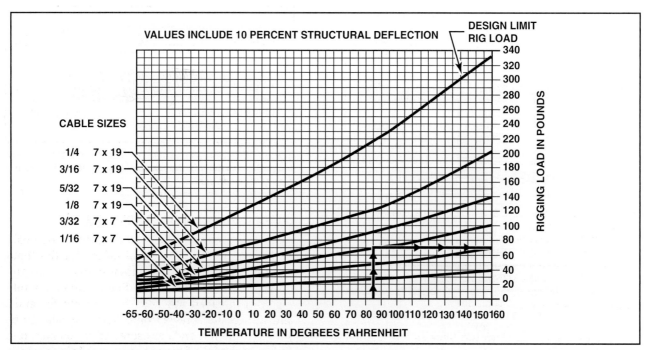

Figure 1-86. A control cable tension chart allows the technician to determine the correct tension for various cable sizes based on the ambient temperature.

proper amount of thread showing and then safetied.

In order for a turnbuckle to develop its full strength, there must be no more than three threads exposed at either end of the turnbuckle barrel, and the turnbuckles should never be lubricated. Turnbuckle barrels are available in two lengths, so if the cable tension cannot be adjusted without there being an excessive number of threads exposed, a long barrel may be used in the system. The use of a long barrel will probably require several of the turnbuckles in the system to be readjusted.

Most of the primary control cables are at least 1/8-inch diameter and are safetied with either 0.040 stainless steel wire, single wrapped, or 0.040 brass wire, double wrapped. In both the single- and double-wrap methods of safetying, you have the option of straight or spiral wiring. In either method, the wiring must be terminated with at least four wraps around the shank of the turnbuckle. [Figures 1-87, 1-88]

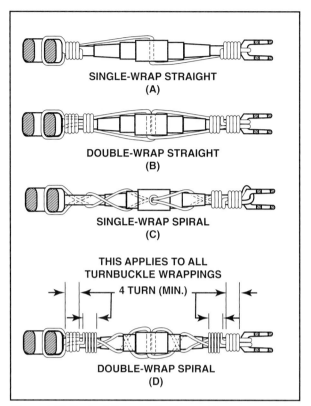

Figure 1-88. Each of these methods provides an acceptable means of safety wiring turnbuckles.

Cable Size	Type of Wrap	Diameter of Safety Wire	Materials (annealed condition)
1/16	Single	0.040	Copper, brass[1]
3/32	Single	0.040	Copper, brass[1]
1/8	Single	0.040	Stainless steel, Monel and "K" Monel
1/8	Double	0.040	Copper, brass[1]
1/8	Single	0.057 min.	Copper, brass[1]
5/32 and greater	Double	0.040	Stainless steel, Monel and "K" Monel[1]
5/32 and greater	Single	0.057 min.	Stainless steel, Monel and "K" Monel[1]
5/32 and greater	Double	0.051[2]	Copper, brass[1]

[1] Galvanized or tinned steel, or soft iron wires are also acceptable.

[2] The safety wire holes in 5/32-inch diameter and larger turnbuckle terminals for swaging may be drilled sufficiently to accommodate the double 0.051-inch diameter copper or brass wire when used.

Figure 1-87. Turnbuckle safety wire requirements are based on the size of the cable, the type of wrap and the material used for the safety wire.

Many of the more modern turnbuckles are safetied with clip-type locking devices such as we see in figure 1-89. When the tension is properly adjusted, the two-piece clips are inserted into a groove in the turnbuckle body and the terminal end on the cable, and then the two pieces are clipped together through the hole in the barrel. [Figure 1-89]

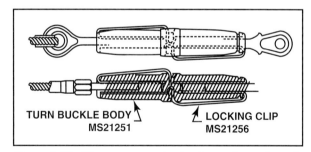

Figure 1-89. Clips are a much simpler and equally safe method for safetying turnbuckles, and make the technician's job much easier.

PUSH-PULL ROD SYSTEM

Many airplanes and almost all helicopters use push-pull rods rather than control cables for the flight control systems. The tubes themselves are usually made of seamless heat-treated aluminum alloy tubing with threaded rod ends riveted into the ends, and fittings are screwed onto these threads. To be sure that the rod ends are screwed far enough into fitting, each of the fittings has a small hole drilled in it. If you can pass a piece of safety wire through the hole, the rod end is not screwed in far enough. A check nut is screwed onto the rod end, and when the length of the rod is adjusted, the check nut is screwed up tight against the end of the fitting to hold it in place. [Figures 1-90, 1-91]

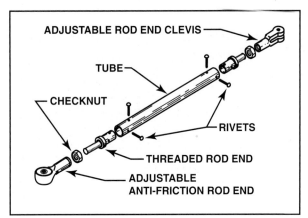

Figure 1-90. Push-pull control rods are frequently used to connect cable-driven bell cranks to the adjacent control surface and may provide a simple means of adjusting the control surface travel. This is a typical push-pull rod using a clevis rod end and a rod-end bearing.

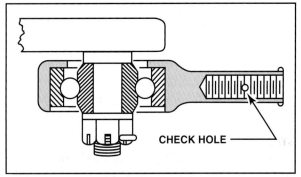

Figure 1-91. The rod must be screwed into the rod end bearing until it covers the check hole.

Helicopters use so many push-pull rods in their control systems that only a small amount of wear in each of the fittings will be amplified and can cause serious control vibration. It is important when checking the rigging of a helicopter that there be no slack in the control rods.

The anti-friction rod-end bearings are usually of the self-aligning type, but when you install one, you should be sure that it is squarely in the center of the groove in which it fits so there will be no tendency for the bearing to contact the housing and rub at the extremes of its travel.

Bell cranks are used extensively in push-pull tubing systems to change direction of travel and to gain or decrease the mechanical advantage of the control movement. [Figure 1-92]

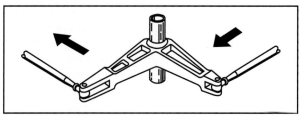

Figure 1-92. Bell cranks are a common component in push-pull control systems.

TORQUE TUBE SYSTEM

Some flaps and ailerons are actuated from the cockpit by means of torque tubes. Tubes made of heat-treated aluminum alloy are rotated by the cockpit control to lower or raise the control surfaces. This type of actuation is the most direct and has the least amount of backlash, but it can be used only on installations where the air loads are not excessive.

Control Surface Balancing

As previously discussed, control surfaces must remain in a balanced condition to prevent aerodynamic flutter. In most situations, the airframe manufacturer provides detailed information for the balancing procedure, and may also offer special balancing tools and equipment. The following provides one method of control surface balancing, but each manufacturer prescribes specific procedures to follow. [Figure 1-92a]

Figure 1-92a. Tools that are typically used for checking the balance of control surfaces include mandrels, a balance beam, scales, spirit level, and measuring equipment. Some manufacturers also provide technical specifications that allow a technician to fabricate any required special tools.

Proper balance must be maintained on any flight control that deflects upward and downward, such as an elevator or aileron, and on surfaces that move left to right, such as a rudder or similarly designed component. Balance is not, however, required on structures such as flaps, spoilers, slats, etc. where the surface is movable to a fixed alternate position.

Typical checks involve verifying that the static weight condition of the component balances within a specified range. If the control surface has either too much weight ahead of or behind the hinge line, the control surface may flutter during flight. In extreme cases, flutter causes oscillations that progressively increase in amplitude, which ultimately can cause the pilot to lose control of the aircraft, or could cause a catastrophic structural failure.

The balance check is done with the control surface removed from the aircraft and assembled as specified in the manufacturer's instructions. Generally, all mounting hardware and components contained within the control surface should be installed in their relative positions. Trim tab assemblies, for example, are usually installed along with mounting hardware and actuating assemblies such as torque tubes or chain sprockets.

Once assembled, the control surface is placed on balancing mandrels within a draft-free room. The mandrels are positioned under the hinge points in a way that allows the control surface to pivot freely without dragging or binding. A balance beam or jig is then installed to determine the moment arm required to balance the control surface.

As seen in Figure 1-92a, the beam consists of a graduated measuring scale in which a weight can be moved to bring the control surface to a horizontally level position or other position specified by the manufacturer. Once in the proper position, the distance that the weight is located ahead of or behind the hinge line is recorded and checked against the manufacturer's specifications. If the moment arm (weight x distance) exceeds the manufacturer's specifications, weight is added or removed from the control surface so that the balance beam weight falls within the manufacturer's specified range.

The manufacturer usually provides an area on the control surface specifically designed to hold balance weight, and prescribes actions to take to correct an out-of-balance condition. For example, some manufacturers require the application of solder or the installation or removal of lead weight to a part of the control surface. Regardless of what technique is used, the technician must make sure any weight that is applied will not become lose or fall off during flight.

BIPLANE ASSEMBLY AND RIGGING

At one time in the history of aviation, a battle raged between biplanes and monoplanes, but it is now obvious which type won out, and it is quite likely that a person can spend their entire career as an aviation maintenance technician and never touch a real live biplane. But it is well to spend a few moments reviewing some of the nomenclature that is unique to this special configuration of airplane. [Figure 1-93]

Figure 1-93. Many home-builders choose to build biplanes for their nostalgia factor. These aircraft are often used for aerobatics and, are capable of withstanding higher "G" loads.

One of the main advantages of a biplane over a monoplane in the early days of aviation was the ease with which a biplane could be braced; thus it could have the required wing area and strength without an excessively heavy internal structure. Most biplanes had struts between the wings and streamlined wires that joined the wings in a strong truss.

BIPLANE TERMINOLOGY

Before we look at the components of a biplane, let's review a few terms that are unique to biplanes:

STAGGER

Normally the upper wing of a biplane is not directly above the lower wing. When the upper wing is ahead of the lower, the airplane is said to have **positive stagger**. If the lower wing is ahead, as was true of the classic "Staggerwing" Beechcraft, it is said to have **negative stagger**.

DECALAGE

The difference in the angle of incidence of the two wings of a biplane is called the decalage. It is positive if the upper wing has the greater angle of incidence and negative if the angle of incidence of the lower wing is greater.

BIPLANE COMPONENTS

CENTER SECTION

If the upper wing is mounted above the fuselage, a center section is normally used, to which the upper wing panels are attached. By using a center section which is rigged true to the fuselage, the upper wing panels may be removed and replaced without disturbing the stagger or decalage.

CABANE STRUTS

The struts between the center section and the fuselage are called the cabane struts. These may be either straight struts, or struts in the form of a V or an N.

CABANE OR STAGGER WIRES

Streamlined wire is used for the stagger wires, those that run between the rear and forward cabane struts to adjust the fore-and-aft position of the center section. By adjusting the stagger wires, you may adjust the stagger and the symmetry of the center section. Additionally, the upper wing can be rigged square with the fuselage using these wires. This is why stagger should be checked at each outer strut, to be sure the assembly is square. Sometimes stagger wires are also called drag and anti-drag wires, because they also function in that capacity.

ROLL OR TRANSVERSE WIRES

Roll or transverse wires are the streamlined wires that tie the cabane struts and the center section to the fuselage laterally. By changing the length of these wires, you adjust the center section laterally above the fuselage.

LANDING WIRES

Landing wires are heavy streamlined wires that support the lower wing from the center section, bracing the wings against loads imposed by landing. More importantly, they set the dihedral for the lower wings and, through the interplane struts, the upper wings as well.

FLYING WIRES

The heavy streamlined wires that attach to the fuselage at the lower wing root and to the outboard portion of the upper wing are the flying wires, opposing the loads of flight. Additionally, they provide the cross bracing through the interplane struts to form a closed bay for the wing structures. Normally both the flying and landing wires are double.

INTERPLANE STRUTS

Interplane struts, usually made of streamlined steel tubing in the form of the letter N, are placed between the upper and lower wing out near the tip. The flying and landing wires attach to the wing spar at the interplane struts and together form the rigid truss structure of the wing. On some installations, the rear strut is adjustable in length and provides a way to change wash-in and wash-out, allowing some correction for wing heaviness. Some airplanes use a single-piece streamlined I-strut rather than the more conventional N-struts.

ASSEMBLY AND RIGGING PROCEDURE

No one should attempt to assemble and rig a biplane without a complete set of specifications and the proper tools, but the procedure for this is fairly straightforward. Level the fuselage both laterally and longitudinally, as all of the measurements are made from the longitudinal and lateral axes of the fuselage. For obvious reasons, the landing gear should be adjusted as closely as possible before hoisting the airplane.

For safety reasons, use chocks, blocking and jacking procedures that are safe and solid, and exercise extreme care. To rig a biplane, you must climb on it while it is "up in the air."

1. Install the center section and adjust the stagger wires and transverse wires to get the center section perfectly square above the fuselage and positioned with the proper amount of stagger.

2. Install the lower wing panels and support them with the landing wires.

3. Adjust the length of the landing wires to get a rough dihedral angle.

4. Install the interplane struts and then the upper wing panels.

5. Check the wing stagger using plumb bobs. Measure from the upper wing leading edge to the lower wing leading edge at the interplane strut point.

6. Install the flying wires and adjust all of the wires to the proper tension, then recheck the dihedral angle. Adjust as required.

7. The wash-in or wash-out required for the wing is adjusted by changing the length of the rear interplane strut. Some airplanes have a special leveling fixture for this purpose.

8. When all of the flying and landing wires are adjusted and safetied, a wooden vibration damper, called a bird or javelin, is normally required to be fastened at the intersection of the wires.

9. Install the tail surfaces and rig all of the movable control surfaces, adjusting their travel and control cable tension as you do for a monoplane.

10. Be sure to check all wire terminal check nuts and clevis pin cotters. Each terminal should be checked for adequate thread length insertion with a piece of safety wire inserted in the terminal safety hole. This is a go/no-go check. If the wire goes in the safety hole, the terminal is not safe.

LARGE AIRCRAFT ASSEMBLY AND RIGGING

There are many reasons for major disassembly of larger aircraft that will subsequently require assembly and rigging of the flight control systems. Among these are repair of structural damage incurred during operations, aging aircraft inspection and repair programs, and removal of components for time or cycle limited inspections, overhaul or replacement.

TYPICAL REPAIR OPERATIONS

The inboard flap of a transport aircraft frequently suffers major damage when a tire blows on takeoff or landing. Airlines generally find it more cost-effective to remove the damaged flap, install a new or overhauled unit, and send the damaged unit to a structural repair shop. There the flap is inspected and repaired and returned to inventory until needed on another airplane.

When a jet transport takes off, is pressurized and de-pressurized, and subsequently lands, it completes a "cycle." Each time the pressure vessel is cycled, it expands and contracts, and after many thousands of cycles, metal fatigue may occur. The wing structure flexes and bends repeatedly under normal flight loads, and these structures are also subject to metal fatigue. Fatigue generally begins as tiny cracks around rivet holes, at corners or bends or at openings such as windows or doors. If not found and repaired, a disastrous failure may occur.

Corrosion can further weaken the structure. The air surrounding a jet transport in cruising flight is generally between 30 and 60 degrees below zero. The outer surfaces of the airplane become very cold and the fuselage is well insulated to keep the interior at a comfortable temperature. People exhale a great deal of moisture when they breathe. Some of this moisture condenses on the cold skin and is trapped between the skin and the insulation blankets. Over time, corrosion can form where the wet insulation contacts the metal structure. Corrosion also forms at other locations where liquids can be trapped by the structure. The belly of the aircraft and the areas beneath the galleys and lavatories are common problem locations.

The aging-aircraft program requires each transport aircraft to undergo a series of tests and inspections after it has been in service for a specified number of years and/or has accumulated a certain number of flight cycles. These inspections are generally a combination of visual and non-destructive testing methods such as ultrasound, eddy current and x-rays. Corrosion control and structural repairs are made when necessary. Since major disassembly of the aircraft is required to complete the required inspections, rigging of the control systems will be necessary.

Some manufacturers may require major structural disassembly after a certain number of flight hours. For example, some Lear Jet models require removal of the wing from the fuselage after approximately 5,000 flight hours for a series of inspections. The manufacture of an aircraft may specify this type of disassembly and inspection after a period of months or years, after a certain number of flight hours or after a specific number of cycles.

Air carrier aircraft undergo a series of routine maintenance inspections called "letter checks." A, B, C and D checks are performed at various intervals. A D Check is the most extensive and involved operation and amounts to a complete disassembly and overhaul of the airplane. Among other things, each flight control surface will be removed, inspected, repaired if necessary and re-installed.

REMOVAL AND INSTALLATION REQUIREMENTS

Due to their large size and weight, the control surfaces of transport aircraft and many business aircraft require special tools, handling fixtures and procedures for their removal and installation. Repair of control surfaces is critical because they must be as carefully balanced after repair as they were when new. Special jigs, fixtures and tooling may be required to ensure that the repaired surface conforms exactly to its original design. Because of their critical requirements and the special nature of the repair operations, these procedures are usually performed only by the manufacturer or specially designated repair stations.

Each aircraft manufacturer usually publishes a structural repair manual in addition to the general maintenance manual. A structural repair manual provides the technician with the detailed instructions, procedures and methods used to safely accomplish repairs to the aircraft structure. In addition, transport aircraft manufacturers usually provide "job cards" for most maintenance and repair operations. Job cards contain the detailed step-by-step instructions for performing specific tasks and provide places for the technician and inspector to sign off the work.

As for any other aircraft, the Type Certificate Data Sheets for a large aircraft will contain the rigging specifications. However, you must refer to the procedures specified in the manufacturers' approved manuals for the details of each assembly and rigging task. This will be a team effort and many technicians and inspectors will be involved. Be certain that you always use the required tools and procedures. In the past, accidents have resulted when technicians tried to develop their own "better, faster" methods.

SECTION C

FUNDAMENTALS OF ROTARY-WING AIRCRAFT

HISTORY OF ROTARY-WING FLIGHT

Long before fixed-wing flight became a reality, the concept of rotary-wing aerodynamics was known. Records indicate that winged toys resembling helicopter rotors were made in the 14th century, and Leonardo da Vinci (1452–1519) proposed the helix. The helix was a spiral-shaped airfoil design that resembled the threads of a screw.

History is replete with attempts to make rotary-wing machines fly. Men such as Ellehammer, Breguet, Cornu, and DeBothezat all contributed to the birth of the helicopter, though their efforts were unsuccessful. Problems with both power and control plagued rotorcraft developers until 1923, when the Spanish engineer Juan de la Cierva solved some control problems. Rather than driving the rotor with power, he used the aerodynamic principle of autorotation to spin his four-bladed rotor to lift his **gyroplane** into the air. Gyroplanes are quite inefficient when compared with airplanes, but they have some desirable features, one of which is slow flight without danger of stalling. The gyroplane's greatest contribution to aviation has been in its serving as a test bed for rotor systems, breaking some of the developmental bottlenecks, and allowing the practical **helicopter** to be born.

The Focke-Achgelis FW-61 flew in Germany in 1936, and test pilot Hanna Reitsch established records for both duration and altitude. But this helicopter was too heavy to carry any payload. In 1939, Igor Sikorsky flew the VS-300, the first helicopter considered truly successful. This machine was in the direct lineage of the first U.S. military helicopter, the R-4, and of all of the helicopters that bear the famous Sikorsky name.

In Korea and Vietnam, the helicopter was given its acid test, and because of these difficult assignments, it grew to be the truly unique machine it is today. As designers continue to develop new materials and manufacturing techniques, the helicopter will continue to grow in popularity as the workhorse of the aviation industry.

CONFIGURATIONS OF ROTARY-WING AIRCRAFT

GYROPLANE

The machine in which Cierva opened the door to successful rotary-wing flight was not a helicopter, it was a **gyroplane**. A gyroplane has no power to the main rotor except, in some cases, to start it spinning. It is turned in flight by an aerodynamic force called the **autorotational force**, which is discussed later.

Many attempts have been made to build a successful gyroplane and a number of them have been on the market. Both the U.S. Army and the U.S. Marines have used gyroplanes, and Eastern Airlines at one time flew the mail from the airport in Philadelphia to the downtown post office by gyroplane.

A gyroplane has many of the problems associated with rotary-wing flight, but since it cannot rise vertically nor hover in still air, its development has been almost brought to a halt in favor of the helicopter. [Figure 1-94]

Figure 1-94. A gyroplane is supported by a rotating rotor that is driven by aerodynamic forces rather than by power from its engine.

SINGLE-ROTOR HELICOPTER

The most popular design, which has become almost universally adopted, has a single main rotor that provides lift and thrust and a small vertical rotor on the tail that compensates for main rotor torque. When the engine spins the rotor, according to Newton's third law, an equal and opposite force tends to rotate the fuselage in the opposite direction. The main transmission drives the **anti-torque rotor**, and the pilot controls its pitch with foot pedals to vary the amount of torque correction. [Figure 1-95]

Figure 1-95. The torque of the single main rotor is countered by thrust from the tail rotor.

DUAL-ROTOR HELICOPTER

The effect of the torque created by the main rotor system has always plagued helicopter engineers. One solution uses two rotors turning in opposite directions. Though each rotor system still produces torque, the torque cancels itself out because they are turning in opposite directions. These rotors can be mounted one above the other and driven by coaxial shafts, or one rotor can be at the front end of the fuselage and the other at the rear end. The Focke-Achgelis FW-61 had two rotors mounted laterally on outriggers, and the Kaman helicopter has two intermeshing rotors mounted angularly on either side of the fuselage. [Figure 1-96]

Figure 1-96. Counter-rotating rotors cancel out torque.

TYPES OF ROTOR SYSTEMS

MAIN ROTOR SYSTEM

Main rotor systems are classified according to how the main rotor blades move in relation to the main rotor. There are three basic classifications: fully articulated, semirigid, and rigid. Some modern rotor systems may use a combination of these types.

FULLY ARTICULATED SYSTEM

In a fully articulated rotor system, each rotor blade is attached to the rotor hub through a series of hinges, allowing the blade to move independently of the others. These rotor systems usually have three or more blades. The horizontal hinge, called the **flapping hinge**, allows the blade to move up and down. This movement is called **flapping** and is designed to compensate for dissymmetry of lift. The flapping hinge may be located at varying distances from the rotor hub and there may be more than one hinge. The vertical hinge, called the **lead-lag** or **drag hinge**, allows the blade to move back and forth. This movement is called **lead-lag**, **dragging**, or **hunting**. Dampers are usually used to prevent excess back and forth movement around the drag hinge. The drag hinge and dampers compensate for the acceleration and deceleration that Coriolis effect causes. Each blade can also be feathered, that is, rotated around its spanwise axis. Feathering the blade means changing the pitch angle of the blade, which controls the thrust and direction of the main rotor disc. [Figure 1-97]

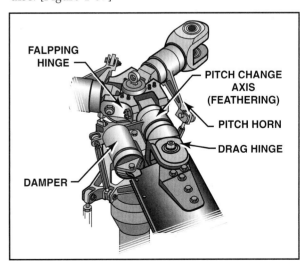

Figure 1-97. Each blade of a fully articulated rotor system can flap, drag, and feather independently of the other blades.

SEMIRIGID ROTOR SYSTEM

A semirigid rotor system uses two blades rigidly mounted to the main rotor hub. There is no vertical drag hinge. The main rotor hub, however, is free to rock and tilt, with respect to the main rotor shaft, on what is known as the **teetering hinge**. This allows the blades to flap together as a unit; as one blade flaps up, the other flaps down. Although there is no vertical drag hinge, lead lag forces still apply, but the lead-lag is usually absorbed through blade bending. [Figure 1-98]

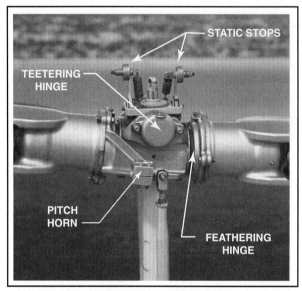

Figure 1-98. A teetering hinge allows the rotor hub and blades to flap as a unit. A static flapping stop located above the hub prevents the hub from excess rocking when the rotors are stationary. As the blades begin to turn, centrifugal force pulls the static stops out of the way.

RIGID ROTOR SYSTEM

In a rigid rotor system, the blades, hub, and mast are rigid with respect to each other. There are no vertical or horizontal hinges so the blades cannot flap or drag, but they can be feathered. Blade bending absorbs flapping and lead/lag forces.

Modern rotor systems may use the combined principles of the rotor systems mentioned above. Some rotor hubs incorporate a flexible hub, which allows for blade bending (flexing) without the need for bearings or hinges. These systems, called **flextures**, are usually constructed from composite material. **Elastomeric bearings** may also be used in place of conventional roller bearings. Elastomeric bearings are constructed from a rubber-type material and have limited movement that is perfectly suited for helicopter application. Flextures and elastomeric bearings require no lubrication and therefore require less maintenance. They also absorb vibration, which means less fatigue and longer service life for the helicopter components. [Figure 1-99]

FORCES ACTING ON THE MAIN ROTOR

Figure 1-99. Modern rotor systems use composite materials and elastomeric bearings, which reduce complexity and maintenance while increasing reliability.

GRAVITY

The weight of the rotor blades causes them to droop when they are not turning. On fully articulated rotors, a droop-stop that is built into the hub prevents the blade from drooping too low and striking the fuselage as it slows down. Droop stops are not required on semirigid rotor heads, as there is no horizontal drag hinge that allows the grip and blade to droop. Instead, a static flapping stop located above the hub prevents the hub from excess rocking when the rotors are stationary.

CENTRIFUGAL FORCE

The rotor blades must be turning for a helicopter to generate lift. The rotation of the rotor system creates **centrifugal force**, which tends to pull the blades straight out from the main rotor hub. The faster the rotation, the greater the centrifugal force. This force gives the rotor blades their rigidity, and in turn, the strength to support the weight of the helicopter. The centrifugal force generated determines the maximum operating rotor r.p.m. due to structural limitations on the main rotor system. The blade grips must be strong enough to hold the blade against these tremendous forces.

LIFT

A helicopter rotor produces lift in the same way that the wing of an airplane does. As the airfoil moves through the air, it creates a low pressure on top that pulls air down and deflects it behind the blade. The action that produces this downwash pushes up on the rotor and lifts the helicopter.

As you recall from your studies of aerodynamics, the center of pressure of an asymmetrical airfoil moves as the angle of attack changes. This is of no real concern on an airplane because the change in angle of attack occurs relatively slowly, so the pilot can easily compensate for center of pressure changes. But in a helicopter rotor, the angle of attack is continually changing. On one side of the rotor disc, the angle of attack is high, while it is quite low on the opposite side. If the center of pressure moves with changes in angle of attack, it creates forces that cause vibration and may eventually damage the rotor system.

The center of pressure of a symmetrical airfoil does not move as the angle of attack changes, and for this reason, most helicopter rotors have a symmetrical airfoil section. [Figure 1-100]

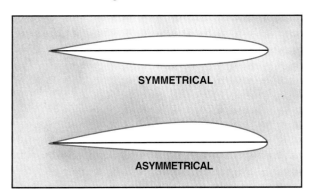

Figure 1-100. The upper and lower curvatures are the same on a symmetrical airfoil and vary on an asymmetrical airfoil.

The airflow velocity over a wing is relatively constant from root to tip in normal flight, and except to achieve a specific stall progression over it, the wing can have a constant airfoil section along its span. A helicopter rotor blade cannot have a constant airfoil section, because its speed through the air varies drastically. The airflow velocity over the blade is quite low at the root and increases as we move out toward the tip. To even out the lift produced by the blade, the airfoil section or the blade pitch angle

must change. To accomplish this, the blade is designed with washout or twist in it so that the tip section always has a lower angle of attack than the root section. Other design features may include using a thinner or smaller blade section toward the blade tips.

Two major forces are acting at the same time when making a vertical takeoff: centrifugal force acting outward and perpendicular to the rotor mast, and lift acting upward and parallel to the mast. These two forces result in the blades assuming a conical path instead of remaining perpendicular to the mast. This is called **coning**. [Figure 1-101]

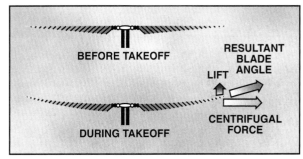

Figure 1-101. Rotor blade coning occurs as the rotor blades begin to lift the weight of the helicopter. In semirigid and rigid rotor systems, coning results in blade bending. In an articulated rotor system, the blades assume an upward angle through movement about the flapping hinges.

GYROSCOPIC FORCES

RIGIDITY IN SPACE

A spinning object, such as a bicycle wheel, a propeller, or a helicopter rotor, has some of the characteristics of a gyro. One of these characteristics is that of rigidity in space. When a wheel or rotor spins in space, it develops an inertial force that causes it to tend to remain rigid, and it will resist any force that tries to move it.

This force is quite pronounced on a rotor blade, but it has an even more obvious application in the stabilizing systems used on some helicopters, specifically some of the older Bell helicopters. A weighted **stabilizer bar** turned together with the rotor tends to remain rigid in space, and as the helicopter pitches or rolls, the mast moves relative to the bar and changes the pitch of the rotor blades. This is discussed in more detail later.

PRECESSION

The gyroscopic characteristic of **precession** explains the cause and effect action that takes place when the cyclic pitch on a helicopter rotor is changed. Gyroscopic precession is the resultant action or

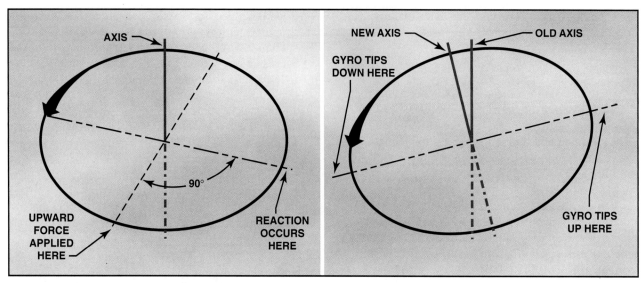

Figure 1-102. Gyroscopic precession principle — When a force is applied to a spinning gyro, the maximum reaction occurs approximately 90° later in the direction of rotation.

deflection of a spinning object when a force is applied to it. This action occurs approximately 90° in the direction of rotation from the point where the force is applied. [Figure 1-102]

Consider a two-bladed rotor system to see how gyroscopic precession affects the movement of the tip-patch plane. Moving the cyclic pitch control increases the angle of attack of one rotor blade, resulting in a greater lifting force applied at that point in the plane of rotation. This same control movement simultaneously decreases the angle of attack of the other blade by the same amount, thus decreasing the lifting force applied at that point in the plane of rotation. The blade with the increased angle of attack tends to rise; the blade with the decreased angle of attack tends to lower. However, because of the gyroscopic precession property, the blades do not rise or lower to maximum deflection until a point approximately 90° later in the plane of rotation. As shown in figure 1-103, the retreating blade angle of attack is increased and the advancing blade angle of attack is decreased, resulting in a tipping forward of the tip-path plane since maximum deflection takes place 90° later when the blades are at the rear and front, respectively. In an articulated rotor system, the movement of the cyclic pitch-control changes the angle of attack of each blade to an appropriate amount so that the end result is the same. [Figure 1-103]

CORIOLIS EFFECT (CONSERVATION OF ANGULAR MOMENTUM)

Coriolis effect, sometimes referred to as conservation of angular momentum, might be compared to

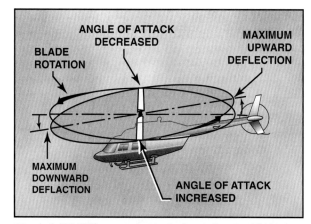

Figure 1-103. With a counterclockwise main rotor blade rotation, the maximum increase in angle of attack occurs as each blade passes the 90° position on the left. The maximum decrease in angle of attack occurs as each blade passes the 90° position to the right. Maximum deflection takes place 90° later — upward deflection at the rear and maximum downward deflection at the front — and the tip-path plane tilts forward.

spinning skaters. When skaters extend their arms, their rotation slows down because the center of mass moves farther from the axis of rotation. When their arms are retracted, the rotation speeds up because the center of mass moves closer to the axis of rotation. The reason for this is that the distance between the center of mass and the axis of rotation, times the rotational velocity of the mass, will try to remain constant.

When a rotor blade of an articulated rotor system flaps upward, the center of mass of that blade moves closer to the axis of rotation and blade acceleration takes place in order to conserve angular momentum.

Conversely, when that blade flaps downward, its center of mass moves further from the axis of rotation and blade deceleration takes place. Keep in mind that due to coning, a rotor blade will not flap below a plane passing through the rotor hub and perpendicular to the axis of rotation. Either dampers or the blade structure itself absorbs the acceleration and deceleration actions of the rotor blades, depending upon the design of the rotor system. [Figure 1-104]

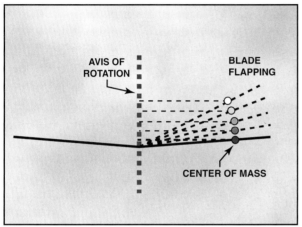

Figure 1-104. The tendency of a rotor blade to increase or decrease its velocity in its plane of rotation due to mass movement is known as Coriolis effect, named for the mathematician who made studies of forces generated by radial movements of mass on a rotating disc.

Two-bladed rotor systems are normally subject to Coriolis effect to a much lesser degree than are articulated rotor systems since the blades are generally underslung with respect to the rotor hub, and the change in the distance of the center of mass from the axis of rotation is small. The blades absorb the hunting action through bending. If a two-bladed rotor system is not "underslung," it will be subject to Coriolis effect comparable to that of a fully articulated system.

HELICOPTER FLIGHT CONDITIONS

HOVERING FLIGHT
For standardization purposes, this discussion assumes a stationary hover in a no-wind condition. During hovering flight, a helicopter maintains a constant position over a selected point, usually a few feet above the ground. For a helicopter to hover, the lift and thrust produced by the rotor system act straight up and must equal the weight and drag which act straight down. While hovering, main rotor thrust can be changed to maintain the desired hovering altitude. Changing the angle of attack of the main rotor blades and varying power does this, as needed. In this case, thrust acts in the same vertical direction as lift. [Figure 1-105]

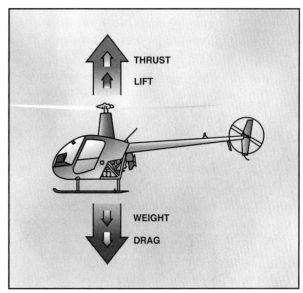

Figure 1-105. To maintain a hover at a constant altitude, enough lift and thrust must be generated to equal the weight of the helicopter and the drag produced by the rotor blades.

The weight to be supported is the total weight of the helicopter and its occupants. If the amount of thrust is greater than the actual weight, the helicopter gains altitude; if thrust is less than weight, the helicopter loses altitude.

TORQUE
Torque is an important consequence of producing thrust. As stated before, for every action there is an equal and opposite reaction. Therefore, as the engine turns the main rotor system in a counterclockwise direction, the helicopter fuselage will turn clockwise. The amount of torque is directly related to the amount of engine power being used to turn the main rotor system. Remember, as power changes, torque changes.

To counteract this torque-induced turning tendency, an antitorque rotor, or tail rotor is incorporated into most helicopter designs. The amount of thrust can be varied produced by the tail rotor in relation to the amount of torque produced by the engine. As the engine supplies more power, the tail rotor must produce more thrust. The use of antitorque pedals achieves this.

TRANSLATING TENDENCY OR DRIFT
During hovering flight, a single main rotor helicopter tends to drift opposite to the direction of

antitorque rotor thrust. This drifting tendency is called **translating tendency**. [Figure 1-106]

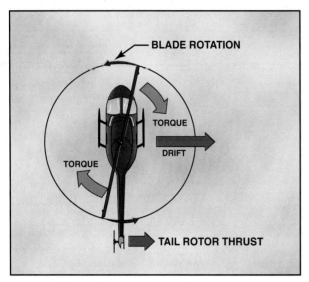

Figure 1-106. A tail rotor is designed to produce thrust in a direction opposite torque. The thrust produced by the tail rotor is sufficient to move the helicopter laterally.

To counteract this drift, one or more of the following features may be used:

- The main transmission is mounted so that the rotor mast is rigged for the tip-path plane to have a built-in tilt that is opposite the tail thrust, thus producing a small sideward thrust.

- Flight control rigging is designed so that the rotor disc is tilted slightly opposite the tail rotor thrust when the cyclic is centered.

- The collective pitch control system is designed so that the rotor disc tilts slightly opposite the tail rotor thrust when in a hover.

DENSITY ALTITUDE

Performance of both the rotor system and the engine depends on the density of the air. The engine must mix a given mass of air with its fuel in order to release energy, and the less dense the air, the greater the volume needed to release the required energy. Lift, as you remember from basic aerodynamics, is determined by the lift coefficient, the area of the surface, and the dynamic pressure. Dynamic pressure is the product of one-half of the density of the air and the square of the velocity of the blade through the air.

Air density has an important effect on helicopter performance, and the pilot is concerned with density altitude, which is the altitude in standard air that is the same as the existing pressure altitude. The pilot can determine the maximum load to carry by knowing the density altitude and referring to appropriate charts.

GROUND EFFECT

When hovering near the ground, a phenomenon known as **ground effect** takes place. This effect usually occurs less than one rotor diameter above the surface. As the surface friction restricts the induced airflow through the rotor, the lift vector increases. This allows a lower rotor blade angle for the same amount of lift, which reduces induced drag. Ground effect also restricts the generation of blade tip vortices due to the downward and outward airflow, producing lift from a larger portion of the blade. When the helicopter gains altitude vertically, with no forward airspeed, induced airflow is no longer restricted and the blade tip vortices increase with the decrease in outward airflow. As a result, a higher pitch angle is required to move more air down through the rotor, and more power is needed to compensate for the increased drag. Specifications for helicopter performance provide the hover ceiling in two ways, IGE and OGE. These figures provide the maximum altitude at which the helicopter will hover at its rated gross weight in ground effect and out of ground effect. [Figure 1-107]

Ground effect is at its maximum in a no-wind condition over a firm, smooth surface. Tall grass, rough terrain, revetments and water surfaces alter the airflow pattern, causing an increase in rotor tip vortexes.

VERTICAL ASCENT AND DESCENT

For a helicopter to hover in still air, the thrust produced by the rotor must exactly equal the combined weight and drag of the helicopter. When pilots wish to rise vertically, they increase the pitch angle of all of the rotor blades with the collective pitch, and at the same time, add engine power. Moving one control does all this, as we will see later. The additional power increases the amount of torque, and the pilot must adjust tail rotor pitch with the pedals to prevent the nose of the helicopter from turning. When the pilot wants to descend vertically, blade pitch angle and the power are reduced, and the pedals are adjusted again to correct for any tendency of the nose to rotate. The helicopter will descend vertically when the thrust produced by the rotor system becomes less than the weight.

FORWARD FLIGHT

DISSYMMETRY OF LIFT

As the helicopter moves forward through the air, the relative airflow through the main rotor disc is

Aircraft Structural Assembly and Rigging

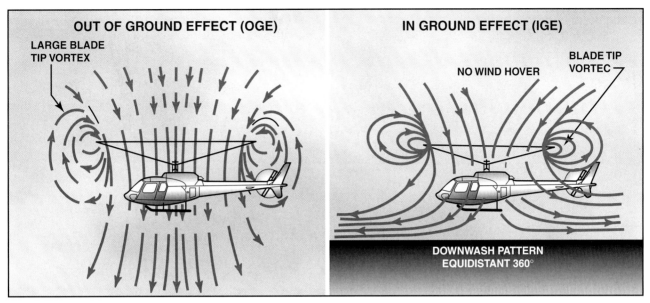

Figure 1-107. Air circulation patterns change when hovering out of ground effect (OGE) and when hovering in ground effect (IGE).

different on the advancing side than on the retreating side. The relative wind encountered by the advancing blade is increased by the forward speed of the helicopter, while the relative wind speed acting on the retreating blade is reduced. Therefore, as a result of the relative wind speed, the advancing blade side of the rotor disc produces more lift than the retreating blade side. This hypothetical situation is defined as dissymmetry of lift. [Figure 1-108]

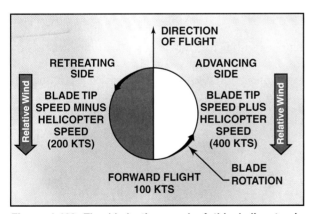

Figure 1-108. The blade tip speed of this helicopter is approximately 300 knots. If the helicopter is moving forward at 100 knots, the relative wind speed on the advancing side is 400 knots. On the retreating side, it is only 200 knots. This difference in speed causes a dissymmetry of lift.

If this condition were allowed to exist, a helicopter with a counterclockwise main rotor blade rotation would roll to the left because of the difference in lift. In reality, the main rotor blades flap and feather automatically to equalize lift across the rotor disc. Articulated rotor systems, usually with three or more blades, incorporate a horizontal hinge to allow the individual rotor blades to move, or flap up and down as they rotate. A semirigid rotor system (two blades) utilizes a teetering hinge, which allows the blades to flap as a unit. When one blade flaps up, the other flaps down.

In figure 1-109, as the rotor blade reaches the advancing side of the rotor disc (A), the increase in lift causes it to flap up until it reaches its maximum upflap velocity. When the blade flaps upward, the angle between the chord line and the resultant relative wind decreases. This decreases the angle of attack, which reduces the amount of lift produced by the blade. At position (C), the decreased lift causes the blade to flap down until it is at its maximum downflapping velocity. Due to downflapping, the angle between the chord line and the resultant relative wind increases. This increases the angle of attack and thus the amount of lift produced by the blade.

The combination of blade flapping and slow relative wind acting upon the retreating blade normally limits the maximum forward speed of a helicopter. At a high forward speed, the retreating blade stalls because of a high angle of attack and slow relative wind speed. This situation is called **retreating blade stall** and is evidenced by a nose pitch up, vibration, and a rolling tendency usually to the left in helicopters with counterclockwise blade rotation. The pilot can avoid retreating blade stall by not violating the never-exceed speed. This speed is designated V_{ne} and is usually marked on the airspeed indicator by a red line and by placards in the cockpit.

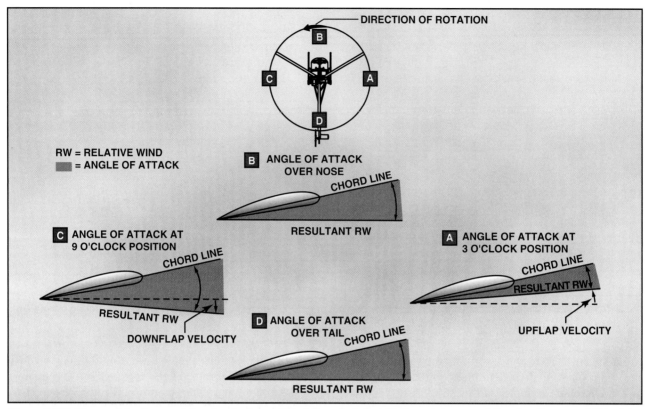

Figure 1-109. The combined upward flapping (reduced lift) of the advancing blade and downward flapping (increased lift) of the retreating blade equalizes lift across the main rotor disc, counteracting the dissymmetry of lift.

RETREATING BLADE STALL

In a fixed-wing airplane, we normally associate stalls with low airspeed, but in a helicopter, the opposite is true. Retreating blade stall is one of the factors that limit a helicopter's top speed.

Stalls on both airplanes and helicopters are caused by exactly the same thing: an excessive angle of attack. As the forward speed of the helicopter increases, the velocity of the air over the retreating blade decreases; and as it flaps downward, its angle of attack increases. At some given airspeed, the angle of attack becomes excessive and a stall occurs. The exact location of the stall on the rotor disc depends upon many variables, but it normally occurs when the retreating blade is between the 270° and 300° position in its rotation, and usually only about the outer 20 or 30 percent of the blade stalls. When a blade stalls, lift is lost and drag builds up. The effect of an action on a rotor disc is felt, not at the point at which it occurs, but at a point some ninety degrees away in the direction of rotation due to precession. Then, when a stall develops, it causes the helicopter to pitch nose up.

TRANSLATIONAL LIFT

Translational lift is present with any horizontal flow of air across the rotor. This increased flow is most noticeable when the airspeed reaches approximately 16 to 24 knots. As the helicopter accelerates through 16 to 24 knots, it moves out of its own downwash and vortices into relatively undisturbed air. The airflow is now more horizontal, which reduces induced flow and drag with a corresponding increase in angle of attack. The additional lift available at this speed is referred to as "effective translational lift" (ETL). [Figure 1-110]

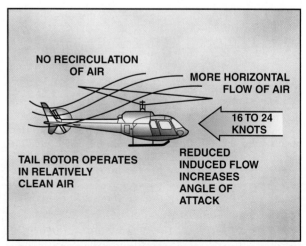

Figure 1-110. Effective translational lift is easily recognized in flight by a transient induced aerodynamic vibration and increased performance of the helicopter.

Aircraft Structural Assembly and Rigging

Translational lift can also be present in a stationary hover if the wind speed is approximately 16 to 24 knots. In normal operations, the pilot should always utilize the benefit of translational lift, especially if maximum performance is needed.

AUTOROTATION

Autorotation is the state of flight where the main rotor system is being turned by the action of relative wind rather than engine power. It is the means by which a helicopter can be landed safely in the event of an engine failure. All helicopters must have this capability to be certified. Autorotation is permitted mechanically because of a freewheeling unit, which allows the main rotor to continue turning even if the engine is not running.

In normal powered flight, air is drawn into the main rotor system from above and exhausted downward. During autorotation, airflow enters the rotor disc from below as the helicopter descends, creating an aerodynamic force that drives the rotor and keeps it turning. This is known as the **autorotative force**. [Figure 1-111]

Figure 1-112 demonstrates the way this autorotative force comes about. Air flowing upward through the rotor produces an angle of attack such as we see here. From our study of basic aerodynamics, we know that the lift produced by an airfoil acts perpendicularly to the relative wind, while the induced drag acts parallel, but in opposite direction, to the relative wind. If we consider the components of lift and drag that act along the plane of rotation of the rotor, we see that there is a net force acting in the direction of rotor rotation along this plane. This is the autorotative force, and it will accelerate the rotor. This autorotative force is found in the portion of the rotor from approximately 25% to 70% of the blade span. This part of the rotor disc is called the autorotation region. [Figure 1-112]

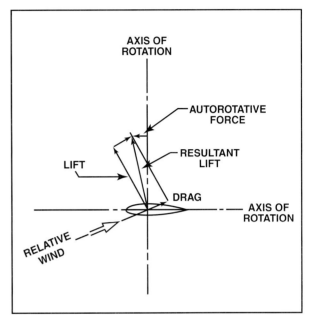

Figure 1-112. When the angle between the relative wind and the axis of rotation is high, the resultant lift will be ahead of the axis of rotation, and there will be an autorotative force that pulls the rotor in its direction of rotation.

From about 70% of the blade span to the tip, the velocity is high and the angle of attack is low. The drag force that acts along the plane of rotation is greater than the component of the lift that acts along this plane. The result is a force that tries to decelerate the rotor. This is called an **anti-autorotative**

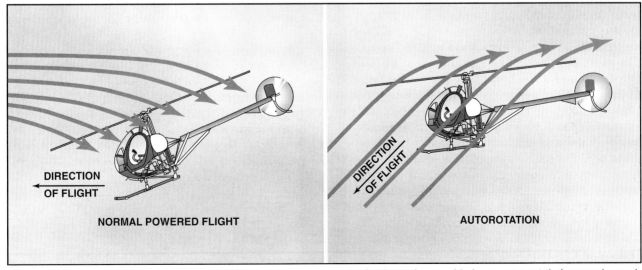

Figure 1-111. During autorotation, the upward flow of relative wind permits the main rotor blades to rotate at their normal speed. In effect, the blades are "gliding" in their rotational plane.

force, and the portion of the rotor disc that produces this force is called the **driven region**. [Figure 1-113]

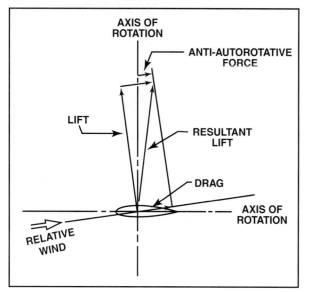

Figure 1-113. When the angle between the relative wind and the axis of rotation is low, the resultant lift is behind the axis or rotation, and a retarding force is produced on the rotor. This occurs in the driven region of the rotor.

ROTORCRAFT CONTROLS

DIRECT ROTOR HEAD TILT

Controlling a rotorcraft is basically done by tilting the lift produced by the rotor. If the rotor disc is tilted forward, the aircraft will fly forward; if it is tilted back, it will fly backward; and tilting it to the side will cause it to fly sideways. The early gyroplanes and some of the simplest rotary wing aircraft, mainly amateur-built, have a control bar that directly tilts the rotor head. This form of direct control usually requires the control forces to be quite light.

SWASH PLATE CONTROL SYSTEM

The direct rotor head tilt system has so many limitations that a far more effective system has been devised and is used in some form on almost all helicopters. The swash plate is used to transmit control inputs from the cockpit controls to the main rotor blades. It consists of two main parts: the stationary swash plate and the rotating swash plate. The stationary swash plate is mounted to the main rotor mast and connected to the cyclic and collective controls in the cockpit by a series of pushrods. It is restrained from rotating but is able to tilt in all directions and move vertically. The rotating swash plate is mounted to the stationary swash plate by means of a bearing and is allowed to rotate with the main rotor mast. Both swash plates tilt and move up and down as a unit. The rotating swash plate is connected to the main rotor grips by the pitch links. [Figure 1-114]

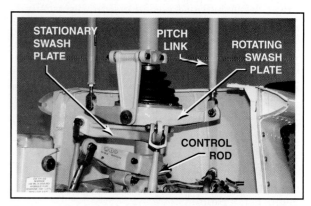

Figure 1-114. Control rods transmit collective and cyclic control inputs to the stationary swash plate, causing it to tilt or move vertically. The pitch links, attached from the rotating swash plate to the pitch arms on the rotor blades, transmit these movements to the blades.

COLLECTIVE PITCH CONTROL

Moving the collective control lever in the cockpit causes the swash plate assembly to slide up and down on the swash plate support, which is mounted on top of the transmission or mast assembly. This causes all the rotor blades to increase or decrease blade pitch angle by the same amount, or collectively as the name implies. As the collective pitch control is raised, there is a simultaneous and equal increase in pitch angle of all main rotor blades, and lift increases. As it is lowered, there is a simultaneous and equal decrease in pitch angle, and lift is decreased. This is done through a series of mechanical linkages, and the amount of movement in the collective lever will determine the amount of blade pitch change. A friction control is adjusted by the pilot to prevent inadvertent collective pitch movement. [Figure 1-115]

Changing the pitch angle on the blades changes the blade angle of attack and lift. With a change in angle of attack and lift comes a change in drag, and the speed or r.p.m. of the rotors will be affected. As the pitch angle of the blades is increased, the angle of attack and drag increase, while the rotor r.p.m. decreases. Decreasing pitch angle decreases both angle of attack and drag, and rotor r.p.m. increases. To maintain a constant rotor r.p.m., which is essential in helicopter operation, a proportionate change in power is required to compensate for the change in drag. A correlator and/or governor is the most common way to accomplish this.

A correlator is a mechanical connection between the collective lever and the engine throttle. When

Aircraft Structural Assembly and Rigging

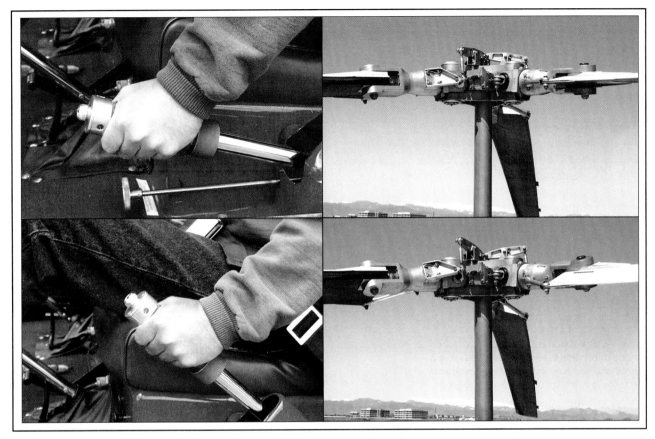

Figure 1-115. Raising the collective pitch control increases the pitch angle by the same amount on all blades.

the collective lever is raised, power is automatically increased, and when lowered, power is decreased. This system maintains r.p.m. close to the desired value, but still requires pilot adjustment of the throttle for fine-tuning.

A governor is a sensing device that detects rotor and engine r.p.m. and makes the necessary adjustments to keep the rotor r.p.m. constant. Once the rotor r.p.m. is set, the governor will keep it steady and there is no requirement of the pilot to make any throttle adjustments. Governors are common on all turbine helicopters.

Some helicopters do not use correlators or governors and require the pilot to coordinate all collective and throttle movements together. When the collective is raised, the throttle must be increased; when the collective is lowered, the throttle must be decreased.

THROTTLE CONTROL

The throttle is usually mounted on the end of the collective lever in the form of a twist grip. Some turbine helicopters have the throttles mounted on the overhead panel or on the floor of the cockpit. [Figure 1-116]

Figure 1-116. A twist grip throttle is usually mounted at the end of the collective pitch lever. Some turbine helicopters have throttles mounted on the overhead panel or on the floor of the cockpit.

The function of the throttle is to regulate engine r.p.m. If the correlator system does not maintain the desired r.p.m. when the collective is moved, or if the system does not have a correlator at all, the throttle will have to be moved manually with the twist grip to maintain r.p.m.

The throttle is used to set normal r.p.m. in helicopters with governors, and then is left alone throughout the rest of the flight. The governor maintains desired r.p.m. when the collective is moved. However, if the governor fails, the pilot will have to use the throttle to maintain r.p.m.

CYCLIC PITCH CONTROL

The cyclic control, mounted vertically between the pilot's knees, tilts the main rotor disc by tilting the swash plate assembly, thus changing the pitch angle of the rotor blades in their cycle of rotation. When the main rotor disc is tilted, the horizontal component of lift will move the helicopter in the direction of tilt. [Figure 1-117]

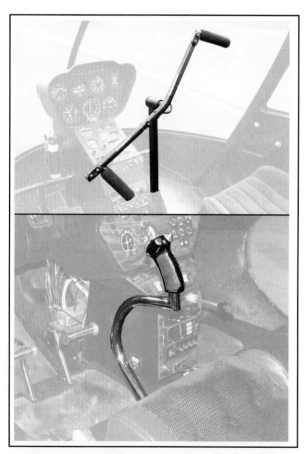

Figure 1-117. The cyclic control may be mounted vertically between the pilot's knees or on a teetering bar on a single cyclic located in the center console. The cyclic can pivot in all directions.

The rotor disc tilts in the direction that pressure is applied to the cyclic. If the cyclic stick is moved forward, the rotor disc will tilt forward, and if the cyclic is moved aft, the disc will tilt aft and so on. Because of gyroscopic precession, the mechanical linkages for the cyclic control rods are rigged in such a way that they decrease the pitch angle of the rotor blade 90° before it reaches the direction of cyclic displacement and increases the pitch angle of the rotor blade 90° after it passes the direction of displacement. An increase in pitch angle will increase angle of attack; a decrease in pitch angle will decrease angle of attack. For example, if the cyclic is moved forward, the angle of attack decreases as the rotor blade passes the right side of the helicopter and increases on the left side. Because of gyroscopic procession, this results in maximum downward deflection of the rotor blade in front of the helicopter and maximum upward deflection behind it, causing the rotor disc to tilt forward. This assumes that the direction of rotor blade rotation is counterclockwise as viewed from above, which is the case of single rotor helicopters manufactured in the United States.

SYNCHRONIZED ELEVATORS

Some helicopters have horizontal tail surfaces that are synchronized with the cyclic pitch control to create a down load that increases when the cyclic control tilts the lift forward. Without this provision, the fuselage will assume an excessive nose-down pitch at high airspeeds, but the elevator holds it at a more level attitude.

Helicopters that do not have synchronized elevators instead may have a fixed horizontal stabilizer, that has a very pronounced airfoil section, mounted on the tail boom in a position whereby its down load increases with increased forward airspeed.

BOOSTED CONTROLS

Most helicopters, other than smaller piston-powered helicopter, incorporate the use of hydraulic actuators on the flight controls. This serves two main purposes: to overcome high control forces and to prevent vibrations from the rotor system from being fed back into the controls. The hydraulic system consists of actuators, also called servos, on each flight control, a pump that is usually driven by the main rotor gearbox, and a reservoir to store the hydraulic fluid. A switch in the cockpit can turn the system off, although it is left on under normal conditions. A pressure indicator in the cockpit may be installed to monitor the system. [Figure 1-118]

When a control input is made, a valve inside the servo directs hydraulic fluid under pressure to the piston to change the rotor pitch. This type of servo uses a system of check valves to make the system irreversible; that is, to make it so the pilot can control the pitch of the rotor blades, but vibration from the rotor cannot feed back into the cockpit controls. The controls can still be moved in the event of a hydraulic system failure, but the control forces will be very heavy.

Aircraft Structural Assembly and Rigging

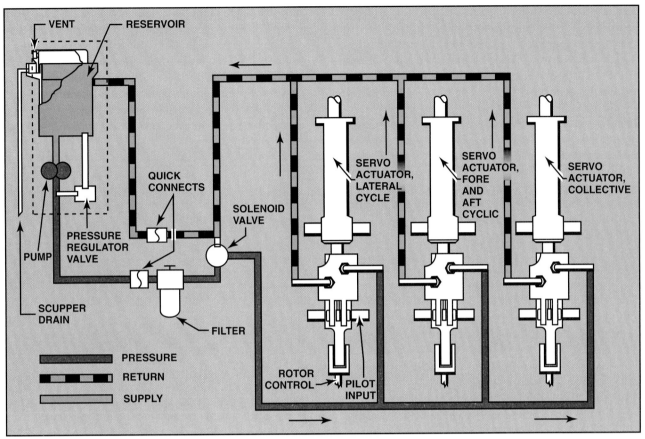

Figure 1-118. A typical hydraulic system for helicopters in the light to medium range is shown here.

Two or more independent hydraulic systems may be installed in those helicopters where the control forces are so high that they cannot be moved without hydraulic assistance. Some helicopters use hydraulic accumulators to store pressure, which can be used for a short while in an emergency if the hydraulic pump fails. This gives the pilot enough time to land the helicopter with normal control.

TORQUE COMPENSATION

One of the most basic laws of physics that affects helicopter flight is that of each force being countered with an equal and opposite force. When the engine drives the rotor, the same force that spins the rotor tries to spin the fuselage. If the helicopter is to be controlled about its vertical axis, there must be some way to counteract this torque.

There have been many approaches to the solution for this problem. Some designers have driven the rotor with a jet of air from the rotor blade tip, thus eliminating any torque from a fuselage-mounted transmission drive. Others used two rotors, either on coaxial shafts, or mounted side by side or fore and aft. Despite all of the different attempts to compensate for torque, the one system that has survived with almost universal use is the tail rotor or anti-torque rotor. [Figure 1-119]

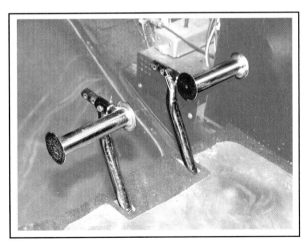

Figure 1-119. The pilot changes the pitch of the tail rotor to correct for torque by moving the pedals in the cockpit.

The main rotor transmission drives the tail rotor, so in the event of engine failure, the tail rotor will still be operational to provide the pilot with directional control for an autorotative landing. It is driven through a drive shaft supported by ball bearings, and incorporates universal joints or intermediate gearboxes where there is any change in direction of the shaft. The drive shaft normally terminates in a right-angle gearbox to which the rotor is attached.

The tail rotor, like the main rotor, has an advancing and retreating blade in forward flight, and to compensate for dissymmetry of lift, they either have a flapping hinge or are free to teeter. Some of the flapping hinges incorporate a Delta-three hinge, whose hinge line is angled with respect to the rotor span. When the blade flaps, it also changes its effective pitch angle and can correct for the dissymmetry of lift without a severe flapping angle.

The pitch of the tail rotor is controlled by the pedals that correspond to the rudder pedals in an airplane. The normal torque from the counterclockwise-turning rotor will try to rotor the nose to the right, and to compensate for this, the tail rotor produces a thrust to the right thus pushing the nose to the left. The pilot varies the thrust to rotate the fuselage while hovering or to compensate for any changes in the torque as power and speed are changed.

Another form of antitorque rotor is the **fenestron**, or "fan-in-tail" design. This system uses a series of rotating blades shrouded within a vertical tail. Because the blades are located within a circular duct, they are less likely to come in contact with people or objects. [Figure 1-120]

Figure 1-120. The fenestron antitorque system provides an improved margin of safety during ground operations.

The **NOTAR®**, or **no tail rotor** system is an alternative to the antitorque rotor. This system uses low-pressure air, which is forced into the tailboom by a fan mounted within the helicopter. The air is forced through horizontal slots, located on the right side of the tailboom, to a controllable rotating nozzle, providing antitorque and directional control. The low-pressure air coming from the horizontal slots, in conjunction with the downwash from the main rotor, creates a phenomenon called **Coanda effect**, which produces a lifting force on the right side of the tailboom. [Figure 1-121]

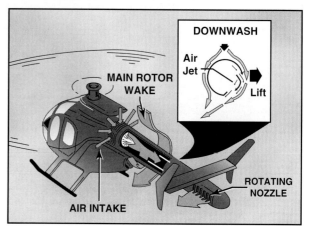

Figure 1-121. While in a hover, Coanda effect supplies approximately two-thirds of the force necessary to maintain directional control. Directing the thrust from the controllable rotating nozzle creates the rest.

STABILIZER SYSTEMS

BELL STABILIZER BAR SYSTEM

The inherent instability of a helicopter was one of the problems that delayed its development for so long. An airplane has both positive static and dynamic stability about all three of its axes, but the helicopter is not so fortunate.

One of the developments pioneered by Bell Helicopter was the **stabilizer bar**. This is a long bar with weights at each end, mounted on a pivot and driven by the rotor mast. The control rods from the swash plate attach to the stabilizer bar, and pitch control links connect the blade pitch arms to the stabilizer bar. In flight, the stabilizer bar acts as a very effective gyro, having rigidity in space; that is, it does not want to depart from its plane of rotation. If the helicopter tilts, the stabilizer bar remains in its original plane of rotation and an angular difference is formed between the bar and the mast. This is transmitted to the rotor blades as a pitch change in the correct direction to right the helicopter. [Figure 1-122]

Too much effectiveness of the stabilizer would lead to control difficulties. This is prevented by limiting the angular difference possible between the mast and the stabilizer bar and by installing hydraulic

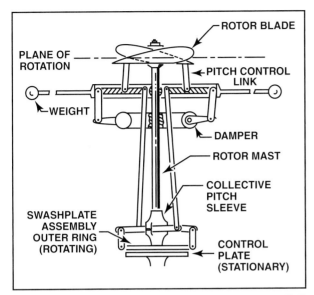

Figure 1-122. The stabilizer bar used on many older Bell helicopters remains rigid in space and changes the pitch on the rotor blade to create a restorative force any time the plane of the rotor is different from the plane of the stabilizer bar.

dampers that cause the stabilizer bar to follow the mast, though at a highly damped rate.

OFFSET FLAPPING HINGE

By simply offsetting the flapping hinge, it is possible to give stability to a helicopter rotor system. If the rotor system in figure 1-123 tilts to the right in flight, the angle of attack of the descending rotor blade will increase and the blade will flap up. As it flaps, the offset hinge will cause its pitch angle to increase and automatically produce lift in the direction needed to right the helicopter. It has the same effect as the pilot applying corrective use of the cyclic control, but this is automatic.

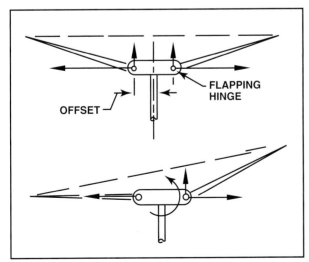

Figure 1-123. Offset hinges on the main rotor provide more corrective action than a semirigid rotor with a teetering hinge.

STABILITY AUGMENTATION SYSTEM (SAS)

Some helicopters incorporate stability augmentation systems (SAS) to aid in stabilizing the helicopter in flight and in a hover. The simplest of these systems is a force trim system, which uses a magnetic clutch and springs to hold the cyclic control in the position where it was released. More advanced systems use electric servos that actually move the flight controls. These servos receive control commands from a computer that senses helicopter attitude. The SAS may be overridden or disconnected by the pilot at any time.

Stability augmentation systems allow the pilot more time and concentration to accomplish other duties. It improves basic aircraft control harmony and reduces outside disturbances, thus reducing pilot workload. These systems are useful when pilots are required to perform other duties such as sling loading and search and rescue operations.

AUTOPILOTS

Helicopter autopilot systems are similar to stability augmentations systems except they have additional features. The autopilot can actually fly the helicopter and perform certain functions selected by the pilot. These functions depend on the type of autopilot and systems installed in the helicopter.

The most common functions are altitude and heading hold. Some more advanced systems include a vertical speed or indicated airspeed (IAS) hold mode, where a constant rate of climb/descent or indicated airspeed is maintained by the autopilot. Some autopilots have navigation capabilities such as VOR, ILS, and GPS intercept and tracking which is especially useful in IFR conditions. The most advanced autopilots can fly an instrument approach to a hover without any additional pilot input once the initial functions have been selected.

The autopilot system consists of electric actuators (servos) connected to the flight controls. The number and location of these servos depend on the type of system installed. A two-axis autopilot controls the helicopter in pitch and roll; one servo controls fore and aft cyclic and another controls left and right cyclic. A three-axis autopilot has an additional servo connected to the antitorque pedals and controls the helicopter in yaw. A four-axis system uses a fourth servo, which controls the collective (power). These servos move the respective flight controls when they receive control commands from a central computer, which receives data input from the flight instruments for attitude reference and from the navigation equipment for navigation and tracking reference. A control panel in the cockpit has a number of switches that allows the pilot to select the desired functions and engage the autopilot.

An automatic disengage feature is usually included for safety purposes, which disconnects the autopilot in heavy turbulence or when extreme flight attitudes are reached. Even though all autopilots can be overridden by the pilot, there is also an autopilot disengage button located on the cyclic or collective which allows you to completely disengage the autopilot without removing your hands from the controls. Because autopilot systems and installations differ from one helicopter to another, it is important that you refer to the autopilot operating procedures located in the Rotorcraft Flight Manual.

HELICOPTER VIBRATION

Vibration has been one of the main problems plaguing helicopter designers ever since rotary-wing flight became possible. Although not all vibration can ever be completely eliminated, irreversible controls and special vibration-absorbing engine and transmission mounts have minimized its effects.

TYPES OF VIBRATION

Helicopter vibration is a complex subject depending on the depth of analysis. To keep it simple and useful, we categorize them basically into two frequency ranges, two modes, and two conditions.

FREQUENCY RANGES

Some texts divide vibrations into three ranges: low frequency, medium frequency, and high frequency. The break between low and medium, and between medium and high is rather nebulous, so in this text we lump them all into either low or high frequencies.

Low-Frequency Vibration

Regardless of the actual number of cycles per second of a vibration, we classify a vibration in which we can feel the beat as a **low-frequency vibration**. These are normally associated with the main rotor system. They may have a ratio of 1:1, 2:1, or 3:1 with the main rotor, depending on the rotor configuration, and may be caused by either a static or a dynamic unbalance condition or by aerodynamic forces acting upon the rotor.

High-Frequency Vibration

Rather than feeling these as a thump or a beat, **high-frequency vibrations** can be thought of as any that are felt as a buzz. Any component that turns at a high r.p.m. usually causes them, such as the engine, cooling fan, or some portion of the tail rotor system.

Vibration Modes

Vibrations can cause the helicopter to jump up and down in what is called a **vertical vibration**, or to shake sideways, which is called **lateral vibration**. Vertical vibrations are generally caused by some disparity in lift as the rotor spins and generally points to an out-of-track condition. Lateral vibrations are most often caused by an out-of-balance condition of the rotor. [Figure 1-124]

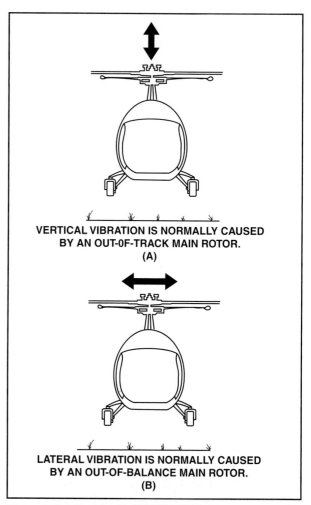

Figure 1-124. Modes of helicopter vibration are shown here.

CONDITIONS OF VIBRATION

We must know where vibration occurs to fully analyze it. A vibration that is felt only on the ground would have one fix, while a similar vibration that occurs only in flight would require a different procedure to correct it.

MEASUREMENT OF VIBRATION

Since vibration is such an important aspect of helicopter maintenance, much study has been made in analyzing and measuring it so its isolation and cure can be efficiently accomplished. With advances in technology, there are now many types of vibration analysis equipment, which use different approaches and techniques to determine vibration and correction.

Earlier equipment uses a combination of hand-held strobe and analog computer to determine track and balance. Once an observation is made, it is plotted on a chart, which then directs us to make a move to correct the track or balance. Generally, only one corrective move is made at a time.

Some of the newer systems use infrared sensors together with computers that direct us to which corrective moves to make. In these systems, more than one move can be made each time, thus accelerating the track and balance procedure.

In almost all of these systems, an electronic pickup is used to key the system each time a rotor blade passes a reference point. An accelerometer or velometer measures the amplitude of the vibration and marks the position in the blade path where this vibration occurs. By plotting this information on special graphs furnished with the equipment and tailored for the specific helicopter, the technician can determine the amount of weight that needs to be added to one of the blades to correct for balance, or the amount of change that should be made in the pitch change linkage to correct for track.

CORRECTION OF VIBRATION

BLADE BALANCING

Both main rotor blades and tail rotor blades are usually statically and dynamically balanced. Static balance refers to balancing a blade while it is not moving and is generally removed from the helicopter. Dynamic balance is accomplished with the blades mounted on the helicopter and turning.

Before a helicopter rotor is mounted on the mast, it must be balanced both chordwise and spanwise. Align semirigid rotors by adjusting the drag links or blade latches until an aligning wire, which is stretched between alignment points on each blade, passes over the correct point on the hub. After this alignment, or chordwise balance, is accomplished, balance the rotor spanwise by adding weight to the light blade until the balance specification is met. Some helicopters, specifically those with three or more blades, may require the hub to be balanced without the blades, and then the blades are installed that have been balanced against a master blade. [Figure 1-125]

BLADE TRACKING

Rotor blades must first be statically balanced, the rotor installed on the helicopter, and then the blades tracked to determine that each blade follows directly in the path of the blade ahead of it. Old methods made use of a marking stick, which was raised until it

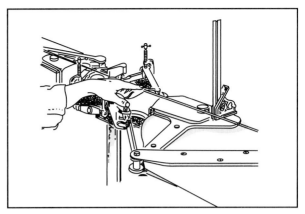

Figure 1-125. Chordwise balance of a two-bladed rotor consists of aligning the blades so that they are straight across the hub.

just contacted the bottom of the rotor blade at the tip, or a flag that was moved into the tip plane until the tip just touched the flag. With the marking stick method, the blades were in track when each blade was marked the same. Before checking the track using the flag method, each blade tip was marked with a different color crayon. They left a mark of their respective color when the blade tips touched the flag. If the rotor was in track, the colored marks were superimposed. The drawbacks to these two methods is that the helicopter can only be tracked on the ground and not in the air where it is most important.

Modern tracking systems either use a strobe light held by the technician or an infra-red light mounted on the helicopter to determine blade track. The rotor is turned at the proper r.p.m. and the light is triggered each time one of the blades passes a pickup that is mounted on the stationary swash plate. A special reflector is installed on the tip of each blade, and they form a distinctive pattern as the strobe illuminates each reflector. If the blades are in track, the images will be all in line, but if they are not in track, the images will be staggered up and down. The distinctive mark on each reflector tells the technician which blade is out of track, in which direction, and how much. [Figure 1-126]

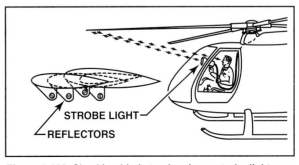

Figure 1-126. Checking blade track using a strobe light may be done on the ground and in flight.

TRACK ADJUSTMENT

Blade track adjustments are usually made two ways, depending on where the out-of-track condition exists. Generally, if a blade is found to be out of track on the ground, the pitch links are adjusted to change the pitch angle of that blade. If the blade is out of track, as seen during an in-flight track check, it is usually corrected by slightly bending one of the rotor blade trim tabs.

HELICOPTER POWER SYSTEMS

POWERPLANT

The piston engine and the turbine engine are the two most common types of engines used in helicopters. Piston engines are generally used in smaller helicopters because they are relatively simple and inexpensive to operate. Turbine engines are more powerful and are used in a wide range of helicopters. They produce a tremendous amount of power for their size but are more expensive to operate.

PISTON ENGINE

The first helicopters and some of the smaller ones today use piston engines similar to those in airplanes. While these engines are made by the same manufacturers and carry the same basic designation as those used in airplanes, they differ in operational details.

In some helicopters, the engine is mounted vertically so the transmission and the rotor mast can mount directly on the engine crankshaft. When this is done, the lubrication system must be changed. They are generally converted from a wet sump system, in which the oil is carried inside the engine itself, to a dry sump system, where the oil is carried in an external tank. The oil is pumped through the engine for lubrication and cooling and returned to the tank. Other helicopters have the engine mounted horizontally and are connected to the transmission with several V-belts.

The lack of a propeller to serve as a flywheel and to provide cooling air gives a helicopter two problems that the airplane does not have. Since the engine carries the load of the rotor system, it must be operated at a much higher idling speed, so there is not nearly as wide a range of engine r.p.m. as there is with an airplane. Since there is no propeller, a fan blows air over the engine cooling fins to remove the heat.

Engine power control is entirely different on a helicopter compared to that on an airplane. Rather than having complete control of the engine speed, the collective pitch control governs the helicopter power output. The rotor r.p.m. and that of the engine are held relatively constant. As more load is placed on the rotor by increasing the blade angle with the collective pitch control, the throttle must be increased to maintain r.p.m. and prevent the rotor from slowing down. This is either done automatically by correlators or governors, or manually accomplished by the pilot. The pilot determines the power that the engine is producing primarily by the relationship between the tachometer and the manifold pressure gauge.

TURBINE ENGINES

The turbine engine and the helicopter are ideal companions. The turbine operates best at a constant speed, which is perfect for the helicopter, but the most outstanding advantage is its tremendous power output for its small size and light weight.

Helicopters use turboshaft engines, which differ from turbojets in that they have extra stages in their turbine section to extract a maximum amount of energy from the expanding gases. Very little thrust is obtained from the exiting gases in a turboshaft engine. [Figure 1-127]

There are two types of turboshaft engines used: the direct-shaft engine and the free turbine engine. The direct-shaft engine uses a set of turbine wheels mounted on a single shaft that drives both the compressor in the gas generator and the output shaft. Reduction gearing reduces the r.p.m. of this shaft to a speed that is usable for the helicopter transmission. The free turbine engine has one set of turbines to drive the compressor for the gas generator and a separate set of turbines for the transmission and rotor. The exhaust gases do not leave a turboshaft engine in a straight line from the turbine section as they do in a turbojet engine, so it is possible to drive the output shaft from either the hot or the cold end of the engine.

The power produced by a turboshaft engine is indicated to the pilot by a torquemeter and a tachometer that is calibrated in percent of r.p.m. There are two tachometers on free turbine engines: N1 shows the speed of the compressor and N2 the speed of the power turbine. The exhaust gas temperature or turbine inlet temperature is also critical for the operation of a turboshaft engine. It is essential that the temperature of the gases entering the turbine stage never becomes excessive, as this can damage the turbine.

Aircraft Structural Assembly and Rigging

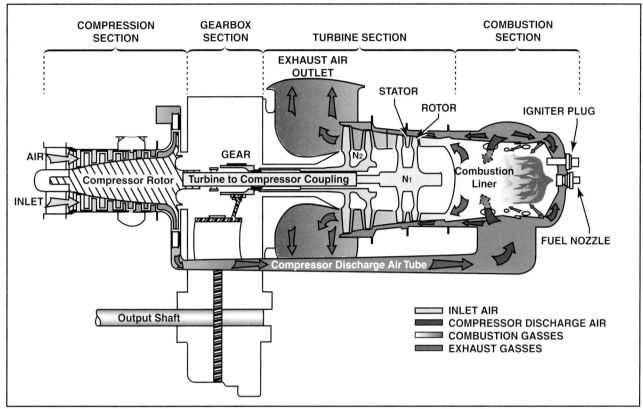

Figure 1-127. The main difference between a turboshaft and a turbojet engine is that most of the energy produced from the expanding gasses is used to drive a turbine rather than producing thrust through propulsion of exhaust gasses.

TRANSMISSION SYSTEM

The transmission system transfers power from the engine to the main rotor, the tail rotor, and other accessories. The main components of the transmission system are the main rotor transmission, tail rotor drive system, clutch, and freewheeling unit. Helicopter transmissions are normally lubricated and cooled with their own oil supply. A sight gauge is provided to check the oil level. Some transmissions have **chip detectors** and warning lights located on the pilot's instrument panel that illuminate in the event of an internal problem.

MAIN ROTOR TRANSMISSION

The primary purpose of the main rotor transmission is to reduce engine output r.p.m. to optimum rotor r.p.m. This reduction is different for the various helicopters, but as an example, suppose the engine r.p.m. of a specific helicopter is 2,700. Achieving a rotor speed of 450 r.p.m. would require a reduction of 6 to 1. A 9 to 1 reduction would mean that the rotor would turn at 300 r.p.m.

A dual-needle tachometer shows both engine and rotor r.p.m. The rotor r.p.m. needle normally is used only during clutch engagement to monitor rotor acceleration, as well as in autorotation to maintain r.p.m. within prescribed limits. When the needles are superimposed or married, the ratio of the engine r.p.m. is the same as the gear reduction ration. [Figure 1-128]

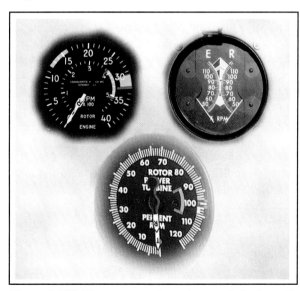

Figure 1-128. There are various types of dual-needle tachometers. When the needles are superimposed or married, the ratio of the engine r.p.m. is the same as the gear reduction ratio.

In helicopters with horizontally mounted engines, changing the axis of rotation from the horizontal axis of the engine to the vertical axis of the rotor shaft is another purpose of the main rotor transmission.

TAIL ROTOR DRIVE SYSTEM

The tail rotor drive system consists of a tail rotor drive shaft powered from the main transmission and a tail rotor transmission mounted at the end of the tail boom. The drive shaft may consist of one long shaft or a series of shorter shafts connected at both ends with flexible couplings. This allows the drive shaft to flex with the tail boom. The tail rotor transmission provides a right angle drive for the tail rotor and may also include gearing to adjust the output to optimum tail rotor r.p.m. [Figure 1-129]

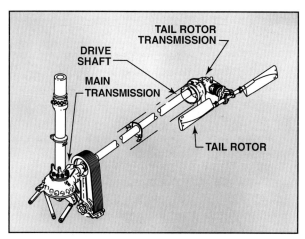

Figure 1-129. The typical components of a tail rotor drive system are shown here.

CLUTCH

In a conventional airplane, the engine and propeller are permanently connected. However, in a helicopter there is a different relationship between the engine and the rotor. Because of the greater weight of a rotor in relation to the power of the engine (as compared to the weight of a propeller and the power in an airplane), the rotor must be disconnected from the engine when the starter is engaged. A clutch allows the engine to be started and then gradually pick up the load of the rotor.

On free-turbine gas engines, no clutch is required as the gas producer turbine is essentially disconnected from the power turbine. There is little resistance from the power turbine when the engine is started. The drive turbine is able to accelerate to normal idle speed. The power turbine then begins to rotate as a gas coupling is formed between the two turbines.

The two main types of clutches are the centrifugal clutch and the belt drive clutch.

Centrifugal Clutch

The centrifugal clutch is made up of an inner assembly and an outer drum. The inner assembly, which is connected to the engine driveshaft, consists of shoes lined with material similar to automotive brake linings. Springs hold the shoes in at low engine speeds, so that there is no contact with the outer drum, which is attached to the transmission input shaft. As engine speed increases, centrifugal force causes the clutch shoes to move outward and begin sliding against the outer drum. The transmission input shaft begins to rotate, causing the rotor to turn, slowly at first, but increasing as the friction increases between the clutch shoes and transmission drum. As rotor speed increases, the rotor tachometer needle indicates this by moving toward the engine tachometer needle. The engine and the rotor are synchronized when the two needles are superimposed, indicating the clutch is fully engaged and the shoes are no longer slipping.

Belt Drive Clutch

Some helicopters utilize a belt drive to transmit power from the engine to the transmission. A belt drive consists of a lower pulley attached to the engine, an upper pulley attached to the transmission input shaft, a belt or a series of V-belts, and some means of applying tension to the belts. The belts fit loosely over the upper and lower pulley when there is no tension on them. This allows the engine to be started without any load from the transmission. Once the engine is running, moving the idler pulley gradually increases tension on the belts. This is accomplished either electrically or mechanically. When the rotor and engine tachometer needles are superimposed, the rotor and the engine are synchronized and the clutch is fully engaged. Advantages of this system include vibration isolation, simple maintenance, and the ability to start and warm up the engine without engaging the rotor.

Since rotating airfoils provide lift in a helicopter, the rotor system must be free to rotate if the engine fails. The freewheeling unit automatically disengages the engine from the main rotor in the event the engine stops providing power to the main rotor. This allows the main rotor to continue turning at normal in-flight speeds. The most common freewheeling unit assembly consists of a one-way sprag clutch located between the engine and main rotor transmission. This usually is located in the upper pulley in a piston helicopter, or mounted on the engine gearbox in a turbine helicopter. When the engine is driving the rotor, inclined surfaces force rollers against an outer drum. This prevents the upper pulley from exceeding transmission input shaft r.p.m. If the engine fails, the rollers move inward, allowing the outer drum to exceed the speed of the inner portion. The transmission input shaft and tail rotor drive shaft can then exceed the speed of the upper pulley. In this condition, engine speed is less than that of the drive system, and the helicopter is in an autorotative state. [Figure 1-130]

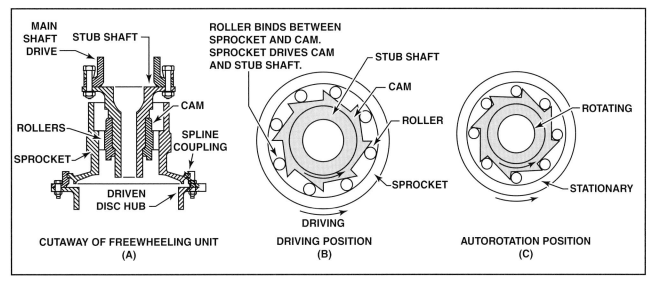

Figure 1-130. Here is an example of a typical freewheeling unit.

CHAPTER 2

SHEET METAL STRUCTURES

INTRODUCTION

Materials used in aircraft construction have changed significantly since the Wright brothers built the first practical airplane. The Wright Flyer was constructed from wood and fabric, as were most early aircraft. Later, to increase strength and durability, manufacturers replaced wood substructures with welded steel tubing. However, metallic materials such as aluminum and stainless steel were eventually used not only in the substructure, but also as the outer covering. Today, most aircraft are primarily fabricated from metallic components, although advanced composite materials are being widely used mostly on control surfaces and nonstructural components. In the future, composite materials will constitute a greater percentage of an aircraft's structure, but metallic materials will certainly continue to be used for many years. This chapter serves as an introduction to the design, construction, inspection, and repair of basic airframe structures made from sheet metal. For detailed instructions regarding maintenance practices for specific airframe components, refer to the manufacturer's publications such as the maintenance manual (MM) or structural repair manual (SRM).

SECTION A

METALLIC AIRCRAFT CONSTRUCTION

During their career, nearly all aviation maintenance technicians are involved to some degree in the fabrication or repair of sheet metal aircraft structures. To fully understand the theories and procedures used in sheet metal work, a technician must recognize the construction techniques incorporated into the design of an airframe that allow it to handle different types of structural loads and stresses. In addition, a technician must also be aware of how different physical properties of sheet metal materials increase the durability and strength of an aircraft. With an understanding of these items, a technician can maintain an aircraft to its original condition, or properly alter its design to improve aircraft performance or safety.

STRESSES AND STRUCTURES

Aircraft structures must be strong, lightweight, streamlined, and durable. However, in most cases, all these requirements are unobtainable without compromise. Truss-type structures can be made both lightweight and strong; but with a resulting angular form, require a superstructure to make them streamlined. A wooden monocoque design does provide a streamlined form with high strength, but with compromises due to the limited useful life of wood, as well as the high cost involved with fabricating a laminated wood structure.

Today, the high volume of aircraft production has caused riveted or bonded sheet metal designs to become the most common method of construction. This is because sheet metal fabrication has been simplified over the years by developing and standardizing the tooling and equipment necessary for assembling most structures. Consequently, sheet metal aircraft are relatively easy to fabricate in a production facility as well as to repair in the field. [Figure 2-1]

The type of metal most widely used for aircraft structures comes from aluminum alloys, which account for as much as 90% of the metals used for

Figure 2-1. Modern sheet metal construction techniques produce airplanes that are strong, streamlined, and economical to assemble or repair.

civil aircraft. Heat-treated aluminum alloys have the advantage of being lightweight with the ability to carry high structural loads, while being comparably inexpensive with regard to other similar strength metals. These assets make aluminum alloy an excellent choice to use for the construction of most modern civil aircraft. The remaining 10% of metals used include titanium, stainless steel, and assorted exotic metals that are predominantly used on military or large transport category aircraft.

TYPES OF SHEET METAL STRUCTURES

There are two basic types of sheet metal structures used for aircraft; monocoque and semimonocoque. Both of these are forms of stressed skin structures, meaning that the greatest part of the load is carried in the external skin. However, these structures differ in the amount of internal components that they use. While both designs typically incorporate formers and bulkheads, a semimonocoque airframe also utilizes stringers and longerons to add rigidity and strength to the structure.

A thin metal beverage can is an excellent example of **monocoque** construction. The can has two thin metal ends attached to an even thinner body, but as an assembly, can support a large load if a force is applied evenly across the ends of the container. However, a problem with a monocoque structure is that just a small dent or crease in a side wall can destroy its ability to support a load.

Semimonocoque structures minimize the problem of failures caused by dents and creases by supporting the external skin on a framework of formers, stringers, and longerons. The internal structure stiffens the skin so it is far less susceptible to failure caused by deformation. Because of its increased strength, a semimonocoque design is used in the construction of most modern aircraft.

Even the skin of an aircraft may be made more rigid across any large unsupported panels by riveting stiffeners or by laminating honeycomb material to the panels. Some modern high-speed jet aircraft have skins that are milled for increased rigidity. With these skins, stiffeners are formed on the inner surface by either conventional machining with a computer-controlled mill or by chemical or electrochemical milling.

STRUCTURAL LOADS

An airplane manufacturer must consider all of the loads to which the structure will be subjected in order to design each component properly. A safety factor is then built in to provide for any unusual or unanticipated loads that may conceivably be encountered. In addition to meeting all of the strength requirements, the structure must be as lightweight as possible and reasonably easy to construct.

As a maintenance technician it is your responsibility to be sure that any repair you make to an aircraft restores both the original strength and stiffness to the structure, while maintaining the original shape of the part. These considerations must be met while keeping the weight of the repair to an absolute minimum. In most cases of major structural damage, to guarantee that a repair meets minimum standards for strength, an aeronautical or structural engineer must develop and approve the design of the repair. These engineers have the credentials and experience to analyze the structural loads on the damaged area. From their analysis, these engineers can produce drawings and instructions that the technician must follow closely to accomplish a satisfactory repair. However, for some repairs, an aircraft technician can refer to the manufacturer's Structural Repair Manual or Maintenance Manual to determine if preapproved information is already available to meet FAA requirements. Even in cases where the damage is slight, it is permissible for a technician to use reference materials such as the Advisory Circular *AC 43.13-1B, Acceptable Methods, Techniques and Practices/Aircraft Inspection, Repair, and Alterations*. With this publication, acceptable procedures for performing some relatively common sheet metal repairs can be found. Although the advisory circular information is not FAA-approved data in itself, it can be used to substantiate an approved repair technique.

STRESSES

In order to determine that a repair will restore full strength to a damaged sheet metal structure, you should understand how various stresses act on an airframe. Five types of stress are of major concern. Two of these are primary, and the other three, for all practical purposes, can be expressed in terms of the first two. Tension and compression are the two basic stresses, and the other three, bending, torsion, and shear, are essentially different arrangements of tension and compression working on a body at the same time.

TENSION

Tension is a primary stress that tries to pull a body apart. When a weight is supported by a cable, the cable is subjected to tension or, as it is often expressed, to a tensile stress. The weight attempts to pull the cable apart. [Figure 2-2]

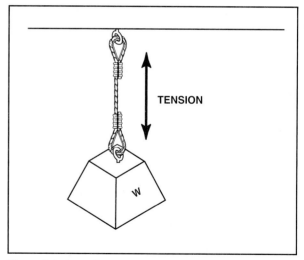

Figure 2-2. A tensile stress exerted on a cable tends to pull the cable apart.

COMPRESSION

Compression, another primary stress, tries to squeeze a part together. For example, a weight supported on a post exerts a force that tries to squeeze the ends of the post together, or to collapse it. This is called a compressive stress. [Figure 2-3]

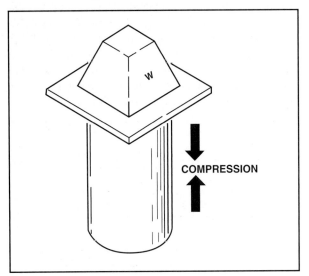

Figure 2-3. An example of a compressive stress is realized by analyzing the effect of a column that is supporting a weight. The weight tends to squeeze the ends of the column together.

BENDING

A bending force tries to pull one side of a body apart while at the same time squeezing the other side together. When a person stands on a diving board, the top of the board is under a tensile stress while the bottom is under compression. The wing spars of a cantilever wing or the section of a wing spar outboard of a strut is subjected to bending stresses. In flight, the top of a spar is being compressed while the bottom is under tension, but on the ground, the top is pulled and the bottom is compressed. Therefore, the major stresses imposed on wing struts reverse in flight as compared to on the ground. For this reason, it is important that spars be able to withstand both compressive and tensile stresses. [Figures 2-4 and 2-5]

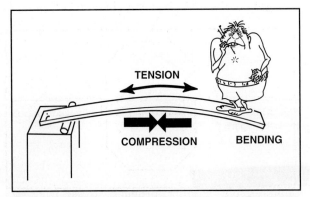

Figure 2-4. When a diving board is under a bending stress, tension stresses occur on the top of the board, while the bottom is under compression.

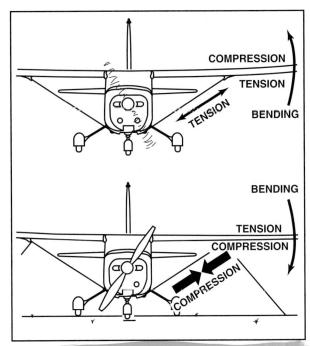

Figure 2-5. Wing struts are predominantly under tension in flight, but under compression when an airplane is on the ground. Outboard of a wing strut, the remaining portion of a wing is subjected to bending stresses.

TORSION

Torsion is a twisting force. When a structural member is twisted or placed under torsion, a tensile stress acts diagonally across the member and a compressive stress acts at right angles to the tension. For example, the crankshaft of an aircraft engine is under a torsional load when the engine rotates the propeller. [Figure 2-6]

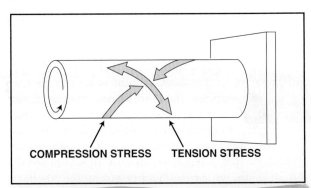

Figure 2-6. A torsional stress consists of tension and compression acting perpendicular to each other, with both acting diagonally across the body.

SHEAR

Shear loads are created when opposing forces are applied on opposite sides of a body. For example, a rivet is primarily designed to withstand shear loads from overlapping sheets of metal that are subjected to being pulled in opposite directions. Rivets hold pieces of aircraft skin together, and in a properly designed joint, the rivets support more bearing or tensile load than shear load. [Figure 2-7]

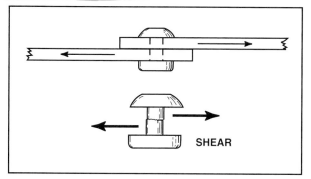

Figure 2-7. A shear stress on a rivet attempts to slide through the rivet shank.

RIVET JOINT CONSIDERATIONS

The design of an aircraft repair is complicated by the requirement that it be as lightweight as possible. If weight were not critical, all repairs could be made with a large margin of safety so there would never be a concern about the strength of the repair. However, in actual practice, repairs must be strong enough to carry all of the loads with the required safety factor, but also as lightweight as possible. On the other hand, a joint must also be manufactured in a way that if it is subjected to extreme loads, the fasteners will fail instead of the base metal. For these reasons, a joint that is too weak cannot be tolerated, but neither can one that is too strong.

BEARING STRENGTH

Bearing strength can be characterized by a sheet of metal being able to withstand being torn away from the rivets in a joint. The bearing strength of a material is affected by both its thickness and by the size of the rivet in the sheet. A joint is said to bear up when, for example, the landing gear makes contact with the runway upon landing and the wings droop downward (with the skins pulling against the rivets), and then springs back into normal position. The spring-back recovery results from the tensile stress produced by the bearing strength of the metal against the shear strength of the rivets. In other words, the joint is manufactured to have a certain amount of elasticity.

SHEAR VERSUS BEARING STRENGTH

Most aircraft structures are held together by the clamping action of either rivets or bolts. When fabricating a riveted joint, consider both the shear strength of the rivet (the amount of force that is needed to cut it in two) and the bearing strength of the sheet metal (the amount of force that will cause the rivet to tear out from the metal). In a properly designed joint, the bearing strength and shear strength should be as near the same as possible, with the shear strength being slightly less. When this is provided, the joint will support the maximum load, but if it does fail, the rivet will shear. It is much less costly to replace a rivet than it is to repair a hole torn in the metal. [Figure 2-8]

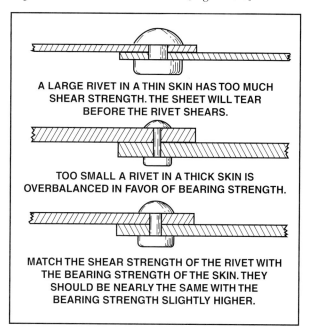

Figure 2-8. Rivet selection must be matched to skin thickness.

TRANSFER OF STRESSES WITHIN A STRUCTURE

An aircraft structure must be designed in such a way that it will accept all of the stresses imposed on it by flight or ground loads without any deformation. When repairs are made to the structure, they must be made to accept the stresses, carry them across the repair, and then transfer them back into the original structure. In this manner, the original integrity of the part is restored. [Figure 2-9]

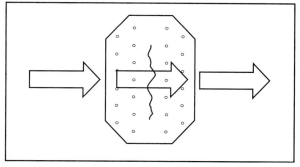

Figure 2-9. Any repair to an aircraft structure must accept all of the loads, support the loads, and then transfer them back into the structure.

Stresses can be thought of as flowing through a structure, so there must be a complete path for them to travel with no abrupt changes in the cross-sectional area along the way. Abrupt changes in area will cause the stresses to concentrate, and it is at such a point that failures occur. A scratch or gouge in the surface of a highly stressed piece of metal will obstruct the flow of stresses and concentrate them to the point that the metal will fail at the defect. [Figure 2-10]

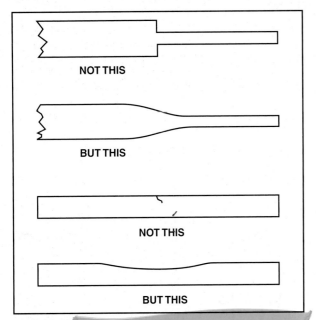

Figure 2-10. Abrupt changes in the cross-sectional area of a part must be avoided. An abrupt area change will concentrate the stresses and could cause the part to fail. To prevent gouges and scratches from becoming stress concentration points, burnishing is often accomplished. Burnishing involves using special tools called burnishing knives or spoons to taper and smooth the edges of the defect.

The thin metal that most aircraft structures are made of is subject to vibration and stresses, which in time, can cause cracks to form. These cracks may start out to be reasonably small, but if they form in the edge of a sheet that is being subjected to a tensile stress, the stresses will concentrate at the end of the crack. Eventually, the crack will propagate, causing the sheet to tear completely through. For example, if a small crack starts in the edge of a piece of 0.032 inch sheet aluminum alloy which has a tensile strength of 64,000 pounds per square inch (psi), it will take a stress of just over two pounds to extend the crack. Normal vibrations can produce stresses that far exceed this amount.

Metal thickness = .032 inch

Width at end of crack = .001 inch

Area subjected to tensile stress = .000032 sq. inches

Ultimate tensile strength of metal = 64,000 psi

Stress needed to cause the crack to extend = 2.048 pounds.

A standard fix for a crack developing in sheet metal is to stop-drill the end of the crack and rivet a patch over the entire damaged area. By drilling a hole with a number 30 bit at the end of the crack, the area on which the stresses concentrate increases from the .000032 sq. inches to .0129 sq. inches. This causes the stress needed to extend the crack to increase to about 819.2 pounds.

Diameter of number 30 drill = .128 inch

Circumference of number 30 drill = .402 inch

Area subjected to tensile stress = .0129 sq. inches

Stress needed to extend crack = 819.2 pounds.

To complete the repair, a small patch can be riveted over the crack to stiffen the edge of the material so the crack will not be subject to vibration. [Figure 2-11]

MATERIALS FOR SHEET METAL AIRCRAFT CONSTRUCTION

In Chapter Seven of the *A&P Technician General Textbook*, aircraft metal classifications and designations were discussed in detail. In this section, those metals used in the repair of aircraft structures are covered, including a brief review of the physical characteristics of these materials. By recognizing these characteristics, you will better understand how to select the ideal materials for a specific repair.

ALUMINUM ALLOYS

Pure aluminum lacks sufficient strength to be used for aircraft construction. However, its strength increases considerably when it is alloyed or mixed with other compatible metals. For example, when mixed with copper or zinc, the resultant aluminum alloy is as strong as steel, with only one-third the weight. Furthermore, the corrosion resistance possessed by the aluminum carries over to the newly formed alloy.

ALLOYING AGENTS

Aluminum alloys are classified by their major alloying ingredient. The elements most commonly used for aluminum alloying are copper, magnesium, manganese, and zinc. Wrought aluminum and wrought aluminum alloys are identified by a four-digit index system. The first digit of a designation

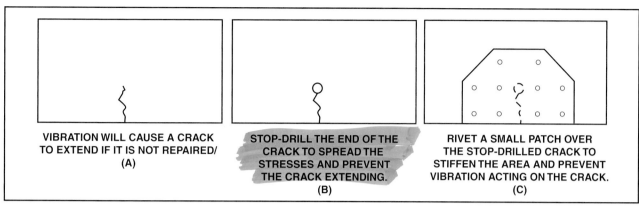

Figure 2-11. A crack in the edge of a sheet metal panel must be repaired to prevent the crack from expanding. However, cracks can also develop in the middle of a sheet due to large vibrations. For example, a common location for cracks to develop is ahead of control surfaces that are improperly balanced or that have extreme wear in the hinges. The air loads imposed on the control surface can cause it to aerodynamically flutter, generating extreme vibrations and stresses that are carried through the airframe structure. Eventually, the vibrations can cause cracks to form, which then expand at an accelerated rate.

identifies the major alloying element used in the formation of the alloy. The most common alloying elements used are as follows:

1xxx — aluminum
2xxx — copper
3xxx — manganese
4xxx — silicon
5xxx — magnesium
6xxx — magnesium and silicon
7xxx — zinc
8xxx — other elements

The second number represents a specific alloy modification. For example, if this digit is zero, it indicates there were no special controls over individual impurities. However, a digit of 1 through 9 indicates the number of controls over impurities in the metal.

The last two numbers of the 1xxx group of alloys are used to indicate the hundredths of 1% above the original 99% pure aluminum. For example, if the last two digits are 75, the alloy contains 99.75% pure aluminum. However, in the 2xxx through 8xxx groups, the last two digits identify the different alloys in the group. [Figure 2-12]

The 1xxx series of aluminum alloys represents commercially pure aluminum, of 99% or higher purity. Pure aluminum offers high corrosion resistance, excellent thermal and electrical properties, and is easily worked. However, pure aluminum is very low in strength.

Alloys within the 2xxx series utilize **copper** as the principle alloying agent. When aluminum is mixed with copper, certain metallic changes take place in the resultant alloys grain structure. For the most part, these changes are beneficial and produce greater strength. However, a major drawback to aluminum-copper alloys is their susceptibility to intergranular corrosion when improperly heat-treated. Most aluminum alloy used in aircraft structures is an aluminum-copper alloy. Two of the most com-

PERCENT OF ALLOYING ELEMENTS ALUMINUM AND NORMAL IMPURITIES CONSTITUTE REMAINDER								
ALLOY	**COPPER**	**SILICON**	**MANGANESE**	**MAGNESIUM**	**ZINC**	**CHROMIUM**	
1100							Seldom used in aircraft
3003			1.2				Cowling and non-structural parts
2017	4.0		0.5	0.5			Obsolete — superseded by 2024
2117	2.5			0.3			Most aircraft rivets
2024	4.5		0.6	1.5			Majority of structure
5052				2.5		0.25	Gas tanks and fuel lines
7075	1.6			2.5	5.6	0.3	High strength requirements

Figure 2-12. A variety of elements are used to produce aluminum alloys.

monly used in the construction of skins and rivets are 2017 and 2024.

The 3xxx series alloys have **manganese** as the principle alloying element, and are generally considered nonheat treatable. The most common variation is 3003, which offers moderate strength and has good working characteristics.

The 4xxx series aluminum is alloyed with **silicon**, which lowers a metal's melting temperature. This results in an alloy that works well for welding and brazing.

Magnesium is used to produce the 5xxx series alloys. These alloys possess good welding and corrosion-resistance characteristics. However, if the metal is exposed to high temperatures or excessive cold-working, its susceptibility to corrosion increases.

If silicon and magnesium are added to aluminum, the resultant alloy carries a 6xxx series designation. In these alloys, the silicon and magnesium form magnesium silicide, which makes the alloy heat treatable. Furthermore, the 6xxx series has medium strength with good forming and corrosion-resistance properties.

When parts require more strength and little forming, harder aluminum alloys are employed. The 7xxx series aluminum alloys are made harder and stronger by the addition of **zinc**. Some widely used forms of zinc-aluminum alloys are 7075 and 7178. The aluminum-zinc alloy 7075 has a tensile strength of 77 thousand pounds per square inch (KSI) and a bearing strength of 139 KSI. However, the alloy is very hard and is difficult to bend. An even stronger zinc alloy is 7178, which has a tensile strength of 84 KSI and a bearing strength of 151 KSI.

CLAD ALUMINUM ALLOY
Most external aircraft surfaces are made of clad aluminum. **Alclad** consists of a pure aluminum coating rolled onto the surface of heat-treated aluminum alloy. The thickness of this coating is approximately 5% of the alloy's thickness on each side. For example, if an alclad sheet of aluminum is .040 inch thick, then 5%, or .002 inches of pure aluminum, is applied to each side. This results in an alloy thickness of .036 inch.

This clad surface greatly increases the corrosion resistance of an aluminum alloy. However, if it is penetrated, corrosive agents can attack the alloy within. For this reason, sheet metal should be protected from scratches and abrasions. In addition to providing a starting point for corrosion, abrasions create potential stress points.

HEAT TREATMENT
Heat treatment is a series of operations involving the heating and cooling of metals in their solid state. Its purpose is to make the metal more useful, serviceable, and safe for a definite purpose. By heat-treating, a metal can be made harder, stronger, and more resistant to impact. Heat-treating can also make a metal softer and more ductile. However, one heat-treating operation cannot produce all these characteristics. In fact, some properties are often improved at the expense of others. In being hardened, for example, a metal may become brittle.

All heat-treating processes are similar in that they involve the heating and cooling of metals. They differ, however, in the temperatures to which the metal is heated and the rate at which it is cooled.

There are two types of heat-treatments used on aluminum alloys. One is called solution heat-treatment, and the other is known as precipitation heat-treatment. Some alloys, such as 2017 and 2024, develop their full properties as a result of solution heat-treatment, followed by about four days of cooling, or **aging**, at room temperature. However, other alloys, such as 2014 and 7075, require both heat-treatments.

Solution Heat Treatment
When aluminum is alloyed with materials such as copper, magnesium, or zinc, the resultant alloys are much stronger than aluminum. To understand why this happens, it is necessary to examine the microscopic structure of aluminum. Pure aluminum has a molecular structure that is composed of weakly bonded aluminum atoms and, therefore, is extremely soft. Aluminum alloys, on the other hand, consist of a base metal of aluminum and an alloying element that is dispersed throughout the structure. In this configuration, when the aluminum alloy is subjected to stress, these alloying particles adhere to the aluminum molecules and resist deformation. However, special processes must be used to allow the base metal and alloy to mix properly. For example, when aluminum is alloyed with copper through conventional processes, approximately .5% of the copper dissolves, or mixes, with the aluminum. The remaining copper takes the form of the compound $CuAl_2$. However, when the aluminum alloy is heated suffi-

ciently, the remaining copper enters the base metal and hardens the alloy.

The process of heating certain aluminum to allow the alloying element to mix with the base metal is called solution heat-treating. In this procedure, metal is heated in either a molten sodium or potassium nitrate bath or in a hot-air furnace to a temperature just below its melting point. The temperature is then held to within about plus or minus 10°F of this temperature and the base metal is soaked until the alloying element is uniform throughout. Once the metal has sufficiently soaked, it is removed from the furnace and cooled, or quenched. It is extremely important that no more than about 10 seconds elapse between removal of an alloy from the furnace and the quench. The reason for this is that when metal leaves the furnace and starts to cool, its alloying metals begin to precipitate out of the base metal. If this process is not stopped, large grains of alloy become suspended in the aluminum and weaken the alloy. Excessive precipitation also increases the likelihood of intergranular corrosion.

To help minimize the amount of alloying element that precipitates out of a base metal, a quenching medium is selected to ensure the proper cooling rate. For example, a water spray or bath provides the appropriate cooling rate for aluminum alloys. However, large forgings are typically quenched in hot water to minimize thermal shock that could cause cracking. Thin sheet metal normally warps and distorts when it is quenched, so it must be straightened immediately after it is removed from the quench. After the quench, all metals must be rinsed thoroughly, since the salt residue from the sodium or potassium nitrate bath can lead to corrosion if left on the alloy.

Precipitation Heat Treatment

Heat-treatable aluminum alloys are comparatively soft when first removed from a quench. With time, however, the metal becomes hard and gains strength. When an alloy is allowed to cool at room temperature, it is referred to as **natural aging** and can take several hours or several weeks. For example, aluminum alloyed with copper gains about 90% of its strength in the first half-hour after it is removed from the quench, and becomes fully hard in about four or five days.

Alloy aging times can be lengthened, or shortened. For example, the aging process can be slowed by storing a metal at a subfreezing temperature immediately after it is removed from the quench. On the other hand, reheating a metal and allowing it to soak for a specified period can accelerate the aging process. This type of aging is referred to as **artificial aging**, or **precipitation heat treatment**, and develops hardness, strength, and corrosion resistance by locking a metal's grain structure together.

Naturally aged alloys, such as the copper-zinc-magnesium alloys, derive their full strength at room temperature in a relatively short period and require no further heat-treatment. However, other alloys, particularly those with a high zinc content, need thermal treatment to develop full strength. These alloys are called artificially aged alloys.

Annealing

Annealing is a process that softens a metal and decreases internal stress. In general, annealing is the opposite of hardening. To anneal an aluminum alloy, the metal temperature is raised to an annealing point and held there until the metal becomes thoroughly heat soaked. It is then cooled to 500°F at a rate of about 50°F per hour. Below 500°F, the rate of cooling is not important.

When annealing clad aluminum metals, they should be heated as quickly and as carefully as possible. The reason for this is that if clad aluminum is exposed to excessive heat, some of the core material tends to mix with the cladding. This reduces the corrosion resistance of the metal. [Figure 2-13]

Heat-Treatment Identification

Heat-treatable alloys have their hardness condition designated by the letter "T", followed by one or more

ALLOY	ANNEALING TREATMENT		
	METAL TEMPERATURE °F	APPROX. TIME OF HEATING HOURS	TEMPER DESIGNATION
1100	550		-0
2017	775	2-3	-0
2024	775	2-3	-0
2117	775	2-3	-0
3003	775		-0
5052	650		-0
6061	775	2-3	-0
7075	775	2-3	-0

Figure 2-13. Aluminum alloys are heat-treated to increase their strength and improve their working characteristics. Heat-treatment temperatures and soak times are critical to disperse the alloying elements.

numbers. A listing of the more popular temper designation codes includes the following: [Figure 2-14]

HEAT-TREATABLE ALLOYS	
TEMPER DESIGNATION	DEFINITION
-0	Annealed recrystallized (wrought products only) applies to softest temper of wrought products.
-T3	Solution heat-treated and cold-worked by the flattening or straightening operation.
-T36	Solution heat-treated and cold-worked by reduction of 6 percent.
-T4	Solution heat-treated.
-T42	Solution heat-treated and aged by user regardless of prior temper (applicable only to 2014 and 2024 alloys).
-T5	Artificially aged only (castings only).
-T6	Solution heat-treated and artificially aged.
-T62	Solution heat-treated and aged by user regardless of prior temper (applicable only to 2014 and 2024 alloys).
-T351, -T451 -T3510, -T3511 -T4510, -T4511	Solution heat-treated and stress relieved by stretching to produce a permanent set 1 to 3 percent, and depending on product.
-T651, -T851 -T6510, -T8510 -T6511, -T8511	Solution heat-treated and stress relieved by stretching to produce a permanent set 1 to 3 percent, and artificially aged.
-T652	Solution heat-treated, compressed to produce a permanent set and then artificially aged.
-T81	Solution heat-treated, cold-worked by the flattening or straightening operation, and then artificially aged.
-T86	Solution heat-treated, cold-worked by reduction of 6 percent, and then artificially aged.
-F	For wrought alloys, as fabricated. No mechanical properties limits. For cast alloys, as cast.

Figure 2-14. All heat treatment processes are identified from the manufacturer by the use of a standardized temper designation number.

Reheat-Treatment

Material which has been previously heat treated can generally be reheat-treated any number of times. As an example, rivets made of 2017 or 2024 are extremely hard and typically receive several reheat-treatments to make them soft enough to drive. However, the number of solution heat-treatments allowed for clad materials is limited due to the increased diffusion of core material into the cladding. This diffusion results in decreased corrosion resistance. As a result, clad material is generally limited to no more than three reheat-treatments.

NONHEAT-TREATABLE ALLOYS

Commercially pure aluminum does not benefit from heat treatment since there are no alloying materials in its structure. By the same token, 3003 is an almost identical metal and, except for a small amount of manganese, does not benefit from being heat treated. Both of these metals are lightweight and somewhat corrosion resistant. However, neither has a great deal of strength and, therefore, their use in aircraft is limited to nonstructural components such as fairings and streamlined enclosures that carry little or no load.

Alloy 5052 is perhaps the most important of the nonheat-treatable aluminum alloys. It contains about 2.5% magnesium and a small amount of chromium. It is used for welded parts such as gasoline or oil tanks, and for rigid fluid lines. Cold working increases its strength.

STRAIN-HARDENING AND HARDNESS DESIGNATIONS

Both heat-treatable and non-heat-treatable aluminum alloys can be strengthened and hardened through strain hardening, also referred to as cold working or work hardening. This process requires mechanically working a metal at a temperature below its critical range. Strain hardening alters the grain structure and hardens the metal. The mechanical working can consist of rolling, drawing, or pressing.

Heat-treatable alloys have their strength increased by rolling after they have been solution heat treated. On the other hand, non-heat-treatable alloys are hardened in the manufacturing process when they are rolled to their desired dimensions. However, at times these alloys are hardened too much, and must be partially annealed.

Where appropriate, the metal hardness, or temper, is indicated by a letter designation that is separated from the alloy designation by a dash. When the basic temper designation must be more specifically defined one or more numbers follow the letter designation. These designations are as follows:

-F For wrought alloys; As fabricated. No mechanical property limits. For cast alloys; As cast.

-O Annealed, recrystallized (wrought materials only); Applies to softest temper of wrought products.

-H Strain hardened.

-H1 Strain hardened only. Applies to products which are strain hardened to obtain the desired strength without supplementary thermal treatment.

-H2 Strain hardened and partially annealed.

-H3 Strain hardened and stabilized.

When a digit follows the designations H1, H2, or H3, the second number indicates the degree of strain hardening. For example, the number 8 in the designation H18 represents the maximum tensile strength, while in H10, the 0 indicates an annealed state. The most common designations include:

-Hx2 Quarter-hard
-Hx4 Half-hard
-Hx6 Three-quarter hard
-Hx8 Full-hard

MAGNESIUM AND ITS ALLOYS

Magnesium alloys are used for castings, and in their wrought form, are available in sheets, bars, tubing, and extrusions. Magnesium is one of the lightest metals having sufficient strength and suitable working characteristics for use in aircraft structures. It has a density of 1.74, compared with 2.69 for aluminum. In other words, it weighs only about 2/3 as much as aluminum.

Magnesium is obtained primarily from electrolysis of sea water or brine from deep wells, and lacks sufficient strength in its pure state for use as a structural metal. However, when alloyed with zinc, aluminum, thorium, zirconium, or manganese, it develops strength characteristics that make it quite useful.

The American Society for Testing Materials (ASTM) has developed a classification system for magnesium alloys that consists of a series of letters and numbers to indicate alloying agents and temper condition. [Figure 2-15]

Magnesium has some rather serious drawbacks that have to be overcome before it can be used successfully. For example, magnesium is highly susceptible to corrosion, and tends to crack. The cracking con-

ALLOYING ELEMENTS	TEMPER CONDITION
A – ALUMINUM	F – AS FABRICATED
E – RARE EARTH	O – ANNEALED
H – THORIUM	H24 – STRAIN HARDENED AND PARTIALLY ANNEALED
K – ZIRCONIUM	T4 – SOLUTION HEAT-TREATED
M – MANGANESE	T5 – ARTIFICALLY AGED ONLY
Z – ZINC	T6 – SOLUTION HEAT-TREATED AND ARTIFICIALLY AGED

Figure 2-15. Magnesium alloys use a different designation system than aluminum. For example, the designation AZ31A-T4 identifies an alloy containing 3% aluminum and 1% zinc that has been solution heat-treated.

tributes to its difficulty in forming and limits its use for thin sheet metal parts. However, this tendency is largely overcome by forming parts while the metal is hot. Treating the surface with chemicals that form an oxide film to prevent oxygen from reaching the metal minimizes the corrosion problem. When oxygen is excluded from the surface, no corrosion can form. Another important step in minimizing corrosion is to always use hardware such as rivets, nuts, bolts, and screws that are made of a compatible material.

In addition to cracking and corroding easily, magnesium burns readily in a dust or small-particle form. For this reason, caution must be exercised when grinding and machining magnesium. If a fire should occur, extinguish it by smothering it with dry sand or some other dry material that excludes air from the metal and cools its surface. If water is used, it will only intensify the fire.

Solution heat-treatment of magnesium alloys increases tensile strength, ductility, and resistance to shock. After a piece of magnesium alloy has been solution heat-treated, it can be precipitation heat-treated by heating it to a temperature lower than that used for solution heat-treatment, and holding it at this temperature for a period of several hours. This increases the metal hardness and yield strength.

TITANIUM AND ITS ALLOYS

Titanium and its alloys are lightweight metals with very high strength. Pure titanium weighs .163 pounds per cubic inch, which is about 50% lighter than stainless steel, yet it is approximately equal in strength to iron. Furthermore, pure titanium is soft and ductile with a density between that of aluminum and iron.

Titanium is a metallic element which, when first discovered, was classified as a rare metal. However, in 1947 its status was changed due to its importance as a structural metal. In the area of structural metallurgy, it is said that no other structural metal has been studied so extensively or has advanced aircraft structures so rapidly.

In addition to their light weight and high strength, titanium and its alloys have excellent corrosion resistance characteristics, particularly to the corrosive effects of salt water. However, since the metal is sensitive to both nitrogen and oxygen, it must be converted to titanium dioxide with chlorine gas and a reducing agent before it can be used.

Titanium is classified as alpha, alpha-beta, or beta alloys. These classifications are based on specific chemical bonding within the alloy itself. The specifics of the chemical composition are not critical to working with the alloy, but certain details should be known about each classification.

Alpha alloys have medium strengths of 120 KSI to 150 KSI and good elevated-temperature strength. Because of this, alpha alloys can be welded and used in forgings. The standard identification number for alpha titanium is 8Al-1Mo-1V-Ti, which is also referred to as Ti-8-1-1. This series of numbers indicates that the alloying elements and their percentages are 8% aluminum, 1% molybdenum, and 1% vanadium.

Alpha-beta alloys are the most versatile of the titanium alloys. They have medium strength in the annealed condition and much higher strength when heat-treated. While this form of titanium is generally not weldable, it has good forming characteristics.

Beta alloys have medium strength, excellent forming characteristics, and contain large quantities of high-density alloying elements. Because of this, beta titanium can be heat-treated to a very high strength.

The grain size of titanium is refined when aluminum is added to the alloy mixture. However, when copper is added to titanium, a precipitation-hardening alloy is produced. Titanium added to high temperature nickel-cobalt-chromium alloy produces a precipitation-hardening reaction, which provides strength at temperatures up to 1,500°F. [Figure 2-16]

Because of its high strength-to-weight ratio, titanium is now used extensively in the civilian aerospace industry. Although once rare on commercial aircraft, alloys containing 10 to 15 percent titanium in structural areas are utilized on modern jet transports.

STAINLESS STEEL

Stainless steel is a classification of corrosion-resistant steels that contain large amounts of chromium and nickel. Their strength and resistance to corrosion make them well suited for high-temperature applications such as firewalls and exhaust system components. These steels can be divided into three general groups based on the chemical structure: austenitic, ferritic, and matensitic.

Austenitic steels, also referred to as 200 and 300 series stainless steels, contain a large percentage of chromium and nickel, and in the case of the 200 series, some manganese. When these steels are heated to a temperature above the critical range and held there, a structure known as austenite forms. Austenite is a solid solution of pearlite, an alloy of iron and carbon, and gamma iron, which is a non-magnetic form of iron. Only cold-working can harden austenitic stainless steels, while heat treatment serves only to anneal them.

Ferritic steels are primarily alloyed with chromium,

ALLOY	COMPOSITION	TENSILE STRENGTH	ELONGATION
ALPHA	5% AL - 2.5% SN	130 KSI	15%
ALPHA-BETA	6% AL - 4% V	140 KSI	15%
ALPHA-BETA HEAT-TREATED	6% AL - 4% V	180 KSI	7%
BETA	13% V - 11% CR - 3% AL	150 KSI	15%
BETA HEAT-TREATED	13% V - 11% CR - 3% AL	200 KSI	6%
DEFINITIONS OF LETTERS AND PERCENTAGES ARE: AL - ALUMINUM CR - CHROMIUM V - VANADIUM SN - TIN			

Figure 2-16. This table illustrates the composition, tensile strength, and elongation of titanium alloys. The degree of strength is denoted by the smaller hole elongation percentage shown in the last column. The titanium alloy most commonly used by the aerospace industry is an alpha-beta heat-treated alloy called 6Al-4V. This alloy has a tensile strength of 180 KSI, or 180,000 pounds per square inch. It is frequently used for special fasteners.

but many also contain small amounts of aluminum. However, they contain no carbon and, therefore, do not respond to heat-treatment.

The 400 series of stainless steel is a **martensitic steel**. These steels are alloyed with chromium only and therefore are magnetic. Martensitic steels become extremely hard if allowed to cool rapidly by quenching from an elevated temperature.

The corrosion-resistant steel most often used in aircraft construction is known as 18-8 steel because it contains 18% chromium and 8% nickel. One of the distinctive features of 18-8 steel is that cold-working may increase its strength.

ALUMINUM ALLOY-FACED HONEYCOMB

Very thin sheets of aluminum alloy may meet all of the strength requirements for an aircraft structure, but they may not provide enough rigidity or stiffness to withstand the rough handling to which the structure is subjected. To provide this rigidity, some wing skins, fuselage panels, and floorboards are made of aluminum alloy-faced honeycomb.

With aluminum alloy-faced honeycomb, a core material is made up of aluminum foil formed into a cellular structure similar in shape to that used by the honey bee. This form gives a maximum amount of strength for its weight, but it is strong only against loads that act in line with the cells. The core material is cut into slabs of the proper thickness, and face sheets of thin aluminum alloy are bonded to both sides of the core. This stabilizes the core and produces a panel whose strength is greater than both of the face sheets. It has the rigidity of a thick panel, but its weight is far less than that of a solid panel of similar dimensions, or of a panel built up by any other method. [Figure 2-17]

Repair of a simple puncture of an aluminum alloy-faced honeycomb panel can be accomplished by covering with a doubler plate. The plate should be cut from a piece of aluminum the same or up to one and one-half times the thickness of the original skin thickness. Additionally, the doubler should be tapered at a ratio of about 100:1. The repair of honeycomb panels is covered in the section on Aircraft Wood and Composite Structural Repair in Chapter Three, Wood and Composite Structures, in this text.

CORROSION PREVENTION OF SHEET METAL MATERIALS

As previously mentioned, the susceptibility of aluminum alloys to corrosion is one of the limiting factors as to its use as a structural material. In many

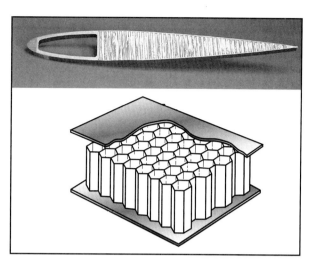

Figure 2-17. Aluminum alloy-faced honeycomb panels are used in aircraft where high strength, light weight, and stiffness are required. Bonded honeycomb is not restricted to straight panels, but may also be used to form complex-shaped control surfaces and is even used for the rotor blades of some helicopters.

cases, this problem has been largely minimized by three methods of protection. These methods are; cladding the alloy with pure aluminum, covering the surface with an impenetrable oxide film, and covering the surface with an organic coating such as primer and paint.

Before looking at each of these methods, a review of how corrosion forms will help you to understand what causes it and what can be done to prevent it. For a more thorough review, this information is covered in much more detail in Chapter Twelve of the *A&P Technician General Textbook*.

Corrosion is an electrochemical action in which an element in the metal is changed into a porous salt of the metal. This salt is the white or gray powder that can be seen on a piece of corroded aluminum. In order for the corrosion to form, three elements must be met. First, there must be an area of electrode potential difference within the metal; second, there must be a conductive path within the metal between these areas; and third, some form of electrolyte must cover the surface between these areas to complete an electrical circuit.

Look at an example of dissimilar metal corrosion to see the way it works. When a steel bolt holds two pieces of aluminum alloy together, there is an opportunity for **galvanic**, or dissimilar metal, corrosion to form. Aluminum is more **anodic**, or more active, than steel and will furnish electrons for the electrical action that occurs when the surface is covered with an electrolyte such as water. When electrons flow from the aluminum to the steel, they leave positive aluminum ions that attract negative hydroxide ions from the water. This results in the formation of aluminum hydroxide, or corrosion, as the aluminum is eaten away.

There are several things that can be done to prevent corrosion. You may have noticed that almost all steel aircraft hardware is plated with a thin coating of cadmium, which has an electrical potential almost the same as aluminum. As long as the cadmium is not scratched through, there will be no contact between the steel and the aluminum and therefore almost no electrode potential difference. To further protect the aluminum from damage, aluminum washers may be placed under both the head of the bolt and the nut, and as even further protection, if the joint is in a corrosive environment, the bolt can be dipped in primer before it is installed. This will exclude water and air from the joint, which will help prevent corrosion from forming. [Figure 2-18]

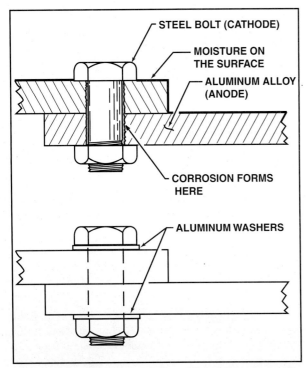

Figure 2-18. Corrosion may be prevented by excluding moisture from the metal and by preventing contact between dissimilar metals.

CLADDING

Pure aluminum will not corrode since there is no electrode potential difference within the metal, but it is too weak for use as an aircraft structural material. However, as previously mentioned, if a thin coating of pure aluminum is rolled onto the surface of the strong aluminum alloy, features of both, the pure metal and the alloy, provide a strong and corrosion-resistant material. For this process, called cladding, there is a penalty by a loss of only about 5% of the strength of the alloy. To help determine if a metal sheet is clad or bare, aluminum alloy can be differentiated from pure aluminum by placing a drop of 10% solution, caustic soda (sodium hydroxide) on the surface. If it is an alloy, it will form a spot.

While pure aluminum does not corrode, it does oxidize; that is, it readily unites with oxygen in the air to form a dull-looking film on its surface. This film is extremely tight and prevents any more oxygen from reaching the metal, so the oxidizing action stops as soon as the film is completely formed on the surface. Clad aluminum may be used as the outside skin for airplanes since it gives a nice silvery appearance without being painted. Care must be taken, however, to prevent any scratches from penetrating through the thin cladding, as the alloy would then be exposed and would corrode. However, it is not unusual for corrosion to form along the edges where the alloy is exposed and between overlapping sheets. These joints are often referred to as fayed edges, and require additional protection. In many cases, fayed edges are sealed with a primer or epoxy filler to exclude moisture.

OXIDE FILM

An oxide film can protect aluminum alloy in the same way it protects pure aluminum; by excluding air and moisture from the metal. Since an aluminum alloy cannot corrode unless an electrolyte is in contact with the metal, the oxide film insulates the surface from the electrolyte, and corrosion cannot form. This protective oxide film may be formed either electrolytically or chemically. Both methods produce a film that not only excludes air from the surface, but also roughens it enough for paint to bond tightly.

PAINT FINISHES

When all-metal airplanes first became popular, they were seldom painted. Instead, the skin was usually of clad aluminum alloy, which had a shiny silver appearance. However, in order to keep the skin shiny, the dull oxide film had to continually be rubbed off, and since this type of surface requires so much care, the modern trend has become to paint the entire aircraft.

With a painted finish, the majority of modern high-volume production aircraft is primed with a two-part wash primer that etches the surface of the metal so paint will adhere. Then, when the primer is completely cured, the entire airplane is sprayed with acrylic lacquer. However, when the cost of the finish allows it and when there is sufficient time in the production schedule, the surface may be primed with epoxy primer and the aircraft finished with polyurethane enamel. These finishes are far more durable than acrylic lacquer and yield a much more attractive finish.

SECTION B

SHEET METAL TOOLS AND FASTENERS

Aircraft constructed from sheet metal require many specialized tools for fabricating the metal and for installing special fasteners. Due to the number of manufacturers of sheet metal equipment and fastener hardware, only a select few items are covered in this section. Continuous work as an aircraft technician, requires learning how to operate additional tools and how to install more types of fasteners. In all cases, refer and adhere to the appropriate manufacturer's instructions regarding the proper use of equipment and hardware that is used for aircraft maintenance.

FABRICATION TOOLS FOR SHEET METAL STRUCTURES

In Chapter 9 of the *A&P Technician General Textbook*, hand tools and measuring devices were discussed in detail. In this section, a review of some of the tools especially designed for sheet metal construction and repair is presented. First, a discussion regarding basic tool selection and operation for sheet metal construction will be covered, followed by an introduction to the use of specialized tools for specific sheet metal fabrications.

LAYOUT TOOLS

When a sheet metal repair is to be made, or a part is to be fabricated, a detailed drawing of the repair or part is sometimes available. For example, a drawing might be available from a structural repair manual. Other times, the technician must draw the repair from scratch, using guidance from a manual or from the FAA Advisory Circular, *AC 43.13-1B, Acceptable Methods, Techniques and Practices/Aircraft Inspection, Repair, and Alterations*. Whether the technician prepares a new drawing or transfers dimensions from a drawing for a pattern, layout tools are necessary to accurately determine dimensions.

SCALES

One of the most commonly used measuring devices for sheet metal layout is a flexible steel scale. These handy tools usually have four sets of graduations on them. For aircraft work, a scale with 32nds and 64ths of an inch on one side, and increments of 50ths and 100ths of an inch on the other, are most useful. For the greatest accuracy when using a scale, do not measure from its end. Instead, start with one of the inch marks. This is recommended because it is quite possible for a scale-end to have been damaged, resulting in an inaccurate measurement. [Figure 2-19]

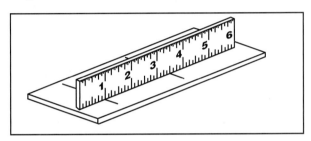

Figure 2-19. Do not make measurements from the end of a steel scale; rather, start with one of its inch marks.

COMBINATION SQUARE

A combination square is another extremely useful tool for sheet metal layout. The blade in this square is removable and is available with many different types of graduations. The blade with graduations of 50ths and 100ths of an inch is the most useful for aircraft layout work. The head of the square has a 45 degree bevel on one side and a 90 degree face on the other, with a spirit level and a steel scribe in the head. [Figure 2-20]

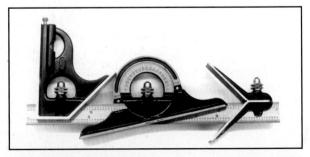

Figure 2-20. A combination square is one of the most useful measuring tools for sheet metal construction and repair. Aside from the stock head, which is commonly found on a combination set, a protractor and center head may be included.

DIVIDERS

A pair of good quality 6 inch spring dividers is essential for layout work. These tools are handy for spacing rivets in a row and for transferring distances when duplicating parts. They may also be used to transfer measurements from a full-scale drawing to a layout on an actual metal part. For example, if an inspection or a lightning hole is needed, dividers are used to scribe the outline of the hole onto the surface of the metal. Dividers are also used to make circles, arcs, and measurement transfers. [Figure 2-21]

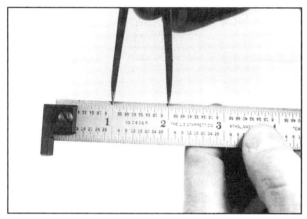

Figure 2-21. Dividers may be set from a steel scale and used to transfer measurements to the sheet metal.

MARKING TOOLS

Once a drawing has been selected, dimensions must be transferred to the surface of the metal for cutting, drilling, shaping, and forming. The marking device will vary according to the surface being marked and the operation to be performed on the sheet metal.

SCRIBES

In sheet metal fabrications, most work is done with soft metal so a carbide-tipped scribe is almost never needed. In many cases, the handiest scribe is made of plain steel and has a removable point that can be reversed in the handle so it will not be dulled by contacting other tools while stored in a toolbox. When using a scribe for marking sheet metal, a colored dye is usually applied to the metal beforehand, and the scribe is then used to scratch through the dye.

Scribes should be used in sheet metal work only for marking the cut-off lines of a part. They should never be used for marking dimension lines for bending metal. Any mark on a piece of sheet aluminum or magnesium that is scratched into the metal with a scribe can cause the part to crack when it is bent. [Figure 2-22]

Figure 2-22. Steel scribes should be used only to mark the cut-off lines on sheet metal. As a result of scratches in soft sheet metal, stresses can concentrate from vibrations at the scratch, eventually causing the metal to crack. To prevent this from occurring, lines that will remain on the metal should be made with a soft felt tip marker, or soft pencil.

PENCILS

Most of the layout marks made on sheet metal can be made with a sharp, soft lead pencil. These marks can be more easily seen if the metal is wiped clean with lacquer thinner or toluol and then sprayed with a light coat of zinc chromate primer. The primer takes pencil marks very well, which show up without eyestrain. Since the film of the primer is generally so thin, it is not likely to interfere with accurate layout measurements and marks.

When using a lead pencil, use caution to avoid making any mark on the hot section of a turbine engine, or on the exhaust system of a reciprocating engine. When heated, these marks can cause the carbon from the pencil to infuse the metal, eventually causing the part to weaken and crack.

Common lead pencils can cause scratching while also inducing graphite into the material. To prevent this from occurring, use only pencils that are acceptable for sheet metal layout marking. Three examples of acceptable commercial pencils are **Stabilo "8008," Dixon "Phano,"** and **Blaisdell**. These pencils are made of soft wax-charcoal, instead of graphite.

FELT MARKING PENS

Felt marking pens are becoming widely used by sheet metal technicians because their marks are more visible than those of other marking tools. To obtain sharp and clear lines, it is best to use a felt marking pen with a fresh, sharp tip. Once a part is fabricated, the lines made by the pen can be easily removed by wiping them with a rag soaked in alcohol.

PUNCHES

The lines drawn on sheet metal indicate where different operations such as bending and drilling are to be performed. This is not always sufficient for some operations. Drilling, for instance, requires some physical indentation on the surface in order to keep the drill from wandering when first starting. To make indentations in metal, different types of punches are available for accurately transferring marks. Other types of punches are designed to impose forces on the metal without damaging adjacent areas.

PRICK PUNCH

Once layout marks have been made with a soft pencil, permanent marks can be made in locations for holes with a sharp prick punch. It takes only a light tap with a small hammer to mark these locations, which can be enlarged later with a center punch.

CENTER PUNCH

Enlarging a prick punch mark with a center punch allows a drill to be centered so that it will start cutting the metal in the proper location. The center punch has a blunt point, ground to an angle of about 60 degrees, which approximates the tip angle of a twist drill. The marks should be deep enough for the drill to start cutting, but the blow used on the punch must not be hard enough to distort the metal.

An automatic center punch is one of the handiest tools a sheet metal technician can have, allowing punch marks to be made quickly and uniformly. This punch is spring-loaded; all that is needed is to center it in the prick punch mark, and depress the punch. When the punch is pressed hard enough, a spring-loaded plunger inside the handle is released and hits the point with a solid blow. [Figure 2-23]

Figure 2-23. An automatic center punch makes uniform indentations for aligning tools such as drills and hole cutters.

TRANSFER PUNCH

Much sheet metal repair work consists of replacing damaged skin with new skins. If the old metal is in sufficient condition, layout marks can be transferred directly to the new material by the use of a transfer punch. A transfer punch is used to locate the exact center of drilled holes, such as those used for rivets. This punch has a shank the diameter of the hardware that will be installed in the hole, and has a sharp point in its exact center. The point will make a mark similar to a prick punch mark in the center of the hole. These punches are made as both solid steel, and automatic punches. [Figure 2-24]

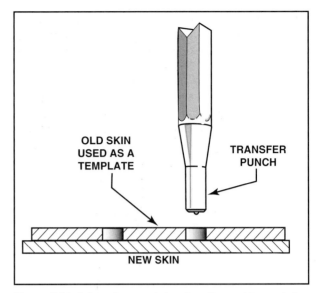

Figure 2-24. A transfer punch is primarily used to mark the center of rivet holes when using an old skin as a pattern.

PIN PUNCH

One of the most useful punches for a sheet metal worker is the pin punch because it is ideal for removing rivets. To do this, begin by drilling through the rivet head down to the base. Then, by using a pin punch of the same diameter as the rivet shank, snap the head off by prying sideways. Once the rivet head is removed, the rivet shank is then punched out of its hole with the pin punch, leaving the original-size rivet hole. When punching the rivet shanks out of thin materials, back up the metal before tapping the pin punch to help prevent distorting the metal. [Figure 2-25]

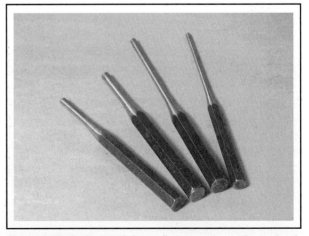

Figure 2-25. A pin punch has a flat tip and a uniform shank that is available in several different sizes.

Pin punches are available in sizes which correspond to standard rivet diameters. They range in sizes from 1/16 inch to 1/4 inch. The drill holes are

Sheet Metal Structures

.002 inch to .004 inch larger than the pin punches. For example, a 1/8 inch (.125 inch) diameter rivet uses a No. 30 (.1285 inch) drill. The pin punch used would be the 1/8 inch size, easily fitting into the rivet hole with over a .003 inch clearance.

CUTTING TOOLS

Metal cutting tools can be divided into hand-operated or floor-mounted types. The list of these tools is rather lengthy. However, in this text, the ones most commonly used in the maintenance of aircraft will be covered. Again, when using any fabrication equipment, always consult the tool manufacturer's information for use and care instructions.

METAL-CUTTING POWER TOOLS

Many of the cutting operations done on sheet metal require the use of hand tools because the repair is done on the aircraft rather than being done on a workbench. Though most operations can be performed with non-powered hand tools, powered tools make the repair task go much faster and often much smoother.

Ketts Saw

The electrically operated portable circular-cutting Ketts saw uses blades of various diameters. The head of this saw can be turned to any desired angle, which makes it very useful for removing damaged sections on stringers and other intricately designed parts. Advantages of a Ketts saw include:

- The ability to cut metal up to 3/16 inch thick.
- A starting hole is not required.
- A cut can be started anywhere on a sheet of metal.
- The saw provides the capability to cut an inside or outside radius.

Although the tool is fairly easy to operate, some basic operating precautions are required. To prevent the blade from grabbing and kicking back, keep a firm grip on the saw handle at all times. In addition, before installing any cutting blade on the tool, the blade should be checked carefully for cracks. A crack could cause the blade to fail during operation, thereby causing parts to fly out, possibly causing serious injury to the operator or bystanders. [Figure 2-26]

Reciprocating Saws

A reciprocating saw is an electrically powered tool used for rough cutting of large damaged sections such as portions of spars or stringers. With these saws, a variety of cutting blades is available to be used on different types and thickness of materials. The cutting blades used to cut through metal are

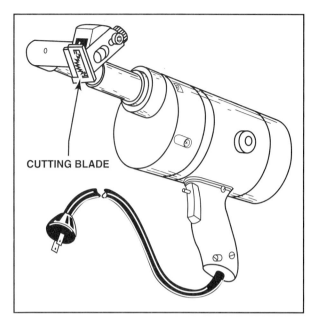

Figure 2-26. A Ketts saw uses various circular metal-cutting blades to easily remove damaged sections of sheet metal.

primarily made of good quality steel and are available with different numbers of teeth per inch of blade. For metal .250 inch or thicker, use a coarse cutting blade. However, if a coarse blade is used on thin metal, the blade can hang up, dull rapidly, or break. [Figure 2-27]

The **saber saw** is another tool that is commonly used for sheet metal repair work. These tools are electrically powered and similar in operation to the reciprocating saw. Often, the saber saw is used to cut holes in flat sheets of metal such as on wings or control surfaces, or for rough cutting the edges of sheet metal parts. One advantage of using the saber saw is that its shoe plate can be tilted, allowing for bevel-edged cuts. Like the reciprocating saw, saber saws can be adapted to cut materials other than metal by using different styles of cutting blades.

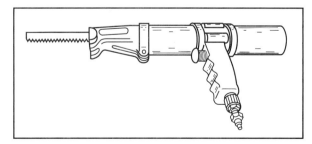

Figure 2-27. When using a reciprocating saw for metal cutting, make certain to use a blade with the proper number of teeth for the particular operation. In all cases, the manufacturer's information regarding the correct blades to use, and proper tool operation, should be consulted and followed.

Nibblers

A nibbler is a tool used for rough cutting small- to medium-size holes in skins, radio chassis, and instrument panels. These tools may be electrically or pneumatically powered, but are also available in a non-powered hand version. Regardless of the operating power, each tool produces a similar style cut.

The main advantage of a nibbler, aside from its simplicity of operation, is the ability to use the tool for making detailed inside cuts. To make a cut, the edge of the sheet metal is placed in a slot in the face of the tool. By pulling a trigger, a cutting blade moves down to shear out a small rectangular section of metal about 1/8 inch deep by 3/16 inch wide. As the metal is fed into the tool, the action repeats until the opening is made to the desired size. One disadvantage of the nibbler however, is that it tends to leave a rough edge requiring the use of a file to smooth the metal once the cut is complete. For this reason, the initial cut should be made leaving excess material, permitting filing to be used to achieve the exact finished dimension. [Figure 2-28]

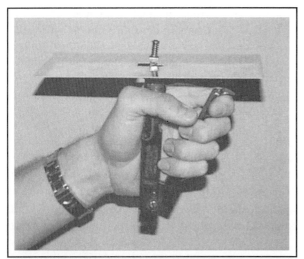

Figure 2-28. Manual and powered nibblers are used to cut out rough-open cutouts in sheet metal. To make inside cuts, it is necessary to drill a pilot hole in the metal large enough to allow the nose of the tool to pass through the hole.

NON-POWERED HAND CUTTING TOOLS

Non-powered hand cutting tools are useful since no outside power is required for their operation, making them ideal for field repairs. These tools also offer control over the amount of cut, which is often lacking when using power tools. Since there are so many types of non-powered cutting tools available, only a few of the most common types will be discussed in this section.

Sheet Metal Shears

One of the most common hand cutting tools used by sheet metal technicians is a pair of tin snip cutters. These cutters are considered useful for cutting metal to a rough shape, but for more accurate and detailed cuts, aviation snips are used. [Figure 2-29]

Figure 2-29. Hand shears and aviation snips are used for cutting thin sheet metal.

Aviation Snips

Aviation snips are compound action shears which have a serrated cutting edge that holds the metal that is being cut. Aviation snips come in sets of three and are manufactured to perform different cuts as indicated by a color-code on the handles. Snips with yellow handles cut straight, those with green handles cut to the right, and snips with red handles cut to the left.

Files

When aircraft sheet metal skins or other parts with close tolerances need to fit together, a file is often used to provide a finished edge or surface. A detailed description of the file is given in Chapter 9, Hand Tools and Measuring Devices of the *A&P Technician General Textbook*, but is briefly discussed here as a review of basic file use and care as it pertains to sheet metal work.

Due to the tolerances and detailed shapes sometimes required when fabricating sheet metal parts, a wide variety of files may be required to complete a sheet metal repair or alteration. For most sheet metal work, the common files include the standard rectangular, half round, three-square, round or rat-tail, and knife-edge shapes. With most of these shapes, the file is available in either a single- or double-cut variety.

Sheet Metal Structures

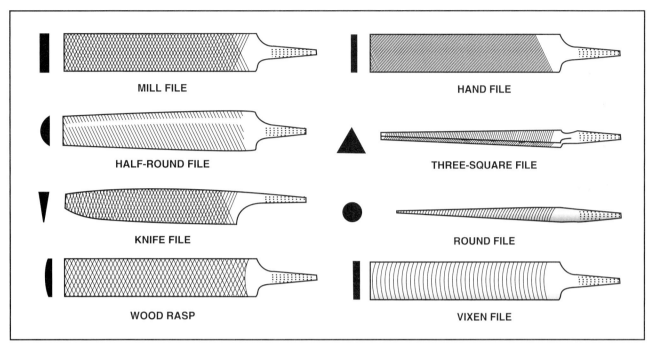

Figure 2-30. Files are available in all sorts of shapes, sizes, and styles. The most common file shapes are shown here.

Keep in mind that single-cut files are primarily used to provide a smooth finish by removing a small amount of material with each draw, while double-cut files provide faster cutting action but leave a rougher finish. Each of these files is also usually available in a variety of coarseness of cuts including coarse cut, bastard cut, second cut, smooth cut and dead smooth cut. As with any hand tool, it is important to select the tool that is designed for the specific task. [Figure 2-30]

File use is also an important consideration. Improper use not only can cause damage to the aircraft part or the file, but also can cause excessive time to be spent on accomplishing a fabrication. To obtain the best results, observe the basic rules of proper file use.

First, remember that a file is a cutting tool with sharp edges in one direction. Generally, when filing metal, do not drag the file backward across the metal with any downward force. Drawing the file backward while pushing down can dull the teeth. However, to help remove soft metal from the teeth, it is acceptable to draw the file backward without any significant downward force to dislodge metal particles. Secondly, before and after using a file, make certain that no metal fragments remain in the teeth by using a file card to brush the teeth out before and after use. It is also advisable to coat a file with light oil before storing the tool for long periods.

Deburring Tools

Once a hole is drilled in sheet metal, it is not uncommon to find that a sharp edge or burrs, is left around the circumference of the hole. A drill several sizes larger than the drilled hole, or a standard countersink cutter held in a file handle, makes a good tool for removing the burrs from the edges of holes. However, a common mistake made by inexperienced technicians is to remove too much material. Remember, a deburring tool should be used to remove burrs and to smooth edges. If too much pressure is applied while deburring, it is possible the hole will become undesirably countersunk. [Figure 2-31]

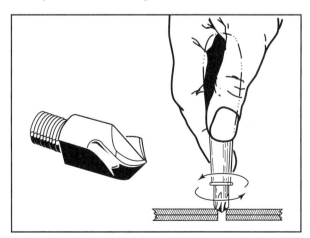

Figure 2-31. When removing the burrs from the edges of a drilled hole, do not cut too deeply. If too much material is removed, the metal will be weakened and any hardware installed in the hole will not have the same shear area to act against.

In addition to sharp edges around holes, when sheet metal is cut, the edge is also left sharp, and with burrs. A file can be used to remove the burrs from the edges of a sheet, but grinding a sharp V-shaped notch in the end of a small file can create a tool that quickens the job. To use the tool, just pull it along the edge of the sheet, and the sharp edges of the "V" will cut the burrs from both sides of the metal at the same time.

SHOP TOOLS

Shop tools, both powered and non-powered, make large-scale and repetitive metal cutting and shaping operations more convenient. Some operations, such as bends in long sheets of metal, can be accomplished only with larger shop equipment. Although most shop tools operate in a similar fashion, consult the tool manufacturer's care and use instructions before using any new equipment.

SQUARING SHEAR

Of all shop tools, a squaring shear is one of the most commonly used tools for sheet metal work. A typical size shear will accept a full 4 foot-wide sheet with the capacity to handle up to a 14-gauge soft metal, but capacity varies widely depending on manufacturer. Most shops use foot-operated shears, but when thick metal must be cut, the shears are operated by electric motors. [Figure 2-32]

Squaring shears have a guide edge that is adjustable to keep the material exactly perpendicular to the blade. Since the guide is adjustable, it is important to periodically check the alignment by making a sample cut, and checking the finished edge with a combination square. This is especially important to check before making cuts in wide sheets.

In addition to the alignment guide, squaring shears also have an adjustable stop fence for setting the depth of a cut, as well as a clamp for holding the metal tight against the table. The clamp allows fingers to be kept away from the blade during operation. When the metal is in place, stepping on a treadle, or foot pedal, causes the clamp to drop down on the metal to hold it firmly as the blade shears across the material. When properly used and maintained, squaring shears will cut metal smoothly, and leave a minimum of burrs on the edges of the material.

However, it should be remembered that a squaring shear can be a dangerous piece of equipment.

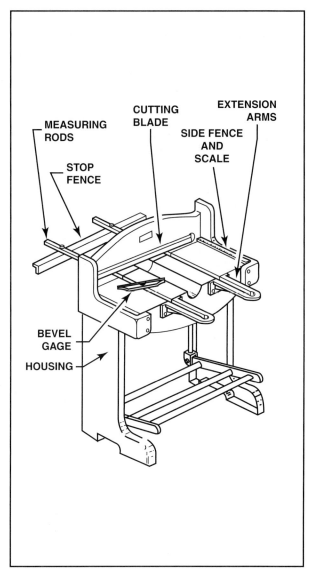

Figure 2-32. A squaring shear is used to make straight cuts across sheet metal.

Exercise care whenever using a squaring shear to verify that all fingers are clear before depressing the treadle. This is especially true when working with another person who may be helping with positioning large metal sheets.

THROATLESS SHEARS

Throatless shears are best used to cut 10-gauge mild carbon sheet metal, or up to 12-gauge stainless steel. The shear gets its name from its construction because it actually has no throat in the frame of the tool. Without this throat, there are no obstructions during cutting, which allows for more mobility in making detailed cuts. In effect, a sheet of any length can be cut, and the metal can be turned in any

Sheet Metal Structures

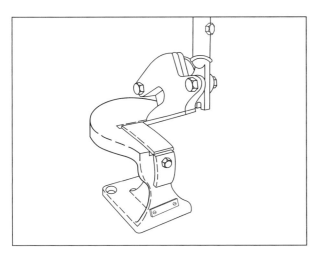

Figure 2-33. The action of throatless shears resembles a large pair of scissors. A hand lever that causes it to shear across a fixed lower blade operates the cutting blade (top blade).

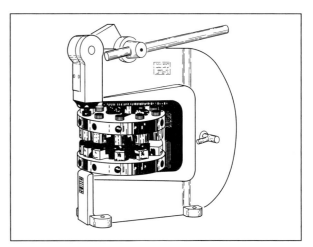

Figure 2-34. The hand-operated rotary punch press is used to punch holes or make radius cuts in metal parts.

direction to allow for cutting irregular shapes. [Figure 2-33]

ROTARY PUNCH PRESS

A rotary punch press is used in an airframe repair shop to punch holes in metal parts, for cutting radii in corners and for many other jobs where circular cuts are required. The machine is composed of two cylindrical turrets, one mounted over the other and supported by the frame. Both turrets are synchronized so that they rotate together, with index pins ensuring correct alignment at all times. The index pins may be released from their locking position by rotating a lever on the right side of the machine. This action withdraws the index pins from the tapered alignment holes and allows an operator to run the turrets to any size punch desired. [Figure 2-34]

To operate the machine, select the desired punch by looking at the stamped size on the front of the die holder. Then place the metal to be worked between the die and punch by positioning the desired center of the hole over a raised teat on the turret. Pulling the lever on the top side of the machine toward you actuates a pinion shaft, gear segment, toggle link and ram, forcing the punch through the metal. When the lever is returned to its original position, the metal is released from the punch.

BAND SAW

When sheet metal must be cut along curved lines, or when the metal is too thick to shear, a band saw can be used to cut the metal. One of the most versatile band saws found in aircraft sheet metal shops is a "Do-all" saw. This saw has a variable speed drive that allows the correct cutting speed to be adjusted for any metal. In addition, the table can be tilted so bevels and tapers can be cut on thick metal.

However, one of the most useful features of this saw is a blade cutter and welding machine. If a large opening or inside hole needs to be cut, a starting hole can be drilled, the saw blade cut in two, and then inserted through the hole. Then, with a resistance welder that is built into the saw, the blade can be welded back together. With just a few minutes of effort, the saw is ready to cut the inside of the hole. [Figure 2-35]

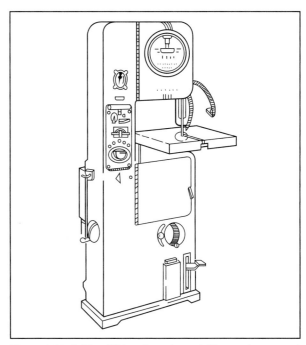

Figure 2-35. A "Do-all" band saw is useful for cutting curved lines in metal, wood, or plastic because the blade speed is adjustable to vary for different types of materials.

DISC SANDER

Wood, plastic materials, and sheet metal can be cut by the band saw to almost the correct size and then finished very accurately with a heavy-duty disc sander. With this tool, the material can be milled right up to a scribed line. [Figure 2-36]

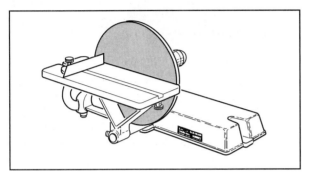

Figure 2-36. A disc sander is used to trim curved cuts in sheet metal, wood, or plastic after they have been rough-cut by a band saw.

SCROLL SHEARS

Scroll shears are used for cutting irregular lines on the inside of a sheet without cutting through from the edge. With this tool, the upper cutting blade is stationary while the lower blade is movable. By moving a handle connected to the lower blade, the shear can pierce the center of a piece of sheet metal, and cut in a similar fashion as a can opener. [Figure 2-37]

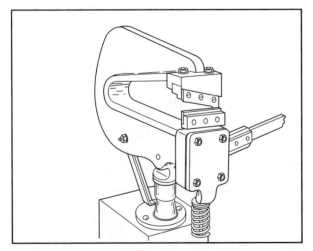

Figure 2-37. Scroll shears are used to make irregular cuts on the inside of a sheet without cutting through to the edge.

DRILLS

Aviation maintenance technicians use drills and associated attachments almost more than any other tool when fabricating sheet metal components. Drills can be either hand-operated or shop mounted. Again, always become familiar with the tool manufacturer's operating and safety instructions, and with specific operating instructions, before using any equipment for the first time.

DRILL MOTORS

The vast majority of holes drilled in aircraft sheet metal structure is small and drilled in relatively soft metal. For this reason, there is seldom a need for a drill motor larger than one with a 1/4 inch chuck. Recall that the chuck is the part of the motor that holds the cutting drill in place, and comes in a variety of sizes depending on the power of the drill motor.

Electric Drill Motors

The convenience of electric outlets in the shop and the relatively low cost of electric drill motors, as compared with air drills, make them useful tools. In addition, a variable speed control makes these tools even more useful. However, an electric drill motor is larger and heavier than an air drill and has the potential for producing an electric spark or shock when being used on an aircraft structure. For these reasons, air drills, rather than electric drills, are generally more accepted for sheet metal work. [Figure 2-38]

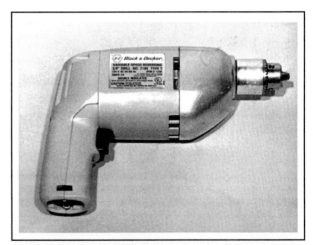

Figure 2-38. An electric drill motor is often used for sheet metal repair work when a supply of compressed air is not readily available. Rechargeable battery-powered drills are also commonly used for small sheet metal repairs because of the convenience of use, but should not be used around compartments containing flammable fluids such as fuel cells.

Pneumatic Drill Motors

The availability of compressed air to operate rivet guns makes pneumatic, or air drill motors, a logical choice for aircraft structural repair. These drills are lightweight, have good speed control, do not overheat regardless of use-frequency, and are available in a number of shapes that allows them to be used in difficult locations. [Figure 2-39]

Sheet Metal Structures

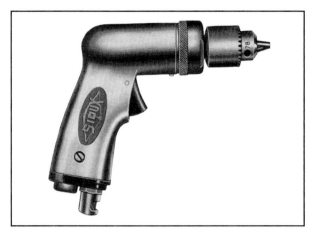

Figure 2-39. A pneumatic, or air drill motor, is the most widely used drilling tool for aircraft sheet metal repair work.

The most popular air drill motor is the pistol grip model with a 1/4 inch chuck. The speed of these drills is controlled by the amount of pull on the trigger, but if it is necessary to limit the maximum speed, a regulator may be installed at the air hose where it attaches to the drill. The regulator can then be adjusted for the maximum amount of air entering the drill to limit the maximum speed, even with the trigger fully depressed.

For drilling holes where the structure interferes, a right-angle drill attachment is available. In addition, if the chuck and the length of the twist drill prevent getting the drill motor in where it is needed, a right-angle drill motor equipped to use short-threaded twist drills can be used. [Figure 2-40]

DRILL ATTACHMENTS AND SPECIAL DRILLS

Drill jigs are used to assist in drilling accurate holes in skins and structural subassemblies. These are held in place by drilling one hole and anchoring the jig so it can be used as a template to drill numerous additional holes. The alignment of the jig makes it possible to obtain holes that are round, straight, and free from cracks. This is especially true when the metal is thick where holes drilled freehand have a tendency to be made crooked. Drill jigs are most commonly used during the assembly process while an aircraft is being built. For example, drill jigs are very useful for installing anchor nuts or anything else that requires holes to be made with a high degree of repetitive dimensional accuracy.

A drill attachment used a great deal by sheet metal technicians during the disassembly of a damaged aircraft, is a rivet removal tool. A rivet removal tool is available with interchangeable twist drills that correspond to standard rivet sizes. Drilling out rivets is made easier because the tool can be adjusted to cut only to the depth of the manufactured rivet head. The procedure is then the same as for the freehand rivet removal technique. Once the head is drilled, simply tap or snap off the drilled head, and tap out the rivet shank with a hammer and pin punch. Again, the material should be backed up with a bucking bar or other similar device to prevent damaging the base metal while the shank is driven out.

Right-Angle Drills And Attachments

Angled drill motors are designed for operation in tight locations where there is limited access to the structure being drilled. Angled drill motors are available in two standard head angles of 45 degrees and 90 degrees. However, for access into even tighter locations, angled drill motor attachments are also available. [Figure 2-41]

A right-angled drill motor attachment is primarily used to open holes in close quarters where even an angled drill motor cannot work. The attachment is chucked into a straight pistol-type motor. The twist drill used on a right angle attachment is installed in a collet, which can hold a standard twist drill. When the twist drill becomes too short in the collet,

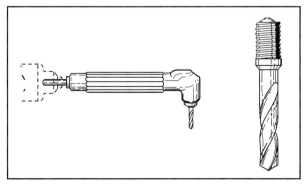

Figure 2-40. Threaded shank twist drills are used in drill motors for a number of reasons. One of the primary reasons is that the twist drill will not slip in the chuck when subjected to an excessive amount of torque.

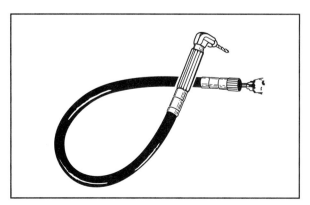

Figure 2-41. The right-angle attachment allows access into locations that are inaccessible with a standard drill motor.

it can be replaced with a broken straight drill with a newly sharpened tip. The twist drill is pressed into the collet and held in place by pressure exerted by the compressed wall of the collet when it is tightened in the attachment's holder.

Snake Attachments

A flexible snake attachment may be used in limited access areas where an angle drill motor or angle attachment cannot be used. The snake attachment basically performs the same function that a right-angle attachment does, except it can be snaked in to drill a hole much farther away than can a right-angle drill. These are excellent tools for getting into locations that a regular motor can't, because the handle will not permit a straight entry on the part being drilled. One use for a snake attachment is for back-drilling through holes in original members, into new, undrilled skins. Back-drilling is done to open holes in new skins through preexisting holes drilled in ribs, stringers, or spars, which were previously made during the original installation of sheet metal parts. [Figure 2-42]

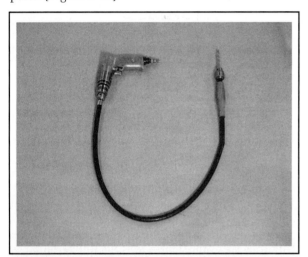

Figure 2-42. A snake attachment with its right angle and threaded twist drill permits access through lightening holes to back-drill skin repairs.

Extension Drills

There are applications in aircraft maintenance where it is necessary to reach through a part of the structure to drill a hole that is beyond the reach of an ordinary twist drill. When this type problem occurs, there are two types of extension drills available to use. One is simply a long drill that must have a piece of aluminum tubing slipped over the shank to prevent it from whipping during use. The other has a heavy shank with a small drill fixed into its end and needs no protective cover, as it is too rigid to whip. [Figure 2-43]

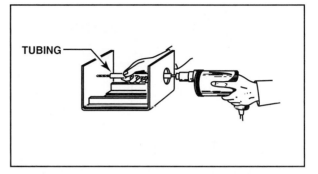

Figure 2-43. With thin shank extension drills, when pressure is applied during the drilling operation, the twist drill can bend and cause whipping. To prevent whipping, the twist drill is placed in a hollow tube to provide support for the drill.

SPRING DRILL STOPS

Although it is not uncommon to need a drill that is reasonably long, it is also important to be conscientious of the fact that a twist drill may be too long for a given application. In fact, it is often necessary to limit the amount that a twist drill can penetrate through a part to prevent damaging components on the other side of the drilled structure. A device that is often used to limit twist drill penetration is a spring drill stop. These stops come in assorted sizes and use a set screw to anchor the stop to the twist drill at any desired position. In this fashion, the exposed length of the twist drill end can be adequately controlled to prevent the drill from penetrating too far through the structure. [Figure 2-44]

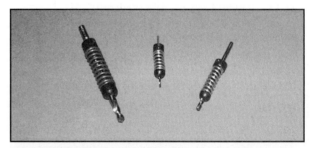

Figure 2-44. Drill stops are typically color coded to match the drill size. Silver stops are used with No. 40 drills; copper stops are used with No. 30 drills; and black stops are used with No. 21 drills.

DRILL PRESSES

Drill presses are available in a variety of styles with the most common being the upright variety. When the upright drill press is in use, the height of the drill table is adjusted to accommodate the height of the part to be drilled. When the height of the part is greater than the distance between the drill and the table, the table is lowered. When the height of the part is less than the distance between the table and the drill chuck, when it is at its full extension, the table is raised to permit the drill to penetrate completely through the part.

Once the table is properly adjusted, the part is placed on the table, and the drill is brought down to aid in positioning the part so that the hole to be drilled is directly beneath the drill point. The part is then clamped to the table to prevent it from slipping during the drilling operation. Parts not properly clamped may bind on the drill and start spinning, causing the operator to suffer serious cuts or the loss of fingers. To prevent injuries, always make sure the part is properly clamped before starting the drilling operation. Never attempt to hold a part by hand.

Another consideration when using a drill press is to make certain to never leave a chuck key in the drill chuck. Failure to remove the key before turning the drill press on will cause the key to fly out, possibly causing serious injury to the operator or bystanders. [Figure 2-45].

TWIST DRILLS

Twist drills are used for opening holes in metal, wood, and other materials. A twist drill has three main parts consisting of the tip, body, and shank. The tip includes two cutting lips that are normally sharpened to a 59 degree angle from the center-line. This produces an included angle of 118 degrees while the heel of the tip is normally ground to an angle of about 12 to 15 degrees. However, depending on the type material being cut, twist drills can have different angles to provide optimum performance.

The shank of a drill is the portion that is chucked into the drill motor, whereas the body of the drill includes hollow flutes and reamer lands. The flutes aid in carrying material out of the hole while also providing a method for cooling-oil to be delivered to the cutting surface. The reamer lands, which are also part of the body, serve to provide a finished dimension to the hole. [Figure 2-46]

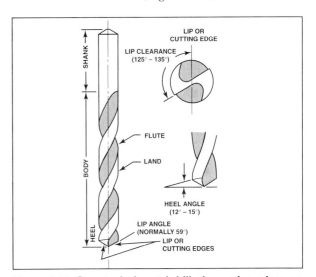

Figure 2-46. On a typical metal drill, the cutting edges on the point perform the actual cutting. The lands and flutes remove the cut material and carry lubricant to the cutting edges as well as provide a finished cut for proper dimensions.

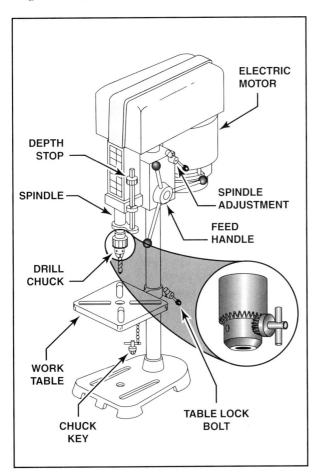

Figure 2-45. To increase the accuracy and straightness of a drill press when using a twist drill, it is recommended to tighten the chuck by using each of the key holes in unison around the chuck. This tends to provide an even tension on the shank of the twist drill and helps to ensure straightness.

Some materials require different included and heel angles to be ground on the tip, while motor speeds and pressures may also need to be varied. To cut holes in aircraft aluminum, an included angle of 118 degrees should be used with a high tip speed and steady pressure on the drill. For soft materials such as plastics, an included angle of 90 degrees should be used with drill motor speeds and pressures being adjusted for the particular density. Stainless steels, on the other hand, require an included angle of 140 degrees with a slow tip speed and reasonably heavy drill motor pressure.

Drill diameters are distinguished in one of four ways by either a number, fraction, letter, or decimal. For sheet metal work, number and letter drills are the types most widely used. Numbered and letter drills are identified in place of fractional drills in order to provide a clearance to accept fractional sized rivets and other hardware. For example, a No. 30 drill has a diameter of .1285 inches whereas a -4 rivet has a diameter of 1/8 or .125 inches. Since the No. 30 drill is .0035 inches larger, the rivet will easily fit into the hole without an excessive clearance. Although fractional drills come in 1/64 inch increments in sizes less than one inch in diameter, if a fractional drill is used, the clearances will either be too close, or too extreme to permit proper hardware clearance. [Figure 2-47]

Millimeter	Dec. Equiv.	Fractional	Number	Millimeter	Dec. Equiv.	Fractional	Number	Millimeter	Dec. Equiv.	Fractional	Number	Millimeter	Dec. Equiv.	Fractional	Number	Millimeter	Dec. Equiv.	Fractional	Number	Millimeter	Dec. Equiv.	Fractional	Number
.1	.0039			1.45	.0570			3.2	.1260			5.4	.2126				.3230		P	14.5	.5709		
.15	.0059			1.5	.0591			3.25	.1279				.2130		3	8.25	.3248			14.68	.5781	37/64	
.2	.0079				.0595		53		.1285		30	5.5	.2165			8.3	.3268			15.0	.5906		
.25	.0098			1.55	.0610			3.3	.1299			5.56	.2187	1/32		8.33	.3281	21/64		15.08	.5937	19/32	
.3	.0118			1.59	.0625	1/16		3.4	.1338			5.6	.2205			8.4	.3307			15.48	.6094	39/64	
	.0135		80	1.6	.0629				.1360		29		.2210		2		.3320		Q	15.5	.6102		
.35	.0138				.0635		52	3.5	.1378			5.7	.2244			8.5	.3346			15.88	.6250	5/8	
	.0145		79	1.65	.0649				.1405		28	5.75	.2263			8.6	.3386			16.0	.6299		
.39	.0156	1/64		1.7	.0669			3.57	.1406	9/64			.2280		1		.3390		R	16.27	.6406	41/64	
.4	.0157				.0670		51	3.6	.1417			5.8	.2283			8.7	.3425			16.5	.6496		
	.0160		78	1.75	.0689				.1440		27	5.9	.2323			8.73	.3437	11/32		16.67	.6562	21/32	
.45	.0177				.0700		50	3.7	.1457				.2340		A	8.75	.3445			17.0	.6693		
	.0180		77	1.8	.0709				.1470		26	5.95	.2344	15/64		8.8	.3465			17.06	.6719	43/64	
.5	.0197			1.85	.0728			3.75	.1476			6.0	.2362				.3480		S	17.46	.6875	11/16	
	.0200		76		.0730		49		.1495		25		.2380		B	8.9	.3504			17.5	.6890		
	.0210		75	1.9	.0748			3.8	.1496			6.1	.2401			9.0	.3543			17.86	.7031	45/64	
.55	.0217				.0760		48		.1520		24		.2420		C		.3580		T	18.0	.7087		
	.0225		74	1.95	.0767			3.9	.1535			6.2	.2441			9.1	.3583			18.26	.7187	23/32	
.6	.0236			1.98	.0781	5/64			.1540		23	6.25	.2460		D	9.13	.3594	23/64		18.5	.7283		
	.0240		73		.0785		47	3.97	.1562	5/32		6.3	.2480			9.2	.3622			18.65	.7344	47/64	
	.0250		72	2.0	.0787				.1570		22	6.35	.2500	1/4	E	9.25	.3641			19.0	.7480		
.65	.0256			2.05	.0807			4.0	.1575			6.4	.2520			9.3	.3661			19.05	.7500	3/4	
	.0260		71		.0810		46		.1590		21	6.5	.2559				.3680		U	19.45	.7656	49/64	
.7	.0276				.0820		45		.1610		20		.2570		F	9.4	.3701			19.5	.7677		
	.0280		70	2.1	.0827			4.1	.1614			6.6	.2598			9.5	.3740			19.84	.7812	25/32	
	.0292		69	2.15	.0846			4.2	.1654				.2610		G	9.53	.3750	3/8		20.0	.7874		
.75	.0295				.0860		44		.1660		19	6.7	.2638				.3770		V	20.24	.7969	51/64	
	.0310		68	2.2	.0866			4.25	.1673			6.75	.2657	17/64		9.6	.3780			20.5	.8071		
.79	.0312	1/32		2.25	.0885			4.3	.1693			6.75	.2657			9.7	.3819			20.64	.8125	13/16	
.8	.0315				.0890		43		.1695		18		.2660		H	9.75	.3838			21.0	.8268		
	.0320		67	2.3	.0905			4.37	.1719	11/64		6.8	.2677			9.8	.3858			21.03	.8281	53/64	
	.0330		66	2.35	.0925				.1730		17	6.9	.2716				.3860		W	21.43	.8437	27/32	
.85	.0335				.0935		42	4.4	.1732				.2720		I	9.9	.3898			21.5	.8465		
	.0350		65	2.38	.0937	3/32			.1770		16	7.0	.2756			9.92	.3906	25/64		21.83	.8594	55/64	
.9	.0354			2.4	.0945			4.5	.1771				.2770		J	10.0	.3937			22.0	.8661		
	.0360		64		.0960		41		.1800		15	7.1	.2795				.3970		X	22.23	.8750	7/8	
	.0370		63	2.45	.0964			4.6	.1811				.2811		K		.4040		Y	22.5	.8858		
.95	.0374				.0980		40		.1820		14	7.14	.2812	9/32		10.32	.4062	13/32		22.62	.8906	57/64	
	.0380		62	2.5	.0984			4.7	.1850		13	7.2	.2835				.4130		Z	23.0	.9055		
	.0390		61		.0995		39	4.75	.1870			7.25	.2854			10.5	.4134			23.02	.9062	29/32	
1.0	.0394				.1015		38	4.76	.1875	3/16		7.3	.2874			10.72	.4219	27/64		23.42	.9219	59/64	
	.0400		60	2.6	.1024			4.8	.1890		12		.2900		L	11.0	.4330			23.5	.9252		
	.0410		59		.1040		37		.1910		11	7.4	.2913			11.11	.4375	7/16		23.81	.9375	15/16	
1.05	.0413			2.7	.1063			4.9	.1929				.2950		M	11.5	.4528			24.0	.9449		
	.0420		58		.1065		36		.1935		10	7.5	.2953			11.51	.4531	29/64		24.21	.9531	61/64	
	.0430		57	2.75	.1082				.1960		9	7.54	.2968	19/64		11.91	.4687	15/32		24.5	.9646		
1.1	.0433			2.78	.1094	7/64		5.0	.1968			7.6	.2992			12.0	.4724			24.61	.9687	31/32	
1.15	.0452				.1100		35		.1990		8		.3020		N	12.30	.4843	31/64		25.0	.9843		
	.0465		56	2.8	.1102			5.1	.2008			7.7	.3031			12.5	.4921			25.03	.9844	63/64	
1.19	.0469	3/64			.1110		34		.2010		7	7.75	.3051			12.7	.5000	1/2		25.4	1.0000	1	
1.2	.0472				.1130		33	5.16	.2031	13/64		7.8	.3071			13.0	.5118						
1.25	.0492			2.9	.1141				.2040		6	7.9	.3110			13.10	.5156	33/64					
1.3	.0512				.1160		32	5.2	.2047			7.94	.3125	5/16		13.49	.5312	17/32					
	.0520		55	3.0	.1181				.2055		5	8.0	.3150			13.5	.5315						
1.35	.0531				.1200		31	5.25	.2067				.3160		O	13.89	.5469	35/64					
	.0550		54	3.1	.1220			5.3	.2086			8.1	.3189			14.0	.5512						
1.4	.0551			3.18	.1250	1/8			.2090		4	8.2	.3228			14.29	.5625	9/16					

Figure 2-47. Fractional twist drills come in 1/64 inch increments in sizes less than 1 inch and can be compared with equivalent decimal and metric measurements by the use of this chart. In addition, the actual decimal equivalent sizes of numbered and lettered drills can be determined from the chart. This can be useful when selecting drills for the installation of rivets where it is common to use a numbered drill that is approximately .003 inch larger than the shank diameter of the rivet.

FORMING TOOLS

Very few sheet metal skins are perfectly flat. In fact, nearly all require bends or curves that must be shaped in some manner. Forming tools include tools that create straight bends as well as those that create compound curves. Some of these tools are manually powered while others are electrically, pneumatically or hydraulically powered.

PRESS BRAKES

The secret of economical mass production of airplanes lies in the ability of the designer's skill in specifying fabrication methods that require only skilled workers to set up machines, and then having workers of far less skill produce the parts. The press brake needs only die installation and adjustment, and the stops properly set by a skilled worker; then any number of pieces can be formed with relatively unskilled labor. [Figure 2-48]

Figure 2-49. The cornice brake, or leaf brake, as it is sometimes called, is one of the most commonly used bend fabricating tools found in many aircraft maintenance shops.

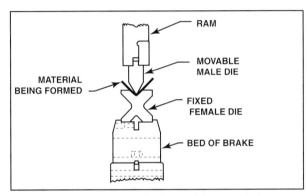

Figure 2-48. The press brake is used for production work where many similar bends must be made in sheet metal.

With press brakes, a female die is fixed and a male die is driven by energy stored in a heavy flywheel by an electric motor. The material is moved over the female die until it rests against the stop and the male die is lowered into it. As the dies come together, they form an accurate bend that can be duplicated many times.

The number and types of dies available for press brakes allow them to be used to make almost any kind of bend in sheet metal. For example, dies are available that bead the edges so wire can be installed to make hinges, or they can also be used to form lock-seams in thin sheet steel, or to form channels and boxes.

CORNICE BRAKES

The cornice brake is one type of widely used bending brake found in most maintenance shops because it will accommodate a wide range of metal thickness. These brakes normally have a rather sharp nose bar, around which bars of any desired milled radius may be placed. The bend radius blocks may be moved back away from the edge of the bending leaf to accurately adjust the setback or the distance of the radius from the bending leaf. This is necessary to take into account the thickness of the metal so the metal will hold a tight contour around the radius blocks. [Figure 2-49]

When the correct radius block is in place, the metal to be bent is slipped in the brake and positioned so the bend line is exactly below the beginning of the radius in the radius block. However, since the bend begins under the radius block, it is not possible to see where the bend line begins. To facilitate placing the metal in the correct position, a sight line is marked forward from the bend line approximately at the same distance as the size of the radius. [Figure 2-50]

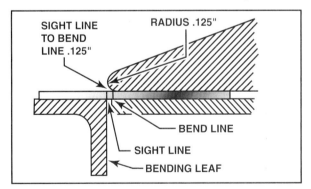

Figure 2-50. By marking a sight line forward of the desired bend line, a technician will be able to align the metal in the nose of the brake so that the bend begins at the correct location. For example, when bending a piece of metal using a 1/8 inch radius in the brake, mark the sight line 1/8 or .125 inches ahead of the desired bend line. Then, when positioning the metal in the brake, look straight down on the nose of the radius block to place the sight line at the front of the bar. A more thorough discussion regarding the setup of bending brakes is covered in the sheet metal bending section later in this chapter.

Once the metal is positioned properly in the brake, a handle is pulled to lower the nosepiece onto the metal. This action holds the metal securely in place while being bent. Bending is accomplished by lifting up on a counterweighted bending leaf, which causes a plate to pivot to force the metal to bend. Raising the leaf higher causes the angle of the bend to increase. When making bends to a specified number of degrees, it is generally necessary to bend past the desired angle. This is because the metal tends to spring back toward a smaller bend angle once the bending leaf is returned to the idle position. Experience facilitates familiarity with how much additional bending is necessary to achieve a finished angle with various metals.

The manufacturer determines the bending capacity of a cornice brake. Standard capacities of this machine are from 12- to 22-gauge sheet metal, and lengths are from 3 to 12 feet. The maximum bending angle of the brake is determined by the bending edge thickness of the radius bars. To provide an increased bending angle, most radius bars are tapered toward their front edge to allow the metal to be bent a maximum amount over the bar. [Figure 2-51]

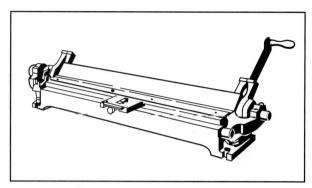

Figure 2-52. The bar folding machine is used to bend the edges of relatively light stock.

Before using the bar folder, several adjustments must be made for thickness of material, as well as the width, sharpness, and angle of fold. Adjusting the screws at each end of the folder makes the adjustment for thickness of material. As this modification is made, place a piece of metal of the desired thickness in the folder and raise the operating handle until a small roller rests on a cam follower. Hold the folding blade in this position and adjust the setscrews so that the metal is clamped securely and evenly through the full length of the folding blade. After the folder has been adjusted, test for uniformity by actually folding small pieces of metal at each end of the machine and comparing the folds.

To make the fold once the machine is adjusted, insert the metal between the folding blade and jaw. Hold the metal firmly against the gauge and pull the operating handle toward you. As the handle is brought forward, the jaw automatically raises and holds the metal until the desired fold is made. When the handle is moved back to its original position, the jaw and blade will return to their original positions and release the metal.

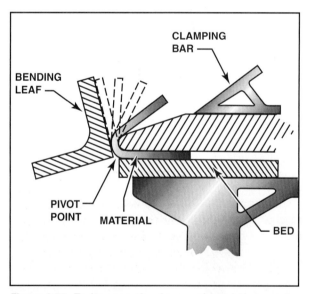

Figure 2-51. To increase the degree of bend that can be achieved with a cornice brake, the radius bars are tapered to allow the bending leaf to have a greater amount of travel.

BAR FOLDING MACHINE

The bar folder is designed for use in making bends or folds along edges of sheets. This machine is best suited for folding small hems, flanges, seams, and edges to be wired. Most bar folders have a capacity for metal up to 22-gauge in thickness and 42 inches in length. [Figure 2-52]

BOX BRAKE

The box brake, sometimes referred to as a pan or finger brake, is probably one of the most widely used forming tools. This extremely handy brake is very similar to the cornice brake except that its top leaf is fitted with a number of radius bars, or fingers. These bars of varying widths can be selected to the dimensions of a box interior in which four sides are bent. Two opposite sides of a box may be formed on a leaf brake, but the last two sides must be formed with a box brake. The fingers are selected to fit just between the two sides that have been formed, and when the box is clamped in place and the leaf is raised, the sides of the box will slip between the

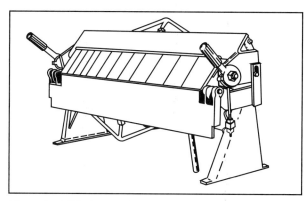

Figure 2-53. The box, or finger brake, is similar to a cornice brake, except that it has split leaves to allow bending of the multiple sides of four-sided box objects.

fingers. This allows the last two sides of the box to be formed. [Figure 2-53]

SLIP ROLL FORMER

All of the machines discussed to this point are used to make rather sharp bends in sheet metal, but sometimes a gentle curve in a part is needed to form a metal tube, or to form a skin for a fuselage. To make these curves, a slip roll former can be used.

A slip roll former is a simple machine consisting of three hard steel rollers in a framework. A drive roller is turned with a hand crank, while a clamp roller is adjustable up or down to provide a tight clamping action to aid in pulling the metal through the machine. In addition to the drive and clamp rollers, a radius roller can be adjusted in or out to increase or decrease the radius being formed in the metal as it passes through the rollers. [Figure 2-54]

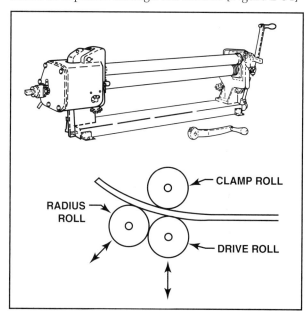

Figure 2-54. A slip roll former is used to make gentle bends to fabricate parts such as tubing or contoured fuselage skins

To use the slip roll former, the metal is placed between the drive and clamp rollers, and the hand crank is turned to pull the metal through the machine. As the metal passes through, it moves over the radius roller. On the first pass, the radius roller should be adjusted to just touch the metal to form a very slight curve. By subsequently passing the metal through the machine with the radius roller adjusted to a tighter fit, the metal will take on a greater curved surface. Each time the metal is rolled through the machine, the radius roller is moved up a bit so it will decrease the radius of the curve in the metal. The metal is passed through the former several times with the radius roller being progressively adjusted to obtain the desired radius.

COMPOUND CURVE TOOLS

In the modern aircraft factory, large compound curved skins are produced on stretch presses where the sheet of metal is grasped in two large sets of jaws, and the sheet is pulled across a male die until it stretches to the desired shape. Once formed, the metal is then trimmed to the proper size.

Stretch Press

Stretch presses are usually found in an aircraft factory. However, variations of this tool can be rented for use in small repair shops or by aircraft homebuilders. These tools are used to form compound curved parts by pulling the sheet across a male die. When formed in this manner, the metal obtains a certain amount of strength and rigidity by being left in a cold-worked condition.

Drop Hammer

A process that has been used longer than the stretch press is drop hammer forming. In this process, large matching metal male and female dies are used. By placing sheet metal over the female die and dropping or slamming the male die onto the female die, the metal will be forged into the contoured shape. This method of forming tends to make a uniform grain pattern in the metal, causing the strength of the material to increase.

Hydropress

Smaller components such as fuselage formers, wing ribs, and all types of compound curved brackets are formed in a hydropress. With a hydropress, a blank for the part is punched out of sheet metal on a punch press. The blank is then placed over a metal male die and held in place with tooling pins sticking through holes in the blank. The die is placed on the bed of a hydropress; then a ram, which carries a thick rubber blanket, is lowered over the die. A

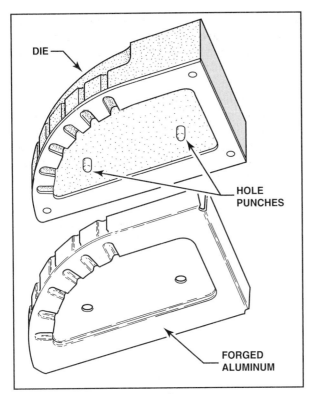

Figure 2-55. A hydropress uses water pressure to force a piece of sheet metal to assume the shape of the die.

Figure 2-56. Sheet metal shrinkers and stretchers work on the principle of clamping the sheet metal between two jaws while moving the jaws together or apart slightly, to compress or expand the material, respectively.

pressure of several thousand tons is applied to press the rubber down over the metal, making it conform to the shape of the die. [Figure 2-55]

Shrinkers And Stretchers

Shrinking and stretching tools are used to form contours in parts by expanding or compressing metal to make it form a curved surface. For example, when the edge of sheet metal is worked in a stretcher, the edge will expand, causing it to form an outside curve. Conversely, a shrinker causes the metal to contract, which causes the metal to form an inside curve. Each tool is constructed in a similar fashion in that they both consist of two pairs of heavy jaws that are operated by a hand lever or foot pedal. In each tool, gripping jaws are opened and the edge of the material is placed between them.

With a shrinker, pulling down on a lever or hydraulically operating the tool by a foot pedal causes the jaws to grip the metal and then move inward to compress a small portion of the edge. With a stretcher, the opposite is true in that the jaws grip the metal and then spread apart. By progressively working the metal over a certain distance along its edge, the metal will eventually begin to take on a contoured radius. The jaws do not move enough to buckle or tear the metal, but just enough

to compress or stretch it somewhat. The material is worked back and forth across the full width of the curve, shrinking or stretching it just a little with each movement of the jaws. [Figure 2-56]

Sandbags

When only one part is to be formed, a heavy canvas bag filled with a good grade of washed sand can serve as a mold. An impression is made in the sand that approximates the desired shape, and the metal is carefully formed into the depression with a round-face plastic mallet. This is strictly a trial and error method of forming, and its results depend upon the skill of the worker and the care that is taken. [Figure 2-57]

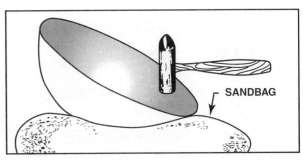

Figure 2-57. A sandbag can be used to form a one-of-a-kind compound curved part.

SPECIAL ASSEMBLY TOOLS

The job of fabricating sheet metal assemblies is made much easier by the use of tools that have been created specifically for working with sheet metal. Since there are so many specialty tools available for sheet metal work, only a select few of the most common types are discussed in this text. However, these tools are representative of those most likely to be encountered when the technician begins performing basic sheet metal repairs to aircraft.

CLAMPS AND SHEET FASTENERS

There are many devices used by technicians to assist them when fabricating sheet metal aircraft components and structures. Some of the most important tools are those used to verify that parts remain in proper alignment during the assembly process. Three of the most common holding tools include Cleco fasteners, C-clamps, and wingnut clamps.

Cleco Fasteners

Before an aircraft sheet metal structure is riveted, it should be temporarily assembled to be sure that all of the parts fit together properly. To provide the closest tolerance fit for rivets, it is standard practice in many operations to drill all of the rivet holes in the individual parts with a pilot drill. The pilot drill is typically smaller than the nominal size of the rivet shank. Eventually, when the parts are mated together, another drill is passed through the pilot holes to open them up to the proper dimensions of the rivet shank. To help prevent the parts from shifting during the final drilling and assembly process, it is common to use clamping fasteners to hold the parts together until the rivets are installed. This helps ensure alignment of the holes so the rivets seat properly.

One of the most widely used clamping devices is the Cleco fastener, a patented product developed by the Cleveland Pneumatic Tool Company. Although there are other manufacturers of similarly designed clamping devices, the name Cleco is generally associated with these types of tools.

Cleco fasteners consist of a steel body, a spring loaded plunger, two step-cut locking jaws, and a spreader bar. To install or remove a Cleco requires the use of a specially designed pair of pliers. With these pliers, the fastener body is held while pressure is applied to the spring loaded plunger on the top of the Cleco. This forces the pair of locking jaws away from the body and past the spreader bar.

As the jaws pass beyond the bar, they come together, decreasing in diameter. This allows the jaws to be inserted into a rivet hole and then when the pliers are released, the jaws draw back in toward the body past the spreader bar. When the jaws spread apart, the diameter increases, causing the steps in the jaws to grab the underside of the metal. When the jaws retract as far as they can into the body, they apply spring pressure between the locking jaws and the sheet metal, as well as filling the hole diameter with the jaws. Once the pliers are removed from the fastener, a tight grip is formed to help prevent slippage of the material. [Figure 2-58]

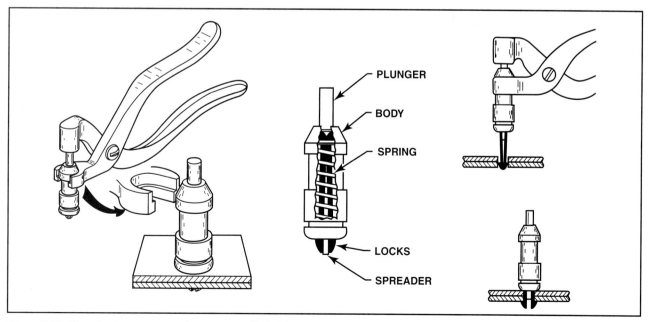

Figure 2-58. Cleco fasteners are used to temporarily hold sheet metal parts together until they are riveted.

Clecos are available in sizes for all the commonly used rivets and even in one larger size. To help identify the designed hole size of Clecos, the body is color coded in one of the following colors:

3/32 inch	(-3 diameter rivet)	Silver
1/8 inch	(-4 diameter rivet)	Copper
5/32 inch	(-5 diameter rivet)	Black
3/16 inch	(-6 diameter rivet)	Brass
1/4 inch		Copper

Wing Nut Fasteners

Wing nut fasteners are used prior to final assembly of aircraft parts that need to be held extra tight before riveting is started. For example, sheet metal parts under tension around a bend tend to spring apart and may need more pressure to hold the metal together than a Cleco can provide. The wing nut fastener, when hand tightened, will clamp the metal together with more pressure than the spring tension of a Cleco, thus ensuring against any possible slippage. However, a major drawback to using wing nut fasteners is the amount of time required to install and remove them. [Figure 2-59]

C-Clamps

The C-clamp is a tool primarily used by machinists, but has been adapted by technicians working with sheet metal for holding work together on aircraft. It is useful for holding sheet metal in place before beginning the drilling operation. C-clamps are available in many sizes. However, smaller sizes are generally preferred for sheet metal applications to prevent damage to the metal.

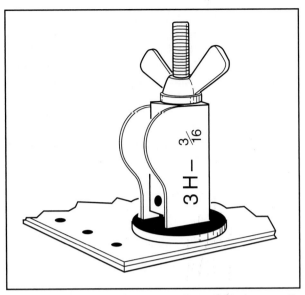

Figure 2-59. Wing nut fasteners are used for temporarily holding sheet metal parts together with more pressure than a Cleco fastener can provide. These fasteners also have the same color coded bodies as Clecos to identify the diameter hole they are designed for.

The C-clamp looks like the letter "C"; hence, its name. The C-frame has a fixed rest on its lower end and a threaded end at the top. The threaded end has a shaft that runs through it with a tee handle running through the shaft, and a floating pad on the end. Before using one of these clamps on sheet metal, it is advisable to place masking tape over each of the pads to help prevent marring of the sheet metal's finish. In addition, before using these clamps, check to make certain the floating pad on the threaded shaft is free to swivel and turn to help prevent marring as the clamp is tightened. [Figure 2-60]

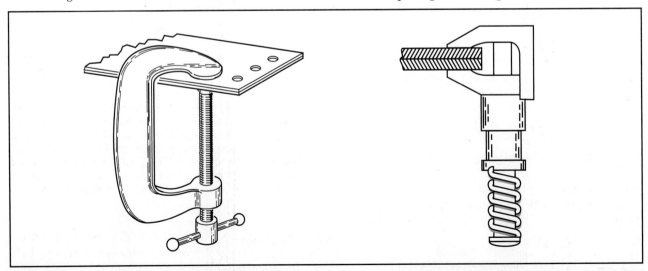

Figure 2-60. C-clamps are useful for holding sheet metal parts together to drill the initial rivet holes as shown on the left. Another device similar to a C-clamp is a side grip clamp, resembling the one shown here on the right. These clamps are spring loaded in the same fashion as a Cleco fastener and are installed and removed using Cleco pliers. Because these clamps are small, they are ideal in tight fitting locations.

HOLE FINDERS

Hole finders are used for locating rivet holes in undrilled skins where a pre-existing hole is hidden by the metal sheet. For example, when an aircraft sheet metal structure is disassembled for a repair, damaged skins may be removed and replaced with new skins while some pre-drilled parts remain on the aircraft. When the new skin is positioned, it may not be possible to see where the holes are located to drill to match the new skin to the pre-drilled parts.

A hole finder consists of two metal straps that are brazed or riveted together at one end. At the opposite end, a pin or pilot extends out from one of the straps while the other strap either has a drill bushing or a center punch type plunger. To use the hole finder, separate the straps and slide the one with the pilot in between the undrilled and drilled parts. When the pilot drops into a rivet hole the second strap with the drill bushing or center punch plunger will be in alignment with the center of the hole. If a drill bushing is used, drill straight through the hole finder to make the new hole in the skin.

On the other hand, if a center punch plunger is used on the hole finder, tap the plunger lightly with a mallet to mark the center of the hole. Once marked, remove the hole finder and drill the part in the normal fashion using the center punch mark as a guide. [Figure 2-61]

CHIP CHASERS

It is sometimes impossible to disassemble skins after drilling a hole, and as a result, there are metal chips that can lodge between the skins that will prevent them from fitting tightly together when a rivet is installed. A tool used to remove these metal chips is commonly called a chip chaser and can be purchased or made from a strip of feeler gauge stock.

When making a chip chaser, use a piece of stock that is thin enough to get between the parts and yet stiff enough to pull out the chips. A strip of .010 inch thick stock is generally considered adequate. Cut a notch near the end of the strip and fasten some sort of handle to the opposite end to make it easy to hold. To use it, just reach in between the skins with the chaser and rake the chips out from between the metal. However, try to pull the parts back so that the chips or the tool does not scratch the finish of the metal. [Figure 2-62]

STRUCTURAL FASTENERS

The integrity of an aircraft joint depends upon the fasteners selected and used to secure its parts together. However, not all aircraft joints are made

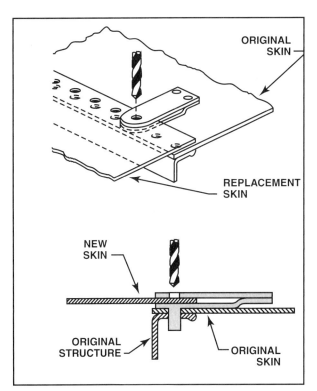

Figure 2-61. A hole finder is used to match holes in an undrilled sheet of metal with existing holes in a hidden part. These tools come equipped with different sized pilots for various rivet diameters. Make sure you do not use a hole finder with too small a pilot, or the drilling accuracy will be diminished and could cause elongation of the pre-existing hole.

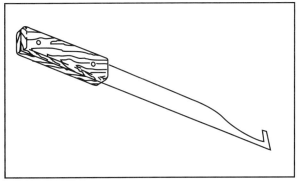

Figure 2-62. A chip chaser can be fashioned to remove drill chips from between sheet metal skins. Another resource for a longer chip chaser is to take an old band saw blade and shear the cutting teeth off its edge. Once cut and shaped, file the edges of the band smooth so that it does not scratch or damage the surface of the metal when used.

using fasteners. Some joints on newer aircraft are made with composite materials that are held together by adhesives. Although this construction technique is gaining popularity, this method of construction will probably never completely take the place of using fasteners in aircraft assemblies. It is therefore important for an aircraft technician to be thoroughly familiar with the different types of fasteners that are encountered in industry. Although this section provides general guidelines in the selection and installation of various types of hardware and fasteners, it is always advisable to get acquainted with the fastener manufacturer's technical information before using its product on an aircraft.

SOLID SHANK RIVETS

The solid shank rivet has been used since sheet metal was first utilized in aircraft, and remains the single most commonly used aircraft fastener today. Unlike other types of fasteners, rivets change in dimension to fit the size of a hole during installation. [Figure 2-63]

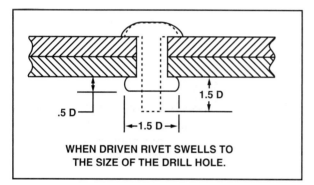

Figure 2-63. Before a rivet is driven, it should extend beyond the base material at least one and a half times the rivet's diameter. Once driven, the rivet shank expands to fill the hole, and the bucktail expands to one and a half times its original diameter. Once the bucktail expands to the appropriate diameter, it should extend beyond the base material by at least one-half the original rivet diameter.

When a rivet is driven, its cross sectional area increases along with its bearing and shearing strengths. Solid shank rivets are available in a variety of materials, head designs, and sizes to accommodate different applications.

RIVET CODES

Rivets are given part codes that indicate their size, head style, and alloy material. Two systems are in use today: the **Air Force - Navy, or AN** system; and the **Military Standards 20 system, or MS20**. While there are minor differences between the two systems, both use the same method for describing rivets. As an example, consider the rivet designation, AN470AD4-5.

The first component of a rivet part number denotes the numbering system used. As discussed, this can either be AN or MS20. The second part of the code is a three-digit number that describes the style of rivet head. The two most common rivet head styles are the universal head, which is represented by the code 470, and the countersunk head, which is represented by the code 426. Following the head designation is a one- or two-digit letter code representing the alloy material used in the rivet. These codes will be discussed in detail later.

After the alloy code, the shank diameter is indicated in 1/32 inch increments, and the length in increments of 1/16 inch. Therefore, in this example, the rivet has a diameter of 4/32 inch and is 5/16 of an inch long. [Figure 2-64]

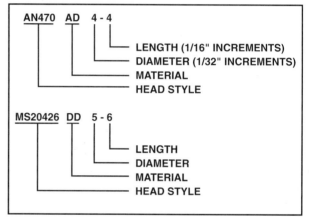

Figure 2-64. Rivet identification numbers indicate head style, material, and size.

The length of a universal head (AN470) rivet is measured from the bottom of the manufactured head to the end of the shank. However, the length of a countersunk rivet (AN426) is measured from the top of the manufactured head to the end of the shank. [Figure 2-65]

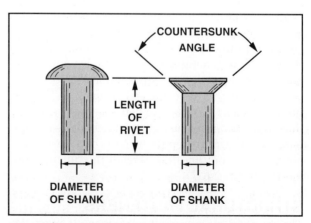

Figure 2-65. Universal and countersunk rivet diameters are measured in the same way, but their length measurements correspond to their grip length.

RIVET HEAD DESIGN

As mentioned, solid shank rivets are available in two standard head styles, **universal** and **countersunk**, or flush. The AN470 universal head rivet now replaces all previous protruding head styles such as AN430 round, AN442 flat, AN455 brazier, and AN456 modified brazier. [Figure 2-66]

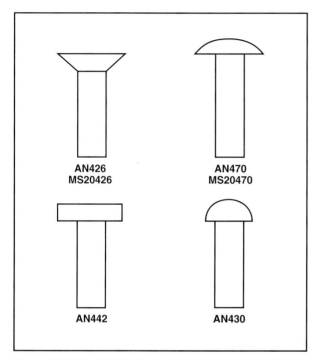

Figure 2-66. The AN470 rivet now replaces almost all other protruding head designs. The round head rivet (AN430) was used extensively on aircraft built before 1955, while the flat head rivet (AN442) was widely used on internal structures. Flat head rivets are still used for applications requiring higher head strength.

AN426 countersunk rivets were developed to streamline airfoils and permit a smooth flow over an aircraft's wings or control surfaces. However, before a countersunk rivet can be installed, the metal must be countersunk or dimpled. Countersinking is a process in which the metal in the top sheet is cut away in the shape of the rivet head. On the other hand, dimpling is a process that mechanically "dents" the sheets being joined to accommodate the rivet head. Sheet thickness and rivet size determine which method is best suited for a particular application.

Joints utilizing countersunk rivets generally lack the strength of protruding head rivet joints. One reason is that a portion of the material being riveted is cut away to allow for the countersunk head. Another reason is that, when riveted, the gun set may not make direct contact with the rivet head if the rivet hole was not countersunk or dimpled correctly, resulting in the rivet not expanding to fill the entire hole.

To ensure head-to-gun set contact, it is recommended that countersunk heads be installed with the manufactured head protruding above the skin's surface about .005 to .007 inch. This ensures that the gun set makes direct contact with the rivet head. To provide a smooth finish after the rivet is driven, the protruding rivet head is removed using a microshaver. This rotary cutter shaves the rivet head flush with the skin, leaving an aerodynamically clean surface. [Figure 2-67]

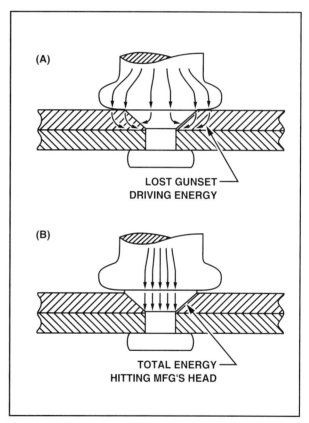

Figure 2-67. (A) — If a countersunk rivet is set with the rivet head flush with the metal's surface, some of the gun set's driving energy is lost. (B) — However, if the rivet head is allowed to protrude above the metal all of the gun set's energy hits the head, resulting in a stronger joint.

An alternative to leaving the rivet head sticking up slightly is to use the Alcoa crown flush rivet. These rivets have a slightly crowned head to allow full contact with the gun set. To drive these rivets, a mushroom-type rivet set is placed directly on the crown flush head. When the rivet is driven, the gun drives the countersunk head into the countersink, while simultaneously completing rivet expansion.

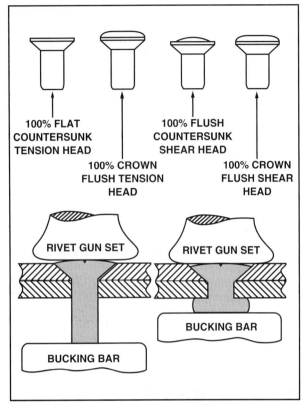

Figure 2-68. The raised head of a crown flush rivet allows greater contact area with a rivet set. This results in a stronger countersunk joint.

This results in a fully cold-worked rivet that needs no microshaving. [Figure 2-68]

RIVET ALLOYS

Most aircraft rivets are made of aluminum alloy. The type of alloy is identified by a letter in the rivet code, and by a mark on the rivet head itself. [Figure 2-69]

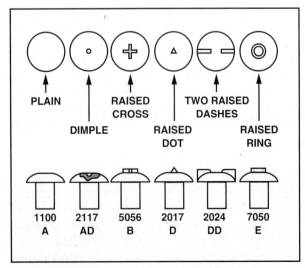

Figure 2-69. Head markings indicate the alloy used in common aircraft rivets.

1100 Aluminum (A)

Rivets made of pure aluminum have no identifying marks on their manufactured head, and are designated by the letter A in the rivet code. Since this type of rivet is made out of commercially pure aluminum, the rivet lacks sufficient strength for structural applications. Instead, 1100 rivets are restricted to nonstructural assemblies such as fairings, engine baffles, and furnishings. The 1100 rivet is driven cold, and therefore, its shear strength increases slightly as a result of cold-working.

2117 Aluminum Alloy (AD)

The rivet alloy 2117-T3 is the most widely used for manufacturing and maintenance of modern aircraft. Rivets made of this alloy have a dimple in the center of the head and are represented by the letters AD in rivet part codes. Because AD rivets are so common and require no heat treatment, they are often referred to as **"field rivets."**

The main advantage for using 2117-T3 for rivets is its high strength and shock resistance characteristics. The alloy 2117-T3 is classified as a heat-treated aluminum alloy, but does not require reheat-treatment before driving.

5056 Aluminum Alloy (B)

Some aircraft parts are made of magnesium. If aluminum rivets were used on these parts, dissimilar metal corrosion could result. For this reason, magnesium structures are riveted with 5056 rivets, which contain about 5% magnesium. These rivets are identified by a raised cross on their heads and the letter B in a rivet code. The maximum shear strength of an installed 5056-H32 rivet is 28,000 pounds per square inch.

2017 Aluminum Alloy (D)

2017 aluminum alloy is extremely hard. Rivets made of this alloy are often referred to as D rivets, and have been widely used for aircraft construction for many years. However, the introduction of jet engines placed greater demands for structural strength on aircraft materials and fasteners. In response to this, the aluminum industry modified 2017 alloy to produce a new version of 2017 aluminum called the "crack free rivet alloy." The minimum shear strength of the older 2017-T31 rivet alloy is 30 KSI, while that of the new 2017-T3 alloy is 34 KSI.

D-rivets are identified by a raised dot in the center of their head and the letter D in rivet codes. Because

Sheet Metal Structures

ALLOY	LETTER	HEAD MARKING	DRIVEN CONDITION	POUNDS IN KSI
1100	A	PLAIN	1100-F	9.5
2117	AD	DIMPLE	2117T3	30
5056	B	RAISED CROSS	5056H32	28
2017	D	RAISED DOT	2017T31	34
2017	D	RAISED DOT	2017T3	38
2024	DD	TWO RAISED DASHES	2024T31	41
7050	E	RAISED RING	7050T73	43

NOTE: $KSI = \dfrac{psi}{1000}$ (e.g. 30 KSI = 30,000 psi)

Figure 2-70. Different rivet alloys produce different shear strengths in their driven condition.

D-rivets are so hard, they must be heat-treated before they can be used. [Figure 2-70]

Recall from the study of heat treatments, that when aluminum alloy is quenched after heat treatment, it does not harden immediately. Instead, it remains soft for several hours and gradually becomes hard and gains full strength. Rivets made of 2017 can be kept in this annealed condition by removing them from a quench bath and immediately storing them in a freezer. Because of this, D-rivets are often referred to as icebox rivets. These rivets become hard when they warm up to room temperature, and may be reheat-treated as many times as necessary without impairing their strength.

2024 Aluminum Alloy (DD)
DD-rivets are identified by two raised dashes on their head. Like D-rivets, DD-rivets are also called **icebox rivets** and must be stored at cool temperatures until they are ready to be driven. The storage temperature determines the length of time the rivets remain soft enough to drive. For example, if the storage temperature is -30°F, the rivets will remain soft enough to drive for two weeks. When DD-rivets are driven, their alloy designation becomes 2024-T31 because of the work hardening achieved during installation.

7050-T73 Aluminum Alloy (E)
A new and stronger rivet alloy was developed in 1979 called 7050-T73. The letter E is used to designate this alloy, and the rivet head is marked with a raised circle. 7050 alloy contains zinc as the major alloying ingredient, and is precipitation heat-treated. This alloy is used by the Boeing Company as a replacement for 2024-T31 rivets in the manufacture of the 767 wide-body aircraft.

Corrosion-Resistant Steel (F)
Stainless steel rivets are used for fastening corrosion-resistant steel sheets in applications such as firewalls and exhaust shrouds. They have no marking on their heads.

Monel (M)
Monel rivets are identified with two recessed dimples in their heads. They are used in place of corrosion-resistant steel rivets when their somewhat lower shear strength is not a detriment.

SPECIAL RIVETS
A rivet is any type of fastener that obtains its clamping action by having one of its ends mechanically upset. Conventional solid shank rivets require access to both ends to be driven. However, special rivets, often called blind rivets, are installed with access to only one end of the rivet. While considerably more expensive than solid shank rivets, blind rivets find many applications in today's aircraft industry.

POP™ RIVETS
Pop rivets have limited use on aircraft and are never used for structural repairs. However, they are useful for temporarily lining up holes. In addition, some "home built" aircraft utilize Pop rivets. They are available in flat head, countersunk head, and modified flush heads with standard diameters of 1/8, 5/32, and 3/16 inch. Pop rivets are made

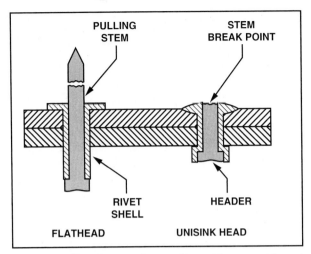

Figure 2-71. Pop rivets are frequently used for assembly and non-structural applications. They must not be used in areas that are subject to moderate or heavy loads.

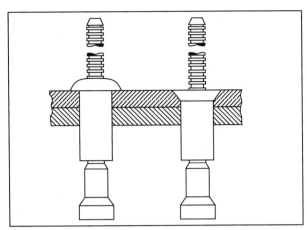

Figure 2-72. The friction-lock rivet assembly consists of a shell and mandrel, or pulling stem. The stem is pulled until the header forms a buck-tail on the blind side of the shell. At this point, a weak point built into the stem shears and the stem breaks off. After the stem fractures, part of it projects upward. The projecting stem is cut close to the rivet head, and the small residual portion remaining in the head is filed smooth.

from soft aluminum alloy, steel, copper, and Monel. [Figure 2-71]

FRICTION-LOCK RIVETS

One early form of blind rivet that was the first to be widely used for aircraft construction and repair was the **Cherry friction-lock rivet**. Originally, Cherry friction-locks were available in two styles: hollow shank pull-through, and self-plugging types. The pull-through type is no longer common. However, the self-plugging Cherry friction-lock rivet is still used for repairing light aircraft.

Cherry friction-lock rivets are available in two head styles: universal and 100 degree countersunk. Furthermore, they are usually supplied in three standard diameters: 1/8, 5/32, and 3/16 inch. However, larger sizes can be specially ordered in sizes up to 5/16 inch. [Figure 2-72]

A friction-lock rivet cannot replace a solid shank rivet, size for size. When a friction-lock is used to replace a solid shank rivet, it must be at least one size (1/32 inch) larger in diameter. This is because a friction-lock rivet loses considerable strength if its center stem falls out due to damage or vibration.

MECHANICAL-LOCK RIVETS

Mechanical-lock rivets were designed to prevent the center stem of a rivet from falling out as a result of the vibration encountered during aircraft operation. Unlike the center stem of a friction-lock rivet, a mechanical-lock rivet permanently locks the stem into place, and vibration cannot shake it loose.

HUCK-LOKS

Huck-Lok rivets were the first mechanical-lock rivets and are used as structural replacements for solid shank rivets. However, because of the expensive tooling required for their installation, Huck-Loks are generally limited to aircraft manufacturers and some large repair facilities.

Huck-Loks are available in four standard diameters: 1/8, 5/32, 3/16, and 1/4 inch, and come in three different alloy combinations: a 5056 sleeve with a 2024 pin, an A-286 sleeve with an A-286 pin, and a Monel 400 sleeve with an A-286 pin. [Figure 2-73]

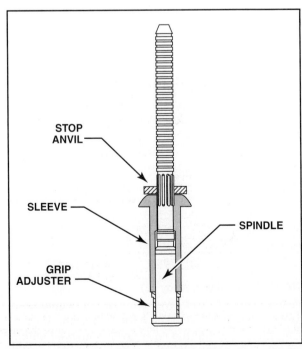

Figure 2-73. Unlike friction-lock rivets, Huck-Loks utilize a lock ring that mechanically locks the center stem in place.

CHERRYLOCKS™

The Cherry mechanical-lock rivet, often called the bulbed CherryLOCK, was developed shortly after the Huck-Lok. Like the Huck-Lok, the CherryLOCK rivet is an improvement over the friction-lock rivet because its center stem is locked into place with a lock ring. This results in shear and bearing strengths that are high enough to allow CherryLOCKS to be used as replacements for solid shank rivets. [Figure 2-74]

CherryLOCK rivets are available with two head styles: 100 degree countersunk, and universal. Like most blind rivets, CherryLOCKs are available with diameters of 1/8, 5/32, and 3/16 inch, with an oversize of 1/64 inch for each standard size. The rivet, or shell, portion of a CherryLOCK may be constructed of 2017 aluminum alloy, 5056 aluminum alloy, Monel, or stainless steel. Installation of CherryLOCK rivets requires a special pulling tool for each different size and head shape. However, the same size tool can be used for an oversize rivet in the same diameter group.

One disadvantage of a CherryLOCK is that if a rivet is too short for an application, the lock ring sets prematurely, resulting in a malformed shank header. This fails to compress the joint, leaving it in a weakened condition. To avoid this, always use the proper rivet length selection gauge and follow the manufacturer's installation recommendations.

OLYMPIC-LOKS

Olympic-lok blind fasteners are lightweight, mechanically-locking, spindle-type blind rivets. Olympic-loks come with a lock ring stowed on the head. As an Olympic-lok is installed, the ring slips down the stem and locks the center stem to the

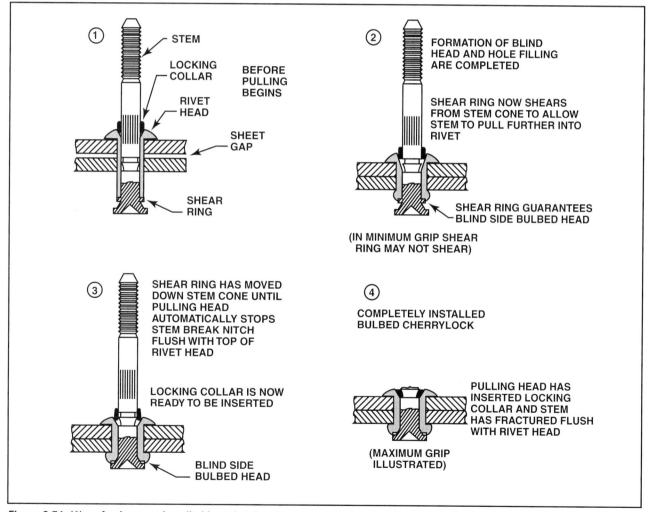

Figure 2-74. (1) — As the stem is pulled into the rivet sleeve, a bulb forms on the rivet's blind side that begins to clamp the two pieces of metal together and fill the hole. (2) — Once the pieces are clamped tightly together, the bulb continues to form until the shear ring shears and allows the stem to pull further into the rivet. (3) — With the shear ring gone, the stem is pulled upward until the pulling head automatically stops at the stem break notch, and the locking collar is ready to be inserted. (4) — When completely installed, the locking collar is inserted and the stem is fractured flush with the rivet head.

outer shell. These blind fasteners require a specially designed set of installation tools. [Figure 2-75]

Olympic-lok rivets are made with three head styles: universal, 100 degree flush, and 100 degree flush shear. Rivet diameters of 1/8, 5/32, and 3/16 inch are available in eight different alloy combinations of 2017-T4, A-286, 5056, and Monel.

When Olympic-loks were first introduced, they were advertised as an inexpensive blind fastening system. The price of each rivet is less than the other types of mechanical locking blind rivets, and only three installation tools are required. The installation tools fit both countersunk and universal heads in the same size range.

CHERRYMAX™

The CherryMAX rivet is economical to use and strong enough to replace solid shank rivets, size for size. The economic advantage of the CherryMAX system is that one size puller can be used for the installation of all sizes of CherryMAX rivets. A CherryMAX rivet is composed of five main parts: a pulling stem, a driving anvil, a safe-lock locking collar, a rivet sleeve, and a bulbed blind head. [Figure 2-76]

Available in both universal and countersunk head styles, the rivet sleeve is made from 5056, monel, and inco 600. The stems are made from alloy steel, CRES, and inco X-750. The ultimate shear strength of CherryMAX rivets ranges from 50KSI to 75KSI. Furthermore, CherryMAX rivets can be used at temperatures from 250°F to 1,400°F. They are available

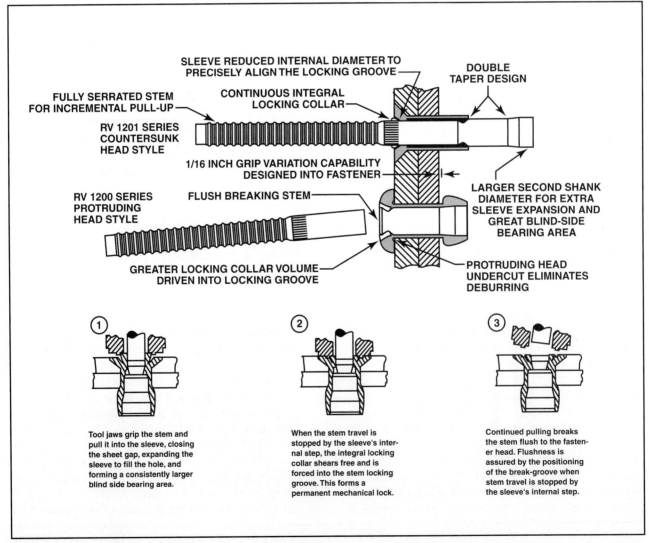

Figure 2-75. (1) — Once an Olympic-Lok rivet is inserted into a prepared hole, the stem is pulled into the sleeve, closing any gap between the materials being riveted: filling the hole, and forming a bearing area. (2) — When the stem travel is stopped by the sleeve's internal step, the locking collar shears free and is forced into the locking groove. (3) — Continued pulling breaks the stem flush with the rivet head.

Sheet Metal Structures

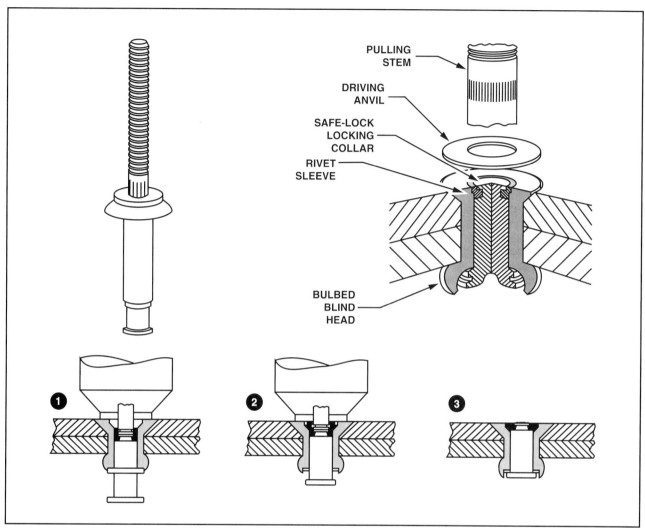

Figure 2-76. (1) — As the stem pulls into the rivet sleeve, it forms a large bulb that seats the rivet head, and clamps the two sheets tightly together. (2) As the blind head is completed, the safe-locking collar moves into the rivet sleeve recess. (3) As the stem continues to be pulled, the safe-lock collar is formed into the head recess by the driving anvil, locking the stem and sleeve securely together. Further pulling fractures the stem, providing a flush, burr-free installation.

in diameters of 1/8, 5/32, 3/16 and 1/4 inches, and are also made with an oversize diameter for each standard diameter listed.

Removal Of Mechanical-Lock Rivets

To remove mechanical-lock rivets, first file a flat spot on the rivet's center stem. Once this is done, a center punch is used to punch out the stem so the lock ring can be drilled out. With the lock ring removed, tap out the remaining stem, drill to the depth of the manufactured head, and tap out the remaining shank. All brands of mechanical-lock blind rivets are removed using the same basic technique.

HI-SHEAR RIVETS

One of the first special fasteners used by the aerospace industry was the Hi-Shear rivet. Hi-Shear rivets were developed in the 1940s to meet the demand for fasteners that could carry greater shear loads.

The Hi-Shear rivet has the same strength characteristics as a standard AN bolt. In fact, the only difference between the two is that a bolt is secured by a nut and a Hi-Shear rivet is secured by a crushed collar. The Hi-Shear rivet is installed with an interference fit, where the side wall clearance is reamed to a tolerance determined by the aircraft builder. When properly installed, a Hi-Shear rivet has to be tapped into its hole before the locking collar is swaged on.

Hi-Shear rivets are made in two head styles: flat and countersunk. As the name implies, the Hi-Shear rivet is designed especially to absorb high shear loads. The Hi-Shear rivet is made from steel alloy

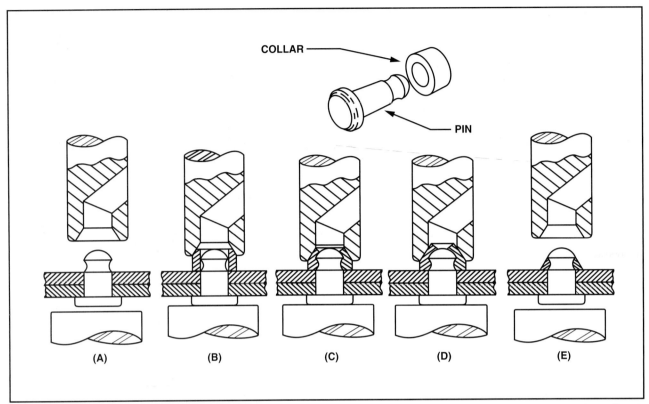

Figure 2-77. (A) — A bucking bar and rivet gun are used to install Hi-Shear rivets. (B) — A collar is placed over the pin's small end. (C) — The rivet gun forces the collar over the pin. (D) — The gun set drives the collar onto the rivet pin and cuts off excess material. (E) — When the collar is fully driven, excess collar material is ejected from the gun set.

having the same tensile strength as an equal size AN bolt. The lower portion of its shank has a specially milled groove with a sharp edge that retains and finishes the collar as it is swaged into the locked position. [Figure 2-77]

SPECIAL FASTENERS

Many special fasteners have the advantage of producing high strength with light weight and can be used in place of conventional AN bolts and nuts. When a standard AN nut and bolt assembly is tightened, the bolt stretches and its shank diameter decreases, causing the bolt to increase its clearance in the hole. Special fasteners eliminate this change in dimension because they are held in place by a collar that is squeezed into position instead of being screwed on like a nut. As a result, these fasteners are not under the same tensile loads that are imposed on a bolt during installation.

LOCK BOLTS

Lock bolts are manufactured by several companies and conform to Military Standards. These standards describe the size of a lock bolt's head in relation to its shank diameter, as well as the alloy used. Lock bolts are used to permanently assemble two materials. They are lightweight, and as strong as standard bolts.

There are three types of lock bolts used in aviation: the pull-type lock bolt, the blind-type lock bolt, and the stump-type lock bolt. The pull-type lock bolt has a pulling stem on which a pneumatic installation gun fits. The gun pulls the materials together and then drives a locking collar into the grooves of the lock bolt. Once secure, the gun fractures the pulling pin at its break point. The blind-type lock bolt is similar to most other types of blind fasteners. To install a blind-type lock bolt, it is placed into a blind hole and an installation gun is placed over the pulling stem. As the gun pulls the stem, a blind head forms and pulls the materials together. Once the materials are pulled tightly together, a locking collar locks the bolt in place and the pulling stem is broken off. Unlike other blind fasteners that typically break off flush with the surface, blind lock bolts protrude above the surface.

The third type of lock bolt is the stump-type lock bolt, and is installed in places where there is not enough room to use the standard pulling tool. Instead, the stump-type lock bolt is installed using

Sheet Metal Structures

an installation tool similar to that used to install Hi-Shear rivets. [Figure 2-78]

Lock bolts are available for both shear and tension applications. With shear lock bolts, the head is kept thin and there are only two grooves provided for the locking collar. However, with tension lock bolts, the head is thicker and four or five grooves are provided to allow for higher tension values. The locking collars used on both shear and tension lock bolts are color coded for easy identification. [Figure 2-79]

HI-LOKS

Hi-Lok bolts are manufactured in several different alloys such as titanium, stainless steel, steel, and aluminum. They possess sufficient strength to withstand bearing and shearing loads, and are available with flat and countersunk heads.

A conventional Hi-Lok has a straight shank with standard threads. Although wrenching lock nuts are usually used, the threads are compatible with standard AN bolts and nuts. To install a Hi-Lok, the hole is first drilled with an interference fit. The Hi-Lok is then tapped into the hole and a shear collar is installed. A Hi-Lok retaining collar is installed using either specially prepared tools or a simple Allen and box end wrench. Once the collar is tightened to the appropriate torque value, the wrenching

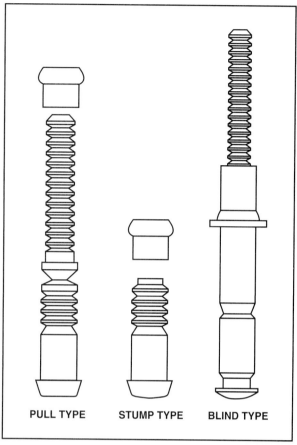

Figure 2-78. Lock bolts are classified as pull type, stump type, and blind type. The bolt used for a particular application depends primarily on access to the work area.

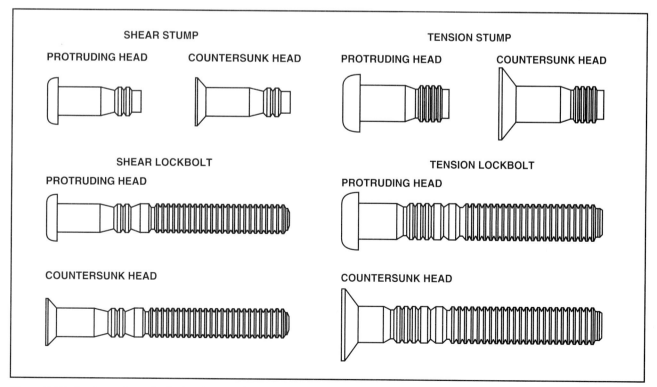

Figure 2-79. As shown here, both the shear and tension type lock bolts come in a variety of sizes with multiple head styles.

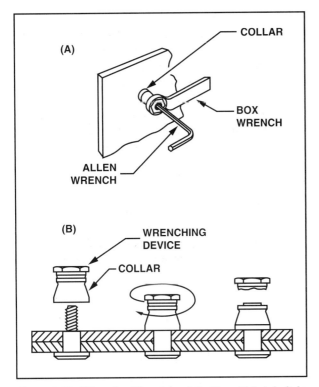

Figure 2-80. (A) — An Allen wrench holds a Hi-Lok bolt in place while a combination wrench is used to tighten the shear nut. (B) — Once a Hi-Lok is installed, the collar is installed and tightened. When the appropriate torque value is obtained, the wrenching device shears off, leaving the collar.

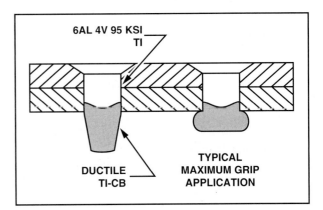

Figure 2-81. CherryBUCK fasteners combine two titanium alloys to produce a one-piece fastener with 95 KSI shear strength.

device shears off leaving only the locking collar. [Figure 2-80]

HI-LITE FASTENERS

The Hi-Lite fastener is similar to the Hi-Lok except that it is made from lighter materials and has a shorter transition from the threaded section to the shank. Furthermore, the elimination of material between the threads and shank yields an additional weight saving, with no loss of strength. The Hi-Lite's main advantage is its excellent strength to weight ratio.

Hi-Lites are available in an assortment of diameters ranging from 3/16 to 3/8 inch. They are installed either with a Hi-Lok locking collar, or by a swaged collar such as the lock bolt. In either case, the shank diameter is not reduced by stretch torquing.

CHERRYBUCK RIVETS

The CherryBUCK is a one-piece special fastener that combines two titanium alloys which are bonded together to form a strong structural fastener. The head and upper part of the shank of a CherryBUCK is composed of 6AL-4V alloy, while

Ti-Cb alloy is used in the lower shank. When driven, the lower part of the shank forms a buck-tail.

An important advantage of the CherryBUCK is that it is a one piece fastener. Since there is only one piece, CherryBUCKs can safely be installed in jet engine intakes with no danger of foreign object damage. This type of damage often occurs when multiple piece fasteners lose their retaining collars and are ingested into a compressor inlet. [Figure 2-81]

TAPER-LOK FASTENER

Taper-Loks are the strongest special fasteners used in aircraft construction. Because of its tapered shape, the Taper-Lok exerts a force on the conical walls of a hole, much like a cork in a wine bottle. To a certain extent, a Taper-Lok mimics the action of a driven solid shank rivet, in that it completely fills the hole. However, a Taper-Lok does this without the shank swelling.

When a washer nut draws the Taper-Lok into its hole, the fastener pushes outward and creates a tremendous force against the tapered walls of the hole. This creates radial compression around the shank and vertical compression lines as the metals are squeezed together. The combination of these forces generates strength unequaled by any other type of fastener. [Figure 2-82]

HI-TIGUE FASTENERS

The Hi-Tigue fastener has a bead that encircles the bottom of its shank and is a further advancement in special fastener design. This bead preloads the hole it fills, resulting in increased joint strength. During installation, the bead presses against the side wall of the hole, exerting a radial force, which strengthens the surrounding area. Since it is preloaded, the joint

Sheet Metal Structures

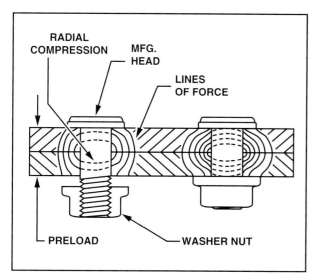

Figure 2-82. The hole for a Taper-Lok is made with a special tapered drill. Once a Taper-Lok is installed and a washer nut is tightened, radial compression forces and vertical compression forces combine to create an extremely strong joint.

is not subjected to the constant cyclic action that normally causes a joint to become cold-worked and eventually fail.

Hi-Tigue fasteners are produced in aluminum, titanium, and stainless steel alloys. The collars are also composed of compatible metal alloys and are available in two types: sealing and non-sealing. As with Hi-Loks, Hi-Tigues can be installed using an Allen and combination wrench. [Figure 2-83]

JO-BOLTS

Jo-Bolts are patented high-strength structural fasteners that are used in close-tolerance holes where strength requirements are high, but physical clearance precludes the use of standard AN, MS, or NAS bolts.

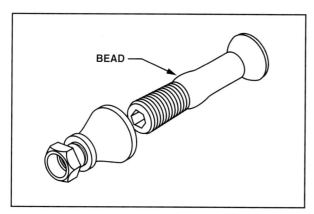

Figure 2-83. A Hi-Tigue fastener features a subtly shaped bead at the threaded end of the shank. This bead preloads the hole it is inserted into, thereby strengthening the joint.

The hole for a Jo-Bolt is drilled, reamed, and countersunk before the Jo-Bolt is inserted, and held tightly in place by a nose adapter of either a hand tool or power tool. A wrench adapter then grips the bolt's driving flat and screws it up through the nut. As the bolt pulls up, it forces a sleeve up over the tapered outside of the nut and forms a blind head on the inside of the work. When driving is complete, the driving flat of the bolt breaks off. [Figure 2-84]

Removal Of Special Fasteners

Special fasteners that are locked into place with a crushable collar are easily removed by splitting the collar with a small cape chisel. After the collar is split, knock away the two halves and tap the fastener from the hole. Fasteners that are not damaged during removal can be reused using new locking

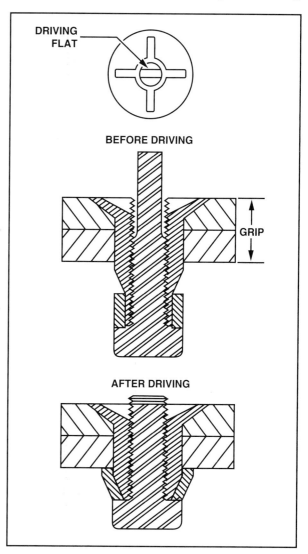

Figure 2-84. Once a Jo-Bolt is inserted into a hole, the bolt is rotated, causing the nut to pull up to the metal. As the nut moves upward, a sleeve is forced over the tapered end of the bolt. This creates a blind head that holds the joint together.

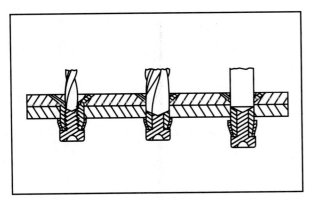

Figure 2-85. Remove flush-type Jo-Bolts by first drilling a pilot hole into the bolt slightly deeper than the inside of the head of the nut. Then, using a twist drill of the same size, drill to the depth of the pilot hole. Drive the shank and blind head from the hole using a pin punch of the proper size.

THREADED RIVETS/RIVNUTS

Goodrich Rivnuts were developed by the B.F. Goodrich Company to attach rubber deicer boots to aircraft wing and tail surfaces. To install a Rivnut, a hole is drilled in the skin to accommodate the body of the Rivnut, and a special cutter is used to cut a small notch in the circumference of the hole. This notch locks the Rivnut into the skin to prevent it from turning when it is used as a nut. A Rivnut of the proper grip length is then screwed onto a special puller and inserted into the hole with a key, or protrusion, aligned with the key-way cut in the hole. When the handle of the puller is squeezed, the hollow shank of the Rivnut upsets and grips the skin. The tool is then unscrewed from the Rivnut, leaving a threaded hole that accepts machine screws for attaching a de-icer boot. [Figure 2-86]

collars. The removal techniques of certain special fasteners are basically the same as those used for solid shank rivets. However, in some cases, the manufacturer may recommend that a special tool be used.

Removal of Taper-Loks, Hi-Loks, Hi-Tigues, and Hi-Lites requires the removal of the washer-nut or locking collar. Both are removed by turning them with the proper size box end wrench or a pair of vise-grips. After removal, a mallet is used to tap the remaining fastener out of its hole.

To remove a Jo-Bolt, begin by drilling through the nut head with a pilot bit, followed by a bit the same size as the bolt shank. Once the nut head is removed, a punch is used to punch out the remaining portion of the nut and bolt. [Figure 2-85]

ACCESS PANEL AND COWLING FASTENERS

Turn-lock fasteners are used to secure inspection plates, doors, cowlings, and other removable panels on aircraft. The most desirable feature of these fasteners is that they permit quick and easy removal or opening of access panels for inspection and servicing purposes. Turn-lock fasteners are produced and supplied by a number of manufacturers under various trade names. Some of the most commonly used are the Dzus, Airloc, and Camlock.

DZUS FASTENERS

Cowling and other inspection access doors, that must be opened or removed frequently, can be secured with Dzus fasteners that require only a quarter of a turn to lock or unlock. With a Dzus fastener, a hard spring-steel wire is riveted across an opening on a fixed part of the fuselage, while a stud

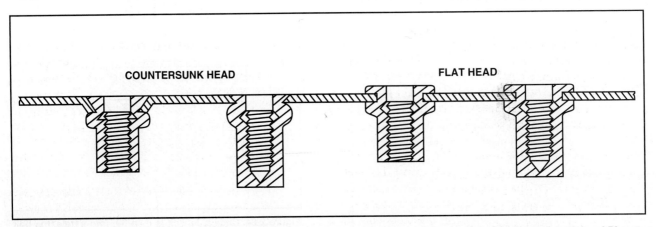

Figure 2-86. Rivnuts are commonly available with flat heads and with 100 degree countersunk heads. Countersunk head Rivnuts are made with both .048 inch and .063 inch head thickness, with the thinner head used when it is necessary to install a Rivnut in a machine countersunk hole in thin material. Closed-end Rivnuts are available for installation in a pressurized structure or sealed compartment, such as a fuel tank.

Sheet Metal Structures

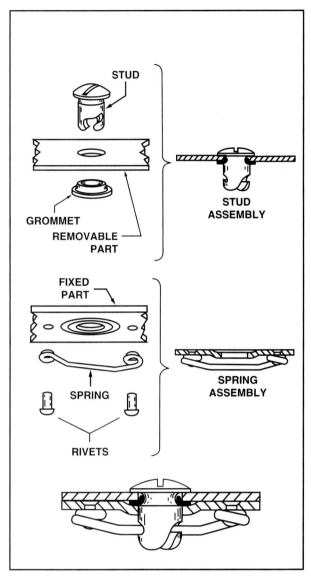

Figure 2-87. With a standard Dzus fastener, a slotted stud engages a spring mounted to the fuselage. As the stud is turned one quarter turn, the fastener locks into place by straddling and securing to the spring.

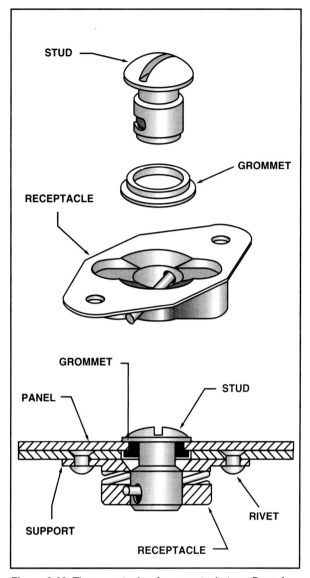

Figure 2-88. The receptacle of a receptacle-type Dzus fastener guides the stud to the exact location it needs to be prior to engaging the spring.

is mounted on the access panel and secured by a metal grommet. Turning the stud one-quarter pulls the stud down by straddling the spring into a beveled slot cut into the stud. As the stud reaches a locked position, the spring drops into a recess in the slot. [Figure 2-87]

When something is fastened with Dzus fasteners, care must be taken that the stud in every fastener straddles each of the springs rather than passing beside them. To ensure that all of the fasteners are properly locked, the heads of the fasteners should all line up once secured. Furthermore, when a Dzus fastener is fastened, a distinct click is heard when the spring drops into the recess of the slot in the

locked position. To aid in ensuring that no stud misses the spring. Special receptacle-type Dzus fasteners are available that guide the stud over the spring. [Figure 2-88]

To aid in identifying the size and head style of a Dzus fastener, the manufacturer provides a cast letter and number in the head of each stud. The letters designate the head style as being a raised oval head, flush head, or winged head. The winged head style allows for turning the fastener without the use of hand tools. On the other hand, the raised oval and flush head styles, usually require a slotted or Phillips style screwdriver to be used to turn the fastener. The numbers following the head style letter

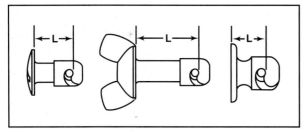

Figure 2-89. Stud lengths of raised oval head and winged Dzus fasteners are measured from the bottom of the head to the bottom of the locking slot, whereas flush head styles are measured from the top of the head to the bottom of the locking slot.

designate the diameter and length of the stud. The first number indicates the diameter of the stud body measured in 1/16 inch increments, while the second number designates the length of the stud measured in hundredths of an inch. [Figure 2-89]

AIRLOC FASTENER

An Airloc fastener consists of a steel stud and cross-pin in a removable cowling or door and a sheet spring-steel receptacle in the stationary member. To lock this type of fastener, the stud slips into the receptacle and is rotated one-quarter. The pin drops into an indentation in the receptacle spring and holds the fastener locked. [Figure 2-90]

CAMLOCK FASTENER

The stud assembly of a Camlock fastener consists of a housing containing a spring and a stud with a steel pin. This assembly is held onto the removable portion of the cowling or access door with a metal grommet. The stud fits into a pressed steel receptacle, and a one-quarter turn locks the steel pin in a groove in the bottom of the receptacle. [Figure 2-91]

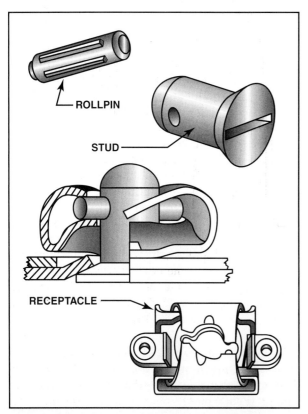

Figure 2-90. Airloc cowling fasteners are similar to Dzus fasteners and are used in many of the same applications.

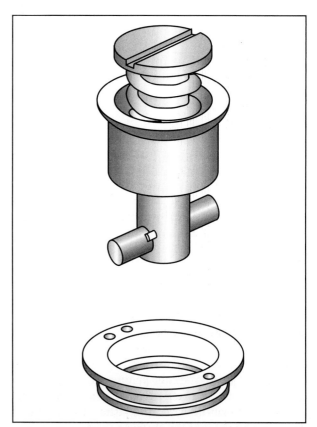

Figure 2-91. With a Camlock cowling fastener, the stud assembly can be inserted when the pin is aligned with the slot in the receptacle.

SECTION C

SHEET METAL FABRICATION

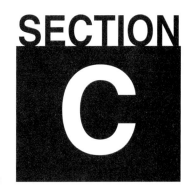

The selection of sheet metal materials, and proper fabrication techniques is crucial during initial construction of a sheet metal structure, or for its restoration of strength during a repair. In most cases, selection of materials and fabrication processes can be determined by consulting the manufacturer's publications, such as the structural repair or maintenance manuals. However, if the manufacturer does not identify specific materials and fabrication procedures, the technician must use standard industry acceptable techniques to perform sheet metal work.

To aid the technician, many of these standard material identification and fabrication processes are addressed in Advisory Circular 43.13-1B, *Acceptable Methods, Techniques and Practices*. By learning and applying standard practices, the technician can perform satisfactory work to many types of sheet metal structures, even when the manufacturer's information is not available to directly cover the task to be performed.

INSTALLATION OF SOLID RIVETS

A primary task performed by any aircraft maintenance technician working with sheet metal is to install solid shank rivets in structures or components. This involves selecting the proper rivets and installing them in such a way that the maximum structural integrity of the product is attained, as determined by engineering experience with properly installed rivets. Although the concept of installing rivets is a straightforward process, there are a number of aspects that the technician must understand and observe to achieve the optimum performance from the structure or component design.

RIVET SELECTION

When initially fabricating or making a repair to an aircraft structure or component, the technician's primary objective is to obtain the maximum structural integrity and aerodynamic shape. When an aircraft is originally certificated, the acceptance of the design is based on each subsequent aircraft being manufactured to the original design specifications. If an aircraft is damaged while in service, it is the technician's responsibility to verify that the aircraft is repaired to those original specifications and certification criteria. With any sheet metal structure, the proper preparation and installation of solid shank rivets is paramount in achieving these goals. One of the most critical aspects of sheet metal work is to be able to select the proper rivet for a given application.

As previously discussed in the beginning of this chapter, when a rivet is used for a particular installation, the shear strength of the rivet, as compared to its bearing strength, must be considered. This is especially important when replacing rivets of different types of alloys. For example, if replacing a 2024 rivet with a 2117 rivet, the relative bearing and shear strengths of each rivet should be considered. These strengths can be found in rivet charts. In this case, due to the varying strength characteristics of the two different types of alloys used in the rivets, the smaller 2024 rivet must be replaced with one size larger 2117 rivet to produce the same strength qualities. In many cases, the only course is to use the same type and size rivet originally specified. Bearing and shear strength should usually be nearly the same. However, the bearing strength should be slightly higher to enable the part to fail by shearing the rivets rather than tearing out sections of sheet metal. [Figures 2-92 and 2-93]

The single-shear strength of aluminum alloy rivets is shown in Figure 2-92. From this figure it can be determined that a 1/8 inch diameter AD rivet will support a load of 331 pounds, and a 1/8 inch DD rivet will support a load of 429 pounds before shearing. For double-shear loads, that is, a joint in which three pieces of material are being held together, the values in the chart are doubled. On the other hand, Figure 2-93 shows the bearing strength of 2024-T3 clad aluminum alloy sheets. From the chart, you can determine what the bearing strength of the aluminum sheet will be, given different sheet thickness and rivet diameters. For example, a .040 inch thick 2024-T3 clad sheet will provide 410 pounds bearing strength using a 1/8 inch diameter rivet. If an AD

SINGLE-SHEAR STRENGTH OF ALUMINUM ALLOY RIVETS — POUNDS									
COMPOSITION OF RIVET (ALLOY)	ULTIMATE STRENGTH OF RIVET METAL (POUNDS PER SQUARE INCH)	DIAMETER OF RIVET (INCHES)							
		1/16	3/32	1/8	5/32	3/16	1/4	5/16	3/8
AD 2117	27,000	83	186	331	518	745	1.325	2,071	2.981
D 2017	30,000	92	206	368	573	828	1,472	2,300	3.313
DD 2024	35,000	107	241	429	670	966	1,718	2,684	3,865

DOUBLE-SHEAR IS FOUND BY MULTIPLYING THE NUMBERS IN THE CHART BY TWO

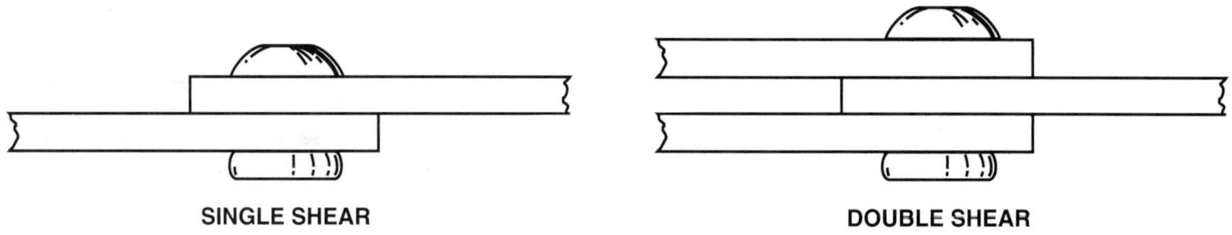

SINGLE SHEAR   DOUBLE SHEAR

Figure 2-92. The shear strength of a rivet is dependent on its alloy and the diameter of the rivet.

rivet is used, the shear strength of the rivet will be 331 pounds, which is less than the 410 pounds of bearing strength. In this example, the rivet and metal combination would provide a satisfactory joint, in that the rivet would shear in two before damaging the sheet metal.

When charts are not readily available for determining the shear strength of rivets and bearing strength of sheet metal, a general formula can be used to determine the proper diameter rivet for single lap joints. In this formula, the proper diameter rivet is equal to three times the thickness of the thickest sheet of metal in the joint. For example, if a joint consists of one sheet of aluminum .032 inch thick and another .040 inch thick, the rivet diameter would be equal to 3 × .040 or .120 inch. The closest rivet diameter to this dimension is a -4-diameter rivet, which is .125 inch. Once the proper diameter rivet has been determined, the correct rivet length can be calculated.

To fabricate a properly driven rivet, the width of the bucked head must be equal to one and a half times the rivet's original shank diameter and the height must be one half the original shank diameter. In

THICKNESS OF SHEET (INCHES)	BEARING STRENGTH OF 2024-T3 CLAD ALUMINUM (POUNDS)							
	DIAMETER OF RIVET (INCHES)							
	1/16	3/32	1/8	5/32	3/16	1/4	5/16	3/8
.016	82	123	164	204	246	328	410	492
.020	102	253	205	256	307	410	412	615
.025	128	192	256	320	284	512	640	768
.032	164	145	328	409	492	656	820	984
.040	205	307	410	512	615	820	1,025	1,230
.051	261	391	522	653	784	1,045	1,306	1,568
.064	....	492	656	820	984	1,312	1,640	1,968
.072	....	553	738	922	1,107	1,476	1,845	2,214
.091	....	699	932	1,167	1,398	1,864	2,330	2,796
.125	....	961	1,281	1,602	1,922	2,563	3,203	3,844
.250	....	1,921	2,562	3,202	3,843	5,125	6,405	7,686
.500	....	3,842	5,124	6,404	7,686	10,250	12,090	15,372

Figure 2-93. The bearing strength of sheet metal is dependent on the alloy, thickness of the sheet, and the size of the installed rivet.

order for the bucked head to develop these finished dimensions, the rivet needs to protrude through the metal approximately one and a half times the shank diameter before being driven.

To calculate the proper length, it is necessary to multiply one and a half times the diameter of the rivet and then add the grip length of the material. For the proper dimension rivet in the previous example, the rivet length is determined by multiplying the diameter of the rivet (.125) by one and a half, or 1.5. This provides an answer of .1855. To this number, add the grip length of the metal (.032 + .040) to obtain an overall length of .2575 inches. Once this number is determined, convert the answer into the closest 1/16 inch increment, which in this case, is equal to 1/4 inch for a -4 length rivet. If the rivets in this sample project had universal head styles, the final rivet identification would be MS20470AD4-4. [Figure 2-94]

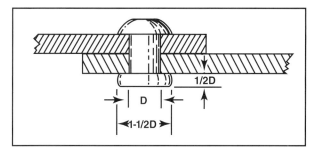

Figure 2-94. A properly formed shop or bucked rivet head will have a diameter of one and a half times the original shank diameter wide, by one half the shank diameter high, as shown in this sketch.

RIVET CUTTERS

To avoid carrying a large assortment of rivet lengths, many shops only purchase longer rivets. For shorter sizes, a rivet cutter can be used to trim the rivets to the desired length. [Figure 2-95]

RIVET LAYOUT PATTERNS

It is important when making a riveted repair that the rivets be installed in such a way that they will develop the maximum strength from the sheet metal. To obtain this strength, not only the rivet and sheet strength must be determined, but the rivet pattern is also a critical factor so the drilled holes do not weaken the joint. This means the spacing between rivets and the distance they remain from the edge of the material cannot be closer than minimum specifications.

RIVETED SHEET METAL STRENGTH

From the previous discussion of sheet metal alloys, it was presented that sheet metal can withstand spe-

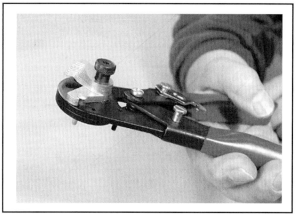

Figure 2-95. Rivet cutters have holes to cut common-sized rivet diameters, and a series of leaves that are rotated into position to shim under the rivet head to vary the shank length.

cific amounts of tensile stress depending on the type and thickness material. In aircraft construction, it is important that sheet metals be able to withstand the minimum tensile stresses when the sheets are used in stressed skin structures. For example, semimonocoque designs rely heavily on the ability of sheet metal skins to carry stresses into a substructure. When an aircraft is designed, engineers determine the amount of load that the sheet metal must carry to provide the desired strength from a structure. If sheet metal skins or components have riveted joints or seams, there must be an adequate number of rivets to carry the load. To determine the proper number of rivets, charts are available to determine the quantity of rivets that must be used for various types and thickness of alloy sheets. [Figure 2-96]

For example, when fabricating a seam in 2024-T3 alclad aluminum that is .040 inch thick, it is necessary to use a minimum of 6.2 1/8 inch diameter rivets per inch of seam width. If the seam is 8 inches wide, it will require 49.6 or 50 rivets to fabricate the seam to withstand the same tensile stresses as the original sheet metal. Additional information regarding the use of these types of charts is covered in the sheet metal repair section of this chapter.

EDGE DISTANCE

It is important when installing rivets that they be placed a certain distance from the edge of the material. If rivets are installed too close to the edge, the sheet metal will tear out instead of shearing the rivet when extreme loads are encountered. Conversely, if the rivets are placed too far away from the edge, the metal sheets can separate, allowing foreign contaminates to enter the joint, ultimately causing corrosion.

THICKNESS "T" IN INCHES	NO. OF 2117-T4 (AD) PROTRUDING HEAD RIVETS REQUIRED PER INCH OF WIDTH "W"					NUMBER OF BOLTS
	Rivet size					
	3/32	1/8	5/32	3/16	1/4	AN-3
.016	6.5	4.9	--	--	--	--
.020	6.5	4.9	3.9	--	--	--
.025	6.9	4.9	3.9	--	--	--
.032	8.9	4.9	3.9	3.3	--	--
.036	10.0	5.6	3.9	3.3	2.4	--
.040	11.1	6.2	4.0	3.3	2.4	--
.051	--	7.9	5.1	3.6	2.4	3.3
.064	--	9.9	6.5	4.5	2.5	3.3
.081	--	12.5	8.1	5.7	3.1	3.3
.091	--	--	9.1	6.3	3.5	3.3
.102	--	--	10.3	7.1	3.9	3.3
.128	--	--	12.9	8.9	4.9	3.3

NOTES:

   a. For stringers in the upper surface of a wing, or in a fuselage, 80 percent of the number of rivets shown in the table may be used.
   b. For intermediate frames, 60 percent if the number shown may be used.
   c. For single lap sheet joints, 75 percent of the number shown may be used.

ENGINEERING NOTES:

   a. The load per inch of width of material was calculated by assuming a strip 1 inch wide in tension.

   b. Number of rivets required was calculated for 2117-T4 (AD) rivets, based on a rivet allowable shear stress equal to 50 percent of the sheet allowable tensile stress, and a sheet allowable bearing stress equal to 160 percent of the sheet allowable tensile stress, using nominal hole diameters for rivets.

   c. Combinations of sheet thickness and rivet size above the underlined numbers are critical in (i.e., will fail by) bearing on the sheet; those below are critical in shearing of the rivets.

   d. The number of AN-3 bolts required below the underlined number was calculated based on a sheet allowable tensile stress of 55,000 psi and a bolt allowable single shear load of 2,126 pounds.

Figure 2-96. Charts are available to provide information regarding the number of rivets required in a joint or seam to endure the same amount of tensile strength as sheet metal. The chart in this figure provides information regarding the minimum number of rivets required per inch of width of a joint or seam to restore the strength of stressed sheet metal structures.

An accepted practice is to place the center of a rivet hole no closer than two rivet shank diameters from the edge and no further back than four diameters. With this in mind, it is common to place the center of the rivet hole about two and a half shank diameters away from the edge of the sheet. However, if possible, the rivet edge distance should be the same as the original by matching the previous pattern, or by consulting the manufacturer's design specifications.

PITCH

The distance between adjacent rivets in a row is called the pitch. To prevent the joint from being weakened by too many holes in a row, the adjacent rivets should be no closer than three diameters to one another. In contrast, to prevent the sheets from separating between rivets, the rivet holes should be no further apart than ten to twelve times the rivet shank diameter. [Figure 2-97]

RIVET GAUGE OR TRANSVERSE PITCH

The distance between rows of rivets in a multi-row layout should be about 75% of the pitch, provided that the rivets in adjacent rows are staggered. If the rivets are not staggered, then the pitch will be the

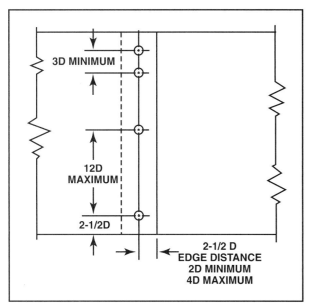

Figure 2-97. Edge distance and rivet pitch are critical to obtaining maximum strength from a riveted repair.

same between rows as it is between rivets in a single row. For most layout patterns, it is most practical to stagger the placement of rivets to reduce the amount of sheet metal that has to be overlapped. In addition, multiple rivet rows are often used to prevent rivets in a single row from becoming too close together, or to improve the cosmetics of a repair.

SAMPLE LAYOUT PATTERN

For a sample layout pattern, assume that you want to join two 3-3/4 inch straps of .040 inch 2024-T3 clad aluminum alloy sheets with 11 MS20470AD4-4 rivets. To install 11 rivets, more than one row will be required, because a single row would cause the rivets to be so close that the joint would be weakened. For this example, assume two rows are used with six rivets in the first row and five in the second. To make a layout pattern, first make two lines to mark off the edge distance of two and a half rivet shank diameters. Since the rivets are 1/8 inch (4/32), the distance will be 5/16 inch from the edge of the sheets. Mark these lines with a soft pencil or felt tip marker so the marks will not scratch the metal. Measure another 5/16 inch up from the ends of these lines to locate the position of the end rivets so they will have the proper edge distance from both the end and the sides of the sheet. Mark these locations with a center punch to enable starting a twist drill in the correct position. Then, with a pair of dividers, separate the distance between the two end rivets into five equal spaces to find the location of the six rivets for the first row. These rivets will be 5/8 inch apart, or five diameters, which is well within the allowable spacing of between 3 and 12 times the rivet shank diameter (3D to 12D).

The gauge, or distance, between the rows should be about 75% of the pitch and, in this case, will be 0.468 inch. For practical purposes, 1/2 inch (.5) is adequate. Mark a row across the strap 1/2 inch from the first row of rivets and locate the five holes needed on this line. These holes should be centered between the rivets in the first row. Again, mark the rivet locations with a center punch. [Figure 2-98]

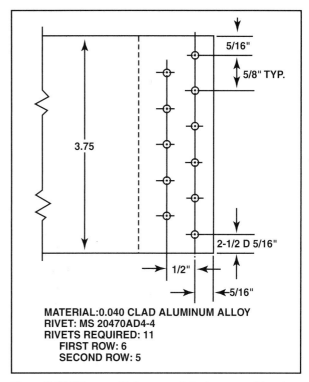

Figure 2-98. When multiple rows of rivets are used in a layout, the transverse pitch should be approximately 75% of the rivet pitch, and the rivets should be staggered.

HOLE PREPARATION FOR RIVETS

The hole in which a rivet is installed is critical to the strength of the finished repair. The hole must be slightly larger, but not so large that the expanded rivet does not fill the hole. Sizes of holes for all conventional rivets can be found in charts included in most aircraft technician handbooks. It is important that the hole be properly drilled and finished prior to a rivet being driven.

DRILL SIZE

The twist drills used for aircraft sheet metal work are most generally of the number and letter sizes, rather than the fractional sizes commonly used in other forms of mechanical work. Most of the rivets used in sheet metal work are between 3-3/32 inch, which is the smallest rivet generally allowed in aircraft structure, and 3/8 inch diameter. Rather than using rivets larger than 3/8 inch, some other form of fastener is normally used. [Figure 2-99]

RIVET DIAMETER	PILOT SIZE	FINAL SIZE
3/32	3/32 (.0937)	#40 (.098)
1/8	1/8 (.125)	#30 (.1285)
5/32	5/32 (.1562)	#21 (.159)
3/16	3/16 (.1875)	#11 (.191)
1/4	1/4 (.250)	F (.257)
5/16	5/16 (.3125)	O (.316)
3/8	3/8 (.375)	V (.377)

Figure 2-99. The final hole for a particular rivet size can be prepared by drilling a hole the size of the rivet and then reaming the hole to the final dimension. Where less critical applications are allowed, the final dimensions can be drilled using a number or letter twist drill.

The number drill size for each diameter rivet is slightly larger than the rivet diameter. As previously mentioned, the holes made by these drills are usually three- or four-thousandths of an inch larger than the diameter of the rivet. This allows the rivet to be slipped in place without forcing it and scraping any protective oxide coating off the rivet shank. The clearance is small enough that, during driving, the shank will swell to take up any excess clearance. [Figure 2-100]

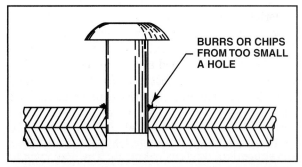

Figure 2-100. Too small a hole will destroy the protective oxide coating on a rivet shank and may also cause the sheet metal to buckle once the rivet is driven.

RIVET HOLE PREPARATION

Since rivet holes drilled in sheet metal are usually small and the metal is soft, lightweight drill motors can be used to produce rivet holes. For safety, most sheet metal drilling is accomplished using pneumatic drills, so the chances of electric shock and arcing are reduced. In addition, pneumatic drill motors are usually equipped with triggers that permit controlling the motor speed by varying the amount the trigger is depressed. This is very handy as it allows the hole to be started with a slow speed, which provides better drill control. As the hole begins to form, a faster speed can then be used to permit production that is more efficient.

When selecting a twist drill for rivet hole preparation, examine the point to be sure that it is properly ground and sharp. To prevent the twist drill from wobbling during drilling, the point of the tip should be in the exact center of the drill. To help ensure uniform tension on the shank of the twist drill, it is recommended that the chuck of the drill motor be tightened using each of the chuck key holes with the proper size chuck key.

Proper drill motor use is an important consideration for preparing acceptable rivet holes as well as for safety. When using a drill motor, hold the tool in a manner similar to that shown in Figure 2-101. This allows for maximum tool control and aids in preventing the drill from pushing through the metal so far as to possibly cause damage to structures behind the drilled surface.

Figure 2-101. When using a drill motor, hold the drill as shown in this figure. Pointing the index finger parallel to the drill chuck will aid in aligning the twist drill with a 90 degree angle to the drilled surface.

To prevent injuries and damage to aircraft structures when using a drill motor, observe the following safety considerations:

1. Be sure there are no burrs on the twist drill shank that could prevent it from fitting properly in the chuck.

2. When changing twist drills, it is advisable to disconnect the air source before tightening the chuck with a chuck key. If the air source is left connected, injuries may occur if the trigger is inadvertently activated. Also, never attempt to

tighten a drill chuck by holding the chuck stationary by hand and operating the drill motor.

3. Before drilling a hole, run the drill motor and watch the end of the drill. It should not appear to wobble. If it does, remove it from the chuck and check to see if it is bent, by rolling the twist drill over a flat surface. Also, check the twist drill shank for burrs that would prevent it from centering in the chuck. Bent drills and worn chucks will cause oversize holes that can ruin a repair.

4. Be sure to wear eye protection when drilling. Fine chips of metal are propelled from the rapidly spinning twist drill and by the air exhaust from pneumatic drill motors. [Figure 2-102]

Figure 2-102. Be sure to wear adequate eye protection when drilling sheet metal. Impact resistant safety glasses with side shields should be worn at all times while working around sheet metal fabrication areas in the shop or on an aircraft.

When the drill and drill motor are ready, prepare the metal. Mark the location for drilled holes with a center punch and make the indentation just large enough for the twist drill to start cutting. Too small a mark will allow the drill to walk, while too heavy a blow with the center punch may distort the metal. Also, to prevent distorting the metal, it is advisable to use a scrap piece of wood to backup the material to oppose the force of the drill motor.

For right handed individuals, hold the handle and the trigger with the right hand while using the left hand to rest on the work. Push back against the drill motor with the left hand to provide a balance to the pressure applied in the drilling operation. Center the point of the drill in the center punch mark and start the drill motor slowly until the twist drill begins to cut the metal with the full face of the tip. Once the tip makes full contact, increase the speed of the drill to a faster speed, but not so fast as to cause loss of control. Use enough force to keep the drill cutting smoothly, but as soon as the tip begins emerging through the metal, relax the pressure on the drill motor. Allow the tip and a portion of the twist drill body to go through the metal, but do not allow the chuck to touch the work. If the chuck contacts the surface, it will cause scratches, damaging the finish of the metal. This is especially critical if the metal is clad with a protective finish.

When two or more sheets of metal are being drilled together, once the first hole is drilled, use a Cleco or similar temporary fastener to secure the metal together. Failure to use a temporary fastener may cause the metal to slip during subsequent drilling, causing misalignment of holes when the drilling is complete.

DEBURRING

When aluminum alloys are drilled sharp burrs can remain on the edge of the hole. If these burrs are not removed before riveting, the sheets of metal will not fit tightly together. This not only causes cosmetic flaws, but also prevents the rivets from providing maximum shear strength. In addition, when sheet metal is cut, burrs can also form along the edges of the metal. To remove these sharp edges and burrs, a process referred to as deburring is accomplished.

Deburring consists of using tools to remove excess material from around the edges of the sheet metal. As previously discussed, examples of tools that can be used for hole deburring include fabricating a handle made from wood and installing a twist drill or countersink cutter that is larger than the drilled hole. Lightly turning the twist drill or cutter in a hole tends to remove the excess material from around the edges. However, when using these tools, make certain to apply only a light pressure to avoid countersinking the hole. Deburring should remove only the material that is above the surface of the sheet metal.

Specialty tools are also available and may be used to increase the speed and accuracy of deburring. For hole deburring, a tool is available that has a cutter that can be rotated by a handle, wherein the handle is swiveled to cause the cutter to rapidly remove sharp edges and burrs without countersinking. For edge deburring of sheet metal, tools are available that have a notch cut in them so the tool can be pulled across the edge to deburr both sides of the metal in a single operation. Make certain all holes and edges are deburred before installing any rivets. When two or more sheets have been drilled together, separate the sheets and deburr the holes on each side of the metal. [Figure 2-103]

Figure 2-103. Specialty deburring tools are available commercially and come in a variety of styles and sizes.

HOLE PREPARATION FOR FLUSH RIVETS

It is extremely important that the skin on high-speed aircraft be as smooth as possible. On many of these aircraft, airfoils are assembled using flush head rivets to increase the laminar flow of air over the surface of the skin. In order for the rivets to seat as flush as possible, the skin must be contoured or machined to accept the head of the rivet to a depth that places the rivet head flush with the surface. To produce a countersunk hole, a number of processes may be used. In some situations, the surface material is machined away by a countersink cutter while in others the rivet head is pressed or dimpled into the skin. In either case, modern countersinking for rivets involves forming a recess with a 100 degree taper. [Figure 2-104]

COUNTERSINKING

When the top sheet of metal being joined is thicker than the tapered portion of the rivet head, the sheet can be countersunk to produce a smooth riveted surface. If the sheet is too thin, the rivet head will protrude past the surface of the metal and cause the shearing zone of the joint to be applied to the rivet head rather than across the shank. This not only reduces the effectiveness of the rivet, but also weakens the skin by removing too much material. [Figure 2-105]

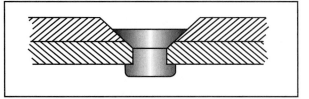

Figure 2-105. The depth of the rivet head limits countersinking a recess in sheet metal. If the metal is too thin, the countersink will enlarge the hole and cause the metal to shear across the head instead of the shank.

Milling a countersunk hole is accomplished by using a countersink or micro-stop cutter after a hole has been drilled to the desired shank diameter. A standard countersink for aircraft rivets has a cutting angle, which matches the 100 degrees of the countersunk rivet head. A standard countersink can be used in a drill motor, but the difficulty in cutting the hole to the correct depth makes this tool impractical when you have multiple holes to countersink.

A **micro-stop**, or **stop countersink**, as it is sometimes called, uses a cage that can be adjusted to limit the depth that the cutter penetrates into the sheet metal. With this tool, the depth of the cutter is adjusted in increments of .001 inch to obtain a uniform countersunk depth in multiple holes. Sample countersinking tools are shown in Figure 2-106.

Figure 2-106. Countersink cutters and the cutters used in micro-stops are manufactured with a 100 degree cutting angle. With some cutters, a pilot is milled onto the end while others have interchangeable pilots. Pilots are available in various sizes to match the diameter of the rivet shank and help position the cutter in the center of the hole.

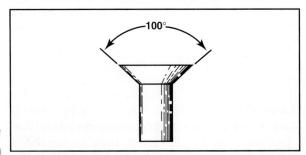

Figure 2-104. Flush rivets used on modern aircraft include the AN426 or MS20426 rivet. These rivets have a head angle of 100 degrees and are measured as shown here.

A micro-stop countersink is used far more often than a standard countersink cutter. When using a micro-stop, the cage portion of the tool is held stationary against the metal, and the countersink is pushed into the hole until the cutter reaches a stop, limiting the depth of the countersink. With a micro-stop, a shaft fits into the chuck of a 1/4 inch air or electric drill motor, and the cutter screws onto this shaft.

When adjusting the stop, set the countersink depth to cut the proper amount by using a piece of scrap metal the thickness of the top sheet being riveted. Drill some holes the size used for the rivet, and adjust the stop by screwing the cage up or down on the body and locking it into position with the locknut. Hold the stop with one hand and position the cage collar tightly against the sheet metal. Then, press the trigger on the drill motor and push the cutter into the metal until the stop is reached. Once milled, slip the countersunk rivet into the hole and check the flushness of the head across the surface of the sheet. Depending on the operation, the rivet may be raised a few thousandths of an inch above the surface, to be milled smooth with a microshaver.

A **microshaver** is a pneumatic tool that turns a flat-faced milling cutter at high speed. The tool is equipped with a stop that is adjustable to limit the amount the cutter extends out from a guard. By regulating the depth of the cutter, the rivet head can be milled perfectly flush with the metal. [Figure 2-107]

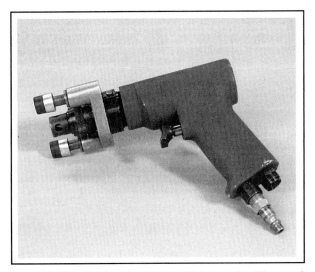

Figure 2-107. A microshaver is a high-speed milling tool that is useful in shaving the heads of rivets, to obtain an accurate flush fit to the surface of the sheet metal.

DIMPLING

When the top sheet of metal is too thin to countersink, the edges of the hole may be formed to accommodate the head of the rivet by using a set of dimpling dies. There are two methods of dimpling sheet metal: coin dimpling, which forges, or coins, the metal into the dies, and radius dimpling, which folds the material down to form the dimple. Although both techniques are commonly used, coin dimpling generally provides a slightly tighter fit but tends to leave a sharper bend around the rivet head. Radius dimpling may not produce as tight a fit, but has the advantage of leaving a more gradual radius bend around the rivet head, helping to prevent cracking during service.

Coin Dimpling

In coin dimpling, a male die fits through the rivet hole, and a coining ram in a female die exerts pressure on the underside of the hole. By forcing the male die into the female die, the metal contours to the shape of the coin. The pressure on the dies forges the edges of the hole to exactly fit the shape of the dies. Coin dimpling gives the hole sharply defined edges that almost resemble machine countersinking. Both the top and the bottom of the dimple are formed to a 100 degree angle, so multiple sheets can be dimpled and stacked, or nested. [Figure 2-108]

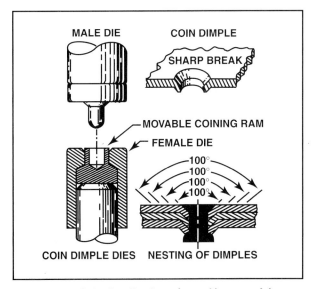

Figure 2-108. Coin dimpling is performed by a special pneumatic machine or press, which has, in addition to the usual dies, a "coining ram." The ram applies an opposing pressure to the edges of the hole so the metal is made to flow into all the sharp contours of the die, giving the dimple greater accuracy and improving the fit.

Radius Dimpling

Radius dimpling is a form of cold dimpling in thin sheet metal in which a cone-shaped male die is forced into the recess of a female die, with either a hammer blow or a pneumatic rivet gun. In some instances, a flush rivet is used as the male die. The male die is forced into the female die. In this form of dimpling, a rivet gun is fitted with a special female dimpling die, and the rivet head is set into the sheet metal by rapid impact blows of the rivet gun. The dimple formed in this way does not have parallel sides, as the lower side has an angle greater than 100 degrees. For this reason, radius dimpling is not usually considered acceptable to stack or nest multiple sheets. [Figure 2-109]

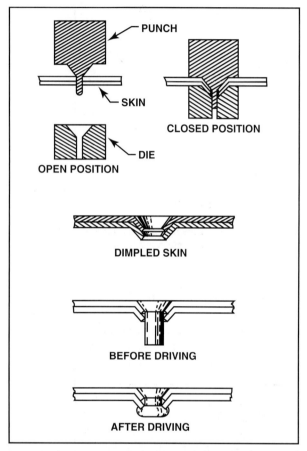

Figure 2-109. Radius dimpling does not allow the sheets to be nested unless the bottom sheet is radius dimpled. Radius dimpling is done because its equipment is smaller than that needed for coin dimpling, and can be used in locations where access with coin dimpling tools is not practical.

Hot Dimpling

Magnesium and some of the harder aluminum alloys, such as 7075, cannot be successfully cold dimpled, because the material is so brittle it will crack when the dimple is formed. To prevent cracking, these materials are heated before dimpling is accomplished. The equipment for hot dimpling is similar to that used for coin dimpling, except that an electrical current heats the dies.

To perform hot dimpling, the dies are preheated and then the metal is positioned between the dies. When the technician presses a pedal, the dies are pneumatically pressed together until they both just make contact with the metal. Once the dies make contact, a dwell time allows sufficient heat to soften the metal before the dies are fully squeezed together to form the dimple. The dwell time for heating is automatically controlled by a timer to prevent destroying the temper condition of the metal. The operator of the machine must be familiar with how to adjust the machine for the various time limits and temperatures for the types of metal being formed.

MULTIPLE SHEET FLUSH RIVETING

The proper preparation of holes for flush riveting depends upon the thickness of the sheets being joined. If the top sheet is thick enough to be countersunk, the substructure, or lower skins, need nothing more than to have the holes drilled for the rivet. But if the top skin is too thin to be countersunk, it must be dimpled and the bottom skin either countersunk or dimpled. In this situation, the top skin must be coin dimpled so the bottom of the dimple will fit into the 100 degree inside angle of the dimple or countersink in the lower skin. [Figure 2-110]

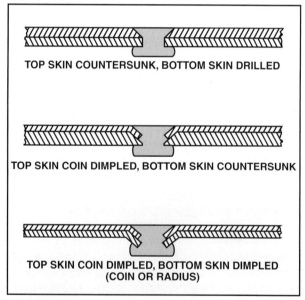

Figure 2-110. Care must be taken to ensure compatibility of countersunk, coin dimpled, and radius dimpled skins.

RIVET INSTALLATION

Because of the many thousands of rivets used to hold an aircraft structure together, it is easy to get complacent with the riveting process and not be concerned about less than perfect riveting techniques. However, each rivet must carry its share of the total load in an aircraft structure. If a rivet is not properly installed, it can force the adjacent rivets to carry more load than they are designed to take, ultimately causing a structure to fail.

In addition to the proper preparation of a hole for a rivet, the strength of a riveted joint is determined by the way the rivets are driven. When installing rivets, it is important to install the rivet with as few impacts as possible so the materials will not work-harden and crack. The shop head of the rivet should be concentric with the shank and flush with the surface without tipping. In addition, the formed, or bucked head, should be fabricated to proper dimensions.

HAND RIVETING

Almost all rivets are driven with either a rivet gun or squeeze riveter, but there are times when building small components, or when working in areas without air or electricity, that it is necessary to drive a rivet by hand. The process used for hand driving aircraft rivets is not the same as that used by some other commercial sheet metal processes. Aircraft rivets driven in flat sheets are never peened over. Instead, the shank is collapsed with a hand set in much the same manner as other aircraft riveting techniques.

To drive a rivet by hand, the material to be joined is prepared by drilling and deburring a hole, and a rivet is inserted to extend one and a half shank diameters through the metal. A special metal bar that has a recessed contour approximately the shape of the rivet head is then mounted in a vise with the recess facing upward. The rivet is placed through the hole in the metal and the rivet head is put in the recess of the metal bar. A draw set is slipped over the rivet shank and tapped lightly with a hammer to draw the sheets of metal tightly together. A hand-set is then placed on top of the rivet shank and struck with a hammer to force the rivet to compress to the proper dimensions. It is important to strike the set hard enough that the rivet compresses with only three or four strikes of the hammer to prevent work-hardening the rivet. Be sure to use the hand-set rather than striking the rivet directly with the hammer, as the hand-set is machined to provide a smooth surface to the formed head. [Figure 2-111]

COMPRESSION RIVETING

When there are a large number of easily accessible rivets to be installed, a compression, or squeeze riveter, can be used instead of hand or gun riveting. These riveting tools reduce the time required to install the rivets and produce a far more uniform shape than can be driven by hand or with a rivet gun.

A squeeze riveter consists of a pair of jaws; one stationary and the other moved by a piston in an air cylinder. A dolly, milled with a recess similar to the shape of the rivet head, is put into the stationary jaw, and a flat dolly is placed in the movable jaw. When a handle or trigger is depressed, air flows into the cylinder and squeezes the jaws together to compress the shank of the rivet in a uniform motion. [Figure 2-112]

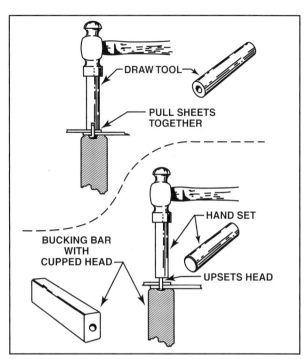

Figure 2-111. Hand riveting is performed where it is inconvenient to use conventional riveting equipment. However, the standards for acceptable hand driven rivets are the same as for all other riveting methods.

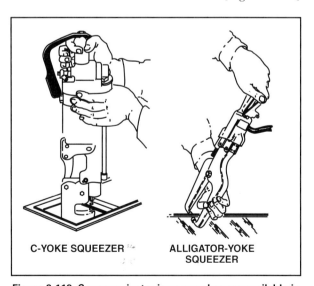

Figure 2-112. Squeeze riveter jaws or yokes are available in different lengths and configurations. Although these tools are extremely useful, the yoke length limits the use of the squeezer to rivets that are located reasonably close to the edges of sheet metal structures. For this reason, these tools are used primarily in production rather than repair processes.

Shims placed between the jaws and the dollies control the separation of the dollies at the end of the piston stroke, and this determines the height of the shop head formed on the rivet. The number of shims needed is determined by trial and error, using scrap material of the same thickness as that to be riveted. Once the dollies are adjusted, all of the compressed rivets will have exactly the same height and diameter. In addition, the smooth compressive pressure used to upset the rivet will have a minimum strain-hardening effect on the rivet shank. [Figure 2-113]

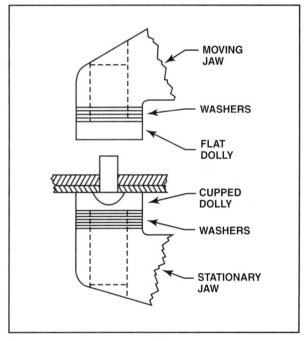

Figure 2-113. To adjust the height of the shop head formed by a squeeze riveter, add or remove washers between the dollies and the jaws.

GUN RIVETING

Hand riveting and compression riveting are used for special conditions, but a rivet gun drives most rivets used in aircraft construction. These tools look and operate in a similar fashion to a reciprocating air hammer, but the number of strokes and force of the impact are considerably different. As such, only guns designed for riveting should be used on aircraft structures.

Rivet Gun Types

There are a number of different types of rivet guns used for aircraft fabrication and repair, but these can be divided into two basic categories: fast-hitting, short-stroke guns, which produce light blows, and guns with long strokes that produce heavy blows. The fast-hitting guns are usually used for 3/32 inch or 1/8 inch rivets. These guns have bodies made from aluminum alloy castings so they are light enough that the user will not be fatigued after using the gun for a prolonged period. The long-stroke gun may be either a slow-hitting reciprocating type, or a one-shot gun that drives the rivet set only one blow each time the trigger is pulled. These guns are used to drive the larger rivets and are much heavier than fast-hitting guns.

Handle styles also vary with different types of guns. The pistol grip and offset handle are the most popular styles, with a push-button type available for special applications where neither of the other styles of guns will fit because of clearance problems. [Figure 2-114]

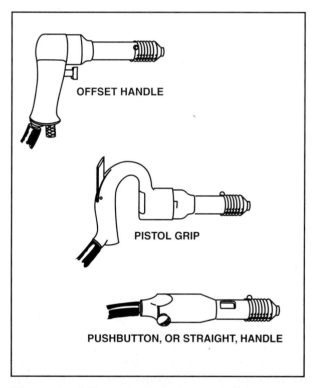

Figure 2-114. Different handle styles are available on rivet guns to fit various locations and operator preferences.

When the trigger, or throttle, as it is sometimes called, is pulled, air enters a sliding valve and drives a piston forward against the stem of a rivet set. When the piston reaches the end of its stroke, a port is uncovered by the valve that directs air to the forward end of the piston, and moves it back so it can get air for another driving stroke. As long as the trigger is held down, the gun will reciprocate, or hammer, on the rivet set. A regulator, built into the handle of the gun or attached to the air hose, restricts the flow of air into the gun. If the regulator is wide

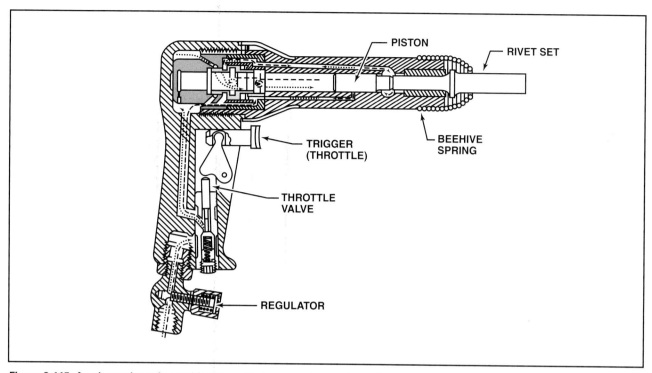

Figure 2-115. An air regulator, in combination with the gun trigger, allows the operator to vary the impact speed and intensity for various sizes and alloy rivets.

open, the gun will hit hard and fast, or the regulator can be adjusted to restrict the air flow to cause the gun to hit slower and softer. [Figure 2-115]

The gun has a provision whereby different sets can be installed for different sizes and head styles of rivets. If the gun is operated without a set being installed, the piston can be severely damaged if the trigger is pulled, or if the set is not pressed tightly against either a rivet or piece of scrap wood. The rivet set is held in place by a retaining spring, sometimes called a beehive because of its appearance. Without the retaining spring, the set can fly out from the gun if the trigger is pulled without the set being against an object. As a safety precaution, the gun should never be pointed at a person and the trigger pulled, even with a retaining spring installed. If the spring were to break, the set could fly out and cause injury to the person.

Rivet Sets

Rivet sets are manufactured differently for various head styles and sizes of rivets. In the past, the numerous styles of raised-head rivets required a set that was designed for each of the styles, causing a technician to acquire a large assortment of sets. Fortunately, with newer sheet metal fabrications, universal head rivets can be used to replace almost any protruding head rivet. This means that a technician rarely needs more rivet sets than those that fit the various sizes of universal head rivets. Flush head rivets also require a special set, but generally, any flush head rivet can be driven with a flush head set.

It is important when selecting a universal rivet set for a particular job that its size and shape be correct for the type rivet being driven. When selecting rivet sets, the radius of the depression in the set must be larger than that of the rivet, but not so large that the set contacts the sheet metal during driving. In addition, damage on the recess face of the set may cause it to slip off the rivet during driving, or the set may leave unacceptable marks on the rivet head.

The recess in a universal style rivet set is machined with a slightly larger radius than the rivet head. By having a greater radius, the set concentrates more of the rivet gun's driving energy to the center of the rivet head. However, if the set is too large, it will produce small indentations in the sheet metal around the rivet head. These indentations are commonly referred to as smiles because of the shape they leave in the metal. In most cases, these indentations are unacceptable because they damage the protective coatings of the metal, or create stress concentration points that can cause the metal to fail around the rivet. On the other hand, if the set is too small, it will produce a similar type mark on the rivet head, which is also unacceptable. [Figure 2-116]

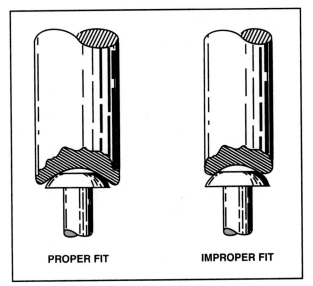

Figure 2-116. The radius of the cup of the rivet set must be slightly larger than the radius of the rivet head, but not so large that the edges of the set contact the surface of the metal.

Not only must the rivet set have the correct size and shape of depression, but it must also fit squarely on the rivet head. Because the structure inside an aircraft sometimes makes it difficult to align the gun exactly with the rivet, rivet sets are made in many lengths and shapes. Some sets are manufactured with a straight shank while others have one or even two offsets. When selecting a rivet set, make certain that its shape concentrates the blows of the rivet gun as close to in-line with the rivet as possible. [Figure 2-117]

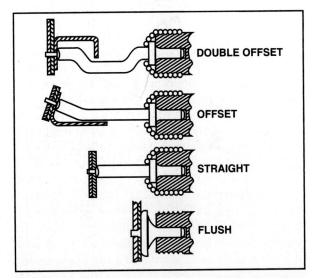

Figure 2-117. Rivet sets are available with different offsets to allow the blows from the gun to be directed straight in-line with the rivet. These sets are useful in locations where structures interfere with a straight shank set.

Bucking Bars

When a rivet is driven, the actual compression of the rivet is not performed by the action of the rivet gun. Instead, the rivet is backed up by a metal bar that reciprocates in response to the beats of the rivet gun. This reciprocating action causes the rivet to be compressed in successive actions. These metal bars are referred to as bucking bars because of the method in which the bar bucks, or vibrates, on the shank of the rivet.

The driving face of a bucking bar is machined smooth and polished so that no marks are left on the rivet shank during driving. This hardened and polished steel surface is held against the end of the rivet shank, and pressure is applied as the gun vibrates the rivet against the bar. The position of the bucking bar is critical in the formation of the shop, or bucked head. If the bar is tipped slightly, the rivet will dump over and not form a concentric head. If too much pressure is held on the bar, or if the bar is too large, the shop head will be driven too thin, or the manufactured head may be forced up from the surface of the metal. On the other hand, if the bar is too small or is not held tight enough, the hammering of the rivet gun may distort the skin.

There are many sizes and shapes of bucking bars used in aircraft maintenance. One of the challenging tasks of performing structural fabrications by riveting is being able to select a bucking bar that will clear the structure and fit squarely on the end of the rivet shank. In the same fashion as a rivet set, when a bucking bar is selected, its mass should be as close to being in-line with the rivet as possible. In addition, the weight of the bar should be heavy enough to allow driving the rivet in a few short beats of the rivet gun, yet not so heavy that the operator has difficulty holding the bar in the correct position. Before using any bucking bar, inspect the face of the bar to make certain no damage is present that will leave an indentation on the formed head of the rivet. [Figure 2-118]

Rivet Gun Set-Up and Adjustment

One of the most difficult tasks for an aircraft technician to learn is how to properly and efficiently install rivets using a rivet gun. The skills necessary to perform gun riveting are acquired over time and must be practiced on a regular basis in order to remain proficient. One task that must be learned before trying to install rivets is setting up and adjusting a rivet gun for a particular operation. Since improper riveting can cause irreparable damage to an aircraft structure, experience should first be gained by developing riveting techniques on practice projects.

Sheet Metal Structures

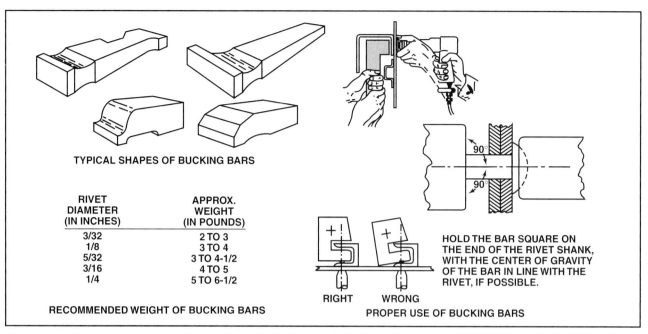

Figure 2-118. Bucking bars are available in different weights and shapes to allow riveting of solid rivets in almost any location.

Begin a practice project by trimming two sheets of aluminum to a manageable dimension. Sheets trimmed to six inches square are often used because of the ease in clamping the material in a vise. Align the edges of the material and drill a number of holes for the correct diameter rivet, deburr the edges, and clamp the material in a padded-jaw vise.

Select a rivet set that is appropriate for the head style and size rivet to be installed and insert it in the rivet gun. Once the set is in place, install a retaining spring to keep the set from being propelled out of the gun during operation. Once the set has been installed, connect an air hose to the gun and place the set against a soft piece of wood to perform an initial adjustment of the air regulator.

With the set in place against the wood, depress the trigger to cause the gun to pound against the wood. If the gun is set properly, the wood will be left with an indentation approximately the depth of the recess in the set, with only a few gun raps. If the gun strikes the wood too hard, adjust the air regulator for a lower pressure, or limit the amount that the trigger is depressed. This is usually a good starting adjustment, but will probably have to be adjusted further when driving a rivet. [Figure 2-119]

Rivet Installation

Once the metal has been drilled, put a rivet of the correct length through a hole and hold the rivet gun set against the manufactured head of the rivet. The set must be directly in line with the rivet, and not

Figure 2-119. Adjust the airflow to the gun so it will indent a scrap of wood but will not shatter it.

tipped or it will contact the sheet metal during driving. [Figure 2-120]

Figure 2-120. Hold the rivet set square with the rivet head and support the gun so the set will not slip off during driving.

Hold a bucking bar flat against the end of the rivet shank and develop a good feel of the balance between the gun and the bucking bar. A proper balance is obtained by applying just enough extra force on the gun to hold the rivet head firmly against the skin. Once the gun is in position, pull the trigger to provide a short burst of raps on the rivet head. The rivet should be driven with the fewest blows possible so it will not work-harden the rivet, which can cause cracks to form. It may be necessary to further adjust the rivet gun airflow to fine-tune it for the installation of subsequent rivets.

Evaluating Driven Rivets

In the process of developing riveting skills, it is inevitable that some rivets will be driven improperly. One of the requirements of a technician is to be able to quickly identify rivets with unacceptable characteristics. In addition, the technician must be able to remove damaged or improperly installed rivets without adversely affecting the base sheet metal, and then be able to correct the problem with the installation of replacement rivets. [Figure 2-121]

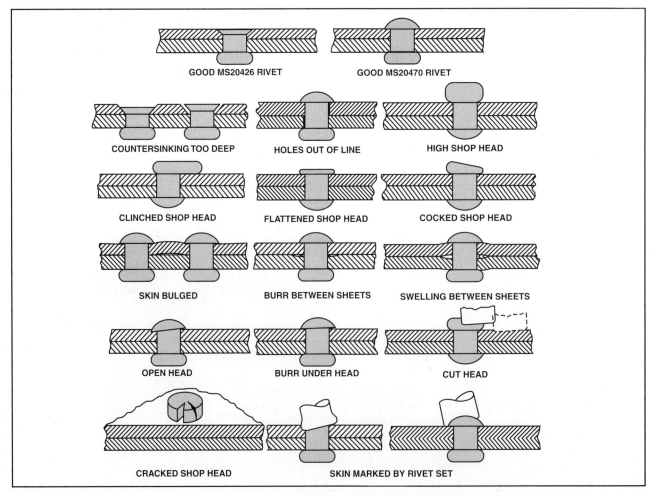

Figure 2-121. A technician must be able to identify properly driven rivets. Until the technician has gained enough experience to evaluate driven head dimensions, special gauges may be fabricated to check shop head shapes and sizes.

As previously mentioned, a properly formed shop head will be one-half the shank diameter in height with a diameter that is one and a half times that of the shank diameter. In addition, the rivet should be uniform with adjacent rivets and each rivet should be driven concentric with the hole. Placing a straightedge on top of shop heads that have been driven in a row can check the uniformity of rivet heights. Each surface of the shop heads should touch the straightedge without gaps.

The rivet shank should also be checked for concentricity. Concentricity is achieved when the shop head of the rivet compresses in an even fashion around the shank. It is detected by looking closely for a circular mark left on the driven head. A rivet that has been properly driven will have an even margin between the circular pattern and the edge of the shop head. If the margin is uneven, it usually indicates the bucking bar slipped across the rivet during driving, causing the shop head to form an oval shape or to be off center with the shank of the rivet.

The manufactured head of the rivet must also be perfectly flat against the metal. If a thin feeler gauge blade can be slipped between the manufactured head and the skin, the rivet must be removed and the cause of the improper fit determined. The rivet may be tipped in the hole by a small burr, or the hole may have been drilled at an angle. If the hole was improperly drilled, the only satisfactory repair is to re-drill the metal to accept the next larger diameter rivet.

Rivet Removal

When it has been determined that a rivet has been improperly installed, the rivet must be removed and replaced without damaging the base metal. To remove a rivet, lightly indent the center of the manufactured head with a center punch. Be sure to back-up the shop head with a bucking bar when center punching so as not to distort the skin. Use a drill the same diameter as the hole or one that is one number size smaller to drill down to the base of the rivet head. Once drilled, use a pin punch with a diameter the same size as the rivet shank diameter to pry the head off, or tap the head lightly with a cape chisel to break it off from the shank. If a chisel is used, be sure that it does not scratch the skin around the rivet head. [Figure 2-122]

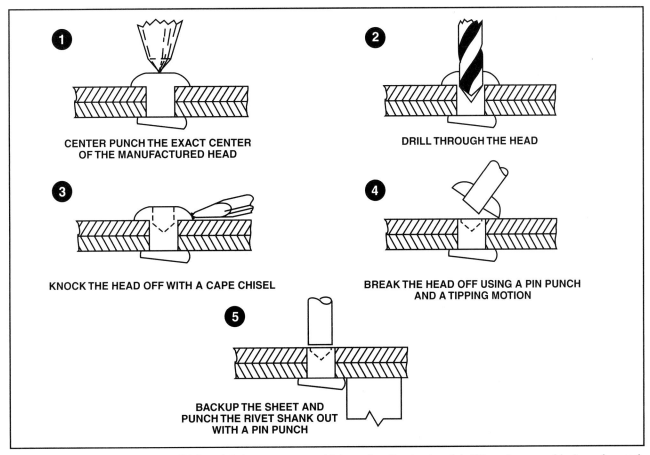

Figure 2-122. When removing a solid rivet, it is important to avoid damaging the sheet metal. Although a cape chisel may be used, removing the rivet head with a pin punch is preferred to help avoid damaging the base metal.

After the rivet head has been removed, back-up the underside of the skin with a bucking bar or piece of wood, and use a pin punch to gently drive the rivet shank from the sheet metal. When the rivet is out, examine the hole, and if it is not elongated, another rivet of the same size may be used as a replacement. If the hole is damaged, use a twist drill for the next larger diameter rivet to re-drill the hole. When using a larger diameter rivet, be sure that the pitch, gauge, and edge distance values are all satisfactory for the pattern.

NACA Flush Riveting

It is possible to drive a rivet in such a way that the shop head will be flush with the outside skin and the protruding manufactured head will be on the inside of the structure. This technique is necessary when a shop head would extend too far beyond a surface and interfere with an adjacent part or cause an unacceptable effect on the aerodynamic characteristics of the structure. To aid manufacturers in standardizing riveting techniques, the **National Advisory Committee for Aeronautics**, or **NACA**, established a set of standards for riveting aircraft and aerospace vehicles. When riveting is conducted to these standards, the process is referred to as NACA riveting. [Figure 2-123]

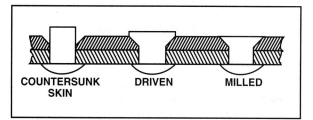

Figure 2-123. The NACA method of riveting produces the smoothest surface possible on a riveted structure.

To perform **NACA riveting**, a hole is drilled and countersunk as it would be for the installation of any flush rivet; however, the rivet is installed from the inside of the structure so its shank sticks out through the countersunk side. When the rivet is driven, the shank is upset to fill the countersunk hole and is allowed to stick up a few thousandths of an inch above the surface. When all the rivets have been driven, a microshaver is adjusted to mill the excess material from the shank to obtain a flush surface. [Figure 2-124]

Team Riveting

When sheet metal parts are riveted onto a complete airframe or a large component, it is often necessary for a person to be on one side of the structure to hold the bucking bar while another person operates

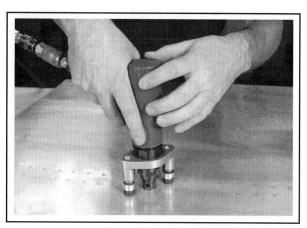

Figure 2-124. A microshaver is used to smooth the upset ends of rivets driven using NACA techniques.

the rivet gun from the opposite side. This operation is called team riveting and requires a great deal of coordination between the technicians. Since there are often many teams working on the same structure, the noise produced in the shop may make it impossible for team members to verbally communicate with one another. To improve communications between the riveter and the bucker, a method of standardized tapping codes has been developed, whereby the person holding the bucking bar can advise the gun operator as to the condition of a driven rivet.

The gun operator, or driver, is responsible for putting the rivets into the holes and driving each of them the same amount. Once a rivet is placed in a hole, the gun is positioned on the rivet head to prevent it from falling out. The bucker then taps the shank one time as the bucking bar is brought into contact with the rivet. After the driver feels or hears the tap, the gun is operated for a short period to upset the rivet shank. As soon as the driver stops the gun, the bucker removes the bucking bar and examines the rivet. If the rivet is sufficiently driven, the bucker taps on the rivet twice with the bucking bar, and the driver goes on to the next rivet. However, if the rivet needs to be driven more, the bucker will tap only one time and immediately place the bucking bar back on the rivet for subsequent driving. If the rivet is dumped over, or for any other reason is not satisfactory, the bucker taps the rivet three times and the driver circles the rivet head with a grease pencil so it can be identified for later replacement.

In the high-speed production needed to build modern aircraft, good, well-coordinated teams of riveters can keep a production line moving effectively.[Figure 2-125]

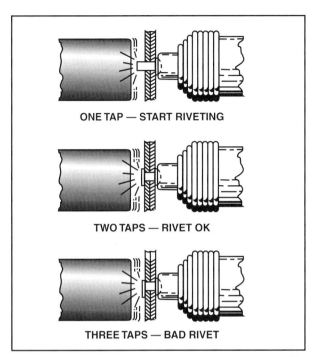

Figure 2-125. A tapping code allows the riveter on one side of the skin to communicate with the bucker on the other side of the structure.

LAYOUT AND FORMING

The fabrication of sheet metal parts for an aircraft requires the technician to have a fundamental knowledge of the physical characteristics of the metal being used, and a working knowledge of applied geometry. In many cases, parts are fabricated from blueprint drawings or are constructed from templates created from pre-existing components. In other cases, the technician must use industry acceptable practices and information from publications such as Advisory Circular 43.13-1B and 2A, *Acceptable Methods and Techniques for Aircraft Alterations and Repairs*, for conducting minor repairs to sheet metal structures. In these situations, the technician must consider the physical characteristics of the metal and perform computations to fabricate parts to the desired dimensions.

When an aircraft is manufactured, fabrication engineers compute all of the bends and cutting dimensions and create dies that can be used in forming machines to speed the manufacturing process. The objective is to allow the metal to be cut to size in one department and formed in another, and yet when the parts are assembled, they all fit properly. However, in field repairs and alterations, it is the technician's responsibility to perform the computations to produce a layout on the sheet metal. A layout is simply the process of placing lines on the metal to distinguish the locations of cuts and bends.

FABRICATION TERMINOLOGY

Fabrication processes require an understanding of the wrought physical characteristics of metals as well as the characteristics of metals when they are shaped or bent. For example, when sheet metal is bent, it must be formed around a radius to allow the metal to gradually change direction. If sheet metal is bent around a sharp corner, the stresses developed will cause fractures during fabrication or while the part is in service. However, when metal is formed around a radius, the amount of material required in the bend will be less than the amount required to form around a sharp corner. Also, since a bend begins and ends at different locations depending on the radius size, it is necessary to compute and layout the location of the bend's starting and ending points to properly position the metal in the forming machine. To understand the methods used in developing sheet metal layouts, it is necessary to consider the physical characteristics of the metal as well as knowing the meaning of various terms used in the fabrication process. [Figure 2-126]

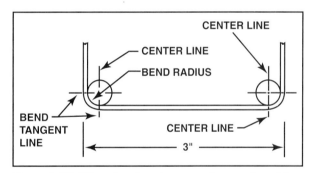

Figure 2-126. When bending sheet metal, a number of factors must be considered. For example, depending on the radius size, the point where the bend begins and ends will vary and must be determined to obtain a desired overall dimension on the flats of the metal.

SHEET METAL GRAIN

Sheet metal used for aircraft construction and repair is formed from ingots of aluminum alloy that are passed through a series of rollers until the metal is reduced to a desired thickness. In the rolling process, the molecules in the metal are elongated in the direction that the metal is passed through the roll. This elongation causes the metal to develop a grain pattern that presents certain physical characteristics that should be considered before forming sheet metal.

Looking closely for small lines that run in one direction through the material distinguishes the grain direction. When bending sheet metal, the strength of the material varies depending on the direction of the bend in relation to the grain. When laying out a pattern, it is better to orient bends to run across, or perpendicular, to the grain of the metal. If the bend is formed parallel to the grain, the grain boundaries tend to separate and cause cracks. In some cases, the metal will initially fabricate without showing signs of fatigue, but the part may prematurely develop cracks while in service.

BEND RADIUS

Non-aviation sheet metal construction does not require the high strength and light weight that is necessary for aircraft, and so for economy of construction, many metal fabrications have sharp bends. In thin sheet steel, this usually gives no problem, but when working with hard aluminum alloy sheets for aircraft parts, sharp bends must be avoided to prevent cracking. To prevent cracks, a minimum bend radius is recommended for different types of alloys and metal thickness. A radius is measured on the inside of a bend and is generally measured in fractions of an inch. For instance, a common radius for aircraft sheet metal bends is 1/8 inch. This is the distance from the external edge to the center of the radius. [Figure 2-127]

As an example, notice that in the -O, or annealed temper, the metal can be bent over a very small radius, but as the metal's hardness and thickness increases, so does the minimum allowable bend radius. If the metal is bent with too small a radius, the outside of the bend, which is stretched, will pull the grain boundaries apart, causing the metal to crack. On the other hand, if the metal is bent around a radius that is larger than the minimum size, the area of the bend may be too large to fit in conjunction with adjacent parts.

NEUTRAL AXIS

When bending a piece of metal around a radius, the metal on the outside of the bend stretches, while the metal toward the inside tends to compress or shrink. Within the metal, a portion neither shrinks nor stretches, but retains its original dimension. The line along which this occurs is called the neutral axis of the metal. This line is not located exactly in the center of the sheet, but is actually about 44.53% of the sheet thickness from the inside of the bend.

Occasionally it is necessary to know the exact length of the neutral axis, but for most practical purposes, you can assume the neutral axis is located in the center of the metal. The slight error from this approximation is usually too small for consideration when constructing a layout pattern.

MOLD LINE

Mold lines are used to designate the dimensions of a piece of metal on a drawing or layout pattern. These are formed by extending a line from the exter-

Alloy and temper	APPROXIMATE SHEET THICKNESS (t) (inch)					
	0.016	0.032	0.064	0.128	0.182	0.258
2024-O[1]	0	0-1t	0-1t	0-1t	0-1t	0-1t
2024-T3[1,2]	1 1/2t-3t	2t-4t	3t-5t	4t-6t	4t-6t	5t-7t
2024-T6[1]	2t-4t	3t-5t	3t-5t	4t-6t	5t-7t	6t-10t
5052-O	0	0	0-1t	0-1t	0-1t	0-1t
5052-H32	0	0	1/2t-1t	1/2t-1 1/2t	1/2t-1 1/2t	1/2t-1 1/2t
5052-H34	0	0	1/2t-1 1/2t	1 1/2t-2 1/2t	1 1/2t-2 1/2t	2t-3t
5052-H36	0-1t	1/2t-1 1/2t	1t-2t	1 1/2t-3t	2t-4t	2t-4t
5052-H38	1/2t-1 1/2t	1t-2t	1 1/2t-3t	2t-4t	3t-5t	4t-6t
6061-O	0	0-1t	0-1t	0-1t	0-1t	0-1t
6061-T4	0-1t	0-1t	1/2t-1 1/2t	1t-2t	1 1/2t-3t	2 1/2t-4t
6061-T6	0-1t	1/2t-1 1/2t	1t-2t	1 1/2t-3t	2t-4t	3t-4t
7075-O	0	0-1t	0-1t	1/2t-1 1/2t	1t-2t	1 1/2t-3t
7075-T6[1]	2t-4t	3t-5t	4t-6t	5t-7t	5t-7t	6t-10t

[1] Alclad sheet may be bent over slightly smaller radii than the corresponding tempers of uncoated alloy.
[2] Immediately after quenching, this alloy may be formed over appreciably smaller radii.

Figure 2-127. The type of alloy, temper, and thickness of sheet metal determines the minimum bend radius.

Sheet Metal Structures

nal side of the metal, out beyond the radius of a bend. Although the mold lines themselves do not include dimensions, they are used as reference points from where dimensions can be established.

MOLD POINT
The point where two mold lines intersect in a bend is referred to as a mold point. When a piece of sheet metal is laid out, all of the measurements are made from one mold point to the next. However, the actual corner of the formed metal does not reach the mold point because of the bend radius. On the other hand, by measuring from the mold point, the position where the bend should start can be located.

BEND TANGENT LINE
Bend tangent lines are generally shown on blueprints and drawings to designate the location where the sheet metal begins to form around the bend radius. When positioning metal into a forming brake, the metal is inserted so that the bend tangent line is located under the back edge of the bend radius. Since the line is positioned under the radius, it is difficult to see the tangent line's location.

To enable more accurate positioning, a sight line is marked off from the bend tangent line. The distance of the sight line from the bend tangent is the same dimension as the radius. For example, if a 1/8 inch radius is used, the sight line is marked 1/8 inch or .125 inch ahead of the bend tangent line. When the metal is placed in the brake, the sight line is positioned directly under the nose of the radius forming block. This will cause the bend tangent line to be located at the approximate position of the beginning of the radius curvature.

SETBACK
The distance between the mold line and the bend tangent line inside the bend area is referred to as the setback dimension. When determining where a bend will begin, it is necessary to subtract the setback amount from the desired dimension of the flat. For a 90 degree bend, the setback is found by adding the dimension of the bend radius to the thickness of the metal. For example, the setback for a 90-degree bend in .040 inch thick material with a 1/8 inch radius would be equal to .165 inches (.125 +.040). If a flat with a developed width of 3 inches were desired with one end bent, the bend tangent line would be located at 2.835 inches. However, where two bends are formed on each end of a flat, it is necessary to subtract the setback twice from the desired developed width of the flat. [Figure 2-128]

On the other hand, for bends greater or less than 90 degrees, it is necessary to compensate the setback amount by applying an additional multiplier to the

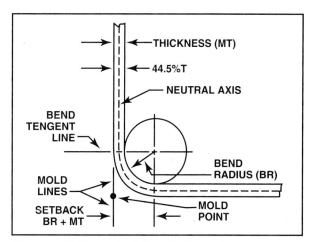

Figure 2-128. The setback for a 90 degree bend is equal to the dimension of the radius plus the metal thickness (Setback = BR + MT).

formula, which is commonly referred to as a K-value. The K-value takes into account that when the metal is bent to an angle ± 90 degrees, the mold point of the bend will move a certain distance. The direction the mold point moves is dependent on whether the bend has an open or closed angle. [Figure 2-129]

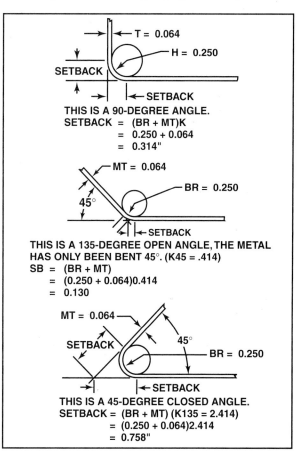

Figure 2-129. As shown in these diagrams, a closed angle bend requires the metal to be bent more than 90 degrees, whereas bending the metal less than 90 degrees develops an open angle.

For bends other than 90° Setback = k [R+T]

Ang [deg]	K value	Ang [deg]	K value	Ang [deg]	K value
1	0.00873	61	0.58904	121	1.7675
2	0.01745	62	0.60086	122	1.8040
3	0.02618	63	0.61280	123	1.8418
4	0.03492	64	0.62487	124	1.8807
5	0.04366	65	0.63707	125	1.9210
6	0.05241	66	0.64941	126	1.9625
7	0.06116	67	0.66188	127	2.0057
8	0.06993	68	0.67451	128	2.0503
9	0.07870	69	0.68728	129	2.0965
10	0.08749	70	0.70021	130	2.1445
11	0.09629	71	0.71329	131	2.1943
12	0.10510	72	0.72654	132	2.2460
13	0.11393	73	0.73996	133	2.2998
14	0.12278	74	0.75355	134	2.3558
15	0.13165	75	0.76733	135	2.4142
16	0.14054	76	0.78128	136	2.4751
17	0.14945	77	0.79543	137	2.5386
18	0.15838	78	0.80978	138	2.6051
19	0.16734	79	0.82434	139	2.6746
20	0.17633	80	0.83910	140	2.7475
21	0.18534	81	0.85408	141	2.8239
22	0.19438	82	0.86929	142	2.9042
23	0.20345	83	0.88472	143	2.9887
24	0.21256	84	0.90040	144	3.0777
25	0.22169	85	0.91633	145	3.1716
26	0.23087	86	0.93251	146	3.2708
27	0.24008	87	0.94890	147	3.3759
28	0.24933	88	0.96569	148	3.4874
29	0.25862	89	0.98270	149	3.6059
30	0.26795	90	1.0000	150	3.7320
31	0.27732	91	1.0176	151	3.8667
32	0.28674	92	1.0355	152	4.0108
33	0.29621	93	1.0538	153	4.1653
34	0.30573	94	1.0724	154	4.3315
35	0.31530	95	1.0913	155	4.5107
36	0.32492	96	1.1106	156	4.7046
37	0.33459	97	1.1303	157	4.9151
38	0.34433	98	1.1504	158	5.1455
39	0.35412	99	1.1708	159	5.3995
40	0.36397	100	1.1917	160	5.6713
41	0.37388	101	1.2131	161	5.9758
42	0.38386	102	1.2349	162	6.3137
43	0.39391	103	1.2572	163	6.6911
44	0.40403	104	1.2799	164	7.1154
45	0.41421	105	1.3032	165	7.5957
46	0.42447	106	1.3270	166	8.1443
47	0.43481	107	1.3514	167	8.7769
48	0.44523	108	1.3764	168	9.5144
49	0.45573	109	1.4019	169	10.385
50	0.46631	110	1.4281	170	11.430
51	0.47697	111	1.4550	171	12.706
52	0.48773	112	1.4826	172	14.301
53	0.49858	113	1.5108	173	16.350
54	0.50952	114	1.5399	174	19.081
55	0.52057	115	1.5697	175	22.904
56	0.53171	116	1.6003	176	26.636
57	0.54295	117	1.6318	177	38.138
58	0.55431	118	1.6643	178	57.290
59	0.56577	119	1.6977	179	114.590
60	0.57735	120	1.7320	180	Infinite

Figure 2-130. A K-chart is used to simplify the problem of determining the setback value for bends other than 90 degrees. To determine the setback with a K-chart, use the number of degrees the metal is bent from the flat layout position.

Sheet Metal Structures

K-values for determining the setback of bends are usually available on charts. These charts make the computations of setbacks for other than 90 degrees rather simple. Once the K-value for the bend is known, the setback is easily found by adding the radius plus the metal thickness and then multiplying by the K-value (S.B. = K × (R + T)). However, if a chart is not available, the K-value can be determined mathematically with most calculators that are capable of performing trigonometric functions, or by using a trigonometric table. To determine the K-value with a calculator or table, find the tangent of one-half the degrees of bend. [Figure 2-130]

For example, to determine the K-value for a closed angle bend of 50 degrees in .040 inch thick material using a 1/8 inch radius, use the following procedures. First, determine the number of degrees the metal will actually be bent from the flat layout position. In this situation, the metal will be bent a total of 130 degrees (180 degrees − 50 degrees = 130 degrees). Next, find the K-value for the degrees of bend. In this example, the K factor is equal to the tangent of one half the bend angle (65 degrees), which is equal to 2.144 inches.

Keep in mind that if a K-chart is used, merely refer to the K-value for the number of bend angle degrees. To complete the setback problem, add the radius (.125 inch) to the metal thickness (.040 inch) and multiply that value by the K factor (2.144 inch × .165 inch = .35376 inch).

FLATS
The distance between inside bend tangent lines from one bend to another, or from the end of a piece of metal to the first bend tangent line is called a flat. This is the amount of metal that is not bent. The length of the flats will always be less than the desired developed width because the bend is setback into the flat.

BEND ALLOWANCE
Bend allowance is the amount of material that is actually involved in the bend and is equal to the length of the neutral axis. When determining the total developed length of a layout pattern, the bend allowance values are added to the lengths of the flats. One method of computing a bend allowance assumes the neutral axis to be located in the exact center of the metal. While this may not be the most accurate method, the final computation is reasonably close and works well for most applications.

To compute a bend allowance, begin by finding the circumference of a circle with a radius equal to the bend radius, plus the thickness of the metal. Since the circumference of a circle is equal to π (3.1415) times diameter (πD), begin by finding the diameter by doubling the radius and then add the thickness of the metal. For example, for a 1/8 inch radius and .040 inch thick metal, the diameter of the circle would be .25 inch (.125 × 2) plus .040 inch for a total diameter of .290 inch.

Next, find the circumference by multiplying the diameter by π (.290 × 3.1415), which for this purpose can be rounded to .911 inch. This would be the amount of metal necessary to make a bend to form an entire circle. To determine the bend allowance for the desired bend angle, it is necessary to divide the circumference by 360 degrees to find the amount of material required for each degree of bend. The amount of material in the example would be found by dividing .911 inches by 360 degrees, which is equal to .0025 inches/degree. This answer multiplied by the number of degrees of the desired bend will be the total bend allowance. For instance, for a 45 degree bend, the bend allowance will be equal to .0025 inches times 45 degrees, which is equal to .1125 inches. This is the approximate amount of material used in the bend. [Figure 2-131]

Another method of finding the bend allowance, which produces accurate results, is found by using an empirical formula. This formula provides for the bend allowance to be located in the actual position

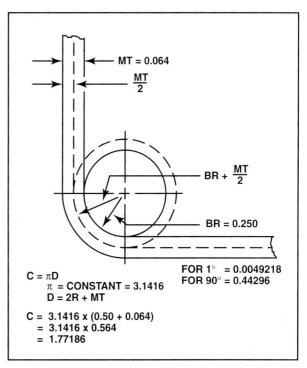

Figure 2-131. One method of finding the bend allowance assumes the neutral axis to be in the center of the metal.

of the neutral axis. The formula for determining the bend allowance for one degree of bend is:

Bend Allowance = (.0078T + .01743R)

The bend allowance for the previous example by the use of this formula is:

B.A. = (.0078 × .040 + .01743 × .125) × 45

= (.000312 + .00218) × 45

= .002492 × 45 = .11214

The empirical formula has been used to compile a table that is found in many aircraft technician and sheet metal fabrication handbooks. With most of these charts, the bend allowance for a 90 degree bend is shown along with the amount of bend allowance for each degree of bend. [Figure 2-132]

Notice that the computation using the first formula versus using a bend allowance chart is within a few ten thousandths inch of each other. For practical purposes, this degree of accuracy is acceptable for most sheet metal fabrications, so either method may be used.

COMPUTATIONS FOR LAYOUTS

To fabricate a sheet metal part with bends, a pattern is first drawn on flat metal to determine the locations of the bend tangent and sight lines, as well as the locations for cutting the metal to the proper length and width. To begin the process, computations are made to determine the dimensions of the flats and locations of the bend tangent lines. For example, to form 2024-T3 aluminum that is 0.032 inch thick into a 4 inch long by 2 inch deep and 2 inch wide U-shaped channel, a layout pattern is cal-

METAL GAUGE	RADIUS OF BEND IN INCHES														
	1/32 .032	1/16 .063	3/32 .094	1/8 .125	5/32 .156	3/16 .188	7/32 .219	1/4 .250	9/32 .281	5/16 .313	11/32 .344	3/8 .375	7/16 .438	1/2 .500	
.020	.062 .000693	.113 .001251	.161 .001792	.210 .002333	.259 .002874	.309 .003433	.358 .003974	.406 .004515	.455 .005056	.505 .005614	.554 .006155	.603 .006695	.702 .007795	.799 .008877	
.025	.066 .000736	.116 .001294	.165 .001835	.214 .002376	.263 .002917	.313 .003476	.362 .004017	.410 .004558	.459 .005098	.509 .005657	.558 .006198	.607 .006739	.705 .007838	.803 .008920	
.028	.068 .000759	.119 .001318	.167 .001859	.216 .002400	.265 .002941	.315 .003499	.364 .004040	.412 .004581	.461 .005122	.511 .005680	.560 .006221	.609 .006762	.708 .007862	.805 .007862	
.032	.071 .000787	.121 .001345	.170 .001886	.218 .002427	.267 .002968	.317 .003526	.366 .004067	.415 .004608	.463 .005149	.514 .005708	.562 .006249	.611 .006789	.710 .007889	.807 .008971	
.038	.075 .000837	.126 .001396	.174 .001937	.223 .002478	.272 .003019	.322 .003577	.371 .004118	.419 .004659	.468 .005200	.518 .05758	.567 .006299	.616 .006840	.715 .007940	.812 .009021	
.040	.077 .000853	.127 .001411	.176 .001952	.224 .002493	.273 .003034	.323 .003593	.372 .004134	.421 .004675	.469 .005215	.520 .005774	.568 .006315	.617 .006856	.716 .007955	.813 .009037	
.051		.134 .001413	.183 .002034	.232 .002575	.280 .003116	.331 .003675	.379 .004215	.428 .04756	.477 .005297	.527 .005855	.576 .006397	.624 .006934	.723 .008037	.821 .009119	
.064		.144 .001595	.192 .002136	.241 .002676	.290 .003218	.340 .003776	.389 .004317	.437 .004858	.486 .005399	.536 .005957	.585 .006498	.634 .007039	.732 .008138	.830 .009220	
.072			.198 .002202	.247 .002743	.296 .003284	.436 .003842	.394 .004283	.443 .004924	.492 .005465	.542 .006023	.591 .006564	.639 .007105	.738 .008205	.836 .009287	
.078			.202 .002249	.251 .002790	.300 .003331	.350 .003889	.399 .004430	.447 .004963	.496 .005512	.546 .006070	.595 .006611	.644 .007152	.745 .008252	.840 .009333	
.081			.204 .002272	.253 .002813	.302 .003354	.352 .003912	.401 .004453	.449 .004969	.498 .005535	.548 .006094	.598 .006635	.646 .007176	.745 .008275	.842 .009457	
.091			.212 .002350	.260 .002891	.309 .003432	.359 .003990	.408 .004531	.456 .005072	.505 .005613	.555 .006172	.604 .006713	.653 .007254	.752 .008353	.849 .009435	
.094			.214 .002374	.262 .002914	.311 .003455	.361 .004014	.410 .004555	.459 .005096	.507 .005637	.558 .006195	.606 .006736	.655 .007277	.754 .008376	.851 .009458	
.102				.268 .002977	.317 .003518	.367 .004076	.416 .004617	.464 .005158	.513 .005699	.563 .006257	.612 .006798	.661 .007339	.760 .008439	.857 .009521	
.109				.273 .003031	.321 .003572	.372 .004131	.420 .004672	.469 .005213	.518 .005754	.568 .006312	.617 .006853	.665 .008394	.764 .008493	.862 .009575	
.125				.284 .003156	.333 .003697	.383 .004256	.432 .004797	.480 .005338	.529 .005678	.579 .006437	.628 .006978	.677 .007519	.776 .008618	.873 .009700	
.156					.355 .003939	.405 .004497	.453 .005038	.502 .005579	.551 .006120	.601 .006679	.650 .007220	.698 .007761	.797 .008860	.895 .009942	
.188						.417 .004747	.476 .005288	.525 .005829	.573 .006370	.624 .006928	.672 .007469	.721 .008010	.820 .009109	.917 .010191	
.250								.568 .006313	.617 .006853	.667 .007412	.716 .007953	.764 .008494	.863 .009593	.961 .010675	

Figure 2-132. A bend allowance chart provides the bend allowance for a 90-degree bend and for individual degrees of bend for various thickness metal and bend radii. The bend allowance for 90 degree bends is shown on top while the bend allowance for each degree of bend is shown in the bottom of each box.

Sheet Metal Structures

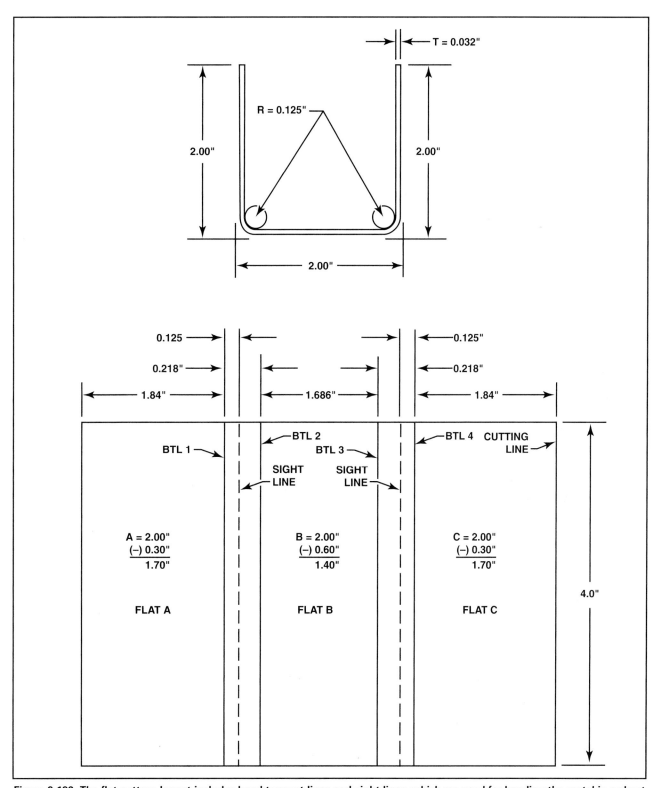

Figure 2-133. The flat pattern layout includes bend tangent lines and sight lines, which are used for bending the metal in a sheet metal brake.

culated and marked in the following manner. [Figure 2-133]

First, begin by finding the minimum radius that will be used to fabricate the bends. In many applications, blueprints or other manufacturing information will specify the radius to use. However, if it is not given, the minimum radius can be determined by referring to a chart similar to the one previously shown in Figure 2-127. For this example, it is found that the minimum radius for 2024-T3 aluminum that is .032 inch thick is between 2 to 4 times the metal thickness, or .064 inch to .128 inch. Converted to the nearest fractional equivalent, the minimum radius would need to be at least 3/32 inch, but 1/8 inch provides for a larger radius to help prevent cracking. For this example, 1/8 inch will be used for fabricating the bends.

Once the minimum bend radius has been determined, it is necessary to find the setback value for the bends. This value is used numerous times in establishing the dimensions in the layout pattern since both bends for the U-shaped channel will be fabricated to 90 degrees. For 90-degree bends, the setback is equal to the bend radius (.125) plus the metal thickness (.032), which equals .157 inch. By subtracting the setback value from each of the mold line dimensions, the lengths of the flats and locations for the bend tangent lines can be determined. To layout the pattern, begin from one end of the metal and progressively mark the bend tangent lines as each location is calculated. For most layout patterns, computations carried out to the nearest hundredth inch are adequate. Mark the metal with a fine tipped marker, orienting it, if practical, so the bends will be made across the grain of the metal.

The first bend tangent line is marked at a distance from the end of the metal that is equal to the mold line dimension, minus the setback. For this example, the mold line dimension for flat A is 2 inches, which will mean the first bend tangent line will be marked at 1.843 inches or 1.84 inches from the end. Next, a second bend tangent line is marked off from the first line, at a distance equal to the bend allowance.

By referring to the chart shown in Figure 2-132, the bend allowance for .032 inch thick aluminum using a 1/8 inch radius is found to equal .218 inch for 90 degree bends. Mark the second bend tangent line on the layout pattern approximately .218 inch away from the first bend tangent line on the side toward the next flat.

From the second bend tangent line, it is necessary to determine the length of flat B to locate the position of the third bend tangent line. The length of flat B is found by subtracting the setback amount twice from the desired developed width of flat B. The two setback values are subtracted to account for the bends at each end of the flat. In this example, the developed width of flat B is 2 inches, so the length of flat B would be 1.686 inches. The third bend tangent line is then marked 1.686 inches from the second bend tangent line.

The final bend tangent line is located at a distance that is equal to the bend allowance away from the third bend tangent line. Since both bends are the same, the bend allowances are also the same. This means the final bend tangent line is marked .218 inch from the third line.

To complete the layout, it is necessary to locate where the metal needs to be cut for length, and then sight lines must be established if a bending brake will be used to fabricate the channel. Since the length of flat C is the same as flat A, the cutting line should be marked 1.84 inches from the fourth bend tangent line. Once the metal is cut on a squaring shear, deburr the edges before bending the metal. As a last step, sight lines are marked depending on how the metal will be positioned in the brake for bending. The sight lines are marked off from the bend tangent line that will be placed under the nose of the brake, at a distance equal to the bend radius used. For this example, the sight lines are marked .125 inch from the bend tangent line.

FORMING BENDS

To form bends using a Cornice or finger brake, begin by verifying that the proper size radius blocks are installed in the upper jaw. For most brakes, the blocks are stamped with the size. However, if the block is not marked, it may be necessary to use a radius gauge to determine the size.

Once the radius has been verified, adjust the nose of the blocks so they are back from the edge of the bending leaf by a distance equal to the thickness of the metal. Once adjusted, open the jaws of the brake and slip the metal in place, lining it up so that looking straight down reveals that the sight line is even with the nose of the radius block. Clamp the jaws of the brake and raise the leaf to the desired angle. Since all sheet metal has some spring-back, the leaf will need to be brought through the desired angle by a few degrees to achieve the properly finished dimension. [Figure 2-134]

When making more than one bend, consider the possibility of the upper jaw of the brake interfering with flats that have already been made. In some cases, the radius blocks may need to be spaced apart so that bent flats can come up along the sides of the blocks. However, if the bend cannot be completely formed because of interference, bumping the metal down with a plastic or rubber mallet over a hardwood block can do the finished shaping.

Sheet Metal Structures

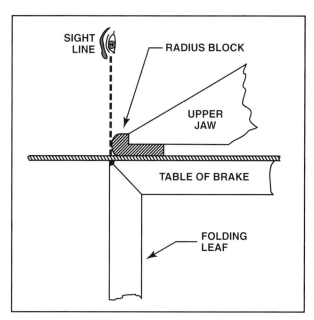

Figure 2-134. The sheet metal being bent should be placed in the brake so that the sight line, which is one radius from the bend tangent line in the brake, is directly beneath the nose of the bend radius blocks.

To perform shaping in this manner, clamp the metal in a vise between a hardwood block that has a radius formed on one edge, and a second hardwood block to prevent damage to the metal from the clamp. Once secured, bump the metal down over the radius block by using a mallet and a piece of hardwood. Use as few blows as possible and spread the force out over as large a distance as possible with the wood block. [Figure 2-135]

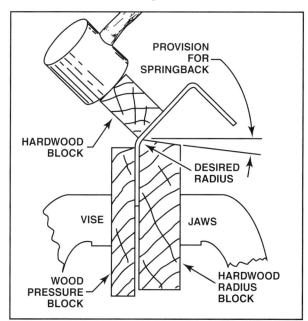

Figure 2-135. When it is not possible to complete a bend in a brake, blocks of wood and a mallet are used to finish the bend.

FOLDING A BOX

One of the most common sheet metal parts that a technician is required to fabricate is a four walled box. Although bending these boxes is similar to forming a U-channel, a few additional processes must be performed. As a project, examine the procedure needed to lay out and form a box of .051 inch 2024-T3 clad aluminum alloy. The box is to be 4 inches square and have sides that are 1 inch high. All of the dimensions are mold line dimensions, and the bend radius will be 5/32 inch. [Figure 2-136]

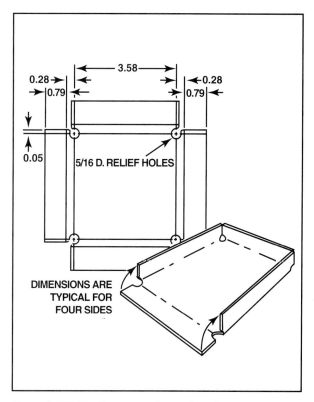

Figure 2-136. The flat pattern layout for a box is similar to a U-channel except that relief holes must be drilled at the inner bend tangent lines to prevent the metal from cracking at the intersection of the bends. In addition, cutting lines must also be marked for material to be removed so the corners can be fabricated.

Determining setback

All of the bends used in this box are 90 degrees, so the setback will be the sum of the bend radius and the metal thickness, or:

$$\text{Setback} = \text{Bend Radius} + \text{Metal Thickness}$$
$$= .156 \text{ inch} + .051 \text{ inch}$$
$$= .207 \text{ inch}$$

Round this off to .21 inch.

Find the length of the flats for the sides
The mold line lengths for the sides are all 1 inch, so the flats of the sides will be this amount less one setback:

Side = 1.00 inch − .21 inch
= .79 inch

Find the length of the flat for the bottom
The mold line length for the bottom is 4 inches, and since there is a setback dimension at each end of this, the flat for the bottom will be:

Bottom = 4.00 inches − .42 inches ×2
= 3.58 inches

Find the bend allowance
Use the chart of Figure 2-132 on page 2-74 and follow the 5/32 inch bend radius column down to the .051 inch thickness line. For a 90 degree bend, .280 inch of material is needed.

Find the length of the developed width for the material
Since the box is to be square, a square piece of metal equal to the sum of the three flats and two bend allowances is needed. For this example, the developed width is found as follows:

Developed Width = .79 inch + .28 inch + 3.58 inches
+ .28 inch + .79 inch
= 5.72 inches.

Lay out the box
Cut a square 5.72 inches on each side, remove all of the burrs, and mark the bend tangent lines in from the edges, equal to the flat dimensions for the walls of the box, or at .79 inch. Make these marks with a fine tipped marker, being careful not to scratch the metal. From these marks, make a second line inside each at a distance equal to the bend radius, or at .28 inch. These lines form the inside and outside locations where the bends begin and end.

Since the bends intersect in the corners, the metal will tend to crack if relief holes are not drilled. A good rule of thumb for relief holes is to use a diameter of twice the bend radius, with the holes centered at the intersection of the inner bend tangent lines. The sight line and the ends of each of the sides will be tangent to these holes.

Cut the material
Drill the relief holes and then use a deburring tool to smooth the edges of the holes. Then cut out the material at each of the corners. Once the metal has been cut to shape, smooth all the edges and corner cuts using a fine toothed file.

Bend the box
A box, or pan brake, should be used to bend the sides. These brakes resemble a leaf brake except that the upper jaw is made up of a series of segmented blocks. These blocks vary in width and can be positioned to accommodate the bent walls of the box as previously mentioned. Once two opposite sides of the box are bent, position the upper jaw of the brake to fit between the formed sides. When the leaf is raised, the two sides will ride up in the slots between the fingers. When using a conventional leaf brake, form two opposite sides on the brake and then bend the remaining sides over a forming block, using a block of hardwood and mallet as previously described in the procedure for forming a channel.

COMPOUND CURVES AND CONTOURS
A part made of thin sheet metal can be considerably stiffened without any increase in its weight by designing it to have compound curves. In addition, sheet metal parts can be made into shapes that allow a part to be used for aerodynamic streamlining. These curves can be fabricated by a number of different methods including hand or machine forming. With hand forming, considerable experience is necessary to make some of the more complex parts that have unique shapes. However, for simple forming, patterns can be used to fabricate contours and curves.

When a repair must be made that requires a compound curved part, it is more economical to buy the part from the factory than it is to fabricate it in the shop. But sometimes parts are not available and the technician must resort to hand forming a replacement.

When a straight bend is made in a piece of sheet metal, the inside of the bend is shrunk and the outside is stretched, but no metal is actually moved, or displaced. When a compound curve is formed, however, metal must be displaced, and the shrinking and stretching take place in more than one plane.

An example of a part that is occasionally hand formed, is an extruded bulb angle stringer. These stringers are used in semimonocoque structures because of the resistance of the metal to distortion. In Figure 2-137, if a load is placed on a bulb angle, it is opposed by the entire width of flat A, which is stiffened by flat B and the bulb. In order to bend, the top side of flat A will have to stretch while flat B and the bulb will have to shrink.

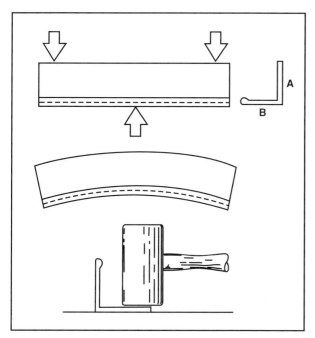

Figure 2-137. Extruded bulb angles are used for stringers in many aircraft. To form a convex curve in a stringer of this type, the metal on the flat with the bulb angle must be stretched. This can be accomplished by bumping the metal with a mallet against a round hardwood block.

To shape a bulb angle to fit the contour of a skin, the thickness of flat A must be reduced by hammering it. Although no metal is removed by hammering, the material that is moved when the thickness is reduced goes to increasing the length of the part. When the outside of flat A is lengthened, flat B, and the bulb, will bend. This method of forming can be used only for very gentle bends, but stringers seldom require more than a slight curvature.

It is more likely required to form a bulb angle into a convex curve than into a concave shape. To do this, the edge of flat A must be shrunk. This can be done, but requires more care than stretching the metal. To form a concave shape, use a V-block made of wood, and hammer on the edge of the flat. Do not hit it hard enough to buckle it. Move the angle back and forth over the V-block as the metal is hit so it will shrink uniformly over the entire bend. Only through practice will the skill be developed to use blows hard enough to minimize work-hardening and yet not too hard to buckle the flat. [Figure 2-138]

One of the best projects that can be performed to develop the skills needed in hand forming aluminum is done by forming a compound curved

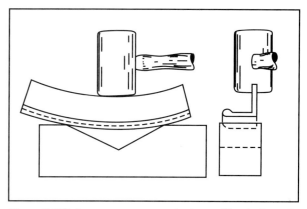

Figure 2-138. A concave curve can be formed in a bulb angle by shrinking the flange over a V-block made of hardwood.

channel. Forming these channels requires considerable stretching on one side of the part and shrinking on the other.

To form a compound curved channel, begin by making a hardwood forming block. Cut the block to the exact shape and size of the inside of the desired channel dimension. Then round the edges with a radius that is greater than the minimum radius allowed for the material that will be used for the part. Taper the edges of the block back about three degrees to allow for spring-back, so the final bend will be a true right angle. Once the first block is made, cut a second block of the proper size and shape so it comes just to the beginning of the bend radius. Drill three holes through both pieces to be used for bolts or other devices to keep the forming blocks and the material from shifting during forming. [Figure 2-139]

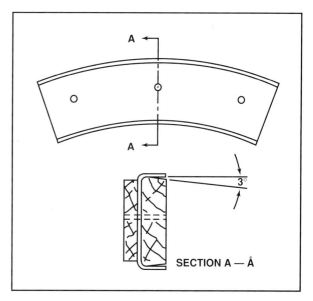

Figure 2-139. A compound curved channel can be formed using two hardwood blocks clamped in a vise.

Lay out the flat pattern for the metal to be bent. Consider the bend allowance when laying out the material, but leave it slightly larger than needed so it can be trimmed to the final size after forming. However, do not allow too much extra material as this will also have to be shrunk and stretched, and too much material can make the project far more difficult. Drill the tooling holes in the metal to match those in the forming blocks. Once drilled, put the metal between the blocks, securing them with the proper size pins or bolts, and clamp the assembly in a vise with the concave side up. Start near the ends of the curve and back the metal with a piece of tapered hardwood. Strike the metal with a plastic mallet as near the bend tangent line as possible. Work from each end toward the middle, folding the metal over just a little each time it is hit. Some buckling will occur, but it can be worked out as the bend progresses across the entire concave side. [Figure 2-140]

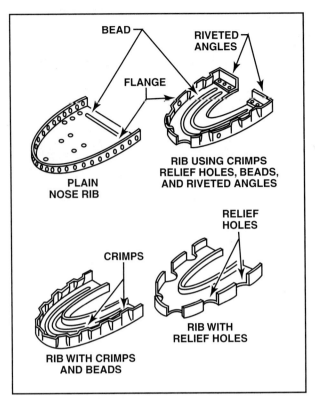

Figure 2-141. Many types of components and aircraft structures can be formed using hand-shrinking and stretching. For example, several different versions of nose ribs can be reproduced using hand-shrinking and stretching, as shown here.

Figure 2-140. When forming a compound curve, the metal should be backed up with a hardwood forming bar for better control of the bend.

Skill in making this kind of bend comes only from practice. A technician must develop a feel for the material so it can be formed with as few blows as possible, to prevent work-hardening the metal, while not creating any big buckles that cannot be worked out.

Once the flats have been formed, turn the assembly over and form the convex side. Back up the material with the tapered block and strike the metal near the bend, but this time start at the center and work outward toward each end. When stretching, as when shrinking, use as few blows as possible so the material will not crack or create wrinkles that cannot be worked out. [Figure 2-141]

BUMPING

It may sometimes be necessary to form a streamline cover for some component that must protrude into the air stream. These parts are usually nonstructural and are much more easily made of fiberglass reinforced resins, but occasionally they are made from aluminum alloy sheet metal.

An example of a part that can be formed by bumping is used to cover bellcrank parts that protrude through an aircraft structure and into the airstream. To form one of these parts, make a forming block of hardwood, hollowed out to the shape of the finished cover. The inside of the depression should be exactly the size and shape as the outside of the cover. Make a hold-down plate of metal or heavy plywood that will hold the edges of the metal, yet allow a mallet to be used to bump the metal on the inside of the form. [Figure 2-142]

Cut a sheet of annealed material, usually 3003-O or 5052-O aluminum, large enough to form the part. Clamp the metal between the forming block and the hold-down plate tight enough to prevent it from wrinkling, yet loose enough that it can slip as the

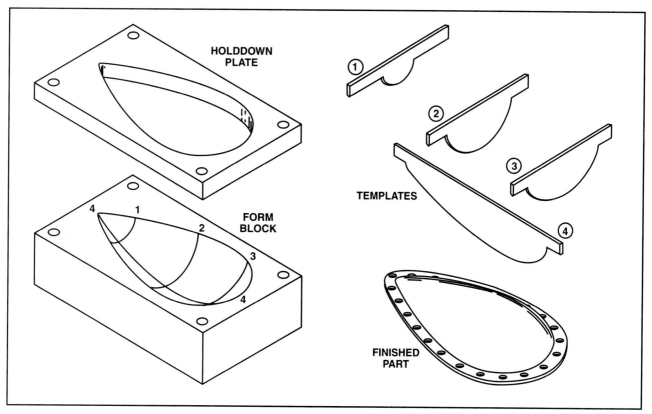

Figure 2-142. A female forming block allows compound curved parts to be made by bumping.

material is forced down into the depression. Begin forming by striking around the edges of the depression with a wedge-shaped plastic mallet. Stretch the material slowly and evenly as it goes into the depression.

In the process of forming deep parts, the material usually work-hardens and becomes difficult to form. When this happens, remove the material and anneal it. It should be annealed in a furnace, but if one is not available, and if the part is strictly nonstructural, a rather rough procedure can be followed that will soften the material enough to finish bumping it to shape.

To perform this annealing process, remove the material from the forming block, and use a welding torch to coat it with a thin layer of carbon by using an extremely rich acetylene flame from a large tip. Then, using a large but very soft neutral flame, carefully heat the metal just enough to burn the carbon off. When the part cools, put it back between the forming block and the hold-down plate and finish bumping it into shape.

FLANGING LIGHTENING HOLES

The thin metal from which aircraft are made usually has ample strength, but its basic structural limitation lies in its lack of stiffness and rigidity. Two benefits can be gained, when making pieces such as fuselage bulkheads and wing ribs, by removing some of the metal to decrease weight and flanging the edge of the cutout to increase the rigidity of the part. These flanged cutouts are called lightening holes.

When a large number of lightening holes are to be cut, it is economical to make or purchase a two-piece flanging die made of steel. These dies are usually made in various sizes ranging from 1/2 inch to 6 inch diameters. [Figure 2-143]

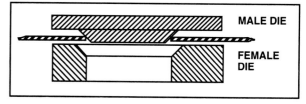

Figure 2-143. Using male and female flanging dies can stiffen the edges of a lightening hole.

To use lightening hole flanging dies, cut the proper size hole in the component with a punch press, hole saw, or fly cutter. Deburr the edges to be sure there are no rough spots that could cause the material to tear when it is stretched. Slip the male die through the hole in the metal and position it over the female die. Once the dies are secured, put the assembly in an arbor press and press the male die into the female die. This forms the metal smoothly and uniformly without work-hardening. When the part is removed, very little finishing is required, making the process fairly quick and easy.

JOGGLING

When a sheet metal structure is built up, there are often locations where the metal is stacked into multiple layers where the parts are joined together. In order for the sheet metal pieces to be flat against the skin and yet have one on top of the other at the joining intersection, a process known as joggling is used. In joggling, the end of one of the pieces is bent up just enough to clear the other, and then it is bent back so it will be parallel to the original piece. Parts should be joggled to fit, rather than attempting to pull them into position with rivets. These joggles may be fabricated by pounding with a soft hammer against a block of wood, or they may be formed in a hydraulic press or with joggling dies. [Figure 2-144]

Joggling dies are often used when many parts are to be joggled with the same dimensions. If dies are not available, joggles may also be formed by stacking sheet metal in a similar fashion to that shown in Figure 2-145, and forming the joggle in a hydraulic or arbor press. [Figure 2-145]

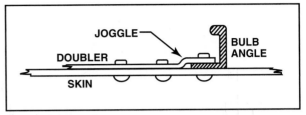

Figure 2-144. Joggles are used primarily to place parts on the same plane so there are no contours in the metal where the parts overlap. One example is where a doubler must fit over a bulb angle. Rather than attempting to pull the doubler against the skin with rivets, a joggle should be formed to allow the two metal pieces to be on the same plane.

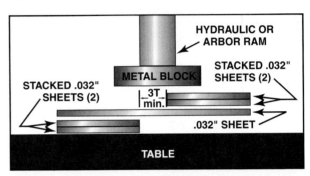

Figure 2-145. Clamping the metal between joggle dies or by placing metal sheets in position and applying force with a press can form joggles. When positioning the metal to form a joggle in this manner, it is important to leave enough distance between the sheets to prevent the bends from exceeding the minimum bend radius of the material being joggled.

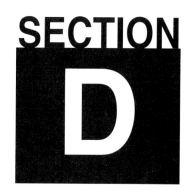

SECTION D

INSPECTION AND REPAIR OF METALLIC AIRCRAFT STRUCTURES

Whenever sheet metal repairs become necessary, it is important to do a thorough inspection for damage that is not immediately apparent. For example, damage that is initially thought to be just a skin repair can have further damage to the underlying structure. Although a visual inspection can usually disclose damaged parts, distortion is occasionally difficult to detect by merely looking at the parts. In these situations, it is advisable for the technician to feel for deformation, such as wrinkles and buckling, by running a hand across the material. In other cases, it may be necessary to make measurements of the aircraft structure to assure symmetry between components when the damage is gradual over a large area. In addition, although inspecting for corrosion should be an on-going process, a more thorough inspection for corrosion can be accomplished whenever skins are removed for repairs.

REPAIR OF SHEET METAL STRUCTURES

The most important part of any repair to a sheet metal structure is to restore the integrity and strength of the component. If a repair procedure is not covered in a structural repair manual, the technician must use various tables to determine the number and sizes of rivets, as well as the best method for restoring the original quality of the structure.

REPAIRABILITY OF SHEET METAL STRUCTURES

Not too many years ago, major sheet metal repairs were done in most aircraft maintenance shops. However, today, with the high cost of labor, most repairs consist of removing the damaged component and replacing it with a new part from the factory. The complex construction and design of many newer sheet metal parts require forming methods beyond the economical capability of most smaller shops.

Major repairs to stressed skin aircraft should never be attempted unless the proper jigs are available to hold the structure in place when the skins are removed. The manufacturer of the aircraft can normally furnish drawings that locate the critical jig points so the jigs can be made accurately. Also, in some cases, prefabricated jigs may be available from the manufacturer for lease. Examples of components that may need to be assembled in jigs include the fuselage, wings, doors, control surfaces, and flaps.

One of the big advantages of sheet metal construction over the formerly used welded steel structures is the ease with which it can be repaired. If, for example, there is major damage to the aft section of a fuselage, the rivets that hold the damaged area can be drilled out and the entire section removed. A new section can then be mated to the undamaged portion in a jig, and with a minimum of man-hours, the aircraft can be restored to its original condition and structural integrity.

ASSESSMENT OF DAMAGE

The difference between making a profit and losing money on a repair job is largely in the assessment of the damage. An intelligent bid must be made, one that includes the repair of every bit of the damage, yet one that is not so large that it disqualifies the bid. Damage that is visible from the surface is usually easy to evaluate, but it is the damage that is not readily apparent that can make the difference between profit and loss.

When examining a damaged structure, use the illustrated parts catalog to determine what types of components are not visible. Consider every piece of skin, rib, former, stringer, and fitting. Some of these parts which have only superficial damage may be quickly repaired, but once the labor of a repair is considered, it may be more economical to use new parts on larger or severely damaged components.

It may also be more economical to exchange a damaged component such as a wing or fuselage from a repair station that specializes in rebuilding these components. A repair station's specialized skills and equipment will allow it to make the repair with a far smaller labor cost than a shop that does the work only occasionally. By exchanging for a component that has already been repaired, the damaged aircraft can be returned to service in far less time than if the work is done in a smaller, less specialized facility.

INSPECTION OF RIVETED JOINTS

Hidden damage may extend beyond the area of visible deformation, and any riveted joint that shows an indication of damage should be inspected well beyond the last deformed rivet. When inspecting the rivets, check both the manufactured heads and the shop heads.

One method of determining if the rivet has been tipped from excessive loads is to try slipping a .020-inch feeler gauge under the rivet head. If it goes under, the rivet may have been stretched. When rivet damage is suspected, drill out the rivet and examine the hole for any indication of elongation or tearing. If the structure has been damaged, the skin will shift when the rivet is taken out. All of the stresses caused by the stretching will have to be removed by drilling out rivets in the seam until there is no more shifting. If the holes are sufficiently out of alignment to require the next size larger rivet, be sure that the edge distance and rivet spacing will allow the use of the larger rivet. Otherwise, the skin will have to be replaced.

INSPECTION FOR CORROSION

Many times, aircraft structures that are enclosed will develop corrosion that will not be detected unless the structure is opened for repair. Also, if a damaged component is improperly stored for a long period of time, corrosion may develop. One of the most important aspects of sheet metal repair is to detect, remove, and treat corroded structural components before the corrosion has a chance to progress too far. If corrosion is found, every trace must be removed and the metal treated to prevent its recurrence. After the treatment, the part should be primed with either epoxy or zinc chromate primer. A more complete coverage of the removal and control of corrosion is found in Chapter Twelve of the *A&P Technician General Textbook*.

REPAIR OF NEGLIGIBLE DAMAGE

Smooth dents in a structure, free from cracks and sharp corners that do not interfere with any structure or mechanism, are considered negligible damage. These may be left as they are, or, if the structure is painted, they may be filled with a resin-type filler, filed smooth, and refinished to match the rest of the surface. However, fillers should not be used on control surfaces, because if the filler becomes dislodged, a severe imbalance could cause aerodynamic fluttering of the control. In fact, most repairs to control surfaces are classified as major because of the critical balance considerations.

Other forms of damage that may be considered negligible include scratches in aluminum alloy skins. However, scratches may harbor corrosion and concentrate stresses enough that they may cause the part to crack. If the scratch is not too deep, it can be burnished with a smooth, rounded piece of steel to force the metal back into the scratch. Work the metal back in smoothly and evenly, but do not allow it to lap or fold over and form an inclusion that will trap moisture and cause corrosion. [Figure 2-146]

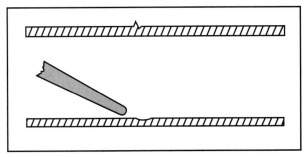

Figure 2-146. A scratch in a piece of aluminum may be burnished. The displaced metal is moved back into the scratch to prevent corrosion from forming. In addition, the blending of the metal reduces the possibility of cracks forming at the bottom of a scratch.

SPECIFIC SHEET METAL REPAIRS

Before discussing any type of specific repairs that could be made on an aircraft, remember that the methods, procedures, and materials mentioned in the following pages are only typical and should not be used as the authority for the repair. When repairing a damaged component or part, consult the applicable section of the manufacturer's **Structural Repair Manual (SRM)** for the aircraft. These manuals may also be referred to by some manufacturers as **Structural Inspection and Repair Manuals (SIRM)**, and essentially contain the same type of information as an SRM.

An SRM differs from the manufacturer's service or repair manual in that it is devoted entirely to the inspection and repair of the aircraft structure. On some smaller aircraft, the manufacturer will include a chapter on structural repair in the service manual rather than in a separate, dedicated SRM. Some of the areas covered in a structural repair manual (or a single chapter) will include the criteria that all repairs must restore a damaged aircraft to its original design strength, shape and alignment.

Specific areas of repair and adjustment will usually include the equipment and tools necessary for repair, control balancing, and setting of the angle of incidence of the wings or stabilizers. Also included will be specifications for materials to be used and the procedures to follow in the repair of the fuselage, wings, ailerons, fin, stabilizer, elevator, rudder, engine mounts, baffling, and cowling. Instructions for all types of materials used in the structure will be included. Normally, a similar repair will be illustrated, and the types of material, rivets and rivet spacing, and the methods and procedures to be used, will be listed. Any special additional information needed to make a repair will also be detailed. If in doubt about any part of a repair detailed in the SRM, the aircraft manufacturer should be consulted.

REPAIR OF STRESSED SKIN STRUCTURE

When repairing damage to a stressed skin structure, it is important to make a repair which fully restores the original strength of the panel. The repair will be required to assume any loads transferred to it and pass them through to the rest of the structure.

APPROVAL OF REPAIRS

For an aircraft to remain legally airworthy, it must continue to meet all of the requirements for its original certification. This means that any repair must retain all of the strength, rigidity, and airflow characteristics of the original structure, and it must be protected against damage from the environment in the same way as the original, or better. Any repair that affects any of these factors must be approved by the FAA. The easiest way of assuring that the repair will meet the required standards is to verify that all data used is FAA approved.

Approved data may be in the form of the repairs that are described by the aircraft manufacturer's service manuals or structural repair manuals, which are usually FAA approved. Before using the manufacturer's repair information, check to see that there is a statement in the manual that designates the manual is FAA approved. Approved data can also be in the form of an Airworthiness Directive, or a Supplemental Type Certificate. If the repair is unusual, the aircraft manufacturer's engineering department may need to be consulted to obtain the instructions on how to perform the repair.

When a repair is made according to approved data, the aircraft can be approved for return to service by an A&P technician holding an Inspection Authorization (IA). An IA must examine the repair to make certain it conforms to the approved data and then sign a statement of conformity on an FAA Form 337. Once the IA signs the 337 Form and other appropriate maintenance record entries are made, the aircraft can be returned to service.

If the repair cannot be made in accordance with approved data, a technician can detail the methods that will be used to conduct the repair on a 337 Form and submit the information for approval from the FAA. It is important that the repair procedure receive approval prior to any work being accomplished on the aircraft. In most cases, the 337 Form can be submitted to the local Flight Standards District Office for approval. However, if the repair is extensive, the local office submits the information to the engineers in the aircraft certification department for approval.

When detailing the methods that will be used to conduct a repair, information to substantiate that the procedure meets FAA criteria can come from a number of sources. One source that is commonly used is to reference procedures outlined in Advisory Circular AC 43.13-1B, *Acceptable Methods, Techniques, and Practices*. However, the information contained in this circular is not FAA approved, but is considered acceptable to be used to meet airworthiness standards. Once the AC is used as a reference, the FAA must verify that the information meets approved criteria.

Upon completion of the repair, in most cases, an IA can perform an inspection to verify that the work has been done in accordance with the approved procedures, and then permit the aircraft to be returned to service.

If a similar repair has been made on another aircraft and written up on a 337 Form and approved, that 337 Form may not be used as approved data because its approval is for a one-time-only specific repair. A similar repair can be made, but the repaired aircraft will have to be approved for return to service by the FAA via another 337 Form.

CRITERIA OF A REPAIR

Any repair made on an aircraft structure must allow all of the stresses to enter, must sustain these stresses, and must then allow them to return into the structure. The repair must be as strong as the original structure, but not different enough in strength or stiffness to cause stress concentrations or alter the resonant frequency of the structure.

All-metal aircraft are made of very thin sheet metal, and it is possible to restore the strength of a repair without restoring its rigidity. All repairs should be made using the same type and thickness of material that was used in the original structure. If the original skin had corrugations or flanges for rigidity, these must be preserved and strengthened. If a flange or corrugation is dented or cracked, the material loses much of its rigidity and must be repaired in such a way that will restore its rigidity and stiffness as well as its strength.

There also must be no abrupt changes in the cross-sectional area of a repair. If a crack is reinforced in a stressed skin, it should not be made with a rectangular patch that causes an abrupt change in the strength of the skin, as the stresses enter, and then another abrupt change as they leave. Rather, use an octagonal patch that gives a more gradual change as the stresses enter and leave the repair area. [Figure 2-147]

Think ahead when considering a repair. Remember that every rivet that is removed must be reinstalled, and to drive a solid rivet, access must be made to both the manufactured head and the shop head. If there are areas that cannot be reached to buck a solid rivet, it may be necessary to use blind rivets. A friction-lock Cherry rivet usually requires one size larger diameter than a solid rivet it replaces, and if one is used, be sure to verify that sufficient edge distance exists. On the other hand, a mechanical-lock Cherry rivet normally allows the use of the same size diameter as the solid rivet it replaces and also may be available in slightly oversized diameters.

Both Hi-Shear rivets and Huck Lockbolts are types of high-strength pin rivets that are used in aircraft factories for applications where high strength is needed, but where light weight and ease of installation make them a preferred choice over bolts and nuts. If one of these special fasteners need to be replaced in a repair and the proper tooling is not available, bolts and self-locking nuts are generally allowed as replacement hardware. However, it is still necessary to verify that the aircraft manufacturer allows the substitution.

REPLACEMENT OF A PANEL

In accordance with FAR Part 43, Appendix A, if additional seams are cut in a stressed skin panel, these types of repairs constitute a major repair, and must have FAA approval before the aircraft can be approved for return to service. In addition, repairs to stressed skin panels that exceed 6 inches in any direction also constitute a major repair, and also

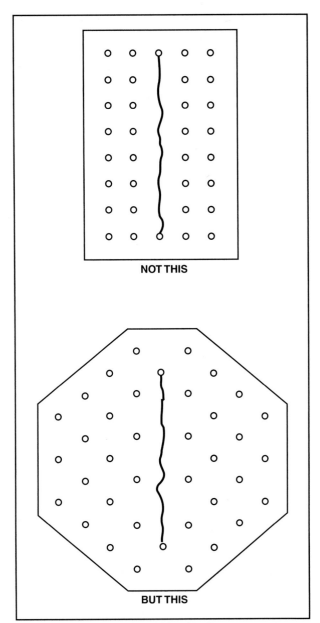

Figure 2-147. A repair should be constructed to allow the stresses to be imparted on the repair area in a gradual manner. Increasing the cross section of a repair over a wider area helps reduce the stress concentrations caused by an abrupt change.

must have FAA approval. On the other hand, if a skin on a wing or tail surface has been extensively damaged, it may be more economical to replace an entire panel rather than making a repair within the panel.

To perform these repairs, remove the damaged panel from one structural member to the next, cutting a generous radius in all of the corners. Usually a radius of 1/2 inch is considered to be the minimum size. Remove all of the rivets immediately sur-

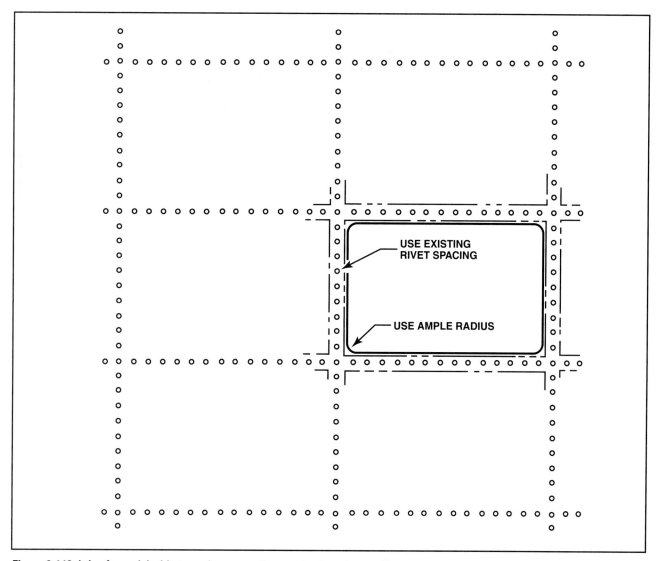

Figure 2-148. It is often advisable to replace an entire panel rather than making a patch within the panel.

rounding the cutout. When installing the new panel, use the same rivet size and spacing as was used on the original structure, unless there is different rivet spacing along a sheet splice for the panel being repaired. If different rivet spacing is used, apply the same spacing as was used on the splice inboard or ahead of the panel being replaced. Be sure when cutting the new panel that sufficient edge distance exists for all of the rivets to be installed. [Figure 2-148]

Cut the panel to size and drill one of the corner rivet holes. Fasten the panel to the structure with a Cleco fastener, aligning the panel with the structure. Now, using a hole finder, locate and drill the other corner holes and temporarily fasten these holes with Cleco fasteners. If possible, back-drill the rest of the holes through the structure or, if not possible, use the hole finder to locate the remaining holes. Then remove the panel and deburr all of the holes and the corners of the panel. Finish the panel by gently crimping the edges so it will fit tightly against the skin, and spray the inside with a light coat of zinc chromate primer. Once the panel is prepared, fasten it back in position with Cleco fasteners, and begin riveting the panel in place.

DESIGN OF PATCHES FOR STRESSED SKIN

When there is damage to a stressed skin, first determine the amount of strength that has been lost, and then design a patch that will restore the strength. A typical example of this kind of repair restores the original strength by placing an octagonal patch over a crack. For example, assume a repair is needed to a .040-inch aluminum alloy skin with a 2-inch crack across the material.

To begin the process of repairing the panel, begin by finding out how much strength the damaged section needs to provide. To determine this value, tables are available that specify the number of rivets that are necessary to impart the same strength through the repair as through an undamaged sheet of aluminum. These tables specify a particular type and size of rivet in various thickness aluminum alloy sheets.

Using one such table for 1/8-inch diameter, 2117 AD rivets in .040 inch 2024-T3 skin, 7.7 rivets per inch of crack width are called for on each side of the damaged area. This falls below the line in this column, which means that the joint is critical in shear, or that it will fail by the rivets shearing rather than by the sheet tearing at the holes. [Figure 2-149]

Begin the repair by drilling a number 30 stop-drill hole in each end of the crack. Clean a piece of .040-inch aluminum alloy with lacquer thinner and spray it with a light coat of zinc chromate primer so the layout lines will show up. Start the layout with the locations of the stop-drill holes, and draw a line between them. Now, draw a parallel line on each side of this line, two and one-half rivet diameters from it. This is the edge distance between the center of the rivet hole and the edge of the crack.

Since the damage to the skin is 2 inches, there will need to be a minimum of 16 rivets on each side of the repair. However, 16 rivets on either side of the crack will not allow a symmetrical patch. On the other hand, if 18 rivets are used on each side, a pat-

Tabe 4-9. Number of rivets required for splices (single-lap joint) in bare 2014-T6, 2024-T3, 2024-T36, and 7075-T6 sheet, clad 2014-T6, 2024-T3, 2024-T36 and 7075-T6 plate, bar, rod, tube, and extrusions,2014-T6 extrusions.

Thickness "t" in inches	No. of 2117-T4 (AD) protruding head rivets required per inch of width "W" (on each side of crack or splice)					No. of Bolts
	Rivet Size					
	3/32	1/8	5/32	3/16	1/4	AN-3
.016	6.5	4.9	--	--	--	--
.020	6.9	4.9	3.9	--	--	--
.025	8.6	4.9	3.9	--	--	--
.036	11.1	6.2	3.9	3.3	--	--
.036	12.5	7.0	4.5	3.3	2.4	--
.040	13.8	7.7	5.0	3.5	2.4	3.3
.051	--	9.8	6.4	4.5	2.5	3.3
.064	--	12.3	8.1	5.6	3.1	3.3
.081	--	--	10.2	7.1	3.9	3.3
.091	--	--	11.4	7.9	4.4	3.3
.102	--	--	12.8	8.9	4.9	3.4
.128	--	--	--	11.2	6.2	3.2

NOTES:
- a. For stringers in the upper surface of a wing, or in a fuselage, 80 percent of the number of rivets shown in the table may be used.
- b. For intermediate frames, 60 percent of the number shown may be used.
- c. For single lap sheet joints, 75 percent of the number shown may be used.

ENGINEERING NOTES:
- a. The load per inch of width of material was calculated by using a strip 1 inch wide in tension.
- b. Number of rivets required was calcuated for 2117-T4 (AD) rivets, based on a rivet allowable shear stress equal to 40 percent of the sheet allowable tensil stress, and a sheet allowable bearing stress equal to 160 percent of the sheet allowable tensile stress, using nominal bolt diameters for rivets.
- c. Combinations of sheet thickess and rivet size above the underlined numbers are critical in (i.e., will fail by) bearing on the sheet; those below are critical in shearing of the rivets.
- d. The number of AN-3 bolts required below the underlined number was calculated based on a sheet allowable tensile stress of 70,000 psi and a bolt allowable single shear load of 2,126 pounds.

Figure 2-149. Tables are available that specify the number of rivets required for each side of a single lap joint repair to restore the original strength to sheet aluminum.

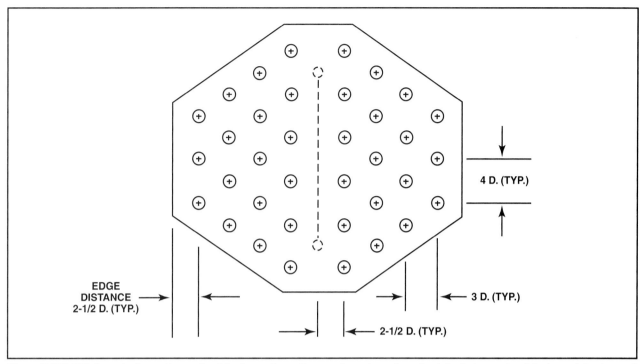

Figure 2-150. This layout of an octagonal patch shows the pattern and number of required rivets to repair a 2-inch crack in stressed .040-inch thick 2024-T3 aluminum alloy skin.

tern can be formed consisting of seven rivets in the first row, six in the second, and five in a third row. Although the pattern will use more than the minimum number of rivets, this is allowed. The determining factor for the maximum number of rivets is that the rivets cannot be spaced closer together than minimums allow, while also maintaining edge distances. By using four rivet diameters for the pitch, a rivet spacing of 1/2-inch can be used and the correct distance between the rows will be 3/8-inch. [Figure 2-150]

Use a fine tipped marker to lay out all the lines and mark the 36 rivet holes. Draw a line, two and one-half diameters from the center of the rivets in the outside rows. Cut the patch along this outside line, deburr all of the edges, and cut a radius in all of the corners. Center punch each rivet location, and drill a number 30 hole for each rivet. Once drilled, deburr each of the holes.

After the patch has been prepared, center it over the stop-drilled crack and drill two holes through the skin to match the holes in diagonal corners of the patch. If the damage is on a curved portion of the skin, the patch will first need to be run through a slip roll former to form its contour to match the skin. Do not try to form the metal by holding the patch by hand while drilling. In most cases, the amount of tension on the metal will cause it to slip. When the rivets are installed, the metal will shift, causing the drilled holes to misalign. The patch must fit smooth and tightly before riveting. Also, slightly crimping the perimeter of the patch prevents the edges from lifting. [Figure 2-151]

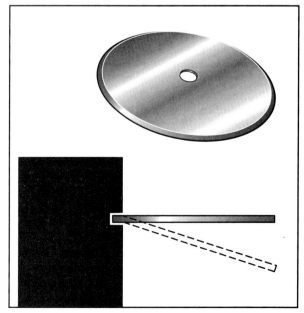

Figure 2-151. A simple tool for beveling the edges of a patch can be fabricated from a phenolic block. Using a band saw, notch the end of the block to the desired depth of the bevel.

Fasten the patch in place with Cleco fasteners and drill all of the remaining rivet holes. After drilling, remove the patch, deburr the holes in the skin, and remove any metal chips. Once deburred and cleaned, verify that the protective coating is in good condition and rivet the patch in place.

FLUSH PATCH STRESSED SKIN REPAIRS

Small damage in a stressed skin may be repaired by making a circular patch that has uniform strength in all directions. For maximum streamlining, a flush patch may be installed on the inside of the skin, using flush rivets, and a plug of the same thickness as the skin. The plug is riveted in the hole where the damaged skin is removed, and contours the surface to the original shape.

Designing a flush patch repair is similar to any other type of patch repair. For example, if there is damage to a .025-inch aluminum alloy stressed skin made from 2024-T3 material, a flush patch can be fabricated in the following fashion.

Before installing a patch, it is necessary to remove any damaged material. For this example, assume that a 2-inch hole has been cut to remove the damaged material. Begin designing the patch by determining the minimum number of rivets required to restore the strength of the sheet metal. By referring to the chart in figure 2-149, it is found that the repair requires 8.6 3/32-inch diameter rivets per inch on each side of the damage in a doubler plate. Although the material will be removed in a circular pattern, the damage still has a 2-inch diameter. This means that the repair must have at least 18 rivets on each side of the damage. [Figure 2-152]

Lay out the pattern on a piece of .025-inch material, making a circle with a 1inch radius, representing the removed damage in the skin. Then draw a circle with a radius of 1.23 inches to locate the first row of rivet holes, which is approximately two and one-half rivet diameters from the edge of the removed damage. A second circle with a radius of 1.61 inches will locate the second row of rivets, approximately four diameters from the circle on which the first row is located. Finally, a third circle with a radius of 1.84 inches is marked to locate the outside perimeter of the doubler. This will provide the minimum edge distance from the outside row of rivets to the edge of the doubler.

Using a protractor, mark off 20 degree increments on the first circle to locate 9 rivets on each half of the

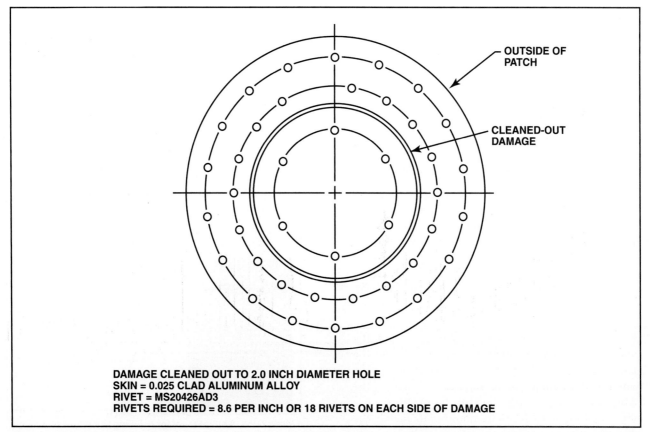

Figure 2-152. A layout of a flush patch shows the rivets in circular rows on a doubler plate, and a non-structural flush patch riveted to fill the hole created by the removed skin.

circle. Now, mark between the center of each of these rivets on the second circle to locate the center of the rivets in the second row. Mark all of these locations, and then center punch them for drilling. Use a number 40 drill to make the rivet holes at each of the locations.

Once the doubler is fabricated, position it on the inside surface of the skin and center it under the cleaned-out damage. Drill one hole, and fasten the doubler in place with a Cleco fastener. Once one hole is fastened, continue to drill for the remaining rivets, securing the doubler with a sufficient number of Cleco fasteners as the drilling progresses.

Once the drilling is complete, remove the doubler, deburr all holes, and clean out any metal chips. Because the repair is to be flush, instead of using universal head rivets, flush head rivets will be installed. Since the skin has insufficient thickness to allow counter-sinking, the rivet holes must be dimpled to accept MS20426AD3 rivets. Once the doubler plate has been dimpled, spray the surface with zinc chromate primer and rivet it in place.

The final step in completing the repair involves cutting a circular patch of .025-inch material that will fill in the removed skin. Once fabricated, place the patch in the hole and drill for rivets through the patch and doubler. Since the patch does not offer any strength to the stressed skin, only enough rivets are required to hold the patch securely in place to prevent the edges from lifting. After the patch is prepared, dimple the metal, and finish installing the rivets.

STRINGER REPAIRS

The most critically loaded stringers in an aircraft structure are those found in the lower surface of a cantilever wing. They are under a tensile load in flight and under a compressive load upon landing. Stringers must receive the greatest amount of care when being repaired to enable them to withstand high stress loads.

When a bulbed stringer must be spliced, first determine the amount of material in the cross-section, and use material having the same or greater area to attach a doubler with the number of rivets specified in figure 2-149. For example, assume a stringer has a 1 inch wide flange on one side, and a 1/2-inch wide flange on the other, and is made of .040-inch 2024T-3 aluminum alloy. In this example, the stringer would have a total cross section of 1.5 inches. According to figure 2-149, there will need to be 7.7 rivets per inch of width, or 12-1/8-inch diameter rivets on each side of the splice (7.7 × 1.5). Putting an extra 3/32-inch diameter rivet at each end will help avoid an abrupt cross-sectional area change. Allowing for an edge distance of two and one-half rivet diameters, and a spacing of three diameters, the splice will have to extend 5.05 inches on either side of the joint. [Figure 2-153]

For splice material, use a piece of the same type of bulb angle. Taper the ends of the top flange back to provide a gradual change in the cross-sectional area and rivet the two pieces together. For formed stringers and extrusions other than bulb angles, the repair is similar. Form a reinforcement piece with

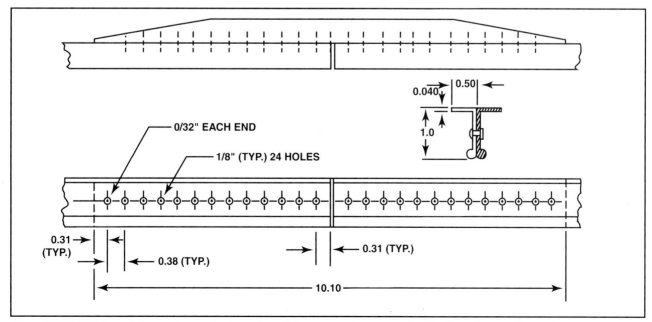

Figure 2-153. This layout shows the required number of rivets spaced out on each side of a stringer splice.

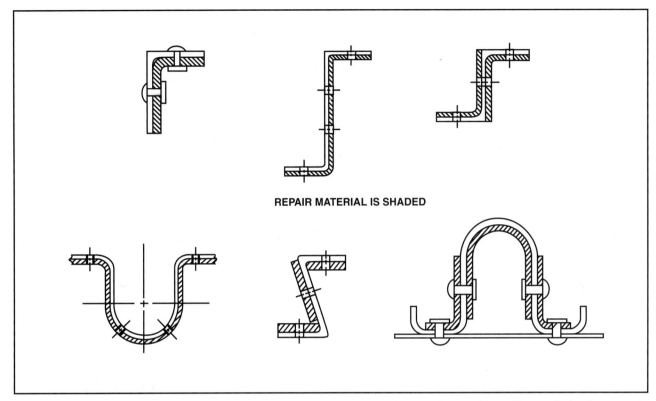

Figure 2-154. The cross-sectional area of the repair material must be greater than the damaged area. On a typical stringer repair, this requires reinforcing more than one angle of the stringer.

more cross-sectional area than the damaged stringer, and rivet it in place with the number of rivets specified in figure 2-149. Remember, the function of a stringer is to provide stiffness as well as strength. [Figure 2-154]

REPAIRS FOR WATERCRAFT

In addition to providing adequate strength with the proper streamlining, any repair to a float or boat hull must be waterproof, and adequately protected from corrosion. The strength requirements and layout procedures are similar to other repairs previously discussed, but when a patch is prepared, coat it with a rubber-like sealant, and then put in place. Also, dip all rivets in the sealant and install them while the sealant is wet.

TRAILING EDGE REPAIRS

The trailing edge of a truss-type wing is usually made of aluminum alloy formed into a V-section. When these areas are distorted or cracked, they should be straightened out as much as possible and an insert formed of the same thickness material, slipped inside and riveted in place. These rivets will be difficult to install, and will have to be put in with the manufactured head on the inside, and bucked with a bar ground to fit in the confines of the

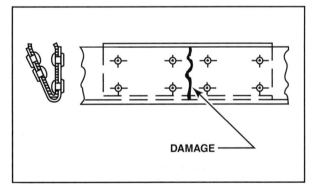

Figure 2-155. A typical trailing edge repair shows the repair material inserted into the inside of the angle and riveted in place.

V. The rivet shank can then be upset with a flush rivet set. [Figure 2-155]

CORRUGATED SKIN REPAIRS

The control surfaces on most light, all-metal airplanes are made from thin sheet metal, which is corrugated to give additional stiffness. When a corrugation is dented or cracked, it can no longer withstand the loads imposed on it, and must be repaired. To perform repairs to corrugated surfaces, remove the damaged area and rivet a new piece of skin in

Sheet Metal Structures

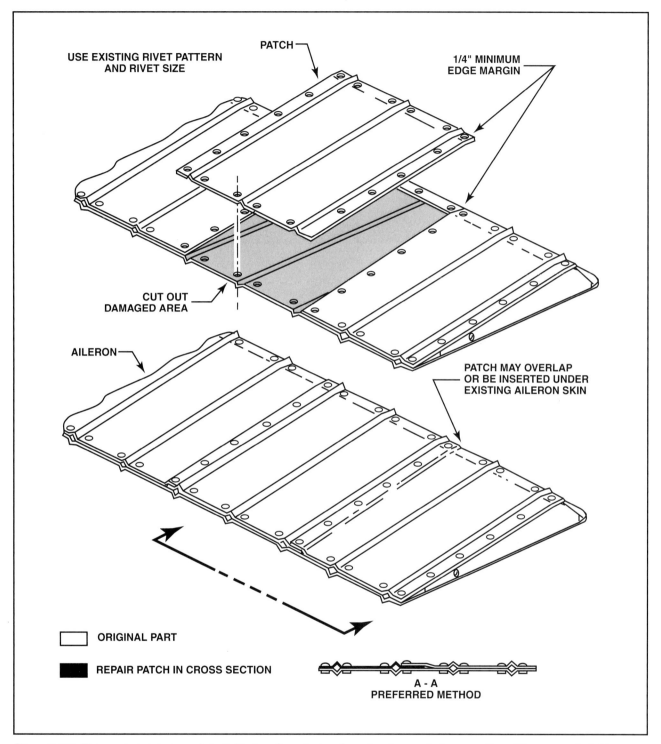

Figure 2-156. The most common repair to light aircraft control surfaces involves replacing a portion of the corrugated skin.

place. Corrugated skins are available from the aircraft manufacturer, or can be fabricated by forming in a press. [Figure 2-156]

It is extremely important when making any repair to a control surface that the repair does not add weight behind the hinge line. In the SRM, the aircraft manufacturer generally specifies the balance conditions for the surfaces. After repairs have been made to a control surface, it must be checked to determine that its balance falls within specifications. In addition, any repair to monocoque or semimonocoque control surfaces constitutes a major repair. [Figure 2-157]

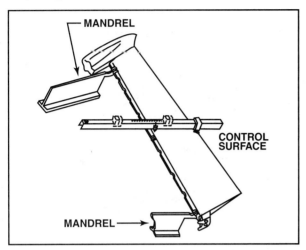

Figure 2-157. The static balance of control surfaces must be checked after a repair has been made to be sure that it is within the tolerances allowed by the manufacturer.

INSPECTION OPENINGS

Many times, a repair must be made to a metal structure that requires access to the inside where the manufacturer has provided no openings. In these situations, inspection hole kits are often available so holes can be cut wherever access is needed. These kits usually include doublers and plates for reinforcing and covering the access opening. Generally, the doubler is riveted to the inside of the structure, and screws are used to secure the cover. Be sure, before cutting a hole in the structure, that it is not located in a highly stressed area. If there is any doubt about the location of the hole, check with the manufacturer or obtain approval from the FAA before cutting a hole. [Figure 2-158]

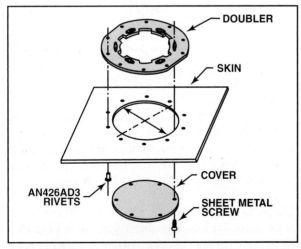

Figure 2-158. Inspection holes may be cut in an aircraft skin to allow access to an area that must be repaired. Be sure that the location for such a hole is approved before cutting the opening.

SPECIALIZED REPAIRS

The high stresses encountered by modern aircraft require that every repair to a structure be carefully considered and made only in accordance with data that is furnished by the manufacturer, or that is specifically approved by the FAA. This is especially true for major load carrying members such as bulkheads, formers, and spars. In addition, special considerations must be used when repairing pressurized aircraft because of the high stress loads encountered in the pressurized sections of the airframe.

FORMER AND BULKHEAD REPAIRS

Bulkheads are primary load carrying members of a fuselage. However, they are also found in wing construction. Bulkheads are usually perpendicular to the longerons, keel beams, or stringers. Some bulkheads also run fore and aft in the fuselage, especially around passenger and cargo doors. On the other hand, formers are often called forming rings, body frames, circumferential rings, or belt frames, because of the manner in which they provide the shape to a fuselage structure. Formers are usually riveted to longerons and stringers in order to carry primary structural loads. Since formers and bulkheads may be subject to high loads, repairs to these components should be conducted in strict accordance with the aircraft manufacturer's repair instructions. [Figure 2-159]

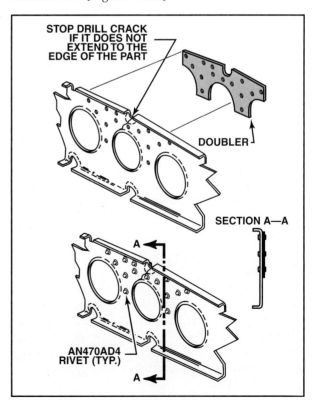

Figure 2-159. A wing rib or fuselage former must be reinforced with a doubler over the damaged area.

Sheet Metal Structures

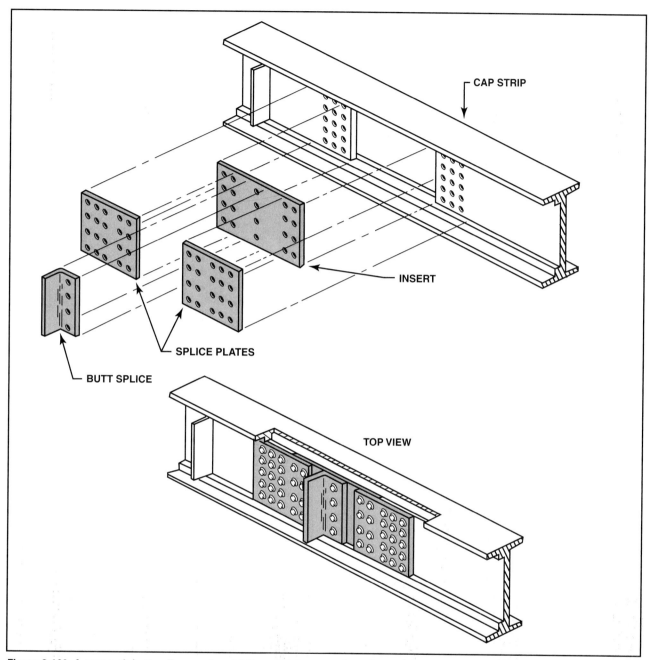

Figure 2-160. A spar web butt splice requires adding an insert to replace the cutout damaged area. Any material used for a splice should be of the same type and thickness as the original spar, and reinforced with doubler plates.

SPAR REPAIRS

The spar, being a primary load carrying member of the wings, usually requires repairs to be made according to the aircraft manufacturer's instructions. However, spar repairs of a general nature are also covered in AC43.13-1B, *Acceptable Methods, Techniques and Practices*. Again, these repairs are not approved for any specific aircraft, but the procedures may be used to obtain FAA approval.

Some spar repairs involve placing an insert in place of a damaged section. Before riveting the insert in place, give all contacting surfaces a coat of zinc chromate primer. The rivets used for attaching the insert section to the spar flange are in addition to those calculated for attaching the splice plates. One typical spar repair is shown in Figure 2-160.

LEADING EDGE REPAIRS

A damaged leading edge usually involves nose ribs, skins and the spar. Repairs made to these components are usually outlined in the manufacturer's SRM or maintenance manual. Since the leading edges are laid out in sections from the wing butt to the tip, the manufacturer often recommends that an entire leading edge section be replaced during the repair procedure.

One difficulty encountered when replacing a leading edge is trying to maintain the contour of the structure during installation. In many cases, it is advisable to use cargo straps or other similar devices to hold the leading edge in position before drilling any holes. Once the leading edge has been initially contoured and positioned, wrap the straps around the wing and tighten them to pull the leading edge into position. Once secured, it is less likely to have problems with the leading edge shifting, which causes misaligned rivet holes. [Figure 2-161]

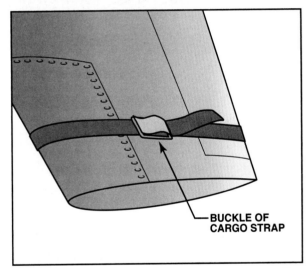

Figure 2-161. Cargo straps are useful for pulling a leading edge section into position before drilling and riveting.

PRESSURIZED STRUCTURE REPAIRS

High performance aircraft with pressurized cabins are becoming more common in aviation, and repairs to this type of structure must take into account the need for additional strength. The increased strength is especially critical due to the flexing of the structure during pressurization and depressurization cycles. When repairs are necessary on these aircraft, it is especially critical to follow the aircraft manufacturer's procedures. In many cases, the manufacturer's engineering department must design and issue specific instructions to be followed when conducting repairs to pressurized airframe structural components. [Figure 2-162]

MISCELLANEOUS REPAIRS

In many situations, it is possible that more than a single sheet metal component will suffer damage, making it necessary to use a combination of repair techniques. One such repair involves damage to the

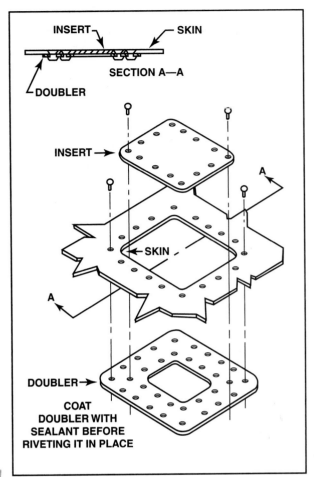

Figure 2-162. A typical repair of a pressurized structure requires that sealant be used to restore air tightness to the pressure vessel of the structure.

skin as well as substructure components. An example of a typical repair for this type damage is shown in figure 2-163.

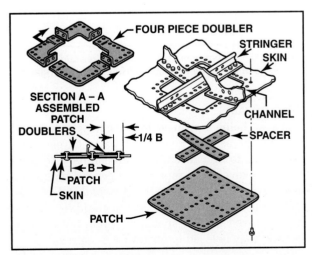

Figure 2-163. If the substructure and skin have been damaged, the substructure must first be repaired and then the skin patched.

Although many repair procedures have been discussed in this chapter, there are countless sheet metal structures in an aircraft. In some cases, damage to these components is minor, and only requires general practices to construct an acceptable repair. However, it is important to use good judgement in determining if a repair may affect the integrity of the aircraft. Always remember to consult the aircraft manufacturer, or a local FAA inspector before proceeding with any questionable procedures.

WOOD, COMPOSITE, AND TRANSPARENT PLASTIC STRUCTURES

CHAPTER 3

INTRODUCTION

Since the time the Wright brothers built their first airplane of wood and fabric, there have been major advances in aircraft construction. Metals, first as steel tubing, and later as aluminum in monocoque-type construction, were a quantum leap forward in terms of the ability to manufacture aircraft quickly and economically. However, wood was used on early aircraft because of its availability and relatively high strength-to-weight ratio. Because wood is a resilient material when properly maintained, many older wooden aircraft still exist, while a few modern designs continue to use wood in select components. This information is only presented as an overview, and as such, information on wood or composite repairs for a specific aircraft should be referenced from the applicable aircraft's structural repair manual. In the event such manuals do not exist, consult Advisory Circular 43.13-1B, *Acceptable Methods, Techniques, and Practices/Aircraft Inspection and Repair*, and approve all major structural repairs through the FAA via Form 337.

SECTION A

AIRCRAFT WOOD STRUCTURES

Although wood was used for the first airplanes because of its favorable strength-to-weight ratio, it is primarily the cost of the additional hand labor needed for wood construction and maintenance that has caused wood aircraft to become almost entirely superseded by those of all-metal construction. However, there are still many home-built airplanes that feature wood construction, and occasionally, commercial designs intended for low-volume production appear using some degree of wood in their structures. [Figure 3-1]

Figure 3-1. This Bellanca Viking incorporates wooden spars in its airframe structure.

This section will provide information on the materials, inspection, and repair of wood structures. For a detailed description of the components and function of aircraft structures, refer to Chapter One of this textbook, Aircraft Structures and Assembly & Rigging.

QUALITY MATERIALS
Wood and adhesive materials used in aircraft repair should meet aircraft (AN) quality standards and be purchased from reputable distributors to ensure such quality. Strict adherence to the specifications in the aircraft structural-repair manual will ensure that the structure will be as strong as the original.

WOOD
Sitka spruce is the reference wood used for aircraft structures because of its uniformity, strength, and excellent shock-resistance qualities. Reputable companies that sell wood for use in aircraft repairs, stringently inspect and verify that the wood product meets the appropriate FAA specifications. To meet the "Aircraft Sitka Spruce" grade specification, the lumber must be kiln-dried to a government specification known as AN-W-2. This specification requires that the specific gravity shall not be less than .36, the slope of the grain shall not be steeper than 1 to 15, the wood must be sawn vertical-grain (sometimes called edge-grained), and shall have no fewer than six annular rings per inch. Each of these specification characteristics is discussed in detail later in this section. Most Sitka spruce now comes from British Columbia and Alaska due to the depletion of old growth spruce forests in the United States, thus making quality spruce valuable and occasionally, limited in supply.

WOOD SUBSTITUTION
Other types of wood are also approved for use in aircraft structures. However, the wood species used to repair a part should be the same as the original wood whenever possible. If using a wood substitute, it is the responsibility of the person making the repair to ensure that the wood meets all of the requirements for that repair. If a substitute wood product meets the same quality standards as the original wood, it is considered an acceptable alternative. For example, you may substitute laminated wood spars for solid-rectangular wood spars as long as they are manufactured from the same quality wood and they are produced under aviation standards.

AC 43.13-1B outlines information regarding acceptable wood species substitutions. If there is any question about the suitability of a specific piece or type of wood for a repair, it would be wise to get the approval of the aircraft manufacturer or local FAA inspector before using it on the aircraft. [Figure 3-2]

PLYWOOD
Structural aircraft-grade plywood is more commonly manufactured from African mahogany or American birch veneers that are bonded together in a hot press over hardwood cores of basswood or

Species of Wood	Strength Properties as Compared to Spruce	Maximum Permissible Grain Deviation (Slope og Grain)	Remarks
Spruce	100%	1:15	Excellent for all causes. Considered as standard for this table.
Douglas Fir	Exceeds spruce	1:15	May be used as substitute for spruce in same sizes or in slightly reduced sizes providing reductions are substantiated. Difficult to work with hand tools. Some tendency to split and splinter during fabrication. Large solid pieces should be avoided due to inspection difficulties. Gluing satisfactory.
Noble Fir	Slightly exceeds spruce except 8 percent deficient in shear	1:15	Satisfactory characteristics with respect to workability, warping and splitting. May be used as direct substitute for spruce in same sizes providing shear does not become critical. Hardness somewhat less than spruce. Gluing satifactory.
Western Hemlock	Slightly exceeds spruce	1:15	Less uniform in texture than spruce. May be used as direct substitute for spruce. Gluing satisfactory.
Pine, Northern White	Properties between 85 percent and 96 percent those of spruce	1:15	Excellent working qualities and uniform in properties but somewhat low in hardness and shock-resisting capacity. Cannot be used as substitute for spruce without increase in sizes to compensate for lesser strength. Gluing satisfactory.
White Cedar, Port Orford	Exceeds spruce	1:15	May be used as substitute for spruce in same sizes or in slightly reduced sizes providing reductions are substantiated. Easy to work with hand tools. Gluing difficult but satisfactory joints can be obtained if suitable precautions are taken.
Poplar, Yellow	Slightly less than spruce except in compression (crushing) and shear	1:15	Excellent working qualities. Should not be used as a direct substitute for spruce without carefully accounting for slightly reduced strength properties. Somewhat low in shock-resisting capacity. Gluing satisfactory.

Figure 3-2. Only certain species of wood are suitable for aircraft structures. This figure outlines the different types of wood approved for aircraft structural repair along with the characteristics and properties of each type in comparison to the standard, Sitka spruce.

poplar. Basswood plywood is another type of aviation-grade plywood that is lighter and more flexible than mahogany and birch plywood but has slightly less structural strength. All aviation-grade plywood is manufactured to specifications outlined in MIL-P-6070, which calls for shear testing after immersion in boiling water for three hours to verify the adhesive qualities between the plies meets specifications.

LAMINATED WOOD

Laminated wood is constructed of two or more layers of solid wood that are bonded together. The lamination process differs from the plywood process in that each layer of laminated wood is bonded with the grain running parallel with each other. Plywood, on the other hand, is constructed of wood layers that are bonded with the grain direction at a 90° angle to the previous layer.

Laminated wood is stronger but less flexible than a piece of solid wood of the same type and size. However, laminated wood is much more resistant to warping than solid wood, making it a good substitute for solid wood components such as laminated spars in place of solid spars. Laminated wood is most commonly utilized for components that require curved shapes such as wing-tip bows and fuselage formers.

WOOD ASSESSMENT

Aircraft technicians who take on a wooden structural repair must be able to properly assess the wood used. Familiarity with the quality and condition of the wood along with the types of defects inherent to wood products is essential to competent wood assessment. The technician must make certain that the quality of wood meets the original specifications outlined in the aircraft's repair manual. The following information describes wood characteristics that the maintenance technician must consider for proper wood assessment, not only for the initial use of a wood product, but also in the inspection phase of wooden structures.

The cut of the wood, slope of the grain, and the number of growth rings are factors to examine when determining quality. The way wood is cut affects its shrinkage characteristics and strength qualities. Aviation-quality wood is usually quarter-sawed to reduce the amount of shrinkage over the life of the component. Quarter-sawn wood is cut from quartered logs so that the annual growth rings are at 90° angles to the wide face.

The slope of the grain is another factor to consider when assessing wood. The maximum slope of the grain for aviation-grade lumber is 1:15. The slope of the grain is the amount of grain rise over the grain length. In other words, the grain may not rise more than one inch in a 15-inch section of wood. [Figure 3-3]

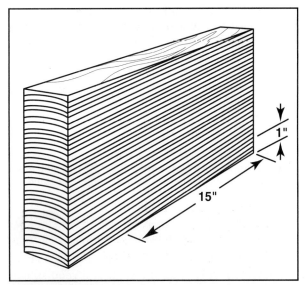

Figure 3-3. According to FAA standards, a grain slope of 1:15 is the maximum allowable slope allowed in aviation-grade wood.

Another factor to consider when assessing wood is the number of growth rings per inch. To accurately calculate the number of rings, look at the end of the board and count the number of growth rings in one inch. The minimum grain count for softwoods is six rings per inch. Port Oxford white cedar and Douglas fir are exceptions. The minimum grain count for these woods is eight rings per inch.

Certain defects are allowed and others disallowed when choosing the appropriate species of wood, which stresses the importance of accurate identification of wood defects. Following are several wood defects the technician must be able to identify to properly assess wood quality. [Figure 3-4]

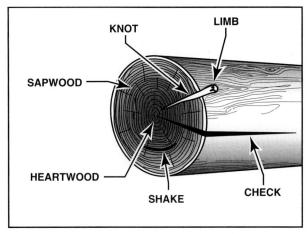

Figure 3-4. This figure illustrates several wood defects that a technician must be able to identify when evaluating wood condition and quality.

- **Brown rot** – Any decay in wood that produces a light to dark brown, easily crumbled residue. An advanced stage of brown is referred to as "cubical rot" that splits the wood along rectangular planes.

- **Checks** – A lengthwise separation or crack of the wood that extends along the wood grain. It develops during drying and is commonly caused by differences in radial and tangential shrinkage or because of uneven shrinkage of the tissues in adjacent portions of the wood.

- **Compression failure** – Characterized by a buckling of fibers that appear as streaks on the surface of the wood that are at right angles to the grain. Compression failures vary from pronounced failures to very fine hairlines that require close inspection. This defect is caused from the wood being overstressed in compression due to: natural forces during the growth of the tree, felling trees on rough or irregular ground, or rough handling of logs or lumber.

- **Compression wood** – Characterized by high specific gravity, it has the appearance of an excessive growth of summerwood. Compression shows little or no contrast in color between springwood and summerwood, making it a difficult defect to identify. If you have any doubt whether a piece of wood is compression wood or not, reject it.

- **Cross grain** – Wood in which the direction of the fibers or grain deviate from a line parallel to the sides. Crossed grain may look like diagonal grain, spiral grain, or a combination of the two.

- **Curly grain** – Wood with distorted fibers resulting in a curly appearance as in bird's-eye wood. The area covered by each curl may vary up to several inches in diameter.
- **Decay** – The destruction and eventual reduction of wood to its component sugars and base elements through attack by organisms such as fungi and certain insects such as termites; may also be referred to as "**dote**." Red heart and purple heart are also forms of decay.
- **Dry rot** – A term loosely applied to any dry, crumbly rot but especially a wood easily crushed to dry powder in its advanced stage.
- **Hard knots** – A knot that is solid across the surface, at least as hard as the surrounding wood, and shows no indication of decay.
- **Heartwood** – The inner core of a woody stem or log, extending from the pith to the sap, which is usually darker in color. This part of the wood contains dead cells that no longer participate in the life processes of the tree.
- **Interlocked grain** – Grain in which the direction of the fibers first follow a left- then a right-handed spiral, then alternate in a spiral direction every few years. Such wood is very difficult to split radially, although it may split easily in the tangential direction.
- **Knot** – That portion of a branch or limb that is embedded in the wood of a tree trunk, or that has been surrounded by subsequent stem growth.
- **Mineral streaks** – An olive to greenish-black or brown discoloration believed to show regions of abnormal concentrations of mineral matter in some hardwoods. Mineral staining is common in hard maple, hickory, and basswood.
- **Pin knot clusters** – Pin knots are knots with diameters less than or equal to 1/2 inch. Several pin knots in close proximity to each other make up a cluster.
- **Pitch pocket** – Lens-shaped opening extending parallel to the annual growth rings in certain coniferous woods. May be empty or may contain liquid or solid resin.
- **Spike knots** – Knots that run completely through the depth of the wood perpendicular to the annual rings. Spike knots appear most frequently in quarter-sawn lumber.
- **Split** – Longitudinal cracks produced by artificially induced stress.
- **Spiral grain** – Wood in which the fibers follow a regular spiral direction (right-handed or left-handed) around the trunk of the tree instead of the normal vertical course. Spiral grain is a form of a cross grain.
- **Shakes** – A separation or crack along the grain, the greater part of which may occur at the common boundary of two rings or within growth rings
- **Wavy grain** – Wood in which the collective appearance of the fibers presents a regular form of waves and undulations.

ACCEPTABLE DEFECTS

Certain types of wood defects are permitted in aviation-grade lumber. The following list of permissible wood defects applies to the species of wood listed in figure 3-2.

- **Cross grain.** Spiral grain, diagonal grain, or a combination of the two is acceptable providing the grain does not diverge from the longitudinal axis of the material more than specified in column 3 of figure 3-2.
- **Wavy, curly, and interlocked grain**. Acceptable if local irregularities do not exceed limitations specified for spiral and diagonal grain.
- **Hard knots.** Sound, hard knots up to 3/8 inch in diameter are acceptable providing: (1) they are not projecting portions of I-beams, along the edges of rectangular or beveled un-routed beams, or along the edges of flanges of box beams except in lowly stressed portions; 2) they do not cause grain divergence at the edges of the board or in the flanges of beams more than specified in column 3 of figure 3-2; and 3) they are in the center third of the beam and are not closer than 20 inches to another knot or other defect.
- **Pin knot clusters.** Small clusters are acceptable providing they produce only a small deviation of grain direction.
- **Pitch pockets.** Acceptable in the center portion of a beam providing they are at least 14 inches apart when they lie in the same growth ring and do not exceed 1-1/2 inches in length by 1/8 inch in depth.
- **Mineral streaks.** Acceptable providing that there is no decay indicated anywhere on the wood.

NON-ACCEPTABLE DEFECTS

While there are certain defects that are allowed in wooden aircraft structures, there are more that are not acceptable. Choosing a section of wood with non-acceptable defects increases the chance of future structural failure. The following is a list of non-acceptable wood defects.

- **Cross grain.** Not acceptable unless they are within the limitations specified in the description of acceptable cross-grain defects listed previously.

- **Wavy, curly, and interlocked grain.** Not acceptable unless they are within the limitations specified in the description of acceptable defects listed previously.

- **Hard knots.** Not acceptable unless they are within the limitations specified in the description of acceptable defects listed previously.

- **Pin knot clusters.** Not acceptable if they produce a large effect on the direction of the grain.

- **Spike knots.** Reject wood that contains this type of defect.

- **Pitch pockets.** Not acceptable unless they are within the limitations specified in the description of acceptable defects listed previously.

- **Mineral streaks.** Not acceptable if any decay is found.

- **Checks, shakes, and splits.** Reject wood containing these defects.

- **Compression wood.** Reject wood that indicates compression wood.

- **Compression failures.** Reject wood that contains an obvious compression failure. If there is a question as to whether wood indicates compression failure, perform a microscopic inspection or toughness test.

- **Decay.** Reject wood that indicates any form of decay or rot including indications of red heart or purple heart.

AIRCRAFT ADHESIVES/GLUES

The adhesive used in aircraft structural repair plays a critical role in the overall finished strength of the structure. The maintenance technician must only use those types of adhesives that meet the performance requirements necessary for use in aircraft structures. Not every type of glue is appropriate for use in all aircraft repair situations. Because of its importance, use each type of glue in strict accordance with the aircraft and adhesive manufacturer's instructions.

TYPES OF ADHESIVES

Most older airplanes were glued with **casein glue**, which was a powdered glue made from milk. Casein glue deteriorates over the years after it is exposed to moisture in the air and to wide variations in temperature. Many of the more modern adhesives are incompatible with casein glue. If a joint that has been glued with casein is to be re-bonded with a different type of glue, scrape all traces of the casein away before applying the new glue. The alkaline nature of casein glue may prevent the new glue from curing properly, thereby compromising the structural integrity. The performance of casein glue is considered inferior to other available products and should be considered obsolete for all aircraft repairs.

Plastic resin glue is a urea-formaldehyde resin that is water-, insect-, and mold-proof. This type of glue usually comes in a powdered form. Mix it with water and apply it to one side of the joint. Apply a hardener to the other side of the joint, clamp the two sides together and the adhesive will begin to set. Mix plastic resin glue in the exact proportions specified by the manufacturer, otherwise the adhesive properties may be impaired.

Plastic resin glue rapidly deteriorates in hot, moist environments, and under cyclic stresses, making it obsolete for all aircraft structural repairs. Any use of this type of glue for aircraft repair should be discussed with the appropriate FAA representative prior to use on certificated aircraft.

Resorcinol glue is a two-part synthetic resin glue consisting of a resin and a hardener and is the most water-resistant of the glues used. The glue is ready for use as soon as the appropriate amount of hardener and resin has been thoroughly mixed. Resorcinol adhesive meets the strength and durability requirements of the FAA, making it one of the most common types of glue used in aircraft wood-structure repair. Again, follow the aircraft manufacturer's recommendations when choosing the type of glue for any structural repair.

Phenol-formaldehyde glue is most commonly used in the manufacturing of aircraft-grade plywood. Phenol-formaldehyde glue requires high curing temperatures and pressures making it impractical for use in the field.

Epoxy resins are two-part synthetic resins that generally consist of a resin and a hardener mixed together in specific quantities. Epoxies have excellent working properties and usually require less attention to joint quality or clamping pressures as compared to other aircraft adhesives. They penetrate evenly and completely into wood and plywood structures. However, varying degrees of humidity and temperature affects the joint durability in different epoxies. Only use the recommended epoxy as outlined in the aircraft's repair manual.

THE BONDING PROCESS

The bonding process is critical to the structural strength of an aircraft wooden structure. To ensure the structural integrity of a wood joint, the bonding process must be carefully controlled. It is imperative to follow the manufacturer's repair procedures in detail when producing a wood joint.

Following are the three most important requirements for a strong and durable structural bond.

- **Preparation** of the wood surface prior to applying the adhesive.

- **Utilization** of a good quality aircraft-standard adhesive that is properly prepared.

- **Performing** a good bonding technique consistent with the manufacturer's instructions.

WOOD PREPARATION

It is imperative to properly prepare the wood surface prior to applying any adhesive. The wood surface must be clean, dry, and free of any oil, grease, or wax, otherwise the adhesive will not penetrate the wood evenly. Without proper adhesive penetration, you will not gain a proper glue line, which will weaken the joint. The glue provides the strength of a properly prepared wood splice-joint. In addition, wood changes in dimension according to its moisture content, therefore, the pieces of wood to be joined should be kept in the same room for a minimum of 24 hours to equalize the moisture content.

Cut the wood to the required bevel with a fine-toothed saw, then plane or scrape the surface until it is smooth and true. Planer marks, chipped or loosened grain, and other surface irregularities are not permissible. The wood joint must join evenly over the entire bonded surface to produce a strong and durable bond. **Do not use sandpaper to smooth the surface**. Sanding may round corners and change the flatness of the wood surface resulting in a joint that does not properly meet. Sanding also produces dust that fills the wood pores and causes a weak glue line. Roughening the wood surface is also not recommended because it will prevent uniform contact of the wood surface, which is necessary for strong and durable glue joints. Before applying adhesive to the joint surfaces, vacuum them to remove anything that remains which may prevent glue penetration.

When wood surfaces cannot be freshly machined before bonding, such as plywood or inaccessible members, sand them lightly using a very fine grit such as 220. Very light sanding improves the penetration quotient of the adhesive in these cases only. However, heavy sanding will change the flatness of the wood and deposit sawdust in its pores. Again, make sure the surfaces are clean and dry before applying adhesive.

APPLYING THE ADHESIVE

When the wood surfaces are prepared and ready to be glued, apply a smooth, even coat of glue to each surface. Then, **following the adhesive manufacturer's recommended procedures**, join the surfaces together. It is important to observe the orientation of the wood grain to avoid applying glue to the end grain. End grain is wood that is cut at a 90° angle to the direction of the grain. An acceptable cut of wood has been cut nearly parallel to the direction of grain. [Figure 3-5]

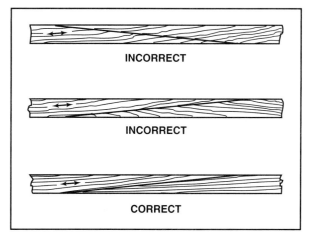

Figure 3-5. Avoid end-grain joints when gluing wood scarf joints. Make sure the wood is cut with the grain of both pieces as close to parallel as possible. Using end-grain joints increases the chance of future warping.

Almost all types of adhesives have **four time-periods** that are critical to the bonding process. **Pot life** is the useable life of the glue from the time it is mixed until the time it must be used. Discard the glue once the pot life has expired. Using glue after the pot life has expired or adding thinners to the adhesive will not extend its life.

The **open-assembly time** is the allowable time between the application of the glue and the time the joint is assembled. If the open-assembly time is too long, the glue will begin to set up on the joint surfaces and the glue line will weaken. Different types of adhesives have varying open-assembly times. Follow the adhesive manufacturer's procedures explicitly when bonding a structure.

The **closed-assembly time** is the allowable length of time between the assembling of the joint and the application of the clamping pressure. Closed-assembly time allows for the movement of parts to place them in the proper alignment.

The **pressing time** is the period during which the parts are pressed or clamped together and is essentially the adhesive curing period. Pressing time must be sufficient to ensure that the joint is strong enough to withstand manipulation or the machining process. The temperature of the bond line also affects the cure rate of the glue. Each type of glue requires a specific temperature during the curing cycle. [Figure 3-6]

CONDITION	TEMPERATURE		
	70° F (21° C)	80° F (27° C)	90° F (32° C)
Mixture pot life (hours)	4 - 5	2 - 1/2 - 3 1/2	1 - 2
Maximum assembly time			
Open (minutes)	15	10	5
Closed (minutes)	25	15	8
Pressure period (hours)	14	8	5
Assembly must be maintained at a temperature of 70° F or above to assure a satisfactory cure of the glue line.			

Figure 3-6. This chart outlines an example of a manufacturer's recommended bonding times, pressure period, and assembly temperature for a specific type of plastic resin glue. Each type of adhesive has specific time-periods and procedures to be followed absolutely. If you waver from the manufacturer's procedures, you must either discard the wood parts or remove the adhesive, clean the bond line, and start over. The structural integrity of the joint will be compromised if the manufacturer's procedures are not followed to the letter.

CLAMPING PRESSURE

When the joint is connected and properly aligned, apply pressure to spread the adhesive into a thin, continuous film between the wood layers. The strength of a glue line is partially dependent upon the correct pressure applied during the curing process. Clamping forces air out of the joint and brings the wood surfaces together evenly. Too little clamping pressure results in thick glue lines and weak glue joints. Too much clamping pressure can squeeze out too much glue weakening the joint. Clamping pressure is accomplished using clamps, presses, or by other mechanical means. Each type of adhesive requires a specific amount of clamping pressure. Therefore, follow the adhesive manufacturer's gluing procedures in detail. For example, the recommended clamping pressure for soft-woods is between 125 and 150 psi and between 150 and 200 psi for hardwoods when using resorcinol glue.

METHODS OF APPLYING PRESSURE

In addition to the amount of clamping pressure, the method used to apply pressure is also important. Different methods range from the use of brads, nails, small screws, and clamps, to the use of hydraulic and electrical presses. The choice of clamping method is important to achieving a strong and durable joint.

Hand nailing is one method of applying pressure using small nails or screws in the bonding of ribs, attachment of plywood skins to the wing, control surfaces, and fuselage frames. However, both nails and screws can produce adverse effects such as splitting small parts and creating points where moisture may enter the wood structure causing decay. If you decide to utilize the hand nailing method, **nailing strips** are often used to spread the pressure over a larger area and to help in the removal of the nails after the glue has cured. To prevent the nailing strip from sticking to the wood structure, place a piece of waxed paper between the strip and the structure.

The nails or screws used in the hand nailing method may or may not be removed after the adhesive has cured. Nails used for clamping pressure are not intended to hold the structure together for strength purposes. When using nails, be careful not to crush the wood with heavy hammer blows and do not penetrate all the way through the wood structure. [Figure 3-7]

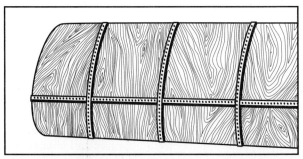

Figure 3-7. Nail strips may be used for clamping pressure on plywood skin during the bonding process. Remove nail strips once the glue is cured. Before applying the finish, fill the nail holes with a manufacturer's recommended wood filler to prevent any areas at which moisture may enter the structure.

Another common method is the use of **screw clamps or "C" clamps** in conjunction with pressure blocks. Pressure blocks distribute the clamping pressure and protect the members from local crushing. Clamps and pressure blocks apply pressure evenly over the entire glue joint to form a thin, even glue line, which produces a strong joint. [Figure 3-8]

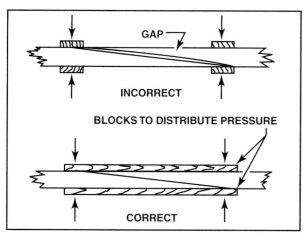

Figure 3-8. Apply pressure evenly over the entire joint to avoid gaps between the mating surfaces. An even clamping pressure ensures that the adhesive is squeezed out of the glue-joint uniformly. Insufficient or uneven pressure usually results in thick bond lines that weaken the joint.

Apply pressure to a joint for the time recommended by the glue manufacturer. When the clamping pressure is removed, clean and inspect the joint and remove any glue that has been squeezed from the joint.

INSPECTION OF WOOD STRUCTURES

To effectively inspect wood structures, be familiar with methods of inspection and the equipment used to examine them. Also, be able to identify the types of defects that are common to wood structures, as well as the failure modes that are unique to them.

Most wood damage is caused by conditions such as moisture, temperature, and sunlight. Because wood is an organic material, it is subject to mildew and rot unless protected from moisture. Keep wood airplanes in well-ventilated hangars and take special care to ensure that all of the drain and ventilation holes remain open. If a ventilation hole becomes obstructed, changes in air temperature will cause moisture to condense inside the structure, which will cause the wood to deteriorate.

TYPES OF DETERIORATION

The maintenance technician must be able to identify wood deterioration to determine the airworthiness of a wood structure. Along with the list under *Wood Defects* discussed earlier in this section, the following are several of the more common types of wood deterioration.

- Wood decay results from the attack and growth of fungus upon wood products. Decay is indicated by softness, swelling when wet, excessive shrinkage when dry, cracking, and discoloration. Musty or moldy odors also indicate wood decay.

- Splitting or cracking of a wood member may occur due to the varying shrinkage rates of bonded wood members, or due to an outside force applied to the structure. Wood splits often result when different types of woods are bonded together. For example, bonding a mahogany plywood doubler to a spruce member may produce a split. As the spruce dries, it tries to shrink. However, the mahogany plywood, which shrinks at a lower rate, holds the spruce firmly in place. The induced stress in the spruce member exceeds its cross-grain strength, thus resulting in a split.

- Bond failure is most commonly due to an improper bonding process or prolonged exposure to moisture. Using the wrong type of glue, not following the manufacturer's bonding procedures or improper wood preparation can all lead to bond failure of the wood joint.

- Finish failure is the breakdown of the protective finish applied to the wood structure to prevent decay. Finish failure results from long-term exposure to water, wood splitting, ultra-violet light, and surface abrasion.

- Stress damage is caused by excessive impact, mechanical, or aerodynamic loads imposed upon the wood structure. Over-tightening of fittings can also cause wood crush and possible bending of the metal fittings. Some applications use steel bushings to prevent the bolts from being tightened to a point where the wood is crushed. Such bushings also add bearing strength to the assembly.

INSPECTION METHODS

When inspecting a wood structure aircraft, move it into a dry, well-ventilated hangar. Before beginning the inspection, remove all of the inspection and access panels to facilitate the drying of the wooden structures. One of the first steps is to check the moisture content of the wood using a moisture meter. If the moisture content is high, dry the wood structures before inspecting further. Wooden structures of the aircraft need to be dry to be able to effectively determine the condition of the bonded joints. The following are several inspection methods and associated equipment employed for inspecting wooden structures.

MOISTURE METERING

Use moisture meters to determine the moisture content of the wood structure. The moisture content of any wooden member is an important factor in its structural integrity. Wood that is too wet or too dry may compromise the strength and integrity of the structure. A moisture meter reads the moisture content through a probe that is inserted into a wooden member. Use a correction card to correct for temperature and the type of wood being tested.

TAPPING

The wood structure may be inspected for structural integrity by **tapping** the suspect area with a light plastic hammer or screwdriver handle. Tapping should produce a sharp, solid noise from a solid piece of wood. If the wood area sounds hollow or feels soft, inspect further.

PROBING

If soft, hollow wood is found during the tap test, probe the suspect area with a sharp metal tool to determine whether the wood is solid. Ideally, the wood structure should feel firm and solid when probed. If the area feels soft and mushy, wood has rotted and disassembly of the structure is necessary to repair or replace the damaged area.

PRYING

Use prying to determine whether a bonded joint shows signs of separation. When prying a joint, be cautious not to use too much force, otherwise you may forcibly separate it. Light prying is sufficient to check the integrity of a joint. If there is any movement between the wood members of the joint, a failure of the bond is confirmed. Repair or replace the bonded structure if a failure has occurred.

SMELLING

Smell is a good indicator of musty or moldy areas. When removing the inspection panels, be aware of any odors that may indicate damage to the wood structure. Odor is an essential indicator of possible wood deterioration. Musty and moldy odors reveal the existence of moisture and possible wood rot.

VISUAL INSPECTION

Visual inspection techniques are used to determine any visible signs of damage. Both internal and external visual examinations are imperative to a complete inspection of the wood structure.

External Visual Inspection

Many airplanes that have an external skin made of thin mahogany plywood are covered with lightweight cotton or polyester fabric to increase both the strength and smoothness of its surface. A thorough inspection is required to ensure that the fabric covering has not pulled loose or torn away from the wood. A split or tear in the fabric could be an indication of internal damage to the wooden structure. Subsequently, any known surface damage requires a careful inspection of the internal structure.

Minor bulging in the panels of a very light plywood structure may be acceptable. Refer to the aircraft manufacturer's repair manual for detailed specifications. However, large bulges or any indication of the skin loosening requires careful examination to determine the source and extent of the damage. It is possible for the layers of plywood to separate, or delaminate, which is indicated by a slight hump in an otherwise smooth skin. Tap the suspected area with a coin. If the tapping produces a dull thudding noise rather than a solid ringing sound, it is possible that the plywood has delaminated. Determine the extent of the damage and repair or replace the skin.

Internal Visual Inspection

The most likely place for wood deterioration to begin is the lowest point inside an aircraft's structure while the airplane is in its normal ground attitude. Dirt collects at these low points and holds the moisture against the wood until the protective coating is penetrated, wetting the wood fibers. Since wood is an organic material, it is subject to mildew and rot unless it is adequately protected from moisture. For the best protection, treat wood structures with a rot-inhibiting sealer, then, after the sealant has dried for a specified length of time, cover the entire structure with good quality varnish.

Open and examine the internal structure if there is any reason to suspect glue failure or wood rot. This may entail creating inspection openings or even removing part of the skin. If any opening must be made, use procedures that are approved by the aircraft manufacturer or by a local FAA inspector. When the area of suspected damage is accessible, carefully scrape away all of the protective coating and examine the wood and glue lines. Be suspicious of any stains in the wood. Stains usually accompany decay and wood rot. Perform a probe test in the suspected area with a sharp point, dental probe, or other similar tool. If the wood pulls up in a chunk, it is rotten. However, if the wood splinters, it is usually an indication that the wood is sound. Remove and replace any wood that shows signs of decay.

Carefully check all of the glue lines for any indication of separation. Inspect glue lines with a magnifying glass, and then try to slip a thin feeler-gauge blade into any portion of the glue line that seems to be separated. If the blade inserts into the crack, the joint is not sound and must be repaired using the methods recommended by the aircraft manufacturer. To determine whether the glue failed or if the joint was forced apart, examine the surfaces of the damaged joint. If the joint separated and the glue surface showed an imprint of the wood but no wood fibers attached to the glue, the adhesive failed. However, if something physically forced the joint apart, pieces of wood would be attached to the glue surface.

If there are any wood screws in the area where decay is suspected, remove them and check to see if they show any signs of corrosion or water stains. Replace the screw if the old one shows no indication of corrosion and the wood shows no sign of decay in or around the screw hole. Replace it with a screw of the same length but of the next larger size. Be sure that the replacement screw is made of the material specified in the aircraft's illustrated-parts manual.

Wood spars utilize reinforcement plates made of birch plywood that are glued to the ends of any splices, under the butt-end fittings, and the strut attachment fittings. Carefully inspect these plates to ensure that they have not separated from the spar. If a glue line failure is indicated between the spar and the plate, remove the plate and all traces of the glue then install a new plate.

Shake the wing to detect any looseness between the struts and the wing spar. Any movement indicates possible elongation or wear at the bolthole. In this case, remove the bolts and carefully examine them and the boltholes for wear, cracking, or elongation. An elongated bolthole or any cracks near them require that you splice in a new section of spar or replace the entire spar. The manufacturer's repair manual outlines acceptable tolerances.

If a wooden structure has been subjected to any unusual strain or extreme loads, carefully inspect the main load-carrying members for any indication of compression failure on the side that carried the compressive load. A **compression failure** usually appears as a fine line across the grain, indicating that the fibers in the wood have actually been ruptured. Replace any wood that shows this type of failure.

WOOD STRUCTURE REPAIR

The basic criterion for any aircraft repair is that the repaired structure must not only be as strong as the original structure, but the rigidity of the structure and the aerodynamic shape must also be equivalent. Materials used for the repair of a wooden structure should be the same as the original unless they have become obsolete. If substitutions are made, they must produce a repair that meets the basic requirements of the manufacturer and the FAA.

WING SPAR REPAIRS

There are several types of wooden spars that are likely to be encountered in aircraft construction. Each type of spar is unique in design and requires specific repair procedures. Reference the aircraft manufacturer's repair manual for specific repair requirements. Some of the most common wood spars include solid spars, laminated spars with rectangular cross sections, and externally routed spars with cross sections resembling I-beams. The I-beam spar is routed to reduce weight while still providing adequate strength requirements. You may also encounter built-up box spars that utilize upper and lower flanges of solid spruce with webs of plywood. Other types include built-up I-beam spars with spruce webs and flanges, as well as internally routed box-spars made of two rectangular pieces of spruce glued together then routed to reduce weight. [Figure 3-9]

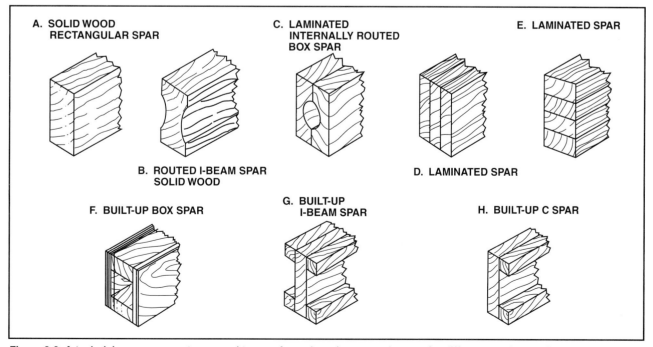

Figure 3-9. A technician can encounter several types of wooden wing spars that require different repair techniques.

SOLID WOOD SPAR REPAIRS

If an inspection reveals a **longitudinal crack in a solid wood spar**, repair it by carefully scraping away the finish on both sides of the spar and gluing reinforcing plates of spruce or plywood on each side of it. Reinforcing plates should be one-fourth as thick as the spar and extend beyond each end of the crack for at least three times the thickness of the spar. Bevel the ends of the reinforcing plates with a 5:1 taper to within 1/8-inch of the thickness of the plate, and attach the plates with glue, using no nails. Nails compromise the structure and produce moisture collection points, thus increasing the chance of wood decay. [Figure 3-10]

Splice or reinforce a wing spar at any point except under the attachment fittings for the wing root, landing gear, engine-mount, lift, or inter-plane struts. None of these fittings may overlap any part of a splice. If a splice will interfere with any of these fittings, you will have to change the design of the repair so that the spar can be repaired without interfering with the fittings. Regardless of the spar type, allow no more than two splices on a single spar.

Attachments for minor fittings, such as those for drag or anti-drag wires or compression members, are allowed to pass through a spar splice with certain restrictions. One restriction is that the reinforcement plates for the splice are not allowed to interfere with the proper attachment or alignment of fittings. These fittings include pulley support brackets, bellcrank support brackets, or control surface support brackets. Do not alter the location of these fittings in any way. A second restriction dictates that the reinforcement plates may overlap drag or anti-drag wire or compression member fittings if the reinforcement plates are located on the front face of the front spar or on the rear face of the rear spar. In these situations, use new, longer length bolts to attach the components.

If a **solid, laminated, or internally routed spar is damaged on either its top or bottom edge**, repair them, providing that all of the damage can be removed without exceeding certain limits. Clean out the damaged material to a depth of no more than one-fourth of the spar thickness. Once the damage is removed, taper the ends of the area to a 5:1 slope then insert and glue a spruce block. Finally, glue spruce or plywood reinforcing-plates to each side of the spar, making them one-fourth the thickness of the spar and tapered to a 5:1 slope. [Figure 3-11]

Splice solid or rectangular wood spars using a **scarf repair** that requires a taper of 1:10 or 1:12. Glue reinforcement plates to the end of the splice. If you decide to splice the spar without completely disassembling the wing, take special care to prepare the spar and the repair material. Cut the spar and the new material to the proper scarf angle.

Once the cut is prepared, put the two scarfed ends together and clamp them to a back-up board that is

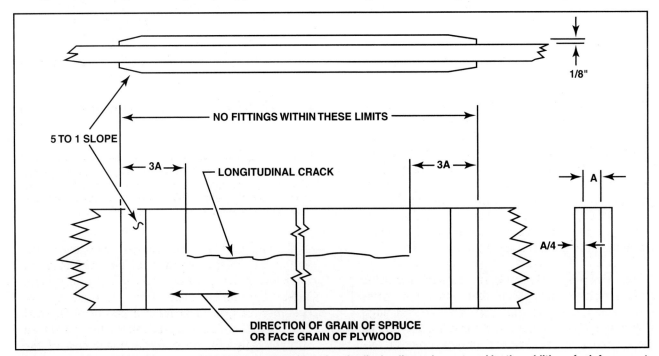

Figure 3-10. A lack of strength in a solid wood spar caused by a longitudinal split can be restored by the addition of reinforcement plates on each side of the spar.

Wood, Composite, and Transparent Plastic Structures

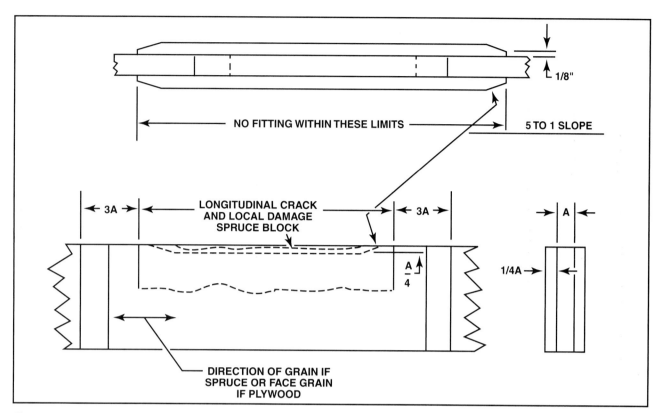

Figure 3-11. Repair of a damaged edge of a solid wood wing-spar requires replacement of the damaged material. After the plug is placed in the damaged area, glue reinforcement plates to both sides of the spar to increase its strength.

the same width as the spar and thick enough to give good, solid support. Be sure that the new material is perfectly straight and aligned with the original spar, then clamp it securely with cabinetmaker's parallel clamps or "C" clamps. Once secured, pass a fine-toothed crosscut saw through the scarf joint to remove material that does not match properly. Blow out all of the sawdust, loosen the clamps on the new material, and then tap the ends of the spar pieces to butt the two tightly together. Tighten the clamps again and make another cut to straighten both sides of the scarfed joint so the two pieces of wood match exactly. The strength of a scarf joint depends heavily on making sure the bevel cuts match precisely. To ensure a tight glue joint, use a very sharp plane or chisel to make a perfectly smooth surface with open pores. Do not use sandpaper to smooth the surface because sawdust will clog the pores and not allow the glue to properly adhere to the wood, weakening the joint. [Figure 3-12]

Spread the properly mixed glue on each prepared surface, join the pieces, and then apply even pressure, being sure that the spar is in correct edge alignment. When the glue has cured for the proper time, remove the clamps and pressure blocks and inspect the glue line, carefully cutting away any glue that squeezed from the joint. Once inspected

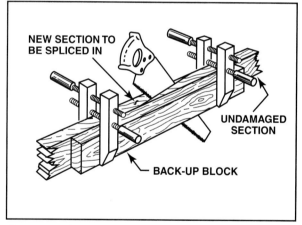

Figure 3-12. Clamping the two scarfed spar pieces together and making another scarf cut ensures a perfectly matched joint. If the initial cut produces rough or chipped wood, use a planer to smooth the surface; never sandpaper.

and cleaned, glue reinforcement plates over each end of the scarf. Make these plates one-fourth the thickness of the spar from solid spruce or plywood. Extend the reinforcement plates across the spar at six times the spar thickness on each side of the scarf line. Taper the ends of the plates to prevent an abrupt change in the cross sectional area of the repaired spar. Because the alignment of the fittings and attachments to a spar are critical, it is important

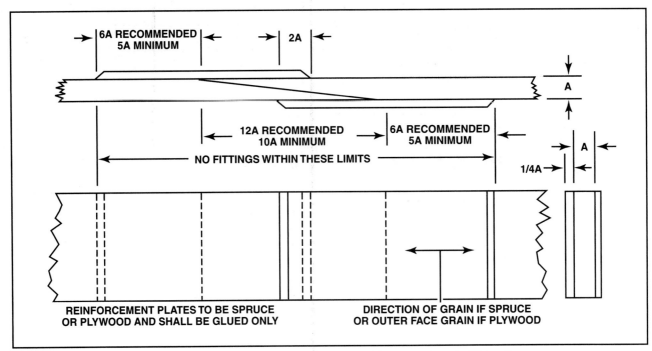

Figure 3-13. A splice for a solid wood wing-spar requires reinforcement plates on each end of the splice.

not to drill the new boltholes in the spar until the splice is completed. [Figure 3-13]

Routed I-beam spars are spliced in much the same manner as solid rectangular spars. The exception is that the reinforcement plates installed on a routed I-beam spar must be one-half the thickness of the spar web and contoured to fit into the routed portion of the spar.

A **built-up I-beam spar** repair requires a 10:1 to 12:1 scarf joint between the original spar and the new material. This type of repair requires that you place solid spruce filler-blocks in between the spar flanges for added support. It also requires plywood reinforcement plates, one-half the spar web thickness, to be glued to the spar flanges and filler blocks to make a box-type repair at the splice. [Figure 3-14]

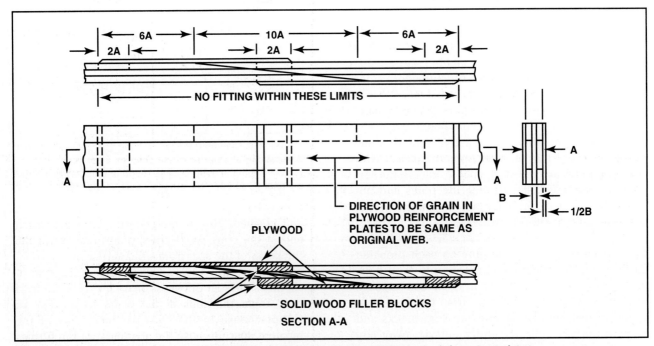

Figure 3-14. Repair to built-up I-beam wood spars includes filler blocks and plywood reinforcement plates.

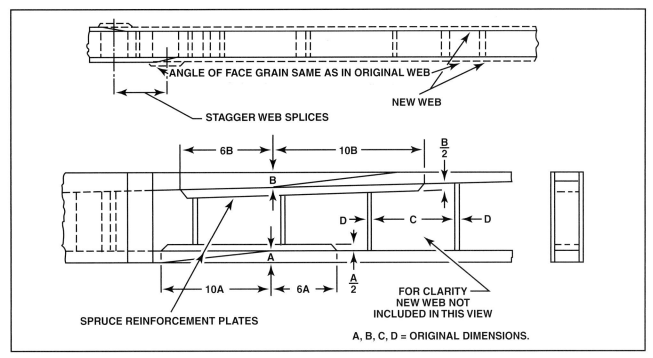

Figure 3-15. The splice for a built-up wood-box spar is critical. This type of spar carries the heaviest loads of all wood spars.

Built-up box spars carry the greatest loads of any of the wooden spars. For this reason, the built-up box spar repair is the most critical, and therefore requires the use of approved drawings as proper guidelines. The typical built-up box spar repair consists of removing portions of the webs from both sides of the spar and cutting the flanges to a 10:1 to 12:1 taper. Then spliced in new flanges and spruce reinforcement plates, one-half the thickness of the flanges, and install them on the inside of the spar. [Figure 3-15]

When replacing the webs on repaired box-spars, install spruce filler blocks that are the same thickness as the flanges between the flanges. Scarf the undamaged portion of the web to a taper of 10:1 and install a filler-block at a point centered under each scarf joint. Stagger the scarf cuts in the two webs along the spar rather than directly across from each other to improve the strength of the webs. At this point, glue and nail the filler-block in place. Cut the new web section to an exact fit, then glue and nail it in place. When the glue has cured and all of the excess glue is removed, glue and nail a plywood cover strip over the end of the splice. [Figure 3-16]

WING RIB REPAIRS

Wood wing ribs are usually made of spruce strips that have a cross section of approximately 1/4 to 5/16-inch. These small strips of wood accept the air loads from the covering of the wing and transmit them into the spars.

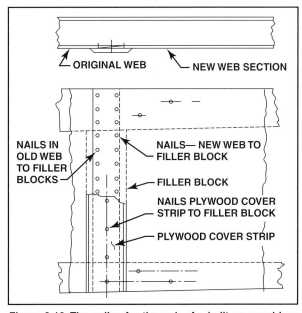

Figure 3-16. The splice for the web of a built-up wood box spar incorporates both filler blocks and plywood cover strips.

When manufacturing wing ribs, soften the upper and lower cap strips with steam before bending them over a form. Cap strips are the upper and lower surfaces that attach to the supporting web. These strips carry the bending loads of the wing and provide a surface for attaching the wing skin. When dry, place them in a jig and cut all of the cross members to fit between them. Cover each intersection between a vertical member and a cap strip with a gusset made of mahogany plywood. Glue the gussets to the strips and secure them with brads to provide the pressure needed to make strong glue joints. Slip the completed ribs over the spars and assemble and square up the wing truss with the drag and anti-drag wires adjusted to the proper tension.

CAP STRIP REPAIRS

If a cap strip is broken between two of the upright members, cut the strip to a taper 10 to 12 times its thickness. Then cut a new piece of the same type material with a matching taper. Cut a reinforcing block of spruce the same size as the cap strip and 16 times as long as its width, and glue it to the inside of the cap strip. Then cover both the cap strip and the reinforcement with plywood faceplates that are glued to the strip and held with brads. [Figure 3-17]

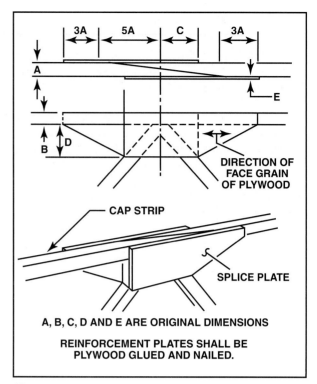

Figure 3-18. Add splice plates to repair a cap strip broken over an upright member. When cutting splice plates, ensure that the grain is parallel to the grain of the cap strips to reduce the chance of warpage.

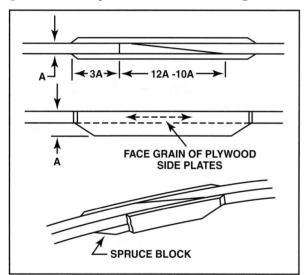

Figure 3-17. Use a rib cap-strip splice to repair a cap strip broken between two upright members.

If the damage is located above one of the upright members, cut the cap strip with a 10:1 to 12:1 taper with the center of the cut over the upright member. Then splice a new piece of cap-strip material into the structure. The upright member serves as the reinforcement so no block is needed under the splice. Put splice plates of thin plywood on each side of the splice so that none of the joints depends upon end-grain gluing. [Figure 3-18]

It is sometimes necessary to replace only the leading or trailing edge portion of a rib, so cap strips may be cut at a spar. When the cap strips are cut over a spar, use a 10:1 to 12:1 taper, and glue gussets of plywood the same size as the original to each side of the rib. [Figure 3-19]

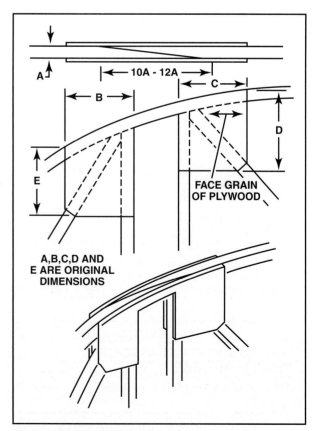

Figure 3-19. Use original size gussets on a rib-cap splice that is located over a spar.

TRAILING EDGE RIB REPAIR

The trailing edge of a wing is the area most likely to be damaged by moisture collecting and causing the wood to rot. All wings must incorporate drainage grommets at the lowest part of each rib bay to drain accumulated moisture. Drainage grommets also ventilate the compartment to prevent condensation. Occasionally, grommets will clog with dirt and not allow adequate drainage. Subsequently, moisture will collect around the wood structures producing an environment ripe for decay. If there is any movement when you flex the trailing edge, cut away the trailing edge fabric and examine the edge structure. If the rear end of the rib has rotted, cut it away and cut a spruce block to fit the removed rib section. Then cut reinforcing plates of plywood, glue them into place, and fasten new trailing edge finishing materials to the repaired rib. [Figure 3-20]

Be sure to treat the wood repair with a rot-resistant sealer before re-covering the structure. When replacing corroded metal sections of the support structure, protect the new metal parts with a corrosion inhibiting primer such as zinc chromate.

Most wooden wings utilize metal compression members. However, wooden compression members are used on certain aircraft. When it is necessary to repair wooden compression members, use a 10:1 to 12:1 scarf joint. Glue reinforcing plates to each side of the splice that are made of the same material as the strip and 12 times its thickness. Then cover the entire repair with plywood to form a boxed rib. [Figure 3-21]

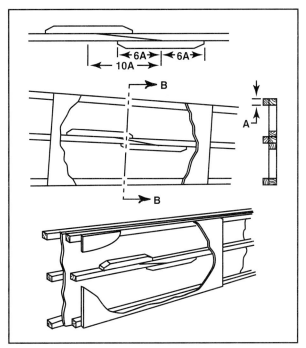

Figure 3-21. Compression rib repairs require reinforcement of the entire splice with plywood to restore the original strength.

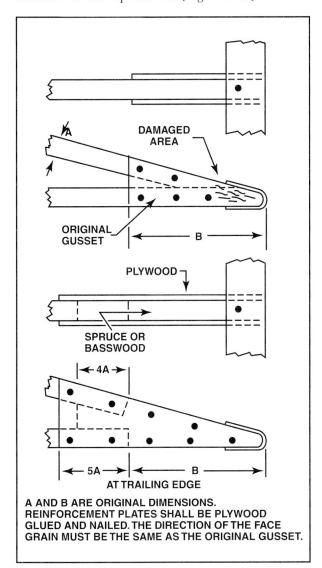

Figure 3-20. Repair to the trailing edge of a wood wing rib requires replacement of the damaged portion with a wood block. Install a larger gusset to reinforce the structure.

PLYWOOD SKIN REPAIRS

Aircraft that incorporate plywood skins normally carry a large amount of stress from the flight loads. Therefore, make repairs to plywood skins in strict accordance with the recommendations of the aircraft manufacturer. If you repair plywood skin exactly as described by the manufacturer or by Advisory Circular 43.13-1B, *Acceptable Methods, Techniques, and Practices, Aircraft Inspection and Repair*, the FAA will most likely approve the repair. If the repair cannot be made according to the approved data, contact the district agent of the FAA for approval of the proposed method before beginning the work.

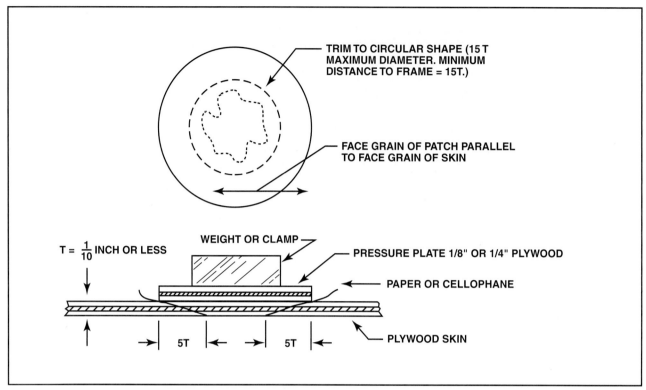

Figure 3-22. Use a splayed patch to repair small holes in thin plywood skin.

Use circular or elliptical plywood patches in plywood skin repair to avoid the stress concentrations developed by abrupt changes in the cross-sectional areas of square or rectangular patches. Following are several types of plywood patches approved for aircraft applications.

SPLAYED PATCH

Small holes in thin plywood skin may be repaired by a splayed patch. Use this type of patch if the skin is less than or equal to 1/10-inch thick and the hole can be cleaned out to a diameter of less than 15 thicknesses (15T). [Figure 3-22]

To fabricate a splayed patch, tape a small piece of scrap plywood over the center of the damage. Use it as a rest for the point of a drafting compass, and draw two circles. Draw one circle to form the trim size of the hole, which can be no more than 15T. For the other circle, the size of the outside of the patch can be no more than 5T beyond the edge of the hole.

To produce the patch, remove the inner circle with a sharp knife, and then, using a chisel, taper the edges evenly from the outer circle to the edge of the hole. Cut the patch plug from the same material as the original skin and taper it to fit the hole exactly. Apply glue to the tapered edges of the hole and to the taper cut on the patch. Put the patch in place, aligning the face grain of the patch with the face grain of the skin. Once installed, place a piece of vinyl plastic or waxed paper over the patch. With a pressure plate cut from scrap plywood that is just slightly larger than the patch, apply pressure and allow the glue to cure. After the glue has cured, remove the pressure plate, fill, sand, and finish the repair to match the rest of the surface.

SURFACE PATCH

If an airplane's plywood skin is damaged, repair it with a surface patch covered with aircraft fabric and finish it to match the rest of the airplane. This does not produce the best looking repair, but its simplicity and economy of time and labor make it a suitable repair for most working-type airplanes. [Figure 3-23]

PLUG PATCH

Make a perfectly flush patch in a section of plywood skin by trimming the damage to a round or oval shape. Put a doubler inside the structure for support, then glue the plug patch to the doubler. [Figure 3-24]

SCARFED PATCH

The most difficult type of patch to make on plywood skin is the scarfed patch. However, because it makes the least change in skin thickness or rigidity, it is preferred for most stressed wood skin repairs. [Figure 3-25]

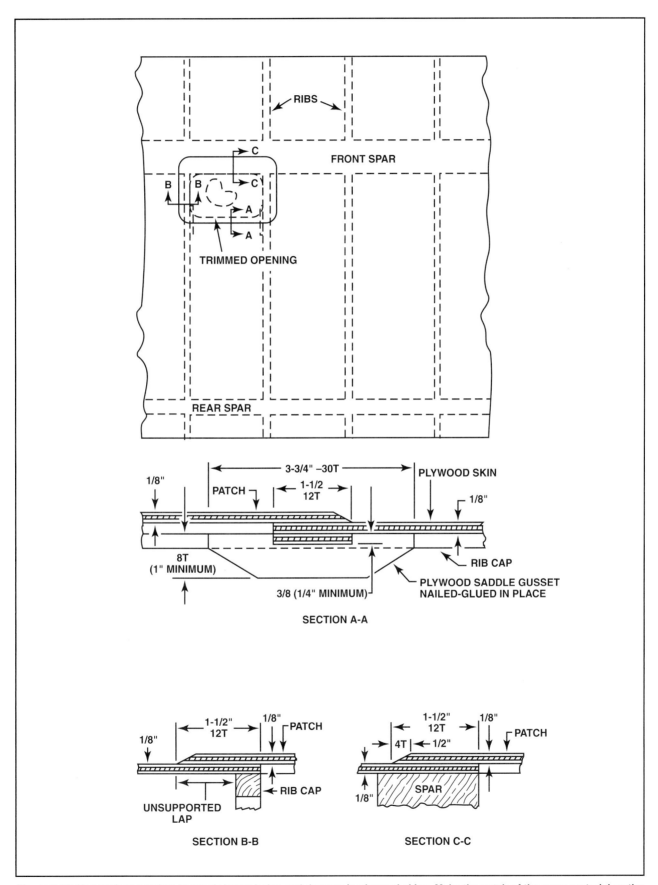

Figure 3-23. Use surface patches to repair larger holes and damage in plywood skins. Make the patch of the same material as the damaged skin and run its face grain in the same direction as that of the skin.

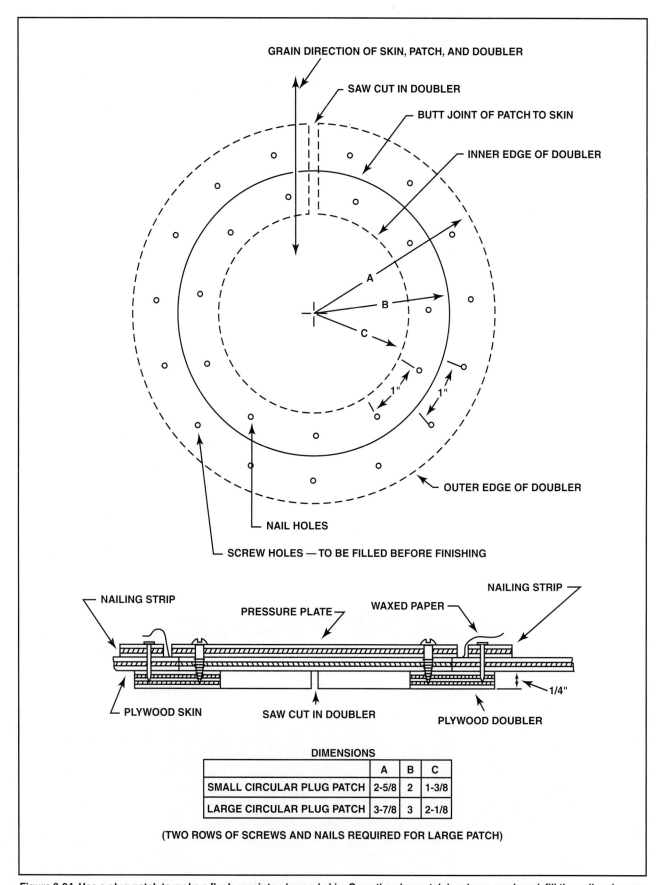

Figure 3-24. Use a plug patch to make a flush repair to plywood skin. Once the plug patch has been produced, fill the nail and screw holes with wood filler, as well as any space between the patch and skin for protection against wood decay.

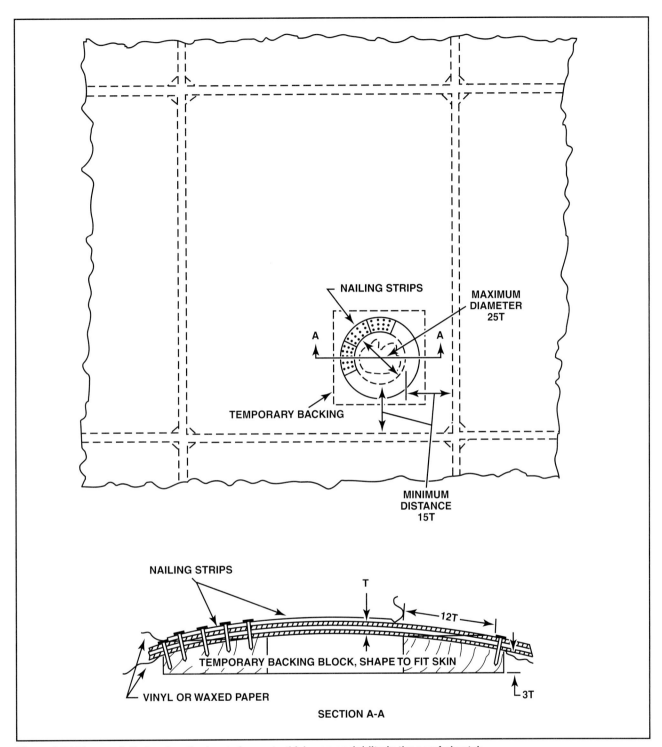

Figure 3-25. The repair that makes the least change to thickness or rigidity is the scarfed patch.

SECTION B
COMPOSITE STRUCTURES

Composites are combinations of two or more materials that differ in composition or form. The constituents or elements that make up the composite retain their individual identities. In other words, the individual elements do not dissolve or otherwise merge into each other. Each can be physically identified, and exhibits a boundary between each other.

Composite structures differ from metallic structures in several ways: excellent elastic properties, ability to be customized in strength and stiffness, damage tolerance characteristics, and sensitivity to environmental factors. Consequently, composites require a vastly different approach from metals with regard to their design, fabrication and assembly, quality control, and maintenance.

One main advantage to using a composite over a metal structure is its high strength-to-weight ratio. Weight reduction is a primary objective when designing structures using composite materials. In addition, the use of composites allows the formation of complex, aerodynamically contoured shapes, reducing drag and significantly extending the range of the aircraft. Composite strength depends upon the type of fibers and bonding materials used, and how the part is engineered to distribute and withstand specific stresses.

COMPOSITE ELEMENTS

In aircraft construction, most currently produced composites consist of a reinforcing material to provide the structural strength, joined with a matrix material to serve as the bonding substance. In addition, adding core material saves overall weight and gives shape to the structure. The three main parts of a fiber-reinforced composite are the fiber, matrix, and interface or boundary between the individual elements of the composite.

REINFORCING FIBERS

Reinforcing fibers provide the primary structural strength to the composite structure when combined with a matrix. Reinforcing fibers can be used in conjunction with one another (hybrids), woven into specific patterns (fiber science), combined with other materials such as rigid foams (sandwich structures), or simply used in combination with various matrix materials. Each type of composite combination provides specific advantages. Following are the five most common types of reinforcing fibers used in aircraft composites.

FIBERGLASS (GLASS CLOTH)

Fiberglass is made from small strands of molten silica glass that are spun together and woven into cloth. Many different weaves of fiberglass are available, depending on a particular application.

One of the disadvantages of fiberglass is that it weighs more and has less strength than most other composite fibers. In the past, fiberglass was limited to nonstructural applications. The weave was heavy and polyester resins were used, which made the part brittle. However, with newly developed matrix formulas, fiberglass is an excellent reinforcing fiber currently used in advanced composite applications.

The two most common types of fiberglass are S-glass and E-glass. E-glass, otherwise known as "electric glass" because of its high resistivity to current flow, is produced from borosilicate glass and is the most common type of fiberglass used for reinforcement. S-glass is produced from magnesia-alumina-silicate, and is used where a very high tensile strength fiberglass is needed. [Figure 3-26]

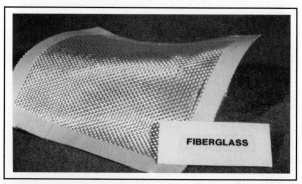

Figure 3-26. Fiberglass is usually a white gleaming cloth. The widespread availability of fiberglass and its low cost make it one of the most common reinforcing fibers utilized in aircraft non-structural composites.

ARAMID

In the early 1970s, DuPont® introduced aramid, an organic aromatic-polymide polymer, commercially known as Kevlar®. Aramid exhibits high tensile strength, exceptional flexibility, high tensile stiffness, low compressive properties, and excellent toughness. The tensile strength of Kevlar® composite material is approximately four times greater than alloyed aluminum. Aramid fibers are non-conductive and produce no galvanic reaction with metals. Another important advantage is its strength-to-weight ratio; it is very light compared to other composite materials. Aramid-reinforced composites also demonstrate excellent vibration-damping characteristics in addition to a high degree of shatter and fatigue resistance. [Figure 3-27]

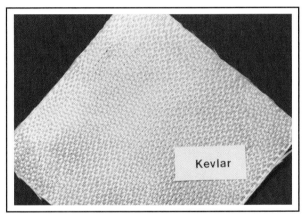

Figure 3-27. Aramid fiber is usually characterized by its yellow color, and as with most reinforcing fibers, comes in various grades and weaves for different uses. Kevlar 49® is predominantly used in aircraft composite reinforced plastics; both in thermoplastic and thermosetting resin systems.

Aramid is ideal for use in aircraft parts that are subject to high stress and vibration. For example, some advanced helicopter designs have made use of aramid materials to fabricate main rotor blades and hub assemblies. Flexibility of the aramid fabric allows the blade to bend and twist in flight, absorbing much of the stress. In contrast, a blade made of metal develops fatigue and stress cracks more frequently under the same conditions.

A disadvantage to aramid is that it stretches, which can cause problems when it is cut. Drilling aramid can also be a problem if the drill bit grabs a fiber and pulls until it stretches to its breaking point. When cutting aramid fabrics, the material will look fuzzy if inappropriate tools are used. Fuzzy material left around fastener holes or seams may act as a wick and absorb moisture or other liquid contaminants such as oil, fuel, or hydraulic fluid. Liquid contaminates may deteriorate the resin materials in the composite structure, producing delamination. It is important to cut aramid cloth correctly because even a slight amount of moisture will prevent aramid from bonding properly. Fuzz around the drilled hole may also prevent a fastener from seating properly, which may cause joint failure.

CARBON/GRAPHITE

Carbon fibers are produced in an inert atmosphere by the pyrolysis of organic fibers such as rayon, polyacrylonitrile, and pitch. The term **carbon** is often interchangeable with the term **graphite**. However, carbon fibers and graphite fibers differ in the temperature at which they are produced. Carbon fibers are typically carbonized at approximately 2400° F and composed of 93% to 95% carbon, while graphite fibers are produced at approximately 3450° to 5450° F and are more than 99% carbon. [Figure 3-28]

Figure 3-28. In general, Americans refer to carbon fibers as "graphite" fiber, while Europeans refer to it as carbon fiber. Carbon actually describes the fiber more correctly, because it contains no graphite structure. Carbon/graphite is a black fiber that is very strong, stiff, and used primarily for its rigid strength characteristics. Fiber composites are used to fabricate primary structural components such as the ribs and skin surfaces of the wings.

Advantages to carbon/graphite materials are in their high compressive strength and degree of stiffness. However, carbon fiber is cathodic while aluminum and steel are anodic. Thus, carbon promotes galvanic corrosion when bonded to aluminum or steel, and special corrosion control techniques are needed to prevent this occurrence. Carbon/graphite materials are kept separate from aluminum components when sealants and corrosion barriers, such as fiberglass, are placed at the interfaces between composites and metals. To further resist galvanic corrosion, anodize, prime, and paint any aluminum surfaces prior to assembly with carbon/graphite material.

BORON

Boron fibers are made by depositing the element boron onto a thin filament of tungsten. The resulting fiber is approximately .004 inch in diameter, has excellent compressive strength and stiffness, and is extremely hard. However, boron is not commonly used in civil aviation because it can be hazardous to work with, and is extremely expensive. In designing components that need both the strength and stiffness associated with boron, many civil aviation manufacturers are utilizing hybrid composite materials of aramid and carbon/graphite instead of boron.

CERAMIC

Ceramic fibers are used where a high-temperature application is needed. This form of composite will retain most of its strength and flexibility at temperatures up to 2,200° F. Tiles on the Space Shuttle are made of a special ceramic composite that dissipates heat quickly. Some firewalls are also made of ceramic-fiber composites. The most common use of ceramic fibers in civilian aviation is in combination with a metal matrix for high-temperature applications.

FIBER SCIENCE

The strength of a reinforcing material within a composite is dependent upon the weave of the material, the wetting process (how the matrix is applied), filament tensile strength, and the design of the part. The tensile strength of fabrics as it is reported in many articles and books is usually the strength of the raw fabric only. However, aircraft composites incorporate a resin material. This decreases the overall strength, because resins tend to make the structure more brittle and lessen the tensile strength. Arranging the fibers in various orientations helps to ensure adequate component strength partially corrects this reduction in strength.

The strength and stiffness of a composite buildup depends upon the orientation of the plies relative to the load direction while a sheet metal component will have the same strength no matter in which direction it is tested. For example, a helicopter rotor blade has high stress along its length because of the centripetal forces emanating from the rotating mass of the blade. If the blade is made of metal, its strength is the same in all directions. In the case of a composite blade, the majority of fibers run the length of the blade to give more strength in the direction of the greatest stress.

In another example, if a wing in flight bends as well as twists, the part can be manufactured so that fibers will run the length of the wing to reduce its tendency to bend. By adding a layer of fibers that run at 45° and at 90°, twisting forces can also be limited. In this manner, each layer may have the major fibers running in a different direction. The strength of fibers runs parallel to the direction that the threads run, allowing designers to customize the strength objective for the type of stress that the part might encounter.

FABRIC ORIENTATION

When working with composite fibers, it is important to understand the construction and orientation of the fabric because all design, manufacturing, and repair work begins with the orientation of the fabric. Unlike metallic structures, the strength of a composite structure relies on the proper placement and use of the reinforcing fibers. Some of the terms used to describe fiber orientation are warp, weft, selvage edge and bias. [Figure 3-29]

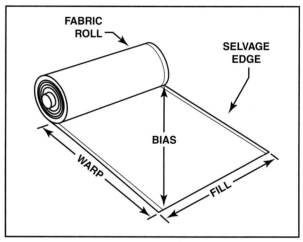

Figure 3-29. All design, manufacturing, and repair work begins with the orientation of the fabric. Unlike metallic structures, the strength of a composite structure relies on the proper placement and use of the reinforcing fibers.

Warp

The warp of threads in a section of fabric run the length of the fabric as it comes off the roll or bolt. Warp direction is designated as 0°. There are typically more threads woven into the warp direction than the fill direction, making it stronger in the warp direction. Because warp is critical in fabricating or repairing composites, insertion of another color or type of thread at periodic intervals identifies the warp direction. Marked plastic backings on the underside of **pre-impregnated** fabrics also identify the orientation of the warp threads. Pre-impregnated fabrics are pre-impregnated with resins by the manufacturer and later cured in the field.

Weft/Fill
Weft or fill threads of the fabric are those that run perpendicular (90°) to the warp fibers. The weft threads interweave with the warp threads to create the reinforcing cloth.

Selvage Edge
The selvage edge of the fabric is the tightly woven edge parallel to the warp direction, which prevents edges from unraveling. The selvage edge is removed before the fabric is utilized. The weave of the selvage edge is different from the body of the fabric and does not have the same strength characteristics as the rest of the fabric.

Bias
The bias is the fiber orientation that runs at a 45° angle (diagonal) to the warp threads. The bias allows for manipulation of the fabric to form contoured shapes. Fabrics can often be stretched along the bias but seldom along the warp or fill.

FABRIC STYLES
Fabrics used in composite construction are manufactured in several different styles: unidirectional, bi-directional, multidirectional, and mat. Component designers can use any or all of these fabric styles, depending on the strength and flexibility requirements of the component part.

Unidirectional
Unidirectional fiber orientation is one in which all of the major fibers run in one direction, giving the majority of its strength in a single direction. This type of fabric is not woven together, meaning that there are no fill fibers. Occasionally, small cross threads are used to hold the major fiber bundles in place. However, the cross threads are not considered woven fibers. [Figure 3-30]

Carbon/graphite materials are sometimes formed into 12-inch wide tapes, which are unidirectional and usually pre-impregnated with resin. Tapes are less expensive than fabric and create a smoother surface. However, for repair work, fabric is usually utilized in place of tapes.

Bi-directional/Multi-directional
Bi-directional or multi-directional fabric orientation calls for the fibers to run in two or more directions. Bi-directional fabrics are woven with the warp threads usually outnumbering the weft, so there is usually more strength in the warp direction than the fill. When using bi-directional fabrics, it is important to align the warp threads when performing a repair, due to the differences in the strength properties of the warp and weft directions.

Another type of bi-directional material is the **intraply hybrid** fabric. This type of fabric is woven from different types of fibers. Intraply hybrid fabrics give composites specific strength, flexibility, and durability characteristics, depending on the combination and proportion of the fibers woven together. A particular structural design may call for different combinations of fibers. Carbon/fiberglass intraply-hybrid fabrics, for example, provide increased stiffness due to the carbon and increased heat resistance due to the fiberglass. Specific characteristics of the final fabric are dependent on the types and proportions of the fibers woven together.

Mats
Mat fabrics consist of chopped fibers compressed together and typically used in combination with woven or unidirectional fabrics. A mat is not as strong as a unidirectional or bi-directional fabric, and is therefore is not commonly used alone in repair work. [Figure 3-31]

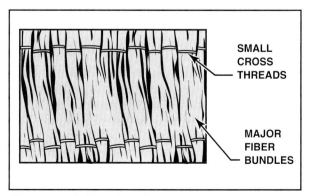

Figure 3-30. Unidirectional fabrics are not woven together. Warp fibers run parallel to each other and are kept in place by small cross threads. The strength of this type of fabric lies in the warp direction, making it imperative to properly lay out the fabric.

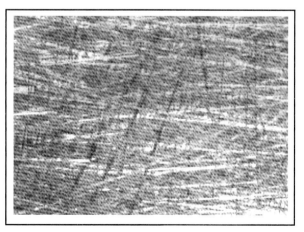

Figure 3-31. Fiberglass mat provides the high strength of glass to reinforce thermosetting resins without the expense of woven cloth.

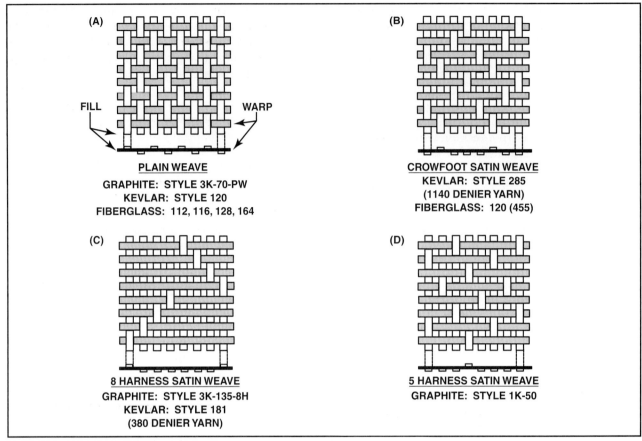

Figure 3-32. The most common weave styles employed in aircraft applications are the plain, crowfoot satin, five-harness satin, and eight-harness satin weaves. While each type of fabric is not necessarily available in all weaves, each fabric that is produced with its own style number. For instance, Kevlar® and carbon/graphite plain weaves utilize different style numbers for the same type weave. Therefore, make sure you use the correct type of material called for by the aircraft manufacturer's structural-repair manual.

FABRIC WEAVES

Woven fabrics are more resistant to fiber breakout, delamination, and damage than unidirectional materials. Because of the wide variety of uses and strength requirements, composite fabrics are available in many weaves. [Figure 3-32]

For a given type of fiber, the fabric's physical stability decreases while its drapability increases progressively from the plain weave to the eight-harness satin weave. To maintain a satisfactory level of stability, more fibers per inch must be added progressively toward the eight-harness satin weave fabric. Thus, plain weaving fibers produce the most stable, least drapable, and lightest fabrics. [Figure 3-33]

MATRIX SYSTEMS

The function of the matrix in a composite is to hold the reinforcing fibers in a desired position. It also gives the composite strength and transfers external stresses to the fibers. The ability of the matrix to transfer stress is the key to the strength of a composite structure.

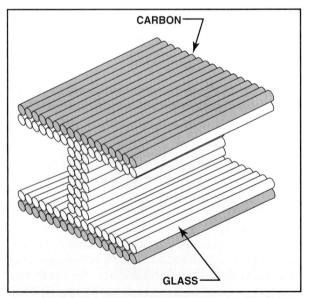

Figure 3-33. Selectively placing different types of fibers in a composite structure may provide greater strength, flexibility, and overall reduced cost. An I-beam, for example, may use carbon/graphite fibers where stiffness is desired, and selectively blend in fiberglass to reduce the cost of the structure.

A wide range of resin systems are used for the matrix portion of fiber reinforced composites. **Resin** is an organic polymer used as a matrix to contain the reinforcing fibers in a composite material. Polyester resin, an example of an earlier matrix, used in conjunction with fiberglass has been used in many nonstructural applications such as fairings and spinners. The old polyester/fiberglass formulas did not offer sufficient strength to fabricate primary structural members. Newer matrix materials display remarkably improved stress distributing characteristics, heat resistance, chemical resistance, and durability.

Resin matrix systems are a type of plastic and include two general categories: thermoplastic and thermosetting. Thermoplastic and thermosetting resins by themselves do not have sufficient strength for use in structural applications. However, plastic matrixes reinforced with other materials form high-strength, lightweight structural composites.

THERMOPLASTIC RESINS

Thermoplastic resins use heat to form the part into the desired shape. However, this shape is not necessarily permanent. If a thermoplastic resin is reheated, it will soften and could easily change shape. One example of a thermoplastic resin is Plexiglas®, which is used to form light aircraft windshields. The shape of the windshield remains fixed after it is cooled at the factory. However, if the windshield is sufficiently reheated, the plastic will melt, which in turn changes its shape.

THERMOSETTING RESINS

Thermosetting resins use heat to form and irreversibly set the shape of the part. Thermosetting plastics, once cured, cannot be reformed even if they are reheated. At this time, most structural airframe applications are constructed with thermosetting resins.

Polyester Resin

Polyester resin, an early thermosetting matrix formula, is mainly used with fiberglass composites to create nonstructural applications such as fairings, spinners, and aircraft trim. While fiberglass possesses many virtues, its greatest limitation lies in its lack of structural rigidity. Polyester resins give fiberglass cohesiveness and rigidity. However, polyester resin/fiberglass composites do not offer sufficient strength to fabricate primary structural members.

Like other plastics, polyester shrinks when cured. While this inherent characteristic helps in some ways, it hurts in others. For instance, when bonding a metal structure to a fiberglass structure, the shrinkage can be helpful. As the polyester resin shrinks, it produces an increasingly tight grip on the embedded metal. However, when installing metal hinges on top of a large, flat surface using long strips of fiberglass and polyester resin, shrinkage may warp the surface. Bonding shorter strips of fiberglass to the fasteners or using a resin with a smaller shrinkage factor such as an epoxy resin usually prevents this type of warpage.

Epoxy Resins

Most of the newer aircraft composite matrix-formulas utilize epoxy resins, which are thermoset plastic resins. Epoxy resin matrices are two-part systems consisting of a resin and a catalyst. The catalyst acts as a curing agent by initiating the chemical reaction of the hardening epoxy. Epoxy resin systems are well known for their outstanding adhesion, strength, and resistance to moisture and chemicals. They are also useful for bonding nonporous and dissimilar materials, such as metal parts to composite components. The manner in which the joints are designed, and how the surfaces are prepared, determines the quality of the bond.

Each epoxy composite system is designed for a specific purpose. For example, a cowling may use an epoxy resin that will withstand high temperatures, while an aileron may require one made to withstand bending stresses. Both use epoxy resin systems but are very different in their chemical makeup, producing structures with different characteristics. Thus, not every type of epoxy resin is suitable for every type of structure or repair. Make sure to use the proper resin called for in the manufacturer's repair manual. Never use an epoxy resin that is not approved for aircraft use, as the strength, flexibility, and moisture resistance qualities cannot be guaranteed.

Some of the properties of epoxy which make it useful for bonded structures are its low shrinkage percentage, high strength-to-weight ratio, exceptional chemical resistance, and ability to adhere to an almost endless variety of materials. Epoxy forms an extremely tight bond between glass and metal. However, if epoxy is used to bond glass to a metal window frame, the glass will crack from temperature changes because of the different expansion rates of the metal and the glass.

Epoxies may be used in place of polyester resins for almost any application. They also have a long shelf life. Unmixed, epoxies generally keep for almost a year at 72° F. Once they are mixed, however, they have a very short **pot life**, which is the amount of time a catalyzed resin remains in a workable state.

ADHESIVES

Resins come in different forms. Resins used for laminating are generally thinner, to allow proper saturation of the reinforcing fibers. Others are used for bonding and are typically known as adhesives because they glue parts together. Adhesive resins and catalysts are available either in pre-mixed quantities or in separate containers.

One of the most unique forms of adhesive is the **film adhesive**. This type of adhesive pre-blends the resin and catalyst on a thin film of plastic. Refrigeration of the film is required to slow the cure rate (the rate of change to its permanent form) of the resin. If left out at room temperature, the resin and catalyst will start to cure. In the freezer, the curing process slows down, lengthening the shelf life of the film. Adhesive films are often used to help bond patches to a repair area. [Figure 3-34]

Another form of adhesive is available in foam, which is primarily used to splice replacement honeycomb core segments to existing cores. When heat is applied to the adhesive, it foams up and expands into crevices. These types of foaming adhesives can also be used to permanently install fasteners.

PRE-IMPREGNATED MATERIALS

Pre-impregnated fabrics, commonly known as "pro-pregs," are fabrics that have the resin system already saturated into the fabric. Because many epoxy resins have high viscosity, it is often difficult to mix and work epoxy resins into the fabric to completely encapsulate the fibers. Fabrics are pre-impregnated with the proper amount and weight of a resin matrix to eliminate the mixing and application details such as proper mix ratios and application procedures. [Figure 3-35]

In addition to woven fabrics, manufacturers also pre-impregnate unidirectional materials. Pre-impregnating unidirectional fabrics involves saturating the fibers with resin directly from individual spools of thread. [Figure 3-36]

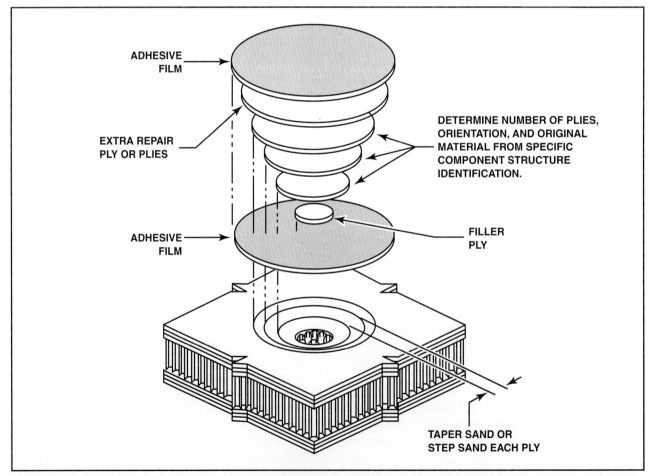

Figure 3-34. When performing a repair to a composite structure, an adhesive film is used many times to help bond laminate patches to the existing part. After installing the first layer of film, apply heat with a heat gun to soften the film. Remove the plastic backing and place the repair plies over the adhesive layer. Finally, apply a final layer of adhesive film over the repair plies to glue the patch in place.

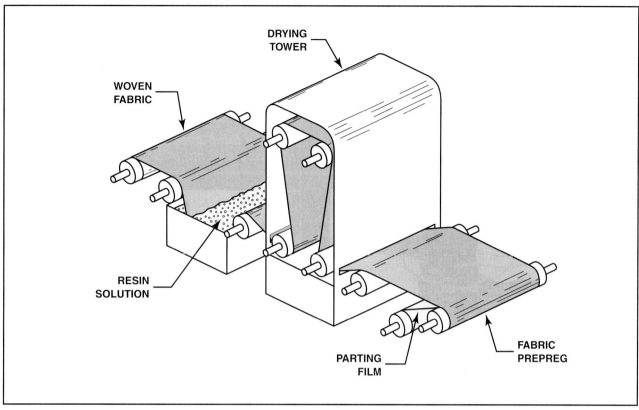

Figure 3-35. Manufacturers often prepare these fabrics by dipping the woven fabric into a resin solution containing the proper amount of resin and catalyst, weighed and mixed together. A catalyst is a substance that changes the rate of a chemical reaction. The wet fabric is then placed into a drying tower, where any excess resin is removed and the fabric is allowed to dry to a somewhat sticky consistency. One or both sides of the resin receives a parting film that prevents the fabric from sticking when it is rolled for storage and shipping. Parting films also help prevent contamination of the resin.

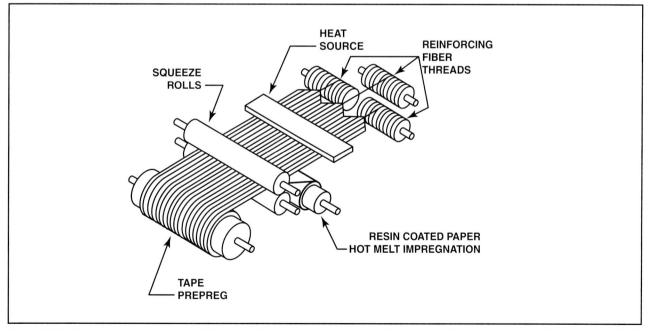

Figure 3-36. When pre-impregnating unidirectional fabrics, each reinforcing fiber is placed in the correct orientation from individual spools of thread. The fibers are then heated while a resin-coated paper is applied to one side of the threads. The heat melts the resin from the paper and soaks the fibers. The paper and threads are then squeezed together to ensure thorough saturation of the threads. The finished material is then wound onto a roll and is ready for use.

Pre-preg materials offer convenience over raw fabrics in many ways. As previously discussed, these materials contain the proper amount of matrix. It does not produce a resin-rich or resin-lean component if cured properly. Before curing, pre- impregnated fabrics generally contain about 50% resin. During the curing process, some of this resin bleeds out of the reinforcing fibers, producing a structure that contains about 40% resin and 60% fibers by weight.

Another advantage of pre-impregnated materials is that the matrix completely encapsulates the reinforcing fibers. During hand lay-up, if a resin system has a high viscosity, it is sometimes difficult to incorporate the resin into and around each individual fiber. This is not a problem with pre-impregnated fabrics. In addition, the technician does not have to worry about distorting the weave while working the resin into the fabric. These fabrics also eliminate the manual weighing and mixing requirements of the resin and catalyst, which not only saves time during the lay-up process, but also helps ensure the quality of the lay-up. In many cases, pre-impregnated materials produce stronger components and repairs, assuming the lay-up processes are strictly followed.

One limitation to pre-impregnated materials is that they must be stored in a freezer to prevent the resin from curing. Pre-preg fabrics cannot be left out of the cold for prolonged periods, and must be warmed slightly before use to achieve better workability. Accordingly, with many of these materials, it is necessary to maintain an accurate log of the "out-of-freezer life limit" (how long the material has been exposed to temperatures above freezing). Keep in mind that, although cold storage of a pre-impregnated fabric helps increase its life limit, most materials still have a shelf life regardless of the storage condition.

Another disadvantage associated with pre-impregnated materials is that they are usually purchased in full roll quantities. Unless a large number of repairs are made, the roll may exceed its shelf life before being used. Although the material may appear to be in good condition, it cannot be used for aircraft applications once the shelf life has expired.

FILLERS
Fillers, also known as **thixotropic agents**, are materials added to resins to control viscosity and weight, to increase pot life and cured strength, and to make the application of the resin easier. Fillers increase the volume of the resin, making it less dense and less susceptible to cracking, as well as lowering the weight of the material. Most fillers are inert and will not react chemically with the resin. Microballoons, chopped fibers, and flox are common types of fillers used in composite construction.

Microballoons are small spheres manufactured from plastic or glass. **Plastic microballoons** must be mixed with a compatible resin system that will not dissolve the plastic. **Glass microballoons**, on the other hand, are not affected by resin mixtures, making them the primary thixotropic agent used in composite construction. The advantages to using microballoons are that they provide greater concentrations of resin in the edges and corners of the structure, they are less dense, which reduces the overall weight, and they provide lower stress concentrations throughout the structure. However, microballoons do not add strength to the composite structure.

Chopped fibers and flox can also be added as fillers and have the advantage of adding strength to the cured mixture. Chopped fibers are made from any type of fiber cut into certain lengths, commonly 1/4 to 1/2-inch lengths. Flox is the fuzzy fiber taken from the fabric strands. Both chopped fibers and flox may be used when added strength is desired. For example, if a hole is accidentally drilled in the wrong place in a composite structure, filling the hole with a mixture of resin and flox provides more strength than pure resin. Using pure epoxy resin produces brittle and heavy plugs.

METAL MATRIX COMPOSITES
A metal matrix composite incorporates metal in the matrix, instead of a plastic resin or ceramic material. Metal matrices provide specific stiffness, improved fatigue, strength, and wear resistance, along with improved thermal characteristics. However, they are not readily used in the aviation field presently; they are still in the experimental stage. Several considerations being analyzed are the potential reinforcement, matrix corrosion reactions, and thermal stresses due to the mismatch of thermal expansions between the reinforcements and the matrix. A large difference in melting temperature between the matrix and the reinforcements may result in matrix creep while the reinforcement remains elastic.

CORE MATERIALS
Core materials are the central members of an assembly and are used extensively in advanced composite construction. When bonded between two thin face sheets, a component can be made rigid and lightweight.

Composite structures manufactured in this manner are sometimes referred to as **sandwich construction**.

A core material gives a great deal of compressive strength to a structure. For example, the sheet metal skin on a rotor blade has a tendency to flex in flight. This constant flexing causes metal fatigue. A composite blade with a central core material provides uniform stiffness throughout the blade and eliminates most of the flexing associated with metal blades. [Figure 3-37]

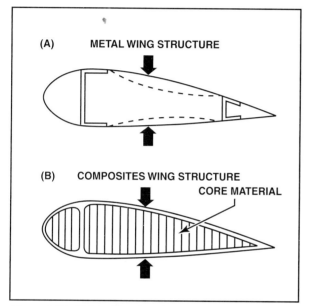

Figure 3-37. Metal skins bend and flex when forces are applied to it in flight. The use of composite construction keeps the structure from flexing, eliminating fatigue.

eycomb. It is important to line up the ribbon direction of the replacement honeycomb core with that of the original when performing a repair honeycomb core repair to ensure consistent structural strength along with uniform compressive strength. [Figure 3-38]

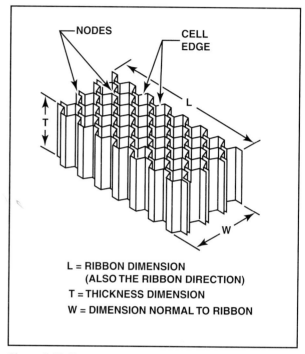

Figure 3-38. Honeycomb comes in a variety of core configurations. Some are more flexible than others and may be bent to form a curve. Crimping the core material into ribbons and joining them together forms the honeycomb shape.

The two most common types of core materials utilized in sandwich construction are honeycomb and foam cores. In addition, wood cores are also occasionally used in composite construction.

HONEYCOMB CORES
Honeycomb core materials consist of the six-sided shape of a natural honeycomb, which provides a core with a very high strength-to-weight ratio. Manufacturers construct honeycomb cores from aluminum, Kevlar®, carbon, fiberglass, paper, and steel. Nomex®, a paper impregnated material, is also widely used as an advanced composite core material.

The **ribbon direction** of a honeycomb core is the direction in which the honeycomb can be pulled apart. Pulling one side of the honeycomb that is perpendicular to the ribbon direction separates it, revealing the ribbon direction. If the pull is parallel to the ribbon, it is nearly impossible to tear the hon-

FOAM CORES
There are many different types of foam core materials available, depending on the specific application. Foam core materials offer different densities and temperature characteristics for high-heat applications and fire resistance. When using foams in a repair operation, it is important to use the proper type and density. Always refer to the manufacturer's repair guidelines for recommended materials and procedures. Styrofoam, urethane foam, poly vinyl chloride (PVC), and strux are several common types of foam cores used in aircraft composite construction.

Styrofoam is commonly used on home-built aircraft and should only be used with an epoxy resin. Polyester resins dissolve Styrofoam. Do not confuse aircraft-quality Styrofoam with the type of Styrofoam used to make Styrofoam cups. Styrofoam cups use foam with large cell configurations that can not be used for structural applications Aircraft-quality

Styrofoam is comprised of smaller cells, which produce a much stronger core material. [Figure 3-39]

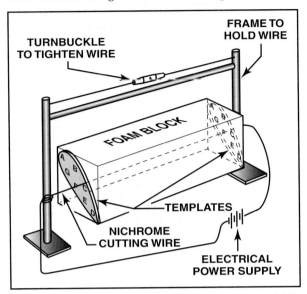

Figure 3-39. A heated cutting wire can be used to cut an airfoil-shaped part from a block of Styrofoam. A hot wire cutter consists of a nichrome wire that is heated electrically. Technicians typically make the tool by stretching the wire in a frame. Attach a template to each end of the foam to provide a uniform cut. Monitor the progress of the cut by making sure the increment marks, placed at each end of the foam, are crossed simultaneously, resulting in a smooth and even surface.

Urethane foam can be used with epoxy or polyester resins. However, urethane cannot be cut with a hot wire. Subjecting urethane foam to high heat produces a hazardous gas. Instead of a hot wire cutter, urethane is cut with a number of common tools. Knives are typically used to rough out the shape, and another piece of urethane foam is used to sand the piece to its desired size and shape.

Other foam core materials include **poly vinyl chloride (PVC)**, and **strux** (cellular, cellulose acetate) foam. PVC foam can be used with either polyester or epoxy resins and cut with a hot wire. Strux foam is commonly used to build up ribs or other structural supports.

WOOD CORES
Balsa wood or laminations of hard wood which are bonded to laminates of high-strength materials are occasionally used for other types of composite sandwich construction. Wood core materials provide high compressive strength to composite structures.

TYPES OF FIBER-REINFORCED COMPOSITES
Now that you are familiar with the basic materials used for composite construction, consider some of the different forms of composites that can be made with these materials. Several methods of fiber-reinforced composite construction are in use today. The two main methods used are laminated and sandwich construction.

LAMINATED COMPOSITES
The use of structural laminations of fiberglass, paper, and linen have been around for some time. As knowledge of composites has evolved, materials made of aramid (Kevlar®), carbon/graphite, boron, and ceramic fibers have been developed. Laminate composites consist of two or more layers of reinforcing material bonded together and embedded in a resin matrix. Laminated composites are built up to desired thicknesses by using multiple layers of reinforcing fabrics.

INTERPLY HYBRID LAMINATES
Interply hybrid composites consist of two or more layers of different reinforcing material laminated together. Blending different advanced composite fabrics in a laminate can achieve the proper balance of stiffness, strength, and weight for a particular application. For example, Kevlar® may be combined with carbon to produce a structure that merges the flexibility, lightweight, and impact resistance of Kevlar® with the stiffness of carbon. [Figure 3-40]

SANDWICH COMPOSITES
Sandwich construction consists of two or more laminated face sheets bonded to each side of a relatively thick, lightweight core. Sandwich composites offer high strength to weight ratios as compared to solid laminated structures. Face materials can be stainless steel, titanium, magnesium, plywood, glass, or nylon depending on the strength requirements and intended usage. [Figure 3-41]

As previously discussed, there are several types of core materials available for use in sandwich construction. Each offers its own advantage and unique strength and rigidity characteristics. In general, the strength of the sandwich composite varies with the thickness and type of core material. [Figure 3-42]

WORKING WITH RESINS AND CATALYSTS
The matrix formula for most advanced composites is very exacting. A slightly improper mix ratio can make a tremendous amount of difference in the strength of the final composite. Because the mixing

Wood, Composite, and Transparent Plastic Structures

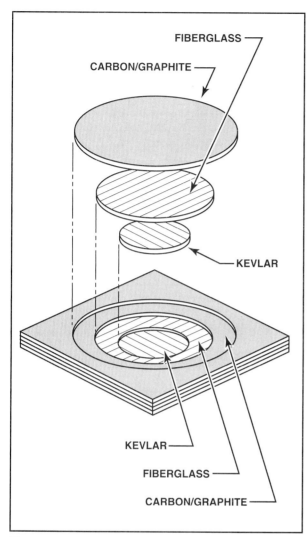

Figure 3-40. Interply hybrid laminates provide the technician the opportunity to customize the composite with specific characteristics. This figure illustrates an interply-hybrid laminate that consists of layers of carbon, fiberglass, and Kevlar®. This type of laminate has the toughness of Kevlar®, the stiffness of carbon, and the heat resistance of fiberglass.

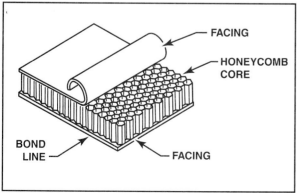

Figure 3-41. The use of core materials in composite construction dramatically increases the strength of a structure. The core material is essentially sandwiched between two or more face sheets.

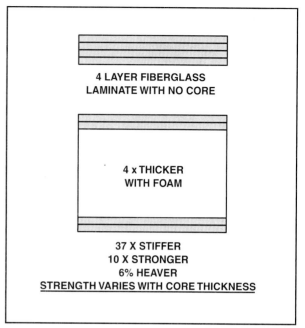

Figure 3-42. The advantages of a sandwich structure can be shown by comparing a four-layer solid fiberglass laminate to that of a foam-core sandwich structure that is four times as thick. The sandwich composite incorporates two layers of fiberglass on the top and bottom of a foam core. In this arrangement, the part becomes 37 times stiffer than the solid fiberglass laminate, and ten times stronger, with only a 6% increase in weight.

procedures are so important, they are always included with the resin containers. The aircraft structural-repair manual also outlines proper mixing procedures. [Figure 3-43]

```
TYPE OF RESIN, TYPE OF CATALYST
DATE OF MFG. 05/04/88
POT LIFE   30 MIN.   @   70   DEG. F
MIX RATIO 100/16. A TO B BY WEIGHT
CURE TEMP 007 DAYS @ 077 DEG. F
SHELF LIFE 12 MO. AT 40 DEG. F
```

Figure 3-43. This figure illustrates a typical resin label that gives all of the pertinent information about the mix ratio and proper curing instructions for that particular material. These instructions should coincide with the aircraft manufacturer's maintenance manual.

Manufacturers often produce pre-measured matrix packages. The advantage to using prepackaged resin systems is that they eliminate the weighing process and therefore remove the possibility of a mixing ratio error. Disposable cartridges that store, mix, and apply two-component materials are also available and convenient to use. They are available in many sizes and can be tailored to specific uses. Like the pre-measured packages described before, cartridges also eliminate mixing ratio errors. [Figure 3-44] [Figure 3-45]

Figure 3-44. The resin and catalyst are divided into separate containers that are attached on one end. When ready for use, the partition, which separates the resin from the catalyst, is broken to allow the two to mix. Still within the package, the resin/catalyst combination is mixed together by squeezing and kneading the package to thoroughly blend the mixture. When completely mixed, the package is cut with scissors and the resin dispensed.

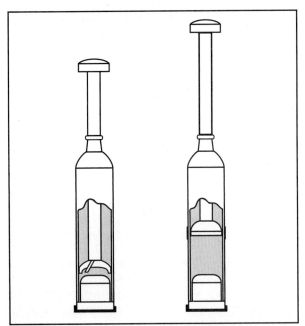

Figure 3-45. To use epoxy cartridges, the seal that separates the two components must be broken with a plunger. The materials are then mixed together by moving the plunger in a twisting and up-and-down motion to thoroughly mix the resin and catalyst. The label describes how many strokes are required to give a thorough mix. A needle or syringe may then be installed onto the end of the cartridge, and the resin dispensed. Be sure to check the cartridge part number, shelf life expiration date, and any special instructions.

In addition to prepackaged resin units, epoxy ratio pumps reduce mixing errors. Epoxy ratio pumps enable the technician to precisely measure varying epoxy ratios. They eliminate the sticky, messy, and in some cases, inaccurate hand proportioning of epoxy resins. [Figure 3-46]

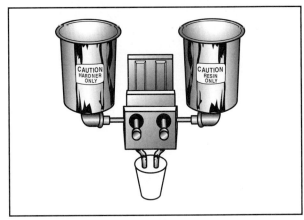

Figure 3-46. This type of epoxy ratio pump offers the advantage of being able to supply adjustable resin ratios. The resin and catalyst are dispensed through separate tubes so there is no mixed material in the pump. Mixing resins using an epoxy ratio pump also helps to increase safety since the user neither touches the resin or inhales the epoxy resin fumes during production.

Some resins systems are weighed verses measured to determine the proper mix ratio. Precision scales are used to weigh the two parts of the resin. The scale surfaces should always be clean, and calibrated periodically to ensure accuracy. If the type of resin system used requires refrigerated storage, allow each part to warm up to room temperature before weighing and mixing. When cold, a resin will weigh more than an equal quantity of the same resin at room temperature.

The resin and catalyst must be mixed thoroughly in order to achieve maximum strength. Mix resin systems together in a wax-free container. If a waxed container is used, the solvents in the resin and catalyst dissolve any wax on the inside of the container, which then contaminates the mixture. Though wax is used for heat control purposes in some resins, the wax from a container may cause the repair to cure incorrectly, or possibly not cure at all. Follow the manufacturer's mixing instructions, which often entails three to five minutes stirring or agitation time to completely mix the components.

Resins that are not mixed properly will not cure to the maximum strength obtainable. If resins are mixed too quickly, small bubbles may rise into the air and could get on your skin or in your hair. Do not be concerned if you have bubbles in the cup because they will be worked out with a roller or squeegee during the lay-up process. Vacuum bagging, which is a process of applying pressure to a lay-up, further ensures that no air bubbles remain in the final composite. If more resin is mixed than necessary to complete a project, any unused amount is wasted. If too much resin is mixed, allow it to cure before throwing it away. In most cases, resins in a cured condition are not considered hazardous materials for disposal.

A large volume of resin and catalyst causes an acceleration of the chemical reaction when curing. When this happens, it starts to cure in the mixing cup, possibly becoming too thick to work completely into the fabric. The pot life is also reduced if large amounts are mixed at one time. If a resin exceeds its pot life, it must not be used. Smaller mixed quantities are generally easier to work with and are more cost effective. One of the best ways to ensure that a properly prepared batch of matrix resin has been achieved is to mix enough for a test sample.

If the work is extensive and takes a long time, the pot life of the resin mixture may be exceeded if too much is initially prepared. Find out the length of the pot life, or working life, of the resin before preparing a resin batch. Some resin systems have very short pot lives (15 minutes), while others have long pot lives (4 hours).

The **shelf life** of a resin is the time that the product is still good in an unopened container. Like the pot life, the shelf life varies from product to product. If it has expired, the resin or catalyst must be discarded. Using a resin that has exceeded its shelf life does not produce the desired chemical reaction, and the strength of the finished product may be insufficient.

If too much resin is applied to the part, it is called **resin rich**. Traditional fiberglass work is used in nonstructural applications, so extra resin is not that critical. However, the use of excessive resin in advanced composite work used for structural applications is very undesirable. Excessive resin affects the strength of the composite by making the part brittle in addition to adding extra weight, which defeats the purpose for using composites for their lightweight characteristics.

A **resin starved** part is one where not enough resin was applied, which weakens the part. The correct amount of fiber to resin ratio is important to provide the structure with the desired strength. In advanced composite work, a 50:50 ratio is generally considered acceptable. However, a 60:40 fiber to resin ratio for advanced composite lay-ups is generally considered the best for strength characteristics. Actual ratios utilized should be in accordance with the manufacturer's instructions.

When working the resin into the fibers, care should be taken not to distort the weave of the fabric. If too much pressure is applied when using a brush or squeegee, the fibers could pull apart, altering the strength characteristics of the fabric. The curing of the resins must also be accomplished correctly to achieve the maximum strength. Be sure to follow the manufacturer's directions concerning curing requirements.

CAUTION: If two batches of resin and catalyst are mixed equally, leaving one batch in a jar and spreading the other in a thin layer, the one in the jar will harden rapidly because of the heat trapped by the glass. This cure rate takes place so quickly that it will cause minute fractures within the plastic. The thin sheet, on the other hand, will not cure as fast because it has a large surface area exposed to the air, which allows the heat to escape.

SAFETY CONSIDERATIONS

Safety is always important when working with composite materials. Many accidents have occurred because of the improper usage and handling of composite materials. Before working with any composite resin or solvent, it is important to know exactly what type of material you are using and exactly how to use it.

MATERIAL SAFETY DATA SHEETS (MSDS)

Material safety data sheets contain information on hazardous ingredients, health precautions, flammability characteristics, ventilation requirements, spill procedures, information for health professionals in case of an accident, along with transportation and labeling requirements. MSDS must be available for review in the shop where the specific material is stored and used. Review the MSDS and become familiar with the specific types of materials you come in contact with before you begin working with the materials.

PERSONAL PROTECTION

Some materials cause allergic reaction and some people are more sensitive to certain materials than others. Therefore, it is imperative to protect your skin from contact with composite matrix materials. The most effective way to provide skin protection is by the use of protective gloves, respirators, face shields, safety glasses, and shop coats. [Figure 3-47]

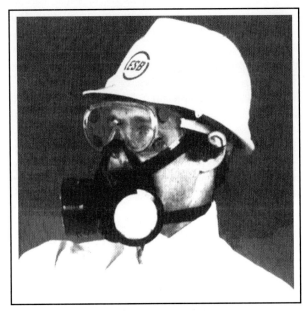

Figure 3-47. The use of proper skin, eye, and respiratory protection cannot be overstressed. Composite chemicals are often caustic to skin and to the respiratory tract, making the use of a respirator, goggles, and gloves a must when working with composite materials.

Protective gloves are produced from many types of material. Make sure you review the MSDS to find out what materials do not react with the composite materials you are using. For example, natural rubber gloves disintegrate when exposed to certain types of epoxy resins. In any case, do not reuse; replace safety gloves after heavy use.

If any materials come in contact with your skin, remove the material immediately and be sure to wash the area thoroughly. In addition, wash your hands before and after working with the materials, and before eating or smoking. Many composite chemicals are irritants and may cause serious skin inflammation and irritation. There are special types of epoxy cleaners available that break down resins without drying or reacting with the skin. Again, be sure to check the MSDS before working with materials that may be potentially hazardous.

Always work in well-ventilated areas when working with resins or solvents. Some resins are toxic enough to cause difficulty breathing and, in some cases, severe allergic reactions. Wear a respirator when working with, mixing, and applying resins, solvents or any other hazardous chemical, and keep contaminated hands away from your mouth. Do not ever ingest any composite chemical, because some are fatally poisonous.

Again, it is important to review and be familiar with MSDS information on each type of material you work with in the shop. Some composite materials can cause severe eye irritation and/or permanent blindness. In addition, plastic contact lenses may craze from resin fumes. Therefore, it is recommended to wear only glasses, not contact lenses, when working with resins. In all events, safety goggles are required to reduce the chance of eye injury. Goggles provide eye protection from front and side impact hazards, chemical splashes, and dust. Face shields also provide protection when working with resins. However, do not wear face shields that pull down over the face when working at an up-draft table. Up-draft tables draw fumes up through an exhaust vent. Face shields that pull down over the face trap fumes and may cause respiratory problems. Face shield-respirator combinations that open at the forehead do not trap up-draft fumes.

FIRE PROTECTION

Many solvents and resins are flammable, making it important to always work with and store these types of materials in well-ventilated areas. Keep all resins away from heat and open flames. Follow the safety recommendations of the manufacturer explicitly to reduce the chance of a fire. The following fire-safety requirements will reduce the fire danger in the shop.

- Eliminate all flames, smoking, sparks, and other sources of ignition from areas where solvents are used.

- Use non-spark-producing tools.

- Ensure that all electrical equipment meets the applicable electrical and fire codes.

- Keep flammable solvents in closed containers.

- Provide adequate ventilation to prevent the buildup of vapors.

- Statically ground the aircraft and any repair carts in use.

- Never unroll bagging films or other materials around solvents, to reduce the chance of static electricity.

- Never store or use solvents in any area when sanding.

SOLVENT SAFETY TIPS

There are many types of solvents used in composite construction today. Solvents are mainly used for cleaning purposes in composite construction. However, most solvents are flammable and must be used with the highest degree of safety in mind. Methyl-Ethyl-Keytone (MEK) and acetone are two common solvents used in composite construction.

MEK is mainly used for cleaning dust, grease, and mold release agents from composite components. Always use protective gloves and goggles when using it. MEK is an excellent cleaner but also a carcinogen. It can be absorbed directly into the bloodstream through the skin and the eyes. Acetone is used for general equipment and tool cleanup, in addition to cleaning the composite parts after sanding as a pre-bond preparation. Follow the manufacturer's recommendations when choosing the proper solvent.

These safety guidelines should be followed when using all solvents and matrices:

- Do not use solvents in any area that may create a static charge.

- Do not pour solvents onto the part. Instead, use a solvent-soaked soft cloth to apply solvents to the working surface.

- Use solvents in a well-ventilated area and avoid prolonged breathing of the vapors.

- Wear gloves when applying solvents to protect the skin from drying out.

- Never use solvents to clean skin. Use suitable epoxy cleaners that are less dangerous.

- Wear goggles when pouring and working with solvents.

- Store solvents in the original containers so they can be readily identified.

MANUFACTURING PROCESSES

The subject of composite manufacturing is addressed with the purpose of familiarizing you with the more common manufacturing methods used today. Being familiar with the composite manufacturing methods helps in subsequent repairs of the composite structure.

In general, most composite manufacturers augment the strength of the finished product by applying heat and pressure to the matrix/fiber mix as it cures, which accomplishes several things:

1. The heat and pressure facilitates the complete saturation of the fiber material.

2. The heat serves to accelerate the curing process of the matrix. In some instances, a high temperature is required to effect a proper cure of the matrix formula.

3. The pressure squeezes out excess resin and air pockets from the reinforcing fibers, which helps to produce a more even blend of fiber and matrix.

COMPRESSION MOLDING

Compression molding is a manufacturing process that uses a male and female mold to form the part. The impregnated reinforcement fabric is laid into a female mold. The male mold is then pressed into the female mold to form the part. If a core material is used, the fabric is wrapped around the core of the desired shape and the two sides of the molds apply pressure giving final shape to the part. [Figure 3-48]

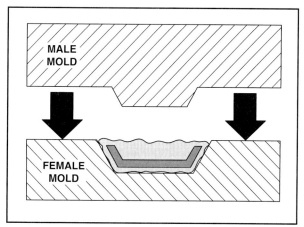

Figure 3-48. Compression molding is normally used to manufacture a large number of precision-formed parts.

Heating the molds at specific temperatures and time intervals cures the part. Two common heating methods used with compression molding are the circulation of heated oil through the mold or the use of electric heating elements imbedded in the mold. Another option for heating the molded part is to place the entire mold assembly into an oven to ensure an even, carefully controlled distribution of heat.

Because composites start out as reinforcing fibers (cloth) and a liquid (uncured matrix material), the only limitations on the shape of molded components are those limitations associated with the mold itself. Once a mold has been produced, it can economically turn out a very large number of precision-formed parts.

VACUUM BAGGING

Vacuum bagging is the most commonly used method to apply pressure to composite repairs. With this technique, the assembly is placed into a plastic bag and the air is then withdrawn by the use of a vacuum source. When the air is removed, pressure is applied by the surrounding atmosphere. The vacuum bag technique can be used in combination with molds, wet lay-up, and autoclave curing. This method applies a very uniform pressure to somewhat complicated shapes and can accommodate moderately large objects. [Figure 3-49]

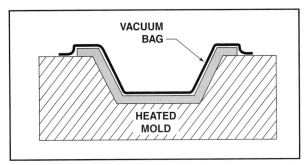

Figure 3-49. Vacuum bag molding can apply pressure to very large and complicated shapes. It is used in both manufacturing and repairing of composites.

Both compression molding and vacuum molding have the advantage of distributing the matrix evenly throughout the composite structure. This helps to eliminate air bubbles and results in a seamless structure that is easier to fabricate and usually stronger than a metallic counterpart.

FILAMENT WINDING

Another manufacturing method that produces incredibly strong structures is the filament winding method. A continuous thread of reinforcing fiber is wound around a mandrel in the same shape of the desired part. In order to provide the precision required in placing the thread, a filament-winding machine, or robot, is used. Some filament wound parts use pre-impregnated threads; others dip the threads into a resin, and use a drying area to dry off extra resin. Once the filament is wound in the desired pattern, the composite mixture is cured. [Figure 3-50]

At this time, very few repairs are approved for filament-wound parts. If the component part is dented and has broken a few of the strands, the damage should not be cut out. Cutting out the damage would weaken the fibers, causing the structure to weaken even more. If the manufacturer of the damaged filament-wound component does not specify a repair, the component must be replaced.

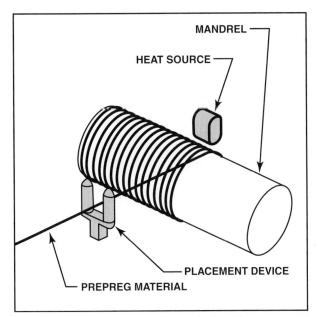

Figure 3-50. The filament winding method produces some of the strongest composite structures. It is commonly used in the fabrication of helicopter rotor blades, propellers, and even entire fuselages.

WET LAY-UP

A manufacturing technique that is less precise than compression, vacuum bag molding, or filament winding is referred to as wet lay-up or hand lay-up. This technique simply involves the mixing of the fiber reinforcement with the matrix, then laying the wet fabric over a surface for curing. [Figure 3-51]

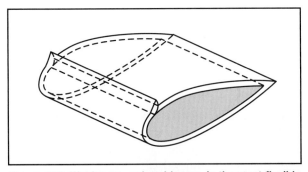

Figure 3-51. Wet lay-up, or hand lay-up, is the most flexible method of laminated construction and is the most frequently used for repairs. The best fiber to resin ratio for an advanced composite wet lay-up is 60 parts fiber to 40 parts resin (60:40 mix).

Although this technique is less precise than other manufacturing methods, it is the most flexible procedure available. The simplicity and flexibility of the wet lay-up has made this technique a favorite of builders of home-built aircraft. Furthermore, the materials and methodology associated with wet lay-up are the same as those often used to make repairs to composite structures.

FIBERGLASS LAY-UP

In the case of fiberglass hand lay-up, flat laminated sheets are laid up by first spreading a heavy coat of resin on a glass or aluminum plate. A sheet of fiberglass fabric is placed onto the resin and a roller is used to work the fiberglass down into the resin. The roller also forces all of the air bubbles out to ensure the complete encapsulation of every fiber. Another layer of resin is then placed on top of the first layer of fiberglass followed by a second sheet of glass cloth. Again all of the air is worked out and the process is continued until the assembly is built up to the desired number of layers. A final coat of resin is then placed on the lay-up and the entire assembly is cured.

ELECTRICAL BONDING

Manufacturers use varying methods to dissipate the electrical charge on composite structures. Several common bonding methods are:

- Aluminum wires woven into the top layer of composite fabric. Fiberglass and Kevlar® are two types of composites that routinely utilize this method. However, carbon/graphite does not use this method due to the galvanic reaction of aluminum and carbon.

- A fine aluminum screen laminated under the top layer of reinforcing fabric. This method may be used with carbon composites if the screen is sandwiched between two layers of fiberglass or some other non-metallic material to prevent galvanic corrosion.

- A thin sheet of aluminum foil bonded to the outer layer of composite.

- Aluminum flame sprayed onto the component, which is molten aluminum sprayed on like a paint finish.

- Metal bonded to the composite structure and another metal structure and out through a static wick.

COMPOSITE FINISHES

The final finish of a composite structure seals the surface and creates a barrier from moisture and ultra-violet light. Moisture is so destructive to a composite, that some manufacturers, such as Boeing, apply a plastic coating on the composite before painting. This plastic coating serves as an additional moisture barrier. Most new-generation aircraft paint can be used on composites as well as metallic structures, providing excellent flexibility and wear-resistance characteristics. In addition, aircraft finishes do not deteriorate as readily as some other types of composite finishes, such as gel coats.

A **gel coat** is a colored polyester resin applied to the outer surface of a composite structure. It is not structural and looks more like a paint coat. Gel coats are susceptible to sun damage and may crack if left in the sun for long periods. They cannot be rejuvenated like dope and fabric aircraft. The gel coat must be sanded off and reapplied. Many aircraft owners today repair damaged gel coats by sanding off and re-finishing with new-generation paints.

COMPOSITE INSPECTION

Today's composite inspection techniques and non-destructive testing (NDT) methods typically involve the use of multiple methods to accurately determine the airworthiness of the structure. Fortunately, many metal inspection and NDT methods transfer to composite applications. Composite structures require ongoing inspection intervals along with non-scheduled damage inspection and testing.

When a composite structure is damaged, it must first be thoroughly inspected to determine the extent of the damage, which often extends beyond the immediate apparent defect. Proper inspection and testing methods help determine the **classification of damage**, which is, whether the damage is repairable or whether the part must be replaced. In addition, classifying the damage helps to determine the proper method of repair. The manufacturer's structural repair manual outlines inspection procedures, damage classification factors, and recommended repair methods.

Today's composite inspection and nondestructive testing procedures typically involve more than one inspection method. Some of the more common composite inspection and testing methods are visual inspection, tap testing, and ultrasonic testing along with several other more advanced NDT methods.

VISUAL INSPECTION

Visual inspection is the most frequently used inspection method in aviation. Ideally, pilots, ground crew, and maintenance technicians visually inspect the aircraft on a daily basis. This method of inspection is generally used to detect resin-rich areas, resin starvation, edge delamination, fiber break-out, cracks, blistering, and other types of surface irregularities. A strong light and magnifying glass are useful tools for visual inspection. In extremely critical cases, a small microscope is helpful in determining whether the fibers in a cracked surface are broken, or if the crack affects the resin only.

Shining a strong light through the structure, called **backlighting**, helps in the identification of cracked or broken fibers, and, in some cases, delamination. The delaminated area may appear as a bubble, an indentation in the surface, or a change in color if viewed from the side opposite the light. However, backlighting does not detect entrapped water. In addition, to properly inspect a composite using the backlight method, you must strip the surface of all paint.

Many times, visual inspection alone is not adequate to accurately determine the soundness of a composite structure. In the case of visually inspecting a sandwich structure, many times core crush is not evident from the surface. The surface may not show any residual damage and may have sprung back to its original location, which is one of the main problems with inspecting composite materials. Internal damage is not always evident from the surface, which further necessitates the use of additional, more advanced methods of inspection when damage is suspected.

The maintenance technician is generally the first person to assess damage using visual inspection techniques. After this initial inspection, more advanced forms of inspection and testing may be required to determine the extent of the damage.

TAP TEST

The tap test is one of the simplest methods used to detect damage in bonded parts. The laminated part is tapped with a coin or small metallic object, such as a tap hammer, to detect delamination. The tap test is an **acoustic test**, one in which you listen for sound differences in the part, and is not the most accurate test method. The tap test detects delaminations close to the surface in addition to transitions to different internal structures. A properly prepared, undamaged laminated area produces a sharp, even pitch as compared to a delaminated area, which produces a dull sound.

However, changes in the thickness of the part, reinforcements, fasteners, and previous repairs may give false readings when using the tap test. Tap testing will not indicate delamination well below the surface in thick parts. In general, in sandwich structures, if the first laminate is over 1/4 inch down to the bond line, the tap test should not be used; it may not produce an indication of damage. The tap test should be limited to near surface inspection of bondline defects. Tap testing bondlines of thicker laminates becomes less and less effective as the thickness increases. [Figure 3-52]

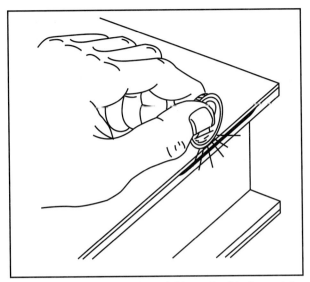

Figure 3-52. The tap test is a viable method to inspect for delamination. However, it should always be followed by another NDT method, such as ultrasonic, or radiography inspection methods to accurately determine damage parameters.

ULTRASONIC INSPECTION

Ultrasonic inspection is the most common instrumental NDT method used on composites today. An ultrasonic tester is useful for detecting internal damage such as delaminations, core crush, and other subsurface defects. Two common methods of ultrasonic testing include the **pulse echo** and **through transmission** methods.

In the pulse echo method, the tester generates ultrasonic pulses, sends them through the part, and receives the return echo. The echo patterns are displayed on an oscilloscope. An advantage to the pulse echo method is that it only requires access to one side of the structure. However, near-surface defects do not readily allow sound to pass through them, making it difficult to detect defects located under the first defect. The pulse echo method works well on laminates because they do not reduce the magnitude of sound waves as much as a bonded core structure.

The "through transmission" method uses two transducers. One transducer emits ultrasonic waves through the part and the other receives them. Defects located at multiple levels throughout the structure are more easily detected because the receiver, located on the backside of the part, receives the reduced amount of sound waves that pass through the defects. The ratio of the magnitudes of sound vibrations transmitted and received determines the structure's reliability. Testing bonded-core structures usually requires the through transmission method due to the fact that sound waves reduce in magnitude as they travel through the sandwich structure. To effectively test this type of structure, the use of a receiver on the backside of the part dramatically increases the likelihood of detecting a defect. [Figure 3-53] [Figure 3-54]

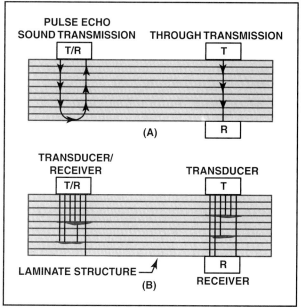

Figure 3-53. View A of this figure compares pulse echo sound transmission with that of "through transmission." As shown in view B, the pulse-echo method does not effectively detect multiple layers of defects. Sound waves do not travel through damage as readily as they do through an intact structure, which makes it difficult for the transducer/receiver to receive the pulse echo back from the lower defects. "Through transmission" inspection provides a more accurate view of the integrity of the composite structure. The receiver that is placed on the backside of the part picks up the low-intensity sound waves that emanate from multiple defects.

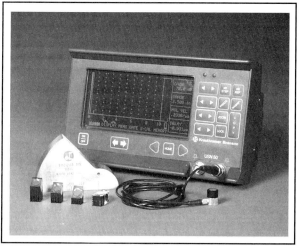

Figure 3-54. This pulse-echo ultrasonic tester uses high-frequency sound waves to inspect for internal damage. The defects are shown on the oscilloscope screen.

RADIOGRAPHY

Radiography or x-ray inspection is used to detect differences in the thickness or physical density when compared to the surrounding material of a composite. It can be used to detect surface as well as internal cracks. Radiography also detects entrapped water inside honeycomb core cells. In addition to detecting the actual defect, it can also detect the extent and size of the damage, unlike ultrasonic or tap testing. X-ray inspection will also detect foreign objects in the composite structure if the object's density is different from the composite structure.

THERMOGRAPHY

Thermography locates flaws by temperature variations at the surface of a damaged part. Heat is applied to the part and the temperature gradients are measured using an infrared camera. Thermography requires knowledge of the thermal conductivity of the test specimen and a reference standard for comparison purposes.

DYE PENETRANT

Dye penetrant successfully detects cracks and other defects in metallic surfaces, but should not be used on composite structure unless called for by the manufacturer. If a dye penetrant is used on the composite structure and allowed to sit on the surface, the wicking action of the fibers may absorb the penetrant. Absorbed penetrant does not allow fibers to bond to new material. The entire area affected by the dye penetrant would have to be removed before new patches could be applied, which could extend the damage to a size that would make the part non-repairable.

ACOUSTIC EMISSION TESTING

Another nondestructive testing technique used to detect composite defects is acoustic emission testing. Presently, this type of test is more commonly found in a production facility versus a maintenance facility. Acoustic emission testing is a comparison test. You must have a good test sample to compare the composite structure test results with to accurately detect flaws. It measures the sounds of a structure and any subsequent defects. Basically, acoustic emission testing picks up the "noise" of the defect and displays it on an oscilloscope. This type of test detects entrapped water, cracks, delamination, and other subsurface flaws.

MACHINING COMPOSITES

The machining process consists of drilling, cutting, sanding, and grinding of material. Composite materials act differently than traditional aluminum when machined. Therefore, it is important to understand the differences in machining techniques commonly used in composite repair.

CUTTING FABRICS

Before a fiberglass or carbon/graphite fabric is combined with a matrix and cured, it can be cut with conventional fabric scissors. Aramid fabric in its raw state is more difficult to cut. Scissors with special steel or ceramic blades and serrated edges are used to cut through aramid. Ceramic scissors cut through aramid with ease and last many times longer than steel. The advantage to ceramic scissors is that the serrated edges hold the fabric, while the blades cut fabric without fraying the edges. A conventional pair of scissors separates the weave, but does not cut the fabric unless very sharp. Scissors designed to cut specific types of fabrics should only be used on those fabrics, to keep the blades sharp.

Pre-impregnated materials can be cut with razor blade and utility knives, using a template or straight edge. The resin tends to hold the pre-impregnated fibers in place while the razor edge cuts through the fibers. Very sharp, defect-free cutting edges are necessary to work with composite fabrics.

DRILLING COMPOSITES

Drilling holes in composite materials presents different problems from those encountered in drilling metal. Composites are more susceptible to failures of the material when being machined. However, proper selection and application of cutting tools produce structurally sound holes.

Delamination, fracture, breakout and separation are types of failures that may occur while drilling composites. Delamination most often occurs as the drill pushes the last layer apart rather than drilling through it. A fracture occurs when a crack forms along one of the layers, usually due to the force of the drill. Breakout occurs when the bottom layer splinters as the drill completes the hole and separation occurs when a gap opens between layers as the drill passes through successive layers.

To reduce the possibility of damage, the composite should be backed with a wood block. A very sharp drill bit is required to prevent the delamination of the last ply. When a blind fastener is to be used with the composite part, and the backside is inaccessible, a wood backup is not possible. In this case, a drill stop is useful to limit the depth of the drill bit. By limiting the depth of the drill passage, fiber breakage on the backside can be eliminated. [Figure 3-55]

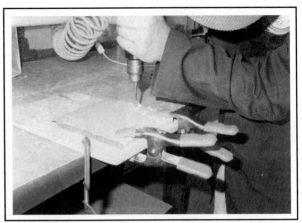

Figure 3-55. A wooden backup block should be used with little or no pressure on the drill, to prevent damage to the part.

Do not use a cutting coolant when drilling holes into bonded honeycomb or foam core structures. The coolant may seep into and remain in the structure after a patch has been bonded over the repair. In addition, the laminate fibers may absorb the coolant liquid, creating an unbondable surface. If a cooling solution is necessary, use clean water only and completely dry the part before applying any lay-up material.

Carbide drills work on all types of composites and have a longer life than standard steel bits. Diamond-coated cutters perform well on fiberglass and carbon; however, they produce excessive fuzzing around the cut on aramid components.

Drill speed is also an important consideration. High speeds work best for most types of materials being drilled, but do not use excessive pressure. The best drill bit to use on composites has an included angle of 135°.

While sanding, drilling, or trimming composite structures, very fine dust particles contaminate the air therefore, respirators must be worn when performing these operations. Use a dust collector or downdraft table in conjunction with a respirator to help pull fine particles out of the air.

DRILLING ARAMID

Machining and drilling materials reinforced with aramid or Kevlar® fibers requires different tools than those made of fiberglass or carbon/graphite fibers. The physical properties of aramid fibers are unique. Using a conventional sheet-metal twist drill on aramid tends to produce fuzz around the holes. Due to the flexibility of the aramid fiber, the drill pulls and stretches the fiber to the point of breaking instead of directly cutting it, which produces the fuzzy appearance around the edge of the drilled hole.

Holes drilled in aramid often measure a smaller diameter than the twist drill that is used, because of the fiber fuzzing. The fuzzing around the hole may not produce a problem in itself. However, if a fastener is installed, it may not seat properly in the hole, which may produce mechanical failure of the bearing surface. To resolve this problem, apply a quick curing epoxy to the fuzzed area. Once the epoxy is cured, use a file or rasp to remove the fuzz. If the composite has just been cured and the peel ply or release fabric layer has not been removed, drill the hole through the peel ply to eliminate the fuzzing.

Special drills designed to machine aramid consist of a brad point and C-shaped cutting edge to cut the fiber without stretching and fraying the material. These drills last longer than conventional twist drills, and produce a cleaner hole. Although carbide drills were specifically designed for aramid composites, they also produce good holes in fiberglass and carbon/graphite. Each drill should be reserved for use on just one type of material. [Figure 3-56]

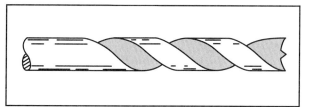

Figure 3-56. Brad point drills with a special C-shaped cutting edge are used to machine aramid composites without causing the fuzzing left by normal twist drills.

DRILLING FIBERGLASS OR CARBON/GRAPHITE

Drilling fiberglass or carbon/graphite materials can be accomplished with most conventional tools. However, the abrasiveness of these particular composite materials reduces the quality of the cutting edge and drastically shortens the life of the drill. Carbide or diamond-coated tools produce better results and longer tool life. Diamond-coated tools are usually steel drills coated with diamond dust. This type of drill works well on both carbon/graphite and fiberglass components.

If a dull or improper drill is used, the fiber may break inside the composite structure, causing a larger hole than wanted. In addition, dust chips allowed to remain in the holes during the drilling process can also enlarge the diameter of the hole. A single cutting-edge dagger or spade bit is used when cutting fiberglass and carbon materials to reduce the tendency of fiber breakage.

SANDING

Sanding is used to remove fabric one layer at a time during the repair process and to smooth the outer surfaces of some composites. Proper sanding materials must be used in order to produce an airworthy repair. Aluminum oxide, for instance, should not be used when sanding carbon fibers. Small particles of aluminum may become lodged in the carbon fibers, which can cause galvanic corrosion. To prevent such corrosion, only silicon-carbide or pure carbide papers should be used.

Because the fabric layers of composites are usually very thin, they sand off very quickly. Therefore, **hand sanding** is effective when removing only one layer, or a very thin coat of paint. If removing a layer of paint, do not sand into the top layer of fabric, or an additional repair may have to be performed. Wet sanding is preferred, using a fine-grit sandpaper of about 240 grit.

When step-sanding laminates to remove damage, a right-angle sander or drill motor is preferred. The tool should be capable of 20,000 r.p.m. and be equipped with a one, two, or three-inch sanding disc. Additionally, a sanding disc can also be used in combination with a drill motor or sander. A smaller one or two-inch disc gives you more control when step-sanding or scarfing the composite structure.

Each composite material sands differently, and various techniques can be used with each. When sanding aramid, expect the material to fuzz and start to gloss when almost through the layer. During the sanding process, it is important to look carefully for the gloss area, which indicates that the next layer of laminate is near.

Carbon/graphite material produces a very fine powder when sanded. It is also usually easier to see the layers of carbon than of aramid. Another way to tell if sanding through one layer has been completed is to look for the change in weave orientation, since most composites are produced with several layers of reinforcing fabrics laid in different directions.

The layers of a composite laminate are very thin, and a common problem is to sand too quickly or with too much pressure. For example, this may present a problem if there are only three layers in the laminate over a core structure and the repair calls for sanding down to the core. If the first two layers are sanded down and counted as one, the honeycomb core may become exposed when the next layer is sanded. In this situation, there may not be enough surface area to laminate new plies to the repair area.

ROUTERS

The most common types of routers operate at 25,000–30,000 r.p.m. and are used to trim composite laminates and to route out damaged core material. When routing Nomex® honeycomb, carbon/graphite, or fiberglass laminates, a carbide blade diamond-cut router bit works best. A diamond-cut router bit refers to the shape of the cut on the flutes. [Figure 3-57]

Because of the flexibility of aramid fibers, a herringbone router bit works best because the flutes change direction. As the bit starts to pull out an aramid fiber, the flute changes direction to cut the fiber, which produces a clean non-fuzzing cut. In order to rout out damaged core material, a circular or oval area of the top laminate skin over the damaged core must first be routed, using a pointed router bit. [Figure 3-58]

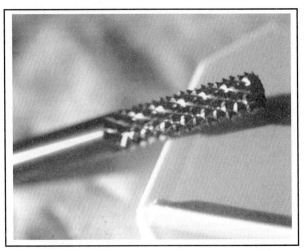

Figure 3-57. A router bit cut in a diamond-shaped pattern works best for cutting Nomex® honeycomb, carbon/graphite, or fiberglass laminates.

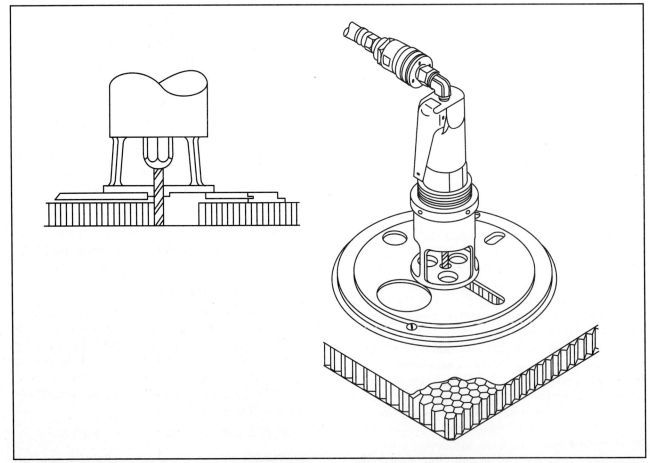

Figure 3-58. If the damage penetrates one skin and the core, care should be taken not to rout into the opposite laminate. A diamond-cut router bit works well to clean out honeycomb core. If the damage requires all of the core material to be removed, adjusted the router to a depth just inside the bottom laminate surface. To prevent the router from cutting into the bottom layers, adjust the depth so that some honeycomb can still be seen. A flush bit is then used to clean out the remaining core material.

If the routed area is tapered such as in the case of a trailing edge, place shims under the template to produce the desired cut. If the part is curved, do not adjust the router depth to cut to the bottom skin. This extra precaution is to prevent laminates on the opposite side from being damaged. Any excess honeycomb in this area can be removed by hand sanding. [Figure 3-59]

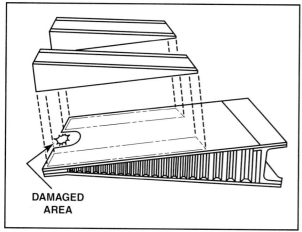

Figure 3-59. Tapered shims should be used to hold the routing template parallel with the skin when removing the core from a tapered control surface.

HOLE SAWS

Holes in some types of composites may also be cut by the use of a hole saw. However, hole saws tend to tear out the honeycomb core of aramid sandwich structures because their teeth do not cut, but rather, fray the edges. Because of this, hole saws are not recommended on aramid laminates. For carbon/graphite cutting, the hole saws may be fitted with a diamond-coated blade to produce a cleaner cut.

WATER-JET CUTTING

Water-jet cutting systems use a fine stream of water pumped at 30,000 to 50,000 psi through a pinhole nozzle. Water-jet cutting does not produce dust or fumes, nor cause delamination or fuzzing of aramid laminates. Water-jet cutting is used most often during the manufacturing process, and is not commonly used for repair applications. A water jet uses such a fine spray of water that water wicking into aramid is not a problem. However, if some water is absorbed, it can quickly be removed by applying heat from a heat lamp to evaporate any water.

BAND SAWS

A band saw may be used to cut composites if the blade has at least 12 to 14 teeth per inch. Blades made for composite sawing are available in carbide or with diamond dust on the cutting edges. Band saws produce some fuzzing when cutting aramid, but this can be cleaned up by hand-sanding the edges.

HYDRAULIC PRESS CUTTING

During the manufacturing process, raw or pre-impregnated fabric can be cut using a large hydraulic press. Pattern pieces are made from metal with sharp cutting edges and are used to stamp out the fabric pieces.

LASER CUTTING

Another cutting device used in manufacturing composites is the laser cutter. A laser uses a highly focused light beam to cut composite materials. Laser cutting can be performed on both cured and uncured composite materials.

COMPOSITE REPAIR

The newer advanced composites use stronger fabrics and resin matrices, which cannot be repaired in the same way as fiberglass. A common misconception of advanced composites is that they can be repaired in the same way as the older fiberglass structures. To repair an advanced composite structure using the materials and techniques traditionally used for fiberglass repairs may result in an unairworthy repair. Such traditional fiberglass repairs allow for excessive weight, increased susceptibility to material fatigue and decreased flexibility.

Depending on the manufacturer of the aircraft, **classification of damage** is usually placed in one of three categories: negligible, repairable, or non-repairable damage. Negligible damage may be corrected by a simple procedure with no flight restrictions. Repairable damage is damage to the skin, bond, or core that cannot be repaired without placing restrictions on the aircraft or structure. A composite structure that is damaged beyond limits must be replaced unless a structurally sound repair can be designed by a structural engineer.

TYPES OF REPAIRS

The exact procedures for repair of various laminated composite structures depends partly on the type of damage incurred. The damage can range from a relatively simple surface scratch, to damage completely through all internal plies and core honeycomb material. There are four basic types of composite repairs:

1. Bolted metal or cured composite patches

2. Bonded metal or cured composite patches

3. Resin injection

4. Laminating new repair plies to the damage

Bolted and bonded surface patches are not usually recommended due to the fact that these types of patches do not restore the strength characteristics of the original structure. A bolted or bonded patch that is attached to the surface also causes undesirable aerodynamic changes.

Resin injection repairs are used to fill holes or voids. They are accomplished by injecting resin into the hole of a damaged area using a needle and syringe. This type of repair is usually done on non-structural parts. The injected resin does not restore the original strength, and, in some cases, expands the delamination.

The most desirable type of permanent repair to composite structure is to laminate new repair plies in the damaged area. This type of repair involves removing the damaged plies, and laminating on new ones.

ASSESSMENT AND PREPARATION

All repairs must be performed correctly, based on the type and extent of damage, in addition to the function of the damaged structure. Several of the more common reasons that composite repairs fail are poor surface preparation, and contamination of the reinforcing fabric or matrices used in the repair. Additionally, repairs may fail if the measuring and mixing requirements of the matrices are not explicitly followed. Failure may also occur if the cure times or temperatures are not adhered to explicitly. Finally, repairs may fail if inadequate pressure is applied to the repair during the curing process.

In order to ensure that composite repairs do not fail from the items described above, proper damage assessment and repair preparation are a must. Steps such as the initial damage assessment and classification, materials preparation, surface preparation, damage removal, cleaning, and water removal are essential to achieve an airworthy composite repair.

DAMAGE ASSESSMENT

Before starting any repair, a complete and total assessment of the damage must be made. It is important to evaluate the damage to determine such information as the type of defect, depth, size and location on the aircraft. With any type of assessment, the most important tool an aircraft maintenance technician has is his or her critical judgment, based on experience. A general guideline to follow when assessing damage includes:

1. Visually examine the part to determine the type and extent of damage.

2. Check the damaged area for water, oil, fuel, dirt, or other foreign matter contamination.

3. Check for delamination around the damaged area.

4. Check for subsurface damage, if warranted, with other forms of advanced nondestructive testing methods.

5. Determine the repairable damage limits as found in the manufacturer's repair manual.

6. Determine the proper repair procedure, if approved, as outlined in the manufacturer's repair manual.

Several of the more common types of composite damages include cosmetic defects, impact damage, cracks, and holes. It is important to understand and be able to identify each of these types of damage to properly classify and subsequently repair them.

Cosmetic Defects
A cosmetic defect is a flaw on the surface of the skin that does not involve damage to the structural reinforcing fibers. Improper handling is a main cause for chipping and scratching. Cosmetic defects do not usually affect the strength of the part, and are primarily repaired for aesthetic reasons. In some cases, surface damage to aramid structural components with a top layer of fiberglass is considered negligible or cosmetic damage.

Impact Damage
One of the most common cause of impact damage results from careless handling during transportation or storage such as standing parts on edge without adequate protection. Improper handling can cause nicking, chipping, cracking, or breaking away of pieces of the edges or corners. Because the face sheets on sandwich panels are extremely thin, they are especially susceptible to impact damage. An area that has been subjected to impact damage should also be inspected for delamination around the impacted area. [Figure 3-60]

Cracks
Cracks can occur in advanced composite structures, just as in metallic ones. Sometimes they can be detected visually while other times require more advanced methods of nondestructive inspection. At times, cracks may appear in the top paint or matrix layer, and not penetrate into the fiber material at all. A crack may also extend into the fiber material and into the core, but appear to be located in the top surface only. A thorough inspection should be made to determine the extent of any cracks.

damaged during the sanding process, the repair must be enlarged to include this area.

Once the surface finish has been removed from the damaged area, the repair zone should be masked off to protect the surrounding areas. This portion of the repair preparation defines the area that will be removed to accommodate the repair patch.

DAMAGE REMOVAL

When preparing for a composite repair, the damaged area must be completely removed to provide a strong adhesion area. If damage has occurred to the core material of a sandwich structure, it must be removed prior to step cutting the laminate face sheets. Routers are routinely used to remove honeycomb core damage. In the case of laminated composite structures, step and scarf cutting are done to remove specific plies.

Routing

Air-driven routers are specially suited for repairing bonded honeycomb structure. Hand-held routers spin the cutting blade 10,000–20,000 r.p.m. The collar and support bracket allow for the adjustment of the cutter so it can cut through the core material without damaging the opposite face sheet. [Figure 3-62]

Figure 3-63. Step sanding will provide the most control when removing laminate plies.

The proper amount of space for each step cut can be marked off by using the following procedures:

1. Outline the entire damaged area that must be cut out and removed.

2. Expand the repair radius (assuming the repair is circular) by a one-half inch for each ply that must be repaired.

3. Extend the radius by an additional one-inch if an overlap patch is called for. [Figure 3-64]

Step cutting is accomplished by sanding away approximately one-half inch of each layer as you taper down to the center of the repair. Initially, aramid (Kevlar®) will fuzz and carbon will produce a fine powder as each layer is sanded through. Eventually, the materials will show a gloss area for each ply removed. The gloss indicates the next layer. When you see the gloss effect, stop sanding, otherwise the next layer may be damaged. Another way to determine if one layer has been sanded is to look for a change in fiber direction, which is only possible when the warp has been placed in alternating positions.

Scarf Cutting

Scarf cutting is used to remove damaged material with a tapered cutout. Dimensions of the scarf are based on the ratio of the total height of the plies to a given length. The scarf should taper down to the center of the repair. By shining a light on the surface of a scarf cut, you can identify the layer transitions. [Figure 3-65]

CLEANING

All repairs must be cleaned after sanding in order to remove all dust, dirt, oils, and any other sanding remnant. The strength of the bond is directly related

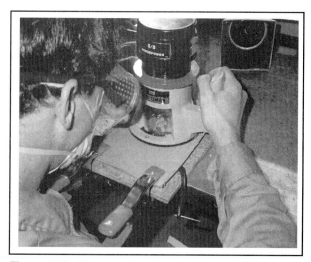

Figure 3-62. High-speed routers are commonly used to remove core material from sandwich structures.

Step Cutting

To properly step cut a laminated structure, each successive layer of fiber and matrix must be removed without damaging the underlying layer. Great care must be exercised during this process to avoid damaging the surrounding fibers. Using sanding as the method of step cutting provides the most control. [Figure 3-63]

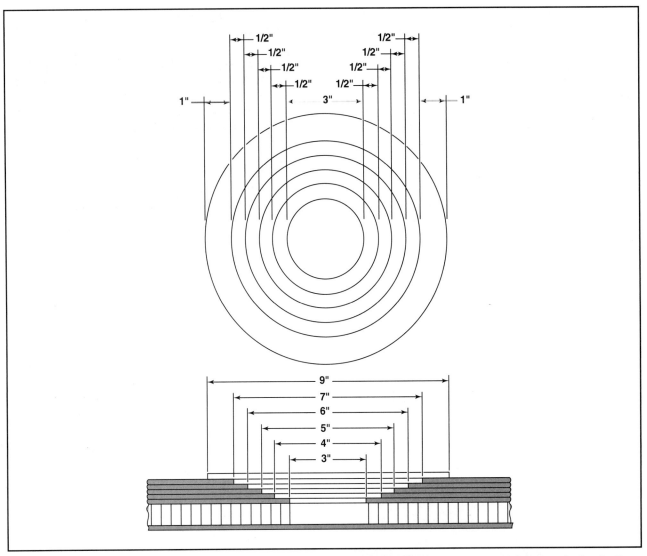

Figure 3-64. This figure illustrates a typical step cut outline regarding damage to the skin and core, which is three inches in diameter and runs through five surface plies. The three inch damaged area is removed and each ply is step-sanded 1/2 inch per ply in addition to a final one-inch overlap. The finished diameter for this repair is nine inches.

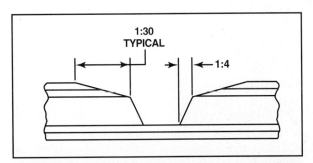

Figure 3-65. A typical scarf ratio is 1:30, which means that the cut tapers 1 increment down for every 30 increments in length. Remember that cores may be scarfed at different ratios than the laminates, as shown here. Read the repair manual carefully and fully understand how to make the appropriate cuts. If the structural repair-manual calls for a scarf cut, you may use the sanding step cut method as long as there is enough bonding area to provide adequate adhesion for the patch.

to the condition of the adhesion surface. If the surface is not properly cleaned, the repair may not bond.

A vacuum cleaner is routinely used to remove the dust from sanding. A solvent wash of MEK, acetone, butyl alcohol, or other approved cleaner removes dirt and oils. Always let the part completely dry before beginning the repair. Once the part is clean, it is important not to touch the surface or any of the repair materials with bare hands, or the entire cleaning process must be repeated. Wear gloves during the repair process, to make sure you protect yourself and keep the structure clean. In addition, do not use compressed air to blow dust from the surface. The introduction of compressed air may cause delamination.

Prior to the bonding process, some manufacturers may require a **water-break test** to make sure all oil and grease has been removed from the surface. The water-break test is accomplished by flushing the repair with room temperature water. If the water beads on the surface, the cleaning process must be repeated. If the surface is properly cleaned, the water will run off the surface in a sheet. Before continuing with the bonding process, make sure the surface is completely dry. A heat lamp is commonly used for this process.

WATER REMOVAL

Moisture trapped within a composite structure can be very dangerous. It may expand when heated, which would build up pressure and cause delamination. Entrapped water that freezes may also cause delamination. Water can enter the structure around improperly sealed edges and fastener holes, and any other area that is not properly sealed.

Moisture can be detected using X-ray, laser holography, and acoustic emission testing. Once moisture is detected, it must be removed. The following actions will help remove water from a composite structure.

1. Remove any standing water with a wet/dry vacuum.

2. Vacuum bag the surface to pull the water out. The bleeder, a barrier normally used to absorb excess resin, in the vacuum bag system helps soak up additional water and the heat blanket evaporates any remaining water. [Figure 3-66]

3. Use a heat lamp to dry out composite structures.

4. Use a reduced temperature rise during the repair process to eliminate blisters and voids.

5. If the entrapped water cannot be removed, remove and repair the affected area.

GENERAL REPAIR PROCESSES

This section describes the more common repairs that you, as the technician, may be called upon to perform in the field. The following procedures are intended for training purposes to familiarize you with basic repair procedures. Always consult the manufacturer's repair manuals to determine damage tolerance specifications, specific repair procedures, and other information pertaining to a specific repair.

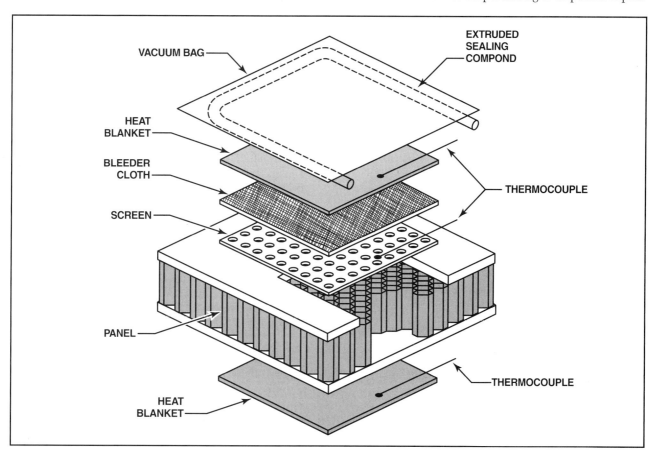

Figure 3-66. This figure illustrates a typical vacuum bag setup to remove water from a composite structure. The combination of the negative pressure, bleeder, and heat blanket helps remove any entrapped water.

The technician must understand how to identify the fabric orientation, fabric-impregnating techniques, vacuum bagging, and the methods of applying pressure and curing before an effective composite repair can be accomplished. For more detailed information regarding other repair processes, refer to Jeppesen's *Advanced Composites* textbook.

FIBER ORIENTATION

During the preparation phase of any composite repair, you must determine the fiber orientation prior to beginning the repair. As discussed previously, you can find the fiber orientation by referencing the structural repair manual. When repairing composite structures, the fiber orientation of the repair patch must be in the same direction as the original to provide uniform and consistent stress diffusion throughout the part.

If the fiber orientation is not correctly applied, the strength is dramatically reduced increasing the chance of failure. On a bolt of fabric, the warp direction is found by looking at the selvage edge. However, on the finished part, the selvage edge is removed making it more difficult to identify the warp direction, which is 0°. A **warp clock** or **compass** tool is used to reference the orientation of the warp direction. [Figure 3-67]

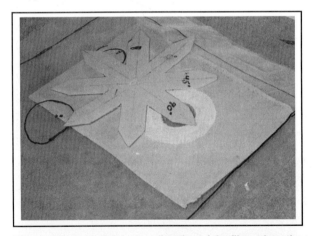

Figure 3-67. To find the warp direction of the fiber, place the warp compass on the repair and align the warp direction of each successive patch in the same orientation as the corresponding layer. Each layer may have a different warp orientation such as 0°, 15°, 45°, 90°, etc. Manufacturers typically have the warp of the top layer of a part running at zero degrees. The structural repair manual defines the zero degree reference orientation.

A general procedure to cut replacement patches to the correct size and shape is described below:

1. Mark the fiber zero degree reference point on the part as found in the structural repair manual.

2. Lay the warp compass on the repair area and orient the zero degree mark on the compass with the zero degree mark on the structure.

3. Lay a clear plastic sheet on top of the repair area and trace the shape of the repair cutout on it starting with the bottom cutout.

4. Note the warp orientation on the plastic for each layer traced.

5. Remove the plastic from the repair and place it over the replacement fabric, being careful to orient the warp as marked on the plastic with the warp of the fabric.

6. Cut the impregnated fabric to the correct shape as outlined on the plastic using a sharp razor knife. [Figure 3-68]

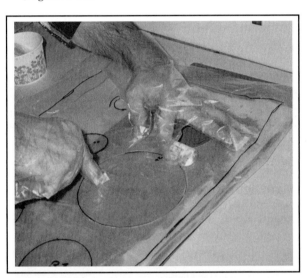

Figure 3-68. Once the fabric is impregnated, the marked plastic sheet is placed over it and the patches cut out with a sharp razor knife. Be sure to properly orient the warp direction of the fabric with the warp reference marks on the plastic sheet.

IMPREGNATING RAW FABRIC

Because pre-impregnated fabrics may not be available for a repair, it is often necessary to impregnate the fabric at the time of the repair. The impregnating process should be done as close to the time you plan to use the material as possible. Remember the pot life when estimating the time needed to properly complete the repair.

Once the shape of the bonding patches has been determined, the repair fabric is placed on a clean work surface such as another sheet of plastic. Be sure to weigh and mix the resins according to the manufacturer's requirements. The liquid resin is poured onto the fabric and worked into it using a squeegee. Keep in mind that the resin must fully saturate the fabric to produce a uniformly solid structure.

When working with a squeegee be sure not to damage the fiber orientation or fray the fabric. Generally, the fabric/resin ratio is 60:40, which produces a matrix that provides complete fiber support. A resin-rich fabric is more susceptible to cracking, due to the lack of fiber support. A resin-starved fabric does not provide the proper stiffness and strength needed for the repair. [Figure 3-69]

Figure 3-69. Impregnating fabric with a catalyzed resin is done frequently in the field. Be sure not to damage the fiber orientation or fray the fabric while using the squeegee. Damaging the fiber may compromise the structure decreasing its strength and uniformity.

APPLYING PRESSURE

When performing a composite repair, pressure should be applied to the surface during the curing operation. Applying pressure assists in the removal of the excess resin that is squeezed out. In addition, sustained pressure compacts the fiber layers together, removes trapped air, and maintains the contour of the repair relative to the original part. It also holds the repair securely to prevent shifting.

Several methods of applying pressure commonly used in composite repair are described next. [Figure 3-70]

- **Shot bags** are effective on large contoured surfaces that cannot be clamped. A plastic sheet must be used to keep the bag from sticking to the repair. A disadvantage to this type of pressure application is that it cannot be used on the underside of any part.

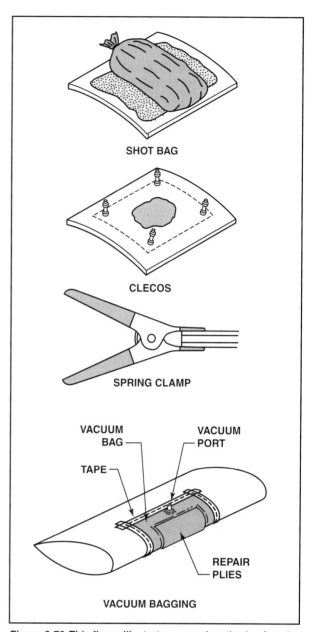

Figure 3-70. This figure illustrates several methods of applying pressure to composite repairs. Each method has advantages and disadvantages specific to their application. Make sure you understand these limitations before beginning the repair.

- **Clecos** are used in conjunction with pre-shaped caul (pressure) plates to support the backside of the repair. This method is not generally recommended due to the added problems of producing holes for the clecos in the structure.

- **Spring clamps** are used in conjunction with caul plates to distribute pressure evenly over the repair area. This method is also not recommended because they do not compensate for pressure changes when the resin begins to flow. Spring clamps also increase the chance of compression damage.

- **Peel ply** may be used to apply pressure in places that are inaccessible to vacuum bagging or that have varied contours and shapes. Peel ply is a section of nylon fabric that is generally used during the vacuum bagging process to facilitate the removal of the bleeder material from the repair area.
- **Vacuum bagging** works by using atmospheric pressure to provide an even pressure over the surface of the repair. This method is the most effective pressure application method used in today's composite repair. It is used on large surfaces and for most repair work.

METHODS OF CURING

Composite matrix systems cure by chemical reaction. Some matrix systems can cure at room temperature while others require heat to achieve maximum strength. Failure to follow the proper curing requirements and improper use of the curing equipment can cause defects in the repair.

Room Temperature Cure

Some types of composite repairs may be cured at room temperature (65–80° F) over a time period of 8–24 hours depending on the type of resin used. In some cases, room temperature curing can be accelerated by applying low heat (140–160° F). Full cure strength is not usually achieved for five to seven days. Room temperature cures are used on nonstructural or lightly loaded parts.

Heat Curing

Most advanced composites utilize resins that require high temperatures during the curing process in order to develop full strength. Consequently, the repair of parts that use these types of resins must also cure at high heat settings (250–750° F) to restore the original strength. The amount of heat applied must be controlled by monitoring the surface temperature of the repair. Although heat curing may produce a stronger repair, overheating can cause severe damage. Do not exceed recommended curing temperatures, so as to avoid disintegration or further delamination of the part.

Heat curing can be accomplished using several different methods, described below.

- **Heat lamps** are not recommended due to the uncontrolled heating of the part. Heat lamps may localize the heat in one spot causing uneven curing.
- **Heat guns** must be controlled with a temperature monitor. Heat guns can produce heat up to 750° F when left on continuously. Excessive heat can evaporate resins, leaving dry areas in the part.
- **Oven curing** offers controlled and uniform heating of all repair surfaces. Some ovens incorporate vacuum ports to provide pressure while curing. Disadvantages to oven curing are that the part must be removed from the aircraft and must be small enough to fit into the oven.
- **Autoclaves** are customarily used in the manufacturing of composites, rather than in repair.
- **Hot patch bonding** utilizes a flexible silicon heating-blanket that incorporates a temperature control. This is the preferred method of curing, due to the controlled even heating of the part. Most hot patch bonding machines also incorporate a vacuum pump to apply pressure during the curing process. [Figure 3-71] [Figure 3-72]

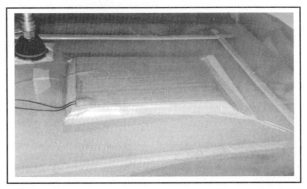

Figure 3-71. Hot patch bonding is accomplished with a heat blanket and is usually used in conjunction with vacuum bagging.

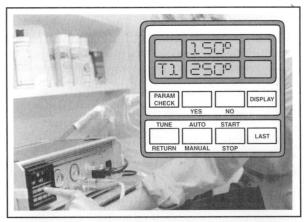

Figure 3-72. Controller devices provide uniform heating during the curing process. They are usually simple machines with easy programmable settings.

Repairs must also be allowed to cool at the proper rate, because they gain much of their strength in the cooling process. A slow temperature rise and gradual cooling period is desirable. Two common methods of heat curing are the **step cure** method and **ramp and soak** method.

Step curing is used in conjunction with a manually operated controller. It requires that the technician make the temperature adjustments manually at specific time intervals. The temperature is brought up slowly to a specific temperature then held for a certain period, then brought up and held again. This step-up process is continued until the cure temperature is reached. After the cure time has elapsed, the temperature is stepped down by slowly reducing the temperature in increments. [Figure 3-73]

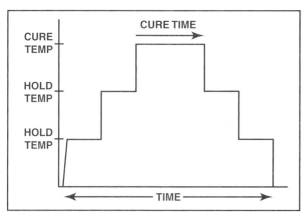

Figure 3-73. This figure illustrates the basic concept of step curing. The part is heated and held at certain temperatures in a step-up process until it reaches its cure temperature. The part is cooled in the same step process. It is cooled to certain temperatures and held in a step down process until the part reaches full strength.

A more sophisticated and accurate curing process may be accomplished using a programmable controller. This controller may be programmed in a **ramp and soak** mode, which is used to heat or cool a repair at a specific rate. This heating process is called the "ramp" time.

For example, a structural repair manual may specify that a repair be heated to a temperature of 400° Fahrenheit and that the temperature must be reached at a slow, constant rate of eight° per minute. If room temperature was 70° F, it would take approximately 41 minutes to reach the 400° mark (400 -70 = 330° F; 330° \ 8° each minute = 41.25 minutes).

Once the repair has been heated to 400° F, the structural repair manual may require that the part must be held at this temperature for a specific amount of time. This specific hold time is referred to as the "soak" period.

Following the soak period, the repair manual may specify that the temperature be ramped (cooled) down to room temperature at a specific rate. Again, citing the above example, a five degree per minute cool down rate would take an hour and six minutes (400° - 70° = 330°; 330° \ 5° each minute = 66 minutes). A ramp and soak profile depicts the entire heating and cooling cycle. Structural repair manuals do not generally provide ramp times because the starting temperatures may not always be the same. If you are working in a cold climate, the ramp up-time will be substantially longer than if you were working in a warmer climate. [Figure 3-74] [Figure 3-75]

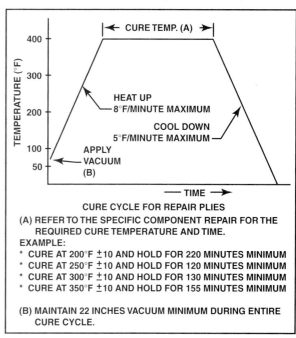

Figure 3-74. This figure illustrates a typical ramp and soak profile used to cure a composite.

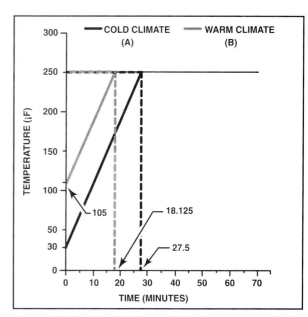

Figure 3-75. Part A of this figure illustrates a ramp profile for a composite cured in a cold climate. If the outside temperature is 30° F, the cure ramp up-time is eight° F per minute (250° - 30° = 220° \ 8° = 27.5 minutes). In this case, it will take 27.5 minutes to climb to the cure temperature of 250° at a rate of 8° per minute. Part B illustrates a ramp profile for a composite cured in warm climate. If the outside temperature is 105° F, the cure ramp up-time is 8° per minute (250° - 105° = 145° \ 8° = 18.125). It would take 18 minutes to climb to the cure temperature 250° at a rate of 8° per minute.

VACUUM-BAGGING PROCESS

Vacuum bagging is one of the most common techniques used to apply pressure to a composite repair. A general description of the vacuum-bagging process follows, to provide you with a basic knowledge of this type of repair process. Always refer to the manufacturer's repair manual for specific instructions for each repair.

Once the composite repair is produced and the patches are in place, the repair area is covered with a perforated parting film (peel ply), which allows any excess matrix to flow through to the upper surface. The parting film also "feathers" the seams or any overlap of fabric to produce a smooth surface. Release fabrics are used in place of parting films if a rough surface is desirable in order to prepare the surface for painting.

Next, a bleeder material is placed over the repair to absorb the excess resin that is squeezed out during the pressure application. Another parting film is then placed over the bleeder to keep it from sticking to the pressure plate, which is placed over the last layer of parting film. A surface breather material is placed over the pressure plate to allow airflow up through the vacuum valve. Bleeder and breather materials can be made of the same material and, in many cases, are used interchangeably.

If a thermocouple or temperature sensing device is used, it should be laid next to the repair area and sealed with sealant tape. In addition, if a heat blanket is used to cure the part, a parting film should be installed to keep it from sticking to the repair. Sealant tape is attached around the edges of the repair to produce an airtight seal. [Figure 3-76]

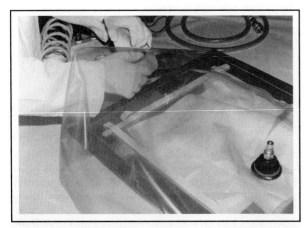

Figure 3-76. Sealant tape and vacuum bagging film is being applied to the repair area.

Vacuum-bagging film is laid over the repair and the edges sealed with sealant tape to produce an airtight seal. Commonly, bagging films are made of nylon to resist tears and punctures. The vacuum bag should be lightly pressed into the tape until it forms an airtight seal. An X is cut into the bagging film and slipped over the valve, making sure the hole is not too big. The film should go down past the threads to the base of the valve. The area around the valve is then sealed using a rubber grommet. The valve is then attached to a vacuum hose, and a vacuum leak test performed. [Figure 3-77] [Figure 3-78]

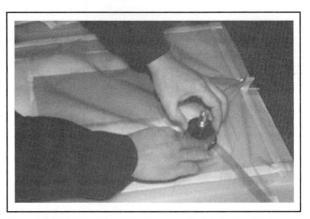

Figure 3-77. The valve is being installed into the X-shaped cut made in the vacuum bag material.

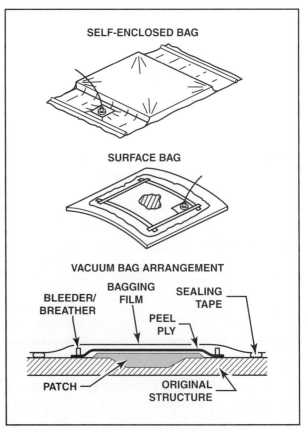

Figure 3-78. There are several different types of vacuum bag set-ups, such as the self-enclosed bag and surface bag.

MECHANICALLY FASTENED REPAIRS WITH PRE-CURED PATCHES

At times, the facilities and bagging equipment are not available to produce a proper composite repair. In this case, a temporary repair made of a pre-cured patch inserted with blind fasteners may be used. However, this type of repair does not produce a structure with the same strength as the original, and it may cause vibration because it is not a flush repair. If composite patches are required, kits with pre-cured patches may be available. Pre-cured patches come in several sizes and are produced to have the fibers of each layer in the correct orientation.

In addition, some manufacturers offer various sizes of core materials that are bonded to pre-cured laminates. The technician can route out the damaged area and simply insert this type of core and laminate patch. This type of repair may have a type of adhesive pre-applied to help it bond. These types of patches are usually stabilized using some type of mechanical blind fastener, which is drilled through the patch and into the original part. The problem with using any type of rivet is that they have a tendency to crush the core and produce delamination.

These types of repairs must be performed using the correct type of fasteners. Hole expanding fasteners such as MS20470 rivets should not be used in composite structures because of the possibility of causing damage. Impact damage and delamination may occur due to the pressure of the rivet gun and bucking bar and the expansion of the rivet. In addition, you must also make sure metallic fasteners will not react with the composite and cause galvanic corrosion. For example, metal fasteners used with carbon/graphite composites must be made of corrosion resistant steel or titanium to prevent this electrolytic action. [Figure 3-79]

POTTED REPAIRS

Potted repairs use a filler to complete the composite repair process. They provide an alternative to installing a core material plug but do not provide as much strength as a core material. Filling a hole with a resin and filler mixture adds weight and decreases the flexibility of the part. However, many structural repair manuals still list the potted repair as a viable repair for composite structures.

Most potted repairs are appropriate for honeycomb core sandwich structures with damage up to one-inch in diameter. In some cases, it is permissible to

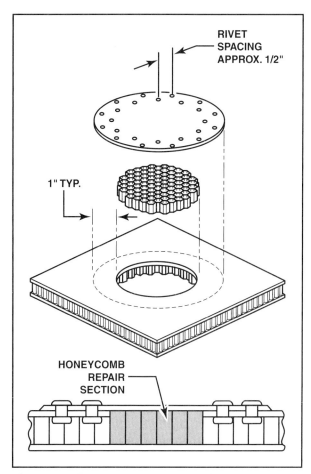

Figure 3-79. Pre-cure patches using blind fasteners and sheet metal plates are usually considered temporary repairs due to the inherent problems associated with them.

drill a small hole into a delaminated area and inject resin to strengthen the part. A typical potted repair procedure requires the technician to:

1. Clean the damaged area.

2. Sand out the delaminated area.

3. Fill the core area with a resin and microballoon mixture.

4. Prepare and install repair patches.

5. Apply pressure and cure.

6. Refinish the part.

Undercut Potted Repair

A composite structure that sustained puncture damage through one face sheet and the core material may be repaired using a potted repair and surface patch. This type of repair differs slightly from the old fiberglass repair in that the technician undercuts the core material and installs a surface patch to help retain the plug. [Figure 3-80]

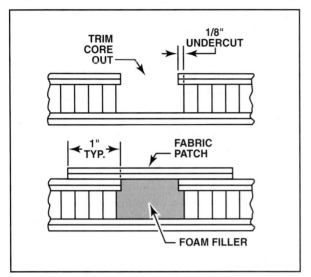

Figure 3-80. The undercut and surface patch of this type of repair helps retain the plug. In earlier fiberglass potted repairs, no surface patch was required and the core material was routed out with no undercut, which allowed the plug to pop out if the structure flexed.

Following is a general guideline to an undercut potted repair.

1. Open the puncture with a drill or router to remove the ragged edges and broken fibers.

2. Clean out the crushed core material.

3. Undercut the core approximately .125 inch and mark the outline of the undercut on the surface of the part.

4. Prepare the surface for bonding.

5. Clean out and vacuum the hole.

6. Fill the hole with a foam filler and allow it to cure. Make sure all air pockets are displaced.

7. Cut the repair patches to size allowing an overlap of approximately one-inch from the edge of the hole.

8. Prepare and install the bonding patches.

9. Apply pressure and cure.

10. Refinish the part.

Mislocated Potting Compound

In some cases, the manufacturer supplies a component with potting compound installed to accommodate a fastener. At times, this potting compound is not correctly positioned. Following is a general procedure for repositioning the potting. [Figure 3-81]

1. Ascertain the correct location of the fastener.

2. Drill a 1/8-inch hole at the correct position through one skin only.

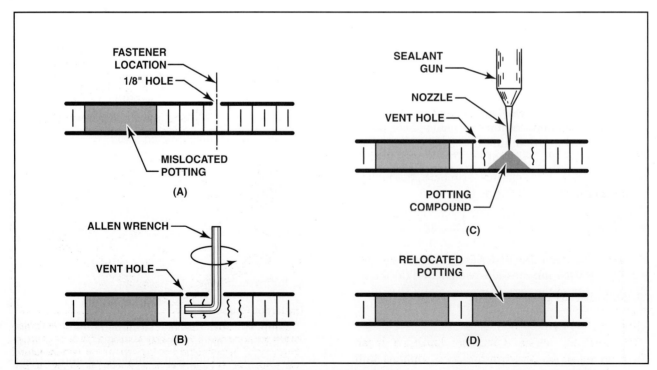

Figure 3-81. When the potting compound for a fastener is mislocated, these steps may be followed to relocate the fastener.

3. Drill a small vent hole approximately one-inch from the center hole.

4. Insert a small Allen wrench through the hole and rotate to break the honeycomb cell walls.

5. Vacuum out the debris.

6. Inject the potting compound using a sealant gun or syringe.

7. Cure in accordance with the manufacturer's instructions.

8. Re-drill the fastener hole and install fitting.

LAMINATE STRUCTURE REPAIR

Laminated structures are susceptible to several different types of damage such as surface scratches, delamination, impact, and puncture damage. Each type of damage requires a different repair procedure. Following are several basic laminate composite repair procedures.

Laminate Cosmetic Repair

A cosmetic defect is a surface scratch that does not penetrate the first structural ply. This type of damage is classified as negligible damage. Superficial scratches, abrasions, or rain erosion can generally be repaired by applying one or more coats of resin to the surface. The general process for repairing a cosmetic defect follows.

1. Clean the repair area with MEK or other approved solvent.

2. Remove the paint from the repair area and feather the edges.

3. Scuff-sand the damaged area to provide a good bond surface.

4. Clean the repair area with solvent to remove all sanding residue.

5. Mix resin and filler or approved surface putty.

6. Apply resin/filler mixture to repair area using a squeegee, brush, or fairing tool.

7. Cover repair area with cellophane and work out all air bubbles.

7. Cure the repair according to the manufacturer's instructions.

8. Re-finish the part.

Delamination Repair

Delamination occurs when the laminate layers become separated or when the plies separate from the core material. It can be caused by sonic vibration, entrapped moisture, and manufacturing defects. Delamination is sometimes referred to as unbonding, or disbonding and, in some cases, can be detected by shining a light over the part and looking at the damaged area at an angle. Delamination may look like a bubble or indentation on the surface of the part.

In some cases, internal delamination is minor enough to repair using a potting compound. It can sometimes be repaired by simply injecting resin into the cavity that was caused by the ply separation. If the delamination is severe enough, it must be removed and repaired or replaced; always check the manufacturer's repair limitations. A typical delamination injection repair procedure for **minor delamination** follows. [Figure 3-82] [Figure 3-83]

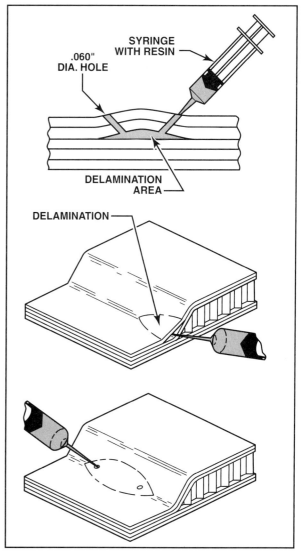

Figure 3-82. Resin injection repairs for small-delaminated areas are considered temporary repairs in the case of structural parts. Delaminated potting repairs on non-structural parts such as fairings or other aesthetic parts may be considered permanent. Check the manufacturer's repair manual for acceptable repair limitations.

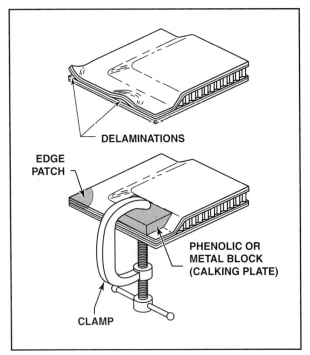

Figure 3-83. Minor edge delamination can sometimes be repaired by injecting resin into the delamination, clamping the edge and allowing the resin to cure. In addition, edges that have been damaged by crushing or puncture can be repaired by scarf cutting, and installing new plies.

1. Clean the surface with an approved solvent.
2. Outline the void area and mark the injection hole locations.
3. Drill two .060-inch holes into the disbonded area taking care not to drill through the part.
4. Inject mixed resin into one hole allowing air to vent from the other.
5. Clean excess resin from the surface of the part.
6. Cure according to the manufacturer's instructions.

Laminate Damage to One Surface
This type of repair calls for the removal and replacement of the damaged laminate plies. Fiber damage to one side of the surface that does not completely penetrate the part may be repaired as follows:

1. Prepare the surface by removing the paint and cleaning.
2. Remove the damage by scarf or step-cutting the plies.
3. Mix the resin and prepare the bonding patches.
4. Apply pressure and cure the part.
5. Re-finish the part. [Figure 3-84]

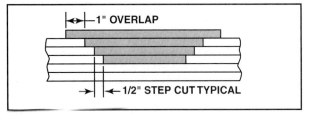

Figure 3-84. This figure illustrates a step-cut repair to a laminate structure with damage to one side only. This repair entails removing the damage and replacing the plies with new composite materials. As discussed previously, each ply is step-cut 1/2-inch and the surface patch is installed with a one-inch overlap.

Laminate Damage Through the Part
Damage that runs through all of the laminate layers can be repaired in several ways depending on the number of plies, the location of the damage, and the size of the damage. Check the manufacturer's repair manual for specific repair limitations regarding each type of damage.

Repairable damage can be fixed in several different ways. The damage can be repaired using a step-cut that starts from one side of the part to the other, or, in the case of thicker laminate structures, repaired by step-cutting from both sides and meeting in the middle. [Figure 3-85]

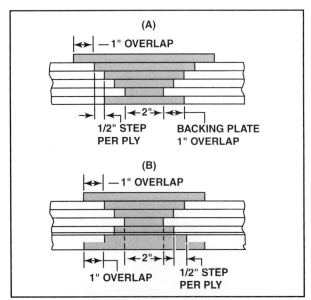

Figure 3-85. View A illustrates a step-cut repair that runs from one side of the part to the other. In addition, a surface patch and backing plate with a one-inch overlap are applied to both surfaces of the repair. View B illustrates a modified step-cut repair. The step-cuts are started from both sides and meet in the middle. This type of repair reduces the size of the patch when performed on thicker laminates. Using view B as an example, the modified two-sided step-cut repair results in a patch that is four inches in diameter. If the one sided step-cut repair was performed on this same five-layer laminate, the patch would be five inches in diameter.

SANDWICH STRUCTURE REPAIRS

Sandwich structures are vulnerable to impact and puncture damage primarily because these types of structures usually incorporate relatively thin face sheets. Because the face sheets of sandwich structures are relatively thin, delaminations commonly occur at the point where the face sheet bonds to the core material. Puncture damage may be repaired in several different ways depending on the size, extent, and location of the damage. Two of the more common types of sandwich structure repair are described below.

Puncture Repair

Small punctures that penetrate one side and into the core material may be repaired using a resin and filler mixture. Check the repair manual damage limitations before proceeding with this type of repair. Generally, small punctures can be repaired using the following procedure.

1. Determine the extent of the damage and check the repair limitations.
2. Vacuum out the hole.
3. Prepare the resin and filler (milled glass fibers).
4. Work the resin/filler mixture into the hole.
5. Cure the resin in accordance with the manufacturer's instructions.
6. Sand the surface with fine sandpaper.
7. Prepare the surface for finishing using an approved solvent.
8. Re-finish the part.

Honeycomb Core Repairs

As discussed previously, potted repairs may be made to areas of damage of up to one-inch diameter. If the damaged area is larger than an inch or in a more critical area, a balsa wood or composite honeycomb plug may be cut and bonded in place. If balsa is used, cut the plug so the grain is perpendicular to the skin. If honeycomb material is used, it should be the same density as the original. [Figure 3-86]

Aluminum alloy-faced honeycomb repairs of a simple puncture of an aluminum alloy-faced honeycomb panel can be accomplished by covering with a doubler plate. The plate should be cut from a piece of aluminum the same or up to one and one-half times the thickness of the original skin thickness. Additionally, the doubler should be tapered at a ratio of about 100:1

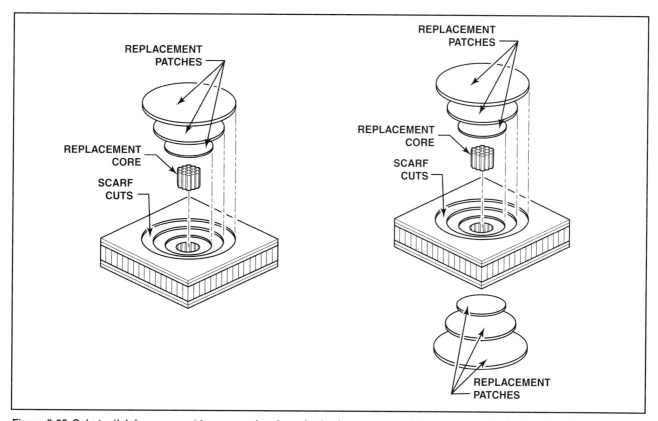

Figure 3-86. Substantial damage must be removed and repaired using a more sophisticated repair technique. In this case, a router is used to remove all of the damaged material. The router may be adjusted to remove one of the face skins only, a face skin and part of the core, a face skin and all of the core, or both the face skins and the core. A plug of core material is then inserted into the structure and held in place by the replacement plies and surface patch.

MAINTENANCE ENTRIES

After performing any repair, a proper maintenance entry is required before the job is complete. FAR Part 43.9 outlines the requirements for a proper maintenance entry. Several items specific to composite repair should be included in the description of work performed such as the time, temperature, and pressure used in the cure cycle. In addition to the type of fabric, core material, matrix, and adhesives used.

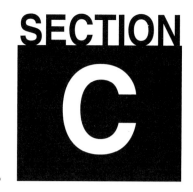

TRANSPARENT PLASTIC MATERIALS

The increasing use of modern lightweight plastics in aircraft construction has reduced aircraft weight and in most cases, improved performance. Plastics have been used in windows, windshields, interior furnishings, and various light covers for a long time. Since transparent materials have unique characteristics, they will continue to be used because of low cost and ease of use. Thermoplastic materials that soften with the application of heat are types that a maintenance technician will likely be required to maintain.

TYPES OF TRANSPARENT PLASTICS

Two types of transparent thermoplastic materials are used for aircraft windows and windshields: one has a cellulose acetate base, and the other is a synthetic resin, acrylic. Cellulose acetate was primarily used in the past. Since it is dimensionally unstable and turns yellow after it has been installed for a time, it is rarely used today, and is not considered an acceptable substitute for acrylic.

Acrylic plastics, known by the tradenames of Lucite® or Plexiglas®, and by the British as Perspex®, meet the military specifications of MIL-P-6886 for regular acrylic, MIL-P-5425 for heat-resistant acrylic, and MIL-P-8184 for craze-resistant acrylic.

One way to distinguish between acrylic and acetate is to rub a bit of acetone on a small piece of the material. If it turns white but does not soften, it is acrylic. Acetate softens without changing color. Perform a positive check by burning a scrap of the material. Acrylic burns with a steady, clear flame, while acetate burns with a sputtering flame and gives off a dark smoke. The odor of burning acrylic is somewhat pleasant but acetate is definitely unpleasant.

STORAGE AND HANDLING

Since transparent thermoplastic sheets soften and deform when heated, store them in a cool, dry location, away from heating coils, radiators, or steam pipes. Keep the fumes found in paint spray booths or paint storage areas away from plastics. Most acrylics have adhesive papers on both sides of the sheets to protect them and prevent scratches. Keep plastic sheets out of the direct rays of the sun since deteriorated adhesive paper can bond to the surface, making it difficult to remove.

Store plastic sheets with the masking paper in place, in bins which are tilted at a ten-degree angle from vertical. This helps prevent buckling. If the sheets must be stored horizontally, avoid getting dirt and chips between them. Stacks of sheets should never be over 18 inches high, with the smallest sheets stacked on top of larger ones, eliminating unsupported overhang. Leave the masking paper on the sheets as long as possible, and be careful not to scratch or gouge the sheets by sliding them against each other or across rough or dirty tables. Store formed sections with ample support to retain their original shape and avoid vertical nesting. Protect formed parts from temperatures higher than 120°, and leave the protective coating in place until they are installed on the airplane.

FORMING PROCEDURES AND TECHNIQUES

Transparent acrylic plastics get soft and pliable when heated to their forming temperatures. When soft, acrylics can be molded to almost any shape, and after cooling, they retain the shape to which they were formed.

Acrylic plastic may be cold-bent into a single curvature if the material is thin and the radius around which it is bent is at least 180 times the thickness of the sheet. Cold bending beyond these limits will impose so much stress on the surface of the plastic that tiny fissures or cracks will form.

HEATING

Before heating any transparent plastic material, remove all of the masking paper and adhesive from it. If the sheet is dusty or dirty, wash it with clean soap and water and rinse it well. Dry it thoroughly by blotting it with soft, absorbent paper towels.

For best results when hot-forming acrylics, use the temperatures recommended by the manufacturer. Use a forced-air oven capable of operating over a temperature range of 120° to 347° F (49° to 190° C). For uniform heating, it is best to hang the sheets vertically. [Figure 3-87]

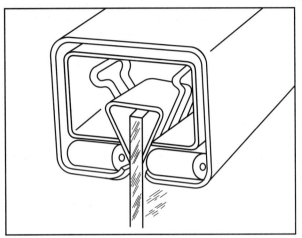

Figure 3-87. While heating acrylic sheets, hang them with spring clips attached to the trim area.

If the piece is too small to hold with clips, or if there is not enough trim area, lay the sheets on shelves or racks covered with soft felt or flannel. Be sure there is enough open space to allow the air to circulate around the sheet and heat it uniformly.

Wear cotton gloves when handling heated plastic to keep finger marks off of the soft surface. You may heat small forming jobs such as landing light covers in a kitchen-type oven. You may also use infrared heat lamps if they are arranged on seven- to eight-inch centers and enough are utilized to heat the sheet evenly. Place the lamps about 18 inches from the material. Never use hot water or steam directly on the plastic to heat it since it may become milky or cloudy. If the part gets too hot during the forming process, surface bubbles may form that will impair the optical qualities of the sheet.

FORMS
Heated acrylic plastic will form with almost no pressure, so the molds can be of very simple construction. Molds made of pressed wood, plywood, or plaster are adequate to form simple curves, but you may need reinforced plastic or plaster to shape complex or compound curves. The mold must be completely smooth. To ensure this, sand the mold and cover it with soft cloth such as flannel or billiard felt. Make the mold large enough to extend beyond the trim line of the part, and make provisions for holding the hot plastic snug against the mold as it cools.

A mold can be made for a complex part by using the damaged part itself. If the part is broken, tape the pieces together, wax or grease the inside so the plaster will not stick to it, and support the entire part in sand. Fill the part with plaster and allow it to harden, then remove it from the mold. Smooth out any rough parts, cover it with soft cloth, and it will be ready to form the new part.

SIMPLE CURVE FORMING
The temperature required for forming sheet materials depends on the type of material and its thickness. Generally, the more complex the shape, the higher the temperature required. [Figure 3-88]

THICKNESS OF SHEET	0.125		0.250		0.125		0.250	
Type of Forming	Regular acrylic plastic. MIL-P-6886				Heat-resistant acrylic plastic, MIL-P-5425, and craze-resistant acrylic plastic, MIL-P-8184			
	C°	F°	C°	F°	C°	F°	C°	F°
SIMPLE CURVE	113	235	110	230	135	275	135	275
STRETCH FORMING (DRY MOLD COVER)	140	284	135	275	160	320	150	302
MALE AND FEMALE FORMING	140	284	135	275	180	356	170	338
VACUUM FORMING WITHOUT FORM	140	284	135	275	150	302	145	293
VACUUM FORMING WITH FEMALE FORM	145	293	140	284	180	356	170	338

Figure 3-88. To select a temperature for forming sheet material, consider the complexity of the shape, the type of material, and its thickness.

After heating the plastic, remove it from the oven and drape it over the prepared mold. Carefully press the hot plastic to the mold, and either hold or clamp the sheet in place until it cools. This process may take from ten minutes to half an hour, but do not force the plastic to cool.

COMPOUND-CURVE FORMING
Compound-curve forming is normally used for such parts as canopies, complex wing tips, or light covers. Four methods, each requiring specialized equipment, are commonly used. These methods include stretch, male and female die, vacuum, and vacuum forming with female dies.

STRETCH FORMING
With stretch forming, preheated acrylic sheets are stretched mechanically in a manner similar to the method for a simple curved piece. Take special care to preserve uniform thickness of the material since certain parts will have to stretch more than others.

MALE AND FEMALE DIE FORMING
Male and female die forming requires expensive male and female dies. The heated plastic sheet is placed between the dies, which are then pressed together. This produces a part that is precisely formed on both inner and outer surfaces and is removed from the dies after cooling.

VACUUM FORMING WITHOUT FORMS
Many aircraft canopies are formed by vacuum pressure without molds. In this process, a clamp with an opening of the desired shape is placed over a vacuum box and the heated sheet of plastic is clamped in place. When air in the box is evacuated, outside air pressure will force the hot plastic through the opening and form a concave bubble that becomes the canopy. It is the weight of the plastic sagging beneath the form that shapes the bubble.

VACUUM-FORMING WITH A FEMALE FORM
If a shape is needed other than that which will be formed by the weight of the plastic, a female mold must be used. In this situation, the mold is placed below the plastic sheet and a vacuum pump is connected. When the air is evacuated, the outside air pressure will force the hot plastic sheet into the mold.

SAWING
Several types of saws can be used with transparent plastics, but circular saws are the best for straight cuts. To prevent binding, the blades should be hollow-ground or have some teeth set. By setting (bending) the teeth, the saw cuts a wider slot, minimizing binding and allowing the blade to run cooler. After the teeth are set, they should be side-dressed to produce a smooth edge on the cut.

To prevent overheating of the acrylic sheet, feed it into the saw blade slowly. If the plastic begins to smoke or the edges of the cut appear smeared, the feed may be too fast for the thickness of the material. The saw blade may also be dull or the wrong type.

Band saws are recommended for cutting acrylic sheets when the cuts must be curved. Close control of size and shape may be accomplished by band sawing a piece to within 1/16-inch of the desired size, as marked by a scribed line on the plastic. Once it is rough cut, sand the part to the correct size with a drum or belt sander.

DRILLING
Unlike soft metal, acrylic plastic is a very poor conductor of heat, and you must make provisions for removing the heat when it is drilled. Deep holes need cooling, and water-soluble cutting oil is a satisfactory coolant, since it tends not to attack the plastic.

Carefully grind the twist drill used to make holes in transparent plastics to be free from nicks and burrs which would affect the surface finish. Grind the drills with a greater included angle (usually 150°) than would be used for soft metal. The rake angle should be zero. When drilling plastics, a back up material such as a scrap of wood should be used to prevent damage when the drill breaks through the underside of the part. [Figure 3-89]

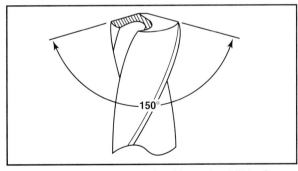

Figure 3-89. Drill acrylic plastic with a twist drill having an included angle of approximately 150°.

The patented Unibit is good for drilling small holes in aircraft windshields and windows. Four sizes enable the technician to cut smooth 1/8- to 7/8-inch diameter holes with no stress cracks around the edges. [Figure 3-90]

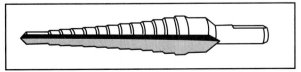

Figure 3-90. A Unibit drill is useful for cutting stress-free holes in acrylic sheet material.

CEMENTING

Cementing is the process of joining two materials to form a structure as strong as the original material. With plastics, materials are softened with cement and held under pressure until the two pieces have joined at a molecular level.

APPLICATION OF CEMENT

Acrylic plastics may be joined by using ethylene dichloride, a clear liquid solvent. It softens the material and forms a cushion between the two pieces, allowing a thorough intermingling of the molecules between the surfaces.

There are two methods of cementing plastics: soaking and gluing. In the soaking method, soak one of the parts to be joined for about 10 minutes, or until a cushion forms. This cushion should be deep enough to take care of any discrepancies in the fit of the parts. After the cushion forms, press the wet surface against the opposite dry surface, and allow it to set for about half a minute. Then apply enough pressure to the joint to squeeze out any air bubbles and assure complete intermingling of the cushions. [Figure 3-91]

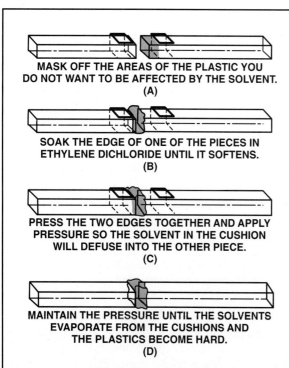

Figure 3-91. The soaking method of bonding acrylic materials. When it is convenient, soak one of the surfaces in the solvent.

Use the glue method when it is not convenient to soak the edges of either of the pieces. Dissolve shavings of acrylic plastic in ethylene dichloride, and mix for a thick syrup or glue consistency. Spread the glue on one of the surfaces and assemble the parts. Allow the glue to set for a few seconds, then apply pressure to mingle the cushions.

When applying any type of cemented patch to an acrylic plastic, make certain that the surface areas around the repair are adequately protected from the solvent. Use pressure-sensitive aluminum tape to mask this area.

APPLICATION OF PRESSURE

It is extremely important, once the parts to be joined have been assembled and the cushions adequately formed, to hold an even pressure on the joint. When acrylic absorbs the solvent and forms the cushions, it expands. However, when the solvent evaporates from the intermingled cushion, acrylic shrinks. Maintain a constant pressure during the cementing process. Spring clamps, weights, or elastic straps are acceptable for this, but parallel clamps or C-clamps will loosen as the joint shrinks. The pressure must be great enough to force out any air bubbles, but not so much as to break down the cushions or to cause localized stresses resulting in fissures and cracking. [Figure 3-92]

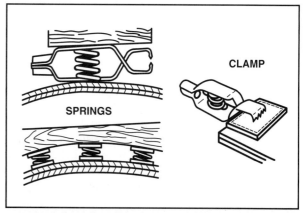

Figure 3-92. Apply pressure to acrylic plastics by a method that will retain the pressure as the joint shrinks from the evaporation of the solvent.

CURING

Solvent in a cemented joint will never completely evaporate from the acrylic or the cushions as they form, and since the cushions are expanded, they will be weaker than the original material. If the cemented material is heated to around 122° F (50° C) and held at this temperature for about 48 hours, the cushions will further expand and the solvent contained in them will be diffused into a larger volume of plastic and become less concentrated. Subsequently, the cemented joint develops more strength. Curing is often called heat treating, and it hardens the surface enough that machining and polishing can be done without soft spots causing an uneven surface. [Figure 3-93]

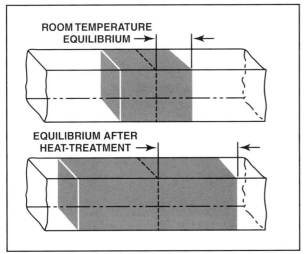

Figure 3-93. Heat treatment of a joint between acrylic pieces will disperse the solvent through a greater volume of the plastic and will increase the strength of the joint.

REPAIRS

When windshields and side windows made of acrylic plastics are damaged, they are usually replaced unless the damage is minor and not in the line of vision. Repairs usually require a great deal of labor, and replacement parts are readily available, so replacement is normally more economical than repair.

TEMPORARY REPAIRS

There are times when a windshield is cracked and must be put in good enough condition to fly to a location where it can be replaced. In this situation, make temporary repairs by stop-drilling the ends of the crack with a number 30 drill to prevent the crack from growing larger. Drill a series of number 40 holes a half-inch from the edge of the crack and about a half-inch apart. Lace through these holes with brass safety wire. Another way to make a temporary repair is to stop-drill the ends of the crack, and then drill number 27 holes every inch or so throughout the crack. Use AN515-6 screws and AN365-632 nuts, with AN960-6 washers on both sides of the plastic. This will hold the crack together and prevent further breakage until the windshield can be properly repaired or replaced. [Figure 3-94]

PERMANENT REPAIRS

Windshields or side windows with small cracks that affect only the appearance rather than the airworthiness of a sheet may be repaired by first stop-drilling the ends of the crack with a number 30 drill to relieve the stresses. Then use a hypodermic syringe and needle to fill the crack with ethylene dichloride, and allow capillary action to fill the crack completely. Soak the end of a 1/8-inch acrylic rod in ethylene dichloride to soften it and insert it in the stop-drilled hole. Allow the repair to dry for about half an hour, then trim the rod flush with the sheet.

POLISHING AND FINISHING

Scratches and repair marks, within certain limitations, can be removed from acrylic plastic. Do not sand any portion of a windshield that could adversely affect its optical properties and distort the pilot's vision. In addition, do not reduce the thickness of windows on pressurized aircraft to the point where the window would be weakened. The manufacturer's service manual will specify the minimum allowable thickness.

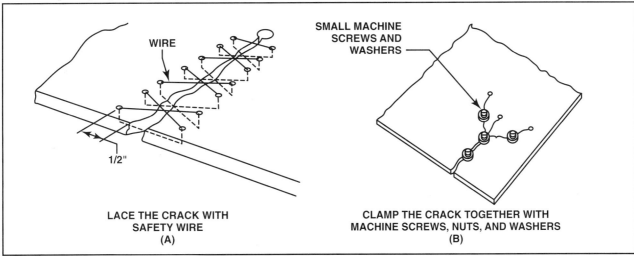

Figure 3-94. When an acrylic windshield is cracked, it can be fixed temporarily with brass safety wire, or with small machine screws and washers.

If there are scratches or repair marks in an area that can be sanded, use 320- or 400-grit abrasive paper wrapped around a felt or rubber pad and hand sand in a circular motion. Use light pressure and a mild liquid soap solution as a lubricant. After sanding, rinse the surface thoroughly with running water. Then, using progressively finer grit paper, continue to sand lightly. Sand and rinse until all of the sanding or repair marks have been removed. After using the finest grit abrasive paper, use rubbing compound and buff in a circular motion to remove all traces of sanding.

CLEANING

Clean acrylic windshields and windows by washing them with mild soap and running water. Rub the surface in the stream of water with bare hands. After all the dirt has been flushed away, dry the surface with a soft, clean cloth or tissue and polish it with a windshield cleaner specially approved for use on transparent aircraft plastics. These cleaners may be purchased through aircraft supply houses. A thin coating of wax fills any minute scratches that may still be present and causes rain to form into droplets, which are easily blown away by the wind.

PROTECTION

Acrylic windshields are often called lifetime windshields to distinguish them from those made of the much shorter-lived acetate material. However, even acrylic must be protected from the ravages of the elements. When an airplane is parked in the direct rays of the sun, its windshield will absorb heat and actually become hotter than either the inside of the airplane or the outside air. The sun will cause the inside of a closed airplane to become extremely hot. The intense heat also affects plastic side windows.

To protect against damage, it is wise to keep the airplane in a hangar. If this is not possible, provide some type of shade to keep the sun from coming in direct contact with the windshield. Some aircraft owners use a close fitting, opaque, reflective cover over the windshield, but in many cases, this has done more harm than good. The cover may absorb moisture from the air and give off harmful vapors, and if it touches the surface of the plastic, it can cause crazing or minute cracks to form in the windshield. Another hazard in using such a cover is that sand can blow up under the cover and scratch the plastic.

WINDSHIELD INSTALLATION

Aircraft windshields may be purchased either from the original aircraft manufacturer or from any of several FAA-PMA (Federal Aviation Administration - Parts Manufacturing Approval) sources. These windshields are formed to the exact shape required, but are slightly larger than necessary so they may be trimmed to the exact size.

After removing the damaged windshield, clean all of the sealer from the grooves and cut the new windshield to fit. New windshields are covered with either protective paper or film to prevent damage during shipping and installation. Carefully peel back just enough of this covering to make the installation. Fit the windshield in its channels with about 1/8-inch clearance to allow for expansion and contraction. If any holes are drilled in the plastic for screws, oversize them about 1/8-inch to allow for expansion.

Place the proper sealing tape around the edges of the windshield and install it in its frame. Tighten the screws that go through the windshield and then back out a full turn, allowing the plastic to shift as it expands and contracts. Do not remove the protective paper or film until the windshield is installed and all of the securing screws are in place.

CHAPTER 4

AIRCRAFT WELDING

INTRODUCTION

Metallic aircraft structures are composed of many individual pieces that must be securely fastened together to form a complete structural unit. Fusion welding, non-fusion welding, hardware fasteners, and adhesives are the principal methods used in the construction and repair of metal aircraft joints. This chapter describes the equipment and basic procedures needed to join metals using fusion and non-fusion welding techniques, and the methods used to repair welded aircraft structures. Since welding has become so highly specialized, the FAA no longer requires a person to perform welding to a high skill level to become a certified aircraft technician. If you have an interest in specializing in this area, you will need to pursue advanced training. One source of additional information on advanced aircraft welding techniques is the book *Welding Guidelines*, which is available from Jeppesen.

SECTION A

WELDING PROCESSES

Aircraft maintenance technicians are no longer required to be highly proficient at performing weld repairs or fabricating welded parts for aircraft. In fact, in most cases the technician should delegate the responsibility for welding aircraft parts to FAA-certified repair stations that specialize in welding processes. However, aircraft technicians are expected to have sufficient knowledge to enable them to identify defective welds to determine the airworthiness of an aircraft.

The two most prominent methods of welding aircraft structures and components are fusion and non-fusion. **Fusion** welding is the blending of compatible molten metals into one common part or joint. Fusing of metals is accomplished by producing sufficient heat for the metals to melt, flow together and mix. The heat is then removed to allow the fused joint to solidify. **Non-fusion** welding is the joining of metals by adhesion of one metal to another. The most prominent non-fusion welding processes used on aircraft are brazing and soldering, which are covered in detail later in this section.

FUSION WELDING PROCESSES

The three principal methods of fusion welding are gas, electric arc, and electrical resistance. Fusion welding has, for some time, been the method of choice for constructing the structural framework of aircraft. It still is used extensively on the framework of many modern small airplanes, particularly in aerobatic, agricultural and homebuilt machines, but is used to some degree in virtually all aircraft. Fusion welding results in superior strength joints because the metal parts are melted together into a single solid object. Since fusion-welded joints are used extensively in high-stress applications, their failure is likely to have catastrophic consequences. To fully appreciate the level of detail that must be exercised when inspecting welded components, you must be aware of the characteristics that define a quality fusion-welded joint.

GENERAL EVALUATION OF WELDS

A good weld is uniform in width, with even ripples that taper off smoothly into the base metal. It shows good penetration, or depth of fusion. In fact, penetration is the most important characteristic of a good weld. To obtain the proper amount of penetration and proper weld dimensions, a welder must use the correct type and size of filler rod and appropriate welding technique for the thickness and type of the material to be joined. A good weld also is well rein-

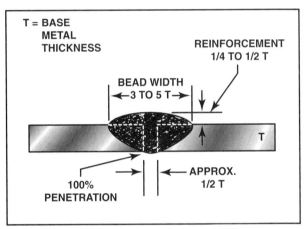

Figure 4-1. Proper penetration is the single most important characteristic of a good weld.

forced. This means there is enough filler material across the joint to provide sufficient strength, as shown in Figure 4-1. A good weld is also free of excessive oxidation.

Poor welds display certain characteristics. A cold weld has irregular edges and considerable variation in depth of penetration, while a weld from excessive heat shows pitting along its edges and long, pointed ripples. If cracks appear adjacent to a weld, it means a part may have cooled too quickly after being welded. If a welded joint displays any of these defects, all of the old weld must be removed and the joint rewelded. In other cases, the part may need additional reinforcement repairs, or it may be necessary or more economical to scrap and replace the part.

OXIDATION

Oxidation is a primary concern to a welder. Metal oxides are formed in the welding zone of most metallic alloys when sufficient heat is applied. Oxygen chemically reacts with the heated surface of the metal and forms metal oxides such as iron oxide (rust) or aluminum oxide. When excessive oxide is present, it often results in porous pockets, causing a weak joint. Limiting the effects of oxidation is critical to maintaining strong weld joints.

Each type of fusion welding controls the formation of oxides differently. The gas welding processes generate carbon dioxide (CO_2), which shields the welding zone from oxygen. CO_2 is a natural by-product of oxygen and acetylene combustion. In welding processes that use an electric arc for heat, other methods of shielding the weld are used, including coating electrodes with flux, or using gases to flood the area around the arc to shield the weld from oxygen.

OXYACETYLENE WELDING

Oxyacetylene welding, often referred to as gas welding, gets its name from the two gases, oxygen and acetylene, that are used to produce a flame. Acetylene is the fuel for the flame and oxygen supports combustion and makes the flame hotter. The combination of these two gases results in sufficient heat to produce molten metal. The temperature of the oxyacetylene flame ranges from 5,600° to 6,300° F.

Most aircraft gas welding is done on thin-gauge steel that ranges from 16- to 20-gauge, or about .027- to .050-inch thick. Welding thicker metals requires larger equipment; however, welding techniques remain much the same. Aircraft structures of steel tubing are usually fabricated by welding the tubes together into a strong, lightweight structure. [Figure 4-2]

Figure 4-2. Oxyacetylene welding was once used almost exclusively in the fabrication of welded-tube aircraft fuselage structures.

Section C of this chapter introduces the procedures for proper handling, setup and safe use of oxyacetylene equipment. It also explains the fundamentals of oxyacetylene-torch cutting. Although you are not required to become proficient in performing welds, a background in basic oxyacetylene welding skills may prove useful in the maintenance of tools and shop equipment.

ELECTRIC ARC WELDING

Electric arc welding includes shielded metal arc welding, gas metal arc welding, and tungsten inert gas (TIG) arc welding. Although TIG welding is the method that is predominantly used in aircraft fabrication and repair, a technician is also required to understand the other methods.

When electricity has sufficient voltage to arc across the space between an electrode to an area of different electrical potential, heat is produced from the movement of electrons. The amount of heat is predominantly determined by the amount of current (amperage) flowing across the gap. For all types of arc welding, a transformer/rectifier (TR) unit is used to produce and control electrical power. An additional feature is that the TR unit has the capability of changing the polarity of electricity that flows out of its terminals.

Electric arc welding produces a blinding light, with infrared and ultraviolet rays, which can burn both skin and eyes. Before being exposed to a welding arc, you must wear an arc-welding helmet, gloves and proper clothing. Anyone observing electric arc welding also needs appropriate protective gear. Heavy clothing or leather aprons should be worn to cover as much skin as practical. The use of lightweight fabric materials does not provide adequate protection to prevent burns to the skin.

SHIELDED METAL ARC WELDING

Shielded metal arc welding (SMAW), or stick welding, is the most common type of arc welding. You may find stick welding useful for fabricating tools and shop equipment, but it is not generally used for the fabrication or repair of aircraft.

In SMAW welding, a metal wire rod, which is composed of approximately the same chemical composition as the metal to be welded, is clamped in an electrode holder. This holder, in turn, is connected to one terminal of the TR power supply by a heavy gauge electrical cable. The metal to be welded is attached to the other terminal of the power supply through another electrical cable usually equipped with a spring clamp. The technician strikes an arc between the wire rod and the metal, which produces heat capable of exceeding 10,000° F. This intense heat melts both the metal and the rod.

As previously mentioned, the welding process accelerates oxidation, a harmful chemical reaction between the metal and the oxygen in the air. It is a major concern, and must be controlled to obtain a quality weld. To accomplish this, the arc-welding rod is coated with flux, which protects the welding zone from the air as it burns. As the flux burns from the rod, it releases an inert gas that shields the molten metal. When the weld is complete, an airtight slag covers the weld bead as it cools. This slag must be chipped off to examine the weld. [Figure 4-3]

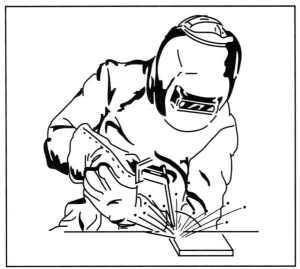

Figure 4-3. Shielded metal arc welding normally is used for welding heavy gauge steel. It seldom is used for aircraft construction or repair.

GAS METAL ARC WELDING

Gas metal arc welding (GMAW), formerly called Metal Inert Gas (MIG) welding, is used primarily in large volume production work. An advantage of GMAW over stick welding is that no slag is deposited on the weld bead. An uncoated filler wire acts as the electrode. It is connected to one terminal on the power supply, and fed into the torch. An inert gas such as argon, helium or carbon dioxide flows out around the wire to protect the weld zone from oxygen. The metal to be welded is connected to the other terminal of the power supply. When power is supplied to the electrode, and it is brought into contact with the work, it produces an arc, which melts the metal and the filler wire.

TUNGSTEN INERT GAS WELDING

Tungsten inert gas welding (TIG) is the form of electric arc welding that is used most in aircraft maintenance. It also is known as gas tungsten arc welding (GTAW) and by the trade names of Heliarc® and Heliweld®. These trade names were derived from the fact that the inert gas originally used was helium. [Figure 4-4]

Figure 4-4. Tungsten inert gas (TIG) welding can be used in most fabrications that can be done by oxyacetylene welding.

Unlike SMAW and GMAW, which use consumable electrodes, TIG welding uses a tungsten electrode that does not act as filler rod. The electrode is connected to an AC or DC electrical power supply to form an arc with the metal being welded. The arc is concentrated on a small area of the metal, raising its temperature to as high as 11,000° F, without excessively heating the surrounding metal. The base metal melts in the area of the arc and forms a puddle into which the filler rod is added. [Figure 4-5]

There are two methods of TIG welding using DC: straight polarity and reverse polarity. The most common method is straight-polarity DC where the metal to be welded is connected to the positive side of the power supply, and the electrode is connected to the negative side. This produces most of the heat in the metal where it is most needed. [Figure 4-6]

With reverse-polarity DC, the metal to be welded is connected to the negative side of the power supply, and the electrode is connected to the positive side. While this method does not generate enough heat in the metal to be used for welding heavy work, it does have an important advantage. Reverse polarity breaks up the surface oxides from the material. This is especially important with aluminum, because the surface oxides have a much higher melting point than the base metal, and these oxides must be loosened from the surface before you can satisfactorily complete a weld. Without loosening the oxides, the temperature control of the weld is difficult to maintain, resulting in insufficient penetration of the weld or excessively burning away the base metal.

Aircraft Welding

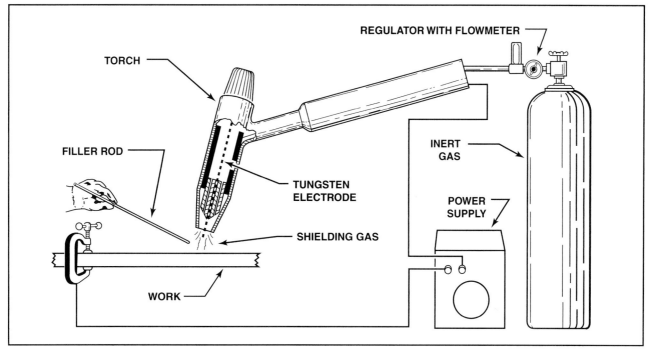

Figure 4-5. A TIG welding setup consists of a power supply, a TIG torch with a tungsten electrode, and a supply of inert gas.

AC electricity is a reasonable compromise. Half the time the arc is putting most of its heat in the metal, and the other half it is blasting the oxides away from the surface. A disadvantage is that 60 Hz AC does not alternate rapidly enough to result in a stable arc. This can be overcome by superimposing on the welding current, a high-voltage, low-current signal operating at a high frequency.

High frequency AC also is used for starting arcs without touching the material. On DC arcs, this starting arc normally is disengaged once the arc is stable. On AC arcs, it normally is left on to augment the arc's stability. When welding aluminum, the stabilizing signal should be on continuously whether using AC or DC.

The torch used with TIG welding is essentially a handle with a chuck to hold the tungsten electrode. A ceramic, high-temperature plastic, or Pyrex® glass nozzle, surrounds the electrode and directs a continuous flow of gas around the arc. An air-cooled

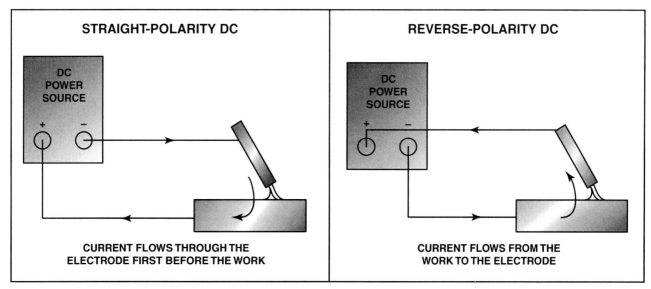

Figure 4-6. Straight- and reverse-polarity DC each has advantages.

torch is appropriate for light-duty welding such as that needed on most aircraft structures. The flow of gas over the electrode carries enough heat away from the electrode to adequately cool it. For heavier applications, water-cooled torches are required.

The tungsten electrodes for TIG welding must be selected according to the size and type of metal to be welded. Three types of electrodes are pure tungsten, zirconium tungsten, and thorium tungsten. Pure tungsten is the least expensive and is satisfactory for most welding; however, it is consumed faster than the others are. Zirconium tungsten lasts longer and provides a more stable arc. Thorium tungsten produces a cleaner weld and its arc is easiest to start. The diameters of the electrodes that are most used for aircraft maintenance welding are 0.040 inch, 1/16-inch, and 3/32-inch. When choosing and preparing the electrode, it is important to follow the manufacturer's instructions for the particular equipment being used. [Figure 4-7]

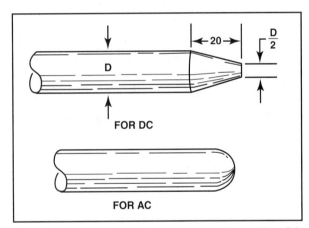

Figure 4-7. When preparing a TIG electrode for welding, follow the manufacturer's instructions. Manufacturers sometimes require the electrodes to be ground to a rounded shape for welding with AC power, while a pointed tip is sometimes called for when using straight-polarity DC.

ELECTRIC RESISTANCE WELDING

Many thin sheet metal parts for aircraft, especially stainless steel parts, are joined by one of the forms of electric resistance welding; either spot welding or seam welding. [Figure 4-8]

SPOT WELDING

When spot welding, two copper electrodes are held in the jaws of a vise-like machine and the pieces of metal to be welded are clamped between them. Pressure is applied to hold the electrodes tightly together while electrical current passes between the electrodes. The current flow through the metal between the electrodes generates enough heat to melt the metal. The pressure on the electrodes

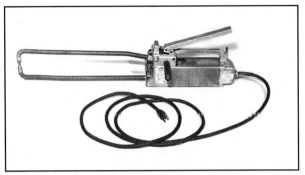

Figure 4-8. A spot welder is an electrical resistance welder used for joining thin sheets of metal.

forces the molten spots in the pieces of metal to flow together. The pressure on the electrodes is then held while the current stops flowing, long enough for the metal to solidify. The duration that the electric current flows is referred to as the dwell time. The amount of current, pressure and dwell time are carefully controlled and matched to the type of material and its thickness to produce the correct spot welds.

SEAM WELDING

While it would be possible to create a seam with a series of closely spaced spot welds, a better method is to use a seam welder. This equipment is commonly used to manufacture fuel tanks and other components where a continuous weld is needed. Instead of the electrodes of a spot welder, the metal in a seam welder is drawn between two copper wheels. Pressure is applied to the wheels and timed pulses of current flow through the metal between them. The pulses create spots of molten metal that overlap to form the continuous seam.

TYPES OF WELDED JOINTS

The result of a weld is a joint. To evaluate the quality of an aircraft welding repair, it is essential to understand the various types of joints and be able to recognize whether proper techniques were used to create these joints. The five basic types of weld joints are shown in figure 4-9.

BUTT JOINTS

Butt joints are used to join metal forms such as sheet, bar, plate, tube and pipe. In aircraft applications, butt joints generally are not used for joining tubing because they are too weak for aircraft structures. In uses other than tubing, butt joints are suitable for most load stresses if the welds are made with full penetration, adequate fusion and proper reinforcement. A typical butt weld should penetrate 100 percent of the thickness of the base metal. [Figure 4-10]

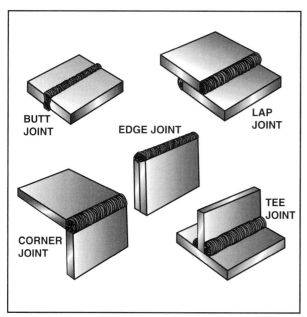

Figure 4-9. A technician must be familiar with the techniques for fabricating these basic welding joints.

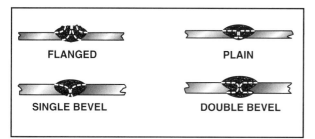

Figure 4-10. Butt joints are simple but require special preparation considerations to provide adequate weld penetration. Bevels are often required before welding thick metals.

TEE JOINTS

Tee joints are quite common in aircraft work, particularly in tubular structures. The plain tee joint is suitable for most aircraft metal thickness. Thicker metals require the vertical member to be either single or double beveled to permit the heat to penetrate deeply enough. [Figure 4-11]

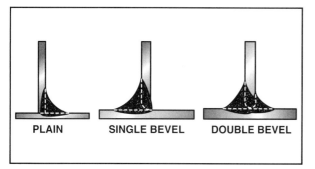

Figure 4-11. Tee joints are suitable for most metal thickness, although beveled edges may be required to achieve sufficient penetration in thicker metals.

LAP JOINTS

A lap joint is seldom used in aircraft structures when welding with gas, but is commonly used when spot welding. The single lap joint has very little resistance to bending and will not withstand shearing stresses. The double lap joint is stronger, but requires twice the welding of the simpler, more efficient, butt weld. [Figure 4-12]

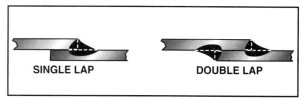

Figure 4-12. In aircraft applications, lap joints are more common than butt welds when sheets of metal are to be joined together.

CORNER JOINTS

A corner joint results when two pieces of metal are brought together so that their edges form a corner of a box or rectangle. This joint can only be used where load stresses are not significant. [Figure 4-13]

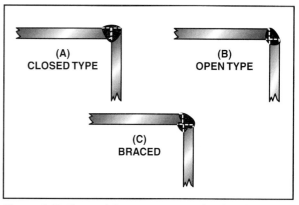

Figure 4-13. Corner joints are common in aircraft components such as airbox assemblies made from aluminum.

EDGE JOINTS

Again where load stresses are not significant, edge joints may be used to join two pieces of sheet metal. To form an edge joint, bend the edges of one or both parts upward and place the two ends parallel to each other. Weld along the outside of the seam formed by the two edges. In some situations, the bent-up edges may provide enough material to form the bead so that a filler rod is not required. [Figure 4-14]

EXPANSION AND CONTRACTION OF METAL

Controlling the expansion and contraction of metals during welding is necessary to reduce distortion,

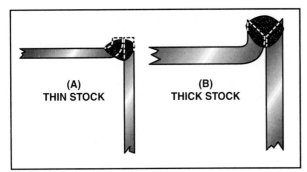

Figure 4-14. In thin metal, edge welded joints can be performed without the use of filler material since the metal will melt to form the weld reinforcement.

complete the weld, the two pieces will warp in towards each other. [Figure 4-15]

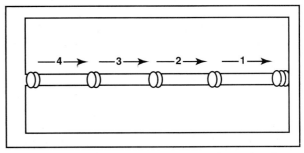

Figure 4-15. Tacking the metal at one-inch intervals prior to completing the weld is one way to compensate for warping (A). Spacing the sheets of metal so that they are closer at one end is another (B).

warping and residual stress. These effects are more noticeable when welding long sections of thin sheet metal. You can control these distortions in the following ways:

1. Distributing heat more evenly.
2. Reducing the amount of heat applied to the metal.
3. Using jigs to hold the metal firmly in place.
4. Allowing for space between the edges of the joint.

In aircraft construction, most welding requires the finished pieces to be free of warping and distortion. To accomplish this, the metal is held in heavy jigs that prevent the metal from warping out of shape. Stresses will remain in the joints after they are welded in a restraining jig, which must be normalized by heating the entire joint to a cherry red and then allowing it to cool to room temperature in still air.

You can readily see warping by butt welding two pieces of sheet metal without anchoring them. As the weld progresses along the joint, one piece will warp over the top of the other. Several methods can be used to prevent this. One way is to space the sheets apart about the thickness of the material and then tack-weld them together at about one-inch intervals. A tack weld is simply a small round weld that temporarily holds the work together. It is created by melting a puddle in the metal and dabbing in a piece of filler rod. Once the sheets are tacked together, a continuous weld can be completed over the tack welds.

Another method of allowing for expansion and contraction is to place the metal pieces to be butt welded in a slight V. The pieces will be farther apart at the end where the weld is to be finished. As you

EVALUATING WELDED JOINTS

In order to evaluate the quality of the welds on an aircraft, it is important to understand the attributes of a properly welded joint. It is necessary to examine the parts of a weld, proportions of the weld, and its formation.

PARTS OF THE WELD

The parts of a weld are shown in Figure 4-16. They are listed below:

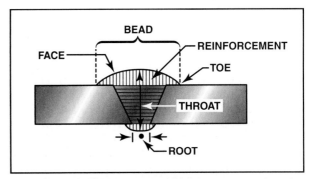

Figure 4-16. Weld bead nomenclature is necessary in order to compare welds with printed standards.

1. Bead is the metal that is deposited as the weld is made.
2. Face is the exposed surface of the weld.
3. Root is the depth that fusion penetrates into the base metal.
4. Throat is the distance through the center from the root to the face.
5. Toe is the edge formed where the face of the weld meets the base metal.
6. Reinforcement is the quantity of weld metal added above the surface of the base metal.

PROPORTIONS OF THE WELD

The three most important proportions of a weld are depth of penetration, width of the bead, and the height of reinforcement.

A finished bead has uniformity when its width is three to five times the thickness of the metal (T), with its edges flowing evenly into the base metal. To ensure proper fusion on lap and butt joints, penetration should be to a depth of 100 percent of the thickness of the base metal. The height of the bead above the surface of the base metal should be approximately 1/4 T to 1/2 T. For other joints, penetration must be adequate to impart enough strength that the metal sheets will fail before the joint fails. [Figure 4-17]

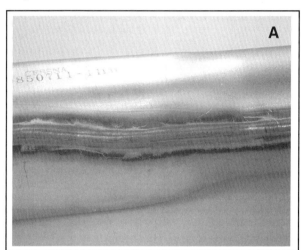

Figure 4-17. A cleanly formed bead should have uniform width, good penetration, adequate reinforcement and uniform ripples as shown in A. In contrast, a poorly formed weld, as shown in B, often fails in service.

FORMATION OF THE WELD

A weld must be formed correctly to achieve the intended strength and to resist joint fatigue. If a joint is not made properly, the strength can be reduced as much as 50 percent. A properly completed weld should exhibit the following attributes:

1. The bead should be smooth and uniform in thickness.
2. The weld should be built up to provide extra thickness at the seam.
3. The bead should taper off smoothly into the base metal.
4. No oxide should be formed on the base metal further than 1/2-inch from the weld.
5. There should be no blowholes, porosity or protruding globules.
6. There should be no signs of pitting, cracking, burning or warping.

A good weld will have a uniform rippled appearance. To achieve this, it must have been completed with the proper amount of heat and the correct filler rod size for the thickness of the material. [Figure 4-18]

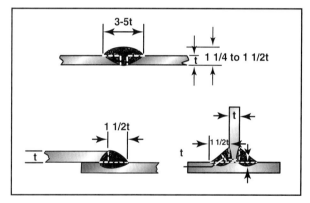

Figure 4-18. Examples of properly formed joints show adequate weld penetration into the base metal and proper reinforcement height. Penetration and bead height are measured relative to thickness of the base metal (t).

Approving an aircraft for return to service when welding joints are weak or fatigued could have serious consequences. Improperly formed welds typically are caused by:

1. Undercutting the base metal at the toe of the weld.
2. Lack of adequate penetration.
3. Poor fusion of the weld metal with the base.
4. Oxides, slag or gas pockets in the weld.
5. Improper heat.

If a welded joint is pitted, rough, dirty or uneven, the weld is almost always defective on the inside. However, not all defects can be observed by visual

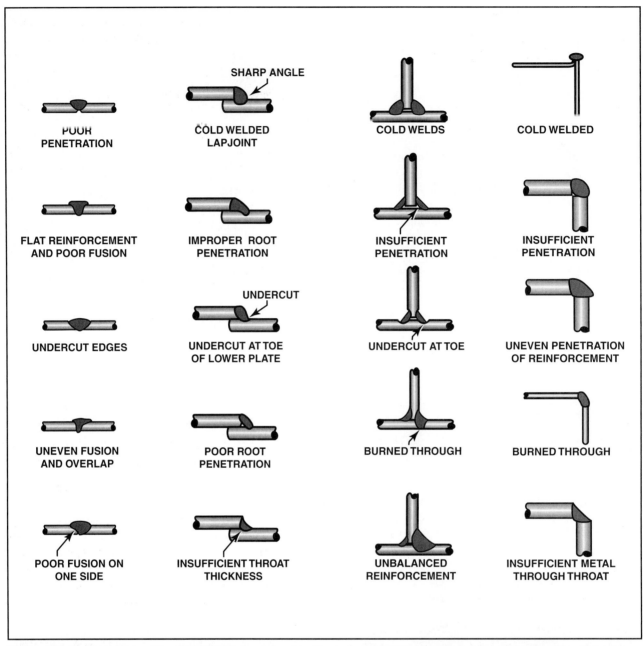

Figure 4-19. These are examples of improperly formed welds. Most are caused by poor technique, inexperience or simply carelessness. If you observe welds with these defects, reject them and arrange to have them repaired by an FAA-approved welding facility.

inspection. Cracks and internal defects may be difficult or impossible to see. [Figure 4-19]

If you suspect a defective joint, but cannot confirm it visually, it will be necessary to perform a more detailed non-destructive inspection. Non-destructive inspection (NDI) techniques include dye-penetrant, magnetic-particle, x-ray, ultrasonic and eddy-current. The dye-penetrant inspection is the most cost-effective and easy test to perform in the field, but does not identify internal flaws. To use other forms of NDI, the inspection equipment must be certified for accuracy and the personnel conducting the inspection must have received special training.

BRAZING AND SOLDERING

Non-fusion joining occurs when two or more pieces of metal are held together by a softer metal such as brass, bronze or silver. The molten metal sticks to the surface by adhesion rather than fusion. A brazing rod or solder used with the proper flux will act like glue when holding a metal joint together. Brazing or soldering a joint is achieved at a much lower temperature than fusion welding. Since brazed or soldered joints are not as strong as fusion-welded joints, they are not typically used in aircraft structural applications. Instead, brazing and soldering are used widely in electrical connectors, fuel and hydraulic line fittings, and many other low stress applications.

TORCH BRAZING

Brazing is a form of metal joining in which an iron-free metal is used as a cohesive material. The nonferrous material, usually brass or bronze, is melted with an oxyacetylene torch at a temperature below that of the base metal, but above 800° F. There are actually two operations that may be performed by brazing. One operation is referred to as braze welding, in which two metals are joined together in much the same manner as fusion welding, except that the filler material does not fuse into the base metal. Because braze welding does not provide the strength of fusion welds, it is seldom used in aircraft structures. The other operation is simply referred to as brazing, and involves melting cohesive brazing metal near closely fitting parts. The cohesive metal, once melted, flows between close-fitting metal surfaces by capillary action to form a tight seal or joint. When the joint cools, it forms a bond between the two metals. For this action to occur properly, the metal must be absolutely clean.

Flux is used to clean the surfaces of the metal being joined. Brazing flux has a caustic base which, when heated, turns to liquid. It washes away oxides and impurities that may have formed on the surfaces of the metals. A flux coating on a filler rod allows it to flow when melted and to adhere to a base metal surface. The term adhesion, as applied to brazing, means the molten filler material will flow into the pores of (but not fuse to) a joint. [Figure 4-20]

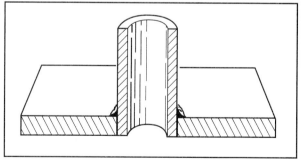

Figure 4-20. Brazing uses capillary action to draw molten metal into close fitting joints.

Braze welding also is sometimes used to join dissimilar metals that cannot be joined by fusion welding. Braze welding also can join metals where the heat of welding would distort the part or destroy its heat treatment. For braze welding, the material must be mechanically cleaned to remove all oxidation and scale. The metal is then positioned, as it would be for regular welding. The end of the brazing rod is heated for about two inches and dipped into a borax-type flux. The flux sticks to the rod and will flow off as the rod melts on the base metal. Covering the joint with flux before welding also helps prevent the formation of oxides and produces a cleaner, stronger weld.

When braze welding, the torch should be adjusted to a neutral flame the same size as would be used for welding. Flame types and adjustment procedures are described later in section C of this chapter. The base metal should be heated to a cherry red. Keep the end of the brazing rod near the torch flame to keep it hot. When the metal is sufficiently heated, touch the rod to it so that the rod melts and flows onto the base metal. The width of a good braze bead should be about the same as a weld bead on the same thickness of material. Bead width is determined by the distance the torch tip is held from the work — the closer the tip, the wider the bead. However, the tip should be held at a sufficient distance to prevent the base metal from becoming molten.

A good brazed joint will be smooth with an even deposit of brazing metal on the top. The brazing material should penetrate the joint and be visible on the bottom. Brazing material should be the same color as the original rod. If the joint is overheated, zinc may burn out of the metal, and it will leave the weld with a copper appearance. [Figure 4-21]

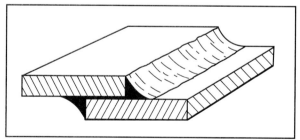

Figure 4-21. Braze welding is similar to brazing, but the filler rod is deposited on the surface rather than depending only upon capillary action to produce the joint.

TORCH BRAZING OF ALUMINUM AND MAGNESIUM

Aluminum and magnesium may be torch brazed in much the same way as steel. This procedure requires the material to be completely clean with no oxides on the surface. Apply fresh flux to the surface, and then use a soft, neutral flame to heat the material until the flux melts. As soon as the flux starts to flow, touch the rod to the material. The rod will melt and flow into the joint. Allow the joint to cool, and scrub every trace of the flux with an acid wash. Rinse away all traces of the acid with plenty of hot water. It is especially important to thoroughly remove all flux from magnesium parts because it can cause severe corrosion.

TORCH SOLDERING

Soldering is similar to brazing, using many of the same techniques and devices as when making brazed joints. Brazing materials normally melt at temperatures above 800° F, while solders melt at temperatures considerably lower than this. As with brazing, the oxyacetylene welding torch is sometimes used for soldering as well as brazing.

SOFT SOLDERING

While there is very little use for soldered joints in aircraft structures, a technician may need to solder joints in an air-conditioning system or other non-structural application.

The basic principle of soldering is similar to that of brazing in that a joint is made with a close fit between the parts. The joint is fluxed and uniformly heated with a soft flame. The solder is melted by touching it to the metal at the edge of the joint, and it is pulled by capillary action into the joint.

Fluxes used for soldering are an important part of the system. The flux dissolves any oxides that are on the surface and prevents the formation of additional oxides until the joint is covered with the solder. Each different type of metal to be soldered has a recommended flux, which can be found in manufacturers' guides or in soldering reference handbooks.

Soft solder is primarily a mixture of tin and lead. The melting point varies considerably with the percentage of each. The solder is identified by two numbers, indicating the percentage of tin followed by the percentage of lead. 40–60 solder must be heated to 460° F to flow, while 50–50 solder flows at 420°F. The mixture that melts at the lowest temperature has 63% tin and 37% lead (63–37), which flows at 361°F. The most commonly used soft solder is of the 50–50 composition, and is sold in the form of a hollow wire with the center filled with flux. [Figure 4-22]

HARD SOLDERING

Hard and silver soldering are actually forms of brazing in which the filler material melts at a temperature more than 800° F. The oxyacetylene flame for silver soldering should be neutral, but may have a slight excess of acetylene. The flame must be soft, not harsh. To conform to commonly accepted terminology in aircraft maintenance, silver brazing will be called silver soldering in this text.

Stainless steel oxygen lines often have their end fittings attached by silver soldering. To do this, clean and assemble the fitting and the end of the tube. Mix the flux with water, apply it to the joint and carefully heat it. The flux will dry as the water boils out of it and will then begin to bubble. As it approaches the correct temperature for applying the

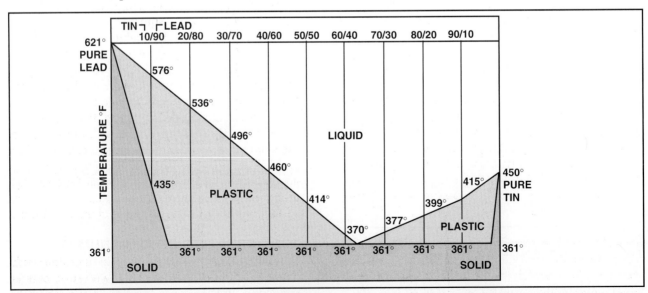

Figure 4-22. The melting temperature of solder depends on the tin-lead mixture and is supplied on a spool marked with an identification number. Examples of where each solder type may be used are shown in this figure with the tin/lead content, melting temperature and the temperature required for each solder to flow.

rod, it will turn into a clear liquid. Touch the rod to the edge of the joint and allow capillary action to draw the molten material into the joint. When a good solid union of the two pieces is formed, withdraw the flame. Because the solder is in a plastic state during the early part of cool down, the joint must be held stationary until the solder has fully solidified. [Figure 4-23]

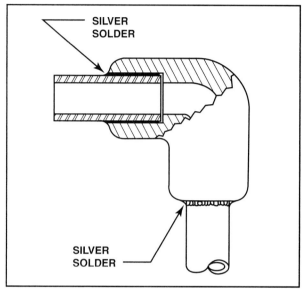

Figure 4-23. Silver soldering is a form of brazing in which the silver solder is drawn into the joint by capillary action.

SOLDERING OF ELECTRICAL WIRES AND CONNECTIONS

In some electrical applications such as wire terminals, soldering is not recommended. The vibrations that occur around wire terminals can cause rigid, soldered connections to crack, resulting in poor electrical contact. In other types of electrical connections such as AN-type connectors, the soldering of pins and sockets is necessary. [Figure 4-24]

Generally, a gas torch should not be used for electrical soldering except that occasionally a propane or oxyacetylene torch may be used on large surface areas such as heavy gauge electrical cables or bonding straps. To control the amount of heat applied to small wire and electrical connecting devices, it is common to use an electric soldering iron of the appropriate wattage.

Soldering irons are available with various wattage ratings to produce specific temperatures, or as soldering stations, which can be adjusted to vary the temperature. Although higher-powered electric soldering guns can be used on some electrical connections, they should not be used on printed circuit

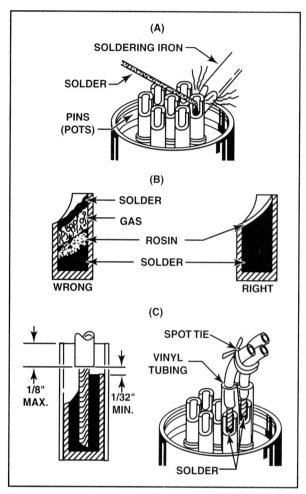

Figure 4-24. AN-type electrical connectors have provisions for installing wires into solder cups, where the application of heat is done by a soldering iron instead of a welding torch.

boards and other intricate components. Regardless of the type iron, each requires special preparation before use. This includes making sure the iron is adequately cleaned and tinned. Tinning involves applying a small amount of solder to the iron tip to produce a uniform coating. Without this coating, heat cannot readily transfer from the iron to the work. Depending on the type iron, tinning may be done by lightly scuffing the tip with sandpaper to remove oxides and then applying a coating of flux and solder. For other types of irons with specially coated tips, do not use abrasive paper for cleaning, as it will destroy the special coating. In all cases, closely adhere to the soldering iron manufacturer's instructions.

For electrical connections, you must use solder with a non-corrosive flux. Wire solder with rosin core is commonly used because the flux is automatically applied as the solder is melted on the joint. The two metal parts being joined by the solder must be heated to the melting point of the solder. Once the metal has been adequately heated, the solder will flow smoothly into the joint and form a solid bond with the metal. Care must be exercised so that heat from the soldering iron does not damage adjacent insulation or electric circuitry. Apply heat sinks to diodes and other temperature-sensitive devices. Once the solder has been applied, the joint must be held stationary until the solder has cooled and solidified. If the part is disturbed before cooling, the solder may fracture, causing an intermittent or high-resistance connection.

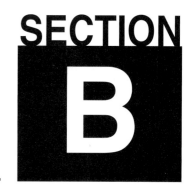

ADVANCED WELDING AND REPAIRS

Repair of aircraft structures on modern aircraft requires more technical welding proficiency than perhaps may have been necessary before. Advances in aircraft materials and in the design of airframe structures have demanded improved welding processes and techniques. The use of basic gas welding to repair structures has generally been replaced by inert gas welding processes, usually Gas Tungsten Arc Welding, or TIG. Improvements in the welding process has also made it necessary to modify welding techniques based on the type of material being welded.

Repair of damaged aircraft structures is also more technically demanding than it may have been in the earlier types of aircraft. Specific rules and FAA requirements now dictate a much more deliberate and planned approach to complicated repairs. A technically simplistic approach to repair of aircraft structures is no longer realistic.

Technicians need to develop an appreciation and understanding of advanced welding techniques and repairs. Although typically not expert welders, aircraft technicians are charged with the ultimate responsibility of inspecting and approving the results of these maintenance activities. General concepts for advanced welding are introduced in this section, however the specific manufacturer's recommendations and guidance should be followed explicitly when welding on any aircraft components.

ADVANCED WELDING

Aircraft construction has migrated from wood structures and fabric covers to aluminum and other strong, lightweight materials. Each type of metal used in constructing or repairing an aircraft has unique properties that should be considered in the welding process.

GAS WELDING ALUMINUM

Aluminum has a reputation for being difficult to weld because it does not change its color as it nears its melting temperature. It can be solid one moment and the next it melts through, leaving a hole. With practice though, aluminum welding can be done as readily as steel.

Most aluminum welding is done with a gas-shielded electric arc, but there are times when aluminum is welded with either acetylene or hydrogen gas. Hydrogen is preferred because of its cleaner flame and less risk of oxidizing the joint. Hydrogen burns with much less heat than acetylene. For an equal thickness of metal, hydrogen requires a larger tip than would be used when welding with acetylene.

Prepare the metal for welding by cleaning it with an aluminum or hog bristle brush. A stainless steel brush or stainless steel wool may also be used. Other types of steel brushes or steel wool should not be used, as tiny steel particles may become embedded in the soft aluminum and contaminate the weld. Support both pieces adequately before welding as the metal at the joint will lose all its strength and rigidity as it melts.

Heat the metal and apply flux to the pieces that will be joined and to the welding rod. Use light blue welding goggles in order to see the very subtle changes in appearance of the metal as it approaches melting. The surface of the metal should be lightly wiped with the welding rod. As soon as the metal feels slightly soft, allow the rod to stay in contact and melt with the metal.

The flame used for welding aluminum should be neutral or slightly carburizing, to reduce the possibility of the metal being oxidized during the welding process. A properly formed bead will be uniform in width and height. It also will have a bright shiny surface, the same color as the filler rod. If the weld is oxidized, it will have a powdery white appearance and a rough surface. If the flame was too carburizing, the weld will have a dirty appearance from the extra carbon. When the weld is complete, scrub off all traces of the welding flux with hot water and a hog bristle brush. Any flux left on the metal will cause it to corrode.

TIG WELDING ALUMINUM

TIG welding is the preferred method for welding aluminum. The heat from the TIG torch is much more concentrated than it is with an oxygen torch. This concentration has less effect on the remainder of the structure and does not affect the heat treatment as extensively as gas welding. In either case, after welding a heat treated aluminum part, the part must be reheat treated before reinstallation on the aircraft. In addition, the inert gas provides shielding of the work area and resists the development of aluminum oxides.

To prepare aluminum for welding, it must first be cleaned as thoroughly as possible. It should be brushed with a hog bristle or stainless steel brush to remove the oxide film that is on the surface. Afterward, it should be scrubbed with an appropriate solvent. This should be done immediately prior to initiating the weld, because bare aluminum will begin oxidizing very quickly.

Use the electrode recommended by the manufacturer of the welding equipment for the thickness of metal being welded. Preparation of the electrode, the amount and type of current, and the flow of gas will also be described in the manufacturer's instructions. Hold the torch over the work and lower it until the arc starts. The superimposed high-frequency signal will allow the arc to start without touching the electrode to the work. With the arc established, the surface of the aluminum must be monitored so that the rod can be added before the metal melts through. When the surface begins to appear frosty and then starts to lift just a little, it is time to touch the rod to the surface. Once initiated, continue the weld by melting the rod along with the surface.

TIG WELDING MAGNESIUM

Magnesium is an extremely active metal whose oxidation rate increases as its temperature is raised. Magnesium burns at a very high temperature and produces oxygen as a byproduct of combustion, therefore, is not extinguishable by normal means. Special agents must be used that cool the magnesium thereby extinguishing the fire. Magnesium melts at temperatures near that of aluminum, but the oxides that form on its surface melt at a far higher temperature. In order to make a successful weld, these oxides must be removed. Before starting the weld, the oxides must be removed with a hog bristle or stainless steel brush and the surface scrubbed with an appropriate solvent. Using AC current with high-frequency arc stabilization and an inert gas keeps the weld area from being contaminated. The filler rod used for magnesium should be the same as the metal itself. The producer of the metal may recommend other types of rods.

TIG WELDING TITANIUM

Titanium, beryllium and zirconium are reactive metals, and welding them presents special problems. These metals have a strong affinity for oxygen and nitrogen when they are heated to a high temperature, and when they combine, they form stable compounds. The oxides that form resist further oxidation. Minor impurities in these metals will cause them to become brittle. Special care must be taken to be sure that all of their oxides and any surface contaminants are removed before they are welded. Titanium and other reactive metals must be welded in an atmosphere that allows no oxygen or nitrogen to contact the hot metal.

When welding titanium, use straight polarity DC current and a thoriated tungsten electrode of the smallest diameter that will carry the welding current. Argon gas, a large gas nozzle and a trailing shield is used to keep air away from the hot metal.

TIG WELDING STAINLESS STEEL

Welding the thin metal in aircraft engine exhaust systems is one of those jobs that creates many problems for a typical aircraft maintenance shop unless TIG welding equipment is available. Exhaust systems are usually constructed of stainless steel, and welding it can be difficult if the metal is not thoroughly cleaned. The exhaust residue becomes impregnated in the metal and can only be removed by abrasive blasting. These exhaust system components should be welded using straight-polarity DC, with the high-frequency arc stabilizer turned on only to get the arc started and then turned off. [Figure 4-25]

Figure 4-25. TIG welding is the best way to repair exhaust systems and exotic metal components.

STRUCTURAL WELDING REPAIRS

Welded repairs to aircraft structures by an A&P technician requires a high degree of professionalism and judgement. As an astute aircraft technician, you should have recognized by now that welding of aluminum, stainless steel and exotic metals requires special tools, training and techniques. The average technician usually will not have the expertise necessary to perform airworthy repairs to these types of structures and components. Recognition of the need to enlist the aid of specialists is one of the hallmarks of a good A & P mechanic. Consider that an improper weld on a structural part of the aircraft could cause failures with serious results. Welded repairs made directly to the airframe must be correctly done in order to retain the design strength of the aircraft. The airframe along with the engine mounts and landing gear assemblies absorb the highest mechanical loads and vibrations in the aircraft.

Before attempting any welding repair of an aircraft structure, it is a good policy to consult with the service department of the aircraft manufacturer. The complexity of modern aircraft and the responsibility placed on a technician requires obtaining the best information prior to making a repair decision.

Although you may be reasonably qualified to perform the repair, it is advisable to consider another option. While the integrity of the repair is the responsibility of a licensed A&P technician, using the skill and experience of a welding specialist may be a wise practice. You may choose to design the repair, cut and fit the pieces, and then deliver the job to a professional welder for completion.

The repair must be designed so it will conform to approved data. If such data does not exist that exactly fits the needed repair, the proposed design must be submitted to the FAA Flight Standards District Office for approval. The FAA must approve any proposed repair that is not listed in the manufacturer's structural repair manual.

Any structural repair decision a technician makes should be based on a clear understanding of specific welded repairs and techniques. A closer look at specific types of repairs and techniques is covered in the next section.

STRUCTURAL REPAIRS

Most repairs requiring welding will be outlined in detail in a structural repair manual for the part being repaired. Some will require design and approval for a repair. This section outlines the general procedures that will be employed in specific types of repairs. This is only an orientation to these types of repairs and any actual repair must be performed according to approved data and procedures.

DENTS AT A CLUSTER WELD

If a truss has been dented at a cluster, the joint will be weakened. It must be reinforced with a patch over the dent and extended out along each tube in the cluster. A patch plate of the same thickness and material as the longeron should be cut and shaped. It should extend 1-1/2 diameters of the tube beyond the damage, and up each tube in the cluster by 1-1/2 diameters. The patch is tack welded into place and, using heat, formed to the contour of the tubing. There should be no space between the edge of the patch and the tube. It is then welded in place. The technician should practice lap joint welding before welding the patch in place. [Figure 4-26]

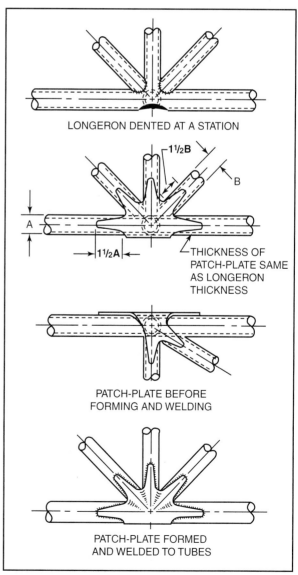

Figure 4-26. A dent at a cluster weld is repaired with a lap-joint welded patch.

DENTS BETWEEN CLUSTERS

If a longeron or cross member is dented or cracked between clusters, it may be repaired by welding a sleeve over the damage. First, straighten the dent as much as possible and then straighten the tube. If the tube is cracked, stop drill the ends of the crack to prevent it from extending any further.

A piece of repair tubing should be selected having the same wall thickness and made of the same material as the damaged member. The repair piece should be one with an inside diameter the same as the outside diameter of the damaged tube. It should be cut to extend 1-1/2 diameters of the damaged tube beyond the ends of the damage. The repair tube should be scarfed from this point with an angle of 30 degrees and split before being welded in place over the damage. This type of repair uses both butt welds and lap welds.

If a piece of tubing that fits correctly is not available, a piece of flat material of the same thickness and material as the damaged member can be wrapped around the tube to form the reinforcement. A pattern of paper should be cut to the correct size and shape, and then transferred to the metal. The repair material should be tack-welded in place and then heated and formed to the tube. Rather than using a scarfed end repair, a fishmouth cut may be used on the ends of the tube weld area. Rosette welds are used on the sides of this repair to strengthen its bond to the damaged tube.

If the dent is no deeper than 1/10 of the tube diameter, does not involve more than 1/4 of the tube circumference and does not have any cracks or sharp edges, it may be repaired by welding a patch over it. The patch of the same thickness and material as the damaged tube should be extended two tube diameters beyond the end of the damage. It should be wrapped around the tube for reinforcement. The two ends of the patch should be close to meeting on the backside of the tube, within 1/4 of the tube circumference. [Figure 4-27]

TUBE SPLICING WITH AN INNER SLEEVE REINFORCEMENT

If a tubing member is damaged enough to require replacement, clamp the structure rigidly to keep its position and alignment correct while it is being cut and welded. Cut across the tube at a 30° angle to remove the damaged portion. Cut a replacement of the same diameter, wall thickness, and material as the original. Cut an inner sleeve for either end. The sleeve should be of the same material and thickness as the original, but should slip inside with a nice

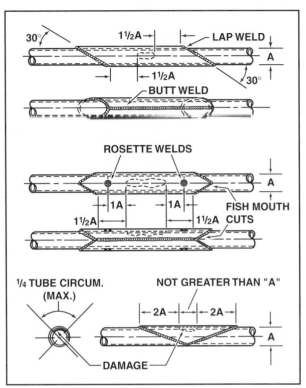

Figure 4-27. Repairs to dents in straight tubing between clusters uses a combination of butt welds, lap welds and rosette welds.

even fit. When the inner sleeve is too tight to slide in easily, heat the outer sleeve and cool the inner sleeve to assemble without difficulty. Each sleeve should be five times as long as the tube diameter and should be centered under the scarf cut. Holes must be drilled in the outer tubing sections to allow for rosette welds. Make Rosette welds through a small hole to hold the inner and outer tubes together, preventing relative movement. Assemble the repair with a gap of 1/8-inch between the ends of the replacement section and the original tube to allow for proper penetration of the weld. Weld the sleeve through the rosette holes, and then run weld bead down the gaps between the tubes. [Figure 4-28]

TUBE SPLICING WITH AN OUTER-SLEEVE REINFORCEMENT

If it does not affect the outside covering of the structure, a tubular member may be spliced using an outer-sleeve type of reinforcement. Secure the surrounding structure with clamps that hold it in position while the damaged material is removed and replaced. Remove the damaged material, using a square cut on the tubing. Cut a replacement tube of the same material, diameter, and wall thickness with a length that fits precisely where the damaged piece was removed.

Cut two pieces of tubing of the same material and wall thickness that fit over the outside of the tube

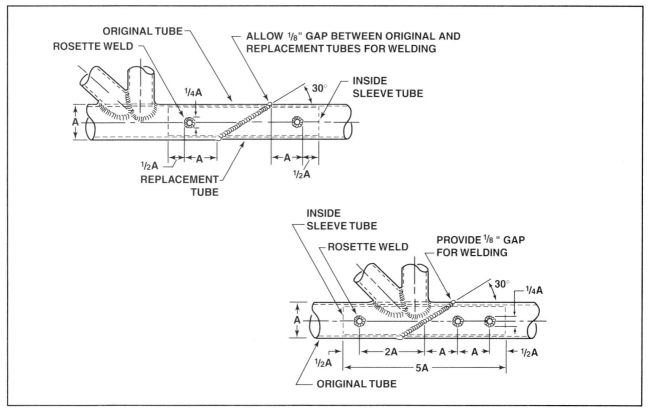

Figure 4-28. Use of a supporting inner-sleeve is one method of repairing a damaged tubular structure.

being repaired. The inside diameter of the reinforcing tubes must be no greater than 1/16-inch more than the outside diameter of the original. Cut the ends of the tubes with a 30-degree scarf or with a fish mouth. The length of the reinforcing section should be three times the diameter of the original tube between the inner ends of the cuts.

Drill rosette holes in the sleeves before installing them over the replacement tube. Center the splice tubes over each end of the replacement section and weld them in place. Weld both ends of one of the sleeves and allow the weld to cool before welding the other sleeve in place. This will minimize the warpage. [Figure 4-29]

If a cluster weld has been damaged too much to be repaired by a patch welded over it, the entire cluster may be cut out and outer sleeves welded over the ends of the tubing to make a new cluster. The procedure is similar to that for an outer-sleeve splice. [Figure 4-30]

EXPANSION AND CONTRACTION

Some allowances must be made for expansion and contraction when designing a welded repair. As a part is heated to welding temperature, it naturally expands. As it cools, it contracts. If a part is clamped rigidly and not allowed to expand during welding, it will, in many cases, be pulled out of alignment. In extreme cases, contraction during cooling can break the repair by creating excessive tension on the part. Structural repair manuals will often specify the procedures to be used in counteracting expansion and contraction.

LANDING GEAR AND ENGINE MOUNT REPAIRS

All welding on an engine mount should be of the highest quality, since vibration tends to accentuate any minor defect. Repaired engine mounts must be checked for accurate alignment. If all members are out of alignment, reject the engine mount and replace it with one supplied by the manufacturer or equivalent.

Aircraft technicians should be aware that certain landing gear assemblies are not repairable. Only assemblies made from steel tubing are candidates for welding repairs. Even some steel-tubing landing gear assemblies are not suitable for repair. Use of heat treating on parts of the gear assembly makes welding repair difficult or impossible. Consult the manufacturer's maintenance manual prior to attempting any welded repairs on aircraft landing gear assemblie.

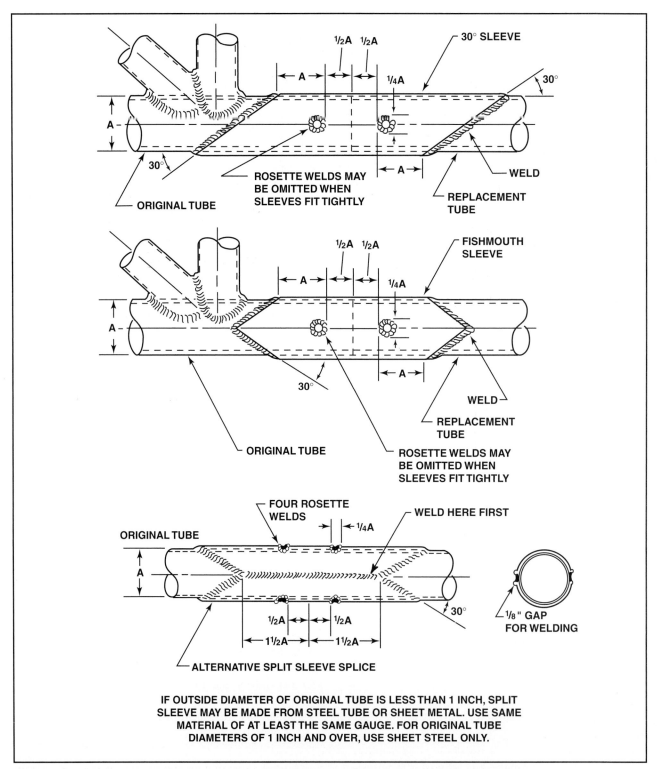

Figure 4-29. Another type of repair to tubular structure is with the use of an outer sleeve as reinforcement.

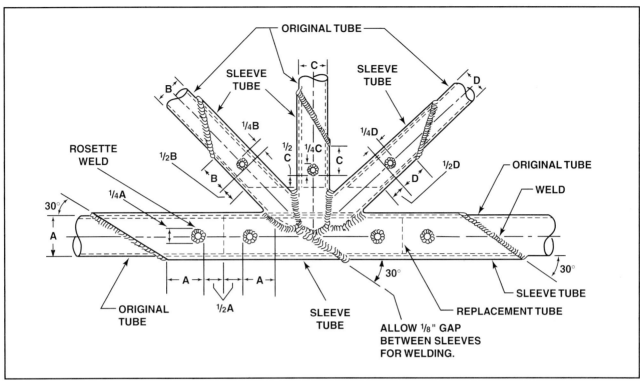

Figure 4-30. An entire cluster welded joint can be replaced using techniques similar to those used for outer-sleeve splices.

SECTION C

BASIC GAS WELDING

Even though tungsten inert gas (TIG) welding is the accepted welding method for modern aircraft, you need to understand basic gas welding before learning these advanced techniques. This includes knowledge of the gases and equipment used, as well as equipment set-up, operation and shutdown. Gas welding, more than most aspects of airframe maintenance, requires constant attention to safety practices and proper use of safety equipment.

GASES

To produce enough heat to melt metal, a combustible gas like acetylene is mixed with oxygen. Other gases, such as propane, natural gas, methylacetylene-propadiene (MAPP) gas, and hydrogen are used primarily for cutting or in the soldering/brazing processes. [Figure 4-31]

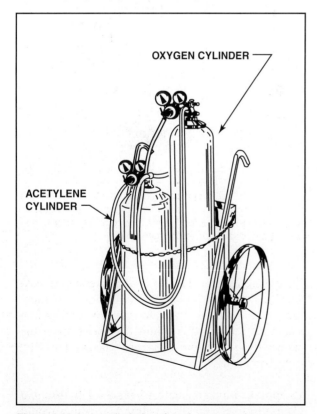

Figure 4-31. In a typical aircraft maintenance shop, oxygen and acetylene cylinders are stored on a rack that can be rolled to the job.

ACETYLENE GAS

A heavy gas, with a disagreeable odor, acetylene becomes unstable at pressures of 15 psi or higher. At 30 psi, the slightest amount of heat generated by friction of acetylene against the atmosphere will cause it to explode. The normal operating pressure for most acetylene welding is 4 to 8 psi.

Acetylene tanks are filled with Fuller's earth, a porous mixture with the characteristics of a sponge. This provides hollow paths for liquid acetone to flow into the bottle. Acetone can absorb up to 25 times its volume in acetylene. Because acetylene is stored in acetone, it can be shipped in bottles without the danger of exploding. The quantity of a freshly charged acetylene bottle is a volumetric measurement calculated from total weight, not psi. However, when an acetylene bottle is connected to a pressure gauge, it can read as high as 250 psi when fully charged. The pressure gauge is not reading acetylene but acetone pressure. It is recommended that acetylene be released slowly from the acetone by opening the tank valve no more than one-quarter to one-half turn. If the acetylene tank valve is opened too far, acetone will flow out of the bottle into the feed line or gas hose.

Acetylene gas is formed when calcium carbide crystals are exposed to water. An acetylene generator utilizes a full water tank, a half-filled carbide container, and a tank filler line connecting the top of the carbide container to the acetylene bottle. As water droplets fall onto the carbide crystals, acetylene gas is released, sort of like a seltzer tablet. As the crystals fizz, released gas builds up pressure at the top of the carbide container where it eventually flows to the acetylene tank and is absorbed by the acetone. The weight of the bottle is monitored, and the tank valve is closed when the proper weight is reached.

OXYGEN

Oxygen is a colorless, odorless, tasteless gas that comprises 21 percent of the volume of the earth's atmosphere. In welding, oxygen is used to support combustion. Oxygen must never be used in the presence of petroleum-based substances because of the danger of a fire or explosion. For example, when high-pressure oxygen comes into contact with petroleum-based engine oil, it lowers its flash point to the level that a spark or minimal friction could set off an explosion. Extreme caution is required when welding on tanks that were once filled with gasoline, alcohol, hydraulic fluid or any other volatile substance. The tanks must be steam-cleaned thoroughly before beginning.

The flame color and pattern from the welding torch changes as the volume of oxygen relative to the acetylene is increased. As the oxygen valve is opened, the flame color changes from a dull orange to a brilliant whitish purple. This **carbonizing flame** is still relatively rich in acetylene. The intensity of the heat continues to increase with additional oxygen until the optimal mixture, and a **neutral flame** is reached. If the amount of oxygen is increased further, the flame changes from neutral to oxidizing. A neutral flame is 6,300° F and is used primarily when welding aircraft steel. A carbonizing flame burns much cooler, and is used for brazing and soldering. An **oxidizing flame** is used for cutting.

EQUIPMENT

Oxyacetylene welding equipment includes regulators, hoses, torches and lighters as well as personal safety equipment like welding goggles and gloves. It is essential to understand the operation and safety practices for this equipment.

PRESSURE REGULATORS

Both oxygen and acetylene are supplied under a pressure far higher than is needed at the torch. A regulator reduces the pressure to a usable value. If the gases are supplied from a manifold system, a single-stage regulator is sufficient to control the pressure. However, most aircraft repair shops use bottled, high-pressure gases, which require two-stage regulators.

OXYGEN

The cylinder pressure for oxygen can be as high as 3,000 psi. Two-stage regulators employ one stage of pressure reduction to a constant intermediate pressure. A final stage further drops the pressure to the specific value needed for welding. Two gauges show the tank pressure and the pressure at the output of the regulator. This two-stage operation ensures that the pressure delivered to the torch will not vary as the cylinder pressure drops. [Figure 4-32]

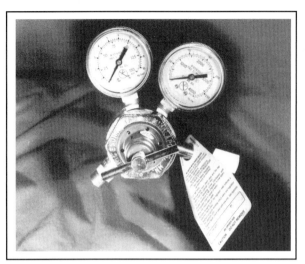

Figure 4-32. Most aircraft maintenance shops use two-stage regulators for controlling the flow of oxygen in oxyacetylene welding.

ACETYLENE

The acetylene regulator also has two gauges. One gauge indicates the pressure of the gas in the cylinder, which can be as high as about 500 psi. The second gauge indicates the pressure of the acetylene gas being fed to the torch. [Figure 4-33]

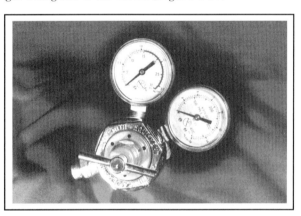

Figure 4-33. A two-stage regulator controls the acetylene gas in oxyacetylene welding. The gauges for pressures in the bottle and at the torch read much lower values than those gauges used for the oxygen gas.

The adjusting screw handle is turned counterclockwise to close the valve. When there is no more opposition from the spring, the gas flow to the torch is shut off. Turning this handle clockwise opens the valve and regulates the pressure of the gas delivered to the torch.

All approved-type regulators have a safety disc that will rupture at a pressure below that which would damage the regulator. This disc is on the low-pressure side of the regulator, and if the regulator should leak, this disc will break before the regulator diaphragm is ruptured.

The oxygen regulator is similar to the one used for the acetylene gas, except that it is built for a much higher pressure.

In addition to a portable welding system, there is another type called the manifold system. The manifold system is a permanent setup with individual welding stations. The manifold system is often used in a welding shop or a welding school. A manifold system has two sets of pipes —one for oxygen and the other for acetylene gas. The gas bottles used in the manifold system are larger than those of the portable system. Gas pressure in the manifold system is set by a main system regulator connected to each gas tank. Each main pressure regulator has two gauges mounted on it, the one near the tank indicates bottle pressure and the other near the line indicates adjusted line pressure.

If the oxygen and acetylene are delivered to a permanently installed welding bench from a manifold, there often is only one pressure gauge. It will be located at the bench and will indicate the pressure being delivered to the torch.

HOSES

Acetylene and oxygen hoses connect the gas bottles to the welding torch. A typical system consists of the bottles with gas shutoff valves, regulators connected to each gas valve, and hoses that run from the regulators to the torch handle.

To prevent interchanging the hoses, the acetylene regulator connection has left-hand threads, while oxygen regulator connection has right-hand threads. The acetylene hose is red and the left-hand fittings can easily be recognized by the universal groove cut into the hex-sided coupling nuts. The oxygen hose

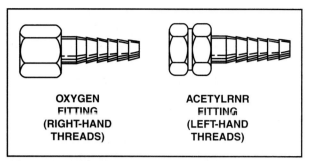

Figure 4-34. The oxygen and acetylene hose couplings have different threads to prevent inadvertently switching the hoses.

is green with right-hand threaded coupling nuts at each end. [Figure 4-34]

TORCHES

There are two types of oxyacetylene welding torches: equal pressure and injector. The equal pressure type is most commonly used in aircraft welding and uses cylinder gases. If the acetylene is supplied to the welding bench from a generator, an injector torch is used.

EQUAL PRESSURE TORCH

An equal pressure torch consists of oxygen and acetylene on/off valves, a mixing chamber and a torch tip. This is the type of torch that is most often used with cylinder gases. The oxygen and acetylene are supplied to the torch at the same pressure, usually between one and five psi, depending upon the thickness of the metal being welded.

Needle type control valves provide fine adjustment of the gas flow. For larger torches these valves are located at the end of the torch where the hoses attach. For light-duty welding of thin-wall aircraft tubing, the torches that give the best control have the valves located at the tip end of the torch. This enables fine adjustments to be made with the fingertips while welding. [Figure 4-35]

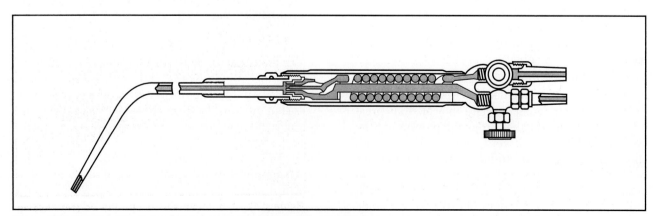

Figure 4-35. Equal pressure torches are used with high-pressure bottled gas.

The mixing chamber for equal pressure torches screws onto the torch body, and the tubes and tips screw into it. The gases flow into the mixing chamber through two separate tubes, where they are thoroughly mixed before they are delivered to the tip.

While some torches utilize separate tips screwed onto a single-torch tube, a combination tip and tube, that screws directly into the mixing chamber, is more common. The manufacturer usually designates the size of the tip. Common orifice sizes are 000, 00, 0, and 1-5, with larger tip sizes being designated by larger numbers. The manufacturer usually lists the number drill size that corresponds to each size tip.

INJECTOR TORCH
Although not commonly used, acetylene generators require a torch that operates differently. The high-pressure flow of oxygen passes through a venturi-like portion of the torch creating a low-pressure area to draw the low-pressure acetylene into the mixing chamber. [Figure 4-36]

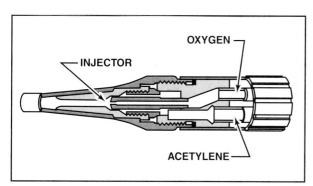

Figure 4-36. Injector torches utilize a venturi to draw acetylene into the mixing chamber.

TORCH LIGHTERS
Torches usually are lit with a flint-and-steel-type lighter. Matches should not be used because of the danger in carrying matches in the welding area. Metal popped from the torch could accidentally ignite them. Many permanently installed welding stations have a combination torch holder and pilot light. When the torch is hung up on the holder, both the acetylene and oxygen are cut off, and the flame goes out. When the torch is lifted, the gases again flow, and a small pilot light on the fixture is used to light the torch.

FILLER RODS
Filler rods are used to reinforce the bead and they also serve as a heat sink. When a filler rod works as a heat sink it pulls heat away from the molten puddle as it melts. The melting of the rod helps the welder to control the flow of metal in the molten puddle.

Most filler rods used for aircraft welding are copper coated to prevent rust from forming on the surface. Oxide on a filler rod is undesirable because it can cause more rapid oxidation while welding. Too many oxides within a molten puddle will result in a porous weld, leading to weak spots in the joint.

OXYACETYLENE WELDING GOGGLES
Although the glow from the white-hot metal will not cause permanent damage to the eyes, the intense flame and molten puddle are bright enough to temporarily blind a welder who looks directly at them. That is why an approved pair of oxyacetylene welding goggles is required equipment. It is important to realize that oxyacetylene welding goggles do not provide adequate protection for electric arc welding; permanent damage to the eyes will likely occur if these oxyacetylene goggles are used for that purpose.

Welding goggle lenses consist of a clear outer cover and shaded filter glass. The inexpensive outer cover is often damaged from weld spatter and needs to be replaced frequently. The filter glass is identified with a number indicating the level of eye protection. Higher numbers denote darker lenses with more protection and lower numbers represent less protection. For aluminum welding, a blue lens filter normally is used, because it allows better observation of the metal surface. For steel, the filters usually are green or brown. Never substitute sunglasses for welding goggles since they do not provide adequate protection. [Figure 4-37]

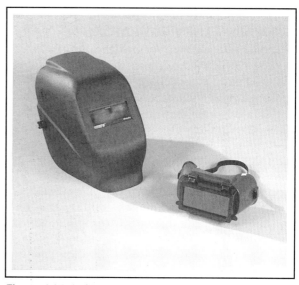

Figure 4-37. It is important to use the correct welding goggles or hood for the type of welding to be done.

WELDING GLOVES

Welding gloves are an important part of the welder's personal safety equipment. They usually are described as gauntlet style because of the extended coverage up the forearm. The welding glove prevents sparks and weld splatters from burning your hands and arms. [Fig 4-38]

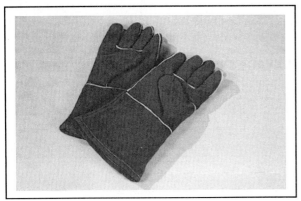

Figure 4-38. Welding gloves are typically made of fire resistant materials such as chemically treated canvas.

EQUIPMENT SETUP

When using welding equipment, it is essential to follow proper safety procedures. The gases can be highly explosive and are under extremely high pressures that can be dangerous if not handled properly.

HANDLING GAS CYLINDERS

The gas cylinders are shipped with protective caps which are threaded to cover the tank's gas valves. The tank caps protect the gas valve from being broken if a tank falls over. A broken valve would turn a fully charged bottle into an uncontrolled rocket. Gas cylinders must always be locked or chained into a holding stand on either portable or stationary bottle stations.

ATTACHING THE REGULATORS

To attach a regulator, remove the protective cap from the cylinder and open the valve slightly to blow out any dirt particles that may be in the fitting. Close the valve and attach the regulator, taking care not to cross-thread the regulator union nut.

CONNECTING THE TORCH

The oxyacetylene torch is connected by two gas hoses to their respective pressure regulators. The oxygen hose (green) and the acetylene hose (red) connect to the back end of the torch. Twist the oxygen hose attachment nut clockwise to tighten it and the acetylene hose attachment nut counterclockwise to tighten.

Check the fitting at the hose and regulator ends for leaks. The recommended method is to brush a commercial leak check fluid on each of the fittings while there is line pressure in the system. Do not use a soap solution, since most soaps contain petroleum compounds. If detecting a leak, gently tighten the fitting. If it still leaks, bleed the system, undo the fitting, check for dirt, reconnect, and check for leaks again.

SELECTING THE TORCH TIP AND ROD SIZES

The thickness of the metal determines the size of the torch tip and filler rod. When welding thicker metal, the torch tip and filler rod sizes increase. Larger torch tips produce a larger flame without increasing temperatures and are required when welding thicker metals. Small tips do not provide enough heat to melt the metal before the heat is dissipated away from the weld area. If a soft flame is required, a larger tip can produce the desired heat output by reducing the amount of oxygen in relation to the acetylene. If the filler rod is too small for the torch tip and metal thickness, the finished bead will be undercut. The bead will not be sufficiently filled, leaving it concave instead of convex. The rod will melt so fast it will not be able to carry heat away from the molten puddle. If the filler rod is too large for the torch tip, it will cool the molten puddle too rapidly. This causes the rod to stick to the puddle, as well as poor penetration and cold edges. Cold edges are roughly formed but not completely fused with the base metal.

USE OF THE OXYACETYLENE TORCH

Lighting and properly adjusting the torch will provide maximum safety and prevent unnecessary wear or damage to the equipment. In all cases, follow the procedures recommended by the manufacturer.

With the equipment properly set up and the correct size tip screwed into the torch, open the oxygen valve on the torch about one turn. Screw in the oxygen regulator handle until the low-pressure gauge indicates the correct pressure for the thickness of metal to be welded. The pressure must be set with the oxygen flowing since the pressure will be lower than it is when there is no flow. Turn off the oxygen valve at the torch and repeat the procedure with the acetylene regulator. Adjust it for the proper pressure with a flow from the torch. After the acetylene pressure is set, shut off the valve at the torch. [Figure 4-39]

To light the torch, open the acetylene valve on the torch handle about 1/6-turn, and ignite the gas with a flint lighter. Continue opening the acetylene valve

METEL THICKNESS	ROD DIAMETER	TIP ORIFICE DRILL SIZE	PRESSURE	
			OXYGEN	ACETYLENE
1/16	1/16	60 - 69	4	4
1/8	1/8	54 - 57	5	5
1/4	3/16	44 - 52	8	8
3/8	1/4	40 - 50	9	9

Figure 4-39. Each different metal thickness requires specific rod diameters, tip sizes, and gas pressures.

until all the heavy black smoke clears from the flame. Then, slowly open the oxygen valve on the torch. Three distinct flames appear; a bright white cone at the orifice tip, a long white middle flame, and a large feathery blue flame. As the oxygen valve is opened further, the flame pattern changes from the heavy acetylene orange to a flame with a whitish-blue hue. As more oxygen is added to the flame, the long white middle cone begins to shrink toward the inner cone at the tip. When the middle flame becomes the same size as the white inner cone, a neutral flame has been formed. From this point it is a simple matter of fine-tuning the flame to the type flame required for a particular welding repair. [Figure 4-40]

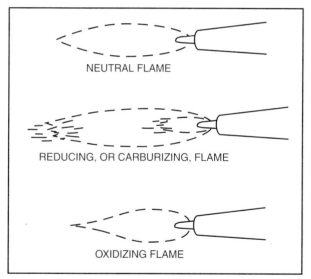

Figure 4-40. The type of oxyacetylene gas flame can be adjusted to a reducing flame by reducing the amount of oxygen and toward an oxidizing flame by adding more oxygen.

THE PUDDLE

A welder must learn to control a molten puddle without using a filler rod. Practice running a puddle without a filler rod before attempting an actual weld with a filler rod.

A beginning welder should practice running puddles on small squares of 18-gauge (.035-inch thick) steel cut to a size of 4"×4". Place the square of steel onto a fire brick (brick which is resistant to high heat) and begin running puddles from one side to the other in a forehand direction. A good puddle should show a rippled penetration on the bottom side of the plate. Occasionally, you might burn a hole all the way through the metal. This usually is caused by not moving the torch fast enough or by a flame that is too hot. If the penetration is not deep enough, it is because the torch is moved too fast or the flame is not hot enough. [Figure 4-41]

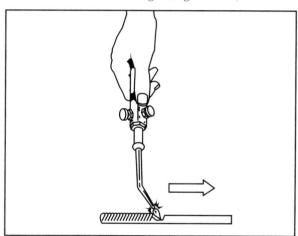

Figure 4-41. Practice initial welds in the forehand direction.

The actual puddle movement is accomplished by keeping the hot cone at the torch tip about 1/4-inch away from the base metal and slowly moving the tip in a small circular motion. When the heat is right and as the tip is directing heat to the puddle, a little glossy eye will appear. When the glossy eye is present it means the metal is melting at a uniform rate. Keeping the torch moving in a circular motion continues this.

After running several puddles on the same sheet, examine the welds. A good puddle should sag uniformly over the full length of the run, the ripples from the underside should be uniform, and there should be no blowholes. After running straight-line puddles, practice curved and side-to-side puddles, along with fore and aft puddle control. Running puddles prepares you to control the action of the welding torch in preparation for future welding jobs. [Figure 4-42]

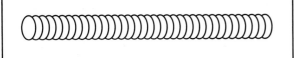

Figure 4-42. A beginning welder must learn to control or run a molten puddle before attempting fusion welding. This skill is developed by practicing running puddle beads across a sheet of steel.

FILLER ROD ADDED TO THE PUDDLE

After mastering the art of molten puddle control, the next step is to begin adding filler rod to the puddle. The operation begins the same as for running a puddle; however, when the puddle forms, start dipping the filler rod into the molten mass while moving the puddle horizontally across the metal at an even rate. [Figure 4-43]

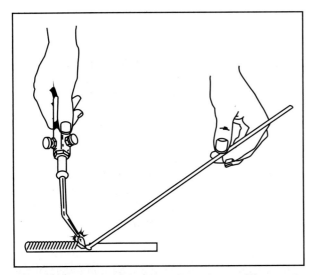

Figure 4-43. Dip the filler rod into the molten puddle so as to add enough metal to produce a slightly crowned bead.

Use a rod of the proper diameter for the metal. For 1/16-inch-thick metal, a 1/16-inch diameter rod is correct. Form the puddle as before, and hold the end of the rod about 1/8-inch above the puddle and just a little ahead of the torch. Momentarily dip the end of the rod into the puddle, and enough of it will melt off to give the puddle a slightly crowned appearance. Continue to move the puddle across the sheet, adding rod by dipping it in as you move.

If the rod, torch tip and metal thickness are correct, a neutral flame should be sufficient to melt both the puddle and the filler rod. As the metal is being heated to form a puddle, the filler rod tip should be resting within the high-heat zone. The filler rod will draw away some of the heat used to melt the puddle. This will give you more control over the molten puddle.

As heat is drawn away from the puddle, and experienced welding operator will slow down the movement and travel of the torch in order to obtain a good solid bead. As the torch moves across the base metal, the front part of the flame preheats the metal and the aft flame after-heats the metal. The heat in the puddle and the cooling of the bead are controlled with precise movement of the flame.

A good bead with proper penetration will have a smoothly crowned surface that fairs evenly into the base metal. This kind of bead can be produced only by keeping the torch motion constant, properly slanted to the work, and by adding the filler metal at regular intervals. [Figure 4-44]

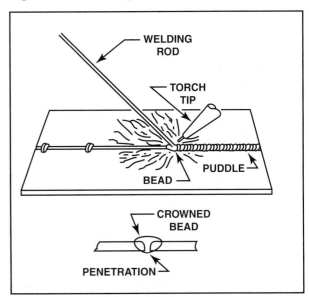

Figure 4-44. Add filler rod into the puddle to get a crowned bead with good penetration.

OXYACETYLENE CUTTING

An oxyacetylene flame can be used to cut steel with unrivaled ease. The cutting torch is similar to an oxyacetylene-welding torch with an additional valve that allows a stream of high-pressure oxygen to flow directly to the work from a hole in the center of the nozzle. Surrounding this oxygen hole is a series of pre-heating holes. [Figure 4-45]

With sufficient oxygen pressure and the right size torch tip, an oxyacetylene cutting flame can be adjusted to cut steel over a foot thick. The cut made by an oxyacetylene cutting torch is known as a kerf.

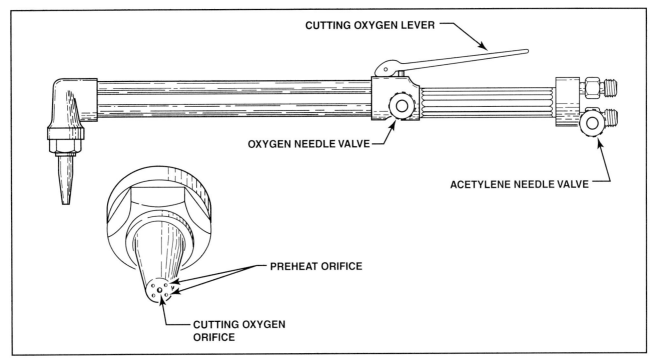

Figure 4-45. An oxyacetylene cutting torch consists of a normal welding torch, with an additional orifice for high-pressure oxygen, and a valve to control it.

A good quality cutting-torch can make as smooth a cut in the metal as a hacksaw.

Different gas pressures and tip sizes are required for cutting various thickness of metal. This data usually is determined by the manufacturer of the cutting instrument. For example, one manufacturer recommends using 15 psi of oxygen, 4 psi of acetylene, and an oxygen jet size of a no. 72 drill to cut a 1/8-inch thick sheet of steel. The same manufacturer recommends using 100 psi of oxygen, 7 psi of acetylene, and an oxygen jet size of a no. 32 drill for cutting a piece of steel 12 inches thick. These pressures will vary according to the make of the cutting instrument. You can find exact procedures on how to operate the equipment in the manufacturer's manual.

When cutting thick materials, higher pressure oxygen is necessary. When oxygen pressure above 100 psi is delivered from the torch tip, it is difficult to hold the torch in place. The high-pressure oxygen stream makes the tip react like a jet engine, pushing away from the cutting surface. To assist in holding the torch a constant distance from the material at high pressures, a cutting torch may be mounted on a trolley that rolls across the metal as it is being cut.

A cutting torch and a regular welding torch connect into the oxyacetylene system the same way. There are gas inlet attachments for both oxygen and acetylene on the torch handles. The cutting torch has an extra oxygen valve that is used to adjust the high-pressure oxygen jet blast for cutting. Just like the regular welding torch, a cutting torch is lit with a striker, then adjusted to a neutral flame. The adjusted neutral flame is used to preheat the metal to a red-hot condition before cutting. The neutral flame should not shift when the jet blast of pure oxygen is applied. If the preheat flame shifts, turn off the torch, clean the tip orifices with a tip cleaner, then re-light the torch.

Squeezing a lever on the torch handle activates the jet of high-pressure oxygen. When this pressurized oxygen hits the red-hot surface of the metal, a chemical and physical reaction occurs. The oxygen accelerates the natural oxidation of the metal to many times the normal rate. The metal disintegrates, or burns, and the resulting residual metal is blown out the bottom of the kerf. Upon examining the remains, you would find ferrous oxides as evidence of this oxidation.

SHUTTING DOWN THE EQUIPMENT

Welding equipment should be shut down when not in use. Shutting off the gas equipment avoids gas leaks and relieves pressure from the gas regulator valves. Here is the procedure (and correct sequence) for shutting down the welding apparatus:

1. Shut off the flame by closing the acetylene needle valve on the torch.

2. Close the oxygen needle valve on the torch.

3. Close the acetylene cylinder valve.

4. Close the oxygen cylinder valve.

5. Open the acetylene valve on the torch to drain the acetylene hose and regulator. This removes pressure on the regulator's working-pressure gauges.

6. Turn the acetylene-regulator adjusting screw to the left to relieve pressure on the diaphragm, and then close the torch acetylene valve.

7. Open the torch oxygen valve, drain the oxygen hose and regulator.

8. Turn the oxygen-regulator adjusting screw to the left to relieve the diaphragm pressure. Then close the torch oxygen valve.

9. Hang up the torch and hose properly to prevent hose or torch damage.

CHAPTER 5

AIRCRAFT FABRIC COVERING

INTRODUCTION
The methods and materials needed to effectively cover aircraft structures with fabric have improved dramatically since the early days of aviation. The first practical covering materials included organic fabrics such as cotton, linen, and even silk. Although these continue to be used today, organic fabrics are being replaced with more resilient synthetic materials. Possessing increased strength and durability, these products have made fabric covering a practical means of aircraft construction that will continue to be used for many years.

SECTION A
FABRIC COVERING PROCESSES

When early airplanes were constructed using organic fabrics, most builders did not use any special processes to increase the strength of the material. These coverings were not airtight, and tended to loosen and wrinkle with changes in humidity. In order to keep humid air from flowing through the fabric, builders began applying a rubberized or varnished finish to the covering surface. Although such finishes were effective, it was found that shrinking the fabric by brushing or spraying it with banana oil and collodion and letting it dry provided better dimensional stability.

Later, an improved cellulose nitrate material consisting of cotton fibers dissolved in nitric acid was used to form a clear dope that could be worked into the fabric to produce a stronger and more durable finish. However, when an aircraft structure was covered with white cotton fabric and coated with the cellulose nitrate dope, the covering became translucent, exposing the interior framework. Over time, it was found that the translucency caused deterioration of the fabric and interior components due to the sun's ultraviolet rays. [Figure 5-1]

To improve durability, manufacturers applied colored enamel over doped fabric to block the sunlight. Although this technique provided increased protection, the paint finish tended to crack and peel due to the lack of adequate adhesion to the dope coating. Further improvements led manufacturers to blend fine aluminum powder directly into the dope before

Figure 5-1. Early aircraft, such as this DeHavilland Moth, used clear dope over the fabric and had no finish coating to protect the fabric or interior structure.

Figure 5-2. Several fabric-covered airplanes are in production today, including this 1995 Maule. New fabric-covered production aircraft are more likely to use synthetic materials such as polyester rather than organic fabrics.

it was applied to the fabric. This technique caused the dope to take on a silver color that blocked the sunlight and also tended to reflect heat away from the fabric.

The process of aluminum pigmenting proved effective, but there were still serious problems with the early style dopes. Besides the problems of organic decay from sunlight, another major problem associated with **cellulose nitrate dope** was its high flammability. In fact, the process of mixing cotton and nitric acid was similar to a technique used to form guncotton, an explosive used for the manufacture of smokeless gunpowder. To alleviate some of the fire hazard, the U.S. Navy developed a cellulose acetate dope for use on their aircraft.

Cellulose acetate dope was later modified to **cellulose acetate butyrate (CAB)** dope, which is still widely used today. This dope is less flammable, but lacks some of the adhesive qualities of cellulose nitrate dope. As a result, cellulose nitrate dope is still often used as a base coat before CAB dope is applied to the fabric. This results in a tradeoff between the benefits of both materials, but is regarded to be one of the best finishing methods for organic fabrics.

Ultimately, manufacturers have continued to look for alternative finishing systems, including new types of fabric that are stronger and safer. In recent years, advances in chemical technology have allowed manufacturers to produce excellent finishes on durable synthetic fabrics that are far superior to organic materials. However, even with these improvements, fabric-covered construction has been largely replaced by sheet metal and advanced composite structures. Although the number of fabric-covered aircraft is relatively small, there are still some airplanes that are being produced with fabric covering, and many fabric-covered vintage aircraft continue to remain in service. [Figure 5-2]

Regardless of the type of fabric-covering process, the application and maintenance practices for fabric covering are reasonably easy, although they are time consuming. To obtain the greatest durability and best appearance, it is important to closely follow the fabric manufacturer's procedures and to only use materials that are specifically approved for use with their product. To maintain the quality and safety of a fabric-covered aircraft, it is the responsibility of the technician to use FAA-approved products and procedures to help maintain the highest safety standards.

FAA APPROVAL CRITERIA

Re-covering an aircraft is considered to be a major repair in accordance with 14 CFR Part 43, Appendix A, and the details of this type of repair must be recorded on an FAA Form 337 (Major Repair or Alteration form). The process of re-covering or repairing a fabric aircraft must match the FAA-approved methods and specifications for both the aircraft type and fabric type.

There are several FAA-approved methods for covering fabric and when one of these methods is used, it is important not to mix or substitute the specified procedures or materials. The manufacturer's materials and application processes have been thoroughly tested and any unauthorized variance could cause serious deficiencies in the strength or durability of the materials.

MANUFACTURER'S SERVICE MANUAL

Before an aircraft is certified to be airworthy by the FAA it must undergo extensive static and flight tests. To remain legally airworthy, the plane must be maintained so that it continues to meet all of its original certification requirements. Most fabric-covered aircraft manufacturers' service manuals provide the necessary information to make sure that the same type and width of fabrics, tapes, threads and stitches are used when maintaining the covering. Also, the original methods of fabric attachment, as well as the finishing processes, are detailed in these manuals.

If an aircraft is re-covered using the materials and procedures that are specified in the manufacturer's service manual, the maintenance is classified as a major repair. As such, the details of the repair must be recorded on FAA Form 337 including the part number of the service manual and specific references to the sections of the manual that were followed. An authorized individual can then approve the aircraft for return to service by verifying that the maintenance was properly accomplished in accordance with FAA-approved data. Some of the individuals authorized to return the aircraft to service include an A&P technician holding an Inspection Authorization (IA), an inspector from an appropriately rated repair station or an FAA Airworthiness Inspector.

SUPPLEMENTAL TYPE CERTIFICATES

If synthetic fabric is to be installed on an aircraft that was originally covered with an organic material, the FAA-approved procedures for making the alteration may be contained in a Supplemental Type Certificate (STC). The FAA issues an STC when an individual or company shows that the alteration of an aircraft from its original design still meets FAA criteria for airworthiness. Once the STC is issued, the holder of the certificate may market their product for installation on any aircraft of the same make and model.

The qualities and advantages of the different fabric covering systems approved with an STC have made them very popular among aircraft owners. However, before starting an alteration using a different type of fabric than the original, verify that the STC approves the use of that specific fabric on the aircraft. An STC may not be approved if the aircraft has been previously changed in a manner that makes a subsequent alteration invalid.

For example, an STC approval for a different type of fabric covering may not be valid if the aircraft has been modified with a higher-powered engine to increase airspeed. Since the airspeed of an aircraft is a determining factor in the certification of the fabric covering, an STC may not be authorized with a change in the original airspeed limits. If there is any question as to the validity of an STC for a specific aircraft, an FAA Airworthiness Inspector should be consulted before beginning the alteration.

An individual or company that provides an STC for fabric covering must provide detailed instructions on how to conduct the alteration. These instructions, usually in the form of a procedures manual, are FAA approved and include information with regard to the method of fabric attachment, the specific types of seams to be used, the widths and placement of tape, and details of the finishing materials, including the methods of application. [Figure 5-3]

Figure 5-3. Several different STCs are available to aircraft owners who wish to use a specially designed covering system. The holder of the STC provides instructions with specific details for attaching the fabric and the methods for applying the finish.

When using an STC process, the details of the alteration must be included on FAA Form 337 along with the STC certificate number and the document number of the FAA-approved procedures manual.

Once the alteration has been completed, an authorized individual approves the aircraft for return to service by inspecting the alteration for conformity to the STC instructions, and then signing the Statement of Conformity on Form 337. Persons that typically approve an aircraft for return to service with an STC alteration include an IA, an inspector from an appropriately rated repair station or an FAA Airworthiness Inspector.

ADVISORY CIRCULAR 43.13-1B

If the manufacturer's service manual is not available or does not provide adequate detail for maintaining the fabric, and an STC is also not available, FAA Advisory Circular 43.13-1B, *Acceptable Methods, Techniques, and Practices—Aircraft Inspection and Repair*, may be consulted. Chapter 2 of this circular describes in adequate detail an acceptable method of re-covering an airplane, using cotton or linen fabric with a cellulose dope finish.

These instructions may be followed with regard to the types of materials, methods of attachment, and application of finishes for airplanes that were originally certified with cotton or linen coverings. When filling out FAA Form 337, list the paragraphs and figures in AC 43.13-1B that were used to perform the repair or alteration. Although the information contained in the circular is not FAA-approved data, the technical information can be used to obtain an FAA field approval for the specific aircraft.

FAA FIELD APPROVALS

If it has been determined that a service manual or STC is not available for a particular fabric covering process, another option for obtaining FAA approval is to request a one-time field approval. In most cases, the closest regional district office of the FAA evaluates and issues these approvals and authorizes the repair or alteration.

An FAA field approval may also be necessary if deviations will be made from a previously approved procedure. For instance, an aircraft may have originally had the fabric attached to the wing ribs with lacing cord as called for in the aircraft manufacturer's service manual. Newer fabric-attaching techniques provide specially designed clips that eliminate the need for rib lacing, making for an easier installation. To deviate from the manufacturer's information and use clips instead of lacing, an FAA field approval would be necessary.

When performing an alteration that requires a field approval, it is best to consult with an FAA Airworthiness Inspector before beginning the work, so that the inspector can assist in determining appropriate procedures for conducting the alteration and identify the required data for obtaining the field approval.

FABRIC-COVERING PRODUCTS

Once it has been decided which approved method would be most appropriate for a fabric-covering repair or alteration, it is necessary to become familiar with the products that will be used in the process. It is the responsibility of the person making the repair or alteration to verify that all of the materials to be used meet the airworthiness requirements for the aircraft. In all cases, fabric-covering materials that are used on FAA-certified aircraft must meet the minimum standards established by the FAA.

To verify that the products meet required specifications, buy them from a reputable aircraft supply dealer, and confirm that the proper identification is present on the materials. In addition, only use products from the supplier of the specific fabric-covering process. If substitutions are made, there is a possibility that the products may be incompatible, causing defects in the fabric that ultimately may compromise safety.

Be aware of old dopes or solvents. Some materials have a recommended shelf life, which if not used in a certain amount of time may not perform to specification. Using products that are substandard or too old can not only affect safety, but may also be very costly in time and labor. Generally, labor is the largest percentage of the total expense of a fabric job, and cutting corners by using old materials can increase costs by having to take the time to correct flaws.

PARTS MANUFACTURER APPROVALS

When a product such as synthetic fabric or glass cloth is developed for aircraft applications, the manufacturer must submit data to substantiate that the item conforms to the standards established by the FAA. Once the FAA evaluates the product and verifies that it meets FAA standards, the manufacturer is granted a **Parts Manufacturer Approval (FAA-PMA)**.

An FAA-PMA designates that a product is approved by the FAA for use on eligible type certificated aircraft. Once the manufacturer inspects the product for quality control standards, it is certified as having met the specifications of the PMA. Upon final inspection, the product is marked to indicate that it meets the standards for an FAA-PMA. [Figure 5-4]

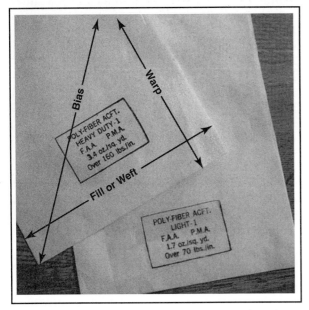

Figure 5-4. A FAA-PMA stamp is located along the selvage edge of aircraft grade fabric, and should be verified to make sure the material is approved for use on certificated aircraft. A selvage edge is located along the length of fabric as it is pulled off a roll, or bolt, and is identified by a specially sewn stitch that helps to prevent the fabric from unraveling.

FABRIC ORIENTATION

It is important to understand the construction and orientation of fabric material because all design, manufacturing, and repair work begins with the orientation of the fabric. Unlike metallic structures, the fabric structure relies on the proper placement and use of the reinforcing fabric to produce a strong covering. Some of the terms used to describe fabric orientation are warp, weft, selvage edge and bias.

The **Warp** of threads in a section of fabric run the length of the fabric as it comes off the roll or bolt. Warp direction is designated as 0°. There are typically more threads woven into the warp direction than the fill direction, making it stronger in the warp direction. The warp is critical in creating or repairing fabric coverings. The fabric must be applied with the warp parallel to the direction of flight.

The **Weft/Fill** threads of the fabric are those that run perpendicular (90°) to the warp fibers. The weft threads interweave with the warp threads to create the reinforcing cloth.

The **Selvage Edge** of the fabric is the tightly woven edge parallel to the warp direction, which prevents edges from unraveling. The selvage edge is removed before the fabric is utilized. The weave of the selvage edge is different from the body of the fabric and does not have the same strength characteristics as the rest of the fabric.

The **Bias** is the fiber orientation that runs at a 45° angle (diagonal) to the warp threads. The bias allows for manipulation of the fabric to form contoured shapes. Fabrics can often be stretched along the bias but seldom along the warp or fill.

ORGANIC FABRIC MATERIALS

As previously mentioned, a common organic fabric covering material is grade-A cotton. This material meets Aeronautical Material Specifications AMS 3806 and is produced under Technical Standard Order TSO-C15. The cloth has between 80 and 84 threads per inch in both the warp (the direction along the length of the fabric) and the weft directions (the direction along the width) and weighs about four ounces per square yard.

In the process of manufacturing grade-A cotton fabric, the natural material is mercerized by dipping the threads in a hot caustic soda solution to give them a sheen and to increase their strength. After the cloth is woven it is calendered by wetting the cloth and passing it through a series of heavy, heated rollers to produce a smooth finish on the fabric.

The minimum tensile strength of new grade-A cotton fabric is 80 pounds per inch of width. This is the minimum strength allowed for new fabric used on aircraft having a never-exceed speed (V_{NE}) of more than 160 miles per hour and a wing loading in excess of nine pounds per square foot. For aircraft operated at these speeds and wing loads, the fabric is considered airworthy until it deteriorates to 70% of its new strength, or 56 pounds per inch of width.

A lighter cotton fabric called **cotton intermediate grade**, meeting specifications AMS 3804 (TSO-C14), was approved for use on aircraft having a V_{NE} of less than 160 miles per hour and a wing loading of nine pounds per square foot or less. Many of the light airplanes built in the 1940s were covered with this fabric. Cotton intermediate-grade fabric has a much finer weave than grade A cotton, with up to 94 threads per inch allowed in both the warp and fill directions. Its minimum tensile strength when new is 65 pounds per inch of width and is considered airworthy on lightweight aircraft until its strength deteriorates to 46 pounds per inch. Because of the tremendous decrease in demand for aircraft fabric, this particular material may be difficult to find. If cotton-intermediate grade fabric is not available, grade-A cotton may be used as an alternate covering material. When grade-A cotton is used on these lightweight airplanes, the fabric strength may also be allowed to deteriorate to 46 pounds per inch before it must be replaced.

A very fine-weave cotton fabric called **glider fabric** has up to 110 threads per inch in warp and fill and meets specification AMS 3802. This fabric is designed for use on gliders and sailplanes having a V_{NE} of 135 miles per hour or less and a wing loading of eight pounds per square foot or less. It is also used to cover plywood surfaces on powered aircraft to protect the wood and give it a smooth finish. Glider fabric has a minimum tensile strength of 50 pounds per inch when new, and for use on gliders, can deteriorate to a strength of 35 pounds per inch before it must be replaced. When used as a cover for plywood, its strength is not a determining factor, and so there is no published minimum strength requirement for these applications.

Irish linen, produced in the British Isles, is another organic fabric. Since this fabric was originally milled by the British, it was designed to meet British specification 7F1 (formerly DTD 540). This fabric is stronger than grade-A cotton, with strength of around 140 pounds per inch when new. It is somewhat heavier than grade-A cotton and requires more dope to get an equivalent finish, but it does meet all of the requirements of grade-A cotton and tends to last longer. However, linen has become increasingly difficult to acquire, and because some synthetic fabrics have such marked advantages over it, linen is seldom used except for restorers who desire the originality enough to pay the difference in time and money to use it. [Figure 5-5]

INORGANIC FABRIC MATERIALS

A man-made inorganic fabric that is produced from synthetic polyester has quickly become one of the most popular aircraft covering materials. Polyester fibers, woven into cloth with different weights, are

MATERIALS	SPECIFICATION	MINIMUM TENSILE STRENGTH NEW (UNDOPED)	MINIMUM TEARING STRENGTH NEW (UNDOPED) (ASTM D 1424)	MINIMUM TENSILE STRENGTH DETERIORATED (UNDOPED)	THREAD COUNT PER INCH	USE AND REMARKS
Airplane cloth mercerized cotton (Grade "A")	TSO-C15d, as amended, references Society Automotive Engineers AMS 3806d, as amended or MIL-C-5646	80 pounds per inch warp and fill	5 pounds warp and fill	56 pounds per inch	80 min., 84 max. warp and fill	For use on all aircraft. Required on aircraft with wing loading of 9 p. s. f. or greater or placarded never exceed speed of 160 mph or greater
Airplane cloth mercerized cotton	TSO-C14d, as amended, references Society Automotive Engineers AMS 3804c, as amended.	65 pounds per inch warp and fill	4 pounds warp and fill	46 pounds per inch	80 min., 94 max. warp and fill	For use on aircraft with wing loading less than 9 p. s. f. and never exceed speed of less than 160 mph.
Airplane cloth mercerized cotton	Society Automotive Engineers AMS 3802, as amended.	50 pounds per inch warp and fill	3 pounds warp and fill	35 pounds per inch	110 max. warp and fill	For use on gliders with wing loading of 8 p. s. f. or less, provided the placarded never-exceed speed is 135 mph or less.
Aircraft linen	British 7F1.					This material meets the minimum strength requirements of TSO-C15.

Figure 5-5. Specifications for cotton and linen fabrics

sold under trade names such as **Ceconite**, **Polyfiber**, and **Superflite**. The fibers used to make the material have been passed through rollers and are woven so that the number of fibers in the warp direction is equal to the number in the fill direction. When the material is finished it is delivered in an unshrunk, or greige, condition. Once heat is applied during the installation process, the unshrunk fabric will constrict back to its original length and size.

Other inorganic fiber-covering systems use fiberglass filaments woven into cloth which will not decay with moisture or mildew and has virtually unlimited life. **Fiberglass cloth** has previously been approved as reinforcement over cotton in sound condition, but treated fiberglass has become an approved direct replacement for grade-A cotton. Treated fiberglass is a loose-weave fabric impregnated with dope that, when the dope is activated, draws the filaments together. Covering processes that utilize fiberglass are sold under STCs and, as with all covering systems, the installation and repair procedures should be followed closely.

FINISHING MATERIALS

Several finish materials that increase the durability and appearance of fabric are used in covering processes. These items provide additional rigidity of the fabric, which helps to transfer the aerodynamic lift provided by the covering into the structure of the aircraft. In addition, inspection hole and drainage grommets, as well as tapes and lacing cords, are vital components to a quality fabric-covered structure.

This section provides a brief overview of finishing products that are used in conjunction with fabric covering. For a detailed description of these products and application procedures, refer to Chapter 6, Aircraft Painting and Finishing, in this textbook.

REINFORCING TAPE

Reinforcing tape is a flat woven cotton material that is available in 1/4-inch, 3/8-inch, and 1/2-inch widths, with a strength of 150 pounds per half-inch of width. This tape is used under rib lacing to act as a reinforcement to prevent the lacing cord or other fabric-attaching devices from pulling through the fabric covering. Another function of this tape includes interrib bracing for wing ribs to hold them in an upright position until the fabric covering is secured to the capstrips.

Reinforcing tapes made from polyester are also available in the same widths as cotton tape. The polyester is less susceptible to decay from moisture and mildew and has more strength than the cotton fiber tape.

SURFACE TAPE

Surface tape is made of the same material as the covering fabric and is used over all seams, ribs, around corners, along leading edges, around the tips and along the trailing edge of all surfaces. The purpose of the tape is to blend the covering around contours and irregularities to make for a smoother surface finish. In addition, the tape aids to prevent the covering material from coming loose in the airstream during flight.

Surface tapes are available in a bias cut or a straight-cut. Straight cut tape has a weave that runs parallel to its edges and is primarily used over flat surfaces such as on top of wing ribs. On the other hand, bias-cut tapes are constructed so that the weave of the fabric runs at a 45° angle to the edge. The bias weave provides for better contouring around curves such as those found on the rudder or wing tip bows. When the tape is stretched during application it can shrink in width by as much as 1/3 its original width. In the unstretched condition, the tape comes in widths ranging from 3/4-inch to six inches.

Both bias- and straight-cut tapes are available with either pinked (serrated) edges or with smooth edges that are sealed to prevent raveling. A pinked edge helps to prevent a continuous thread from pulling out if the edge of the tape becomes loose. Some surface tapes are predoped, or impregnated with dope to aid in its application. [Figures 5-5 and 5-6]

RIB LACING CORD

As the name implies, rib lacing cord is used to secure aircraft covering to the capstrips of ribs. The lacing helps to prevent the fabric from pulling away from the ribs during flight, when the airstream tends to pull the fabric up as a result of the production of lift. The security of the fabric is not only necessary to prevent it from tearing, but also to maintain the shape of the airfoil to prevent the disruption of lift.

Rib lacing cord is available in cotton and polyester fibers. Cotton-fiber rib lacing cord is prewaxed with beeswax to allow the cord to pass more easily through the fabric during lacing and also helps to prevent thread decay. It has a minimum tensile strength of 80 pounds, double. Polyester lacing cord can be purchased in several different strengths and thickness and is usually impregnated with a fungicidal wax to prevent decay.

MACHINE SEWING THREADS

Machine sewing threads are used primarily to sew lengths of fabric to form large blankets or to form an envelope to slip over a wing or other surface. These threads are available in grade-A cotton or polyester.

MATERIALS	SPECIFICATIONS	YARN SIZE	MINIMUM TENSILE STRENGTH	YARDS PER POUND	USE AND REMARKS
Reinforcing tape, cotton	MIL-T-566 1 E, Type 1 MIL-Y-1104H		150 pounds per 1/2 inch width.		Used as reinforcing tape on fabric and under rib lacing cord. Strength of other widths approx. in proportion.
Lacing cord, prewaxed braded cotton	Federal T-C-57 1F		40 pounds	310 minimum	Lacing fabric to structures. Unless already waxed, must be lightly waxed before using
Lacing cord, braided cotton	MIL-C-5648A		80 pounds	170 minimum	Lacing fabric to structures. Unless already waxed, must be lightly waxed before using
Lacing cord thread, high tenacity cotton	MIL-T-5660B	Ticket No. 10.	62 pounds	480 minimum	Lacing fabric to structures. Unless already waxed, must be lightly waxed before using
Machine thread cotton	Federal V-T-276H	20/4 ply	5 pounds	5,000 normal	Used for all machine sewing
Hand-Sewing thread cotton	Federal V-T-276H Type III B	8/4 ply	14 pounds	1,650 normal	Use for all hand sewing. Use fully waxed thread
Finishing (Surface) tape cotton	Same as fabric used.		Same as fabric used		Use over seams, leading edges, trailing edges, outer edges and ribs, pinked, raveled or straight edges.

Figure 5-6. The specifications for common cotton and linen tapes and threads are shown in this table.

Grade-A cotton is available in a 20/4-ply thread with a tensile strength of five pounds, single, whereas polyester sewing thread has as many as 4 plies and has a 10 pound tensile strength.

HAND SEWING THREAD

Cotton thread with an 8/4-ply yarn size and a tensile strength of 14 pounds, single, is used for hand sewing stitches. This thread is generally supplied without any coating but should be lightly waxed with beeswax before being used. Polyester hand sewing thread is commonly uncoated with multiple plies and has a tensile strength of over 15 pounds.

DRAINAGE GROMMETS AND INSPECTION RINGS

Drainage grommets are small doughnut-shaped plastic, aluminum or brass rings that are installed in numerous locations on the aircraft. Typical installation positions include the lowest point on the bottom of the wing and tail surfaces, toward the rear of each rib bay and on the fuselage fabric at the lowest point of each compartment. These items are usually installed when the second coat of dope is applied to the fabric, while the dope is still wet. When all the finishing coats have cured, the center of the grommet is cut out with a sharp knife blade to allow any moisture within the structure to drain out and to ventilate the inside of the structure to minimize condensation.

Larger inspection hole grommets, or inspection rings, are installed on the fabric over any location where access to the interior structure may be needed. These rings add additional strength to the fabric and prevent the fibers from unraveling from around the cut-out opening. Inspection grommets are installed at the same time as drainage grommets during the dope application. When the covering is complete and the final finish coat has cured, the fabric in the center of the ring opening is cut out and a metal cover plate is slipped in place to close over the opening. [Figure 5-7]

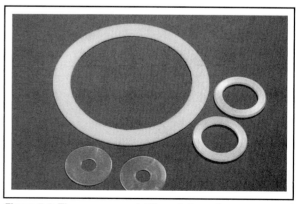

Figure 5-7. There are several different sizes of grommets and inspection rings as well as other types of hardware that are used in fabric covering processes. Additional grommets may be available from the fabric manufacturer for areas such as locations where control cables pass through the fabric.

FINISHING DOPE

As previously discussed, there are two types of dope used with organic fabric coverings: cellulose acetate butyrate and cellulose nitrate. For practicality, these are often simply referred to as butyrate and nitrate dope. Each type of dope has a film base made from a cellulose product such as cotton, which is dissolved in appropriate solvents along with plasticizers to give the film added resilience. The resulting material is thinned to a spraying or brushing consistency with dope thinners which, along with the solvents, evaporate out to leave an airtight film on the surface of the fabric.

Nitrate dope is quite flammable and would not be used at all today except for the fact that it bonds to the fibers of the fabric and to the structure better than butyrate dope. Because of its adhesive quality, nitrate dope is often used to bond the fabric to the structure in much the same way as glue. In addition, since it bonds so well to fabric fibers, it is often used for the first coat or two before applying butyrate dope as a top finish coat. Because nitrate dope has such high flammability, adequate ventilation and care should be taken whenever applying it or when sanding the material once it has cured.

Butyrate, or CAB dope, is one of the most popular dope products used today because, while it will burn, it is more difficult to ignite than nitrate and burns much more slowly. In addition, butyrate dope has a greater tautening action on the fabric than nitrate dope and may be applied over a nitrate film. The solvents used in nitrate dope are less potent than those used in butyrate dope, and for this reason, nitrate dope cannot be put over a butyrate film; it will not soften a butyrate film enough to bond properly.

Nontautening dopes are available that have different types of plasticizers which cause the film to shrink less when the solvents evaporate. These nontautening products are used to provide the fill over the surface of polyester fabrics without producing additional tautening. These dopes are used on many of the synthetic fabric processes to prevent distortion of the airframe structure, since synthetic fabrics are capable of applying extreme pressures if allowed to over-tauten.

Since aircraft-finishing dope is highly toxic, it must be disposed of properly. As a safety precaution, always dispose of waste dope products in a sealed container that is marked to indicate the contents. In addition, as with any hazardous materials, always have the material safety data sheets (MSDS) for the finishing products available for reference in the event of an emergency.

THINNERS

Dope is normally supplied with a viscosity that is proper for brushing, which means it must be thinned before being sprayed. There are thinners that are available for nitrate dope and thinners for butyrate dope, but there are also universal thinners that are compatible with either type of dope. Use the specific thinner recommended by the dope manufacturer, but do not mix brands of finishing products.

DOPE RETARDERS

Properly applied dope should dry with a smooth, clear film. However, if the humidity in the air is high, the temperature drop caused by the rapid evaporation of solvents out of the dope may cause water to condense out of the air and onto the fabric covering. When the water enters the uncured dope, it tends to create a porous surface that takes on a dull, chalky appearance. This effect is commonly referred to as a blushing finish. To prevent blushing when dope is sprayed under high-humidity conditions, a retarder may be added to the thinned dope. Retarder is a special, slow-drying solvent, that prevents rapid evaporation and the accompanying temperature drop.

FUNGICIDAL PASTE

Since cotton and linen are both organic materials, they are subject to deterioration from mildew and fungus. Fungus spores are always present in the air and can penetrate into the fabric before dope has been applied. In the dark, warm, and humid conditions of a hangar, these spores can rapidly multiply and weaken covering fabrics.

To prevent organic fabrics from suffering the effects of rot from living organisms, a fungicidal paste may be mixed in with thinned clear dope that is used for a first coat. The fungicidal paste kills the spores and protects the fabric. To help ensure that the first coat has adequate penetration into the fibers, the paste is normally mixed with a dye so there is better contrast between the fabric and the dope.

ALUMINUM PASTE

Ultraviolet (UV) rays of the sun tend to quickly weaken a cellulose film base of clear dope and the organic fibers of cotton or linen, as well as the synthetic fibers of polyester fabrics. To block these rays, a thin opaque coating of aluminum flakes may be applied over a clear dope finish. The aluminum powder, which is simply tiny flakes of aluminum metal, is mixed with clear dope and sprayed on after the last coat of clear dope, and before any pigmented color coats are applied. The layers of aluminum dope serve as a mechanical barrier and tend to be much more effective than chemical UV blockers that can be added to colored topcoats.

Aluminum powder is difficult to mix with dope, and so, many manufacturers have started selling the powder already mixed in a suitable vehicle, which is more readily blended into the dope. If the aluminum paste is not purchased in this form, it should be added into a container and then the dope mixed thoroughly as it is added. Generally, one pound of paste should be mixed with five gallons of unthinned clear dope.

REJUVENATOR

Over time, if fabric deteriorates or loses its strength, there is nothing that can be done to restore the fabric. However, if the fabric is in good condition and only the dope is cracked or has become brittle, the finish may be treated with rejuvenator and then repainted. A rejuvenator is a mixture of very potent solvents and plasticizers that penetrate into the dope to restore its resilience.

To apply the rejuvenator, the surface of the fabric is cleaned and the rejuvenator is sprayed on in a wet coat. As solvents soften the old finish, new plasticizers become part of the preexisting finish. When the rejuvenator dries, the surface can be sanded and the final finish restored. If cracks have developed in the finish before the rejuvenator is applied, aluminum dope can be used as a fill material before applying the final finish coats.

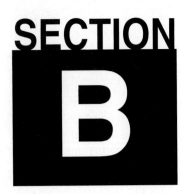

COVERING PROCEDURES

Processes involved in maintaining an aircraft with fabric covering are unique when compared to maintaining a metallic or composite skinned aircraft. Some of the most striking differences are in the materials used, methods of attaching the covering to the airframe, procedures for repairs, and the effects of the sun and weather on the covering. Although fabric covering has less strength and durability when compared to sheet metal, it does have certain advantages. Fabric coverings are generally lightweight, economical, easy to install and repair, and are visually appealing when a proper finish is applied.

DETERMINING FABRIC STRENGTH

One of the disadvantages of a fabric-covered aircraft is that the covering materials deteriorate over time, causing them to lose their strength. Eventually it becomes necessary to perform repairs in areas where the material has degraded below acceptable strength standards or to re-cover the entire aircraft. One of the most important duties that an aircraft technician must perform is to determine the airworthiness of the fabric covering by checking its strength.

Due to the expense and time involved, most aircraft are re-covered only when the strength of the fabric drops below the minimum airworthy value. Since the strength of the fabric is a major factor in the airworthiness of an airplane, its condition is determined during each 100-hour, annual, or other required airworthiness inspection. Fabric is considered to be airworthy until it is evaluated to be less than 70% of its original strength. For example, if grade-A cotton is used on an airplane that requires only intermediate fabric, it may deteriorate to 46 pounds per square inch before it must be replaced (70% of the strength of intermediate fabric). There are a variety of methods available to determine fabric-covering strength. Some of these methods can be done in the field using a simple testing apparatus, but for more thorough and accurate testing, a sample piece of fabric must be checked with more elaborate equipment.

SEYBOTH TESTER

To determine fabric strength, a **Seyboth** tester is often used by maintenance technicians working in the field. These testers are sometimes called a **"punch tester"** because of their method of operation. These tools provide a direct indication of the strength of the fabric. With this instrument, a spring-loaded housing holds a shaft, which has a flared point at one end with a hardened steel tip in its center. The opposite end of the shaft is marked with red, yellow, and green bands. When pressure is applied to the tip, the bands become exposed at the top of the housing. To use the tester, hold it vertically over the covering surface and press straight into it until the tip penetrates the fabric. The point on the instrument must break the fabric and enter far enough to allow the shaft face to make full contact. The color band that is even with the end of the housing just as the instrument penetrates the covering indicates the amount of force needed to punch through the fabric. A small amount of pressure moves the red band out of the housing to indicate that the fabric is weak. The yellow band indicates that the fabric is stronger, and the green band indicates the condition of good quality fabric. After the test is complete, cover the hole in the fabric with a small circular patch. Since the Seyboth tester punches holes in the fabric during each test, another type of tester that does not leave a hole is usually preferred. One such tester is a Maule tester.

MAULE TEST INSTRUMENT

A **Maule tester** is somewhat similar to the Seyboth tester in that it measures the amount of pressure applied directly to the fabric. This tester consists of a tubular housing containing a calibrated spring. When pressed against the fabric, pressure is measured and indicated on a scale. If the fabric fails, the Maule tester penetrates the fabric prior to reaching the specified point on the strength scale. If the fabric has adequate strength, the tester will not penetrate the fabric and a repair is not required once the test is complete. [Figure 5-8]

Aircraft Fabric Covering

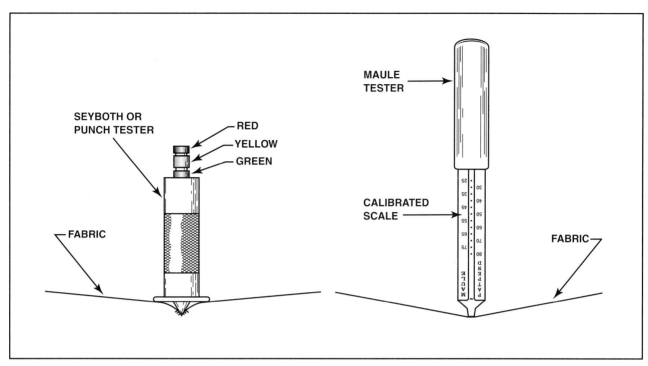

Figure 5-8. The Seyboth, or 'punch tester,' and the Maule tester are popular devices that are used to quickly determine the approximate strength of fabric.

Failing a Seyboth test or a Maule test is not positive proof that the fabric covering is not airworthy. However, when fabric passes the test in the high green (for a Seyboth test) or without penetrating the surface of the fabric (for a Maule test), there is little doubt that the fabric meets good quality strength standards. However, be cautious when using these testers because surface finishes such as dope or paint can effect the results. If the fabric tests are questionable, use a more quantitative method to avoid these errors.

When a higher quality test is required, use a **pull test** or **grab test**. These tests require special equipment that, unless available, usually require a sample piece of fabric to be sent to a specialized testing facility. Regardless of where the test is conducted, a sample of fabric is required. The parts of an aircraft most susceptible to deterioration are locations that are exposed to the sun and finished in a dark color. The sample piece is a 1-inch swath, which is often cut from the upper surface of the fuselage or wings. Once the sample is cut, remove all traces of the finish materials by soaking it in acetone or another reducing solvent, then pull threads from the sample until it is exactly one inch wide. Once prepared, the sample is clamped in the jaws of the tester and pulled until it tears apart. The fabric cannot be considered airworthy if it breaks below the minimum allowable strength. [Figure 5-9]

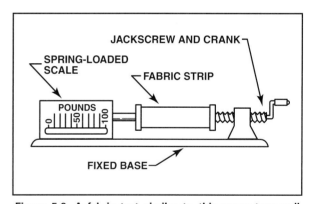

Figure 5-9. A fabric test similar to this screw-type pull tester is acceptable to the FAA in accordance with ASTM method D-5035 (Grab Test) and provides a quantitative measure of the strength of a covering material.

FABRIC-COVERING REMOVAL

When it has been determined that an entire aircraft must be re-covered, the airframe structure must be disassembled to prepare it for re-covering. Obtain a copy of the manufacturer's service manual and follow any special procedures the manufacturer recommends. As the major components are removed, place them where they will not be damaged and protect them from potential corrosion. Do not, for example, store the wings with their leading edge in direct contact with a concrete floor since the lime in the cement will accelerate corrosion of any exposed aluminum sub-structure.

If the aircraft is a high-performance type, the FAA may require the flight control surfaces to be weighed and the balance point checked to prevent aerodynamic flutter. In the case of older aircraft, it may be necessary to record the weight and balance position of the control surface before removing the old fabric. On newer aircraft, the manufacturer usually provides the final weight and balance checks that are required once the re-covering and finishing process is complete. Balance checks are generally performed using jigs, similar to the one shown in figure 5-10.

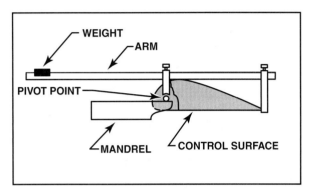

Figure 5-10. Measure the control surface balance before and after re-covering to be sure the balance of the surface has not significantly changed. When finishing the surface, try to keep as much weight off from behind the hinge line to help prevent an out-of-balance condition. This is especially important during the application of finish materials and is covered further in Chapter 6, Aircraft Finishes, in this manual.

For other structures, it is best to leave each major component covered until it is ready to be worked on. An uncovered structure is far more subject to damage than one with the fabric on it. When the component is ready to be worked on, cut the fabric carefully in such a way that it will stay in one piece. Roll up the cover and keep it until the re-covering job is complete. To reduce time and effort, refer to the old cover to locate fabric stitching, inspection openings, and to determine exactly where the control cables penetrate. However, do not assume that everything on the old cover is entirely correct. If there is any conflict between the removed cover and approved data for the types of seams, width of tape, or stitch spacing, then adhere to the approved data.

STRUCTURAL INSPECTIONS

Once an aircraft is re-covered, expect that it will be from five to fifteen years before the complete structure can be fully inspected again. With this in mind, carefully inspect the structure when the covering is removed and replace or repair any component that is questionable. Conduct the inspection according to the manufacturer's specifications and approved methods. While the fabric is removed, take the time to make sure all Airworthiness Directives pertaining to the airframe are complete and consider any manufacturer's recommendations for improving design safety. Evaluate each structure with regard to its anticipated use and environmental exposure.

FUSELAGE AND EMPENNAGE STRUCTURES

When checking the fuselage and empennage, check all steel tube components, especially at the lowest part of the structure, when the aircraft is in a ground storage attitude. The outside of the tubing may look good, but rust can form on the inside of the tubing and not be visually evident. To preclude overlooking bad condition tubing, use an ice pick or an awl to punch the tubing anywhere weakness from internal rust is suspected.

Thoroughly clean the aircraft tubing and steel superstructures. After cleaning, coat the metal structures with an epoxy primer. If the airframe has wood formers and stringers, examine them for any indication of rot. Repair or replace any wood components showing any signs of decay and apply a fresh coat of clear spar varnish to all wooden surfaces.

Carefully check control cables, pulleys, pulley brackets, and fairleads. Replace any damaged components or those showing signs of significant wear. Check all fluid-carrying lines that may be difficult or impossible to reach once the covering is installed. Examine all electrical wiring to be sure that it is in good condition, properly bundled, and adequately secured to the structure.

Inspect landing gear and wing attach fittings to be certain that none of them are cracked or have elongated holes. Consider the use of dye penetrant or other non-destructive testing methods to verify the condition of the fittings. It is much easier to access these areas of the aircraft when the covering is removed, allowing for easier repairs. For example, if a fitting must be replaced or repaired by welding, it is best to complete the repair when the fabric is removed.

WING STRUCTURES

The wing structure is less rugged than the fuselage and must be checked carefully to ensure that it is in condition to last for the entire period it will be covered. Be systematic in the inspection to avoid overlooking anything. First, check the truss. Examine each of the spars, and if they are wood, check them for any indication of rot, cracks, loose fittings, or for

Aircraft Fabric Covering

any signs of damage. Once the wood condition has been verified, apply a fresh coat of clear spar varnish to all wood structures.

Check the squareness of the truss with a bar and trammel points as described in Chapter 1, Structural Assembly and Rigging, of this manual. Make sure the rigging of each wing and control surface is within specifications and that each rib bay is perfectly square. All drag and anti-drag wires should be properly tensioned and locked. Also, to prevent abrasive wear, tie the wires together where they cross, using rib-lacing string. [Figure 5-11]

Check the leading edges of the wings for dents or damage, and inspect all of the ribs that are covered with the leading edge metal. Examine the wing tip bow for security and condition. If it is metal, coat it with epoxy primer, and if it is wood, give it a coat of clear spar varnish. Treat any wood surfaces that come in contact with the doped fabric with a protective coating such as aluminum foil, cellulose tape, or dope-proof paint to protect them against the solvents in the dope. Inspect the aileron and flap wells for any indication of loose metal or damage. Pay particular attention to the control hinge brackets and to their attachment to the spar. Finish out the wing structure with a careful examination of the trailing edge and all of the ribs.

If there are fuel tanks in the wing, examine their cradle and attachments. Examine all electrical wiring and replace any that is questionable. Be sure that the pitot and static tubes are in good condition if they are located in the wing.

Give the aileron and flap control system a careful examination. Check all of the bolts for indication of wear and for proper safety. Lubricate any mechanism as the manufacturer recommends. Pull the control cables taut and secure them at the root rib in their proper location. There should be no interference between the cables and the rib stitching cord when the fabric is attached.

If the wing uses inter-rib bracing, be sure that it is installed in the way the manufacturer specifies and is tied off at the root rib. With cloth tape, cover all the screw heads and edges of metal sheets that will touch the fabric to prevent their wearing through.

INSTALLING THE FABRIC

When the inspection is complete, it will be well worth your time to build and cover a test panel prior to re-covering the aircraft. Practice the installation of the fabric up through the final finish coats of dope to be familiar with the process and the qualities of the materials in each phase of the installa-

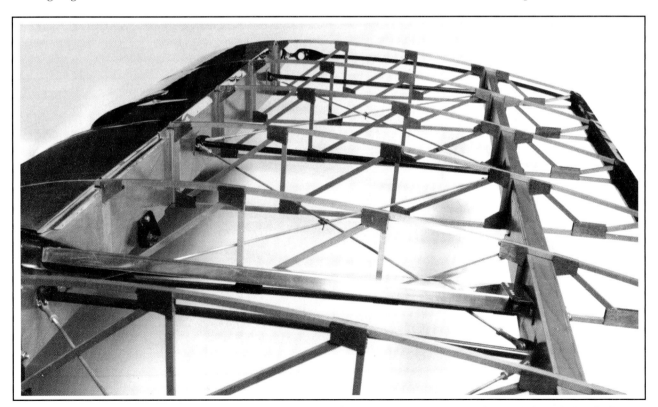

Figure 5-11. Because there are many items to inspect on an aircraft structure, it is important to be meticulous during the inspection. For airplanes, ensure the wing will retain its proper rigging and strength.

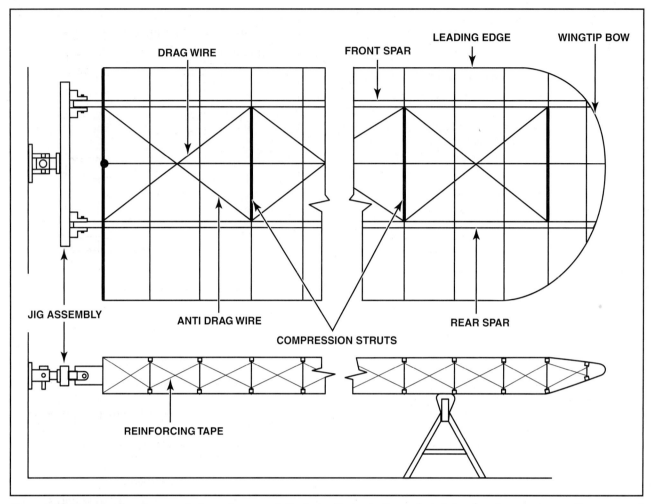

Figure 5-12. The wing may be mounted on a homemade jig that will steady the project. A jig also allows you to easily rotate and turn the wing as each component is being inspected, covered and finished.

tion. Start by building a panel no smaller than a foot square. Build it from wood, sheet metal channel, welded steel tubing, or whichever type structure you will cover first. Have a bracket stick up on the panel similar to a control cable attachment fitting so you can see the way the fabric moves during the initial shrinking.

Envelope Method Of Wing Covering

Most of the vendors of aircraft covering materials can supply fabric envelopes to fit almost any certificated airplane and many of the more popular home-built models. The envelopes are sewed to fit in order to minimize the amount of hand stitching or cementing required in the installation of the fabric.

To cover a wing by the envelope method, support the inspected and prepared wing on two padded sawhorses. Another method is with one sawhorse under the spars near the tip and a jig, such as the one in figure 5-12, attached to the spars at the wing root.

Slip the covering envelope over the wing from the tip and straighten the seam along the trailing edge and around the wing tip bow. If the original fabric was sewed in place at the root and in the control surface wells, duplicate the sewing with approved hand-sewing thread. Fold back about a half-inch of material on both pieces to be joined and use a baseball stitch with at least four stitches per inch. Lock the stitching every six inches with a modified seine knot. [Figures 5-13 and 5-14]

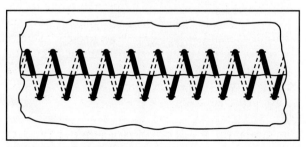

Figure 5-13. The baseball stitch is used for almost all hand sewing of fabric seams.

Aircraft Fabric Covering

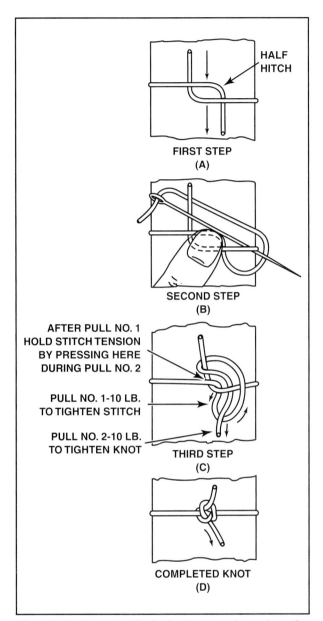

Figure 5-14. Use a modified seine knot, as shown here, for locking baseball stitches and rib stitching.

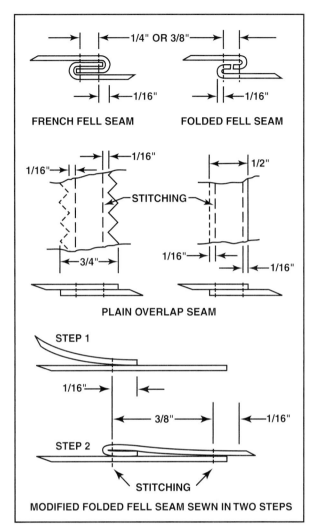

Figure 5-15. There are several types of machine-sewn seams that can be used to join the fabric where it meets when using the blanket method of covering.

Blanket Method Of Wing Covering

If you decide not to use the envelope covering method over the structure, the blanket method may be used, instead. The blanket method refers to the fact that the wing is wrapped or blanket-covered with the fabric. Use a blanket large enough to cover the entire wing, top and bottom. If the blanket is be made of machine-sewn, grade-A cotton, use a **French fell** or **folded fell seam** and two parallel rows of stitches. [Figure 5-15]

For airplanes with a never-exceed speed (V_{NE}) of more than 150 m.p.h., the blanket will have to be sewn together along the trailing edge, using either a plain overthrow or baseball stitch. This type of covering is seldom used for airplanes with a high never-exceed speed because of the large amount of hand labor involved in sewing.

If the V_{NE} of the airplane is 150 m.p.h. or less, the blanket method of covering may be used with a doped seam along the trailing edge.

Prior to attaching the fabric, scrub the trailing edge and the tip bow with an appropriate solvent to remove all traces of grease or other contamination that could prevent the dope from sticking. Wrap the blanket over the wing from the trailing edge, around the leading edge, and back to the trailing edge. Position the blanket so that none of the chordwise seams will fall over a wing rib. Brush a coat of full-bodied nitrate dope or nitrate cement over the trailing edge and around the wing tip, and allow it to dry.

Nitrate cement is similar to dope except that it does not have the plasticizer, and there is a different balance of solvents. Brush on a second coat of dope and work the fabric into it, pressing the fabric down into the dope and working out all of the trapped air. Wrap the fabric around the trailing edge and around the tip bow. Notch the fabric, if necessary, as it goes around the tip bow so that it will not double up or wrinkle.

Pull the blanket smoothly over the wing and take out the slack. Do not pull it too tight; just get it smooth and line up the seams. Clamp the fabric in place with spring clamps or pin it with T-pins to hold it in position. Brush a coat of dope over the attached fabric, and work the over-lapping fabric into it. Press the fabric down into the dope and work out all of the air bubbles. Continue to work the fabric down along the trailing edge and around the tip bow. As you work around the tip bow, it may be necessary to notch the fabric so it will lie smoothly with no wrinkles. Trim the fabric to have a one-inch overlap and very carefully dope it in place.

Now close out the aileron and flap well. Brush on a full-bodied coat of dope or the cement and allow it to dry, then trim the fabric so it will fit smoothly in the well. Brush on a second coat of dope and work the fabric down into it, pressing out any air bubbles and thoroughly saturating the fabric with the dope. Keep the fabric smooth and uniformly taut. It should not be tight, just smooth and wrinkle-free. Dope down the fabric from the bottom first, and when it is completely dry, dope it from the top.

Finally, close the root end of the wing in the same way the aileron well was closed. Allow at least a one-inch overlap of the fabric.

COVERING THE FUSELAGE AND TAIL SURFACES

Most fuselages and tail surfaces are made of steel tubing. After they are thoroughly inspected, prime them, preferably with a good grade of epoxy primer. If the fuselage has any fabric covering on the inside, install it first and then the outside covering.

Cover the fuselage in the same way as the wings, by either the envelope or the blanket method. If the airplane has a V_{NE} of more than 150 m.p.h., all of the seams will have to be sewn, and it is much easier to use the envelope method since a minimum of hand sewing will be needed.

If the V_{NE} is 150 m.p.h. or less, the blanket method may be the easiest to use as the fabric can be doped to the structure. To prepare the structure, clean the tubing to which the fabric will be attached with an appropriate solvent. Wipe the metal clean with a rag, and give these tubes a complete coating of nitrate dope or cement. When this has dried, brush a second coat of dope over the tubing, lay the fabric over the fuselage and work the fabric down into the dope. Attach the fabric to the outside of the tube, and when the dope dries, trim the fabric to go around the tube but not lap up against the fabric on the inside. Now dope this fabric in place, being very careful that no dope runs down or falls through to the inside of the fabric. Continue covering the tail surfaces in the same way as the fuselage.

REMOVING THE WRINKLES

When all of the fabric is properly attached to the structure, wet it thoroughly with either distilled or demineralized water to pull all of the wrinkles out. Either spray the water on or rub it in with a sponge or a piece of terry cloth. The water swells the fibers and shortens them, pulling the fabric very tight almost as soon as the water is applied. The fabric will pull tight around any fittings or brackets that stick out from the structure and will have to be cut for them to come through. However, do not cut the holes at this time. Wait until you have some dope on the fabric before cutting. If cutting is done at this time, the fabric will move around enough as it shrinks so that the hole created will end up in the wrong place. The only correction if this occurs is to install a large, unsightly patch, over the hole.

THE FIRST COAT OF DOPE

Allow the fabric to dry completely, at least overnight, before applying any dope. When the water evaporates, the fabric will return to almost its original tension and most of the wrinkles will be taken out, leaving the fabric in a good condition to receive the dope. When all of the water has evaporated, and within 48 hours if at all possible, brush the first coat of dope into the fabric. In most cases, cotton- or linen-covered airframes are finished with butyrate dope. Butyrate dope has superior tautening properties and is much less flammable than nitrate dope, however, nitrate dope will bond to the fibers better. Nitrate dope is often used, not only to attach the fabric to the structure but for the first couple coats of dope as well. Since the solvents in butyrate dope are more potent than those used in nitrate dope, in latter coats, the butyrate material will open up the nitrate finish enough that the bond will be entirely satisfactory. Especially for cotton or linen fabrics, mix a special fungicidal additive that does not affect adhesive qualities with the first coat of dope. Both cotton and linen are organic fibers and are subject to damage from fungus growth and mildew unless they are properly treated. Mix this fungicidal paste at a rate of roughly four ounces of paste per each gallon of dope. Pour the dope into the paste and mix it thoroughly. Next, thin the dope with equal parts of thinner to get a viscosity that will thoroughly penetrate the fabric and wet every fiber.

Use a good quality animal bristle brush and work the thinned fungicidal dope into the fabric. The application of this first coat of dope requires a technique that you should have perfected on the test panel. Most fungicidal paste has a non-bleeding dye in it that will help you see when you are getting uniform and complete coverage. Work the dope into the fabric, but don't put on so much that it runs down on the inside of the fabric. If any dope does run or drip inside, it will create a wrinkle that will have to be cut out and the fabric patch-repaired.

Sometimes, after the initial shrinking, the water will pull the wrinkles out of the fabric, but after the first coat of dope dries, the fabric becomes loose again and full of wrinkles. Although this may alarm you, don't forget that these wrinkles will disappear and the covering will become taut and smooth as the additional coats of dope shrink the fabric. These coats of dope slowly pull the fabric fibers together and can help prevent damage caused to the airframe if the fabric is initially too tight.

ATTACHING THE FABRIC

Unless the fuselage has large, flat surfaces, the attachment of the fabric to the structure as previously described is all that is needed. But if there are large, flat sides or if the top of the cabin area forms an airfoil section, the fabric will have to be attached to the structure by much the same method as described next for the wing.

For wings, if the airplane has a never-exceed speed in excess of 250 m.p.h., use anti-tear strips under all of the rib-stitching. This is a strip of grade-A cotton, linen, or polyester fabric placed over each of the wing ribs, under the reinforcing tape. In the slipstream, the tape should run from the trailing edge, around the leading edge, and back to the trailing edge. Outside of the slipstream, it needs to extend from the trailing edge, over the top, and back as far as the front spar on the bottom of the wing. Put a layer of dope where the tape should go and then place the tape in position. Press the air bubbles out of it, and then brush on another coat of dope.

Cut lengths of reinforcing tape and saturate them with nitrate dope. Lay the tape over the rib with about a half-inch extending beyond the first and last rib stitches. Pull the tape smooth and rub it down onto the rib with your fingers.

If the fabric was originally attached with any special fasteners such as **Martin clips** or sheet metal screws, use them to attach the fabric. Be sure that the rib capstrips are in good enough condition to receive these fasteners, and that all of the required fasteners are used. If you use sheet metal screws, place a thin plastic washer between the screw head and the reinforcing tape. [Figure 5-16]

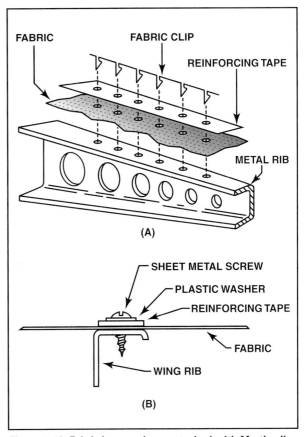

Figure 5-16. Fabric is sometimes attached with Martin clips or sheet metal screws. If the original type of fastener is not used, approved supplemental data must be obtained to use the alternate method.

Duplicate the rib stitch spacing that was used on the original aircraft. Do not copy the spacing from the removed fabric unless you know that the spacing on it was the same as the manufacturer used. If you do not know the original spacing, use the rib stitch spacing chart in AC 43.13-1B, which is also shown in figure 5-17. Notice that the spacing is determined by the V_{NE} of the aircraft and whether the stitching is in or out of the slipstream. For this purpose, the **slipstream** is considered to be the diameter of the propeller plus one rib on each side of the propeller's diameter.

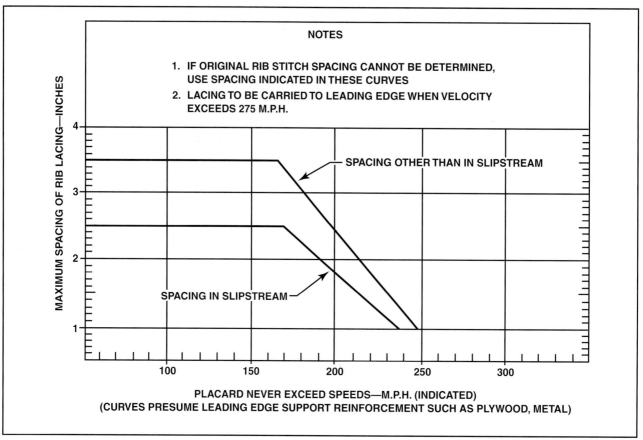

Figure 5-17. A rib stitch spacing chart, such as the one shown here, may be used if the manufacturer does not provide the information.

It is possible to save time and end up with straight stitches if you mark the root rib and the last rib for the in-slipstream spacing. Then, very lightly pop a chalk line between these two ribs to mark the spacing on all the ribs between them. Do the same thing for the ribs outside of the slipstream. When all of the rib-stitch locations are marked, use a rib stitch needle to pierce the fabric on each side of the ribs at these locations, right against the capstrip.

There are no requirements as to which side of the wing that the rib stitch knots must be placed, but the chance of disrupting the airflow over the curved airfoil and reducing lift is less on the bottom. Many times, the stitching is placed on the bottom of a low-wing airplane or on the lower wing of a biplane to keep a clean appearance. They can be placed on the top for a high-wing airplane or upon the top wing of a biplane.

Cut a piece of waxed, rib-stitch cord that is about four times the length of the cord of the wing. Start stitching at the trailing edge in the first holes that have been punched. (For this explanation, assume the knots will be on the top of the wing.) Pass the cord around the rib and tie a square knot in the center of the reinforcing tape on the bottom of the wing. Now, return both ends of the cord through the same holes just used and tie a square knot in the center of the reinforcing tape on top. Lock the square knot with half hitches, using each end of the cord, and cut off the surplus from the short end.

With the starting stitch tied, bring the long end of the cord up to a set of holes halfway between the regular spacing and pass the cord around the rib through these holes. Using the modified seine knot as shown in figure 5-14, make a half hitch around the cord as it goes into the wing with the end of the cord that just came out (figure 5-18). Place the intersection of these cords formed by the half hitch to the side of the reinforcing tape and hold your thumb over them. Pull the free end of the cord with about ten pounds of force in the direction you are stitching, to tighten the stitch. Keep your thumb in place and loop the free end back over the cord that just came out of the wing and under the cord coming from the last stitch. Now, go back over the cord you just went under, and also under the one coming out of the wing. Finish the knot by going over the loop.

Aircraft Fabric Covering

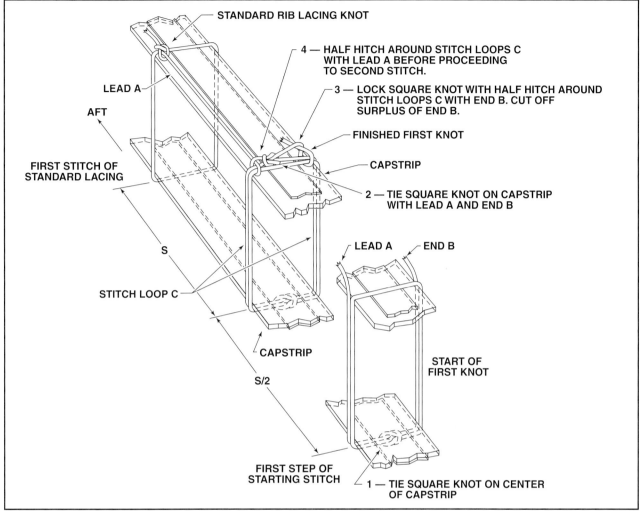

Figure 5-18. The type of rib-stitching knot is determined by its location.

Now, pull the free end with about a ten-pound pull to tighten the knot, and you are ready to go to the next stitch. Place all of the knots right beside the reinforcing tape where they will have the least interference with the airflow. Do not pull the knots back through the holes in the fabric, as they will have a tendency to pull back out and loosen the stitches.

If the rib stitch cord is not long enough, splice on another piece, using the splice knot shown in figure 5-19. Do not use a square knot, as it will not hold with the waxed cord.

When finishing the rib stitching on each rib, place the last stitch at one-half of the normal spacing. Use a double loop and secure the knot with an extra half hitch.

Surface Tape Application

When the rib stitching is complete, brush on a second coat of dope. This time, use a full-bodied butyrate dope. Rather than working the dope into the fabric as with the first coat, lay it on the surface with a fully saturated brush and use even brush

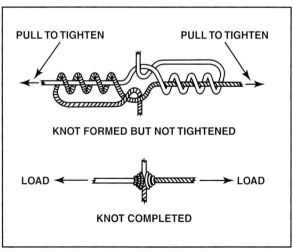

Figure 5-19. When splicing rib stitch cord use a splice knot to join it together.

strokes. Overlap each stroke enough to get a smooth and even coat. As this second coat of dope dries, the fabric will begin to retighten, and when the dope is completely dry, sand the **nap** off of the fabric. The nap is composed of the ends of the fibers that are loosened and stick up like short hairs after they have been stiffened by the dope. If the nap is not removed before applying the tape, the standing ends will prevent the tape from lying smooth.

Remember, that the nitrate dope used for the first coat is extremely flammable. If not careful, the fumes inside the structure can be ignited by static electricity produced from rubbing over the surface with sandpaper. Before sanding, the structure should be electrically grounded to a cold water pipe or to the metal structure of the shop. Lightly sand the surface with 320-grit dry sandpaper, just enough to remove the nap. Be extremely careful when sanding the fabric since it takes only a light pass over the stiffened fabric to cut through.

Cover all seams, rib stitching points where abrasion could occur, and areas along the leading and trailing edges with pinked-edge surface tape, sometimes called finishing tape. Cut lengths of two-inch pinked-edge surface tape long enough to go completely around the wing. Saturate the tape in butyrate dope, brush a coat of full-bodied butyrate dope over the rib, and lay the tape into it. Pull the tape straight and work it down into the dope by either straddling the reinforcing tape with two fingers or using the heel of your hand to work all of the air bubbles out. Lay all of the tapes on one side of the wing, then flip the wing over and dope the rest of the tape down on the other side.

Now you can cut the fabric over all of the protruding fittings. The second coat of dope has given the fabric enough stiffness that it will not move around any more, and yet it will still shrink enough to pull out any remaining looseness. Very neatly slit the fabric and lay it smoothly beside these fittings, then reinforce each opening with surface tape or with a pinked-edge patch, which should be cut to fit.

Cut and fit tape to cover all of the corners and all of the edges of the fabric in the control surface wells. Cut a piece of four-inch-wide tape long enough to go the full length of the leading edge and around the tip bow to the aileron well. Brush on a full-bodied coat of dope and lay the tape straight down the center of the leading edge. Work all of the air bubbles out, and when you get to the tip, brush on a wet coat of dope and pull the tape all the way around the tip. Use enough tension to force the tape to conform to the curvature. Clamp the tape at the aileron well with a spring clamp. Go back and work the tape down into the dope as you brush a coat of dope over it.

Cover the trailing edge with three-inch tape. If the airplane has a V_{NE} of 200 miles per hour or more, notch the trailing edge tape every 18 inches with a notch one inch deep and one inch wide. The reason for the notches is that the edges of the tape face into the wind, and if they should ever start to lift, they will form a very effective spoiler. The ailerons are particularly affected and can cause control problems that could be very serious. If the tape is notched, it will tear off at a notch and not cause as much trouble.

At the same time the tape is being applied, give the entire surface its third coat of dope, which is the second coat of full-bodied dope. This coat and each one from here on will be butyrate because of its better shrinking characteristics and because it is more fire resistant than nitrate. However, do not work the dope in, but rather lay it on with smooth strokes of the brush.

Lay on all of the inspection rings and drainage grommets with this coat. At this stage, having access to the old fabric that was removed from the structure will assist you in placing the rings and grommets in the correct location. These shaped pieces of plastic or metal are laid into the wet dope. You will probably need to put a drainage grommet at the lowest point in each rib bay and an inspection ring every place that access may be needed. Since the inspection openings are not cut unless they are needed, it is much better to have some that will never be used than to later need one that was not installed.

Dope Fill Coats

The quality of the fabric finish is determined by the final steps of applying fill coats of clear dope. The number of fill coats used depends upon the type of finish wanted. If the airplane is a working machine where strength and aerodynamic smoothness are the basic criteria, use a minimum number of coats, just enough to produce a taut and well-filled finish. Normally this will require about two coats of brushed on full-bodied clear butyrate dope, followed by one or two cross-coats that are sprayed on. Restorers of antique airplanes or others who want to regularly display the aircraft they are refinishing will choose to give it a deep gloss, hand-rubbed finish. Spraying on many coats of dope with wet sanding between them, and finally rubbing the pigmented coats down with an abrasive rubbing compound achieve a deep gloss.

Aircraft Fabric Covering

The glass-like finish comes from the fill coats of clear dope and not from the aluminum dope. Spray on a wet cross-coat of clear butyrate dope, then, before the solvents evaporate, cover it by spraying at right angles to this coat.

When the dope is thoroughly dry, wet-sand it with 400-grit paper, wash off all of the sanding residue, and air-dry the surface. Spray on another wet cross-coat of clear dope. Continue this, finally using 600-grit paper, until the surface is at the desired smoothness.

ALUMINUM DOPE COATS

The sun has extremely detrimental effects on the clear dope and on the fabric of an aircraft. Cotton, linen, and even polyester fabric can be affected by the ultraviolet rays of the sun. Sunlight weakens and attacks the cellulose film base of the clear dope. To prevent this damage and to increase the life of the covering, spray an opaque coating of aluminum dope on the surface after the fill coats of clear dope have been sprayed on and sanded. The pigment in this dope is made of aluminum in the form of tiny flakes, some as fine as 325 mesh. These tiny flakes are suspended in clear dope, and when sprayed on the surface, they form a continuous lightproof layer of aluminum, which the ultraviolet rays of the sun cannot penetrate. One of the problems with aircraft finishing comes from the improper application of the aluminum dope coats. If too much powder is used, there will be an adhesion problem and the topcoats can be peeled off of the lower coats. Properly mixing between three and three and one-half ounces of aluminum paste with one gallon of unthinned clear dope will create the best consistency. Mix some thinner with the paste, then pour the dope into it. Agitate the dope as it is being sprayed to prevent the aluminum pigment from settling out.

After the fill coats have been wet-sanded and all of the sanding residue washed off, and with the surface thoroughly dried and free from dust, spray the aluminum dope on with a wet cross-coat. There should be a minimum of aluminum dope on the surface, just enough to form a completely light-tight cover. Properly applied and mixed aluminum dope will dry with a slightly glossy finish. If you can rub across the dried finish and transfer any of the aluminum flakes to your hand, there is too much aluminum powder in the dope and the topcoats will be likely to peel off easily.

FINISH COATS

The final coats of a finishing system are the most visible part of the entire aircraft, and it is not only the design of the paint scheme that is noticeable but the smoothness and uniformity of the finish itself. In the system referred to here, the color coats of finish will be pigmented butyrate dope. Before spraying on the colored dope, the aluminum dope coats should be thoroughly dry and wiped clean of all dust. It is possible for there to be quite a delay between the time the first surface is covered and the time the aircraft is ready for its final assembly. If this is the case, as the surfaces are covered, they should be finished up through the aluminum dope coating then wrapped in polyethylene sheeting such as is used for drop cloths. When the aircraft is all ready for its final finish, spray it all at one time so it will have a consistent color throughout.

For the final coats of dope to adhere to the aluminum dope, it must be treated so the pigmented dope will bond to it. If it has been only a short time since the aluminum dope was sprayed, spray a fine coat of thinner on the surface to soften the film until it becomes tacky, then spray on the colored dope. However, if the surface has set up enough that the thinner does not cause it to become tacky, it may be necessary to spray on a mixture of thinner and rejuvenator. The stronger solvents in rejuvenator will soften the film of the aluminum dope so the colored dope can adhere.

Some dopes can be given a richer appearing color if they are sprayed over a base coat of white dope. These dopes are semi-transparent, and the white undercoating will reflect the light through the pigmented film to produce a more pleasing final color. If a white base coat is used, allow it to dry for at least 24 hours before spraying on the final topcoat.

INORGANIC SYSTEMS

Supplemental Type Certificates using inorganic fabrics have become the most popular method of re-covering aircraft. Most of these systems use a polyester fiber, woven into different weights of cloth. This polyester fabric has proven to be one of the most useful of the synthetic materials for aircraft covering. Other material, like **Orlon**, is made from acrylic resin, and has been used as aircraft covering material, but it is unsatisfactory because of its short life. Nylon, a polyamide resin, has a good lifespan, but tends to be too elastic.

The vendor of the covering system specifies the size of the filament, the number of filaments per thread, the arrangement of the filaments in the threads (either straight or twisted), and the number of threads per inch in both warp and fill. The material is delivered to the vendor in its **greige** state; that is, just as it is removed from the mills in an un-shrunk condition. The fabric that meets specifications is stamped along its selvage edge with the FAA-PMA (Federal Aviation Administration-Parts Manufacturing Approval) number (as shown in figure 5-4 in section A of this chapter).

A few popular systems use Ceconite fabric in weights of 3.7, 2.7 and 1.7 ounces per square yard. The 3.7-ounce fabric has a strength of 135 pounds per square inch, and the strength of the 2.7-ounce material is 100 pounds per square inch. The lightweight fabric is suitable for wing loading less than 9 lbs. per square foot with a strength of over 70 pounds per square inch. The strength of the two heavier fabrics is in excess of that of grade-A cotton, and both may be used for aircraft that were originally covered with cotton.

There are several systems approved for the installation of Ceconite fabric, all using heat to shrink the fabric, instead of water. However, the systems use different finishing methods. Most systems exclusively use nontautening dope for the finish, another STC uses special resins designed to remain dimensionally stable, and yet another uses a factory applied precoating on the fabric and then a water-thinned filler. Regardless of the system used, it is extremely important to follow the FAA-approved instruction manual in detail and that no substitutions are made in materials or application procedures. Remember that re-covering with any of these systems constitutes an alteration of the aircraft, and the approved data for the alteration is the instruction manual that comes with the material. An aircraft technician holding an Inspection Authorization (IA) can certify that the entire job was done according to the approved data and subsequently approve the aircraft for return to service.

SYNTHETIC FABRIC INSTALLATION

Install polyester fabric according to the stipulations of the Supplemental Type Certificate issued by the FAA to the vendor of the covering materials. High-strength adhesives are available to attach this fabric to the airframe. Some adhesives even have enough strength that a never-exceed speed restriction is not mandated for cemented seams as there is with a doped seam for grade-A cotton. Use the fabric, the liquids, and the application methods specified by the Supplemental Type Certificate you are using. Any deviation from the system would prevent the application manual from being used as approved data, and voids the procedure.

With grade-A cotton and linen, wrinkles are pulled out of a freshly installed fabric with water and then shrunk with dope; but polyester fabric is shrunk with heat. When polyester is installed on the structure, use a calibrated iron to heat the fabric. With the iron set at the specified temperature, usually between 250°F to 350°F (roughly between the Wool and the Rayon settings on most irons), move the iron over the entire surface at about the same speed you would use in ironing a shirt. The degree of shrinking is directly related to the temperature, so accurate temperature calibration of the iron is important to prevent damage to the covering. Some procedures use incremental temperature settings and several passes over the fabric at different settings to gradually shrink the entire covering. This is a good practice since excessive shrinking in localized areas can warp the airframe structure.

Continue shrinking the fabric uniformly until it is at the degree of tautness you want without exceeding the maximum temperature allowed for in the STC (usually around 350°F). If a seam is crooked, straighten it by applying a bit of localized heat in the concave portion of the curve. Use a heated blower such as a hair dryer or even a radiant heater to apply light heat, but this type of heat is discouraged in most applications because the structure beneath the fabric absorbs the heat. Ultimately, without accurate temperature control, uneven tautening can result, and direct contact between a calibrated iron and the fabric is the only approved source of heat for some STC procedures.

SEALING AND ATTACHING SYNTHETIC FABRIC

When the fabric has been shrunk, apply the first coat of finishing materials. It is here that the systems can differ greatly. Carefully follow the STC and use the material that is specified, applying it in exactly the manner that the process manufacturer's manual specifies. Many polyester systems prohibit the use of butyrate dope on bare, unsealed fabric because it will not adhere since many systems use a proprietary surface sealer. After the first coat of finish is dry, attach the fabric to the structure with rib stitching cord, screws, rivets, clips, or by whatever method was originally used.

SURFACE TAPE APPLICATION

The width of surface tapes over cemented seams is specified in the STC. Because the width of the tape is important to the strength of the seams and fabric system, be sure to use all of the correct widths of surface tape.

One difference between polyester tape and grade-A cotton tape is in pulling the tape around the wing tip bow. If the tape does not lie smoothly along the bow, use a little bit of heat from the tip of the iron upon the edge of the tape to locally shrink it. It is also common to use bias-cut tape around sharp curves where laying the tape flat against the surface is important. Install all of the drain grommets and inspection rings at the same time the surface tape is laid down.

FILL COAT APPLICATION
Grade-A cotton fabric shrinks with dope, but polyester fabric shrinks with heat. Dope upon polyester fabric should not shrink it. Some polyester systems use a form of resin to fill and finish the surface, while some systems use non-tautening butyrate dope.

FINISH COATS
The finish on polyester fabric should be in accordance with the STC used. One system recommends the use of pigmented, non-tautening butyrate dope, and others recommend special pigmented resins furnished under proprietary names. It is important to emphasize again that anytime a covering is put on an aircraft in any way that differs from the method and materials used by the manufacturer, it constitutes an alteration. The procedure must adhere in all details with the recommendations of the holder of the STC.

GLASS CLOTH SYSTEMS
A third system of covering uses a specially treated glass cloth to cover the airframe. Some of the methods only use the glass to reinforce the fabric already on the structure, which is good only as long as the underlying fabric has a given amount of strength. The third method applies glass cloth directly to the structure, which is the system now used.

Glass fabric is woven of filaments of glass that are inorganic and therefore not weakened by the sun or by most agricultural or industrial chemicals. The cloth has approximately twice the tensile strength of grade-A cotton and weighs about the same. The difficulty in working with glass cloth is primarily because of its loose weave and slickness of the threads. To prevent this problem, pre-treat the glass fabric with a special butyrate dope to assure a good bond. The glass filaments will not shrink, so apply the cloth smoothly and without wrinkles. Attach it to the structure with butyrate dope applied to the surface with a spray gun rather than a brush. Since the glass filaments do not shrink, tautening of the cover depends upon the doped finish bonding tightly to the glass, and as the chemicals evaporate, the dope shrinks, pulling the fabric tightly to the structure. The amount of shrinkage is less than other types of fabric, and because of this, the initial smoothness of the fabric is even more critical.

SECTION C

INSPECTION AND REPAIR OF FABRIC COVERING

Periodic inspections and repair of fabric should be performed as routine procedures during the life of a fabric covering. To increase the accuracy of the results, select the areas for inspection that are known to deteriorate most rapidly. First inspect surfaces exposed to direct sunlight or large temperature fluctuations. Also, areas covered with darker colors, which absorb more heat, are susceptible to more rapid deterioration.

INSPECTION

Open the inspection rings during examination of the fabric. Visually inspect the interior of the wing and fabric using appropriate mirrors and flashlights. Pay special attention to the material around control cable inlets, inspection rings, and drain grommets, looking for areas of physical wear and general deterioration. Make sure that no light penetrates the fabric and illuminates the interior of the wing. If it does, test these areas for deterioration caused by the sun's ultraviolet rays. Continue the inspection by looking in the interior of the wing for condensation, or markings and residue from evaporated water. Carefully inspect these areas for signs of rot or decay. If the inspection reveals excessive deterioration or minor rips and tears, a repair will be necessary to return the fabric covering to its original strength and tautness. [Figure 5-20]

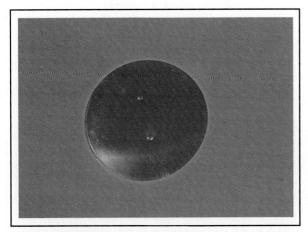

Figure 5-20. Inspection rings are installed at every location where it is necessary to gain access to the inside of the structure for service or inspection.

REPAIR TYPES

The types of repairs needed to restore the fabric to its original strength vary depending on the damage and type of fabric. Much of the damage that requires a patch results from accidental physical damage to the fabric, often referred to as hangar rash. Many of these repairs will be on fabric covering that was applied using new adhesives in accordance with STC procedures, which usually allow patches to be made without hand sewing. However, as with the initial covering, the use of these materials for repair is subject to specific approved data obtained by the manufacturer or distributor of the covering process. The FAA provides acceptable methods of repair in Advisory Circular 43.13-1B, *Acceptable Methods, Techniques, and Practices/ Aircraft Inspection and Repair*. Advisory Circular 43.13-1B is a reliable source of acceptable data. However, only use the **acceptable** repair methods provided within AC 43.13-1B in the absence of the manufacturer's **approved data**. Even then, submit the plan for a major alteration or repair on Form 337 for approval by the local FAA Flight Standards District Office.

L-SHAPED TEARS IN THE FABRIC

A common repair is when the covering is torn in an L shape. An acceptable repair method is to sew the rip together and cover the area with a patch or pinked-edge finishing tape.

Start the repair by removing the existing finish and dope down to the first layer of dope. If the old finish is dope, carefully brush some butyrate dope around the tear and allow it to soften the finish. Be sure that none of the dope runs down on the inside of the structure and drips onto the fabric on the opposite side. This would create a lump in the finish that would require the spot be cut out and patched. When the dope softens, scrape it away with a dull knife blade until all of the old finish is removed down to the clear coats of dope. Before doing the baseball stitching, remove the surrounding dope down to the clear coats for a distance of at least two inches from the tear. With a curved needle, start sewing at the apex of the tear using waxed

CHAPTER 6

AIRCRAFT PAINTING AND FINISHING

INTRODUCTION

Aircraft finishes are important, not only for the attractive appearance they give the aircraft, but for the protection they afford the lightweight, highly reactive composites, metals, and fabrics from which the structure is covered and made. When an airplane leaves the factory, it has been given a finish that is both decorative and protective. It is the responsibility of maintenance personnel to see that the finish is maintained in such a way that it will keep its beauty and continue this protection. If the airplane is to be refinished, the technician must properly prepare the surface and apply a new finish that will protect at least as well as the original. Finishing and refinishing operations consist of many different materials and techniques for applying the protective covering. Some material is sprayed, some brushed or dipped, and some is attracted to the surface of the aircraft via an electrical charge, a process called electroplating. There are almost as many different methods of application as there are materials to apply. Apply each material according to the manufacturer's instructions to obtain the best possible finish and the greatest possible protection.

SECTION A

FABRIC FINISHING PROCESSES

Chapter 5 of this text detailed fabric covering, and this chapter will examine the finishing process for fabric materials. Although fabric-covered aircraft constitute a small percentage of the aircraft that require refinishing, technicians should still be aware of the finishing systems that are available. The types of fabric that are used today and the methods used for their application were discussed in detail in Chapter 5. The discussion centered on how the fabric is installed and protected by layers of dope that contain fungicide and aluminum paste, which are followed by a more decorative finish. This section will concentrate on the finishing system's esthetic properties that give aircraft a pleasing and durable finish, rather than the airworthiness of the covering.

DOPE FINISH

The traditional dope finish for an aircraft is one of the easiest to apply and requires the least experience and equipment to do a satisfactory job, but it has a drawback of being prone to cracking. Except for those striving for an authentic finish on a restored aircraft, cracking causes refinishers to consider some of the newer systems. There is now a wide selection of finishing systems for fabric-covered aircraft, as opposed to being limited only to nitrate or butyrate dope. New systems are far more durable and more attractive. However, some STC systems require the use of compatible paint or finishing products, essentially limiting the choice of topcoats.

Continuous flexing of fabric covering also requires the finish to be flexible or it will crack easily. To obtain flexibility in a dope film, manufacturers use a plasticizer, such as castor oil, tricresyl phosphate, or one of the more modern vinyl resins. Plasticizers have a tendency to migrate out of the dope film, and after a period of time the film becomes brittle. If struck with a rock or if someone pushes against it, the protective film will crack, usually in a ringworm form. Sunlight and moisture can get to the fabric and cause it to lose strength prematurely. While making the film flexible, plasticizers also make it soft enough so that it can be abraded by sand or attacked by chemical fumes.

REJUVENATION

If the dope film becomes brittle and cracks but the fabric is still good, the finish may be rejuvenated and new topcoats applied. The first step is to scrub the old finish with Scotch-Brite™ pads and plenty of water. Any wax on the surface should be scrubbed with a rag that is dampened with an appropriate solvent. Dry the surface completely, then spray on a good wet coat of rejuvenator thinned with some butyrate dope thinner. Rejuvenator is essentially a mixture of potent solvents and plasticizers. Solvents soften the dried dope film, keeping it permeable long enough for the plasticizers to become part of the film.

The first coat of rejuvenator causes the fabric to loosen, but a second coat pulls it back taut. Sand the surface and check carefully to be sure that the dope film has filled the cracks completely. Wash off the sanding residue from the surface and then spray on a good wet cross-coat of aluminum pigmented butyrate dope. Wet sand the aluminum dope, wash off the residue, and apply a new topcoat.

APPLICATION OF DOPE

Spray on the finish coats of dope as heavily as possible without the dope running or sagging. If lighter coats are all that conditions permit, apply several top coats in order to fill the weave of the fabric and produce a smooth finish. In order to build a film with a minimum number of coats, use a pressure pot instead of a suction-cup-type spray gun. More dope can be applied with less thinner. [Figure 6-1]

Figure 6-1. A well-applied finish to a fabric-covered structure is not only pleasing to the eye but protects the fabric from the elements.

Aircraft Painting and Finishing

Mix the final coat of pigmented dope with about 20% clear dope to get a glossier finish. Use less than about 20% of clear dope because the ultraviolet rays of the sun will degrade it if not protected by the pigmented dope and the surface will soon become dull and chalky looking. Prevent UV damage by adding Ultra Violet blocker to the topcoats of paint. Chemical blockers are much less effective than the aluminum dope coats. Substitute retarder for some of the thinner to achieve a sheen or gloss. Spray on a number of very thin coats of the pigmented dope and wet-sand it with 600-grit sandpaper between each coat to get a deep gloss finish. After the finish has dried for at least a month, hand-rub it with rubbing compound and wax.

DIFFICULTIES WITH DOPE

Dope and fabric finishing can be difficult to apply. Changing conditions in humidity and temperature can cause problems with equipment settings and chemical makeup. Quickly diagnose any problems during the finishing process. Unfortunately, in some cases the problem won't manifest itself until a period of time after the finish coats have been applied.

DOPE ADHESION

Proper surface preparation is imperative to good dope adhesion. All paint or enamel must be removed before the application of dope because dope softens them and causes loss of adhesion. Spraying nitrate dope over butyrate dope can sometimes cause the dope film of the topcoat to peel off, coming off in strips. Solvents in the nitrate dope are not strong enough to penetrate the film formed by the butyrate, which results in a weak bond.

Topcoats may also peel off if there is too much aluminum powder in the aluminum dope coats. Use only a thin layer of aluminum dope to keep the sunlight from degrading the dope and the fabric under it. If a trace of aluminum powder can be found on one's fingers when the aluminum doped surface is rubbed, then the layer is too thick and the topcoats may fail to adhere.

BLUSHING

The most common trouble with dope finishing is blushing, which occurs when the humidity in the air is relatively high. The solvents evaporating from the dope cause its temperature to drop. Water condenses out of the air into the dope film, and the nitrocellulose will precipitate out, causing the dope film to be porous and have a chalky appearance. Blushed dope finish is not as strong as a normal dope finish. [Figure 6-2]

Figure 6-2. Moisture condensing from the air onto the surface of the fresh dope can cause the dope to blush. Blushed dope is not as strong as a normal finish.

The cure for blushing is to prevent water from condensing out of the air. If the conditions are not too severe, use some retarder in the dope to slow down the evaporation and prevent the temperature drop that causes blushing. Even better, lower the relative humidity by warming the air so that it will hold the moisture in its vapor form. In practice, if a surface is starting to blush, wait for more favorable conditions.

If the dope has blushed, but not too severely, remove it by spraying on a mixture of one part retarder and two parts thinner in a light mist coat. Allow it to dry and spray on another coat. This should melt the blushed surface film and allow it to re-form in a smooth and glossy manner. If this treatment does not remove the blush, sand off the surface and respray when the weather conditions are more favorable.

PINHOLES

Dope is composed of the dry solids that remain on the surface, as well as the thinners and solvents that evaporate when the liquid changes to a vapor. If the dope film is exposed to too much heat or wind, the surface can become hardened to the point that vapors cannot easily escape from the surface but join to form large bubbles. Bubbles that punch through the film that forms on the surface leave a crater or a pinhole. Excessively atomized air on the spray gun can also cause bubbles to form in the dope film, leaving pinholes when the bubbles burst. [Figure 6-3 on page 604]

SAGS AND RUNS

Carelessness causes sags and runs because of too much dope being applied. Prime causes of sags and runs are moving the gun too slowly over the surface, holding the work too close, or not thinning the dope properly. Trying to cover these imperfections with additional paint only makes the blemish

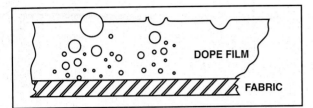

Figure 6-3. Pinholes form in the dope film when bubbles of the solvent break through the surface of the dope which has prematurely hardened. The hardened surface cannot flow back and cover the hole.

more pronounced. Vertical surfaces are especially prone to sags or runs when excess dope is applied. [Figure 6-4]

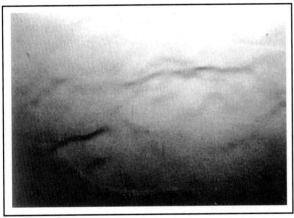

Figure 6-4. Spraying too much dope on the surface causes sags and runs. Essentially, gravity overcomes the surface tension and the material starts to flow.

ORANGE PEEL
Orange peel or localized roughening of the finish is caused by the surface of the dope drying before the dope beneath it does. When the surface dries, it shrinks and wrinkles, giving it the rough appearance of an orange peel. Improper spraying techniques, thinners that evaporate too quickly or an air draft over the surface are usually causes of orange peeling.

FISHEYES
Fisheyes are localized spots within the film of the dope that do not dry and are usually caused by wax, oil, or some silicone product contaminating the surface and preventing the dope from curing as it should. Be sure that the surface to be sprayed is perfectly clean to eliminate fisheyes. Scrub the surface with a rag that is damp with an appropriate solvent.

DOPE ROPING
Dope roping is a condition that occasionally forms through the use of a brush in applying dope. Instead of the dope flowing out of the brush in a nice, wet, smooth coat, the brush sometimes leaves a rough trail. Solvents evaporating out of the dope before they flow out over the surface cause this condition. Use a bit more thinner to prevent roping. Roping is sometimes caused by the dope being too cold to flow out onto a smooth surface. Always allow the dope to warm up to room temperature.

POLYURETHANE FINISH
Polyurethane finishes on fabric-covered aircraft are criticized primarily for two reasons. The first is that polyurethane finishes are designed for use over metal. Since fabric is not as rigid as metal, polyurethane has a tendency to crack as the fabric flexes beneath the finish. The other criticism is that the enameled surfaces are more difficult to repair. Newer polyurethane topcoats are formulated to produce a flexible film that does not crack when it is sprayed over the fabric. Newer finishes are primarily designed to be used over synthetic fabrics and a special flexative modified primer. Primer is sprayed on the surface after all of the fabric is shrunk, rib stitched, and all of the tapes are installed. If the primer is left uncovered for more than a week, scuff it with a light abrasive before applying the topcoats. Aluminum dope is not an absolute necessity on all synthetic fabrics. Some manufacturers specify a special black coat to be sprayed on after the primer and just before the topcoat. Topcoats in this system also have the flexative added so that they will not crack. If repairs are required on fabric that is painted with polyurethane, remove the topcoats by sanding them off. Polyurethanes do not soften in the same way as dope. Use a 280-grit, open-coat sandpaper to cut away the polyurethane film and get down to the clear dope coats underneath.

SECTION B

AIRCRAFT PAINTING PROCESSES

Painting processes vary greatly and often depend on the type of material, the painting surface, and the equipment used. Paint and equipment manufacturers often provide helpful information with their products to assure the appropriate settings for equipment, which is based on conditions and the type of finishing material being applied. Manufacturer's directions should be followed closely to create a finish that is smooth and pleasing to the eye. The smooth surface of a quality finish will help reduce drag as well as protect the base material from corrosion and abrasion.

METAL AND COMPOSITE FINISHING

Although the skills required to finish metal and composite surfaces are similar to fabric, there are a few additional steps in the process that differ. One difference is in the way the existing coating is removed. Normally, stripping dope or paint from fabric aircraft on a large scale is rare. For fabric covered aircraft, if the existing color has faded, simply add a new coat or replace the fabric and apply a new finish. However, on metal and composite aircraft, completely strip the paint to expose a clean surface to which the new finish will adhere.

STRIPPING

Preparation of the surface is one of the most important steps in refinishing an aircraft. Paint will not adhere to metal nor properly protect it from corrosion without a smooth and properly prepared surface. The first step is to remove all of the old finish. There are three basic techniques for stripping paint: chemical, mechanical and pyrolytic. The stripping of metal parts is most commonly done with a chemical paint stripper or by air blasting with an abrasive material. If the material is composite or fiberglass, mechanically remove the paint through careful sanding or scraping because the active ingredients in a chemical paint stripper will soften the resins in most fiberglass components. Pyrolytic methods use high temperatures to expand and extract the paint and are rarely used on aircraft parts due to the probability of damaging the base material. Paint strippers remove the finish by penetrating the surface film; either softening it and causing it to swell (enamels, epoxies, and polyurethanes), or by dissolving it (dope and lacquer). Chemical strippers have a number of potent ingredients, some of which are quite toxic, and must be used with care to prevent personal injury or damage to the aircraft. The Environmental Protection Agency (EPA) has stringent rules on the disposal of some stripping chemicals, which must be strictly followed. In addition, the Occupational Safety and Health Administration (OSHA) governs the use of these chemicals because of their hazards. Products are beginning to arrive on the market that are just as effective as previous strippers but without the environmental and health concerns. As the FAA approves these products, they will become excellent alternatives to the strippers of the past.

While refinishing an aircraft, it is extremely important to follow the instructions furnished by the manufacturer of the finishing materials. Today there are some rather exotic chemicals used in the stripping, priming, and finishing processes, and if misused or modified, the results can be totally unacceptable. In addition, the disposal of these chemicals is highly regulated, and it is important to follow all of the manufacturer and governmental safety and disposal recommendations. The ingredients used in different brands of paint strippers may be quite incompatible, and should never be mixed.

Before stripping an aircraft, put it in an area where the fumes will be filtered and vented away from all personnel. Also, properly dispose of any air-filtering element. If there are any parts of the aircraft that must not have stripper on them, such as windshields or windows, or any plastic components such as wing tips, wheel pants, or cowling, they should be masked with aluminum tape. Another option is to use aluminum foil and polyethylene sheeting that is taped down tightly so that no stripper can run under it. Remove all of the flight control surfaces prior to stripping them to make it easier to reach difficult areas. Flight control surfaces will have to be removed after repainting anyway, in order to balance them.

Apply a thick layer of stripper with a brush without rubbing it in. As soon an area has been completely wet down, cover it with a piece of polyethylene sheeting, such as a drop cloth. This will prevent the solvents from evaporating and keep the stripper working until it thoroughly penetrates the film. After the stripper has properly worked into the surface of the paint, remove it with hot water or a steam cleaner.

Let the stripper remain on the surface for the time specified by the manufacturer. Premature removal of the chemicals will usually provide poor results. Even with multiple applications of paint remover, paint may not easily come off of the structure, especially around rivets and in some of the cracks. Despite the difficulty, remove every bit of the old finish to achieve a good finishing job.

Be careful when working around the windows, windshield and the plastic parts that have been masked off. Remove every bit of the paint from the metal next to them without spreading the chemical onto sensitive areas. Use a small brush to locally apply the stripper here, and be sure that it only contacts the metal parts. When all of the old finish is removed, use warm water, a good detergent, and a Scotch-Brite™ pad to scrub the entire surface and get rid of every trace of the old finish and residue from the stripper. Properly dispose of all the waste sludge and paint. [Figure 6-5]

COMPOUND BLASTING

Mechanically blasting an abrasive material or compound against the paint is the most familiar method of removing old finishes and corrosion. Sand blasting has been one of the most common methods of aircraft paint removal but has several major disadvantages. Foremost is the violent and aggressive nature of the reaction of the sand against the material being blasted. Sand is very hard and course and grinds against the base material called the *substrate*. Blasting for too long in one spot can make the metal thinner. When blasted upon thin walled metal tubing, like engine mounts, the wall thickness can be reduced to the point where the mount is not airworthy. Sand blasting aluminum damages the metal, sometimes beyond repair. This removes any cladding and the soft metal stretches due to the

Figure 6-5. One of the most important parts of a complete refinishing job is the proper stripping of the old finish from the surface. Properly protect all surfaces that can be damaged by the chemical strippers.

hammering of the sand particles, often with disastrous results. Give careful consideration to the results before sand blasting

Media blasting, however, is a different matter. Several methods that use other types of materials have been developed that avoid the dangers of sand blasting. The most popular method uses a plastic material that removes the old finish and the surface corrosion without harming the metal underneath. Most systems use special equipment that not only provides the blasting action but removes the residue as well. In some, the blasting media is separated and recycled, while the removed material is available for proper disposal. Some waste materials cannot be disposed of by simply throwing them away. The used blasting media may contain hazardous materials and may require special handling.

As with any stripping operation, proper preparation is most important. Many items must be masked, not only for the stripping operation, but also for the sake of safety. Stray materials can plug instrument openings, fuel vents, and get into cracks and crevices that could cause problems if not properly cleaned afterward. Most plastic media blasting uses a system approach, in which the equipment, as well as the type and size of media, is specifically designed for the job.

Additional types of blasting media that can be used are wheat starch, sodium bicarbonate, glass beads, nut shells, cornhusks, and fruit seeds. Lightweight and organic products like fruit seeds and wheat starch can be used on thin and sensitive substrate like magnesium and even many composite materials. Carefully follow the manufacturer's instructions and be sure that the removal of the paint does not damage or weaken any of the base material.

CORROSION REMOVAL AND PREVENTION

When all of the finish is removed from the structure, examine it carefully for any indications of corrosion. When polyurethane topcoats are improperly applied, a common problem develops known as filiform corrosion, starting under the dense film of polyurethane and eating away at the metal in a long, thread-like pattern. Sand away any corrosion and if the corrosion is too advanced, the metal may have to be replaced. [Figure 6-6]

When all of the corrosion is removed, acid etch the surface to allow the subsequent coatings to adhere. After the acid treatment, apply a conversion coating

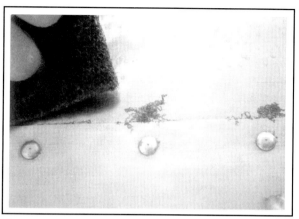

Figure 6-6. Filiform corrosion is first noticed as a puffiness of the finish, generally around the lap joints. This type of corrosion often extends beyond the initial area of indication given by the lifting paint.

to change the active aluminum surface of the metal into an oxide film that is chemically inert. This will prevent the formation of filiform corrosion and allow the primer to adhere. It would also be wise to use other methods of corrosion detection and check the airframe and fasteners for strength with eddy current or ultrasonic testers. Once a surface has been cleaned, do not touch it with bare hands since oils in the skin can easily cause adhesion problems when painting is begun. Corrosion can be minimized by timely touchup of the aircraft finish. Beside improving the overall appearance of the aircraft, touchup helps prevent corrosion from beginning in the first place.

FINISHING MATERIALS

The quality of the materials used to cover the substrate should match the desired durability, the type of material to be covered, and the desired look. Glossy finishes, like enamel, are best for rigid aircraft where there is little chance of surface cracking. Other types of finishes look good in addition to their ability to take punishment from abrasives and high temperatures. Ultimately, picking the correct finishing materials is as much an issue of preference as it is a by-product of need. All of the components must be compatible before the process is begun.

PRIMERS

After the surface has been properly pre-treated, apply a primer to provide a good bond between the metal and the topcoats. For years, zinc chromate has been the standard primer for aircraft because of its good corrosion resistance. But its use is decreasing since it does not provide as good a bond to the surface as some of the new primers, and it is also a hazardous material. Two-component epoxy primer is recommended.

WASH PRIMER

High-volume production of all-metal aircraft has brought about the development of a wash primer which provides a good bond between the metal and the finish, and cures after a half hour. These primers can be used on aluminum, magnesium, steel, or on fiberglass. Apply acrylic or enamel topcoats directly over the wash primer, but for maximum protection, as is required for seaplane or agricultural aircraft, apply an epoxy primer over the wash primer. When wash primer is applied over a properly cured conversion coating, the organic film of the wash primer bonds with the inorganic film, providing excellent adhesion between the topcoat and the surface. It also provides good protection for the metal.

Wash primer is a three-component material. Four parts of primer are mixed with one part acid and four parts thinner and allowed to stand for twenty minutes to begin curing. The primer must be re-stirred to assure a consistent mix and then sprayed on the surface. Adjust the viscosity by adding more thinner to get the extremely thin film that is required, but never more than eight parts of thinner to four parts of primer. Apply wash primers with a film thickness of no more than 0.3 mil (0.0003 in., 0.0076 mm). This can be determined by looking at the surface. A film of proper thickness will not hide the surface but will give a light amber cast to the aluminum.

Wash primers are popular because they may be topcoated shortly after application without the finish sinking in and losing its gloss. The phosphoric acid requires about thirty minutes to convert into the phosphate film, so the topcoat must not be applied until the conversion has finished. However, it must be applied within eight hours or the glaze on the primer will be so hard that the topcoat will not adhere to it. If it is impossible to finish within eight hours, apply another coat of primer. Omit the acid when mixing the primer for the second coat or when it is going to be used over fiberglass or plastic.

Critical to the application of wash primers is the necessity of having sufficient moisture in the air to properly convert the acid into the phosphate film. Proper conversion requires nine-hundredths of a pound of water for every pound of dry air during the primer application. To determine if there is sufficient moisture in the air, use a chart comparing wet and dry bulb temperatures. [Figure 6-7]

This is a modification of a relative humidity chart. To use the chart, compare a wet-bulb thermometer with a dry-bulb thermometer. Wrap a dampened cotton wick around the bulb of one, and with the thermometers placed side by side in the spray booth, blow air from the spray gun across them. This will evaporate the water from the wick and lower the temperature on that thermometer. The temperature of the thermometer without the wick (dry-bulb) is located on the bottom of the chart.

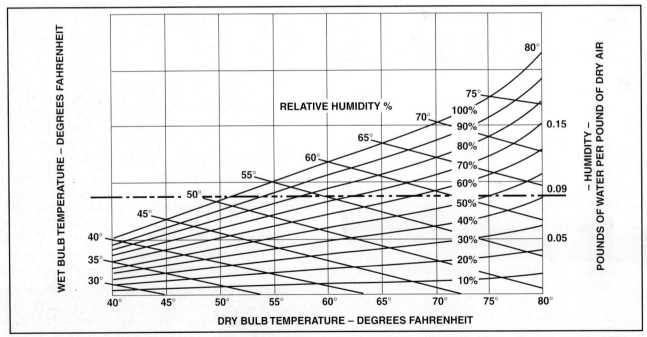

Figure 6-7. If the lines representing the wet-bulb and dry-bulb temperatures intersect above the 0.09-pound line, there is sufficient water in the air. If the intersection is between the 0.09 and 0.05-pound lines, add one ounce of distilled water to one gallon of the thinner. If the intersection is below the 0.05-pound line, add a maximum of two ounces of distilled water to each gallon of thinner to aid the cure of the primer.

Follow this line up until the slanted line representing the temperature of the wet-bulb thermometer crosses it. The amount of water is on the horizontal line through this intersecting point. If the dry-bulb temperature is 70 degrees Fahrenheit and the wet-bulb temperature is 60 degrees, these two lines cross above the horizontal line indicating 0.09 pounds of water per pound of dry air, actually at about 0.095 pounds. There is enough water in the air to properly convert the acid in the primer.

If there is not enough water in the air for proper conversion, the finish will trap active acid against the metal. To prevent this, and the subsequent danger of corrosion, add water to the thinner. The thinner for wash primer is primarily an alcohol and it will accept water. When there is between 0.05 and 0.09 pounds of water to each pound of air, add one ounce of distilled water to each gallon of thinner. If there is less water than 0.05 pounds per pound of dry air, add two ounces to each gallon of thinner; but this is the maximum amount permissible under any circumstances. [Figure 6-8]

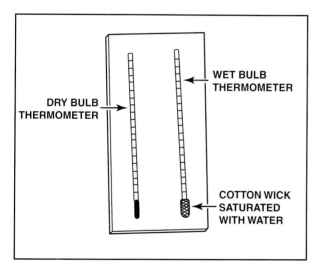

Figure 6-8. Saturate the wick surrounding the wet bulb thermometer with water, and blow air across the two. Compare the two readings on the table in figure 6-7 to determine the amount of water in the air.

Acrylic lacquer applied over an improperly cured wash primer is porous enough to allow the moisture from a heavy dew to penetrate the film, unite with the free acid and convert it. In the process of doing this, the paint will blister from the surface. If blisters appear shortly after the painting has been finished put the airplane in the sunshine to thoroughly warm it. These blisters will subside and the surface will be smooth again. The acid will have received sufficient water for its conversion and the primer will have its proper cure; the finish will not have been damaged, provided the initial condition was not too severe.

SYNTHETIC ENAMELS

Enamel paint is one of the older finishes for metal aircraft and one that has been commonly used for automobiles. The finish can be applied over zinc chromate primer, and cured by the process of oxidation. These finishes have a good gloss and do not require rubbing. Their chemical resistance is nominal and not as resistant to abrasion when compared to some of the more modern finishing systems.

To apply enamel, thin it to the viscosity specified by the manufacturer, spray on a light mist coat and allow the thinners to evaporate. This takes about ten to fifteen minutes and is followed by a full, wet cross-coat. The enamel should be ready to mask and tape in about 48 hours.

ACRYLIC LACQUER

Many aircraft produced on high-volume production lines are finished with acrylic lacquer because of the speed at which paint can be applied. They are primed with a two-part wash primer, and as soon as the primer is entirely dry, they are sprayed with the acrylic lacquer. These lacquers are easy to apply. They have a lower solids content than enamels but they produce a good gloss, especially if they are polished. They are fairly resistant to chemical attack and quite weather resistant.

The low solids content of acrylics makes them somewhat touchy to apply. Thin the lacquer to the viscosity specified by the manufacturer, spray on a very light tack coat, and then follow that with at least three wet cross-coats. Allow about a half-hour drying time between coats. It is much better to use more thin coats than it is to spray on a heavier coat of lacquer because thick coats tend to develop pinholes and orange peel. Mix the final coat by using one part of retarder to three parts of thinner. This will give the coat a little more time to flow out to a nice, smooth finish. If retarder is used, the finish should dry overnight before tape is put on.

POLYURETHANE

One of the most durable and attractive topcoats on high-speed, high-altitude aircraft is produced by the polyurethane system. This hard, chemically resistant finish not only provides a beautiful wet look, but it is the most durable finish for agricultural aircraft, seaplanes, and other flight vehicles that operate in hostile environments. In fact, polyurethane

Figure 6-9. A polyurethane finish is extremely durable and is noted for its "wet" look.

paints are even resistant to Skydrol® hydraulic fluid, which is highly corrosive. [Figure 6-9]

Polyurethane is a two-part, chemically cured finish that has a solids content of up to 60%, twice that of acrylic lacquers. The characteristically high gloss of these finishes is due to the slow-flowing resins used. The thinners evaporate but the resins continue to flow until they form a perfectly flat surface and cure uniformly throughout. Light reflecting from its flat surface gives it the wet look. The drying time of polyurethane paints can be shortened by increasing the percentage of catalyst added to the resin. This additional "drier" speeds up drying time. However, if too much is added it can create a brittle film that is prone to cracking and peeling.

Polyurethane is resistant to abrasion, to most of the chemicals used in agricultural applications, and to the strong action of the phosphate-ester-based hydraulic fluids used in many modern jet aircraft. Even a solvent as potent as acetone has minimal effect on it. To remove a polyurethane finish, hold a solvent-type paint stripper against the surface for a considerable time to give the active ingredients an opportunity to break through the film and degrade the primer. The primer used under a polyurethane topcoat is a critical part of the system. Wash primers may be used, but if they do not cure properly, they can cause filiform corrosion.

Mix polyurethane with its catalyst in the proportions specified by the instructions on the can and allow it to stand for fifteen to thirty minutes to begin curing. This is called the induction time of the material and is an extremely important procedure. After the proper induction time, stir and mix in the proper reducer to get the correct viscosity for spraying. Determine the viscosity by using a viscosity cup. Dip the cup in the thinned polyurethane and lift out. The amount of time in seconds between the moment the cup is lifted and the first break in the stream flowing out of the hole in the bottom of the cup is a measure of the viscosity. The correct viscosity is determined from experience, but once this is known, it can be duplicated for every refinishing job.

Spray on a very light tack coat when the viscosity of the liquid is properly adjusted, and a full, wet cross-coat when the thinner evaporates from the liquid. These finishes can be applied with any of the accepted paint application systems. The main difficulty with a polyurethane finish is that it is easily sprayed on too thick. This causes it to build up at the skin lap joints and possibly crack because it is not as flexible as some of the other finishes. The slow drying and low surface tension of a polyurethane finish allows it to flow out flat, and it will actually require several days before the finish has reached its final smoothness and hardness. The surface of polyurethane is normally dry enough to tape in about five hours, but it is much better if it can be left for at least 24 hours before putting on any tape.

The catalysts used with polyurethanes react to moisture, and the cans of material must not be left open. If a can of catalyst remains open for a period of time and is then resealed, it is possible that it has absorbed enough moisture to start the reaction; the can may swell up and burst. Remove all catalyzed material from the spraying equipment as soon as spraying is complete because if it sets up in the equipment, it is almost impossible to remove. A lack of flexibility has been one of the difficulties with polyurethane, but continued research and development have produced polyurethane topcoats flexible enough that they will not crack, even if applied to a flexible surface. The pot life of a catalyzed material such as polyurethane is the time between the mixing of the material and the time it has set-up too much to use. The pot life for polyurethanes is about six to eight hours. After this time, it must be discarded because it no longer usable.

Repainting an aircraft with a polyurethane paint usually includes picking an entire process system. The complete polyurethane system includes all of the metal preparation chemicals and procedures as well as the primers, thinners, and additives such as fisheye eliminators. When using a system, do not substitute materials or try to modify the times or procedures.

ACRYLIC URETHANES

This system of finishes has the advantages of both the acrylics and the urethanes. It is easy to apply like an acrylic, and is nearly as durable and chemical resistant as a polyurethane finish. Many acrylic urethanes are used as a clear coat on top of color coats.

SPECIAL FINISHES AND FINISHING PRODUCTS

Traditional paints can not always be utilized over all the surfaces of the aircraft. In some cases, special paints and processes should be utilized in areas where a specific requirement or need exists. Wing walks, instrument panels, and battery boxes are good examples of areas that require special finishes and treatments.

HIGH-VISIBILITY FINISHES

The need to make aircraft more visible, both on the ground and in the air, has caused the paint manufacturers to develop a series of vivid color finishes. Normally these finishes are not used for the complete airplane but are used for wing-tips, cowlings, the empennage, or for colored bands around the fuselage. These finishes consist of a coat of transparent pigment applied over a white, reflective base coat. A clear, ultraviolet-absorbing topcoat is sprayed over the colored pigments to help retard fading of the vivid transparent pigments. If the airplane has a good white finish on it, use this for the base.

Light penetrates the transparent topcoats and is reflected off the base; the viewer sees the colored reflection. The application of these finishes is the same as for any sprayed-on finish. The reducer, the pigmented material, and the topcoat material must all be compatible. Paint manufacturers sell all of this material in kit form, so there is usually enough of all the components for the typical small application. It is also available in bulk for larger applications.

WRINKLE FINISH

Instrument panels, electronic equipment, and other aircraft parts subject to considerably rough treatment may be finished with a wrinkled surface. This is essentially a material with fast drying oils. The surface dries first and as the bottom dries, it shrinks, pulling the surface into a wrinkled pattern. The size of the wrinkles is determined to a great extent by the formulation of the material.

Spray all wrinkle-finish materials with a heavy coat and allow them to dry in the way that is recommended by the manufacturer. There are two types of finish: one that cures by heat and is baked to produce the proper wrinkle, and another that dries in the air. Air-dry wrinkle finish that is available in aerosol cans is considerably softer than the baked finish and is not recommended for areas where there will be much handling or wear. It is also not recommended that air-dry finishes be baked because the pigments will discolor.

FLAT BLACK LACQUER

Flat black lacquer is a durable, non-reflective coating for instrument panels and glare shields. Spray this on, either with a gun, or from an aerosol can, and allow it to air dry. Spray flat finishes on thin so they will not flow out and gloss. If they are put on too thick, there will be spotty areas of glossy finish in the predominantly flat coat.

WING WALK COMPOUND

Mix a special sharp grained sand into a tough enamel material to form non-slippery surfaces for wing walks or any part of the airplane where a rough surface is desirable. Apply it either with a coarse brush, or a spray, using a special nozzle for the dense, highly abrasive material. Apply it directly over the regular finish after thoroughly cleaning the surface and breaking the glaze if the finish is old. Thin the wing walk compound with conventional enamel thinners.

ACID-PROOF PAINT

Battery boxes are one of the more corrosion-prone areas of an airplane because of the continual presence of acid fumes and the occasional spilled acid. To prevent damage to the metal, treat battery boxes and the surrounding area with an acid-proof paint. One commonly used material is a black asphaltum that resembles tar. Dilute it with thinner and brush it onto the surface after all traces of corrosion have been removed and treated with a conversion coating such as Alodine. The process of alodining is the chemical application of a protective chromate conversion coating upon aluminum. The alodining process provides good corrosion protection, even when scratched. Paint sticks to it extremely well, and in some cases, it can substitute for primer. Alodine is a brand name for a process that is generically called chromate conversion coating. These coatings are electrically conductive and corrosion resistant but with almost no abrasion resistance. Anodizing is an electrochemical process that converts the surface aluminum into aluminum oxide. It is non-conductive with significant corrosion and abrasion resistance. Both coatings are widely used as adhesion promoters. The final finish must cover the corrosion treatment. Generally, chromate conversion coatings are less expensive than anodizing, while offering better protection.

A good coat of polyurethane enamel is an acid-proof finish that is far superior to the black asphaltum paint. When an airplane is being painted with polyurethane, coat the battery box and the adjacent area with polyurethane. It will provide protection from the fumes of the lead-acid or nickel-cadmium batteries and will not chip or break away from the metal. Polyurethane will not wash away with gasoline or any ordinary solvents.

FLOAT BOTTOM COMPOUND

Seaplane floats take a beating, both from the abrasion of the water and from rocks on the beach or floating debris. In addition to the mechanical damage, they are subject to maximum exposure to corrosive elements. For protection to the bottom of the float, a material similar to acid-proof paint is used. It is an asphaltum product, which is thinned to spraying consistency. If a black finish is not desired on the bottom of the float, suspend aluminum paste in thinner and spray it onto the black compound; the thinner will soften the material and allow the aluminum powder to embed in the finish. Polyurethane enamel provides a good abrasion-resistant finish for floats and may be used instead of the more conventional float bottom compound.

FUEL TANK SEALER

Built-up fuel tanks may develop seeping leaks around rivets and seams. These can be stopped with a resilient, non-hardening tank sealer. Seal the tanks which can be removed from the structure by sloshing them with the sealer; hence the more common name of sloshing compound.

Tank Preparation

a. — Drain the tank and ventilate it thoroughly.

b. — Remove the tank from the airplane.

c. — Remove the gauge sender, vent line fittings, main line screen, and quick drain.

d. — Rinse the tank with clean white gasoline to get rid of all the fuel dyes, then air-dry the tank for at least thirty minutes at room temperature. If the tank has been previously sealed, remove all of the old sealer by pouring about a gallon of acetone or ethyl acetate into the tank and sealing it up for an hour or two. The vapors will soften the sealer and the liquid may then be sloshed around in the tank and dumped out. Repeat the process until the solvents come out clean and inspect that there is none of the old sealant in the tank. Drain it completely and dry the tank with compressed air.

e. — Plug all the threaded holes with pipe plugs and cover the gauge hole with tape or a metal plate.

f. — Pour about a gallon of sloshing or sealing compound into the tank, thinned as recommended on the can. Cover the filler hole and slowly rotate the tank until every bit of the inside is covered. Leave the main line plug slightly loose to relieve the pressure that builds up during sloshing.

g. — Place the tank over a container and remove the quick-drain plug. Allow as much compound to drain out as will. This compound will remain usable if it is covered immediately after draining it from the tank. If it has thickened, it can be thinned with an appropriate thinner or solvent.

h. — Reinstall the drain plug and put on another coat by pouring a gallon of sealer in the tank and rotating the tank as before.

i. — Drain and dry the tank for at least 24 hours; or if low pressure air is circulated through the tank, it can be used after 16 hours.

j. — Clean all of the threaded openings with a bottle brush and then install the fittings, using an appropriate thread lubricant.

Aircraft Painting and Finishing

k. — Coat the float of the sender unit with light grease to prevent its sticking to the compound, and reinstall the sender, using a new gasket.

l. — Reinstall the tank according to the manufacturer's recommendations and fill it with fuel. Check the operation of the sender and if stuck in fresh sealer, free it from the bottom of the tank.

A tank should never be resealed until every bit of the old material has been removed. Don't put new material in the tank if even a trace of the old material remains. Use this type of sealant around rivets and seams in built-up tanks. Be sure the area is perfectly clean and scrubbed with an appropriate solvent. Brush sealant into the seams and around the rivets inside the tank to a thickness about the same as there would be from two coats of sloshed sealant.

SEAM PASTE

Seam paste is a thick zinc-chromate material with organic fibers embedded in it. Use it for making waterproof joints in seaplane hulls or floats and to make leakproof seams in fuel tanks. Use it also as a dielectric for joining dissimilar metals. Put it on with a putty knife or squeegee and smooth it down to the desired thickness then join the seam. Note that this material will not harden.

HIGH TEMPERATURE FINISHES

Pyrolytic paint removal utilizing high temperature, open flames, or hot chemical solutions is an effective method for paint removal. However, certain aircraft components are constantly exposed to such conditions and without paint material designed to withstand them, they would quickly be worn away. Paints designed to take the abuse of heat are important protectors of engine components and structural members. As with all other types of finishes, strictly adhere to the method of application.

Engine Enamel

This enamel has pigments that are colorfast under high temperatures. The special colors used by engine manufacturers are available in this material. Thin it with regular enamel reducer or other conventional thinners.

Heat Resistant Aluminum Paint

This material is specially designed to resist temperatures up to about 1,200 degrees Fahrenheit. It is used on exhaust systems and heater shrouds and has a high content of aluminum shavings, much like an aluminum dope.

ROT-INHIBITING SEALER

Organic materials such as aircraft woods are subject to fungus or mildew, which destroy the strength of the fibers. Mix a special alkyd resin having a very low solids content with fungicidal materials. Treat wood structures with this before they are varnished. Dip large and intricate wood structures into a vat of this material to be sure every portion is protected. Dry for at least 24 hours before covering it with spar varnish. Rot-inhibiting sealer, as any fungicide or mildewcide, is poisonous, but because of its extremely low toxicity, no special safety precautions beyond adequate ventilation and normal hand washing are required for its use.

SPAR VARNISH

Spar varnish is a phenolic modified oil which cures by oxidation rather than evaporation of its solvents. It produces a tough, highly water-resistant film that is not softened by solvents used in the varnish. Use it over the rot-inhibiting sealer for aircraft wood structures. A topcoat of spar varnish in which a fungicide is dissolved often protects electronic components such as circuit boards. This is a transparent coating with a light amber cast. When used over circuit boards it must be completely removed from any point to be soldered.

TUBE OIL

Use this thin, non-hardening, raw linseed oil to protect the inside of the tubular structure in aircraft fuselages, empennage structure, and landing gear. Drill a hole into each tube section and force in the tube oil. Rotate the structure so the oil will fill every portion of the tube then drain the oil out. After the oil has drained, plug the holes with sheet metal screws, drive screws, or by welding.

THINNERS AND REDUCERS

Dopes, enamels, and lacquers are formulated in such a way that the pigments or film materials are suspended in the appropriate solvents. Thin or reduce these to make them less viscous for spraying. Adding the correct type and amount of thinner is of the utmost importance. Use only the thinner or reducer recommended by the paint manufacturer, and thin to the proper viscosity by mixing the material and thinner as specified; or better, by the viscosity called for or known to be correct from experience. Perform this test this using a viscosity cup.

NITRATE DOPE THINNER

Use nitrate dope thinner, some of which meets Federal Specifications TT-T-266C, to thin nitrate dope, nitrocellulose lacquers, or nitrate cement. This thinner, if rubbed on a dry dope film, will determine whether the dope is nitrate or butyrate. If the film softens immediately, it is nitrate; if it does not, it is butyrate.

BUTYRATE DOPE THINNER

Butyrate dope thinners can be used in butyrate or nitrate dope, but nitrate thinners cannot be used in butyrate. Acrylic lacquer thinner may be used in either butyrate or nitrate dope but is not recommended. Butyrate thinner cannot be used to thin acrylics. There is a universal thinner that will thin nitrate, butyrate, and acrylic lacquers, but because of the special requirements of each, this type of material is a compromise and is not generally recommended. Always use thinners and reducers made specifically for the product being thinned.

RETARDER

Retarder is a special type of thinner that has rich solvents. These dry very slowly and prevent the temperature drop that condenses moisture and causes blushing. If dope spraying must be done in times of high humidity, and there is no way to control the amount of moisture in the air, use retarder in place of some of the regular thinner. One part retarder to four or five parts of regular thinner is a good rate of mixture to try. One part retarder to three parts thinner is the absolute maximum that will do any good. A mixture of one part retarder and two parts thinner that is lightly mist-coated over a blushed surface will sometimes remove the blush.

ANTI-BLUSH THINNER

Airshow-quality fabric finishes consist of many coats of dope, sanded with number 600-grit paper between each coat, and rubbed down after the last coat. Spray these coats on wet and thin. Use anti-blush thinner in this type of finish because its slower drying solvents allow each coat more time to flow out and form a smoother film. Anti-blush thinner lies between regular thinner and retarder in its quantity of solvents and length of drying time.

ENAMEL REDUCER

There are several proprietary reducers on the market for enamel, but a good reducer does a satisfactory job of reducing enamels, engine enamels, wing walk compounds, zinc chromate primer, acid-proof black paint, float bottom compound, and white, dope-proof paint. Use it also for washing down a surface to remove the wax after a paint stripper has been used.

ACETONE

Acetone is universally used as a solvent and a cleaner. Use it to remove lacquer finishes and for clean-up after painting. It will soften acrylics or lacquers that have set-up in spray guns or hoses, but it has little effect on polyurethane. Use only virgin acetone since recovered acetone is often so acidic that it can damage anything it is used upon.

REJUVENATOR

Rejuvenator is composed essentially of potent solvents and a plasticizer. Tricresyl-phosphate (TCP) is a permanent, somewhat fire-retardant plasticizer used in many rejuvenators. It softens the old dope and flows the cracks back together.

If a fabric job has been interrupted in the process of finishing, and the aluminum dope has been on the fabric for a considerable time before the topcoats are applied, spray on a coat of rejuvenator to soften the dope and then spray on the color coats.

SPOT PUTTY AND SANDING SURFACER

Use nitrocellulose spot putty to fill cracks or low spots in wood skins before covering them with fabric. When using spot putty, be certain that any defects that may cause a loss of structural strength are not covered up and hidden. If the skin is to be covered only with a film of enamel, use enamel spot putty so that the solvents in the putty will not lift the film. After applying the filler, apply a sanding surfacer over the wood or fiberglass laid-up structure. The sanding surfacer is designed to fill surface irregularities with a material having enough body that it can be sanded smooth.

SECTION C

FINISHING EQUIPMENT AND SAFETY

As always, safety should be one of the primary concerns in the shop. Proper storage of paint and chemical strippers, electrical outlets, personal protection, and fire safety are all major concerns when working with paint. In addition, proper knowledge of the equipment and tools available to do the job will increase safety as well as ease the workload. The maintenance technician should become knowledgeable concerning the paint shop facility and its tools, and be familiar with the equipment required to finish the job safely, quickly, and professionally.

PAINT ROOM

Businesses that do a large volume of aircraft painting normally have a paint hangar where nothing is done except painting and finishing aircraft. Ideally, these buildings are equipped with vapor-proof lights and have their temperature and humidity controlled for the optimum spraying conditions. However, a significant amount of aircraft finishing takes place under conditions that are far less elaborate. In any case, there are certain safety conditions that must be met wherever finishing material is applied. [Figure 6-10]

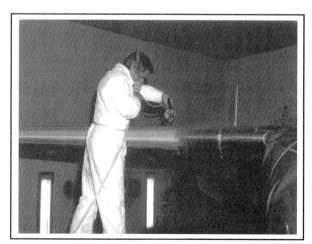

Figure 6-10. If at all possible, use the paint shop for nothing besides painting aircraft.

First, all of the lighting, electrical switches, and outlets must meet not only the electrical code for the building, but the insurance company requirements for electrical equipment in a paint area. In addition, Occupational Safety and Health Administration (OSHA) standards must also be met for the safety of employees. For example, the spray room should provide for moving the air resulting in little more than a slight odor of the finishing material. If painting is being done in a corner of the hangar that has been closed off with polyethylene sheeting, install exhaust fans to rid the area of fumes. Most of the fumes are heavier than air, so place the fan near the floor. The fan should be belt-driven with the motor located in an area that is free of potentially explosive fumes. Wear proper masks and breathing equipment while spraying the paint. Protect the skin from direct contact with chemicals and paints.

Store all of the finishing equipment in an area that is approved by the insurance carrier and by Occupational Safety and Health Administration (OSHA) inspectors. If it is stored outside the painting facility, bring the material inside with plenty of time for it to come to room temperature before using it.

AIR SUPPLY

One of the most important considerations in a paint shop with conventional high-pressure spray guns is the supply of compressed air. There must be an adequate volume with sufficient pressure, and the air must be free of oil and water. Most aircraft maintenance shops have compressors capable of producing air pressure of about 150 psi. The air is stored in an air tank called a *receiver* and then piped to the paint shop. Before it is used, it is filtered and regulated, typically with a wall-mounted filter and regulator unit that contains a water trap. Drain the filter unit and the air receiver every day to keep all water out of the system. Some systems use a chemical desiccant as a final filter to remove any humidity in the compressed air. Check the desiccant regularly and replace it when it becomes saturated. [Figure 6-11]

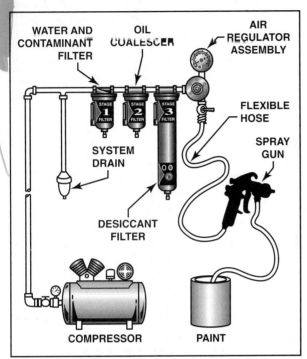

Figure 6-11. A typical system with just one filtered air outlet may look like this. It should provide an adequate supply of clean compressed air essential for a successful paint job. Check the filtering system prior to every operation to assure clean, water-free air.

The important pressure for spray painting is not the pressure at the compressor, or even at the regulator, but at the gun itself. When a gun is connected to the regulator with a long hose, or even a short length of small-diameter hose, there will be an appreciable pressure drop in the flexible line. Air pressure at the gun will be considerably less than at the regulator. In order to get the proper pressure for the material being sprayed, there should be a pressure gauge at the gun. A temporary pressure gauge can be installed at the gun to determine what primary regulator settings achieve the desired pressure at the gun. Pressure required to spray the material will vary depending upon the material and type of application.

PAINTING AND SPRAY EQUIPMENT

There are several systems for spraying liquid or powder finishing materials. Popular systems use compressed air to atomize the liquid and spray it onto the surface and are used almost exclusively in aircraft maintenance shops. Other methods are becoming more popular because of new environmental laws and the need for increased efficiency in paint application. Some of the newer methods include the use of a high volume/low pressure spray system, electrostatic paint spray systems, and finally, powder coating systems.

HIGH VOLUME/LOW PRESSURE (HVLP)

The High Volume/Low Pressure (HVLP) painting system uses a relatively low pressure, resulting in lower paint velocities. Essentially, the system atomizes paint with a high volume of air delivered at low pressure. HVLP systems generally never exceed 10 psi of atomizing air while producing over 20 cubic feet per minute through the paint nozzle. The more common high-pressure systems are set at 45–60 psi and produce 8 cubic feet per minute. In most cases, the use of an HVLP system warrants holding the gun slightly closer to the painting surface due to the lower velocity. Since there is less force in the paint stream, the ability to control the application is much greater than using a high-pressure painting system. In addition, these systems are more environmentally friendly because they create less overspray and use less paint for the same application as compared to high-pressure systems. [Figure 6-12]

Figure 6-12. HVLP or High Volume/Low Pressure systems are becoming more popular due to their efficiency and environmental safety. These systems use a high volume of air at low pressure to atomize and apply paint.

ELECTROSTATIC SYSTEMS

Electrostatic systems are used where overspray must be kept to an absolute minimum. Seldom found in maintenance shops, this type of equipment is generally reserved for larger shops and complex painting jobs. Electrostatic painting can be done effectively on metal and some wood. Paint is applied from a gun and attracted to an electrically grounded workpiece. Spray guns are available in airless, air atomized, rotating discs and bells, and HVLP. When the material leaves the gun, the negatively charged atomized paint is drawn electrostatically to the surface of the substrate. This minimizes overspray due to the attraction that the paint has to the surface of the aircraft or part. Approximately 75% of the paint transfers effectively to the surface, compared to 50 to 65 percent in the HVLP system. However, its potential drawbacks can offset the increased efficiency.

Unfortunately, when painting complex surfaces using an electrostatic system, the Faraday cage effect can cause paint particles to deposit around small cavities or complex curves in the metal. Several problems are created due to excessive paint buildup but it can be overcome by increasing the spray pressure or particle charge. However, the result is slightly less efficiency and more overspray. Another problem with electrostatic painting is that plastic, rubber, ceramic and glass cannot be grounded and will not attract the electrically-charged atomized paint.

POWDER COATING SYSTEMS

Powder coating is generally referred to as a "dry painting" process, which involves melting a dry powdered paint onto the surface of ferrous material. Superior metal protection is achieved but it may not always be as aesthetically pleasing. "Dry painting" is typically reserved for parts of the airframe, like engine mounts or landing gear struts, that will be exposed to stresses, temperatures, and chemicals that would normally reduce the life of conventional "wet paints."

SPRAY GUNS

The type of spray gun used in aircraft painting needs to match the type of material and system that is used during application. Each spray gun manufacturer publishes air cap charts showing the optimal atomization setting for viscosity in conjunction with the specific gun configuration. Making sure these settings are made for each viscosity and temperature combination will help produce the desired finish.

There are basically two types of spray guns used for aircraft finishing, suction feed guns and pressure feed guns. The difference between the two is in the fluid tip and the air cap. [Figure 6-13]

SUCTION GUNS

The suction-feed gun normally has a one-quart cup attached and is used for touch-up painting or other low-volume work. The fluid tip sticks out of the air cap just enough to produce a low pressure, or a suction, which pulls the material out of the cup. As compressed air flows past the needle valve orifice and out the spray tip, it creates a partial vacuum in the paint tube. A small hole in the top of the paint container allows atmospheric pressure to push the paint up to the needle valve. Pressure-feed guns, on the other hand, use the direct force of compressed air to push the paint up into the needle valve.

PRESSURE GUNS

A pressure-feed gun is attached to either a pressure cup or to a pressure pot with the fluid tip flush with the air cap. Airflow around the tip does not create a low pressure since the fluid is delivered to the gun under a slight positive pressure from the air supply. When spraying a considerable volume of material, use a pressure-fed gun with a pressure pot. Pots are

Figure 6-13. A good quality spray gun is essential for an appealing finishing job. There are a variety of spray guns designed for use with HVLP, High Pressure, or other types of paint application.

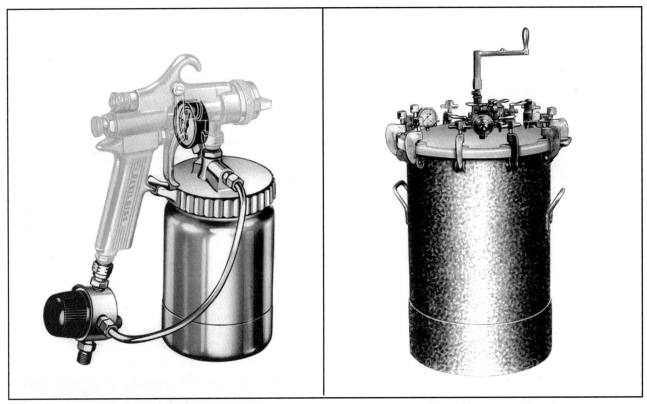

Figure 6-14. Smaller suction cups, front left, are often used for spraying the trim and for touch-up. A smaller two-quart pressure pot or a larger pressure pot, shown in this figure, is used for priming and painting the base coats or for larger areas of accent color.

available in sizes that hold from two quarts to fifteen gallons, with the five-gallon pot being a handy size for finishing aircraft. These pots normally have an agitator to keep the material mixed while spraying and may either be air driven or turned with a crank. [Figure 6-14]

AIRLESS GUNS

Another common spray gun uses a pump to deliver the material under high pressure to the special spray gun where it is released through a small nozzle. Instead of pressurized air, the high fluid velocity tears the material apart, or atomizes it. Paint particles are generally larger when using this system, which allows for a thick coating of paint. Airless spraying is used where large areas must be covered in a short period of time and is normally found only in shops where a large volume of painting is done. No matter what type of gun or system is used, utilize the proper protection from vapors and paint particles.

RESPIRATORS AND MASKS

Many of the solvents and thinners used in modern finishing systems are toxic, and some form of respirator or mask must be worn when spraying the material. Most dust respirators or filter-type masks will remove solids from the air you breathe, but they will not filter out fumes. Removal of fumes and vapors depend upon adequate ventilation in the paint room. [Figure 6-15]

Figure 6-15. Dust respirators or respirator masks are not designed to remove fumes from the air; they can only filter out solids.

If you must stay in a room that has a heavy concentration of fumes, wear an airflow-type of mask that is slightly pressurized with shop air from the feed line to the spray gun. This keeps fumes out of the mask, shields the face and protects the eyes from potentially irritating vapors. [Figure 6-16]

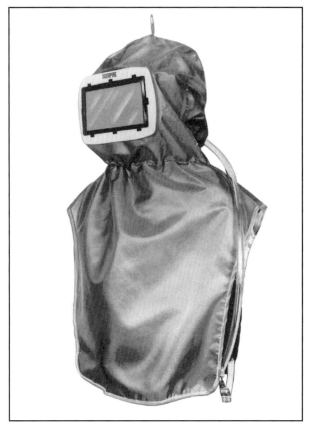

Figure 6-16. Use airflow-type respirators if there are fumes in the paint shop that cannot be removed. This is an example of a visor/hood model that provides extra protection. Other systems supply air only to a mask.

An organic vapor respirator can also be used to remove some vapors and solids from the air. Chemical absorption removes vapors from the breathing air and a mask removes solids with a pre-filter. They are ideally suited for surface preparation such as sanding and paint removal.

MIXING AND VISCOSITY MEASUREMENT EQUIPMENT

If a shop does a considerable amount of finishing, a mechanical shaker is a valuable piece of equipment to have in the shop. A can of finishing material is clamped into the machine that shakes, or agitates, the can for fifteen to twenty minutes, assuring a thorough mixture of paint. If a mechanical shaker is not available, a good mixture of the pigments can be obtained by using a hand agitator or an agitator driven by an air drill motor. An electric drill motor must never be used to drive the agitator since the vapors from many of the finishing materials in use are quite flammable. Electrical arcing from the brushes in these motors can easily ignite vapors that are stirred up.

After the paint is thoroughly mixed, measure its viscosity. Many of the modern finishes are quite sensitive with regard to their viscosity when they are sprayed. Use a viscosity cup to duplicate the consistency of the paint from one batch to the next. Paint laboratories use Zahn or Ford cups, which are precision devices, but a small plastic cup is available at paint supply houses that is entirely adequate.

Dip the viscosity cup into the thinned and mixed paint. Time the flow from the cup until the first break appears. Elapsed time is dependent upon the viscosity of the material. By knowing the viscosity that produces a good finish, each succeeding batch can be mixed the same. [Figure 6-17]

Figure 6-17. A viscosity cup will ensure consistency of viscosity between batches of finishing material.

SPRAY GUN OPERATION

There are many different spray gun manufacturers and each gun is designed, built, and used differently. Since operating procedures are unique for each gun, follow the directions provided by the manufacturer closely. When the material is ready and the tools have been chosen, the next step will be to adjust the spray pattern of the gun. The proper use of the spray gun makes the difference between a truly fine finish and one that leaves something to be desired. Understanding clearly the operation of the paint gun permits the technician to accurately set the various controls.

For this explanation, assume that the gun is a suction-feed gun and has a one-quart suction cup screwed directly to the gun, as shown in figure 6-18. Notice that the air line is attached to the fitting at the bottom of the handle, and the material supply is attached to the fitting at the front end of the gun.

The vent hole in the lid of the suction cup must be open and the gasket properly fit, sealing the lid to the cup so that material will not spill out. Connect the air line to the gun and pull the trigger to open the air valve. Adjust the regulator to get the desired pressure at the gun while the air is flowing. [Figure 6-19]

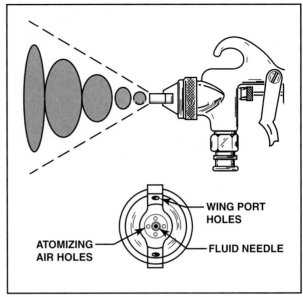

Figure 6-19. The amount that the air valve is open determines the shape of the spray pattern. When there is no airflow through the wing ports, the pattern is round; and as the airflow increases, the pattern flattens out.

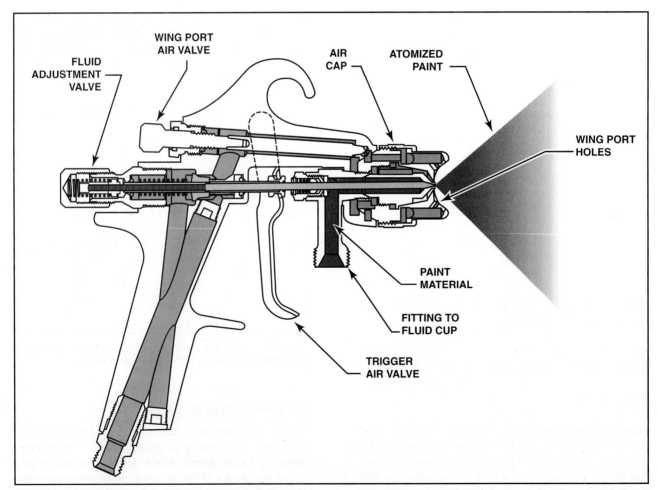

Figure 6-18. When the trigger is pulled, the air valve opens, sending atomizing air to the nozzle and the wing ports. Continued pulling of the trigger lifts the fluid needle off its seat so the material can flow from the nozzle. The fluid adjustment determines the amount of material allowed to flow, and the air valve controls the shape of the spray pattern.

Notice in figure 6-18 that pulling the trigger allows air to flow out the center hole of the air cap, producing a low pressure that will pull fluid from the cup when the fluid adjustment valve is open. Continuing to pull back on the trigger moves the fluid needle back so that material from the cup can be pulled up and sprayed out with the air. As the liquid leaves the air cap, it is broken down into extremely tiny droplets, or atomized. The spray pattern from the gun in this condition should be round.

When the gun is putting out a round pattern of spray, opening up the wing-port air valve allows air to flow out of the holes in the wing ports of the air cap. Air flowing from these holes blows against the stream of atomized material, and determines the shape of the spray pattern. A correctly adjusted spray gun should produce a uniform, fan-shaped spray, with the fan perpendicular to the wing ports.

If the paint pattern is heavy at either the top or the bottom, material has built up on the air cap, or some of the holes may be plugged up. To correct this, remove the air cap, soak it in thinner, and probe the holes with a broom straw or toothpick. Do not use any kind of wire, as wire can damage these precision holes. Afterward, blow dry the cap with compressed air and reinstall it.

If the pattern is too heavy in the center, even with the spreader adjustment valve wide open, either too much material is coming out or the material is too thick. Closing the fluid valve a bit should improve the shape of the pattern. If not, thinning the material should help. If the spray pattern splits or is too thin in the middle, too much air is coming through the wing port holes and the spreader adjustment should be shut down a little. Improper material viscosity can also cause this. A jerky or intermittent spray can be caused by air getting into the fluid line from an air leak or by too little material in the cup. A banana-shaped spray pattern is caused by the wing-port hole being plugged on the convex side of the pattern. Remove the air cap, soak it in thinner, and blow out the holes with compressed air. Too much overspray is caused by too much air pressure on the gun, and can be cut down at the regulator. [Figure 6-20]

Runs and sags on the surface may be caused by either the gun being held too close to the work or by too much material being sprayed out for the speed that the gun is being moved. It is also possible that the gun being held at an angle with respect to the surface or perhaps the material is too thin. If the material on the surface is too dry and is rough rather

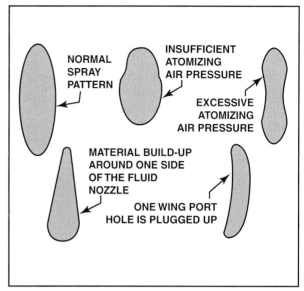

Figure 6-20. Spray gun problems or incorrect settings can often be identified by the pattern of the spray.

than flowing, it is an indication that the gun is being held too far from the work, or there is too much air pressure supplied to the gun. It may also indicate an incorrect fluid tip size or a low fluid pressure for HLVP systems. An orange peel surface clearly indicates that too much material is being applied. [Figure 6-21]

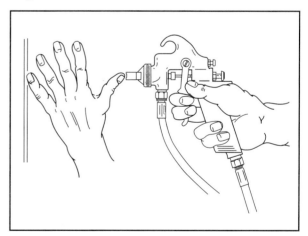

Figure 6-21. Hold the spray gun about one hand span from the surface; this is equal to about eight inches.

APPLYING THE FINISH

To spray the surface, begin with the edges or the corners and move the gun parallel with the surface about a hand span—about eight inches—away. Begin the stroke of the gun before reaching the surface. Pull the trigger just before reaching the surface and keep going until after passing the end. The fan-shaped spray is perpendicular to the wing ports, which can be set at any angle to give a spray in the proper direction for cutting in edges. [Figure 6-22]

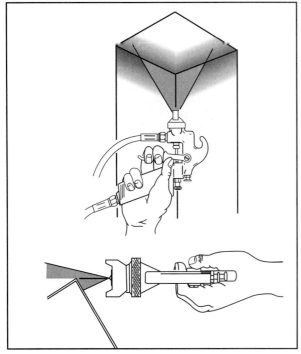

Figure 6-22. When starting to paint a surface, spray the corners and edges first, then the flat surfaces.

When all of the edges and corners are finished, spray the flat portion of the surface with straight passes across the surface. Hold the gun level and at a constant distance from the surface. If the gun is properly adjusted, the passes should be about 10 or 12 inches wide, and each succeeding pass should overlap the previous one by about two-thirds of their width. Proper lapping will give the finish a uniform film thickness. Arcing the gun rather than moving it parallel to the surface can cause non-uniform film thickness. [Figure 6-23]

SEQUENCE FOR PAINTING AN AIRPLANE

Careful planning of the sequence to use in painting an entire airplane will make the work proceed much more easily and will minimize problems from overspray. Position the aircraft in the paint room so that moving air will pass over the aircraft from the tail to the nose. Any overspray should fall on the unpainted portion. Paint the ends and leading edges of the ailerons and flaps first, followed by the flap and aileron wells, the wing tips, and the leading and trailing edges. Spray all of the landing gear, wheel wells, control horns and hinges. Before starting on any flat surfaces, paint all of the difficult areas then proceed in a systematic way. In some areas, it may be difficult to keep a uniform distance from the

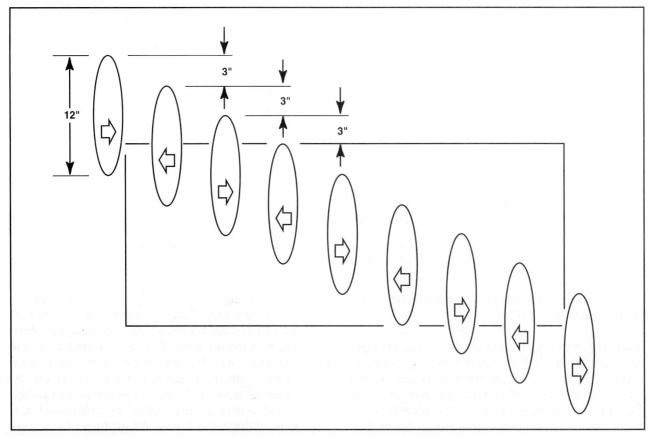

Figure 6-23. Overlap each succeeding pass by about two-thirds of its width to get a film of uniform thickness.

Aircraft Painting and Finishing

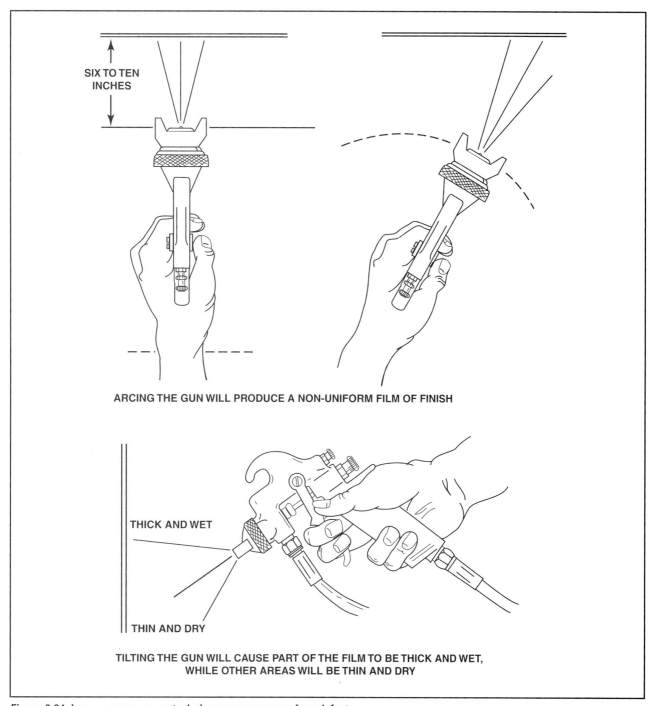

Figure 6-24. Improper spray gun technique may cause surface defects.

work surface, but arcing or tilting the gun may cause defects in the paint. [Figure 6-24]

Paint the bottom of the airplane first, using a creeper for the belly and the bottom of low-wing airplanes. Prime the bottom of the horizontal tail surfaces first, starting at the root and working outward, spraying chordwise. Work up the fuselage, allowing the spray to go up the sides, all the way up to the engine. Spray the bottom of the wing, starting at the root and working toward the tip, spraying chordwise. Jack up the nose of the airplane to lower the tail far enough to allow painting of the top of the fin. When spraying the top of the fuselage, tilt the gun so that overspray will fall out ahead of the area being painted with new material wiping it out. On the upper surfaces, apply the primer across the fuselage and spanwise on the vertical and horizontal surfaces of the tail and wing. Allow the primer to cure and use the same sequence to spray on the finish.

Spray the tack coat and final coat using the same sequence and direction as the prime coat.

It is usually impossible to completely reach across the top of the wing, so spray as far as can while working from the root to the tip, along the trailing edge, then from the tip back toward the fuselage. Tilt the gun back so the overspray does not fall on the rear half of the wing where the paint has hardened.

Use a very thin coat of acrylic lacquer to wash out acrylic overspray. It softens the film enough for the overspray to sink into the finish. Dried overspray from any material other than polyurethane can be worked into the finish by spraying a mixture of one part retarder and two parts thinner upon the surface while the overspray and topcoat are still fresh. The surface softens enough for the overspray to sink in resulting in a glossy surface. Enamel overspray isn't as bad as lacquer or dope because it dries much slower.

Once an aircraft has been painted, control surfaces must be checked for proper balance. The exact requirements and procedures for balancing can be found in the manufacturer's service manual or by contacting the manufacturer.

CLEANING THE SPRAY PAINT EQUIPMENT

As with any precision tool, a spray gun will provide satisfactory results for a long time if it is properly maintained. The spray gun is especially vulnerable and must be kept clean. If using a suction cup or gravity cup, immediately after spraying, dump the material from the cup and clean it. Pour thinner into the cup and spray it through the gun. Pull the trigger repeatedly to flush the passageways and clean the tip of the needle. Spray thinner through the gun until it comes out with no trace of the material.

When cleaning pressure-fed guns, first empty the gun and related hose by loosening the air cap on the gun and the lid of the pressure pot. Pull the trigger after covering the air cap with a rag. Atomizing air backing up through the gun and the fluid line will force all of the material back into the pot. Then place thinner in the pot and replace the lid. Spraying thinner through the hose and gun will clean the entire system.

After cleaning the inside passages of the gun, soak the nozzle in a container of thinner. Do not soak the entire gun in the thinner since the packings may be ruined. Lubricate the air valve stem and all of the packings around the fluid needle with light oil. They will continue to operate smoothly and remain soft and pliable. Tighten the packing nuts finger-tight only. (Figure 6-25)

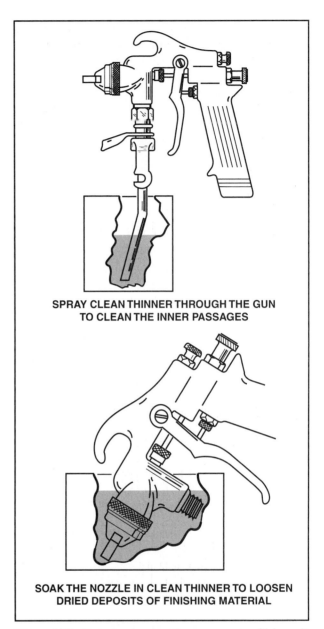

SPRAY CLEAN THINNER THROUGH THE GUN TO CLEAN THE INNER PASSAGES

SOAK THE NOZZLE IN CLEAN THINNER TO LOOSEN DRIED DEPOSITS OF FINISHING MATERIAL

Figure 6-25. Properly cleaning the spray gun will keep it operating correctly and extend the life of the equipment.

Material should never be left in the gun because it will set-up and plug the passages. If the passages become plugged with dope or acrylic lacquer, disassemble the gun and soak the parts in acetone. If not flushed out immediately after use, catalyzed materials such as epoxies and polyurethanes will set-up in the gun and hoses. If this happens, discard the hoses and clean the passages in the gun by digging out the material. This is not only time consuming but there is a good chance the gun will be damaged inside.

COMMON FINISHING PROBLEMS

Because of the number of variables that need to be controlled in order to attain a good finish, closely follow the procedures suggested by the spray gun manufacturer and the paint producer. Before proceeding, it is always helpful to practice painting on a surface with the same qualities as the aircraft. Once ready to begin painting, realize that recognizing errors and problems quickly is a good way to avert a potential repainting job. Some of the most common problems and their characteristics include:

- Rough Finish—Dope or paint is too cold or viscous. The aircraft, dope, and thinner should all be at the same temperature, about 70° F, before spraying.

- Fabric Will Not Tauten—Fabric put on too loosely. Fabric remains undoped for too long. Too much retarder used for thinning dope.

- Blushing—Humidity too high. Moisture in spray system. Dope applied over a moist surface.

- Pinholing or Blisters—Water or oil in spray system or on surface. Undercoat not thoroughly dry. Too fast surface drying. Film coat too heavy.

- Bubbles and Bridging—Dope too cool, or not brushed out properly. Temperature of dope room too high.

- Runs and Sags—Use of improper equipment. Incorrect adjustment of equipment. Improper thinning or faulty spray technique.

- Dope Will Not Dry—Oil, grease, or wax on surface.

- Dull Spots—Porous spot putty or undercoating, allowing dope or lacquer to sink in.

- Bleeding—Organic pigments or dyes used in the undercoats which are soluble in the topcoat solvents.

- White Spots—Water in spray system or on surface.

- Paint or Primer Peeling—Wax from stripper or detergent from cleaning process may still be on surface. If dope is applied over old paint or enamel, the old paint will softened and eventually peel from the surface.

- Brown Spots—Oil in spray system.

- Orange Peel—Spraying with too high pressure. Use of too fast drying thinner. Cold, damp draft over surface.

- Wrinkling—Reaction between solvents and primer or undercoats.

- Overspray—Wrong spray technique. Too fast drying thinner.

- Fisheyes—Silicone, wax, or polish contamination on surface. Oil from air compressor.

MASKING AND APPLYING THE TRIM

After the base coats are complete, install the trim details. Applying the details of the aircraft finish can be the most time consuming portion of the process. When changing colors for trim details, allow the base paint to dry sufficiently before masking off the areas for this work. Fortunately there are time- and cost-saving materials and methods to help, but working with basic materials will usually cost less.

MASKING FOR THE TRIM

Masking an aircraft and getting it ready for the trim may seem to be a minor operation, but there is a lot more to it than meets the eye. Consider that modern finishes will penetrate some of the more porous masking products, so use only the best quality masking paper and tape. Newspapers and masking tape bought at the supermarket will not stop the finishes from seeping through or bleeding under, and the few cents saved using anything less than the best can cost an entire topcoat job. [Figure 6-26]

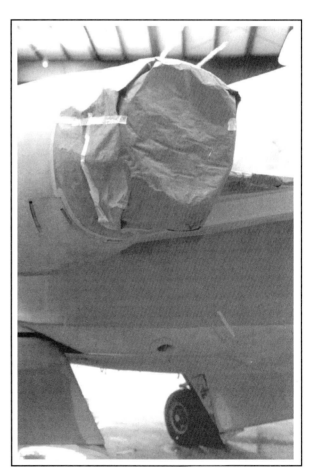

Figure 6-26. Before spraying any finish, mask off all of the parts not to be painted with a material that will not allow any finish to bleed through.

The most important thing about the trim is to attain good, smooth edges and usually the quality of the masking tape determines this. When spraying over the edge of the masking tape, the film will build up to the thickness of the tape edge, and when the tape is removed, a thick edge is left. If the finish dries before the tape is removed, the edge will not only be thick, but rough as well.

To get the sharpest line, use an extremely thin polypropylene tape. A tight edge is formed and the finish cannot bleed under. Do not use this tape for blocking off the entire surface, just for the edge of the trim. Back it up with regular crepe masking tape and cover the large areas with high quality masking paper.

After masking off the surface, spray the finish, and as soon as it is no longer tacky to the touch, remove the tape. Pull it back at an angle of about 170 degrees, just far enough away from the fresh finish that the tape will not get on it. By pulling the tape back over itself, it can be removed without pulling up any of the still partially soft trim, and the finish will flow down smoothly and not leave a thick edge. [Figure 6-27]

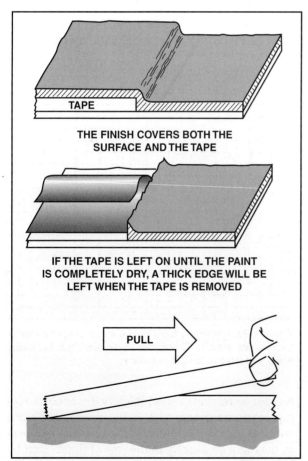

Figure 6-27. Remove the masking tape by pulling it back over itself as soon as the finish becomes tacky to the touch.

When masking an aircraft, allow the finish to dry for the length of time specified by the manufacturer. Allow the primer to cure and use the same sequence to spray on the finish. Spray the tack coat and final coat using the same sequence and direction as the prime coat, and do not allow the tape to remain on the surface any longer than is absolutely necessary. If the tape is put on several days before the trim is sprayed, and especially if the aircraft has been in the sun or has gotten hot, it is probable that the tape will cure to the finish and be very difficult to remove. If this happens, use aliphatic naphtha to try and soak the tape free.

LAYING OUT REGISTRATION NUMBERS

The placement of registration numbers on an aircraft is specified in 14 CFR Part 45 of the Federal Aviation Regulations. All aircraft use a Roman font, but the placement and size of registration numbers varies for each aircraft. For rotorcraft, lay out a set of numbers horizontally on both sides of the fuselage, cabin, boom, or tail. The numbers should have a height of 12 inches. If the surfaces do not allow for the full sized letters, make them as high as practicable, but at least two inches high. Generally, helicopter numbers should be on the side of the fuselage below the window line and as near to the cockpit as possible.

For fixed-wing aircraft, place the numbers between the trailing edge of the wing and the leading edge of the horizontal stabilizer, or on the engine pods if they are located in this area. If this is not a good location, place them on both sides of the vertical tail. Marks on a fixed-wing aircraft must be at least 12 inches high unless the aircraft has a maximum cruising speed of less than 180 knots and is an experimental, exhibition, or amateur-built aircraft. For these aircraft, the numbers may be only three inches high. In addition, if the aircraft was built at least 30 years ago and is thus classified as an antique, it can have two-inch-high registration numbers on its vertical tail or the sides of its fuselage.

DESCRIPTION

Certificated aircraft that are registered in the United States must display the capital Roman letter N followed by the registration number of the aircraft. If the aircraft is registered in a foreign country, it will have its own distinct number, letter, or combination of the two. The number must be two-thirds as wide as it is high, except for the number **1** which is one-sixth as wide as it is high, and the letters **M** and **W** which may be equally wide as they are high. The letters must be formed with a solid line that is one-sixth as thick as the character is high, and the spac-

Aircraft Painting and Finishing

AIRCRAFT NATIONALITY MARKS			
AN	Nicaragua	PK	Indonesia
AP	Pakistan	PP, PT	Brazil
B	China	SE	Sweden
CCCP	Russia	SN	Sudan
CF	Canada	SP	Poland
CU	Cuba	SU	Egypt
D	Germany	SX	Greece
EC	Spain	TC	Turkey
ET	Ethiopia	TF	Iceland
F	France	TG	Guatemala
G	Great Britain	TI	Costa Rica
HB	Switzerland	VH	Australia
HC	Ecuador	VP, VQ, VR	Great Britain - Colonies & Protectorates
HK	Colombia	VT	India
I	Italy	XA, XB, XC	Mexico
JA	Japan	XH	Honduras
LN	Norway	XV	Vietnam
LV, LQ	Argentina	YA	Afghanistan
LZ	Bulgaria	YS	El Salvador
N	United States	YU	Yugoslavia
OB	Peru	YV	Venezuela
OD	Lebanon	ZK, ZL, ZM	New Zealand
OE	Austria	ZP	Paraguay
OH	Finland	S, ZT, ZU	Union of South Africa
OO	Belgium	4X	Israel
PH	Netherlands	5A	Libya

Figure 6-28. International aircraft marks identify the country in which it is registered.

ing between the letters or numbers may not be less than one-fourth of the character width. [Figure 6-28]

APPLICATION

Newer developments make the task of laying out registration numbers much easier. Today there are pressure-sensitive vinyl numbers available that may be stuck on the aircraft and provide a professional looking job with a minimum amount of skill and time. These stickers have also been used on larger commercial aircraft in place of expensive and detailed paint schemes. There are also commercially available stencils, which are simply stuck on and spray painted over. When the finish is no longer tacky, remove the stencil and the registration numbers will be perfectly spaced.

There is always the possibility of having to lay out the numbers without any of these aids, and for this eventuality there are two tools that can easily be made. The template similar to the one in figure 6-29 has been the standard of the industry for years. It is cut from thin sheet aluminum and has the numbers eight and one together. Any letter or number can be laid out with this template.

A tool that is easier to make and one that works so well that it is used by many painters is simply a piece of flexible metal or transparent plastic with some marks scribed on it. For the 12-inch charac-

ters, make the tool about 14 inches long and two inches wide. Scribe a line down its center and mark lines across it that are two, four, six, nine, and twelve inches apart. [Figure 6-29]

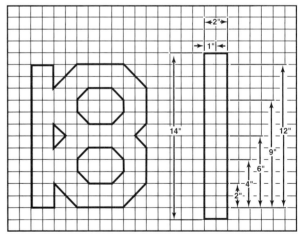

Figure 6-29. A lettering tool may be made of light metal or transparent plastic. These two examples can be easily made and used to lay out any letter.

As soon as the finish is sufficiently dry to allow taping, locate the proper spot on the aircraft and block out the space for the numbers. Lay out the numbers so they will be directly in the line of flight. Start by laying down two strips of thin polypropylene masking tape that are perfectly parallel and are as far apart as the height of the numbers. These strips serve as the top and the bottom of the numbers. If the numbers N46382 are to be laid out, the block will have to be 58 inches long and 12 inches tall.

Using the layout tool, mark out the blocks for each of the characters, which will be eight inches wide and separated by two inches. Since all of the dimensions for the characters will be the same as those marked on the layout tool, it is a simple matter to mark out all of the lines. Lay out all of the blocks with a soft lead pencil. Do not use a ball point pen as the ink will migrate up through almost any finish applied over it. Use narrow polypropylene tape for all of the edges of the characters and the holes. Mask large areas further with a good grade of crepe masking tape and sheets of masking paper. [Figure 6-30]

When masking is complete, check the layout carefully to be certain no letters or numbers have been forgotten or that some area that needs to be painted has not been masked. When the layout is correct, mix the finishing material and spray. When it dries enough so that it is no longer tacky, remove the masking.

DECALS, MARKINGS AND PLACARDS

The final part of any finishing operation is completing the mandatory markings and placards. This includes all of the service markings, safety warnings, capacity and grade markings, and a myriad

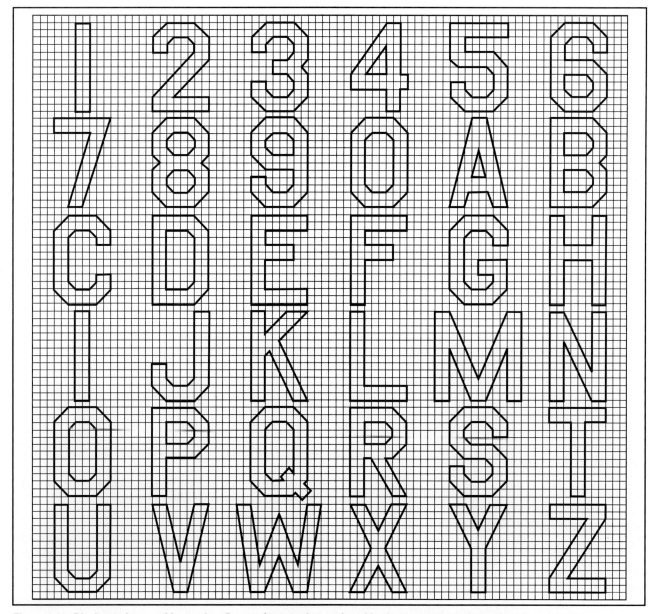

Figure 6-30. Block numbers and letters in a Roman font can be produced in the proportions shown here.

other items. Decals, or decalcomanias, are the most frequently used. They consist of printing or silk screening upon a clear film with a paper backing. Remove the backing by soaking the decal in clean, warm water until it separates, then transfer it to its final position.

Some placards are printed on vinyl and transferred the same way as the vinyl registration numbers. Some registration numbers, placards, and specialized markings are fabricated from High Performance 3M or Calon Vinyl. After application, a 7-year-plus life can be expected. A wide variety of colors and shadowing can be done with these decals. Others are applied to a surface that is flooded with a detergent mixture, which allows for their placement before a squeegee is used and they become permanently applied. Still others are small metal signs that are applied with a MIL-SPEC cement. There are so many different types of decals in use today that a general set of directions is impossible. In all cases, the placards should be applied as specified in their instructions. One word of caution: the airplane is not airworthy until ALL of the required placards, warnings, safety and service instructions have been applied.

SAFETY IN THE PAINT SHOP

The very nature of aircraft finishing has an element of danger. Working with volatile and sometimes toxic chemicals produces a need for extra caution, so it is important to emphasize a few safety precautions. First, never use a dry broom to sweep overspray from the paint room or hangar floor. This material is usually highly flammable and the static electricity from sweeping can ignite it. Always hose it down with plenty of water and sweep it while wet.

When a fabric-covered aircraft is being dry-sanded, ground the aircraft structure, especially if the first coats of finish were nitrate dope. Static electricity from the sanding could cause a spark to jump inside the structure and ignite the fumes.

Do not stir finishing materials with a beater attached to an electric drill motor. The sparking brushes could ignite the fumes that are stirred up. If the material should ever catch fire, simply cover the flame and smother it. Even a piece of cardboard will cut off the oxygen.

Human eyes are very sensitive to airborne chemicals and paints. If any finishing material gets in the eyes, flush them immediately with plenty of water. In addition to the sensitivity of the eye, lungs, nasal passages, open cuts, and even exposed skin can show signs of irritation. When painting around the top of a vertical fin or in any high place, use a good scaffold, rather than trying to reach too far from a shaky stepladder.

AIRFRAME ELECTRICAL SYSTEMS

INTRODUCTION
An aviation maintenance technician must be familiar with aircraft electrical systems, including ways in which electricity is generated and routed to various aircraft components. By understanding the principles of electricity and electrical system designs, a technician can effectively diagnose, isolate, and repair malfunctions. However, because the expense involved in possessing the proper tools, test equipment, and current technical publications is prohibitive for most individuals, FAA certified repair stations — or the component manufacturer — service many major electrical components such as generators, motors and inverters. Although a technician may be required only to remove and replace (R&R) these items with serviceable units, it is still the technician's responsibility to know how these components operate in order to perform system troubleshooting and routine servicing of the electrical system.

SECTION A

AIRBORNE SOURCES OF ELECTRICAL POWER

GENERATORS

Energy for the operation of most electrical equipment on large aircraft and some small aircraft is supplied by a generator. A generator is any piece of equipment that converts mechanical energy into electrical energy by electromagnetic induction. Generators designed to produce direct current are called DC generators and those that produce alternating current are called AC generators.

On many older aircraft, the DC generator is the source of electrical energy. With this type of system, one or more DC generators are driven by the engine(s) to supply power for all electrical equipment as well as for charging the battery. In most cases only one generator is driven by each engine; however, some large aircraft have two generators driven by a single engine.

THEORY OF OPERATION

After the discovery that electric current flowing through a conductor creates a magnetic field around the conductor, there was considerable scientific speculation regarding whether or not a magnetic field could create current flow. In 1831, English scientist Michael Faraday demonstrated that this, in fact, could be accomplished. This discovery is the basis for the operation of the generator.

To show how an electric current is created by a magnetic field, several turns of wire are wrapped around a cardboard tube, and the ends of the conductor are connected to a galvanometer. A bar magnet is then moved through the tube. As the magnet's lines of flux are cut by the turns of wire, the galvanometer deflects from its zero position. However, when the magnet is at rest inside the tube, the galvanometer shows a reading of zero, indicating no current flow. When the magnet is moved through the tube in the opposite direction, the galvanometer indicates a deflection in the opposite direction. [Figure 7-1]

The same results are obtained by holding the magnet stationary and moving the coil of wire. This

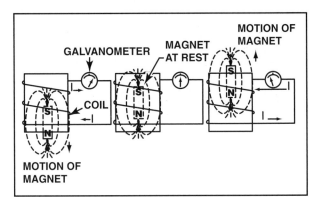

Figure 7-1. Current flow induced into a coil by magnetic flux lines is seen on a galvanometer. The direction of current flow is dependent upon how the magnetic fields cross the conductor.

indicates that current flows as long as there is relative motion between the wire coil and the magnetic field. The strength of the induced current depends on both the strength of the magnetic field and the speed at which the lines of flux are cut.

When a conductor is moved through a magnetic field, an electromotive force (EMF) is induced into the conductor. The direction, or polarity, of the induced EMF is determined by the direction the conductor is moved in relation to the magnetic flux lines.

The left-hand rule for generators is one way to determine the direction of the induced EMF. For example, when the left index finger is pointed in the direction of the magnetic lines of flux (north to south), and the thumb in the direction the conductor is moved through the magnetic field, the second finger indicates the direction of the induced EMF, when extended perpendicular to the index finger. [Figure 7-2]

When a conductor in the shape of a single loop is rotated in a magnetic field, a voltage is induced in each side of the loop. Although the two sides of the loop cut the magnetic field in opposite directions, the induced current flows in one continuous direction within the loop. This increases the value of the induced EMF. [Figure 7-3]

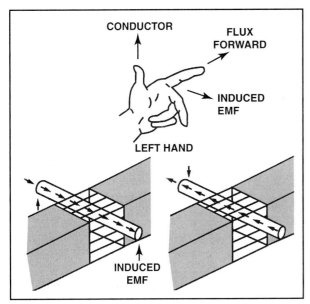

Figure 7-2. When applying the generator left-hand rule, the index finger points in the direction the lines of magnetic flux travel, the thumb indicates the conductor's direction of movement, and the second finger indicates the direction of induced EMF.

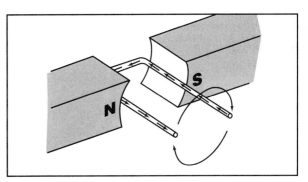

Figure 7-3. When a loop of wire is rotated in a magnetic field, the current induced in the wire flows in one continuous direction.

When the loop is rotated half a turn so that the sides of the conductor have exchanged positions, the induced EMF in each wire reverses its direction. This is because the wire, formerly cutting the lines of flux in an upward direction, is now moving downward; and the wire formerly cutting the lines of flux in a downward direction, is now moving upward. In other words, the sides of the loop cut the magnetic field in opposite directions.

In a simple generator, two sides of a wire loop are arranged to rotate in a magnetic field. When the sides of the loop are parallel to the magnetic lines of flux, the induced voltage causes current to flow in one direction. Maximum voltage is induced at this position because the wires are cutting the lines of flux at right angles. This means that more lines of flux per second are cut than in any other position relative to the magnetic field.

As the loop approaches the vertical position, the induced voltage decreases. This is because both sides of the loop become perpendicular to the lines of flux; therefore, fewer flux lines are cut. When the loop is vertical, the wires momentarily travel perpendicular to the magnetic lines of flux and there is no induced voltage.

As the loop continues to rotate, the number of flux lines being cut increases until the 90-degree point is reached. At this point, the number of flux lines cut is maximum again. However, each side of the loop is cutting the lines of flux in the opposite direction. Therefore, the direction, or polarity, of the induced voltage is reversed. Rotation beyond the 90-degree point again decreases the number of flux lines being cut until the induced voltage becomes zero at the vertical position. [Figure 7-4]

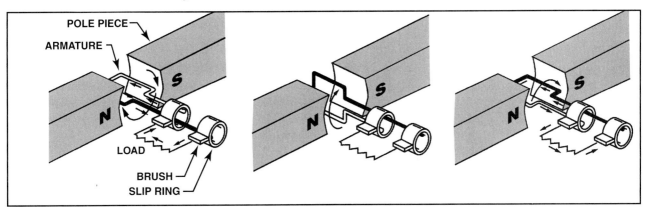

Figure 7-4. In a simple generator, the magnets are called pole pieces and the loop of wire is called the armature. Attached to each end of the loop is a slip ring, on which a set of brushes rides, to complete a circuit through a load. Maximum voltage is induced into the armature when it is parallel with the flux lines. Once the armature is perpendicular with the flux lines, no lines of flux are cut and no voltage is induced. As the armature rotates to the 90-degree point, the maximum number of flux lines is being cut, but in the opposite direction.

When the voltage induced throughout the entire 360 degrees of rotation is shown on an oscilloscope, the curve increases from a zero value at zero degrees, to a maximum positive voltage at 90 degrees. Once beyond 90 degrees the curve decreases until it reaches 180 degrees, where the output is again zero. As the curve continues beyond 180 degrees, the amount of negative voltage produced increases up to the 270-degree point, where it is maximum. The amount of negative voltage then decreases to the zero point at 360 degrees. [Figure 7-5]

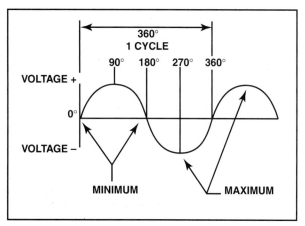

Figure 7-5. The output of an elementary generator is represented by the sine curve, as a loop rotates 360 degrees through the lines of magnetic flux.

As the illustration shows, the output produced by a single loop rotating in a magnetic field is alternating current. By replacing the slip rings of the basic AC generator with two half-cylinders, commonly referred to as a single commutator, the alternating current is changed to direct current. [Figure 7-6]

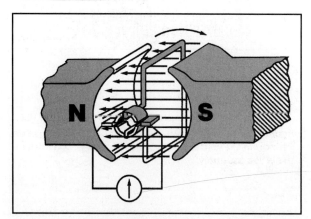

Figure 7-6. In this illustration, the black side of the coil is connected to the black segment, and the white side of the coil to the white segment. The segments are insulated from each other, and two stationary brushes are placed on opposite sides of the commutator. The brushes are mounted so that each brush contacts each segment of the commutator as it revolves with the loop.

To explain how DC current is obtained, start with the armature at zero degrees where no lines of flux are cut and therefore, no output voltage is obtained. Once the armature begins rotating, the black brush comes in contact with the black segment of the commutator, and the white brush comes in contact with the white segment of the commutator. Furthermore, the lines of flux are cut at an increasing rate until the armature is parallel to the lines of flux, and the induced EMF is maximum.

Once the armature completes 180 degrees of rotation, no lines of flux are being cut and the output voltage is again zero. At this point, both brushes are contacting both the black and white segments on the sides of the commutator. After the armature rotates past the 180-degree point, the brushes contact only one side of the commutator and the lines of flux are again cut at an increasing rate.

The switching of the commutator allows one brush to always be in contact with that portion of the loop that travels downward through the lines of flux, and the other brush to always be in contact with the half of the loop that travels upward. Although the current reverses its direction in the loop of a DC generator, the commutator action causes current to flow in the same direction. [Figure 7-7]

The variation in DC voltage is called ripple, and is reduced by adding more loops. As the number of loops increases, the variation between the maximum and minimum values of voltage is reduced. In fact, the more loops that are used, the closer the output voltage resembles pure DC. [Figure 7-8]

As the number of armature loops increases, the number of commutator segments must also increase. For example, one loop requires two commutator segments, two loops require four segments, and four loops require eight segments.

The voltage induced in a single-turn loop is small. Increasing the number of loops does not increase the maximum value of the generated voltage. However, increasing the number of turns in each loop does increase the voltage value. This is because voltage is obtained as an average only from the peak values. The closer the peaks are to each other, the higher the generated voltage value.

DC GENERATOR CONSTRUCTION

Generators used on aircraft differ somewhat in design because they are made by various manufacturers. However, all are of the same general construction and operate similarly. [Figure 7-9]

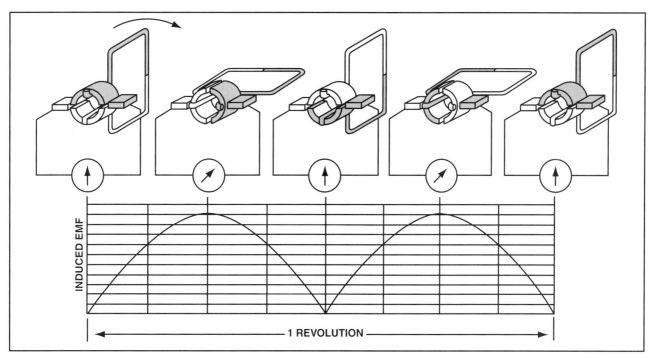

Figure 7-7. As the armature rotates in a DC generator, the commutator allows one brush to remain in contact with that portion of the loop that moves downward through the flux lines and the other brush to remain in contact with the portion of the loop that moves upward. This commutator action produces pulsating DC voltage that varies from zero to a maximum, twice in one revolution.

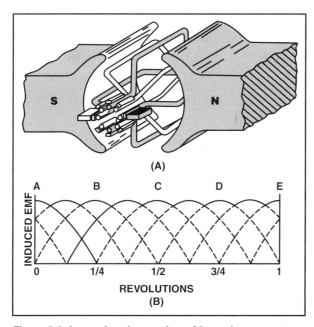

Figure 7-8. Increasing the number of loops in an armature reduces the magnitude of the ripple in DC voltage.

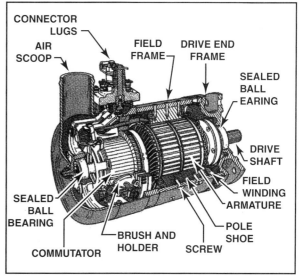

Figure 7-9. The major parts, or assemblies, of a DC generator include the field frame, rotating armature, commutator, and brush assembly.

FIELD FRAME

The field frame, or yoke, constitutes the foundation for the generator. The frame has two primary functions: 1) It completes the magnetic circuit between the poles, and 2) It acts as a mechanical support for the other parts. In small generators, the frame is made of one piece of iron; however, in larger generators, it is usually made up of two parts bolted together. The frame is highly permeable (magnetic lines of flux pass through it freely) and, together with the pole pieces, forms the majority of the magnetic circuit.

The magnetizing force inside a generator is produced by an electromagnet consisting of a wire coil called a field coil, and a core called a field pole, or shoe. The **pole shoes** are bolted to the inside of the frame and are usually laminated to reduce eddy current losses and concentrate the lines of force produced by the field coils. The frame and pole shoes are made from high quality magnetic iron or sheet steel. There is one north pole for each south pole, so there is always an even number of poles in a generator. [Figure 7-10]

The field coils are made up of many turns of insulated wire. The coils are wound on a form that is securely fastened over the iron core of the pole shoes. The current used to produce the magnetic field around the shoes is obtained from an external source or from the current generated by the unit itself. The magnetic field is created by the field coils, and there is no electrical connection between the windings of the field coils and the pole shoes. The pole shoes serve to concentrate the magnetic field produced by the field windings. [Figure 7-11]

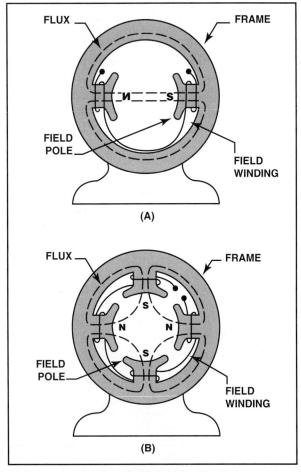

Figure 7-10. Generator field frames typically house either two or four pole shoes. These shoes are not permanent magnets, and rely on the field windings to produce a magnetic field. To try and produce the magnetic field necessary with only permanent magnets would greatly increase the physical size of the generator.

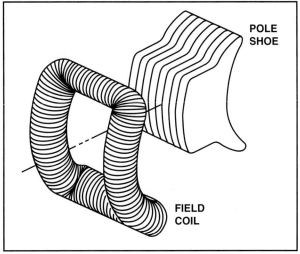

Figure 7-11. Field coils are form fitted around the pole shoes and are connected in such a manner that the north and south poles are in alternate polarity order.

ARMATURE

The armature assembly consists of the armature coils, the commutator and other associated mechanical parts. The armature is mounted on a shaft that rotates in bearings located in the generator's end frames. The core of the armature acts as a conductor when it is rotated in the magnetic field and it is laminated to prevent the circulation of eddy currents. [Figure 7-12]

A drum-type armature has coils placed in slots in the core of the armature. However, there is no electrical connection between the coils and core. The coils are usually held in the slots by wooden or fiber wedges. The coil ends are brought out to individual segments of the commutator.

COMMUTATORS

The commutator is located at one end of the armature and consists of wedge-shaped segments of hard-drawn copper. Each segment is insulated from the other by a thin sheet of mica. The segments are held in place by steel V-rings or clamping flanges

Note that the pole shoes project from the frame. The reason is that since air offers a great deal of resistance to a magnetic field, most generator designs reduce the width of air gap between the poles and the rotating armature. This increases the efficiency of the generator. When the pole pieces are made to project inward from the frame, they are called salient poles.

Airframe Electrical Systems

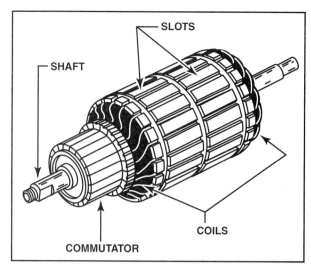

Figure 7-12. The armature rotates within the frame assembly, and current is induced into it by the electromagnetic field created by the field coils and pole shoes.

fitted with bolts. Rings of mica also insulate the segments from the flanges. The raised portion of each segment is called the riser, and the leads from the armature coils are soldered to each riser. In some generators, the segments have no risers. In this situation the leads are soldered to short slits in the ends of the segments. [Figure 7-13]

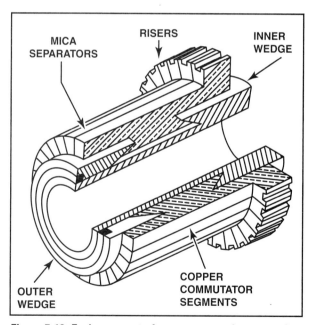

Figure 7-13. Each segment of a commutator is mounted to an inner wedge and separated by thin pieces of insulating mica. The risers on each segment hold the leads coming from the armature coils.

One end of a single armature coil attaches to one commutator segment, while the other end is soldered to the adjacent segment. In this configuration, each coil laps over the preceding one. This is known as lap winding. When an armature rotates at operational speed, the magnetic field that it produces lags behind the speed of rotation. Lap winding is a method for stabilizing the armature magnetic field. [Figure 7-14]

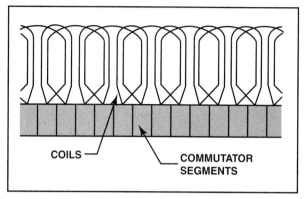

Figure 7-14. Lap winding connects one end of two coils to each commutator segment, and the other ends of these coils to adjacent segments.

BRUSHES

Brushes ride on the surface of the commutator and act as the electrical contact between armature coils and an external circuit. A flexible braided-copper conductor, called a pig-tail, connects each brush to the external circuit. The brushes are made of high-grade carbon and held in place by spring-loaded brush holders that are insulated from the frame. The brushes are free to slide up and down in their holders so they can follow any irregularities in the commutator's surface and allow for wear. A brush's position is typically adjustable so that the pressure on the commutator can be varied, and so the brush position with respect to the risers can be changed as necessary. Low spring tension can result in brush arcing. [Figure 7-15]

Figure 7-15. Carbon brushes connect to an external circuit through pig-tails. The brushes are typically adjustable to allow varied pressure on the commutator and position on the segments.

The constant making and breaking of connections to armature coils necessitate the use of a brush material that has a definite contact resistance. This material must also have low friction to prevent excessive wear. The high-grade carbon used to make brushes must be soft enough to prevent undue wear to the commutator, yet hard enough to provide reasonable brush life. The contact resistance of carbon is fairly high due to its molecular structure, and the commutator surface is highly polished to reduce friction as much as possible. Oil or grease must never be used on a commutator, and extreme care must be used when cleaning a commutator to avoid marring or scratching its surface.

TYPES OF DC GENERATORS

There are three types of DC generators: the series-wound, shunt-wound, and shunt-series, or compound-wound. The difference between each depends on how the field winding is connected to the external circuit.

SERIES-WOUND

The field winding of a series-wound generator is connected in series with the external load circuit. In this type of generator, the field coils are composed of a few turns of large wire because they must carry the full load current. Magnetic field strength in a series-wound generator is created more because of the large current flow than the number of turns in the coil.

Because of the way series-wound generators are constructed, they possess poor voltage regulation capabilities. For example, as the load voltage increases, the current through the field coils also increases. This induces a greater EMF that, in turn, increases the generator's output voltage. Therefore, when the load increases, voltage increases; likewise, when the load decreases, voltage decreases.

One way to control the output voltage of a series-wound generator is to install a rheostat in parallel with the field windings. This limits the amount of current that flows through the field coils, thereby limiting the voltage output. [Figure 7-16]

Since series-wound generators have such poor voltage regulation capabilities, they are not suitable for use in aircraft. However, they are suitable for situations where a constant RPM and constant load are applied to the generator.

SHUNT-WOUND

A generator having a field winding connected in parallel with the external circuit is called a shunt-wound

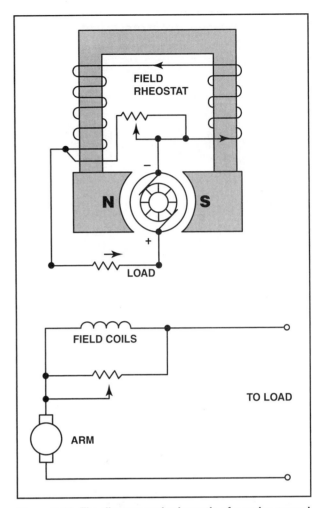

Figure 7-16. The diagram and schematic of a series-wound generator show that the field windings are connected in series with the external load. A field rheostat is connected in parallel with the field windings to control the amount of current flowing in the field coils.

generator. Unlike the field coils in a series-wound generator, the field coils in a shunt-wound generator contain many turns of small wire. This permits the field coil to derive its magnetic strength from the large number of turns, rather than from the amount of current flowing through the coils. [Figure 7-17]

In a shunt-wound generator, the armature and the load are connected in series; therefore, all the current flowing in the external circuit passes through the armature winding. However, due to resistance in the armature winding, some voltage is lost. The formula used to calculate this voltage drop is:

$$\text{IR drop} = \text{current} \times \text{armature resistance}$$

From this formula it can be seen that as the load, or current, increases, the IR drop in the armature also increases. Since the output voltage is the difference

Airframe Electrical Systems

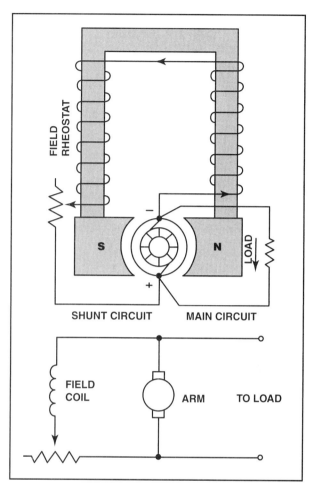

Figure 7-17. In a shunt-wound generator, the field windings are connected in parallel with the external load.

between induced voltage and voltage drop, there is a decrease in output voltage with an increased load. This decrease in output voltage causes a corresponding decrease in field strength because the current in the field coils decreases with a decrease in output voltage. By the same token, when the load decreases, the output voltage increases accordingly, and a larger current flows in the windings. This action is cumulative and, if allowed, the output voltage would rise to a point called field saturation. At this point there is no further increase in output voltage. Because of this, a shunt-wound generator is not desired for rapidly fluctuating loads.

To control the output voltage of a shunt generator, a rheostat is inserted in series with the field winding. In this configuration, as armature resistance increases, the rheostat reduces the field current, which decreases the output voltage. For a given setting on the field rheostat, the terminal voltage at the armature brushes is approximately equal to the generated voltage minus the IR drop produced by the armature resistance. However, this also means that the output voltage at the terminals drops when a larger load is applied. Certain voltage-sensitive devices are available which automatically adjust the field rheostat to compensate for variations in load. When these devices are used, the terminal voltage remains essentially constant.

The output and voltage-regulation capabilities of shunt-type generators make them suitable for light to medium duty use on aircraft. However, DC alternators generally have replaced most of these units.

COMPOUND-WOUND

A compound-wound generator combines a series winding and a shunt winding so that characteristics of each are used. The series field coils consist of a relatively small number of turns made of large copper conductor, either circular or rectangular in cross-section. As discussed earlier, series field coils are connected in series with the armature circuit. These coils are mounted on the same poles as the shunt field coils and, therefore, contribute to the magnetizing force, or magnetomotive force, which influences the generator's main field flux. [Figure 7-18]

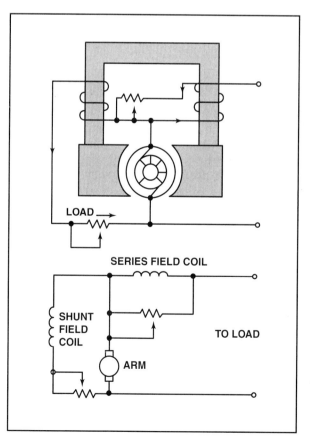

Figure 7-18. Compound-wound generators utilize both series and shunt windings.

If the ampere-turns of the series field act in the same direction as those of the shunt field, the combined magnetomotive force is equal to the sum of the series and shunt field components. Load is added to a compound-wound generator in the same manner as a shunt-wound generator: by increasing the number of parallel paths across the generator. When this is done, the total load resistance decreases, causing an increase in armature-circuit and series-field circuit current. Therefore, by adding a series field, the field flux increases with an increased load. Thus, the output voltage of the generator increases or decreases with load, depending on the influence of the series field coils. This influence is referred to as the degree of compounding.

The amount of output voltage produced by a compound-wound generator depends on the degree of compounding. For example, a flat-compound generator is one in which the no-load and full-load voltages have the same value. However, an under-compound generator has a full-load voltage less than the no-load voltage, and an over-compound generator has a full-load voltage higher than the no-load voltage.

Generators are typically designed to be over-compounded. This feature permits varied degrees of compounding by connecting a variable shunt across the series field. Such a shunt is sometimes called a diverter. Compound generators are used where voltage regulation is of prime importance.

If, in a compound-wound generator, the series field aids the shunt field, the generator is said to be cumulative-compounded. However, if the series field opposes the shunt field, the generator is said to be differentially-compounded. [Figure 7-19]

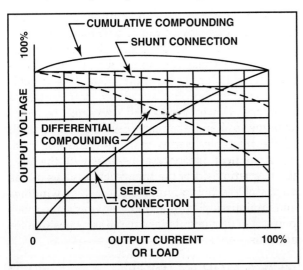

Figure 7-19. The characteristics of shunt-wound, series-wound, and compounded generators can be compared on this chart.

If the shunt field of a compound-wound generator is connected across both the armature and the series field, it is known as a long-shunt connection. However, if the shunt field is connected across the armature alone, it is called a short-shunt connection. These connections produce essentially the same generator characteristics.

STARTER GENERATORS

Many small turbine engines are equipped with starter generators rather than separate starters and generators. This saves appreciably in weight, as both starters and generators are very heavy. A typical starter generator consists of at least two sets of windings and one armature winding. When acting as a starter, a high current flows through both sets of field windings and the armature to produce the torque required to start the engine. However, in the generator mode, only the high resistance shunt-winding receives current, while the series-winding receives no current. The current flowing through the shunt-winding is necessary to produce the magnetic field that induces voltage into the armature. Once power is produced, it flows to the primary bus.

ARMATURE REACTION

Any time current flows through a conductor, a magnetic field is produced. Therefore, when current flows through an armature, electromagnetic fields are produced in the windings. These fields tend to distort or bend the lines of magnetic flux between the poles of the generator. This distortion is called armature reaction. Since the current flowing through the armature increases as the load increases, the distortion becomes greater with larger loads. [Figure 7-20]

Armature windings of a generator are spaced so that during rotation there are certain positions when the brushes contact two adjacent segments on the commutator, thereby shorting the armature windings. When the magnetic field is not distorted, there is no voltage induced in the shorted windings and no harmful results occur. However, when the field is distorted by armature reaction, a voltage is induced in the shorted windings, and sparking takes place between the brushes and the commutator segments. Consequently, the commutator becomes pitted, the wear on the brushes becomes excessive, and the output of the generator is reduced.

To correct this condition, the brushes are set so that the plane of the coils being shorted is perpendicular to the distorted magnetic field. This is accomplished by moving the brushes forward in the direction of rotation. This operation is called shifting the

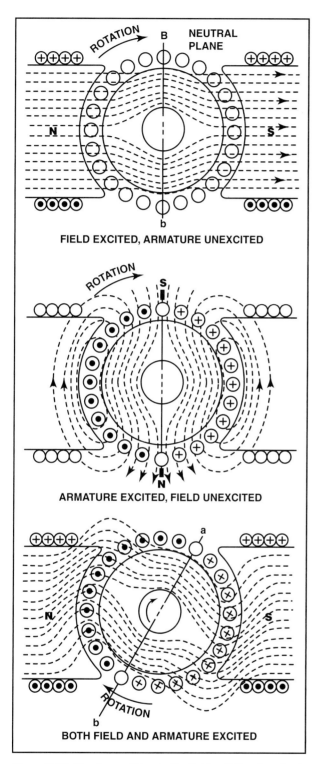

Figure 7-20. The lines of flux in the field coil flow in a horizontal path from north to south and induce voltage into the armature. However, as this is done, magnetic fields are produced in the armature, and tend to distort or bend the lines of flux produced by the field coil.

brushes to the neutral plane, or plane of commutation. The neutral plane is the position where the plane of the two opposite coils is perpendicular to the magnetic field in the generator. On a few generators, the brushes are shifted manually ahead of the normal neutral plane to the neutral plane caused by field distortion. On nonadjustable brush generators, the manufacturer sets the brushes for minimum sparking.

In some generators, special field poles, called **interpoles**, are used to counteract some of the effects of field distortion when the speed and load of the generator are changing constantly. An interpole is another field pole that is placed between the main poles. [Figure 7-21]

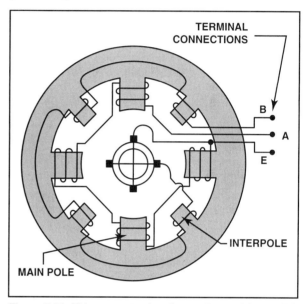

Figure 7-21. This generator has four poles and four interpoles. The interpoles are used to counter the effect of armature reaction.

An interpole has the same polarity as the next main pole in the direction of rotation. The magnetic flux produced by an interpole causes the current in the armature to change direction as the armature winding rotates under the interpole's field. This cancels the electromagnetic fields produced by the armature windings. The interpoles are connected in series with the load and, therefore, the magnetic strength of the interpoles varies with the generator load. Since the field distortion also varies with the load, the magnetic field of the interpoles counteracts the effects of the field around the armature windings and minimizes distortion. In other words, the interpoles keep the neutral plane in the same position for all loads.

GENERATOR RATINGS

A generator is rated according to its power output. Since a generator is designed to operate at a specified voltage, the rating is usually given as the number of amperes the generator can safely supply at its rated voltage. For example, a typical generator rating is 300 amps at 28.5 volts. A generator's rating and performance data are stamped on the name plate attached to the generator. When replacing a generator make sure it is the proper rating.

The rotation of generators is termed as either clockwise or counterclockwise, as viewed from the driven end. If no direction is stamped on the data plate, the rotation is marked by an arrow on the cover plate of the brush housing. To maintain the correct polarity, it is important to use a generator with the correct direction of rotation.

The speed of an aircraft engine varies from idle rpm to takeoff rpm; however, the majority of flight, is conducted at a constant cruising speed. The generator drive on a reciprocating engine is usually geared between 1-1/8 and 1-1/2 times the engine crankshaft speed. Most aircraft generators have a speed at which they begin to produce their normal voltage. This is termed the coming-in speed, and is typically around 1,500 rpm.

GENERATOR TERMINALS

On large 24-volt generators, electrical connections are made to terminals marked B, A, and E. The positive armature lead connects to the B terminal, the negative armature lead connects to the E terminal, and the positive end of the shunt field winding connects to terminal A. The negative end of the shunt field winding is connected to the negative terminal brush. Terminal A receives current from the negative generator brush through the shunt field winding. This current passes through the voltage regulator and back to the armature through the positive brush. Load current, which leaves the armature through the negative brush, comes out of the E lead and passes through the load before returning through the positive brush.

GENERATOR VOLTAGE REGULATION

Efficient operation of electrical equipment in an aircraft depends on a voltage supply that varies with a system's load requirements. Among the factors that determine the voltage output of a generator, the strength of the field current is the only one that is conveniently controlled.

One way to control the field current is to install a rheostat in the field coil circuit. When the rheostat is set to increase the resistance in the field circuit, less current flows through the field coils and the strength of the magnetic field decreases.

Consequently, less voltage is induced into the armature, and generator output decreases. When the resistance in the field circuit is decreased with the rheostat, more current flows through the field coils, and the magnetic field becomes stronger. This allows more voltage to be induced into the armature, which produces a greater output voltage. [Figure 7-22]

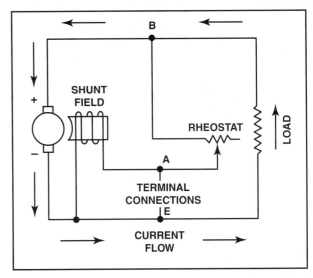

Figure 7-22. When generator voltage is regulated by field rheostat, more resistance results in less output voltage while less resistance results in more output voltage.

One thing to keep in mind: The weaker the magnetic field is, the easier it is to turn the armature. On the other hand, if the strength of the magnetic field is increased, more force is required to turn the armature. This means when the load on a generator increases, additional field current must be supplied to increase the voltage output as well as overcome the additional force required to turn the armature.

This principle is further developed by the addition of a solenoid which electrically connects or removes the field rheostat from the circuit as the voltage varies. This type of setup is found in a vibrating-type voltage regulator. [Figure 7-23]

When the output voltage rises above a specified critical value, the downward pull of the solenoid's coil exceeds the spring tension, and contact B opens. This reinserts the field rheostat in the field circuit. The additional resistance reduces the field current and lowers output voltage. When the output voltage falls below a certain value, contact B closes, shorting the field rheostat, and the terminal voltage starts to rise. Thus, an average voltage is maintained with or without load changes. The dashpot P provides

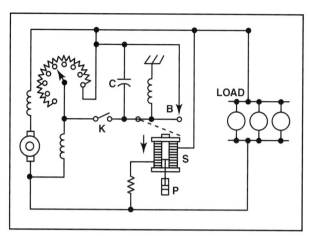

Figure 7-23. With the generator running at normal speed and switch K open, the field rheostat is adjusted so that the output voltage is about 60 percent of normal. At this level, solenoid S is weak and contact B is held closed by the spring. However, when K is closed, a short circuit is placed across the field rheostat. This action causes the field current to increase and the output voltage to rise.

smoother operation by acting as a dampener to prevent hunting, and capacitor C across contact B helps eliminate sparking.

With a vibrating-type voltage regulator, contact B opens and closes several times per second to maintain the correct generator output. Based on this, if the solenoid should malfunction or the contacts stick closed, excess current would flow to the field, and generator output would increase.

Certain light aircraft employ a **three-unit regulator** for their generator systems. This type of regulator includes a current limiter, a reverse current cutout, and a voltage regulator. [Figure 7-24]

The action of the voltage regulator unit is similar to the vibrating-type regulator described earlier. The current limiter is the second of three units, and it limits the generator's output current. The third unit is a reverse-current cutout, which disconnects the battery from the generator when the generator output is lower than the battery output. If the battery were not disconnected, it would discharge through the generator armature when the generator voltage falls below that of the battery. When this occurs, the battery attempts to drive the generator as a motor. This action is called motoring the generator and, unless prevented, the battery discharges in a short time.

Since contacts have a tendency to pit or burn when large amounts of current flow through them, vibrating-type regulators and three-unit regulators cannot be used with generators that require a high field

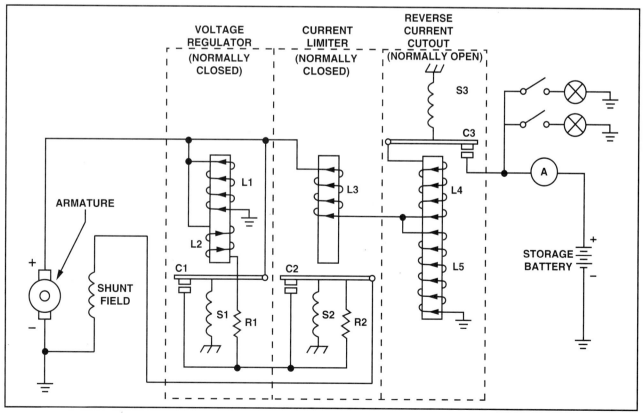

Figure 7-24. A voltage regulator contains three coils: a voltage regulator coil, a current limiter coil, and a reverse current cutout coil.

current. Therefore, heavy-duty generator systems require a different type of regulator, such as the carbon-pile voltage regulator. The carbon-pile voltage regulator relies on the resistance of carbon disks arranged in a pile or stack. The resistance of the carbon stack varies inversely with the pressure applied. For example, when the stack is compressed, less air exists between the carbon disks, and the resistance decreases. However, when the pressure is reduced, more air is allowed between the disks causing the resistance to increase.

Pressure on the carbon pile is created by two opposing forces: a spring and an electromagnet. The spring compresses the carbon pile, and the electromagnet exerts a pull on the spring that decreases the pressure. [Figure 7-25]

Whenever the generator voltage varies, the pull of the electromagnet varies. If the generator voltage rises above a specific amount, the pull of the electromagnet increases, thereby decreasing the pressure exerted on the carbon pile and increasing its resistance. Since this resistance is in series with the field, less current flows through the field winding and there is a corresponding decrease in field strength. This results in a drop in generator output.

On the other hand, if the generator output drops below a specified value, the pull of the electromagnet decreases and the carbon pile places less resistance in the field winding circuit. This results in an increase in field strength and a corresponding increase in generator output. A small rheostat provides a means of adjusting the current flow through the electromagnet coil.

DC GENERATOR SERVICE AND MAINTENANCE

Because of their relative simplicity and durable construction, generators operate many hours without trouble. The routine inspection and service done at each 100-hour or annual inspection interval is generally all that is required to keep a generator in good working order. Generator overhaul is often accomplished at the same time as engine overhaul. This minimizes aircraft down time and increases the likelihood of trouble-free operation when the aircraft is placed back in service.

ROUTINE INSPECTION AND SERVICING

The 100-hour and annual inspection of a generator should include
the following items:

1. Inspect the generator for security of mounting; check the mounting flange for cracks and loose mounting bolts.

2. Inspect the mounting flange area for oil leaks.

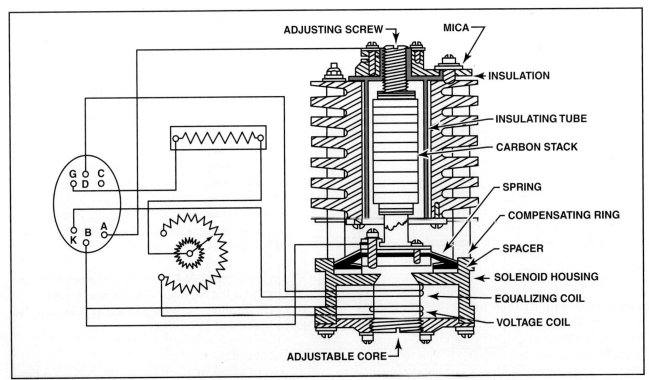

Figure 7-25. A carbon pile voltage regulator relies on the amount of air space within a stack of carbon disks to control generator voltage. Pressure is maintained on the disks by a spring while an electromagnet controls spring tension.

Airframe Electrical Systems

3. Inspect the generator electrical connections for cleanliness and security of attachment.

4. Remove the band covering the brushes and commutator. Use compressed air to blow out accumulated dust. Inspect the brushes for wear, and freedom of movement. Check tension of the brush springs, using a spring scale.

5. Inspect commutator for cleanliness, wear, and pitting.

6. Inspect the area around the commutator, and brush assemblies for any solder particles. The presence of solder indicates that the generator has overheated and melted the solder attaching the armature coils to the risers. When this happens, an open is created in the armature.

If a DC generator is unable to keep an aircraft battery charged, and if the ammeter does not show the proper rate of charge, first check the aircraft electrical system associated with the battery and generator. Physically check every connection in the generator and battery circuit and electrically check the condition of all fuses and circuit breakers. Check the condition of all ground connections for the battery, battery contacts, and the generator control units. When it has been determined that there are no obvious external problems and the generator armature turns when the engine is cranked, check the generator and the voltage regulator.

One of the easiest ways to determine which unit is not operating is to connect a voltmeter between the G terminal of the voltage regulator and ground. This checks the generator's output voltage. However, because this check requires the generator to be turning, it must be accomplished with the engine running, or on an appropriate test stand. In either case, observe proper safety precautions. Even when the field winding is open, or the voltage regulator is malfunctioning, the generator should produce residual voltage. In other words, the action of the armature cutting across the residual magnetic field in the generator frame and field shoes, produces voltage. This should be around one or two volts.

If there is no residual voltage, it is possible that the generator needs only the residual magnetism restored. Residual magnetism is restored by an operation known as flashing the field. This is accomplished by momentarily passing current through the field coils in the same direction that it normally flows. The methods vary with the internal connections of the generator, and with the type of voltage regulator used.

For example, in an "A" circuit generator the field is grounded externally, and the positive terminal of the battery must be touched to the armature. It also may be necessary to insulate one of the brushes by inserting a piece of insulating material between the brush and commutator. To flash the field in an internally grounded "B" circuit generator, the positive battery terminal is touched to the field. Whatever the case, be certain to follow the specific manufacturer instructions. Failure to do so could result in damage to the generator and/or voltage regulator.

If the generator produces residual voltage and no output voltage, the trouble could be with the generator or the regulator. To determine which, operate the engine at a speed high enough for the generator to produce an output, and bypass the voltage regulator with a jumper wire. This method varies with the type of generator and regulator being used, and should be performed in accordance with the manufacturer's recommendations.

If the generator produces voltage with the regulator shorted, the problem is with the voltage regulator. If this is the case, be sure that the regulator is properly grounded, because a faulty ground connection prevents a regulator from functioning properly. It is possible to service and adjust some vibrator-type generator controls. However, due to expense, time involved, and test equipment needed to do the job properly, most servicing is done by replacing a faulty unit with a new one. If the generator does not produce an output voltage when the regulator is bypassed, remove the generator from the engine and overhaul or replace it with a serviceable unit.

GENERATOR OVERHAUL

Generator overhaul is accomplished any time a generator is determined to be inoperative, or at the same time the aircraft engine is overhauled. Although an overhaul can be done in some aircraft repair facilities, it is more often the job of an FAA Certified Repair Station licensed for that operation.

The steps involved in the overhaul of a generator are the same for the overhaul of any unit: (1) disassembly, (2) cleaning, (3) inspection and repair, (4) reassembly, and (5) testing.

DISASSEMBLY

Disassembly instructions for specific units are covered in the manufacturer's overhaul manual and must be followed exactly. Specialized tools are sometimes required for removing pole shoes since the screws holding these in place are usually staked to prevent them from accidentally backing out. Special instructions must also be followed when removing bearings. If the incorrect procedures or tools are used, damage to the bearings or their seating area could result.

CLEANING

Care must be taken when cleaning electrical parts. The proper solvents must be used, and, in general, parts are not submerged in solvent tanks. Using the wrong solvent could remove the lacquer-type insulation used on field coils and armatures, resulting in short circuits after the generator is reassembled.

INSPECTION AND REPAIR

Inspect components for physical damage, corrosion or wear, and repair or replace as required. Testing for proper operation of electrical components is accomplished using a growler and an electrical multimeter. A **growler** is a specially designed test unit for DC generators and motors and a variety of tests on the armature and field coils is performed using this equipment.

Growlers consist of a laminated core wound with many turns of wire that are connected to 110 volts AC. The top of the core forms a V into which the armature of a DC generator fits. The coil and laminated core of the growler form the primary of a transformer, while the generator armature becomes the secondary. Also included on most growlers is a 110-volt test lamp. This is a simple series circuit with a light bulb that illuminates when the circuit is complete. [Figure 7-26]

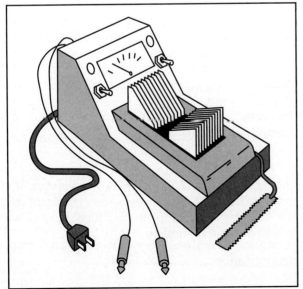

Figure 7-26. A growler is used to test the armature of a DC generator for open circuits.

To test an armature for an open circuit, place it on an energized growler. Using the probes attached to the test lamp, test each armature coil by placing the probes on adjacent segments. The lamp should light with each set of commutator bars. Failure of the test lamp to illuminate indicates an open circuit in that coil, and replacement of the armature is called for.

Armatures are also tested for shorts by placing them on a growler, energizing the unit and holding a thin steel strip, typically a hacksaw blade, slightly above the armature. Slowly rotate the armature on the growler. If there are any shorts in the armature windings, the blade will vibrate vigorously. [Figure 7-27]

Figure 7-27. When a short exists in the armature windings, the hacksaw blade vibrates vigorously.

A third test for armature shorts is accomplished using a 110 volt test lamp to check for grounds. To use a test lamp, one lead is touched to the armature shaft while the second lead is touched to each commutator segment. If a ground exists between any of the windings and the core of the armature, the test lamp illuminates. [Figure 7-28]

A generator's field coil can be tested for shorts by also using a test lamp. To do this, one probe is placed at the field winding and the other probe at the generator frame. If the light illuminates, a short exists. [Figure 7-29]

The field coil can be tested for continuity by using an ohmmeter set to the low-ohms scale. A shunt field coil should indicate between 2 and 30 ohms, depending on the specific coil. A series type field coil shows almost no resistance. In some cases, a current draw test is specified by the manufacturer. This test is accomplished by connecting a battery of proper voltage across the field coils and measuring

Airframe Electrical Systems

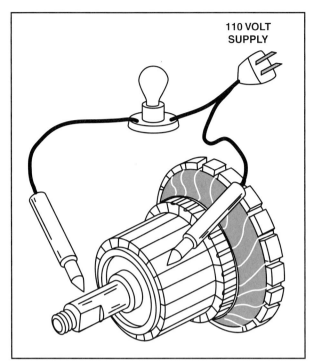

Figure 7-28. When a ground exists between the windings and the armature core, the test light illuminates. This test may also be accomplished using an ohmmeter.

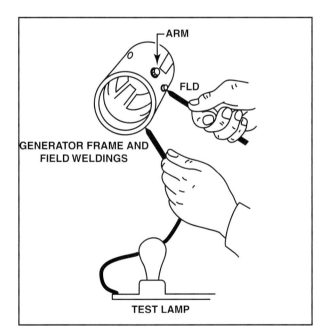

Figure 7-29. In order for the frame and field windings to be considered good, the light should not illuminate.

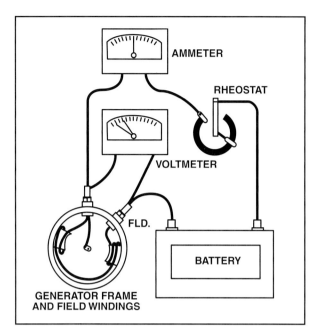

Figure 7-30. To test for shorted turns in a field coil, measure the current drawn by the field at a specific voltage. The current drawn by the field must be within the manufacturer's prescribed range in order for the field windings to be good.

the current flow. This value must be within the limits specified in the manufacturer's test specifications. [Figure 7-30]

To resurface a commutator to remove irregularities or pitting, an armature is turned in a special armature lathe, or an engine lathe equipped with a special holding fixture. When doing this, remove only enough metal to smooth the commutator's surface. If too much material is removed, the security of the coil ends is jeopardized. If the commutator is only slightly roughened, it is smoothed using No. 000 sandpaper. Never use emery cloth or other conductive material, since shorting between commutator segments could result.

With some commutators, the mica insulation between the commutator bars may need to be undercut. However, when doing this, follow the most recent instructions provided by the manufacturer. This operation is accomplished using a special attachment for the armature lathe, or a hack-saw blade. When specified, the mica is undercut about the same depth as the mica's width, or approximately .020 inch.

REASSEMBLY

Prior to reassembly, the painted finish on the exterior of the frame is restored. In certain cases the special insulated coatings on the interior surfaces are renewed. Furthermore, all defective parts are replaced in accordance with the reassembly procedure specified by the manufacturer.

When reassembling a generator, make certain that all internal electrical connections are properly made and secured, and that the brushes are free to move in their holders. Check the pigtails on the brushes for freedom, and make sure they do not alter or restrict the brushes' free motion. The purpose of the pigtail is to conduct current and help eliminate any current in the brush springs that could alter its spring action. The pigtails also eliminate possible sparking caused by movement of the brush within the holder, thus minimizing brush side wear.

Generator brushes are normally replaced at overhaul, or when-half worn. When new brushes are installed they must be seated, or contoured, to maximize the contact area between the face of the brush and the commutator. Seating is accomplished by lifting the brush slightly to permit the insertion of No. 000, or finer sandpaper, rough side out. With the sandpaper in place, pull the sandpaper in the direction of armature rotation, being careful to keep the ends of the sandpaper as close to the commutator as possible to avoid rounding the edges of the brush. When pulling the sandpaper back to the starting point, the brush is raised so it does not ride on the sandpaper. [Figure 7-31]

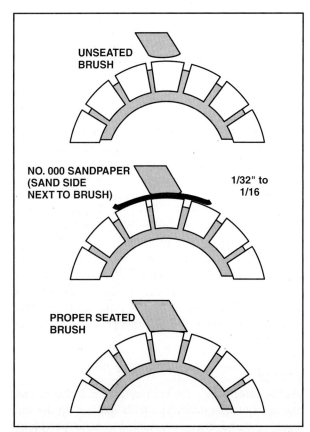

Figure 7-31. When a new brush is installed in a generator, the brush must be contoured to fit the commutator. To do this, insert a piece of No. 000 sandpaper between the brush and the commutator and sand the brush in the direction of rotation.

The brush spring tension is checked using a spring scale. A carbon, graphite, or light metal brush should exert a pressure of only 1-1/2 to 2-1/2 pounds on the commutator. If the spring tension is not within the limits set by the manufacturer, the springs must be replaced. When a spring scale is used, the pressure measurement exerted by a brush is read directly on the spring scale. The scale is attached at the point of contact between the spring arm and the top of the brush, with the brush installed in the guide. The scale is then pulled up until the arm just lifts the brush off the commutator. At that instant, the force on the scale is read.

After the generator has run for a short period, the brushes should be reinspected to ensure that no pieces of sand are embedded in the brush. Under no circumstance should emery cloth or similar abrasive be used for seating brushes or smoothing commutators, since they contain conductive materials.

TESTING
Operational testing of generators is accomplished on test benches built for that purpose. Bench testing allows the technician the opportunity to flash the field, and ensure proper operation of the unit before installation. Generator manufacturers supply test specifications in their overhaul instructions that should be followed exactly.

GENERATOR SYSTEMS
When installed on most aircraft, the output of a generator typically flows to the aircraft's bus bar where it is distributed to the various electrical components. In this type of system, the allowable voltage drop in the main power wires coming off the generator to the bus bar is 2 percent of the regulated voltage when the generator is producing its rated current. As added insurance to make sure that a given electrical load does not exceed a generator's output capability, the total continuous electrical load permitted in a given system is limited to 80 percent of the total rated generator output. For example, if an aircraft has a 60-amp generator installed, the maximum continuous load that can be placed on the electrical system is 48 amps.

In the event a generator quits producing current or produces too much current, most aircraft systems have a generator master switch that allows the generator to be disconnected from the electrical system. This feature helps prevent damage to the generator or to the rest of the electrical system. On aircraft that utilize more than one generator connected to a common electrical system, the Federal Aviation

Regulations require individual generator switches that can be operated from the cockpit. A malfunctioning generator can therefore be disconnected to protect the remaining electrical system.

ALTERNATORS

There are two types of alternators used in today's aircraft: the DC alternator and the AC alternator. DC alternators produce relatively small amounts of current and, therefore, are typically found on light aircraft. AC alternators, on the other hand, are capable of producing a great deal of power and, therefore, are typically found on larger aircraft and military aircraft. Furthermore, since AC electricity can be carried through smaller conductors, AC alternators allow appreciable weight savings.

DC ALTERNATORS

DC alternators have the same function as DC generators. They produce AC that is then converted to DC before it enters an aircraft's electrical system. However, the difference is that, in an alternator the magnetic poles rotate and induce voltage into a fixed, or stationary winding. Furthermore, the AC current produced is rectified by six solid-state diodes instead of a commutator. [Figure 7-32]

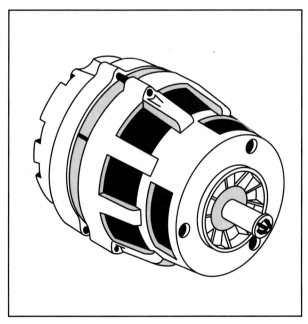

Figure 7-32. DC aternators are used in aircaft that require a low or medium amount of electrical power.

All alternators are constructed in basically the same way. The primary components of an alternator include the rotor, the stator, the rectifier, and the brush assembly.

ROTOR

An alternator rotor consists of a wire coil wound on an iron spool between two heavy iron segments with interlacing fingers. Some rotors have four fingers while others have as many as seven. Each finger forms one pole of the rotating magnetic field. [Figure 7-33]

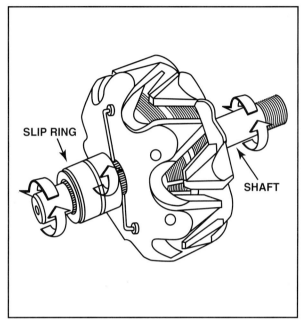

Figure 7-33. Rotors consist of interlacing fingers that form the poles of the rotating magnetic field.

The two coil leads pass through one segment and each lead attaches to an insulated slip ring. The slip rings, segments, and coil spool are all pressed onto a hardened steel rotor shaft which is either splined, or has a key slot to secure the rotor to the shaft. In an assembled alternator, this shaft is driven by an engine accessory pad, or fitted with a pulley and driven by an accessory belt. The slip-ring end of the shaft is supported in the housing with a needle bearing, and the drive end with a ball bearing. Two carbon brushes ride on the smooth slip rings to bring a varying direct current into the field from the DC exciter and the voltage regulator.

STATOR

As the rotor turns, the load current is induced into stationary stator coils. The coils making up the stator are wound in slots around the inside periphery of the stator frame, which is made of thin laminations of soft iron. Most alternators are three-phase alternators. This means that the stator has three separate coils that are 120 degrees apart. To do this, one end of each coil is brought together to form a common junction of a Y-connection. [Figure 7-34]

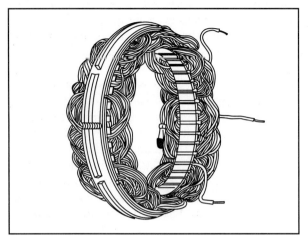

Figure 7-34. Three sets of coils in a stater are typically brought together to form a Y-connection.

With the stator wound in a three-phase configuration, the output current peaks in each set of windings every 120 degrees of rotation. However, after the output is rectified, the DC output becomes much smoother. [Figure 7-35]

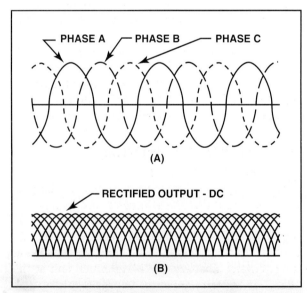

Figure 7-35. (A) The waveform produced by a three-phase stator winding results in a peak every 120 degrees of rotation. (B) Once rectified, the output produces a relatively smooth direct current, with a low-amplitude, high-frequency ripple.

Because an alternator has several field poles and the large number of stator windings, most alternators produce their rated output at a relatively low rpm. This differs from a generator that must rotate at a fairly high speed to produce its rated output.

RECTIFIERS

The three-phase, full-wave rectifier in an alternator is made up of six heavy-duty silicon diodes. Three of the diodes are pressed into the slip-ring end frame, and the other three are pressed into a heat sink that is electrically insulated from the end frame. [Figure 7-36]

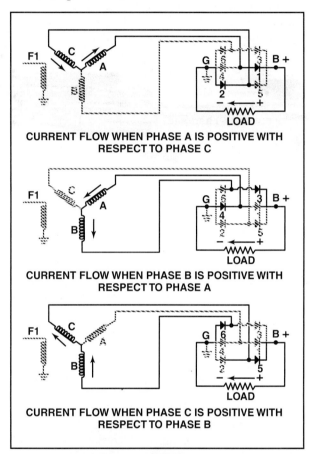

Figure 7-36. This circuit diagram illustrates how the AC produced in each winding is rectified into DC.

Figure 7-36 shows that at the instant the output terminal of winding "A" is positive with respect to the output end of winding "C," current flows through diode 1 to the load, and back through diode 2, which is pressed into the alternator end frame. From this diode, it flows back through winding "C."

As the rotor continues to turn, winding "B" becomes positive with respect to winding "A," and the current flows through diode 3 to the load, and then back through diode 4 and winding "A."

When the rotor completes 240 degrees of rotation, "C" becomes positive with respect to "B." When this occurs, current flows through diode 5 to the load, and back through diode 6. After 360 degrees of rotation, the process begins again.

BRUSH ASSEMBLY

The brush assembly in an alternator consists of two brushes, two brush springs, and brush holders. Unlike a generator that uses brushes to supply a path for current to flow from the armature to the load, the brushes in an alternator supply current to the field coils. Since these brushes ride on the smooth surface of the slip rings, the efficiency and service life of alternator brushes is typically better than on DC generators.

ALTERNATOR CONTROLS

The voltage produced by an alternator is controlled in the same way as in a generator-by varying the DC field current. Therefore, when the output voltage rises above the desired value, the field current is decreased. By the same token, when the output voltage drops below the desired value, the field current is increased.

The process of increasing and decreasing the field current could be accomplished in low-output alternators with vibrator-type controls that interrupt the field current by opening a set of contacts. However, a more efficient means of voltage control has been devised that uses a transistor to control the flow of field current.

The transistorized voltage regulator utilizes both vibrating points and transistors for voltage control. The vibrating points operate the same as they do in vibrator-type voltage regulators. However, instead of the field current flowing through the contacts, the transistor base current flows through them. Since this current is small compared to the field current that flows through the emitter-collector, there is no arcing at the contacts. [Figure 7-37]

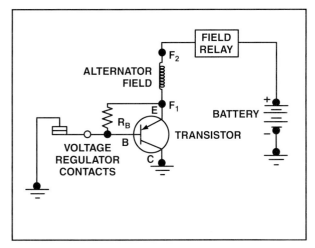

Figure 7-37. Alternator field current flows from the emitter to the collector only when the voltage regulator contacts are closed and current flows to the base.

In a completely solid-state voltage regulator, semiconductor devices replace all of the moving parts. These units are very efficient, reliable, and generally have no serviceable components. Therefore, if a completely solid-state unit becomes defective, it is typically removed from service and replaced.

Alternator control requirements are different from those of a generator for several reasons. For example, since an alternator uses solid-state diodes for rectification, current cannot flow from the battery into the alternator. Therefore, there is no need for a reverse-current cutout relay. Furthermore, since the alternator field is excited by the system bus with a limited voltage, there is no way an alternator can yield enough current to burn itself out. Because of this, there is no need for a current limiter.

With an alternator there must be some means of shutting off the flow of field current when the alternator is not producing power. To do this, most systems utilize either a field switch or a field relay that is controlled by the master switch. Either setup allows isolation of the field current and shuts it off if necessary.

Another control that most aircraft alternator circuits employ is some form of overvoltage protection. This allows the alternator to be removed from the bus should a malfunction occur that increases the output voltage to a dangerous level. This function is often handled by an **alternator control unit**, or ACU. Basically, the ACU drops the alternator from the circuit when an overvoltage condition exists.

DC ALTERNATOR SERVICE AND MAINTENANCE

When an alternator fails to keep the battery charged, first determine that the alternator and battery circuits are properly connected. This includes checking for open fuses or circuit breakers. If everything is connected properly, check for battery voltage at the alternator's battery, or "B" terminal, and at the "Batt," or "+" terminal of the voltage regulator.

There are basically two problems that prevent an alternator from producing electrical power. The most likely is a shorted or open diode in the rectifying circuit. The other problem is the possibility of an open circuit in the field.

To check for a shorted circuit, measure the resistance between the alternator's "B" terminal and ground. To accomplish this, set the ohmmeter on the R × 1 scale and measure resistance. Then, reverse the ohmmeter leads and measure the resistance again. If the diodes are good, there will be a relatively low resistance reading when the diodes are forward biased, and an infinite or very high reading when the diodes are reverse biased. If an infinite or very high reading is not obtained, one or more of the diodes are shorted. [Figure 7-38]

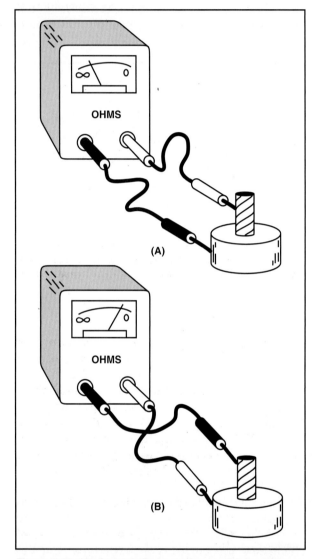

Figure 7-38. (A) A good diode produces a high resistance reading when reverse biased. (B) On the other hand, resistance should be low when forward biased. A shorted diode typically causes a low resistance reading when checked in the forward and reverse biased directions, while an open diode produces high resistance readings in both directions.

Since the diodes in an alternator are connected in parallel, an open diode cannot be detected with an ohmmeter. However, an ohmmeter will identify an open diode, if the diodes are checked individually.

Solid-state diodes are quite rugged and have a long life when properly used. However, they can be damaged by excessive voltage or reverse current flow. For this reason they should never be operated without being connected to an electrical load. Since alternators receive their field current from the aircraft bus and do not rely on residual magnetism to be started, never flash the field or polarize an alternator.

To aid in systematic alternator troubleshooting, some manufacturers have specialized test equipment. This test equipment is usually plugged into the aircraft electrical system between the voltage regulator and the aircraft bus. Through the use of indicator lights, the test equipment tells whether a problem exists in the voltage regulator, the overvoltage sensing circuit, or the alternator field/output circuit. By using this type of test equipment, time can be saved and the unnecessary replacement of good components can be avoided.

To avoid burning out the rectifying diodes during installation, it is extremely important that the battery be connected with the proper polarity. In addition, any time an external power source is connected to the aircraft, ensure correct polarity is applied.

AC ALTERNATORS

Direct current is used as the main electrical power for small aircraft because it is storable and aircraft engines are started through the use of battery power. However, large aircraft require elaborate ground service facilities and external power sources for starting. Therefore, they can take advantage of the appreciable weight savings provided by using alternating current as their primary power source.

In addition to saving weight, alternating current has the advantage over direct current in that its voltage is easily stepped up or down. Therefore, when needed, AC carries current a long distance by raising the voltage through use of a step-up transformer. This promotes additional weight savings since high voltage AC is conducted through relatively small wires. Once the voltage arrives at its destination, it passes through a step-down transformer where voltage is lowered and current is stepped up to the value needed.

In some situations, such as charging batteries or operating variable speed motors, direct current is required. However, by passing AC through a series of semiconductor diodes it is easily changed into DC with relatively little loss. This is another advantage of AC as compared to DC.

TYPES OF AC ALTERNATORS

AC alternators are classified in order to distinguish differences. One means of classification is by the number of output voltage phases. Alternating current alternators can be single-phase, two-phase, three-phase, and sometimes even six-phase or more. However, almost all aircraft electrical systems use three-phase alternators.

In a single-phase alternator, the stator is made up of several windings connected in series to form a single circuit. The windings are also connected so the AC voltages induced into each winding are in phase. This means that, to determine a single-phase alternator's total output, the voltage induced into each winding must be added. Therefore, the total voltage produced by a stator with four windings is four times the single voltage in any one winding. However, since the power delivered by a single-phase circuit is pulsating, this type of circuit is impractical for many applications. [Figure 7-39]

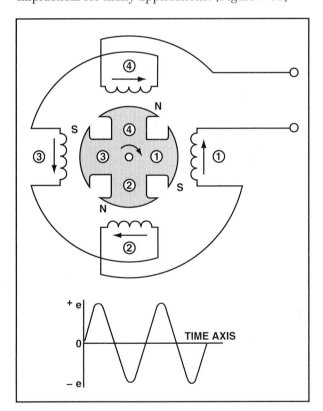

Figure 7-39. In this single-phase alternator the rotating field induces voltage into the stationary stator. The number of stator windings determines the output voltage.

Two-phase alternators have two or more single-phase windings spaced symmetrically around the stator, so that the AC voltage induced in one is 90 degrees out of phase with the voltage induced in the other. These windings are electrically separate from each other so that when one winding is cutting the maximum number of flux-lines, the other is cutting no flux lines.

A three-phase, or polyphase, circuit is used in most aircraft alternators. The three-phase alternator has three single-phase windings spaced so that the voltage induced in each winding is 120 degrees out of phase with the voltage in the other two windings. [Figure 7-40]

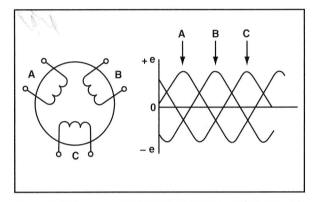

Figure 7-40 Sinewave "A" is 120 degrees out of phase with sine wave "B," and sinewave "B" is 120 degrees out of phase with sine wave "C."

The three individual phase voltages produced by a three-phase alternator are similar to those generated by three single-phase alternators, whose voltages are out of phase by 120 degrees. In a three-phase alternator, one lead from each winding is connected to form a common junction. When this is done, the stator is Y- or star-connected. A three-phase stator can also be connected so that the phases are end-to-end. This arrangement is called a delta connection.

Still another means of classifying alternators is to distinguish between the type of stator and rotor used. When done this way, there are two types of alternators: the revolving-armature type, and the revolving-field type.

The revolving-armature type alternator is similar in construction to the DC generator, in that the armature rotates within a stationary magnetic field. This type of setup is typically found only in alternators with a low power rating and generally is not used.

The revolving-field type alternator has a stationary armature winding (stator) and a rotating-field winding (rotor). The advantage of this configuration is that the armature is connected directly to the load without sliding contacts in the load circuit. Direct connection to the armature circuit makes it possible to use large cross-section conductors that are adequately insulated for high voltage. [Figure 7-41]

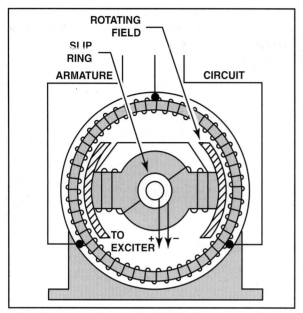

Figure 7-41. In a revolving-field type alternator, the armature is directly connected to the load without the use of sliding contacts. The rotating-field alternator is used almost universally in aircraft systems.

BRUSHLESS ALTERNATORS

The AC alternators used in large jet-powered aircraft are of the brushless type and are usually air-cooled. Since the brushless alternators have no current flow between brushes or slip rings, they are very efficient at high altitudes where brush arcing is often a problem.

As discussed earlier, alternator brushes are used to carry current to the rotating electromagnet. However, in a brushless alternator, current is induced into the field coil through an exciter. A brushless alternator consists of three separate fields, a permanent magnetic field, an exciter field, and a main output field. The permanent magnets furnish the magnetic flux to start the generator, producing an output before field current flows. The magnetism produced by these magnets induces voltage into an armature that carries the current to a generator control unit, or GCU. Here, the AC is rectified and sent to the exciter field winding. The exciter field then induces voltage into the exciter output winding. The output from the exciter is rectified by six silicon diodes, and the resulting DC flows through the output field winding. From here, voltage is induced into the main output coils.

The permanent magnet, exciter output winding, six diodes, and output field winding are all mounted on the generator shaft and rotate as a unit. The three-phase output stator windings are wound in slots that are in the laminated frame of the alternator housing. [Figure 7-42]

The main output stator winding ends of a brushless alternator are connected in the form of a Y, and in the case of the previous figure, the neutral winding is brought to the outside of the housing along with the three-phase windings. These alternators are usually designed to produce 120 volts across a single phase and 208 volts across two phases.

The GCU actually monitors and regulates the main generator's output by controlling the amount of cur-

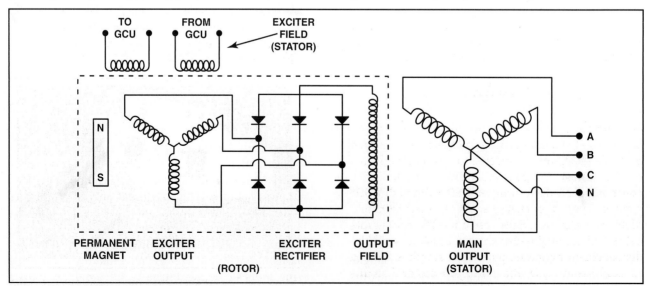

Figure 7-42. In a brushless alternator there are actually three generators: the permanent magnet generator, the exciter generator, and the main generator. The permanent magnet generator induces voltage into the exciter generator, which in turn supplies the field current for the main generator.

rent that flows into the exciter field. For example, if additional output is needed, the GCU increases the amount of current flowing to the exciter field winding that, in turn, increases the exciter output. A higher exciter output increases the current flowing through the main generator field winding, thereby increasing alternator output.

Since brushless alternators utilize a permanent magnet, there is no need to flash the field. In addition, the use of a permanent magnet eliminates the need to carry current to a rotating assembly through brushes.

ALTERNATOR RATINGS

AC alternators are rated in volt-amps, which is a measure of the apparent power being produced by the generator. Because most AC alternators produce a great deal of power, their ratings are generally expressed in kilo-volt amperes, or KVA. A typical Boeing 727 AC alternator is rated at 45 KVA.

FREQUENCY

The AC frequency produced by an AC generator is determined by the number of poles and the speed of the rotor. The faster the rotor, the higher the frequency. By the same token, the more poles on a rotor, the higher the frequency for any given speed. The frequency of AC generated by an alternator is determined using the equation:

$$F = \frac{P}{2} \times \frac{N}{60} = \frac{PN}{120}$$

Where:
F = Frequency
P = thenumberofpoles
N = the speed in rpm

With this formula, the frequency of a two-pole, 3,600 rpm alternator has a frequency of 60 hertz.

$$F = \frac{2}{2} \times \frac{3,600}{60} = 60 \text{ Hz}$$

To provide a constant frequency as engine speed varies, and to maintain a uniform frequency between multiple generators connected to a common electrical bus, most AC generators are connected to a **constant-speed drive unit, or CSD**. Although CSDs come in a variety of shapes and sizes, their principle of operation is essentially the same. The drive units consist of an engine-driven, axial-piston, variable-displacement hydraulic pump that supplies fluid to an axial-piston hydraulic motor. The motor then drives the generator. The displacement of the pump is controlled by a governor that senses the rotational speed of the AC generator. The governor action holds the output speed of the generator constant and maintains an AC frequency of 400-hertz, plus or minus established tolerances. [Figure 7-43]

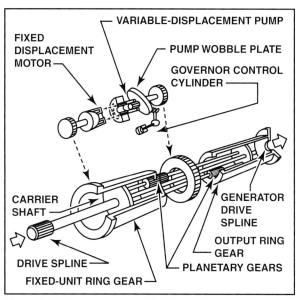

Figure 7-43. A constant-speed drive axial-gear differential, such as the one shown here, is used in the Sunstrand Integrated Drive Generator.

Some modern jet aircraft produce AC with a generator called an integrated drive generator, or IDG. The IDG is a high output generator that uses brushes and slip rings to carry DC exciter current to the rotating field. An IDG comprises both the constant speed drive unit and the generator sealed in the same housing. If a problem occurs with an IDG, it is usually not field repairable and is simply removed and replaced with a new or serviceable unit. When a CSD or IDG malfunctions, they can be automatically or manually disconnected in-flight. Once disconnected, they can not be reconnected in-flight and can only be put back on-line by personnel on the ground. [Figure 7-44]

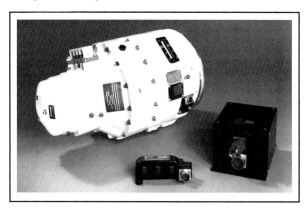

Figure 7-44. The constant speed drive unit is included in the housing with the Sunstrand Integrated Drive Generator above. Also shown is the generator's control unit and a current transformer assembly.

AC ALTERNATOR MAINTENANCE

Maintenance and inspection of alternator systems are similar to that of DC systems. The proper maintenance of an alternator requires the unit to be kept clean and all electrical connections tight and in good repair. Alternators and their drive systems differ in design and maintenance requirements; therefore, specific information is found in the manufacturer's service publications and in the maintenance program approved for a particular aircraft.

STORAGE BATTERIES

Batteries are the standard source of electrical energy for starting engines and supplying power in the event of generator failure. Batteries also stabilize and smooth out the generator output during extreme load changes.

LEAD-ACID BATTERY

The lead-acid battery has traditionally been the most popular type used in smaller aircraft. Today, for systems that require a constant high current output over a relatively long period, such as is needed to start turbine-powered aircraft, lead-acid batteries have been replaced with nickel-cadmium batteries. However, the less expensive and less maintenance intensive lead-acid units will continue as a power source where constant high current output is not required. In fact, improved capacity lead-acid batteries are replacing nickel-cadmium batteries in some usages. [Figure 7-45]

DETERMINING CONDITION OF CHARGE

The open-circuit voltage of a lead-acid battery remains relatively constant at about 2.1 volts per cell and, consequently, does not reflect a battery's state of charge. However, since the concentration of acid in the electrolyte changes as the battery is used, the electrolyte's specific gravity gives a good indication of the state of charge. In a fully charged, new battery, the electrolyte is approximately 30 percent acid and 70 percent water (by volume). This results in a specific gravity of between 1.275 and 1.300 with an electrolyte temperature of 80 degrees Fahrenheit.

As a battery is used, sulfate ions are removed from the electrolyte, leaving water. As a result, the specific gravity reading decreases. When an electrolyte's specific gravity drops below 1.150, the battery is considered to be discharged. At this level, there is not enough chemical strength in the electrolyte to convert the active materials into lead sulfate.

Battery Testing

Specific gravity is checked with a **hydrometer** which measures the depth a calibrated float sinks in a sample of electrolyte. The graduation on the float's stem that is even with the liquid shows the specific gravity of the electrolyte. The more dense the liquid, the higher the specific gravity reading. In other words, the more buoyant the hydrometer bulb, the more dense the liquid. [Figure 7-46]

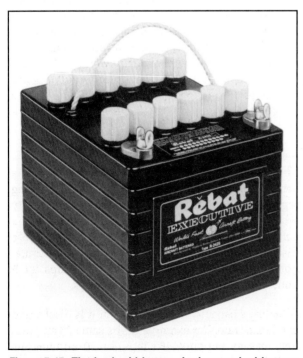

Figure 7-45. The lead-acid battery is the standard battery used in most light aircraft.

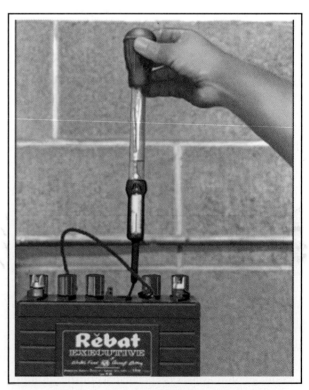

Figure 7-46. The specific gravity of the electrolyte is measured with a hydrometer.

The temperature of the electrolyte affects its specific gravity. Therefore, a standard of 80°Fahrenheit is used as the reference. If an electrolyte's temperature is something other than 80°F, a correction must be applied to the hydrometer reading. The electrolyte is less dense at higher temperatures and more dense at lower temperatures. [Figure 7-47]

ELECTROLYTE TEMPERATURE		CORRECTION POINT ADD OR SUBTRACT
°C	°F	
60	140	+24
55	130	+20
49	120	+16
43	110	+12
38	100	+8
33	90	+4
27	80	0
23	70	−4
15	60	−8
10	50	−12
5	40	−16
−2	30	−20
−7	20	−24
−13	10	−28
−18	0	−32
−23	−10	−36
−28	−20	−40
−35	−30	−44

Figure 7-47. This is an example of a correction chart used when determining an electrolyte's specific gravity. For example, suppose the specific gravity read 1240 at a temperature of 60°F. According to the chart, a correction of −8 points should be made. The corrected specific gravity is 1232 which is equivalent to 1.232.

BATTERY RATINGS

The open-circuit voltage of a lead-acid battery is 2.10 volts per cell when the electrolyte has a specific gravity of 1.265. The physical size of the cell or the number of plates has no effect on this voltage.

When a load is placed on a battery, the active material begins to convert into lead sulfate. As the lead sulfate forms, it increases the battery's **internal resistance** and causes the closed-circuit terminal voltage to drop. A battery's internal resistance can be calculated by using Ohm's law. For example, a lead-acid battery with 12 cells has a no-load voltage of 2.1 volts per cell. If the battery delivers 5 amps to a load having a resistance of 3 ohms, what is the internal resistance?

From the information given, the total no-load voltage is determined to be 25.2 volts (12 cells × 2.1 volts/cell). By applying Ohm's law (E=IR), the voltage under load is the product of amps times resistance. Therefore, the load voltage is 15 volts (5 amps × 3 ohms = 15 volts). The internal resistance in the battery thus causes a voltage drop of 10.2 volts (25.2 volts − 15 volts = 10.2 volts). To determine the battery's internal resistance, apply Ohm's law to the voltage drop (10.2 volts) and the amperes the battery delivers (5 amps). The internal resistance is 2.04 ohms.

Example:
$$E = IR$$
$$10.2 \text{ volts} = 5 \text{ amps} \times R$$
$$R = 2.04 \text{ ohms}$$

CAPACITY

The **capacity** of a battery is its ability to produce a given amount of current for a specified time. Capacity is measured in ampere-hours with one **ampere-hour** equaling the amount of electricity that is put into or taken from a battery when a current of one ampere flows for one hour. Any combination of flow and time that moves this same amount of electricity is referred to as one ampere-hour. For example, a flow of one-half amp for two hours or two amps for one-half hour is one ampere-hour. In theory, a 100 ampere-hour battery can produce 100 amps for one hour, 50 amps for two hours, or 20 amps for five hours.

The capacity of a battery is affected by four things. They are: the amount of active material, the plate area, the quantity of electrolyte, and temperature. An increase in the amount of active material, the plate area, or the quantity of electrolyte results in an increase in capacity. On the other hand, using a battery in cold temperatures effectively decreases its capacity. For example, at 50°F, a fully charged battery may be able to provide power for 5 hours. However, at 0°F, the same battery may only supply power for 1 hour. The reason for this is that as the temperature drops, the chemical reactions within a battery slow.

Five-Hour Discharge

The standard rating used to specify the capacity of a battery is the **five-hour discharge rating**. This rating represents the number of ampere-hours of capacity when there is sufficient current flow to drop the voltage of a fully charged battery to a completely discharged condition over the course of five hours. For example, a battery that supplies 5 amps for 5 hours has a capacity of 25 ampere-hours.

A battery's capacity decreases when it is discharged at a higher rate. For example, if the same 25 ampere-hour battery supplies 48 amps for 20 minutes, its capacity drops to 16 ampere-hours. If the battery is discharged in 5 minutes, the capacity drops to 11.7

BATTERY VOLTAGE	PLATES PER CELL	DISCHARGE RATE					
		5-HOUR		20-MINUTE		5-MINUTE	
		A.H.	AMPS	A.H.	AMPS	A.H.	AMPS
12	9	25	5	16	48	12	140
24	9	17	3.4	10.3	31	6.7	80
* Battery is considered discharged when closed-circuit voltage drops to 1.2 volts per cell.							

Figure 7-48. Relationship between ampere-hour capacity and discharge rate.

ampere-hours. This is due to heat, sulfation of the plates, and a tendency of the electrolyte to become diluted immediately around the plates. [Figure 7-48]

CELL TEST

If a battery's construction is such that the voltage of an individual cell can be measured with a VOM meter, a good indication of the cell condition can be determined while under load. To do this, first verify the electrolyte is at the proper level. Then, apply a heavy load to the battery for about three seconds by cranking the engine with the magneto switch(es) off, or the mixture in the idle-cutoff position. Now, turn on the landing lights and taxi lights to draw about ten amps. While the load current is flowing, measure the voltage of each cell.

A fully charged cell in good condition should have a voltage of 1.95 volts and all cells should be within 0.05 volt of each other. If some of the cells are below 1.95 volts, but all are within 0.05 volt of each other, the cells are in good condition but the battery is somewhat discharged. If any of the cells read higher than 1.95 volts and there is more than a 0.05-volt difference between any of them, there is a defective cell in the battery.

The individual cell voltages of closed cell lead-acid batteries can be checked in a similar manner by using special test equipment. For these measurements, a probe is placed in the electrolyte of each cell to measure the voltage passing through it when the battery is under load. With this equipment, closely follow the manufacturer's instructions and make sure that the probes are rinsed and dried after each use. [Figure 7-49]

SERVICING AND CHARGING

One of the most important aspects of battery servicing is keeping the battery clean and all of the terminals tight and free of corrosion. If any corrosion exists on the battery terminals or within the battery box it should be removed. To do this, scrub the bat-

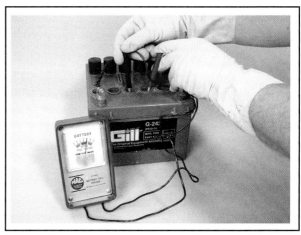

Figure 7-49. The individual cell voltages of a lead-acid battery can be checked by using test equipment that measures the voltage present in each cell's electrolyte.

tery box and the top of the battery with a soft bristle brush and a solution of sodium bicarbonate (baking soda) and water. When washing the top of the battery, avoid getting any baking soda in the cells since it neutralizes the electrolyte. After the battery and box are clean, rinse them with clean water and dry thoroughly. Coat the battery terminals with petroleum jelly or general purpose grease, and touch up any paint damage to the battery box or adjacent area with an acid-resistant paint.

The electrolyte in each cell should just cover the plates. Most batteries have an indicator to show the correct level. If the electrolyte level is low, add distilled or demineralized water. Never add acid to the battery unless it has been spilled, and then, follow the recommendations of the battery manufacturer in detail. The normal loss of liquid in a battery is the result of water decomposing during charging.

Most new batteries are received in a dry-charged state with the cells sealed. When putting a new battery into service, remove the cell seals and pour in the electrolyte that is shipped with the battery. In order to ensure a fully charged battery, the battery must be given a slow freshening, or boost, charge. Once this is done, allow the battery to sit for an hour or so and then adjust the electrolyte level.

It is normally not necessary to mix electrolyte. However, if it should ever become necessary to dilute acid, it is extremely important that the acid be added to the water, and never the other way around. If water is added to acid, the water, being less dense, floats on top, and a chemical action takes place along the surface where they meet. This action can generate enough heat to boil the water and splash acid out of the container causing serious injury if it gets on the skin or in the eyes. If acid should get into your eyes, flush them with generous amounts of clean water and get medical attention as soon as possible.

When acid is added to water, the acid mixes with the water and distributes the heat generated by the chemical action throughout the battery. This action still causes the water temperature to rise, but not enough to cause boiling or a violent reaction.

Automotive and aircraft electrolytes are different and should not be mixed. Automotive electrolyte has a lower specific gravity when charged and, therefore, an aircraft battery may never obtain a full charge with automotive electrolyte.

BATTERY CHARGERS
A storage battery is charged by passing direct current through the battery in a direction opposite to that of the discharge current. Because of the battery's internal resistance, the voltage of the external charging source must be greater than the open-circuit voltage. For example, the open-circuit voltage of a fully charged 12-cell, lead-acid battery is approximately 25.2 volts (12 × 2.1 volts). However, the battery's internal resistance causes a voltage drop of 2.8 volts. Therefore, approximately 28 volts are required to charge the battery. Batteries are charged by either the constant-voltage or constant-current method.

Constant-Current Charging
The most effective way to charge a battery is by inducing current back into it at a constant rate. The amount of current induced is typically specified by the manufacturer. However, in the absence of manufacturer information, you should use a current value of no more than seven percent of the battery's ampere-hour rating. For example, if you are charging a 40-ampere-hour battery and do not have specific information from the battery manufacturer, you should charge it at a rate not exceeding 2.8 amperes (40 ampere-hour × .07 = 2.8 amps).

As a battery begins to charge, the no-load voltage increases. Therefore, the voltage on a constant current charger must be varied in order to maintain a constant current throughout the charge. Because of this, a constant current charger usually requires more time to complete and additional attention.

When charging more than one battery with a constant-current charger, connect the batteries in series. One way to remember this is to recall that current remains constant in a series circuit and, therefore, a constant current charge requires multiple batteries to be connected in series. The batteries being charged can be of different voltages, but they should all require the same charging rate. When charging multiple batteries, begin the charge cycle with the maximum recommended current for the battery with the lowest capacity. Then, when the cells begin gassing freely, decrease the current and continue the charge until the proper number of ampere hours of charge is reached.

Constant-Voltage Charging
The generating system in an aircraft charges a battery by the **constant voltage** method. This method utilizes a fixed voltage that is slightly higher than the battery voltage. The amount of current that flows into a battery being charged is determined by a battery's state-of-charge. For example, the low voltage of a discharged battery allows a large amount of current to flow when the charge first begins. Then, as the charge continues and the battery voltage rises, the current decreases. The voltage produced by a typical aircraft generating system is typically high enough to produce about one ampere of current flow even when a battery is fully charged.

Constant-voltage chargers are often used as shop chargers. However, care must be exercised when using them since the high charging rate produced when the charger is first connected to a discharged battery can overheat a battery. Another thing to keep in mind is that the boost charge provided by a constant-voltage charger does not fully charge a battery. Instead, it usually supplies enough charge to start the engine and allow the aircraft generating system to complete the charge.

Like constant-current chargers, several batteries can be simultaneously charged with a constant-voltage charger. However, since the voltage supplied to each battery must remain constant, the batteries must have the same voltage rating and be connected in parallel.

On-Board Battery Charging

The aircraft electrical generating system is designed to carry most of the electrical loads and to keep the battery fully charged. The battery has an internal resistance that causes a voltage drop when it is being charged, so in order to fully charge a battery, the charging voltage must be slightly higher than the battery's. For example, with a 24-volt system, the alternator or generator must produce a higher voltage, normally around 28 volts. On the other hand, a generator or alternator producing approximately 14 volts will normally charge a 12-volt system.

If a lead-acid battery requires frequent servicing of the electrolyte level, it's possible that the charging voltage is too high, causing the electrolyte to boil. If this occurs, check the voltage output of the charging system in accordance with the aircraft manufacturer's procedures.

CHARGING PRECAUTIONS

Whenever working around lead-acid batteries, there are several precautions that must be observed, especially when charging. When a battery is charging, gaseous hydrogen and oxygen are released by the battery cells. Since these gases are explosive, it is essential for a battery to be charged in a well-ventilated location, isolated from sparks and open flames. To prevent sparking from the battery, always turn off the battery charger before connecting or disconnecting the electrical cables. Furthermore, when removing a battery from an aircraft, always disconnect the negative lead first, and when installing it, connect the negative lead last. When this method is used, it is less likely to cause sparking if the tools should contact the grounding plane of the airframe structure.

Since lead-acid battery electrolyte is extremely corrosive and will burn skin, always wear eye and hand protection whenever working with batteries. If electrolyte is spilled from a battery, it should be neutralized with sodium bicarbonate (baking soda) and rinsed with water.

BATTERY INSTALLATION

Before installing any battery in an aircraft, be sure it's correct for the aircraft. The voltage and ampere-hour ratings must meet the manufacturer's specifications. Some aircraft use two batteries connected in parallel to provide a reserve of current for starting and for extra-heavy electrical loads, while others meet the system voltage by series connecting batteries. When installing batteries in either of these systems, be sure they are the type specified in the aircraft service manual.

Be sure that the battery box is properly vented and that the battery box drain extends through the aircraft skin. Some batteries are of the manifold type, which do not require a separate battery box. With this type of battery, a cover is placed over the cells, and the area above the cells is vented outside the aircraft structure.

The fumes emitted from storage batteries are highly corrosive and toxic. Therefore, they typically must be neutralized before they are released into the atmosphere. To do this, many battery installations vent the fumes inside the battery box through sump jars containing absorbent pads moistened with a solution of sodium bicarbonate and water.

After installing a battery, make certain that it supplies enough current to crank the engine. Also, be sure that the aircraft generating system keeps the battery charged. If an aircraft ammeter shows a full charge rate, but the battery discharges rapidly, it is most likely that the battery is shorted internally.

NICKEL-CADMIUM BATTERIES

Turbine engines require extremely high current for starting. However, high-rate discharges cause the plates of lead-acid batteries to build up sulfate deposits, thereby increasing internal resistance and causing a subsequent voltage drop. This drawback spurred the development of an alkaline battery for aircraft use. The **nickel-cadmium** or **ni-cad** battery has a very distinct advantage that its internal resistance is very low, so its voltage remains constant until it is almost totally discharged. This low resistance is also an advantage in recharging, as it allows high charging rates without damage.

While high discharge and charging rates are favorable, there are dangers involved. These dangers begin with the high temperatures associated with nickel-cadmium batteries. For example, the discharge or charging cycle of a nickel-cadmium battery produces high temperatures that break down the cellophane-like material that separates the plates within the cell. The breakdown of the cell separator creates a short circuit allowing current flow to increase. The increased current flow creates more heat, causing further breakdown of the separator material. This condition is aggravated by the fact that the internal resistance of a ni-cad battery drops as the temperature rises. These factors all contribute to the process known as vicious-cycling, or **thermal runaway**.

New cell separator materials and advanced on-board charging equipment have reduced the likelihood of thermal runaway. However, most nickel-cadmium battery installations are required to have temperature monitoring equipment that enables the flight crew to recognize an overheat condition that can lead to thermal runaway.

CONSTRUCTION
Most nickel-cadmium batteries are made up of individual removable cells. The positive plates are made of powdered nickel (plaque) fused, or sintered, to a porous nickel mesh. This porous mesh is impregnated with nickel hydroxide. The negative plates are made of the same type of porous plaque as on the positive plates, but are impregnated with cadmium hydroxide. Separators of nylon and cellophane keep the plates from touching each other. The cluster of plates and separators is assembled into a polystyrene or nylon cell case and the case is sealed.

A thirty-percent-by-weight solution of potassium hydroxide and distilled water serves as the electrolyte. The specific gravity of this liquid is between 1.24 and 1.30 at room temperature. Since the electrolyte acts only as a conductor during charging and discharging, its specific gravity is no indication of a battery's state of charge.

An individual cell produces an open circuit voltage of between 1.55 and 1.80 volts, depending on the manufacturer. Batteries used in 12-volt aircraft systems use either 9 or 10 cells, while batteries used in 24-volt aircraft systems are made up of 19 or 20 individual cells. [Figure 7-50]

CHEMICAL CHANGES DURING DISCHARGE
As a nickel-cadmium battery discharges, metallic cadmium on the negative plates combines with hydroxide ions in the electrolyte. This releases electrons which flow to the negative terminal. During this process, the cadmium is converted to cadmium hydroxide. At the same time, hydroxide ions leave the positive plates and go into the electrolyte solution. This allows the electrolyte solution to remain

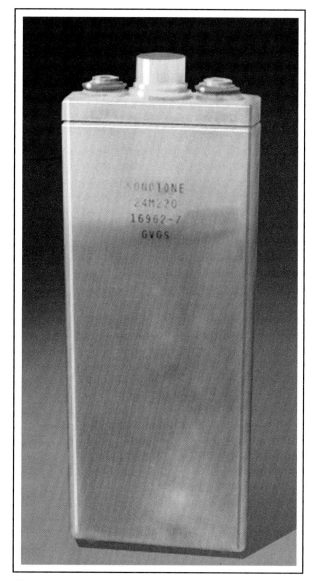

Figure 7-50. Nickel-cadmium batteries are made up of individual cells in transparent cases.

about the same. Therefore, specific gravity readings of electrolyte in nickel-cadmium batteries do not indicate the battery's state of charge.

CHEMICAL CHANGES DURING CHARGE
When charging current flows into a nickel-cadmium battery, oxygen is driven from the cadmium oxide on the negative plate leaving metallic cadmium. The nickel hydroxide on the positive plate accepts some of the released oxygen and becomes more highly oxidized. This process continues until all of the oxygen is removed from the negative plate. If charging is continued, gassing occurs in the cell as the water in the electrolyte is decomposed by electrolysis.

During the charging cycle, electrolyte is driven from the positive and negative plates. Therefore, the electrolyte is at its highest level immediately after a charging cycle. Because of this, the electrolyte level is checked and water added only when a nickel-cadmium battery is fully charged.

CELL IMBALANCE

One characteristic of a nickel-cadmium cell being charged is that the negative plate controls the cell's voltage characteristics. This, coupled with a slightly lower charge efficiency in the positive plates, results in an imbalance between the negative and positive plates in each cell. Constant-voltage charging is unable to recognize this condition. Voltages appear normal and the battery appears to be fully charged.

As long as the battery stays on a constant-voltage charge, the imbalance condition becomes worse each time the battery is cycled. Eventually the imbalance reduces the battery's available capacity to the point where there is not enough power to crank the engines or supply emergency power.

Cell imbalance problems have been greatly reduced by more sophisticated charging techniques. For example, in a pulse charging system, battery voltage is monitored and charging current regulated accordingly.

Another way to reduce the chance of cell imbalance is to terminate a constant-voltage charge prior to the battery obtaining a full charge. Then, complete the charge at a constant current rate equivalent to approximately ten percent of the battery's ampere-hour capacity. This technique drives the negative plates into a controlled overcharge and allows the positive plates to be brought to full charge without generating excessive gas and damaging the battery.

SERVICING NICKEL-CADMIUM BATTERIES

The electrolytes used by nickel-cadmium and lead-acid batteries are chemically opposite, and either type of battery can be contaminated by fumes from the other. For this reason, it is extremely important that separate facilities be used for servicing nickel-cadmium batteries and lead-acid batteries.

The alkaline electrolyte used in nickel-cadmium batteries is corrosive. It can burn your skin or cause severe injury if it gets into your eyes. Be careful when handling this liquid. If any electrolyte is spilled, neutralize it with vinegar or boric acid, and flush the area with clean water.

Nickel-cadmium battery manufacturers supply detailed service information for each of their products, and these directions must be followed closely. Every nickel-cadmium battery should have a service record that follows the battery to the service facility each time it is removed for service or testing. It is very important to perform service in accordance with the manufacturer's instructions, and to record all work on the battery service record.

It is normal for most nickel-cadmium batteries to develop an accumulation of potassium carbonate on top of the cells. This white powder forms when electrolyte spewed from the battery combines with carbon dioxide. The amount of this deposit is increased by charging a battery too fast, or by the electrolyte level being too high. If there is an excessive amount of potassium carbonate, check the voltage regulator and the level of electrolyte in the cells. Scrub all of the deposits off the top of the cells with a nylon or other type of nonmetallic bristle brush. Dry the battery thoroughly with a soft flow of compressed air.

Internal short circuits can occur between the cells of a ni-cad battery and are indicated when the battery won't hold a charge. Check for electrical leakage between the cells and the steel case by using a milliammeter between the positive terminal of the battery and the case. If there is more than about 100 milliamps of leakage, the battery should be disassembled and thoroughly cleaned. [Figure 7-51]

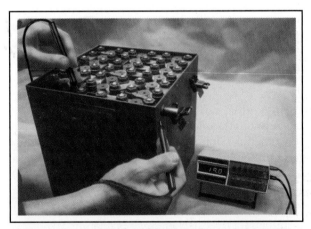

Figure 7-51. Cell-to-case leakage should be measured with a milliammeter. If there is more than 100 milliamps of leakage, the battery should be disassembled and cleaned.

Check the condition of all the cell connector hardware and verify there is no trace of corrosion. Dirty contacts or improperly torqued nuts can cause overheating and burned hardware. Heat or burn marks on nuts and contacts indicates the hardware was torqued improperly.

The only way to determine the actual condition of a nickel-cadmium battery is to fully charge it, and then discharge it at a specified rate to measure its ampere-hour capacity. When charging, use the five-hour rate and charge the battery until the cell voltage is that specified by the manufacturer. When the battery is fully charged, and immediately after it is taken off the charger, measure the level of the electrolyte. Ni-cad cell plates absorb electrolyte as a battery discharges or when it sits for long periods. However, the plates release electrolyte as the cells charge. If the level is not checked immediately after the charge is completed, the level drops and the correct level is difficult, if not impossible, to ascertain. Spewing of water and electrolyte during charge is a good sign that water was added while the battery was partially discharged. When water is added, the amount and cell location must be recorded on the battery service record.

When the battery is fully charged and the electrolyte adjusted, it must be discharged at a specified rate and its ampere-hour capacity measured. If the capacity is less than it should be, it is an indication that an imbalance exists. In this situation, the cells must be equalized through a process known as **deep-cycling**. To deep-cycle a battery, continue to discharge it at a rate somewhat lower than that used for the capacity test. When the cell voltage decreases to approximately 0.2 volts per cell, short across each cell with shorting straps. Leave the straps across the cells for three to eight hours to completely discharge them. This process is known as equalization. [Figure 7-52]

After equalization, the battery is ready to charge. Nickel-cadmium batteries may be charged using either the constant-voltage or constant-current methods. The constant-voltage method results in a faster charge; however, the constant-current is most widely used. For either system, the battery manufacturer's service instructions must be followed.

Monitor the battery during charge, and measure individual cell voltages. The manufacturer specifies a maximum differential between cells during the charging process. If a cell exceeds the specification, it must be replaced. Battery manufacturers specify the maximum number of cells that can be replaced before the battery must be retired.

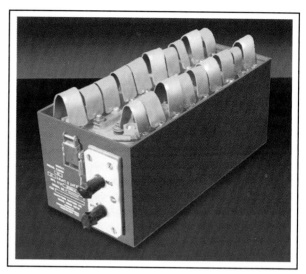

Figure 7-52. When the cell voltage falls to approximately 0.2 volts per cell, shorting straps are used to short the cells and ensure that they are equally discharged.

When a cell is replaced, enter its serial number in the battery service records and its location in the battery.

As a battery nears the completion of a charge, the cells release gases. This is normal, and must occur before the cell is fully charged. It is normal to overcharge a nickel-cadmium battery to 140 percent of its amp-hour capacity. If the battery has been properly serviced and is in good condition, each cell should have a voltage of between 1.55 and 1.80 at a temperature between 70 and 80°F. However, the actual voltage of a charged cell does vary with temperature and the method used for charging. [Figure 7-53]

Figure 7-53. Complete the deep-cycle operation by charging the battery to 140 percent of its ampere-hour capacity.

When working with nickel-cadmium batteries, it is sometimes helpful to have a troubleshooting chart. These charts allow you to associate a probable cause and corrective action to an observed condition. [Figure 7-54]

OBSERVATION	PROBABLE CAUSE	CORRECTIVE ACTION
High-trickle charge — When charging at constant voltge of 28.5 volts(+0.1) volts, current does not drop below 1 amp after 30- minute charge.	Defective cells.	While still charging, check individual cells. Those below .5 volts are defective and should be replaced.Those between .5 and 1.5 volts may be defectve or may be imbalanced, those above 1.5 volts are alright.
High-trickle charge after replacing defective cells, or battery fails to meet amp-hour capacity charge.	Cell imbalance.	Discharge battery and short out individual cells for 8 hours. Charge battery using constant-current method. Check capacity and if OK, recharge using constant-current method.
Battery fails to deliver rated capacity.	Cell imbalance or faulty cells.	Repeat capacity check, discharge and constant-current charge a maximum of three times. If capacity does not develop, replace faulty cells.
No potential available.	Complete battery failure.	Check terminals and all electrical connections. Check for dry cell. Check for high-trickle charge.
Excessive white crystal deposits on cells. (There will always be some potassium carbonate present due to normal gassing.)	Excessive spewage.	Battery subject to high charge current, high temperature, or high liquid level. Clean battery constant-current charge and check liquid level. Check charger operation.
Distortion of cell case.	Overcharge or high heat.	Replace cell.
Forign materials in cell — black or gray particles.	Impure water, high heat, high concentration of KOH, or improper water level	Adjust specific gravity and electrolyte level. Check battery for cell imbalance or replace defective cell.
Excessive corrosion of hardware.	Defective or damaged plating.	Replace parts.
Heat or blue marks on hardware.	Loose connections causing overheating of intercell connector or hardware.	Clean hardware and properly torque connectors.
Excessive water consumption. Cell dry.	Cell imbalance.	Proceed as above for cell imbalace.

Figure 7-54. Nickel-cadmium troubleshooting chart.

SECTION B

AIRCRAFT ELECTRICAL CIRCUITS

SMALL SINGLE-ENGINE AIRCRAFT

Electrical systems for an aircraft are best analyzed by referring to an electrical system schematic. Most aircraft manufacturers provide service manuals that contain schematic diagrams of all the electrical circuits broken down with only one main circuit per diagram. This makes diagnosing problems, or troubleshooting, much easier by isolating the individual circuit wiring and components.

The intent of these diagrams is to show the aircraft technician the relationship between all the electrical components and the associated wiring of a particular system. However, the diagrams do not show the physical location of components, except that some manufacturers use code numbers to help the technician locate the general position of components on the aircraft.

BATTERY CIRCUIT

Small aircraft typically use lead-acid batteries rated at either 12 or 24 volts and, almost without exception, they use a negative-ground, single-wire system. This means that the negative terminal of the battery is connected directly to the metallic portion of the aircraft structure, which places the structure at a negative potential. This allows electrical components to be connected to the structure for a return path to the battery to complete a closed circuit. When the circuit is closed for a path back to the battery, the procedure is commonly referred to as grounding.

Benefits for grounding in this manner include the reduced chance of developing radio frequency interference (RFI) by having fewer wires that can emit RFI energy, and also a lighter weight system from the reduced amount of wire required in the system. The electrical components need only a single wire from the positive voltage source, and a negative or ground connected to the metal structure to provide a return path for the current. In most situations, the electrical component has a metallic case, which serves as the ground source when it is attached to the structure.

The exception to this design occurs when the aircraft is built using composite materials in place of metallic structures. Most composite materials are non-conductive, so a two-wire electrical system must be used. One wire is needed for the positive connection, and one for the negative connection to a grounding source remotely located from the composite material.

A typical battery circuit for light aircraft is shown in figure 7-55. The positive terminal of the battery connects to the battery solenoid, which is a normally-open heavy-duty switch. Keep in mind that some manufacturers refer to a solenoid as a contactor, but both terms are used to describe components that essentially perform the same task; to remotely control a large current source with a small conductor and switch.

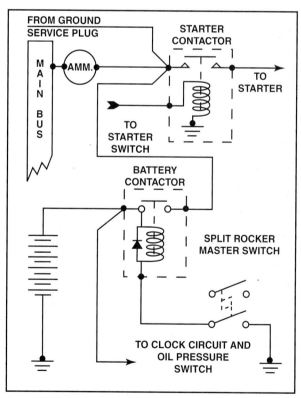

Figure 7-55. With some styles of solenoids, one end of the control coil connects internally to the main power terminal, which always has a voltage when the battery is installed and charged; and the other end of the coil is sent to ground through the battery master switch at the instrument panel. The wire from the master switch to the solenoid coil is relatively small and allows for controlling the battery circuit, which consists of large, heavy gage cable.

When the master switch is closed, it completes a ground path to energize the battery solenoid coil. Once the coil is energized, the main power connection is closed internally in the solenoid to complete the circuit from the battery to the electrical distribution bus. In this manner, the solenoid controls the large electrical supply of the aircraft with wires and a switch that are reasonable size for installation in the instrument panel.

A clipping diode is installed across the coil of the solenoid to eliminate spikes of voltage that are induced when the master switch is opened and the magnetic field from the coil collapses. The resulting spikes of voltage from a solenoid or relay coil may cause damage to sensitive electronic components if they are not controlled by a clipping diode. The diode simply isolates, or blocks, the induced voltage from flowing into the bus side of the electrical system, thereby protecting the electronic components from the high voltage.

The battery and battery solenoid on a light aircraft are often located on the engine-side of the firewall, or may be in the rear baggage compartment, depending upon the weight-and-balance requirements of the aircraft. However, in some installations, such as on multi-engine aircraft, the batteries may be installed in the wings.

Most aircraft have some components that require power all the time, whether the master switch is on or not. These low-current items, such as clocks and flight-hour meters, are connected to the main terminal of the battery solenoid, and supply current to these components anytime a charged battery is installed in the aircraft. These circuits often employ a fuse to protect the circuit, which is either mounted adjacent to the battery or as an in-line installation, located in a special housing in the wiring loom.

To cut down on the length and overall weight of electrical cables, a starter solenoid (contactor) is generally located on the engine-side of the firewall, and serves as a junction point for the heavy cable from the battery solenoid. In a similar fashion to the battery solenoid, the starter solenoid allows for remotely controlling the circuit with a relatively small wire from the starter switch to the solenoid coil.

GENERATOR CIRCUIT

An aircraft generator or alternator is the primary source of electrical energy. It must have sufficient capacity to supply all of the electrical loads and to keep the battery fully charged. Remember that when the generator is functioning properly, the battery should provide power to the system only when intermittent high-current-demand electrical items are energized. These include items such as landing lights, electric landing gear motors, and flap motors. Otherwise, the generator should provide all primary electrical power. This means the generator must have sufficient capacity to carry the load and have enough power remaining to keep the battery charged when all the continuous load items are on.

It is possible for an aircraft to have electrical demands that exceed the rating of the generator. If this occurs for two minutes or less, no flight crew member action is required. However, if the overload condition is continuous, the aircraft must have placards in plain sight of the crew to indicate exactly what combination of equipment can be on at any given time to prevent an overload condition.

To do away with the possibility of exceeding the capacity of the charging system, thereby eliminating limitation placards, many smaller aircraft electrical systems use a current limiter in the generator control to limit the current the generator can produce. For aircraft that typically use a DC alternator, a circuit breaker is used to limit the current.

An ammeter is also usually installed in either the generator output lead or battery positive lead. An ammeter installed in these positions can show the charge or discharge rate of the battery. An ammeter installed in the battery lead should not show discharge except during the momentary operation of high-current devices such as landing gear motors or flaps.

If a continuous discharge is shown, the generator is incapable of supplying enough current to meet the load demands. If the ammeter is installed in the generator output lead, the meter will indicate a positive value whenever the generator is working correctly and the electrical load is below the generator's maximum output.

On some installations, instead of having an ammeter, a loadmeter is provided to monitor the electrical load being placed on the system. These gauges are marked in percentage rather than in amps. The flight crew should never allow the electrical load to exceed the 100% rating indicated on the meter. A shunt resistor for the load-meter is matched to the generator output so the meter will indicate 100%

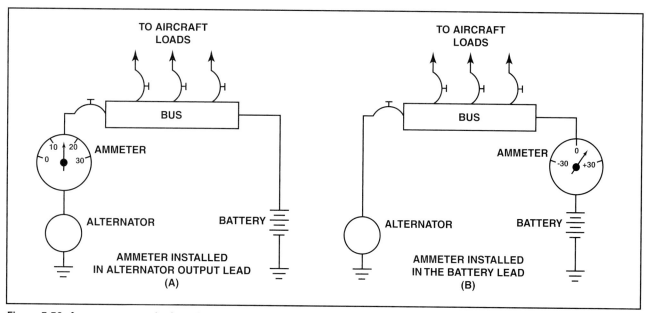

Figure 7-56. An ammeter can be installed in either the alternator output lead or in the battery lead. If installed in the alternator output lead, it is often called a loadmeter. Loadmeters directly indicate the amount of current being delivered from the generator or alternator. The indicator display is usually placarded or labeled to indicate the maximum amount of current that should be allowed to operate continuously, and to show the percentage of generator power being demanded on the system at any given time.

when the generator's rated current is flowing. [Figure 7-56]

In an installation where there is no way to monitor the load, it is important that the continuous duty electrical demands be limited in order for the battery to be kept fully charged. For these types of installations, the continuous duty load must be limited to 80% of the generator's rated output. For example, if a 100 amp generator is installed, the total continuous duty electrical component load should not exceed 80 amps (100 × .80 = 80 amps).

In a typical light-aircraft generator circuit, the generator output goes to the voltage regulator, through the reverse current relay inside the regulator unit, and then to the main bus. The output current must pass through a circuit breaker of adequate capacity to protect the generator in case the current limiter should malfunction. The generator field current flows from the field terminal of the generator, through one side of the master switch to the F-terminal of the voltage regulator, and to ground inside the regulator. [Figure 7-57]

The generator and the battery are in parallel with their common terminal at the main bus. The ammeter measures only the current flowing into or out of the battery, and only indirectly tells what the generator is doing.

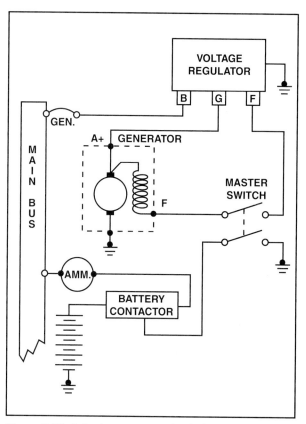

Figure 7-57. A basic generator circuit for a small single-engine aircraft has a generator, battery and a voltage regulator. Most aircraft also include a protective circuit breaker for the generator, a master switch, a battery solenoid (contactor) and an ammeter or loadmeter.

ALTERNATOR CIRCUIT

The advantages of more current for less weight have made the alternator the modern choice for production of DC electrical power on most aircraft. The exception to this is the use of starter generators found on turbine-powered aircraft. The external circuit of a DC alternator is similar to that used by a generator, with a few exceptions. [Figure 7-58]

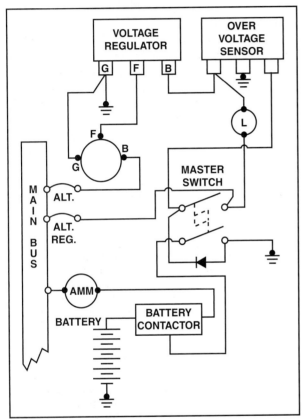

Figure 7-58. A typical DC alternator circuit for a light aircraft is similar to the circuit for that of a generator with the addition of an overvoltage sensor.

The G-terminal of the alternator is connected to ground at the G-terminal of the voltage regulator. The B-terminal of the alternator is its positive output terminal, and it connects to the aircraft main power bus through a circuit breaker that protects the alternator from exceeding its current limits. The F-terminal of the alternator connects directly to the F-terminal of the voltage regulator, and this circuit continues out the B-terminal of the regulator, through the over-voltage sensor to the alternator side of the master switch, and then to the main bus through the alternator regulator circuit breaker.

Some alternator systems incorporate the over-voltage sensor, an under-voltage sensor and the voltage regulator into one unit. This unit, often called an Alternator Control Unit (ACU), is typically a solid-state unit with no moving parts.

The electrical system master switch is an interlocking double-pole single-throw switch. The battery side of the switch can be turned on and off independent of the alternator, but the alternator side of the switch cannot be turned on without also turning on the battery. The alternator can be turned off, however, without affecting the battery side of the switch.

An over-voltage relay in the field circuit senses the output voltage, and if it becomes excessive, the relay will open the field circuit, shutting off the alternator output. The relay also turns on a warning light on the instrument panel, informing the pilot that the alternator is off-line because it has produced an excessively high voltage.

A solid-state diode is placed across the master switch from the field connection of the alternator side to the ground terminal of the battery side. This allows any spikes of voltage, that are induced into the system when the field switch is opened, to pass harmlessly to ground rather than getting to the main power bus.

EXTERNAL POWER CIRCUIT

Because of the heavy drain the starter puts on the battery, many aircraft are equipped with external power receptacles where a battery cart, or external power supply, may be plugged in to furnish power for engine starting.

Power is brought from a battery cart or rectifier through a standard three-terminal external power plug. Two of the pins in the aircraft receptacle are larger than the third, and are also longer. When the cart is plugged in, a solid contact is made with the two larger plugs. The external power relay in the aircraft remains open, and no current can flow from the external source until the plug is forced all the way into the receptacle, and the smaller pin makes contact. This small pin then supplies power through a reverse-polarity diode to the external power relay that closes, connecting the external power source to the aircraft bus.

The reverse-polarity diode is used in the circuit to prevent an external power source with incorrect polarity from being connected to the aircraft's bus. The diode simply blocks current from flowing to the external power relay, if the applied power is connected backwards or is offering reverse polarity.

Depending on the system design, some external power sources can be connected to charge the aircraft battery, while others isolate the battery from the external power. For systems that provide for battery charging, it is quite possible that the aircraft battery can be so completely discharged that the battery contactor cannot get enough current to close. This would prevent the external power source from charging the battery.

To allow for battery charging, a circuit consisting of a diode, a current-limiting resistor and a fuse is connected between the positive terminal of the external power plug and the battery side of the battery contactor. With this arrangement, enough current can flow from the external power source to energize the battery contactor coil, so that it can close and allow the battery to be charged. A diode (D2 in Figure 7-52) is in the circuit to prevent the positive pin in the external power receptacle from having power, or being "hot," when no external power plug is connected. [Figure 7-59]

Today, with many of the smaller aircraft having 12-volt systems while others have 24-volt systems, it is extremely important to connect the correct voltage when using an external power source. Diode D1 prevents the external power relay closing if the power source has the wrong polarity, but there is normally no protection against improper voltage. Connecting the aircraft to the wrong voltage can severely damage sensitive electronic equipment.

STARTER CIRCUIT

In most aircraft the starter is activated through a solenoid by an ignition switch or start button found on the instrument panel. A spring-loaded START position on the switch sends current from the main bus to the coil of the starter solenoid (contactor). When the starter contactor closes, the high current required for the starter motor flows from either the battery or the external power source to the starter. [Figure 7-60]

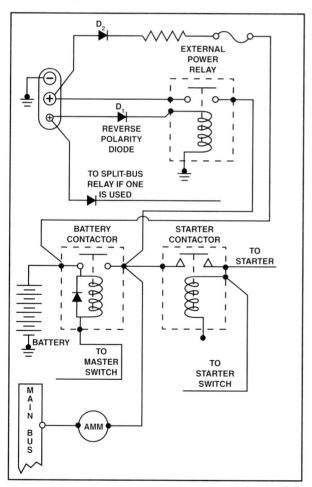

Figure 7-59. The external power circuit is usually connected to the external power source through a standard three-terminal external power plug that ensures a correct hookup of cables.

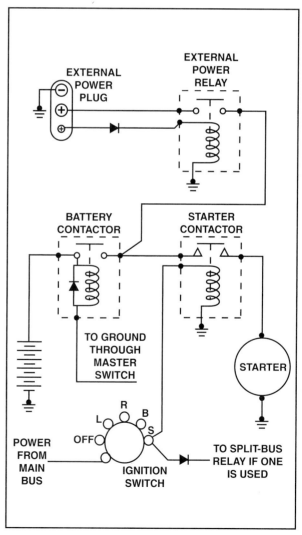

Figure 7-60. Starter circuits usually employ a solenoid to carry the high current loads required by the engine starter.

AVIONICS POWER CIRCUIT

The vast majority of the avionics equipment used today has solid-state components, which can be damaged by voltage spikes such as those produced when a magnetic field collapses and sends its induced voltage into the system. To prevent this type of damage to the avionics equipment, the radios often receive power from a separate bus that may be isolated from the main bus when the engine is being started, or when the external power source is connected.

The avionics bus may be connected to the main bus with a split-bus relay, which is a normally-closed relay that is opened by current from the ignition switch when it's in the START position, and from the control pin of the external power receptacle. Diodes are used in the circuit to both of the sources of current to isolate one when the other is being used. [Figure 7-61]

master switch is either turned on or off, and any time the engine is started or external power is connected to the aircraft. [Figure 7-62]

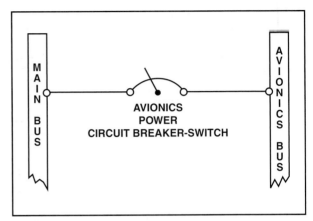

Figure 7-62. Another means of isolating the avionics bus is to have it controlled by a separate circuit-breaker switch.

Using a circuit-breaker switch to isolate the avionics bus has a couple of very definite advantages. The circuit breaker-type switch can be used as a master switch for all of the avionics equipment, so their individual switches will not have to be used. And in the case of a fault in any of the avionics equipment, this circuit breaker will open to protect the avionics wiring.

LANDING GEAR CIRCUIT

The landing gear circuit for a typical twin-engine general-aviation airplane is hydraulically operated by a reversible DC electric motor. It turns in one direction to lower the gear and in the opposite direction to raise it. [Figure 7-63]

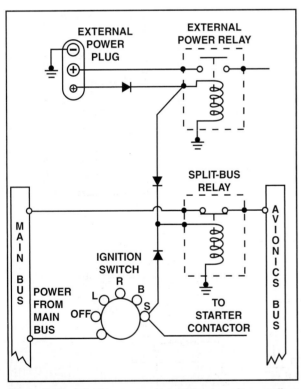

Figure 7-61. Avionics are very sensitive to voltage spikes and are protected by being powered from a bus separate from the main bus. One method of doing this is by means of a split-bus relay that de-energizes the avionics bus during the start cycle.

The avionics bus of some aircraft is isolated from the main bus by a circuit breaker switch rather than with a relay. When an avionics power switch is installed, it is the responsibility of the aircraft operator to be sure this switch is turned off before the

In figure 7-63(A) the airplane is in the air with the landing gear down and locked, but the landing gear selector switch has been moved to the GEAR UP position. The down-limit switches in each of the three landing gears are in the DOWN position, and the up-limit switches are all in the NOT UP position. Current flows through the NOT UP side of the up-limit switches through the FLIGHT side of the squat switch, and through the hydraulic pressure switch to the gear up relay.

The squat switch in the circuit is mounted on one of the struts so that it's in one position when the airplane weight is on the landing gear, and in the other position when there is no weight on the wheels. The purpose of the squat switch in this circuit is to prevent the gear from being retracted while the aircraft is still on the ground. The hydraulic pressure switch

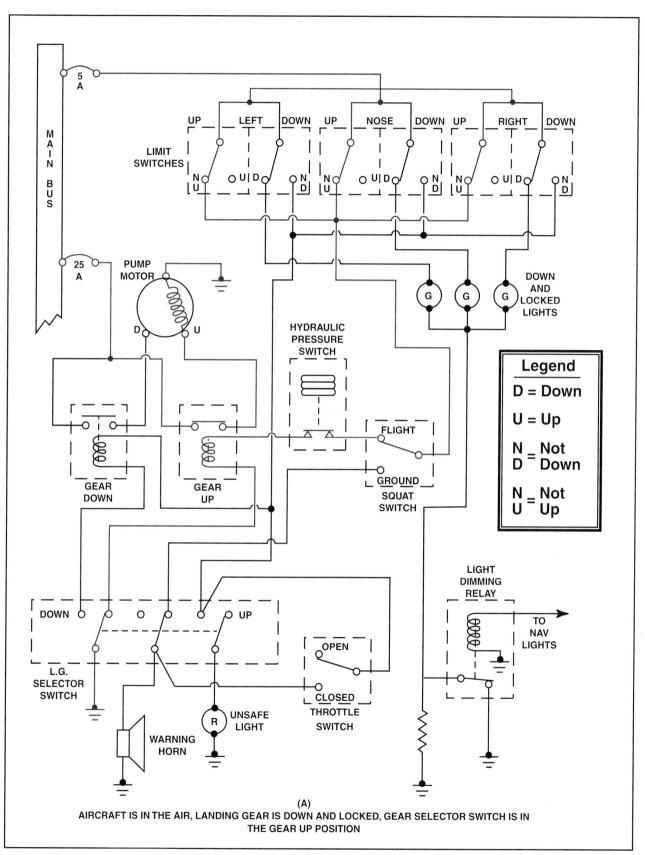

Figure 7-63(A). Landing gear circuit (1 of 3).

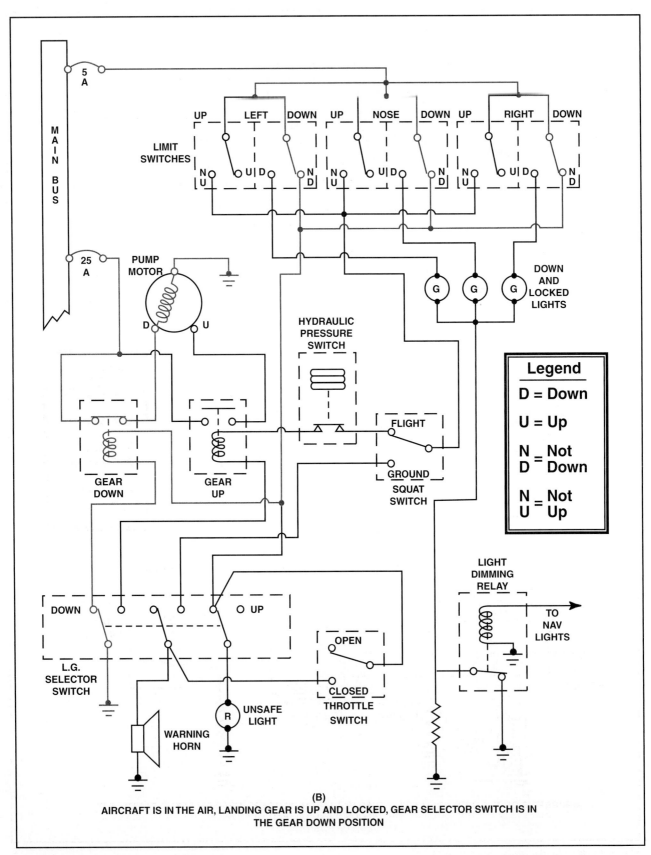

Figure 7-63(B). Landing gear circuit (2 of 3).

Airframe Electrical Systems

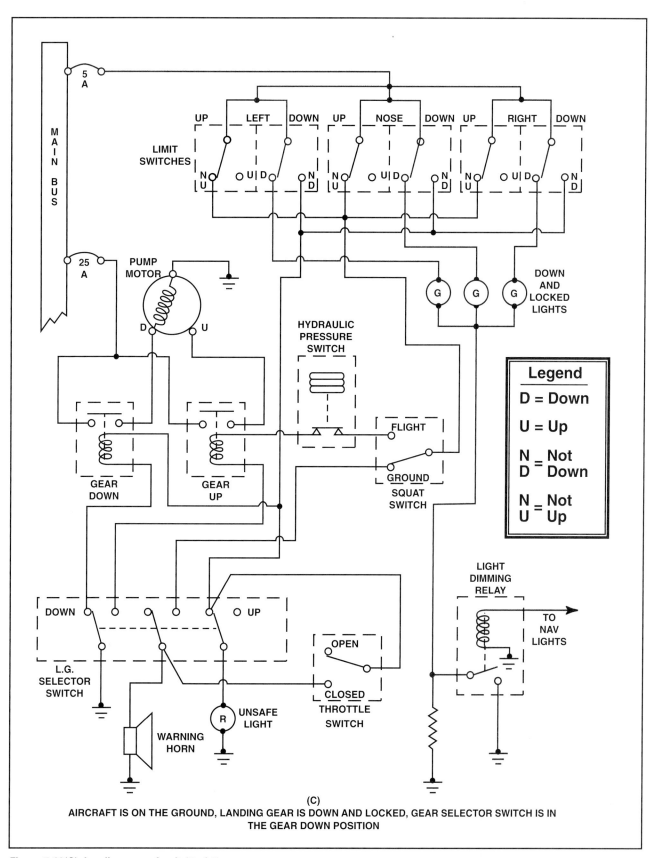

Figure 7-63(C). Landing gear circuit (3 of 3).

is used to turn off the pump motor in the event hydraulic pressure exceeds its limits.

From the pressure switch, the current goes through the coil of the LANDING GEAR UP relay and to ground through the landing gear selector switch. This flow of current creates a magnetic field in the LANDING GEAR UP relay and closes it, so that current can flow from the main bus through the relay contacts and the windings that turn the motor in the direction to raise the landing gear.

As soon as the landing gear is fully up, the up-lock switches in each gear move to the UP position, and current is shut off to the landing gear motor relay, and the motor stops. The down-limit switches have moved to the NOT DOWN position, and the three down-and-locked lights go out. If the throttle is closed when the landing gear is not down and locked, the warning horn will sound.

The landing gear may be lowered in flight by moving the landing gear selector switch to the GEAR DOWN position as in figure 7-63(B). Current flows through the NOT DOWN side of the down-limit switches, through the coil of the LANDING GEAR DOWN relay, and to ground through the landing gear selector switch. The motor turns in the direction needed to produce hydraulic pressure to lower the landing gear.

Figure 7-63(C) shows that when the landing gear is down and locked, and the landing gear selector switch is in the down position, current flows through the DOWN sides of the down-limit switches, and the green GEAR DOWN AND LOCKED lights come on. In the daytime, current from these lights goes directly to ground through the closed contacts of the light dimming relay, but at night when the navigation lights are on, current from this light circuit energizes the relay, and current from the indicator lights must go to ground through the resistor. This makes the lights illuminate dimly so they will not be distracting at night.

If any of the limit switches are in a NOT UP or a NOT DOWN position, a red UNSAFE light will illuminate. But when the landing gear selector switch is in the LANDING GEAR DOWN position and all three gears are down and locked, the light will be out. Also, if the selector switch is in the LANDING GEAR UP position and all three gears are in the up position, the light will be out. If the airplane is on the ground with the squat switch in the GROUND position, and the landing gear selector switch is moved to the LANDING GEAR UP position, the warning horn will sound, but the landing gear pump motor will not run.

ALTERNATING CURRENT SUPPLY

Small aircraft for which DC is the primary source of electrical power have little use for alternating current except for certain instruments that require 26-volt, 400-hertz AC, and some lighting circuits known as Electro Luminescent Panels. Most aircraft needing this type of power are equipped with solid-state, or static, inverters. These units consist of a solid-state sine-wave oscillator followed by a transformer that produces the required power. Some instruments that use 26-volt AC have built-in inverters and can be operated with a DC input voltage. [Figure 7-64]

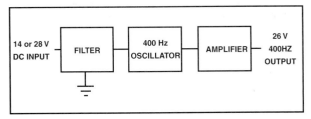

Figure 7-64. A solid-state inverter uses an oscillator to produce 400-hertz alternating current that is amplified so it can drive alternating current instruments.

Before solid-state electronics became such an important factor in aircraft instrumentation, most DC-powered aircraft got their AC for instruments from small single-phase rotary inverters, which had a small DC motor driving an AC generator. The frequency of the AC is dependent on the RPM of the DC motor, and the voltage output of the AC generator is dependent on the DC current supplied to the field of the generator. These units may still be found in some older aircraft, but because they are noisy, subject to mechanical problems and are electrically inefficient, in most cases solid-state inverters are replacing rotary inverters. [Figure 7-65]

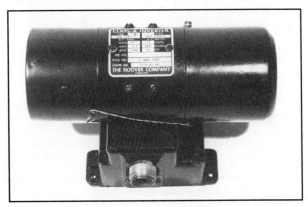

Figure 7-65. A rotary inverter is used in some older aircraft to produce alternating current from direct current.

SMALL MULTI-ENGINE AIRCRAFT

The vast majority of light twin-engine aircraft uses two generators or alternators, and theses aircraft have their voltage controlled so that when they are connected together on the power bus, the regulators work together to keep the output voltages of the two sources the same. This is called **paralleling**.

PARALLELING WITH VIBRATOR-TYPE VOLTAGE REGULATORS

Some of the lower-output electrical systems use vibrator-type voltage regulators with provisions for paralleling the two generators. These generator controls work in the same way as the three-unit controls for a single-engine installation, except that the voltage regulator relay has an extra coil that is connected through a paralleling switch or relay, to a similar coil in the voltage regulator of the other engine. [Figure 7-66]

The paralleling circuit for a twin-engine system using vibrator-type regulators is quite simple in its operation. With the paralleling switch open, each generator acts in exactly the same way it would in a single-engine system. The generator with the highest voltage will produce the most current. If the paralleling switch is closed when the left generator is producing a higher voltage than the one on the right engine, current will flow from the left-engine regulator to the regulator on the right engine, through both paralleling coils.

The flow through the left coil produces a magnetic field that assists the voltage coil, and the contacts open sooner, inserting the field resistor more often and lowering the output voltage of the left generator. This same current flows through the paralleling coil in the right voltage regulator, except that it flows in the direction that will cause its magnetic field to oppose the field from the voltage coil. This allows more field current to flow through the right generator and increases its voltage.

The only current that flows in the paralleling coils is that caused by the difference in the output voltages of the two generators, and this small current will produce just enough magnetic field difference to keep the generators putting out the same voltage and thus sharing the load equally. The theory of operation applies to vibrating-type voltage regulators; however, transistorized regulators can provide the same paralleling functions using solid-state devices.

PARALLELING WITH CARBON-PILE VOLTAGE REGULATORS

Light twin-engine aircraft with generators of greater output than can be controlled with a vibrator-type

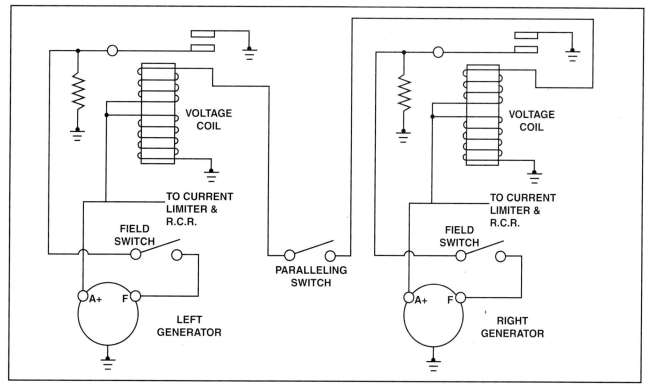

Figure 7-66. A paralleling switch activates the paralleling circuit for a light twin using vibrator-type voltage regulators

voltage regulator use carbon-pile voltage regulators. The field current is controlled by varying the pull of an electromagnet on the voltage regulator armature. The higher the output voltage, the more current through the electromagnet and the more pull on the armature. This loosens the carbon pile, increasing its resistance and decreasing the field current.

The only difference between a carbon-pile voltage regulator used on a single-engine installation and that used for a twin-engine system is a paralleling coil. This coil is connected in the same way as the paralleling coil of a vibrator-type regulator; the basic difference being that the current that flows through it is produced by the voltage drop across paralleling resistors in the ground wire leads to the armatures of both generators. All of the generator output flows through these resistors.

If the left generator yields more current than the right, the voltage drop across the left paralleling resistor will be higher than that across the right, and electrons will flow through the paralleling coils of both voltage regulators. The magnetic field of the left paralleling coil will assist the left voltage coil and loosen the carbon pile. This will decrease the field current in the left generator and lower its voltage. At the same time, the field of the right paralleling coil will oppose that of its voltage coil, and the spring will compress the carbon pile and increase the output voltage of the right generator. [Figure 7-67]

PARALLELING TWIN-ENGINE ALTERNATOR SYSTEMS

Most modern light twin-engine aircraft contain DC alternator systems for electrical power generation. These alternators are typically controlled through two relatively complex solid-state **alternator control units (ACU)**. The alternator control units receive input from both alternators in order to provide paralleling. The ACU also provides voltage and current regulation as well as over-and-under voltage sensing and protection. In the event of a fault condition, the ACU would automatically isolate the defective alternator and alert the flight crew.

LARGE MULTI-ENGINE AIRCRAFT

There are two basic types of large aircraft power distribution systems—the **split-bus** and the **parallel system**. The split-bus system is typically found on twin-engine aircraft such as the Boeing 757 and 767, the McDonnell Douglas MD-80 and the Airbus A320. The parallel system is typically used on aircraft containing three engines such as the DC-10 and Boeing 727. Most four-engine aircraft employ a modified split-bus system.

A simplified version of a split-bus power distribution system has AC generator power from the right engine connected to the right distribution bus and isolated from the left bus by **bus tie breakers (BTB)**. The left AC generator supplies power only to the left bus.

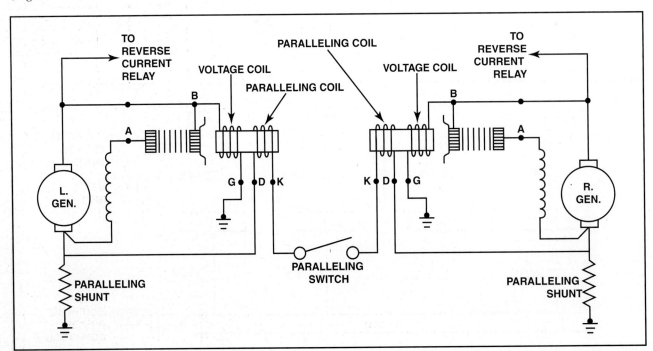

Figure 7-67. The paralleling circuit for use with carbon-pile voltage regulators is similar in operation to that used for vibrator-type voltage regulators.

Airframe Electrical Systems

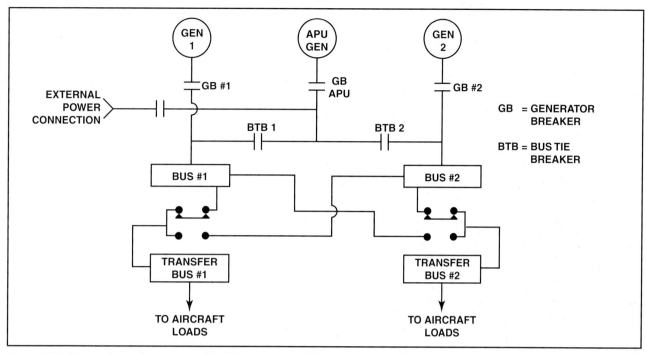

Figure 7-68. Most twin-engine commercial airliners use a split-bus power distribution system.

In the event of a generator failure, the failed generator is isolated by the generator breaker (GB), and BTB 1 and 2 close to connect the isolated bus to the operating generator. When this occurs, it is often necessary to limit the electrical load to essential components to prevent overloading the operating generator. [Figure 7-68]

On some aircraft, an Auxiliary Power Unit (APU) equipped with a generator can be started during flight and may be used to carry the load of the failed generator. In that case the left and right busses would once again be isolated.

In a parallel power distribution system all three generators are connected to a common bus and share the electrical loads equally. In the event of a generator failure, the failed unit would be isolated from the bus by its generator breaker, and the flight would continue with two generators supplying the electrical power. [Figure 7-69]

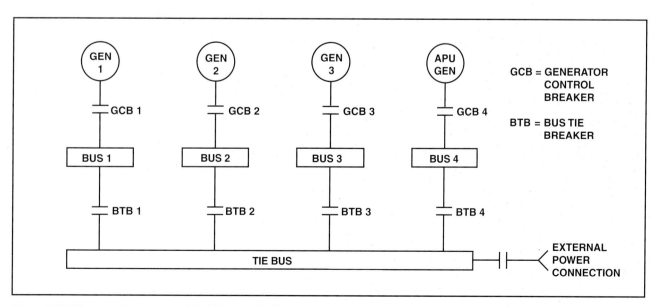

Figure 7-69. Three-engine commercial airliners typically use a parallel power distribution system.

A modified split-bus system found on some four-engine aircraft connects the two right-side generators in a parallel configuration. The two left-side generators are also paralleled. The right- and left-side busses are kept isolated by a split-system breaker. In the event of a generator failure the associated paralleled generator will carry the entire load for that bus. If both generators on the same side should fail, the split-system breaker would close and send power to the inoperative bus from the working generators.

The Boeing 727 is one of the most popular jet airliners, and its electrical system is typical for this type of aircraft. The B727 uses a parallel power distribution system where each of the three engines drives a three-phase generator through a hydraulic constant-speed-drive (CSD) unit. The CSD provides for frequency control by matching the RPM of each generator to produce three-phase, 115 volts AC at 400 cycles per second (Hz.).

Single phase transformers reduce some of this power to 28 volts AC to operate most of the lights, and there are three transformer-rectifier (T-R) units that convert the 115 volts AC to 28 volts DC for emergency lights and system control circuits. The 28-volt DC system is also used for charging the battery. [Figure 7-70]

A gas turbine-powered **auxiliary power unit (APU)** located in the wheel well drives a three-phase, 115-volt alternator identical to the ones driven by the engines. The APU is used to supply all of the needed electrical power when the aircraft is on the ground. When the aircraft is operated on the

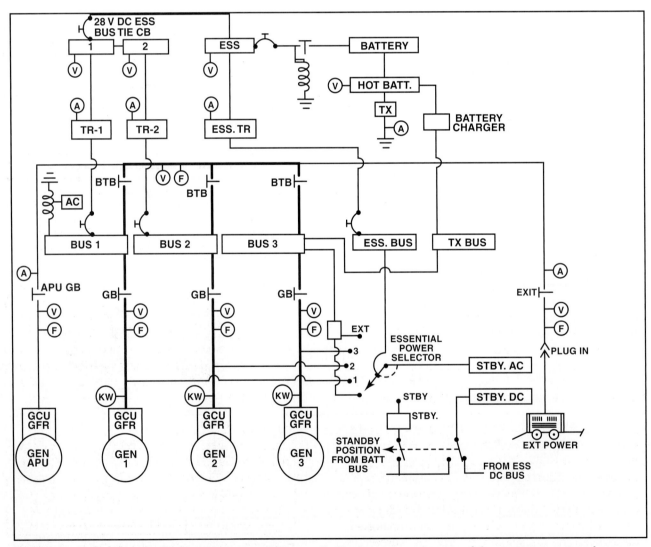

Figure 7-70. The Boeing 727 aircraft employs a parallel power distribution system. Any one of the generators, external power, or an APU can power an essential bus. The essential bus, as the name implies, has equipment connected to it that is essential for safe flight operations in the event of a major electrical system fault or failure.

ground, an internal fan cools the APU. Unlike some aircraft, the 727's APU cannot be operated when the aircraft is in the air. [Figure 7-71]

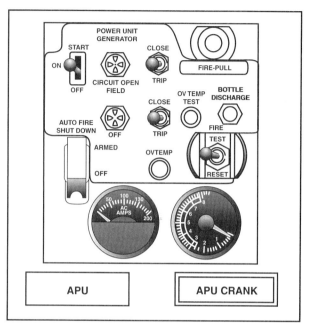

Figure 7-71. The APU can be started and the output monitored from the APU control panel.

Each engine-driven generator feeds its own bus through a generator breaker, and all three buses can be tied to a common sync, or tie, bus through bus tie breakers, when the generators are synchronized. In this way the generators can divide the electrical load among themselves. [Figure 7-72]

While most of the electrical power used by the 727 comes directly from the AC generators, the aircraft is equipped with a battery to start the APU and to provide power for certain essential lighting, avionics and instrument equipment. In normal operation the battery is disconnected from the DC loads. DC power is supplied by transformer-rectifier units connected to the main AC busses. The battery is kept fully charged by a battery-charger circuit that draws its power from the AC transfer bus. [Figure 7-73]

Equipment that is essential for flight gets its power from an AC and a DC essential bus, and a selector on the upper panel allows the flight engineer to supply the essential busses from any of the three main AC busses, the external power supply or the APU. In the event that all three AC generators fail, the engineer can select the STANDBY position of the selector switch, and the standby AC and DC busses are supplied directly from the battery bus with a standby inverter providing the AC.

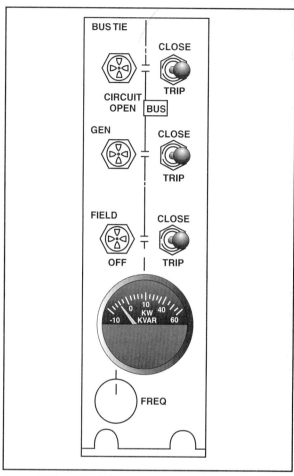

Figure 7-72. Each of three generators on a Boeing 727 has its own generator control panel. The switches control the bus tie breaker, the generator breaker, and the field relay. A kilo-watt/kilo-volt-amp-reactive (KW-KVAR) meter indicates the power the generator is producing, and the frequency control varies the output of the CSD unit to manually adjust the generator frequency when automatic phase syncing is not available.

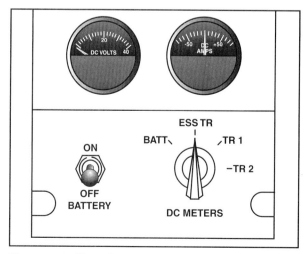

Figure 7-73. The DC electrical panel of the Boeing 727 allows the flight engineer to select the source of DC power and to monitor the voltage and current being supplied.

AC ALTERNATOR DRIVE

Each of the generators is driven through a hydraulic constant-speed drive unit, and the flight engineer has controls that allow disconnecting the CSD in case of a generator or CSD malfunction. The CSD has its own lubrication system, and the flight engineer has a low oil pressure warning light and an oil temperature gauge for the unit. The temperature gauge can either indicate the temperature of the oil entering the CSD unit, or it can show the amount of temperature rise as the oil passes through the drive unit. Since the generator is cooled by oil spray, it gives an indication of the load that is being carried by the CSD. [Figure 7-74]

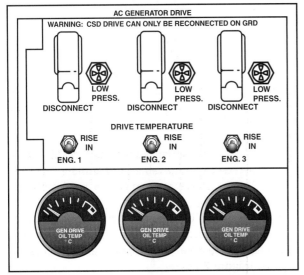

Figure 7-74. The constant-speed-drive control panel allows for individual CSD units to be disconnected in flight in case of a malfunction. Once disconnected, the CSD unit can be reconnected only on the ground. The generator drive oil temperature indicator shows the temperature of the drive oil in the CSD. In the RISE position, it indicates the rise in temperature of the oil as it passes through the CSD.

The paralleling process of two or more generators requires that the flight engineer connect them to a tie bus. If there are no other power sources connected to the tie bus, the first generator can be connected without complications by simply closing its BTB. But, before the second or third generator can be placed on the tie bus, they must be synchronized with regard to voltage output, phase and frequency to match the generator(s) already connected to the bus, or on-line.

The phase rotation is determined when the generator is installed and will not change in flight. The voltage is controlled by the generator control unit to hold the output to 115 volts, plus or minus 5 volts, and the frequency is maintained by the CSD. There is a control on the flight engineer's panel that allows the frequency to be manually adjusted for paralleling or paralleling may be performed automatically on newer aircraft.

The CSD maintains the frequency at 400 hertz, plus or minus 8 hertz. It is the flight engineer's job to determine that all generators are producing equal values before they are connected to the tie bus. If a generator is paralleled improperly, damage may occur to the generator or the CSD. [Figure 7-75]

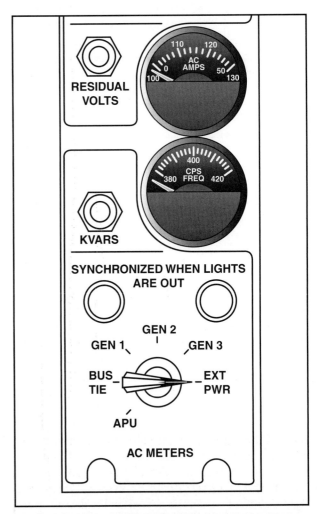

Figure 7-75. The voltmeter and frequency meter on the AC control panel of the Boeing 727 allow the flight engineer to monitor voltage and frequency of selected AC sources. In modern aircraft, the frequency is measured very accurately through use of an integrated circuit chip that has an internal clock circuit. Depressing a residual voltage button causes the voltmeter to indicate the voltage output of the selected generator with no field current flowing. This voltage is produced by the residual magnetism of the genera-

The KVAR button causes the selected generator's KW-KVAR meter to read in kilo-volt-amps-reactive. This indicates how hard the generator is working to supply power for both the resistive and reactive loads. The synchronizer lights are lit when the selected generator and the tie bus are out of phase. The bus tie breaker should be closed only when these two lights are out.

GENERATOR INSTRUMENTATION AND CONTROLS

A frequency meter and an AC voltmeter may be selected to monitor the voltage and frequency of the APU generator, the external power source or any of the three engine-driven generators, as well as that of the tie bus. A push-button beside the voltmeter allows it to read residual voltage, which is the voltage produced by the selected generator when no field current is flowing. If there is residual voltage, the generator is turning, but when no voltage is present it indicates the CSD has been disconnected or the residual magnetism has been lost.

Two synchronizing lights on the panel illuminate when the selected generator is not synchronized with the power on the tie bus, and must not be connected until it is synchronized. The generator frequency should be adjusted until the lights blink at their slowest rate and the bus tie breaker closed when both lights are out.

The generator control unit used with each of the engine-driven generators and with the APU generator contains all of the circuitry needed to maintain a constant voltage as the load current varies. In addition, the control unit will open the bus tie breaker, the generator breaker or the generator field relay if a fault occurs in either the system or the generator. The control unit will also alert the flight engineer to the defect in order that corrective action can be taken to divert power to the affected bus.

Each of the three main generator control panels has a control for the bus tie breaker (and a light that's on when the breaker is open), a generator breaker switch and the generator field switch. There is a light that is on when the generator breaker is open, and one when the field is off. The generator field switch connects voltage to the generator field through the voltage regulator. The generator breaker connects the generator output leads to the correct AC power distribution bus. The bus tie breaker connects the number 1, 2 or 3 AC bus to the tie bus for paralleled operation.

Each alternator is equipped with a KW-KVAR meter that normally indicates the amount of electrical power being produced by the alternator in kilowatts. When a button on the instrument panel is pressed, the indicator shifts circuits and indicates the reactive power in kilo-volt-amps. This indication is to inform the flight crew of the true power of the generator. True power is the measure of power with respect to both the resistive and reactive loads.

AUTOMATED AC POWER SYSTEMS

Many of the latest generation commercial aircraft use automated systems for controlling the various AC power distribution functions previously mentioned. The use of automated systems has made it possible to reduce flight crew workloads enough that most modern commercial airliners require only two flight crewmembers with the flight engineer's position having been replaced by automated electronic systems.

The automated systems not only are designed to lighten flight crew loads, they also incorporate built-in test circuitry to ease maintenance troubleshooting. The **Built In Test Equipment (BITE)** circuitry is typically contained in the control units located in the aircraft's electrical and electronic equipment (E&E) bay.

The control units, known as **Line Replaceable Units (LRU)** are easily removable to facilitate maintenance. The BITE system employs a light emitting diode(LED) display, which can be viewed by maintenance personnel during system troubleshooting. The technician then refers to a code book to determine the defective component and necessary repairs. [Figure 7-76]

It should be noted that the complex electronic circuitry which makes the automated systems possible

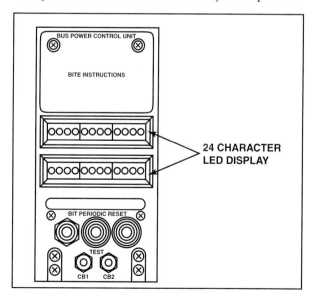

Figure 7-76. The generator control unit from a Boeing 757 contains a 24-character LED display for visual display of system faults.

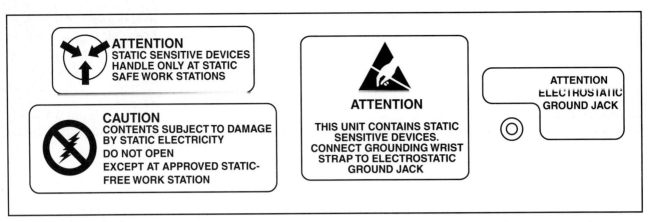

Figure 7-77. A&P technicians should be certain to use an approved wrist strap when working on or around equipment labeled with Electro Static Discharge Sensitive (ESDS) symbols and placards.

is very sensitive to stray electrical currents. The current produced from the static electricity, which is commonly produced by a technician's movements around the aircraft, could be harmful to the LRU. It is therefore very important to ground one's self using an approved wrist strap before touching any electrical component labeled as **Electro Static Discharge Sensitive (ESDS)**. [Figure 7-77]

The generator control unit monitors the AC systems and performs most of the fault analysis and resolution that was required to be done by a flight engineer on older aircraft. With the newer automated systems, the electrical controls are condensed into a smaller area in the control cabin, which permits the electrical system to be operated by the pilots. [Figure 7-78]

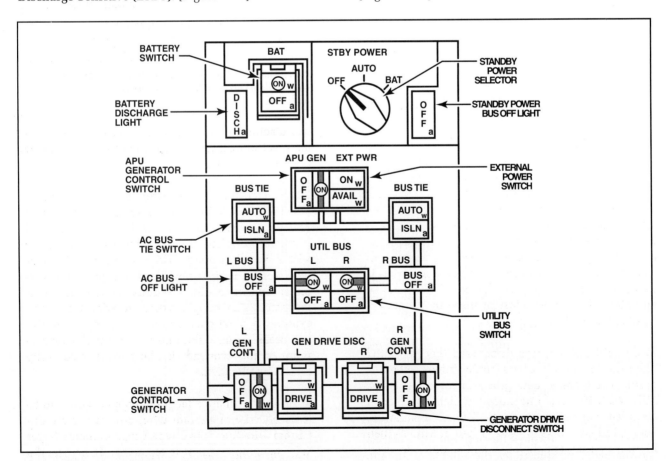

Figure 7-78. The simplicity of automated AC power systems, such as this one on the Boeing 757, helps reduce the flight crew workload.

SECTION C

WIRING INSTALLATION

The satisfactory performance of any modern aircraft depends on the continuing reliability of electrical systems and subsystems. One of the most critical factors for obtaining a high degree of reliability involves the quality of workmanship that a technician uses when installing electrical connectors and wiring. Although the concept of assembling and installing wiring may initially seem rather basic, rigid quality standards require the technician to be highly trained. Improperly or carelessly installed and maintained electrical wiring can be a source of both immediate and potential danger.

WIRE

When choosing the wire for an electrical system, there are several factors that must be considered. For example, the wire selected must be large enough to accommodate the required current without producing excessive heat or causing an excessive voltage drop. In addition, the insulation must prevent electrical leakage and be strong enough to resist damage caused by abrasion.

WIRE TYPES

The majority of wiring in aircraft is made from stranded copper. In most cases, the wiring is coated with tin, silver, or nickel to help prevent oxidation. Prior to the mid-1990s, the wire used most often was stranded copper wire manufactured to MIL-W-5086 standards. This wire is made of strands of annealed copper covered with a very thin coating of tin. For insulation purposes, a variety of materials is used, including polyvinyl chloride (PVC), nylon, and glass cloth braid. Most of these insulators are rated to 600 volts. [Figure 7-79]

In the mid-1990s it was discovered that polyvinyl chloride insulation emits toxic fumes when it burns. Therefore, a new group of mil spec wires, MIL-W-22759, was introduced which consists of stranded copper wire with Teflon® insulation. Today, MIL-W-22759 is used in lieu of MIL-W-5086 in new wiring installations and should be used when replacing original aircraft wiring. However, prior to replacing any aircraft wiring, consult the

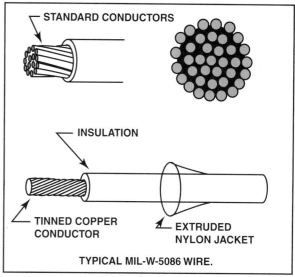

Figure 7-79. MIL-W-5086 wire consists of stranded copper conductors that are covered with tin and wrapped with PVC insulation. For added protection, most of this type of wiring is also covered in a nylon jacket.

manufacturer's service manual and service bulletins for specific instructions and information regarding both wire type and size.

Where large amounts of current must be carried for long distances, MIL-W-7072 aluminum wire is often used. This wire is insulated with either fluorinated ethylene propylene (FEP Fluorocarbon), nylon, or with a fiberglass braid. While aluminum wire does save weight, it has a few disadvantages. For example, it can carry about two-thirds as much current as the same size copper wire. In addition, when exposed to vibration, aluminum wire can crystallize and break. In fact, aluminum wire smaller than 8-gauge is not recommended because it is so easily broken by vibration.

In situations where a length of copper wire is to be replaced with aluminum wire, always use an aluminum wire that is two wire gauge numbers larger than the copper wire it is replacing. Therefore, if a piece of 8-gauge copper wire is to be replaced with aluminum, use at least a 4-gauge wire.

WIRE SIZE

Aircraft wire is measured by the **American Wire Gage (AWG)** system, with the larger numbers representing the smaller wires. The smallest size wire normally used in aircraft is 22-gauge wire, which has a diameter of about .025 inch. However, conductors carrying large amounts of current are typically of the 0000, or 4-aught size, and have a diameter of about .52 inch. Each gauge size is related to a specific cross-sectional area of wire.

The amount of current a wire is capable of carrying is determined by its cross-sectional area. In most cases, a wire's cross-sectional area is expressed in **circular mil** sizes. A circular mil is the standard measurement of a round conductor's cross-sectional area. One mil is equivalent to .001 inches. Thus, a wire that has a diameter of .125 is expressed as 125 mils. To find the cross-sectional area of a round conductor in circular mils, square the conductor's diameter. For example, if a round wire has a diameter of 3/8 inch, or 375 mils, its circular area is 140,625 circular mils (375 × 375 = 140,625).

The **square mil** is the unit of measure for square or rectangular conductors such as bus bars. To determine the cross-sectional area of such a conductor in square mils, multiply the conductor's thickness by its width. For example, the cross-sectional area of a copper strip 400 mils thick and 500 mils wide is 200,000 square mils.

Note that one circular mil is .7854 of one square mil. Therefore, to convert a circular mil area to a square mil area, divide the area in circular mils by .7854 mil. Conversely, to convert a square mil area to a circular mil area, multiply the area in square mils by .7854. [Figure 7-80]

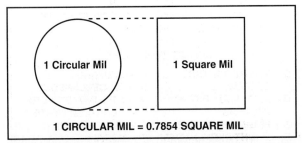

Figure 7-80. One circular mil is slightly smaller than one square mil. Circular mils are used as an area measurement for small, round items, since to measure in square inches would be unwieldy.

When replacing wire to make a repair, the damaged wire is normally replaced with the same size and type of wire. However, in new installations, several factors must be considered in selecting the proper wire. For example, one of the first things to know is the system operating voltage. It is also necessary to know the allowable voltage drop and whether the wiring operates on a continuous or intermittent basis. [Figure 7-81]

NOMINAL SYSTEM VOLTAGE	ALLOWABLE VOLTAGE DROP	
	CONTINUOUS OPERATION	INTERMITTENT OPERATION
14	0.5	1.0
28	1.0	2.0
115	4.0	8.0
200	7.0	14.0

Figure 7-81. Before wiring a new component in an aircraft, the system voltage, the allowable voltage drop, and the component's duty cycle must be known. All of these factors are interrelated.

As an example, a cowl flap motor is to be installed in a 28-volt system. Since cowl flaps are moved infrequently during a flight, a cowl flap motor is considered to operate on an intermittent basis. Reference to the chart in the previous figure indicates that the allowable voltage drop for a component installed in a 28-volt system, operated intermittently, is 2 volts. Once these three factors are known, an electric wire chart is used to determine the wire size needed for installation. [Figure 7-82]

The three curves extend diagonally across the chart from the lower left corner to the right side. These curves represent the ability of a wire to carry the current without overheating. Curve 1 represents the continuous rating of a wire when routed in bundles or conduit. If the intersection of the current and wire length lines is above this curve, the wire can carry the current without generating excessive heat.

If the intersection of the current and wire length lines falls between curve 1 and 2, the wire can be used to carry current continuously in free air. If the intersection falls between curves 2 and 3, the wire can be used to carry current intermittently. Intermittent is for two minutes or less.

WIRE MARKING

There is no standard system for wire identification among the manufacturers of general aviation aircraft. However, some form of identification mark is required every 12 to 15 inches along a wire. The identification marking should identify the wire with regard to the type of circuit, location within

Airframe Electrical Systems

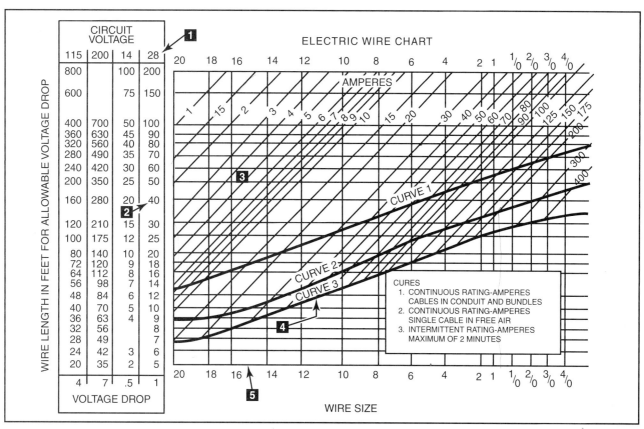

Figure 7-82. In this example, assume a 28-volt cowl flap motor draws 6 amps, and 40 feet of wire must be used for the installation. To determine the correct wire size, locate the column on the left side of the chart representing a 28-volt system (item 1). Move down this column to the horizontal line representing a wire length of 40 feet (item 2). Follow this line to the right until it intersects the diagonal line for 6 amps (item 3). Because the wire carries an intermittent current, item 3 must be at or above curve 3 on the chart (item 4). In this case, the intersection is above curve 3, and dropping down vertically to the bottom of the chart indicates a wire size between 14 and 16 (item 5). Whenever the chart indicates between two sizes, the larger wire must be selected. In this case, size 14 wire is required.

the circuit, and wire size. Each manufacturer will have a code letter to indicate the type of circuit.

For example, wire in flight instrumentation circuits is often identified with the letter "F" while the letter "N" is sometimes used to identify the circuit ground. Easy identification of each wire greatly facilitates troubleshooting procedures and saves time. [Figure 7-83]

A typical wire identification code would be F26D-22N. In this case, the letter "F" indicates the wire is in the flight instrumentation circuit, while the "26" identifies the wire as being the 26th wire in the circuit. The "D" indicates the fourth segment of the No. 26 wire. The "22" is the gauge size of the wire, and the "N" indicates that the wire goes to ground.

Another wire marking system that is sometimes used employs a two-letter identification code. This marking system provides much more detail about

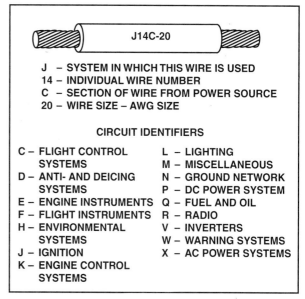

Figure 7-83. The marking J14C-20 indicates that this wire is part of an ignition circuit (J) and is the 14th wire in that circuit (14). In addition, it is the third individual segment of wire No. 14 (C), and it is a 20-gauge wire.

```
A - ARMAMENT
B - PHOTOGRAPHIC
C - CONTROL SURFACE
    CA - AUTOMATIC PILOT
    CC - WING FLAPS
    CD - ELEVATOR TRIM
D - INSTRUMENT (OTHER THAN FLIGHT OR
    ENGINE INSTRUMENT)
    DA - AMMETER
    DB - FLAP POSITION INDICATOR
    DC - CLOCK
    DD - VOLTMETER
    DE - OUTSIDE AIR TEMPERATURE
    DF - FLIGHT HOUR METER
E - ENGINE INSTRUMENT
    EA - CARBURETOR AIR TEMPERATURE
    EB - FUEL QUANTITY GAUGE & TRANSMITTER
    EC - CYLINDER HEAD TEMPERATURE
    ED - OIL PRESSURE
    EE - OIL TEMPERATURE
    EF - FUEL PRESSURE
    EG - TACHOMETER
    EH - TORQUE INDICATOR
    EJ - INSTRUMENT CLUSTER
F - FLIGHT INSTRUMENT
    FA - TURN AND BANK
    FB - PITOT STATIC TUBE HEATER & STALL
         WARNING HEATER
    FC - STALL WARNING
    FD - SPEED CONTROL SYSTEM
    FE - INDICATOR LIGHTS
G - LANDING GEAR
    GA - ACTUATOR
    GB - RETRACTION
    GC - WARNING DEVICE (HORN)
    GD - LIGHT SWITCHES
    GE - INDICATOR LIGHTS
H - HEATING, VENTILATING, & DEICING
    HA - ANTI-ICING
    HB - CABIN HEATER
    HC - CIGAR LIGHTER
    HD - DEICING
    HE - AIR CONDITIONERS
    HF - CABIN VENTILATION
J - IGNITION
    JA - MAGNETO
K - ENGINE CONTROL
    KA - STARTER CONTROL
    KB - PROPELLER SYNCHRONIZER

L - LIGHTING
    LA - CABIN
    LB - INSTRUMENT
    LC - LANDING
    LD - NAVIGATION
    LE - TAXI
    LF - ROTATING BEACON
    LG - RADIO
    LH - DEICE
    LJ - FUEL SELECTOR
    LK - TAIL FLOODLIGHT
M - MISCELLANEOUS
    MA - COWL FLAPS
    MB - ELECTRICALLY OPERATED SEATS
    MC - SMOKE GENERATOR
    MD - SPRAY EQUIPMENT
    ME - CABIN PRESSURIZATION EQUIPMENT
    MF - CHEM $O_2$-INDICATOR
P - DC POWER
    PA - POWER CIRCUIT
    PB - GENERATOR CIRCUITS
    PC - EXTERNAL POWER SOURCE
Q - FUEL & OIL
    QA - AUXILIARY FUEL PUMP
    QB - OIL DILUTION
    QC - ENGINE PRIMER
    QD - MAIN FUEL PUMPS
    QE - FUEL VALVES
R - RADIO (NAVIGATION & COMMUNICATIONS)
    RA - INSTRUMENT LANDING
    RB - COMMAND
    RC - RADIO DIRECTION FINDER
    RD - VHF
    RE - HOMING
    RF - MARKER BEACON
    RG - NAVIGATION
    RH - HIGH FREQUENCY
    RJ - INTERPHONE
    RK - UHF
    RL - LOW FREQUENCY
    RM - FREQUENCY MODULATION
    RP - AUDIO SYSTEM & AUDIO AMPLIFIER
    RR - DISTANCE MEASURING EQUIPMENT (DME)
    RS - AIRBORNE PUBLIC ADDRESS SYSTEM
S - RADAR
U - MISCELLANEOUS ELECTRONIC
    UA - IDENTIFICATION-FRIEND OR FOE
W - WARNING AND EMERGENCY
    WA - FLARE RELEASE
    WB - CHIP DETECTOR
    WC - FIRE DETECTION SYSTEM
X - AC POWER
```

Figure 7-84. A two-letter identification code is used in some wire marking systems to provide more detail on a wire's function and location.

the location of a given wire, making it especially helpful in large aircraft having many systems. [Figure 7-84]

In addition to marking individual wires, wire bundles may have a specific identification number. In this case, pressure-sensitive tape or flexible sleeves are stamped with identification codes and wrapped around a wire bundle. Individual wires within each bundle are usually hot-stamped for easy identification. Typically, wire bundles are marked near the points at which they enter and leave a compartment. [Figure 7-85]

WIRING INSTALLATION

As a general rule, electrical wiring is installed in aircraft either as open wiring or in conduit. With open wiring, individual wires, or wire bundles, are routed inside the aircraft structure without protective covering. On the other hand, when installed in conduit, electrical wiring is put inside either a rigid or flexible tubing that provides a great deal of protection.

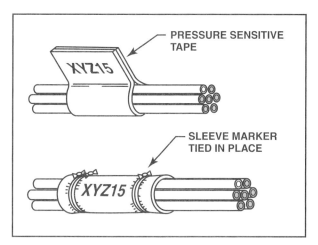

Figure 7-85. Several methods of identifying wire bundles include pressure-sensitive tape or sleeve markers tied in place.

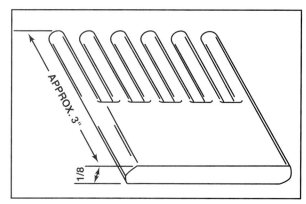

Figure 7-86. A plastic comb is a helpful tool used to keep wires straight and parallel when making up new wire bundles.

OPEN WIRING

The quickest and easiest way to install wiring is to install it as open wiring. In addition, open wiring allows easy access when troubleshooting or servicing individual circuits.

To help provide a more organized installation, electrical wiring is often installed in bundles. Several methods of assembling wires into bundles may be utilized, depending on where a bundle is fabricated.

For example, in the shop, wires stamped with identification markings may be lined up parallel to one another on a bench and tied together. However, inside an aircraft, wires may be secured to an existing bundle, or laid out to form a new bundle and tied together. In some cases, it may be helpful to connect one end of the wires to their destination terminal strip or connector before securing them together in a bundle.

Wiring harnesses fabricated at a factory are typically made on a jig board prior to installation in an aircraft. This method allows the manufacturer to preform the wire bundles with the bends needed to fit the bundle into an aircraft. Regardless of the method of bundle assembly, a plastic comb can be used to keep individual wires straight and parallel in a bundle. [Figure 7-86]

When possible, limit the number of wires in a single bundle. This helps prevent the possibility of a single wire faulting and ultimately damaging an entire bundle. In addition, no single bundle should include wires for both a main and back-up system. This helps prevent the possibility of neutralizing a system if a bundle were damaged. Additionally, it is better to keep ignition wires, shielded wires, and wires not protected by a fuse or circuit breaker separate from all other wiring.

Once all the wires in a bundle are assembled, the bundle should be tied together every 3 to 4 inches. Typically, wire bundles that are assembled on a jig are tied together with waxed linen or nylon cord using a clove hitch secured with a square knot. [Figure 7-87]

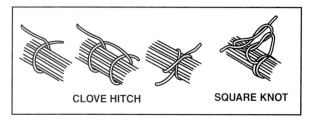

Figure 7-87. When tying a wire bundle together with either linen or nylon cord, it is best to use a clove hitch that is secured by a square knot.

In the field, nylon straps called wire ties are often used to hold wire bundles together. These small nylon straps are wrapped around the wire bundle and one end is passed through a slot in the other end and pulled tight. Once tight, the excess strapping is cut off. When cutting the ends, it is best to use a pair of flush cutting diagonal pliers or other tool that will provide a flush cut. Nylon straps can leave extremely sharp edges if they are not cut smooth against the locking portion of the strap.

Another method that is used to hold wire bundles together is with either single- or double-lacing. However, lacing should not be used with wire bundles installed around an engine because a break anywhere in the lacing cord loosens an entire section of the bundle.

Once tied, some wire bundles are covered with a heat-shrinkable tubing, a coiled "spaghetti" tubing or various other types of coverings made of Teflon, nylon, or fiberglass. Such materials give added abrasion protection.

ROUTING AND CLAMPING

All of the wire bundles installed in aircraft should be routed so they are at least three inches away from any control cable and will not interfere with any moving components. If there is any possibility that a wire or wire bundle could touch a control cable, some form of mechanical guard must be installed to keep the wire bundle and cable separated.

If possible, it is better to route electrical wiring along the overhead or the side walls of an aircraft rather than in the bottom of the fuselage. This helps prevent the wires from being damaged by fluids that may leak into low areas. In addition, electrical wiring must be routed where it cannot be damaged by persons entering or leaving the aircraft or by any baggage or cargo.

When electrical wires are routed parallel to oxygen or any type of fluid line, the wiring should be at least 6 inches above the fluid line. However, the distance can be reduced to 2 inches as long as the wiring is not supported in any way by the fluid line, and the proper mechanical protection is provided. Acceptable "secondary" protection typically includes additional clamps, approved sleeving, and the use of conduit.

Wire bundles must be securely clamped to the aircraft structure using clamps lined with a non-metallic cushion. In addition, the clamps should be spaced close enough together so the wiring bundle does not sag or vibrate excessively. [Figure 7-88]

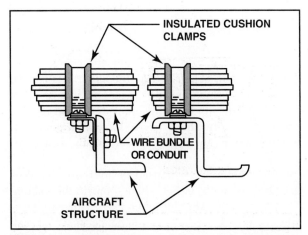

Figure 7-88. Cushion clamps, or Adel clamps, are used to support and secure wire bundles to the aircraft structure.

Another point to consider when securing wire bundles is that, once installed, the bundle should be able to be deflected about a half inch with normal hand pressure applied between any two supports. In addition, the last support should allow enough slack that a connector could be easily disconnected and reconnected for servicing. In some cases, a service loop in the wiring is the preferred method for providing adequate slack to enable equipment removal and replacement.

Any bundle that passes through a bulkhead should be clamped to a bracket which centers the bundle in the hole. In addition, if there is less than 1/4-inch clearance between the bundle and the hole, a protective rubber grommet must be installed. The grommet prevents wiring from being cut by the sharp edges of the metal hole should the centering bracket break or become bent. [Figure 7-89]

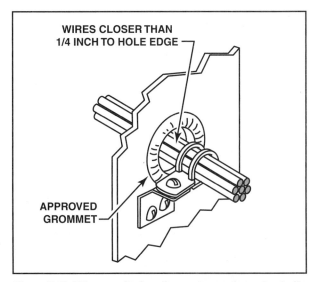

Figure 7-89. When a wire bundle must pass through a bulkhead or frame, a supporting bracket is required. If less than 1/4-inch clearance exists, a rubber grommet is installed to prevent possible wiring damage.

Where bundles must make a bend, use a bend radius that does not cause the wires on the inside of the bend to bunch up. Using a bend radius of approximately 10 times the diameter of the wire bundle is a good practice.

CONDUIT

The method of installation that provides the best mechanical protection for electrical wiring is to enclose the wiring in either a rigid or flexible metal conduit. This is the preferred method of installation in areas such as wheel wells and engine nacelles, where wire bundles are likely to be chafed or crushed.

The inside diameter of the conduit must be 25% larger than the maximum diameter of the wire bundle. When the technician figures the conduit size needed, the nominal diameter of a conduit represents the conduit's outside diameter. So twice the wall thickness must be subtracted from the conduit's outside diameter to determine the inside diameter.

When the technician installs a wire bundle inside a conduit some soapstone talc should be blown through the tubing first. The talc will act as a lubricant between the wiring and conduit wall to make it easier to install the wires and to reduce the possibility of damage. Another installation tip is to attach a long piece of lacing cord to the bundle and blow it through the tubing with compressed air. Once through the tubing, the lacing cord can be used to help pull the wire bundle through. Flaring the ends of the conduit will also make it easier to insert wiring without damaging the wire. Flaring conduit into a "bell mouth" shape also helps prevent fraying from occurring after installation.

Like open wiring, conduit must be supported by clamps attached to the aircraft structure. In addition, a 1/8-inch drain hole must be made at the lowest point in each run of the conduit. This is done to provide a means of draining any moisture that condenses inside the conduit.

Fabricating conduit for electrical wiring requires removing all burrs and sharp edges from the conduit. In addition, bends must be made using a radius that will not cause the conduit tube to kink, wrinkle, or flatten excessively.

SHIELDING

Anytime a wire carries electrical current, a magnetic field surrounds the wire. If strong enough, this magnetic field could interfere with some of the aircraft instrumentation. For example, even though the tiny light that illuminates the card of a magnetic compass is powered with low-voltage direct current, the magnetic field produced by the current flow is enough to deflect the compass. To minimize this field, a two-conductor twisted wire is used to carry the current to and from this light. With twisted wire, the magnetic fields cancel each other.

Alternating current or pulsating direct current has an especially bad effect on electronic equipment. To help prevent interference, wires that carry AC or pulsating DC are often shielded. Shielding is a method of intercepting electrical energy and shunting it to electrical ground.

Most shielding consists of a braid of tin-plated or cadmium-plated copper wire that surrounds the insulation of a wire. Typically, the braid is connected to aircraft ground through a crimped-on ring terminal. When a wire is shielded, the radiated energy from the conductor is received by the braided shielding and passed to the aircraft's ground where it cannot cause interference.

Shielding in aircraft electrical systems is typically grounded at both ends of the wiring run. However, shielding in electronic circuit wiring is usually grounded only at one end to prevent setting up a loop that could cause electromagnetic interference.

WIRING TERMINALS

Electrical wiring used in aircraft is generally terminated with **solderless terminals** that are staked, or crimped, onto the wire. Crimping is a term used to describe the squeezing of a terminal around a wire to secure the wire and provide a high quality electrical connection. There are several different types of terminals used in aircraft including the ring terminal, slotted terminal, and hook type terminal. However, the ring terminal is used most often because it virtually eliminates the possibility of circuit failure due to terminal disconnection.

Typically, solderless terminals used on 10-gauge and smaller wire use color-coded insulated terminals. For example, terminals with red insulation are used on wire gauge sizes 22 through 18, while blue insulation identifies a terminal used on 16- and 14-gauge wires. If a terminal has yellow insulation, it is used for 12- and 10-gauge wires. [Figure 7-90]

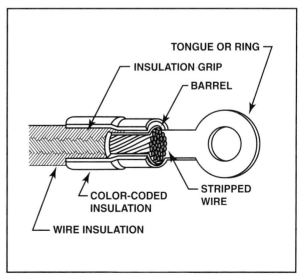

Figure 7-90. Wire gauge sizes 10 through 22 are typically terminated with preinsulated solderless terminals that are color-coded to identify various sizes.

Wires larger than 10-gauge typically use **uninsulated terminals**. In this case, a piece of vinyl tubing or heat-shrinkable tubing is used to insulate the terminal. However, the insulating material must be slipped over the wire prior to crimping. Then, after the crimping operation, the tubing is pulled over the terminal barrel and secured. [Figure 7-91]

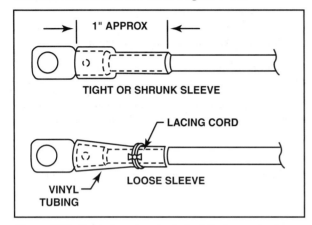

Figure 7-91. Solderless terminals installed on wire larger than 10-gauge require a separate piece of insulation. Heat-shrinkable tubing may be used or vinyl tubing may be secured over the terminal's barrel with lacing cord.

When choosing a solderless terminal, it is important that the materials of the terminal and wire are compatible. This helps eliminate the possibility of dissimilar metal corrosion.

Before a wire can be attached to a terminal, the protective insulation must be removed. This is typically done by cutting the insulation and gently pulling it from the end of the wire. This process is known as stripping the wire. Whenever stripping a wire, expose as little of the conductor as necessary to make the connection. In addition, care must be taken not to damage the conductor beyond allowable limits. Aluminum wiring must not be damaged at all because individual strands will break easily after being nicked. [Figure 7-92]

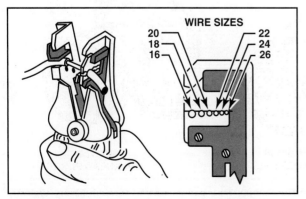

Figure 7-92. When stripping electrical wire, it is best to use a quality stripper that is designed for a specific size of wire.

The FAA specifies limits for the number of nicked or broken strands on any conductor. For example, a 20-gauge copper wire with 19 strands may have two nicks and no broken strands. In general, the larger the number of strands or the larger the conductor, the greater the acceptable number of broken or nicked strands.

Special crimping tools are needed to crimp a terminal onto a wire. A properly crimped terminal should provide a joint between the wire and the terminal that is as strong as the tensile strength of the wire itself. The preferred crimping tool is a ratchet-type crimper that meets military specifications and that is periodically calibrated to ensure a consistent and proper crimp. When a ratchet-type crimping tool is used, its handles will not release until the jaws have moved close enough together to properly compress the terminal barrel. [Figure 7-93]

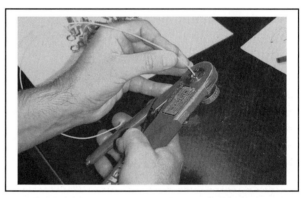

Figure 7-93. Ratchet-type crimping tools are periodically calibrated to provide a quality crimp on every terminal.

Pneumatic crimping tools are often used on wire gauge sizes 0 through 0000 because of the force required to properly crimp the wire. Like the ratchet-type crimper, a pneumatic crimper must be calibrated periodically.

Aluminum wire presents some special problems when installing terminals. Aluminum wire oxidizes when exposed to air, and the oxides formed become an electrical insulator. Therefore, to ensure a good quality electrical connection between an aluminum wire and terminal, the inside of a typical aluminum terminal is partially filled with a compound of petroleum jelly and zinc dust. This compound mechanically grinds the oxide film off the wire strands when the terminal is compressed, and then seals the wire from the air, preventing the formation of new oxides. [Figure 7-94]

CONNECTORS

Connectors are usually installed on wiring that is frequently disconnected. As an example, the wiring that

Airframe Electrical Systems

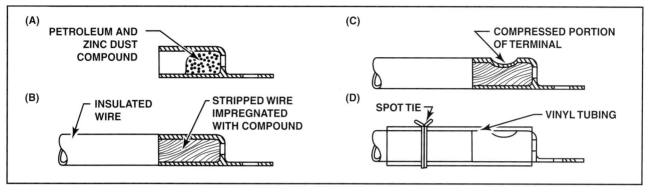

Figure 7-94. (A) To help prevent oxidation, special terminals filled with a compound of petroleum jelly and zinc dust are required for terminating aluminum wiring. (B) When a stripped wire is inserted into this type of terminal, the wire strands become impregnated with the petroleum jelly and zinc compound. (C) As the terminal is crimped, the zinc dust cleans the wire while the petroleum jelly creates an airtight seal. (D) Once crimped, a piece of vinyl tubing is tied over the terminal end to protect it from moisture.

is used to supply power to avionic components typically utilize multi-pin connectors that can be easily removed to perform maintenance. The most common **connector** used in aircraft electrical circuits are the **AN (Army-Navy) and MS (Military Specification)** available in a variety of sizes and types.

These connectors are designed to meet military specifications and have been adapted for use in most aircraft today. When it is necessary to use an electrical connector in an area where it may be exposed to moisture, special moisture proof connectors should be used. Mating connectors consist of one connector with female contacts, or sockets, and another connector with male contacts, or pins.

The ground side of an electrical power conductor is typically connected to a male connector while the power side of the conductor is attached to the female connector. This is done to reduce the chance of an accidental short between the power side of a circuit and any conductive surface when the mating connectors are separated. [Figure 7-95]

There are two ways the wires are typically connected to the contacts in a connector. Newer plugs

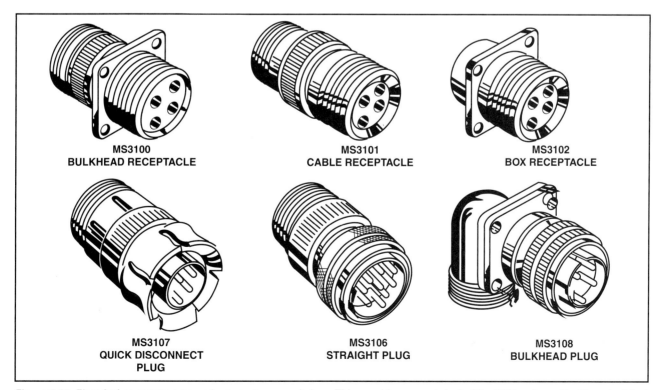

Figure 7-95. Electrical connectors come in many types and sizes. This provides a great deal of flexibility when attaching electrical wiring to various components.

use tapered pins that are crimped onto the wire end. The pin is then slipped into a tapered hole in the pin or socket end of a connector. A special tool is used to remove or replace the pin if it is ever necessary.

Another, less common, type of connector plug requires the wires to be soldered into each end of a connector. To install a typical soldered connector, begin by stripping enough insulation from the end of each wire to provide approximately 1/32-inch of bare wire between the end of the insulation and the end of a pin when inserted.

Next, slip a specified length of insulated tubing over the end of each wire. Using an appropriate soldering iron, solder, and flux, apply a small amount of solder to the stripped portion of each wire. This is a process known as Tinning the wires and greatly facilitates their insertion into the connector pins.

Once the wires are tinned, fill the end of each pin, or solder pot, with solder. While keeping the solder in a solder pot molten, insert the appropriate wire. Once inserted, remove the soldering iron and hold the wire as still as possible until the solder solidifies. If the wire is moved before the solder solidifies, the solder will take on a granular appearance and cause excessive electrical resistance. Repeat this process until all pins and wires have been soldered in place. After all of the wires are installed, clean any remaining flux residue from the connector with an approved cleaner. [Figure 7-96]

Inspect the soldered connector before final assembly. If excess solder is bridging two pins, or if some pins do not have enough solder, re-solder the connection. The solder in a properly soldered connection should completely fill each pot and have a slightly rounded top. However, the solder must not wick up into the strands of each wire, or the wire may become brittle and break when bent. When the soldered joints are satisfactorily completed, slip the insulation tubing on each wire over its respective solder pot. Once all solder pots are covered, spot tie the bundle just above the insulation tubing to keep it in place.

SPLICING REPAIRS

In general, splices in electrical wire should be kept to a minimum and avoided completely in areas of high vibration. Splices in bundles should be staggered with no more than one splice in any one segment of wire. Splices are not to be used to salvage scrap wire and cannot be installed within 12 inches of a termination device. In certain instances, these

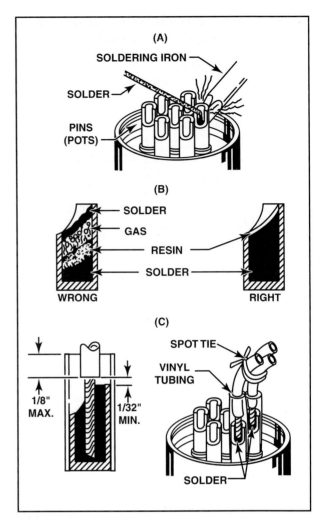

Figure 7-96. (A) After the wires have been tinned, fill a solder pot with solder. (B) When doing this, allow enough time for the pot to heat up sufficiently. Otherwise, resin could become trapped in the solder producing a weak connection. (C) Once installed, there should be a minimum of 1/32-inch clearance between the wire insulation and the top of the solder joint. In addition, a piece of vinyl tubing should be slipped over each connection and the wires tied off.

rules can be amended if an approved engineer has authorized the splice installation. In some cases, emergency splices have been approved for use in commercial aircraft, with the understanding that the wire would be repaired with a permanent installation within a limited number of flight hours.

TERMINAL STRIPS

A terminal strip consists of a series of threaded studs that are mounted on a strip of insulating material. A typical terminal strip is made of a plastic or a paper-based phenolic compound that provides a high mechanical strength as well as good electrical insulation properties. The terminal studs should be secured to prevent rotation. In addition, most terminal strips have barriers between adjacent studs to keep the wires properly separated. [Figure 7-97]

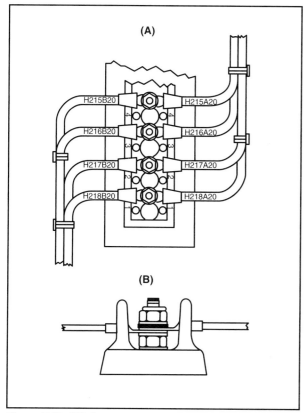

Figure 7-97. (A) A common terminal strip used in aircraft consists of several studs set in a nonconductive base. (B) To help prevent shorting between terminal connectors, most terminal strips have some sort of barrier between each stud.

The size of the studs on a terminal strip usually range from size 6 up to about 1 inch in diameter, with the smaller sizes suitable only for low current control circuits. The smallest terminal stud allowed for electrical power systems is a No. 10.

When attaching the terminal ends of a wire bundle to a terminal strip, fan the wires out from the bundle so the wires align with the terminal studs. A typical installation centers the wires and arranges them at right angles to the terminal strip. A short distance from the terminal strip, the wires are then turned 90 degrees to form a wire bundle.

Ideally, there should be only two wires attached to any one stud. However, the FAA allows up to four terminals to be stacked on any single stud. When attaching multiple terminals to one stud, or lug, the wires must be angled out from the terminal stud and stacked to avoid damaging the tongue of each terminal. [Figure 7-98]

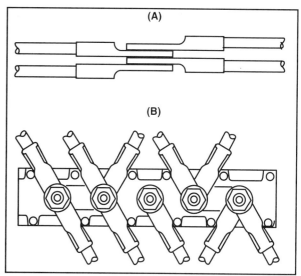

Figure 7-98. (A) Proper stacking of multiple terminals requires terminal tongues to be alternately placed up and down. (B) A typical method of attaching multiple wires to a terminal stud stacks and angles the wires out from each stud.

JUNCTION BOXES

Junction boxes installed in aircraft powerplant compartments or other high heat/fire areas are typically made of stainless steel or a heat resistant aluminum alloy. Junction boxes for aircraft use must be constructed in a way that no oil canning exists since this is considered to be a shorting hazard.

Oil canning occurs when the metal of a junction box is pushed inward and remains in the bent condition when the pressure is removed. In addition, junction boxes are mounted in a way that minimizes the possibility of water getting into the box and causing electrical shorts or corrosion. Most junction boxes in powerplant compartments are mounted vertically to help prevent small hardware from becoming lodged between terminals and causing a short circuit or electrical fire.

Junction boxes should be isolated from electrical power whenever the cover is removed for maintenance. On occasions where it is necessary to apply power to circuits in a junction box for performing maintenance, be very cautious. A metal tool carelessly placed across several terminals could cause severe damage to the aircraft electrical system and components. In addition, remove hand jewelry to avoid accidental shocks or burns while working in the box.

BONDING

Bonding is a process that grounds all components in an aircraft together electrically. This prevents a difference in potential from building to the point that sparks jump from one component to another. For example, all the control surfaces on an aircraft are electrically grounded to the main aircraft structure with braided bonding straps. This helps prevent the inherent resistance in a control surface's hinge from insulating the control surface from the main structure. In the powerplant compartment, shock-mounted components are also bonded to the main structure with bonding straps.

Any electrical component which uses the aircraft structure as the return path for its current must be bonded to the structure. When selecting a bonding strap, it must be large enough to handle all the return current flow without producing an unacceptable voltage drop. An adequately bonded structure requires bonding straps that hold resistance readings to a negligible amount.

For example, the resistance between any component and the aircraft structure must be 5 milliohms (.005 ohms) or less and the resistance of the bonding strap should not exceed 3 milliohms (.003 ohms).). A bonding strap must be long enough to allow free movement of a component, but not so long as to increase resistance too much. Bonding straps should be made from a material that does not produce galvanic corrosion. Use of appropriate washers can minimize the chance of corrosion to the aircraft structure itself. For example, installing a sacrificial aluminum washer between a copper strap and an aluminum stringer allows all potential galvanic corrosion between the copper and aluminum to be concentrated in the washer rather than the structure.

Aluminum alloy jumpers are recommended for most cases, but copper jumpers should be used to bond together parts made of stainless steel, cadmium plated steel, copper, brass, or bronze. The bonding strap should have enough mechanical strength to withstand constant flexing and should be easily installed to allow removal of the component being bonded. [Figure 7-99]

The large flow of current from an electrical engine starter requires a heavy bonding strap between the engine and airframe. This is especially necessary on engines that are mounted in rubber shock mounts that have high electrical insulation properties. If a bonding strap is not installed between the engine and airframe, or if a break or extensive corrosion exists in an installed bonding strap, the risk of fire is greatly increased. This occurs because much of the starter return current will be forced to pass through metal fuel and primer lines to reach ground.

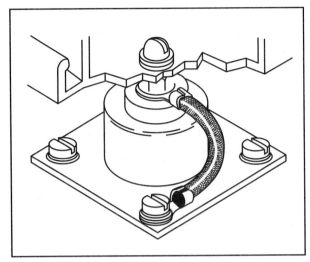

Figure 7-99. A braided bonding strap of adequate length and electrical capacity provides a current return path for otherwise insulated shock-mounted components.

COAXIAL CABLE

Wiring circuits that connect various pieces of electronic equipment to an antenna are especially prone to interference from outside sources of **radio frequency interference (RFI)**. To preclude RFI, special types of shielded wire are available that both hold in RFI and protect the circuit from outside RFI. The most common type of RF cable is referred to as coaxial cable. The installation and maintenance of coaxial cable are critical because of the possibility of causing erroneous electronic equipment indications and interference.

Despite the shielding of coaxial cable, cable runs should be as direct and as short as possible to further minimize RFI hazards. Coaxial cable has an inner conductor, a layer of insulation and a braided outer conductor encased in a vinyl jacket. The braided outer conductor is grounded at one end or the other, or at both ends, to protect the electronic circuitry.

There are a number of sizes and types of coaxial cable used for electronic installations, and each type must be terminated in the way specified by the manufacturers of the cable and the equipment being wired. The **BNC connector** is the type of connector most widely specified. The proper installation of the BNC connector is critical to the integrity of the installation. [Figure 7-100]

When installing coaxial cable, if the cable must be routed around corners, its minimum bend radius should be 6 times the diameter of the cable to prevent kinking. If the cable becomes kinked, dented,

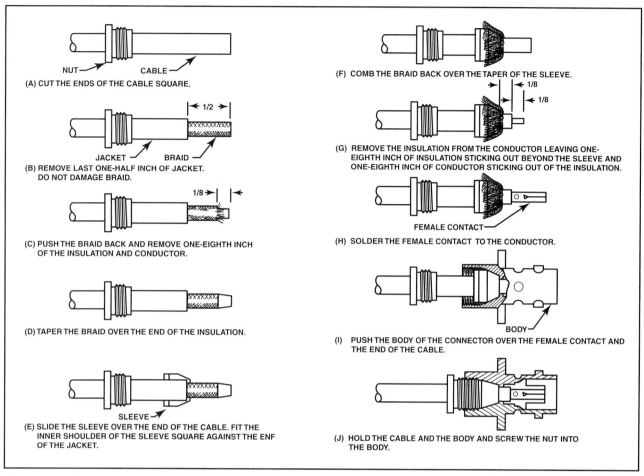

Figure 7-100. Coaxial cable is usually terminated in a BNC connector. Proper installation of the connector provides a shielded installation that meets the stringent requirements of electronic installations.

or crimped, the reduction in the distance from the center conductor to the shielding may cause electrical interference. To prevent damage, the cable should be supported at a minimum of 24-inch intervals and located in a position that precludes it from being stepped on or used as a handhold.

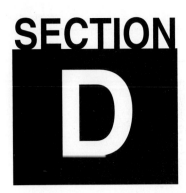

SECTION D

ELECTRICAL SYSTEM COMPONENTS

SWITCHES

The purpose of a switch is to interrupt the flow of current to the component it controls. Each switch is rated with regard to the voltage it can withstand and the current it can carry. However, the continuous voltage and current in a circuit are not typically the limiting factors for switch ratings. Instead, voltage and amperage ratings must take into account that the switch's contacts may be subject to high voltage arcing as the switch is turned off, or high current may flow through the contacts when the switch is initially closed. Depending on the types of components installed in a circuit, it's necessary to derate switches, which simply means that a switch must be rated to sustain more voltage and current than what the circuit normally carries continuously.

For example, inductive loads can create a large voltage spike that can cause a switch to fail if it's not derated. Inductive loads, such as those produced by relays and solenoids, will produce a high voltage as the circuit is turned off due to the inductance generated as magnetic flux lines collapse back into the control coil. If not suppressed with a diode or capacitor, the high voltage may arc across the contacts as the switch is turned off, which could cause the contacts to weld together, or, in severe cases, burn away entirely.

Other components that must have their control switch derated include motors, which have a high in-rush of current when first energized. Once the motor develops its full rotational speed, it draws its normal rated current, but the initial current draw may be sufficiently high to damage the switch if its rating is inadequate. On the other hand, when a motor is turned off, the collapse of the motor's field causes a high voltage to be induced into the circuit, which can also damage the switch in the same manner as relays and solenoids.

In another inductive circuit, the resistance of the filament in a lamp is very low when it is cold. Therefore, when a switch is first closed in a lamp circuit, the initial surge of current is as much as 15 times as high as the current the lamp uses for continuous operation. Once the filament heats up, its resistance increases significantly, which reduces the amount of current flowing through the lamp.

Because a switch must be able to carry its rated current continuously, and at the same time be able to operate the type of load in which it is installed, a derating factor has been devised that allows matching the switch to the type of load. A derating factor table, such as that found in AC43.13-1B, should be consulted when choosing a switch for any circuit. [Figure 7-101]

NOMINAL SYSTEM VOLTAGE	TYPE OF LOAD	DERATING FACTOR
28 VDC	Lamp	8
28 VDC	Inductive (relay-solenoid)	4
28 VDC	Resistive (heater)	2
28 VDC	Motor	3
12 VDC	Lamp	5
12 VDC	Inductive (relay-solenoid)	2
12VDC	Resistive (heater)	1
12 VDC	Motor	2

NOTES:
1. To find the nominal rating of a switch required to operate a given device, multiply the continuous current rating of the device by the derating factor corresponding to the voltage and type of load
2. To find the continuous rating that a switch of a given nominal rating will handle efficiently, divide the switch nominal rating by the derating factor corresponding to the voltage and type of load

Figure 7-101. A switch-derating chart is used to select the proper nominal switch rating when the continuous load current is known. It can also be used to determine the continuous load a particular switch can handle.

For example, a typical aircraft toggle switch may be rated at 24 volts for 35 amps. If this switch is to be installed in a lamp circuit, it should have a derating factor of 8 applied to the switch rating. This means that the switch should not have more than approximately 4.3 amps flowing through it continuously (35 amps ÷ 8=4.375 amps).

Once the derating factor has been calculated, determine that the switch rating is sufficient for the circuit. This is easily accomplished by referring to the rating information that is indicated on the case of the switch. If the information is not legible or is missing, most manufacturers have reference manuals that can be checked using the switch's part number.

SWITCH INSTALLATION

Hazardous errors in switch operation can be avoided by logical and consistent installation. For example, two-position "on-off" switches should be mounted so that an upward or forward movement of the toggle obtains the "on" position. In addition, switches that control movable aircraft components such as landing gear or flaps should be installed so the switch moves in the same direction as the desired motion of the component.

Certain circuits must be operated only in an emergency. For these circuits, the switch is normally enclosed in a cover that must be lifted before the switch can be actuated. These switches are said to be guarded. If the switch is one that could create a hazard if it's operated inadvertently, the guard cover may be wired shut with a lightweight safety wire that can be broken if it becomes necessary to gain access to the switch.

TOGGLE AND ROCKER SWITCHES

Toggle switches have been one of the more popular type switches for older aircraft electrical systems, but in the past few years, rocker-type switches that allow for easier operation and yield a more attractive installation have replaced toggle switches. Rocker switches are also safer should the pilot or passenger accidentally strike a switch during flights in rough air, a crash or hard landing. [Figure 7-102]

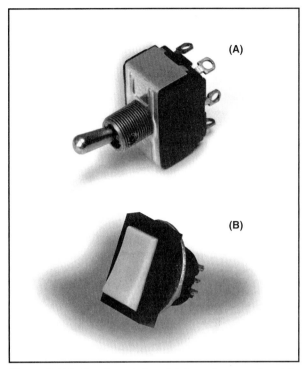

Figure 7-102. Toggle switches (a) have been replaced in newer aircraft by rocker switches (b), since they are less hazardous to crewmembers who might bump into them.

Electrically, both toggle and rocker switches operate the same way and are available with a number of different contact arrangements. These switches are named according to the number of circuits they control and the number of sets of poles they have.

A single-pole, single-throw (SPST) switch has two contacts, and controls only one circuit. It can only be either on or off. A single-pole, double-throw (SPDT) switch selects two conditions for a single circuit. A flap motor switch would, for example, have two positions for sending current through the flap motor, driving the motor in the direction either to raise or lower the flaps. [Figure 7-103]

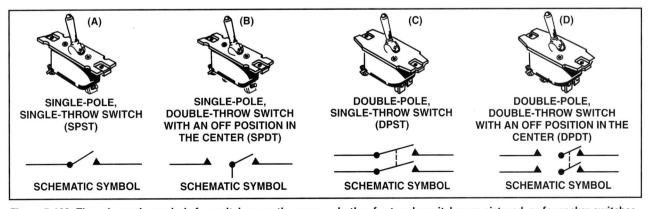

Figure 7-103. The schematic symbols for switches are the same, whether for toggle switches as pictured, or for rocker switches.

A **double-pole, single-throw (DPST)** switch controls two circuits but with only an open and closed position. The advantage of a double-pole switch is that both circuits are controlled together, but they can be fused separately. If one circuit fails it would not effect the other. An example of this type of switch might be used to control navigation and panel lights. Both circuits could be turned on with one switch, and yet the circuits are completely independent.

Double-pole, double-throw (DPDT) switches control two circuits in two conditions and may have two or three positions. A two-position switch controls the circuit(s) in two conditions only, such as open or closed. A three-position switch would also have a center position for turning the circuit off. If the switch must be held in a particular position to actuate a circuit, it is usually called a momentary switch. Another description of these types of switches is "normally open" or "normally closed." For example, a switch that has only two positions for a single wire circuit would be called a single pole, single throw (SPST), two-position, normally open switch if it must be held in the "on" position.

ROTARY SWITCHES

When it is necessary to select several conditions for a circuit, a rotary switch may be used. These switches are made up of wafers with contacts arranged radially around the central shaft and contact arm. Any number of wafers can be stacked onto the shaft to control as many circuits as are needed. [Figure 7-104]

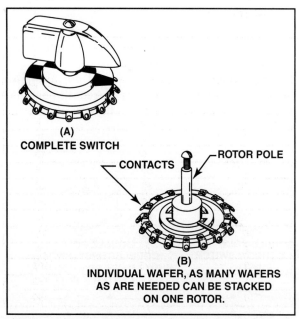

Figure 7-104. Rotary switches allow selection of any of several conditions within a circuit. Stacking wafers allows more than one circuit to be controlled simultaneously.

PRECISION (MICRO) SWITCHES

Precision switches, generally known by the trade name *Microswitches*™, are used in many applications in an aircraft. Most commonly, these switches are actuated when some mechanical device reaches a particular position. These switches require only a slight movement of the operating plunger to cause the internal spring to snap the contacts open or closed.

When precision switches are used to limit the movement of a mechanism, they are typically referred to as limit switches. For example, the electric motor of a flap actuator would be turned off by a limit switch when the flaps reach their up or down limit. In a similar manner, the up-and-locked and down-and-locked lights of a landing gear system are controlled through a precision switch when the gear is firmly locked in the appropriate position. [Figure 7-105]

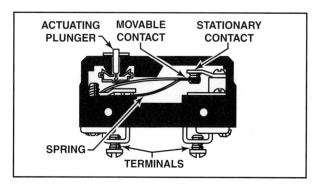

Figure 7-105. A precision switch is actuated when a mechanical device moves into position to depress an actuating plunger. These switches may be designed to be normally open or normally closed, or the switch may have three terminals that allow it to be used either way.

RELAYS AND SOLENOIDS

One of the features of an electrical system is the ability to remotely control components which are located in some far corner of the aircraft. By using a solenoid, a very small switch can be used to control the current needed to operate an aircraft engine starter or other high-current device.

Relays and solenoids are quite similar, with only a mechanical difference. Normally a relay has a fixed soft-iron core around which an electromagnetic coil is wound. Movable contacts are closed by the magnetic pull exerted by the core when the coil is energized, and are opened by a spring when the coil is de-energized (normally open relay). A relay can also be normally closed with the opposite functions of the spring and electromagnet.

A solenoid has a movable core that is pulled into the center of an electromagnetic coil when the coil is

Airframe Electrical Systems

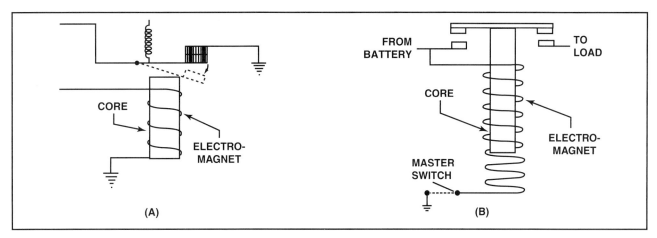

Figure 7-106. Electrical circuits can be controlled remotely by use of relays (a), or by solenoids (b). Both are sometimes referred to as contactors.

energized. Due to the movable core, solenoids respond quicker and are stronger than relays. Both relays and solenoids are used for electrical controls. Solenoids are typically used for high current applications and also find important use as mechanical control devices; for example, to move locking pins into and out of mechanically actuated devices. To eliminate the confusion between relays and solenoids, many aircraft manufacturers refer to all magnetically activated switches as contactors. [Figure 7-106]

Two very commonly used magnetically operated switches are the battery contactor and the starter contactor used on most light aircraft. The basic difference between these two devices is the operation of the coil. One end of the coil in the battery contactor is internally connected to the main terminal that connects to the battery. The other end of the coil comes out of the contactor case through an insulated terminal and goes to the battery master switch, through which it is grounded when the master switch is turned on.

One end of the control coil in the starter solenoid is grounded inside of the housing, and the other end comes out through an insulated terminal and goes to the ignition switch. When the ignition switch is in the START position, current flows through the starter contactor coil and closes it, allowing the high current to flow to the starter. It should also be noted that starter contactors are intended for intermittent duty cycles. The electromagnetic coil may overheat if activated for long periods of time. On the other hand, a battery contactor is rated for continuous operation.

Although both contactors often look the same externally, it's important to verify that the unit has the correct part number. If a starter contactor is inadvertently used in place of a battery contactor, it may become overheated and fail during extended operation.

CURRENT LIMITING DEVICES

Electrical circuit protection devices are installed primarily to protect circuit wiring. To do this adequately, the protection devices should be located as close as possible to the electrical power source bus. Two types of circuit protection devices that are used on aircraft are circuit breakers and fuses.

A current limiting device should open an electrical circuit before its conductors become hot enough to emit smoke. In order to achieve this, a circuit breaker or fuse is selected with time/current characteristics that fall below that of the associated conductor. Circuit protection should be matched to conductor requirements in order to obtain maximum utilization of the equipment operated by the circuit. [Figure 7-107]

WIRE AND GAUGE COPPER	CIRCUIT BREAKER AMPERAGE	FUSE AMP
22	5	5
20	7.5	5
18	10	10
16	15	10
14	20	15
12	30	20
10	40	30
8	50	50
6	80	70
4	100	70
2	125	100
1		150
0		150

Figure 7-107. This chart is an example of the type used to select circuit breaker and fuse protection for copper conductors. Such charts are usually applicable to a specific set of ambient temperatures and wire bundle sizes.

FUSES

A fuse is simply a piece of low-melting-point alloy encased in a glass tube with metal contacts on each end. It is placed in the circuit, usually at the bus bar, so all of the current flowing to a circuit must pass through it. If too much current flows, the link will melt and open the circuit. To restore the circuit, the blown fuse must be replaced with a new one.

There are two types of fuses used in aircraft circuits—the regular glass tubular fuse and the slow-blow fuse. The regular fuse has a simple narrow strip of low-melting-point material that will melt as soon as an excess of current flows through it. The slow-blow fuse has a larger fusible element that is held under tension by a small coil spring inside the glass tube. This fuse will pass a momentary surge of high current such as that when the switch in a lighting circuit is closed, but it will soften under a sustained current flow in excess of its rating. The spring will pull the fusible link in two, opening the circuit. Aircraft fuses are rated according to the maximum continuous amperage they can carry. [Figure 7-108]

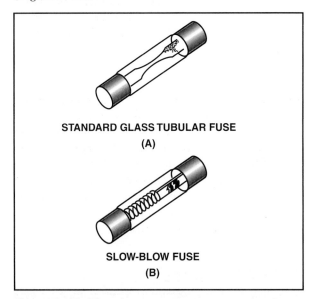

Figure 7-108. Though not common, some aircraft are equipped with fuses for circuit protection.

For certification of a normal, utility, acrobatic, or commuter category aircraft, CFR 14 Part 23 requires that a spare fuse of each rating, or 50% of each rating of accessible fuses, whichever is greater, be carried and readily accessible in flight. CFR 14 Part 91 requires the operator of an aircraft, using fuses as protective devices and flying at night, to carry a spare set. The spare set may consist of a complete replacement set of fuses or a quantity of three replacement fuses for each kind required.

Fuses that are accessible by maintenance personnel are found on some aircraft. These fuses, commonly called **current limiters**, are often used to isolate a complete distribution bus in the event of a short to that bus. In the event the current limiter opens, the pilot would continue the flight without use of the isolated bus and have the problem corrected upon landing. [Figure 7-109]

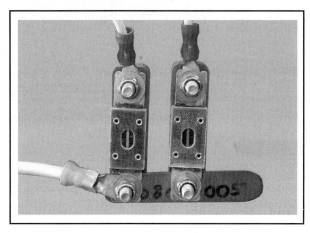

Figure 7-109. A current limiter operates in the same manner as a fuse and is often used to protect an entire electrical bus. The center section of the limiter has a clear inspection window where the condition of the fusible material can be visually inspected.

CIRCUIT BREAKERS

Circuit breakers are used rather than fuses because it is so much easier to restore a circuit in flight by simply resetting the circuit breaker than it is to remove and replace a fuse. There are two basic types of circuit breakers used in aircraft electrical systems: those that operate on heat, and those that are opened by the pull of a magnetic field. Most aircraft breakers, however, work on the principle of heat. When more current flows than the circuit breaker is rated for, a bimetallic strip inside the housing warps out of shape and snaps the contacts open, disconnecting the circuit. [Figure 7-110]

There are three basic configurations of circuit breakers used in aircraft electrical systems: the push-to-reset, the push-pull type and the toggle type. The button of the push-to-reset type of circuit breaker is normally in, and it pops out only when the circuit has been overloaded. This type of circuit protection device cannot be used as a switch, as there is no way to grip the button to pull it out.

The push-pull-type circuit breaker has a small lip that can allow the breaker to be pulled to open the circuit. Normally these circuit breaker buttons are in, but if a breaker is overloaded, the button pops

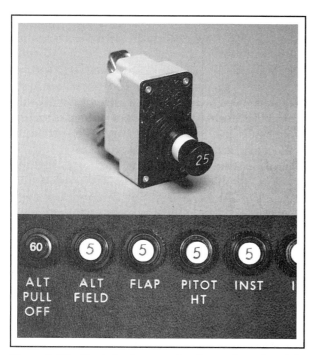

Figure 7-110. Most circuit protection for modern aircraft is provided by circuit breakers that can be reset in flight. The push-pull type is shown in the upper panel, with the push-to-reset type in the lower panel.

out and is easily identified. Many of these circuit breakers have a white band around the button that is visible when the breaker has popped. This type of circuit breaker should be used as a switch only for isolating a circuit for maintenance purposes.

The toggle-type circuit breaker is normally used as a control switch as well as a circuit breaker. When the toggle is up, the circuit is closed, but if the circuit is overloaded, the breaker will pop partially down. To restore the circuit, the toggle is moved all the way down and then back up.

All aircraft circuit breakers must be of the trip-free type, which simply means that the breaker contacts will remain open as long as a short-circuit exists, regardless of the actuating control position. This prevents the circuit breaker from being closed and manually held closed if a fault exists that could generate enough heat to cause a fire. Automatic reset circuit breakers are not allowed in aircraft installations.

Anytime a circuit breaker opens a circuit, it should be allowed to cool before being reset. If it tripped because of a transient condition in the electrical system, it will remain closed and the electrical system should operate normally. If it was opened by an actual fault, it will pop open again and should be left open until the cause of the excessive current draw is found.

On many large commercial aircraft, electronic circuit breakers are being used to control overloaded power distribution circuits. The current flowing through a conductor is measured using inductive pickups called current transformers. The inductive pickups generate a signal that is monitored by a control unit computer, and if an overcurrent condition exists, the computer opens the circuit breaker.

ELECTRICAL CONTROL PLACARDS

When electrical control components such as switches or circuit breakers are installed, it is important to make sure that the device is placarded to indicate its function. These placards, or labels, should be readily visible to the flight crew in day or nighttime conditions, and should also indicate any special operating considerations. For example, when flashing anticollision lights are installed, a placard is usually located near the control switch to advise the flight crew to turn the lights off when operating in clouds or haze. This precaution is necessary to prevent the crew from becoming disoriented due to the light being reflected back into the control cabin.

AIRCRAFT LIGHTS

As aircraft have become more complex and airspace has become more crowded, the number of lighting systems required for safe operations and passenger comfort has increased. Interior lighting systems are needed to illuminate most areas of the aircraft, from cargo compartments to the pilot's instrument panel, while exterior lights are required to ensure safety by illuminating landing and taxi environments, and to provide recognition or position indications to other aircraft.

EXTERIOR LIGHTS

Most aircraft use a variety of exterior lighting systems. Some examples include position, landing, taxi, anticollision and wing inspection lights. Each of these systems is designed to power a lamp, or bulb, that is ideal for the application. For external lighting there is a number of different types of lamps used including incandescent, halogen, and xenon.

INCANDESCENT LAMPS

Incandescent lamps are the most popular types of lights for exterior use and are also found extensively in the interior of an aircraft. These lamps use a small coil of tungsten wire, called a filament, which is surrounded by a sealed bulb containing an inert gas. When electrical current is applied to the lamp, it causes the filament to illuminate. However, as the lamp ages, the filament slowly erodes until it finally breaks, or "burns out."

In most cases the filament can be seen through the clear glass bulb; thus, many technicians rely on a visual inspection to determine if the lamp is good. However, it's sometimes possible for the filament to appear good, when in fact, it's burned out.

The best way to check an incandescent lamp is to conduct a continuity test of the filament, using an ohmmeter: Set the ohmmeter to a low resistance range and then hold the meter leads on each of the lamp's terminals. If the meter shows resistance, the filament has continuity. However, the resistance read when conducting this test will not be the same value as when the lamp is operating because the filament heats up when the lamp is on, which causes the resistance to significantly increase.

Incandescent lamps come with various reflector designs and in a wide variety of sizes to suit the application. In addition, most are available with different voltage and amperage ratings. Because lamps may look identical, yet have different ratings, it's important to verify that a replacement lamp is correct for the light. Check the part number on the lamp with the one indicated in the aircraft manufacturer's illustrated parts catalog. Never assume the removed lamp has the correct part number, in the event the wrong one was previously installed.

HALOGEN LAMPS

Halogen lamps are often used on the exterior of an aircraft for such items as flashing beacons and illumination lights. These are similar to incandescent lamps in that they use a tungsten filament. However, instead of being filled with an inert gas, the bulb is filled with a halogen gas such as iodine-vapor. The halogen gas allows the filament to burn much brighter, but also increases the lamp's useful life by reducing the decomposition of the tungsten filament.

Halogen lamps are checked and maintained in essentially the same manner as incandescent lamps. Since a filament is used in a halogen lamp, its operation can be checked with an ohmmeter. However, one major precaution when working with these lamps is to avoid bare-hand contact with the bulb, since oil from the skin can burn into the bulb during operation. This decreases the lamp's intensity and can also lead to premature failure.

XENON LAMPS

Xenon lamps are used primarily on the exterior of an aircraft for flashing anticollision lights, commonly referred to as strobes. A strobe is a compact unit that develops a flash of white light that has a much greater intensity than can be developed with an incandescent lamp. Because of the intensity, strobes not only increase flight safety at night, but also permit an aircraft to be seen more readily in daylight conditions.

Xenon lamps are available in a variety of shapes and sizes to fit small areas such as wingtips and tail cones. Unlike an incandescent lamp, xenon lamps do not have a filament. Instead, they use a sealed, clear glass flash tube that is filled with xenon gas, which illuminates when high voltage is passed through the flash tube between electrodes mounted at each end of the tube.

Because a xenon lamp does not have a filament, it is difficult to determine if the lamp is burned out. Without a filament, a continuity test cannot be performed to test the lamp. Also, the bulb tends to discolor as it ages, which can cause the lamp to appear burned out, yet still be functional. About the only way to test these lamps is to swap a questionable lamp with a good one, or to use special test equipment from the manufacturer.

Xenon lamps are sensitive to oil in the same manner as a halogen lamp. If oil from the skin is left on the flash tube, it may cause the glass to concentrate heat in that area during operation and cause cracking. Avoid touching a xenon lamp by handling it with a soft, clean cloth. If the flash tube is inadvertently touched, clean it with isopropyl alcohol and allow it to dry before installation.

POSITION LIGHTS

Position lights are used to indicate the location and movement of an aircraft during night operations. By noting different colors of position lights, the operator of an aircraft can detect if an aircraft is stationary or moving, and then safely navigate around that aircraft. Hence, position lights are often referred to as navigation lights.

Position lights are installed in a manner that allows different colored lights to be seen in various locations around the aircraft. A red colored light is located on the left side of the aircraft (from the pilot's seat perspective), while a green light is located on the right side. An additional white light is used to indicate the back of the aircraft. By noting the light colors, aircraft operators can identify the

Airframe Electrical Systems

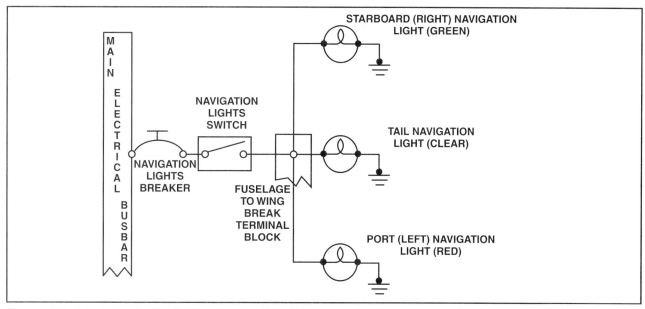

Figure 7-111. A typical navigation light system has all the lights connected in parallel, controlled by a single switch, and protected by a circuit breaker or fuse.

positions and movement of other aircraft. For example, when a green and red light are seen, the colors indicate that the aircraft is heading toward the observer. On the other hand, if a white light is seen, the aircraft is moving away from the observer.

Although the lamp of a position light has a clear glass, different colors are achieved by placing a red or green lens over the bulb. While most navigation lights operate in one mode only, some older or larger aircraft have flasher circuits incorporated in the electrical circuit to allow the lights to pulse. [Figure 7-111]

ANTICOLLISION LIGHTS

Anticollision lights are found in two basic styles depending on the age of the aircraft. Older aircraft were originally equipped with rotating beacons either on top of the vertical stabilizer or on top or bottom of the fuselage. Newer systems often utilize solid state electronics to create a flashing-, or strobe-type anticollision light. The rotating beacon system typically contains a stationary light bulb and a rotating reflector covered by a red glass lens. [Figure 7-112]

As previously mentioned, **strobe-type flashing anticollision lights** have an extremely bright flash produced by a xenon tube that requires a high voltage to operate. The high voltage is produced by a strobe power supply that uses a capacitor charging system. Modern aircraft equipped with a strobe-type anticollision light system typically consist of two white wing tip strobe lamps. Different configurations can be used, depending on the type of aircraft, with all

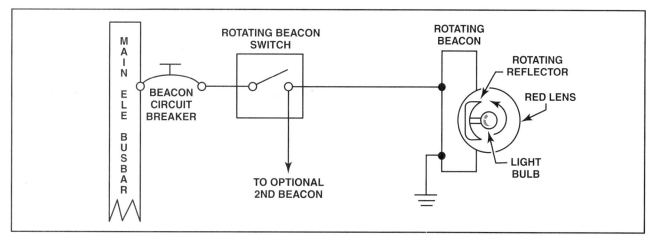

Figure 7-112. A rotating beacon consists of a stationary light bulb, a rotating reflector driven by a small motor, and a red lens.

systems meeting lighting criteria as outlined in 14 CFR Part 23. [Figure 7-113]

To prevent radio frequency interference (RFI), the power wires going to motorized rotating beacons and flashing strobe lights are shielded, and the components of the system are bonded to the structure with jumper wires, or by directly mounting the light unit to the structure. Isolating the source of RFI is easily accomplished by operating each lighting system while listening to the audio from a communications radio. Once the RFI source is isolated, a thorough visual inspection of the lighting system should be performed to locate broken shielding and improper electrical bonding of the light unit to the airframe structure.

LANDING AND TAXI LIGHTS

In order to provide night visibility for landing and taxiing, sealed beam lamps are mounted on the aircraft, facing forward. These lights can be mounted in the leading edge of the wing, in the cowling or on landing gear struts. Some aircraft employ a retractable light that extends from the wing during use. Since the airplane is in a nose-high attitude during landing, landing lights must point forward and slightly down. On the other hand, taxi lights point mostly straight forward from the airplane while in a level attitude.

Since both landing and taxi lights are relatively high powered circuits, they are often controlled, or switched, by a solenoid or relay. The pilot activates the appropriate control cabin switch, which engages the relay or solenoid coil to pull the switch contacts into the "on" position. In some cases of an inoperative landing or taxi light, the relay or solenoid may be the faulty component and should be considered when isolating a malfunction.

WING INSPECTION LIGHTS

Many aircraft are equipped with anti-ice or deicing equipment. The purpose of this equipment is to remove or prevent the buildup of ice on critical surfaces such as the wing's leading edge. Wing inspection lights are typically flush mounted on the fuselage or engine nacelle and are directed toward the leading edge of the wing. If the pilot suspects the formation of ice, the wing can be illuminated to allow a visual inspection from inside the aircraft.

INTERIOR LIGHTS

There is a variety of interior lights found on modern aircraft including instrument lights, overhead lights, step lights, reading lights, and many others. In general, these lights can be divided into two basic categories: incandescent and fluorescent.

INTERIOR INCANDESCENT LIGHTING

Incandescent lamps used inside the aircraft operate on the same principles as exterior incandescent lamps, but are generally much smaller. Even with the smaller size, the interior lamps may be suffi-

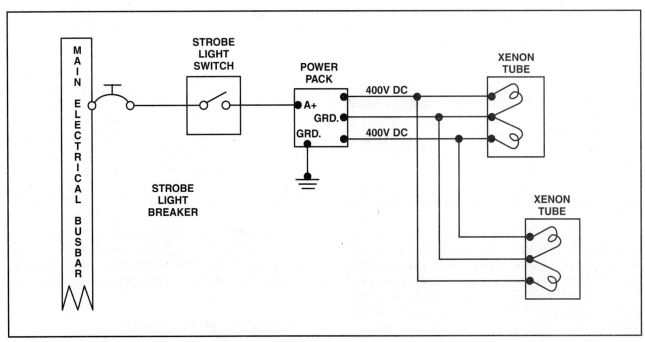

Figure 7-113. A typical strobe-type anticollision light system consists of a power supply to provide the high voltage required, and one or more xenon lamps.

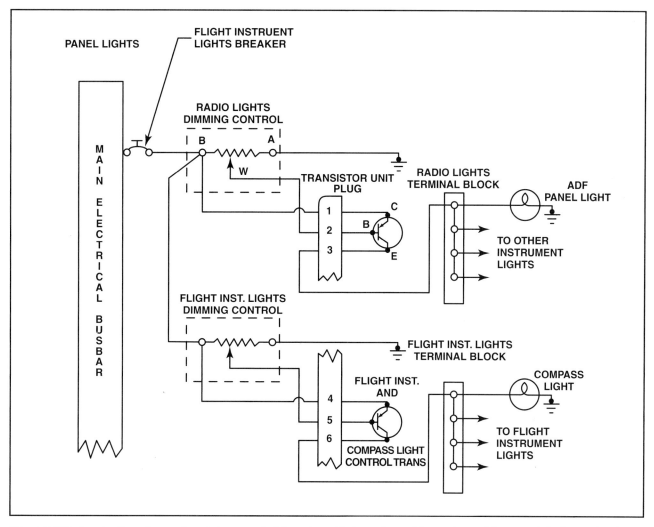

Figure 7-114. A potentiometer, or rheostat, connected in series with incandescent lamps is used to control the input signal to a transistor, thus controlling the current to the light. By using transistors, the size and electrical current through the control is greatly reduced. This design allows smaller lighting controls on the instrument panel, which also helps eliminate excessive heat around the instruments and radios.

ciently bright to require a dimming circuit. These lamps are often made dimmable by using a solid-state, transistorized circuit, to control the current in the lighting circuit. [Figure 7-114]

The transistors for dimming circuits are usually located remotely from the dimmer control. They are normally installed in a heat sink, which aids in dissipating the heat from the transistors. To increase the effectiveness of the heat sink, the transistors have a white heat sink paste applied to the mounting surface of the case. When installing a new transistor, it's important to make sure that the heat paste is applied to the transistor. In addition, when a transistor is replaced, it's important to properly position any insulators. If an insulator is bad, the transistor case may come into contact with the heat sink and short out, causing it to fail.

Fluorescent lights are made of a gas-filled glass tube that glows when a high AC voltage is applied to electrodes at each end. Applying current causes electrodes to emit free electrons. The free electrons strike atoms of mercury vapor in the tube and this produces ultraviolet light. The invisible ultraviolet light strikes the phosphorous coating on the inside of the tube and it glows with a visible white light. The conversion from ultraviolet to visible light is called fluorescence. Fluorescent lamps are much more efficient than incandescent lamps, however they require the use of transformers and AC voltage. Therefore, fluorescent lamps are usually found only on larger aircraft.

FLUORESCENT LIGHTS

Fluorescent lights can operate in a bright or dimmed position. The fluorescent tube is in the dim position

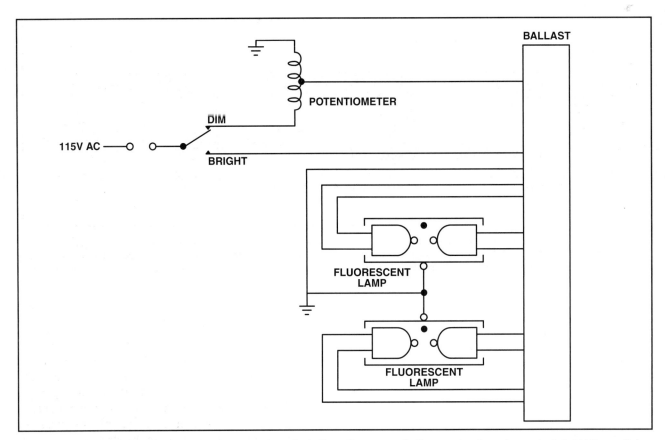

Figure 7-115. A fluorescent-type lamp system consists of a ballast, fluorescent bulbs, a potentiometer, and a bright/dim switch.

when a lower voltage is provided to the ballast transformer through a potentiometer. In the bright position, the potentiometer is bypassed, providing more current to the fluorescent tube. [Figure 7-115]

Another type of interior lighting system that is primarily used on aircraft instrument panels is called an Electro Luminescent (EL) panel. An EL panel contains a fluorescent paste sandwiched between two layers of plastic. The paste glows when an AC voltage is applied to the panel. The light glows through the unpainted areas of the plastic, typically displaying the etched-in lettering. Since Electro Luminescent panels operate only with alternating current, most light aircraft with EL systems use a solid-state static inverter specifically installed for the panel.

MAINTENANCE AND INSPECTION OF LIGHTING SYSTEMS

Most lighting circuits are relatively low maintenance items. Periodic inspections of the wire for chafing and hardware security, corrosion of components, and general condition of the circuit should be performed during routine inspections. Lamp replacement is generally the most needed repair for lighting systems, for which the most critical task is the installation of the correct bulb.

One important consideration when dealing with lighting circuits is to avoid electric shock. This is especially true of any high-intensity flashing lamp or strobe system. These systems operate at such a high voltage that human contact with the voltage source can be fatal. To help avoid shock when working with these systems, always allow the lighting circuit to sit in the OFF position for approximately five minutes prior to performing any maintenance.

Other maintenance and inspection considerations exist for each type of lighting circuit. When malfunctions occur, the best resource for troubleshooting information is obtained from the aircraft manufacturer's maintenance manual. However, some lighting circuits may have been installed after the aircraft was manufactured. For these systems, it may be necessary to locate and use the lighting manufacturer's service information.

MOTORS

Many aircraft functions require an application of force greater than what a pilot can perform manu-

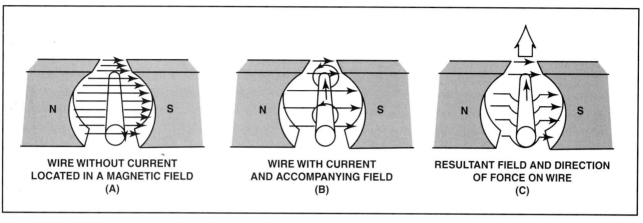

Figure 7-116. (A) When no current flows through a wire that is between two magnets, the lines of flux flow from north to south without being disturbed. (B) When current flows through the wire, magnetic flux lines encircle it. (C) The flux lines from the magnet and the flux lines encircling the wire react with one another to produce a strong magnetic field under the wire and a weak magnetic field above it.

ally. For example, raising and lowering the landing gear with a hand crank, or extending and retracting the flaps, would take a great deal of time and effort on larger, high performance aircraft. Electric motors can perform these and many other operations quickly and easily.

DC MOTORS

Many devices in an airplane, from the starter to the autopilot, depend upon the mechanical energy furnished by direct-current motors. A DC motor is a rotating machine that transforms direct-current electrical energy into mechanical energy.

MOTOR THEORY

The lines of flux between two magnets flow from the north pole to the south pole. At the same time, when current flows through a wire, lines of flux set up around the wire. The direction that these flux lines encircle the wire depends on the direction of current flow. When the wire's flux lines and the magnet's flux lines are combined, a reaction occurs.

For example, when the flux lines between two magnetic poles are flowing from left to right, and the lines of flux, that encircle a wire between the magnetic poles flow in a counterclockwise direction, the flux lines reinforce each other at the bottom of the wire. This happens because the lines of flux produced by the magnet, and the flow of flux lines at the bottom of the wire, are traveling in the same direction. However, at the top of the wire the flux lines oppose, or neutralize, each other. The resulting magnetic field under the wire is strong and the magnetic field above the wire is weak. Consequently, the wire is pushed away from the side where the field is the strongest. [Figure 7-116]

Using this same principle, if the current flow through the wire were reversed, the flux lines encircling the wire would flow in the opposite direction. The resulting combination of the magnetic flux lines and wire flux lines would create a strong magnetic field at the top of the wire and a weak magnetic field at the bottom. Consequently, the wire is pushed downward away from the stronger field.

PARALLEL CONDUCTORS

When two current-carrying wires are in the vicinity of one another, they exert a force on each other. This force is the result of the magnetic fields set up around each wire. When the current flows in the same direction, the resulting magnetic fields encompass both wires in a clockwise direction. These fields oppose each other and, therefore, cancel each other out. [Figure 7-117]

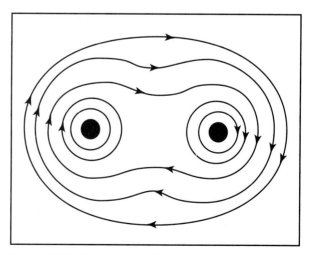

Figure 7-117. When two parallel wires have current flowing through them in the same direction, they are forced in the direction of the weaker field, which is toward each other.

When the electron flow in the wires is opposite, the magnetic field around one wire radiates outward in a clockwise direction, while the magnetic field around the second wire rotates counterclockwise. These fields combine, or reinforce each other, between the wires. [Figure 7-118]

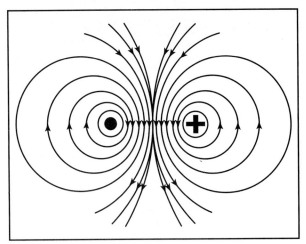

Figure 7-118. When the flow of electrons in two wires is opposite, the resulting magnetic fields force the wires apart.

DEVELOPING TORQUE

When a current-carrying coil is placed in a magnetic field, the magnetic fields produced cause the coil to rotate. The force that produces rotation is called torque. [Figure 7-119]

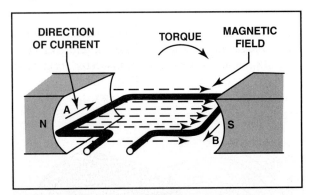

Figure 7-119. The coil above has current flowing inward on side A and outward on side B. Therefore, the magnetic field encircling wire A is counterclockwise, while the magnetic field encircling wire B is clockwise. The resultant force pushes wire B downward, and wire A upward. This causes the coil to rotate until the wires are perpendicular to the magnetic flux lines. In other words, torque is created by the reacting magnetic fields around the coil.

The amount of torque developed in a coil depends on several factors, including the strength of the magnetic field, the number of turns in the coil, and the position of the coil in the field.

The right-hand motor rule is used to determine the direction a current-carrying wire moves in a magnetic field. If the right index finger points in the direction of the magnetic field and the middle finger in the direction of current flow, the thumb indicates the direction the wire moves. [Figure 7-120]

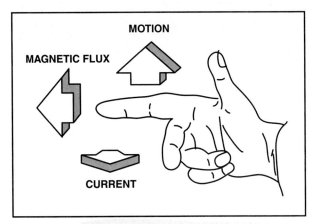

Figure 7-120. When using the right-hand motor rule, the right index finger indicates the flow direction of the magnetic flux, while the middle finger indicates the direction of current flow, and the thumb, the direction the coil rotates.

BASIC DC MOTOR

Torque is the technical basis governing the construction of DC motors. The torque causing a coil to rotate is strongest when it is at a 90-degree angle to the magnetic field produced by two magnets. Therefore, when a coil lines up with the magnetic field, it does not rotate because the torque at that point is zero. Since the coil of a motor must rotate continuously in order for the motor to operate efficiently, it is necessary for a device to reverse the coil current just as the coil becomes parallel with the magnet's flux lines. When the current is reversed, torque is again produced and the coil rotates. When a current-reversing device is set up to reverse the current each time the coil is about to stop, the coil rotates continuously.

One way to reverse the current in a coil is to attach a commutator similar to what is used on a generator. When this is done, the current flowing through the coil changes direction continuously as the coil rotates, thus preserving torque. [Figure 7-121]

A more effective method of ensuring continuous coil torque is to have a large number of coils wound on an armature. When this is done, the coils are spaced so that, for any position of the armature, a coil is near the magnet's poles. This makes torque both continuous and strong. However, it also means that the commutator must contain several segments.

Airframe Electrical Systems

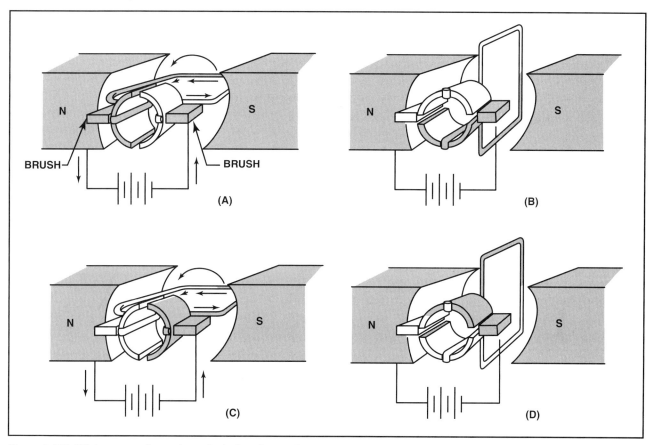

Figure 7-121. (A) As current flows through the brushes to the commutator and coil, torque is produced and the coil rotates. (B) As the coil becomes parallel with the magnetic lines of flux, each brush slides off one terminal and connects the opposite terminal to reverse the polarity. (C) Once the current reverses, torque is again produced and the coil rotates. (D) As the coil again becomes parallel with the flux lines, the commutator again reverses current, and torque continues to rotate the coil.

To further increase the amount of torque generated, the armature is placed between the poles of an electromagnet instead of a permanent magnet. This provides a much stronger magnetic field. Furthermore, the core of an armature is usually made of soft iron that is strongly magnetized through induction.

DC MOTOR CONSTRUCTION

The major parts in a practical motor are the armature assembly, the field assembly, the brush assembly and the end frames. This arrangement is very similar to a DC generator.

ARMATURE ASSEMBLY

The armature assembly contains a soft-iron core, coils, and commutator mounted on a rotatable steel shaft. The core consists of laminated stacks of soft iron that are insulated from each other. Solid iron is not used because it generates excessive heat that uses energy needlessly. The armature windings are made of insulated copper wire that is inserted into slots and protected by a fiber paper, sometimes called fish paper. The ends of the windings are physically connected to the commutator segments with wedges or steel bands. The commutator consists of several copper segments insulated from each other and the armature shaft by pieces of mica. Insulated wedge rings hold the segments in place. [Figure 7-122]

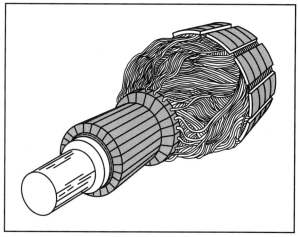

Figure 7-122. The armature of a typical DC motor is similar to that of a DC generator.

FIELD ASSEMBLY

The field assembly consists of the field frame, a set of pole pieces, and field coils. The field frame is located along the inner wall of the motor housing and contains the laminated steel pole pieces on which the field coils are wound. The field coils consist of several turns of insulated wire that fit over each pole piece. Some motors have as few as two poles, while others have as many as eight.

BRUSH ASSEMBLY

The brush assembly consists of brushes and their holders. The brushes are usually made of small blocks of graphitic carbon because of its long service life. The brush holders permit the brushes to move somewhat and utilize a spring to hold them against the commutator. [Figure 7-123]

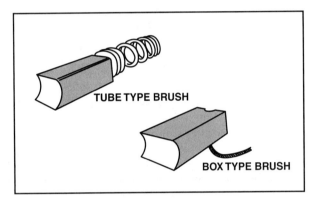

Figure 7-123. Two of the most common brushes used in DC motors are the tube type and box type.

END FRAME

The end frame is the part of the motor that the armature assembly rotates in. The armature shaft, which rides on bearings, extends through one end frame and is connected to the load. Sometimes the drive end frame is part of the unit driven by the motor.

MOTOR SPEED, DIRECTION, AND BRAKING

Some applications call for motors whose speed or direction is changeable. Some applications also require a motor to be stopped very quickly. For example, a landing gear motor must be able to both retract and extend the gear and stop when the gear has reached a particular position, while a windshield wiper motor must have variable speeds to suit changing weather conditions. Internal or external changes need to be made in the motor design to allow these operations.

CHANGING MOTOR SPEED

A motor in which the speed is controlled is called a variable speed motor, and can be either a shunt or series motor. Motor speed is controlled by varying the current in the field windings. For example, when the amount of current flowing through the field windings is increased, the field strength increases, causing the armature windings to produce a larger counter electromotive force (EMF) that slows the motor. Conversely, when the field current is decreased, the field strength decreases, and the motor speeds up because the counter EMF is reduced. [Figure 7-124]

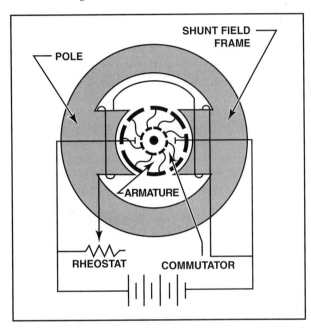

Figure 7-124. A shunt motor with variable speed control uses a rheostat to control the motor's speed.

In a shunt motor, speed is controlled by a rheostat that is connected in series with the field winding. Therefore, the speed depends on the amount of current flowing through the rheostat to the field windings. To increase motor speed, the resistance in the rheostat is increased. This decreases the field current, which decreases the strength of the magnetic field and counter EMF. This momentarily increases the armature current and torque which, in turn, causes the motor to speed up until the counter EMF increases and causes the armature current to decrease to its former value. Once this occurs, the motor operates at a higher fixed speed.

To decrease motor speed, resistance in the rheostat is decreased. This action increases the current flow through the field windings and increases the field strength. The higher field strength causes a momentary increase in the counter EMF, which decreases the armature current. As a result, torque decreases and the motor slows until the counter EMF

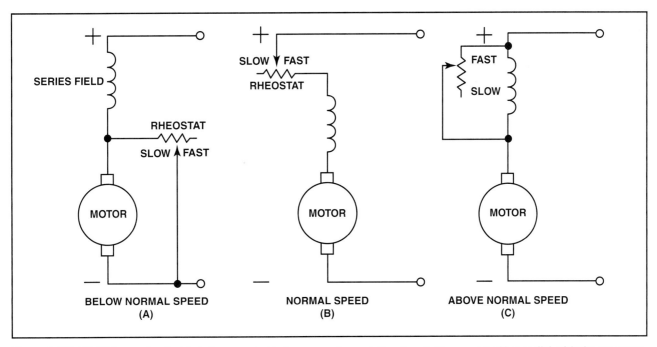

Figure 7-125. (A) With a motor that is to be operated below normal speed, the rheostat is connected in parallel with the armature, and decreasing the current increases the motor speed. (B) When a motor is operated in the normal speed range, the rheostat is connected in series with the motor field. In this configuration, increasing the voltage across the motor increases motor speed. (C) For above normal speed operation, the rheostat is connected in parallel with the series field. In this configuration, part of the voltage bypasses the series field, causing the motor to speed up.

decreases to its former value. Once the counter EMF and armature current are balanced, the motor operates at a lower fixed speed than before.

In a series motor, the rheostat speed control is connected in one of three ways. The rheostat is either connected in parallel or in series with the motor field, or in parallel with the motor armature. Each method of connection allows for operation in a specified speed range. [Figure 7-125]

REVERSING MOTOR DIRECTION

The direction of a DC motor's rotation is reversed by changing the direction of current flow in either the armature or the field windings. In both cases, this reverses the magnetism of either the armature or the magnetic field the armature rotates in. If the wires connecting the motor to an external source are interchanged, the direction of rotation is not reversed, since these wires reverse the magnetism of both the field and armature. This leaves the torque in the same direction.

One method for reversing the direction of rotation employs two field windings wound in opposite directions on the same pole. This type of motor is called a split field motor. A single-pole, double-throw switch makes it possible to direct current to either of the two windings. [Figure 7-126]

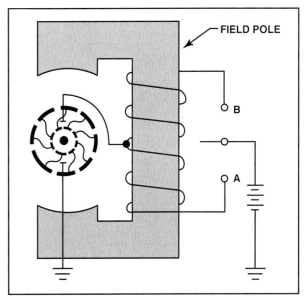

Figure 7-126. When the switch is in the lower position, current flows through the lower field winding, creating a north pole at the lower field winding and at the lower pole piece. However, when the switch is placed in the "up" position, current flows through the upper field winding. This reverses the field magnetism and causes the armature to rotate in the opposite direction.

Some split field motors are built with two separate field windings wound on alternate poles. An example of this is the armature in a four-pole reversible motor. In this configuration, the armature rotates in one direction when current flows through one set of windings, and in the opposite direction when current flows through the other set of windings.

Another method of reversal is called the switch method. This type of motor reversal employs a double-pole, double-throw switch that changes the direction of current flow in either the armature or the field. [Figure 7-127]

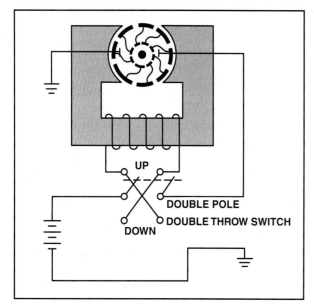

Figure 7-127. In the illustration above, a double-pole, double-throw switch is used to reverse the current through the field. When the switch is in the "up" position, current flows through the field windings. This establishes a north pole on the right side of the motor. When the switch is moved to the "down" position, polarity is reversed and the armature rotates in the opposite direction.

MOTOR BRAKING

Many motor-driven devices, such as landing gear, flaps, and retractable landing lights must be stopped at precise positions. These devices are usually very heavy and a lot of inertia is generated when they are being positioned. Brake mechanisms are used to stop the motor and gear train under these circumstances.

The most common brake mechanism consists of a drum or disc, which is attached to the gear mechanism, and friction pads, or shoes, which are controlled by an electromagnet. When the motor is energized, an electromagnet pulls the friction pad away from the disc or drum. When the motor is turned off, the electromagnet is deenergized and allows a spring to pull the friction material into contact with the disc or drum.

TYPES OF DC MOTORS

DC motors are classified by the type of field and armature connection used and by the type of duty they are designed for. For example, there are three basic types of DC field-armature connections: series, shunt, and compound.

SERIES DC MOTOR

In a series motor, the field windings consist of heavy wire with relatively few turns that are connected in series with the armature winding. This means the same amount of current flows through the field windings and the armature windings. In this configuration, an increase in current causes a corresponding increase in the magnetism of both the field and armature. [Figure 7-128]

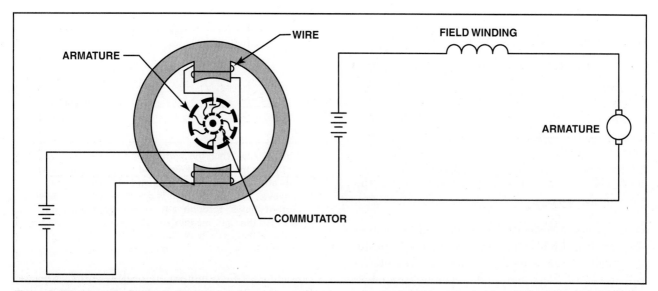

Figure 7-128. Since the field windings and armature in a series motor are connected in series, an increase of current through the field windings results in an increase of current in the armature.

The series motor is able to draw a large starting current because of the winding's low resistance. This starting current passes through both the field and armature windings, therefore, producing a high starting torque. For this reason, series motors are often used in aircraft as starters and for raising and lowering landing gear, cowl flaps, and wing flaps. However, as the speed of a series motor increases, the counter EMF builds and opposes the applied EMF. In turn, this reduces the current flow through the armature and reduces the current draw.

The speed of a series motor depends on the load applied. Therefore, any change in load is accompanied by a substantial change in speed. In fact, if the load is removed entirely, a series motor will operate at an excessively high speed, and the armature could fly apart. In other words, a series motor needs resistance to stay within a safe operating range.

SHUNT DC MOTOR

In a shunt motor, the field winding is connected in parallel with the armature winding. To limit the amount of current that passes through the field, the resistance is high in the field winding. In addition, because the field is connected directly across the power supply, the amount of current that passes through the field is constant. In this configuration, when a shunt motor begins to rotate, most of the current flows through the armature, while relatively little current flows to the field. Because of this, shunt motors develop little torque when they are first started. [Figure 7-129]

As a shunt-wound motor picks up speed, the counter EMF in the armature increases, causing a decrease in the amount of current draw in the armature. At the same time, the field current increases slightly, causing an increase in torque. Once torque and the resulting EMF balance each other, the motor will be operating at its normal, or rated, speed.

Since the amount of current flowing through the field windings remains relatively constant, the speed of a shunt motor varies little with changes in load. In fact, when no load is present, a shunt motor assumes a speed only slightly higher than the loaded speed. Because of this, a shunt motor is well-suited for operations in which a constant speed is desired and a high starting torque is not.

COMPOUND DC MOTOR

The compound motor is a combination of the series and shunt motors. In a compound motor there are two field windings: a shunt winding and a series winding. The shunt winding is composed of many turns of fine wire and is connected in parallel with the armature winding. On the other hand, the series winding consists of a few turns of large wire and is connected in series with the armature winding.

The starting torque is higher in a compound motor than in a shunt motor, and lower than in the series

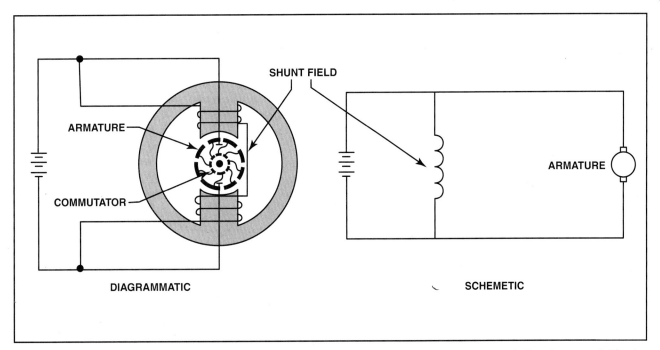

Figure 7-129. In a shunt-wound motor, the field winding is connected in parallel with the armature winding. Because of this, the amount of current that flows to the field when the motor is started is limited, and the resulting torque is low.

motor. Furthermore, variation of speed with a load is less than in a series-wound motor but greater than in a shunt motor. The compound motor is used whenever the combined characteristics of the series and shunt motors are desired. [Figure 7-130]

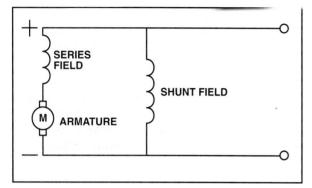

Figure 7-130. In a compound motor, one field winding is connected to the armature winding in series, while the other is connected in parallel.

TYPE OF DUTY

Electric motors must operate under various conditions. For example, some motors are used for intermittent operations, while others operate continuously. In most cases, motors built for intermittent duty may be operated for only short periods of time before they must be allowed to cool. On the other hand, motors built for continuous duty are operable at their rated power for long periods.

ENERGY LOSSES IN MOTORS

When electrical energy is converted to mechanical energy in a motor, some losses do occur. By the same token, losses also occur when mechanical energy is converted to electrical energy. Therefore, in order for machines to be efficient, both electrical and mechanical losses must be kept to a minimum. Electrical losses are classified as either copper losses or iron losses, while mechanical losses originate from the friction of various moving parts.

Copper losses occur when electrons are forced through the copper armature and field windings. These losses occur because some power is dissipated in the form of heat due to the inherent resistance possessed by copper windings. The amount of loss is proportional to the square of the current, and is calculated with the formula:

$$\text{Copper Loss} = I^2 \times R$$

Iron losses are divided into hysteresis and eddy current losses. Hysteresis losses result from the armature revolving in a magnetic field, which causes the current in the armature to alternate, thus magnetizing it in two directions. Since some residual magnetism remains in the armature after its direction is changed, some energy loss does occur. However, since the field magnets are always magnetized in one direction by DC current, they produce no hysteresis losses.

Eddy current losses occur because the armature's iron core acts as a conductor revolving in a magnetic field. This sets up an EMF across portions of the core, causing currents to flow within the core. These currents heat the core and, when excessive, can damage the windings. To keep eddy current losses to a minimum, a laminated core made of thin, insulated sheets of iron is used. The thinner the laminations, the greater the reduction in eddy current losses.

INSPECTION AND MAINTENANCE OF DC MOTORS

The inspection and maintenance of DC motors should be in accordance with the guidelines established by the manufacturer. The following is indicative of the types of maintenance checks typically called for:

1. Check the unit driven by the motor in accordance with the specific installation instructions.

2. Check all wiring, connections, terminals, fuses, and switches for general condition and security.

3. Keep motors clean and mounting bolts tight.

4. Check the brushes for condition, length, and spring tension. Procedures for replacing brushes, along with their minimum lengths, and correct spring tensions are given in the applicable manufacturer's instructions. If the spring tension is too weak, the brush could begin to bounce and arc, causing commutator burning.

5. Inspect the commutator for cleanliness, pitting, scoring, roughness, corrosion, or burning. Check the mica between each of the commutator segments. If a copper segment wears down below the mica, the mica will insulate the brushes from the commutator. In some instances, the mica is undercut to prevent it from wearing against the brushes. However, in some newer motor designs, the mica is left flush with the surface of the commutator. Always check the manufacturer's specifications to determine the limitations regarding the depth of the mica between commutator segments before attempting to cut the mica down.

Clean dirty commutators with the recommended cleaning solvent and a cloth. Polish rough or corroded commutators with fine sandpaper (000 or finer) and blow out remaining particles with compressed air. Never use emery paper or crocus cloth because they contain particles that can cause shorts between the commutator segments. Replace the motor if the commutator is burned, badly pitted, grooved, or worn to the extent that the mica insulation is out of acceptable tolerances

6. Inspect all exposed wiring for evidence of overheating. Replace the motor if the insulation on the leads or windings is burned, cracked, or brittle.

7. Lubricate the motor only if called for by the manufacturer's instructions. Most motors used today do not require lubrication between overhauls.

8. Adjust and lubricate the gearbox or drive unit in accordance with the applicable manufacturer's instructions.

Troubleshoot any problems and replace the motor only when the trouble is due to a defect in the motor itself. In most cases, motor failure is caused by a defect in the external electrical circuit or by mechanical failure in the mechanism driven by the motor.

AC MOTORS

AC motors have several advantages over DC motors. For example, in many instances AC motors do not use brushes or commutators, and therefore cannot cause arcing like a DC motor can. Furthermore, AC motors are well suited for constant-speed applications, although some are manufactured with variable speed characteristics. Other advantages some AC motors have include their ability to operate on single or multiple phase lines as well as at several voltages. In addition, AC motors are generally less expensive than comparable DC motors. Because of these advantages, many aircraft are designed to use AC motors.

Because aircraft electrical systems typically operate on 400 hertz AC, an aircraft AC motor operates at about seven times the speed of a 60-hertz commercial motor with the same number of poles. In fact, a 400-hertz induction type motor typically operates at speeds ranging from 6,000 to 24,000 rpm. This high rotation speed makes AC motors suitable for operating small high-speed rotors. Furthermore, through the use of reduction gears, AC motors are made to lift and move heavy loads such as wing flaps and retractable landing gear, as well as produce enough torque to start an engine. There are three basic types of AC motors; the universal motor, the induction motor, and the synchronous motor. Each type represents a variation on basic AC motor operating principles.

UNIVERSAL MOTORS

Fractional horsepower (less than one horsepower) AC series motors are called universal motors. A unique characteristic of universal motors is that they can operate on either alternating or direct current. In fact, universal motors resemble DC motors in that they have brushes and a commutator. Universal motors are used extensively to operate fans and portable tools such as drills, grinders, and saws. [Figure 7-131]

Figure 7-131. An electric drill uses a universal motor that is similar in construction to a series-wound DC motor.

INDUCTION MOTORS

The most popular type of AC motor is the induction motor. In an induction motor there is no need for an electrical connection between the motor housing and the rotating elements. Therefore, there are no brushes, commutators, or slip rings to contend with. Induction motors operate at a fixed rpm that is determined by their design, and the frequency of AC applied. In addition, an induction motor can be operated on either single-phase or three-phase alternating current.

A single-phase induction motor is used to operate devices such as surface locks, intercooler shutters, oil shutoff valves, and places where the power requirements are low. Single-phase induction motors require some form of starting circuit that automatically disconnects after the motor is running. Single-phase induction motors operate well in either rotational direction, with the direction determined by the starting circuit.

Unlike single-phase induction motors, three-phase induction motors are self-starting and are commonly used when high power is needed. Common applications for three-phase induction motors include engine starting, operating flaps and landing gear, and powering hydraulic pumps.

CONSTRUCTION

The two primary parts of an induction motor are the stator and the rotor. The stator is unique because instead of having field poles that extend outward, windings are placed in slots around the stator's periphery. These windings make up a series of electromagnets that produce a magnetic field.

The rotor of an induction motor consists of an iron core made of thin circular laminations of soft steel that are keyed to a shaft. Longitudinal slots are cut into the rotor's circumference and heavy copper or aluminum bars are embedded in them. These bars are welded to a heavy, highly conductive ring on either end. [Figure 7-132]

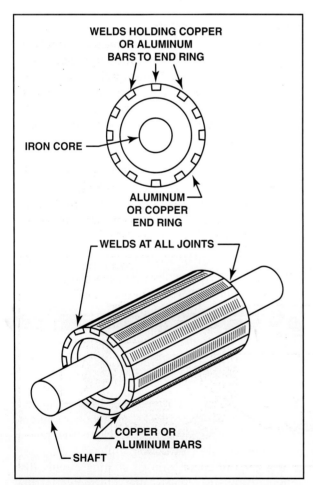

Figure 7-132. The complete rotor in an induction motor is sometimes called a squirrel cage. For this reason, motors containing this type of rotor are often called squirrel cage induction motors.

When AC is applied to the stator, the strength and polarity of the electromagnets changes with the excitation current. Furthermore, to give the effect of a rotating magnetic field, each group of poles is attached to a separate phase of voltage.

When the rotor of an induction motor is subjected to the revolving magnetic field produced by the stator windings, a voltage is induced in the longitudinal bars. This induced voltage causes current to flow through the bars, which produces its own magnetic field that combines with the stator's revolving field. As a result, the rotor revolves at nearly a synchronous speed with the stator field. The only difference in the rotational speed between the stator field and the rotor is that necessary to induce the proper current into the rotor to overcome mechanical and electrical losses.

If a rotor were to turn at the same speed as the rotating field, a resonance would set up. When this happens, the rotor conductors are not cut by any magnetic lines of flux, so no EMF is induced into them. Thus, no current flows in the rotor, resulting in no torque and little rotor rotation. For this reason, there must always be a difference in speed between the rotor and the stator's rotating field.

The difference in rotational speed is called motor slip, and is expressed as a percentage of the synchronous speed. For example, if the rotor turns at 1,750 rpm and the synchronous speed is 1,800 rpm, the difference in speed is 50 rpm. The slip is therefore equal to 50/1,800, or 2.78 %.

SINGLE-PHASE INDUCTION MOTOR

A single-phase motor differs from a multiphase motor in that the single-phase motor has only one stator winding. In this configuration, the stator winding generates a field that pulsates. This generates an expanding and collapsing stator field that induces currents into the rotor. These currents generate a rotor field opposite in polarity to that of the stator.

The field opposition exerts a turning force on the upper and lower parts of the rotor that try to turn it 180 degrees from its original position. Since these forces are exerted in the center of the rotor, the turning force is equal in each direction. As a result, the rotor will not begin turning from a standing stop. However, if the rotor starts turning, it continues to rotate. Furthermore, since the rotor momentum aids the turning force, there is no opposition to rotation.

SHADED-POLE INDUCTION MOTOR

The first effort in the development of a self-starting, single-phase motor was the shaded-pole induction motor. Like the generator, the shaded-pole motor has field poles that extend outward from the motor housing. In addition, a portion of each pole is encircled with a heavy copper ring. [Figure 7-133]

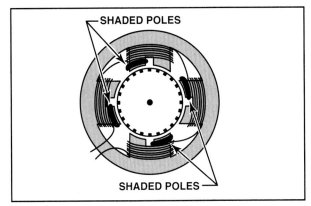

Figure 7-133. In a shaded-pole induction motor, a copper ring encircles a portion of each salient pole.

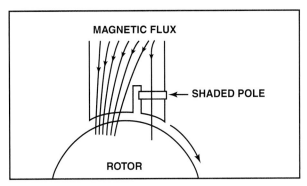

Figure 7-134. The portion of magnetic flux lines that passes through the shaded pole lags behind the opposite pole, thereby creating a slight component of rotation.

The presence of the copper ring causes the magnetic field through the ringed portion of the pole face to lag appreciably behind that of the other half of the pole. This results in a slight component of rotation in the field that is strong enough to cause rotation. Although the torque created by this field is small, it is enough to accelerate the rotor to its rated speed. [Figure 7-134]

SPLIT-PHASE MOTOR

Another variety of self-starting motor is known as a split-phase motor. Split-phase motors have a winding that is dedicated to starting the rotor. This "start" winding is displaced 90 degrees from the main, or run, winding and has a fairly high resistance that causes the current to be out of phase with the current in the run winding. The out of phase condition produces a rotating field that makes the rotor revolve. Once the rotor attains approximately 25 % of its rated speed, a centrifugal switch disconnects the start winding automatically.

CAPACITOR-START MOTOR

With the development of high-capacity electrolytic capacitors, a variation of the split-phase motor was made. Motors that use high-capacity electrolytic capacitors are known as capacitor-start motors. Nearly all fractional horsepower motors in use today on refrigerators, oil burners, and other similar appliances are of this type.

In a capacitor-start motor, the start and run windings are the same size and have identical resistance values. The phase shift between the two windings is obtained by using capacitors connected in series with the start winding. [Figure 7-135]

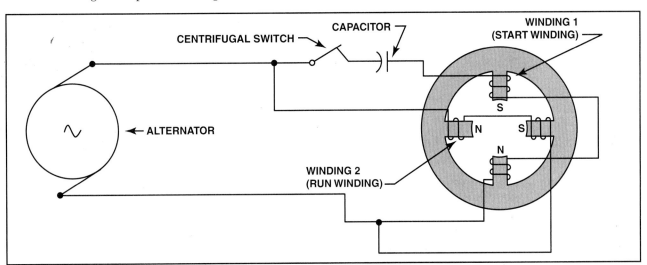

Figure 7-135. A single-phase motor with capacitor start windings has a capacitor connected in series with an alternator and the start winding.

Capacitor-start motors have a starting torque comparable to their rated speed torque and are used in applications where the initial load is heavy. Again, a centrifugal switch is required for disconnecting the start winding when the rotor speed is approximately 25 % of the rated speed.

DIRECTION OF ROTATION

The direction of rotation for a three-phase induction motor is changed by reversing two of the motor leads. The same effect is obtained in a two-phase motor by reversing the connections on one phase. In a single-phase motor, reversing the connections to the start winding reverses the direction of rotation. Most single-phase motors that are designed for general application are built to enable reversal of the connections to the start winding. On the other hand, a shaded-pole motor cannot be reversed because its rotational direction is determined by the physical location of the copper shaded ring.

SYNCHRONOUS MOTORS

Like the induction motor, a synchronous motor uses a rotating magnetic field. However, the torque developed by a synchronous motor does not depend on the induction of currents in the rotor. Instead, the principle of operation of the synchronous motor begins with a multi-phase source of AC applied to a series of stator windings. When this is done, a rotating magnetic field is produced. At the same time, DC current is applied to the rotor winding, producing a second magnetic field. A synchronous motor is designed so that the rotor is pulled by the stator's rotating magnetic field. The rotor turns at approximately the same speed as the stator's magnetic field. In other words, they are synchronized.

To understand the operation of a synchronous motor, assume that poles A and B in figure 7-136 are physically rotated clockwise in order to produce a rotating magnetic field. These poles induce the opposite polarity in the soft-iron rotor between them, creating an attraction between the rotating poles and the rotor. This attraction allows the rotating poles to drag the rotor at the same speed. [Figure 7-136]

When a load is applied to the rotor shaft, its axis momentarily falls behind that of the rotating field. However, the rotor catches up and again rotates with the field at the same speed, as long as the load remains constant. If the load is too large, the rotor pulls out of sync with the rotating poles and is unable to rotate at the same speed. In this situation, the motor is said to be overloaded.

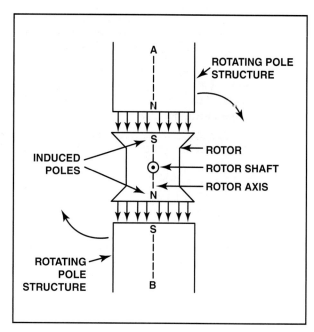

Figure 7-136. In a synchronous motor, a rotating magnet induces opposite magnetic fields in a soft iron rotor. The rotor then turns at the same speed as the magnet.

The idea of using a mechanical means to rotate the poles is impractical because another motor would be required to do this. Therefore, a rotating magnetic field is produced electrically by using phased AC voltages. In this respect, a synchronous motor is similar to the induction motor.

The synchronous motor consists of a stator field winding that produces a rotating magnetic field. The rotor is either a permanent magnet or an electromagnet. If permanent magnets are used, the rotor's magnetism is stored within the magnet. Conversely, if electromagnets are used, the magnets receive power from a DC power source through slip rings.

Since a synchronous motor has little starting torque, it requires assistance to bring it up to synchronous speed. The most common method for this is to start the motor with no load, allow it to reach full speed, and then energize the magnetic field. The magnetic field of the rotor then locks with the magnetic field of the stator, and the motor operates at synchronous speed. [Figure 7-137]

The magnitude of the induced rotor poles is so small that sufficient torque cannot be developed for most practical loads. To avoid the limitation on motor operation, a winding is placed on the rotor and energized with direct current. To adjust the motor for varying loads, a rheostat is placed in

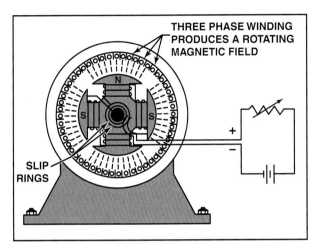

Figure 7-137. The weak field induced in the rotor poles limits the ability of the poles to produce torque. This problem is overcome by applying DC current through a rheostat to vary the field strength of the poles.

series with the DC source to provide a means of varying the pole's strength.

A synchronous motor is not self-starting. Since rotors are heavy, it is impossible to bring a stationary rotor into magnetic lock with a rotating magnetic field. As a result, all synchronous motors have some kind of starting device. One type of simple starter used is another AC or DC motor that brings the rotor up to approximately 90 % of its synchronous speed. The starting motor is then disconnected and the rotor locks in with the rotating field.

Another method utilizes a second squirrel-cage type winding on the rotor. This inducton winding brings the rotor almost to synchronous speed before the direct current is disconnected from the rotor windings, and the rotor is pulled into sync with the field.

CHAPTER 8

HYDRAULIC AND PNEUMATIC POWER SYSTEMS

INTRODUCTION

Early aircraft were equipped with flight controls and systems that were connected directly to the cockpit controls. As aircraft became more complex, it became necessary to operate systems remotely, and the first of these was probably the brake system. Instead of cables or pushrods operating the brakes, hydraulic pressure was used to solve routing problems and multiply force on the braking surfaces. While small aircraft continue to use cables or pushrods for operating flight controls, larger aircraft are equipped with hydraulic or pneumatic control systems at least for the primary system. Today's aviation maintenance technician must be familiar with the principles of hydraulic and pneumatic systems as well as how the different aircraft systems utilize these principles.

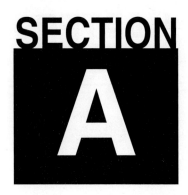

SECTION A

PRINCIPLES OF HYDRAULIC POWER

The word hydraulics is based on the Greek word for water, and originally meant the study of physical behavior of water at rest and in motion. Today the meaning has been expanded to include the physical behavior of all liquids, including hydraulic fluids.

Although some aircraft manufacturers make greater use of hydraulic systems than others, the hydraulic system of the average modern aircraft performs many functions. Among the units commonly operated by hydraulic systems are landing gear, wing flaps, speed and wheel brakes, and flight control surfaces.

Hydraulic systems have many advantages as a power source for operating various aircraft units. Hydraulic systems combine the advantages of lighter weight, ease of installation, simplification of inspection, and minimum maintenance requirements. Hydraulic operations are almost 100% efficient, with only a negligible loss due to fluid friction.

Aircraft hydraulic systems belong to that branch of physics known as fluid power systems. Work is done by moving an incompressible fluid. Pneumatic systems work in much the same way, obeying many of the same laws, but the fluid used (air) is compressible. Pneumatic systems will be covered later in this chapter.

To better understand how a hydraulic system accomplishes its task, a brief review of the physics involved is necessary. Basic knowledge of how fluids behave under certain conditions will better enable you to evaluate and troubleshoot hydraulic system problems.

STATIC FLUID PRESSURE

A column of liquid produces pressure that is directly proportional to the height of the column, and it in no way depends upon either the shape of the container or the amount of liquid the container holds. For example, one cubic inch of water weighs 0.036 pound, and if a tube is 231 inches tall with a cross section of one square inch, it will hold one gallon of water (one gallon = 231 cubic inches). If the tube is standing straight up, the one gallon of water will exert a pressure of 8.34 pounds per square inch at the bottom of the tube.

If the tube were 231 inches high and had an area of 100 square inches, it would hold 100 gallons of water, but the pressure at the bottom would still be 8.34 pounds per square inch. The force exerted by the column of water is equal to the pressure acting on each square inch times the number of square inches, or 834 pounds. [Figure 8-1]

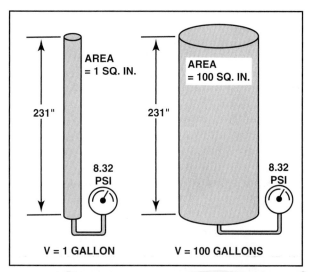

Figure 8-1. The pressure exerted by a column of liquid is determined by the height of the column and is independent of its volume.

It makes no difference as to the shape or size of the vessel that contains the liquid; it is the height of the column that is the critical factor. The pressure read by the gauges will be the same in all instances, since the height is the same. Naturally, all of the vessels must be filled with the same liquid. [Figure 8-2]

PASCAL'S LAW

This is the basic law of transmitting power by a hydraulic system. The French mathematician Blaise Pascal observed that any increase in the pressure on a confined liquid was transmitted equally and undi-

Hydraulic and Pneumatic Power Systems

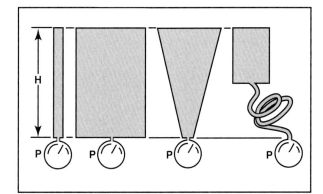

Figure 8-2. Neither the shape nor the volume of a container affects the pressure. Only the height of the column has any effect.

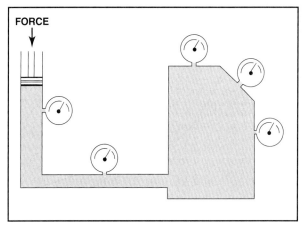

Figure 8-3. Pressure exerted on a fluid in an enclosed container is transmitted equally and undiminished to all parts of the container and acts at right angles to the enclosing walls.

minished to all parts of the container. Hydraulic pressure acts at right angles to the enclosing walls of the container. This means that if an enclosed vessel is full of liquid, and a force is applied to a piston in the vessel to raise the pressure, this increase in pressure will be the same anywhere in the system. Each of the gauges attached to the container will have the same reading. [Figure 8-3]

RELATIONSHIP BETWEEN PRESSURE, FORCE, AND AREA

Pressure is a measure of the amount of force that acts on a unit of area. In most American hydraulic systems, pressure is measured in pounds per square inch (psi), and in the metric system it is expressed in kilograms per square centimeter.

The relationship between force, pressure, and area may be expressed by the formula:

$$\text{Force} = \text{Area} \times \text{Pressure}$$

This may be visualized by looking at a segmented circle. The bottom half represents the area in square inches and the pressure in pounds per square inch. This product equals the amount of force in pounds, which is represented by the top half of the circle. [Figure 8-4]

The area needed to produce a given amount of force with the available pressure can be found by using the formula:

$$\text{(C) Area} = \text{Force}/\text{Pressure}$$

In order to find the amount of pressure needed for a piston to produce a given amount of force, the force required in pounds is divided by the area of the piston in square inches:

$$\text{Pressure} = \text{Force}/\text{Area}$$

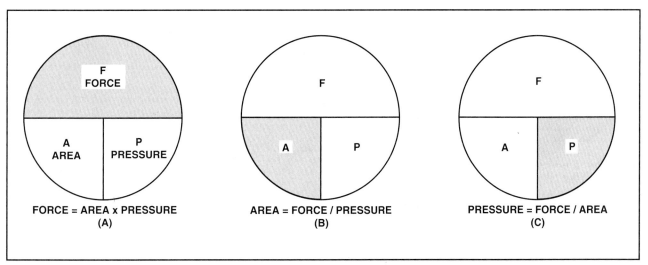

Figure 8-4. The relationship between area, pressure, and force is represented by these subdivided circles.

Some examples of this relationship:

How much piston area is required to produce a force of 250 pounds if a pressure of 3,000 psi is used?

$$\text{Area} = \text{Force/Pressure}$$

$$= 250/3,000$$

$$= 0.083 \text{ square inch}$$

A piston having an area of only 0.083 square inch will produce this amount of force with 3,000 psi hydraulic pressure.

How much pressure will be needed to produce a force of 1,000 pounds if the piston has an area of 2.5 square inches?

$$\text{Pressure} = \text{Force/Area}$$

$$= 1000/2.5$$

$$= 400 \text{ psi}$$

A pressure of 400 psi when acting on a piston with an area of 2.5 square inches, will produce a force of 1,000 pounds.

RELATIONSHIP BETWEEN AREA, DISTANCE, AND VOLUME

Another relationship in hydraulics that must be understood is the one between the area of the piston, the distance it moves, and the volume of the fluid displaced. This relationship can be represented by a segmented circle. The bottom half represents the area in square inches and the distance the piston moves in inches. This product equals the amount of volume in cubic inches, represented by the top half of the circle. If using the metric system, the area will be in square centimeters, the distance in centimeters, and the volume in cubic centimeters. [Figure 8-5]

The following are example problems of this relationship.

How many cubic inches of fluid is needed to move a piston having an area of 2.5 square inches a distance of six inches?

$$\text{Volume} = \text{Area} \times \text{Distance}$$

$$= 2.5 \times 6$$

$$= 15 \text{ cubic inches of fluid}$$

What is the area of the piston required to move 1,000 cubic inches of fluid as it travels a distance of five inches?

$$\text{Area} = \text{Volume/Distance}$$

$$= 1,000/5$$

$$= 200 \text{ square inches}$$

How many centimeters will 250 cubic centimeters of fluid move a piston whose area is 20 square centimeters?

$$\text{Distance} = \text{Volume/Area}$$

$$= 250/20$$

$$= 12.5 \text{ centimeters}$$

MECHANICAL ADVANTAGE IN A HYDRAULIC SYSTEM

A hydraulic system has two major advantages over other types of mechanical systems. One is the ease with which force can be transmitted over large dis-

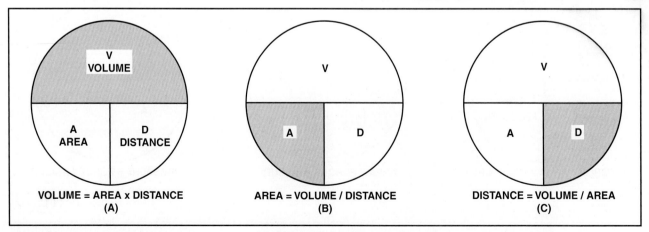

Figure 8-5. The relationship between volume, area, and distance can also be represented by a segmented circle.

tances. The other is the large gain in mechanical advantage made possible by varying the size of pistons.

Mechanical advantage is achieved in a hydraulic system by having an output piston that is larger than the input piston. If a piston whose area is one square inch is pressed down with a force of one pound, it will produce a pressure of one pound per square inch, and for every inch it moves, it will displace one cubic inch of fluid.

A cylinder containing the piston described above is connected to one having a piston with an area of 20 square inches. Every square inch will be acted on by the same one psi pressure, and a force of 20 pounds will be produced. The one cubic inch of fluid displaced when the small piston moves down one inch spreads out under all 20 square inches of the large piston, and so it will move up only $\frac{1}{20}$-inch. [Figure 8-6]

By looking at Figure 8-6 the relationship is:

A (small) × D (small) = A (large) × D (large)

$$1 \times 1 = 20 \times \frac{1}{20}$$

$$1 = 1$$

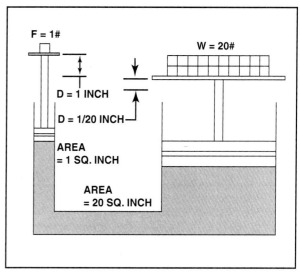

Figure 8-6. The product of the force times the distance moved of the small piston is equal to the product of the force times the distance moved of the large piston.

It is possible to have an application in an aircraft hydraulic system that requires a large amount of movement but only a small amount of force. When this is needed, a large piston can be used to drive a smaller one. All of the fluid moved by the large piston will enter the cylinder with the small piston and move it a distance equal to the volume of fluid divided by the area of the small piston.

SECTION B

HYDRAULIC SYSTEM COMPONENTS AND DESIGN

The basic physics of hydraulics apply to any hydraulic system. All hydraulic systems are essentially the same, whatever their function. Regardless of application, each hydraulic system has a minimum number of components, and some type of hydraulic fluid.

HYDRAULIC FLUID

While we may not normally think of fluid as being a component, the fluid used in aircraft hydraulic systems is one of the system's most important parts. This fluid must be able to flow through all of the lines with a minimum of opposition, and it must be incompressible. It must have good lubricating properties to prevent wear in the pump and valves. It must inhibit corrosion and not chemically attack any of the seals used in the system. Above all, it must not foam in operation, because air carried into the components will give the system a spongy action.

Manufacturers of hydraulic devices usually specify the type of liquid best suited for use with their equipment. These recommendations are made in view of the working conditions, the service required, temperatures expected inside and outside the systems, pressures the liquid must withstand, and the possibilities of corrosion. The major characteristics that must be considered when selecting a satisfactory liquid for a particular system are viscosity, chemical stability, flash point, and fire point.

VISCOSITY

One of the most important properties of any hydraulic fluid is its viscosity. Viscosity is internal resistance to flow. A liquid such as gasoline flows easily (has a low viscosity) while a liquid such as tar flows slowly (has a high viscosity). Viscosity increases as temperature decreases.

A satisfactory liquid for a given hydraulic system must have enough body to give a good seal at pumps, valves and pistons. However, it must not be so thick that it offers excessive resistance to flow, leading to power loss and higher operating temperatures. Excessive viscosity will add to the load and to excessive wear of parts. A fluid that is too thin will also lead to rapid wear of parts which move or are subject to heavy loads, due to excessive friction.

The viscosity of a liquid is measured with a viscosimeter or viscometer. There are several types, but the instrument most often used by engineers in the U.S. is the **Saybolt universal viscosimeter**. This instrument measures the number of seconds it takes for a fixed quantity of liquid (60 cc.) to flow through a small orifice of standard length and diameter at a specific temperature. Time-of-flow measurements are taken in seconds, and the viscosity reading is expressed as SSU (seconds, Saybolt universal). [Figure 8-7]

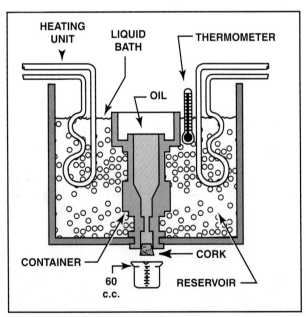

Figure 8-7. The Saybolt viscosimeter is used to measure the viscosity (thickness) of liquids.

CHEMICAL STABILITY

Chemical stability is another property which is exceedingly important in selecting a hydraulic liquid. It is the ability of the liquid to resist oxidation and deterioration for long periods. All liquids tend to undergo unfavorable chemical changes under severe operating conditions. This is the case, for example, when a system operates for a considerable period of time at high temperatures.

Excessive temperatures greatly affect the life of a liquid. It should be noted that the temperature of the liquid in the reservoir of an operating hydraulic system does not always represent a true state of operating conditions. Localized hot spots occur on bearings, gear teeth, or at the point where liquid under pressure is forced through a small orifice. Continuous passage of a liquid through these points may produce local temperatures high enough to carbonize the liquid. The fluid could become like sludge, yet the liquid in the reservoir may not indicate an excessively high temperature. Liquids with a high viscosity have a greater resistance to heat than do light- or low-viscosity liquids that have been derived from the same source. The average hydraulic liquid has a low viscosity. Fortunately, there is a wide choice of liquids available for use within the viscosity range required of hydraulic liquids.

Liquids may break down if exposed to air, water, salt, or other impurities, especially if they are in constant motion or subject to heat. Some metals, such as zinc, lead, brass, and copper have an undesirable chemical reaction with certain liquids and can result in the formation of sludge, gums, and carbon or other deposits. Sludge and other deposits clog openings, cause valves and pistons to stick or leak, and give poor lubrication to moving parts. As soon as small amounts of sludge or other deposits are formed, their rate of formation generally increases. As they are formed, certain changes in the physical and chemical properties of the liquid take place. The liquid usually becomes acidic, darker in color and higher in viscosity.

FLASH POINT

Flash point is the temperature at which a liquid gives off vapor in sufficient quantity to ignite momentarily (flash) when a flame is applied. A high flash point is desirable for hydraulic fluids because it indicates a good resistance to combustion and a low degree of evaporation at normal temperatures.

FIRE POINT

Fire point is the temperature at which a substance gives off vapor in sufficient quantity to ignite and continue to burn when exposed to a spark or flame. As with flash point, a high fire point is desirable in hydraulic fluids.

TYPES OF HYDRAULIC FLUID

To assure proper system operation and to avoid damage to nonmetallic components of the hydraulic system, the correct fluid must be used.

When adding fluid to a system, the only fluid that should be used is the type specified in the aircraft manufacturer's maintenance manual or on the instruction plate affixed to the reservoir or unit being serviced. There are three types of hydraulic fluids currently being used in civil aircraft: vegetable-base, petroleum-base and phosphate ester-base.

INTERMIXING OF FLUIDS

Due to the difference in composition, vegetable-base, petroleum-base and phosphate ester-base fluids will not mix. Neither are the type of seals for any one fluid usable with or tolerant of any of the other fluids. Should an aircraft hydraulic system be serviced with the wrong type of fluid, it should immediately be drained and flushed. The seals should be replaced or serviced according to the manufacturer's specifications.

VEGETABLE-BASE FLUID

MIL-H-7644 fluid was used in the past, when hydraulic system requirements were not as stringent as they are today. This fluid is essentially castor oil and alcohol, and is used primarily in older aircraft. Although similar to automotive brake fluid, it is not interchangeable, and is dyed blue for identification. Natural rubber seals are used with vegetable-base fluid. If a vegetable-base fluid system is contaminated with petroleum-base or phosphate ester-base fluids, the seals will swell, break down and block the system. The system may be flushed with alcohol. Vegetable-base fluids are flammable.

MINERAL-BASE FLUID

MIL-H-5606 is the most widely used hydraulic fluid in general aviation aircraft today. It is basically a kerosene-type petroleum product, having good lubricating properties and additives to inhibit foaming and prevent the formation of corrosion. It is quite stable chemically and has very little viscosity change with temperature. MIL-H-5606 fluid is dyed red for identification, and systems using this fluid may be flushed with naphtha, varsol, or Stoddard solvent. Neoprene seals and hoses may be used with MIL-H-5606 fluid. This type of fluid is also flammable.

SYNTHETIC FLUID

Non-petroleum-base hydraulic fluids were introduced in 1948 to provide a fire-resistant hydraulic fluid for use in high-performance piston engine and turbine-powered aircraft. These fluids were tested for fire resistance by being sprayed through a welding torch flame (6,000 degrees). There was no burning, and only occasional flashes of fire. These and other tests proved that these non-petroleum-base fluids would not support combustion. Even though they might flash at exceedingly high temperatures, they could not spread a fire because burning was localized at the source of heat. Once the heat source was removed or the fluid flowed away from the source, no further flashing or burning occurred.

The most commonly used fluid of this type is **SKYDROL**®(a registered trade name of the Monsanto Chemical Co.) This fluid is dyed light purple for identification and is slightly heavier than water. It sustains operation at a wide range of operating temperatures, from approximately -65°F to more than 225°F. Currently there are three grades of Skydrol in use: Skydrol 500B4, Skydrol LD-4, and Skydrol 5. Skydrol LD-4 has a lower density and offers some advantage in jumbo jet transport aircraft where weight is a prime factor. Skydrol 5 is more compatible with painted surfaces than the other two.

Skydrol is not without its problems for the A&P technician, however. It is quite susceptible to contamination by water from the atmosphere and must be kept tightly sealed. When servicing a system using Skydrol, extreme care must be taken to use only seals and hoses having the proper part number. Skydrol systems may be flushed out with trichlorethylene.

COMPATIBILITY WITH AIRCRAFT MATERIALS
Aircraft hydraulic systems designed around Skydrol fluids should be virtually trouble-free if properly serviced. Skydrol does not appreciably affect common aircraft metals—aluminum, silver, zinc, magnesium, cadmium, iron, stainless steel, bronze, chromium, and others—as long as the fluid is kept free of contamination.

Due to the **phosphate ester base** of synthetic hydraulic fluids, thermoplastic resins may be softened chemically by these fluids. Thermoplastic resins include vinyl compositions, nitrocellulose lacquers, oil base paints, linoleum and asphalt. Skydrol 5 has less effect on painted surfaces than the other types of Skydrol, but manufacturer's instructions should be followed closely. Skydrol will attack polyvinyl chloride, and must not be allowed to drip on to electrical wiring, as it will break down the insulation. However, this chemical reaction usually requires longer than just momentary exposure; and spills that are wiped up immediately with soap and water do not harm most of these materials. Skydrol is compatible with natural fibers and with a number of synthetic fibers (including nylon and polyester) which are used extensively in many aircraft.

Note: Petroleum oil hydraulic seals of **neoprene** or **Buna-N** are not compatible with Skydrol. They must be replaced with seals of butyl rubber or ethylene-propylene elastomers for units that are intended for use in systems utilizing phosphate ester-base hydraulic fluid. These seals are readily available from suppliers.

HEALTH AND HANDLING
Skydrol fluid does not present any particular health hazard when used as recommended. Skydrol has a very low order of toxicity when taken orally or applied to the skin in liquid form. It causes pain on contact with eye tissue and other areas of sensitive skin, but animal studies and human experience indicate that it causes no permanent damage. First aid treatment for eye contact includes flushing the eyes immediately with large volumes of water and the application of an anesthetic eye solution. If pain persists, the individual should see a physician as soon as possible.

If mist or fog form, Skydrol is quite irritating to nasal or respiratory passages and generally produces coughing and sneezing. Such irritation does not persist after exposure is terminated. Silicone ointments, rubber gloves, and careful washing procedures should be utilized to avoid excessive repeated contact with Skydrol in order to avoid solvent effect on skin.

BASIC HYDRAULIC SYSTEMS
A hydraulic system is much like an electrical system. It must have a source of power, a means of transmitting this power, and finally some type of device to use the power. Hydraulic systems can be open or closed. Only closed systems are of use in aviation applications.

OPEN HYDRAULIC SYSTEMS
The most basic form of an open hydraulic system is that used by hydroelectric power plants. Large dams block streams of water to form lakes that store billions of gallons of water. This stored water represents the potential energy in the system. This potential energy is converted to kinetic energy as the water flows downward through penstocks, or pipes, to the turbine. The kinetic energy of the flowing water is converted to mechanical energy as it turns

the turbine. This mechanical energy is used to drive the generator. [Figure 8-8]

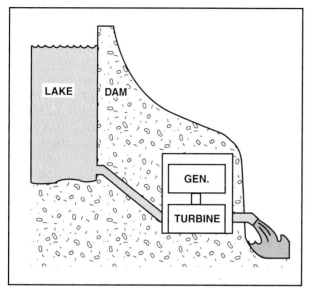

Figure 8-8. The potential energy stored water is converted into mechanical energy to drive the electrical generators in this type of open hydraulic system.

This type of hydraulic system works well to operate a grist mill, or for the production of electrical energy, but has no practical application to airborne systems.

CLOSED HYDRAULIC SYSTEMS

To apply hydraulic power to aircraft systems, the fluid must be enclosed and moved through a system of rigid lines and flexible hoses. The resulting energy is then put to use in various types of actuators and hydraulic motors.

Among the first hydraulic systems used on airplanes was the hydraulic brake. In its simplest form, this hydraulic system consisted of a rubber expander tube similar to an inner tube in a tire. This tube was slipped over the body of the brake and connected to a fluid-filled brake housing in the airplane cockpit. When the pilot pressed on a diaphragm in the brake housing, fluid was forced into the expander tube. The tube expanded and forced blocks of brake lining material against the rotating brake drum and produced enough friction to slow the wheel. [Figure 8-9]

When the pilot released the brake pedal, the diaphragm in the brake housing moved back, and springs between the brake blocks in the wheel pressed the fluid out of the expander tube and back into the housing. This simple system worked quite

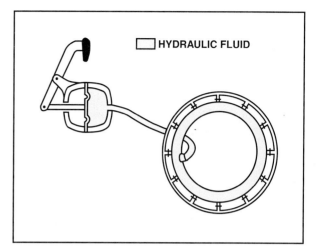

Figure 8-9. Expander-tube brakes used on early aircraft are an example of a basic closed hydraulic system.

well, but it had limitations. If the totally enclosed fluid expanded due to heat, it caused the brakes to drag. If any fluid was lost, the brakes could not function since there was no way to automatically replenish the fluid.

These problems have been solved quite well in a more modern brake system similar to that used on automobiles. Most modern airplanes use disc brakes, but the hydraulic systems for both disc and shoe brakes are similar.

The hydraulic cylinder inside the wheel has two rubber cups which act as pistons. They are both pushed into the cylinder as the brake return spring pulls the shoes away from the drum. The piston in the master cylinder is pushed back by a spring so it just uncovers a compensator port and opens the passage between the vented reservoir and the inside of the master cylinder. [Figure 8-10]

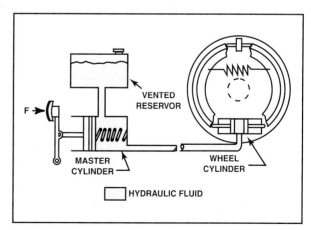

Figure 8-10. Hydraulic brakes are one of the simplest hydraulic systems used on modern airplanes.

When the pilot applies the brakes, the first movement of the piston in the master cylinder covers the compensator port and traps the fluid in the line to the wheel cylinder. As the piston continues to move, the fluid is forced into the wheel cylinder where it moves the pistons out and pushes the brake shoes against the drum. When the brake pedal is released, the spring in the wheel pulls the shoes away from the drum. At the same time, the spring inside the master cylinder moves the piston back, uncovering the compensator port. If any fluid is lost, it will be automatically replaced from the reservoir. If the fluid in the brake line expands due to heat, the expanded fluid will back up into the reservoir and not cause the brakes to drag.

SECTION C

HYDRAULIC POWER SYSTEMS

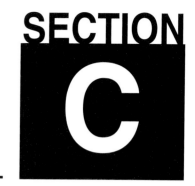

The systems discussed to this point are the most elementary of aircraft hydraulic systems, but they are limited because all they can do is apply the brakes. As aircraft have become more complex, the demand for hydraulically operated equipment has increased. Retractable landing gear, wing flaps, engine cowl flaps, passenger doors and stairs, windshield wipers, and even motors to operate air conditioning compressors are but a few of the functions that can be efficiently handled with hydraulics.

As a typical hydraulic system evolves from the simple to the complex, the systems need the same basic components. But as the systems grow in complexity, the components themselves become more elaborate, and auxiliary devices must be added to make the basic components operate more effectively.

EVOLUTION OF THE HYDRAULIC SYSTEM

A simple hydraulic system might consist of a vented reservoir, a hand-operated pump with a check valve at both its inlet and outlet, and a selector valve that will direct fluid either from the pump into the actuator or from the actuator back into the reservoir. This actuator is a simple single-action unit that uses fluid to force the piston out, but a spring returns the piston when the fluid is released back into the reservoir. [Figure 8-11]

This basic system has two limitations. First, it requires manual operation of the pump, and second, it can apply hydraulic pressure to only one side of the actuator.

DOUBLE-ACTING ACTUATOR AND TWO-WAY SELECTOR VALVE

If, as shown in figure 8-12a, the actuator is replaced with a double-action cylinder that uses hydraulic pressure on both sides of the piston, the actuator can operate under hydraulic pressure in both directions—a big improvement in the system. The figure also shows that the selector valve has been replaced. This selector valve has four ports, and in one position it directs fluid from the pump to the

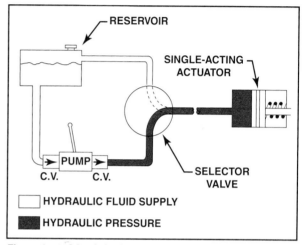

Figure 8-11. A basic hydraulic system consists of a supply of hydraulic fluid, a pump, an actuator, and a selector valve.

upper side of the piston while the lower side of the piston is connected to the reservoir. When the selector valve is rotated 90 degrees, as shown in figure 8-12b the lower side of the piston is connected to the pump outlet and the fluid from the upper side is returned to the reservoir.

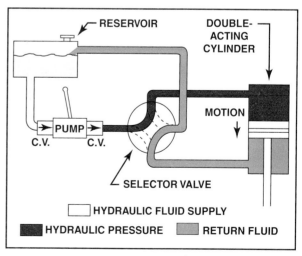

Figure 8-12a. A double-acting actuator allows the actuator to be powered in both directions. A two-way valve allows selection of either side of the actuator.

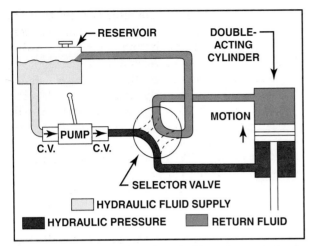

Figure 8-12b. Movement of a double-acting actuator is acomplished by redirecting fluid-flow with the two-way valve.

ENGINE-DRIVEN PUMP AND PUMP-CONTROL VALVE

The next big improvement in this system is to replace the hand pump with an engine-driven pump. This design can have problems, though, because the engine is robbed of power by maintaining hydraulic pressure on the system when it is not needed. The pump is coupled directly to the engine and there is no way to disconnect it, but it needs very little power when it is not moving the fluid against an opposition. If a pump-control valve is installed between the pump outlet and the return line into the reservoir, the pump-control valve can be opened when pressure is not needed, and fluid flows from the bottom of the reservoir, through the pump, back into the top. This fluid circulates freely with almost no opposition, and the pump requires very little engine power. The selector valve holds fluid in the actuator by maintaining pressure on both sides of the piston.

To actuate this system, the pilot puts the selector valve in the desired position and then closes the pump-control valve. The pump now directs fluid into one side of the actuator while the fluid on the opposite side of the piston returns to the reservoir. The pump continues to move fluid after the piston reaches the end of its travel, and the pressure rises enough to unseat the relief valve. This valve allows fluid to return to the reservoir and maintain a pressure below the bursting pressure of the lines, or below the pressure that could damage the pump.

The pilot now returns the selector valve to its neutral position and opens the pump-control valve to relieve the engine of the load. Many selector valves and pump-control valves used in this type of system are automatic in their action. The pilot needs only to select the desired position of the selector valve and then close the pump-control valve. When the actuator reaches the end of its travel, the selector valve automatically returns to the neutral position and the pump-control valve opens.

A hydraulic system using a pump-control valve can operate both the landing gear and the flaps or even more actuators, if needed. The pilot must select the system wanted and then close the pump-control valve each time a system is to be actuated. A better system design would be to maintain the pressure at all times. In this way the pilot would always have pressure for raising or lowering the landing gear, and the brakes could operate from the main hydraulic system rather than having to use independent master cylinders. [Figure 8-13]

UNLOADING VALVE AND ACCUMULATOR

In order to maintain pressure on the system without causing the engine to continually work against the pressure, two new types of components are required: an unloading valve and an accumulator.

A hydraulic pump does nothing more than move the non-compressible fluid through the hydraulic system. There is no pressure generated until this flow of fluid is opposed. The fluid can move the piston of an actuator from one end of its travel to the other with only enough rise in pressure to overcome the friction. When the piston reaches the end of its

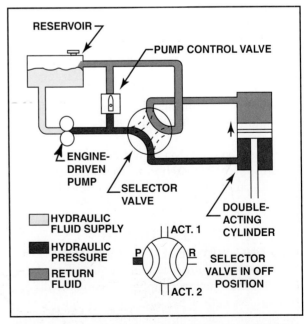

Figure 8-13. A hydraulic system with an engine-driven pump is similar to the two-way system driven by a hand pump. It incorporates a bypass valve (pump-control valve) to unload the pump when it is not being used to power the actuator.

travel, the pressure rises immediately because there is no longer a place for the fluid to go. If there is no relief valve in the system, this sudden rise in pressure may cause damage.

If an unloading valve is installed in the line between the pump and the selector valves, the pump-control valve can be dispensed with. The unloading valve has a return line to the reservoir. There is a check valve after the unloading valve to prevent any reverse flow of fluid from the pressure manifold back to the reservoir. This check valve, however, is normally built into the unloading valve.

An accumulator is a device having two compartments separated by a movable partition; either a piston, a diaphragm, or a bladder. One compartment is connected directly to the pressure manifold, and the other is sealed and filled with either compressed air or nitrogen. The air (or nitrogen) pressure is initially about one-half of the system operating pressure. When the pump forces fluid into the pressure manifold, some of it flows into the accumulator and moves the partition, further compressing the air, giving it the same pressure as that of the fluid. When the system pressure reaches the desired value, the unloading valve automatically "kicks out," which means that it opens the return line to the reservoir and unloads the pump. No fluid can flow back from the pressure manifold through the unloading valve because of the check valve, and the air in the accumulator holds pressure on the fluid in the manifold. As soon as any component connected to the pressure manifold is actuated, the pressure will drop, and the unloading valve will detect this drop in pressure through the sensor line. This pressure sensing, like the check valve, is actually built into the unloading valve and causes the valve to "kick in," putting the pump online until the pressure rises to the "kick-out" pressure. The shock-absorbing action of the accumulator will prevent the pressure from changing so rapidly that it could damage the system. [Figure 8-14]

HAND PUMP AND STANDPIPE

The next item to be added to the hydraulic system is a hand pump to provide pressure for the brakes before the first engine is started. A slight change in the reservoir is also made at this point. The engine-driven pump no longer receives its fluid from the bottom of the reservoir, but now takes it from a standpipe. If one of the lines breaks, the engine-driven pump can exhaust the reservoir only down to the top of the standpipe. If the broken line can be isolated, there will still be enough fluid available to

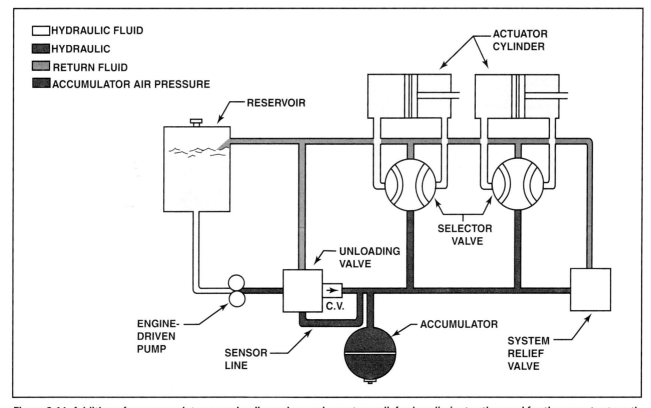

Figure 8-14. Addition of an accumulator, an unloading valve, and a system relief valve eliminates the need for the operator to activate the pump control valve every time pressure is required by the system.

the hand pump to lower the landing gear and operate the brakes. [Figure 8-15]

FILTERS AND THERMAL RELIEF VALVES

To keep the fluid in the system clean, a filter is needed. A typical location for the filter is in the return line to the reservoir. Here, it will catch all fluids used to actuate any of the cylinders and all fluids circulated by the pump when it is unloaded. Most filters are equipped with a bypass valve so that if the filter plugs up, the fluid will open the valve and return to the reservoir without doing any damage. Even though the filter is in the return line where there is normally no pressure, if it should clog it will cause enough opposition to the flow of the return fluid that pressure will build up across it, causing the bypass valve to open.

If the fluid is trapped in the line to the actuator by the selector valves in the neutral position and should expand due to heat, thermal relief valves will offseat just enough to relieve the pressure. Since there is no compressible fluid in these lines, it takes but a few drops to decrease the pressure enough to prevent damage. [Figure 8-16]

When any part of a hydraulic system requires removal, there is a possibility of system contamination. The best way to minimize contamination is to cap all lines and fittings as soon as they are opened. Special pairs of caps are designed to fit both sides of the connection. Care must be taken to ensure

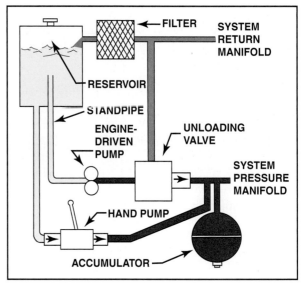

Figure 8-15. Addition of a hand pump and a standpipe provide emergency capability in the event of fluid exhaustion to the engine-driven pump.

removal of all plugs and caps when the system is reconnected. The high pressures and flow rates in typical hydraulic systems can quickly draw any installed plugs into the system and can cause expensive damage to delicate components.

SPECIAL TYPES OF AIRCRAFT HYDRAULIC SYSTEMS

The hydraulic system described above can be expanded with the addition of selector valves and

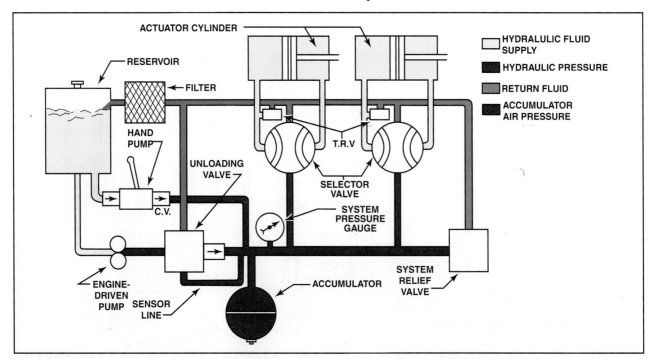

Figure 8-16. Adding a filter keeps the fluid clean and the addition of thermal relief valves (T.R.V.) guards against overpressure due to temperature rise.

actuating units to accommodate all of the systems of a large aircraft. To allow smaller aircraft to take advantage of hydraulic power, but utilize a less complex system, open-center hydraulic systems and hydraulic power-pack systems have been developed.

OPEN-CENTER SYSTEM

Many of the lighter aircraft use an open-center hydraulic system that performs the functions of the more complex systems, but with relatively simple components. The reservoir, pumps, filter, and system relief valve all function in the same way as in the closed-center system, except that in the open-center system the two selector valves are in series, while those in the closed-center system are in parallel. [Figure 8-17]

The open-center valves serve the functions of the selector valve and unloading valve. When the selector handle is in the neutral position, fluid flows straight through the valve. Since all of the valves in the system are in series, the fluid flows out of the tank, through the pump, through both selector valves, and back into the reservoir through the filter.

When the pilot moves the landing gear selector to either the GEAR UP or GEAR DOWN position, fluid flows from the pump to the appropriate side of the landing gear actuator and moves the piston over. Fluid that is forced out of the other side of the landing gear actuator flows through the flap selector valve back to the reservoir. When all of the landing gears are in the up or down position, the landing gear selector valve shifts to the neutral position so that fluid can flow straight through it.

HYDRAULIC POWER PACK SYSTEM

To continue the simplification of the hydraulic systems, many manufacturers use an electric motor to drive the hydraulic pump. They incorporate the reservoir, control valve, and many of the auxiliary

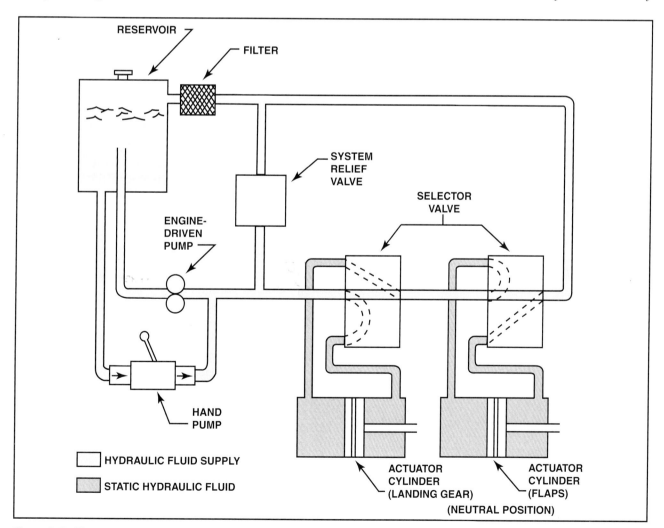

Figure 8-17. The open-center hydraulic system has the selector valves in series as opposed to the parallel valves in the closed-center system.

valves into a single unit, normally called a power pack. In this system the pump is driven by a reversible DC motor in the appropriate direction to lower the landing gear. When the gear reaches its limits, the power-pack motor stops, ceasing the flow of hydraulic fluid and relieving the system pressure. When the selector is placed in the GEAR DOWN position as illustrated in Figure 8-18, the pump turns in the direction indicated. Fluid comes from the reservoir down through the right check valve, around the outside of the gears, and down to the shuttle valve. The shuttle valve is pushed over, compressing the spring and opening the passage to the down-side of the three landing gear cylinders. This same pressure moves the gear-up check valve over and opens the valve so that fluid returning from the up-side of the cylinders can flow back through the pump and to the down-side of the cylinders. [Figure 8-18]

When the selector is placed in the GEAR UP position as illustrated in Figure 8-19, the pump turns in the opposite direction. Fluid is drawn from the reservoir through the filter and around the gears in the pump, and through the gear-up check valve, to the up-side of the three cylinders. The return fluid passes through the shuttle valve that is now moved over by the spring, and back up into the reservoir. When all three gear cylinders are up, pressure will build up and open the pressure switch that shuts off the electrical power to the pump motor. There are no mechanical up-locks in this system, and the gear is held up by hydraulic pressure. If the pressure bleeds off, the pressure switch will start the pump motor and restore the pressure before the gear has a chance to fall out of the wheel wells. [Figure 8-19]

HYDRAULIC SYSTEM COMPONENTS

Most hydraulic systems are made up of the components discussed so far. Even though most systems have all of these components, the components themselves vary according to the type of system or the action needed from the hydraulic system.

RESERVOIRS

There is a tendency to envision a reservoir as an individual component; however, this is not always true. There are two types of reservoirs:

1. Integral — This type has no housing of its own but is merely a space set aside within some major component to hold a supply of operational fluid. A familiar example of this type is the reserve fluid space found within most automobile brake master cylinders.

2. In-Line — This type has its own housing, is complete within itself, and is connected with other components in a system by tubing or hose.

In all reservoirs, a space is provided above the normal level of the fluid for fluid expansion and the escape of entrapped air. Reservoirs are never intentionally filled to the top with fluid. Most reservoirs are designed so the rim of the filler neck is somewhat below the top of the reservoir to prevent overfilling during servicing. Many reservoirs are equipped with a dipstick or a glass sight gauge by which fluid level can be conveniently and accurately checked. Reservoirs are either vented to the atmosphere or closed to the atmosphere and pressurized.

UNPRESSURIZED RESERVOIRS

Aircraft that fly at lower altitudes normally have hydraulic systems supplied with fluid from unpressurized reservoirs. These reservoirs must be large enough to hold all of the fluid required for any position of the actuating cylinders.

The fluid return to the reservoir is usually directed in such a way that foaming is minimized, and any air in the fluid is swirled out, or extracted. Some reservoirs have filters built into them at the return line so that all of the fluid entering the tank is strained. The filler cap on an un-pressurized reservoir may be opened while the system is operating. However, since the fluid is out in the system, the level should only measured with the system at rest. If fluid is added while the system is operating, the maximum fluid level will likely be exceeded when the system is shut down.

Many reservoirs have two outlets. One is located in the bottom and the other is either part way up the side, or is connected to a standpipe that sticks up inside the reservoir. A standpipe outlet feeds the engine-driven pump. In the event of a break in the system that causes the engine pump to lose all its fluid, the hand pump can still pick up enough fluid to lower the landing gear and flaps and actuate the brakes. [Figure 8-20]

PRESSURIZED RESERVOIRS

Jet aircraft that operate at altitudes where there is not enough air pressure to assure a positive feed of fluid to the pump have pressurized hydraulic reservoirs. This ensures that an adequate supply of fluid, free from foaming, is always available at the pump inlet. There are three ways of pressurizing these reservoirs: using variable displacement hydraulic pumps,

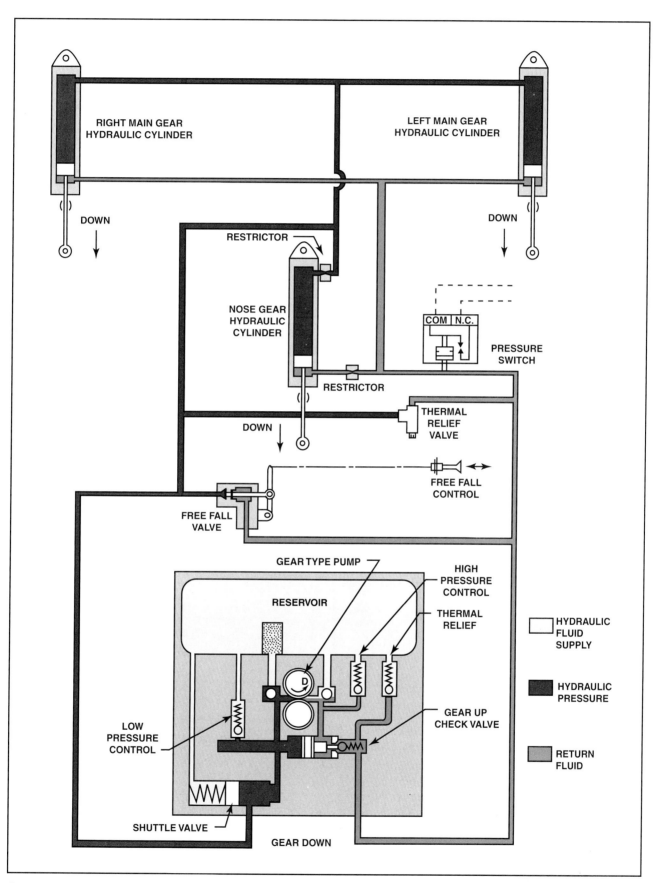

CAP: Figure 8-18. A power pack is a hydraulic system powered by a reversible DC motor. The direction of rotation determines the direction and action of the hydraulic fluid and actuating cylinders. In this drawing, the landing gear is being lowered.

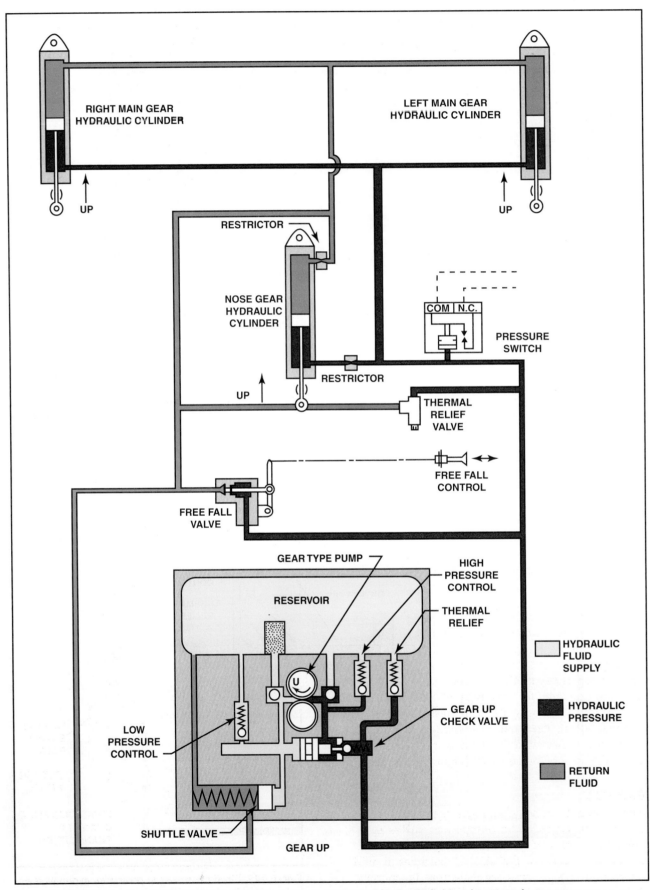

Figure 8-19. Reversing the direction of motor rotation reverses the flow of hydraulic fluid and retracts the gear.

Hydraulic and Pneumatic Power Systems

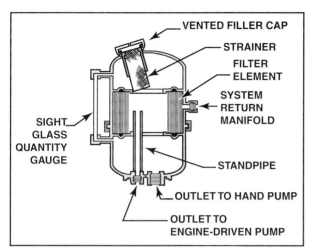

Figure 8-20. Non-pressurized hydraulic reservoirs are found on airplanes that fly only at lower altitudes.

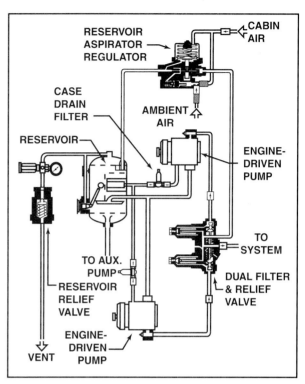

Figure 8-21. One type of pressurized hydraulic reservoir utilizes an aspirator in the return line to pressurize the fluid. The aspirator operates similarly to a jet pump found in some fuel systems.

engine bleed air, or by the use of hydraulic system pressure acting on a small piston in the reservoir.

One method is observed in aircraft that use variable displacement hydraulic pumps. Since there is always some fluid flow back to the reservoir, an **aspirator** is installed in the return line to the reservoir. Fluid flowing through the aspirator or venturi-tee draws either cabin or ambient air pressure by jet action into the reservoir thereby pressurizing it. A pressure regulator maintains a pressure of between 30 and 35 psi in the reservoir. [Figure 8-21]

Another system uses bleed air from the aircraft's turbine engines to maintain a pressure in the main hydraulic reservoir of 40 to 45 psi. All of the pressurizing air must be released before removing the reservoir cap. [Figure 8-22]

A third type of reservoir may be pressurized by hydraulic system pressure acting on a small piston. The resulting force, produced by a larger piston attached to the smaller one, causes a reduction of pressure on reservoir fluid. Pressure ratios of near 50:1 are common for this type of reservoir. This means that a 3,000-psi system can pressurize the fluid to about 60 psi. The quantity of fluid in this type of reservoir is indicated by the amount the piston sticks out of the body of the reservoir. [Figure 8-23]

FILTERS

The extremely small clearances between components in many hydraulic pumps and valves make effective filtering of the fluid extremely important. A filter is rated by the size of particles it will remove, and these sizes are measured in microns, with one micron equal to one millionth of a meter or 0.000039 inch. To get a good idea of how small this is, the unaided eye can see something only as small

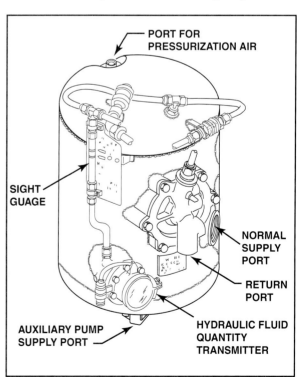

Figure 8-22. Bleed air is used to pressurize hydraulic reservoirs on turbine-powered aircraft.

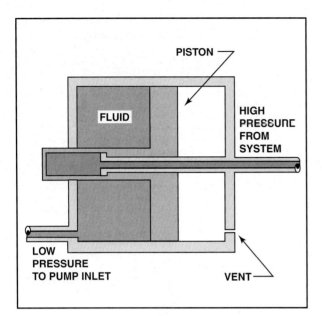

Figure 8-23. The pressure of the hydraulic system can be used to pressurize the reservoir. A pressure-reduction mechanism is used to lower the high system pressures to a low pressure suitable for the reservoir. A 50:1 reduction occurs if the area of the input portion of the piston (A) is one square inch and the area of the piston portion bearing on reservoir fluid (B) is 50 square inches.

as 40 microns, and white blood cells are about 25 microns. [Figure 8-24]

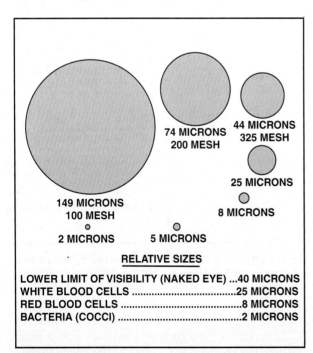

Figure 8-24. The relative size of particles may be used to visualize the effectiveness of a hydraulic filter. The large circle represents the size of a particle that would be stopped by a 100-mesh screen (100 threads per inch). A filter that will remove particles down to 10 microns will provide fluid that is clean enough for almost any hydraulic system.

One of the more efficient types of filters used in aircraft hydraulic systems is made of specially treated paper folded into pleats to increase its surface area. This pleated paper micronic element, as it is called, is wrapped around a spring steel wire coil to prevent it from collapsing. These filters often have a bypass valve across the filtering element so that if the filter ever plugs up, the fluid will bypass the element. This type of filter is usually installed in the fluid return line. Micronic filters are replaced, rather than cleaned, periodically in accordance with manufacturer's instructions. [Figure 8-25]

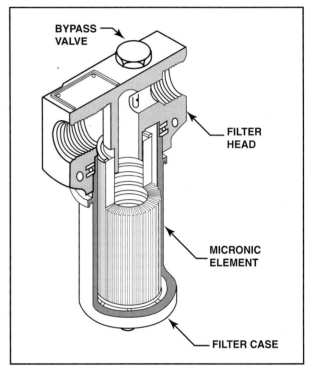

Figure 8-25. A micronic-type filter uses a folded paper element to increase the effective area of the filtering medium.

Filters which have elements made of stainless wire woven into a mesh and wrapped around a wire frame have proven to be quite durable. Mesh with openings as small as .0055 inch are available. [Figure 8-26]

A special two-stage filter is used in the return line for some of the large aircraft hydraulic systems in place of the standard single element unit. This type of filter allows the use of an extremely fine element at low flow rates without causing an excessive pressure drop. [Figure 8-27]

The first stage element has a filter rating of 0.4 to 3 micron and the second stage filter has a rating of 1.5 to 15 microns. All flow less than five gallons per minute, which is adequate for normal cruise flight

Hydraulic and Pneumatic Power Systems

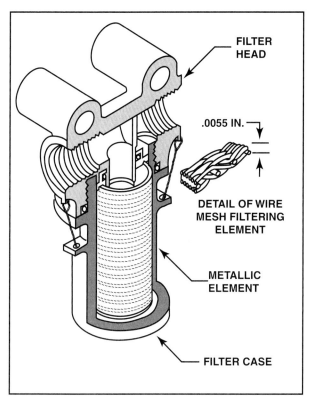

Figure 8-26. Some filters use a woven wire mesh around a central frame. In the detail is seen a close-up of the wire mesh used in the filtering element.

operation, passes through both the first- and second-stage elements and back into the reservoir. During the higher flow operations, such as during actuation of the landing gear or flaps, the flow is divided with up to five gallons per minute passing through both elements, while all fluid in excess of this amount bypasses the first-stage element and passes through the second-stage element only. This allows the pressure drop across the filter to be held to a reasonable value during conditions of high flow rate.

There are differential pressure indicators on top of the filter that indicate when the element is contaminated and needs to be replaced. There are also relief valves across both elements to prevent over-pressurizing the return line if, for any reason, the pressure drop across the elements becomes excessive.

Some filters are built strong enough that they can be used on the pressure side of the system. One of the popular types of pressure filters is the **Cuno filter**. The Cuno filtering element is made up of a stack of discs and spacers mounted on a rod, with a cleaner blade between each of the discs. This entire assem-

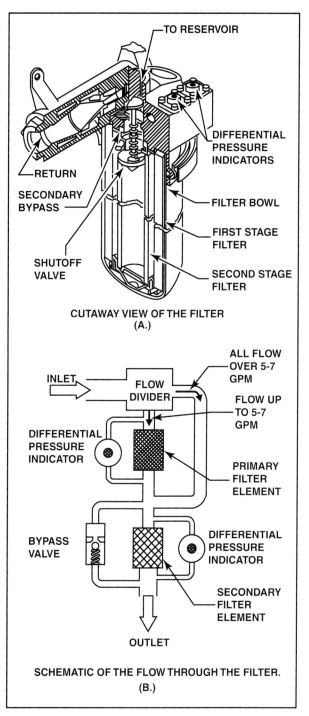

Figure 8-27. Two-stage filters are used to filter out extremely small particles and still maintain an acceptable flow rate.

bly is mounted inside a bowl, and fluid flows from the outside of the stack between the discs to the inside and out to the system. The rod should be turned with the handle that protrudes through the filter housing. In this way, contaminants that have been trapped between the discs will be scraped out by the cleaner blades and will fall to the bottom of the bowl where they can be removed during the next inspection. [Figure 8-28]

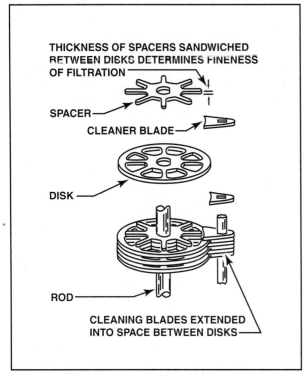

Figure 8-28. The Cuno filter is sturdy enough to be used on the pressure side of the system. A handle can be turned to clean the disks. Contaminants then fall to the bottom of the filter where they can be removed.

PUMPS

Hydraulic power is transmitted by the movement of fluid by a pump. The pump does not create the pressure, but the pressure is produced when the flow of fluid is restricted. A knowledge of how an electrical circuit works helps in understanding hydraulic power.

The flow of fluid in a line is equivalent to the flow of electrons in a wire, the current. The pressure that causes the flow is the same as the voltage, and the opposition to the flow of fluid is the same as the resistance. If there is very little friction in the line, very little pressure is needed to cause the fluid to flow.

In Figure 8-29(A) is a very simple electrical system, consisting of a battery, an ammeter, a voltmeter, and a resistor. The ammeter measures the flow of electrons in the circuit, and the voltmeter measures the voltage (pressure) drop across the resistor. The hydraulic system in Figure 8-29(B) is very similar in its operation. The pump moves the fluid through the system and may be compared to the battery that forces electrons through the circuit. The flowmeter measures the amount of flow, the valve acts as a variable opposition to the flow, and the pressure gauge measures the pressure drop across the valve. [Figure 8-29]

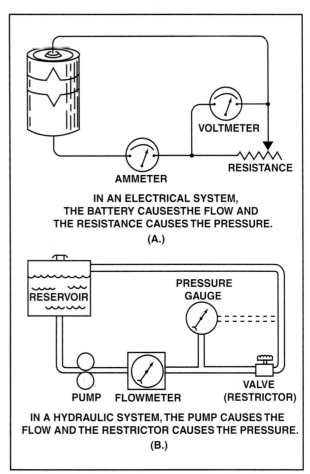

Figure 8-29. A basic hydraulic system is similar to a basic electrical circuit.

When the variable resistor is set to its minimum resistance, the current will be maximum and there will be a minimum voltage drop across the resistor. In the same way, when the valve is fully open, there will be a maximum flow of fluid and a minimum pressure drop across the valve.

When the resistance in the electrical circuit is increased, the voltage drop across the resistor will increase and the current will decrease. In the hydraulic system, as the valve is closed, the flow will decrease and the pressure will increase. When the valve is fully closed, there will be no flow and the pressure will increase to a value as high as the pump can produce. As we will see very soon, if the pump is of the constant displacement type, there must be some provision in the system to relieve the high pressure; otherwise the pump will be damaged, or something in the system will be broken.

HAND PUMPS

Single-action pumps move fluid only on one stroke of the piston, while double-action pumps move it on both strokes. Double-action pumps are the ones most commonly used in aircraft hydraulic systems because of their greater efficiency.

One of the more commonly used types of double-action hand pumps is called a piston rod displacement pump, because its pumping action is caused by the difference in area between the two sides of the piston. One side of the piston has less surface area because of the piston rod. [Figure 8-30]

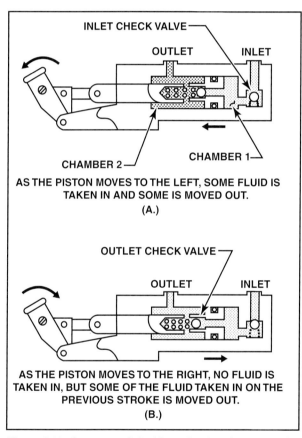

Figure 8-30. One type of double-action hand pump is the piston rod displacement pump.

In view A, the handle is moving in the direction to pull the piston to the left, and fluid is drawn into the pump through the inlet check valve. When the piston is at the end of its stroke, chamber 1 is full of fluid, and the spring closes the check valve. In view B, as the handle is moved to the right, the piston is forced into chamber 1, and fluid flows through the outlet check valve into chamber 2. The volume of chamber 2 is smaller than that of chamber 1 because of the piston rod, and so the fluid fills chamber 2, and the excess leaves the pump through the outlet port. On the return stroke of the piston, the remainder of the fluid in chamber 2 is forced out, while a fresh charge is being brought into chamber 1.

By assigning some values to the area of the piston, it can be seen how this pump moves the fluid. Assume the right side of the piston has an area of two square inches and the piston rod an area of one square inch. When the piston moves two inches with each stroke, four cubic inches of fluid will be pulled into chamber one when the piston moves its full travel. The piston rod decreases the area of the piston on the left side to one square inch so that the volume of chamber two is only two cubic inches. When the piston moves to the left, four cubic inches of fluid flow in, and two cubic inches is forced out. On the next stroke, the piston moves to the right, and no fluid is taken in, but two cubic inches is forced out.

If a force of 500 pounds is exerted on the piston as it moves to the left, the one square inch of area will produce a pressure of 500 pounds per square inch. But as the piston moves to the right, this same 500 pounds of force is spread out over two square inches of piston area, and the pressure will be only 250 pounds per square inch.

POWERED PUMPS

The only function of a pump is to move fluid through the system, and there are two basic types of pumps—those having a constant displacement and those having variable displacement. Pumps may be powered by an electric motor or by direct drive from an engines accessory section. Pumps driven by the engine are usually protected by a shear shaft that disconnects the pump from the engine if the pump should stall.

Constant-Displacement Pumps

A constant-displacement pump moves a specific volume of fluid each time its shaft turns. It must have some form of regulator or relief valve in the system to relieve the pressure that builds up when the pump moves more fluid than the system can use. Constant-displacement pumps come in several different types, each with its own characteristics, and include gear pumps, gerotor pumps, piston pumps and vane pumps.

The **gear pump** is one of the most generally used types of constant-displacement pumps for medium-pressure hydraulic systems. Gear pumps are rugged and dependable and are relatively inexpensive to manufacture. In a typical spur gear-type hydraulic pump the upper gear is driven by the engine through a splined shaft. This gear rides in a close-fitting housing and drives the lower gear. As the teeth of the two gears separate, the volume of the inlet chamber increases and lowers the pressure so that fluid will flow into the pump from the reservoir. This fluid is trapped between the teeth and the wall of the pump body, and the fluid is moved around the outside of the gears to the outlet side of the pump. [Figure 8-31, view (A)]

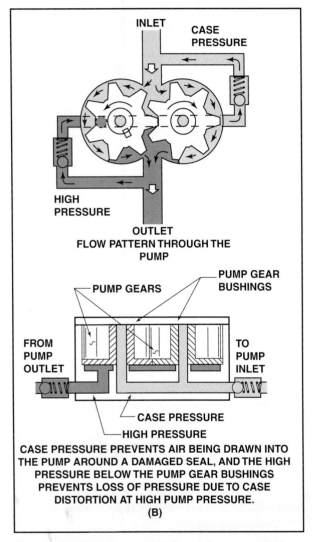

Figure 8-31. The most commonly used constant-displacement pump is the gear pump.

A small amount of fluid leaks past the gears and around the shaft for lubrication, cooling, and sealing. This fluid drains into the hollow shafts of the gears where it is picked up by the low pressure at the inlet side of the pump. A weak relief valve holds the oil in the hollow shafts until it builds up a pressure of about 15 psi. This is called case pressure, and is maintained so that, in the event the shaft or seal becomes scored, fluid will be forced out of the pump rather than air being drawn in. Air in the pump would displace some of the fluid needed for lubrication, and the pump would be damaged. [Figure 8-31, view (B)]

The inside of the gear cavity of some of the pumps is fitted with a bushing and flange, to minimize the problem of the case distorting when the output pressure is high. Distortion will increase the leakage and cause a loss of pressure. Fluid from the output side of the pump is fed back through a check valve into a cavity under the bushing flange. As the output pressure rises, it forces the flange tight against the gears. This minimizes the leakage and compensates for wear.

The **gerotor pump** is a combination internal-external gear pump. The six-tooth spur-type drive gear is turned by an accessory drive from the engine, and as it turns, it rotates a seven-tooth internal-gear rotor. [Figure 8-32]

By following the relationship between the two gears, it can be seen that in view A the two marked teeth are meshed, and the tooth of the spur gear almost completely fills the cavity in the rotor. As the drive gear rotates and pulls the driven gear around, the volume of the cavity increases until in view C it is at a maximum. During the rotation from view A to view C, the expanding cavity is under the inlet port and fluid is drawn into the pump. As the gears continue to rotate, the cavity formed by the marked teeth moves under the outlet port. As the drive gear meshes with the cavity next to the marked cavity in the rotor, its volume decreases. The fluid in this cavity is forced out of the pump through the outlet port.

High-pressure hydraulic systems often use a fixed-angle piston-type pump. These pumps usually have either seven or nine axially-drilled holes in a rotating bronze cylinder block. Fitted into each of these holes are close-fitting pistons, attached by a ball-jointed rod to the engine driven pump drive plate. The housing is angled so that the pistons on one side of the cylinder block are at the bottom of their stroke, while those on the other side are at the top of theirs. As the pump rotates one half turn, half of the pistons move from the top of their stroke to the bottom, while the pistons on the other side move from the bottom of their stroke to the top. A valve plate with two crescent-shaped openings covers the end of the cylinders. One of the openings is above the pistons moving up, and the other opening is above the pistons that are moving down. As the pistons move down, they pull fluid into the pump, and as they move up, they force the fluid out. [Figure 8-33]

Some hydraulic systems require a pump to move a relatively large volume of fluid, but do not need to

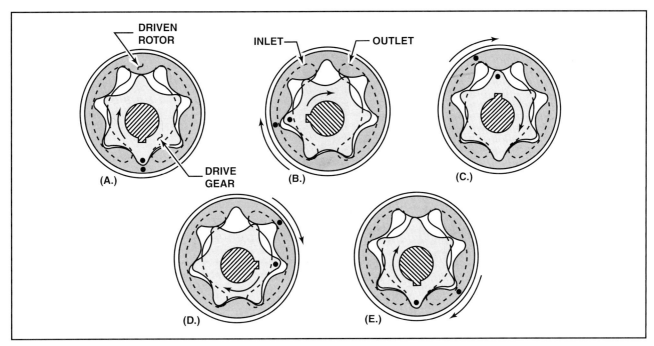

Figure 8-32. By following the marked tooth of the drive gear, it can be seen how the gerotor pump draws in fluid during A and B, and then discharges the fluid during C, D, and E.

produce a very high pressure. For these applications the **vane pump** may be used.

The vanes in the pump are free-floating in the rotor and are held against the wall of the sleeve by a spacer. As the rotor turns in the direction shown by the arrow, the volume between the vanes on the inlet side increases, while the volume between the vanes on the outlet side decreases. This change in volume draws fluid into the pump through the inlet port and forces it out through the outlet port. [Figure 8-34]

Variable-Displacement Pump

A variable-displacement pump does not move a constant amount of fluid each revolution, but only the amount the system will accept. By varying the pump output, the system pressure can be maintained within the desired range without the use of regulators or relief valves. Variable-displacement pumps can turn without any fluid being forced into the system. So, to prevent overheating, these pumps usually bypass some fluid back to the reservoir so there will always be some flow of fluid to cool the pump.

An unloading valve of some sort is needed when a constant-displacement pump is used. But the same force used to control this valve may be used to control the output of a variable-displacement pump with no need for a separate control valve. One of the more popular variable-dis-

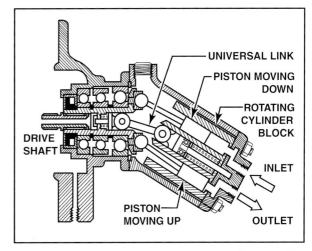

Figure 8-33. A fixed-angle piston pump is often used in high-pressure hydraulic systems.

placement pumps used for high-pressure aircraft hydraulic systems is the Stratopower demand-type pump. This pump uses nine axially-oriented pistons and cylinders. The pistons are driven up and down in the cylinders by a wedge-shaped cam, and the pistons bear against the surface of the cam with ball-joint slippers. When the thick part of the cam is against the piston, it is at the top of the stroke. As the cam rotates, the piston moves down the cylinder until, at the thin part of the cam, it is at the bottom. The stroke is the same, regardless of the amount of fluid demanded by the system, but the effective length of the stroke controls the amount of fluid pumped. [Figure 8-35]

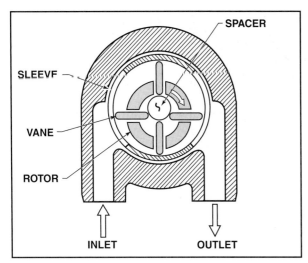

Figure 8-34. Vane pumps are used more generally for moving fuel and air than they are for moving hydraulic fluid.

The balance of forces that controls the pump pressure on the system is between the compensator spring and the compensator stem piston. In Figure 8-36(A) a passage from the discharge side of the pump directs output fluid pressure around the compensator stem. This stem is cut with a shoulder that serves as a piston. As the system pressure rises, the fluid pushes the stem up, compressing the compensator spring. The spider, Figure 8-36(A), which moves the sleeves up or down the pistons, is attached to the stem. [Figure 8-36]

When the pressure is high, as shown in Figure 8-36(B), it acts on the stem piston to raise the spider against the compensator spring and the relief holes near the bottom of the pistons are uncovered during the entire stroke. The pistons now stroke up and down, but no fluid is forced out of the pump, since it is all relieved back into the pump. Near the top of the stroke, a bypass hole in the piston aligns with a passage in the pump housing so that a small amount of fluid bypasses back into the reservoir just enough for lubricating and cooling the pump. When the pressure is low, as in Figure 8-36(C), the compensator spring forces the spider and sleeves down the piston, covering the relief hole when the piston is near the bottom of its stroke. In this way, the full stroke of the piston is utilized to move the fluid. Fluid is forced out

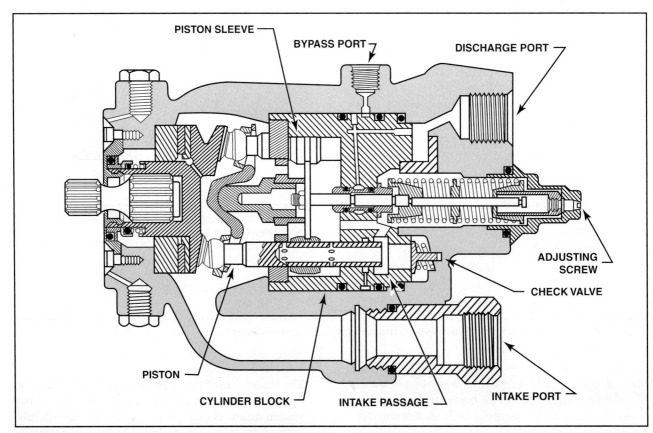

Figure 8-35. The variable-displacement axial-piston-type hydraulic pump utilizes a compensator assembly to effectively lengthen or shorten the stroke of the pistons. This variable displacement maintains the preset system pressure by varying the amount of fluid pumped.

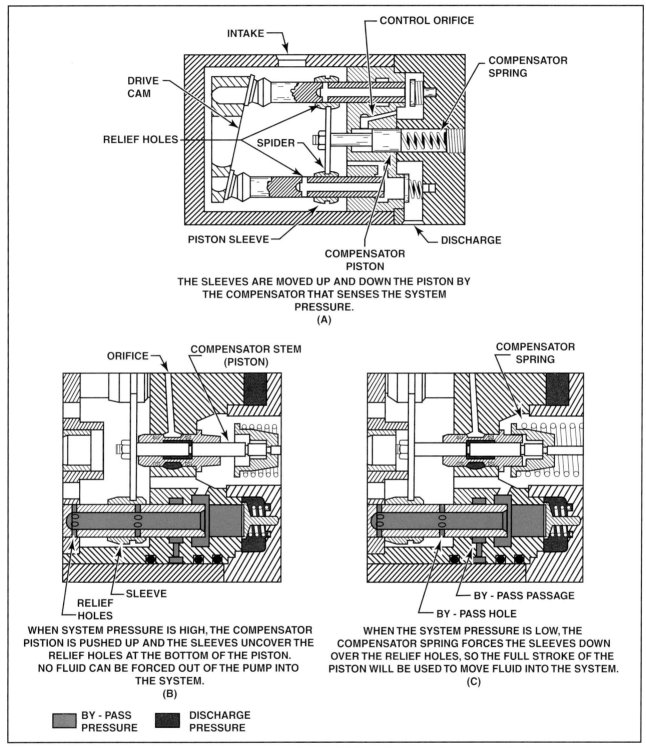

Figure 8-36. This figure details the control characteristics of the variable-displacement piston-type hydraulic pump.

through the check valves into the pump discharge line. In any condition of intermediate pressure, the sleeve closes the relief holes at some point along the stroke of the piston. This allows just enough fluid to be pumped to maintain the system pressure at that level for which the compensator spring is set.

VALVES

The valves used in hydraulic systems may be divided into flow-control and pressure-control valves. A flow-control valve selects the route of flow of the fluid through the system, and is not normally concerned with the pressure. Pressure-control valves, on the other hand, adjust, regulate, or limit the amount of pressure in the system, or in any portion of the system.

FLOW CONTROL VALVES

For hydraulic components and systems to operate as intended, the flow of fluid must be rigidly controlled. Fluid must be made to flow according to definite plans. Many kinds of valve units are used for exercising such control. Examples include selector valves, check valves, sequence valves, priority valves, quick disconnects, and hydraulic fuses.

Selector Valves

One of the most familiar flow-control valves is the selector valve, which determines the direction of flow of fluid to retract or extend the landing gear, or to select the position of the wing flaps. There are two commonly used types of selector valves. The open-center valve directs fluid through the center of the valve back to the reservoir when a unit is not being actuated. The closed-center valve stops the flow of fluid when it is in its neutral position. Both valves direct fluid from the pump to one side of the actuator and vent the opposite side to the reservoir.

The open-center valve is an open-center poppet-type selector valve utilized in an open-center hydraulic system (refer back to Figure 8-17). When the selector handle is in the neutral position, the appropriate poppet is off of its seat. Fluid flows straight through the valve from the pump to the next selector valve and on to the reservoir. [Figure 8-37, view (A)]

Moving the gear selector handle to the GEAR DOWN position causes the cams to open valves 1 and 4. Fluid can now flow from the pump to the actuator around poppet 1, while the return fluid from the vent side of the actuator flows around poppet 4 back to the reservoir as in Figure 8-37, view (B). When the actuator reaches the end of its travel, the pump continues to produce a flow of fluid. A system relief valve (refer back to Figure 8-17) must off-seat to allow a flow back to the reservoir until the selector valve is moved to its neutral position. Some open-center valves have an automatic feature that causes them to move to the neutral position when the pressure rises to a specified value.

When the gear selector is moved to the GEAR UP position, poppets 2 and 5 open, and fluid from the pump flows around poppet 2, while the return fluid flows around poppet 5 as in Figure 8-37, view (C).

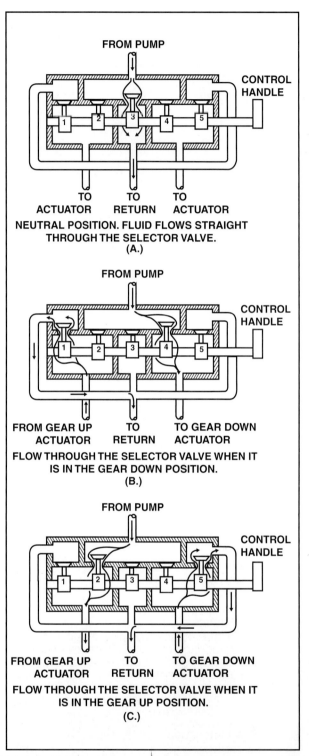

Figure 8-37. The open-center poppet-type valve routes the fluid straight through the valve when no hydraulic action is currently selected. This is done in this example by opening poppet 3.

Plug-type closed-center valves are often used in aircraft hydraulic systems. In one position, the pressure port and one actuator port are connected, while the other actuator port is connected to the return port. Rotating the selector handle ninety degrees reverses the connection between the actuator ports and the pressure and return lines. [Figure 8-38]

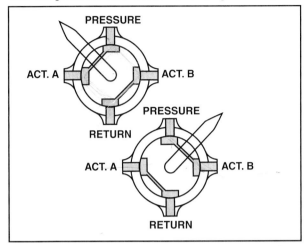

Figure 8-38. The closed-center valve has no provision for a return line as a part of the valve. Systems utilizing this type of valve have a pressure regulator that maintains a constant system pressure.

A more positive shutoff of fluid may be provided by using a poppet-type selector valve, such as the one shown in Figure 8-39. In the GEAR UP position (view A), poppets 2 and 3 are off of their seats, and poppets 1 and 4 are seated. Fluid flows from the pump around poppet one to one side of the actuator piston and raises the landing gear. The fluid from the opposite side of the actuator piston is pushed out of the cylinder, around poppet three, and back to the reservoir.

When the gear selector handle is placed in the GEAR DOWN position (view B), poppets 1 and 4 are off-seated, and poppets 2 and 3 are seated. Fluid flows from the pump around poppet 1 into the actuator, while the displaced fluid flows around poppet 4 back into the reservoir.

In a closed-center hydraulic system, the selector valve may shut off the flow of fluid without causing any rise in the system pressure. This is because the system uses a pressure regulator to maintain the system pressure independent of the position of any selector valve.

Check Valves

There are many instances in an aircraft hydraulic system where it is desirable to allow fluid to flow in one direction but prevent its flow in the opposite direction. This is done with check valves.

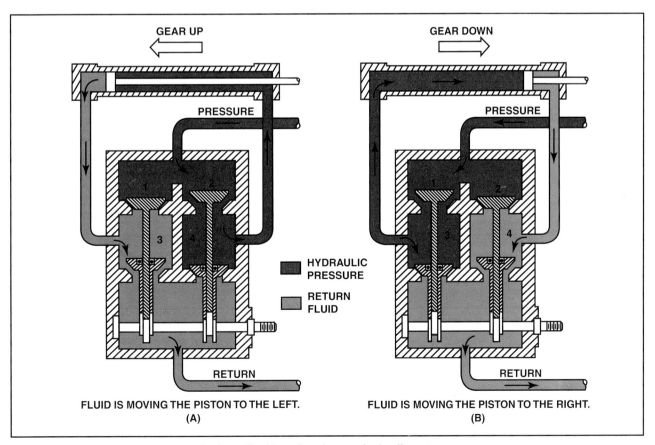

Figure 8-39. A poppet-type valve can also be utilized in a closed-center hydraulic system.

Check valves are made in two general designs to serve two different needs. In one, the check valve is complete within itself. It is interconnected with other components by means of tubing or hose. Check valves of this design are commonly called in-line check valves. In the other design, the check valve is not complete within itself because it does not have a housing exclusively its own. Check valves of this design are commonly called integral check valves. This valve is actually an integral part of some major component and, as such, shares the housing of that component.

There are several types of in-line check valves in use. The ball-type valve is perhaps the most familiar. It allows fluid to flow in one direction, but it cannot flow in the opposite direction. [Figure 8-40, view (A)]

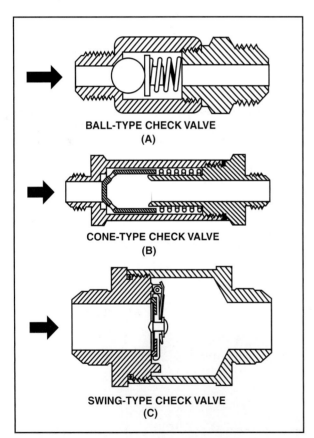

Figure 8-40. In-line check valves allow fluid to flow in one direction, but not the other.

The cone-type check valve, Figure 8-40, view (B), behaves in much the same way except that the seal is provided by a cone rather than by a ball. Neither of these check valves allows an entirely free flow in the direction of the arrow since the flow is opposed by the spring that holds the valve seated. A valve of this type is used to maintain case pressure in a gear-type hydraulic pump. The spring force determines the pressure needed to open the check valve.

Some applications for a check valve cannot tolerate the opposition to the flow of fluid, and if this is the case, a swing-type valve may be used, as in Figure 8-40, view (C). A disk is held with a very weak spring over the opening to the passage inside the check valve. Fluid flowing in the direction of the arrow can easily force the valve open and flow through with almost no opposition. But, fluid attempting to flow in the opposite direction will force the disc over the opening and prevent any flow.

Certain applications require full flow of fluid in one direction and restricted flow in the opposite direction. An example of this is in a landing-gear system where air loads and the weight of the gear cause the extension to be excessively fast. The weight of the gear against the air load requires every bit of pressure possible to get the gear up. An orifice check valve is an example of an integral check valve and is installed in such a way that fluid flowing through the gear-up lines is not restricted, while fluid leaving the gear-up side of the actuator is restricted by the orifice in the check valve. [Figure 8-41]

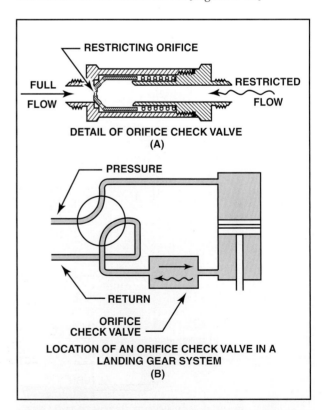

Figure 8-41. Orifice check valves are used where it is desired that flow is unrestricted in one direction, but restricted in the other. Landing gear systems often use these so that the gear does not extend too rapidly.

Sequence Valves

Modern aircraft with retractable landing gear often have doors that close in flight to cover the wheel well and make the airplane more streamlined. To be sure the landing gear does not extend before the doors are opened, sequence valves may be used. These are actually check valves which allow a flow in only one direction, but may be opened manually to allow fluid to flow in either direction. [Figure 8-42]

In a landing gear system, the wheel well doors must be fully open before the landing gear is extended. In this system, the gear door actuating piston opens the sequence valve when the doors are fully open, allowing fluid to flow into the main landing gear cylinder. The return fluid flows unrestricted through the sequence valve on its way back into the reservoir. [Figure 8-43]

Priority Valves

Priority valves are similar to sequence valves, except that they are opened by hydraulic pressure rather than by mechanical contact. They are called priority valves because such devices as wheel well doors, which must operate first, require a lower pressure than the main landing gear, and the valve will shut off all the flow to the main gear until the doors have actuated and the pressure builds up at

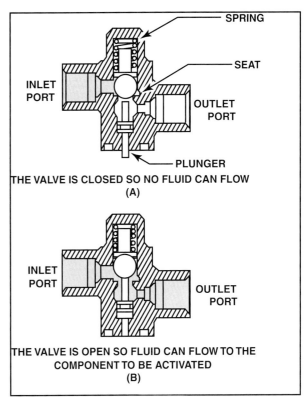

Figure 8-42. Sequence valves are used to permit actuation only when specific prior events have occurred.

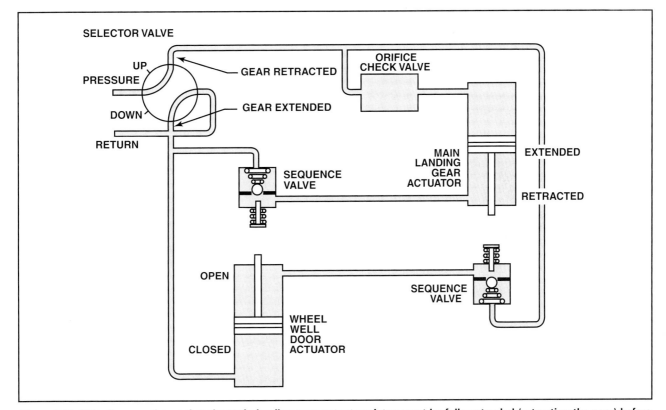

Figure 8-43. This diagram shows that the main landing gear actuator piston must be fully extended (retracting the gear) before the wheel well door actuator piston can retract and close the doors. In the GEAR EXTENDED position, the wheel well door actuator piston must be fully extended (door open) before the landing gear piston can retract (extending the gear).

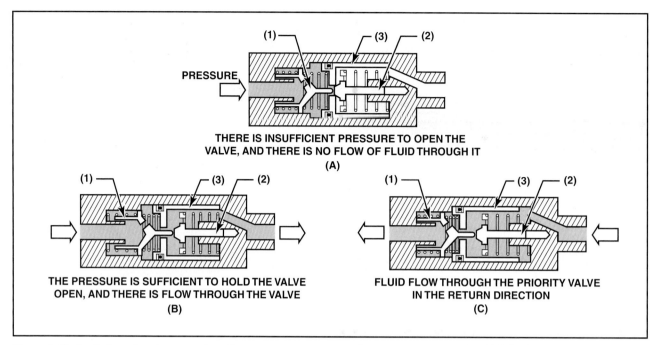

Figure 8-44. A priority valve does not allow flow until a preset pressure is reached. When this pressure is reached, pin (1) pushes valve (2) open and allows flow through the valve. When flow is reversed, the valve seat (3) is pushed back against pin (1) which allows flow in the reverse direction.

the end of the actuator stroke. When this buildup occurs, the priority valve opens so fluid can flow to the main gear. [Figure 8-44]

Quick-Disconnect Valves

Quick-disconnect, or line-disconnect valves, are installed in hydraulic lines to prevent the loss of fluid when units are removed. Such valves may be installed in the pressure and suction lines of the system just in front, and immediately behind the power pump. These valves can also be used in ways other than just for unit replacement. A power pump can be disconnected from the system and a hydraulic test stand connected in its place.

Quick disconnect valve units consist of two interconnecting sections coupled by a nut when installed in the system. Each valve section has a piston and poppet assembly. These are spring loaded to the CLOSED position when the unit is disconnected. [Figure 8-45]

The top illustration of Figure 8-45 shows the valve in the LINE-DISCONNECTED position. The two springs (a and b) hold both poppets (c and f) in the CLOSED position as shown. This prevents loss of fluid through the disconnected line. The bottom illustration of Figure 8-45 shows the valve in the LINE-CONNECTED position. When the valve is

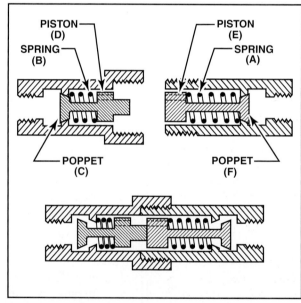

Figure 8-45. Quick-disconnect valves are designed to be closed when the two lines are disconnected and open when the two lines have been fully joined.

being connected, the coupling nut draws the two sections together. The extension (d or e) on one of the pistons forces the opposite piston back against its spring. This action moves the poppet off its seat and permits fluid to flow through that section of the valve. As the nut is drawn up tighter, one piston hits a stop; then the other piston moves back against its spring and, in turn, allows fluid to flow. Thus, fluid is allowed to continue through the valve and on through the system.

Bear in mind that the above disconnect valve is only one of many types presently used. Although all line-disconnect valves operate on the same principle, the details will vary. All manufacturer have their own design features.

A very important factor in the use of the line-disconnect valve is its proper connection. Hydraulic pumps can be seriously damaged if the line disconnects are not properly connected. If in doubt about the line disconnect's operation, the aircraft maintenance manual should be consulted.

The extent of maintenance to be performed on a quick-disconnect valve is very limited. The internal parts of this type valve are precision built and factory assembled. They are made to very close tolerances; therefore, no attempt should be made to disassemble or replace internal parts in either coupling half. However, the coupling halves, lock-springs, union nuts, and dust caps may be replaced. When replacing the assembly or any of the parts, follow the instructions in the applicable maintenance manual.

Hydraulic Fuses

Modern aircraft depend on their hydraulic systems not only for raising and lowering the landing gear and flaps, but for control system boosts, thrust reversers, brakes, and many auxiliary systems. For this reason, most aircraft use more than one independent hydraulic system, and provisions are made in these systems to block a line if a serious leak should occur. This blocking is done with hydraulic fuses.

There are two basic types of hydraulic fuses in use. One of these operates in such a way that it will shut off the flow of fluid if a sufficient pressure drop occurs across the fuse. Fluid flows as long as the spring holds the piston away from any of the holes, as in Figure 8-46, view (A). If a break should occur in a line beyond the fuse, the pressure on one side will drop, and the pressure on the other will force the piston over to cover the holes in the body and stop all flow of fluid. Only when pressure on the normal flow inlet side is greater than it is on the outlet side will the fuse close the line, so a reverse flow is not restricted in any way. [Figure 8-46, view (B)]

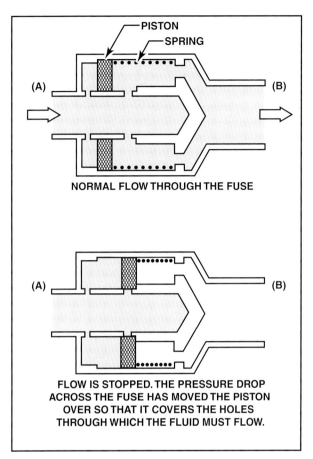

Figure 8-46. A hydraulic fuse will remain open as long as the pressure on side (A) and side (B) are close to equal. If pressure on side (B) drops, the piston is moved to a cutoff point closer to side (B). Reverse flow is not hampered.

Another type of hydraulic fuse shuts off the flow when a specified volume of fluid passes through the fuse. [Figure 8-47]

The fuse in Figure 8-47(A) is in its static condition, with the pressure at sides (A) and (B) the same, and there is no flow through the fuse.

In Figure 8-47(B), fluid is flowing normally through the fuse. Some of it passes through the metering orifice and drifts the piston to the right. The fluid has pushed the sleeve valve back and opened the passage for fluid to flow out of the fuse.

Figure 8-47(C) illustrates what happens when the limiting amount of fluid has passed through the fuse. Enough fluid has passed through the metering orifice to drift the piston over the holes in the housing and shut off the flow through the fuse, resulting in a pressure drop across the fuse. The piston holds the spring in the sleeve valve compressed. While the sleeve valve holds the piston over the holes, preventing any fluid flow.

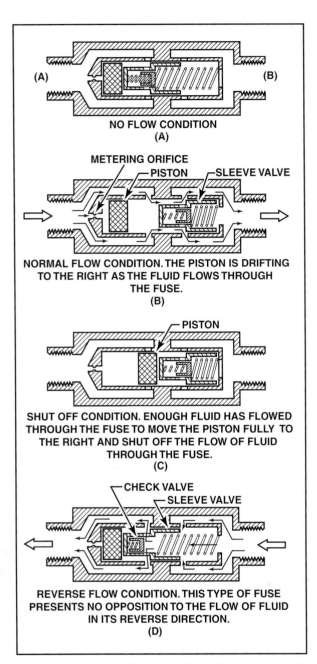

Figure 8-47. This type of hydraulic fuse allows a preset volume of flow before shutting down forward flow. Reverse flow is not hampered.

A volume limiting fuse has no appreciable opposition to the reverse flow of the fluid. Figure 8-47(D) illustrates what happens when the flow is reversed. Both the sleeve valve and the check valve move to the left and the sleeve valve forces the piston away from the holes, allowing the fluid to follow the paths shown by the arrows.

PRESSURE-CONTROL VALVES

While it is essential for the flow of hydraulic fluid to be directed to the proper actuator, it is also important that the pressure of that fluid be proper for the particular application. The devices used to control pressure include relief valves, pressure regulators, and pressure reducers.

Relief Valves

The simplest type of pressure-control valve is the relief valve. It is used primarily as a backup rather than a control device because of the heat generated and the power dissipated when the valve relieves pressure. The main system pressure-relief valve is set to relieve any pressure above that maintained by the system pressure regulator, and only in the event of a malfunction of the regulator will the relief valve be called into service.

The relief valve pictured in Figure 8-48 operates on simple mechanical principles. Increasing pressure attempts to unseat a ball and allow the pressure to escape. Counteracting this force is a spring which holds the ball on the seat. A small adjustment screw can be set to increase or decrease the downward pressure on the ball. When a hydraulic system incorporates several relief valves, they should be adjusted in sequence, which will permit each valve to reach its particular operating pressure. Thus, the highest pressure-valves should be adjusted first and the lowest pressure valves last. [Figure 8-48]

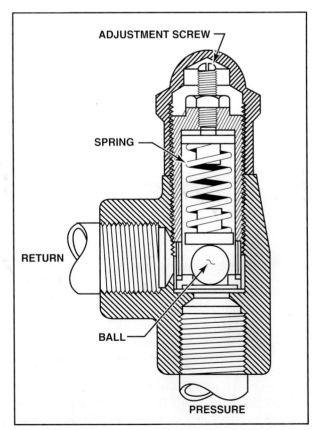

Figure 8-48. Pressure-relief valves are used as backups to the main pressure regulator.

In systems where fluid may be trapped in a line between the actuator and its selector valve, there is the problem of pressure buildup by heat expansion of the fluid. Thermal-relief valves are installed in these lines to prevent damage by releasing a small amount of fluid back into the return line.

Pressure Regulators

A closed-center hydraulic system whose pressure is supplied by an engine-driven hydraulic pump needs a regulator to maintain the pressure within a specified range and to keep the pump unloaded any time no unit is being actuated. The simplest pressure regulator is the balanced type. [Figure 8-49]

Starting with a discharged, or flat, system, the pump forces fluid through the check valve into the system and the accumulator. When no fluid is required for actuation, the accumulator fills and pressure builds up. This pushes up on the piston and down on the ball, and a condition is soon reached where there is a balance of forces. The fluid pressure pressing down on the ball and the spring pressing down on the piston are both opposed by the fluid pressure forcing the piston up.

Assuming that the piston has an area of one square inch, the ball has a seat area of $1/3$ square inch, and the spring exerts a force of 1,000 pounds on the top of the piston, this system balances when the hydraulic system pressure is 1,500 psi. There is an upward force of 1,500 pounds, and the downward force on the ball is one-third of this, or 500 pounds, and the spring exerts the other 1,000 pounds. Just as soon as the system pressure rises above 1,500 psi, the piston is slowly forced up, and the pin barely separates the ball from its seat. As soon as the ball is unseated, 500 pounds of downward force is lost, and the piston then forces the pin up enough to raise the ball completely off its seat. Fluid flows from the pump, around the ball, and out the return to the reservoir. Since the pressure on the inlet side

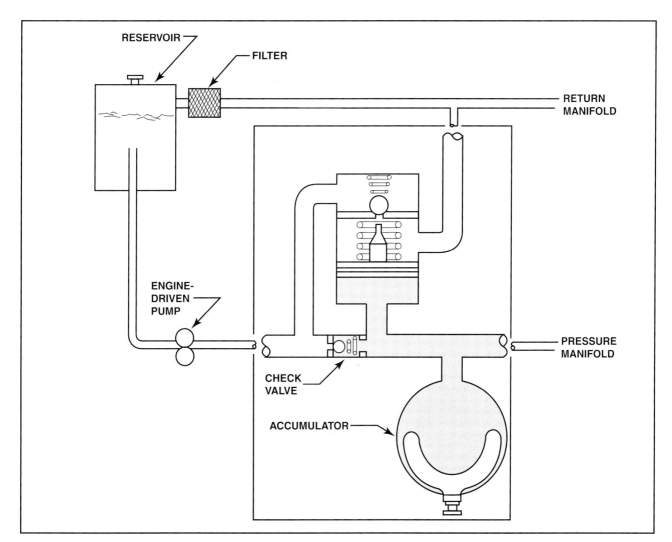

Figure 8-49. The pressure regulator maintains a balance of pressure between the pump and the accumulator.

of the check valve has dropped, the system pressure closes the check valve. The pressure is trapped and held in the system by the air pressure in the accumulator. This is the unloaded condition of the valve. The pump remains unloaded until the system pressure drops to less than 1,000 psi, which is called the kick-in pressure because the pump is placed on-line at that time. The spring then forces the piston down so the ball can reseat. With the ball seated, the pressure builds back to 1,500 psi, which is called the kick-out pressure because the pump is unloaded at that time.

Pressure Reducers

Sometimes, it is necessary to operate some portion of a hydraulic system at a pressure lower than the normal system pressure. This can be done by using a pressure reducer. [Figure 8-50]

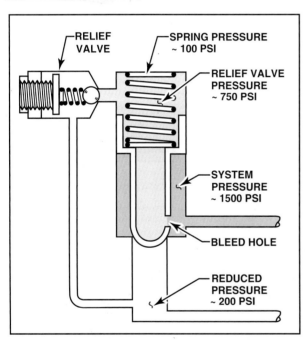

Figure 8-50. A pressure reducer is used to lower the pressure in targeted portions of the hydraulic system.

A balance of forces must be reached to maintain the pressure at a predetermined value below the regulated system pressure. In order to understand how this works, the following values are assumed: the system pressure is held constant at 1,500 psi by the system pressure regulator, and a reduced pressure of 200 psi is required to operate the hydraulic servos in an automatic pilot system. The relief valve in the reducer maintains the pressure above the piston at 750 psi, and the spring presses down on the piston with a force of 100 pounds. The area of that portion of the piston that is exposed to system pressure is one-half square inch, and the area of the ball seat exposed to the 200 psi reduced pressure is one-half square inch. Fluid from the main system bleeds through the hole in the piston, where it builds up pressure above the piston until the relief valve unseats to maintain this pressure at 750 psi. The downward force of 750 pounds caused by the pressure above the piston acting on the entire one square inch of area is added to the 100 pounds of spring force, to give us a total downward force of 850 pounds. This is opposed by an upward force of 850 pounds. Seven hundred and fifty pounds of this force is caused by the system pressure acting on the one-half square-inch piston shoulder area. We also have 100 pounds of upward force caused by the reduced pressure acting on the one-half square-inch ball seat area.

The relief valve maintains 750 psi inside the piston cavity, by a balance of forces. The pressure inside the piston is trying to move the relief valve ball off its seat. Reduced pressure and the force of the spring acting on the opposite side hold the ball seated. When the reduced pressure drops, the seating force decreases, and the ball moves off its seat and drops the pressure above the piston. This enables the system pressure to raise the piston enough for some fluid to enter the reduced pressure portion of the system and bring its pressure back up to 200 psi. The very small bleed hole in the piston prevents the piston chattering as it maintains the reduced pressure. The ball end of the piston remains off of its seat, just enough to maintain the reduced pressure as fluid is used in this portion of the system.

ACCUMULATORS

Hydraulic fluid is non-compressible, and pressure may be stored only with compressible fluids. The effect of compressibility for hydraulic fluid can be created by using an accumulator.

All accumulators consist of a high-strength container divided by some form of movable partition into two sections, or compartments. One compartment is connected to the hydraulic pressure manifold, and the other compartment is filled with either compressed air or with nitrogen. There are three types of accumulators commonly found in aircraft hydraulic systems: the piston type, the bladder type, and the diaphragm type. Before disassembling any accumulator, make sure that the **preload pressure** has been discharged.

The **piston-type accumulator** is cylindrical and a free-floating piston divides the cylinder into the two compartments. A high-pressure air valve allows one compartment to be charged with a **preload** of air or nitrogen of approximately one-half to one-third of the normal system operating pressure. When there is no system pressure, the piston is forced all the way up until it contacts the cylinder head. As soon as fluid is moved into the pressure manifold by a pump, some fluid enters the accumulator and forces the piston down against the compressed air. [Figure 8-51]

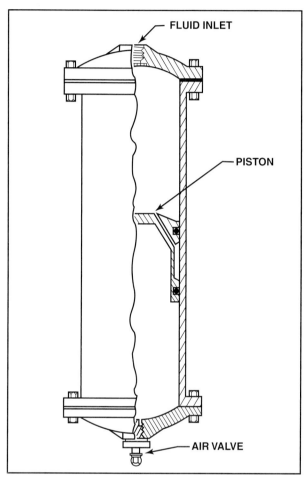

Figure 8-51. An accumulator is used to maintain a preset system pressure. The piston-type accumulator has a piston separating the hydraulic fluid from a compressible gas. As fluid flows into the top of the accumulator, it compresses the gas until a predetermined pressure is reached. This pressure is then maintained in the system by action of the accumulator.

This movement increases the pressure of the air and holds this pressure on the fluid. When the system pressure rises to the kick-out pressure of the unloading valve, the pump unloads, and a check valve traps the fluid in the pressure manifold. With the pump no longer forcing fluid into the system, the pressure is maintained by the air in the accumulator forcing the piston against the fluid.

Some hydraulic systems have the pressure gauge connected to the air side of the accumulator. When there is no hydraulic pressure, the gauge will indicate the air preload. If the system pressure gauge is connected into the fluid side of the system, the preload air pressure can be found by watching the gauge as the hand pump is cycled. No pressure will be shown on the gauge as fluid begins to move into the accumulator, but as soon as the piston moves, the air will oppose it and create a pressure on the fluid equal to the pressure of the air. At this point, the fluid pressure is indicating the preload pressure of the gas.

Bladder- and diaphragm-type accumulators are both spherical, one having a synthetic rubber bladder and the other a diaphragm. In both these types of accumulators, compressed air or nitrogen fills one compartment, and as hydraulic fluid is pumped into the other chamber, the flexible partition allows the compressed air to hold pressure on the fluid. [Figure 8-52]

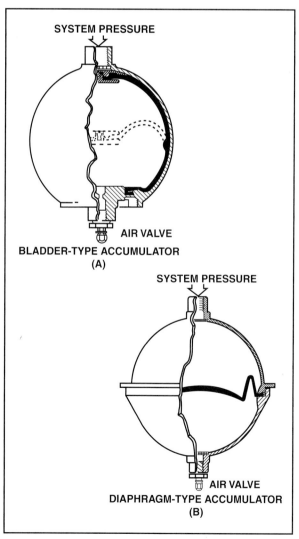

Figure 8-52. Bladder-type and diaphragm-type accumulators function similarly to a piston-type accumulator, but differ in the type of mechanism separating the air from the fluid.

AIR VALVES

There are three types of air valves that may be used in accumulators: the AN812, the AN6287-1 and the MS28889-1.

The AN812 is the simplest valve and it screws directly into the air chamber of the accumulator. It holds the air with a high-pressure valve core. This core is similar in appearance to the one used in an inner tube or with a tubeless tire, but it is definitely not interchangeable. High-pressure valve cores are identified by the letter "H" embossed on the end of their stem.

To charge an accumulator, the air hose is connected from the high-pressure regulator on the air or nitrogen cylinder to the valve to fill the air chamber. (Note: there must be no hydraulic fluid trapped in the accumulator while it is being charged.) To release air from the accumulator, the valve housing is loosened until air can flow through the hole in the side of the valve and escape around the threads. [Figure 8-53]

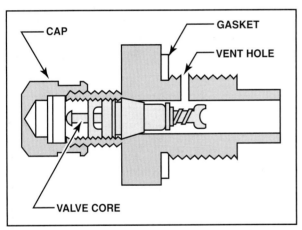

Figure 8-53. The AN812 valve is similar in appearance to the valve used on tires. It differs in having the letter "H" stamped on the end of the stem.

The AN6287-1 valve does not depend upon the valve core to provide the seal. The seal is provided by metal-to-metal contact between the stem and the body of the valve. To charge the accumulator, the high-pressure air hose is attached to the valve and the swivel nut is loosened about one turn. Air can flow through the core and between the loosened metal-to-metal seal into the air chamber. When the correct preload is reached, the swivel nut is tightened snugly, but not tight enough to damage the seal. To release air from the accumulator, the swivel nut is loosened approximately one turn and the stem of the valve core is depressed. [Figure 8-54]

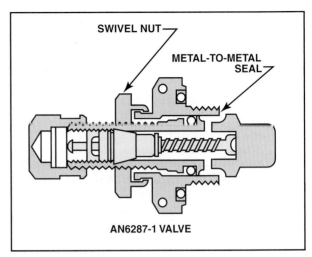

Figure 8-54. The AN6287-1 valve maintains pressure through a metal-to-metal seal. The valve core is used for inflation and controlled deflation only.

CAUTION: When using the AN6287-1 valve, the air in the accumulator is under a very high pressure. Dirt particles may be blown into the skin or eyes and cause serious injury. To prevent this danger, the valve core stem should always be depressed with a special tool to deflect the escaping air away from the body. [Figure 8-55]

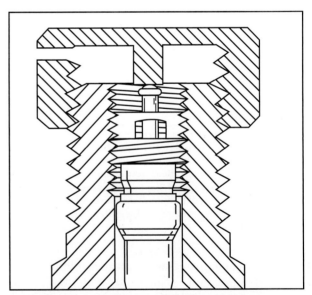

Figure 8-55. A special tool is used to deflate air pressure through the AN6287-1 valve. This tool protects the technician from the high-velocity air escaping from the valve.

The MS28889-1 valve is used in many high-pressure systems and is somewhat similar to the AN6287-1, but has three features that make it different. First, the swivel nut is the same size as the hex on the valve body, while the swivel nut on the

AN valve is smaller than the body. Second, unlike the AN valve, the stem is retained in the body of the valve with a roll pin. The stem can be backed out of the body far enough for air to flow, but not far enough to allow the stem to drop out of the body. The third and most important difference between these two valves is the lack of any valve core in the MS valve. To charge the accumulator, the air hose is attached to the valve and the swivel nut is loosened until the proper charge is reached. The swivel nut is then tightened to provide a good metal-to-metal seal between the stem and the body. The accumulator is discharged by removing the cap from the valve and loosening the swivel nut just enough to allow the air to bleed out. [Figure 8-56]

When any of the three types of valves is used in an accumulator, it must be capped with the proper type of high-pressure valve cap. This will prevent dirt or other contaminates getting into the opening of the valve.

ACTUATORS

The ultimate function of any hydraulic or pneumatic system is to convert the pressure in the fluid into work. The portion of a hydraulic system that does the work is the actuators. In order to perform their work, actuators must move. Actuators are classified as either linear or rotary, depending on the type of movement required.

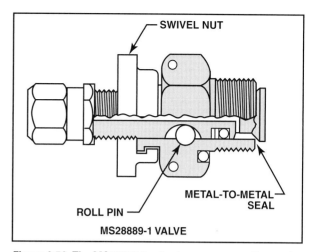

Figure 8-56. The MS28889-1 valve has no core, so it must be attached directly to the inflation hose instead of using an inflation nozzle. Once the proper pressure has been reached, the valve must be closed before removing the hose.

LINEAR ACTUATORS

Linear actuators produce straight-line movement and consist of a cylinder and piston. The cylinder is usually attached to the aircraft structure, while the piston is attached to the component being moved. [Figure 8-57]

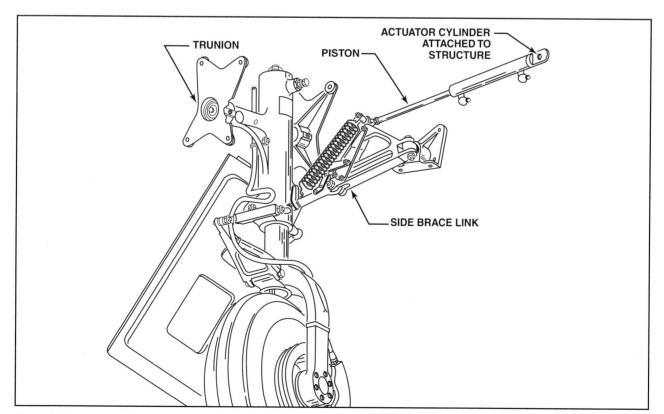

Figure 8-57. In a typical landing gear actuator installation, the cylinder attaches to the wing spar and the piston to the landing gear side-brace link. To retract the landing gear, the piston is pulled into the cylinder.

There are three basic types of linear actuators: single-acting, double-acting unbalanced, and double-acting balanced. The piston in a single-acting cylinder is moved in one direction by hydraulic pressure, and it is returned by a spring. The wheel cylinders in shoe-type brakes are good examples of single-acting cylinders. Hydraulic pressure moves the pistons out to apply the brakes, but when the pedal is released, springs pull the shoes away from the drum and move the pistons back into the cylinder. [Figure 8-58]

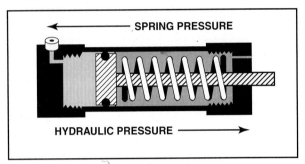

Figure 8-58. In this single-acting linear hydraulic actuator the hydraulic pressure moves the piston to the right, and a spring returns it when the pressure is released.

Double-acting unbalanced actuators are normally used for such applications as raising and lowering the landing gear. The fluid entering the up-port acts on the entire area of the piston, while the fluid entering the down port acts only on that portion of the piston not covered by the actuating rod. Because of this difference in effective piston area, there is a much greater force produced to raise the landing gear than is used to lower it. [Figure 8-59]

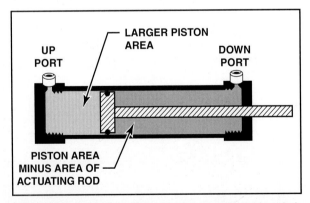

Figure 8-59. A double-acting unbalanced actuator is unbalanced because the area of the actuating rod is subtracted from one side of the actuating piston. On landing gear the side with the larger area is usually used to raise the landing gear since this requires more force.

Double-acting balanced actuators are useful in applications that require the same amount of force in both directions of piston movement. For example, a double-acting balanced actuator might be used as an automatic pilot servo actuator. [Figure 8-60]

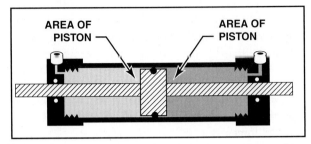

Figure 8-60. A double-acting balanced actuator provides equal force in both direction.

There are many special applications for linear actuators. One is a cushioned gear actuator in which the piston starts its movement slowly, accelerates to full speed, and then is cushioned at the end of its movement. Fluid enters the actuator through the gear-down port, and it must flow around the metering rod to move the piston out of the cylinder. As soon as the piston travels far enough to remove the metering rod from the orifice, the fluid flow increases and moves the piston out at its full speed. As the piston nears the end of its travel, the piston head contacts the poppet and compresses the poppet spring to bring the piston to a smooth stop at the end of its travel. When the selector is placed in the gear-up position, the fluid enters the gear-up port and moves the piston rapidly until the metering pin enters the orifice. The travel is then slowed until it reaches the full-up position. [Figure 8-61]

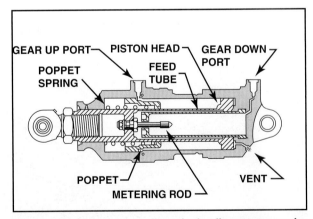

Figure 8-61. A cushion-type linear hydraulic actuator can be used to start slowly, then operate at normal speed until it is brought to a cushioned stop at the end of its travel. This allows for properly controlled actuation of devices such as landing gear.

Some actuators incorporate a means for locking the piston at the end of its travel. When the piston is in its retracted position, the landing gear is down and locked. The locking pin holds the locking ball in the groove in the piston so the piston cannot move out of the cylinder. [Figure 8-62, view (A)]

To raise the gear, fluid under pressure enters the gear-up port and moves the locking pin back, which allows the ball to drop out of the groove in the piston and release it so the fluid can move the piston out of the cylinder and raise the landing gear. The collar holds the ball down in the step of the locking pin and prevents its extending until the gear is lowered. The piston forces the collar back, allowing the ball to release the locking pin. This in turn locks the ball into the groove in the piston. [Figure 8-62, view (B)]

ROTARY ACTUATORS

Perhaps one of the simplest forms of a rotary actuator is the rack-and-pinion type that is used to retract the main landing gear in the popular high-performance single-engine Cessna aircraft. The piston has a rack of teeth cut in its shaft, and these teeth mesh with those in a pinion gear that rotates as the piston moves in or out. Rotation of the pinion shaft raises or lowers the landing gear. [Figure 8-63]

If a continuous rotational force is needed, a hydraulic motor may be used. Fluid under pressure from the system enters the motor through the inlet port and forces the pistons to the bottom of the cylinder block. As they move down the cylinder bore, they force the driveshaft to rotate. By the time

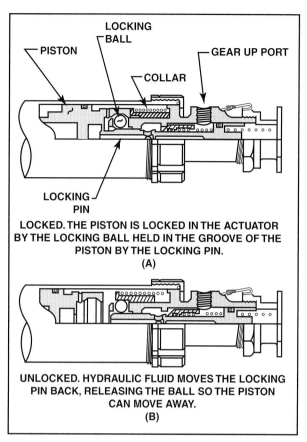

Figure 8-62. This actuator incorporates a locking ball that allows the cylinder to be locked in the extended or retracted positions.

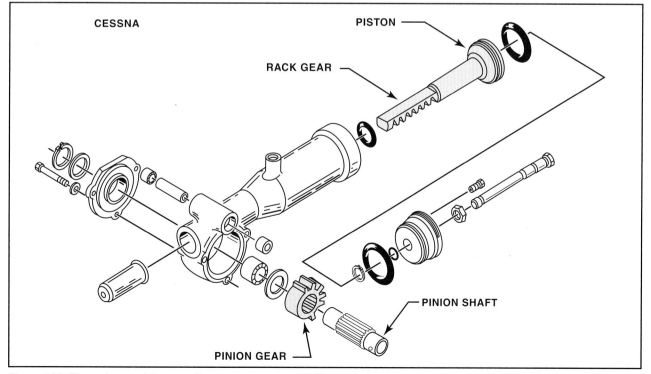

Figure 8-63. The rack-and-pinion type linear actuator converts linear motion into rotary motion. A good example is this exploded view of the mechanism from popular high-wing Cessna retractables.

the pistons reach the bottom of the bore, the cylinder block has rotated until the cylinders whose pistons are moving upward are under the outlet port, and as they move up they force fluid out into the return manifold. Piston-type hydraulic motors have many applications on larger aircraft where it is desirable to have a considerable amount of power with good control, the ability to instantaneously reverse the direction of rotation, and no fire hazard if the motor is stalled. [Figure 8-64]

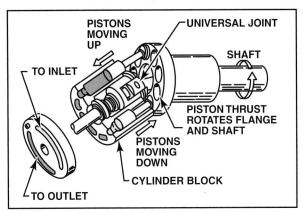

Figure 8-64. The piston-type rotary hydraulic motor is similar in operation to the piston-type hydraulic pump. It converts the linear motion of individual pistons into rotary motion of the assembly.

Where less torque is needed, a vane-type motor may be used in which pressure is directed to vanes on opposite sides of the rotor to balance the load on the shaft. Fluid under pressure enters the inlet chambers of the motor and pushes the vanes around to the outlet chambers. The vanes are free to slide back and forth in the slots in the rotor, and centrifugal force holds them against the outside of the chambers. [Figure 8-65]

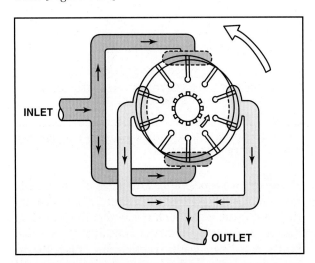

Figure 8-65. The vane-type hydraulic motor works in the reverse manner to the vane-type hydraulic pump. Hydraulic pressure and flow is converted to rotary motion.

SEALS

Seals are used throughout hydraulic and pneumatic systems to minimize internal leakage and the loss of system pressure. There are two types of seals in use: gaskets are used where there is no relative movement between the surfaces, and packings are used where relative movement does exist.

ONE-WAY SEALS

Chevron or **V-ring packings, U-ring packings,** and **D-ring packings** all get their name from their shape and all are one-way seals. This means that the seal will stop the flow of fluid in one direction only. [Figure 8-66]

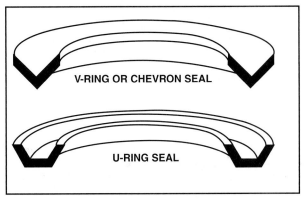

Figure 8-66. Chevron and U-ring seals are two examples of one-way seals.

To prevent a flow from both directions, two sets of seals must be installed, each having their open end facing the direction from which the pressure is applied. The apex, or point, of the seal rests in the groove of a metal backup ring on the shaft, and a spreader ring having a triangular cross section fits into the groove of the seal. When both seals are assembled on the shaft, the adjusting nut is tightened to spread the seals and hold them tight against the wall of the actuating cylinder. [Figure 8-67]

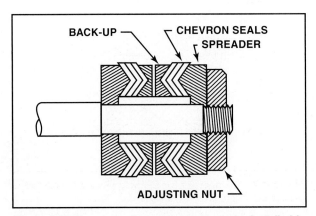

Figure 8-67. Two sets of one-way seals can be installed in opposite directions in order to prevent fluid flow in both directions.

TWO-WAY SEALS

The most commonly used two-way seal is the **O-ring**, which may be used as either a gasket or a packing. This type of seal fits into a groove in one of the surfaces being sealed. The groove should be about 10% wider than the width of the seal and deep enough that the distance between the bottom of the groove and the other mating surface will be a little less than the width of the O-ring. This provides the squeeze, or pinch, necessary to seal under conditions of zero pressure. [Figure 8-68]

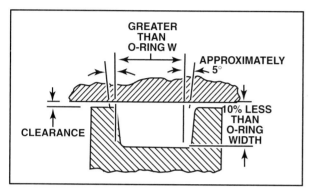

Figure 8-68. The groove in which an O-ring seal fits would have a width approximately 10% greater than the width of the O-ring, but a depth slightly less than the width of the seal.

Referring to Figure 8-69, view (A), when the O-ring fits properly the ring will fit into the groove with the proper pinch. If there is no pinch, the O-ring cannot seal, as in view (B). [Figure 8-69]

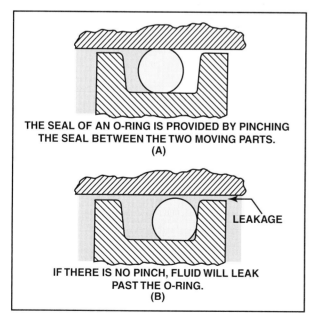

Figure 8-69. The depth of the O-ring groove is critical to the proper sealing action of the O-ring.

The mouth of the cylinder into which an O-ring-equipped piston fits must be chamfered in such a way that the pinch will be applied to the O-ring gradually as the piston enters the cylinder. If there is no chamfer, there is danger that the O-ring will be damaged as the piston is slipped into the cylinder. [Figure 8-70]

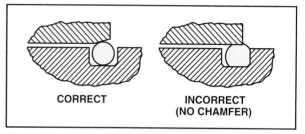

Figure 8-70. When placing an O-ring equipped piston into a cylinder, the cylinder must have a chamfer in order to avoid damaging the O-ring.

BACKUP RINGS

An O-ring of the appropriate size can withstand pressures up to about 1,500 psi without distortion, but beyond this, there is a tendency for the ring to extrude into the groove between the two mating surfaces. To prevent this, an anti-extrusion, or backup, ring should be used. [Figure 8-71]

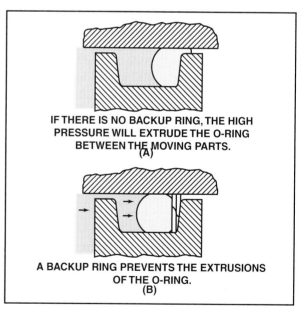

Figure 8-71. Backup rings are used to prevent O-rings from extruding under high pressure.

There are two types of backup rings in use: One is made of leather and the other of Teflon. Leather rings are installed in such a way that the hair side, the smooth side, of the ring is against the O-ring. When installing a leather backup ring, it is softened by soaking it in the fluid with which the ring will be used. Teflon rings are scarfed and spiraled, and if installed incorrectly, the scarfs will be on the wrong side, and the O-ring will be damaged. [Figure 8-72]

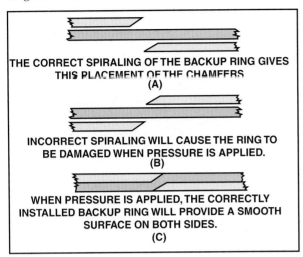

Figure 8-72. Backup rings are scarfed in a certain direction. Care must be taken to install them as in (A) so that when pressure is applied as in (C), the ring will provide a smooth surface on both sides. If installed as in (B), the O-ring can be damaged.

SEAL MATERIALS

There is perhaps no other component as small as a hydraulic seal upon which so much importance must be placed. Seals may look alike, and it is highly possible that the wrong seal may be installed and appear to work. The material of the seal, its age, and its hardness all are important when making the proper replacement. The rule for replacing seals in a hydraulic system is to use the exact part number of the seal specified in the manufacturer's service information. Seals should be purchased from a reputable aircraft parts supplier, and they should be in individual packages marked with the part number, composition of the ring, manufacturer, and cure date. The cure date is the date of ring manufacture and is given in the year and quarter. If, for example, a ring is marked 2Q91, the ring was manufactured sometime during April, May, or June, the second quarter of 1991. Normally, rubber goods are not considered fresh enough for installation if they are more than 24 months old.

The part number is the only sure way of knowing that the correct O-ring is being used, but most rings are also marked with a series of colored dots or stripes to indicate the type of fluid with which they are compatible:

Blue dot or stripe:	Air or MIL-H-5606 hydraulic fluid
Red dot or stripe:	Fuel
Yellow dot:	Synthetic engine oil
White stripe:	Petroleum-base engine oil or lubricant
Green dash:	Skydrol hydraulic fluid

It is important to buy hydraulic seals from a reputable supplier, since it is possible for out-of-date rings to be repackaged by an unscrupulous dealer and stamped with a fresh date. The ring could be installed in good faith by an A&P technician and still fail because of deterioration. Yet, the technician is the one held liable for the failure, because it is only when an improper part is installed in an airplane that a violation of regulations is committed.

O-RING INSTALLATION

When installing O-rings, extreme care must be used to prevent the ring being nicked or damaged by either the sharp edges of the threads or by the tool. A number of special tools are available for installing O-rings. If these types of tool can not be found commercially, tools can be made of brass and polished so there will be no sharp edges to nick or cut the seal. [Figure 8-73]

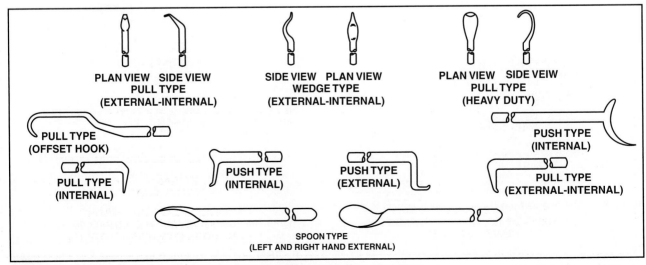

Figure 8-73. Specialty tools are used to remove and install O-rings without damage.

There are many ways to remove O-rings from internal and external grooves. When using removal tools, the technician must use tools very carefully to avoid damaging the rings. [Figure 8-74]

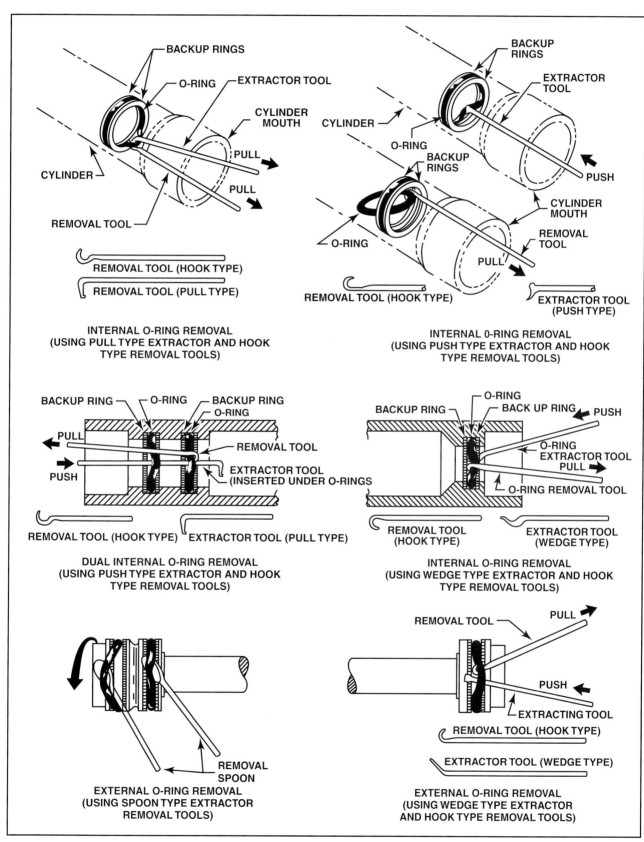

Figure 8-74. With the proper tools and a little care, most O-rings can be removed without damage.

When installing an O-ring over a sharp edge or threads, the sharp portion should be covered with paper, aluminum foil, brass shim stock, or with a piece of plastic. [Figure 8-75]

WIPERS

Neither O-rings nor chevron seals eliminate all seepage past the shaft, and there is usually enough leakage to lubricate the shaft. But this lubrication attracts dust, and in order to prevent the seals being damaged when the shaft is retracted into the cylinder, a felt wiper is usually installed in a counterbore around the shaft. This wipes off any dust or dirt without restricting the movement of the shaft.

LARGE-AIRCRAFT HYDRAULIC SYSTEMS

Large jet transport aircraft usually have a utility system and an auxiliary system. The fluid for the utility system is supplied from the utility reservoir, and the pressure is produced by two engine-driven pumps. These pumps are of the variable displacement type, and therefore they do not require a pressure regulator; but they do have a return line back into the reservoir. An accumulator holds the pressure on the system. A relief valve is installed to bypass fluid back into the reservoir if there should ever be a malfunction in the pumps that

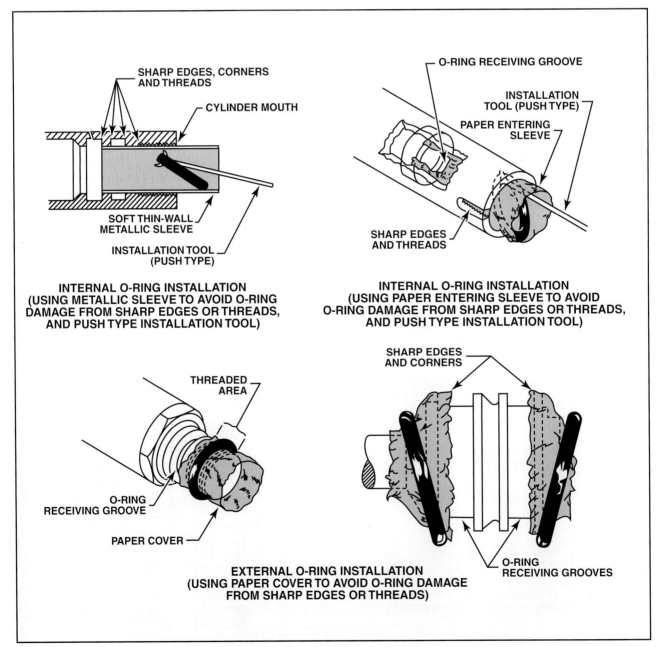

Figure 8-75. Sharp edges and threads should be covered to avoid damaging O-rings during installation.

Hydraulic and Pneumatic Power Systems

would produce too much pressure in the system. For example, a relief valve called a wing flap overload valve is used to prevent the flaps from being lowered above a certain airspeed. [Figure 8-76]

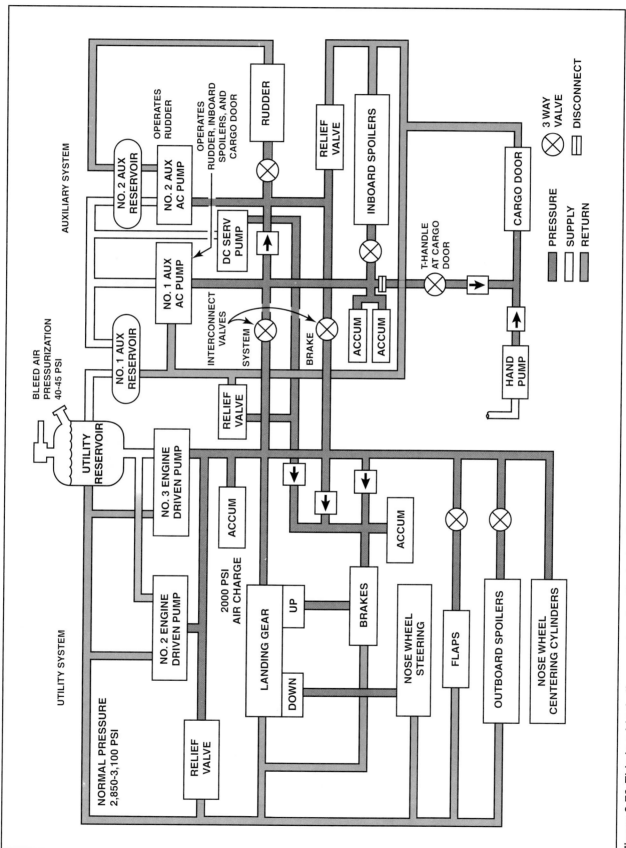

Figure 8-76. This is a block diagram showing the hydraulic system of a large jet transport airplane.

This utility system provides the pressure to raise and lower the landing gear, to operate the brakes, the flaps, and the outboard spoilers, and to provide for centering the nose wheel as it is being retracted. When the landing gear is down, pressure for nose wheel steering is provided by this system. When the landing gear is retracted, the brakes are automatically applied to stop the wheels spinning before they enter the wheel wells.

The auxiliary system consists of two reservoirs and two AC electric motor-driven hydraulic pumps and one DC electric motor-driven service pump. Both AC pumps can supply pressure to the rudder, and pump number one also supplies pressure to the inboard spoilers and the cargo door. Through an interconnect valve, it can supply pressure to the landing gear. Pump number two can supply the brakes through an interconnect valve.

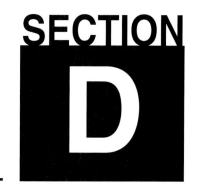

AIRCRAFT PNEUMATIC SYSTEMS

Modern aircraft may use compressed air, or pneumatic, systems for a variety of purposes. Some use pneumatics rather than hydraulics for the operation of the landing gear, flaps, brakes, cargo doors, and other forms of mechanical actuation. Other aircraft using hydraulics for these major functions may have a cylinder of compressed air or nitrogen as a backup source of power in the event of a failure of the hydraulic power. Still other aircraft use pneumatics only for de-icing and for the operation of various flight instruments. Finally, some aircraft use pneumatic systems only to provide a positive air pressure in the cabin for flight at high altitude, where pressurization is used to supply the passengers and crew with the environment needed to fly without supplemental oxygen.

Some of the advantages of using compressed air over hydraulics or electrical systems are:

- Air is universally available in an inexhaustible supply.

- The units in a pneumatic system are reasonably simple and lightweight.

- Compressed air, as a fluid, is lightweight and since no return system is required, weight is saved.

- The system is relatively free from temperature problems.

- There is no fire hazard, and the danger of explosion is minimized by careful design and operation.

- Installation of proper filters minimizes contamination as a problem.

The type of unit used to provide pressurized air for pneumatic systems is determined by the system's air pressure requirements.

HIGH-PRESSURE SYSTEMS

For high-pressure systems, air is usually stored in metal bottles at pressures ranging from 1,000 to 3,000 psi, depending on the particular system. This type of air bottle has two valves: a charging valve, to which a ground-operated compressor can be connected to add air to the bottle; and a control valve, which acts as a shutoff valve to keep air trapped inside the bottle until the system is operated. [Figure 8-77]

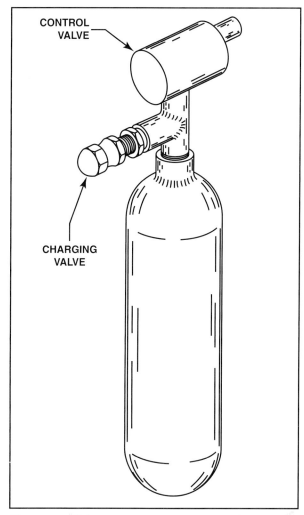

Figure 8-77. High-pressure air is usually stored in steel bottles. One disadvantage of these bottles is that they cannot be charged in flight.

Although the high-pressure storage cylinder is light in weight, it has a definite disadvantage. Since the system cannot be recharged in flight, operation is limited by the small supply of bottled air. Such an arrangement cannot be used for the continuous operation of such systems as landing gear or brakes. The usefulness of this type of system is increased, however, if other air-pressurizing units are added to the aircraft.

On some aircraft, permanently installed air compressors have been added to recharge air bottles whenever pressure is used for operating a unit. Several types of compressors are used for this purpose. Some have two stages of compression, while others have three. [Figure 8-78]

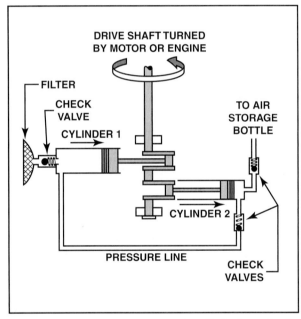

Figure 8-78. This diagram shows a simplified schematic of a two-stage compressor. The pressure of the incoming air is boosted first by cylinder No. 1 and again by cylinder No. 2.

The compressor in Figure 8-78 has three check valves. Like the check valves in a hydraulic hand pump, these units allow air to flow in only one direction. Some source of power, such as an electric motor or aircraft engine, operates a drive shaft. As the shaft turns, it drives the pistons in and out of their cylinders. When piston No. 1 moves to the right, the chamber in cylinder No. 1 becomes larger, and outside air flows through the filter and check valve into the cylinder. As the drive shaft continues to turn, it reverses the direction of piston movement. Piston No. 1 now moves deeper into its cylinder, forcing air through the pressure line and into cylinder No. 2.

Meanwhile, piston No. 2 is moving out of cylinder No. 2 so that cylinder No. 2 can receive the incoming air. Cylinder No. 2 is smaller than cylinder No. 1; thus, the air must be highly compressed to fit into cylinder No. 2.

Because of the difference in cylinder size, piston No. 1 gives the air its first stage of compression. The second stage occurs as piston No. 2 moves deeper into its cylinder, forcing high-pressure air to flow through the pressure line and into the air storage bottle.

MEDIUM-PRESSURE SYSTEMS

A medium-pressure pneumatic system (100 - 150 psi) usually does not include an air bottle. Instead, it generally takes bleed air from the turbine engine compressor section. Engine bleed air will first be routed to a pressure-controlling unit and then to the operating units.

Some jet aircraft use compressor bleed air from the engines to provide a relatively large volume of compressed air at a low pressure. Compression of bleed air provides heated air for the leading edge of the wing to help prevent the formation of ice, and for cabin heat. Compressor bleed air also provides air flow for starting engines and pressurized air for controlling cabin pressure. [Figure 8-79]

LOW-PRESSURE SYSTEMS

Many aircraft use low-pressure, air-driven gyro instruments as either the primary gyro instruments or as backup instruments when the primary gyros are electrically driven. For many years all air-driven gyro instruments used an engine-driven vacuum pump to evacuate the instrument case, and filtered air was pulled into the instrument to spin the gyro. This was done because it was much easier to filter air being pulled into the instrument than it was to filter the air after it had been exposed to an engine-driven pump lubricated by engine oil. The output of such pumps always contained some particles of oil.

Pressurized aircraft created extra problems for suction-operated instruments, and the latest generations of air-driven gyros now almost all use pressure. Turbine-powered aircraft bleed some of the pressure from the engine compressor, regulate and filter it, and then direct it over the gyros. Aircraft with reciprocating engines use vane-type air pumps driven by electric motors or by the aircraft engine to provide the airflow for the gyros. This air is regulated and filtered before it is ready for the instrument. [Figure 8-80]

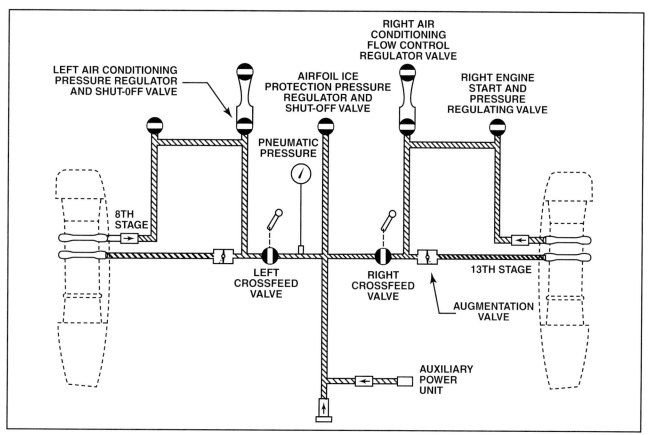

Figure 8-79. Bleed air from turbojet engines can be used to provide pressurized air for medium-pressure systems.

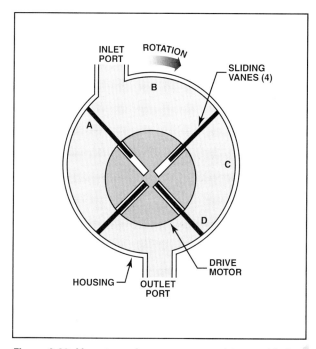

Figure 8-80. Vane-type air pumps are used to provide low pressure air on aircraft fitted with reciprocating engines. Sliding vanes are rotated by the driveshaft and as the shaft turns, the chambers located at positions A and B become larger, while those at positions C and D decrease in size. Air is pulled into the pump at the position the chambers enlarge, and it is moved out as they decrease

There are two types of air pumps used to provide instrument airflow, and both are vane-type pumps. One is called a **"wet" pump** and the other a **"dry" pump**, based on their method of lubrication.

"Wet" vacuum pumps use steel vanes moving in a sealed cast-iron housing and are lubricated by engine oil metered into the inlet air port. This oil is discharged with the air and is removed with an oil separator before the air is either used for inflating de-icer boots or is pumped overboard. A schematic drawing of this type of system is shown in figure 8-81 on the following page.

The more modern instrument air systems use "dry" pumps that have carbon vanes and rotors and require no external lubrication. These pumps may be used to drive the instruments by producing a vacuum and pulling air through them or by using the output of the pump to force the air through the instruments. [Figure 8-82]

PNEUMATIC SYSTEM COMPONENTS

Pneumatic systems are often compared to hydraulic systems, but such comparisons can only hold true

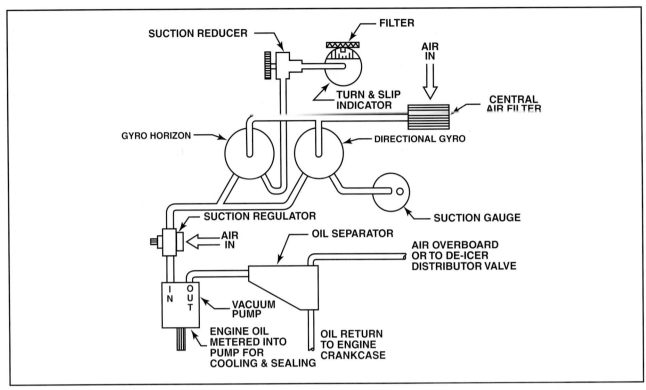

Figure 8-81. Wet-type vacuum systems utilize an air-oil separator to remove lubricating oil from the system.

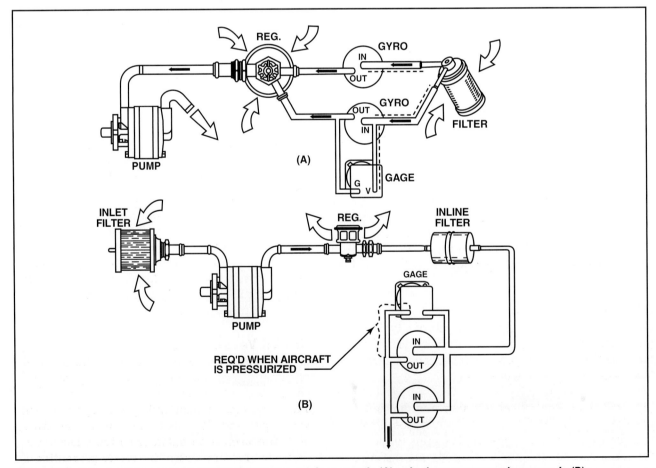

Figure 8-82. A dry-type pump can be used in the vacuum mode as seen in (A) or in the pressure mode as seen in (B).

in general terms. Pneumatic systems do not utilize reservoirs, hand pumps, or accumulators. Similarities, however, do exist in some components.

RELIEF VALVES

Relief valves are used in pneumatic systems to prevent damage. They act as pressure-limiting units and prevent excessive pressures from bursting lines and blowing out seals. [Figure 8-83]

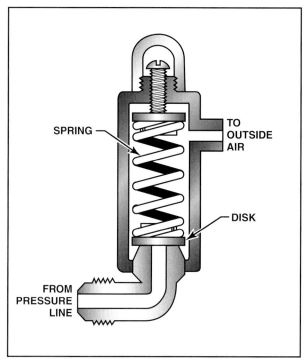

Figure 8-83. A pneumatic-system relief valve is utilized to prevent excess pressures from damaging the system.

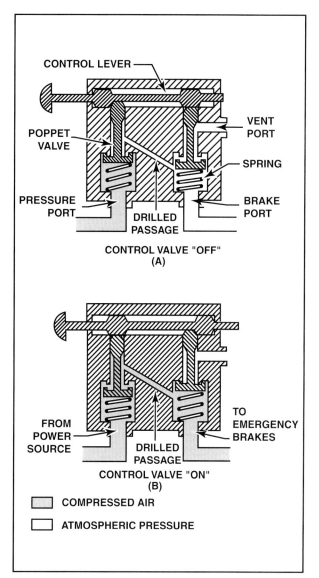

Figure 8-84. This flow diagram of a pneumatic control valve shows how a valve is used to control emergency air brakes.

At normal pressures, a spring holds the valve closed, and air remains in the pressure line. If pressure grows too high, the force it creates on the disk overcomes spring tension and opens the relief valve. Then, excess air flows through the valve and is exhausted as surplus air into the atmosphere. The valve remains open until the pressure drops to normal.

CONTROL VALVES

Control valves are also a necessary part of a typical pneumatic system. The control valve consists of a three-port housing, two poppet valves, and a control lever with two lobes. [Figure 8-84]

In view (A), the control valve is shown in the "off" position. A spring holds the left poppet closed so that the compressed air entering the pressure port cannot flow to the brakes. In view (B), the control valve has been placed in the "on" position. One lobe of the lever holds the left poppet open, and a spring closes the right poppet. Compressed air now flows around the opened left poppet, through a drilled passage, and into a chamber below the right poppet. Since the right poppet is closed, the high-pressure air flows out of the brake port and into the brake line to apply the brakes.

CHECK VALVES

Check valves are used in both hydraulic and pneumatic systems. In a flap-type pneumatic check valve, air enters one port of the check valve and compresses a light spring, forcing the check valve open and allowing air to flow out the other port. However, if air enters from the other direction, air pressure closes the valve, preventing a flow of air out the intake port. Thus, a pneumatic check valve is a one-direction flow-control valve. [Figure 8-85]

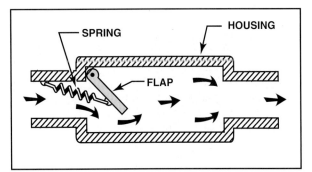

Figure 8-85. A pneumatic check valve utilizes a weak spring to restrict the direction of air flow.

RESTRICTORS

Restrictors are a type of control valve used in pneumatic systems. One type of orifice restrictor has a large inlet port and a small outlet port. The small outlet port reduces the rate of airflow and the speed of operation of an actuating unit. [Figure 8-86]

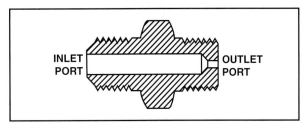

Figure 8-86. A large inlet combined with a small outlet impedes the flow of air through a restrictor.

Another type of speed-regulating unit is the variable restrictor. It contains an adjustable needle valve, which has threads around the top and a point on the lower end. Depending on the direction turned, the needle valve moves the sharp point either into or out of a small opening to decrease or increase the size of the opening. Since air entering the inlet port must pass through this opening before reaching the outlet port, this adjustment also determines the rate of airflow through the restrictor. [Figure 8-87]

FILTERS

Pneumatic systems are protected against dirt by means of various types of filters. A micronic filter consists of a housing with two ports, a replaceable filter cartridge, and a relief valve. Normally, air enters the inlet, circulates around the cellulose cartridge, then flows to the center of the cartridge and out the outlet port. If the cartridge becomes clogged with dirt, pressure forces the relief valve open and

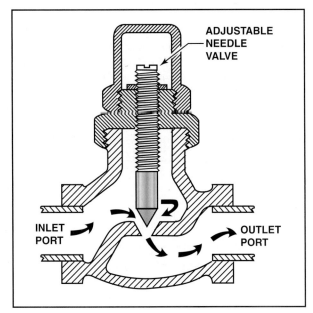

Figure 8-87. A variable orifice enables the variable pneumatic restrictor to be set for a wide range of airflow, from the maximum provided by the air source down to no flow at all.

allows unfiltered air to flow out the outlet port. [Figure 8-88]

A screen-type filter is similar to the micronic filter, but contains a permanent wire screen instead of a replaceable cartridge. In the screen filter, a handle extends through the top of the housing and can be used to clean the screen by rotating it against metal scrapers. [Figure 8-89]

DESICCANT/MOISTURE SEPARATOR

Moisture in a compressed air system will condense and freeze when the pressure of the air is dropped for actuation and, for this reason, every bit of water must be removed from the air. A moisture separator collects the water that is in the air on a baffle, and holds it until the system is shut down. When the inlet pressure to the separator drops below a preset value, a drain valve opens and all of the accumulated water is blown overboard. An electric heater built into the base of the separator unit prevents the water from freezing.

After the air leaves the moisture separator with about 98% of its water removed, it must pass through a desiccant, or chemical dryer, to remove the last traces of moisture. This unit consists of a tubular housing with inlet and outlet ports and contains a desiccant cartridge. These replaceable cartridges consist of a dehydrating agent (MIL-D-3716) and incorporate a bronze filter at each end. Any moisture not removed by the separator will be absorbed by the dehydrating agent.

Hydraulic and Pneumatic Power Systems

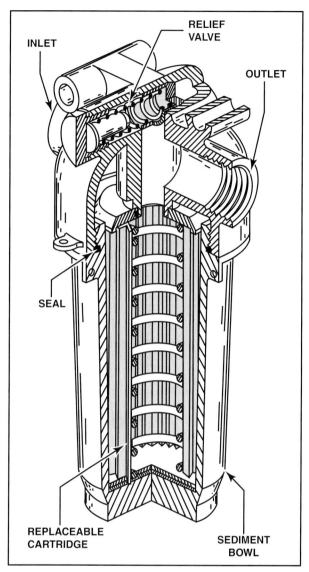

Figure 8-88. The micronic filter is equipped with a replaceable paper element.

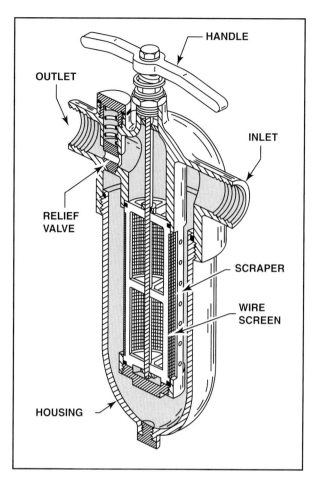

Figure 8-89. The screen-type filter does not have a replaceable element, but is equipped with a scraper used to clean the wire screen.

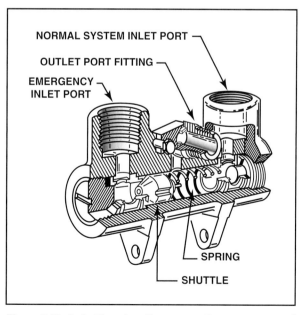

Figure 8-90. A shuttle valve allows more than one source of air pressure to be attached to a system, but only allows the one with greater pressure to actually power the system. The inoperative side is isolated from the system.

SHUTTLE VALVES

Shuttle valves may be installed to allow a pneumatic system to operate from a ground source. When the pressure from the external source is higher than that of the compressor, as it is when the engine is not running, the shuttle slides over and isolates the compressor. The pneumatic systems may then be operated from the ground source. Shuttle valves may also be used to provide an emergency pneumatic backup for hydraulically operated landing gear or brake systems. [Figure 8-90]

EMERGENCY BACKUP SYSTEM

All aircraft with retractable landing gear must have some method of ensuring that the gear will move down and lock in the event of failure of the main extension system. One of the simplest ways of lowering and locking a hydraulically actuated landing gear is by using compressed air or nitrogen stored in an emergency cylinder. The gear selector is placed in the gear down position to provide a path for the fluid to leave the actuator and return into the reservoir. Compressed air is then released from the emergency cylinder, and it enters the actuator through a shuttle valve. This valve is moved over by air pressure to close off the hydraulic system so no air can enter. The air pressure is sufficient to lower and lock the landing gear against the flight loads. [Figure 8-91]

Emergency operation of the brakes is also achieved in many airplanes by the use of compressed air. When the pilot is sure he has no hydraulic pressure to the brakes, he can rotate the pneumatic brake handle. Clockwise rotation of this handle increases the brake pressure, and when the handle is held stationary, the pressure is constant. Nitrogen pressure released by this control handle forces hydraulic fluid in the transfer tube into the main wheel brakes through shuttle valves. When the brake handle is rotated counterclockwise, pressure is released and the nitrogen is exhausted overboard.

TYPICAL PNEUMATIC POWER SYSTEM

Some popular European-built twin-engine commuter transport airplanes utilize a fairly complex pneumatic system. Examination of the components illustrates how they work together to create a reliable system. [Figure 8-92]

COMPONENTS

Each of the two compressors is a four-stage piston-type pump driven from the accessory gearboxes of the two turboprop engines. Air is taken into the first stage through an air duct and is compressed, then passed successively to the other three stages. The discharge air from the fourth stage is routed through an intercooler and a bleed valve to the unloading valve. The bleed valve is kept closed by engine oil pressure and, in the event of a loss of the engine lubricating oil, the valve will open and relieve the pump of any load.

The unloading valve maintains pressure in the system between 2,900 and 3,300 psi. When the pressure rises to 3,800 psi, a relief valve dumps the output of the pump overboard. When the system pressure drops to 2,900 psi, the output of the pump is directed back into the system.

Shuttle valves in the lines between the compressors and the main system make it possible to charge the system from a ground source. When the pressure from the external source is higher than that of the compressor, as it is when the engine is not running, the shuttle slides over and isolates the compressor.

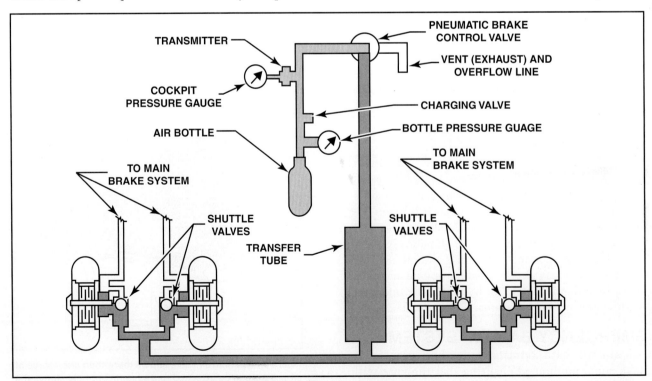

Figure 8-91. The emergency backup system on some aircraft utilizes high-pressure air to operate the brakes in an emergency. Shuttle valves prevent the air from entering the hydraulic system.

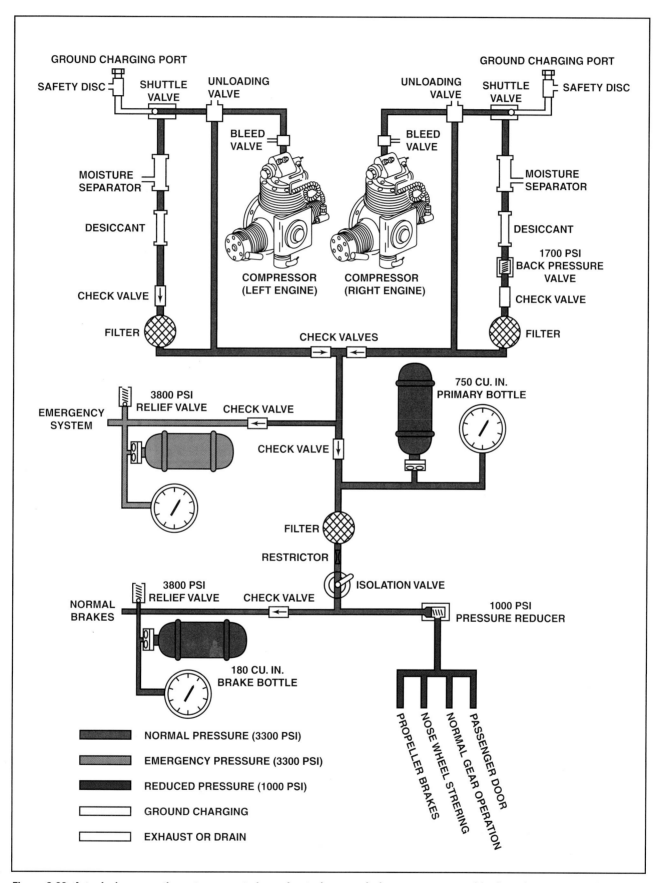

Figure 8-92. A typical pneumatic system on a twin-engine turboprop airplane powers normal brakes, the passenger door, normal gear operation, nosewheel steering, and propeller brakes.

A **moisture separator** collects the water that is in the air and holds it until the system is shut down. When the inlet pressure to the separator drops below 450 psi, a drain valve opens and the accumulated water is blown overboard. After the air leaves the moisture separator it passes through a desiccant, or chemical dryer, to remove the last traces of moisture. Then, the air is filtered through a 10-micron sintered metal filter before it enters the actual operating system. After drying and filtering, air is provided for system use free of contaminates and water vapor.

A back pressure valve is installed in the right engine nacelle. This is essentially a pressure-relief valve in the supply line that does not open until the pressure from the compressor or ground charging system is above 1,700 psi, ensuring that the moisture separator will operate most efficiently. To operate the system from an external source of less than 1,700 psi, it can be connected into the left side where there is no back pressure valve.

There are three air storage bottles in a twin-engine turboprop: a 750-cubic-inch bottle for the main system, a 180-cubic-inch bottle for the normal brake operation, and a second 180-cubic-inch bottle for emergency operation of the landing gear and brakes. A manually operated isolation valve allows a technician to close off the air supply so the system can be serviced without having to discharge the storage bottle.

The majority of the components in this system operate with pressure of 1,000 psi, so a pressure-reducing valve is installed between the isolation valve and the supply manifold for normal operation of the landing gear, passenger door, drag brake, propeller brake, and nose wheel steering. This valve not only reduces the pressure to 1,000 psi, but it also serves as a backup pressure-relief valve.

The emergency system stores compressed air under the full system pressure of 3,300 psi and supplies it for landing gear emergency extension.

PNEUMATIC POWER SYSTEM MAINTENANCE

Maintenance of a pneumatic power system consists of servicing, troubleshooting, removing and installing components, and operational testing.

The air compressors' lubricating oil level should be checked daily in accordance with the applicable manufacturers' instructions. The oil level is indicated by means of a sight gauge or dipstick. After oil is added, the filler plug should be properly torqued and safety wire should be installed as required.

The pneumatic system should be purged periodically to remove contamination, moisture, or oil from the components and lines. Purging the system is accomplished by pressurizing it and removing the plumbing from various components throughout the system. Removal of pressurized lines will cause a high rate of airflow through the system, causing foreign matter to be exhausted. If an excessive amount of foreign matter, particularly oil, is exhausted from any one system, the lines and components should be removed and cleaned or replaced.

Upon completion of pneumatic system purging and after re-connecting all the system components, the system air bottles should be drained to exhaust any moisture or impurities that may have accumulated there.

AIRCRAFT LANDING GEAR SYSTEMS

CHAPTER 9

INTRODUCTION
The landing gear of the very first airplanes was not very complex. The Wright Flyer, for instance, took off from a rail and landed on skids. However, soon after the basic problems of flight were solved, attention was turned to providing better control and stability of the aircraft while it was operated on the ground. Bicycle and motorcycle wheels were first used, which in turn, gave way to specially designed landing gear and wheels that absorbed the extreme loads imparted during takeoffs and landings. In addition, braking systems were installed to provide safer and more efficient control for slowing an airplane after landing. In later years, as aircraft designs improved to increase speed and efficiency, retraction systems were provided to allow the landing gear to be stowed during flight to reduce aerodynamic loads, or drag. With continued improvements in technology, landing gear systems on modern aircraft are highly reliable and capable of handling extreme conditions, enabling safe transitions between flight and ground mobility.

SECTION A

LANDING GEAR SYSTEMS AND MAINTENANCE

LANDING GEAR TYPES

The first airplanes with wheels used a tricycle landing gear. In a tricycle gear configuration, the main wheels are located behind the center of gravity and an auxiliary wheel (nose wheel) is located at the front of the aircraft. When airplanes with aft or "pusher" propellers were replaced with those having the propeller up front, the tailwheel landing gear was used in order to keep the propeller further above the ground. This configuration became so popular that it is called "conventional" landing gear, even though it has the disadvantage of being difficult for ground maneuvering.

When engines were developed that produced their power at speeds that allow for shorter propellers, the ground handling advantages of the tricycle landing gear eventually came back as the most popular landing gear configuration.

Slow airplanes gain flight efficiency by using fixed landing gear. The weight savings more than make up for the additional drag inefficiencies, especially when the wheels are enclosed in streamlined wheel fairings, or "pants." However, as the speed of the airplane increases, the parasitic drag from a fixed landing gear offsets the advantage of its light weight. Hence, it has become advantageous to retract the wheels into the wing or the fuselage to decrease drag. [Figure 9-1]

LANDING GEAR ARRANGEMENT

Landing gear arrangement is determined by the manufacturer. The most prevalent arrangement on modern aircraft is the tricycle gear configuration. However, it is important that the maintenance technician be familiar with other arrangements, particularly the tailwheel or conventional gear arrangement. Occasionally, a technician will encounter some other form of landing gear such as the tandem arrangement in which the wheels are located down the centerline of the longitudinal axis of the aircraft, such as in the case of some gliders.

TAILWHEEL-TYPE LANDING GEAR

The majority of modern aircraft do not utilize conventional landing gear, resulting in a generation of pilots who have never flown an airplane with a tailwheel arrangement. Tailwheel aircraft are configured with the two main wheels located ahead of the aircraft's center of gravity and a much smaller wheel at the tail. Moving the rudder pedals that are linked to the tailwheel steers the aircraft on the ground. The rudder pedals are connected to the tailwheel steering by a control cable system and spring. The spring provides steering dampening. Some aircraft have no provision for steering the tailwheel. In that case, the

Figure 9-1. Decreased drag can be achieved by the use of streamlined fairings or by retracting the gear.

wheel is locked in-line with the fuselage for takeoff and unlocked for landing allowing it to swivel freely for taxiing. Control on the ground is then achieved by the use of differential braking. [Figure 9-2]

Figure 9-2. The configuration of an airplane having a tailwheel-type landing gear is often called a "conventional" landing gear.

Tailwheel aircraft are highly susceptible to ground loop; the abrupt, uncontrolled change in direction of an aircraft on the ground. The center of gravity of a tail wheel aircraft is located behind the main gear making them prone to ground loop. The pilot must be careful to keep the airplane in control and rolling straight or the center of gravity may uncontrollably swing around ahead of the wheels, causing the airplane to spin around on the ground.

Prior to World War II, almost all airplanes used the tailwheel-type landing gear. During WWII such airplanes such as the Lockheed Lightning, the Consolidated Liberator, and the Boeing Superfortress, as well as the commercial Douglas DC-4, proved that the tricycle gear configuration was superior in ground handling ease. The tricycle gear configuration has since become the most widely used landing gear arrangement.

TRICYCLE-TYPE LANDING GEAR
Nearly all currently produced aircraft use the tricycle landing gear configuration in which the main gear are located behind the airplane's center of gravity and the nose of the airplane is supported by the nose gear. Steering the nose wheel through connections to the rudder pedals provides control on the ground for small airplanes, while large airplanes utilize hydraulic steering cylinders to control the direction of the nose gear. [Figure 9-3]

TANDEM LANDING GEAR
The tandem wheel arrangement is seldom used on civilian aircraft other than gliders. Some of the

Figure 9-3. The tricycle landing gear configuration substantially improved the ground handling characteristics of modern airplanes.

heavy bombers of the past have used the tandem gear arrangement. The main wheels are located in line under the fuselage and outrigger wheels support the wings.

FIXED OR RETRACTABLE LANDING GEAR
All aircraft must contend with two types of aerodynamic drag, parasite drag, which is produced by the friction of the airflow over the structure and induced drag which is caused by the production of the lift. Parasite drag increases as the speed increases, while induced drag decreases as the speed increases because of the lower angle of attack required to produce the needed lift. Slower aircraft lose little efficiency by using the lighter-weight fixed landing gear. Faster aircraft retract the landing gear into the structure and thus gain efficiency even at the cost of slightly more weight.

Fixed landing gear decreases parasitic drag markedly by enclosing the wheels in streamlined fairings, called wheel pants. Many light airplanes utilize fixed landing gear that consist of spring or tubular steel landing gear legs with small frontal areas that produce minimum drag. [Figure 9-4]

Figure 9-4. Streamlined wheel fairings called "wheel pants" are used to decrease the wind resistance of fixed landing gear.

SHOCK ABSORBING AND NON-ABSORBING LANDING GEAR

Some aircraft landing gear absorb landing shock and some do not. Non-absorbing gear include spring steel, composite, rigid, and bungee cord construction. Shock absorbing gear incorporate shock absorbers that converts motion into some other form of energy, usually heat.

SPRING STEEL AND COMPOSITES

Most aircraft provide for absorbing the landing impact and shocks of taxiing over rough ground. Some aircraft, however, do not actually absorb these shocks but rather accept the energy in some form of elastic medium and return it at a rate and time that the aircraft can accept. The most popular form of landing gear that does this is the spring steel gear used on most of the single-engine Cessna aircraft. These airplanes use either a flat steel leaf or a tubular spring steel strut that accepts the loads and returns it in such a way that it does not cause the aircraft to rebound. [Figure 9-5]

Figure 9-6. Fabric enclosed rubber bungee cords in this landing gear accept both landing impact and taxi shocks.

Figure 9-5. The thin spring-steel landing gear struts of this airplane do not truly absorb the shocks, but rather accept them and return them to the aircraft at a rate that will not cause the aircraft to bounce.

RIGID GEAR

Certain older types of aircraft use rigid landing gear that transmit all the loads of landing touchdown directly to the airframe's structure. Some of the shock is absorbed by the elasticity of the tires. However, this type of landing gear system is not only hard on the aircraft's occupants, but can cause structural failure during a hard landing. Some aircraft, such as helicopters, that normally land very softly utilize rigid landing gear.

BUNGEE CORD

Some aircraft use rubber to cushion the shock of landing. This may be in the form of rubber doughnuts or as a bungee cord, which is a bundle of small strands of rubber encased in a loosely woven cloth tube. Rubber bungee cords accept both landing impact and taxi shocks. [Figure 9-6]

SHOCK STRUTS

The most widely used shock absorber for aircraft is the air-oil shock absorber, more commonly known as an oleo strut. The cylinder of this strut is attached to the aircraft structure, and a close fitting piston is free to move up and down inside the cylinder. It is kept in alignment and prevented from coming out of the cylinder by torsion links, or scissors. The upper link is hinged to the cylinder and the lower link to the piston. The wheel and axle are mounted to the piston portion of the strut. [Figure 9-7]

Figure 9-7. The oleo strut has become the most widely used form of shock absorber on aircraft landing gear.

Shock Strut Operation

The cylinder of a shock strut is divided into two compartments by a piston tube. The piston itself fits into the cylinder around the tube. A tapered metering pin, which is a part of the piston, sticks through a hole in the bottom of the piston tube. To fill the strut, the piston is pushed all of the way into the cylinder. The strut is then filled with hydraulic fluid to the level of the charging valve. With the weight of the aircraft on the wheel, enough compressed air or nitrogen is pumped through the charging valve to raise the airplane until the piston sticks out of the cylinder a distance specified by the manufacturer. [Figure 9-8]

When the weight is removed from the landing gear, the piston extends the full amount allowed by the torsion links and the fluid drains past the metering pin into the fluid compartment in the piston. Some shock struts are equipped with a damping or snubbing device consisting of a recoil valve on the piston or recoil tube. This reduces the rebound during the extension stroke, and prevents the shock strut from extending too rapidly. A sleeve, spacer, or bumper ring is incorporated to limit the extension stroke of the strut. Then, when the wheels contact the ground on landing, the piston is forced up into the cylinder. The metering pin restricts the flow of fluid into the cylinder, and forcing the fluid through this restricted orifice absorbs much of the energy of the impact. As the fluid is forced through the orifice, the motion of the fluid is converted to heat. The taper of the metering pin provides a graduated amount of opposition to the fluid flow and smoothly absorbs the shock.

Servicing Shock Struts

The air-oil type oleo strut should be maintained at proper strut tube extensions for the best oleo action. Both the nose and main gear struts will have a specific length of piston tube exposed. These measurements should be taken with the airplane sitting on a level surface under normal fuel loading conditions. Whenever servicing any part of the gear, wheels, and tires, the shock strut should be inspected for cleanliness, evidence of damage, and proper amount of extension. Manufacturer's repair manuals should be consulted for proper specifications. [Figure 9-9]

The manufacturer's current service manual should always be checked before servicing the shock struts. The exact procedures for each aircraft will differ, but some generic procedures outlined here give some idea of the steps involved. Dirt and foreign particles collect around the filler plugs of the landing gear struts, therefore, before attempting to

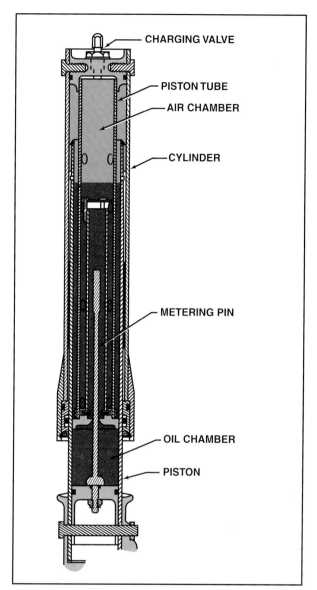

Figure 9-8. The oleo, or air-oil, strut absorbs the landing impact as oil transfers from the oil chamber into the air chamber around the tapered metering pin. Taxi shocks are dampened by the compressed air in the upper chamber.

remove these plugs, the tops of the struts should be cleaned. The next step is to place the airplane on jacks. A drip pan should be placed under the gear to catch spillage of the fluid used in the strut. At the filler plug, air pressure is relieved from the strut housing chamber by removing the cap from the air valve and depressing the valve core. The filler plug is then removed at the top or side of the gear strut housing. The strut piston tube is raised until it is fully compressed by pushing up on the tire. Fluid should be poured from a clean container through the filler opening until it reaches the bottom of the filler plug hole. The filler plug is then installed finger-tight, and the strut extended and compressed two or three times to remove air from the housing.

Figure 9-9. For proper action, the strut tube must be in a position to travel in both directions. The exact resting position for strut exposure can be found in the manufacturer's service manual.

The filler plug is then removed again, and the strut raised to full compression and filled with fluid if needed. The fill plug is then reinstalled and tightened. After making certain that the oleo strut has been sufficiently filled with fluid, remove the aircraft from the jacks. The oleo strut is then inflated until the correct distance is exposed on the strut piston tube. The airplane should be rocked to actuate the strut to determine if the gear returns to the correct strut position. Then the valve core should be checked for leakage before replacing the cap on the air valve core. These procedures are an example only. The manufacturer's service manual should be checked for exact procedures and measurements.

AIRCRAFT WHEELS

Wheels have never appeared to be a spectacular component in aircraft development. Wheels do their job, without attracting much attention. This dependability has come about because the design of these vital and highly stressed components has kept abreast of the development of the aircraft itself.

The wheels used on many early aircraft were designed as one-piece units. The tires were flexible enough that they could be forced over the wheel rim with tire tools in much the same way we force tires on automobile wheels today. However, modern aircraft tires are normally so stiff they cannot be forced over the rims, and, as a result, almost all modern aircraft wheels are constructed of two-piece units.

The development of tubeless tires promoted the development of two-piece wheels that are split in the center and made airtight with an O-ring seal placed between the two halves. Today, this form of wheel is the most popular for all sizes of aircraft, from small trainers up to large jet transports.

WHEEL CONSTRUCTION

Aircraft wheels must be lightweight and strong. Most wheels are made of either aluminum or magnesium alloys and, depending upon their strength requirements, may be either cast or forged. The bead seat area is the most critical part of a wheel. To increase wheel strength against the surface tensile loads applied by the tire, bead seat areas are usually rolled to pre-stress their surface with a compressive stress.

INBOARD WHEEL HALF

The inboard wheel half is the half of a two-piece wheel that houses the brake. Rotating brake disks are driven by tangs on the disk which ride in steel-reinforced keyways, or by steel keys bolted inside the wheel that mate with slots in the periphery of the disk.

A polished steel bearing cup is shrunk into the bearing cavity of the wheel, and a tapered bearing cone slides over the landing gear axle to support the wheel. A grease seal covers the bearing to hold grease in the bearing and prevent any dirt or water from getting to its surfaces.

One or more fusible plugs are installed in the inboard half of the main wheels of jet aircraft to release the air from the tire in the event of an extreme overheat condition, such as heavy braking that is required during an aborted takeoff. Rather than allowing the heat to increase the tire pressure so high that the tire blows out, the low-melting-point alloy in the center of the plug will melt and deflate the tire without further damage.

OUTBOARD WHEEL HALF

The outboard half of the wheel bolts to the inboard half and holds a shrunk-in bearing cup in which a tapered bearing cone rides. A seal protects the roller and bearing surfaces from water and dirt and retains the lubricant in the bearing.

A cap is held in place with a retaining ring to cover the end of the axle shaft and the bearing. If the aircraft is equipped with an antiskid system, this cap is fitted with a bracket to drive the wheel-speed sensor that is mounted in the axle. An inflation valve in the outboard wheel half allows air to be added to a tubeless tire. If a tube-type tire is installed, a hole in this half of the wheel is provided for the valve stem of the tube.

WHEEL INSPECTION

Whenever the maintenance technician comes in contact with the wheels, they should be inspected. When the wheels are on the aircraft, inspect for general condition and proper installation, which includes checking for proper axle torque. When the wheel is off the aircraft, more extensive checks can be and are performed. These on and off aircraft checks include the following checks and procedures.

CHECK FOR PROPER INSTALLATION

It is possible with some types of wheel and brake assemblies that the wheel can be installed with the disk drive tangs between the drive slots, rather than mating with the slots. When inspecting the wheel the technician must make certain that the brake is correctly installed and everything is in its proper place.

CHECK FOR AXLE NUT TORQUE

If too little torque is used on the axle nut, it is possible for the bearing cup to become loose and spin, enlarging its hole and requiring a rather expensive repair to the wheel. If the torque is too high, the bearing can be damaged because the lubricant will be forced out from between the mating surfaces. The amount of torque required varies with the installation. Follow procedures used for installing and securing the axle nut that are recommended by the airframe manufacturer.

LOOSENING THE TIRE FROM THE RIM

Before a wheel can be inspected, the tire must be removed. While this admonition is so basic that it can almost go unsaid, it is tragic that it is sometimes overlooked. Before loosening the wheel half retaining bolts, BE SURE THE TIRE IS COMPLETELY DEFLATED. It is also advisable to deflate the tire before removing the wheel/tire assembly from the axle. In the event that the bolts holding the wheel halves have failed, the only thing holding the assembly together is the axle nut.

For safety, let the air out of the tire by using a deflator cap screwed onto the valve and, after most of the air is out, remove the valve core. This procedure is not as vital for small tires that carry low inflation pressures. It is possible with some jet aircraft that carry more than 100 psi in their tires for the valve core to be ejected with such velocity that it can injure anyone it strikes.

After the tire has been deflated, the bead of the tire should be broken from the wheel by applying an even pressure to the tire as close to the wheel as possible. Screwdrivers or any type of tire tool should never be used to pry the bead away from the rim, because it is easy to nick or damage the soft wheel in the critical bead area. Any damage here will cause a stress concentration that can lead to wheel failure. [Figure 9-10]

Figure 9-10. The bead of the tire should be broken away from the bead seat of the wheel with a steady pressure as near the rim of the wheel as possible.

DISASSEMBLY OF THE WHEEL

The wheel should be placed on a clean, flat surface and the bearing seals and cones removed from both wheel halves. The nuts from the wheel bolts are removed to separate the wheel halves. Impact wrenches are not used on aircraft wheels. Even though it is common practice with automotive wheels to use impact wrenches for speed, the uneven torque produced by these wrenches creates stresses these lightweight wheels are not designed to take.

CLEANING THE WHEEL ASSEMBLY

Stoddard solvent or similar cleaning fluid should be used to remove any grease or dirt from the wheel. A soft bristle brush will aid in removing stubborn deposits. Do not use scrapers that will remove any of the protective finish from the wheel. After all of the parts have been cleaned, they should be dried with compressed air.

CLEANING THE BEARINGS

Clean solvent should be used to wash the wheel bearings. Soak them to soften the grease and any hardened deposits in the bearings, then brush them with a soft bristle brush to remove all of the residue. Dry the bearings by blowing them out with low-pressure dry compressed air. DO NOT SPIN THE BEARINGS AS YOU DRY THEM. Rotating dry metal against dry metal will damage both the rollers and the races. Bearings should never be cleaned with steam, because the heat and excess oxygen will cause a premature breakdown of the bearing surface.

BEARING INSPECTION

If a bearing is difficult to remove from the axle shaft, it should be removed with a special puller. It should never be driven from the shaft with any form of drift. Bearings that have been difficult to remove from the shaft often have indications of galling on their inner bore, which is cause for rejection of the bearing cone.

Water stains on a bearing may not look bad, but they are an indication of intergranular corrosion in the surface of the rollers or the races. Any bearing showing signs of water marks should be rejected. [Figure 9-11]

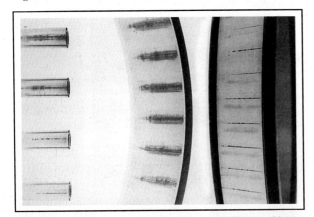

Figure 9-11. Water stains on a bearing are evidence of intergranular corrosion.

Any damage to the large end of the rollers is a reason for rejection, but minor flattening at the small end will usually cause no problem because this end does not provide rolling contact. Any spalling, which is a slight chipping of the rolling surface of the bearing is a reason to reject the bearing. Any grooves on the roller surface that are deep enough to feel with the fingernail will also require replacement of the bearing. Any indication of a bearing overheating that shows up as discoloration of the rollers is cause for rejection of the bearing. An indication of dry operation, which is indicated by rust on the rolling surface, is also cause for rejection. [Figure 9-12] [Figure 9-13]

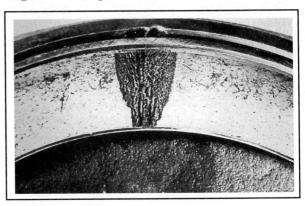

Figure 9-12. Spalling, or failure of the bearing surface, is reason to reject the bearing.

Figure 9-13. These bearings have been overheated.

Bearing cages hold the rollers spaced from each other and aligned on the races. If there are any bends or distortion of the cage, it is likely that there is some hidden damage in the bearing. Rather than risk bearing failure, the bearing should be replaced. If the cage, which is not a wearing surface, shows any signs of wear or of peripheral scratches it is an indication of bearing damage.

The bearing cup should be carefully inspected for any signs of water marks, discoloration, rust, or brinell marks, which are shallow, smooth depressions made by the rollers forced against the cup by excessive pressure. Any of these indications are a cause for rejection.

If it is necessary to remove the bearing cup, the wheel should be heated in boiling water for no more than one hour, or in an oven at a temperature no higher than 225° F for thirty minutes. When the wheel is at the proper temperature, the cup is tapped out of its cavity with a fiber drift. A new cup is installed by again heating the wheel and chilling the cup with dry ice. The outside of the cup is coated with zinc chromate primer and tapped into place with a fiber drift. When heating the wheel, care must be taken not to overheat the assembly, which will impair its heat treatment.

BEARING LUBRICATION
Bearings should be packed using the grease specified by the aircraft manufacturer. MIL-G-81322D is the most recently developed type of lubrication and has superior qualities to previously developed lubrications. Not all brands of MIL specification grease are the same color, but those having the same specification number are compatible and interchangeable, regardless of their color.

If equipment is available for pressure-packing the bearings, it should be used. Pressure packing uniformly covers all of the rollers. If the bearings must be packed by hand, the grease must be worked up around each of the individual rollers and inner cone until they are all uniformly covered. After the bearing is thoroughly greased, the inner cone and roller assembly should be packed in clean waxed paper and stored until ready for installation. A light film of grease should be applied to the bearing cups in the wheel halves, which protects the cup from dirt or damage until the wheel is reassembled.

INSPECTING THE WHEEL HALVES
The most difficult area of an aircraft wheel to inspect is the bead seat region. This area, which is highly stressed by the inflated tire, can be distorted or cracked by a hard landing or a seriously overinflated tire. When all of the forces are removed and the tire dismounted, these cracks may close up so tightly, especially on forged wheels, that penetrant cannot enter the crack. This makes any form of penetrant inspection useless for examination of the bead seat area. [Figure 9-14]

Eddy current inspection should be used for the bead seat area. It measures the amount of current necessary to induce a current flow into the material under the probe. If the conductive characteristics of the material change due to the presence of a crack, the indicator will show that a different amount of current is induced. Eddy current inspection equipment is relatively easy to use, small enough in size, and low enough in cost that it is found in most

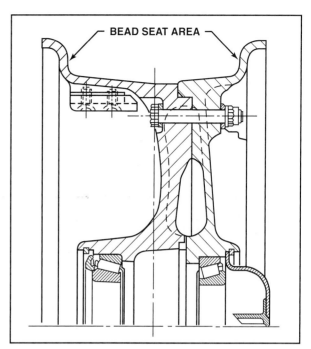

Figure 9-14. The bead seat area of a wheel assembly is the most difficult area of a wheel to inspect. Eddy current inspection is a type of inspection that can reliably find cracks in this area.

maintenance shops. Eddy current inspection is a comparison system rather than an absolute measurement, and its effectiveness depends to a large extent on the experience and judgment of the operator. Dye penetrant inspection procedures can be used to check for cracks in the key slot areas because cracks in this area have no tendency to close up. [Figure 9-15]

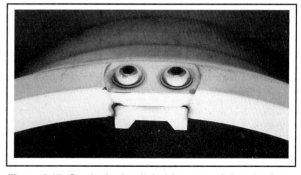

Figure 9-15. Cracks in the disk drive area of the wheel can be inspected by penetrant type inspection because cracks in this area have no tendency to close.

The entire wheel should be inspected for indications of corrosion likely to form at any place where moisture is trapped against the surface of the wheel. The rim area where moisture could be trapped between the tire and the wheel is also a point of possible corrosion. The surface of the wheel that is inside the tire assembly is an area open for inspection only when the tire has been removed. [Figure 9-16]

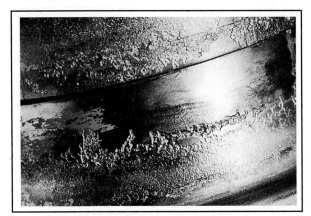

Figure 9-16. Corrosion is likely to occur anywhere that water can be trapped against the surface of the wheel.

Any corrosion pits that are found must be completely dressed out without removing more metal than the manufacturer's service manual allows. After all traces of the damage have been removed, the surface of the metal must be treated to prevent further corrosion.

A good visual inspection is one of the best ways to locate a defect in a part. All suspect areas in the wheel should be carefully inspected with a ten or fifteen-power magnifying glass. All of the dimensions specified in the overhaul manual must be checked to be sure that there are no parts worn to the extent that they will require repair or replacement.

WHEEL BOLT INSPECTION
The wheel bolts should be inspected by magnetic particle inspection. Pay particular attention to the junction of the head and shank, and to the end of the threaded area. Because the cross-sectional area of the shank changes at these two locations, these are the most likely locations for cracks to form.

KEY AND KEY SCREW INSPECTION
The disk drive keys are subject to some of the most severe forces acting on a wheel because they try to rotate the disks against the friction in the brake. Absolutely no play can be tolerated between the keys and the wheel half so they must be checked carefully for looseness, cracks, and excessive wear. Each of the key attachment screws is staked to prevent its becoming loose in service. Therefore, stakes must be checked for proper tolerances.

FUSIBLE PLUG INSPECTION
Carefully examine the condition of the fusible plugs in the wheel to be sure that none of them show any sign of the core melting. Even if only one of the plugs indicates any deformation, all of the plugs must be replaced. [Figure 9-17]

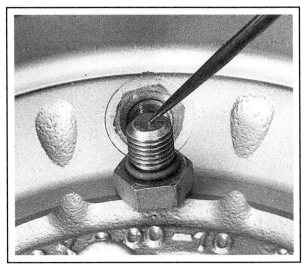

Figure 9-17. Fusible plugs should be inspected for signs of softening due to excessive heat.

BALANCE WEIGHTS
Almost all wheels having a diameter of more than ten inches are statically balanced when they are manufactured. If the weights, installed by the manufacturer, have been removed for any reason, they must be put back in their original position. The final balancing of the wheel is done after the tire is mounted. The weights for this final balancing are usually installed around the outside of the rim of the wheel or at the wheel bolt circle. [Figure 9-18]

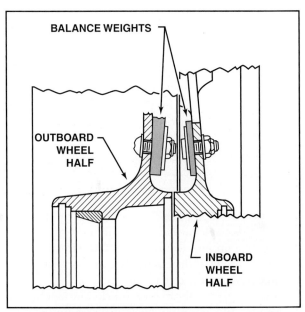

Figure 9-18. Wheel balancing is done statically at manufacture and the weights must be returned to their original positions if removed for any reason.

NOSE WHEEL STEERING SYSTEMS

Nose wheel steering is found on most tricycle gear aircraft. On small aircraft the nose wheel is usually controlled by a direct connection between the rudder pedals and the nose gear. Large aircraft steering is usually activated by a hydraulic actuator that is controlled by the rudder pedals or by a separate steering mechanism. A shimmy damper is also a part of most nose wheel steering systems.

SMALL AIRCRAFT

Almost all airplanes with tricycle landing gear utilize some type of nose-wheel steering on the ground by controlling the nose wheel. Some of the smallest airplanes, however, have a castering nose wheel. In these cases, differential braking does the steering. Other small airplanes link the nose wheel to the rudder pedals directly.

LARGE AIRCRAFT

Large aircraft are steered on the ground by directing hydraulic pressure into the cylinders of dual shimmy dampers. A control wheel operated by the pilot directs fluid under pressure into one or the other of the steering cylinders. The actual control of the fluid can be transmitted from the pilot's control to the hydraulic control unit mechanically, electrically or hydraulically. Fluid from the opposite side of the piston in these cylinders is directed back to the system reservoir through a pressure relief valve that holds a constant pressure on the system to snub any shimmying. An accumulator located in the line to the relief valve holds pressure on the system when the steering control valve is in its neutral position or when pressure to the steering damper system is lost. [Figure 9-19]

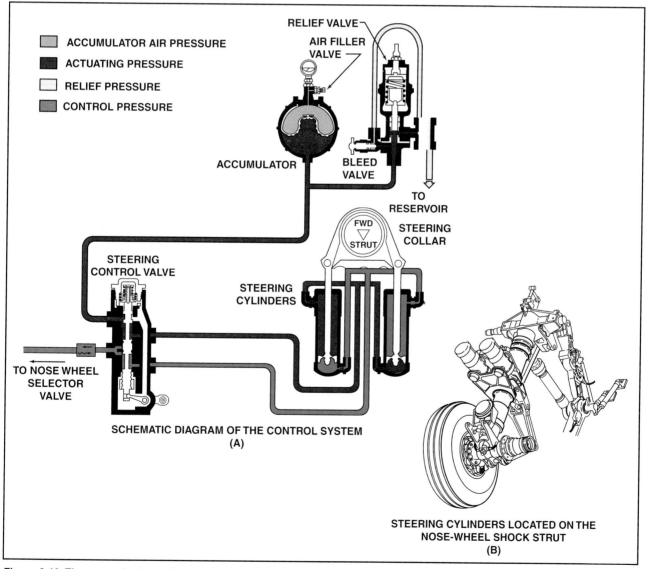

Figure 9-19. The nose wheel steering system for a large aircraft consists of actuators, control valves and other associated components typically found in most hydraulic systems.

SHIMMY DAMPERS

The geometry of the nose wheel makes it possible for it to shimmy, or oscillate back and forth, at certain speeds, sometimes violently. To prevent this highly undesirable condition, almost all nose wheels are equipped with some form of hydraulic shimmy damper as a part of the nosewheel steering system. The shimmy damper is a small hydraulic shock absorber that is installed between the nose-wheel fork and the nose-wheel cylinder.

Shimmy dampers are normally small piston-type hydraulic cylinders that control the bleed of fluid between the two sides of the piston. The restricted flow prevents rapid movement of the piston, but has no effect on normal steering. [Figure 9-20]

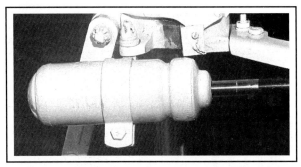

Figure 9-20. A shimmy damper reduces the rapid oscillations of the nose wheel, yet it allows the wheel to be turned by the steering system.

STEERING DAMPERS

In many cases, the steering actuators serve as the steering dampers because they are constantly charged with hydraulic fluid under pressure. As the nose wheel attempts to vibrate or shimmy, these cylinders prevent movement of the nose gear. This type of system is used on large aircraft while a piston type shimmy damper is usually used on small aircraft.

LANDING GEAR ALIGNMENT, SUPPORT AND RETRACTION

In order for the wheels to do their part in supporting the aircraft, there must be a structure that connects the wheels to the aircraft. This structure is the landing gear. The landing gear must be accurately aligned, provide adequate support for the aircraft at any design gross weight, and allow the wheels to retract if necessary.

WHEEL ALIGNMENT

Alignment of the main gear wheels is very important in that misalignment adversely affects landing and take-off, roll characteristics, tire wear, and steering during ground operations. Severe misalignment can cause malfunction and failure of some of the major components of the landing gear system. Alignment consists of checking and adjusting the toe-in or toe-out configuration and the camber of the gear. The aircraft maintenance manual normally specifies the amount of toe-in and camber the landing gear should have. The torque links are also very important in the alignment of the landing gear.[Figure 9-21]

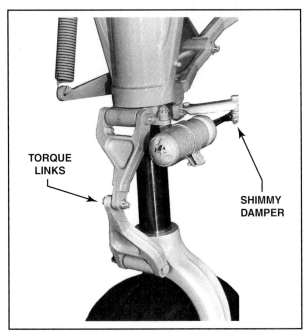

Figure 9-21. The torque links, sometimes called scissors, limit the extension of the oleo strut and keep the wheel in alignment.

The amount of toe-in or toe-out of aircraft wheels is outlined in the aircraft's service manual. An aircraft's wheels are configured in a toe-in position if lines drawn through the center of the two wheels, perpendicular the axles, cross ahead of the wheels or toe-out if the lines cross behind the wheels. As an aircraft moves forward in a toe-in arrangement, the wheels try and move closer together. A toe-out configuration causes the wheels to try and move apart.

In order to measure toe-in, a carpenter's square is held against a straightedge placed across the front of the main wheels. The straightedge should be perpendicular to the longitudinal axis of the aircraft. If this is correct, then the distance between the blade of the carpenter's square and the front and rear flanges of the wheel will indicate toe-in or toe-out. [Figure 9-22] [Figure 9-23] [Figure 9-24]

Aircraft Landing Gear Systems

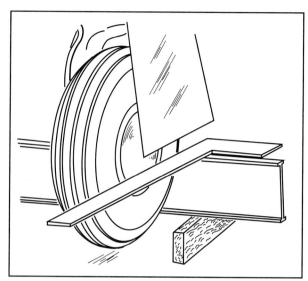

Figure 9-22. One method of checking the toe-in or toe-out of an airplane landing gear utilizes a straightedge and a carpenter's square.

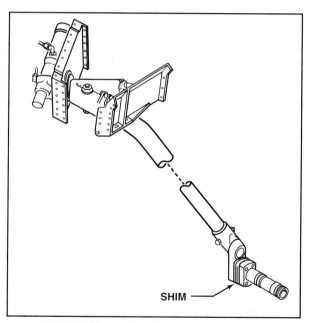

Figure 9-23. This illustration shows the shims used to align the main landing gear on a spring steel landing gear strut. Toe-in is adjusted on spring steel landing gear by placing shims between the axle and the gear leg.

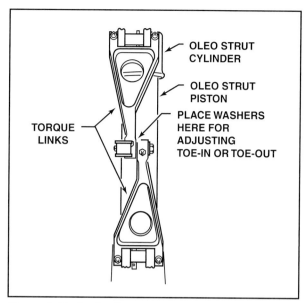

Figure 9-24. On landing gear using an oleo-type shock absorber, toe-in is adjusted by adding or removing washers from between the torque links.

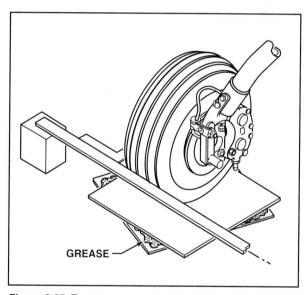

Figure 9-25. To get an accurate wheel alignment check on spring steel gear, the wheels should be rolled onto a sandwich of two metal plates separated by a layer of grease, which allows the aircraft to settle into its normal alignment position.

A spring-steel landing gear moves so much, as the weight of the aircraft is placed on the gear, that an alignment check is difficult unless special procedures are used. The recommended method of checking for toe-in or toe-out is to roll each wheel onto a pair of aluminum plates with grease between them. If the aircraft is rocked back and forth a bit before the measurement is taken, the greased plates will allow the wheels to assume their true position of alignment. [Figure 9-25]

Camber is a measure of the amount the wheel leans, as viewed from straight ahead. If the top of the wheel leans outward, the camber is positive. If it leans inward, the camber is negative. Camber on a spring-steel-type landing gear is affected to a great extent by the operational weight of the aircraft and should be adjusted by the use of shims between the axle and the gear leg. A zero-degree camber at the weight at which the aircraft is most generally operated is the camber usually recommended in the service manual of most small aircraft. [Figure 9-26]

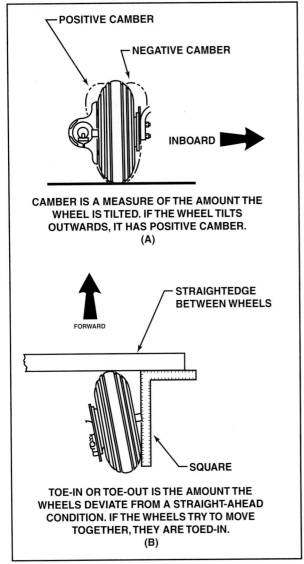

Figure 9-26. Camber is the measure of the amount the top of the wheel tilts inward or outward. Toe-in and toe-out are measures of the amount the front of the wheel points inward or outward.

SUPPORT

The landing gear is generally supported by the aircraft's structure. The wings spars, along with additional structural members, support and attach the main landing gear to the wings on larger aircraft. Non-retractable landing gear is generally attached to the aircraft structure by bolting the landing gear struts to the structure directly. Retractable landing gear systems must provide for the landing gear to move, so the upper shock strut is attached to the airframe using trunnion fittings, which are extensions or shafts attached to the shock strut that mount into fittings bolted to the airframe.

SMALL AIRCRAFT RETRACTION SYSTEMS

When the design speed of an aircraft becomes high enough that the parasite drag of fixed landing gear is greater than the induced drag caused by the added weight of the retracting system, retractable landing gear becomes practical. Some smaller aircraft use a simple mechanical retraction system, incorporating a roller chain and sprockets operated by a hand crank. Many aircraft use electric motors to drive the landing gear retracting mechanism and some European-built aircraft use pneumatic systems.

The simplest hydraulic landing gear system uses a hydraulic power pack containing the reservoir, a reversible electric motor-driven pump, selector valve, and sometimes an emergency hand pump along with other special valves. [Figure 9-27]

To raise the landing gear, the gear selector handle is placed in the GEAR UP position. This starts the hydraulic pump, forcing fluid into the gear-up side of the actuating cylinders to raise the gear. The initial movement of the piston releases the landing gear down-locks so the gear can retract. When all three wheels are completely retracted, a pressure switch stops the pump. There are no mechanical up-locks, and the gear is held in its retracted position by hydraulic pressure in the actuators. A pressure switch stops the pump at a predetermined pressure. If the pressure drops enough to allow any one of the wheels to drop away from its up-limit switch, the pump will start and restore the pressure.

To lower the landing gear, the selector switch is placed in the GEAR DOWN position, which releases the pressure on the up-side of the cylinders. The shuttle valve moves over and fluid flows through the power pack, allowing the gear to free fall and lock down. The pump operates to build up pressure and ensure that all of the gears will lock down. When they are all locked, the limit switches shut off the pump motor.

All retractable landing gear systems must have some means of lowering the gear in the event the main extension system should fail. This simple system depends on the gear free-falling and locking in position. To actuate the emergency extension, a control in the cockpit is actuated to open a valve between the gear-up and gear-down lines that dumps fluid from one side of the actuators to the other. This allows the gear to fall down and lock in place.

An additional feature of this particular landing gear system is the automatic extension system that

Aircraft Landing Gear Systems

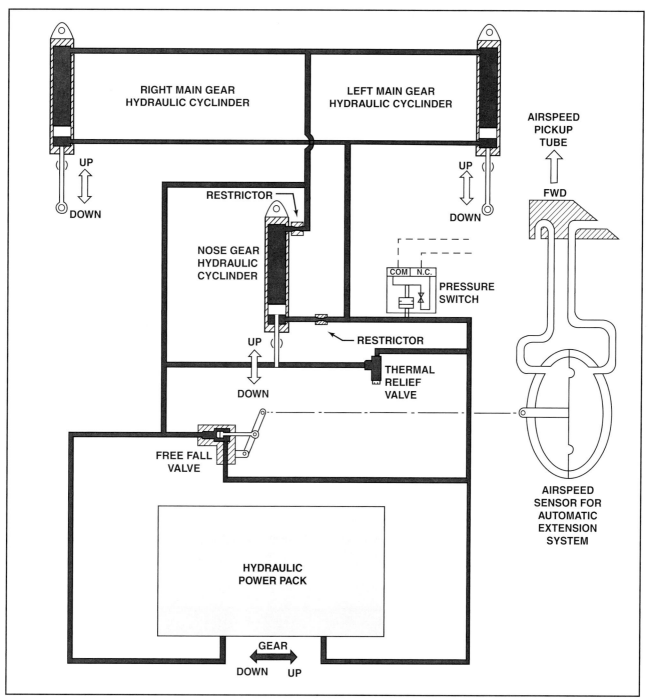

Figure 9-27. This simple landing gear system for small aircraft has an added feature of an airspeed controlled automatic landing gear extension system. At a certain airspeed, the landing gear will automatically extend regardless of gear handle position.

will lower the landing gear when the airspeed slows below a specified value, regardless of the position of the landing gear selector. A diaphragm actuated by the difference in pitot, or ram, air pressure and static, or still air, pressure controls the free-fall valve. An airspeed pickup tube on the side of the fuselage brings pitot and static pressure into the automatic extension valve. When the airspeed is low, the diaphragm holds the free-fall valve open. Regardless of the position of the landing gear selector, the actuating cylinders cannot receive hydraulic pressure to unlock the downlocks and retract the wheels. When the airspeed reaches the predetermined value, the pressure differential across the diaphragm is great enough to close the free-fall valve so that pressure from the hydraulic pump can unlock the landing gear and retract it.

If the airspeed in flight drops below the preset value, the pressure across the diaphragm will be low enough to open the free-fall valve and allow the wheels to free-fall and lock in place.

A manual override on this valve allows the pilot to retract the landing gear when the airspeed is below the preset value if this is necessary to reduce drag after a short-field takeoff, and also to hold the gear up during slow flight for certain flight training maneuvers. This override system simply allows the pilot to hold the free-fall valve closed against the force produced by the airspeed valve.

LARGE AIRCRAFT RETRACTION SYSTEMS

The actual system for retracting and extending the landing gear on large aircraft is similar to that just described. However, there are several additional features and components used because of the size and complexity of the system.

Normally, large aircraft have wheel-well doors that are closed at all times the landing gear is not actually moving up or down. Sequence valves are used in the system to ensure the doors are opened before the landing gear is actuated.

Most large aircraft use mechanical locks to hold the landing gear in its UP or DOWN position. There must be a provision in these systems for the hydraulic pressure to release the locks before fluid is directed into the actuating cylinders.

The brakes are usually applied when the landing-gear selector is placed in the GEAR UP position. This prevents the fire hazard that would exist if the wheels were spinning while in the wheel wells as well as preventing the possibility of damaging the aircraft due to a spinning wheel.

Most of the large aircraft landing-gear systems use an orifice check valve in the fluid lines to the actuators. The weight of the landing gear dropping out of the wheel well could cause it to fall so fast that damage to the structure is a possibility. Therefore, the return flow from the actuator is restricted which prevents uncontrolled free fall. Unrestricted flow, however, is allowed into and out of the actuator when the gear is being retracted. [Figure 9-28]

EMERGENCY EXTENSION SYSTEMS

Retractable landing gear systems must have a means of lowering the landing gear if the primary method of lowering the gear fails. Because there are many methods used to actuate the landing gear, this dis-

Figure 9-28. Because of the size and weight of the gear on large transport aircraft, an orifice valve is necessary to prevent the gear from coming down too fast.

cussion will be general in nature. Emergency extension systems generally use a variety of methods to lower the gear. Some of the methods can include mechanical, alternate hydraulic, compressed air or free-fall techniques to lower the gear. In all cases, the emergency extension system's purpose is to release the up-locks and move the gear to the down and locked position.

LANDING GEAR SAFETY DEVICES

Retractable gear aircraft must have gear safety devices to prevent unwanted actions from taking place. Among these are devices to prevent raising the gear on the ground, devices to lock the gear down when the aircraft is sitting on the ramp, indicators to tell the pilot what position the gear is in, and provisions for centering the nose gear before it is raised into the wheel well.

SAFETY SWITCH

Most aircraft with retractable landing gear are equipped with a means of preventing the retraction of the landing gear while the aircraft is on the ground. If not, the landing gear would retract if the aircraft's hydraulic system was powered and the gear handle was moved to the up position. To prevent this from happening, a squat switch with a lever attached prevents the gear control handle from being placed in the up position when there is weight on the aircraft wheels. A switch connected to the aircraft's landing gear senses whether the aircraft is on the ground or in the air. If the aircraft is in the air, the switch pulls the lever away from the gear handle so it can be placed in the up position. Some aircraft are equipped with an override trigger that will manually pull the lever clear so the gear handle can be moved to the up position. The flight crew can use this provision in case the sensor switch malfunctions when the gear needs to be raised in flight.

GROUND LOCKS

Ground locks are used to secure the landing gear in the down position. These locks are generally removed manually by ground personnel. Ground locks are placed into position after the aircraft lands and are kept engaged until the aircraft is ready for the next flight. The locks generally consist of a pin inserted into the retraction mechanism in such a manner to block the retraction of the landing gear.

GEAR INDICATORS

Position indicators, generally located close to the landing gear lever, include green gear down and locked lights, a red gear door open light, and red a gear disagreement light, but may use a red gear unsafe/in transit light. Smaller airplanes do not use gear door open lights. Generally when the gear is up and locked, all the lights will go out signaling that the gear is up and locked. Switches or proximity probes at each gear position control the lights in the cockpit.

A warning system is used to inform the pilots if the gear is not down during a landing attempt. As the throttles are retarded toward closed, a switch is closed. If the gear is not down, a warning horn will sound until the throttles are opened or the gear is lowered. Although there are some variations from system to system, most are based on this basic idea of warning the pilot that the gear is not in the down and locked position while in a landing configuration.

NOSEWHEEL CENTERING

The nose wheel is equipped with centering cams located in the nose wheel shock strut. These centering cams center the nose wheel when the strut is extended after take-off. The nose gear will remain centered until the weight of the aircraft, upon landing, compresses the strut moving the centering cams away from their slots. This allows the wheel to turn as commanded by the steering tiller or the rudder pedals.

LANDING GEAR RIGGING AND ADJUSTMENT

The purpose of having retracting gear is to reduce parasite drag. If the system is out of adjustment so that one or more components are extended into the slipstream, there is more parasite drag than there should be. Therefore, proper rigging is essential

GEAR LATCHES

The landing gear latches, or up and down locks, must be adjusted so the gear will lock in the full up and down positions without binding. The gear must remain locked after the gear handle is moved to the off or neutral position. Many gear latches are mechanical devices that either lock the gear in place or lock the gear over center so it cannot become unlocked until the retraction system causes it to unlock and move to a new gear position.

GEAR DOOR CLEARANCES

To set or check the gear door clearances, the airplane is first placed on jacks. Once the main gear is properly adjusted in the retracted position, the gear doors can be adjusted. This is usually done by setting the retraction rod end at the gear door so the door will pull up tightly when the gear is fully up. Over-tightening may result in door buckling. However, if the door is too loose, it will gap in flight. All rod ends should be checked for adequate thread engagement, for safety, and tightness of jam nuts.

DRAG AND SIDE BRACE ADJUSTMENT

The drag and side brace adjustments must be set so the gear stops in the correct position and that the gear's alignment is set properly. Since landing-gear systems are so different in construction always consult the manufacturer's maintenance manual before performing any work on these components.

LANDING GEAR RETRACTION CHECK

Landing gear retraction checks are performed as a part of hundred-hour and annual inspections. They are also performed after replacement of landing gear components or after an event that could damage the gear such as a hard landing. To perform a retraction test, the airplane is placed on jacks. Next the main gear is inspected for full travel from gear-down to gear-up, which is accomplished by cycling the gear up and down while watching for looseness, binding, and chafing of the gear or related parts. The gear doors should be adjusted so that they pull up tightly when the gear is fully retracted. If the doors are too loose, they will gap in flight. Check all gear components for damage, for safety, and tightness.

During a retraction test, the emergency extension system (alternate extension system) should also be operated to ensure it is functioning properly. The gear-up warning horn can also be checked during a retraction test. With the gear in the up position and the throttles are pulled to the idle position, the horn should sound until the throttles are moved forward. The manufacturer's service manual should always be followed when performing a retraction test on an aircraft.

SECTION B

AIRCRAFT BRAKES

There are two basic types of disc brakes in use today. For smaller aircraft, on which brakes are used primarily as a maneuvering device and do not require the dissipation of great amounts of kinetic energy, the single-disc brake using spot-type linings has proven very effective. [Figure 9-29]

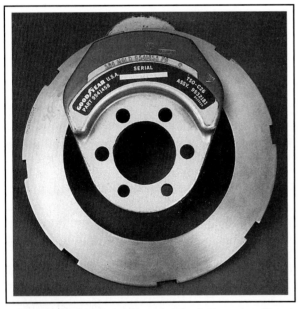

Figure 9-29. Most small general-aviation aircraft are equipped with single-disc brakes.

Large aircraft, whose brakes must dissipate tremendous amounts of kinetic energy at braking, use multiple-disc brakes. Brake rotors and stators are stacked together. The rotors turn with the aircraft wheel, while the stators are fixed to the wheel hub. When the brakes are applied, actuators extend, pushing the rotors and stators against each other. The frictional forces generated by the rotors and stators slow the aircraft. [Figure 9-30]

TYPES OF BRAKES

SINGLE-DISC BRAKES

Aircraft that must dissipate a relatively small amount of kinetic energy may use single-disc brakes with linings made of an organic or metallic friction materials. Many smaller general-aviation aircraft

Figure 9-30. Multiple-disc brakes are used in the wheels of large jet-transport aircraft.

use a brake disc that is bolted rigidly to the wheel and rotates between two brake linings mounted in a caliper that is free to move in and out but is restrained from rotating. When the brake is applied, a hydraulic piston moves out and clamps the linings to the rotor disc. [Figure 9-31]

The linings used in these brakes are usually made of an organic material that has a high coefficient of friction and good thermal characteristics. In the past, an asbestos compound was used, but because of the manufacturing and health hazards associated with asbestos, modern linings use other forms of friction material. When it is necessary to increase the capacity of these single-disc brakes, additional cylinders and lining pucks may be added. [Figure 9-32]

It is important to know the difference between organic and metallic friction materials. Pads that have visible metal in them are not necessarily metallic. Organic linings may or may not have metal incorporated in their construction. Organic linings are solid all the way through and are also known as "semi-metallic" linings. Metallic linings are linings

Figure 9-31. Most light general-aviation aircraft are equipped with a fixed-disc and a single-disc actuator.

Figure 9-32. Multiple-piston, single-disc brakes are used on some medium-size general-aviation aircraft.

made of metal which are bonded to a metal backing plate. The friction material is essentially a powdered metal that is melded through **sintering**. Sintering is a process in which heat and pressure weld the powdered metal together and to the backing plate, then are held in place with two rivets on the pressure plate, or back plate, of the brake. The other term that is used for metallic linings is "sintered" linings. There are two distinct layers when looking at a metallic brake lining: the pad and the plate.

MULTIPLE DISC BRAKES

When the tundra tire, with its large outside dimension and small inside diameter, became popular just before World War II, an effective brake was needed that would fit into the small diameter wheel. To gain the maximum amount of friction area, stacks of discs were keyed to a torque tube bolted to the wheel strut. Between each of these discs were thin bronze- or copper-plated steel discs keyed to rotate with the wheel. Hydraulic pressure forced an annular piston out against a pressure plate that clamped the discs together, thereby slowing the wheel. This type of brake had a very smooth action and provided high torque, but the thin discs had a tendency to warp, requiring frequent manual adjustment to compensate for disc wear. [Figure 9-33]

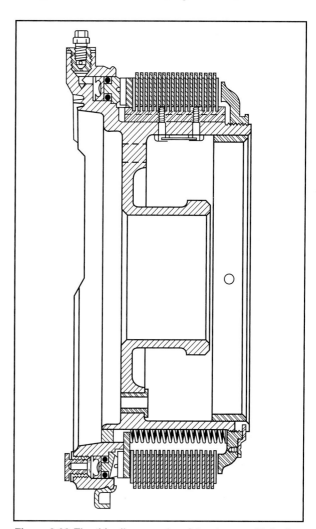

Figure 9-33. The thin-disc, annular-piston type multiple disc brake was one of the first types of multi-disc brakes.

The requirements for the dissipation of vast amounts of kinetic and heat energy brought about the use of multiple-disc brakes that use fewer but thicker discs. Instead of bronze or copper plating on the rotating disc, modern brakes use a sintered copper and iron base friction-material, bonded to the steel rotating discs. Rather than a singular annular piston, a series of small circular pistons force the pressure plate against the disc stack to apply the clamping force. Automatic adjusters compensate for disc wear each time the brake is applied, which eliminates the need for manual brake adjustment.

SEGMENTED ROTOR-DISC BRAKES
The segmented rotor-disc brake is a multiple-disc brake in which the rotating plates or rotors are made up of several segmented plates. The space between the rotor segments and brake linings allow this type of brake to provide improved brake cooling. Because this type of brake cools more efficiently than other types of multi-disc brakes, it can provide more efficient and longer braking action before the temperature limit of the brake is reached.

CARBON BRAKES
Many aircraft employ multiple-disc brakes that use discs made of pure carbon. The weight saving of carbon brakes is tremendous, and the heat-dissipating properties of the relatively thick carbon discs eliminate the need for extensive cooling periods after heavy braking. At present, these brakes are used primarily on higher-performance military aircraft and jet transport aircraft.

BRAKE CONSTRUCTION
The aviation maintenance technician can better maintain brakes by understanding the differences between single-disc, multiple-disc, segmented-rotor, carbon, and expander-tube braking systems.

SINGLE-DISC BRAKE
Brake systems utilizing a single disc have the disc attached rigidly to the wheel or floating with the attachment to the wheel by way of drive keys. The popular Cleveland brake for general-aviation aircraft uses a **fixed brake disc** bolted to the inside wheel half. The caliper for this brake consists of a cylinder assembly holding one or more pistons that are actuated by hydraulic pressure to squeeze the disc between two brake linings, one of which is riveted to the pressure plate and the other to the back plate. [Figure 9-34]

The entire caliper is free to move back and forth on two anchor bolts that ride in holes in the torque plate that is bolted to the landing-gear axle. [Figure 9-35]

A **floating-disc brake system** utilizes a forged-steel disc with a smooth ground surface that is keyed to rotate with the wheel. The disc is free to move in

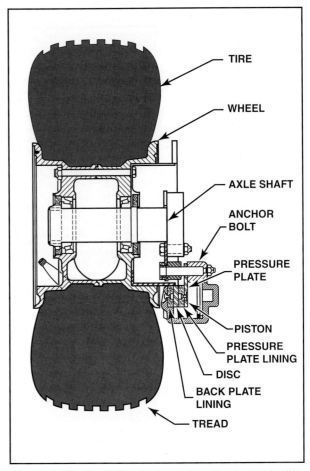

Figure 9-34. A sectional view of a fixed, single-disc hydraulic brake with a movable caliper, as found on Cleveland brakes. This type of brake is very common on light aircraft.

and out enough to prevent binding as the brakes are applied. Anti-rattle clips apply a spring pressure to the outside of the disc to hold it centered so it will not rattle against the wheel as it rotates.

A cast-aluminum or magnesium alloy housing is bolted to a flange on the aircraft landing-gear strut, and lining cavities in the housing hold the fixed anvil lining on one side of the rotating disc and the movable piston-side lining on the opposite side.

A piston cavity, or cylinder, holds a machined aluminum alloy piston that is sealed in the cavity with an O-ring packing around its circumference. The cavity is closed with a cylinder head that is either threaded into the housing or held in place by snap-ring or machine screws. An O-ring seal keeps the cylinder head from leaking. When hydraulic fluid is forced into the cylinder, the piston pushes the movable lining out and clamps the rotating disc between the two linings.

Aircraft Landing Gear Systems

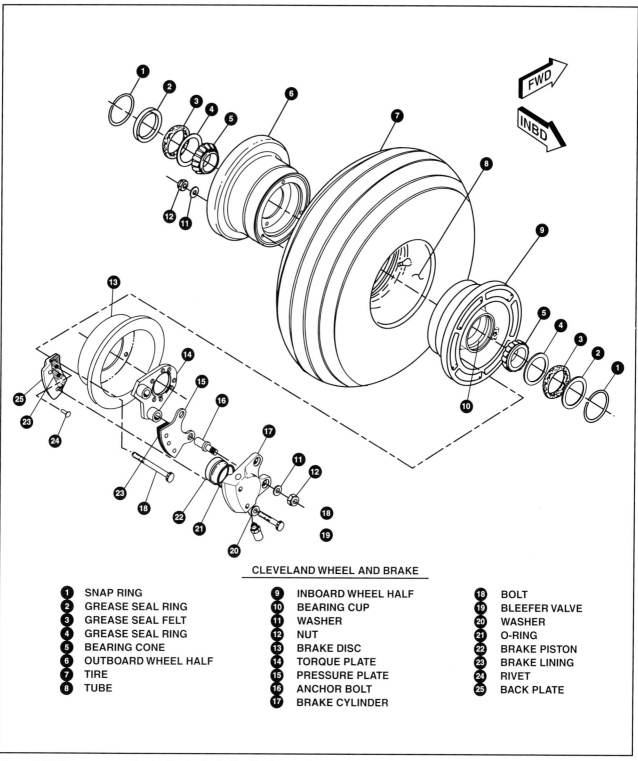

Figure 9-35. The components of a fixed, single-disc brake are shown here. This Cleveland brake assembly utilizes a single piston.

Modern organic linings are made without the use of asbestos. New materials are being developed that have high coefficients of friction, good wearing capabilities, and excellent thermal characteristics, and also cause no health problems during manufacture.

Particles of brass, copper, or copper wool may be embedded in the brake lining material to provide the exact friction characteristics needed. To give the lining the required strength, either a piece of steel mesh may be embedded near the back, or the lining may be bonded into a steel cup. When installing the linings, they must be installed with the smooth side next to the disc, and with any lettering or the steel cup away from the disc. [Figure 9-36]

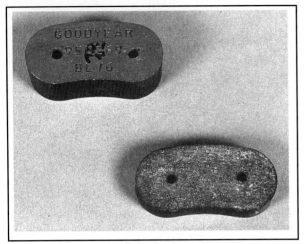

Figure 9-36. When installed on an aircraft, the lettered side of the brake lining must be installed facing away from the disc.

Many models of single-disc brakes incorporate automatic adjusters to maintain a constant clearance between the lining and the disc when the brake is released. Pistons on brakes that include automatic adjusters use a pin with a large head held centered in the piston cavity by a return spring and a spring retainer. The end of the pin extends through the cylinder head where it is held by a friction grip collar. [Figure 9-37]

When the brake is applied, hydraulic pressure forces the piston against the lining, resulting in the two linings clamping the disc. When the piston moves out, the return spring collapses enough for the spring retainer to press against the head of the pin and pull the pin stem through the grip collar. The more the lining wears, the more the pin pulls through the collar. When the pressure is released from the brake, the return spring pushing against the head of the pin moves the piston back until the bottom of the piston cavity rests against the head of the pin. This spacing will be maintained throughout the life of the brake, automatically adjusting itself each time the brake is applied.

MULTIPLE-DISC BRAKE

A high-strength steel torque tube bolts to the center of the landing-gear strut, and the brake housing bolts to the torque tube. Keyed teeth on the outside circumference of the torque tube engage slots in the pressure plate and the stationary discs to prevent rotation. The back plate bolts rigidly to the torque tube to form the clamp for the stack of discs. [Figure 9-38]

The brake housing is usually made of cast aluminum or magnesium alloy and attaches to the strut by bolts through the torque tube. Cavities in the

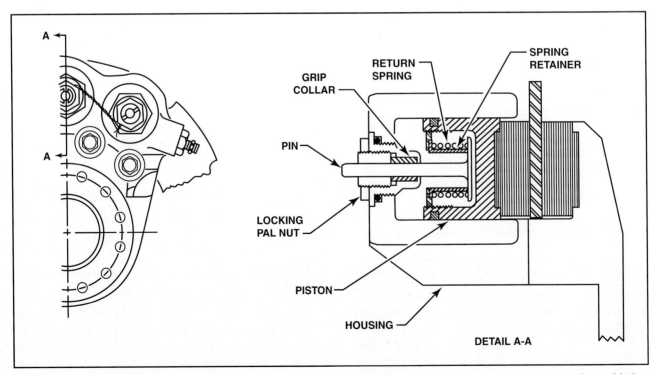

Figure 9-37. Automatic adjusters are used on many models of single-disc brakes. This device maintains the proper disc to friction puck spacing as the puck wears with use.

Figure 9-38. Multiple-disc brakes are used in larger aircraft, such as this executive jet aircraft.

housing hold the pistons that provide the clamping action as they force the pressure plate against the stack of discs. [Figure 9-39]

Figure 9-39. The brake pucks on a multiple-disc brake are enclosed in a cast metal housing that contains multiple brake pistons.

Drilled passages within the housing connect the cylinders to provide hydraulic pressure to the brakes. Some housings have each alternate cylinder connected to one hydraulic system and a backup system connected to the other cylinders. A brake with this arrangement is said to have "A" and "B" systems. This arrangement supplies adequate pressure from either the main or the backup hydraulic system for brake application.

Space is provided in the housing for a series of return springs and automatic adjusters which pull the pressure plate back from the disc stack each time the brake is released. Each of the cavities, or cylinders, in the housing is fitted with a machined-aluminum alloy piston. These pistons are sealed in the cylinder with an O-ring packing, backed up with a spiraled Teflon® backup ring. A composition insulator is attached to the face of each piston, where it presses against the pressure plate. This insulator minimizes the transfer of heat from the disc stack to the piston, preventing adverse effects on the fluid and the seals.

Almost all multiple-disc brakes use some form of return system to pull the pressure plate back from the disc stack when the brake is released. These return systems also serve as automatic adjusters to maintain a constant clearance between the discs as they wear.

A machined-steel spring housing slips through a hole in the brake housing to provide a base for the return spring. A heavy coil spring and spring holder is located inside the spring housing and held in place by a retaining ring when the brake is released.

The head of the adjusting pin engages a slot in the pressure plate, allowing its stem to pass through a hole in both the spring housing and the spring holder. A grip and tube subassembly slips over the pin and is held in place with a nut.

When the brake is applied, the pistons force the pressure plate toward the discs and clamp the disc stack against the back plate. The grip around the tube inside the spring holder forces the holder to compress the return spring until the holder bottoms against the spring housing. As the discs wear, the pressure plate moves further away from the bottom of the spring housing, and the tube is forced to slip through the grip.

When the brakes are released, the return spring forces the spring holder back against the retaining ring, and the grip around the tube is sufficiently tightened to pull the pin back, moving the pressure plate away from the disc stack so the clearance will remain the same as the stack wears. [Figure 9-40]

The pressure plate for multiple-disc brakes is a special stationary disc made of high-strength steel. It has steel wear pads riveted to its inner surface where it contacts the rotating disc. The back plate is also made of high-strength steel, and also serves as a stationary disc. But instead of being keyed to the torque tube so it can move back and forth, it is bolted to the housing through the torque tube, and it serves as the rigid member of the clamp.

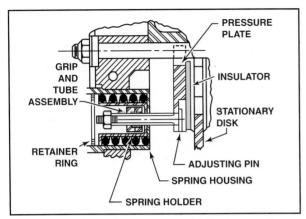

Figure 9-40. Automatic adjusters maintain disc spacing as the pucks wear on multiple discs.

When the brakes are applied, the pistons act against the pressure plate and force the disc stack over against the back plate. Wear pads are riveted to the face of the back plate where it contacts the rotating disc.

The stationary discs are keyed to the torque tube so they are free to slide back and forth as the brakes are applied, but they cannot rotate. Many brakes have steel wear pads riveted to each side of the stationary disc to extend their life. These pads may be replaced rather than replacing the entire disc. [Figure 9-41]

Figure 9-41. The stationary discs are keyed to a torque tube that allows them to move laterally, while not allowing them to rotate with the wheel and rotating disc assembly.

Narrow slots are often cut into the stationary discs to allow for expansion as they get hot. These expansion slots prevent the disc from warping, which could cause the brakes to drag.

A rotating disc is placed between each of the stationary discs. These discs are driven by a tang (projecting prong) on the disc, which fits into a steel-reinforced slot in the wheel. They can also be driven by a steel key attached to the inside of the wheel that engages a slot in the outer rim of the disc. [Figure 9-42]

Figure 9-42. The rotating discs are keyed into the wheel by disc-drive tangs.

The rotating discs are constructed of steel and have a special friction surface made of sintered material bonded to their surface by heat and pressure. Brake disks have expansion slots cut into them to prevent warping from excessive heat. However, if the parking brake is set when the brakes are hot, the disks may warp. Setting the parking brake clamps the disks together, which does not allow heat to dissipate.

CARBON BRAKE

This is a multiple-disc brake that uses a thick carbon disc as the brake rotor. These brakes can absorb tremendous amounts of kinetic energy and yet have relatively low weight. The thick, black carbon disc dissipates heat extremely well, is very light, and has excellent wear properties. [Figure 9-43]

EXPANDER TUBE BRAKE

One of the early types of non-servo brakes used on airplanes as small as the Piper Cub and as large as the Boeing B-29 was the expander tube brake. This type of brake incorporates a flat synthetic rubber tube installed around the brake body on the axle. The tube was filled with hydraulic fluid under pressure from the brake master cylinder or from the power brake-control valve. As the fluid filled the tube, it forced asbestos-compound blocks against the inner surface of a rotating iron drum. When the

Aircraft Landing Gear Systems

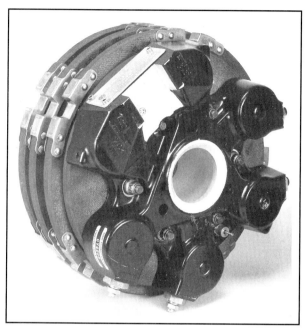

Figure 9-43. A typical carbon brake setup appears similar to a multiple-disc system of steel discs. The carbon discs are thicker than the steel discs but have better wear and heat dissipation characteristics.

brakes were released, flat steel springs in the brake body pressed the blocks back against the expander tube and away from the drum. [Figure 9-44]

BRAKE ACTUATING SYSTEMS

Independent master cylinder, boosted, power, and emergency brake systems each include unique design and operating characteristics. The technician should be aware of these differences when servicing brake systems and follow the manufacturer's service recommendations.

INDEPENDENT MASTER CYLINDERS

As brake design has evolved, so has brake actuation. Most of the early drum-and-shoe brakes were mechanically operated by a flexible steel cable pulling on a lever inside the brake. This lever actuated a cam to move the lining against the drum. The cables were pulled by a "Johnson bar," a long lever which, if pulled straight back, applied both brakes. If the Johnson bar was pulled back and to one side, it applied only the brake on that side. This system gave the pilot some degree of independent braking.

In order to increase the pressure applied to the brake linings, hydraulic cylinders soon replaced the mechanical cams, and individual master cylinders were used to apply pressure to the cylinders inside the wheels when the pilot pulled back on the brake lever, or pushed on the brake pedals.

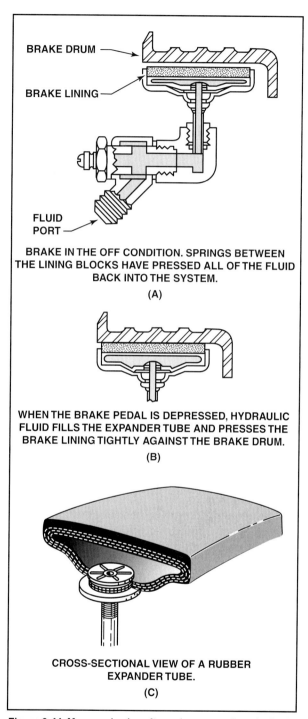

Figure 9-44. Many early aircraft used an expander tube braking system. Hydraulic fluid pressure was used to inflate the tube, thereby pressing it against the rotating drum.

In the quest for simplicity, hydraulic brakes for some smaller aircraft use a sealed hydraulic system consisting of a diaphragm-type master cylinder that is connected via the appropriate tubing to the actuator in the wheel and is filled with hydraulic fluid. When the brake pedal is depressed, pressurized fluid flows from the master cylinder to the wheel cylinder to apply the brakes. This type of system has been used on many of the smaller aircraft for independent braking, and the pilots operate the pedals with their heels. [Figure 9-45]

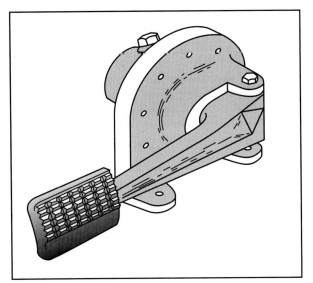

Figure 9-45. Early aircraft used heel operated brake pedals to operate the master cylinder.

Later, this same system was used on light, nose wheel-type airplanes with both brakes connected to one master cylinder. The brakes were operated by a cable pulled from a handle under the instrument panel. In order to apply the parking brake with this type of system, a shutoff valve is located between the master cylinder and the wheel unit. The brakes are applied and the shutoff valve is closed to trap the pressure in the line.

Larger aircraft require higher hydraulic pressure and more fluid for their brakes. In order to prevent brake dragging due to the thermal expansion of the fluid, there is a need to vent this fluid to the atmosphere when the brakes are not applied. There are many types of vented master cylinders, but all of them have the same basic components. One of the more popular vented master cylinders attaches directly to the rudder pedal. The body of the master cylinder serves as the reservoir for the brake fluid, and it is vented to the atmosphere through a vent hole in the filler plug. The piston is attached to the rudder pedal so that when the pilot pushes on the top of the pedal, the piston is forced down into the cylinder. When the pedal is not depressed, the return spring forces the piston up so the compensator sleeve will hold the compensator port open. Fluid from the wheel unit is vented to the atmosphere through the compensator port. When the pedal is depressed, the piston is pushed away from the compensator sleeve, and a special O-ring and washer (the Lock-O-Seal) seals fluid in the line to the brake. The amount of pressure applied to the brake is proportional to the amount of force the pilot applies on the pedals. When the pedal is released, the compensator port opens and vents the brake line fluid into the reservoir. If fluid passes by the piston seal rather than being forced out of the system under pressure, the brakes will not hold pressure, which could result in fading brakes. [Figure 9-46]

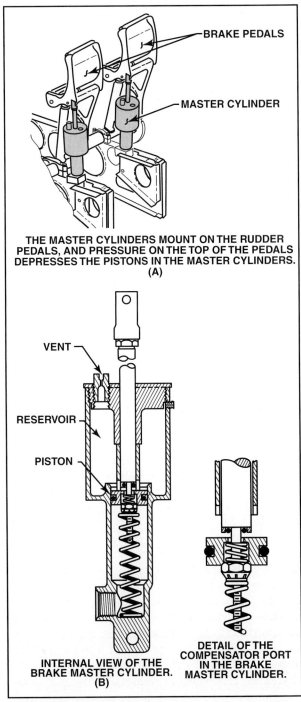

Figure 9-46. Master cylinders must be vented to allow for the expansion of the fluid as it heats. A compensating port allows venting without admitting air into the system.

Aircraft Landing Gear Systems

The parking brake for this type of master cylinder is a simple ratchet mechanism that holds the piston down in the cylinder. To apply the parking brake, the pedal is depressed and the handle pulled, which locks the piston. To release the brake, the pedal is depressed, then the ratchet can be released.

BOOSTED BRAKES

Midsize airplanes require more braking force than can be applied with an independent master cylinder, yet do not require the complex system of a power brake. The boosted brake was designed to fill this need.

The boosted brake master cylinder is typical for this type of operation. The boosted brake cylinder is mounted on the rudder pedal and attached to the toe-brake pedal in such a way that depressing the pedal pulls on the rod and forces fluid out to the brake cylinder. If the pilot needs more pressure on the brakes than can be applied with the pedal, the pilot continues to push. As the toggle mechanism straightens out, the spool valve is moved over so it will direct hydraulic system pressure behind the piston to assist the pilot in forcing fluid to the brake. When the pedal is released, the spool valve moves back to its original position and vents the fluid on top of the piston back to the system reservoir. At the same time, the compensator poppet unseats and vents the fluid from the brakes to the reservoir. [Figure 9-47]

POWER BRAKES

Most large aircraft use brakes operated by pressure from the main hydraulic system. This cannot be done by simply directing part of the system pressure into the brake-actuating unit, because of the high pressures used in large aircraft hydraulic systems. The brake application must be proportional to the force the pilot exerts on the pedals, and the pilot must be able to hold the brakes partially applied without there being a buildup of pressure in the brake lines. Because power brakes are used on airplanes so large that the pilot has no way of knowing when one of the wheels is locking up, there must also be some provision to prevent wheel skidding. The pressure supplied to the wheel must be lower than the pressure of the main hydraulic system, so a pressure-reducing or deboosting system must be incorporated in the system. Because the wheels are susceptible to damage, provision should also be made to lock off the fluid from a wheel in the event a hydraulic line is broken. Finally, there must be an emergency brake system that can actuate the wheel units in the event of a failure of the hydraulic system. [Figure 9-48]

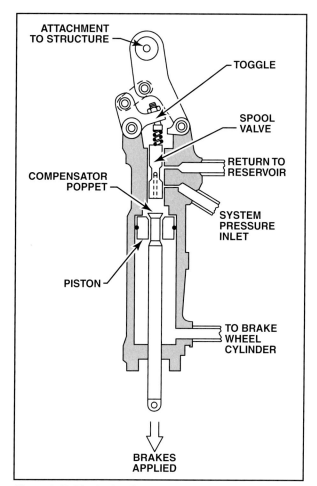

Figure 9-47. The hydraulic system pressure boosted brake master cylinder has a toggle that vents boost pressure to the backside of the piston to assist the pilot in applying the brakes.

In Figure 9-48, a simplified schematic of the brake system of a large jet transport-type aircraft is shown. The brakes get their fluid from the main hydraulic system through a check valve. An accumulator stores pressure for the brakes in the event of a hydraulic system failure. The pilot and copilot operate power brake control valves through the appropriate linkages. These valves are actually pressure regulators, which provide an amount of pressure to the brakes that is proportional to the force the pilot applies to the pedals. Once the desired pressure is reached, the valve holds it as long as that amount of force is held on the pedals. In large aircraft, the pilot does not have a feel for every wheel, so an anti-skid system is installed to sense the rate of deceleration of each wheel and compare it with the maximum allowable rate of deceleration. If any wheel attempts to slow down too fast, as it does at the onset of a skid, the anti-skid valve will release the pressure from that wheel back into the system return manifold, thus preventing a skid by that wheel.

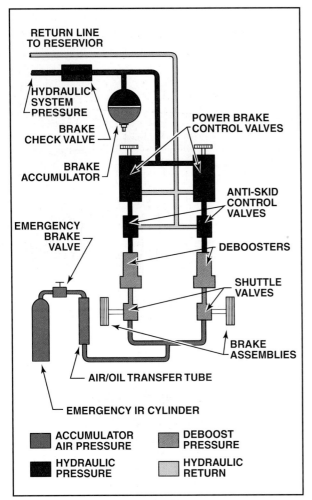

Figure 9-48. Large jet transports utilize boosted brakes that incorporate anti-skid into the system. Anti-skid is essential because the pilot cannot feel the wheels locking up.

The pressure applied by the brake control valve is too high for the proper brake application, so a debooster is installed in the line between the anti-skid valve and the brake. This lowers the pressure and increases the volume of fluid supplied to the wheel units.

Power brake control valves are used on many large aircraft. When the brake is applied, the plunger depresses and moves the spool over to connect the pressure port to the brake line. Fluid under pressure is also directed behind the spool to move it back when the desired brake pressure has been reached; thus, pressure to the brake will not increase, regardless of how long the pilot depresses the pedal. If more pressure is required, the pilot presses on the brake pedal harder, further compressing the plunger spring and allowing more fluid to flow to the brake. When the pedal is released, the return spring forces the spool back, and fluid flows from the brake into the system return line. [Figure 9-49]

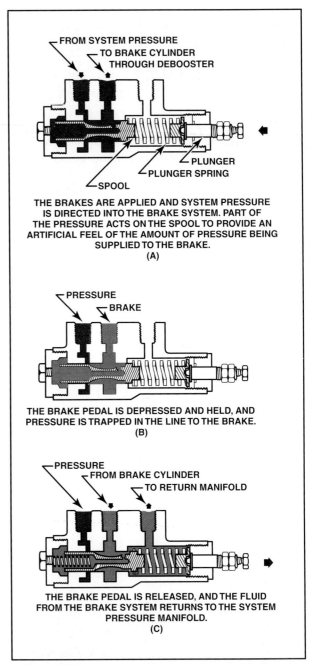

Figure 9-49. Large aircraft use internal spring-type power brake control valves.

Optimum braking is obtained when there is just enough pressure supplied to the brake to hold the wheel on the verge of a skid, but not allow the skid to develop. This is done by the anti-skid system that is discussed in Chapter 10 of this text.

Because hydraulic system pressure is normally too high for brake action, **deboosters** are installed between the anti-skid valve and the wheel cylinders. Deboosters are primarily pressure-reducing valves that operate on the basis of a pressure differential being produced by varying the area of the pistons within the deboosters. [Figure 9-50]

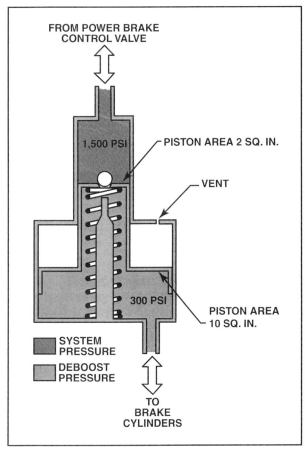

Figure 9-50. Brake debooster valves are used to decrease system pressure and increase system volume of hydraulic fluid flow.

The principle of the debooster is illustrated in Figure 9-50. System pressure of 1,500 psi is applied to a piston having an area of two square inches, which generates 3,000 pounds of force. The other end of the piston has an area of ten square inches, which produces a pressure of 300 psi, thereby reducing or deboosting system pressure. The other function of the debooster is to increase the volume of fluid going to the brakes. When the 1,500 psi system pressure moves the small piston down one inch, two cubic inches of fluid is used, but this same travel of the larger piston moves ten cubic inches of fluid to the brakes. The debooster has a pin-operated ball valve that allows brake fluid to be replenished if there should be a leak in the line. If the debooster piston should move down far enough for the pin to push the ball off its seat, fluid under system pressure will flow into the lower chamber and replenish the lost fluid. As soon as enough fluid enters the chamber, the piston will rise and the ball will reseat.

In the event of a fluid leak downstream of the debooster valve, a lockout debooster will allow the piston to go all the way to the bottom. The pin pushes the ball off its seat, but the spring-loaded valve prevents fluid from entering the lower chamber until the reset handle is lifted. This prevents pumping all of the hydraulic system fluid overboard through a brake line leak. [Figure 9-51]

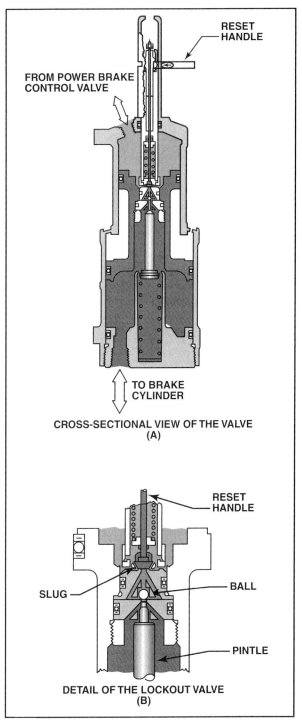

Figure 9-51. The lockout debooster valve prevents loss of the entire hydraulic system in case of a brake line rupture downstream of the debooster.

EMERGENCY BRAKE SYSTEM

In the case of a total failure of the hydraulic system, the pilot of most large aircraft can operate a pneumatic valve on the instrument panel and direct compressed air or nitrogen into the brake system. [Figure 9-52]

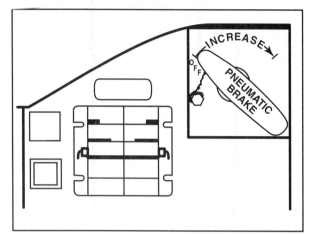

Figure 9-52. In the event of the loss of hydraulic pressure to the brake system, most aircraft have a pneumatic emergency system as a backup.

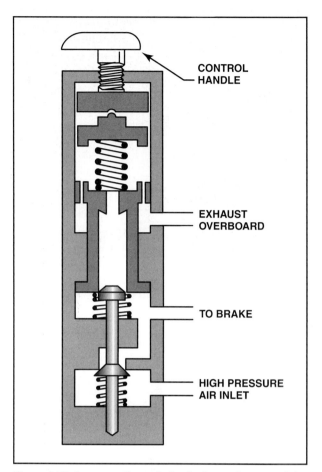

Figure 9-53. A pneumatic control valve is used to control the amount of air pressure applied to the brake system.

When the handle is turned, a regulator that controls air pressure to the brake is adjusted. As pressure on the control spring is increased, the piston moves downward and opens the inlet valve. When sufficient pressure reaches the brake line, the piston moves up against the force of the control spring and shuts off the inlet valve. The compression of the spring determines the amount of pressure supplied to the brake. When the brake handle is rotated in the direction to release the brakes, the air is exhausted overboard. [Figure 9-53]

Rather than allow compressed air to enter the wheel cylinders, which would require the entire brake system to be bled, the emergency air may be directed into a transfer tube. The air forces hydraulic fluid from this tube into the brake system, thereby producing braking action.

BRAKE INSPECTION AND SERVICE

Inspection and servicing of brakes is done both with the brakes installed on the aircraft and with the brakes removed. One of the major brake servicing jobs done by a technician is the replacement of brake linings. This job differs from system to system, and the manufacturer's instructions should always be followed.

ON AIRCRAFT SERVICING

Aircraft brakes are designed and built to provide excellent service, and they normally require little attention while they are on the airplane. Several important inspections can be performed with the brakes installed. One of the most important inspections is the wear of the brake lining material.

The amount of **lining wear** on Goodyear single-disc brakesis usually indicated by the length the return pin of the automatic adjuster sticks through the bushing. The manufacturer's service manual specifies the minimum extension allowed before the brake should be disassembled and the linings replaced. [Figure 9-54]

If the brake does not have automatic adjusters, the lining wear can be measured by applying the brake and measuring the distance between the disc and the housing. If this measurement is greater than that allowed, the linings have worn excessively and must be replaced. [Figure 9-55]

Cleveland brake linings should be replaced when they have worn to a thickness of 0.100 inch or less. Part of the lining is usually visible and its thickness can be compared with the diameter of a number 40

Aircraft Landing Gear Systems 9-31

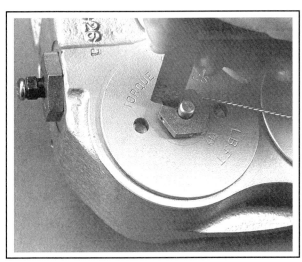

Figure 9-54. On Goodyear single-disc brakes, the amount of wear on the linings is determined by the amount of extension of the return pin of the automatic adjuster.

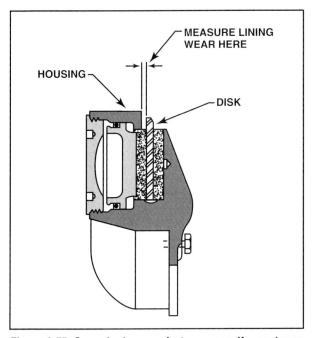

Figure 9-55. Some brake manufacturers specify maximum lining wear in relation to the distance between the disc and the housing with the brakes applied.

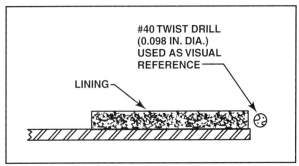

Figure 9-56. Cleveland brake linings are replaced when they get down to 0.100 inch thick. A number 40 twist drill is .098 inch in diameter and serves as a good gauge when inspecting.

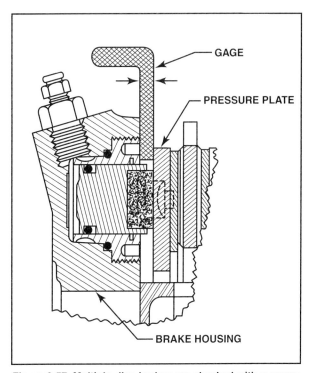

Figure 9-57. Multiple-disc brakes are checked with a gauge measuring the distance between the housing and the pressure plate with the brakes applied. Excess space dictates disassembly of the brakes in order to inspect individual discs for wear.

twist drill. The linings must never be allowed to wear to this thickness (0.098 inch). [Figure 9-56]

The total stack wear of the discs in a multiple-disc brake may be checked by applying the brake and measuring the amount of clearance between the back of the pressure plate and the edge of the housing. Excessive clearance indicates that the total stack wear exceeds the allowable limit, and the brakes must be disassembled and each disc checked for condition and the amount of wear. [Figure 9-57]

Air in the system or **deterioration of the flexible brake lines** cause spongy braking action. Bleeding the brakes removes air from the lines, which restores proper brake function. When bleeding brakes, the airframe manufacturer's service information must be followed in detail.

When the brake master cylinder is fully released in aircraft not equipped with power brakes, there is a direct passage from the brake cylinder through the compensator port to the reservoir. This prevents any pressure buildup from heat, which may cause the brakes to drag. To bleed this type of brake, the screw is removed from the bleeder valve, and a brake bleeding pressure pot is connected to the brake with a flexible hose. Before connecting the line to the bleeder valve, all air should be bled from the line. [Figure 9-58]

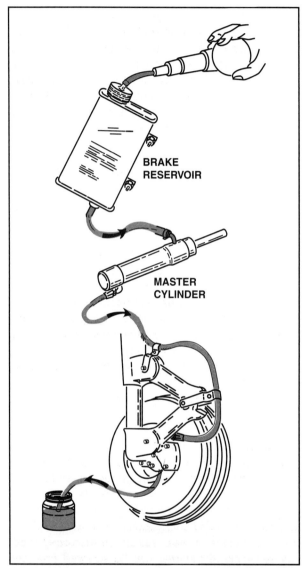

Figure 9-58. The pressure method of bleeding the brakes involves using brake fluid under pressure. The fluid is introduced at the brake end of the system and flows through the system to the brake reservoir.

CAUTION: Be sure that the pressure pot is filled with the proper type of hydraulic fluid. Improper fluid can cause serious damage to the system.

A flexible hose should be attached to the brake reservoir vent, and the other end of the hose placed in a clean container. With pressure applied from the pot, open the bleeder valve to force fluid up through the brake, the master cylinder, and the reservoir into the container. Allow the flow to continue until there are no more air bubbles. The bleeder valve is then closed, the valve on the pressure pot is shut off, and the line is removed from the brake. The bleeder screw is replaced, and the line removed from the reservoir. Finally, the reservoir is checked for the proper amount of fluid.

Aircraft without power brakes may also be bled by forcing trapped air out of the system at the wheel cylinder. One end of a section of flexible tubing is placed over the bleeder valve on the brake, and the other end of the tubing immersed in a container of clean hydraulic fluid. After making certain that the reservoir is full of fluid, the pedal on the master cylinder is depressed. The piston should be held down and the bleeder valve cracked open on the brake unit at the wheel. The piston in the master cylinder should be allowed to go all the way down, and then the bleeder valve closed before releasing the pedal at the master cylinder. This procedure should be continued, while being certain to keep the reservoir full until fluid flows from the brake with no bubbles. [Figure 9-59]

On aircraft equipped with power brakes, the bleeder screw must be removed from the brake's bleeder valve and a length of flexible hose installed. The end of the hose is placed in a clean container of hydraulic fluid. Then, the bleeder valve is opened, and with the hydraulic system pressure in the proper operating range, the brakes are applied very carefully. The fluid is allowed to flow from the brake until there are no more air bubbles. Then the bleeder valve is closed, the hose removed, and the bleeder screw replaced.

It is especially important that the airframe manufacturer's service manual be followed in detail when power brakes are bled, due to the many installed devices such as anti-skid components, deboosters and hydraulic fuses. The bleeding procedure is not complete until both the main and the emergency, or backup, systems are free of air.

The **quantity of fluid and type** must be noted and followed. The hydraulic system of modern aircraft may use either a phosphate-ester-base synthetic fluid such as Skydrol®, or mineral-base, MIL-H-5606, fluid that is commonly called "red oil." The proper fluid must be used, because use of the wrong type will damage the seals it contacts and will contaminate the entire system. If improper fluid is ever used, the system will have to be drained and flushed and all of the seals replaced.

Aircraft Landing Gear Systems

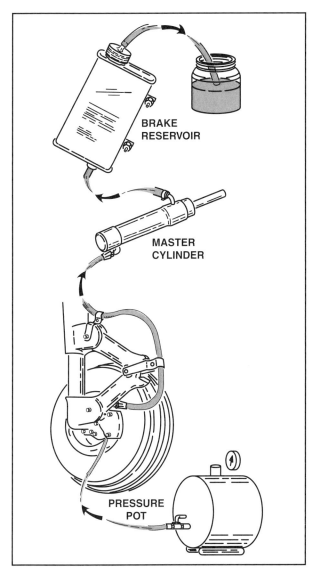

Figure 9-59. As an alternative, the brakes can be bled by forcing the air out of the system at the wheel cylinder. In this method, the master cylinder is used to produce the pressure to force the fluid through the system.

The entire system must be **checked for leaks**. Brake system leakage can be an indication of hidden or impending damage, and the cause of the leakage must be found and corrected.

Leaking fluid-line connections deserve special care, as tightening the fitting is not necessarily a sure way to stop the leak. If a fitting is leaking, the pressure should be removed from the system and the fitting checked for proper torque. If it is obviously loose, it should be tightened and re-checked for proper torque under pressure.

If the connection is made with an MS-type flareless fitting, it is important that no attempt be made to stop any leak by tightening the fitting. Over-tightening can cause the leak to increase. If the fitting is leaking, the pressure should be removed from the system, the fitting loosened, and the sealing surfaces checked. The fitting should be replaced if it shows any indication of damage.

Almost all brake housings are made of either cast-aluminum or cast-magnesium alloy, and it is possible that the porosity of the casting could cause a seep-type leak. If it is determined that this is the cause of fluid on the outside of the brake, the casting may be returned to the manufacturer where it can be impregnated with a resin that will seal all of the pores and stop the seeping.

All of the components of an aircraft landing gear are subject to hammering action, and it is vital that all of the bolts have the **proper torque**, because any looseness will accelerate wear and damage. After a bolt has been properly torqued, a small touch of paint should be placed across the end of the bolt and the nut or across the bolt head and the adjacent structure. A break in the paint will show if there has ever been any motion between the fastener and the structure. If this paint line is unbroken, it is reasonably certain that the torque is still adequate. If there is no paint mark, the brake attachment bolts should be checked for the proper torque. A torque wrench of known accuracy and recent calibration should be used to be sure that the bolt has been tightened to the torque recommended in the aircraft service manual.

OFF AIRCRAFT SERVICING

When a brake has been removed from an aircraft and brought into the shop for inspection and repair, a thorough inspection of all components can be conducted. The first items to be inspected are the **bolts and other threaded connections** for condition. Bolts that show signs of wear should be discarded, and any self-locking nut that can be screwed onto its bolt with less than approximately six inch-pounds of torque should be replaced.

When replacing any hardware, it is possible that some of the nuts and bolts may have a manufacturer's part number and yet closely resemble a standard AN or MS part. Using any parts other than those specifically called for in the parts manual can jeopardize the safety of the aircraft. While parts may look alike, those parts bearing a special parts number have either been manufactured to a closer tolerance or made of a material different from that of a standard part. It is also possible that parts having a special number may have had a more intensive inspection than that of a similar-appearing standard part.

The condition of the threads in the housing where the inlet adapter and the bleeder adapter are installed should be checked very carefully. There must be no burrs in the threaded area, or in the chamfer where the O-ring packing seals.

The entire **disc stack should be inspected** for evenness of wear. If there is an indication that the stack has worn more in one area than in another, there is a good possibility that some of the automatic adjusters have not been pulling back on the pressure plate as they should, and these adjusters should be carefully examined.

The stationary discs should be checked for indications of cracks. The cracks are most likely to form around the end of the relief slot, and any crack is cause for rejection of the disc.

The wear pads should be checked for excessive or uneven wear. The wear pads heat up from the brake action and have a tendency to curl away from the disc. The edges between the rivets will wear more than any other part of the pad. The total thickness of the pad and the disc should be measured. If it is less than that allowed, but the disc is in otherwise good condition, all of the pads may be removed and replaced. When riveting the new pads onto the disc, the manufacturer's recommendations should be followed in detail regarding the tools to use and the tightness of the pad to the disc. [Figure 9-60]

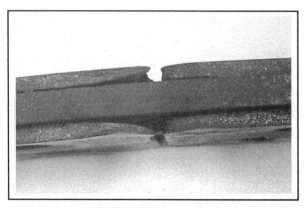

Figure 9-60. Heat will sometimes cause the pads on multiple-disc brakes to pull away from the disc and wear along their edges.

The disc slots should be checked for wear. If the disc has been allowed to hammer against the torque tube, the key slots may be battered until their width is excessive. The widened slot will allow more battering and will accelerate wear to the disc, which can damage the torque tube. If the slots are wider than allowed by the service manual, the disc must be replaced.

The disc should slide freely on the torque tube, and replaced it if it binds in any of the key slots or anywhere along its inner circumference. The end of the relief slots in the rotating discs should be inspected for any indication of cracks. If cracks are found, the disc(s) must be replaced.

It is possible for the sintered material on the rotating disc to become glazed if the brake surface has become locally overheated without the interior of the disc dissipating the heat. This can occur when taxiing a turbine-powered aircraft with the engine operating at high rpm and the brakes applied in short jab-like applications. Glaze is a hard, shiny, glass-like surface on the sintered material that will cause the brakes to chatter or squeal. It may be removed by abrasive blasting the surface until it is restored to its original appearance.

The disc should not have any key slot or drive tang wear. If the width of the slot or tang is worn beyond the tolerance allowed in the manufacturer's service manual, it must be replaced. The pressure plate is inspected in the same way as a stationary disc. It must be replaced if it is cracked, if the key slots are worn more than is allowed, or if it does not slide freely on the torque tube. [Figure 9-61]

Figure 9-61. The pressure plate should be inspected for cracks, worn key slots and freedom of movement.

The back plate is inspected in much the same way as the pressure plate. Worn wear pads may be replaced by riveting on an entire new set, and dishing may be straightened if it is within the allowable limits. The back plate serves as a stationary member against which the pressure plate forces the disc stack. [Figure 9-62]

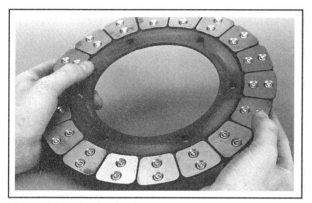

Figure 9-62. The back plate should be inspected for the same items as the regular discs and the pressure plate. Wear pads should be replaced as necessary with a new set of pads. Warped discs and backing plates may be straightened if within limits set by the manufacturer.

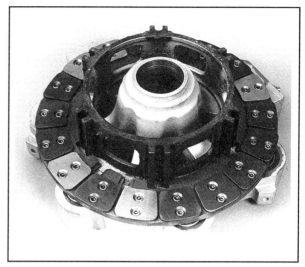

Figure 9-63. The torque tube must be checked for proper dimensions, wear, and for cracks before being returned to service.

The **automatic adjuster return pin** should be inspected for condition. The return pin must not be bent or damaged in any way, or have any damaged threads, nor may there be any nicks or scratches on the stem or under its head. It should be magnetically inspected for cracks, and the pin rejected if any are found.

The grip and tube assembly should be inspected to be sure the grip will move through its full travel with the force specified in the overhaul manual. In addition, inspect to be sure the grip is positioned the proper distance from the end of the tube.

The technician should **inspect the torque tube** for indication of wear or burrs, and scratches. The tube should be inspected for cracks using magnetic particle inspection, and the tube rejected if cracks are found that exceed the limits set in the manufacturer's service manual. The key sections of the torque tube are dimensionally checked, and replaced if any of the keys are worn beyond the allowable limit. [Figure 9-63]

The entire housing should be inspected for cracks using a fluorescent penetrant inspection method. The housing should be replaced if cracks are found. A visual check of the housing should be made for scratches, tool marks or evidence of corrosion. If any damage of this nature is found, it must be dressed out, removing as little material as possible. The cleaned and dressed area should be treated with Alodine or some similar chemical to form a protective film on the surface. The piston cavity diameters should be measured and the housing replaced if any of the cavities are worn beyond the allowable limits. The pistons should be visually checked for evidence of corrosion, scratches or burrs, and dimensionally checked to be sure they have not worn beyond the limits allowed.

The insulators on the bottom of the pistons should be checked to be sure they are not cracked or damaged and that they have sufficient thickness. Minor blisters in the material may be smoothed with a file.

The condition of the seals is very important to the operation of a braking system. Seals are some of the smallest, but most important, components in an aircraft brake system. In modern brake construction, most of the seals used are of the O-ring type, and much research and engineering have gone into the composition of the material used in their construction and in all aspects of their use.

An O-ring fits into a groove in such a way that the diameter of the ring is about ten percent more than the depth of the groove. When the two parts are assembled, the O-ring will be compressed by approximately ten percent. This ensures a good seal and at the same time prevents excessive drag or wear of the ring.

When a packing is exposed to high pressure, there is a tendency for the seal to extrude into the space between the fixed and moving part. In this case, a backup ring, usually made of Teflon or similar material, is installed. Backup rings are always placed on the side of the O-ring away from the pressure. [Figure 9-64]

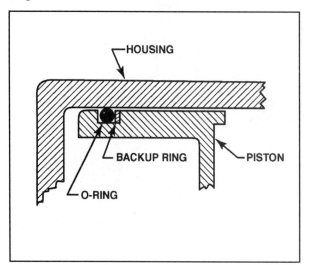

Figure 9-64. O-ring backup rings are used to keep the O-ring from extruding between the moving parts. They should be placed on the side away from the pressure.

Excessive heat, contamination and improper fluid are all enemies of brake seals. Any time a component has been subjected to any of these, it must be disassembled and all of the seals replaced. When a seal is exposed to high pressure and temperature, it will mold itself to its cavity and harden. When the temperature drops, the seal will not have enough resilience to press back against the surface and will allow fluid to leak.

It is impossible for a technician to identify the material of which a seal is made by looking at it, therefore, it is vital that only the correct seal be used in each application. All seals used in aircraft brakes should be received in a sealed container marked with the part number of the seal. If you do not know for sure what the seal is made of, do not use it. If the seal is made of rubber, it should also have the cure date. Dated seals should normally be installed within twelve quarters, or three years, of the cure date. However, this time can be considerably shortened if the seals are not stored properly. All seals should be stored in a cool, dark area away from electrical equipment such as motors or other devices that produce ozone because it is highly detrimental to rubber.

Because it is so difficult to tell the condition of a seal by its physical appearance, unscrupulous operators obtain out-of-date surplus seals, repackaged them with a bogus cure date, and offer them on the market at greatly reduced prices. The use of these seals can seriously jeopardize the safety of flight and must never be used. When buying seals, the technician must be sure to buy only from a reputable vendor so the seals are as advertised and of fresh stock.

REPLACEMENT OF BRAKE LININGS

Replacement of disc linings is one of the most common repairs a maintenance technician performs. Two common types of systems are the Goodyear single-disc brakes and the Cleveland single-disc brakes.

The linings used on **Goodyear single-disc brakes** are free-floating pucks of an organic material that fit into cavities located in the housing. To replace these lining pucks, the anti-rattle clips that center the disc in the wheel are removed, then the wheel is removed from the axle. The disc will remain between the linings. Now, the disc is carefully removed from between the linings, and the lining pucks can be lifted out. The new linings are slipped into the cavities with their smooth sides next to the disc. The disc is inserted between the linings and the wheel reinstalled on the axle. The disc is centered in the wheel with the anti-rattle clips and the axle nut adjusted so there is no end play and the wheel is free to rotate. Then the nut is saftied to the axle with a cotter key. [Figure 9-65]

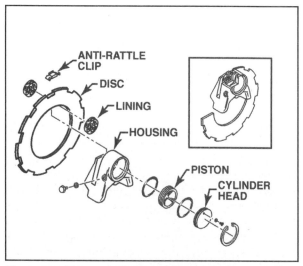

Figure 9-65. This exploded view of the brake disc and caliper assembly shows the relationship of the various components.

It is not necessary to remove the wheel to replace the brake linings on a **Cleveland single-disc brake**. The cylinder assembly is removed from the back plate. When the bolts are backed out, the back plate will drop down, which allows the entire assembly to drop away from the torque plate. The pressure plate can then be pulled off the anchor bolts. [Figure 9-66]

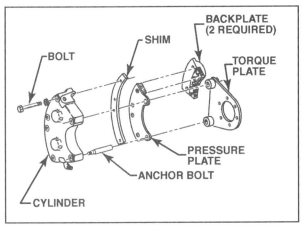

Figure 9-66. The linings on Cleveland brakes can be replaced without removing the wheel from the aircraft.

One of the linings is riveted to the pressure plate and the other to the back plate. A special riveting tool is used to punch the old rivets out of the linings. The plates should be inspected and new linings installed with the rivets furnished in the brake re-lining kit or purchased per Cleveland specifications. [Figure 9-67]

The entire brake should be checked for condition and the pressure plate reinstalled. The two anchor bolts are then inserted into the holes in the torque plate and the back plate reinstalled. Some of these brakes require a shim between the back plate and the pressure plate for proper fit. The bolts are then tightened to the required torque, and safetied as required.

After the linings have been replaced, the new linings should be conditioned, or "burned in," by the method recommended in the manufacturer's service manual. Basically this consists of taxiing the aircraft straight ahead at a fair speed and then stopping it with smooth, even applications of the brake. The brakes should be allowed to cool for about a minute and the procedure repeated. The number of stops, the taxi speed and the cooling interval will vary with the specific type of brake.

MALFUNCTIONS AND DAMAGE

Brakes are subject to a wide variety of operating conditions, which can cause a number of malfunctions and can even permanently damage the brake system. Among the most common malfunctions encountered by the maintenance technician are overheating, dragging brakes, and chattering or squealing brakes.

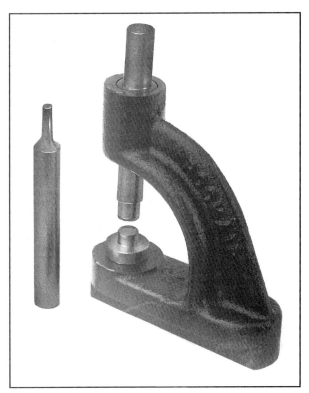

Figure 9-67. A special tool is used to remove the old rivets and install new ones that secure the lining to the backing plate. Use caution when riveting linings to the backing plate. Too much force used during the riveting process can crack the braking linings.

OVERHEATING

One of the most demanding requirements for the modern aircraft brake is its ability to dissipate tremendous amounts of heat energy. If overheating is allowed to continue, the brakes can lose their structural strength or become distorted to the point of causing other problems. Any time a brake shows signs of overheating, or if it has been involved in an aborted takeoff, it must be removed from the aircraft, given a complete inspection, and rebuilt.

The housing must be carefully examined for cracks or warpage, and it must be given a hardness test at the points specified in the service manual. This hardness test will show any loss of heat treatment that would seriously affect the strength of the brake. All of the seals must be replaced because excessive heat will destroy their ability to seal.

The discs must all be checked to see that they have not warped or that there is no transfer of friction mix (the surface material) from the rotating disc to the stationary discs. A specified amount of mix may be lost from the rotating discs. The mix. must be checked to be sure it does not exceed the allowable amount.

After the brake has been rebuilt, it must be bench-checked for leakage and for proper operation before it is reinstalled on the aircraft.

DRAGGING

When a brake fails to completely release after the pressure is removed, it is said to be dragging. If this condition is not immediately corrected, it will lead to excessive heating and wear of the discs.

Dragging may be caused by discs that have become warped from overheating, by weak or missing return springs, or by a slipped pin in the return system. Air trapped in the system will expand as it is heated and keep the brakes partially applied, which will also cause them to drag.

It is vitally important when building up the brakes that the proper components be used, because wrong components in the adjusters, in the disc stack, or in any wear area may cause the brake to fail to release properly and cause the discs to drag.

CHATTERING OR SQUEALING

An extremely annoying condition can exist when the brakes are applied and released many times a minute, rather than exerting a smooth and even friction between the surface of the discs. This produces a loud chattering noise, and, if the frequency of this chattering is quite high, produces a loud squeal. Chattering and squealing brakes are not only annoying to hear, but the vibration generated is harmful to the landing gear and brake structure. Warped or glazed discs will cause chattering, along with any unparalleled condition on the surface of the disc stack. If the discs have been overheated, there is a possibility that some of the mix has been transferred from a rotating disc to a stationary disc. The uneven friction caused by this transfer will also produce chattering.

ANTI-SKID BRAKE CONTROL SYSTEMS

Modern, high-speed jet aircraft usually have more than one wheel on each side, and all of the brakes on one side are controlled with one pedal. With this arrangement, the pilot has no way to know when one of these wheels begins to skid so that corrective action can be taken. But, if corrective action is not taken within a few seconds to release a locked-up wheel, the tire is likely to blow out and control of the aircraft can be lost.

The essence of aircraft braking control is to continuously adjust brake pressure to maintain optimum brake torque. This optimum level balances tire and runway friction at its peak level yielding maximum braking deceleration. Automatic braking provides a pilot with a valuable safety feature. Upon rejected take-off, the brakes are activated immediately at a maximum level, thereby eliminating the normal transition time and runway field length requirements.

For maximum brake effectiveness, the friction between the tire and the runway surface should closely relate to the friction in the brake so that the peripheral speed of the tire is slightly less than the speed of the aircraft. When this is true, the tire will grip the runway surface and slip just a little. This produces the maximum tire drag.

Maintaining this optimum friction is no easy matter, because if the brake pressure is held constant after the slip starts and the wheel begins to decelerate, the brake friction will rapidly increase to the point that the wheel will lock up. The tire will skid over the runway and produce very little effective braking.

A very simple form of manual anti-skid control in an automobile that does not have anti-lock brakes (ABS) installed when driving on ice is to pump the brakes. For the most effective stopping, the brakes are pumped, applying them only enough to slow the wheel, but releasing them before the wheel decelerates enough to lock up. This same on-and-off type of operation was employed in some of the early aircraft anti-skid systems. However, it can have a major drawback if the control valves do not operate fast enough. Modern aircraft are equipped with anti-lock brake systems that sense an impending slip of a tire and adjust the brake pressure to keep the tire from slipping fully. The description and explanation of the operation of the anti-skid system is covered in *Chapter 10, Position and Warning Systems*.

SECTION C

AIRCRAFT TIRES AND TUBES

TIRE CLASSIFICATION

Aircraft tires are classified according to their type, size and ply rating, and whether they are tubeless or tubed. The United States Tire and Rim Association has established nine types of aircraft tires, but only three of these types are of primary concern.

TIRE TYPES

The **type III** tire is the most popular low-pressure tire found today on piston-powered aircraft. The section width is relatively wide in relation to the bead diameter. This allows lower inflation pressure for improved cushioning and flotation. The section width and rim diameter are used to designate the size of the tire. For example, a tire having a section width of 9 1/2 inches that fits a 16-inch wheel would be identified as a 9.50-16 tire. [Figure 9-68]

Figure 9-68. The type III low-pressure tire is found on most small general aviation aircraft.

Type VII extra-high pressure tires are the standard for jet aircraft. They have exceptionally high load-carrying ability and are available in ply ratings from 4 to 38. The tire sizes are designed by outside diameter and section width, with a designation such as 38×13. [Figure 9-69]

Figure 9-69. Jet aircraft are equipped with type VII extra-high pressure tires.

Type VIII tires are used for high performance jet aircraft with their extremely high takeoff speeds. They use extra high inflation pressure and have a low profile. Their size designation includes the outside diameter, section width, and rim diameter. An example of a tire designation for a Type VIII tire would be 30×11.50-14.5.

PLY RATING

In the past, tires were rated for strength by the number of fabric plies used in the construction of the carcass or body of the tire. Newer materials have much greater strength than the cotton originally used. Fewer actual layers of the new materials are needed to get the same strength. Today, tires are given a ply rating, rather than specifying the actual number of layers of fabric material used in the carcass. The ply rating of a tire relates to its maximum static load and its inflation pressure.

TUBE OR TUBELESS TIRES

Aircraft tires are manufactured as both tube-type and tubeless, with the basic difference between the two being the inner liner. Tubeless tires have an inner liner that is approximately one-tenth of an inch thick and serves as a container for the air. Tube-type tires have no such liner, but are somewhat smoother on the inside so the tube will not be damaged by chafing against the inside of the tire. Tubeless tires are identified by the word TUBELESS on their sidewall, and the lack of identification signifies that a tube should be used in the tire.

TIRE CONSTRUCTION

Automobile and truck tires are built to completely different specifications from those used for aircraft tires. They are required to operate for long periods of time, carrying a relatively large but steady load at reasonably high rotational speeds. Because of this, they are allowed to have only a relatively small amount of deflection. For example, passenger car tires are designed for a continual deflection of only about 12 to 14%.

Airplane tires, on the other hand, must be strong enough to absorb the tremendous loads exerted on them at touchdown, and, while they must operate at very high speeds, the ground roll duration is limited. Because of these requirements, aircraft tires are allowed a deflection of between 32 and 35%, more than twice that allowed for automobile tires. [Figure 9-70]

THE BEAD

The most important part of a tire is the bead. It is the bead that anchors the carcass and provides a firm mounting surface for the tire on the wheel. Ultimately, all of the ground forces on the tire terminate in the bead. The beads are made of bundles of high-strength carbon-steel wire, with one, two, or three of these bundles used in each side of the tire. [Figure 9-71]

Rubber apex strips streamline the round bead bundles, so the fabric will fit smoothly around them with no voids. Layers of rubber and fabric called

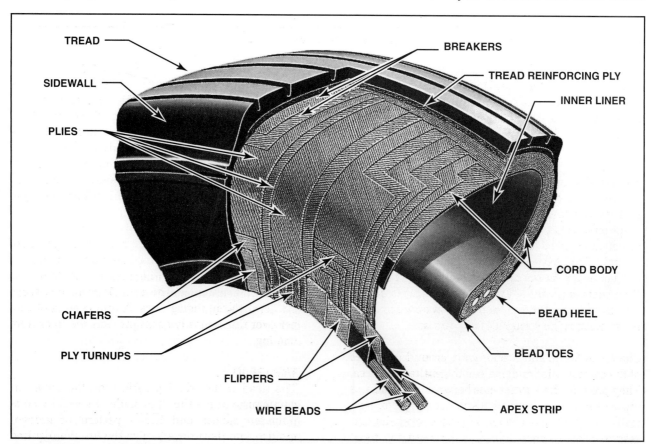

Figure 9-70. This cutaway shows the basic construction of an aircraft tire.

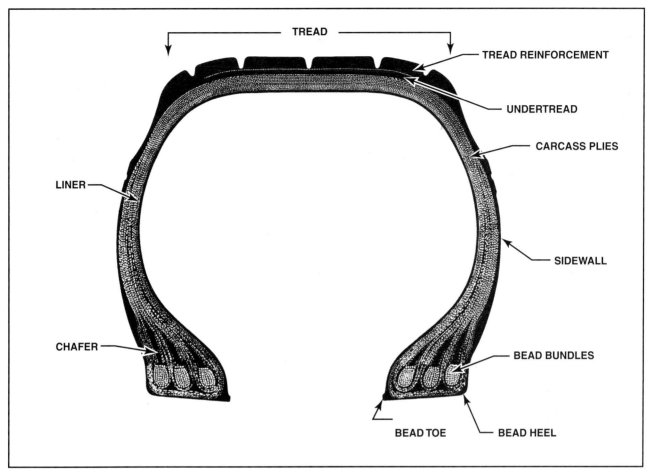

Figure 9-71. This cutaway shows the tread, sidewalls, beads, and other components of a typical low-pressure tire.

flippers enclose the bead bundles to insulate the carcass plies from the bead wires. Because the greatest amount of heat in the tire is concentrated in the bead area, this insulation increases the durability of the tire.

THE CARCASS

The carcass of the tire is made up of layers or plies of rubber-coated nylon-cord fabric. This fabric is cut into strips on the bias, meaning that the cords of the fabric run at an angle of approximately 45° to the length of the strip. The strips extend completely across the tire and lap back over the beads to form the ply turn-ups. Each successive ply of the fabric is placed in such a way that the cords cross each other at approximately 90° to balance the strength of the carcass.

Chafers of fabric and rubber wrap around the edges of the carcass plies and enclose the entire bead area. They provide chafe resistance between the bead and the wheel. An undertread, which is a layer of specially compounded rubber, is placed over the carcass to provide good adhesion between the tread and the carcass.

The tread reinforcement, made of one or more plies of nylon fabric, strengthens the tread and opposes the centrifugal forces that try to pull the tread from the carcass during high-speed operation. It also stabilizes the tread on the cord body and prevents it from squirming or moving. This reinforcement does not enter into the ply rating of the tire, but it is used as a guide for the retreaders when buffing the tire to remove the tread.

Tubeless tires are lined with a special compound of rubber that is less permeable than the rubber used in the rest of the tire. This liner acts as a container for the air and minimizes the amount of air that seeps out through the casing plies. A thin coating of rubber over the inner ply cords protects the liner from chafing.

THE TREAD

The tread is the wearing surface on the outer circumference of the tire. It is made of specially compounded rubber and has a pattern of grooves molded into its surface to give the tire the required traction characteristics according to the type of runway surface the aircraft will encounter. There have been a number of basic tread patterns used on aircraft tires. The most familiar are the plain tread, all-weather tread, rib tread, and the chine.

The **plain, or smooth, tread** was popular for tires used on airplanes with no brakes, or for aircraft whose brakes were used primarily as a taxi aid, rather than for slowing the aircraft in its landing roll. Today, this type of tread is found only on some helicopters and on very light airplanes.

All-weather treads have a rib tread in the center and diamonds molded into the shoulders. The diamond-shaped tread pattern is effective for aircraft operating on grass or hard-packed dirt, and the rib tread in the center gives good traction on hard surfaced runways. [Figure 9-72]

Figure 9-73. Most aircraft tires are equipped with a ribbed tread.

Figure 9-72. All weather treads consist of a center rib tread for hard surface runways and a diamond-shaped pattern on the shoulders for dirt and grass.

The **rib tread** is the most popular tread pattern found on aircraft today. It is designed especially for use on hard-surfaced runways and gives long tread wear, good traction, and exceptionally good directional stability. [Figure 9-73]

The width and depth of the grooves and their placement on the tread are factors determined by the operating conditions of the aircraft for which they are designed.

Jet aircraft with aft-mounted engines have a problem with water or slush being thrown up by the nose wheel and entering the engines, causing damage or flame-out. To prevent this, tires used on the nose wheels of these aircraft have **chines**, or deflectors molded into the upper sidewall to deflect the water or slush away from the engine intakes. Tires for dual nose wheel installations have chines on one side only, while single nose wheel installations have dual chines (deflectors), one on either side of the tire. [Figure 9-74]

Figure 9-74. When it is necessary for water or slush to be deflected away from engines, a tire with special deflectors or chines is installed. This tire is one of a pair of nose wheel tires. A single nose wheel tire will have a chine molded on both sides of the tire.

THE SIDEWALL

The sidewall is a rubber covering that extends from the tread down to the bead heel to protect the carcass from injuries such as cuts or bruises and from exposure to moisture and ozone.

Tubeless tires have an inner liner that is designed to hold air, but some air does seep through. If the sidewall trapped this escaping air in the body plies, it could expand when the tire was heated and cause ply separation and possibly allow the tread to be thrown from the tire. To prevent this, tubeless tires have vent holes in their lower sidewall to allow this air to escape. These holes are marked with paint and must be kept open when the tire is retreaded. Tube-type tires are also vented to release air that is trapped between the tire and the tube when the tire is mounted.

TIRE INSPECTION ON THE AIRCRAFT

Most inspection of aircraft tires occurs with the tires and wheels attached to the aircraft. In addition to scheduled inspections of the tires, the technician should take a look at the tires any time work is being done in the vicinity of the gear.

INFLATION

The greatest enemy of an aircraft tire is heat, either the heat generated within the tire as it rolls over the ground, or heat from external sources such as the brakes or hot runway surfaces. It is the internally generated heat which causes damage that is not likely to be discovered until it results in a tire failure. As the tire rolls over the ground, the sidewalls flex and cause internal heat. Aircraft tires are designed to withstand the heat generated by normal flexing for a reasonable amount of time. Because the air in the tire supports the weight of the airplane, the inflation pressure is critical. The pressure should be checked daily and before each flight. Differences in air pressure between dual mounted tires, whether main or nose, should cause concern. Pressure differences allow one tire to carry more of the load than the other, which defeats the purpose of dual mounted tires. Any pressure difference should be corrected and noted in the aircraft log. Consult the aircraft service manual to determine the exact tire pressure tolerances.

Over-inflation causes accelerated centerline wear on the tread while leaving rubber on the shoulder. When a tire is worn in this way, it has much less resistance to skidding than when its tread wears uniformly.

While over-inflation is bad, under-inflation is even worse. Under-inflation causes excess heat within the tire. If a tire is allowed to deflect as much as 45%, about three times as much heat will build up in the tire as it is designed to withstand. This over-deflection can cause internal carcass damage, which may not be visible and could easily result in premature failure of the tire.

Tires that have been operated with low inflation pressure will have their tread worn away on the shoulders more than in the center. Tires showing this pattern of wear should be carefully examined for evidence of hidden damage.

The importance of maintaining the proper inflation pressure in a tire makes the tire pressure check one of the most important parts of routine preventive maintenance. The proper tire inflation pressure is specified in the aircraft service manual. The pressure in the aircraft service manual should be used, rather than that listed in the tire manufacturer's product manuals due to the fact that tire pressures vary from one aircraft design to another.

The pressure specified in the airframe manufacturer's manual is for a loaded tire; that is, for the tire supporting the weight of the aircraft. When the tire is subjected to this load, it is deflected the designed amount, and the volume of its air chamber is decreased enough to raise the pressure by about four percent. For example, if the service manual specifies a pressure of 187 psi, the proper inflation pressure, if serviced and inflated off the aircraft, is 180 psi; four percent less than 187 psi.

Inaccurate pressure gauges are one of the major causes for chronic inflation problems. To be sure that the gauge is accurate, it should be periodically calibrated. The best gauges for this purpose are dial-type indicators, because they are subjected to more careful handling and easier to read in the small increments needed to accurately determine the tire pressure. [Figure 9-75]

Figure 9-75. For accurately measuring tire pressure, the preferred gauge is one with a dial-type indicator.

Inflation pressure should always be measured when the tire is cold. For this reason, two to three hours should elapse after a flight before the pressure is measured. Inflation pressure of a tire varies with the ambient temperature by about one percent for every five degrees Fahrenheit. For example, if a tire is inflated and allowed to stabilize with a pressure of 180 psi in a shop where the temperature is 70° F, and the airplane is rolled outside where it remains overnight with a temperature of 0° the pressure will drop by 14% to about 155 psi. Obviously, it would then be seriously under-inflated and should be reinflated.

If an airplane is to fly into an area where the temperature is much lower than that of the departing point, theoretically, the pressure should be adjusted before the airplane leaves. If, for example, the temperature is 100° F and the airplane is to land where the temperature is 40° F, the pressure should be increased before takeoff. If the tire requires 187 psi, the 60-degree temperature drop will require the pressure to be 12% greater, or 210 psi. The airframe manufacturer's manual should be consulted before the pressure is changed to see if the maximum allowable safe-inflation pressure will not been exceeded.

Nylon tires will stretch when they are first inflated and will increase their volume enough to cause a pressure drop of about five to ten percent of the initial pressure in the first 24 hours. Nylon tire pressures should therefore be adjusted 12 to 24 hours after installation.

TREAD CONDITION
Because the basic strength of the tire is in its carcass, a tire loses none of its strength as long as the tread does not wear down to the body plies. When the tread is worn away, the traction characteristics of the tire are seriously affected.

A tire that has been properly maintained and operated with the correct inflation pressure will wear the tread uniformly. It should be removed for retreading while there is still at least 1/32-inch of tread left at its most shallow point. [Figure 9-76]

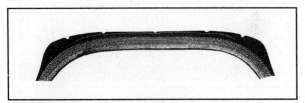

Figure 9-76. The tread wear pattern for a tire that has been properly maintained and operated with the correct inflation pressures will be uniformly worn across the tread.

If the center ribs are worn away while the shoulder ribs still have an appreciable depth, the tire has been operated in an over-inflated condition and is highly susceptible to cuts and bruises. It should be carefully checked for this type of damage. [Figure 9-77]

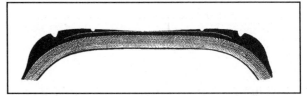

Figure 9-77. Excessive wear of the center portion of the tread indicates that the tire has been operated in an overinflated condition.

Under-inflation will cause the shoulder ribs to wear more than those in the center. Any tire showing this wear pattern should be carefully inspected for signs of bulges that could indicate ply separation. [Figure 9-78]

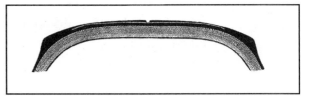

Figure 9-78. Excessively worn tread shoulders are an indication of underinflation.

Tread that has been worn until the body plies are visible indicates poor maintenance. If it is worn only to the point that the tread reinforcement is showing, it is possible that retreading can salvage the tire. But if it is worn into the body plies, it must be rejected and cannot be retread. [Figure 9-79]

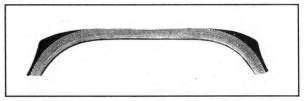

Figure 9-79. If wear has progressed into the body plies, the tire is beyond its serviceable life and must be rejected.

Uneven tread wear can indicate that the landing gear is out of alignment. Wheel alignment should be checked and adjusted so it conforms to the specifications of the aircraft service manual. If the uneven wear is slight and the landing gear is not adjusted, the tires may be removed, carefully inspected, and reversed so the wear will even out.

Tread wear in spots can be caused by malfunctioning brakes, improperly torqued bearings, worn strut parts, or landing with brakes engaged.

Any time the tread is cut more than halfway across a rib, or any of the carcass plies are exposed, the tire should be removed from service. [Figure 9-80]

Figure 9-80. A tire should be removed from service if the tread is cut across more than one-half of a rib width.

Figure 9-81. Hydroplaning combined with a locked brake causes intense heat buildup that actually burns the rubber on the tire.

If there are bits of glass, rock or metal embedded in the tread, they should be carefully removed with a blunt awl or a small screwdriver. The technician must be very careful not to puncture the tire when using these tools.

When a wheel locks up on a water-covered runway and rides on the surface of the water, a tremendous amount of heat builds up at the point of contact and actually burns the rubber. Tires showing this type of damage should be removed from service. [Figure 9-81]

Operating on grooved runways can cause chevron-shaped cuts across the ribs of a tire. If they extend across more than one-half of the rib, the tire should be removed from service. [Figure 9-82]

Any damaged or suspect area of the tire should be marked with a light-colored marker before deflating the tire. When the tire is deflated, these areas are almost impossible to locate.

Any tread damage should be carefully evaluated by a licensed retreading agency to determine whether or not the tire is repairable.

Figure 9-82. These shallow cuts are caused by operations on a grooved runway. The maximum length of the cut is one-half of the width of the rib. Cuts that exceed one half the width of the rib warrant tire rejection.

SIDEWALL CONDITION

The main purpose of the sidewall of a tire is to protect the carcass plies from damage, either from mechanical abrasion, from deterioration by chemicals, or by the sun. Small snags, cuts, or weather cracking in the sidewall rubber that do not expose the cords are not normally considered a cause for removal of the tire; but if any of the ply material is exposed, the tire must be removed.

The liner of a tubeless tire is utilized to hold the air, but some of the air diffuses into the body plies. The sidewalls of these tires are vented to allow this air to escape, but if the vents do not adequately relieve the pressure, ply separation may occur. Be sure to check for open vent holes when inspecting the tires.

TIRE REMOVAL

To remove a tire, the aircraft is jacked up according to the aircraft service manual procedure. When the weight is off the tire, it is deflated using a deflator cap. The high pressure in some aircraft tires can eject the valve core with enough velocity to injure anyone it might hit. After all the air is let out, the core can be safely removed.

When removing the wheel, following the aircraft manufacturer's instructions, bearings should be protected from damage and stored in a safe place until they can be cleaned, inspected, and repacked with proper grease.

The wheel should be placed on a flat surface and the bead separated from the bead seat area of the wheel using a straight push as near to the rim as possible. No kind of tire tool should be used to pry the bead from the wheel, because the wheel can become nicked or scratched. This type of damage will cause stress concentrations at the point of the nick or scratch that could lead to wheel failure. [Figure 9-83]

When the bead has been broken from both sides of the wheel, remove the wheel bolts, then remove the wheel halves from the tire. If the tire is tubeless, the O-ring seal between the wheel halves must be inspected for damage.

TIRE INSPECTION OFF OF THE AIRCRAFT

Any tire that has been involved in an aborted takeoff, severe braking, or has been exposed to enough heat that the fusible plug in the wheel has melted and deflated the tire, should be replaced. Excessive heat causes damage to the tire that, even though it is not obvious, has weakened the tire enough that it will likely fail in service. If one tire in a dual instal-

Figure 9-83. The bead should be broken away from the bead seat of the wheel by pushing straight down with a removal tool. Conventional hand tire removal tools should not be used to pry against the wheel due to the possibility of damage.

lation fails, enough extra stresses have been put on the other tire that it should be discarded also, even though there may be no visible damage.

To inspect the inner liner, the beads must be pried apart very carefully. Force used to spread the beads should not be concentrated in one area, and the beads should not be spread more than the section width of the tire. The use of an improper procedure when breaking or spreading the bead can kink the wire bundles so the bead cannot seat against the wheel when it is reinstalled. A tire with a kinked bead should be scrapped.

The inner liner of tubeless tires is inspected for any bulges or blisters with any suspect areas evaluated by a retreading agency.

Areas that were marked when the tire was inflated should be carefully inspected. Cuts should be opened up enough to check their depth and care must be taken not to puncture the tire. Punctures that do not exceed one-quarter inch on the outside of the tire, one-eighth inch on the inside, and injuries that do not penetrate more than 40% of the actual body plies can be repaired when the tire is retread. Any bulges marked when the tire was inflated should be carefully checked to determine whether they are ply separations or separations between the tread and the carcass. Tread separation may possibly be repaired by retreading, but, if it is a ply separation, the tire must be scrapped.

The sidewall should be carefully inspected for condition, and, if any of the cords have been damaged, the tire cannot be repaired. Exposed cords also make a tire non-repairable, because it is reasonable to suspect that the exposed cords have been weakened by exposure to the elements.

The most important part of a tire is the bead area where all of the forces of the tire are carried into the wheel and where an airtight seal must be maintained with tubeless tires. The bead and the adjacent area should be inspected for indications of damage from tire tools or from chafing against the rim. Any severe damage at the bead would require tire rejection. If the damage is only through the chafer, it can be repaired when the tire is retreaded.

Damage from excessive heat usually shows up on the bead area because the heat can build up faster than it can be dissipated. If the bead area is damaged or has an unusual appearance or texture, the tire cannot be repaired.

The bead surface from the wheel flange to the toe of the bead is the sealing surface for a tubeless tire. If it has been damaged by tire tools or by slipping on the wheel, it will not seal. Bare chafer cords, however, if they are not broken, will not normally cause a tire to leak; and they are not necessarily a cause for removing the tire from service.

TIRE REPAIR AND RETREADING

Aircraft tires are highly stressed, and no repair should be attempted by anyone not adequately equipped for or experienced in this work. General guidelines for repairable tires are given in the FAA Advisory Circular 43.13-1B, but the actual repair should only be made by a repair station equipped and approved for this work. Time and money can be saved, however, if the technician knows what definitely constitutes a non-repairable tire, so that it can be discarded without first being sent to the repair station.

Repair is not recommended for any tire that has the following kinds of damage:

- Any evidence of breaks caused by flexing. This type of damage is often associated with other damage that may not be visible.
- Bead damage to more than the chafers, or any damage that would prevent the bead of a tubeless tire from sealing to the wheel.
- Any evidence of separation between the plies around the bead wire.
- Any injury that would require a reinforcement.
- Kinked or broken beads.

- Weather checks or radial cracks in the sidewall that extend into the cord body.
- Blisters or other evidence of heat damage.
- Cracked, deteriorated, or damaged inner liners of tubeless tires.

Before rejecting a tire that is questionable, a technician should always get expert advice from a certificated retreading agency.

The great amount of abrasion a tire experiences on its tread each time the airplane lands or taxis on a hard surface wears away the tread long before the carcass is worn out. Therefore, it is standard practice for commercial aircraft tires to be retreaded. When a tire is received by the retreading agency, it is thoroughly inspected. The tread, sidewalls and beads are checked for cuts, bruises, other damage, or wear. Air is injected into the sidewall to check for any ply separation. The tire is checked for fabric fatigue and for any indication of contamination by oil, grease, or hydraulic fluid. The tires that pass this inspection then have their old tread rubber removed by contour buffing, which produces a smooth shoulder-to-shoulder surface. New tread rubber and reinforcement are then applied to the buffed carcass. The tire is placed in a heated mold and cured. After it is taken from the mold, balance patches are bonded to the inside of the tire to achieve the proper static balance, and the tire is then given a final inspection.

The tire is then identified as a retreaded tire, and a record is made of the number of times it has been retreaded. There is no specific limit to the number of times a tire can be retread. Retreading is determined by the condition of the carcass.

TIRE STORAGE

All new and retreaded tires should be stored in a cool, dry area, out of direct sunlight and away from any electrical machinery. Fluorescent lights, electric motors, generators, and battery chargers all convert oxygen into ozone, which is very harmful to rubber.

The storage room should not be subjected to extreme temperatures, and should be maintained between 32° and 80° F (0° and 27° C). Whenever possible, the tires should be stored vertically in racks with the tire supported on a flat surface which is at least three or four inches wide. If tubeless tires are stacked horizontally, the bottom tires in the stack may be distorted so much that the beads will not seat on the wheel unless a special bead-seating tool is used. If it is necessary, however, to stack them horizontally, they should not be stacked more than five tires high for tires with a diameter of up to 40 inches, four tires high for those between 40 and 49 inches, and three high for tires larger than 49 inches. [Figure 9-84]

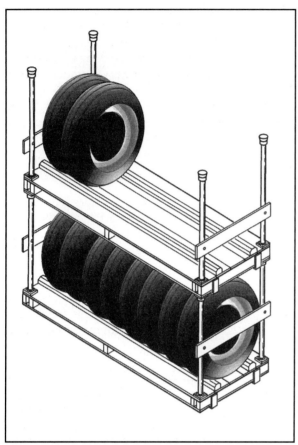

Figure 9-84. Tires should be stored vertically on racks whenever possible.

AIRCRAFT TUBES

Though many tires today are tubeless, a large number of tubed tires may be found. The technician needs to be aware of the procedures for handling tubed tires, as they vary in a number of ways from tubeless tires.

TUBE CONSTRUCTION AND SELECTION

A great number of aircraft tires, ranging from some of the small type III, 5.00-5, up to large type VII, 56 × 16, require tubes. Tubes for these tires are available in either unreinforced rubber for normal applications or in a special heavy-duty reinforced tube. Heavy-duty tubes include a layer of nylon fabric molded to the inside circumference to protect it from chafing against the rim and from heat caused by brake application.

All aircraft tubes are made of a specially compounded natural rubber that holds air with a minimum of leakage. A hole in the tube or a defective valve are two primary causes for an aircraft tube to leak.

It is extremely important that only the tube recommended for a particular tire be used with that tire. If the tube is too small for the tire, its splices will be overstressed, which will weaken the tube.

TUBE INSPECTION

If a tube is suspected of leaking, the valve should be checked by spreading a drop of water over the end of the valve to see if a bubble forms. If a bubble does form, the valve core should be replaced. If the leak is not in the valve, the tire must be deflated, removed from the wheel, and the tube removed. If the tube is not too large, it should be inflated and submerged in water to find the source of bubbles. If the tube is too large for the available water container, water may be poured over the surface of the tube to identify the source of the leak.

It is extremely important when inflating a tube that is not in a tire, that only enough air is used to round it out. Check the tube carefully around the valve stem and the valve pad for any indication of the pad pulling away from the tube. [Figure 9-85]

Figure 9-85. When an inner tube is leaking, the first place to check is the valve stem. The valve core should be checked for leakage and the area around the stem checked for cracks.

The inside circumference of the tube should be examined for evidence of chafing against the toe of the bead. Any tube that is chafed enough to lose some of its thickness in spots should be replaced.

The brakes on modern, high-performance aircraft absorb tremendous amounts of energy, and while some wheels have heat shields, the tires and tubes cannot be completely protected from the heat. Examine the inner circumference of the tube for any indication that it has been heated enough to have lost its smooth, contour-developed square corners. Any tube that is deformed in this way should be replaced. Reinforced tubes should be used on installations where there is enough heat to damage a regular tube.

TUBE INSPECTION AND STORAGE
Tubes should be stored in their original cartons whenever possible. If the original cartons are not available, the tires should be dusted with tire talcum and wrapped in heavy paper. Tubes may also be stored inflated by putting them in the proper size tire and inflating them just enough to round them out. The inside of the tire and the outside of the tube should be dusted with tire talc to prevent the tube sticking to the tire. Tubes should never be stored by hanging them over nails or pegs, or supporting them in any way that would cause a sharp fold or crease which would eventually cause the rubber to crack. Tubes with creases should not be put into service. Tubes, like any other rubber product, should be stored in a cool, dry, dark area, away from any electrical equipment that would produce rubber-damaging ozone.

Tube inspection should include checks for tube elasticity, cracking, evidence of chafing, and a thorough check of the valve stem. The valve core should be checked to see that it is tight and that it does not leak air. Before installing a used tube, it is a good idea to check for leaks by inflating until rounded to its normal shape and placing it in a tank of water.

TIRE MOUNTING
Tire mounting is a critical step in the buildup of tires and wheels for aircraft. The procedures for tubed and tubeless tires are similar in many respects, but technicians must be aware of the differences.

TUBELESS TIRES
Most modern aircraft use the split-type wheel which makes tire mounting far easier than it was with either the single-piece drop-center wheel or with wheels having a removable flange held on with a locking ring.

The fact that the wheel is so highly stressed makes it extremely important that the manufacturer's service information be followed in detail when mounting and removing the tires. This information includes such details as bolt torque, lubrication requirements, and wheel balancing details.

Before the tire is mounted on a wheel, the wheel must be carefully inspected to be sure that there are no nicks or scratches in the bead seat area that could cause the air to leak. For tubeless tires, the O-ring seal area between the wheel halves is critical.

The entire wheel should be inspected for corrosion and for any evidence of the finish being scratched through or worn off. Any balance weights installed when the wheel was manufactured must be securely fastened. The thermal fuse plugs should be checked for security and condition, and the air valve checked for the condition of its O-ring seal. The bead seat area and the O-ring seal area should be cleaned with a cloth dampened with isopropyl alcohol, and the inboard wheel half should be placed on a clean, flat surface.

After checking to be certain that the tire is approved for the aircraft on which it is being mounted and that the word TUBELESS is on the sidewall, the tire should be inspected thoroughly. There should be no foreign material inside the tire, and the bead area should be wiped clean with a rag dampened with isopropyl alcohol. The O-ring is lubricated with the same grease used for the wheel bearings and is carefully placed in the groove without stretching or twisting it.

Because of the tight fit needed between the bead of the tire and the wheel, it may be helpful to apply a little tire talc to the toe, or inner edge, of the bead to help the bead seat when the tire is inflated. However, no powder should be allowed to get between the bead and the wheel flange. The tire should then be placed over the inboard wheel half with the red dot indicating the tire's light point adjacent to the wheel valve, or, if some other mark on the wheel identifies its heavy point, adjacent to that mark. The outboard wheel half is then placed inside the tire and the bolt holes lined up.

An anti-seize compound such as Lubtork® should be applied to the threads of the bolts, to both sides of the washers, and to the bearing surface of the nuts. The bolts and nuts are then installed, and all of the nuts drawn up in a crisscross fashion to one-half of the required torque. Then all of the fasteners are tightened to full torque. Wheel bolts should only be tightened using an accurate hand-torque wrench. An impact wrench should never be used on any bolt where the torque is critical. The torque of an impact wrench is applied in a series of blows or jerks which create stresses that are considerably greater than the bolt is designed to take.

The wheel and tire assembly should be placed in a safety cage, the air pressure regulator adjusted to the recommended tire pressure, and the tire inflated gradually. While the tire is inflating, the technician should watch the tire to be sure the bead seats against the wheel flange.

All nylon tires stretch when they are initially inflated and should be allowed to remain for 12 to 24 hours with no load applied. This stretch may cause a five- to ten-percent decrease in pressure. The pressure should be adjusted after this period. Continue to monitor the inflation pressure daily. There will be some pressure loss, but it should not exceed 5% in any 24-hour period.

TUBE-TYPE TIRES
When mounting a tube-type tire on a wheel, the tire and tube must both be correct for the installation. The wheel should be inspected for any indication of damage or corrosion. If any corrosion is found, all traces of the damage must be removed and the protective oxide film restored. Two coats of zinc chromate primer should be applied and the finish restored to match the rest of the wheel. To ensure cleanliness, wipe the bead seat area with a rag dampened with isopropyl alcohol before mounting the tire.

The inside of the tire must be clean and free of all foreign matter, then dusted with an approved tire talcum powder. The inner tube is then folded, dusted with talc, and slipped inside the tire with the valve sticking out on the side of the tire having the serial number. The tube is inflated just enough to round it out and is adjusted inside the tire so the yellow mark indicating the heavy point of the tube aligns with the red dot on the tire indicating its light point. If there is no balance mark on the tube, it is assumed that the valve is the heavy point.

The tire and tube are then installed on the outboard wheel half so the valve stem sticks out through the hole in the wheel. A bit of tire talc may be rubbed on the toe of the bead to help it slide over the wheel and seat itself.

The inboard half of the wheel is then placed in the tire without pinching the tube. The bolts should be lubricated with anti-seize compound and the nuts tightened in a crisscross fashion to one-half of the required torque. Then, all of the nuts are tightened to the recommended torque value with a good smooth pull on the handle of the torque wrench. [Figure 9-86]

The tire is placed in a safety cage, and using a clip-on chuck, the air pressure is brought up gradually to

Figure 9-86. When installing the wheel halves, make certain the tube does not get pinched between the halves.

the recommended value to seat the bead; then the tire is deflated. After seating the bead, the tire is reinflated to the correct pressure. This inflation, deflation, and reinflation procedure allows the tube to straighten itself out inside the tire and will remove any wrinkles from the tube. [Figure 9-87]

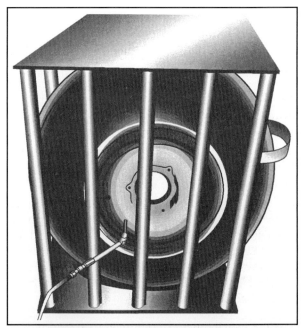

Figure 9-87. Tires should be placed in a safety cage during inflation. A tire, if defective or overinflated, can explode with enough force to cause serious injury or death.

The air pressure in a tube-type tire will drop after its initial inflation because the nylon plies stretch in the same way as a tubeless tire, and there may also be air trapped between the tube and the tire. When the trapped air leaks out around the valve under the beads or through the sidewall vents, the inflation pressure drops. All of this air should be out within the initial 12- to 24-hour period, and the pressure may then be adjusted and the tire put into service. It is advisable to paint a stripe or mark on the side of the tire extending to the rim of the wheel. This mark is called a **slippage mark** and is used detect slippage of the tire on the wheel, which most often results from under-inflation or severe stress from a hard landing or heavy braking.

TIRE BALANCING

As aircraft takeoff speeds increase, the vibration caused by unbalanced wheels becomes annoying. This vibration is especially noticeable on nose wheels, because they extend quite a distance below the airplane on a slender strut. Nose wheels usually do not have a brake to help dampen the vibrations.

After the tire is mounted on the wheel, inflated, and allowed to take its initial stretch, the assembly is mounted on a balancing stand with the cones of the balancing shaft sitting firmly against the bearing cups in the wheel. The shaft is placed on the balancing stand and the wheel allowed to rotate until its heavy point comes to a rest at the bottom. [Figure 9-88]

The wheel is counterbalanced with test weights until the assembly is balanced, and then the correct amount of weight is installed on the wheel at the location identified by the test weights.

Some balance weights are installed on special brackets that mount under the head of the wheel bolts. Others fasten to the wheel rim by a cotter pin through holes that have been drilled in the rim for that purpose. [Figure 9-89]

Many smaller wheels do not have provisions for mechanically attaching balance weights. For these wheels, lead strips having an adhesive backing may be used. Only the type of weight that is approved for the particular wheel being balanced should be used. As always, the instructions in the aircraft service manual should be followed for the installation of these weights. [Figure 9-90]

OPERATING AND HANDLING TIPS
TAXIING
Taxiing is the controlled movement of the airplane under its own power while on the ground. Since an airplane must be moved under its own power for taxi and run-up, the technician must thoroughly understand taxiing procedures and be proficient in maintaining positive control of the airplane's direc-

Figure 9-88. Wheel-balancing stands can be used to find and counterbalance the heavy spot in each tire-wheel combination.

tion and speed. In addition, the operator must be alert and visually check the location and movements of everything along the taxi path.

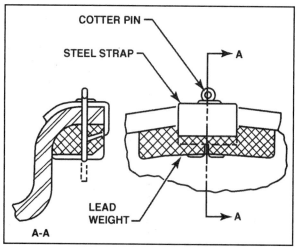

Figure 9-89. Some wheel balance weights are held on the wheel with a cotter pin through a hole drilled in the wheel for this purpose.

Figure 9-90. Strip-type lead balance weights with adhesive backs are used on some wheels that do not have provision for mechanically attached weights.

An awareness of other aircraft that are taking off, landing, or taxiing, and consideration for the right-of-way of others is essential to safety. To really observe the entire area, the operator's eyes must cover almost a complete circle. While taxiing, the technician must be sure the airplane's wings will clear all obstructions and other aircraft. If at any time there is doubt about the clearance of the wingtips, the airplane should be stopped so that someone can check the amount of clearance from the object.

It is difficult to set any rule for a safe taxiing speed. What is safe under some conditions may be hazardous under others. The primary requirement of safe taxiing is positive control, and the ability to stop or turn where and when desired. Normally, the speed should be at the rate where movement of the airplane is dependent on the throttle; that is, slow enough that when the throttle is closed, the airplane can be stopped promptly. Generally, taxiing up to five to ten miles per hour is a good rule of thumb.

BRAKING AND PIVOTING

Initially, taxiing should be done with the heels of the feet resting on the cockpit floor and the balls of the feet on the bottom of the rudder pedals. To brake the aircraft, slide your feet up to the brake pedal, the upper portion of the rudder pedal, and depress the brakes. The feet should be slid up onto the brake pedals only when it is necessary to depress the brakes; otherwise you may have a tendency to ride the brakes, causing undue wear and tear on the brakes and tires. Only after considerable experience is gained, should the feet be positioned with the arches placed on the rudder pedals and the toes near, but not quite touching, the brake portion of the pedals, which permits the simultaneous application of rudder and brake whenever needed. The brakes are used primarily to stop the airplane at a desired point, to slow the airplane, or as an aid in making a sharp controlled turn. Whenever used, they must be applied smoothly, evenly and cautiously to avoid unnecessary wear on the tires and brakes.

FIELD CONDITION

While taxiing over areas that could contain loose gravel and sand, the engine rpm should be kept as low as possible to prevent propeller damage. Rocks can be drawn up into the area of the propeller blades, striking the blades' leading edge and causing serious blade damage. Tires and, if applicable, the wheel fairings should be checked due to the increased amount of friction from the rough surface. Operating an aircraft on muddy fields can also pack mud into the wheel fairings. After drying, the mud can rub on the tire surface creating premature wear patterns.

CHAPTER 10

POSITION AND WARNING SYSTEMS

INTRODUCTION
Pilots of today's complex aircraft can no longer fly by the seats of their pants. The pilot receives indications of what the aircraft is doing through instruments and warning systems. These include airspeed indicators, unsafe system warnings, and remote position indicators. Some systems, such as antiskid brake systems, allow the pilot to obtain maximum performance, which may be impossible without mechanical assistance. This section covers some of these systems and the hardware necessary to operate them.

SECTION A

ANTISKID BRAKE CONTROL SYSTEMS

It is important that a pilot avoid excessive braking to prevent skidding and loss of control. With a tailwheel-type airplane, too much braking could result in a nose-over or ground loop. With large-diameter tires on small wheels, heavy braking could cause the tire to slip on the rim and pull the valve out of the tube.

Modern high-speed jet aircraft usually have more than one wheel on each side, and all of the brakes on one side are controlled with one pedal. With this arrangement, the pilot has no way of knowing when one of these wheels begins to skid. Without prompt corrective action to release a locked-up wheel, the tire is likely to blow out and damage the aircraft, or in severe cases, result in loss of control.

Friction created by the brakes reduces the wheel rotation rate, and friction between the tire and the runway slows the aircraft. If the tire rotation slows too rapidly, the tire will begin to slip on the runway instead of gripping it. Once the tire begins to slip, a skid soon develops and braking effectiveness decreases rapidly to near zero. For maximum brake effectiveness, only enough brake pressure should be applied to cause the tire to reach the point where it just begins to slip. This produces the maximum deceleration rate.

Maintaining this optimum friction is not easy. As the airplane slows, less brake pressure is needed to maintain the correct balance. Contamination such as water, snow or ice on the runway reduces the coefficient of friction between the tire and the runway. This, too, complicates the problem of maintaining the right amount of brake pressure to achieve maximum braking without excessive tire slippage.

SYSTEM OPERATION

You use a simple form of manual antiskid control when driving on ice. For the most effective stopping, you pump the brakes. They are applied only enough to slow the wheel, then released before the wheel decelerates enough to lock up. This same on-and-off type of operation was employed in some of the early aircraft antiskid systems. However, this method only works well when the control valves are capable of operating very quickly. [Figure 10-1]

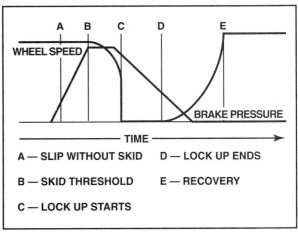

Figure 10-1. This graph shows the wheel speed relative to the amount of brake pressure applied manually by the pilot of an aircraft.

In figure 10-1, the brakes are applied and the pressure rises until the wheel starts to slip, but not skid, at point A. This is the ideal condition, but the pilot, having no indication that a slip has been reached, continues to increase the force on the brake pedal. Sufficient pressure is soon reached to produce enough friction in the brake to cause the tire to start to skid on the runway, as shown at point B. The wheel now decelerates fast enough to be felt, so the pilot reduces pressure on the pedal. Since the braking force that is needed lessens as the wheel slows, the wheel continues to decelerate even though the brake pressure decreases. At point C, the wheel has completely locked up, even though the pressure continues to drop. At point D, the pressure is low enough for the friction between the tire and the runway surface to start the wheel rotating again, and soon after, the brake pressure drops to zero. The wheel then comes back up to speed.

A successful antiskid system requires two features that early on-and-off systems did not have. There must be some form of wheel-speed sensor that can detect a change in the rate of deceleration and send a signal for the pressure to be released before the wheel gets deep into a skid. A valve is also needed that acts quickly enough to prevent all of the pressure from being released before the next application of the brake. This controlled amount of retained pressure prevents the brake-return system from pulling the pressure plate all of the way back, and allows the brakes to reapply almost immediately. The modern modulated antiskid system provides the fastest wheel-speed recovery and produces the shortest stopping distance on any kind of runway surface.

When the pilot wants to stop the aircraft in the shortest distance possible, it is necessary to depress the brake pedals all the way to induce maximum braking. All of the brakes receive the maximum pressure. If any wheel should decelerate at a rate indicating an impending skid, some of the pressure to that brake is dumped into the system-return manifold. The control circuit then measures the amount of time required for the wheel to spin back up and applies a slightly reduced pressure to the brake. This reduced pressure is determined by the time required for the spin-up. If this reduced pressure again causes a skid to develop, the cycle is repeated. Some pressure is maintained in the wheel cylinders to prevent the pressure plate from moving all of the way back. This application and release process continues with progressively decreasing pressure until the wheel is held in the slip area, but not allowed to decelerate fast enough to produce a skid. It produces the proper amount of braking for any runway surface condition, with the pilot having only to apply a hard, steady pressure to the brake pedal.

When the airplane slows down to approximately 20 miles per hour (m.p.h.) and there is no further danger of skidding, the antiskid system automatically deactivates. This gives the pilot full control of the brakes for maneuvering and parking. As with most auxiliary systems in modern aircraft, the antiskid systems have built-in test circuits, and may be deactivated in the event of a malfunction to give the pilot normal braking but no antiskid protection.

Many large jet-transports have an auto-brake feature that works in conjunction with the antiskid system. When the system senses weight on the main wheels, it automatically applies the brakes to produce one of several pilot-selected levels of deceleration. This results in a more immediate application of the wheel brakes and maximizes the use of the antiskid system. The pilot can override and disarm the auto-brake system by applying manual brakes.

SYSTEM COMPONENTS

An antiskid system consists of three basic components: wheel-speed sensors, an antiskid computer, and control valves. [Figure 10-2]

Figure 10-2. A typical antiskid brake system consists of wheel-speed sensors on each main wheel, a control unit, and control valves for each brake.

WHEEL-SPEED SENSORS

There are two types of systems in use, an AC system and a DC system. They are essentially alike except for the wheel-speed sensors and one circuit in the control unit. The AC sensor is a variable-reluctance AC generator in the axle of the landing gear that uses a permanent magnet surrounded by a pickup coil. The outside of this sensor has four equally spaced poles with teeth cut into their periphery.

A soft iron exciter ring with internal teeth is mounted in the hubcap of the wheel so that it rotates around the sensor. The two sets of teeth are separated by a small gap, and as the exciter ring rotates, the teeth approach each other and then move apart. As the distance between the teeth changes, the reluctance of the magnetic circuit is alternately increased and decreased. This causes the amount of magnetic flux cutting across the pickup coil to change and induces an alternating current in the coil. The faster the wheel turns, the higher the frequency of the induced current. [Figure 10-3]

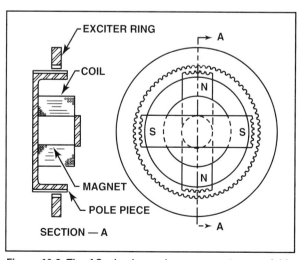

Figure 10-3. The AC wheel-speed sensor creates a variable frequency AC current. The control unit converts the varying frequency AC into a DC signal voltage that is proportional to the frequency of the AC current.

The DC sensor is essentially a small, permanent-magnet direct-current generator, which produces a voltage output directly proportional to the rotational speed of its armature. With this type of sensor, there is no need for a converter in the control unit. There also is less danger of interference with the brakes due to the induction of stray voltage into the sensing system. [Figure 10-4]

Figure 10-4. The DC wheel-speed sensor does not require an AC-DC converter in the control unit because it generates a direct current proportional to wheel speed. The shaft of the armature is fitted with a blade driven by a bracket in the wheel hubcap and rotates with the wheel. The generator output usually is in the range of one volt for each ten m.p.h. of wheel speed.

CONTROL VALVES

A three-port antiskid control valve is located in the pressure line between the brake valve and the brake cylinder, with the third line connecting the control valve to the system-return manifold. During normal operation of the brakes, with no indication of a skid, the valve serves only as a passage and allows the brake fluid to flow into and out of the brake. When a wheel begins to decelerate fast enough to cause a skid, the control unit detects the changing output voltage of the wheel-speed sensor. The control unit sends a DC signal to the control valve, which closes off the pressure port and opens the passage between the brake and the system return. This rapidly operating valve maintains an output pressure that is directly proportional to the amount of signal current from the control unit. [Figure 10-5]

The DC signal from the control unit flows through a coil around the armature of the flapper valve. This armature is free to pivot and is centered between two permanent magnets. [Figure 10-6]

When the signal from the control unit indicates that no skid is impending, and the braking action should be normal, the magnetic field of the coil reacts with the fields of the permanent magnets and holds the flapper centered between the nozzles. [Figure 10-7]

Figure 10-5. The antiskid control unit operates a brake control valve.

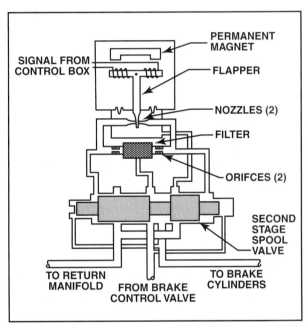

Figure 10-6. A direct-current signal from the control unit energizes the coil on the armature of the flapper valve, and the movement of the flapper changes the pressure drop across the fixed orifices.

Fluid from the brake valve flows through the filter and discharges equally from each nozzle. Since the amount of flow is the same through each orifice, the pressure drop across the orifices will be the same, and the second-stage spool valve will assume a position that allows free passage between the brake valve and the brake.

When the control unit receives a signal from the wheel-speed sensor indicating an impending skid, it sends current through the coil of the armature to polarize it. This causes the flapper to pivot and unbalance the flow from the nozzles. In figure 10-8, the flapper has moved over, restricting the flow

Position and Warning Systems

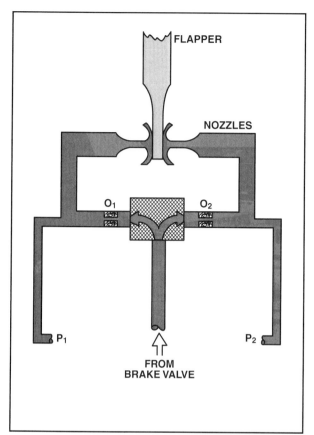

Figure 10-7. When the flapper is centered between the nozzles, the pressure-drops across orifices O_1 and O_2 are equal, resulting in output pressure P_1 equaling P_2.

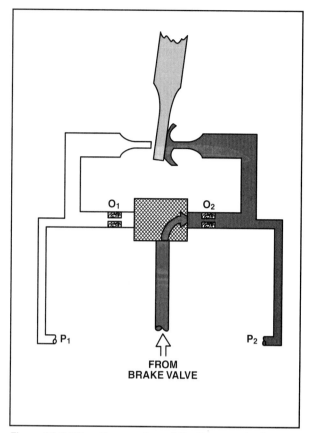

Figure 10-8. When the armature of the flapper valve is energized, the flapper moves over and restricts the flow through orifice O_1 while increasing it through O_2. The increased pressure drop across O_2 causes P_1 to be greater than P_2.

from the left nozzle and opening the flow from the one on the right. There is now more flow through orifice O_2 and therefore a greater pressure drop across it, leaving P_1 greater than P_2. This imbalance of pressures moves the second-stage spool over, shutting off the flow of fluid from the brake valve to the brake, and opening a passage from the brake to the return manifold.

The extremely fast reaction time of this valve allows it to maintain a pressure at the brake that is directly proportional to the amount of current flowing in the armature coil.

CONTROL UNIT

The control unit has three main functions: to generate electrical signals usable by the control valve; to regulate brake pressure to prevent a skid during landing deceleration; and to prevent application of brake pressure prior to touchdown. Before the airplane touches down, the locked-wheel detector sends a signal into the amplifier, which causes the control valve to open the passage between the brakes and the system-return manifold. This pre-

vents the pilot from landing with the brakes applied. [Figure 10-9]

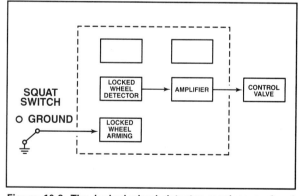

Figure 10-9. The locked-wheel detector receives a signal from the squat switch, which indicates whether the aircraft is airborne or on the ground. If airborne, the circuitry prevents the brakes from being applied before touchdown.

As soon as the airplane touches down, the squat switch registers that weight is on the wheels. The wheels start to spin up, and at approximately 20 m.p.h., generates enough voltage in the wheel-speed sensor to signal the locked-wheel detector. The detector then removes the touchdown control signal from the amplifier. This allows the control valve to apply full pressure to the brakes. [Figure 10-10]

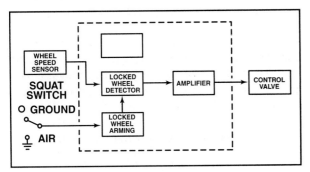

Figure 10-10. On touchdown, the squat switch removes the ground from the locked-wheel arming circuit, and the wheel-speed sensor generates a signal which allows the control valve to send full pressure to the brakes.

When the airplane is on the ground and the wheels are rotating at more than 20 m.p.h., the skid detector and modulator provide almost all of the antiskid control. [Figure 10-11]

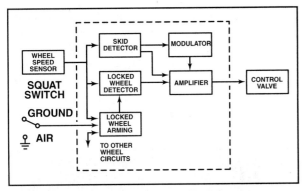

Figure 10-11. When the airplane is on the ground and all wheels are rotating more than 20 miles per hour, the skid detector and the modulator provide signals for the amplifier.

A deceleration threshold is designed into the skid detector circuit. The reference normally is set to about 20 feet per second, with a wheel speed that is at least six m.p.h. below the speed of the airplane. When a wheel decelerates at a rate greater than this threshold value, the skid detector signals the amplifier and then the control valve to reduce the brake pressure. It also signals the modulator, which automatically establishes the amount of current that will continue to flow through the valve after the wheel has recovered from the skid. When the amplifier receives its signal from the modulator, it maintains this current, which is just enough to position the flapper to prevent the pressure from being completely released. The applied current maintains a pressure slightly less than that which caused the skid. A timer circuit in the modulator then allows this pressure to increase slowly until another skid starts to occur, repeating the cycle.

When the aircraft is on a wet or icy runway, the antiskid system holds the wheels in the slip region. However, the locked wheel detector activates whenever one wheel hydroplanes or hits ice and slows down to less than ten m.p.h. while its mated reference wheel still rotates faster than 20 m.p.h. A timer measures the duration of the skid detector signal. If it is more than one-tenth of a second, it sends a "full dump" signal that holds the valve in the full-dump position until the wheel spins back up above ten m.p.h.

When all of the wheels are turning at less than 20 m.p.h., the locked-wheel arming circuit disarms, giving the pilot full braking action for low-speed taxiing and parking. [Figure 10-12]

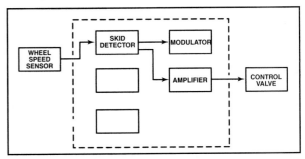

Figure 10-12. When the airplane is on the ground and all three wheels are rotating less than 20 miles per hour, the locked-wheel arming circuit is inoperative and the pilot has full brake control for low speed taxiing and parking.

The control unit for antiskid systems using AC sensors operates in the same way as those using DC generators, the only difference being the addition of a converter circuit. This circuit receives the varying-frequency alternating current and converts it into a varying voltage of direct current. The changes in the DC voltage exactly follow the frequency changes of the AC. [Figure 10-13]

SYSTEM TESTS

Because it is vitally important that a pilot know the exact condition of the brake system before using it, antiskid systems include test circuits and control switches. These allow the pilot to test the entire system, and if any faults are found, disable the system without affecting normal braking action. There is an anti-skid warning light in the flight deck to warn pilots whenever the system is off or has failed.

GROUND TEST

The integrity of the antiskid system can be tested on the ground before flight. The pilot turns on the anti-

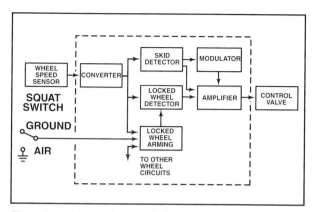

Figure 10-13. The difference between the control unit of an antiskid system using an AC wheel-speed sensor and one using a DC sensor is in the converter between the sensor and the control circuit.

skid control switch and presses the brake pedal. Both the left and right brake lights should illuminate, indicating that all of the pressure from the brake valves is being routed to the brakes.

With the brakes still applied, the pilot presses the test switch and holds it for a few seconds. This sends a signal through the wheel-speed sensors into the control unit to simulate a wheel speed of more than 20 m.p.h. The lights should remain on. When the test switch is released, the two brake lights should go out and stay out for a couple of seconds, then come back on. This simulates a wheel lockup that causes a release of, then restoration of, pressure. This test checks the continuity of all of the wiring and operation of the locked-wheel circuits, amplifiers, and control valves. These procedures vary with aircraft type. Consult the appropriate manuals to determine the correct procedure for your aircraft.

IN-FLIGHT TEST

The antiskid system is included in the pilot's pre-landing checklist. With the airplane configured for landing, the pilot depresses the brake pedals. The brake lights should remain off, which indicates the control valves are holding the brakes in the fully released position.

The pilot then presses the test switch, which should illuminate the brake lights for as long as the switch is held down. The test switch sends a signal through the wheel speed sensors, simulating a wheel speed greater than 20 m.p.h. If the system is operating properly, the control valve will direct normal pressure to the brake.

SYSTEM MAINTENANCE

If a flight crew reports an antiskid or brake malfunction, verify that there is no air in the brake system before condemning the antiskid system. If the brakes are spongy, remove the air by bleeding them. Carefully check for warped disks, malfunctioning return systems, and any indications of damage.

Inspection and maintenance of antiskid systems requires logical troubleshooting to locate faults. Due to the complexity of the components, they are usually returned to the manufacturer or a repair station for any needed repairs. If one of the tests shows a malfunction in the system, the most logical place to start troubleshooting is with the wheel-speed sensor.

WHEEL-SPEED SENSOR

Some DC wheel-speed sensors can be checked on the airplane by removing the wheel hubcap to expose the blade of the sensor. With your finger, give the blade a sharp spin in its normal direction of rotation with the brakes applied and the antiskid switch on. It will not turn more than 180 degrees. It is not the amount of rotation that is important, but the rate at which it is turned. If the system is operating properly, the brakes should momentarily release and then reapply. Watch the brake disk stack for relaxation then tightening, this will confirm proper system operation. If this "tweak" test does not cause the brakes to release, consult the maintenance manual for the specific type of airplane on which you are working to determine the correct test procedures. [Figure 10-14]

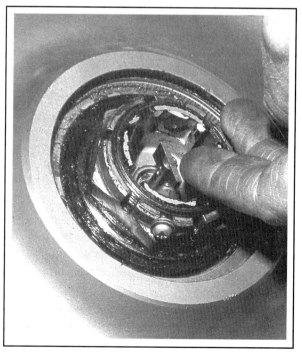

Figure 10-14. When the blade of the wheel-speed sensor is flipped, it should cause the brakes to release and then reapply.

CONTROL UNIT

The control unit, shown in figure 10-15, may be checked using a substitution method. Remove both of the connector plugs from the box and swap them left to right. For example, suppose the trouble indication was originally on the left side of the airplane. If the leads from the box are switched and the indication remains on the left side, the trouble is probably not with the control unit. However, if the indication moves to the right side, the control unit may be defective. Any time you switch the leads, be sure to reinstall them on the proper receptacles and properly secure them before returning the aircraft to service.

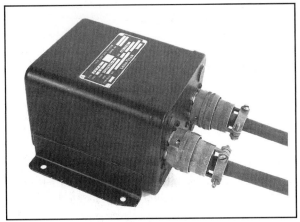

Figure 10-15. The two leads on the antiskid control unit may be switched as a part of the troubleshooting procedure.

CONTROL VALVE

If the trouble remains after checking the two devices that were the easiest to access, all that remains in the antiskid system is the control valve. These valves are electrohydraulic, and the trouble could be in either the electrical or hydraulic section.

The easiest check is the electrical resistance of the coil. Remove the connector plug and measure the resistance of the coil with an accurate ohmmeter. It should measure within the tolerance specified in the service manual. If the trouble is traced to the control valve and is not electrical, the valve must be removed. The problem is probably in the hydraulic portion of the valve.

The extremely close tolerances used in the manufacture of this valve make the use of absolutely clean fluid imperative. A fifteen-micron steel-mesh screen is commonly installed in the line before the orifices to insure that no contaminants reach the inside of the valve. If this screen clogs, the valve may malfunction. Check the manufacturer's service manuals to see if it is possible to replace this filter in the field. If it is allowed, follow the service instructions carefully. If any field servicing is allowed on the valve, it must be done in an area free from contamination. Again, be certain to follow the manufacturer's latest service information.

SECTION B

INDICATING AND WARNING SYSTEMS

STALL WARNING INDICATOR

A stall is a flight condition where the airflow over the upper surface of the wing separates and becomes turbulent. It occurs when the aircraft reaches a critically high angle-of-attack (AOA). If an airplane does not provide sufficient aerodynamic warning of an impending stall, such as buffeting, the pilot must be warned through some other means. Small general aviation aircraft usually use an audible tone or a red light. Many high-performance aircraft use a stick shaker, which vibrates the control column, or which may even force the column forward to reduce the angle-of-attack.

Many stall warning systems, particularly on lower performance aircraft, measure the movement of the stagnation point on the wing. The stagnation point marks the particular location on the leading edge of an airfoil where the air separates, some passing over the top of the surface and the rest passing below it. As the angle-of-attack increases, the stagnation point moves down toward the lower surface. The stagnation point is always in the same location when the airflow over the surface becomes turbulent, indicating the approach to a stall.

ELECTRIC STALL WARNING

An electrically operated stall warning system uses a small vane mounted near the stagnation point in the leading edge of the wing. At flights above the stall speed, the airflow over the vane is downward and the vane is held down. An electrical switch connected to the vane is open while the vane is down. As the angle-of-attack increases toward an impending stall, the stagnation point moves down until the airflow over the vane is upward. The vane is blown up, closing the switch and illuminating a red light or sounding a warning horn. [Figure 10-16]

NON-ELECTRIC STALL WARNING

The reed-type stall warning system operates in a manner similar to a musical instrument reed which produces a tone when air travels through it. The inlet of the small reed-type horn is located on the leading edge of the wing near the stagnation point.

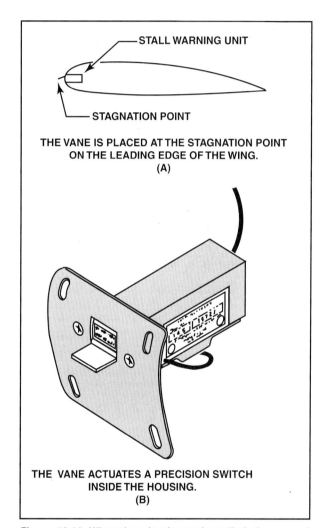

Figure 10-16. When the wing is nearly stalled, the upward airflow moves the vane to activate the stall warning.

As the angle-of-attack increases, the low-pressure air traveling over the wing moves into an area where the reed inlet is located, causing it to sound. By listening to the changing pitch of the horn, the pilot can easily identify the point at which the stall will occur.

On many high-performance aircraft, the margin between the aerodynamically generated pre-stall buffet and the actual stall is insufficient. Using the stagnation point to activate a stall warning system may not provide enough warning. Many corporate jet and transport category aircraft use a stick shaker to provide the pilot with an earlier and more reliable warning of an impending stall. The stick shaker consists of a motor that drives an eccentric weight. This motor is attached to the control column and shakes it to alert the pilot before a stall develops. A stall-warning computer based on airspeed, angle-of-attack, flap configuration, and power setting activates the stick shaker. The system is energized at all times when the aircraft is airborne and is deactivated on the ground by squat switches on the gear.

ANGLE-OF-ATTACK INDICATORS

All stall warning systems provide an indication of an impending stall that is related to the angle-of-attack. For precision flying, the pilot needs to know the actual angle-of-attack during various stages of the flight. One system for measuring and displaying the angle-of-attack uses a slotted probe sticking out of the side of the aircraft fuselage. The slots carry impact air into the housing of the probe where it moves a set of paddles connected to a variable resistor. The change in resistance moves a pointer around the indicator dial, which is calibrated in percent of the stall-speed angle-of-attack, or color-coded with a qualitative indication of angle-of-attack. [Figure 10-17]

Another method of measuring angle-of-attack utilizes a vane-type sensor. A thin, wedge-shaped vane is mounted on a short arm that is free to rotate. In flight, the vane streamlines with the relative wind. As the angle-of-attack changes, the arm pivots and a potentiometer connected to the arm transmits a position signal to the stall warning system. The vane is heated to prevent ice formation. [Figure 10-18]

The pilot can set a reference bug to show the desired ratio of the airspeed to the stall airspeed. For example, if the pilot wants to make an approach to landing at an airspeed of 30% over the stall speed, the reference bug would be set on 1.3. The pilot then maintains the attitude needed to center the angle-of-attack needle on the reference bug and the approach speed will automatically be correct. If the angle-of-attack goes above or below the desired value, the indicator will move away from the bug.

REMOTE POSITION INDICATING SYSTEMS

A pilot needs to know that a control surface has actually moved when commanded. Remote position indicating systems provide feedback about the status of control surfaces, landing gear, control valves,

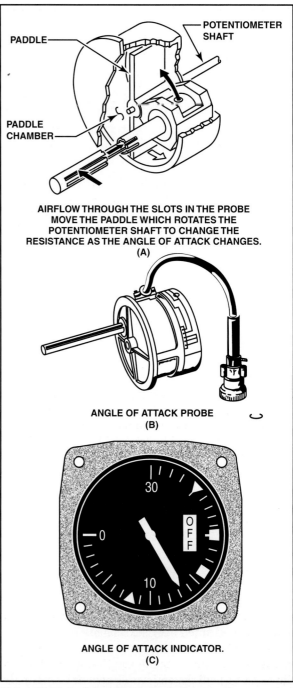

Figure 10-17. As the angle-of-attack changes, the amount of air entering the angle-of-attack sensor changes. This causes the paddles inside to change position. These paddles are attached to a potentiometer that varies the current to an indicator that in turn gives an indication of AOA.

and other mechanically actuated devices.

DIRECT CURRENT

Direct-current remote indicating systems are used in some aircraft to transmit position information so that it can be seen on an instrument dial. The position pickup, or transmitter, is a variable resistor

Position and Warning Systems

Figure 10-18. Many large airplanes utilize a vane-type sensor for angle-of-attack.

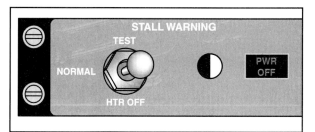

Figure 10-19. When the pilot selects the Test position on a Boeing 747 stall warning system, the air/ground relay is bypassed, the stick shaker operates, the black and white test indicator rotates, and the system checks the angle-of-attack vane and flap position sensor.

made of wire wound around an insulating core in the shape of a cylinder. Two wipers contact bare portions of the wire along one edge of the cylinder, and current flows into the circuit through one of the wipers and out through the other. The cylindrical resistor is tapped at each 120-degree position and is connected to a coil in the indicator that is wound on a ring-shaped core. The indicator coil is also tapped at each 120 degrees and connected to form an electrical delta circuit. The current through each of the three portions of the coil varies depending upon the position of the two wipers in the transmitter. As the current changes, so does the magnetic field. Since a small permanent magnet attached to the pointer always aligns with the composite magnetic field, the indicator is always aligned with the wiper arms in the transmitter. [Figure 10-20]

ALTERNATING CURRENT

Many larger aircraft require greater accuracy than is available from a DC remote position indicating system. For these applications, alternating-current systems of either the Autosyn® or Magnesyn®-type are used.

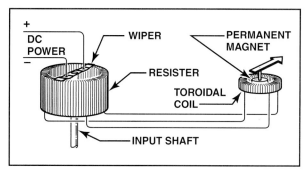

Figure 10-20. A variable resistor provides a variable current to a coil that aligns a permanent magnet with the resistor's wiper.

AUTOSYN® SYSTEMS

One of the more popular remote indicating systems used for all types of mechanical movement is the Autosyn® system. Autosyn® is a registered trade name for a system that uses a single-phase electromagnet for the rotor and a three-phase delta connected coil for the stator. [Figure 10-21]

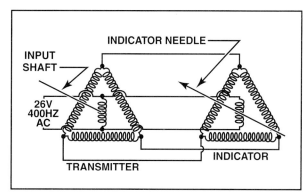

Figure 10-21. The Autosyn®-type alternating-current remote indicating system employs two delta-wound coils. These coils align with each other; one of them attached to an input shaft and the other to a remote pointer.

The synchronous motors in the indicator and transmitter are identical. The rotors are connected in parallel and supplied with 28-volt, 400-hertz AC. The three-phase stators are also connected in parallel, and in most installations, one side of the rotor is connected to one of the terminals of the stator.

Whatever position is being monitored physically moves the rotor of the transmitter. This could be the flap position, landing gear position, or oil or fuel quantity, as well as many of the pressure measurements made with bourdon tubes or pressure capsules.

The AC magnetic field in the rotor induces a voltage in the three windings of the stator, and because the two stators are connected in parallel, the magnetic field in the indicator will be exactly the same as that in the transmitter. The same AC voltage as the rotor in the transmitter excites the rotor in the indicator so their magnetic fields are identical. Since mechanical load on the indicator rotor is nothing more than a small pointer, the rotor will assume the same position inside the indicator as the rotor inside the transmitter. The rotor in the indicator immediately follows any movement of the transmitter rotor.

Many Autosyn® systems use dual indicators. The two synchronous motors are stacked, and the shaft of the rear motor sticks through the hollow shaft in the forward motor. One dial serves both indicators, and the two pointers move in the same way the hands of a clock do.

MAGNESYN® SYSTEMS

Magnesyn® is another remote indicating system bearing a registered trade name and operating on AC. The basic difference between an Autosyn® and a Magnesyn® system is in the rotor. The Magnesyn® system uses a permanent magnet for its rotor rather than the electromagnet used in the Autosyn® system.

The stator of a Magnesyn® system is a toroidal coil: a coil wound around a ring-shaped iron core. The transmitter and indicator are not necessarily the same physical size and configuration, but they are alike in their electrical characteristics.

The coils in both the transmitter and the indicator are supplied with 28-volt, 400-Hertz AC, are tapped each 120 degrees, and are connected in parallel. The voltage generated in the transmitter coil is carried into the indicator coil where it produces magnetic fields in its three sections. The composite field of these coils pulls the permanent magnet in the indicator into exactly the same alignment as the magnet in the transmitter. Any movement of the transmitter magnet causes the magnet in the indicator to mirror the transmitter position. [Figure 10-22]

CONFIGURATION WARNING SYSTEMS

The number and complexity of modern aircraft systems require various warning systems to alert the pilot of malfunctions or incorrect aircraft configuration for a particular flight mode. Most warnings are visual, aural, tactile, or some combination. Warnings alert the aircrew to conditions that require some sort of action to ensure proper and safe operation of the aircraft. The type of signal depends upon the degree of urgency. One type of warning system is the fire warning system, which will be covered in

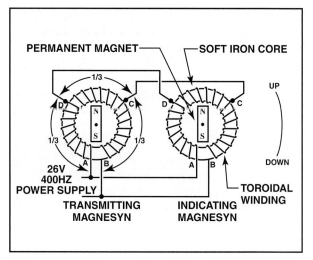

Figure 10-22. A Magnesyn®-type AC remote indicating system uses the paired relationship of two permanent magnets to transfer transmitter position information to an indicator.

depth in Chapter 16. Other types of warning systems include takeoff configuration warning, landing gear configuration warning, Mach/airspeed warning, stall warning, ground proximity warning system (GPWS), and the engine indication and crew alerting system (EICAS).

TAKEOFF CONFIGURATION WARNING SYSTEM

The takeoff configuration warning system is armed when the aircraft is on the ground and one or more thrust levers are advanced to the takeoff power position. A warning light and/or aural warning will sound if the stabilizer trim is not properly set, trailing edge flaps are not in the correct position, any leading edge devices are not properly set, or the speed brake is not properly stowed. The warning signal stops when all monitored devices are properly set.

LANDING GEAR CONFIGURATION WARNING SYSTEM

The landing gear indication lights are activated according to signals from each gear and the landing gear lever. The particular gear indications may vary slightly, but the FAA requires positive indication of "up and locked" and "down and locked" gear positions. A typical system might indicate the landing gear down and locked with an illuminated green light for each individual gear. Another may use a single green light for the entire gear configuration "down and locked" indication. If a single green light is used, the switches at each gear are connected in series so that the "down and locked" light only illuminates when all gear are in the proper position.

When the landing gear is in disagreement with the landing gear lever position, a red light illuminates, meaning that the gear is in transit or in an unsafe condition. When the landing gear is in the proper up position and the gear lever is also in the "UP" position, the gear position lights go out signifying an "up and locked" condition. A technician normally checks the gear warning system during landing gear retraction tests. Problems with the warning system are often caused by the gear position switches. Always consult the manufacturer's service manual for the proper procedures for adjusting the landing gear position switches in addition to any other maintenance performed.

On some aircraft, a steady warning horn is provided to alert the pilot that the airplane is in a landing configuration and the gear is not down and locked. The landing gear warning horn is usually dependent on flap and thrust lever position.

Generally, when a thrust lever is retarded and any landing gear is not down and locked, the landing gear warning horn will sound, but can be silenced using the warning horn cutout switch. Under certain conditions, the landing gear warning horn cannot be silenced. Although the actual flap settings and thrust lever positions will vary from one aircraft type to another, generally some provision is made to remove the pilot's ability to silence the gear warning when specific conditions occur. For example, the warning horn cutout might be disabled if the radar altimeter indicates less than 1,000 feet above ground with the aircraft in a landing configuration and with an unsafe gear.

MACH/AIRSPEED WARNING SYSTEM

Some aircraft are equipped with Mach/airspeed warning systems that provide a distinct aural warning any time the maximum operating airspeed is exceeded. Reducing speed below the limiting value is usually the only way to silence the warning.

The system operates from an internal mechanism inside the Mach/airspeed indicator. Test switches allow an operational check of the system at any time. Maximum operating airspeeds exist primarily due to airplane structural limitations at lower altitudes and airplane handling characteristics at higher altitudes.

GROUND PROXIMITY WARNING SYSTEM (GPWS)

The ground proximity warning system (GPWS) provides warnings and/or alerts to the flight crew when any of the following conditions exist:

- Excessive descent rate.
- Excessive terrain closure rate.
- Altitude loss after takeoff or go-around.
- Unsafe terrain clearance when not in the landing configuration.
- Excessive deviation below an ILS (Instrument Landing System) glide slope.
- Descent below the selected minimum radio altitude.
- Windshear condition encountered.

When one of these conditions is encountered, the computer flashes warning lights and sounds an alarm or warning. Some warnings are computer-generated directions such as "Pull up" or "Windshear."

ENGINE INDICATION AND CREW ALERTING SYSTEM (EICAS)

Older commercial airplanes utilize electromechanical system indicators that employ multiple visual and aural cautions and warnings to alert of hazardous conditions such as engine problems or open cabin doors. Most of these systems use an annunciator that provides a master warning light along with an aural indication to alert the crew that a malfunction has occurred and that corrective action may be required. These indicators do not offer the versatility and redundancy available with modern digital technology.

New generation aircraft use electronic displays and a full-time monitoring system known as EICAS, Engine Indication and Crew Alerting. The use of EICAS requires very little monitoring by the crew and promotes quick, accurate identification and recording of problems.

EICAS reduces flight crew workload by automatically monitoring and recording engine parameters for later review. EICAS also alerts the aircrew of problems when necessary. It is operative through all phases of flight, from power-up through post-flight maintenance. Parameters used to set and monitor engine thrust are displayed full time. The system automatically displays any out-of-tolerance values on a cathode-ray-tube (CRT) or liquid-crystal display (LCD) in an appropriate color. The colored messages are designed to alert the aircrew to any failure and convey the urgency in which to respond. By utilizing electronic displays, EICAS provides accurate, timely information on a single screen rather than multiple engine instruments scattered throughout the panel.

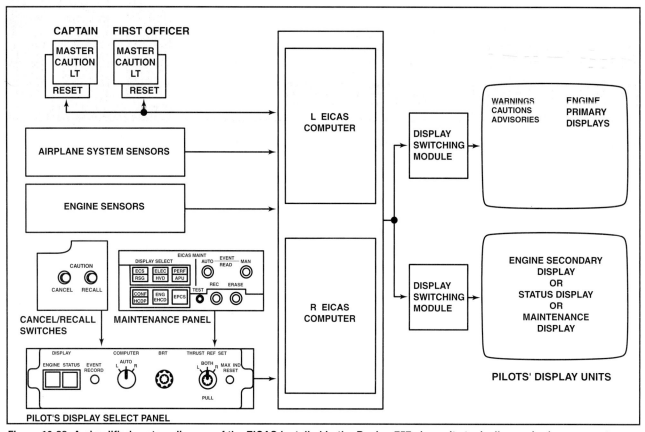

Figure 10-23. A simplified system diagram of the EICAS installed in the Boeing 757 shows its typically required components.

EICAS provides an improved level of maintenance data for the ground crew without causing the flight crew any extra workload. This has been achieved by designing a system that will automatically record subsystem parameters when malfunctions are detected. The system also provides the flight crew with the capability for manual data recording with the push of a single button. This eliminates the need for extensive hand recording of systems and performance data. These features increase the accuracy of maintenance data recordings and improve the communication between the aircrew and ground maintenance crews.

EICAS usually includes two multicolor display units, two computers, and two control panels. These components, together with two display-switching modules, cancel/recall switches, and captain's and first officer's master caution lights, jointly perform the various EICAS functions. [Figure 10-23]

The EICAS computer processes and displays all engine and aircraft system information required by the crew. One computer is used at a time for displaying the data on both display units. Computer selection is done on the display select panel. The upper display unit shows primary engine parameters and crew alerting messages, and the lower display unit shows secondary engine parameters. [Figure 10-24]

EICAS monitors inputs from airplane subsystems and sensors. When an abnormal condition is detected, EICAS will generate and display an alert, status, or maintenance message.

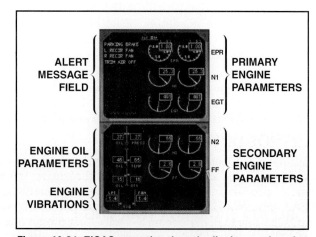

Figure 10-24. EICAS operational mode displays and engine parameters are presented on two displays. A pilot can select status and maintenance readouts on the secondary display using the EICAS maintenance panel.

CHAPTER 11

AIRCRAFT INSTRUMENT SYSTEMS

INTRODUCTION
Maintenance technicians must be familiar with the various types of instruments used to convey information to the pilot. Some are flight instruments that depict the attitude, airspeed, and altitude of the aircraft. Other instruments provide information such as engine operational parameters and electrical system performance. While the FAA prohibits aircraft technicians from performing any maintenance or repairs to the actual instruments, they must be able to inspect them to verify their proper operation and installation. In addition, technicians must maintain the components that support the instruments, such as electrical wiring and fluid-line plumbing.

SECTION A

PRINCIPLES OF INSTRUMENT SYSTEMS

The development of efficient flight instruments is one of the most important factors that contributed to the growth of the present air transportation system. Prior to World War II, few airplanes were equipped for flight without using ground reference navigation or pilotage. Poor visibility or low cloud cover, therefore, required flying at dangerously low altitudes or not flying at all. [Figures 11-1, 11-2]

Figure 11-2. In addition to the basic engine instruments, the panel of this 1929 Cessna only had an airspeed indicator and altimeter.

Figure 11-1. Flight instruments were no problem in the early Curtis airplane — because there were none.

On September 24, 1929, Jimmy Doolittle made a flight without outside visual references. The Consolidated NY-2 airplane he flew had an artificial horizon, which gave him an indication of the pitch and roll attitude of the airplane relative to the earth's surface. He also had a sensitive altimeter that showed the airplane's altitude above the ground within a few feet, and a radio direction finder, which allowed him to determine his position relative to the landing area. With this equipment, Doolittle proved that blind flight indeed was possible.

During the 1940s, instrument flight in zero visibility became commonplace in civilian aviation. With the aid of these newly developed flight instruments, the pilot could keep the airplane straight and level with no outside reference. However, this was of little use without some method of navigating the airplane once it was off the ground. Flight instruments and navigation equipment normally operate in conjunction with each other, particularly with modern auto-flight systems.

Integrated circuits containing microprocessors and other digital electronics have revolutionized flight instrumentation and control systems. New-generation flight instruments show textual and analog information on bright color displays. [Figure 11-3]

PRESSURE-MEASURING INSTRUMENTS

Many aircraft instruments rely on measuring the pressures of fluids such as air, fuel, and oil. Pressures can be measured directly by applying the fluid's force to a movable bellows, or by allowing

Figure 11-3. The instrument panel of a modern twin-engine corporate airplane provides the pilot with all of the information needed to conduct a safe flight under instrument meteorological conditions.

the fluid to enter a transducer that converts the pressure energy directly into an electrical signal. This signal is then used to drive an electrical or electronic display.

PRINCIPLES OF PRESSURE MEASUREMENT

Pressure is the force differential between two points. The force exerted by the air that constantly surrounds us is known as atmospheric pressure, created by the weight of all the air that sits on top of the earth's surface. Because many factors cause atmospheric pressure to vary, scientists have found it necessary to establish an average or "standard day" pressure value. This common reference for temperature and pressure is called the International Standard Atmosphere (ISA). [Figure 11-4]

Many aircraft and engine performance specifications are referenced to standard pressure. Although standard pressure exists at sea level and other elevations, it can change with weather patterns. These changes affect an aircraft in many ways. For example, a wing produces lift because of differential air pressure. The pressure above the upper surface of a wing is lower than the pressure below the underside of the wing. This creates a lifting force. As the atmospheric pressure changes, the amount of lift produced by the wings changes also.

Because pressure changes directly affect the performance characteristics of an aircraft, it is extremely important that an aircraft technician understands them. There are three basic ways of measuring pressure: absolute pressure, gauge pressure and differential pressure.

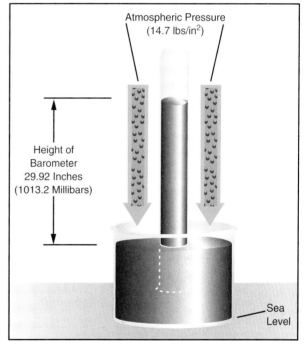

Figure 11-4. Atmospheric pressure can be measured with a mercury barometer. An inverted tube, closed at one end and open at the other, is placed in a bowl of mercury. The weight of the atmosphere causes the mercury to rise up in the tube. The pressure (weight) of the atmosphere balances the height of the column. At sea level in a standard atmosphere, the weight of the atmosphere supports a column of mercury 29.92 inches high.

ABSOLUTE PRESSURE

Absolute pressure is the measurement of pressure relative to a total vacuum. Atmospheric pressure is measured in absolute terms in that it is compared to zero, or a complete vacuum. Pressure is always a comparison between two forces. Atmospheric pressure can be measured in inches of mercury, hectoPascals, or pounds per square inch (psi). HectoPascals (hPa) is the modern term for an older measurement called "millibars." Many jet aircraft have altimeters whose setting windows are labeled in inches of mercury and millibars. Under standard day conditions, atmospheric pressure that equals 29.92 inches of mercury equals 1013.2 hPa or 14.7 psi.

Absolute pressure is often used on the aircraft in comparison to other pressures. In order to make this comparison, a device called an aneroid wafer was devised. This simple device consists of an enclosed chamber, typically made of thin sheet metal. Air in the chamber is evacuated, creating a vacuum that can be used as a reference to measure absolute pressure. [Figure 11-5]

inside the tire and the air pressure outside. This pressure differential is what causes the tire to inflate.

Gauge pressure is measured simply by applying a pressure to a given area and measuring the force it exerts. If a force of 32 psi (absolute pressure) is applied to one side of a bellows and 14.7 psi (atmospheric pressure) is on the other side, the resulting force is 17.3 psi. If the bellows has a total surface area of 0.5 square inches, the force applied is 8.65 pounds. This force can be used to move a mechanical pointer over a dial that would be marked in pounds per square inch. With this amount of applied force, the pointer would be calibrated to indicate 17.3 psi. [Figure 11-6]

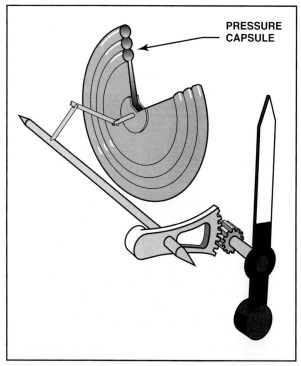

Figure11-5. The aneroid wafer measures the difference in pressure between the vacuum inside a sealed chamber and the ambient pressure around it. Expansion and contraction of the pressure capsule or chamber mechanically moves a needle on a gauge.

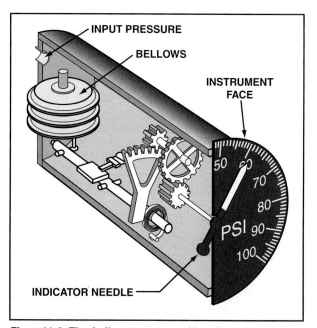

Figure11-6. The bellows-type aneroid wafer measures the difference between the pressure inside a bellows and the pressure outside of the bellows.

The aneroid wafer compares the pressure outside the aneroid to the inside pressure, which is zero. This difference is absolute pressure.

GAUGE PRESSURE

The most commonly used type of pressure is gauge pressure. It is the difference between atmospheric pressure and the pressure being measured. For example, when measuring tire pressure, it is important to know the difference between the air pressure

To measure high pressure, a Bourdon tube-type instrument is used. This instrument consists of a hollow brass or bronze elliptical-shaped tube formed into a semi-circle. One end of the tube is open and connected to the fluid to be measured; the opposite end is sealed. When pressure is applied, the elliptical tube changes shape, which tends to straighten the semi-circular curve. A useful instrument for measuring pressure can be created by attaching the Bourdon tube to a mechanical linkage and pointer. [Figure 11-7]

DIFFERENTIAL PRESSURE

Differential pressure is the comparison between two different pressures. The most common differential

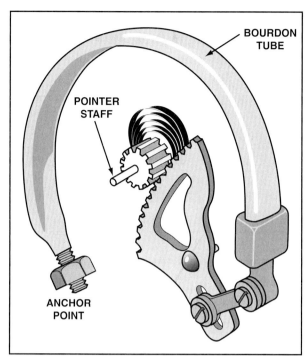

Figure 11-7. A Bourdon tube is connected to a mechanical linkage in order to create a useful pressure-measuring instrument.

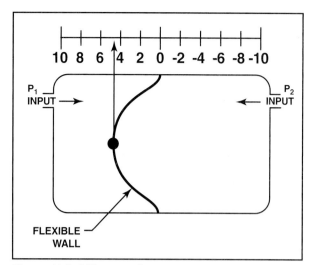

Figure 11-8. Differences in pressure between P1 and P2 cause the flexible wall to bend and to move a pointer on a calibrated scale.

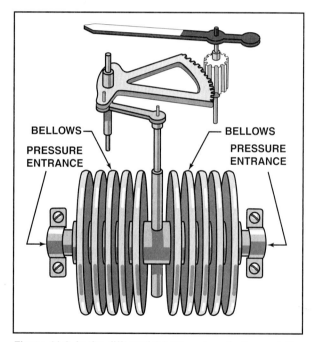

Figure 11-9. In the differential bellows-type instrument, the pressure differential creates a linear movement of the bellows assembly. The indications are the result of pressure differential, but the mechanical linkage allows the presentation of the pressure to be displayed on a round dial.

pressure gauge found on an aircraft is the airspeed indicator. This instrument measures the difference between ram air or pitot pressure, and the static or ambient air pressure. Ram pressure is created by the aircraft's forward motion and static pressure is the atmospheric pressure outside the aircraft. This special type of differential pressure instrument is typically called a pitot-static instrument.

In order to measure differential pressure, two sealed chambers must be used with each chamber connected to its respective pressure. The two pressure forces oppose each other along the common wall of the two chambers. The common wall is flexible so it can bend into or out of the chambers. If pressure P1 exceeds P2, the flexible wall moves toward P2, and vice versa. A pointer mechanism can be attached, and calibrated to measure the differential pressure. [Figure 11-8]

A more useful type of differential-pressure instrument is the differential bellows type. This instrument uses two enclosed bellows, each filled with the associated pressures to be measured. The bellows with the greatest pressure compresses the other bellows and moves the pointer mechanism. [Figure 11-9]

SPECIAL PRESSURE INSTRUMENTS

Although the measurement of various pressures is important to the operation of the aircraft and its powerplant(s), the specific type of pressure being measured is generally irrelevant. Therefore, special pressure instruments have been developed to measure the specific values that provide the flight crew with the information needed to operate the aircraft safely and efficiently. Special pressure instruments include manifold pressure gauges, engine-pressure ratio indicators, pressure switches, altimeters, airspeed indicators, Mach meters, and vertical speed indicators.

MANIFOLD PRESSURE GAUGE

The absolute pressure inside the induction system of an engine is an important indicator of the power the engine is developing. Although this manifold pressure is not directly proportional to horsepower, it is related qualitatively. In a normally aspirated (non-supercharged) engine, the maximum differential between manifold pressure and atmospheric pressure will occur when the engine is at idle. As the throttle is opened, manifold pressure rises and approaches atmospheric pressure. When the throttle is wide open, the engine is producing maximum manifold pressure and horsepower. The aircraft flight manual includes charts showing the various combinations of r.p.m. and manifold pressures that produce desired engine performance. [Figure 11-10]

Figure 11-10. The manifold pressure gauge gives the pilot an approximation of the power the engine is producing.

For normally aspirated engines, the manifold pressure gauge usually has a range from 10 to 40 inches of mercury. For turbocharged engines, the range extends high enough to cover the highest manifold pressure for which the engine is rated.

ENGINE PRESSURE RATIO INDICATOR (EPR)

Thrust produced by a turbojet engine using a centrifugal compressor usually is calculated using the indications of the tachometer and the exhaust gas temperature gauges. In an axial flow turbine engine, the best indication of the thrust produced is the engine pressure ratio. This is measured with a differential pressure gauge that senses the pressure difference between the tail pipe total pressure, usually abbreviated as Pt7, and the compressor inlet total pressure, or Pt2. These gauges usually have a range from 1.1 to 2.5 or 3.0.

PRESSURE SWITCHES

A pressure switch is simply a microswitch activated by the movement of a bellows under the pressure applied by a fluid. The pressure is applied to one side of the bellows and a spring force is applied to the opposite side. If the fluid pressure is strong enough to collapse the spring, the microswitch is actuated. In some cases the bellows may have the spring force replaced by a second pressure monitored by the pressure switch unit. [Figure 11-11]

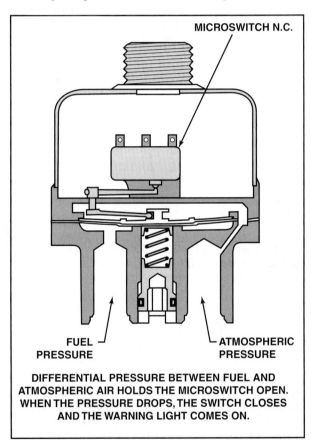

Figure 11-11. Pressure switches are set to react to a given pressure. The switches are used to activate warning lights or to sequence system activities.

Pressure switches are often used to illuminate warning lights if the pressure being monitored exceeds limits. A normally open microswitch is used for this purpose. A normally closed microswitch is used to turn on a warning light when pressure falls below a certain limit.

ALTIMETERS

An altimeter is simply a barometer that measures the absolute pressure of the air. This pressure is caused by the weight of the air above the instrument. As an aircraft climbs, there is less atmosphere above the aircraft and the absolute pressure decreases. The instrument is calibrated to indicate higher altitude with this decrease in pressure and is usually referenced to sea level. As the air circulates above the earth's surface, the pressure at a given location and altitude changes. For this reason, an aircraft altimeter can be adjusted to indicate the correct altitude for a given sea-level pressure.

Development
The altimeter is one of the oldest flight instruments. Early balloon flights carried some form of primitive barometer, which indicated altitude indirectly. The altimeter in many of the early airplanes contained a simple evacuated bellows in which expansion and contraction were measured by an arrangement of gears and levers. These gears and levers translated changes in the bellows' dimensions into movement of the pointer around a dial. The dial was calibrated in feet, and since a change in the barometric pressure changes the pointer position, the dial could be rotated so the instrument read zero when the aircraft was on the ground. This form of operation was adequate for aircraft that seldom flew cross-country and for flights that had little need for accurate altitude information.

Today's flights require accurate altitude indications. The pilot needs to be able to quickly read the altitude within a few feet. This requirement is complicated by a pressure lapse rate (the decrease in pressure with altitude) that is not linear. The pressure change for each thousand feet is greater in the lower altitudes than it is in the higher altitudes. For this reason, bellows are designed with corrugations that allow linear expansion with a change in altitude, rather than a change in pressure. This allows the use of a uniform altitude scale and multiple pointers.

Types of Altitude Measurement
An altimeter can measure height above almost any convenient reference point. Airplanes flying below 18,000 feet measure their altitude with reference to the pressure at mean sea-level (MSL). Indicated altitude is the altitude above mean sea level, which can be read directly from the altimeter when the current altimeter setting is set in the barometric window. All elevations on aeronautical charts are measured from mean sea level. From indicated altitude, a pilot can calculate the aircraft's height above any charted position, also referred to as absolute altitude. When the airplane is on the ground with the local altimeter setting in the barometric window, the altimeter should indicate the surveyed elevation of the aircraft's position on the airport..

The barometric window is an opening in the face of the instrument through which a scale, calibrated in either inches of mercury, or millibars, is visible. A knob on the instrument case, usually located at the seven o'clock position, rotates the scale. Through a series of gears, the knob also rotates the mechanism to which the indicator needles are attached. This allows the instrument reading to be corrected for variations in atmospheric pressure. Airport control towers and air traffic control sites along a flight route give the pilot the altimeter setting, corrected to sea level, for their local barometric pressure. Altimeters with barometric correction capability are often referred to as sensitive altimeters.

Indicated altitude provides for terrain clearance below 18,000 feet MSL. Aircraft operating at 18,000 feet and above maintain vertical separation, using another type of altitude called pressure altitude. When the barometric pressure scale is adjusted to standard sea-level pressure (29.92 inches of mercury or 1013.2 millibars), the altimeter measures the height above this standard pressure level. Although this is a constantly changing reference, it is not a problem because all aircraft in the upper level airspace have their altimeters set to the same reference level. The height above sea-level of an airplane flying at a constant 30,000 feet pressure altitude will vary, but since this affects all aircraft equally, the vertical separation will remain the same. When an aircraft is flying at or above 18,000 feet with the altimeter set to indicate pressure altitude, it is operating at what is called a flight level (FL). FL 310 means a pressure altitude of 31,000 feet.

Until recently, aircraft flying in upper level airspace maintained a 2,000 foot vertical separation with all useable flight levels being odd altitudes. To accommodate the large increase in air traffic over certain, primarily trans-oceanic routes, Reduced Vertical Separation Minimums (RVSM) airspace has been, or will soon be, introduced in many areas. Aircraft flying in RVSM airspace maintain only 1,000 feet of vertical separation. This has placed new and more stringent demands on the maintenance and testing of altimeter systems. Aircraft approved to operate in RVSM airspace must have altimeters, altitude alerting systems, and automatic altitude control systems that are capable of controlling the airplane's altitude within ±150 feet.

Aircraft and engine performance is determined by the air density, which is affected by temperature as well as pressure. Density altitude is pressure altitude corrected for non-standard temperature. It cannot be measured directly, but must be computed

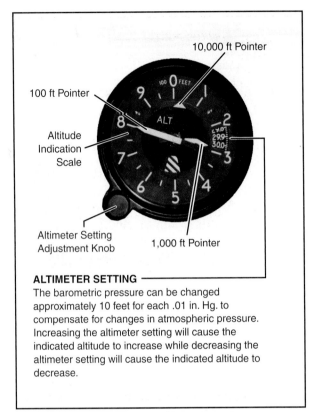

ALTIMETER SETTING
The barometric pressure can be changed approximately 10 feet for each .01 in. Hg. to compensate for changes in atmospheric pressure. Increasing the altimeter setting will cause the indicated altitude to increase while decreasing the altimeter setting will cause the indicated altitude to decrease.

Figure 11-12. The "three pointer" altimeter is a commonly used altimeter in light, general aviation aircraft.

Figure 11-13. The addition of a drum readout of altitude makes the altimeter easier to read.

from pressure altitude using a chart or with a flight computer.

Types of Altimeters
For many years, all sensitive altimeters had three pointers. The long pointer indicates hundreds of feet, a short, fat pointer indicates thousands of feet, and a third, smaller pointer is geared to indicate tens of thousands of feet. The range of these altimeters usually is 20,000; 35,000; 50,000 or 80,000 feet.[Figure 11-12]

To make the altimeter easier to read, and to reduce the chance of misreading the altitude, modern instruments combine a drum scale with a single pointer. The drum indicates thousands of feet digitally, while the pointer indicates hundreds of feet. Since there are five divisions between each number, each mark on the dial represents twenty feet. [Figure 11-13]

Encoding altimeters contain an electronic pressure sensor that sends a digital code to the aircraft's transponder. When the transponder replies to an interrogation from air traffic control radar, it sends altitude data, which appears on the radar screen as a numerical readout next to the radar return for the aircraft.

Position error is inherent with static systems, because the static port is often in disturbed air. The error varies with aircraft design, airspeed and altitude. The **servo altimeter** utilizes a built-in compensation system, tailoring the instrument to the particular aircraft and minimizing this error for the full range of flight speeds and altitudes.

A **radio altimeter** displays the aircraft's altitude as measured by a radio signal, instead of by atmospheric pressure. It sends a high-frequency signal toward the ground, which is reflected back to the aircraft's radio altimeter receiver. Typically, this instrument is used at altitudes within 2,500 feet of the ground, and provides a digital display of the aircraft's absolute altitude above ground level (AGL). These instruments are sometimes referred to as radar altimeters.

Altimeter Tests
The altimeter is one of the most important instruments used on an aircraft, especially when the aircraft is operated in instrument meteorological conditions. FAR 91.411 requires that the aircraft static pressure system, including all altimeters and automatic pressure-altitude reporting systems be tested and inspected every 24 calendar months and found to comply with Appendix E of FAR Part 43. The regulation requires that the inspection be conducted either by the manufacturer of the airplane or a certificated repair station equipped and approved to perform these tests.

Aircraft Instrument Systems

FAR Part 43, Appendix E requires that the following tests be conducted before certifying an altimeter for instrument flight:

- **Scale error**. The altimeter must indicate the same altitude shown on the master indicator or manometer within a specified allowable tolerance.

- **Hysteresis**. The reading taken with the altitude increasing must agree with the readings at the same pressure level when the altitude is decreasing. A specified tolerance is allowed for this test.

- **After effect**. The altimeter must return to the same indication, within tolerance, after the test as it had when the test began.

- **Friction**. Two altitude readings are to be taken at each pressure level, one before and one after the instrument is vibrated. There should be no more than a specified difference between the two readings.

- **Case leak**. When the pressure inside the altimeter case corresponds to an altitude of 18,000 feet, it should not leak down more than a specified amount in a specified period of time.

- **Barometric scale error**. The correlation between the barometric scale and the indication of the altimeter pointers must be correct within an allowable tolerance.

The altimeter must be tested to the highest altitude the aircraft will be flown. After the test is completed, the results of the test must be recorded in the aircraft records. These records include an altimeter correction form that indicates the errors at thousand-foot intervals.

AIRSPEED INDICATORS

An airspeed indicator is a differential pressure gauge that measures the difference between the pitot, or ram air pressure, and the static, or ambient, air pressure. It consists of an airtight case that is vented to the static source, and a thin metal diaphragm vented to the pitot air source. The diaphragm expands in proportion to the pressure difference between the pitot and static air pressure sources. The diaphragm is also mechanically linked to a pointer on the instrument face, which indicates airspeed. [Figure 11-14]

No matter what type of airspeed indicator is used on the aircraft, its accuracy relies heavily on the correct operation of the pitot-static system. The airspeed indicator must be connected into a pitot tube that senses impact air pressure, and a static port that picks up undisturbed ambient air pressure. The pitot static system will be discussed later in this section.

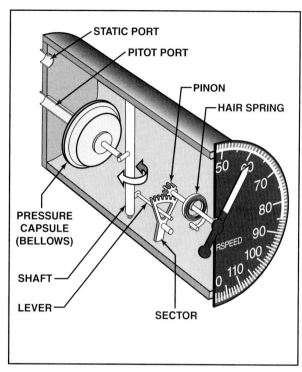

Figure 11-14. As the differential pressure changes, the diaphragm expands or contracts, causing the pointer to indicate changes in airspeed.

The uncorrected, or direct, reading of an airspeed indicator is called **indicated airspeed**. It is almost impossible to find a location for the static port that is entirely free from airflow distortion that produces errors in airspeed indication. To compensate, a flight test is conducted and the airspeed indicator that is connected to the aircraft's static system is calibrated against one that is connected to a trailing static pickup with no airflow disturbance. A calibration card is included with the aircraft flight manual that can be used to calculate calibrated airspeed from indicated airspeed. In most modern production aircraft, the error between indicated and calibrated airspeed is so small that, for practical purposes, it is often ignored.

While pilots are concerned about calibrated airspeed as it relates to the aircraft stall speed, this speed is not very useful for calculating the time that it will take to fly from one location to another. To create a useful flight plan, a pilot must convert calibrated airspeed to true airspeed.

Calibrated and true airspeed are the same at sea level under standard atmospheric conditions.

Under non-standard conditions, the pilot can convert true airspeed into calibrated airspeed by entering the non-standard pressure and temperature into a circular slide rule or electronic flight computer. However, many airspeed indicators have a movable dial that a pilot can adjust to provide a true airspeed readout. [Figure 11-15]

Figure 11-15. The dial on this airspeed indicator can be rotated similar to the scale on a circular flight computer. When the air temperature is lined up with the pressure altitude, the true airspeed can be read from the outside scale.

TRUE AIRSPEED INDICATORS

A true airspeed indicator internally corrects indicated airspeed to true airspeed using a temperature sensor and altitude bellows. This relieves the pilot from computing true airspeed. [Figure 11-16]

MAXIMUM ALLOWABLE AIRSPEED INDICATOR

Airplanes with maximum speeds that are limited by structural considerations have their never-exceed speed marked by a fixed red radial line on the dial of the airspeed indicator.

If the maximum speed is limited by a critical Mach number, the fixed red line is replaced by a movable red pointer or red and white barber pole that is driven by an altimeter bellows. When an airplane flies at or near the speed of sound, parts of the airframe experience airflow that is greater than the speed of sound. If the aircraft is not designed to accommodate this supersonic airflow, destructive shock waves can form, and loss of control or damage can occur. Airplanes that are not designed to fly at supersonic airspeeds must never be allowed to reach their critical Mach number. This is the speed that results in supersonic velocity airflow over parts of the aircraft. If, for example, the airplane is limited to Mach 0.75, the pointer is set for standard sea-level conditions at 497 knots. This is 75% of 662 knots; the speed of sound under standard sea-level conditions. As the speed of sound decreases with altitude, the pointer moves counterclockwise to a lower airspeed. [Figure 11-17]

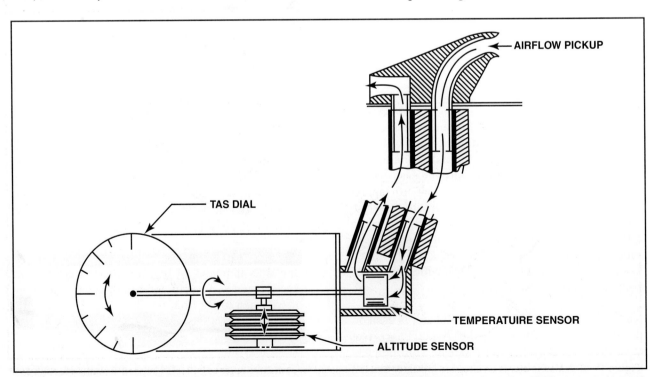

Figure 11-16. A true airspeed indicator contains altitude and air temperature correction devices in order to indicate true airspeed.

ALTITUDE FT.	DENSITY RATIO σ	PRESSURE RATIO δ	TEMPER-ATURE °F	SPEED OF SOUND α KNOTS
0	1.000	1.000	59.00	661.7
1000	0.9711	0.9644	55.43	659.5
2000	0.9428	0.9298	51.87	657.2
3000	0.9151	.08962	48.30	654.9
4000	0.8881	0.8637	44.74	652.6
5000	0.8617	0.8320	41.17	650.3
6000	0.8359	0.8014	37.60	647.9
7000	0.8106	0.7716	34.04	645.6
8000	.07860	0.7428	30.47	643.3
9000	0.7620	0.7148	26.90	640.9
10000	0.7385	0.6877	23.34	638.6
15000	0.6292	0.5643	5.51	626.7
20000	0.5328	0.4595	−12.32	614.6
25000	0.4481	0.3711	−30.15	602.2
30000	0.3741	0.2970	−47.98	589.5
35000	0.3099	0.2353	−65.82	576.6
40000	0.2462	0.1851	−69.70	573.8
45000	0.1936	0.1455	−69.70	573.8
50000	0.1522	0.1145	−69.70	573.8
55000	0.1197	0.0900	−69.70	573.8
60000	0.0941	0.0708	−69.70	573.8
65000	0.0740	0.0557	−69.70	573.8
70000	0.0582	0.0438	−69.70	573.8
75000	0.0458	0.0344	−69.70	573.8
80000	0.0360	0.0271	−69.70	573.8
85000	0.0280	0.0213	−64.80	577.4
90000	0.0217	0.0168	−56.57	583.4
95000	0.0169	0.0134	−48.34	589.3
100000	0.0132	0.0107	−40.11	595.2

Figure 11-17. The speed of sound decreases as altitude increases due to the lower air temperature.

Machmeter

Because the indicated airspeed (IAS) that corresponds to a particular Mach number varies with altitude, aircraft that operate at transonic and supersonic speeds need a Machmeter to measure their speed relative to the speed of sound. Examples of Machmeter indications are Mach .75 at 75 percent of the speed of sound, Mach 1.0 at the speed of sound, and Mach 1.25 at 125 percent of the speed of sound. A Machmeter is somewhat similar to an airspeed indicator. Both have a pointer that is moved by an expanding bellows. Both compare pitot pressure with static pressure. But in the Machmeter, the mechanical advantage of the gears from the airspeed measuring system is varied by the action of an altimeter mechanism. [Figure 11-18]

Combination Airspeed Indicator

A combination airspeed indicator combines the airspeed indicator with a Machmeter, and also shows the maximum allowable operating airspeed. Small indicators (sometimes called "bugs") may appear around the periphery of the dial. The pilot can set these bugs to indicate the correct speed for certain flight conditions, such as during takeoff or during approach to landing. [Figure 11-19]

VERTICAL SPEED INDICATOR

The rate-of-climb indicator, more properly called a vertical speed indicator (VSI), helps a pilot establish a rate of climb or descent to allow arrival at a spec-

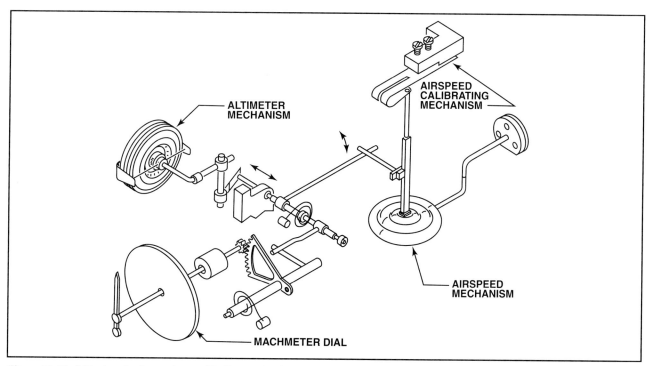

Figure 11-18. A Machmeter is an airspeed indicator that is adjusted for altitude in order to indicate airspeed in relation to the speed of sound.

Figure 11-19. A combination airspeed indicator indicates airspeed, Mach, and maximum allowable airspeed. The airspeed indicator shown here is used in early models of the Boeing 747 and combines airspeed and mach indications with a barber-pole pointer to show the maximum allowable airspeed. The instrument also displays calibrated airspeed on the lower digital counter. Five white bugs can be manually positioned by the pilot to provide visual reminders of critical flight speeds for various flight regimes such as takeoff or landing.

ified altitude at a given time. The VSI also backs up other instruments, such as the altimeter, by providing early indications of changes in pitch.

The vertical speed indicator contains a bellows, or pressure capsule, which is connected to the static source and vented to the inside of the instrument case through a diffuser, which provides an accurately calibrated leak. [Figure 11-20]

When the aircraft climbs, the pressure inside the capsule, which matches the outside static pressure, decreases to a value less than that inside the instrument case. The capsule compresses, causing the levers and gears to move the pointer so it indicates a climb. The pressure inside the case now begins to decrease by leaking through the diffuser. This leak is calibrated to maintain a difference between the pressure inside the capsule and that inside the case, which is proportional to the rate of change of the outside air pressure. As soon as the aircraft levels off, the pressure inside the case and inside the capsule equalizes, and the indicator then shows a zero rate of change.

INSTANTANEOUS VERTICAL SPEED INDICATOR

A disadvantage of the vertical speed indicator (VSI) is that its indication lags behind the actual pressure change. The instantaneous vertical speed indicator (IVSI) was developed to correct this deficiency. An

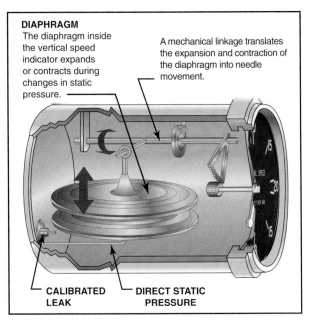

Figure 11-20. A vertical speed indicator measures the rate of change of static pressure and indicates the result as a rate of climb or descent.

IVSI supplements the conventional VSI mechanism with an accelerometer-operated dashpot that pumps air across the capsule. One dashpot compensates for climbs and the other for descents. For example, when the aircraft noses over to begin a descent, the inertia of the accelerometer piston causes it to move upward. This instantaneously increases the pressure inside the capsule and lowers the pressure inside the case, which causes an immediate indication of a descent. After a brief interval, the normal VSI mechanism catches up with the descent rate, and it begins to indicate the descent without assistance from the pumps. When established in a stabilized descent, there is no more vertical acceleration; the accelerometer piston centers and ceases assisting the ordinary VSI mechanism. [Figure 11-21]

TEMPERATURE-MEASURING INSTRUMENTS

All temperature indicators measure the amount of temperature increase or decrease, which occurs in a material being measured. This change in temperature, known as Delta T, can be measured electrically or monitored through the expansion of different materials.

Temperature-measuring instruments can be non-electrical or electrical. Non-electrical instruments depend on expansion or contraction of liquids, solids, or gases to indicate a rise or fall in temperature. Electrical instruments measure changes in electrical resistance, which are proportional to changes in temperature.

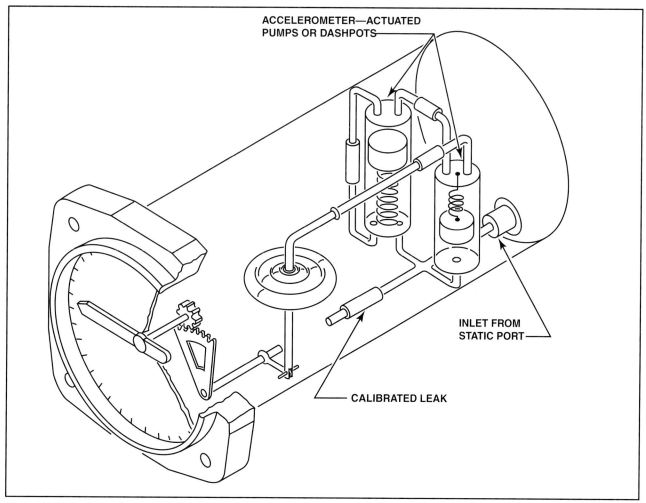

Figure 11-21. An instantaneous vertical speed indicator incorporates accelerometers to help the instrument immediately indicate changes in vertical speed.

NONELECTRICAL TEMPERATURE INSTRUMENTS

Most materials expand when subjected to an increase in temperature and contract when the temperature decreases. This characteristic is used to monitor temperatures in all types of non-electrical temperature instruments. If the amount of expansion or contraction of the material is constant for each degree of temperature change, the temperature indicator can utilize a simple linear scale.

LIQUID EXPANSION

A good example of a liquid expansion temperature measuring instrument is a common household or medical thermometer. Since it consists of an enclosed breakable glass tube, it is seldom used in aircraft. One end of the tube is enlarged and filled with mercury or colored alcohol. As an increase in temperature causes the liquid to expand, it is pushed up in the tube where its height can be measured on a linear scale next to the tube.

SOLIDS EXPANSION

A bimetallic strip thermometer utilizes a sandwich of two metals with different expansion coefficients. Many aircraft are equipped with a bimetallic thermometer to measure the outside air temperature (OAT). Common metals in this type of thermometer include brass and iron. Brass expands nearly twice as fast as iron as its temperature increases. If these two metals are bonded together, the difference in expansion rates causes the bimetallic combination to warp as temperature increases. This action causes deflection of a needle that indicates temperature on a circular scale.

GAS EXPANSION

Another type of temperature-sensing instrument works on the principle of an expanding gas. Typically, this type of instrument consists of a liquid-filled temperature bulb connected to a capillary tube. The capillary tube is then connected to a curved Bourdon tube pressure-sensing device. As the temperature bulb is subjected to a higher temperature, the liquid inside the bulb vaporizes. The vapor acts through the capillary tube and produces a higher pressure in the Bourdon tube. The Bourdon tube senses the change in pressure and straightens out proportionally to the increase in pressure in the tube. This action moves a pointer across a calibrated pointer scale. The Bourdon tube portion of this instrument functions the same as a Bourdon tube pressure instrument, except that the pointer scale is marked in temperature units. [Figure 11-22]

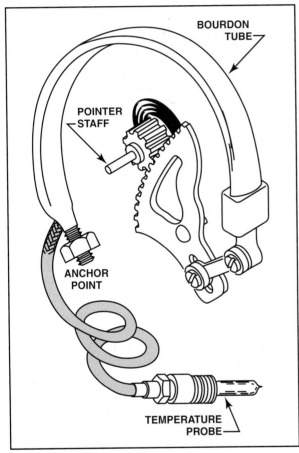

Figure 11-22. Expanding gases cause the Bourdon tube to straighten, indicating a temperature increase.

ELECTRICAL TEMPERATURE INDICATORS

Temperature can be determined electrically either by measuring resistance or voltage and using the result to move an electrical temperature indicator. Resistance instruments measure the change in resistance produced when a temperature-sensing bulb is subjected to heat. Voltage instruments measure voltage produced when certain metals are subjected to heat.

Resistance Instruments

Resistance change-type temperature instruments use a fine nickel-wire coil placed into the end of the temperature-sensing bulb. As the bulb is heated, the resistance of the nickel wire increases in proportion to the change in temperature. This moves an electrical temperature indicator.

The nickel wire is connected to a Wheatstone bridge or to a ratiometer circuit. The Wheatstone bridge contains three resistors with the temperature probe, formed into a bridge circuit with the temperature indicator. The current flow through the indicator changes when the resistance through one of the legs of the bridge is varied, in this case, when the nickel wire (shown here as R4) is subjected to a temperature change. [Figure 11-23]

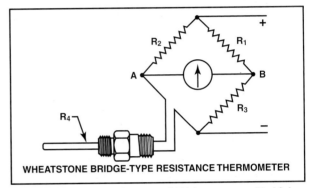

Figure 11-23. The resistance of R4 increases with higher temperature, causing increased current flow through the temperature indicator.

For example, if the resistance of legs R1/R2 is equal to the resistance of legs R3/R4, there is no current flow through the indicator. If the temperature at the probe changes, the resistance of R4 changes. This causes current to flow through the indicator, in one direction or the other, and the pointer indicates a change in temperature.

Another, more accurate, means of monitoring the resistance change of a temperature probe is a ratiometer. This circuit measures the resistance of the sensing bulb by comparing the current flow through the bulb with the flow through a set of resistors inside the instrument case. By using a ratio of current, the effect of variations in system voltage is minimized. [Figure 11-24]

A ratiometer-type temperature indicator may be used with either a 14- or 28-volt electrical system. On the 28-volt system, the power is connected through pin A so the current flows through a resistor that drops the voltage to that required by the instrument.

When the temperature (and bulb resistance) is low, most of the current flows through the low-end coil and the bulb, rather than through the high-end coil.

Aircraft Instrument Systems

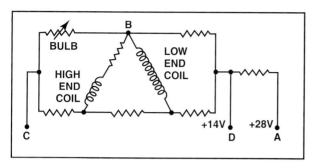

Figure 11-24. The ratiometer is more accurate than the Wheatstone bridge because it compares ratios, which eliminates indicator errors due to small changes in source voltage.

When the temperature (and bulb resistance) is high, more current flows through the high-end coil. The pointer is mounted on a small permanent magnet that aligns itself with the magnetic field produced by the low- and high-end coils.

Thermocouple Instruments

Thermocouples, are typically used to measure relatively high temperatures. They work on the principle that certain dissimilar metals produce a voltage when subjected to heat. This extremely small voltage (in the millivolt range) is measured between the two junctions (ends) of a twisted pair of dissimilar metals. The thermocouple is connected to a sensitive voltmeter with a scale calibrated in degrees Celsius or Fahrenheit. Common applications include measurement of exhaust gas temperature, cylinder head temperature, and turbine inlet temperatures. The dissimilar metals consist of either iron and constantan, or chromel and alumel. Chromel/alumel thermocouples are generally used in the high heat environment of turbine engine exhaust sections. [Figure 11-25]

There are two common types of thermocouple probes classified according to resistance: two-ohm and eight-ohm. Thermocouple instruments are extremely sensitive to resistance within an electrical circuit. It is essential that all connections are clean and properly torqued. When replacing a thermocouple probe, be careful to install the correct replacement.

The circuit's sensitivity to variations in resistance also requires that the wiring connecting the thermocouple be carefully matched to the system. These thermocouple leads are color-coded for easy identification and come in matched pairs that are secured together by a common braid. Some thermocouples are permanently attached to the leads; others are removable. A technician should always use the correct lead for the system, and never

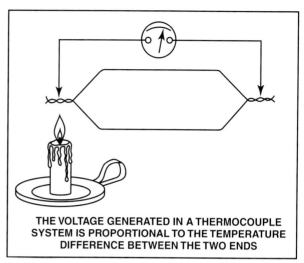

Figure 11-25. A thermocouple generates voltage that is proportionate to the temperature being measured. The resulting current can drive an indicator that is calibrated in degrees.

change the length since that would change the system's resistance.

MECHANICAL MOVEMENT MEASUREMENT

Instruments that measure mechanical movement include accelerometers, tachometers, and synchroscopes. Accelerometers measure forces on the airframe, and the other instruments indicate engine operation.

ACCELEROMETER

Aircraft structures are designed to withstand certain G-loads. Aerobatic and other aircraft that experience severe aerodynamic loads are equipped with an accelerometer (G-meter) so the pilot can determine when the aircraft is approaching its design limit load factor.

Panel-mounted accelerometers use a weight suspended on two guide shafts, centered by a spring-loaded control cord. Three pointers are attached to a pulley driven by the control cord. When the aircraft pitches up or down, inertia acts on the weight and causes it to ride up or down on the guide shafts. This movement turns the pulley and moves one main and two auxiliary pointers. The main pointer indicates the load factor at that moment. The auxiliary pointers have ratchet mechanisms that hold them at the farthest point of their movement. One pointer remains at the maximum positive G indication, while the other indicates maximum negative Gs. Both auxiliary pointers may be returned to their at-rest condition by turning the reset knob in the front of the accelerometer, which releases the pawl on the ratchet mechanism. [Figure 11-26]

Figure 11-26. An accelerometer indicates the present G force acting on the aircraft, plus the maximum negative and positive Gs experienced since the accelerometer was last reset.

Before installing an accelerometer in an instrument panel, place it on a flat surface and verify all the pointers indicate +1 G. Next, turn the instrument on its side. All of the pointers should indicate zero. When inverted, the main pointer should indicate −1 G. When the instrument is moved smoothly up and down the pointers should show positive and negative indications, with one auxiliary pointer stopping at the maximum positive excursion and the other at the maximum negative travel. When the reset knob is rotated, the two auxiliary pointers should return to the same position as the main pointer.

SYNCHROSCOPES

A synchroscope is an instrument that helps the pilot of a twin-engine airplane synchronize the r.p.m. of the propellers. The indicator is a small rotating disc that turns toward the left to indicate the left engine is faster, and toward the right to indicate the right engine is faster. A synchroscope monitors the electrical output of each tachometer generator to determine the rotational speed of each engine. If the engines are turning at the same speed, the output voltage from both tachometer generators is equal. If one engine is at a higher r.p.m., a differential voltage is applied to the synchroscope indicator, causing the indicator to rotate.

TACHOMETERS

A tachometer is a mechanical or electrical instrument that displays engine r.p.m. A pilot uses this instrument to determine how much power an engine is producing at a given throttle setting, and to ensure the engine is operating within its limits. The tachometer face is color coded to indicate the limits, as determined from the aircraft's type certificate data sheets (TCD). Red indicates the maximum r.p.m. allowed, and green indicates the normal operating range. Occasionally, a yellow range may be included, indicating a cautionary range. (See Chapter 2 in the Jeppesen *A&P Technician Powerplant Textbook* for a detailed description of the operation and types of tachometers.) [Figure 11-27]

Figure 11-27. The tachometer is a primary engine instrument. A typical tachometer instrument face is calibrated in hundreds of r.p.m. and has a green arc and red radial line.

GYROSCOPIC INSTRUMENTS

Gyroscopes, or gyros, have made it possible to fly an aircraft without outside visual reference. A gyro is simply a rotating mass; a familiar example is a child's top. Gyroscopes have unique behaviors which make them useful in aircraft instruments.

GYROSCOPIC THEORY

In 1851, the French physicist Leon Focault devised a small wheel with a heavy outside rim. When spun at a high speed, the wheel demonstrated the strange characteristic of remaining rigid in the plane in which it was spinning. Focault deduced that because the wheel remained rigid in space, it could show the rotation of the earth. Accordingly, he named the device the gyroscope, a name that translated from Greek means "to view the earth's rotation."

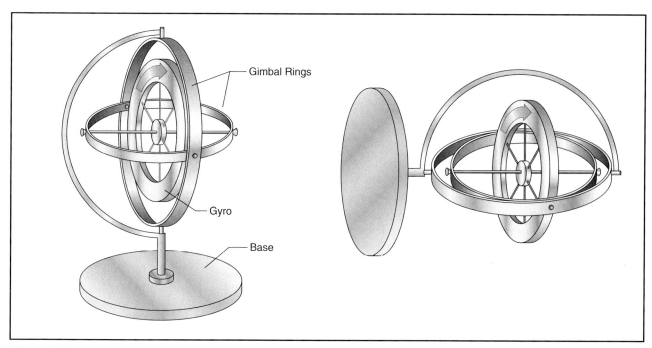

Figure 11-28. Regardless of the position of its base, a gyroscope tends to remain rigid in space, with its axis of rotation pointed in a constant direction. A gyroscope's rigidity in space makes it useful for determining the attitude of an aircraft and the direction the aircraft is pointing.

A spinning gyroscope possesses two characteristics that make it useful for aircraft instruments: rigidity in space and precession. **Rigidity in space** refers to the principle that a wheel with a heavily weighted rim will, when spun rapidly, remain in a fixed position in the plane in which it is spinning. By mounting this wheel, or gyroscope, on a set of gimbal rings, the gyro is able to rotate freely in any direction. Even though the gimbal rings are tilted, twisted, or otherwise moved, the gyro remains in the plane in which it was originally spinning. [Figure 11-28]

When an outside force tries to tilt a spinning gyro, the gyro responds as if the force had been applied at a point 90 degrees further around in the direction of rotation. This effect is called **precession**, because the cause precedes the effect by 90 degrees. [Figure 11-29]

Unwanted precession is caused by friction in the gimbals and bearings of an instrument, causing slow drifting in heading indicators and occasional small errors in attitude indicators. Other instruments, like turn indicators, depend on precession for their operation.

GYROSCOPIC INSTRUMENTS

Gyroscopic principles were first employed in aircraft instruments when Elmer A. Sperry invented the turn and bank indicator in 1918, followed by the gyrocompass and artificial horizon. The gyrocompass, or heading indicator, indicates an aircraft's heading relative to a fixed horizontal reference, such

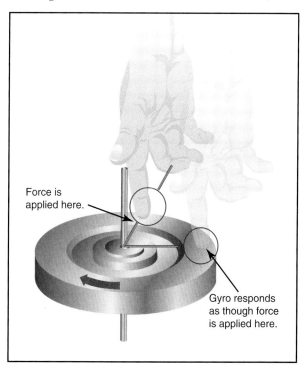

Figure 11-29. Gyroscopic precession can be observed with any rotating mass. In this figure, a rotating bicycle wheel is placed on a gimbal. When a force is applied to the wheel while it is rotating, the effect of the force will cause the wheel to pivot, or precess, around the gimbal in a horizontal direction.

as magnetic north. An artificial horizon indicates an aircraft's attitude relative to the earth's horizon. They are essential for instrument flight. A turn indicator is an example of a rate gyro, which measure the rate of rotation about an aircraft's vertical axis.

HEADING INDICATOR

A gyroscope's rigidity in space can be used to display a stable indication of an aircraft's heading. This is needed because the primary direction instrument, the magnetic compass, bounces around so much that it is difficult to read during flight. Since a gyroscope has no north-seeking tendency, it must be set to agree with the magnetic compass during times when the magnetic compass is providing a stable indication.

Early heading indicators resembled magnetic compasses. The gyro rotor was suspended in a double gimbal, with its spin axis in a horizontal plane inside a calibrated scale. The rotor was spun by a jet of air directed toward buckets cut into its periphery. Pushing in on the caging knob in the front of the instrument leveled the rotor and locked the gimbals. The knob could then be turned to rotate the entire mechanism and bring the desired heading under the reference mark, or lubber line. Pulling the knob out unlocked the gimbals so the rotor could maintain its position in space while the aircraft turned about the gyro. This gave the pilot a reference between the heading of the aircraft and magnetic north. One problem with these early instruments is excessive friction in the gimbals, which caused the gyro to precess. The heading had to be reset frequently to keep the instrument agreeing with the magnetic compass. [Figure 11-30]

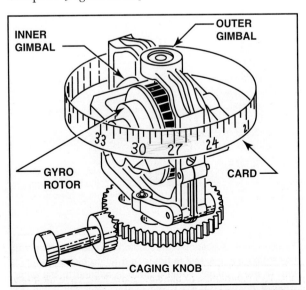

Figure 11-30. Early heading indicators indicated heading on a drum-type card which could be viewed through a window.

In this drum-type heading indicator, as on a magnetic compass, the indications on the directional scale are reversed from what a pilot might expect to see. For example, when a pilot is on a heading of north and wants to turn to a heading of 330°, the number 33 (for 330 degrees) appears to the right of the zero mark, even though the aircraft needs to turn left. This can be confusing to the pilot. [Figure 11-31]

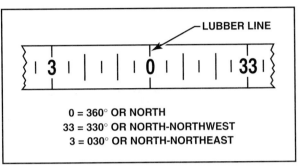

Figure 11-31. The dial of a drum-type heading indicator is marked opposite the way a pilot would normally visualize a change in heading.

To provide more intuitive heading indications, the vertical card compass was invented. Instead of a simple lubber line in front of the card, this instrument has an airplane symbol on its face, in the center of the dial. The airplane nose points up, representing straight ahead, and the circular dial is connected to the gyro mechanism, which remains pointing in one horizontal direction when the gyro is spinning. As the airplane turns about the gimbal, the dial rotates. The pilot pushes and turns the knob on the front of the instrument to turn the dial, matching the indicated heading to the heading shown on the magnetic compass. When the knob is released, a spring pushes it back out, disengaging the knob from the instrument mechanism. [Figure 11-32]

The bearings in modern gyroscopic heading indicators have much lower friction than the original heading indicators. This minimizes precession; however, a pilot still should check the gyroscopic heading indicator against the magnetic compass at regular intervals.

ATTITUDE INDICATOR

The attitude indicator, or artificial horizon, is a mechanical substitute for the natural horizon. The heart of this instrument is a heavy brass rotor that spins in the horizontal plane, mounted on dual gimbals that allow it to remain in that plane regardless of aircraft movement. Before the gyro can spin in the horizontal plane, it must erect itself. While the aircraft is taxiing, gravity provides the force to level

Aircraft Instrument Systems

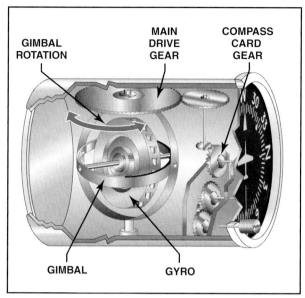

Figure 11-32. Modern heading indicators use a vertical compass card connected to the gyro gimbal by a series of gears.

the gyro. On a vacuum-driven attitude indicator, this is accomplished through the action of pendulous vanes. Air exits the gyro assembly through four ports at right angles to each other located near the base of the assembly. These ports can be individually blocked and opened by the pendulous vanes swinging in front of them. The thrust of the exiting air acts to level the gyro. [Figure 11-33]

A moveable face, painted like a horizon and visible to the pilot, is attached through a counterweighted arm to the gyro housing. The bar indicates the attitude of the gyro and therefore, the position of the earth's horizon. A symbol indicating the wings of the airplane is mounted inside the instrument case, to show the relationship between the airplane and the horizon.

One of the big improvements of newer attitude indicators is in the tumble limits of the gyro. Early instruments were limited in the amount of pitch and roll they would tolerate. Beyond this limit, the mechanism would apply a forces to the gyro housing, that, through precession, would cause the gyro to tumble over. Newer instruments are designed so that the aircraft has complete freedom of movement about the gyro. The aircraft can actually loop or roll without tumbling the gyro, so there usually are no caging mechanisms on these instruments.

Another feature of the newer instruments is the way the information is presented to the pilot. Rather than representing the horizon with a bar, these instruments use a two-color movable dial. Above the horizon, the dial is a light color, usually blue to represent the sky. Below the horizon, it is a dark color, usually brown or black to represent the ground. The lower half of the instrument is marked with lines, converging at the center to help the pilot

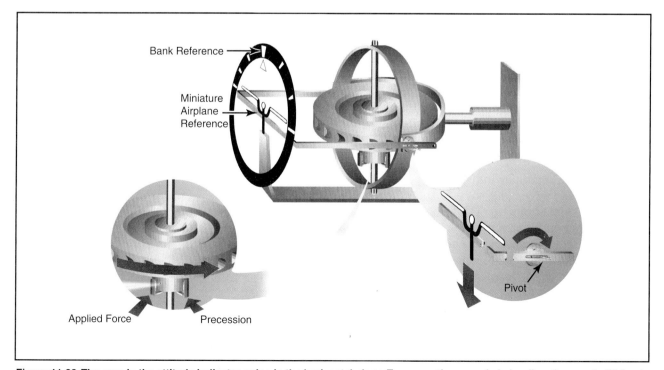

Figure 11-33. The gyro in the attitude indicator spins in the horizontal plane. Two mountings, or gimbals, allow the gyro to tilt freely in the pitch and roll planes. Due to rigidity in space, the gyro remains in a fixed position relative to the horizon as the case and airplane rotate around it.

visualize this as the horizon. (These lines also provide angular references to establish the desired bank angle, although this is not their primary purpose.) Short horizontal lines both above and below the horizon help the pilot to establish pitch angles. Across the top of the instrument, a pointer may be aligned with index marks to establish the desired bank angle. These marks are located at 10, 20, 30, 60 and 90 degrees. [Figure 11-34]

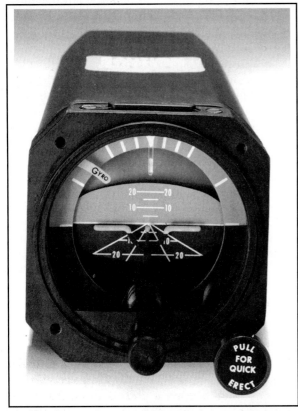

Figure 11-34. Modern attitude indicators do not need caging mechanisms, and have the ability to be quickly erected. They are clearly marked with pitch and bank reference marks, and with a horizon card that resembles an actual horizon.

Many of the newer attitude indicators on larger airplanes use electric motors to spin their gyros. However, on light airplanes, the attitude and heading indicators typically are vacuum powered while the turn indicator is electric. This makes it possible for a pilot to maintain aircraft control in the event one of the systems fails in instrument conditions.

Attitude Director Indicator

A flight director is that portion of an automatic flight control system which provides visual cues to the pilot. Steering commands provided by the flight director allow the pilot to fly the aircraft more precisely. The human pilot receives the same information as an automatic pilot, and can manually fly the airplane in much the same way the servos do when the autopilot is engaged.

The attitude director indicator is similar to that of a normal attitude indicator, except that it has command bars driven by the flight computer. When the pilot complies with a flight director command to fly straight and level, the command bars are aligned with the wing tips of the symbolic aircraft (Figure 11-35, view A). When the flight director commands a climb (view B), the command bars move up and the pilot must raise the nose of the aircraft to place the wingtips on the bars (view C). The symbolic aircraft actually does not move, but since both the command bars and the horizon card behind the aircraft move, it appears to the pilot that the symbolic aircraft has moved to answer the command. The command bars can move to command a climb, descent, turn, or any combination of these maneuvers. [Figure 11-35]

Flight director command bars can take different forms depending on the manufacturer of the system. One popular flight director uses a pair of narrow wedge command bars and a delta, or flattened triangle, to represent the airplane. The command bars can command pitch or roll maneuvers, or both simultaneously. To respond, the pilot maneuvers the airplane so as to fly the delta into the V formed by the command bars. [Figure 11-36]

TURN INDICATORS

Turn indicators measure the rate of rotation of an aircraft about its vertical axis. The basic difference between these rate-of-turn indicators and attitude and heading indicators is in the mounting of the gyro itself. Turn indicator gyros have fewer degrees of freedom because they are mounted in a single gimbal that can move about only one axis. There are two basic types of turn indicators: the turn and slip indicator and the turn coordinator.

The gyroscopic turn and slip indicator was the first "blind flight" instrument invented. It has been known by a number of names including "needle and ball" and "turn and bank indicator."

The turn indicator contains a brass rotor, which is spun either by a jet of air or by an electric motor. This rotor's spin axis is oriented as shown in Figure 11-37, and can tilt left or right, as viewed from the pilot seat. A centering spring holds the gimbal level when there is no outside force acting on it. When the rotor is spinning and the aircraft rotates about its vertical, or yaw, axis, the instrument experiences the turning forces shown in the figure. Because of

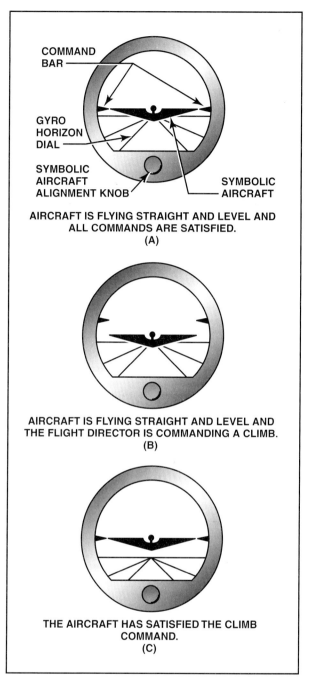

Figure 11-35. The attitude director indicator is similar to a normal attitude gyro, except that it includes command guidance bars that are controlled by the flight director system.

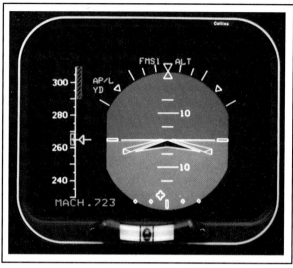

Figure 11-36. The two narrow wedges on this flight director provide pitch and roll information to the pilot. The pilot maneuvers the airplane to "fly" the delta-shaped airplane symbol into the V of the command bars.

precession, this force is felt 90 degrees from where it applied, which causes the rotor to tilt. This tilt is opposed by a dashpot that dampens the force, and by a calibrated spring that restricts the amount the gimbal can tilt. (For simplicity, the spring and dashpot are not shown in the figure.) A pointer is connected to the gimbal in such a way that it indicates the direction and rate of yaw. Note that the pointer is mechanically linked to the gimbal in such a way that it moves in the opposite direction that the gimbal tilts. [Figure 11-37]

The turn coordinator, also shown in Figure 11-37, has the additional advantage of being able to sense roll as well as yaw. It can do this because of the way the gyro is canted. This capability makes the turn coordinator more useful in backing up the bank information provided by the attitude indicator.

The inclinometer is the part of the turn indicator that contains the fluid and the ball. The position of the ball indicates whether a pilot is using the correct angle of bank for the rate of turn, by measuring the balance between the pull of gravity and centrifugal force caused by the turn. In a slip, the rate of turn is too slow for the angle of bank, and the ball falls to the inside of the turn. In a skid, the rate of turn is too great for the angle of bank, and the ball is pulled to the outside of the turn. A pilot steps on the ball, or applies rudder pressure on the side the ball is deflected, to correct an uncoordinated flight condition.

DIRECTION-INDICATING INSTRUMENTS

Navigational maps and charts are based on a grid system of latitude and longitude, with the geographic north and south poles and the equator being the references for this grid. In order to fly from one location on a chart to another, the pilot must have some sort of instrument that will maintain a constant relationship with this grid. The device pilots have relied on for this purpose is the magnetic compass. A limitation of the compass is that it measures

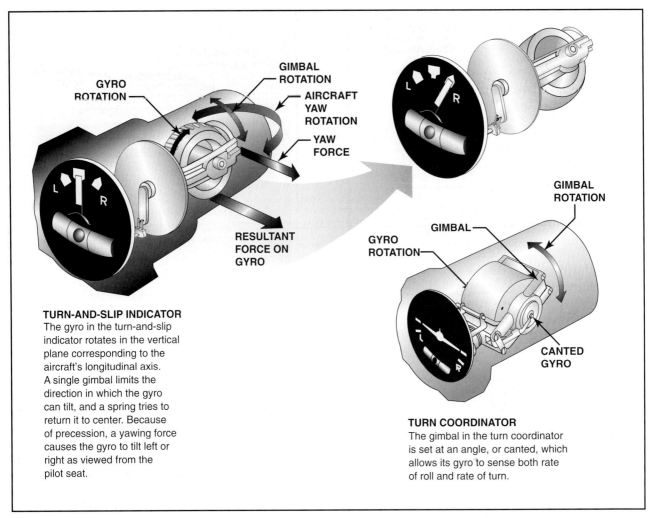

Figure 11-37. Turn indicators translate an aircraft's turning motion into an indication of the rate of turn.

direction relative to magnetic, not geographic, north. However, it is an effective navigation tool (except near the magnetic north pole) once the pilot has corrected for the variation between true and magnetic north. [Figure 11-38]

Because of newer technology like the remote indicating compass and slaved gyro compass, the magnetic compass has become more of a standby instrument in many aircraft. Nonetheless, it is important to understand how the magnetic compass works.

MAGNETIC COMPASS

The magnetic compass is a relatively simple instrument. The main body is a cast aluminum housing with a glass lens. A vertical reference mark called a lubber line is painted across the lens. Inside the housing there is a small brass float, which rides on a low-friction post. A compass card with a dial graduated in 360 degrees is connected to this float. North corresponds to 0 or 360 degrees, east is 90 degrees, south is 180 degrees, and west is 270 degrees. Two small bar-type magnets are soldered to the bottom of the float, aligned with north and south. [Figure 11-39]

The housing is filled with compass fluid, to dampen the oscillations of the float. The fluid is a hydrocarbon product similar to kerosene, but with additives to keep it clear. The housing must be completely full, with no bubbles. To prevent damage to the housing when the fluid expands due to heat, an expansion diaphragm, or bellows, is mounted inside the housing. A set of compensator magnets is located in a slot in the housing outside of the compass bowl. A small instrument lamp screws into the front of the housing and shines inside the bowl to illuminate the lubber line and the numbers on the card. [Figure 11-40]

VARIATION

The geographic north and south poles form the axis for the earth's rotation. These positions are also

Aircraft Instrument Systems

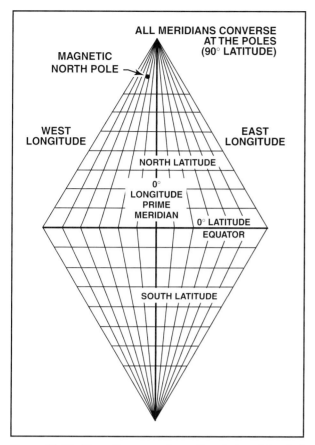

Figure 11-38. Since the magnetic north pole is not located at the geographic north pole, pilots must adjust their heading in order to maintain orientation to the geographic grid.

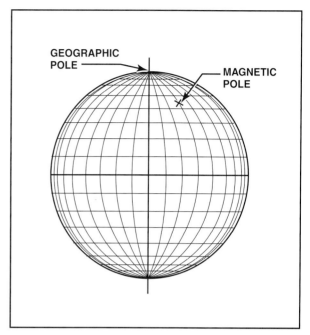

Figure 11-40. The magnetic north and south poles are not located at the geographic poles.

referred to as true north and south. The magnetic north and south poles form another axis. Lines of magnetic force flow out from each magnetic pole in all directions, and eventually return to the opposite pole. A freely mounted bar magnet will align itself

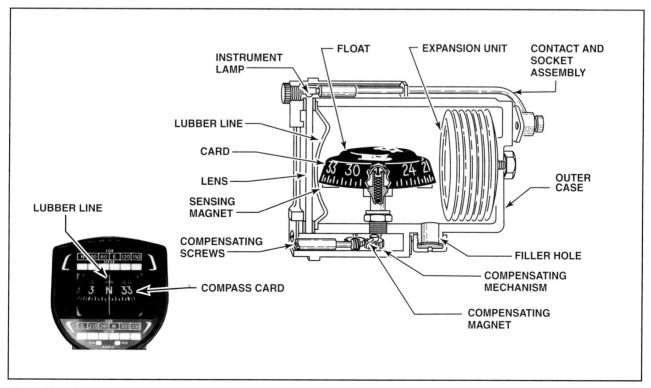

Figure 11-39. The magnetic compass is a simple and reliable navigation instrument.

with the magnetic axis formed by the north/south magnetic field of the earth. [Figure 11-41]

The angular difference between the true and magnetic poles at a given point is referred to as variation. Since most aviation charts are oriented to true north and the aircraft compass is oriented to magnetic north, pilots must convert a true direction to a magnetic direction by correcting for the variation. The amount of variation a pilot needs to apply depends upon the aircraft's location on the earth's surface. [Figure 11-42]

DEVIATION

Deviation refers to a compass error which occurs due to disturbances from magnetic fields produced by metals and electrical accessories within the airplane itself. A magnetic field surrounds any wire carrying electricity, and almost all of the steel parts of an aircraft and the engine have some magnetism in them. Magnetos, alternators, and generators have strong magnetic fields, which are close enough to the compass to influence it. To minimize the electrical interference from the compass light, the wiring consists of twisted two-wire conductors instead of conventional single-wire conductors.

Aircraft compasses are equipped with two or more small compensator magnets in the housing, that may be adjusted to cancel the effect of most of the local magnetic fields in the aircraft. This is called "swinging" the compass. The remaining error is

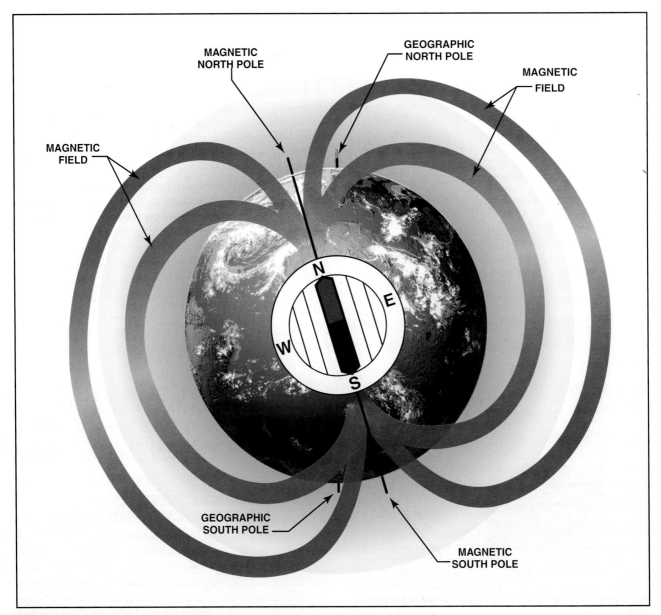

Figure 11-41. A magnetic compass will align itself with the magnetic poles of the earth.

Aircraft Instrument Systems

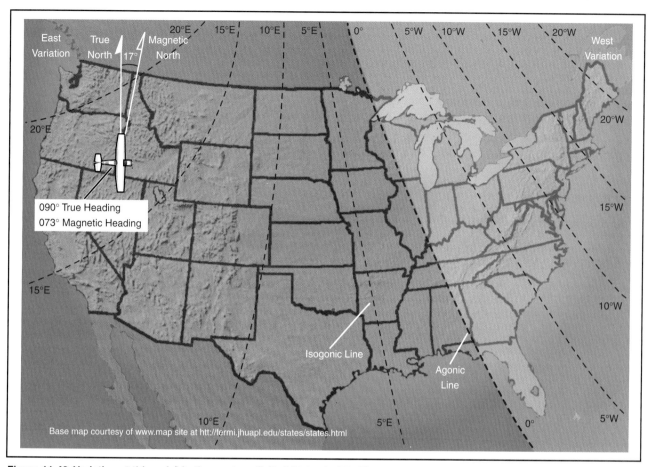

Figure 11-42. Variation at this point in the western United States is 17°. Since the magnetic north pole is located to the east of the true north pole in relation to this point, the variation is easterly. When the magnetic pole falls to the west of the true pole, variation is westerly. Isogonic lines connect points where the variation is equal, while the agonic line defines the points where the variation is zero.

recorded on a chart, called a compass correction card, which is mounted near the compass. [Figure 11-43] Whenever equipment is added to the aircraft that could affect compass deviation, or when the pilot reports that the compass is inaccurate, a technician must perform a compass swing. (This procedure is covered in Section B of this chapter.)

REMOTE INDICATING COMPASS

The instrument panel of the aircraft is the most desirable place for a compass from the pilot's point of view. However, this location is surrounded by a variety of electrical circuits, which create magnetic forces that aggravate compass deviation problems. The remote indicating compass was developed to mitigate this problem. This system uses a remotely-mounted compass transmitter located in an area least likely to be subject to stray magnetism. The rear of the fuselage, the wing tips, and the vertical stabilizer are common locations.

The remote compass transmitter is electrically connected to the magnetic indicator located on the instrument panel. This is called a synchro-type or Magnesyn-type remote indicating system. It is used to transmit the action of the compass to the needle of the indicator.

SLAVED GYRO COMPASS

A heading indicator, combined with direction-sensing instrumentation, overcomes the limitations of either a conventional magnetic compass or a gyroscopic heading indicator without directional input. The resulting instrument is called a slaved gyro. A flux gate, or flux valve, picks up an induced voltage from the earth's magnetic field and, after processing

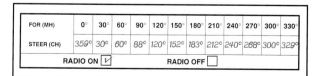

Figure 11-43. When swinging the compass, the technician adjusts the compensator magnets to minimize deviation. Any remaining error is recorded on the compass correction card.

it, directs it to a slaving torque motor in the instrument. This motor precesses the gyro, rotating the dial until the airplane's magnetic heading is under the lubber line on the face of the instrument.

A horizontal situation indicator (HSI) combines the slaved heading indicator with a VOR and glide slope indicator. This provides a navigation display that is easy for a pilot to interpret. Warning flags (NAV, HDG, or GS) warn the pilot whenever the navigation or compass system become inoperative. [Figure 11-44]

Figure 11-45. A radio magnetic indicator (RMI) simplifies ADF and VOR navigation.

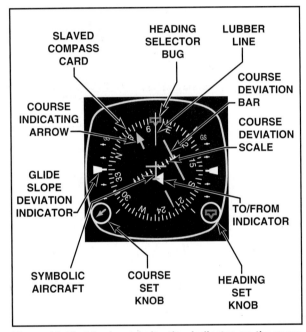

Figure 11-44. A horizontal situation indicator, as the name implies, provides a pilot with excellent horizontal situational awareness.

A radio magnetic indicator (RMI) combines a slaved heading indicator with one or more needles. This simplifies a pilot's calculation of bearings to and from radio beacons and/or VOR stations. [Figure 11-45]

INSTRUMENT PNEUMATIC SYSTEMS

All of the early gyro instruments were powered by a jet of air flowing over buckets cut into the periphery of the gyro rotor. Most light airplanes still use at lease some pneumatic-powered gyroscopic instruments.

VENTURI SYSTEMS

Early airplanes used a venturi mounted on the outside of the aircraft to generate the air for powering pneumatic instruments. The venturi produced a low pressure, or vacuum, which pulled air through the instruments. Air entered the system through a paper filter and then traveled through a nozzle aimed at the rotor. [Figure 11-46]

The primary advantage of venturi systems is that they are simple and require no power from the engine, or electrical system. A major disadvantage is that a venturi is susceptible to ice, and can become unusable at a time when instruments are most needed. Furthermore, the venturi produces no vacuum until the airplane is flying. This means the flight instruments are not available until an airplane has taken off and the instrument gyros have had a few minutes during flight to spin up to operating rpm. Venturis also add a small amount of drag during flight.

As shown in figure 11-46, there are two sizes of venturi tubes: those that produce suction equal to four inches of mercury (in. Hg.) and those that produce two inches Hg. suction. The four inch Hg. venturis are used to drive the attitude and heading indicators, and smaller two inch Hg. tubes are used for turn and slip indicators. Some installations use two of the larger venturi tubes connected in parallel to the attitude and heading indicators, and the turn and slip indicator is connected to one of these instruments through a needle valve.

VACUUM PUMP SYSTEMS

To overcome the venturi tube's susceptibility to ice, most small aircraft are now equipped with engine-driven vacuum pumps. These pumps create suction, which pulls air through the instrument and

Aircraft Instrument Systems

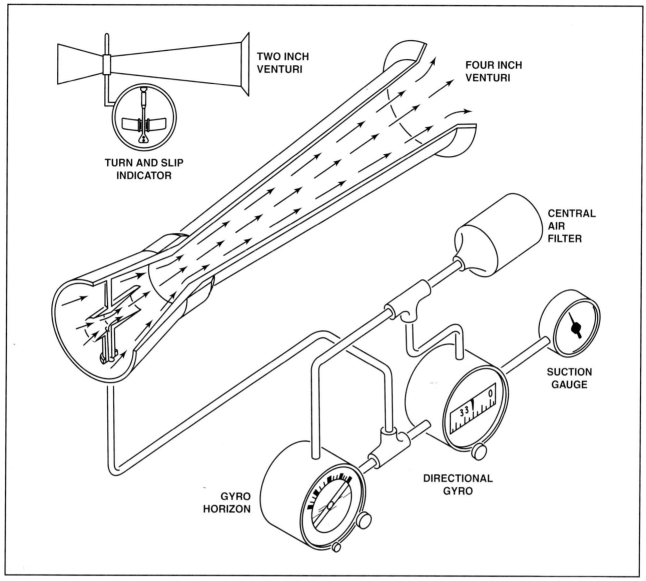

Figure 11-46. Early gyro instruments were powered by venturis located on the outside of the aircraft.

drives the gyro rotor. A suction relief valve maintains the desired pressure (usually about four inches of mercury) on the attitude and heading indicators. Most early instruments used only paper filters in each of the instrument cases, but in some installations a central air filter was used to remove contaminants from the cabin air before it entered the instrument case. On airplanes equipped with vacuum pumps, the turn indicators are normally electric-powered to enable a pilot to maintain aircraft control in case of a vacuum pump failure.

VACUUM PUMPS

The early vacuum pumps were vane-type pumps, also called "wet" pumps, with a cast iron housing and steel vanes. Engine oil is metered into the pump to provide sealing, lubrication and cooling. This oil mixes with the air that the pump draws through the vacuum-powered instruments, and is then blown through an oil separator. Here, the oil collects on baffles and is returned to the engine crankcase, and the air is exhausted overboard. Aircraft equipped with rubber de-ice boots use this discharge air to inflate the boots. In these installations, the air is passed through a second stage of oil separation, then to the distributor valve and to the boots. [Figure 11-47]

Modern vacuum pumps are of the dry type. They do not require oil for lubrication, since they utilize self-lubricating carbon vanes. The vanes lubricate the pump as they wear away at a predetermined rate. Other than not requiring an oil separator, a dry air pump system is almost the same as a wet pump

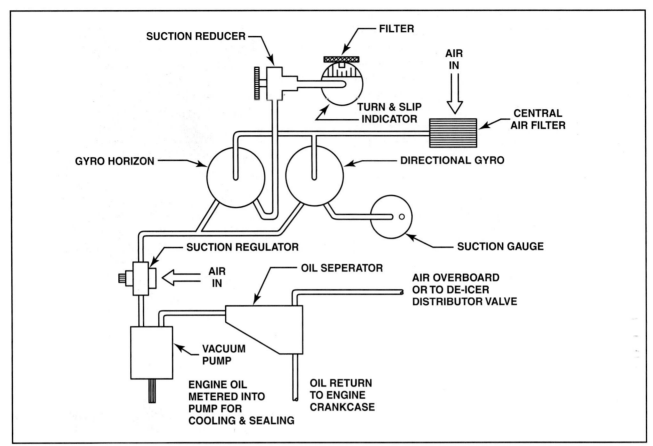

Figure 11-47. Wet-type vacuum pumps require oil separators to recover oil that is used for lubrication and sealing in the pump. Pressurized air from the output side of the pump, with oil removed, is often used to inflate de-icing boots.

system. However, it is essential to keep a dry pump perfectly clean. Solid particles drawn into the system through the suction relief valve can break particles off the carbon vanes. The dislodged particles will damage all of the other vanes and destroy the pump. The air inlet is covered with a filter to prevent particles from entering the system. This filter must be cleaned or replaced at intervals recommended by the aircraft manufacturer. [Figure 11-48]

SUCTION RELIEF VALVES

Vacuum-powered instruments require a specific amount of suction to maintain the correct r.p.m. This is controlled with a pressure regulator

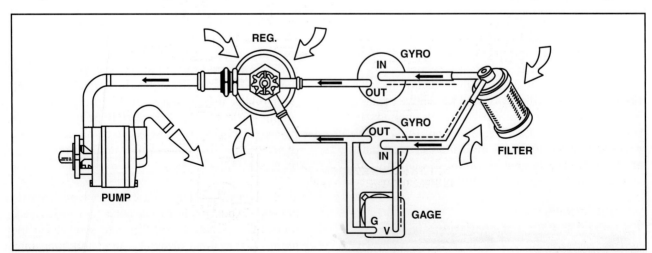

Figure 11-48. Dry-type vacuum pumps require no lubricating oil. However, they cannot tolerate dirt or other particles, because any grit will quickly damage the graphite vanes.

mounted between the pump and the instruments. The regulator has a spring-loaded suction relief valve that allows cabin air to bypass the instruments and enter the pump directly. If the pump is developing too much suction, the relief valve opens, and this maintains the correct negative pressure inside the instrument cases. This valve has a spring-loaded valve that allows cabin air to enter the pump. This maintains the correct negative pressure inside the instrument case. [Figure 11-49]

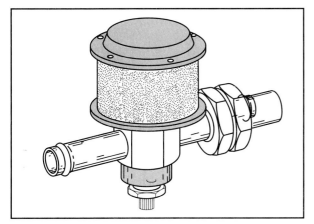

Figure 11-49. This suction relief valve has a removable foam filter.

FILTERS

The life of an air-driven gyro instrument is dramatically affected by the cleanliness of the air that flows over the rotor. In vacuum systems, this air is drawn from the cabin where there usually is dust and sometimes tobacco smoke. Unless all solid contaminants are removed from the air before it enters an instrument, the contaminants will accumulate, usually in the rotor bearings, and slow the rotor. This

Figure 11-50. All particles can damage the rotors in gyros. An air filter ensures that only clean air reaches these instruments.

causes inaccuracy in the instrument and shortens its service life. [Figure 11-50]

Because of the susceptibility to damage from airborne contaminants, all of the filters in the system must be replaced on the schedule recommended by the aircraft manufacturer. If the aircraft is operated under particularly dusty conditions, filters should be replaced more frequently. This is especially true if pilots or passengers regularly smoke while onboard the aircraft. [Figure 11-51]

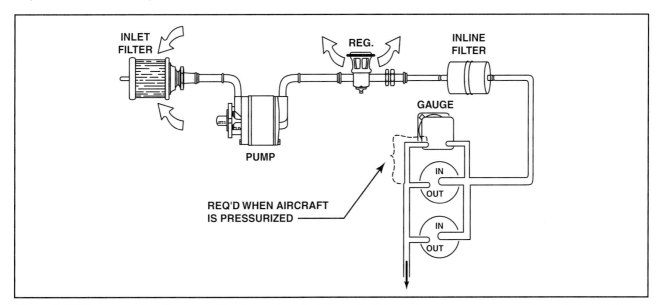

Figure 11-51. Vacuum system filters must be changed regularly to protect the dry type vacuum pumps and gyro instruments.

INSTRUMENT SYSTEM SERVICING

Under the Federal Aviation Regulations, an aircraft technician has very limited authority to repair instruments. Generally, a technician can remove and replace instruments and their related systems; however, the actual repair of the instrument must be performed at an authorized repair station. Regulations do permit routine maintenance or servicing of the system.

Technicians often replace filters, troubleshoot and replace defective instruments, and replace vacuum pumps. The plumbing to and from the various instruments often requires inspection and repair. Many shops are equipped with a ground test vacuum pump that allows the technician to apply vacuum to the system without using the aircraft's engine-driven pump. A ground test vacuum pump is particularly useful when troubleshooting a system with suspected leaks.

POSITIVE PRESSURE SYSTEMS

As an aircraft climbs above 18,000 feet MSL, vacuum pumps can't draw enough air through the system to provide sufficient rotor speed. To remedy this problem, many aircraft that fly at high altitude use positive pressure systems to drive the gyros. These systems use the same type of air pump as vacuum systems, except that the pump is upstream, rather than downstream, of the instruments. The discharged air from the pump is filtered and directed into the instruments through the same fitting that receives the filtered air when a vacuum system is used. A filter also is installed on the inlet of the pump, so the system has two filters instead of one. A pressure regulator is located between the pump and the in-line filter that feeds the air to the instruments.

PITOT-STATIC SYSTEM

Pitot-static instruments were discussed earlier in this chapter; they include the airspeed and/or Mach indicator, the altimeter, and the vertical speed indicator (VSI). The pitot-static system encompasses the pitot tube, static port, and associated plumbing that must be connected to the pitot-static instruments to make them work. Pitot pressure is ram air pressure picked up by a small open-ended pitot tube This tube is about a quarter of an inch in diameter and is mounted facing the oncoming air stream. Ram air entering the tube produces a pressure that increases with the speed of the aircraft through the air. [Figure 11-52]

Static pressure is the pressure of the still or ambient air surrounding the aircraft. It is used to measure the

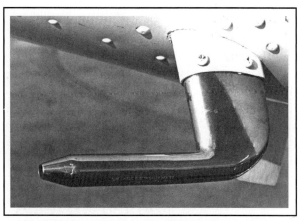

Figure 11-52. The pitot tube is mounted on the outside of the aircraft where it can pick up air flowing past the aircraft. As the speed of the air increases, the pressure inside the closed pitot system also increases.

altitude and as a reference for measuring airspeed. Ambient air pressure is taken from a static port at a location on the aircraft where flight tests have determined there is a minimum disturbance to the airflow. [Figure 11-53]

Figure 11-53. Static ports are typically mounted flush on the exterior of the aircraft.

Only airspeed indicators are connected into the pitot system. The center port of the indicator is connected to the pitot tube, which usually is installed in the wing of a single-engine aircraft, outside of the propeller slipstream, or on the fuselage of a multi-engine aircraft. The pitot head is mounted in such a way that it points directly into the airflow In larger aircraft, a machmeter may also installed in the pitot-static system. [Figure 11-54]

Aircraft Instrument Systems

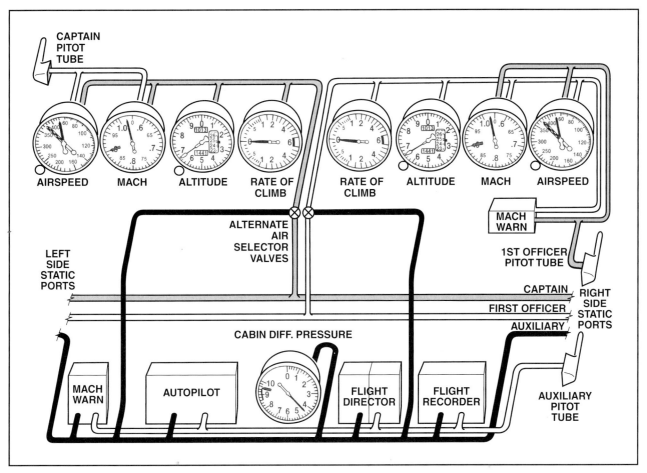

Figure 11-54. Large jet transport airplanes normally contain multiple pitot static systems.

To prevent ice inside the pitot head, many aircraft are equipped with electric pitot tube heaters. The heater may be checked on a preflight inspection by turning it on for a few seconds, and then turning it off, which will leave the pitot tube warm to the touch. However, these heaters produce so much heat that they should not be operated for extended periods of time until takeoff, when adequate airflow will prevent overheating. In flight, a pilot can verify the operation of the heater by checking the ammeter as the switch is turned on. [Figure 11-55]

Some aircraft have a combination pitot-static head, in which the pitot pressure is taken from the open end of the tube, and static pressure from holes or slots around the head, back from the open end. However, most modern aircraft have flush static ports on the sides of the fuselage. Many aircraft have static ports on both sides of the fuselage so side slipping will not cause an inaccurate static pressure. It is extremely important that the static holes are never obstructed, and also important that there is no distortion in the aircraft skin around the

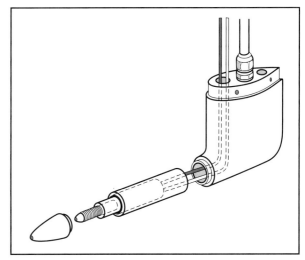

Figure 11-55. A heating element is used to prevent the formation of ice in a pitot tube.

holes. Irregularities on the aircraft surface can cause turbulent airflow, which can produce an inaccurate static pressure. Water traps usually are installed in all of the static lines. These traps should be drained on each routine maintenance inspection. [Figure 11-56]

Figure 11-56. Some aircraft are equipped with a blade probe with a pitot tube in front and a static port in the bottom.

Provisions must be made to keep water and ice out of the pitot system. If water should collect in the pitot line, the airspeed indicator may oscillate as the water moves back and forth in the line. This oscillation can result in an inaccurate reading. If ice should form inside the pitot head, the pitot pressure inside the system will be trapped. As the aircraft descends, the static pressure increases, which can cause the airspeed indicator to show a false decrease in airspeed. To prevent accumulation of water and ice in the system, the plumbing that connects the pitot tube to the airspeed indicator is run as directly as possible. There usually is a T-fitting or sump at a low point in the line to collect any accumulated moisture. [Figure 11-57]

Most aircraft have an alternate static source that a pilot can select in the event the normal ports become plugged with ice. The alternate source valve is located on or under the instrument panel. In unpressurized aircraft, the alternate air is sampled from behind the instrument panel. On pressurized aircraft, the alternate air comes from some portion of the fuselage outside of the pressure vessel. [Figure 11-58]

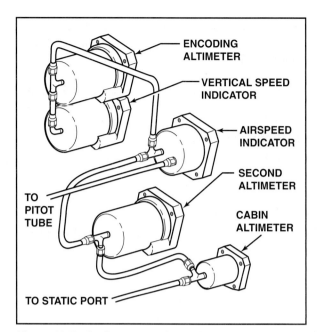

Figure 11-57. Static pressure is provided for the airspeed indicator(s), altimeter, and vertical speed indicator. Some aircraft may also have static pressure routed to a cabin altimeter, standby altimeter or a separate altitude encoder.

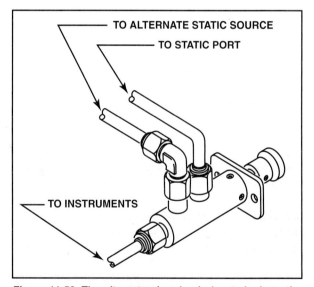

Figure 11-58. The alternate air valve is located where the pilot can select an alternate static air source if the primary source becomes plugged.

STATIC SYSTEM TESTING

FAR 91.411 requires that the static system be checked every 24 calendar months on aircraft that are flown under instrument flight rules (IFR). This check also is required any time the system is opened for service or repair, or may be initiated if a pilot reports a problem. For example, if an instrument attached to a static system were to become disconnected inside a pressurized cabin, the static pressure would be higher than the correct ambient,

Aircraft Instrument Systems

outside pressure. The altimeter would indicate a lower altitude than actual and the airspeed would read lower than the correct airspeed.

The aircraft manufacturer, an authorized technician, or an authorized repair station must perform the system check, and the results must be recorded in the aircraft maintenance records.

FAR 43, Appendix E requires the following:

1. The static system must be free of entrapped moisture and restrictions.
2. The leak check to be within established tolerances.
3. The static port heater, if installed, must be operative.
4. The technician must ensure that no alterations or deformations of the airframe surface have been made that would affect the relationship between air pressure in the static pressure system and true ambient static air pressure for any flight condition.

The procedure is outlined in Section B of this chapter.

FUEL QUANTITY INDICATING SYSTEMS

FAR Part 91 requires a fuel gauge indicating the quantity of fuel in each tank. These gauges, or systems, may be as simple as a wire attached to a cork float. The amount of wire protruding from the cap indicates the amount of fuel in the tank.

MECHANICAL INDICATORS

Other direct-reading fuel quantity indicators move a pointer across the dial by magnetic coupling. A float rides on the top of the fuel which, through a bevel gear, rotates a horseshoe-shaped permanent magnet inside the indicator housing. A pointer attached to a small permanent magnet and mounted on a pivot is separated from the horseshoe magnet by an aluminum alloy diaphragm. The permanent magnet is coupled to the horseshoe magnet by the magnetic fields. The pointer will move around the dial to indicate the level of fuel in the tank. [Figure 11-59]

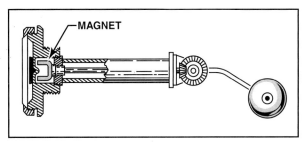

Figure 11-59. The mechanical fuel indicator operates with nearly a direct connection between the fuel float and the indicator. The magnetic coupling dampens the sloshing of the fuel, and reduces wear on the gauge.

DIRECT CURRENT ELECTRICAL INDICATORS

In most cases, a direct connection between the float and the indicator is not possible. A DC electrical indicator solves this problem. It converts mechanical motion of the float into a varying direct current. This current then drives a mechanical indicator or is converted to a digital readout. [Figure 11-60]

The tank unit consists of either a wire-wound resistor or a segment of composition resistance material. A wiper arm driven by the float moves across this resistance material, changing the circuit resistance. Before troubleshooting one of these systems, consult the aircraft service manual to determine how the resistance corresponds to fuel level. Some units

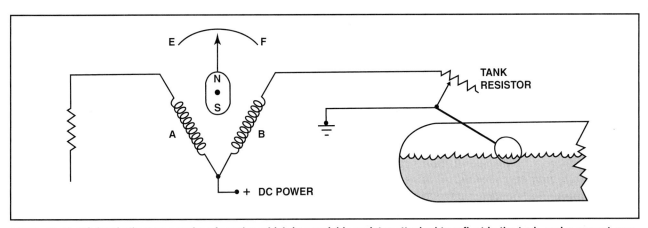

Figure 11-60. DC fuel indicators consist of sender, which is a variable resistor attached to a float in the tank, and a current measuring instrument as the indicator.

signal a full tank with maximum resistance while others do so with minimum resistance.

Most indicators are a ratiometer-type gauge to minimize errors caused by variations in system voltage. Current flows through both coils, the fixed resistor, and the tank unit as shown in Figure 11-60. As the resistance of the tank unit varies, so does the current through coil B, which varies its magnetic strength. The pointer is mounted on a small permanent magnet and moves across the dial in such a way that it indicates the level of fuel in the tank.

CAPACITANCE FUEL QUANTITY SYSTEMS

Many modern aircraft use a capacitance-type fuel quantity measuring system that factors in fuel density, as well as volume. This is an electronic system that measures the capacitance of the probe, or probes, that serve as the tank sender units.

A capacitor is a device that can store an electrical charge. It consists of two conductive plates separated by an insulator called a dielectric. The amount of charge a capacitor can store varies with the area of the plates, the separation between the plates, which equals the thickness of the dielectric, and the dielectric constant of the material between the plates.

The probes in a capacitance fuel quantity indicating system consist of two concentric metal tubes that serve as the plates of the capacitor. The area of the plates is fixed, as well as the separation between them, so the only variable is the material that separates them. [Figure 11-61]

These probes extend from the top to the bottom of the tank. When the tank is empty, the plates are separated by air that has a dielectric constant of one. When the tank is full, the dielectric is fuel that has a constant of approximately two. In any condition between full and empty, part of the dielectric is air, part is fuel, and so the capacitance of the probe varies according to the level of fuel in the tank.

The probes also take into account the density of the fuel. They provide an accurate indication of the mass of the fuel in the tanks, which more accurately reflects the available energy in the fuel than its volume. While the difference is negligible on small airplanes powered by avgas, it can be significant on large airplanes powered by kerosene. Fuel density can change due to variations in temperature or due to substitution of a different grade of jet fuel. The capacitive probes sense this because denser fuel has a higher dielectric constant, which increases the measured fuel quantity at a given volume.

Since a capacitive fuel sensor measures the mass of the fuel, some compensation is required if the desired indication is to be in gallons. To accomplish this, a compensator is built into the bottom of one of the tank units. It is electrically in parallel with the probes which cancels the changes in dielectric constant caused by changes in fuel temperature. To provide accurate indications, the compensator should be calibrated for the grade of fuel normally used in the aircraft. [Figure 11-62]

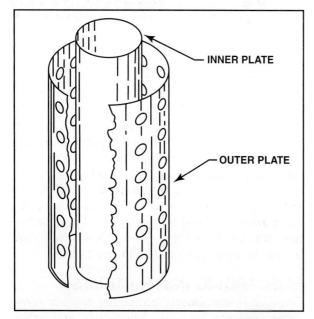

Figure 11-61. A capacitance-type fuel indicator system operates on the principle that the amount of fuel between the plates causes a difference in measurable capacitance.

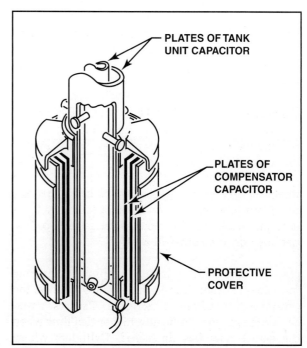

Figure 11-62. A compensator is built into the system in order to adjust the indication for temperature changes.

Aircraft Instrument Systems

Capacitive fuel systems can be tailored for tanks of all sizes and shapes. Another advantage is that all the probes in the aircraft can be connected so the system that integrates their output to show the total amount of fuel on board.

Most capacitance-type systems have a test feature. Actuating the test switch causes the gauges to drive towards Zero or Empty. When the switch is released, the pointer(s) should promptly return to the original quantity indication.

Troubleshooting a capacitance-type fuel-quantity indicating system is quite simple. Following the appropriate instructions, connect a test unit into the system. The tester substitutes known capacitance values for the probes and includes an accurate amplifier and indicator into the system. The testing unit may also provide a technician with the capability to calibrate the system following troubleshooting and/or repairs.

FUEL SYSTEM MONITORING INSTRUMENTS

These instruments include fuel pressure indicators, fuel injection system flow meters, and other types of fuel flow instruments.

FUEL PRESSURE INDICATORS

Fuel pressure instruments usually are found on aircraft with fuel injected engines. They are used to determine how much to prime the engine. Fuel pressure is monitored on many aircraft for starting and to check fuel system operation during flight. The pressure is typically measured in pounds per square inch (psi) using a Bourdon tube type instrument

FUEL INJECTION SYSTEM FLOWMETER

In order to monitor fuel consumption and engine performance, fuel-injected engines use a simple flowmeter. This device operates on the principle that the pressure drop across a fixed orifice is directly proportional to the flow through that orifice. Thus, this type of flowmeter is actually a pressure gauge calibrated in pounds, or gallons, of fuel per hour, or some other flow rate.

One problem with this type of flow measuring instrument is that, if the orifice partially clogs, the pressure drop across it increases. The restriction causes the gauge to indicate a higher fuel flow when the actual flow has decreased. Cylinders whose actual fuel flow is restricted dramatically, will instead show significantly higher fuel flows.

VOLUME FLOW MEASUREMENT

Another type of fuel flow measurement system incorporates a movable vane unit in series with the fuel line to the engine. [Figure 11-63] The spring-loaded vane is displaced in proportion to the amount of fuel flow. Since the vane's movement must be linear to get an accurate flow measurement, the size of the restriction created by the vane in its housing must increase as the vane is displaced. This increase is calibrated into the design of the flowmeter sender unit. The movement of the vane is sent to the flow indicator via an Autosyn transmitter and receiver unit. Refer to Chapter 10 for more information about Autosyn transmitters. [Figure 11-63]

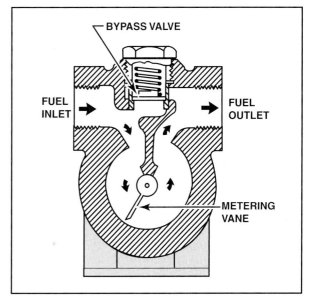

Figure 11-63. A flowmeter measures the amount of a fluid passing a given point by measuring the revolutions of a calibrated vane.

MASS FLOW MEASUREMENT

A mass flow measurement device works on the principle that the viscosity of the fuel changes with the mass of the fuel. A turbine is placed in line with a motor that swirls the fuel. The viscosity of the fuel will determines the force placed on the turbine, which is resisted by a calibrated spring force. The turbine is connected to an AC system, which transmits fuel flow information to the indicator on the instrument panel. While this system is by far the most accurate means of monitoring fuel flow; it also is also the most complicated. [Figure 11-64]

ELECTRONIC INSTRUMENTS

Older style instruments containing rotating mass gyros, delicate bearings, hairsprings and other extremely sensitive components have been replaced with electronic solid-state devices. The new "strap

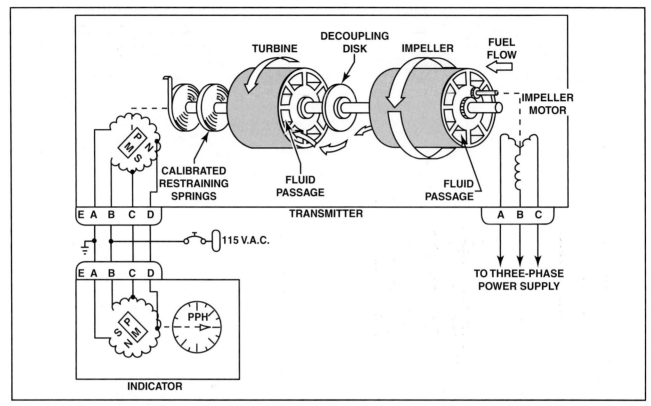

Figure 11-64. Measurement of fuel flow based on mass is accomplished by spinning the fuel toward a turbine with an impeller. The amount the turbine is displaced depends on the quantity and viscosity of the fuel that is spun toward it.

down" systems contain no moving parts and boast of greater reliability and higher time between failures.

Electronic Flight Instruments Systems (EFIS) have several advantages over conventional instruments. Primarily, EFIS offers the advantage of greatly increased reliability, and, ultimately, reduced maintenance costs. EFIS also reduces instrument panel clutter by combining several instruments into one unit. A pilot can customize the information presented on the displays, which can reduce workload. Many EFIS systems offer a moving map display with superimposed weather radar information. Collision avoidance information can even be incorporated into certain EFIS displays. [Figure 11-65]

A typical EFIS system consists of four interchangeable CRT displays and their associated symbol generators. The center symbol generator is used as a redundancy check between the left and right systems. If one or more of the units fail, the other(s) will control the electronic displays. The symbol generator receives signals from the various instruments and navigational sensors located throughout the aircraft. The weather radar receiver/transmitter receives signals from the aircraft's radar antenna and sends information to both the pilot and copilot's symbol

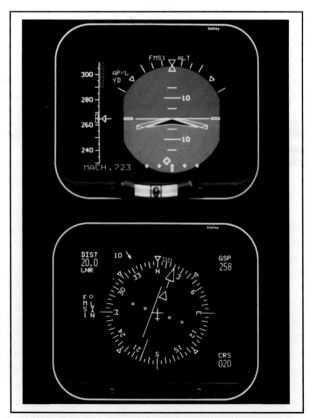

Figure 11-65. This two tube EFIS display is set up to display attitude information on one tube, and navigation information on the other.

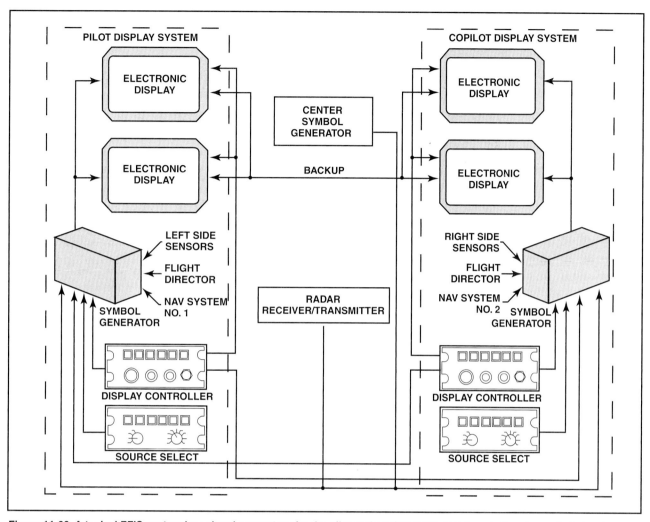

Figure 11-66. A typical EFIS system has signal generators for the pilot and copilot sides of the cockpit and a central signal generator that acts as a control in case of disagreement between the other two.

generators. The display controller allows the pilot to select the appropriate system configuration for the current flight situation. [Figure 11-66]

The use of digitally-based microprocessor electronics allows several mechanical instruments to be replaced with modern electronic flight instruments displaying the information on one or more cathode ray tubes (CRT's). The signals sent to the various components of the systems are typically linked through a digital data bus. A data bus is made up of a twisted pair of insulated wires surrounded by an outer shielding. The electrical signals sent through the data bus consist of short pulses of voltage on or voltage off (binary ones and zeroes). These pulses are extremely short in duration. A typical system is capable of transmitting signal pulses that are only 10 microseconds in length.

One common bus system is known as ARINC 429. ARINC stands for Aeronautical Radio Incorporated, and 429 is a code for a specific digital data standard. This system uses a 32-bit word for all information transmitted over the data bus. The 32-bit word is made up of one parity bit, a sign status matrix, the data, the source destination indicator, and the label or identification of the sending unit. Use of a digital data standard allows computer-processing units to talk to each other. [Figure 11-67]

ELECTRONIC ATTITUDE DIRECTOR INDICATOR (EADI)

The electronic attitude director indicator, like its mechanical predecessor, displays much of the basic flight data needed to maintain a smooth and comfortable flight. From basic pitch and roll information to approach decision height, the information is displayed in a color format that is readable even in full sunlight. There are both full-time and part-time displays on most EADIs. The full-time displays give the pilot information needed for flight control. The part time displays offer information typically

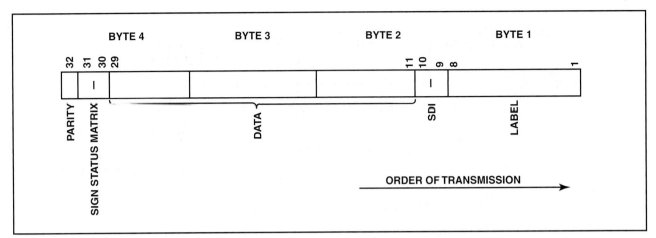

Figure 11-67. The ARINC 429 data bus system is one way that digital components can communicate with each other.

needed for runway approach or basic navigation and are only active during the pertinent portion of the flight. [Figure 11-68]

The full-time displays include the aircraft symbol that is used as a reference for pitch and roll information. To determine the aircraft's attitude the pilot must compare the aircraft symbol to the attitude sphere. The amount of pitch and roll (10, 20, degrees etc) is indicated on the attitude sphere. The attitude source indicator is also displayed to inform the flight crew which symbol generator is currently driving this EADI.

The part-time displays include:

- Rising runway display that appears at 200 feet above ground level and is used during the final portion of an approach.

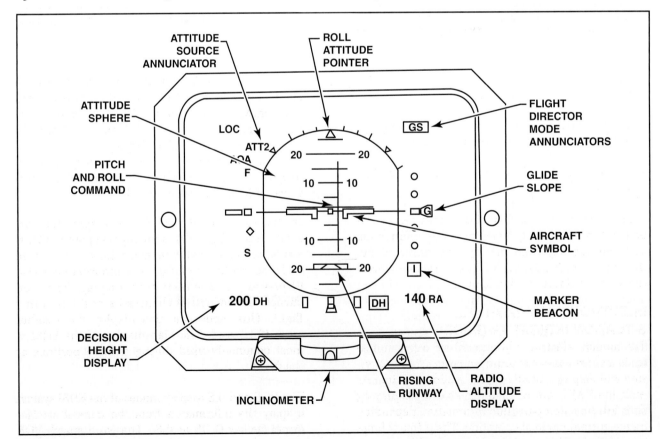

Figure 11-68. The electronic attitude director indicator (EADI) can display most of the flight attitude information required by the pilot for any given phase of flight.

Aircraft Instrument Systems

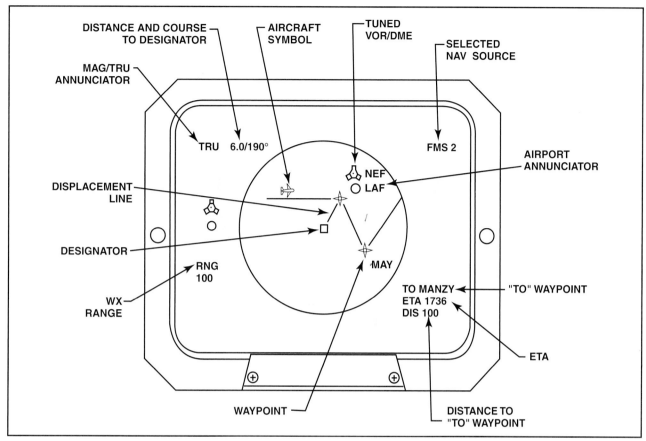

Figure 11-69. The electronic horizontal situation indicator (EHSI) displays most of the horizontal situation information required by the pilot. The exact data displayed will depend on the mode selected.

- Glide slope and localizer indications that guide the pilot to the touchdown point.

- Marker beacon and radio altimeter displays shown in the lower left corner of the display.

- Decision height displayed in the lower right corner.

It should be noted that this EADI is only one of several versions currently available. The information displayed on any particular system may vary, but the basic configuration of the instrument will remain very similar to that shown.

ELECTRONIC HORIZONTAL SITUATION INDICATOR (EHSI)

The modern electronic horizontal situation indicator is modeled after the older electromechanical version and displays much of the same information. As with the EADI, the EHSI is capable of displaying both full-time and part-time information depending on the current mode of operation. The primary function of an EHSI is to display navigational information. [Figure 11-69]

The electronic horizontal situation indicator can be set for one of four modes of operation. These modes are Plan, Map, VOR, and ILS. The flight crew selects the various modes at the EFIS display controller. In the Plan mode, the CRT displays enroute flight information entered into the flight management system to provide the flight crew with a visual portrayal of their recorded flight plan.

In the Map mode, the EHSI will display the currently active flight plan showing waypoints, VORs, airports, etc. A real-time magnetic compass rose and other pertinent navigational information is also displayed on a moving map display. The map changes as the aircraft changes its position during flight. This mode can also display the weather radar if selected. The Map display of the EHSI is most commonly used during enroute portions of the flight.

The VOR and ILS display modes of the EHSI system display the information from the current navigational facility (VOR or ILS). The wind speed, VOR or ILS frequency received, and the current aircraft heading are also displayed.

ELECTRONIC SYSTEMS MONITORING DISPLAYS

In an effort to further reduce instrument panel clutter and pilot workload, many of the traditional engine and system instruments have been replaced with electronic monitors. These monitors rely on computers to receive data inputs from the various systems of the aircraft. The computer analyzes the information and transmits a digital signal to one or more CRT displays that inform the flight crew of various systems conditions. The monitoring systems are also used to alert the flight crew of any system malfunction and, in some case, provide suggested corrective actions. [Figure 11-70]

ELECTRONIC CENTRALIZED AIRCRAFT MONITOR (ECAM)

The electronic centralized aircraft monitor system is comprised of two CRT display units, a left and right symbol generator, an ECAM control panel, discrete warning light display unit, two flight warning computers and a digital-to-analog data converter. The various components communicate through data bus

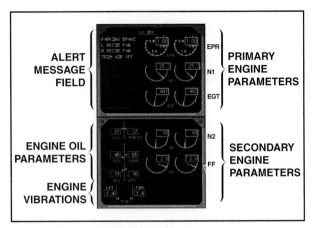

Figure 11-70. Electronic systems monitoring displays replace traditional analog gauges with electronic presentations that are easier to read and more reliable.

systems. The left CRT contains information on systems status, warnings and any associated corrective actions. The right CRT displays information in a pictorial format such as control surface positions. All information displayed on either CRT is shown in a digital format. [Figure 11-71]

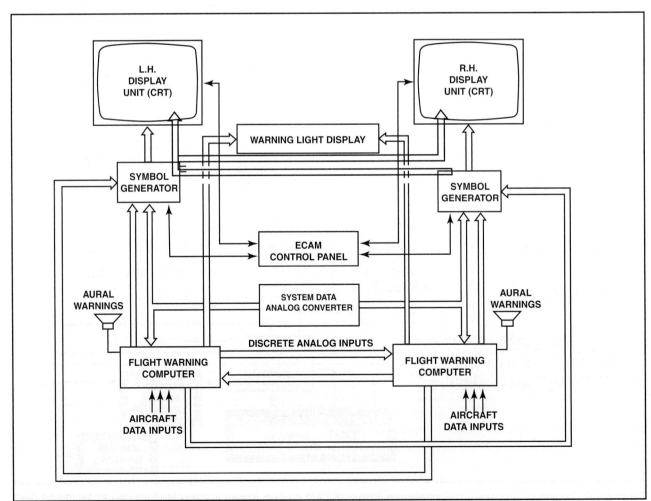

Figure 11-71. The electronic centralized aircraft monitor (ECAM) system displays system information to the pilots.

There are four basic modes of operation for the ECAM system. The manual mode will display pictorial diagrams of various aircraft systems. The other three modes operate automatically and are referred to as the flight phase, advisory and failure modes. The flight mode displays information related to the current phase of flight, such as preflight, take-off, climb, enroute, descent and landing. The flight mode information is displayed on the right-hand CRT. The advisory mode information is displayed on the left CRT and contains information of concern to the flight crew, yet not critical.

The failure mode of operation automatically takes precedence over all other modes. The failure mode displays any information that may be considered critical to the flight safety. The left-hand CRT displays appropriate information and corrective action while the right unit displays the status of the failed system. If a failure occurs, the flight crew is also alerted to the problem through aural and visual warnings.

ENGINE INDICATOR AND CREW ALERTING SYSTEM (EICAS)

This system is used to monitor the engine and various other systems of the aircraft similar to the ECAM, however, the EICAS uses both a digital and analog format to display the information. The engine indicator and crew alerting system consists of two CRT displays, a left and right system computer, a display selector panel, discrete caution and warning displays, and a standby engine indicator. [Figure 11-72]

The CRTs of the EICAS system are arranged one on top of the other and a discrete annunciator is located in front of both the pilot and copilot. The standby engine display is a liquid crystal display unit located in the center area of the instrument panel. The liquid crystal display shows the engine performance parameters.

The upper CRT of the system displays primary engine system information in both analog and digital formats. Warnings, cautions and advisories are also displayed on the upper CRT. The lower CRT is normally blank during flight unless the flight crew selects specific information. The lower CRT displays information such as systems status, aircraft configuration, fluid quantities, various temperatures and maintenance information.

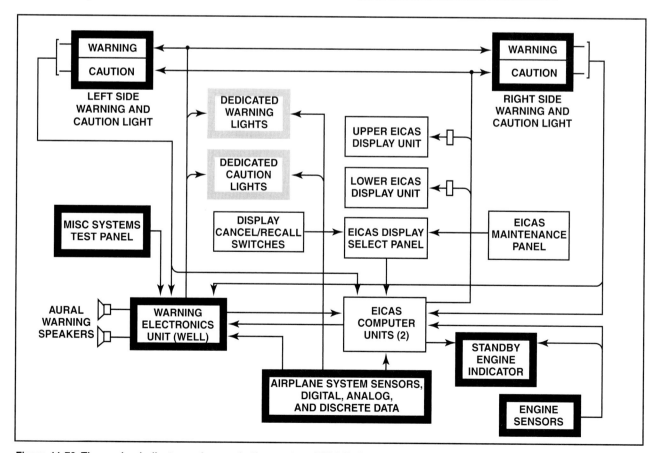

Figure 11-72. The engine indicator and crew alerting system (EICAS) displays aircraft systems similar to the ECAM. EICAS uses both analog and digital displays.

In the event of one CRT failure, all needed information is displayed on the operable CRT. In this case, the analog display of information is removed and the system is automatically converted to the compact mode. In the event the second CRT fails, the flight crew can still find critical engine information on the liquid crystal display.

AUXILIARY INSTRUMENTS

OUTSIDE AIR TEMPERATURE

An outside air temperature gauge is an extremely simple instrument used on most small single-engine aircraft. It consists of a bimetallic-type thermometer in which strips of two dissimilar metals are welded together into a single strip and twisted into a helix. One end is anchored into a protective tube, and the other end is affixed to the pointer that indicates the outside air temperature on a circular dial. As the temperature of the bimetallic strip changes, it causes it to expand or contract. This causes the metal to change the amount of twist and moves the pointer. The bimetallic strip is inside a tube that sticks through a hole in the windshield, with the dial in easy view of the pilot. [Figure 11-73]

CLOCK

A clock is one of the most fundamental of instruments, used for timing flight maneuvers, for navigation and for determining engine functions such as the fuel consumption rate.

Until recently, most aircraft clocks have been of the analog type, all equipped with a sweep second hand. Clocks with digital displays are now becoming popular. These not only display the local time, but also Greenwich mean time (Zulu time), as well

Figure 11-73. The pilot uses outside air temperature (OAT) to calculate true airspeed. The OAT gauge is usually mounted near the top of the windscreen or out the side window beside the pilot.

as the day and date. Most of these clocks are equipped with circuits that allow them to be used as a stopwatch and an elapsed time indicator.

Electrically operated clocks are normally installed in the aircraft with their power lead connected directly to the aircraft battery, so they will always have power regardless of whether the master switch is on or off. They are protected with a low current fuse.

SECTION B

INSTRUMENT SYSTEM INSTALLATION AND MAINTENANCE PRACTICES

Federal Aviation Regulations do not allow an aircraft maintenance technician to repair aircraft instruments. However, a technician is allowed to perform a number of functions including, installation, simple adjustments, and marking of instruments. A technician may install instruments into instrument panels, connect power supplies to instruments, place the range markings on the faces of the instruments, and perform a compass swing on the magnetic compass.

PANEL LAYOUT

Examination of almost any older aircraft confirms that the instrument arrangement in the panel was haphazard at best. The non-standard size of older directional and attitude gyros made it difficult to place these instruments in the panel in a manner that would allow an efficient scan of the instruments for instrument flying. The now-familiar T-arrangement for the flight instruments became standard only after the newer 3 1/8 inch instruments became common. This "Basic T" arrangement is now standard in smaller general aviation aircraft as well as in airliners. As shown in figure 11-74, the basic T-arrangement of instruments is located directly in front of the pilot. The top instrument is the attitude indicator, and directly below this is the directional gyro. To the right of the attitude indicator is the altimeter, and in the left arm of the T is the airspeed indicator. The turn coordinator below the airspeed indicator and the vertical speed indicator below the altimeter supplement the four primary instruments. The clock, airspeed indicator and turn coordinator are all grouped together to permit more efficient scanning of these instruments when the pilot is flying without use of the attitude indicator and directional gyro (partial panel).

The radio equipment is located in the center of the panel and in a row of indicators on the right side of

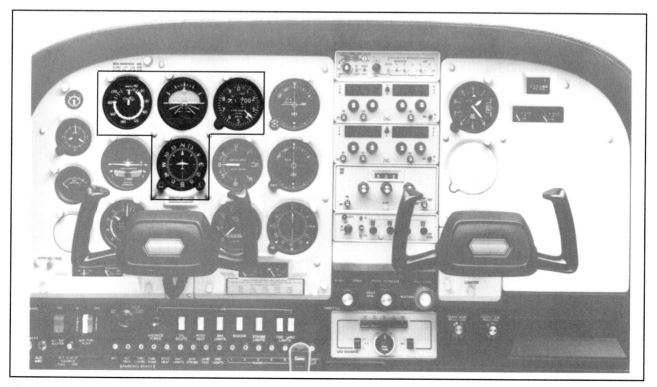

Figure 11-74. Most light aircraft have the flight instruments arranged in a "T" directly in front of the pilot.

the pilot's panel. At the top of the avionics stack are the three marker beacon indicator lights and the audio panel which allows the pilot to select which piece of radio equipment is connected to the speaker and which is connected to the headphones. The next pieces of equipment down from the top are the two communication transceivers and navigation receivers.

The indicators to the left of the radio equipment are operated by the navigation receivers. The upper indicator has a glide-slope needle, while the lower indicator does not. Both indicators, however, can be used to display VOR and localizer information. Below the VHF nav-comm equipment is the automatic direction finder receiver (ADF), and its indicator is just below the VOR/LOC indicators.

Below the ADF is the control head for the radar beacon transponder. The altimeter in front of the pilot is an encoding altimeter, and the one in the right panel is a three-pointer pneumatic altimeter that does not have the encoding feature. To the right of the altimeter in the right panel are the two fuel quantity indicators and the hour meter that records the total time on the engine.

The engine instruments are, in general, located below the flight instruments. The exhaust gas temperature indicator is to the left of and below the turn coordinator. Directly below the turn coordinator is a combination manifold pressure gauge and fuel-flow indicator for the fuel injection system. To the right of this instrument is the tachometer. The bottom row of instruments, almost hidden in this photograph by the control wheel, are the small rectangular ammeter, cylinder head temperature indicator, and oil temperature and oil pressure gauges. The automatic pilot controls are located at the bottom center of the instrument panel.

Small, multi-engine aircraft also reflect the "T" arrangement of flight instruments. Twin-engine aircraft often have dual gauges to reflect the various engine functions for both engines in single gauges. Multi-engine aircraft are usually better equipped with navigation/communication equipment and sometimes even radar. [Figure 11-75]

EQUIPMENT AND INSTRUMENT MOUNTING

Large shock mounts are used to install electronic components to the aircraft structure. Large mount bolts connect a bracket that is riveted to the electronic unit to the aircraft structure. The load must be carried by this type of mount in a direction parallel to the bolt. [Figure 11-76]

Most instruments are mounted in a shock-mounted sub-panel and are covered with a false panel for appearance. The instruments are mounted in the sub-panel according to the case design. Flangeless instruments use an expanding type clamp which is secured to the back of the panel. This clamp is tightened with a screw from the front of the panel.

The shock mount used for instrument panels is simply a pair of threaded studs embedded in a rubber

Figure 11-75. Multi-engine aircraft are usually better equipped than single engine aircraft.

Aircraft Instrument Systems

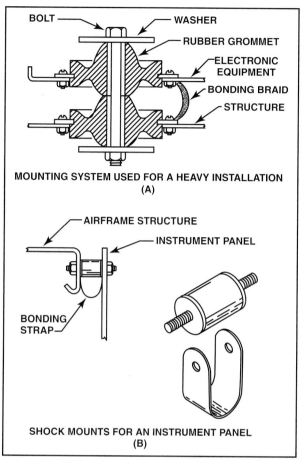

Figure 11-76. Shock mounts isolate instruments and equipment from low frequency, high amplitude vibration that might occur during landing or in rough air. A bonding strap must be used to complete the protective circuit that is broken by using nonconductive shock mounts.

spacer (figure 11-76). The panel is supported from the sub-structure by as many mounts as needed to carry the load. On each inspection, the technician must ensure that the panel cannot be deflected enough for any instruments to hit against the structure or against any of the other instruments. The instrument panel or electrical equipment must be electrically bonded to the structure to provide a current return path. Bonding is accomplished through either a braided bonding strap or with a thin strap of aluminum around the shock mounts. These bonding straps should be checked during every inspection.

POWER REQUIREMENTS

Many instruments found on a modern aircraft require electrical power for operation or lighting. Instruments that are electrically operated are constructed with iron or steel cases in order to control electromagnetic fields. Instrument lighting is typically contained within the instrument or provided by individual post lights. Many instruments, along with the radio and autopilot systems, require electrical power for operation. To ensure proper operation, the power ratings of each component must fall within the aircraft's electrical specifications. Both the voltage and amperage ratings of the instruments must be checked. Whenever installing new or additional electrical equipment to the aircraft, a load check is performed. This check is done by adding up all the continuous electrical loads found on the aircraft. The total is then compared to the output rating of the aircraft's alternator system. The total electrical load must not exceed the alternator output. If the total load is greater than the alternator output, the new equipment should not be added to the aircraft without modification to the electrical system or operating procedures manual.

RANGE MARKINGS

Most aircraft instruments show both the current operating status and any limitations set forth by the manufacturer for the function of that particular instrument. Since equipment limitations can indicate many things such as never-exceed limits, normal operating ranges, caution ranges, etc., color-coding is sometimes used to mark limits on the appropriate instruments. The exact value of each specific range can be found in the aircraft maintenance manual, aircraft flight manual, type certificate data sheet (TCD), or Aircraft Specifications.

In the event that the aircraft has been modified in any way, make sure you check the supplemental type certificate (STC) for any changes to the limitation values. The person installing an instrument is responsible for ensuring that the correct markings are used and placed. The manufacturer or repair station most often places the markings on the instrument dial. However, markings placed on the front face of the instrument glass by maintenance technicians are allowed in some cases. A slippage mark must be placed on all gauges that have range markings on the front glass of the instrument. The **slippage mark** consists of a radial white line that marks both the glass and the case. These marks are not required if the range markings are applied to the inside of the instrument (not on the glass). However, maintenance technicians are not allowed to open the case and apply markings inside the gauge. Only authorized repair facilities are allowed to open the case.

AIRSPEED INDICATOR

- White arc — Used to indicate flap operating range. The lower side of the arc indicates stall speed with flaps extended. The upper end indicates the maximum flap extension speed.

- Green arc — Indicates the normal airspeed operating range of the aircraft. The bottom of the arc is stall speed (flaps up) and the top of the arc is maximum airspeed in rough air.

- Yellow arc — Indicates a limited flight operations range. The aircraft should be operated in the yellow arc airspeed range only in smooth air.

- Red radial line — This is the never-exceed speed for the aircraft. The aircraft should never be operated at speeds at or above this line. [Figure 11-77]

- Blue radial line — On twin-engine aircraft, the blue radial line indicates the best single-engine rate of climb.

CARBURETOR AIR TEMPERATURE INDICATOR

- Red radial line — This line indicates the maximum carburetor inlet air temperature for proper engine operation.

- Green arc — The engine inlet air normal operating temperatures are indicated by the green arc. The upper limit meets the red radial line and the lower limit meets the yellow arc.

- Yellow arc — This is the range at which carburetor ice is most likely to form.

CYLINDER HEAD TEMPERATURE GAUGES

- Red radial line — Maximum permissible cylinder head temperature.

- Yellow arc — Maximum continuous operating temperature.

- Green arc — Normal operating range. [Figure 11-78]

MANIFOLD PRESSURE (MAP) GAUGE

- Red radial line — This line indicates the maximum permissible manifold pressure.

- Yellow arc — This range indicates the maximum manifold pressure for intermittent operations if different from the normal operating range.

- Green arc — This is the normal operating range for the manifold pressure. This arc extends from the minimum acceptable pressure to the bottom of the yellow arc. [Figure 11-79]

FUEL PRESSURE

- Red radial line — Both the maximum and minimum acceptable pressures are indicated by a red line

- Yellow arc — This is a limited operating range. If the system is operating in the yellow arc the fuel system should be inspected for possible restrictions as soon as possible.

- Green arc — Normal operating range.

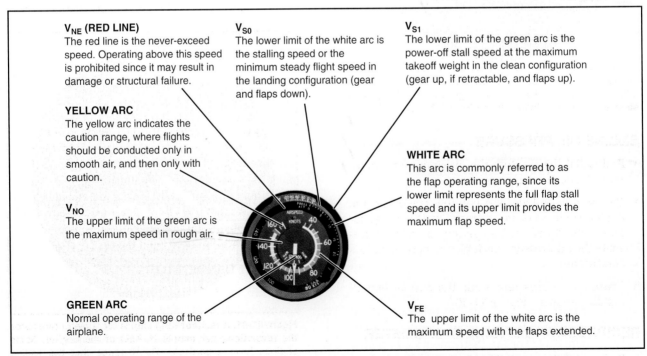

Figure 11-77. The markings on the airspeed indicator provide the pilot with quick, easily interpreted, information on various airspeeds for that aircraft.

Aircraft Instrument Systems

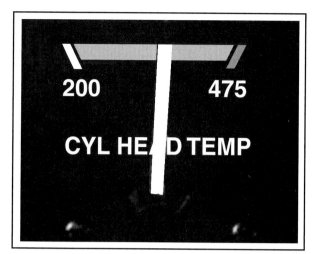

Figure 11-78. Cylinder head temperature markings allow the pilot to quickly determine if the engine is operating within the correct range of temperatures.

Figure 11-80. The oil pressure gauge measures the pressure in pounds per square inch of engine oil pressure.

Figure 11-79. The pilot can determine immediately if manifold pressure is in the correct range. Some MAP gauges show a yellow arc if intermittent high MAP is allowed.

ENGINE OIL PRESSURE

- Red radial line — Both the maximum and minimum acceptable oil pressure.
- Yellow arc — This is a caution range and the engine should not be operated in this range for an extended period. The yellow arc often represents the oil pressure during cold starting or idle conditions.
- Green arc — This represents the normal operating oil pressure. [Figure 11-80]

RECIPROCATING ENGINE TACHOMETER

- Red radial line — Indicates the maximum permissible engine r.p.m.

- Yellow arc — This is the range just above the green arc and below the red radial line. The engine may be operated at this r.p.m. for limited periods only.
- Green arc — Indicates the normal operating r.p.m. range.
- Red arc — This is the range in which the engine should not be operated except to pass through in order to achieve a higher or lower r.p.m. [Figure 11-81]

TURBINE ENGINE TACHOMETER

- Red radial line — Maximum permissible r.p.m.

Figure 11-81. A reciprocating engine tachometer measures the revolutions per minute (R.P.M.) of the engine. Some engines have a particular R.P.M. range that the engine should not be operated in continuously. This range is represented by a red arc.

- Yellow arc — From the maximum continuous operating range to the maximum permissible r.p.m.

- Green arc — This arc extends from the maximum to the minimum r.p.m. for continuous operation. This is the normal operating range.

EXHAUST GAS TEMPERATURE, TURBINE ENGINE

- Red radial line — Maximum permissible temperature for wet or dry operations, whichever is greater.

- Yellow arc — This range extends from the top of the green arc to the red radial line.

- Green arc — This range is the normal operating temperature.

TORQUEMETER

- Red radial line — Maximum permissible torque as set for by the aircraft or engine manufacturer.

- Yellow arc — Caution range extending from the top of the green arc to the red radial line.

- Green arc — Normal continuous operating range.

DUAL TACHOMETER-HELICOPTER

- Red radial line (engine) — Maximum permissible r.p.m. for the engine.

- Red radial line (rotor) — Maximum and minimum r.p.m. for the power-off operation of the rotor.

- Yellow arc (engine) — Caution ranges for the engine r.p.m. These areas of operation are restricted to certain conditions.

- Green arc (engine) — Normal operating range for the engine r.p.m.

- Green arc (rotor) — Normal operating range for the rotor.

GAS PRODUCER (N1) TACHOMETER, TURBOSHAFT HELICOPTER

- Red radial line — Maximum permissible r.p.m.

- Green arc — normal operating r.p.m.

COMPASS SWING

At many larger airports, a compass rose will be laid out on the tarmac according to magnetic directions. It is usually marked with a line every thirty degrees, and is as far as possible from electrical interference. Compass roses are normally on one of the least used taxiways where a technician will be undisturbed while swinging a compass. If there is no compass rose available, you can lay one out, using an accurate compass that does not have any compensating magnets installed.

The aircraft is prepared for the compass swing by removing any material from the instrument panel and glove box that could possibly interfere with the compass. All of the normally installed instruments and radio equipment should be in place and properly functioning. The compensator magnets should be adjusted so the dot on the screw head is aligned with the dot on the instrument case. The aircraft should be taxied onto the center of the north-south line, facing north. If the aircraft is of the tail-wheel type, the tail will have to be supported on some type of wooden stand to bring the aircraft to a level flight condition.

The N-S, north/south, compensator screw should be adjusted with a non-magnetic screwdriver until the compass reads north (0°). The aircraft is then turned until it is aligned with the east-west line and pointed east. The E-W screw is adjusted until the compass reads east (90°). Next the aircraft is turned south (180°) and the N-S screw adjusted to remove one-half of the south heading error. This will throw the north heading off, but the total north-south error should be divided equally between the two headings. The adjustment is completed by turning the aircraft west (270°), and the E-W screw adjusted to remove one-half of the west error.

Once all of the adjustments are completed, the compass should be checked at each 30 degree heading and a calibration card filled out. At the completion of the adjustments, the aircraft is headed west. On this heading, record the compass reading with the radios off and then on. Then, turn the aircraft to align with each of the lines on the compass rose and record the reading. There should not be more than about a ten-degree difference between any of the compass headings and the magnetic heading of the aircraft. If there is, the compass must be adjusted until there is no more than ten degrees difference. Steel screws in the vicinity of the compass, magnetized control yokes, or structural tubing can cause unreasonable compass errors, as can electrical wiring that is routed too close to the compass.

When the compass is adjusted and the calibration card filled out, the technician should sign it with name, A&P number, and the date. The completed card should be placed in a holder in easy view of the pilot.

PITOT-STATIC SYSTEM TEST

A certified pitot-static tester must be used when testing the altimeter and static system as required by FAR part 91.411. In addition, the manufacturer or a certified repair station must perform these tests. However, a certificated airframe mechanic may test and inspect the static pressure system but not the altimeter system. [Figure 11-82]

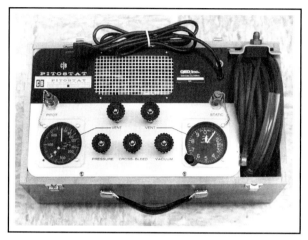

Figure 11-82. A pitot-static tester is used to conduct the static system and altimeter tests required by FAR 91.411.

When conducting tests on the pitot/static system, avoid using tape or other material that might leave anything behind that could plug or block these systems. When it is necessary to plug an opening, it should be done in such a manner that is noticeable during preflight inspection. Additionally, if a pitot or static port becomes clogged, avoid using tools or sharp objects that might alter the shape of the holes, since the shape is aerodynamically critical.

LEAKAGE TEST

The most common type of static check involves the installation of a certified static system check unit. If the aircraft has two flush static ports, cover one of the ports with tape. Use black tape and make a big "X" over the hole to ensure removal of the tape. Never use transparent tape or even masking tape because it may be inadvertently left in place after the test is completed. Attach the test unit to the other static port. Apply vacuum to the static system until the altimeter indicates 1,000 feet above the present altitude. Seal off the unit and let the system sit for one minute.

After one minute, the altimeter indication must not have dropped more than 100 feet. If the leakage of the system exceeds these limits, it must be located. To do this, systematically isolate and test portions of the line. Start with the part of the line that is connected to the instruments and work back towards the static ports themselves. If the system being tested has its static ports in a combination pitot-static head, be sure to cover any drain holes in the head.

ENTRAPPED MOISTURE REMOVAL

This should be completed prior to the leakage test. Moisture can periodically enter the static system, and if it remains, the static air source may become blocked, or the water could freeze and cause complete instrument failure. To remove entrapped moisture, disconnect the line from all instruments and carefully blow out the line from the instrument end, using low-pressure, moisture-free, air. Excess moisture will exit the system through the static port(s). When the line has been carefully blown out, reconnect all instruments, and perform the leakage test.

CHAPTER 12

AIRCRAFT AVIONICS SYSTEMS

INTRODUCTION
The term avionics encompasses the design, production, installation, use, and maintenance of electronic equipment mounted in an aircraft. Avionics systems continue to advance at a more rapid pace than any other part of the aircraft. Additionally, most of these advancements have resulted from the same technology that produced personal computers and the telecommunications industry. Basic VHF communications and navigation systems used in aviation today were developed in the 1940s, but the introduction of newer systems has increased dramatically in recent years. This chapter provides an overview of the types of avionics and autopilot equipment that are in widespread use, and the latest developments that are presently being installed in aircraft.

SECTION A

AVIONICS FUNDAMENTALS

Radio is a term used to refer to the wireless transmission of information from one point to another. Many aircraft avionics systems utilize radio wave technology. Therefore, the operating principles and basic components found in radios are explained here to provide a fundamental knowledge of aircraft avionics. This will make the discussion of more advanced avionics easier to understand.

AVIONICS THAT USE RADIO WAVES

Communications radios were the first avionics systems installed in airplanes. It was only much later that navigational radios and other avionics that use radio waves were developed. Radio wave technology is used in the following avionics systems:

- Communications
- Navigation
- Air Traffic Control (ATC) Radar
- Weather Avoidance Radar
- Approach and Landing Aids
- Altitude Measurement
- Airborne Collision Avoidance

RADIO OPERATING PRINCIPLES

To understand how a radio works you need to be familiar with the basic terminology of electromagnetic waves such as frequency, carrier wave, modulation, and ground, sky, and space waves. In addition, understanding the basic components that make up a radio such as amplifiers, oscillators, modulators, filters, antennas, tuning circuits, transmitters, receivers, and speakers makes troubleshooting and installation much easier.

ELECTROMAGNETIC WAVES

An electromagnetic wave is produced by the synchronized oscillations of electric and magnetic fields. Radio waves are a form of electromagnetic wave. A radio transmits and receives information by means of electromagnetic waves. However, information must be converted into electronic signals before it can be transmitted by a radio wave. The range and diversity of **electromagnetic (EM) waves** is very broad. The entire spectrum of EM waves includes not only radio waves but gamma rays, x-rays, ultraviolet, visible light, and infrared light. [Figure 12-1]

FREQUENCY

Frequency is the number of times something occurs in a given period of time. In the case of radio transmissions, frequency is the number of electromagnetic field oscillations that take place in one second. Each oscillation is called a cycle and the frequency is measure in cycles per second (cps) or hertz (Hz). Electromagnetic waves may be emitted

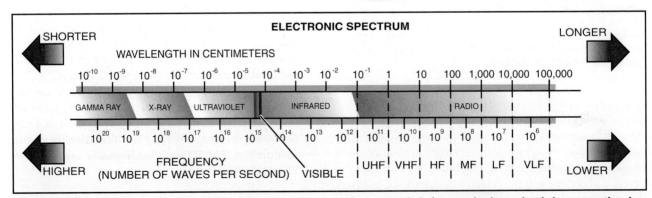

Figure 12-1. All electromagnetic waves are classified according to their characteristic frequencies into what is known as the electromagnetic spectrum. As shown here, visible light has its own distinct frequency and wavelength. While we can perceive some electromagnetic waves, the rest of the electromagnetic spectrum is invisible, and exhibits frequencies that pass through a broad range. Gamma rays, x-rays, and ultraviolet light exhibit the highest frequencies and shortest wavelengths. Infrared radiation and radio waves occupy the lower frequencies of the spectrum. Radio waves, such as FM and AM radio broadcasts, are simply electromagnetic waves with a much lower frequency and longer wavelength than visible light.

in various patterns and in an infinite number of frequencies. Radio waves are classified according to the frequency band they occupy. Starting at the low end, these frequency bands include: very low frequency (VLF), low frequency (LF), medium frequency (MF), high frequency (HF), very high frequency (VHF), ultra high frequency (UHF), super high frequency (SHF) and extremely high frequency (EHF). [Figure 12-2]

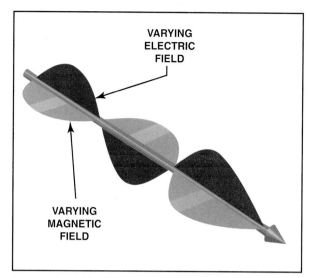

Figure 12-3. This figure shows an electromagnetic wave, with the electric field and magnetic field at right angles to each other and to the direction of wave travel.

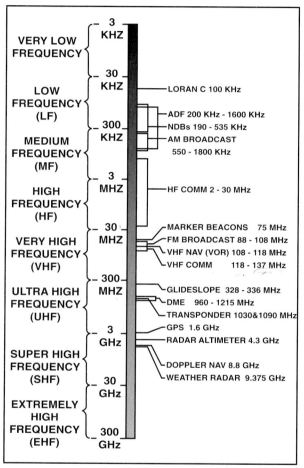

Figure 12-2. This chart illustrates the radio frequency bands on the left side of the chart and operating frequencies of some of the more common aircraft systems on the right side. Notice that there is an aviation application for all of the bands except VLF and EHF.

All radio waves consist of electric and magnetic energy fields that travel through space. The two fields are located at right angles to each other and to the direction of wave travel. They are produced when a radio-frequency electrical signal is sent down a conductor to an antenna, which transforms the signal into radio waves. [Figure 12-3, 12-4]

CARRIER WAVE

A **carrier wave** is necessary to transmit radio wave information. Antenna length is usually one-quarter to one-half the length of the wave it transmits and receives. Low-frequency radio waves directly transmitted without the use of a carrier wave would require an extremely large antenna. For example, audio signals transmitted directly would require an antenna several miles long. However, high-frequency carrier waves with very short wavelengths are received by smaller antennas than those needed for low-frequency long-wavelength transmissions. Hence, a high-frequency carrier wave is employed to deliver information, which in turn reduces the length of the required antenna. In addition, because there is more room at the higher frequencies, many channels can be formed to carry thousands of concurrent transmissions without interference.

MODULATION

To produce a useful transmission with little interference, information is superimposed on a higher frequency carrier wave. If a radio transmitter is designed to send out a steady carrier wave, it will not be able to transmit any data or information. The receiver will only produce a steady hum, or nothing at all, depending on its design. In order to transmit information, it is necessary to vary or alter the carrier wave in some way. This alteration is called **modulation**. The information signal is used to modulate the carrier wave.

Current radio systems use a number of different types of modulation. The two most common are amplitude modulation (AM) and frequency modulation (FM). In amplitude modulation, the amplitude of the carrier wave varies with the change in

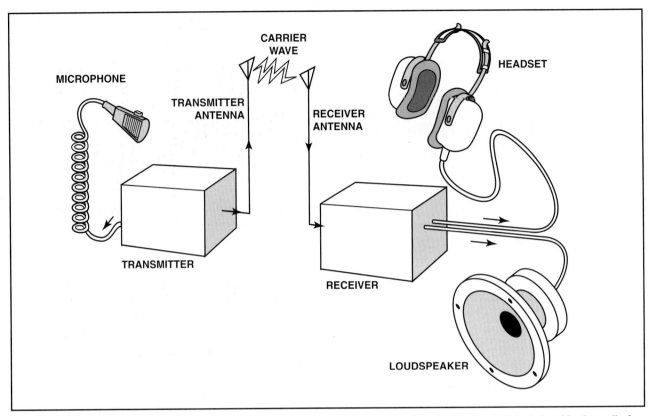

Figure 12-4. This figure illustrates a simplified radio system. Radio waves are produced when an electrical signal in the radio frequency range is sent to an antenna. The antenna converts the electrical signal into radio waves, which travel outward from the antenna through space at the speed of light (186,284 m.p.h). When the radio waves strike a receiving antenna, a voltage and current is induced, which produces an electrical signal that matches the one applied to the transmit antenna. The basic operating frequency of a radio is called the carrier frequency because it carries the transmitted information from one place to another.

amplitude and frequency of the information signal. [Figure 12-5]

In the case of frequency modulation (FM), the carrier amplitude stays constant while the carrier frequency is changed by the modulating or information signal. The frequency of the carrier changes proportionately with a change in the information signal's amplitude. If the amplitude of the information signal decreases, the carrier frequency decreases and vice versa. [Figure 12-6]

Frequency modulation is considered better than amplitude modulation. FM is less affected by atmospheric electrostatic emissions, commonly called noise or static, created by thunderstorms and other disturbances. It rejects interfering signals and improves transmitter effectiveness. However, FM uses an excessive amount of bandwidth space.

GROUND, SKY AND SPACE WAVES

The behavior of radio waves as they travel through the earth's atmosphere and beyond are classified by the terms ground, space and sky waves. Radio waves at frequencies below the HF

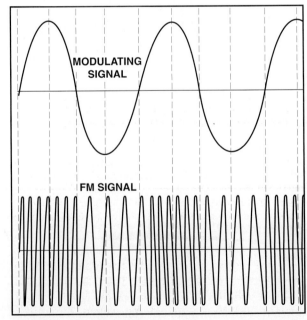

Figure 12-6. This figure illustrates what a carrier wave looks like when it has been modified by frequency modulation (FM). The amplitude of the signal is constant while the varying frequency causes the waveform to compress and expand. This type of waveform can be decoded in a manner similar to that used with an AM signal.

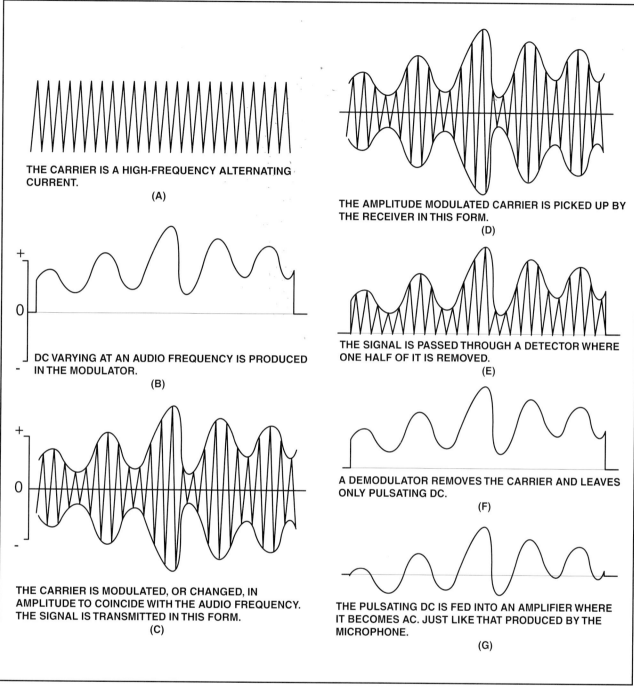

Figure 12-5. Amplitude modulation (AM) is one method of carrying information by EM waves. The captions in this figure briefly describe and illustrate the process of adding information to a carrier wave by amplitude modulation.

band (below 3 MHz) are called **ground waves** because they follow the curvature of the earth. Radio waves that operate in the HF band from 3 MHz to 30 MHz are called **sky waves**. They tend to travel in straight lines and do not follow the curvature of the earth. In addition, sky waves bounce or refract off the ionosphere, which is made up of layers of ionized particles from about 60–200 miles high. When sky waves strike a layer of the ionosphere in the right way, they are refracted back to earth hundreds of miles away. This characteristic can be used to achieve long-range transmission of radio signals. **Space waves** are radio waves transmitted at frequencies above the HF band. They travel in straight lines, but do not bounce off the ionosphere. For example, the radio signals used to communicate with orbiting satellites are above 30 MHz. [Figure 12-7]

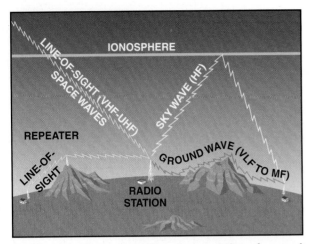

Figure 12-7. The propagation characteristics of ground waves, sky waves and space waves can be seen in this figure. Space waves are considered line-of-sight waves because of their straight-line nature and because they become blocked by terrain and the curvature of the earth. Repeater stations are utilized to extend line-of-sight radio waves. A repeater is a receiver and transmitter unit that picks up the space wave, amplifies it, and retransmits it on another frequency to the end receiver.

BASIC RADIO COMPONENTS

Radio communications equipment transmits and receives, or transceives, radio frequency (RF) signals to transfer audio sounds from one point to another. On the other hand, navigation radios are generally used to receive RF signals that are subsequently converted to an electrical signal in order to drive a graphic display. The display is used to depict specific information. Regardless of whether the radio is used for communication or navigation purposes, every radio type has internal components that serve similar functions.

TRANSMITTERS

A transmitter is the electronic unit that accepts information and converts it into a radio frequency signal capable of being transmitted over long distances. Every transmitter has three basic functions: (1) it must generate a signal of the correct frequency at a desired point in the EM spectrum, (2) it must provide some form of modulation that causes the information signal to modify the carrier signal, and (3) it must provide sufficient power amplification to ensure that the signal level is high enough so that it will carry over the desired distance.

At the heart of radio wave avionics is the **oscillator**. An oscillator generates the carrier signal at the desired frequency. Three common types of oscillators used in avionics are the variable frequency oscillator (VFO), the crystal oscillator, and the phase locked loop oscillator (PLL). A simple oscillator is an inductive-capacitive (LC) circuit made up of a capacitor and inductor connected in parallel. The LC circuit has a resonant frequency that matches the desired frequency. An LC circuit by itself will not continue to oscillate because of the resistance in the components and circuit. In order to maintain oscillations, some energy must be fed back into the circuit. The resonant frequency or oscillation frequency is determined by the values of capacitance and inductance in the LC circuit.

AMPLIFIERS

An amplifier is a component that increases the strength of a signal and is found in both transmitters and receivers. A transmitter must increase the strength of the signal sent to the antenna so that the EM waves will travel a useful distance. A receiver also needs amplifiers because the strength of the signal from the antenna is very low and must be increased to enable the signal to be heard. Until the 1960s, most amplifiers relied on vacuum tubes to increase the strength of signals. Today, transistors and integrated circuits have replaced vacuum tubes for most applications. Three basic types of amplifiers used in transmitters are linear, Class C, and switching amplifiers.

Linear amplifiers, classified as A, AB, or B, provide an output directly proportional to their input but at a higher power level. All audio amplifiers are linear. Class A and B amplifiers are used to increase the power level of changing amplitude radio frequencies such as AM radio signals. However, FM signals use more stable non-linear amplifiers such as Class C or switching amplifiers.

MODULATORS AND DEMODULATORS

Voice communications radios provide an ideal example of the purpose and function of modulators and demodulators. In the radio transmitter, a device is needed to superimpose the audio frequency (AF) signal onto the carrier wave signal before it is sent to the antenna. This is the function of a **modulator**. The output of the modulator is called modulated radio frequency. In order to hear the voice as an output of the receiver, the AF and carrier signal must be separated. The **demodulator** removes the RF component of the modulated RF signal and produces an audio frequency output.

When the AF and RF signals are combined in the modulator, they must have the proper relative strengths for maximum efficiency. The amount of modulation is called the modulation rate. If the AF signal is too weak compared to the RF signal, the modulation rate will be low and the efficiency also low. If the modulation rate is over 100%, distortion can occur.

Most radio transmitters are adjusted to about 90–95% modulation to provide a margin to prevent distortion. One of the most common causes of over-modulation occurs from shouting into a microphone. To obtain the best clarity from the transmitter, use a normal voice tone whenever speaking into a microphone.

FILTERS

A filter is used in a radio circuit to remove or filter out unwanted frequencies. The signals that are processed by the circuits in a radio often have additional frequencies present that are not needed. If the proper filter is installed, it will filter out unwanted frequencies. A filter is usually made up of an arrangement of inductors and capacitors. Without proper filters, the generated audio signal will have so much noise that it may be unusable.

A low-pass filter removes all frequencies above a certain value, and passes the low frequencies. A high-pass filter does the opposite. If a range of frequencies must be blocked, a filter referred to as a band-reject filter may be used. The use of a band-pass filter allows a certain band of frequencies to go through while other frequencies either above or below a specific range are blocked.

ANTENNAS

An antenna is basically an electrical conductor that radiates or receives radio waves. Depending on the particular radio system involved, an antenna may be used for transmit only, receive only, or both. The maintenance, inspection and installation of antennas are usually the responsibility of the airframe technician since they are attached to the structure or skin of the aircraft.

Antenna length is an important factor. It is specific to the wavelength of the electromagnetic waves it receives or sends out. Antenna length is usually a quarter to one-half the length of the EM waves it is set up to transmit or receive. They are most effective when the length is directly related to the wavelength of the transmitted signal.

Antennas often have general names that describe some of their basic characteristics. Common categories are the Hertz dipole, Marconi monopole, wire, and loop antenna. The **Hertz dipole** is a half-wave dipole antenna. It is called a half-wave antenna because the overall length is equal to one half the wavelength of the EM wave it is designed to send out or pick up. Essentially, it is a conductor that is 1/2 wavelength long at the operating frequency. It is divided into two quarter-wavelength sections with the transmission line connected at the center.

A **Marconi monopole antenna** is a single metal conductor with a length of 1/4 wavelength. In order to work properly, the Marconi antenna must have metal surrounding the mounting base. The metal at the base is called the groundplane. [Figure 12-8]

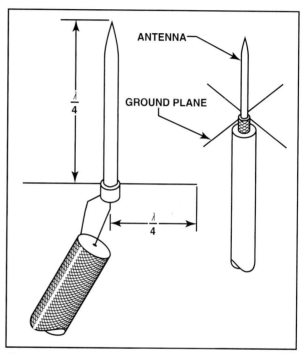

Figure 12-8. Mounting the antenna to the metal skin of the aircraft provides the groundplane for most antennas. In situations where an antenna is mounted to a fabric or composite surface, a metal plate of adequate size must be installed under the surface where the antenna is to be mounted. The metal plate must also have a bonding or grounding wire installed to provide a return path to the primary groundplane.

A **wire antenna** is a length of wire that is supported by masts and attachments above or below the aircraft fuselage. They are found most often on smaller aircraft and older aircraft. Turbine aircraft seldom use wire antennas because of the vibration and increased chance of damage at high speeds. Wire with an outer covering of insulation material is superior to non-insulated wire, because it reduces noise caused by static electricity. An example of a wire antenna is an ADF sense antenna, which is a wire that runs from the upper leading edge of the vertical stabilizer to the forward portion of the fuselage. This type of wire antenna can still be found on older aircraft. Modern aircraft combine the loop and sense antenna together into a single ADF antenna system.

Winding the antenna in the form of a loop may enhance its directional characteristics. . In a **loop antenna**, when a signal is received from a transmitter directly broadside to the loop, the voltage induced into the two sides is of equal magnitude but opposite in polarity, which causes the signals to cancel each other out. This characteristic makes the loop antenna useful for directional finding.

POLARIZATION

Most antennas must also be installed with the correct polarization. Polarization refers to the orientation of the electric field relative to the earth. If the electric field is vertical, it has vertical polarization. The Marconi antenna produces a vertically polarized radiation pattern as illustrated in figure 12-9. Horizontal polarization means that the electric field will be parallel to the earth's surface as illustrated in figure 12-10.

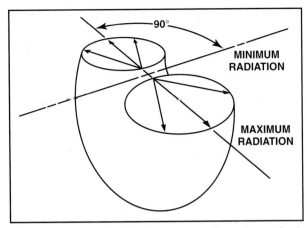

Figure 12-10. A Hertz dipole antenna will produce a horizontally polarized pattern as depicted in this figure. A horizontally polarized antenna receives its signal best from a station on either side and perpendicular to the antenna. Its strength will be the least for a signal from a transmitter in line with its length.

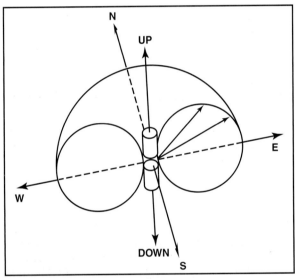

Figure 12-9. A vertically polarized Marconi antenna will produce a vertically polarized radiation pattern as illustrated here. A vertically polarized antenna is omnidirectional, which means that it receives a signal equally well from any direction.

The polarization of the aircraft antenna should normally match the polarization of the ground-based antenna and the EM waves transmitted or received. An aircraft VHF communications antenna is an example of a Marconi antenna with vertical polarization. The common example of a Hertz antenna is the VHF navigation antenna found on small airplanes. It is a V-shaped dipole antenna with horizontal polarization.

TUNING CIRCUITS

An antenna will intercept many different EM waves of varying frequencies; therefore, some method must be used to isolate the desired frequency. A tuning circuit performs this function. A simple tuning circuit consists of a variable capacitor and an inductor connected in parallel. As the tuning knob is rotated on the radio, the variable capacitor changes value until the resonant frequency of the circuit matches the frequency of the desired station. This signal is passed into the radio and the other frequencies are blocked out.

A more efficient type of tuner, which is found in most modern radios, uses a frequency synthesizer that contains a number of crystals that can be combined to match the desired frequency. Each crystal has a particular frequency. By using switches, the crystals can be combined to produce many additional frequencies. When two frequencies are combined, two new frequencies are created that are equal to the sum and the difference of the two original frequencies. By using this technique, hundreds of frequencies can be created using a relatively small number of fixed-frequency crystals.

RECEIVERS

A communications receiver reproduces the information signal received by the antenna. It must have the sensitivity to select the desired frequency from the thousands of frequencies present. In addition, the receiver must provide enough amplification to recover the modulating signal. In other words, a receiver must have selectivity and sensitivity.

In the 1920s, the superheterodyne (also known as a superhet) receiver was invented. This type of receiver reduces the modulated RF signal from the antenna to an audio frequency signal in more than one stage. Virtually all modern radios utilize superhet receivers. In a superhet radio, the RF signal from the antenna is combined with a local oscillator frequency by a mixer to produce a lower intermediate frequency (IF). The basic principle of the mixer is that when two different frequencies are combined, two new frequencies are created; the sum and the difference of the two combined frequencies. In this example, the output of the mixer is the difference between the RF frequency and the local oscillator frequency. The IF signal is amplified and then sent to a detector and demodulator. The detector removes half of each sine wave to produce a varying DC signal from an AC signal. This AF signal is then amplified and used to drive the speaker.

SPEAKERS AND MICROPHONES

Voice transmitters require an audio input from a microphone and speakers, which transform electrical signals into sound waves. The speakers used for aircraft applications are not the same as those used in household or automotive applications. Household speakers typically incorporate large magnets. These magnets can produce magnetic field emissions that would adversely affect a variety of aircraft instrument systems, including the magnetic compass. To alleviate this problem, most speakers used in aircraft installations utilize a metal plate attached over the magnet to shield the field flux and keep it in close proximity to the speaker.

Another type of speaker commonly used in aircraft installation is the dynamic speaker. This type of speaker does not produce a large magnetic field disturbance. When the audio frequency signal is applied to windings in the dynamic speaker, it sets up a magnetic field that expands and contracts at an audio rate. This field causes a metal diaphragm to vibrate at a corresponding rate to produce the movement of air that generates sound waves.

Dynamic microphones are also available. Most microphones used in aircraft applications transform the vibrations of sound waves into varying electrical signals.

AVIONICS SYSTEMS

With improved technology, radios and other avionics systems have become smaller and more reliable. When radio equipment was first installed in aircraft, the equipment was usually large, heavy and produced tremendous amounts of heat. This caused most equipment to be installed in remote locations away from the cockpit. Advances in technology have allowed most electronic components to be miniaturized. Many newer aircraft have all their radio and avionics equipment installed in the instrument panel. However, many larger aircraft still have the equipment installed in remote locations, primarily to make it more accessible for servicing. The control heads allow the flight crew to remotely operate the equipment from the flight deck.

COMMUNICATIONS RADIOS

There are a number of different radio communications systems available for aircraft use. They differ primarily in the frequencies used and the type of communication involved. The most important use of communications radios is for Air Traffic Control (ATC), since the controllers need to be in contact with the pilots to give necessary instructions. The general trend since the 1930s has been the use of higher frequencies and the development of specialized communications for other than ATC purposes.

HF COMMUNICATIONS

Until the 1940s, most aircraft radio communications utilized frequencies in the LF, MF and HF bands because suitable equipment was not available to use higher frequencies. Aircraft high frequency (HF) radios operate on frequencies between 2 and 30 MHz. Currently, modern aircraft that carry HF communications radios are those that operate long distances over water or in the remote regions of the earth. Air carrier and business jets that routinely fly the Atlantic and Pacific oceans usually make use of HF communications radios for air traffic control purposes.

HF communications radios have a maximum reception range of about 1,500 to 2,000 miles compared to a maximum of about 250 miles for VHF communications However, the reception range of VHF radios may be less since it is restricted to line-of-sight distances. Additionally, HF frequencies are generally referred to as "short waves."

The probe and flush mount antennas used for HF communications require a special antenna tuning and coupling device, which automatically repositions each time a new frequency is selected, in order to tune the antenna for that particular frequency. Smaller aircraft with HF radios generally use a long wire antenna that extends from a wing tip up to the vertical fin. Until the 1960s, many aircraft used a long-wire-trailing antenna, which extended out the aft fuselage of the airplane. This antenna could be run in and out to select the proper antenna length. However, it is not suitable for high-speed aircraft.

HF communications radios utilize ground and sky waves to achieve their greater reception range.

Aircraft HF transmitters produce an output power of 80–200 watts, which is much higher than the output power typically found with VHF transmitters. This is necessary to achieve long-distance communication. A disadvantage of HF is that it is affected by atmospheric interference. An aircraft flying over the ocean may lose communication because of thunderstorms or other atmospheric disturbances.

VHF COMMUNICATION
Communication radios in civil aircraft use a portion of the very high frequency (VHF) range, which includes the frequencies between 118.0 megahertz (MHz) and 135.975 MHz. VHF communication radios are classified according to the number of channels they are designed to accommodate. A 360-channel radio uses 50 kHz (.05 MHz) spacing between channels, such as 118.05, 118.10, 118.15, 118.20. A 720-channel radio doubles the frequencies available by using 25 kHz (.025 MHz) spacing, such as 118.025, 118.050, 118.075, 118.100. In some areas of the world, radios with 8.33 kHz spacing are now required to reduce congestion caused by the large number of transmissions.

In recent years, the Federal Communications Commission (FCC) has imposed regulations limiting the use of some older radio designs. In older radios having less than 720 channel frequency spacing, there is a potential problem that the bandwidth may cause interference in adjacent 25 KHz channels. As such, many older radios are not approved for transmitting, but can continue to serve as a receiver. If these types of radios are already installed and working, they can remain in an aircraft providing they are only used for signal reception. However, these radios are not permitted for new installations and if they become inoperative, they must be modified or replaced with a radio that meets current FCC requirements.

VHF communication provides much clearer reception and is much less affected by atmospheric conditions. However, EM waves in the VHF band are space waves, so they are limited to line-of-sight reception. At 1,000 ft., the reception range is approximately 30—39 miles. The maximum reception range using ground-based stations is about 250 miles at altitudes above 35,000 ft. Much less power is required for VHF than for HF communications. Aircraft VHF transmitters have an output power of 5—20 watts. The standard radio communications system in the U.S.A. for ATC purposes is VHF. This is also true for most other countries of the world. The International Civil Aviation Organization (ICAO) has designated VHF as the standard radio communication system for ATC purposes over land. [Figure 12-11]

Most modern VHF radios display two different frequencies. One set of numbers indicates the active frequency while a second set of numbers indicates a standby frequency, which is held in memory. This is a valuable feature, which allows the operator to switch the two frequencies in the displays by simply pushing a transfer button. Technicians should be familiar with other features and the use of aircraft radios for troubleshooting purposes. In addition, technicians might have to taxi an airplane on a controlled airport, which may require the use of the radio.

VHF antennas are usually the bent whip rods or plastic-encapsulated blade types that are mounted on top of the cabin. The range of VHF transmissions is limited to line of sight, which means that obstructions such as buildings, terrain, or the curvature of the earth block the radio waves. [Figure 12-12, 12-13]

INTERCOM AND INTERPHONE SYSTEMS
Intercoms and interphones are not radio systems, but are considered to be avionics equipment. Instead of utilizing RF signals, this equipment uses audio signals to permit communication between various points in and around the aircraft. Both systems operate in a similar manner. An intercom is used for voice communications from one point to another within the aircraft. Large aircraft have intercom systems so that the cockpit crew can communicate with the cabin crew and vice versa. On small airplanes, an intercom is used to communicate within the cockpit area and is generally needed because of high noise levels. On the other hand, an interphone system permits conversation between the cockpit crew and someone outside the aircraft, usually maintenance or service personnel.

The operation of intercom and interphone systems is the same. Phone jacks are available at different locations where a handset or headset can be connected. The handset or headset contains a microphone, a small speaker and a push-to-talk switch (PTT). The phone jacks and wiring are connected to an audio amplifier to control the volume. Switches are available to select the desired system and a bell is used to alert the other party. On larger aircraft, a passenger address (PA) system is included so that the flight or cabin crew can make announcements to the passengers. In some applications, external interphone jacks are located in the nosewheel area, avionics equipment bay area and in the aft fuselage near the auxiliary power unit (APU). These external

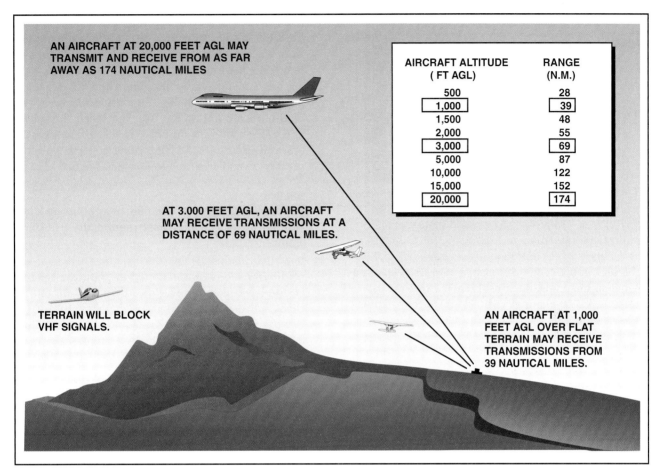

Figure 12-11. Because VHF radio signals are limited to line-of-sight, aircraft flying at higher altitudes are able to transmit and receive at greater distances. This figure illustrates the line-of-sight restriction characteristics that apply to VHF and other space wave transmissions.

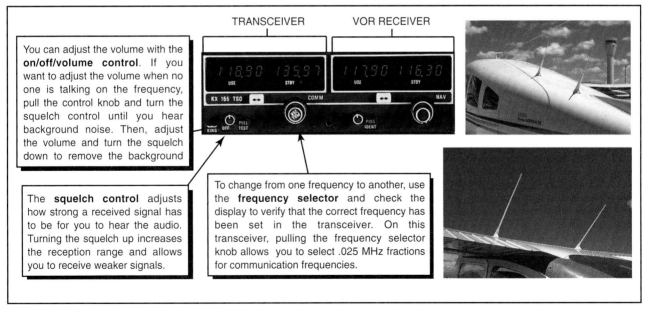

Figure 12-12. The basic components of the VHF communications systems consist of a transceiver, all associated connections, and the antennas. The antennas used with VHF communications are Marconi 1/4-wave monopoles that use vertical polarization. There is usually a separate antenna for each VHF communications radio and they are placed in various locations around the aircraft. An airplane equipped with two transceivers usually has a microphone selector switch for the respective radios, so a second microphone is not required.

Figure 12-13. This illustration shows a complete set of radios that might be found on a typical general aviation airplane. Frequency transfer buttons can be seen on the VHF communications and navigation radios. The transfer button allows a standby frequency to be quickly switched to become the active frequency. The VHF communications and navigation radios show both active and standby frequencies.

jacks permit communication between the cockpit and maintenance personnel working at these locations. Large aircraft such as the Boeing 747 have service interphone connections at each engine, at the refueling stations and other locations where communications with the cockpit may be required during maintenance or servicing.

RADIOTELEPHONE

Aircraft often carry a radiotelephone system, which is somewhat similar to the portable cellular phone available for cars. It employs radio signals to permit telephone calls to and from the aircraft in flight. The frequencies used are 450–500 MHz in the UHF band. The antenna used is a Marconi antenna of a slightly different shape and size compared to a VHF communications antenna.

SATCOM

Satellite Communications, or SATCOM, consists of a UHF radio installed in the aircraft for voice and data communications. SATCOM uses satellites that orbit the earth to relay information from a transmitter to a receiver. To date, it is used for telephone calls from business or air carrier jets, in addition to datalinks from an aircraft in flight to the airline computer system. This permits monitoring of flight progress and aircraft systems status. In some cases, SATCOM is used to replace HF radios for communications and ATC purposes for aircraft over the oceans or remote areas. To enable maximum satellite utilization, the antenna used with SATCOM is a special type that must be installed on the top of the aircraft.

SELCAL

SELCAL is an abbreviation that stands for selective calling, a special communications system for air carrier aircraft. SELCAL is not a separate radio system. It is connected to the existing VHF and HF communications radios on the aircraft. The system is used for communications between aircraft in flight and people on the ground such as the airline managers or dispatch. It is called selective calling because it works somewhat like a telephone system. When a person wants to communicate directly with a specific aircraft crew, they can selectively choose which aircraft to call. This prevents disruptions to the flight crew from unwanted communications.

A typical SELCAL unit incorporates a decoder that connects to the VHF and HF communications radios. Each aircraft is assigned a code number, which is a part of the SELCAL equipment. When the proper code is received, a tone is heard in the cockpit to alert the crew that someone is calling them. The code consists of four tones that are transmitted to the aircraft in series. Each of the four tones has twelve possible frequencies, so that over 20,000 different codes are available.

Assume that a dispatcher for an airline needs to call the flight crew of one of the airline's aircraft to relay a message. According to the flight schedule, the airplane is somewhere between Boston and Atlanta. The dispatcher can pick up the telephone and dial a special access code along with the access code for that particular airplane. The signal is sent out over many different ground transmitters and received by hundreds of airplanes in flight. However, the phone will only ring in the cockpit of the specific airplane called. The SELCAL system is a great help to the airline when they must reroute a flight or pass along important information to the crew in flight.

AIRCRAFT COMMUNICATION ADDRESSING AND REPORTING SYSTEM (ACARS)

ACARS provides an automatic and manual means to transmit and receive operational, maintenance, and administrative information between the airplane and a ground station. Most newer airliners are equipped with ACARS to reduce flight crew workloads and eliminate the need to transmit routine information by voice radio. Messages such as flight plan changes, requests for weather information, engine performance monitoring data, takeoff and landing times, etc. are transmitted by digital data link. Messages to the aircraft are displayed on a screen and printed out on a small cockpit printer. Messages from the airplane are keyed in by the crew, reviewed on the screen and transmitted with the push of a button.

ACARS communicates through either a VHF radio or SATCOM. The third VHF radio is normally reserved for use by the system and it will automatically tune this radio as required. If the VHF radio is used for voice communications, ACARS will still be operational, and if it fails, it may interfere with voice operation of the radio. For this reason, ATC voice communication on the third VHF radio is normally prohibited. Selecting the ACARS prompt on the Flight Management System (FMS) Control Display Unit (CDU) usually accesses ACARS.

AUDIO CONTROL PANELS

When an aircraft has more than one radio, an audio control panel provides an efficient means of switching the microphone and speaker connections from one radio to another. Most audio control panels have a row of toggle switches or push buttons that can be used to connect the audio output of the various radios to the speaker or headphones. It may also have a rotary selector switch to connect the microphone output to the different radio transmitters and intercom systems that may be available. In addition, some audio control panels incorporate three annunciator lights for the marker beacon system. [Figure 12-14]

NAVIGATIONAL SYSTEMS

There is a much wider variety of navigational systems available to aircraft than communications systems. Navigation electronics are used to identify exact location by determining direction and distance from certain points. This section provides a general overview of several common navigation avionics systems that you, as the technician, will maintain.

AUTOMATIC DIRECTION FINDER (ADF)

The Automatic Direction Finder (ADF) system has been in use since the 1930s. Even though it is not as accurate as more modern navigation systems, it is still widely used in general aviation aircraft. Many smaller airports that have no other radio aids for navigation provide transmitters called non-directional beacons (NDB) that can be used with ADF equipment. In less-developed countries, this system may be the only navigational aid available and will be used by even the largest transport aircraft. The term ADF applies to the aircraft equipment, and the NDB is the associated ground-based equipment. For simplification, the term "ADF system" encompass both the ADF and NDB equipment since both the airborne and ground-based components are required to operate the system. The ADF receiver can receive signals transmitted in the 190–1,800 kHz range.

Because ADF signals fall within the low-to-medium frequency range, they are considered ground waves and will follow the curvature of the earth. They are not limited by line-of-sight restrictions, making the ADF system a reliable navigation system at lower altitudes. In addition, ADF equipment also receives AM radio signals, some of which are referenced on navigation charts.

Figure 12-14. An audio control panel performs the switching functions between the various radio microphones and speakers. Marker beacon annunciator lights may also be incorporated in the audio panel. These lights illuminate when the aircraft passes over certain ground radio stations to indicate the aircraft position.

The ADF equipment on the aircraft can receive two different types of transmitted signals. The range 190–500 kHz is used by NDB transmitters that are specifically designed for aircraft use. The range of 550–1,800 kHz is the band used by commercial AM broadcast stations. Again, AM broadcast stations can be used for navigation. However, not all AM transmitters are noted on navigational charts. NDB locations are shown on aeronautical charts, so the location of the transmit antenna can be determined more accurately. The signals transmitted from the ground sites are omnidirectional so the NDB does not provide specific directional information. Instead, the ADF equipment determines only the station direction relative to the aircraft's position.

An ADF requires two antennas: a directional antenna and a sense antenna. On modern aircraft the directional and sensing functions are generally combined into one unit, called a loop-and-sense antenna. The directional antenna on modern aircraft is generally a flat oval or teardrop-shaped antenna that provides the loop functions to the ADF system. These loops receive the radio signals and send them to the ADF receiver. The nondirectional sense antenna receives signals with equal efficiency in all directions.

The strength of the output signal from the antenna depends on the angle between plane of the loop and the direction of travel of the EM wave. When the EM wave is at right angles or perpendicular to the plane of the loop, the signal is at minimum strength, which is commonly called the null. When the EM wave and the loop antenna are parallel, the signal strength is at a maximum. As the loop antenna is rotated, a rise and fall in the signal strength is received.

The null is used to determine station direction, rather than the peak, because there is a greater change in the signal strength when the null is reached versus when the peak is reached. With a loop antenna alone there would be two nulls for each 180° of loop rotation, which means the station could be in one of two directions. Using a second antenna, called the sense antenna, solves this ambiguity problem. [Figure 12-15, 12-16]

VERY HIGH FREQUENCY OMNIRANGE (VOR)

The VERY HIGH FREQUENCY OMNIRANGE (VOR) system has been the standard radio navigation system for cross-country flying in the United States and most of the rest of the world for many years. The VOR system was developed to overcome the problems of the earlier navigational aids. The major advantages of a VOR system are as follows.

1. Provides an infinite number of radials or course indications.

2. Reduces the amount of indication errors from adverse atmospheric conditions.

3. Accurately provides directional information.

VOR systems operate in the VHF range: 108–117.95 MHz. VHF frequencies offer relatively interference-free navigation, but, unlike lower frequency radio waves, which can skip within the atmosphere or travel over the ground for great distances, VOR

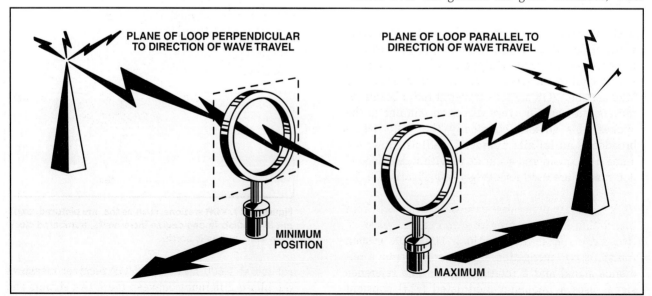

Figure 12-15. This figure illustrates the directional characteristics of the loop antenna used in the ADF system. Without the signals from the sense antenna, ADF indications would be ambiguous, providing indications in one of two possible directions.

Aircraft Avionics Systems

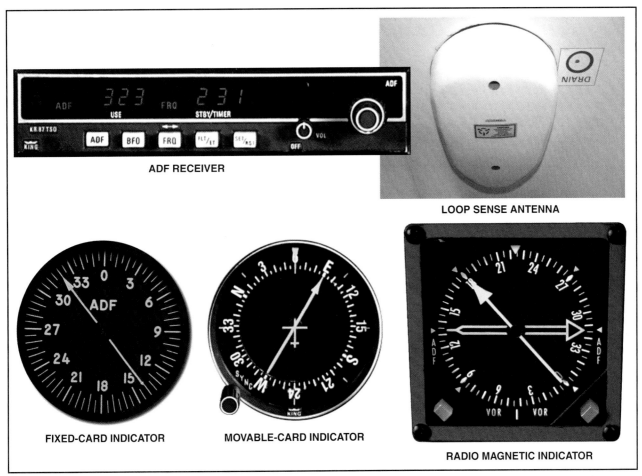

Figure 12-16. This figure illustrates the typical equipment that make up an ADF system: the ADF receiver, a modern loop-and-sense antenna, and three variations of ADF bearing indicators.

reception is strictly line-of-sight. This limits the usable signal range at low altitudes or over mountainous terrain.

Basic VOR systems only provide course guidance, while VOR/DME facilities also provide distance information to aircraft equipped with distance measuring equipment (DME), which is discussed in a later section. VOR stations transmit radio beams, or radials, outward in every direction, similar to the spokes on a wheel. Technically, the VOR station broadcasts an infinite number of radials. However, since it is considered accurate to within one degree, 360 radials are used for navigation. [Figure 12-17]

VOR receivers determine azimuth (direction) from the station by comparing the timing, or phase, of two signals from that station. The VOR system transmits two navigation signal components; a reference signal and a rotating signal. The reference signal uses a frequency-modulated (FM) constant pulse at all points around the VOR. The other signal is amplitude modulated (AM) and electronically

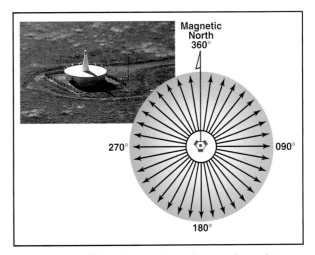

Figure 12-17. VOR stations, such as the one pictured, transmit 360 radials in one-degree increments, numbered clockwise from magnetic north.

rotated at 1,800 r.p.m. The VOR receiver measures the phase difference between these two signals and calculates its direction from the station. The two signals are aligned so that they will be in phase when the receiver is due north of the ground site. [Figure 12-18]

VOR EQUIPMENT CHECK FOR IFR OPERATIONS

According to FARs, VOR equipment must be routinely checked if the aircraft is flown under instrument flight rules (IFR). The specific regulations outlining this equipment check is found in FAR Part 91. Following is a brief summary of the current required VOR equipment check.

No person may operate a civil aircraft under IFR conditions using the VOR system of radio navigation unless the VOR equipment of that aircraft:

1. Is maintained, checked and inspected under an approved procedure; or

2. Has been operationally checked within the preceding 30 days and was found to be within the limits for bearing error set forth below.

The check must use one of the following:

1. An approved FAA or Repair station ground test signal — ±4°.

2. Designated VOR checkpoint on the airport surface — ±4°.

3. Designated airborne checkpoint — ±6°.

4. An airborne check using a VOR radial and prominent ground point that can be seen from the air as established by the person doing the check — ±6°.

5. If two separate VOR receivers are installed, they may be checked against each other — ±4°.

Maintenance record entry. —Each person performing one of the above checks shall enter the date, place and bearing error in the aircraft log or other record and sign it. If a test signal from a repair station is used, the repair station certificate holder must enter the bearing transmitted and date in the aircraft log or other record.

DISTANCE MEASURING EQUIPMENT (DME)

The military services have their own radio navigation system, which operates on principles similar to those of VOR. The system is known as TACAN (Tactical Air Navigation). VOR/DME and VORTAC, a combined VOR and TACAN facility, provide distance information in addition to course guidance. Distance measuring equipment (DME) typically indicates distance in nautical miles (n.m.) to the associated VOR/DME or VORTAC site as well as

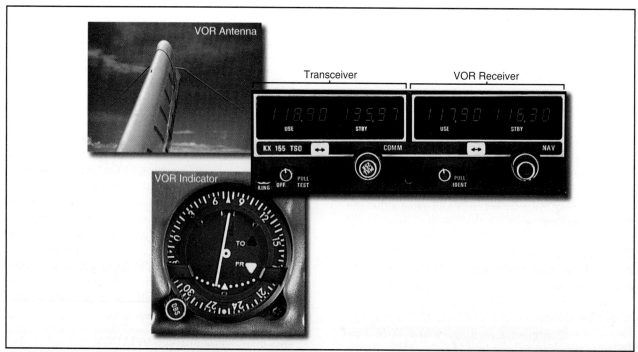

Figure 12-18. VOR equipment usually consists of a dipole antenna, receiver, and bearing indicator. The procedure for tuning in a station with the VOR is similar to that used for tuning an ADF. The pilot locates the station on an aeronautical chart or other reference and determines the frequency. The pilot then tunes in the desired frequency and listens for the Morse code identifier. VOR systems transmit a three-letter identifier in Morse code. When the station has been identified, the pilot is ready to use it for navigation by means of the appropriate flightdeck bearing indicator. Modern airliner flight management systems automatically decode the Morse code signal and present the pilot with the alphabetic station identifier.

groundspeed and time enroute to the station. The frequencies utilized by DME are in the range of 960–1,215 MHz. To obtain a distance from the station, the DME transceiver first transmits an interrogation signal to the station. The ground station then transmits a reply back to the aircraft. The transceiver then measures the round trip time of this signal exchange, computes the distance in nautical miles, and displays it digitally in the flightdeck. Depending on altitude and line-of-sight restrictions, reliable signal can be received up to approximately 200 n.m. from the station. [Figure 12-19]

Although the DME is tuned using a VOR frequency, the DME signal is transmitted from a separate facility. Each VOR frequency is tied to a specific DME

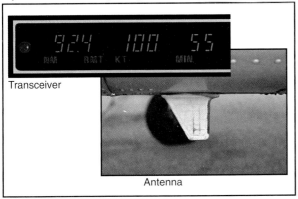

Figure 12-19. The DME system consists of a transceiver located in the instrument panel and a blade or shark-fin antenna, as it is commonly called, located on the bottom of the aircraft.

channel under an arrangement called frequency pairing. When the pilot selects the proper frequency for the VOR that is being used, the DME equipment is tuned automatically to the proper DME channel. When a VOR frequency is tuned in, a VOR identifier is repeated three or four times, followed by the single-coded DME identifier, which is transmitted approximately every 30 seconds and signals that the DME is functioning.

DME can be a very useful navigation aid; however, it does have limitations. For example, since DME measures groundspeed by comparing the time lapse between a series of pulses, flight in any direction other than directly to or away from the station will result in an unreliable reading. While DME normally is accurate to within 1/2 mile or 3% of the actual distance, whichever is greater, DME measures slant range, not horizontal distance to a station. Slant range distance is the result of two components: the horizontal and vertical distance from the station. The difference between the slant range

distance and the horizontal distance is not significant if the aircraft is at least 1 mile from the station for every 1,000 feet of altitude. The error is greatest when the aircraft is directly above the station, where the DME simply indicates the aircraft's altitude in nautical miles. [Figure 12-20]

AREA NAVIGATION

Area navigation allows the pilot to fly direct to a destination without the need to overfly VORs or

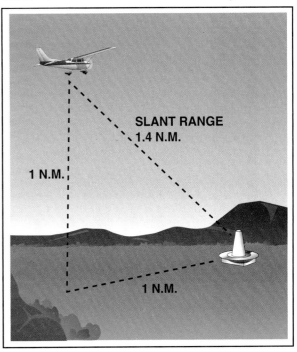

Figure 12-20. If the aircraft is located at an altitude of 1 n.m. and at a horizontal distance of 1 n.m. from the station, the DME will indicate a distance of 1.4 n.m.

other ground-based facilities. Some of the available direct navigation systems include VORTAC-based systems (RNAV), long range navigation (LORAN), inertial navigation systems (INS), and global positioning systems (GPS). Although not discussed in this section, flight management systems (FMS) also provide advanced area navigation. Area navigation courses are defined by waypoints, which are predetermined geographical positions used for route and instrument approach definition or progress reporting purposes. They can be defined relative to VOR/DME or VORTAC stations or in terms of latitude/longitude coordinates.

VORTAC-BASED AREA NAVIGATION

The use of conventional VOR navigation along airways requires that the aircraft be flown directly from one VOR site to the next. Since VOR sites seldom line up directly along a desired flight path, the aircraft ends up flying a zigzag course to get from one place to another over long distances. The use of RNAV equipment permits the aircraft to fly directly to the destination without having to fly straight to and from each of the VOR sites. [Figure 12-21]

RNAV based on VORTAC and VOR/DME facilities has been in use for some time. It is accomplished with a course-line computer (CLC) that creates phantom VOR waypoints at convenient locations for a direct route of flight. The pilot navigates to and from these phantom VORs the same way as actual VOR navigation. The CLC requires DME to calculate the location of the waypoint. To create a phantom VOR waypoint, an actual VORTAC or VOR/DME frequency is tuned into the computer. A waypoint is established as a direction and distance from a VOR and DME site. For example, the waypoint OMN 240/25 would indicate a point that is 25 nautical miles southwest (240°) of the OMN transmitter site. The pilot programs the RNAV computer by desig-

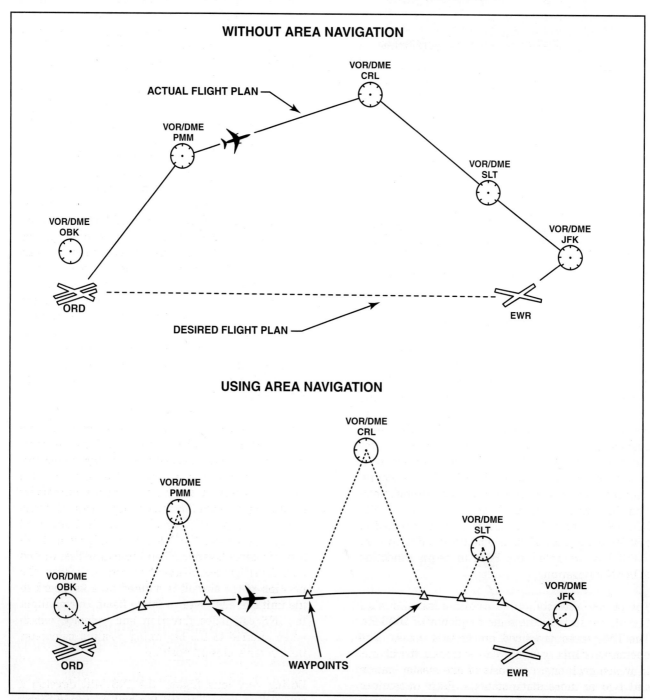

Figure 12-21. The use of RNAV equipment permits direct flights using phantom waypoints instead of requiring the aircraft to fly a specific VOR route.

nating a number of waypoints along the desired flight path. [Figure 12-22]

Even though the RNAV equipment is designed to permit direct routes, the aircraft must be able to

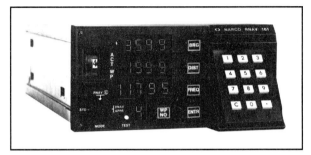

Figure 12-22. A typical VORTAC-based RNAV transceiver is illustrated here. The types of antennas and bearing indicator are the same as the VOR system.

receive usable signals from VORTAC sites. Waypoints cannot be used if they would take the aircraft beyond the line-of-sight reception range of the VOR facility. RNAV does not extend a VOR's standard service range, which means that if a pilot is flying below 14,500 feet using low or high altitude VORs, they must remain within 40 nautical miles of the VORTAC or VOR/DME facilities to receive a reliable navigation signal. Another limitation on the use of RNAV for IFR flight is the ATC system. In congested air traffic areas, air traffic controllers may not approve direct routes of flight.

LONG-RANGE NAVIGATION SYSTEMS (LORAN)

Long range navigation (LORAN) uses a network of land-based radio transmitters originally developed to provide all-weather navigation for mariners along the U.S. coasts and in the Great Lakes. To serve the needs of aviation, LORAN coastal facilities were augmented in 1991 to provide signal coverage over the entire continental U.S. When used for IFR navigation, the current system, **LORAN-C**, is accurate within 0.25 nautical mile and defines aircraft position in terms of latitude and longitude. While LORAN is a reliable and fairly accurate navigation system, the advent of the Global Positioning System (GPS) has prompted the FAA to begin canceling LORAN navigation.

The LORAN-C system uses ground transmitter sites that all transmit on the same frequency of 100 kHz. The EM waves produced are ground waves. The operation of this system involves transmitter chains in which each chain consists of one master station and two or more slave stations. Each transmitter uses a tower about 1,000 ft. tall and has an output power of approximately 4,000,000 watts.

The transmissions are sequenced so that the master transmits first followed by the slaves. The location of the master and slaves are hundreds of miles apart. The signals received by the aircraft will have a time separation that is determined by the aircraft location relative to the transmitter sites. A computer in the LORAN-C receiver performs the calculations that determine location and has an accuracy in the order of 400–1,000 ft. in most cases. The LORAN-C equipment in the aircraft does not have to be tuned since all signals are received on 100 kHz. Each chain can be identified by the time delay between transmission pulses. An advantage of this system is that signals can be received at any altitude, even with the aircraft on the ground.

INERTIAL NAVIGATION SYSTEM

An inertial navigation system (INS) is a self-contained system that uses gyros, accelerometers, and a navigation computer to calculate position. It does not rely on the reception of radio waves. The system is totally self-contained within the aircraft. By programming a series of waypoints, the system provides navigation along a predetermined track. INS is extremely accurate when set to a known position upon departure. Inertial navigation systems can be programmed with complete routes of flight and can be coupled to the aircraft autopilot to provide steering commands. However, without re-calibration, INS accuracy degrades 1 to 2 nautical miles per hour. To maintain precision, many INS systems automatically update their position by incorporating inputs from VOR, DME, and/or other navigation systems. INS systems can be approved as a sole means of navigation or used in combination with other systems.

The key to the operation of INS is the very accurate measurement of acceleration forces. The accelerometer sensors measure acceleration in directions parallel to the surface of the earth. The INS unit can calculate direction and velocity of the aircraft by measuring acceleration forces, but it cannot determine aircraft position when the unit is first turned on. For this reason, the INS must be aligned and calibrated before takeoff. When the INS is first turned on before flight and before the aircraft is moved, the position of the aircraft is entered on a keyboard so the unit can align and calibrate itself. During flight the INS calculates direction and velocity, which when applied to the beginning position calculates the aircraft's present position.

During very long flights, the INS will develop a cumulative error. Toward the end of a long flight, the error might be as much as 20 miles. To eliminate this cumulative error, modern INS units can be updated using radio signals received from ground stations where they are available. Simple types of INS sensors use small weights, which react to acceleration forces by movement about a hinge point.

In order to give accurate readings, the accelerometers must be mounted on a gyro-stabilized platform so that they only measure horizontal forces. In effect, the accelerometers measure north-south and east-west accelerations in order to determine aircraft position.

Another type of inertial navigation system that does not use conventional spinning gyroscopes and does not need a gyro-stabilized platform is the **Inertial Reference System (IRS)**. This type of inertial navigation system contains three accelerometers and three ring-laser gyros (RLGs). The three accelerometers measure acceleration forces along the aircraft's three axes: vertical, lateral and longitudinal. The RLG is a device with no moving parts that replaces a conventional gyro with a spinning rotor. The laser gyro uses a triangular housing and two different laser beams. The mirrors at the corners direct the two laser beams in opposite directions around the triangular course. Sensitive detectors measure the Doppler frequency shift that occurs when the unit is rotated. Three of these RLGs are needed to measure rotation around the three axes of the aircraft. A computer processes the signals from the three accelerometers and the three laser gyro sensors to determine aircraft heading, position and groundspeed.

An IRS is also referred to as a strapdown system because it does not require a gyro-stabilized platform like that required for an INS. The corrections that are needed are calculated by the computer, which means that precession in the indicator is virtually eliminated. Like any inertial navigation system, the strapdown INS must be given the geographical coordinates for present position during the alignment before takeoff.

GLOBAL POSITIONING SYSTEM (GPS)
The global positioning system (GPS) consists of three segments: space, control, and the user. The space segment consists of 24 NAVSTAR satellites, three of which are spares, in circular orbits approximately 10,900 n.m. above the earth. The satellites are positioned so that a minimum of five satellites are always in view to users anywhere in the world. The satellites continually broadcast position and time information, which is used by GPS receivers to calculate extremely accurate position information, including the aircraft's altitude. The GPS receiver communicates with the satellites using frequencies in the 1.6 GHz range. For accurate navigation, the aircraft must be able to communicate with at least four different satellites.

Five monitor stations, three uplink ground antennas, and a master control station located at Falcon Air Force Base in Colorado Springs, Colorado, make up the control segment. The stations track all GPS satellites and calculate precise orbit locations. From this information, the master control station issues updated navigation messages for each satellite, thereby maintaining the most accurate signal information possible.

The user segment, which includes antennas, receivers, and processors, uses the position and time signals broadcast from the GPS constellation to calculate precise position, speed, and time for pilots, mariners, and land-based operators. Similar in principle to LORAN, the GPS receiver establishes a position at the point of intersecting lines-of-position (LOP). By matching timing from the unique coded signal broadcast from each satellite in view, the GPS measures the time delay for each signal to reach the receiver to determine ranges. Measurements collected simultaneously from three satellites can produce a two-dimensional position. To determine a three-dimensional fix to include altitude or vertical position, signals from a minimum of four satellites must be received. [Figure 12-23]

Once a GPS receiver calculates its own position, it can then determine and display the distance, bearing, and estimated time enroute to the next waypoint. Although the GPS unit uses latitude and longitude coordinates to identify a position, it may display waypoint names, fix names, and database identifiers in lieu of latitude and longitude coordinates.

While GPS is an extremely accurate system, small errors are caused by a number of factors, some of which are inherent to the system. Atmospheric factors, satellite position anomalies, and timing errors are relatively small factors in GPS use. The man-made error, called **selective availability (SA)** is potentially larger. The Department of Defense (DOD) introduced SA to degrade the accuracy of GPS for national security reasons. As of May 1, 2000, the government turned off selective availability, resulting in more accurate GPS navigation. The removal of selective availability is estimated to increase the accuracy of GPS from approximately 100 meters (328 ft) to about 20 meters (65 ft).

Figure 12-23. Three components including a satellite constellation, ground monitoring stations, and an airborne GPS receiver act together to provide precise position information. For accurate navigation, the aircraft must be able to communicate with at least four different satellites.

TRANSPONDERS

The transponder equipment found on aircraft is designed to make it easier for air traffic controllers to identify specific aircraft so that they can prevent mid-air collisions and provide guidance to aircraft. Since the transponder is a device that is related to radar, we will begin with a brief discussion of radar with respect to aircraft.

RADAR (radio detection and ranging) uses a synchronized radio transmitter and receiver to emit radio waves and process their reflections for display. The principle used is called **primary radar** or echo location radar. The radar transmitter sends out a brief pulse of EM waves, which travel outward at the speed of light and bounce off the metal parts of an airplane. The reflected energy or echo is received back at the radar site where it produces a spot of light on the radarscope. The range of an aircraft (distance from the antenna) is determined by measuring the time it takes for the radio waves to reach the aircraft and then return to the receiving antenna. The azimuth, or angle of the aircraft from the radar site, is determined by the position of the rotating antenna when the reflected portion of the radio wave is received.

The problem with only using primary radar is that all the blips on the radar scope look the same. During World War II, a system was developed to make it easier to distinguish the friendly aircraft from the enemy aircraft. This type of radar system is now called **secondary radar**. A small radar frequency receiver and transmitter unit is installed in each airplane. When the radar pulse (interrogator signal) from the ground site strikes the aircraft, the secondary equipment sends a coded signal (reply signal) back to the ground site. The coded signal received at the radar site from the aircraft permits it to be identified. In the years since World War II, both primary and secondary radar have been adapted for air traffic control (ATC) purposes. The radio waves used in ASR-9, a radar system commonly used by ATC facilities, operate on a frequency of 2.7 GHz (gigahertz).

The **transponder** is the name of the secondary radar equipment installed on an aircraft. The aircraft transponder system uses two different frequencies, one to transmit and one to receive. The transponder receives on the 1030 MHz frequency and transmits on the 1090 MHz frequency. The ground radar site sends out a coded interrogation pulse, which, in effect, asks the airborne equipment to answer or reply. When the transponder receives a valid interrogation, it sends back the proper reply signal. The coding used in the transponder signals is digital or binary. Each interrogation and reply signal consists of a number of pulses in a pulse train. For each location in the pulse train, a pulse can either be present or absent. The flightdeck controls for the transponder permit the pilot to set one of 4,096 different numerical codes. The numbers set into the transponder represent an octal coding so there are no 8s or 9s in the code setting window. The possible code settings range from 0000 to 7777. The computer in the ground radar site can identify the aircraft by the code its transponder sends out. [Figure 12-24]

The air traffic controller for the entire flight normally assigns only one code. Some transponder codes are reserved for special purposes; 0000 is used by the military, 1200 is used for aircraft that are operating under Visual Flight Rules (VFR) and not in ATC control, and 7500, 7600 and 7700 are reserved for specific types of aircraft emergency situations. Do not cycle the transponder through 0000, 7500, 7600, or 7700 when the transponder is turned on. If the aircraft is operated on the ground set the transponder to the off or standby mode to prevent inadvertent signals being emitted to ATC.

Currently, there are several different operating modes associated with transponder equipment.

1. Mode 3/A — This is the basic transponder mode that can utilize one of 4,096 different codes.

2. Mode C — This mode includes the above capabilities but adds a coded message giving the aircraft's pressure altitude when an altitude encoder is installed.

3. Mode S — This mode has the capability of sending additional messages such as ATC instructions or weather reports that can be viewed on a Cathode Ray Tube (CRT) or printed on paper in the flightdeck. Mode S also increases the number of different identification codes for the aircraft to over one million. Mode S is also used with Traffic Alerting and Collision Avoidance Systems (TCAS). If two aircraft with Mode S transponders and TCAS systems are on a collision course, the systems will communicate with each other and present each pilot with appropriate guidance to resolve the potential conflict.

ATC TRANSPONDER TESTS AND INSPECTIONS

Because transponder returns form the basis for radar separation, regulations require that transponders be tested and inspected every 24 calendar months for operations in controlled airspace. In addition, following any installation or maintenance, which

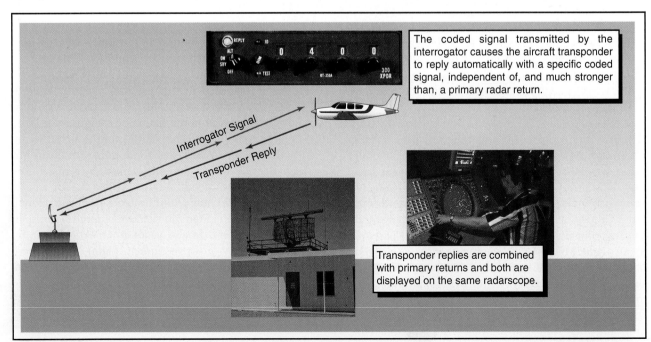

Figure 12-24. The decoder, which is part of the ground-based equipment, enables an ATC controller to assign a discrete transponder code to a specific aircraft. It is also designed to receive altitude information if a Mode C encoding altimeter is installed.

could have introduced transponder errors, the integrated system must also be tested in accordance with FAR 43, Appendix F.

The above tests and inspections must be conducted by:

1. An appropriately rated repair station; or
2. A holder of a continuous airworthiness program, or
3. The manufacturer of the aircraft, if the transponder was installed by that manufacturer.

INSTRUMENT LANDING SYSTEM (ILS)

The instrument landing system (ILS) is a precision approach navigational aid that provides highly accurate azimuth course (horizontal), glide slope (vertical), and distance guidance to a given runway. The ILS can be the best approach alternative in poor weather conditions for several reasons. First, the ILS is a more accurate approach aid than any other widely available system. Secondly, the increased accuracy generally allows for lower approach minimums. Third, the lower minimums can make it possible to execute an ILS approach and land at an airport when it otherwise would not have been possible using another type of approach system.

There are two basic types of instrument approach procedures: precision approaches and non-precision approaches. The difference is that precision approaches give the pilot vertical or descent guidance while non-precision approaches do not. Signals from VORs and NDBs can be used for non-precision approaches. The standard type of precision approach system used in the U.S. and most of the world for civilian aircraft is the ILS.

The ground equipment needed for an ILS system includes four parts.

1. Localizer — A radio beam used for lateral guidance.
2. Glideslope — A radio beam used for vertical guidance.
3. Marker Beacons — Radio signals that give distance indications to the runway and indicated on annunciator lights in the flightdeck.
4. Runway and approach lights.

Guidance information is received from ground-based localizer and glide slope transmitters. In addition, to help determine the distance from the runway, the ILS installation may provide DME fixes or marker beacons located along the ILS approach path. To facilitate the transition from instrument to visual flight as the aircraft approaches the airport, runway and approach lighting systems are installed to aid the pilot in the final portion of the landing. [Figure 12-25]

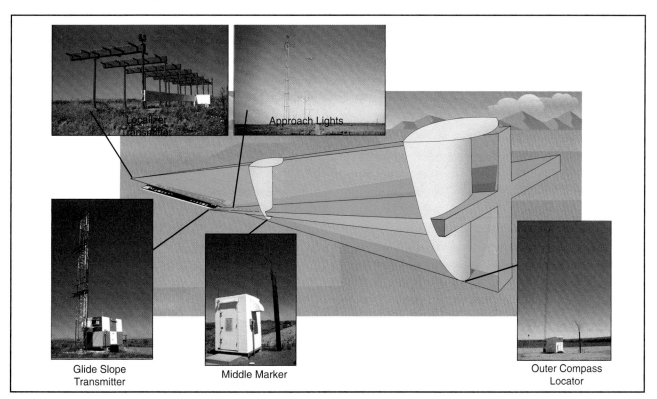

Figure 12-25. A typical ILS system follows this general arrangement. Specific installations vary somewhat in terms of glide slope elevation, localizer width, or marker utilization.

The ILS uses a localizer transmitter to provide information regarding the alignment of the aircraft with the runway centerline (azimuth). The **localizer** transmitter emits a signal from a navigational array located at the far end of the runway, opposite the approach end. The signals transmitted by the localizer are on frequencies between 108 and 112 MHz. A dual beam is transmitted outward from the far end of the runway. The right half of the signal is modulated at 150 Hz and the left half is modulated at 90 Hz. The aircraft receiver measures the relative strength of the 90 and 150 Hz signals. When they are equal, the aircraft is lined up with the centerline of the runway. The cockpit indicator utilizes a vertical needle just like a VOR indicator, and, in some cases; it is often the same one. If the needle swings to the left, the pilot must turn left to get back on course.

The **glideslope** uses a principle similar to the localizer, but it transmits on frequencies of 328–336 MHz in the UHF band. The glideslope signal uses 90 Hz modulation above the glidepath and 150 Hz modulation below the glidepath. The center of the glidepath would produce equal parts of 90 and 150 Hz signal in the receiver. The flightdeck indicator for glideslope is a horizontal needle. [Figure 12-26]

The glideslope and localizer frequencies for an ILS are paired together in set combinations. The glideslope receiver is usually slaved to the localizer receiver so that when a localizer frequency is tuned in, the correct glideslope frequency is automatically set in the glideslope receiver.

The **marker beacons** are low-powered transmitters that transmit a cone-shaped pattern straight up into the air. When the aircraft flies directly over the marker beacon site, an indication is given in the flightdeck to indicate the distance to the approach end of the runway. All marker beacons transmit at 75 MHz so different modulations must be used to identify the inner, middle and outer markers. The outer marker is modulated with a frequency of 400 Hz and causes a blue light to illuminate in the flightdeck when crossed. The middle marker is modulated with a frequency of 1300 Hz and lights an amber annunciator when the crossed. The inner marker is not used with all ILS systems. It uses a modulation of 3000 Hz and illuminates a white light when the aircraft flies over the inner marker. The three marker beacon indicator lights are generally incorporated in modern audio control panels. When a light illuminates, a corresponding audio tone also sounds.

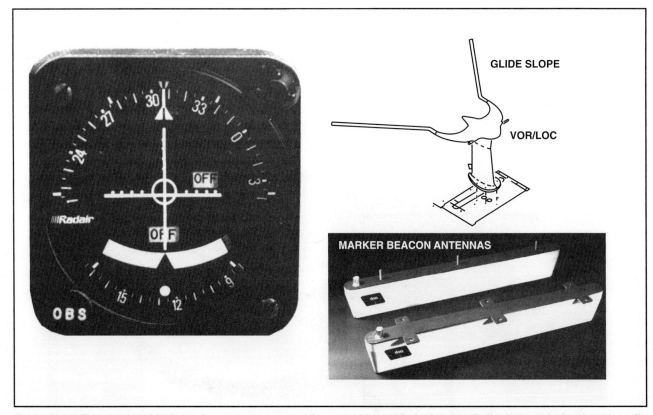

Figure 12-26. This simple ILS indicator incorporates two needles: a vertical needle for localizer indications and a horizontal needle for the glideslope indication. The glideslope antenna is a UHF dipole, which makes this antenna much shorter than the VHF dipole used by the localizer and the VOR. In some cases, as shown here, the glideslope antenna is combined into one unit with the VOR antenna.

EMERGENCY LOCATOR TRANSMITTERS (ELT)

The ELT is a self-contained transmitter that is designed to help locate an airplane after a crash. It is required on most small airplanes, but is not required on air carrier jets and business jets. The ELT is battery powered, and is automatically turned on by crash forces. It transmits a special swept tone for 48 hours on two different emergency frequencies: 121.5 MHz, which is the civilian emergency frequency, and 243.0 MHz, which is the military emergency frequency.

When ELTs were first mandated for installation in the 1970s, they were designed to be detected by overflying aircraft. Today, the primary sensors are spaceborne satellites. Many distress signals by ELTs are not received by the satellite sensors. Because of this, and the numerous inadvertent activations of ELTs, a new digital ELT system was developed using a frequency of 406 MHz. The new system transmits a distress signal that includes a coded identification of the aircraft in distress. This allows for the identification of false signals much more quickly. A new development ties the 406 MHz ELT to a Global Positioning System. This adds location to the position and identification of the distress aircraft. This measurably reduces response time to the crash site.

An acceleration switch activates the transmitter when a rapid deceleration force is applied along the longitudinal axis of the aircraft. The ELT must be installed as far aft as possible but in front of the tail surfaces, since this area has been shown to remain intact in most airplane crashes. The batteries in the ELT must be replaced or recharged at specific intervals as required by the FARs. In addition to checking the expiration dates on an ELT battery, the technician may need to test the ELT operation.

If possible, the ELT should be tested with the antenna disconnected or shielded to prevent the transmission of emergency signals into the air. If this cannot be done, it is still permissible to test the ELT, but only during the first five minutes of the hour and for three audio sweeps maximum. To test the ELT, tune a VHF communications radio to 121.5 MHz. Switch the ELT on manually until the signal is heard on the receiver and then switch it off. Emergency locator transmitter (ELT) tests and inspection regulations are outlined in FAR part 91. [Figure 12-27]

COCKPIT VOICE RECORDERS AND FLIGHT DATA RECORDERS

The cockpit voice recorder (CVR) and the flight data recorder (FDR) are designed to automatically record information in flight that can be used during an investigation following an accident or serious incident. They are installed on all air carrier jets and some commuter airliners and privately owned aircraft. The recorders are installed in the aft fuselage since this area is least likely to be severely damaged in an accident.

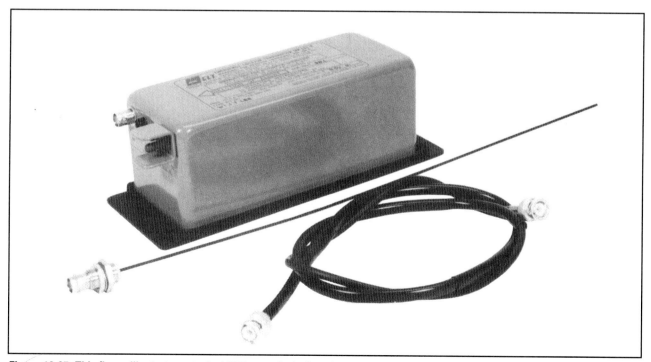

Figure 12-27. This figure illustrates a typical ELT setup including the transmitter/battery, coaxial cable, and antenna.

The CVR is designed to record sounds in the cockpit and communications on the intercom and radio systems. It has a hot microphone in the flightdeck, which is always activated to record voices, warning sounds, engine noise etc. The CVR is connected to the intercom so that conversations between the members of the crew can be recorded. It is also connected to the radios so that communications with ATC are recorded. The CVR has a continuous recording system that holds approximately the last 30 minutes of data. It is waterproof and protected against fire and impact forces.

The flight data recorder (FDR) has many more inputs than the cockpit voice recorder. It has a recording time of 8 hours on smaller aircraft and about 24 hours on larger aircraft. The CVR and FDR are located in the same area of the aft fuselage and have similar protection from water, fire etc. Following is the data that is recorded on the FDR. For a complete list of data, reference FAR Part 91 Appendix E.

The flight recorder required for certain aircraft under FAR 91.609 must record the following items:

1. Indicated airspeed
2. Altitude
3. Magnetic Heading
4. Vertical Acceleration
5. Longitudinal Acceleration
6. Pitch Attitude
7. Roll Attitude
8. Stabilizer Trim Position, or
9. Pitch Control Position
10. Roll Attitude
11. N_1, EPR or Prop RPM and Torque
12. Vertical Speed
13. Angle of Attack
14. Autopilot Engagement
15. TE Flap Position
16. LE Flap Position
17. Thrust Reverser Position
18. Spoiler/ Speedbrake Position

RADAR ALTIMETER

A **radar altimeter** displays the aircraft's absolute altitude above ground level (AGL). The height above terrain is measured by radio signals, instead of by atmospheric pressure receiver. Typically, the usable range for a radar altimeter extends up to 2,500 ft. It is mainly used during instrument approaches in bad weather, and provides a digital display of the altitude AGL. These instruments are also referred to as radio altimeters.

A radar altimeter is a more precise way to measure AGL altitude during IFR approaches. The radar altimeter uses antennas that are installed on the belly of the aircraft. The transmitter sends out radio waves at 4.3 GHz which strike the earth and bounce back to the receive antenna. By measuring the travel time for the radio waves, the system accurately calculates the height above the surface.

GROUND PROXIMITY WARNING SYSTEM (GPWS)

The ground proximity warning system (GPWS) is designed to provide warnings to the flight crew when the aircraft is in danger of striking the ground due to excessive descent rate or rising terrain. This equipment is required on all air carrier jets and it is found on some business jets. The main component in a GPWS system is a computer, which monitors numerous inputs and makes calculations to determine if the aircraft is in danger of hitting the ground. Some of the inputs to the computer are barometric altitude, radar altitude, rate of climb or descent, flap position and landing gear position. The GPWS is one of the few systems on a civil aircraft that gives a spoken voice command to the flight crew. When the system determines that a warning must be given, a recorded voice is activated which tells the flight crew to "PULL UP, PULL UP" or a similar type of message. Enhanced GPWS systems, which have recently become available, incorporate a worldwide terrain database and GPS inputs. These features allow the system to monitor the aircraft's position relative to the surrounding terrain and to provide much greater warning times of impending ground contact.

WEATHER RADAR

Airborne Weather Radar Airborne radar operates under the same principle as ground-based radar. The directional antenna, normally located behind the nose cone, transmits brief pulses of radar frequency EM waves ahead of the aircraft in order to locate and avoid thunderstorms.

There must be something present in a thunderstorm that will reflect the radar pulse. Clouds are invisible to radar, but ice, hail and rain reflect the energy back to the aircraft radar antenna. The size of the raindrops, the rainfall intensity, and the type of radar system installed all affect the strength of the return. Color radar uses different colors for different intensity levels. Green, yellow and red are often used with red indicating the highest intensity of rainfall.

Aircraft weather radar generally uses one of two frequency ranges: X-band and C-band. X-band systems, which are more common in general aviation aircraft, transmit on a frequency of 9.375 GHz and a wavelength of only 0.03 mm. This extremely short wave is reflected by very small amounts of precipitation. Due to the high amount of reflected energy, X-band systems provide a higher resolution and "see" farther than C-band radars. A disadvantage is that very little energy can pass through one storm to detect another, which may be located behind the first. The C-band frequency (5.44 GHz) can penetrate further into a storm, providing a more complete picture of the storm system. This capability makes C-band weather radar systems better for penetration into known areas of precipitation. Consequently, C-band radar is more likely to be found on large commercial aircraft. However, weather radar is prone to many of the same limitations as ground-based systems. It cannot detect water vapor, lightning, or wind shear.

Weather radar, like any other radar, is called a pulse radar system because it transmits very brief pulses of energy. This is necessary in order to use the same antenna for transmit and receive and to produce a usable maximum range. The transmitter sends out a pulse that has a total duration of about one microsecond. Then the antenna is switched to the receiver for about 2,500 microseconds. The receiver must be connected long enough for the pulse to travel out to the maximum range and back again. The use of pulse radar also makes the system more efficient, since the transmitter energy is concentrated in brief pulses. This permits much higher values of peak power than would otherwise be possible. [Figure 12-28]

On some radar systems, a tilt control is included so that the antenna can be tilted up and down to gauge the vertical extent of the storm cell. Radar antennas on air transport aircraft may also be stabilized by signals from the INS or IRS. Stabilization cancels the effects of pitch and roll so that the radar beam remains pointed in the desired direction.

The radar antenna is protected by a plastic or fiberglass radome, which must be carefully maintained

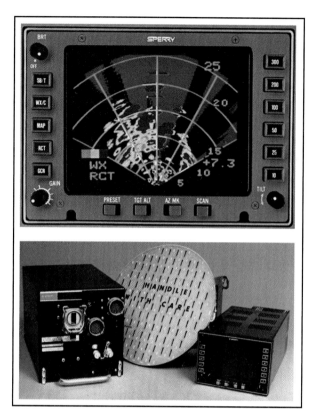

Figure 12-28. The top figure shows a typical weather radar display. The bottom figure shows the three units of a typical weather radar installation, from left to right: receiver/transmitter, antenna, control unit.

to prevent adverse effects on radar performance. The radome generally has conducting strips fastened on the outside to conduct static charges and lightning strikes away from the radome. The radome should only be painted with approved types of paint, which will not interfere with the radar frequency signals that must pass through the radome.

Personal safety is very important when working on aircraft with radar systems. Some of the components in the receiver/transmitter unit can hold very high voltages and should only be serviced by personnel familiar with the necessary safety precautions. The emissions from the radar antenna can be very hazardous to human beings. The radar should never be turned on while on the ground unless special precautions are taken. The manufacturer's maintenance instructions usually include some information on the maximum permissible exposure level (MPEL). The MPEL provides safe distances from aircraft radar antennas. Never walk in front of an aircraft when the radar might be activated.

STORMSCOPE™

The Stormscope is a weather avoidance system that uses completely different methods to locate thunderstorms than a radar system. The Stormscope is designed to receive the radio frequency EM waves produced by lightning discharges. It uses a directional antenna system similar to that used by the ADF equipment. In some cases it is possible to connect the Stormscope to the aircraft ADF antennas with special couplers.

The direction of the lightning is determined using the directional antenna and measuring the relative intensity of the discharge. The intensity is used as a pseudo range on the display. It is not the actual range like that obtained from weather radar, but it does provide useful information to the pilot. The display instrument in the cockpit is normally a small round LCD display that shows a light dot for each lightning strike that is detected. From the patterns on the display the pilot can determine where the worst areas are located and avoid them. [Figure 12-29]

Figure 12-29. This figure illustrates a typical Stormscope display.

The purpose of all weather detection systems is avoidance. A very strong thunderstorm cell has the capability of tearing apart even the strongest of aircraft. Since the Stormscope and the weather radar react to different aspects of thunderstorms, the best weather avoidance system would be to have both installed in the aircraft.

TCAS —AIRBORNE COLLISION AVOIDANCE SYSTEM

The traffic alert and collision avoidance system (TCAS) was designed to supplement ATC collision avoidance prevention. It is generally found on transport category aircraft in addition to a growing number of business jets. TCAS equipment uses some of the same equipment and principles as transponders.

The TCAS equipment generally includes a computer, a display screen in the flightdeck and a directional antenna system. The unit sends out interrogation signals in all directions. Any transponder-equipped aircraft within a specific range will send back a reply and the TCAS calculates its direction, range and altitude. TCAS can only determine altitude if the other aircraft is Mode C equipped. If the other aircraft does not have a transponder, it will not be detected. Each aircraft that has been detected within a certain range will be displayed as a lighted symbol on the display screen. If the other aircraft gets closer and creates a threat, the symbol changes color and shape. [Figure 12-30]

If the TCAS equipment determines that a sufficient danger level is present, it displays a vertical avoidance maneuver command to the pilots. The present equipment is not able to suggest turns as avoidance maneuvers. The display directs the pilot to climb or descend at a certain rate to avoid the threat aircraft. The TCAS system normally uses a Mode S transponder and a special type of directional antenna.

TYPES OF ANTENNAS

Many different types of antennas are used in aircraft avionics systems. Aviation technicians should be familiar with the common types of antennas so that they can properly identify, inspect and maintain them. Some of the common types of aircraft antennas and their basic characteristics will be described in this section. Aircraft antennas usually have a speed rating and should only be installed on aircraft that operate at and below their rated speed. [Figure 12-31]

VOR ANTENNAS

There are two basic types of VOR antennas found on aircraft: the half-wave dipole and the balanced loop types. The half-wave dipole antenna is a "V" shaped antenna that has a figure-eight-shaped reception pattern. The antenna has two metal rods in the shape of the letter "V" or a fiberglass-covered element made of thin sheet metal. It is installed on the aircraft on the vertical fin or on top of the fuselage with the open end of the "V" pointed either forward or aft. The figure-8 reception pattern works well for normal VOR airway. It does not work well for RNAV when the VOR station may be off the side of the aircraft. The dipole VOR antenna requires a special impedance matching device called a "balun." The balun is located at the antenna end of the coaxial

Aircraft Avionics Systems

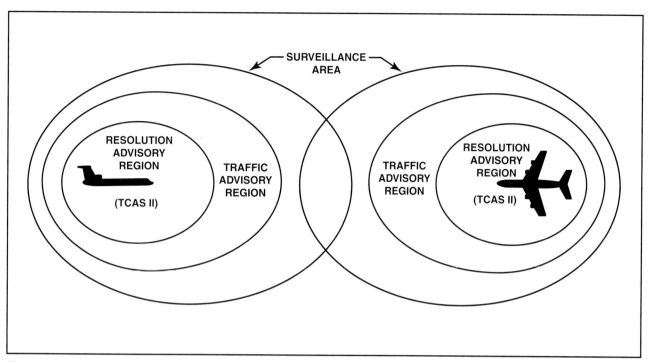

Figure 12-30. This figure outlines a typical surveillance area of a TCAS system.

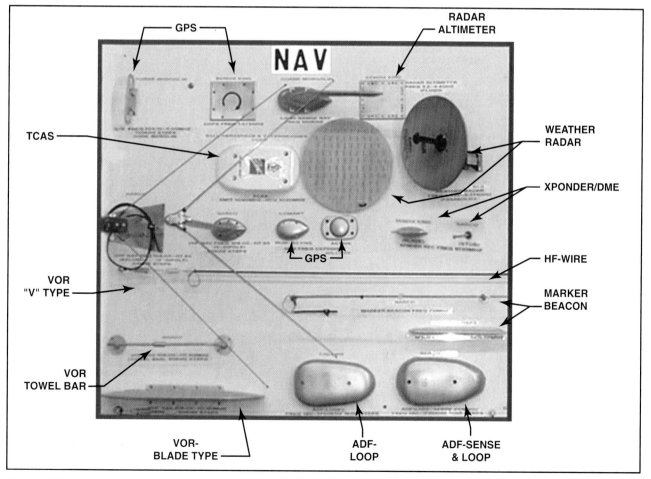

Figure 12-31. Examples of a variety of navigation antennas including VOR, GPS, radar altimeter, weather radar, transponder, marker beacon, and ADF antennas.

cable for more efficient transfer of energy from the antenna to the coax and receiver. [Figure 12-32]

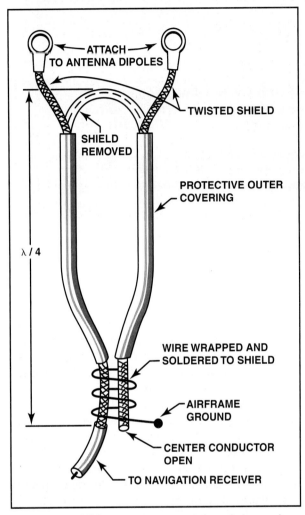

Figure 12-32. VOR antennas may require a balun such as this one.

Figure 12-33. Types of VOR antennas include V-shaped, blade, and towel bar types. The V-shaped is shown here.

The balanced loop VOR antenna has a circular reception pattern and is therefore a more efficient type of antenna for RNAV. There are three types of balanced loop antenna: the open loop towel bar, the blade and the internal mount. These antennas come in two halves that are mounted on opposite sides of the vertical fin on airplanes. On helicopters, they are mounted on each side of the aft fuselage or tail boom. The blade-type, balanced loop VOR antenna has a higher speed rating than the towel bar or V-type and is used on business jets and similar aircraft. Air carrier jets use a VOR antenna that is mounted inside the vertical fin with non-metallic flush covers on each side. [Figure 12-33]

LOCALIZER and GLIDESLOPE ANTENNAS

Small airplanes generally do not utilize a separate localizer antenna. The VOR antenna is used to receive localizer signals. On air carrier jets and similar aircraft, the large fuselage can block the localizer signals, so a separate localizer antenna is installed. This internally mounted antenna is usually installed inside the radome on the nose section of the aircraft.

The signals from glideslope transmitters can be received on a VOR antenna because they operate at a frequency that is approximately the third harmonic of the VOR frequency. Single-engine airplanes commonly use a signal splitter or coupler to supply the glideslope receiver from the VOR antenna. Other general aviation airplanes often use a V-shaped glideslope antenna. This antenna looks a lot like a V-shaped VOR antenna but it is only about 1/3 the size because of the shorter wavelength of glideslope signals. When a separate glideslope antenna is installed on the aircraft, it needs to be located on the front of the aircraft to prevent interference from the fuselage. The loop-type glideslope antenna can be installed either externally or internally on the forward part of an aircraft. The dipole glideslope antenna is designed to be installed inside a radome. [Figure 12-34]

MARKER BEACON ANTENNAS

All modern marker-beacon antennas are installed on the bottom of the aircraft, because the signals are received when the aircraft is directly over the transmitter site. A type of marker beacon antenna found on smaller aircraft is the sled type, which is a bent metal rod about 3-1/2 to 4 ft. long and uses a sliding clip for the lead-in connection. When the antenna is installed on the aircraft, the clip can be loosened and moved to tune the antenna. Another type of marker beacon antenna is the boat-type antenna,

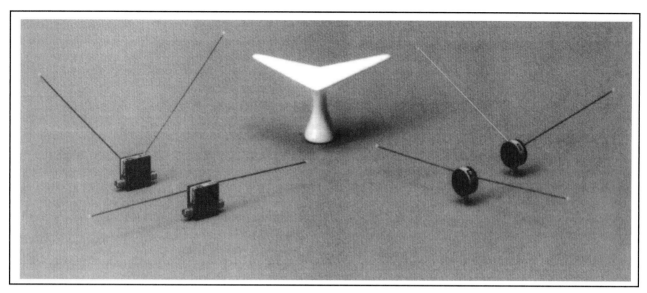
Figure 12-34. This figure shows several examples of localizer and glideslope antennas.

which is smaller and more streamlined than the wire or sled-type antennas.

HF COMMUNICATION ANTENNAS

Older air carrier jets used a probe-type HF antenna similar to the vertical fin antenna. This type of antenna includes a special coupler/tuner that re-tunes the antenna each time the frequency is changed on the HF radio. This kind of antenna can be mounted on the vertical stabilizer, or on a wing tip. Later model air carrier jets use a flush-mounted HF comm. antenna installed inside the vertical stabilizer. This antenna also requires a special tuning device that is installed at the antenna connection point.

VHF COMMUNICATION ANTENNAS

The VHF comm radios on aircraft use a separate antenna for each radio. These antennas are 1/4-wave, monopole antennas that can be mounted on the top or bottom of the aircraft. Lower speed aircraft generally use the thin whip type antennas while faster aircraft utilize blade type antennas that create less drag. The antenna may be straight or bent. The bent antennas have the advantages of less drag and less height for belly mounting. In addition, some blade-type VHF comm antennas have a stainless steel leading edge to prevent damage. [Figure 12-35]

DME/TRANSPONDER ANTENNAS

The same type of antenna can be used for either DME or transponder systems because they operate at similar frequencies and have similar characteristics. These antennas are usually installed on the bottom of the aircraft, but they can be located on the top of a narrow tail boom or other location that does not cause serious blockage. The two common types are the spike and blade antennas. The spike is a short metal rod with a ball on the end. This type is easier to install, but it is more easily damaged and creates more vibration and drag. The blade type is the most common type on modern aircraft. This antenna can be distinguished from the VHF comm blade because it is much smaller, about 2-4" long. These antennas are all 1/4-wave monopoles with vertical polarization, so an adequate groundplane must be provided during installation. [Figure 12-36]

ELT ANTENNAS

A common type of ELT antenna is a thin metal rod that is located close to the ELT itself. The antenna is a Marconi 1/4-wave antenna that requires a groundplane. It should normally be installed as close as possible to the ELT because of the low output power of ELT transmitters. A blade type of ELT antenna is also available for higher speed aircraft. [Figure 12-37]

SATELLITE COMMUNICATIONS ANTENNAS

The SATCOM antenna must be installed on the top of the aircraft to prevent signal blockage. A variety of different designs are produced for this kind of antenna. The antenna in Figure 12-38 is an example of a SATCOM antenna designed for large air carrier jets.

TCAS ANTENNAS

The Traffic Alert and Collision Avoidance system found on air carrier jets requires a special type of

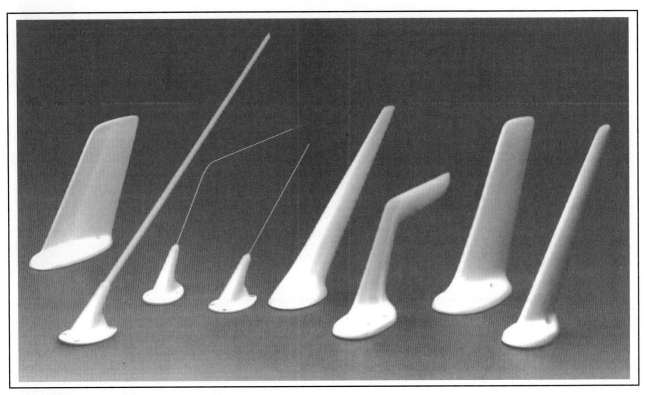

Figure 12-35. This figure illustrates a variety of whip and blade-type VHF comm antennas.

Figure 12-36. These are typical antennas used for DME and transponder.

directional antenna like that seen in Figure 12-39. This TCAS-II antenna is normally located on the top of the fuselage and has three ports for connection to the aircraft's TCAS-II equipment.

RADIOTELEPHONE ANTENNAS

Radiotelephone antennas come in a wide variety of shapes and sizes. These UHF antennas are normally installed on the bottom of the aircraft since they

Aircraft Avionics Systems

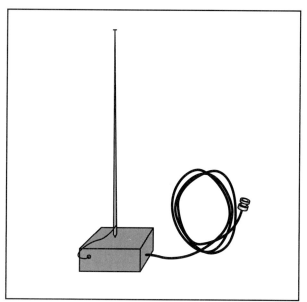

Figure 12-37. This figure illustrates a whip-type ELT antenna.

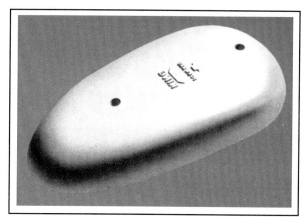

Figure 12-39. This figure illustrates a typical TCAS antenna.

Figure 12-38. This figure illustrates SATCOM antennas for air carrier-type jets.

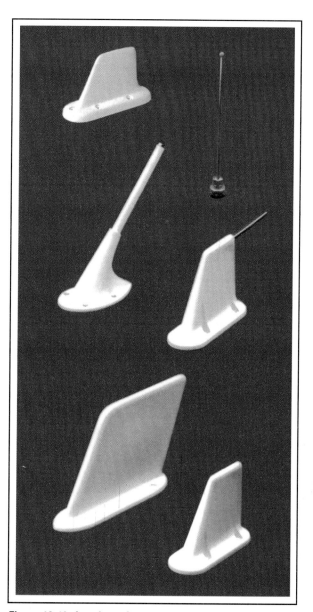

Figure 12-40. A variety of radiotelephone antennas are illustrated including blade, whip, pole, and combination type antennas.

operate in conjunction with ground-based systems using line-of-sight radio waves. A number of different kinds of radiotelephone antennas are shown in Figure 12-40. A major consideration when installing this type of antenna is preventing noise that can be caused by loose joints and poorly bonded surfaces on the aircraft.

SECTION B

AUTOPILOTS AND FLIGHT DIRECTORS

The primary reason that autopilots are installed in aircraft is to relieve the pilot's need to manually control the aircraft during long periods of flight. The FAA classifies autopilots as aircraft instruments, so A&P technicians cannot repair or alter autopilots. There are many tasks related to autopilots that aircraft technicians might perform such as installation, inspection, and troubleshooting.

An autopilot is an expensive and complicated device. It often has various components located in many different areas of the aircraft and many interconnections. The autopilot is connected to the flight control system, and autopilot malfunctions can be very serious.

The FAA must always approve an autopilot system for the specific make and model of aircraft in which it will be installed. A type of autopilot may be approved for a number of different aircraft, but different torque settings and adjustments may have to be made for each application. Always follow the instructions that apply to the specific autopilot installation to insure that the unit is correctly adjusted and tested for the particular aircraft in which it is installed. The basic operating principles for aircraft autopilots will be described here along with some specific examples of aircraft autopilot installations.

TYPES OF AUTOPILOTS

Autopilot systems are categorized according to their complexity and the number of aircraft axes of rotation they control. The autopilot utilizes the same control surfaces that the human pilot does. The three control axes of an airplane are shown in figure 12-41. The rudder controls aircraft rotation about or around the vertical or yaw axis. The elevators control rotation about the lateral or pitch axis. The ailerons control aircraft rotation about the longitudinal or roll axis. Autopilots can be described as single-axis, two-axis or three-axis types. A single-axis autopilot usually operates the ailerons only and is often referred to as a wing leveler. A two-axis autopilot controls the ailerons and elevator to provide pitch and roll stability. A fully integrated, three-axis autopilot operates all three primary control surfaces: ailerons, elevator and rudder. There is a large difference in the capabilities of a three-axis autopilot found on a small general aviation airplane and the three-axis autopilot found on air carrier jets and similar aircraft. For this reason, two other categories of autopilot will be added to the three already mentioned.

The term "Automatic Flight Control System" (AFCS) generally represents the state-of-the-art that was reached a few years ago. The autopilot in the Lockheed L-1011 is an example of an AFCS. This is a three-axis autopilot that can control the aircraft during climbs, descents, cruise flight, and during instrument approaches. It also has an auto-throttle system, which will automatically control engine power or thrust. Some AFCS autopilots have auto-land capability where the autopilot can actually land the airplane on the runway. These types of autopilots require many back-up systems and high levels of redundancy. The AFCS typically includes a flight director function that will be explained later.

The latest types of autopilots are referred to as Flight Management Systems (FMS). These systems include additional computers called Flight Management Computers. Flight Management Computers permit an entire flight profile to be preprogrammed and then automatically controlled from just after takeoff to landing and rollout on the runway. The Flight Management Computer can be thought of as a master computer that monitors flight parameters and controls the autopilot and auto-throttle computers. The computers can store in their memory many different routes and flight profiles, and they can be used to provide maximum economy in fuel consumption or other desired controlling factors. The standard Boeing 767 autopilot systems will be used later as an example of the capabilities of an FMS installation.

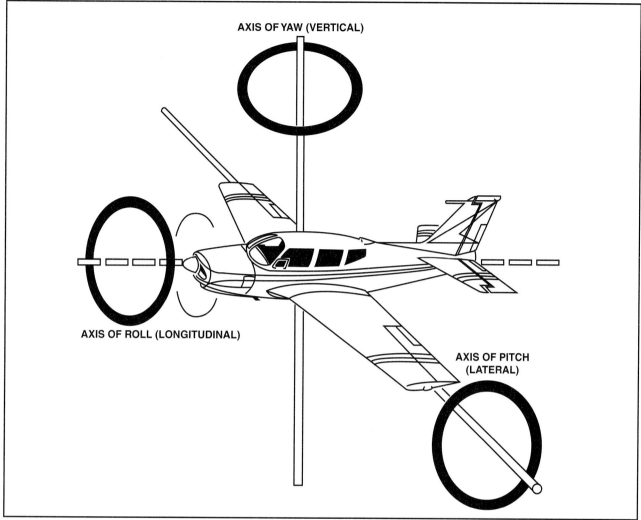

Figure 12-41. These are the control axes for an airplane.

BASIC AUTOPILOT OPERATION

The FAA states in AC 65-15A that the purpose of an automatic pilot system is primarily to reduce the work, strain and fatigue of controlling the aircraft during long flights. The capabilities of a modern autopilot go way beyond simply controlling the aircraft during cruise operations. A sophisticated autopilot system can land the airplane in weather conditions that are so bad that the human pilot could not legally land the airplane. In this section, the basic parts and operation of simple autopilots will be described. Figure 12-42 shows the basic parts of the rudder control channel of an autopilot. The aileron and elevator channels work in a similar fashion.

The basic parts and their functions are:

1. Sensors or Gyros — These detect a change in aircraft attitude using gyros or similar sensing devices.

2. Amplifier or Computer — This component processes the signals from the sensors and sends signals to the servos to correct the attitude.

3. Servos — Servos receive signals from the computer and supply the physical force necessary to move the flight control surface.

4. Feedback — All but the simplest autopilots have a feedback system that sends signals back to the computer to indicate the motion of the flight control surface. Without feedback, control of the aircraft would not be smooth and precise.

5. Controller — Figure 12-43 shows a typical controller. This unit is in the cockpit and contains the actuating switches and the pitch and turn knobs. The pilot can move the pitch knob or turn knob to supply manual commands to the autopilot that change the pitch attitude or command a turn.

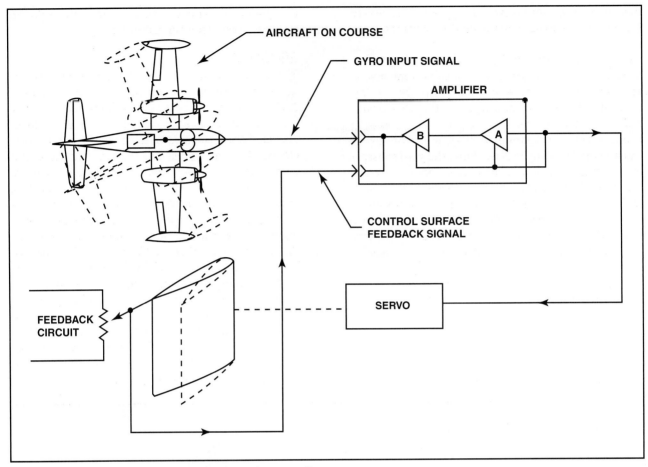

Figure 12-42. This illustrates the basic operation of an autopilot.

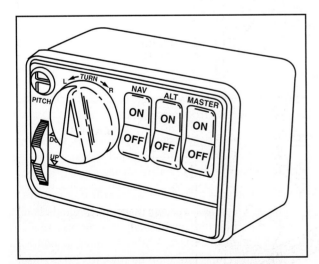

Figure 12-43. This illustrates an autopilot controller.

The operation of any autopilot follows the basic principles of error sensing, correction, follow-up and command, although different types of sensors, servos, etc. may be used. On a modern autopilot, digital computers have replaced earlier analog devices, and there are often separate computers for each of the three control axes.

Some method is required to determine when the flight condition of the aircraft differs from that commanded by the pilot. Almost all modern autopilots use some type of gyro for this purpose, and there are two ways of generating error signals: by attitude gyros or by rate gyros.

Many systems use the directional gyro and attitude indicator to provide a stable reference from which to measure the error signal. The gyros sense relative motion between itself and its supporting system to generate an error signal. Rate gyros sense the rate at which the aircraft has departed from the desired flight condition.

Once it has been established that an error in the aircraft's attitude or altitude exists, the autopilot sets about to correct the situation. The error signal from the sensor needs amplification by mechanical, electrical or hydraulic means. The amplified signal is then sent to the servos, which apply the necessary force or torque to actuate the control surfaces.

Causing the control surface to move in the correct direction is not enough. There must be some means to follow up the correction and stop the motion about the affected axis at the proper time. For example, if the gyro senses that the left wing is low, it commands the servo to move the left aileron down, causing the left wing to rise. When the gyro senses a wings-level condition, it signals the servo to streamline the aileron. However, the inertia of the rising wing could result in an overshoot of wings-level, causing the gyro to sense that the left wing was now high. Without some means to prevent this, a violent oscillation would result. Modern autopilots have a follow-up system that nullifies the input signal so that control-surface movement stops upon reaching the desired deflection. This system allows the surface to be brought back to its streamlined position with no over- or under-shoot.

A command system is incorporated for the autopilot to fly the aircraft as the human pilot wants. The control panel or flight controller allows the pilot to manually command the autopilot.

SENSORS

The gyroscopic sensors used with autopilots are similar to the gyro instruments described in Chapter 11. These sensors detect aircraft pitch, roll, and yaw motions. Deviations in attitude or in the rate of change from a selected attitude are converted into electrical error signals and sent to the autopilot computer. On small aircraft, the autopilot sensors are frequently built in to the attitude and heading gyros. The latest types of autopilots use sensors that employ laser beams instead of a spinning gyro rotor. Figure 12-44 shows one of these laser sensors, called ring laser gyros, or RLGs. The RLG has two laser beams that travel in opposite directions around a triangular course. Sensitive detectors measure the Doppler shift or frequency change whenever the unit is rotated. One RLG is needed for each axis measured. RLGs are much more expensive than an actual gyro, but they do not precess, and they eliminate the moving parts that cause a conventional gyro to gradually wear out.

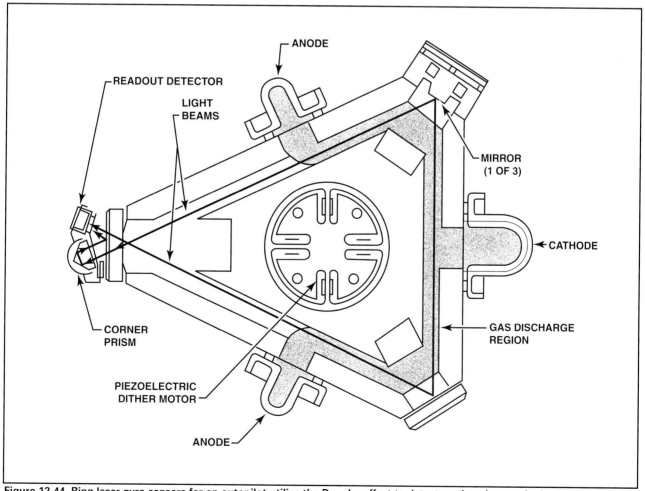

Figure 12-44. Ring laser gyro sensors for an autopilot utilize the Doppler effect to detect motion about a single axis.

SERVOS

Servos supply the force needed to move the flight control surfaces. There are three basic kinds described here: pneumatic servos, electric-motor servos, and electro-hydraulic servos.

Some simple autopilots found on small airplanes use vacuum sources like those used to operate gyro instruments. The vacuum is directed to pneumatic servos that are connected mechanically to the normal flight control system. As seen in figure 12-45, the pneumatic servo is an airtight housing which contains a movable diaphragm. When vacuum is applied to the servo, the diaphragm is displaced, pulling on the bridle cable connected to the main control cable by a bridle clamp. Two of these servos would be needed for each control axis.

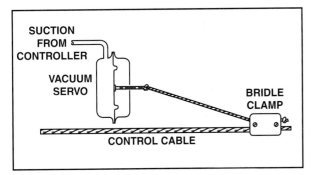

Figure 12-45. This illustrates the pneumatic servo for a small aircraft autopilot.

Servos that utilize electric motors are shown in figures 12-46 and 12-47.

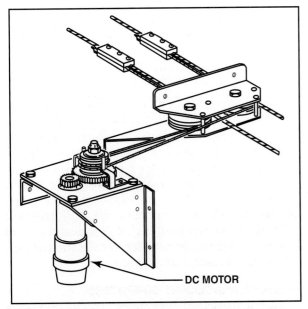

Figure 12-46. This illustrates an autopilot servo with a reversible DC motor, reduction gears, and bridle cables to supply the force to move the control surface in both directions.

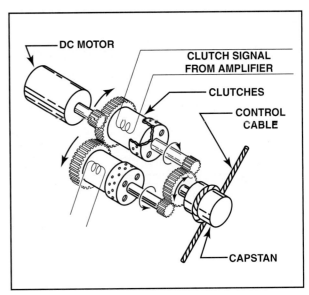

Figure 12-47. This illustrates an autopilot servo with a motor that runs continuously and uses magnetic clutches to engage the mechanism and apply torque to the capstan and control cable. This type's advantage is eliminating the inertial forces of starting and stopping the motor. It can be engaged and disengaged more rapidly and precisely.

Air carrier jets and some of the larger business jets use hydraulically-powered flight controls. The normal flight control system employs mechanical linkages that control hydraulic units called Power Control Actuators, or PCAs. The autopilot servos on these types of aircraft are electro-hydraulic servo valves that utilize electrical signals from the autopilot computers to direct hydraulic fluid under pressure to a hydraulic actuator. The actuator portion of the electro-hydraulic servo valve supplies mechanical force to the normal linkage of the flight control system. Figure 12-48 shows the electro-hydraulic servo valve for a typical large aircraft autopilot system. Figure 12-49 shows an autopilot servo for elevator control in the tail section of an air carrier jet. The servo's mechanical force is transmitted by a push-pull tube to the normal flight-control linkage that activates the PCAs. The level of redundancy in this system is typical for this class of aircraft.

SMALL AIRCRAFT AUTOPILOTS

A single-axis autopilot for a single engine airplane is shown in figure 12-50. This simple autopilot uses pneumatic servos to actuate the ailerons. The source of power is an engine-driven, dry-air vacuum pump. The sensor is a gyro turn coordinator, which controls the pneumatic power applied to the servos. Some of the torque settings and rigging instructions for the autopilot can be seen in this drawing. This is the type of autopilot that is often called a "wing leveler" since it controls only the aileron control surfaces.

Aircraft Avionics Systems

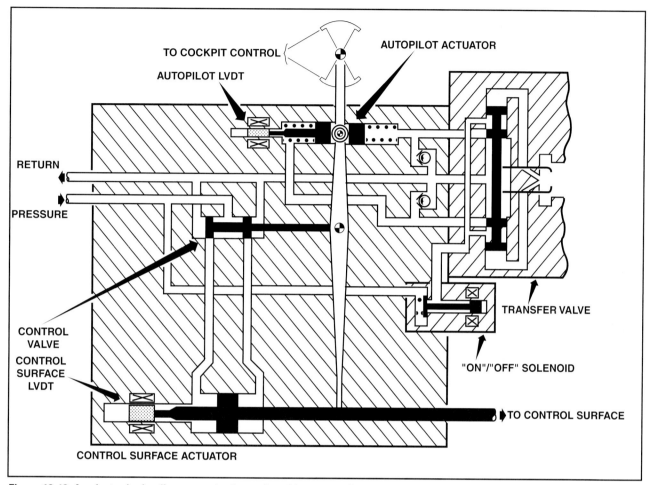

Figure 12-48. An electro-hydraulic servo valve is employed in an autopilot system for aircraft with hydraulically powered flight controls. Linear variable differential transformers (LVDTs) provide feedback signals of the movement of the system's mechanical parts.

A three-axis autopilot with electric motor servos is illustrated in figure 12-51. This system uses gyro sensors and an altitude sensor. The altitude sensor shows that this autopilot has an altitude hold capability.

The autopilot can use radio signals from the aircraft's navigation radios to steer the aircraft along a desired VOR or localizer course. The pitch, roll, and yaw servos receive signals from the computer, which activates the motors that move the control surfaces. A pitch trim servo is included so that the autopilot can apply nose up or nose down pitch trim as required. The aircraft can operate with a wide range of CG positions and the autopilot, like the human pilot, uses pitch trim to reduce the elevator control force to an acceptable level. The autopilot controller has switches to engage the heading, radio NAV and altitude operating modes. It also contains an on/off switch, a pitch control indicator, and the knobs for manual control of autopilot pitch and roll. It should be noted that these autopilot components are located in various parts of the aircraft and some minor components such as bridle cables are not shown. This autopilot system has the ability to guide the aircraft on an ILS approach using both localizer and glideslope signals. This feature is called an "approach coupler" and is required for certain types of instrument approaches.

FLIGHT MANAGEMENT SYSTEM (FMS)

The Boeing 767 will be used as an example of a flight management system or FMS. This system is capable of automatically controlling the airplane from just after takeoff (above 400 ft. AGL) through roll out on the runway after landing. The human pilot must take over to turn off the runway and taxi to the gate. This does not mean that all flights will use all these capabilities, but the autopilot and flight director will be used for some portion of each flight under normal circumstances.

FLIGHT MANAGEMENT COMPUTERS

Flight Management Computers (FMCs) provide a number of advanced features and functions that were not found on earlier autopilot systems. Some

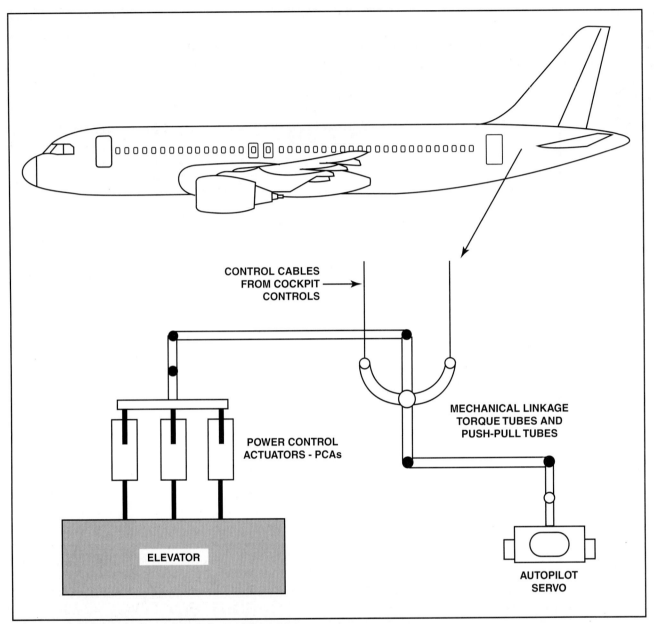

Figure 12-49. The autopilot servos on a large jet airplane provide mechanical force to move the normal control linkage, activating the hydraulic PCAs that move the flight control surfaces.

of the functions of the Flight Management Computer are:

1. Flight Planning — The entire flight can be programmed into the computer using a cockpit keyboard.

2. Performance Management — The system can provide optimum profiles for climb, cruise, descent and holding patterns. A minimum cost flight can be flown automatically by using optimum climb settings, cruise settings, etc.

3. Navigation Calculations — The FMC can calculate great circle routes, climb and descent profiles, etc.

4. Auto Tune of VOR and DME — The FMC can automatically tune the radios to the correct station frequencies.

5. Autothrottle Speed Commands — These are displayed on the Electronic Attitude Deviation Indicator (EADI) as FAST/SLOW indications.

The FMC is a master computer that integrates the functions of the laser sensors, Flight Control Computers, Thrust Management Computers, Air Data Computers, navigation sensors, and EICAS computers. The autopilot sensors are called Inertial Reference Units (IRUs) and are the same in basic operation as the Ring Laser Gyros (RLGs) previously described.

Aircraft Avionics Systems

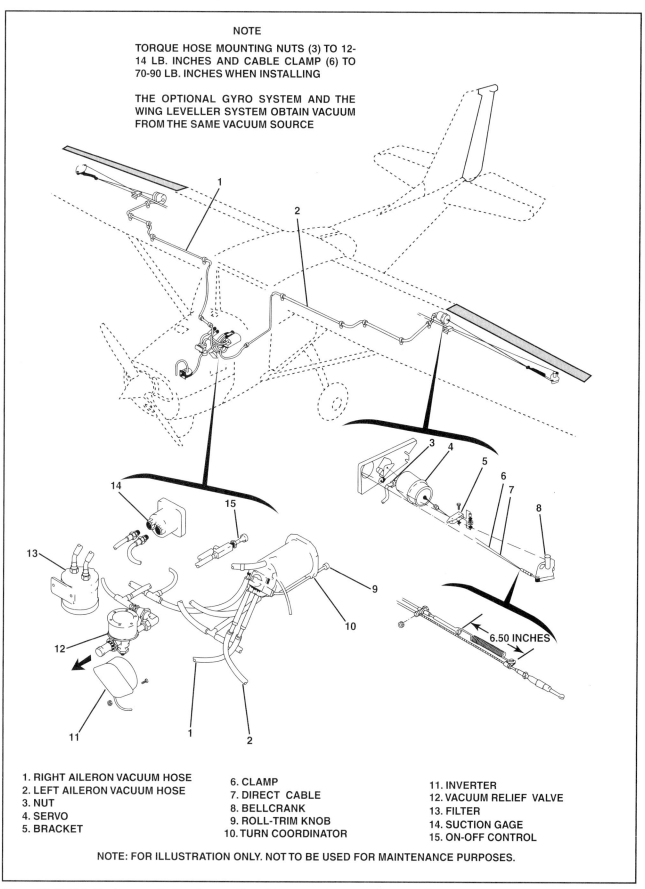

Figure 12-50. This illustrates a single-axis autopilot with pneumatic servos. (Courtesy Cessna Aircraft Corp.)

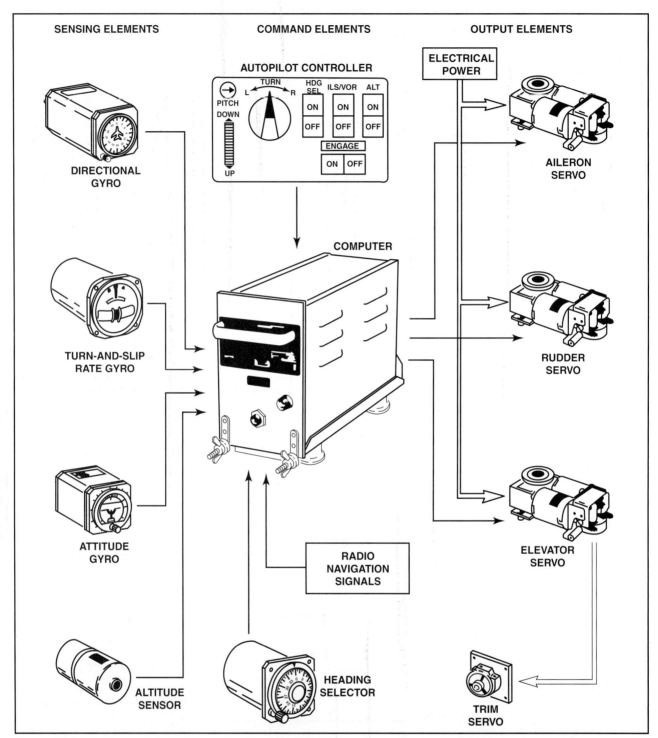

Figure 12-51. This illustrates a 3-axis autopilot that can be coupled to radio navigation receivers.

FLIGHT CONTROL COMPUTERS

There are three flight control computers in the autopilot system. A block diagram of the connections to the three flight control computers is shown in figure 12-52. The three computers are independent so that a failure in one will not affect the other two. These modern computers are faster and more compact than earlier types.

THRUST MANAGEMENT COMPUTER (TMC)

The purpose of the TMC is to automatically set the proper thrust level for the engines. A diagram of the autothrottle system is shown in figure 12-53. The output servo moves the throttle linkage to set the level of engine power that the TMC calculates. The system includes sensors on the engines, which monitor the important engine operating parameters.

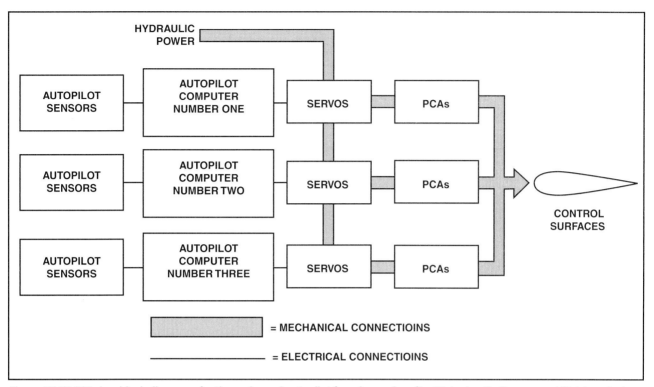

Figure 12-52. This is a block diagram of a three-channel autopilot for a large aircraft with hydraulically powered flight controls.

This is to prevent the pilot from exceeding any engine operating limitation for RPM, EPR, EGT, etc. The autothrottle system can be used to maintain a given climb rate, indicated airspeed, Mach number, or descent rate. Since the 767 has autoland capabilities, the autothrottle system will automatically close the throttles just prior to landing to enable a smooth touchdown. The TMC system also provides a minimum speed protection, which will maintain a safe margin above stall speed for the particular flight configuration. The autopilot system and the autothrottle system can be engaged separately or together using the controls on the flight control panel.

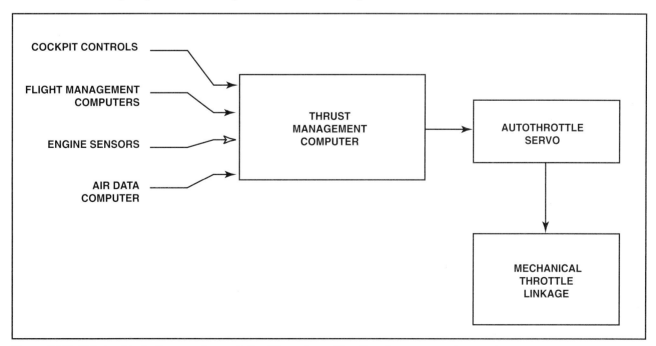

Figure 12-53. This is a block diagram of an autothrottle system with a thrust management computer.

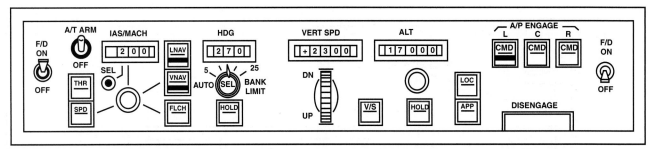

Figure 12-54. The control panel includes switches to control the autopilot, the flight director, and the autothrottle systems.

FLIGHT CONTROL PANEL

The flight control panel contains the switches for activating the various autopilot functions and for adjusting the settings for the desired vertical speed, IAS, Mach number, etc. Also included in the flight control panel are the indicator lights for the different operating modes. This panel is in the glareshield above the center instrument panel. [Figure 12-54]

CONTROL WHEEL STEERING (CWS)

Control wheel steering is an operating mode for the autopilot in addition to the command operating mode. The command mode is the normal autopilot mode in which the pilot does not touch the controls because the autopilot is flying the airplane. In CWS mode, the autopilot maintains the pitch and roll attitude existing at the time of engagement. When the pilot desires to change that attitude, light pressure is applied to the controls as in normal flight. The force upon the controls is measured and used as an input signal to the autopilot computers. In effect, the human pilot is flying the airplane, but the autopilot moves the control surfaces. Figure 12-55 shows the connections between the force transducer and the flight control computer. The operation of a typical force transducer is illustrated by figure 12-56. The three electrical windings and the armature above them make up a special type of variable transformer. The AC input signal is applied to the center winding and the outer windings produce the output signal. The force transducer's housing is flexible so that its length will change based on the force applied to it. When the housing changes in length, it causes relative motion between the armature and the coils. This motion alters the magnetic coupling and therefore produces a change in the output signal.

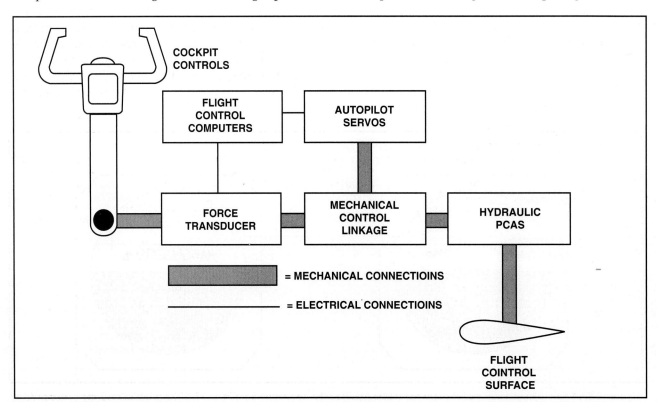

Figure 12-55. This shows the location of the force transducers and servos in the control system of an air carrier jet airplane.

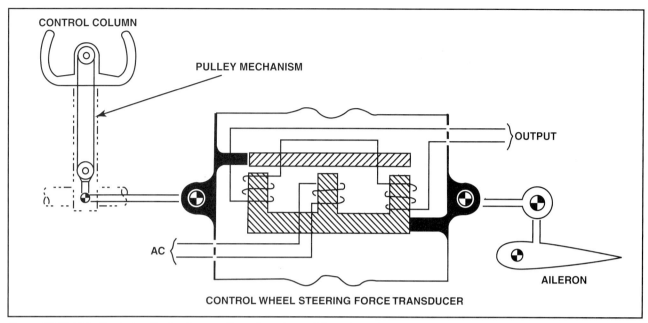

Figure 12-56. This illustrates the basic operation of one type of force transducer.

FLIGHT DIRECTOR

A flight director is a system that uses some of the basic components of an autopilot, but not all of them. A flight director uses sensors and computers, but it does not have servos. The flight director computer uses the signals from sensors to calculate a correction, which is then displayed as a command for the pilot to follow. The commands from the flight director are displayed to the pilot on the EADI by the command bars. Operation of the command bars on the EADI is shown in figure 12-57. On the left, the command bar symbol is above the airplane symbol. The indication is that the pilot needs to raise the nose to satisfy this flight director command. On the right, the nose of the airplane has been raised so that the airplane symbol aligns with the command bar. During flight director operations, the pilot maintains manual control of the aircraft but follows the steering commands directed by the command bars. One of the primary uses for the flight director is during an instrument approach. By using the flight director, the pilot can fly more accurately on an ILS approach because the computer makes rapid calculations to predict the optimum heading and attitude for the approach. Corrections for wind drift are automatic; all the pilot has to do is follow the flight director commands. Another condition when the flight director is helpful is in setting the proper takeoff pitch attitude. [Figure 12-57]

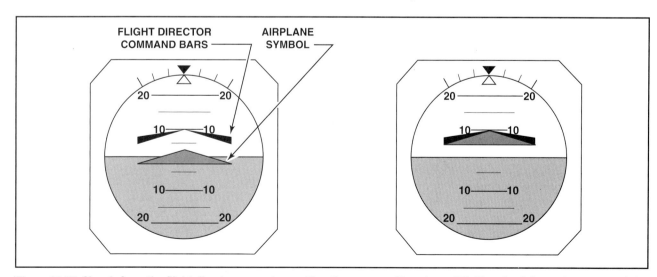

Figure 12-57. Signals from the flight director computer position the command bars in an ADI. The pilot follows the commands by aligning the airplane symbol with the command bars.

ADDITIONAL FEATURES

Some additional features of the Boeing 767 Autopilot and Flight Director System that are typical for this class of aircraft will be described briefly.

The Stability Augmentation System (SAS) involves certain functions of the yaw control system. One purpose of the SAS is to eliminate a potential problem known as Dutch Roll. Many large, swept-wing airplanes display a peculiar type of instability at high altitudes under certain flight conditions. This can result in a continuous pitching and rolling motion known as Dutch Roll. The SAS will automatically make rapid and precise rudder movements to reduce Dutch Roll motions. The system is also referred to as the "yaw damper," which can be engaged separately from the rest of the autopilot.

The runway alignment feature of the Boeing 767 is a part of the autoland system. It will automatically align the longitudinal axis of the airplane with the runway prior to touchdown. This feature is important during crosswind landings because it is designed to prevent the airplane from landing at a crab angle to the runway. Figure 12-58 illustrates this. The runway alignment feature is limited to control surface deflections of 25° for the rudder and 2° for ailerons. A very strong crosswind at 90° to the runway can not be completely counteracted. This same feature will also provide corrections if an engine fails during the landing approach. It would supply the control corrections to counteract the asymmetrical thrust situation.

The Altitude Select and Altitude Hold features allow the crew to select a specific altitude, which the autopilot will capture and subsequently hold. The crew sets the desired altitude on the autopilot control panel. Then, depending upon the system design, the autopilot will either automatically pitch up or down to climb or descend, or the pilot must manually command the climb or descent. Upon reaching the selected altitude, the autopilot will level off and hold that altitude within a few feet.

Autopilots of this type often incorporate a Speed Mode Selection function. The autopilot can be set to maintain a specific indicated airspeed (IAS Hold) or Mach number (Mach Hold). Also available is a Vertical Speed mode that allows the pilot to select a climb or descent rate, which the autopilot will maintain. When a speed mode is selected, the autopilot and autothrottle computers work together to adjust engine thrust as required for the specified flight mode.

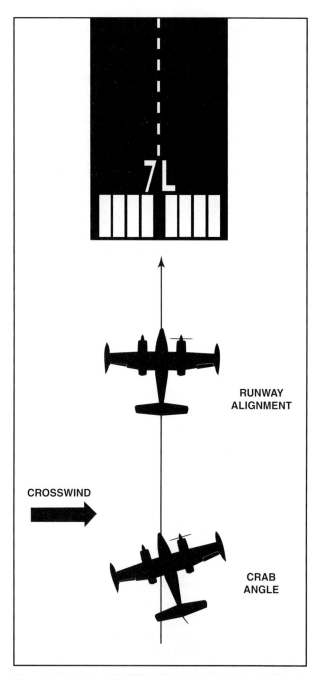

Figure 12-58. The stability augmentation system (SAS) in an aircraft with autoland capability has a special operating mode called "runway alignment." At approximately 500 ft., the runway alignment feature will eliminate the crab angle caused by a crosswind so that the aircraft will point down the runway at touchdown.

AUTOPILOT MAINTENANCE

The information in this section is not meant to relate to any particular aircraft autopilot system. The procedures are general and could be applied to most autopilots as appropriate. Maintenance of autopilots consists of visual inspections, replacement of components, cleaning, lubrication, troubleshooting and operational checkouts of the system.

Perform an operational check of the autopilot whenever one is installed, when components are replaced, and whenever a malfunction is suspected. Many things can be operationally checked on the ground, but some situations may require a test flight with an airborne checkout of the autopilot. Some general procedures follow for a ground checkout of an autopilot.

With the autopilot disengaged, manipulate the flight controls to see if they function smoothly and without excessive drag or interference from autopilot components. Check the autopilot's alignment to the aircraft. This normally involves checking such things as cable tension, torque settings, dimensional adjustments, etc.

Before engaging the autopilot for an operational check, allow the gyros to come up to speed. This normally requires from 2 to 4 minutes. After engaging the autopilot, make the following checks:

1. Rotate the turn knob on the controller to the left. The rudder pedals and control column should move in the proper directions to indicate a left turn. The motion should be smooth and without excessive binding, jerking or hesitation.

2. Rotate the turn knob to the right and watch the controls for proper operation and motion.

3. Rotate the pitch knob up and down and watch for the correct motion of the control column aft and forward.

4. If the autopilot has automatic pitch trim, check the proper motion of the trim control as the control column moves fore and aft. When the control column moves back, the system should apply nose up trim and vice versa.

5. With the autopilot engaged, try to overpower it by grasping the controls and applying force. It should be possible to overpower the autopilot if it is adjusted properly.

6. Check all of the controls and switches for proper actuation and correct indications.

7. It may be desirable to taxi the aircraft to check out some of the operating modes. If you engage the heading hold mode and make a taxi turn to the right, the controls should show motion commanding a turn to the left.

8. Check the autopilot disconnect switches to ensure that the autopilot disconnects rapidly and positively. There may be several ways to disconnect the autopilot; check them all.

9. If the aircraft has a flight director, check for proper indications by the command bars in the ADI or EADI. Check the autopilot mode indicators in the ADI or EADI if so equipped.

If the aircraft has both an autopilot and flight director, check them against each other to aid in troubleshooting. The autopilot and flight director share some components while others are only used by the autopilot. Use this to help locate the source of the problem when malfunctions are suspected. A ground checkout can often help to locate the source of a problem by comparing the indications of the flight director and autopilot. If the flight director is commanding an incorrect control movement and the autopilot is moving the controls in the same incorrect direction, then the fault is most likely a shared component such as the sensors or computer. If the flight director shows a correct command for nose up pitch but the autopilot does not move the controls to agree with this command, then the problem is not likely in the sensors or computer.

Some complaints about autopilot malfunctions result from faults in components other than the autopilot itself. If a pilot reports that the autopilot will not track a VOR radial, the problem could be a fault in the wires that carry radio signals to the autopilot rather than a problem with the autopilot itself. Conditions such as rigging problems or binding of the main control cables can adversely affect the autopilot's operation. Because the interactions and interconnections associated with autopilots can be complex, a good system schematic and a thorough knowledge of the autopilot are necessary for efficient troubleshooting and maintenance of autopilot systems.

SECTION C

INSTALLATION AND MAINTENANCE OF AVIONICS

There are numerous factors that make the installation of avionics equipment more critical than the installation of other types of equipment. Radios and avionics are very sensitive to electromagnetic interference, which can be created by nearby wiring and other electrically operated devices. The installation and maintenance of good bonding jumper connections is important to ensure proper operation of avionics units. Avionics equipment is easily damaged by excessive heat, which requires that provisions be made for adequate air circulation. One of the most common reasons for avionics equipment failure is overheating.

A thorough knowledge of avionics installation practices and other important considerations is necessary to ensure proper avionics performance. Appropriately rated FAA repair stations repair avionics equipment, but maintenance technicians often perform the installation, inspection and routine maintenance on these units. Some important considerations when installing and inspecting avionics equipment are covered in this section. For detailed explanation of specific installation and maintenance procedures, refer to the manufacturers recommended maintenance procedures. In addition, *Acceptable Methods, Techniques, and Practices, Aircraft Inspection and Repair AC 43.13-1B/2A.*, provides additional guidance to installation and inspection of avionics equipment.

CLEANING OF ELECTRONIC EQUIPMENT

It is important to clean electronic equipment to remove dust, dirt and lint that can block cooling holes and cause overheating. Dirt and lint that collects on open terminal strips and other electrical connections can absorb moisture and cause short circuits. Clean open terminal strips regularly to prevent the accumulation of foreign material. [Figure 12-59]

Electrical connections should be kept clean and free of corrosion and oxidation, which can add unwanted resistance. A mild abrasive such as very

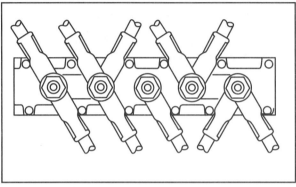

Figure 12-59. Open terminal strips should be kept clean and free of corrosion, dirt, and lint. Check for metal objects that could fall across the terminals and cause shorts.

fine sandpaper is recommended for removing corrosion and oxidation on terminal strips and mating surfaces.

Older electronic equipment made extensive use of rotary selector switches and similar devices with many sets of contacts. Special cleaning solvents are available for cleaning the contacts of these devices. Before using a spray contact cleaner, ensure that it is compatible with the plastic or non-metallic parts of these switches.

ROUTING WIRES

Wiring of all types should be routed above lines that carry fluids and clamped securely to the aircraft structure. Wires are routed to prevent abrasive damage from control cables, mechanical linkages and other moving parts in the aircraft. Clamps and ties should be used to prevent excessive wire movement due to in-flight vibration and other factors. [Figure 12-60]

At wire termination points, leave enough slack to allow for shock mount motions. If wires are tightly clamped too close to the termination point, normal aircraft motion and vibrations place bending loads on the wire connectors and cause premature failure in the wires. Provide sufficient slack at each end of a wire bundle to permit the replacement of termi-

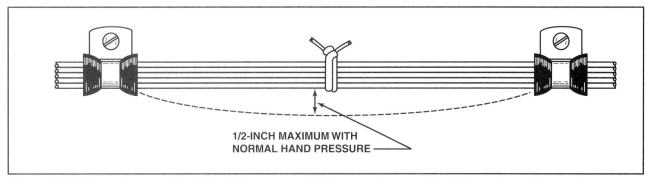

Figure 12-60. Wires and antenna leads should be supported with proper clamps and ties. The maximum distance between support clamps is 24 inches and the recommended distance between ties is 4 to 6 inches. In addition, install wiring with sufficient slack so that the bundles and individual wires are not under tension. Normally, wire groups should not exceed 1/2-inch deflection between support points.

nals to prevent mechanical strain and to permit the shifting of equipment for maintenance and inspection purposes.

Another factor to consider when routing wiring for electronic equipment is electromagnetic interference (EMI). Route antenna leads and other wiring sensitive to EMI away from the wires for inverters, power supplies, strobe lights, motors, and other components that are known to cause interference. When troubleshooting a noise or interference problem in aircraft radios and sensitive electronic equipment, it is often necessary to reroute wires away from the source of the EMI once it has been located.

SWITCHES AND CIRCUIT BREAKERS

Aircraft radios are usually connected to an avionics master switch, which makes it easier for the pilot to ensure that all electronic equipment is turned off when starting and shutting down the engine. This switch is separate from the normal master switch.

Turn off radios during engine starting and shutdown to prevent damage from surge currents and spikes of high voltage. When installing switches in aircraft circuits, the rating of the switch must be adequate to handle both the type and amount of current and voltage for the circuit. [Figure 12-61]

Whether the circuit is AC or DC makes a significant difference in the proper selection of switches. For example, a common aircraft switch is rated for 10 amps at 125-250 volts AC. The same switch is rated at 0.3 amps when it is used in a DC circuit up to 125 volts. If this switch were installed in a 10-amp DC circuit, the points would quickly burn and fail. The reason for this difference in ratings is that the current in an AC circuit drops to zero twice each cycle.

Figure 12-61. Switches used in aircraft circuits should have the appropriate AC or DC rating to prevent premature failures. This figure illustrates a typical aircraft toggle switch used in aircraft electronics systems.

This greatly reduces the problem of arcing as the points in the switch open. To ensure adequate performance and service life, it is important to use a switch with the proper ratings for both amperage and type of current. The condition of switches can be checked during inspection by operating the switch and checking the "feel" during operation. Aircraft switches are required to be of the "snap-action" type, which ensures rapid and positive opening and closing. These switches have a distinct feel when operated. When the switch is worn and ready to fail, it often begins to feel sloppy and loose in operation.

Circuit breakers for aircraft circuits should be the "trip-free" type. This means that the circuit breaker cannot be overridden by holding it in the engaged position. It will open the circuit regardless of the position of the control toggle or push button. Various types of circuit breakers are available, and the correct selection of circuit breaker ratings for the particular circuit is important to prevent dangerous overloads. [Figure 12-62]

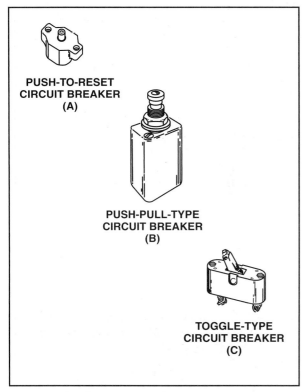

Figure 12-62. Circuit breakers should be the trip-free type and inspected regularly for proper operation.

During inspections, determine the proper operation of the circuit breaker. Most types can be manually opened to interrupt current flow. Even a small, general aviation airplane may have a large number of switches and circuit breakers; inspect each one for proper operation. [Figure 12-63]

BONDING AND SHIELDING

Radio reception can be completely blocked or severely interfered with by improper bonding and shielding in the aircraft. Aircraft radios can be affected by noise interference from sources inside and outside the aircraft. Outside interference comes from precipitation static (p-static) and thunderstorms. Inside interference can be produced by current flow in other circuits and EMI emitters such as ignition systems. The proper installation and maintenance of bonding jumpers is a key factor in preventing radio interference. Both braided wire bonding jumpers and thin metal straps are used for bonding connections. An installation of a braided bonding jumper on a shock mount is shown in figure 12-64.

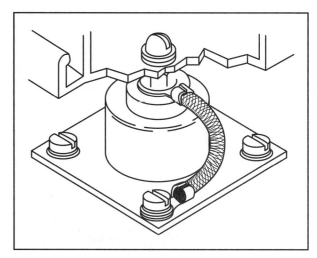

Figure 12-64. Bonding jumpers on shock mounts must allow freedom of movement on the shock mounts, and should be inspected regularly to detect breakage or corrosion.

All parts of the aircraft that could create noise problems should be bonded. Electrical equipment that is shock mounted should have adequate bonding jumpers to carry the ground path current without producing excessive voltage drop. When the bonding jumpers carry ground path currents, always use more than one. If there is only one and it breaks, the radio or other piece of equipment will be inoperative. When attaching bonding jumpers, remove all dirt, grease, paint and anodized coatings around the jumper connection point to ensure a good electrical contact. [Figure 12-65]

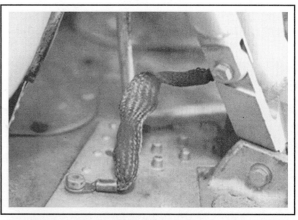

Figure 12-65. Heavy-duty bonding straps are often required for bonding major airframe components.

A general rule is that the maximum resistance for a bonding jumper connection should be no more than .003 ohms. However, the FAA states that if a bonding jumper is only used for static electricity purposes and does not carry ground path currents, 0.01 ohms

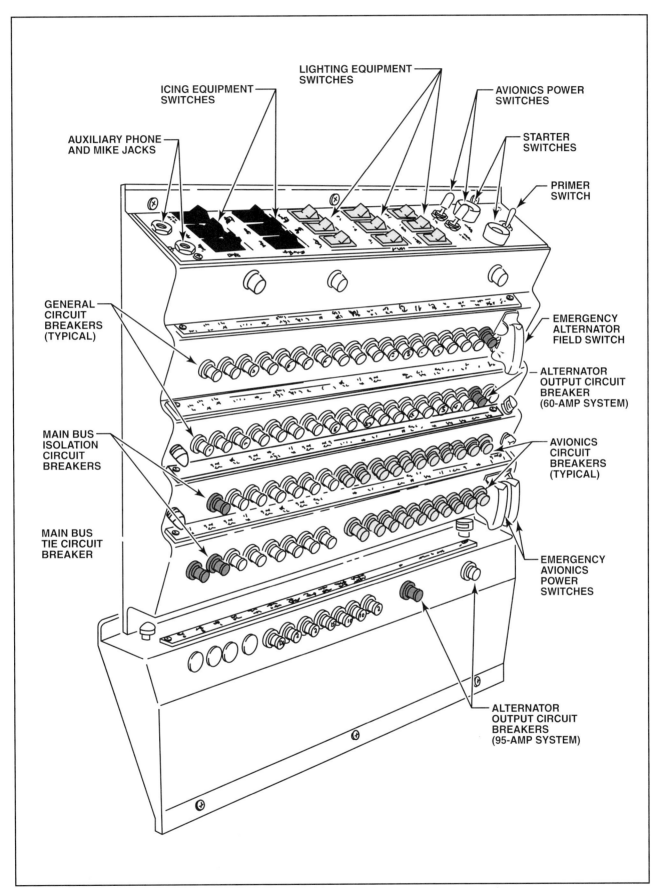

Figure 12-63. This figure illustrates a typical circuit breaker panel for a twin-engine airplane.

is acceptable. Following are several functions that bonding jumpers accomplish on the aircraft.

1. Supply the ground path for current flow for electrical equipment.
2. Reduce radio interference.
3. Decrease the possibility of lightning damage (at control surface hinges, for example).
4. Allow static charges to equalize between different parts of the airframe. This can reduce the fire hazard caused by arcing near fuel tank vents, etc.

In addition, the following are several factors to keep in mind when installing and inspecting bonding jumpers.

1. Bonding jumpers should be as short as possible. However, allow for any necessary motion such as in a control surface area.
2. Do not solder bonding jumpers.
3. Do not paint bonding jumpers. Paint makes them brittle.
4. Ensure good contact by removing dirt, grease, paint and other coatings.
5. Use compatible mounting hardware to prevent corrosion.
6. Use compatible bonding jumpers (aluminum alloy for aluminum alloy structures and copper or brass jumpers for parts made of steel, stainless steel, brass or bronze).

Shielding is an important part of noise suppression for aircraft radios. Shielding can be applied at the source of the noise or at the component or circuit that is sensitive to EMI. Shielding consists of a metal outer cover for a wire or component. Electromagnetic fields that could cause interference are captured in the metal cover and sent to ground. The ignition system of an aircraft engine can produce serious interference. Therefore, all parts of the ignition system need to be shielded. [Figure 12-66]

A metal housing or outer cover provides the necessary shielding for magnetos and spark plugs. Ignition wires use an outer wire braid shielding. The primary or "P" lead is the wire that connects the magneto to the flightdeck ignition switch. It should be a shielded wire to prevent noise. If all parts of the ignition system have been shielded and ignition noise is still present, it may be necessary to install a filter capacitor on the magneto. A capacitor

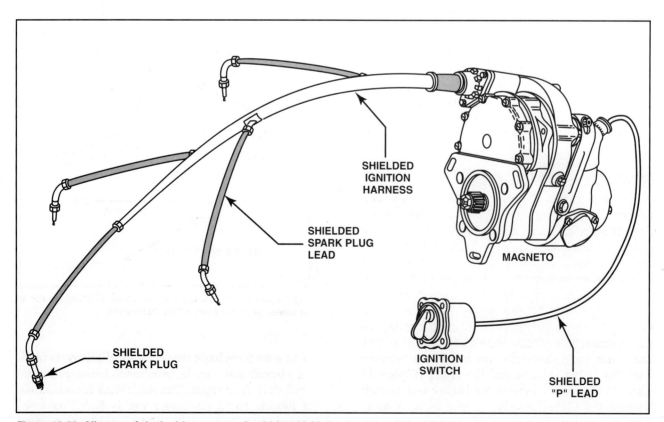

Figure 12-66. All parts of the ignition system should be shielded to prevent radio noise. On a reciprocating engine, for example, the magneto, ignition wires, spark plugs, and "P" lead need shielding.

or condenser of the correct size helps to filter out noise at the source. Other aircraft components, such as certain motors and power supplies, may also require filters.

Under certain circumstances, the shielding on electrical wiring can cause noise and interference. Shielded wires can sometimes result in a phenomenon known as groundloop interference. Depending on the types of electrical signals involved, groundloops can cause interference between different circuits in the aircraft. To prevent groundloop problems, ground the shielding in only one place and leave the other end of it "floating" or ungrounded. Special precautions are recommended in AC 43.13-2A for installing inverters to prevent these kinds of problems. The recommended procedures to prevent inverter interference are:

1. Install inverters in separate areas, away from sensitive electronic circuits.

2. Separate the input and output wires of the inverter.

3. Properly bond the inverter case to the airframe.

4. Use shielded wires for inverter output wires, and ground the shielding at the inverter end only.

A number of items of aircraft equipment can create special interference problems. Examples include inverters, motors, strobe lights, rotating beacon lights, etc. Sometimes trial and error troubleshooting is necessary to eliminate noise and interference problems. The use of shielded wires and physical separation are basic techniques for preventing or eliminating noise and interference between different aircraft systems and equipment.

STATIC DISCHARGERS

A common cause of noise in aircraft radios and related equipment is P-static interference. Precipitation or P-static noise is caused by static electricity that builds up on an aircraft in flight. P-static is commonly caused by friction between the metal skin of the aircraft and particles in the atmosphere. It can build up to 80,000 volts or more under certain conditions. Flying through rain, snow, ice, or even dust particles can result in a static charge on the airframe. The exhaust stream of a turbine engine can cause static electricity due to friction between particles in the exhaust and the metal tailpipe. P-static is a greater problem for high-speed aircraft because the higher speeds produce more friction. High-speed aircraft usually require many static dischargers to bleed off the static charge.

Static dischargers, also called static wicks, are small devices fastened to the trailing edges of the aircraft, usually on the primary control surfaces. They are designed to discharge the static electricity into the atmosphere. Dischargers or wicks are commonly installed on all types of aircraft that operate IFR and require all weather radio reception. [Figure 12-67]

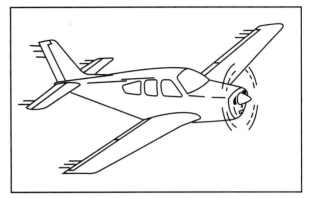

Figure 12-67. Static wicks are installed on the trailing edges of the flight control surfaces to help remove static charges in flight to prevent noise in the radios. High-speed aircraft may have additional static dischargers on the outboard tips of the wing and horizontal stabilizer. This figure shows a typical static wick installation. [Figure 12-68]

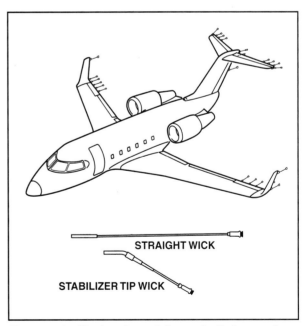

Figure 12-68. The locations of the static dischargers for a business jet are shown in this illustration.

There are three basic types of static dischargers used on aircraft: static wicks, wire braid dischargers, and null field dischargers. The static wick is also called a flexible static discharger and is found on low-speed aircraft. It consists of a plastic tube or outer covering with a fabric braid inside. The inner braid can be cotton, nylon, or some other material. The inner braid extends beyond the plastic covering where it is fanned out to produce the discharge points. The FAA recommends that one-inch of the inner braid extend beyond the outer cover. When they become worn, static dischargers can be trimmed to this dimension until they become too short, at which point they must be replaced. The inner braid of a static discharger is designed to have some built-in resistance to control the discharge current and further reduce noise. Wire braid static dischargers are also called semi-flexible dischargers. [Figure 12-69]

The wire braid discharger does not have built-in resistance, so it is not as effective as the other two types of static dischargers. Jet aircraft normally use the null field discharger, which is more rugged than the others for high-speed use. The null field discharger consists of a rigid shaft made of fiberglass or composite materials with sharp metal points at the aft end. The metal points are sometimes made of tungsten for longer life. [Figure 12-70]

Static dischargers should be maintained properly to ensure that they will perform their intended function. The attachment to the aircraft must be tight and provide a good electrical contact. Any corrosion or looseness at the attachment point can create noise in the radios. Replace damaged or badly worn static dischargers with new ones of the approved type. The noise that P-static produces affects the frequency bands of HF and below more so than the higher frequency bands. Therefore, if the pilot complains of noise on a radio system that operates at HF or below, inspect the static dischargers to determine if the noise is P-static related.

INSTALLATION METHODS

The installation of electronic and radio equipment follows some of the same basic practices that are

Figure 12-69. The semi-flexible wire braid static dischargers do not have any built-in resistance. They are simply a piece of wire braid made of stainless steel.

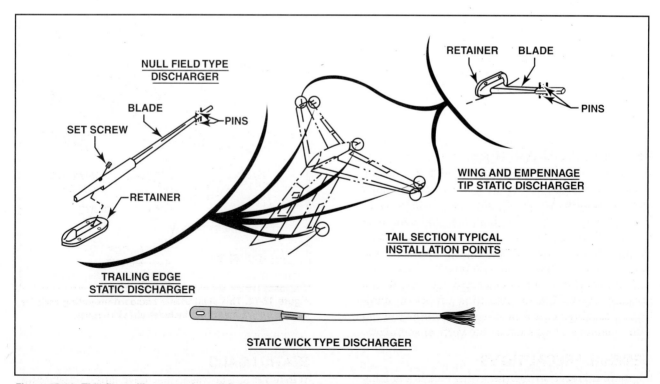

Figure 12-70. This figure illustrates the null field and static wick types of static dischargers. The metal points of the null field dischargers are at right angles to the direction of flight. This feature helps to further reduce noise, compared to the other kinds of static dischargers.

used for other equipment. Avionics systems are especially sensitive to certain types of problems, so special procedures may be required to prevent them. Always follow the specific instructions of the manufacturer when available. Some general recommendations from AC 43.13-1B and -2A will be described here, along with some precautions to observe for all types of installations.

When installing antennas, they must not only be matched to the proper radio, but the conductor that connects the radio and antenna is critical. A type of conductor used to connect radios and antennas is called a **coaxial cable** or coax. It consists of a center conductor covered by insulation and an outer conductor around the insulation. Plain wires cannot be used for radio frequency signals because the energy loss is too great at these frequencies. To ensure proper performance from the radio system, keep antennas and coaxial cables in good condition. [Figure 12-71]

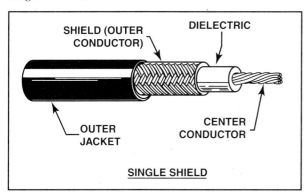

Figure 12-71. Coaxial cable is used to carry radio frequency electrical signals between radios and antennas. When inspecting coax, look for breaks in the shielding and any kinks in the cable.

When selecting an antenna for an aircraft installation, keep in mind that most antennas are speed-rated. For example, an antenna rated at 250 m.p.h. should not be installed on aircraft that fly at higher speeds because the aerodynamic loads may cause the antenna to break or might apply too much force to the aircraft structure. Along with the electronic considerations when installing an antenna, the installer must consider the problems of vibration and flutter. These factors can result in the antenna breaking away from its mounting surface in addition to causing control problems if mounted on an aerodynamic surface such as the vertical stabilizer

GENERAL PRECAUTIONS

If a standard location and mounting rack is available from the aircraft manufacturer, use it to install avionics equipment. If the manufacturer's mounting

rack is not available, the installer will have to determine the best location and means of mounting the equipment. Several factors to consider when making this type of determination are:

1. Sufficient air circulation to prevent overheating. This might require a certain free air space in some cases and the installation of a cooling fan in others.

2. Adequate clearance from high temperatures and flammable materials (next to a combustion heater would not be good place to install a radio).

3. Protection from water, fumes, hydraulic fluid, etc.

4. Protection from damage by baggage or seat deflection.

5. Sufficient clearance to prevent rubbing or striking upon aircraft structures, control cables, movable parts, etc.

6. Preventing interference and noise. Separate sensitive electronic equipment from inverters, power supplies, strobe lights, motors, etc.

7. If shock mounts will be used, ensure that the equipment does not exceed the weight carrying capability of the shock mounts and install adequate bonding jumpers or straps. [Figure 12-72]

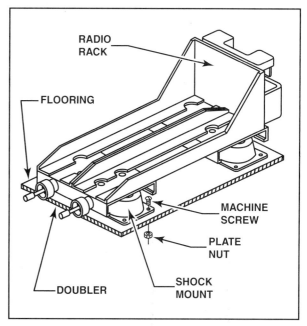

Figure 12-72. This illustrates a standard mounting rack for avionics equipment that includes shock mounts.

STATIC LOADS

Whenever it is necessary for the installer to fabricate a mounting bracket for aircraft equipment, verify the strength of the mounting with a load test. The equipment installed in aircraft must be able to withstand the acceleration forces or "G" loads that are experienced in flight. In a steep turn, for example,

CERTIFICATION CATEGORY OF AIRCRAFT			
DIRECTION OF FORCE APPLIED	NORMAL/ UTILITY	ACROBATIC	ROTOCRAFT
SIDEWARD	1.5 Gs	1.5 Gs	2.0 Gs
UPWARD	3.0 Gs	4.5 Gs	1.5 Gs
FORWARD*	9.0 Gs	9.0 Gs	4.0 Gs
DOWNWARD	6.6 Gs	9.0 Gs	4.0 Gs
*WHEN EQUIPMENT IS LOCATED EXTERNALLY TO ONE SIDE, OR FORWARD OF OCCUPANTS, A FORWARD LOAD FACTOR OF 2.0 G IS SUFFICIENT.			

Figure 12-73. This chart outlines static test load factors that could be used for testing equipment mountings and attachments. The load factors for the test can be obtained from AC 43.13-2A.

the additional "G" load is applied to the equipment in the aircraft as well as the wings and other structures. [Figure 12-73]

To illustrate a static load test, we will use the example of a 5 lb. radio installed in the baggage compartment behind the rear seats of a normal category airplane. The mounting bracket that is fabricated to hold the radio would be tested by applying loads equal to the weight of the equipment multiplied by the appropriate load factor. The sideward test load would be 7.5 lbs., upward load – 15 lbs., forward load – 45 lbs., and the downward load – 33 lbs. If a location was chosen in the nose section of the aircraft and forward of all occupants, a forward test load of 2.0 Gs or 10 lbs. would have been sufficient.

The mountings for aircraft equipment must be able to withstand the appropriate level of acceleration forces or load factors that might be experienced in flight. Standard industry practices for rivets, bolts, screws, etc. would be followed to ensure that the fasteners, mounting brackets and similar parts would provide adequate strength. AC 43.13-2A recommends the use of machine screws and anchor nuts for the removable fasteners to hold aircraft radios in place. In addition, use existing nutplates or install new ones when possible. If this is not practical, then use machine screws and self-locking nuts.

When radios or other equipment are installed in and extend behind the instrument panel, attach a brace or support to the side or back of the equipment to minimize the load on the instrument panel. [Figure 12-74]

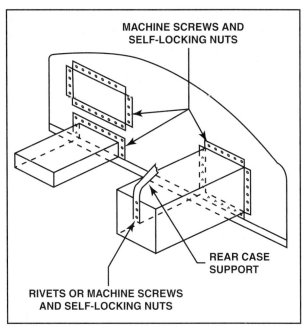

Figure 12-74. This is an example of a rear brace or support for radio equipment installed in an instrument panel.

ANTENNA INSTALLATIONS

The antennas in aircraft radio installations are critical to the proper operation of the radio system. Antennas must be carefully installed and maintained to provide the needed efficiency for good radio reception and transmission. There are many factors that can affect the efficiency of aircraft antennas. Inspection and maintenance of aircraft antennas is part of an aircraft technician's responsibility because, in most cases, the antennas are fastened to the skin or other structure. Some key concepts that affect antenna operation will be covered in this section.

COAXIAL CABLES AND CONNECTORS

Coaxial cables are required for antenna connections on most aircraft radios because of the RF frequencies that are used. The proper installation and maintenance of coaxial cables is important since large signal losses can occur if a fault is present. Coaxial cables should always be routed separately from other wires to avoid radio frequency interference. Reject coaxial cables if they have become dented or if kinks are found. Any distortion or crushing that causes the cable to be oval or flattened is also cause for rejection. If abrasion or rubbing has exposed or damaged the wire braid, replace the cable. Support the coaxial cable with clamps about every two feet to help prevent damage. A good rule of thumb for a coaxial cable bend radius is to use a minimum radius of 10 times the cable diameter. This will help reduce the possibility of kinks from sharp bends.

Special types of end connectors are used with coaxial cable, which come in a number of different styles. Some can be removed and reused while other types are crimped or swaged on and cannot be reused. When installing and removing coaxial cable connections, be careful not to damage to the connectors. If corrosion is found on the connectors, replace them rather than trying to clean them. Even small amounts of corrosion or corrosion pits can cause a signal loss.

When installing a reusable type of coaxial cable connector, carefully spread the wire braid out over the braid clamp and avoid breaking any of the wires. Assemble the connector carefully to provide tight connections with good electrical contact and to avoid distorting the coaxial cable or the connector itself. [Figure 12-75]

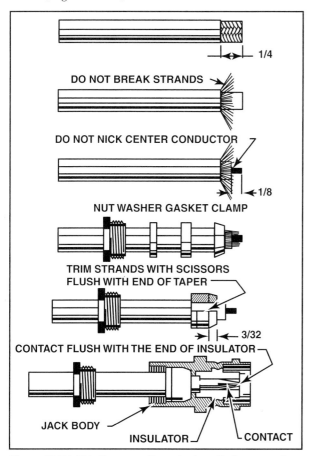

Figure 12-75. This figure outlines the installation procedure for a reusable coaxial cable connector.

If it is necessary to solder a connector pin onto the center conductor, only use an approved electrical solder. Never use acid core solder or acid flux on electrical connectors. An acceptable solder is 60/40 rosin core solder. Use great care when soldering to prevent excessive heat damage to the coaxial cable insulation materials.

Some antenna cables are matched to the radio and antenna and should not be shortened or spliced. This is true for some ADF antenna leads, for example. On other installations, keep the antenna coax as short as possible and route it as directly as possible to reduce signal loss.

Follow the specific antenna or radio manufacturer's installation instructions carefully since there are many different procedures that may apply.

GROUNDPLANE CONSIDERATIONS

When a 1/4-wave, Marconi-type antenna is installed on an aircraft, an adequate groundplane or counterpoise is required for proper operation. The aircraft systems that use 1/4-wave antennas are VHF communications, ATC transponders, DME and UHF radiotelephones. When these antennas are installed on metal skinned aircraft, the metal skin supplies the groundplane. If the antenna is installed too close to fiberglass areas or windshields, the groundplane area is reduced and may result in poor performance. A basic rule of thumb is that the groundplane should extend in all directions outward from the base of the antenna a distance equal to the height of the antenna. If the groundplane is too small, there is a possibility of adversely effecting the signal pattern and strength. For DME and transponders, which use similar frequencies, the groundplane should extend 8–12 inches in all directions from the antenna base. For VHF communications antennas, a groundplane that extends 24 inches in all directions is desirable. It is not always possible to supply a large enough groundplane when installing antennas on aircraft with limited metal skin area, such as small helicopters, but the groundplane area should always be considered.

If it is necessary to install these types of antennas on aircraft with non-metal skin, the installer must provide a groundplane. This usually entails installing metal foil strips or wire mesh fastened on the inside of the aircraft covering. [Figure 12-76]

When installing 1/4-wave antennas, it is recommended that all grease, dirt and paint be removed from the skin area under the base of the antenna. Some avionics experts recommend that a gasket not be used so that the base of the antenna contacts the skin of the aircraft. Whether or not a gasket is used, clean and strip the skin and apply a sealant around the base of the antenna after installation so that moisture can not penetrate the skin of the aircraft.

Installing antennas to the skin of aircraft requires additional reinforcement to preserve the strength of

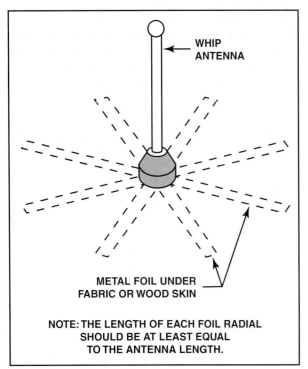

Figure 12-76. When installing Marconi antennas on an aircraft with non-metal skin, a groundplane must be provided.

the aircraft structure. The use of a doubler will reinforce the structure and provide additional support for antenna drag loads. [Figure 12-77]

REDUCING ANTENNA INTERFERENCE

An important factor in the proper performance of aircraft antennas is preventing interference between one system and another. Interference can also occur between a radio system antenna and other components of the aircraft. A basic consideration is that a certain minimum distance must separate antennas to prevent interference for systems that operate on similar frequencies. The possible interactions that can adversely affect aircraft radio systems are many and varied. The more common problems that can occur will be described here. However, a particular interference problem may require trial and error to eliminate the cause of the antenna interaction.

The important factors that affect mutual interference are frequency and wavelength, polarization and type of modulation. The operating frequencies for the various radio systems are listed in the frequency chart in section A of this chapter. The polarization of radio waves is based on the orientation of the electric field relative to the earth's surface. The antenna installed on the aircraft needs to have the proper polarization relative to the ground-based antenna for optimum performance, particularly at frequencies above HF. [Figure 12-78]

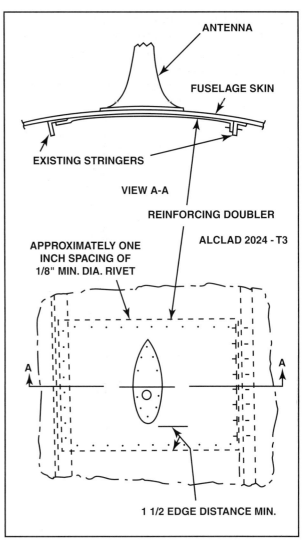

Figure 12-77. A reinforcing doubler should be installed inside the skin at the base of the antenna to provide additional strength.

RADIO SYSTEM	RECEIVE, TRANSMIT, OR BOTH	POLARIZATION
LORAN	RECEIVE	VERTICAL
ADF	RECEIVE	VERTICAL
VHF COM	BOTH	VERTICAL
DME & TRANSPONDER	BOTH	VERTICAL
ELT	TRANSMIT	VERTICAL
VOR & LOCALIZER	RECEIVE	HORIZONTAL
MARKER BEACONS	RECEIVE	HORIZONTAL
GLIDESLOPE	RECEIVE	HORIZONTAL

Figure 12-78. This chart illustrates the polarization for various types of aircraft radio systems. Notice that all systems use vertical polarization except for VOR and the three parts of the ILS instrument approach system.

VHF COMMUNICATIONS ANTENNAS

Aircraft that are equipped for IFR operations commonly have 2 or 3 separate VHF comm radios, which utilize separate antennas. The VHF comm antennas should be separated from each other by at least 5 feet. This is easily accomplished on an air carrier jet, which has plenty of fuselage skin area available, but may be difficult on small aircraft, which have much less. [Figure 12-79]

When two VHF comm antennas are installed on small aircraft, the best coverage is usually obtained with one antenna on the top and the other on the bottom of the fuselage.

The ELT antenna can cause serious interference with VHF communications and should be separated by at least 5 feet from any VHF comm antenna. Radio interference can be caused by parts of the aircraft as well as by other antennas. If a VHF comm antenna is installed too close to the vertical fin of an aircraft, significant signal blockage can occur. A top-mounted VHF comm antenna that is installed closer than 5 feet to the vertical fin will result in blockage and poor radio reception and transmission to the rear of the aircraft. A common mistake is installing a VHF comm too far forward on the upper fuselage. If it is less than 24 inches from the top of the windshield, the lack of a groundplane in the forward direction can distort the signal pattern.

DME AND TRANSPONDER ANTENNAS

These two antennas are treated as equals because they use similar frequencies, polarization and modulation. The antennas used for these two systems are 1/4-wave Marconi antennas with vertical polarization, which both transmit and receive. Since the wavelength is shorter at higher frequencies, the minimum separation is less than that for VHF comm antennas. The DME and transponder antennas should be separated from each other by at least 2 feet. These antennas are normally installed on the bottom of the aircraft to prevent signal blockage by the fuselage. A top-mounted antenna may be used on a narrow portion of the aircraft that will not cause significant blockage. The top of the tail boom on a helicopter can be an acceptable location.

VOR AND LOCALIZER

VOR antennas are most often installed on the vertical fin of the aircraft. This gives good reception

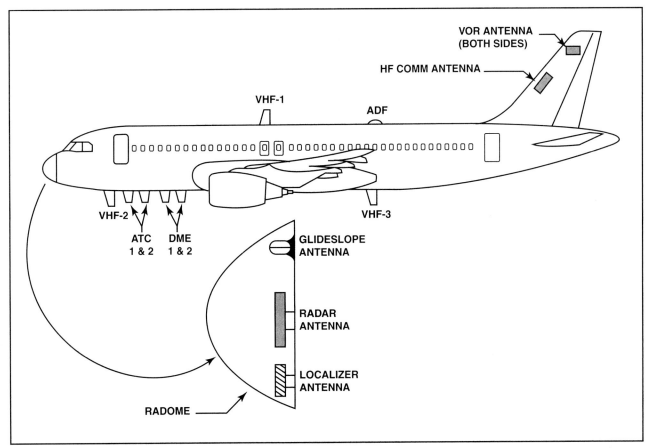

Figure 12-79. Antenna installations on modern air carrier jets often include localizer and glideslope antennas. These are often located inside the radome, with flush mounted VOR and HF comm antennas located in the vertical fin.

characteristics from all directions on most aircraft. On small aircraft, the VOR antenna is sometimes mounted on the top of the fuselage. If the VOR antenna is mounted too far forward, a propeller modulation problem can occur. When receiving signals from the front of the aircraft, the propeller may interfere with the radio wave at certain rpms. The cure for this involves changing propeller rpm or relocating the antenna. Small aircraft often use the same antenna for both VOR and localizer reception. This is practical because the two systems operate on similar frequencies. When the localizer is used for an instrument approach, the signals are always received from the front of the aircraft. On a large aircraft, it is not possible to use the tail-mounted VOR antenna for localizer reception because of fuselage blockage. These aircraft will use a separate localizer antenna or antennas for weather radar that are mounted in the nose section inside the radome.

GLIDESLOPE ANTENNAS

Like the localizer, signals from the ground transmitters for the glideslope are always received from the front of the aircraft. Some small aircraft use the VOR antenna to receive glideslope signals as well as localizer signals. The glideslope operates on frequencies that are the third harmonic of VOR frequencies. This means that the glideslope frequencies are three times the frequencies for VOR. A special antenna coupler is used so that the VOR antenna can supply two separate VOR and localizer receivers and also supply signals for the glideslope receiver.

The same fuselage blockage problems occur on large aircraft for both localizer and glideslope reception, so avoiding these problems is a primary consideration in locating localizer and glideslope antennas. The glideslope antenna or antennas for air carrier jets are often installed inside the radome on the nose of the aircraft. Aircraft that do not have a nose radome can utilize a separate glideslope antenna that is mounted somewhere on the forward fuselage. Interference from other antennas is not as great a problem with these systems as it is for some other radio systems.

ADF ANTENNAS

The primary consideration for locating ADF antennas is to obtain the proper relationship between the loop and sense antenna to ensure accurate indications of station direction. ADF antennas can be installed with both loop and sense antennas on the top of the fuselage, both on the bottom or one each on the top and bottom.

The most common installation on small aircraft is with a wire sense antenna on the top and the loop antenna on the bottom of the aircraft. In any case, the loop antenna must be located in the electrical center of the sense antenna for accurate readings. ADF uses a directional antenna system, and interference from parts of the aircraft can sometimes cause bearing errors. This is one reason that a check of quadrantal error should always be performed when ADF antennas are installed or relocated. Proper bonding jumper and static discharger installations are important to prevent P-static noise in ADF receivers. ADF antennas should be located far enough from aircraft generators and alternators to minimize interference. Filter capacitors can be used to reduce interference from alternators and similar devices.

The sense antennas used with the dual antenna installations are either the wire type or whip type. The whip-type sense antenna is a metal rod about 4 feet long and installed either on the top or bottom of the fuselage. This antenna is still found on some helicopters where there is not enough room for a long wire sense antenna. The long wire sense antenna is about 15 to 20 feet long and most often installed using the vertical fin as the aft anchor point to gain more fuselage clearance. The recommended minimum clearance from the fuselage is 12 inches. The sense wire can be installed on the bottom of the aircraft if adequate ground and fuselage clearance can be obtained. Like the long wire HF antenna, the ADF sense wire uses masts, tension units, and weak links as part of the installation.

Because they are highly directional, ADF antennas must be installed and calibrated correctly to give accurate navigational information. A calibration check or a check for quadrantal error verifies the accuracy of the installation. Whenever an antenna is installed or any change is made that could affect the accuracy of the ADF, a check for quadrantal error should be performed. The checks can be made on the ground but should always be confirmed with a flight check. To perform the ground check, tune in a nearby NDB of known location, then check and adjust the bearing at least every 45° as the aircraft is turned on the ground. The flight check involves locating geographical points on the ground with known bearings from the NDB, and flying the aircraft over those locations to confirm the accuracy of the ADF bearing information. Perform this flight check at low altitude to reduce errors in accurately establishing the aircraft's position.

CHAPTER 13

AIRFRAME ICE AND RAIN CONTROL

INTRODUCTION
Ice is a flight hazard because it destroys the smooth air flow across the aircraft's lifting and control surfaces. Ice on wings and other surfaces decreases their ability to produce lift, increases the weight of the aircraft, and increases induced drag. Removal or prevention of ice is necessary for an aircraft to be safe to fly. Rain is a hazard because it affects the pilot's ability to see clearly. This chapter deals with operating and maintaining aircraft ice prevention and removal systems, as well as procedures and equipment for ground ice and snow removal. Also reviewed are rain control systems and methods of protecting windscreens from the effects of rain.

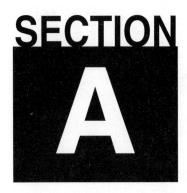

AIRFRAME ICE CONTROL SYSTEMS

Only aircraft that meet the requirements of 14 CFR Part 23.1419 can continue flight into known icing conditions. Aircraft certified for flying into known icing conditions have been proven capable of operating safely in continuous maximum and intermittent maximum icing conditions as described in FAR 25. Many light aircraft have only minimal equipment to deal with icing conditions inadvertently encountered during flight and are not certified to fly into forecasted icing conditions (known icing conditions).

Frost, wing leading-edge ice, horizontal stabilizer ice, carburetor ice, propeller ice, engine intake and windshield ice are all examples of aircraft icing. Each type of icing presents aircraft operators, pilots, and maintenance personnel with unique operational problems prior to or during flight. With the appropriate equipment and adequate procedures, most of these can be effectively managed, allowing an aircraft to take off, or continue flight if airborne.

Prevention of ice is called "anti-icing" and the removal of ice is called "de-icing." A variety of equipment is specifically designed to prevent the formation of ice or remove it when encountered. Different aircraft typically use different combinations of equipment to meet the certification standards of FAR 23. One example of a combined anti-icing and de-icing system is installed on the Learjet 35/36 series. This series uses three different methods of ice protection. The primary anti-ice system uses turbine engine bleed air to prevent ice formation on the wing leading edges, horizontal stabilizer edges, engine nacelle inlets, engines, and the windshield. The second method of protection is in the form of electric heating elements for static ports, stall warning vanes, and pitot tube probes. Finally, methyl alcohol is used to protect the radome and as a backup for the windshield. Aviation maintenance technicians can better deal with the various de-icing and anti-icing systems by understanding both the effects of the various forms of ice and the types of ice detection devices.

ICING EFFECTS

Frost forms on the surface of non-hangared aircraft when the temperature of the air drops at night and moisture is present. If the air is warm, dew will form, but if the temperature is below freezing, water will freeze and form tiny crystals of frost. Frost does not add appreciable weight, but must be removed before flight. Surprisingly, frost is a very effective aerodynamic spoiler that increases the thickness of the boundary layer and adds a significant amount of drag.

As an aircraft flies into clouds with the outside air temperature near freezing, it will likely collect ice. Any exposed surface such as the wings, tail, windshield, propeller, engine intakes and radio antennas can accumulate ice. Ice adds a great deal of weight and changes the aerodynamic shape of the surfaces, destroying much of the lift. For turbine engine intakes, ice will disturb the flow of air into the engine or break off and be ingested into the engine's compressor. Build-up of ice causes propellers to become inefficient and out of balance.

Another dangerous type of icing is one that can occur in the carburetors of reciprocating engines. Float-type carburetors break down liquid fuel into tiny droplets and mix them with air. When fuel changes from a liquid into a vapor, heat is absorbed from the air, causing the surrounding air temperature to drop. Moisture in the form of an invisible water vapor will condense into liquid water. When the temperature is low enough, it will freeze and stick to the walls of the carburetor throat. The flow of air is restricted and the engine will run rough or completely stop. It is not necessary to have visible moisture to experience carburetor ice.

ICE DETECTION SYSTEMS

Ice protection systems should be operated only when necessary so that operational expenses, and unnecessary wear can be minimized. By having some form of ice detection, the prevention or removal system operates only when needed. There are visual detection methods as well as several optical and electronic methods of ice detection.

VISUAL DETECTION

When conditions are favorable for ice formation, flight crews routinely observe the aircraft structure during flight for ice buildup. For example, it is easy to look at the wing leading edge or check the windshield for early signs of ice accumulation. When flying at night, lights are necessary to illuminate aircraft structures. FAR 23 requires some form of lighting that is adequate for the crew to detect ice at night as well as monitor the operation of the de-icing equipment. Some other form of detector is required on those portions of the aircraft where it is not possible for the crew to see. Most often, some form of electronic detection device is used.

ELECTRONIC DETECTION

Ice detectors consist of a microprocessor circuit with an aerodynamic strut and probe extending into the slipstream. The probe vibrates at a pre-determined frequency and when ice attaches to the probe, its frequency decreases. A microprocessor turns on an annunciator light when the probe reaches a preset minimum frequency value.

After detecting ice, a heating element within the probe melts it away and continues to recheck for icing conditions. As long as the probe continues to sense icing at each check, the ICE annunciator remains on. The light will go out when ice is no longer detected. [Figure 13-1]

OPTICAL ICE DETECTORS

Optical detection devices measure ice thickness on airplane wings and transmit the measurements to the pilot. One design produces a measurement range of 15 mm of ice thickness with a resolution of 15 µm. These units mount flush with the airfoil without any protrusions, making it suitable for supersonic flight applications.

CONTAMINANT/FLUID INTEGRITY MEASURING SYSTEM (C/FIMS™)

C/FIMS™ is an aircraft-mounted sensor system developed by Allied Signal Aerospace of Canada for detecting the buildup of ice, frost, and snow on the flight surfaces of airplanes. Strategically located sensors on critical aircraft surfaces interact with processors and a central control unit to provide warning information via a flight deck display. Flush mounted sensors are installed on the aircraft skin and detect the electromagnetic properties of the contaminants. The system can also be used to measure the effectiveness of de/anti-icing fluids and the aircraft's skin temperature.

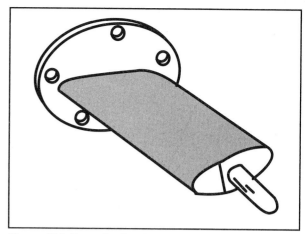

Figure 13-1. Example of an ice detection probe.

ANTI-ICING SYSTEMS

Onboard anti-icing systems are designed to prevent ice formation on certain portions of the aircraft. These are areas most affected by ice build-up, or where damage would occur if accumulated ice were to break free. Typically, anti-ice systems serve the components are around the engine intakes, on locations upstream from the engines, or on components that would not operate properly if allowed to ice over. [Figure 13-2]

Propellers are frequently protected by anti-ice systems because ice can cause an unbalanced condition that makes continued flight difficult or impossible. Anti-icing systems use several different methods to remove the ice, typically by heating the surface or component with hot air, engine oil, or electric heating elements. Another type of system uses chemicals with low freezing points injected at the root of the blade, which flow outward by centrifugal force. This type of system often protects windscreens also.

THERMAL ANTI-ICING

Heated air can be directed through specially designed heater ducts in the leading edges of the wings and tail surfaces to prevent ice formation. Combustion heaters or heater shrouds around the engine exhaust system heat the air in reciprocating engine aircraft. Hot air is also routed to the carburetor when conditions are conducive to carburetor icing. Combustion heaters are controlled by thermocycling switches that turn off the flow of fuel when a certain temperature is reached and turn it back on when the heater has cooled. [Figure 13-3]

Most aircraft that use thermal anti-icing systems today are turbine powered, in which case, a portion

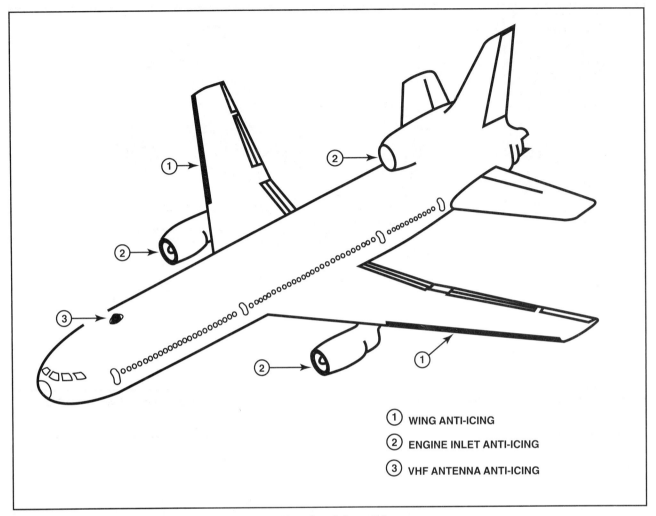

Figure 13-2. Bleed air heats some surfaces on large turbine aircraft for anti-ice purposes.

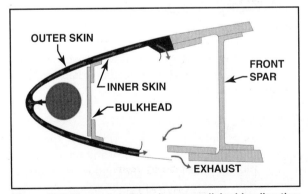

Figure 13-3. Thermal de-icing is accomplished by directing hot air through a duct in the leading edge of the wing.

of the engine's heated compressor bleed air is routed to the leading edges to prevent ice formation. Mixing cold air with the hot bleed air sometimes controls anti-ice air temperature. The Boeing 727 bleeds air from the two outboard engines, directs it through the wing anti-icing control valves to a common manifold and then out into the wing leading edge ducts. As illustrated in figure 13-4, two inboard leading edge flaps and eight leading edge slats are protected with hot air. Overheat sensor switches protect portions of the wing from overheating, which is usually caused by a break in the bleed air duct. If a surface overheats, an overheat warning light illuminates and the anti-icing valves close, shutting off the flow of hot air. When the duct temperature drops to an allowable range, the overheat light will go out and hot air will flow into the duct again.

Turbine engines are susceptible to damage from chunks of ice breaking off from the engine inlet and drawn into the engine's compressor. Compressor bleed air heats most turbine engine intakes by circulating around the intake leading edge to prevent ice. Boeing 727 engines also have hot compressor bleed air directed through the inlet guide vanes, the engine bullet nose, the oil cooler scoop for the constant speed drive, and the inlet duct for the center engine. [Figure 13-5]

Airframe Ice and Rain Control

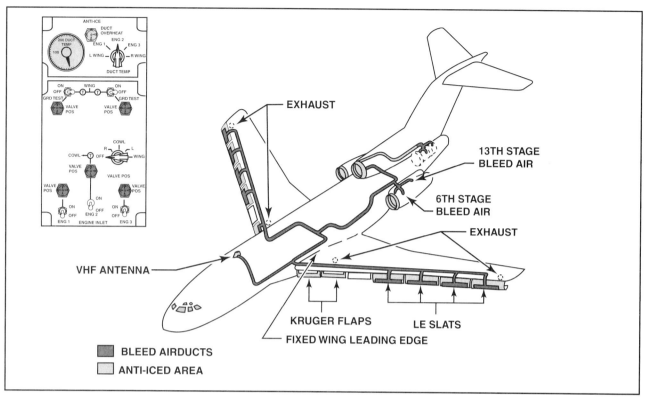

Figure 13-4. Bleed air provides anti-ice protection to wings, control surfaces, antennas, and engine inlets of the Boeing 727.

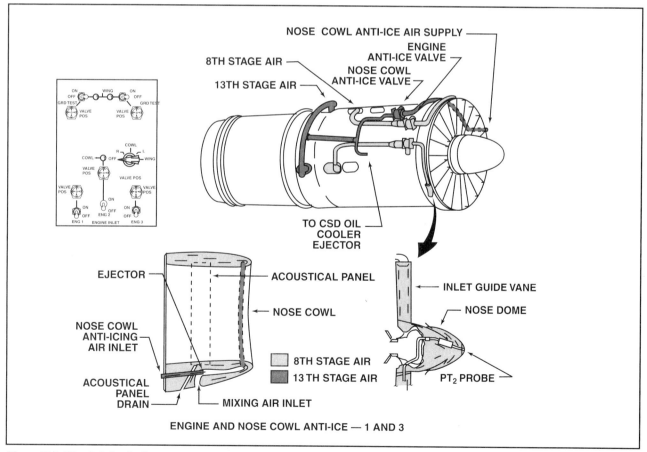

Figure 13-5. Bleed air heats the structures associated with the engine intakes.

A Boeing 727 has the center engine's air intake at the rear top of the fuselage. As a result, some hot air is ducted to the upper VHF radio antenna to prevent ice chunks from breaking off and being ingested into the center engine.

ELECTRIC ANTI-ICING

Aircraft that may possibly encounter icing usually have electrically heated pitot tubes. Pitot heaters are so powerful that they should not be operated on the ground because they may burn out without an adequate flow of air over them. Their in-flight operation is monitored by indicator lights or through the ammeter. Current flow is high enough that the ammeter will deflect noticeably when the heater is on. A heated pitot tube prevents ice from plugging the entry portal. [Figure 13-6]

Figure 13-6. The pitot tube and sometimes the supporting mast are heated electrically to prevent the formation of ice.

Static ports and stall warning vanes on many aircraft are also electrically heated. The static ports on some of the smaller aircraft are not heated. If there is no provision for melting the ice around static ports, the aircraft should be equipped with an alternate static source valve. An alternate valve allows the pilot to switch the flight instruments to a static source inside a non-pressurized aircraft. Large transport aircraft with flush toilets and lavatories often have electric-powered heating elements to prevent the drains and water lines from freezing. Engine intakes of some turboprop aircraft are anti-iced by using electric heating elements.

Windscreens and flight deck windows of many aircraft use electrically heated systems to prevent ice from obscuring the vision of the flightcrew. There are two methods of electrically heating laminated windscreens. One method uses tiny resistance wires embedded inside the windscreen, and the other uses a conductive coating on the inside of the outer layer of glass. Electric current flowing through the wires or the conductive film heats the windscreen. Problems encountered with electrically heated windscreens include delamination, discoloration, scratches, and arcing. If arcing is encountered in an electrically heated windshield, it usually indicates a breakdown in the conductive coating and can lead to localized overheating and possible damage to the windscreen. [Figure 13-7]

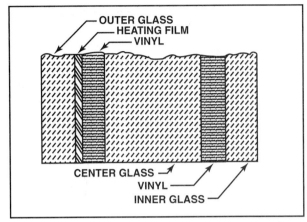

Figure 13-7. Windscreens on most modern jet aircraft are made up of laminations of glass and vinyl. A conductive film is used to heat the windscreen for ice protection.

Windshields on high-speed jet aircraft are highly complex and costly. For all transport category aircraft, windshields must absorb the stresses caused by pressurization, normal abuse, and flight loads. In addition, they must also withstand, without penetration, the impact of a four-pound bird striking the windshield at a velocity equal to the airplane's designed cruising speed. A windshield's strength is derived from its complex layered construction. Windshields can be up to an inch and a half thick, and constructed of three plies of tempered glass with vinyl layers between them. The outer ply of glass has an inner surface with an electrically conductive material that produces enough heat to melt ice. Thermistor-type temperature sensors and an elaborate electronic control system prevent these windshields from becoming overheated. The remainder of the system on these types of heated windshields consist of auto-transformers, heat control relays, heat control toggle switches, and lights to indicate when the system is operating.

In addition to providing anti-ice capabilities, heating the windshield also strengthens them against bird strikes. When the windshield is heated, vinyl layers are less brittle and withstand impacts with much less chance of penetration than when they are cold. Although windscreen defrosters provide some measure of anti-ice capability on light aircraft, they are not acceptable on aircraft approved for flight into known icing.

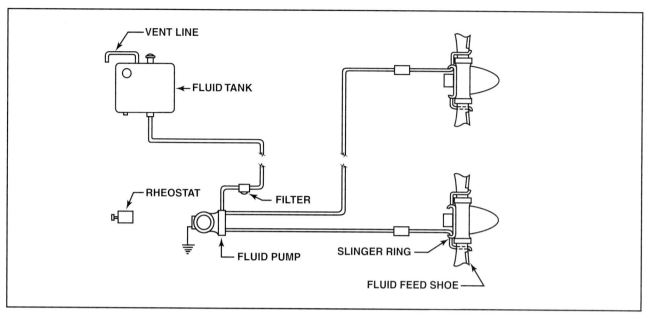

Figure 13-8. Anti-icing fluid is transferred from the feed lines to the propeller through a slinger ring.

CHEMICAL ANTI-ICING

Some aircraft surfaces and components may be coated with either isopropyl alcohol, methyl alcohol, or a mixture of ethylene glycol and alcohol. Chemicals lower the freezing point of water on the surface of the aircraft, and at the same time make the surface too slick for ice to collect upon it. Chemical anti-icing is normally used on carburetors, propellers, and windshields and is stored in a tank on the aircraft.

Propeller anti-icing uses isopropyl alcohol sprayed onto the leading edges of its blades. Alcohol is stored in a tank and pumped to the propeller when needed. A rheostat-controlled electric motor drives the pump. A pilot can control the amount of alcohol flowing to the propeller by controlling pump speed through the rheostat. Each propeller has a slinger ring that uses centrifugal force to distribute alcohol to the blade nozzles. The amount of alcohol the tank can carry limits the system operation. [Figure 13-8]

Windscreen anti-ice on some aircraft utilizes chemicals to prevent the formation of ice. Chemicals are delivered through a fluid spray bar located just ahead of the pilot's windscreen. A pump provides just enough flow to coat the windshield and prevent ice formation. Radome anti-icing is often done with chemicals as well, helping minimize inaccurate readings due to ice accumulation.

WEEPING WING

A weeping wing is an anti-icing system that pumps fluid from a reservoir through a mesh screen embedded in the leading edges of the wings and tail. In this system, fluid coats the wing to prevent ice from accumulating. A switch in the cockpit activates the system and liquid flows all over the wing and tail surfaces, de-icing as it flows. This type of system is also used on propellers and windscreens. [Figure 13-9]

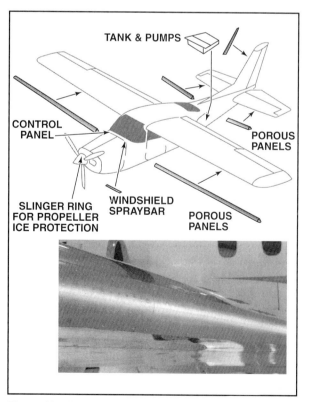

Figure 13-9. The weeping wing system coats the protected surfaces with glycol. The system can both de-ice and provide anti-ice protection.

Flight Ice, Inc. produces a weeping wing system under the trade name TKS. The TKS ice protection system coats the surfaces with a protective film of glycol. The system is designed for prevention in known icing conditions, but it is also capable of de-icing an aircraft. When ice accumulates on the leading edges, the antifreeze solution chemically breaks down the bond between the ice and airframe, allowing the aerodynamic forces on the ice to carry it away. This capability allows the system to clear the airframe of accumulated ice before transitioning to its prevention.

A valuable side effect of TKS ice protection is the reduction of runback icing on the wings and tail. Runback icing is a problem with many leading edge de-ice systems as the water from the melted ice refreezes as it flows rearward. Large water droplets that strike the wings behind the heated section also sometimes freeze to the surface. Once fluid departs the porous panel on the leading edge, it flows aft over the upper and lower surfaces and departs the aircraft at the trailing edge. This effect precludes ice accumulation aft of the panels due to runback or from impact of larger water droplets. There is some increased weight due to the amount of fluid required. As with most chemical systems, if the fluid reservoir runs dry during operation, there is usually no other method to prevent ice from accumulating.

DE-ICING SYSTEMS

De-icing systems remove the ice after it has formed, typically by using pneumatic de-icer boots on the leading edges of the wings and tail. Propeller de-icing normally uses heating elements to melt the accumulated ice and the propeller's centrifugal force to remove it.

Anti-icing systems prevent the formation of ice on the protected component. However, it may be more effective on surfaces of slower airplanes to allow the ice to form and then crack it. The normal airflow over the surface will naturally carry the ice away. This method also avoids runback that can form a ridge on the back edge of the surface, effectively becoming an aerodynamic spoiler.

RUBBER DE-ICER BOOT SYSTEM

Airline flying was hindered in the early days of aviation because of aircraft ice accumulation. With improved instruments and radios, and the introduction of higher performance aircraft, flight into icing conditions could no longer be avoided. To remove the ice, B.F. Goodrich developed a rubber de-icer boot that was installed on the leading edges of the wings and empennage. This allowed aircraft to fully utilize their improved capabilities. De-icer boots are used on many aircraft to this day. [Figure 13-10]

PRINCIPLE OF OPERATION

A rubber boot contains several tubes fastened to the leading edge of the surface to be protected. Low-pressure air passes through a timer-operated distributor valve into the tubes. On a sequentially operated system, the center tube inflates first and any ice formed over it will crack. The center tube deflates and the outer tubes inflate, pushing up the cracked ice. Air flowing over the wing gets under it and blows it off the surface. All tubes then deflate and suction holds them tight against the boot until the ice reforms, and the cycle is repeated. Alternating inflation of the boots keeps the overall disturbance of the airflow to a minimum. [Figure 13-11]

The operating cycle inflates the tubes in a symmetrical manner so the disruption of lift during inflation will be uniform, minimizing any flight control problems. The manufacturer of the aircraft determines the proper cycle time for de-icing boot operation from flight tests.

Larger aircraft using this type of de-icing system have an electric motor-driven timer to operating solenoid valves to continually cycle the system through all of the tubes. The timer then provides the proper duration of rest time, allowing ice to form over the boots before repeating the cycle. Any time the tubes are not inflated, suction is applied to hold them tight against the aircraft surface.

Smaller aircraft do not use an elaborate timer. The pilot turns on the system after observing an accumulation of ice on the leading edges. When the de-icing switch is turned on, the boots will cycle through one, two, or three operating cycles, depending upon the system's design. Similar to more elaborate systems, tubes connected to the vacuum side of the air pump hold the boots tight against the leading edge. [Figure 13-12]

SOURCES OF OPERATING AIR

Low-pressure air is required to operate the boots in the boot de-ice system. Depending on the type of aircraft, several different sources can be utilized for inflation.

1. On most turbine aircraft, pneumatic air is taken from turbine engine bleed air. Air from the turbine engine is at a much higher pressure than required

Airframe Ice and Rain Control

Figure 13-10. This medium twin is equipped with de-icer boots and electrically de-iced propellers. Being able to remove ice accumulations enables it to fully utilize modern navigation and communication equipment. Bottom photos show de-icing boots before and during inflation.

for inflating the boots. A step-down regulator is used to reduce the pressure suitable for de-icing boot operation. Bleed air can also be used through a venturi to create a vacuum. Negative pressure, or vacuum, is used to hold the boots down smoothly to the leading edge during the deflation cycle.

2. Air for inflating boots can also come from the exhaust of an engine-driven air pump (instrument system vacuum pump). Some pumps are the "wet type," which use engine oil taken into the pump through holes in the mounting flange to lubricate and seal the steel vanes. Since oil

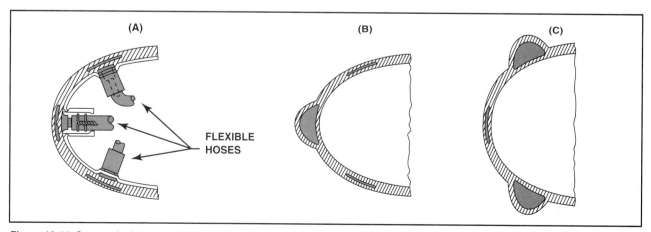

Figure 13-11. Sequential inflation of the boot's tubes breaks up the ice so the airflow can sweep it away.

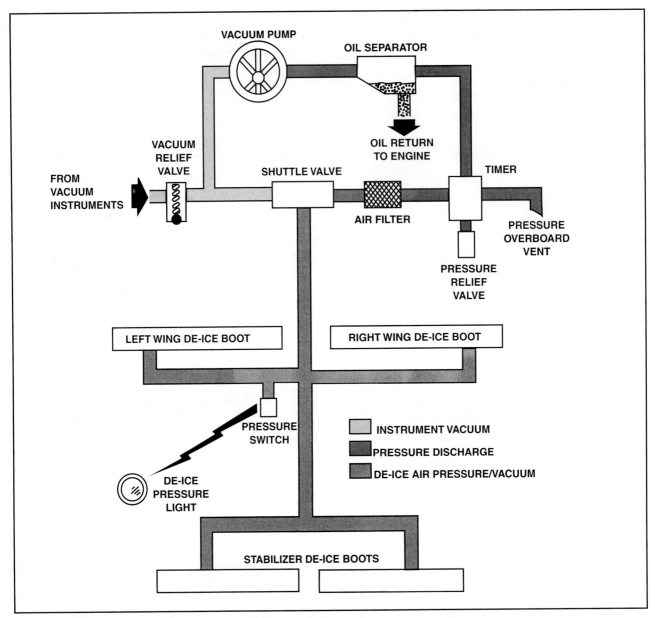

Figure 13-12. The de-icing system for small aircraft does not include a timer.

would damage the boots, an oil separator removes all of it and sends it back into the engine crankcase before the air is used.

3. Newer "dry-type" pumps are used for many installations, and do not require an oil separator since carbon vanes make the pump self-lubricating.

4. Other less common de-icing systems inflate the boots from a cylinder of compressed air that is carried just for this purpose.

DE-ICING SYSTEM COMPONENTS (TYPICAL)

Some of the main components in a pneumatic de-icer system are the air pump (vacuum pump), vacuum regulator, pressure control valve, timer module, and de-icer boots. A vacuum pump is normally used to create a vacuum for operating the flight instruments. The output side of a pump provides air pressure that is used to inflate the de-icer boots. The vacuum regulator controls the amount of vacuum applied to the de-ice boots and the instruments. Likewise, the pressure control valve controls the amount of pressure allowed in the system. Under normal operations, a pressure gauge fluctuates as the boots are alternately inflated while the vacuum gauge remains relatively steady. A switch in the cockpit activates a timer module, which sequences the de-icer boots through one complete de-ice cycle. Normally, the system shuts off after one complete cycle, and applies a vacuum to the

Airframe Ice and Rain Control

cells or tubes until the pilot calls for another cycle. Other components include filters, valves, and miscellaneous tubing and lines.

CONSTRUCTION AND INSTALLATION OF DE-ICER BOOTS

There are several configurations of de-icer boots, but all accomplish their work in the same way. Ice forms then breaks off as the tubes inflate. Some boots use span-wise tubes that inflate alternately, and others inflate simultaneously. Certain configurations of boots have chord-wise tubes that may inflate either alternately or simultaneously. Flight-testing determines the tube configuration and only the specific boot that is approved for the aircraft should be used. [Figure 13-13]

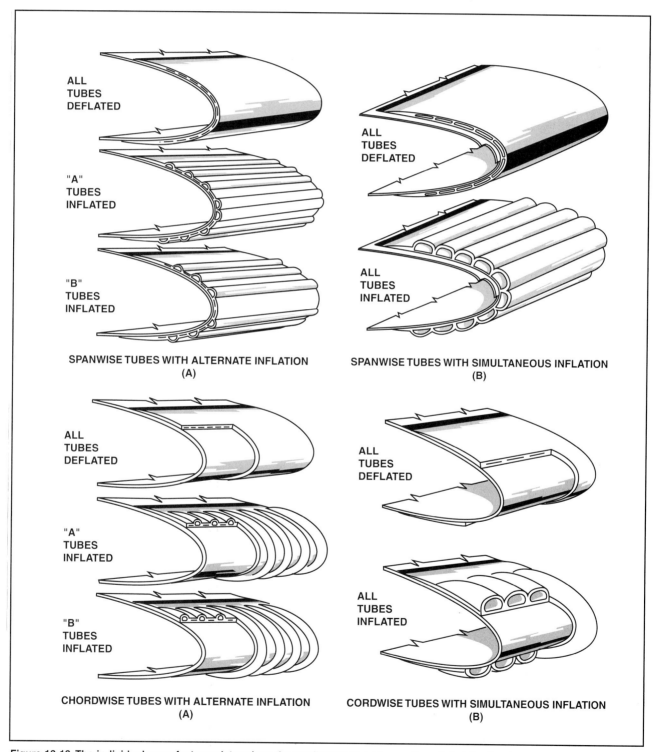

Figure 13-13. The individual manufacturer determines the configuration and operation of de-icing boots.

When rubber de-icer boots were first developed, adhesives had not been developed to the extent that they are today. Boots were installed with machine screws driven into Rivnuts installed in the skin. A narrow metal fairing strip that covers the screw heads at the boot edges can identify this type of installation. Almost all of the newer boot installations fasten the boot to the surface with adhesives, eliminating the need for Rivnuts and screws. When removing or installing a de-icer boot, follow the instructions in the aircraft service manual or the manufacturer's approved information. Do not substitute methods or materials.

To remove de-icer boots, soften the adhesive with the recommended solvent and carefully apply tension to peel back the edges of the boot. Keep the separation area wet with solvent and carefully pull the boot away from the surface.

To begin installation, remove all of the paint and primer from the area where the boot is to be installed. Clean both the surface and the back of the boot thoroughly. Apply adhesive to the back of the boot and leading edge. Secure the hoses to the boot and place the boot in the proper position. Press tightly to the surface with a roller. Obviously, the actual process is considerably more complicated than this, but in all maintenance activities, follow the manufacturer's recommendations carefully.

The only way to know for sure that the boot is properly bonded to the surface is to make a test strip. Apply a trimmed scrap of boot material to the leading edge near the boot installation. Clean the scrap in the same manner as the rest of the boot and apply cement. Attach it to the leading edge by the identical procedures used with the boot. After the recommended cure time, measure the amount of force needed to peel the test strip away. The required force must be within the tolerance specified by the manufacturer.

INSPECTION, MAINTENANCE, AND TROUBLESHOOTING OF RUBBER DE-ICER BOOT SYSTEMS

The most important part of de-icer boot maintenance is keeping the boots clean. Wash the boots with a mild soap and water solution. Remove any cleaning compounds used on the aircraft from the boots using clean water. Remove oil or grease by scrubbing the surface of the boot lightly with a rag that is damp with benzoil or lead-free gasoline. Wipe dry before the solvent has a chance to soak into the rubber. Boots are often sprayed with silicon to give the rubber an extremely smooth surface that the ice cannot adhere to. During inspection, check the surface of the boots for condition and security. Also, inspect the condition of plumbing fittings and lines. Conclude the inspection with a thorough operational check of the system.

A very important part of deicer boot maintenance is to keep the boots clean. Wash the boots with a mild soap and water solution. Remove any cleaning compounds used on the aircraft from the boots using clean water. Remove oil or grease by lightly wiping the surface with a rag that is damp with benzoil or other manufacturers' approved cleaner. Wipe dry before the cleaner has a chance to soak into the rubber. Boots are often sprayed with silicon to give the rubber an extremely smooth surface that the ice cannot adhere to. During inspection, check the surface of the boots for condition and security. Also, inspect the condition of plumbing fittings and lines. Conclude the inspection with a thorough operational check of the system.

A deicer boot repair referred to as a "cold patch repair" includes refurbishing scuff damage, repairing damage to the tube area, and tears in the fillet area. Scuff damage is the most common type of damage that A&P technicians encounter when maintaining deicer boots. For any type of deicer boot damage, refer to the manufacturer's maintenance manual for guidance in making appropriate repairs and follow the approved repair procedures explicitly.

ELECTROTHERMAL DE-ICING

Many modern propellers installed on both reciprocating and turboprop engines are de-iced with an electrothermal de-icer system. Rubber boots with heater wires embedded in them are bonded to the leading edges of the propeller blades. Electrical current passes through the wires to heat the rubber and melt any ice that has formed, while centrifugal force and wind carry the ice away.

In some installations, boots are made in two sections on each blade. Current flows for about a half minute through the outboard section of all blades and then for the same time through the heaters on the inboard section of all blades. Flight tests have determined the amount of time that current flows in each section. The time is sufficient to allow ice to form over the inactive section while loosening it from the heated section. Alternating the de-icing current between the blade sections is the most efficient method for propeller ice removal.

The propeller de-icer system consists of the following components:

1. Electrically heated de-icers bonded to the propeller blades.

2. Slip-ring and brush block assemblies that carry the current to the rotating propeller.

3. A timer to control the heating time and sequence of the de-icing cycle.

4. An ammeter to indicate the operation of the system.

5. Wiring, switches, and circuit breakers necessary to conduct electrical power from the aircraft electrical system into the de-icer system. [Figure 13-14]

The slip-ring assembly is attached to the propeller either through a specially adapted starter gear, the spinner bulkhead, or to the crankshaft flange. The brush block is mounted on the engine so that the three brushes ride squarely on the slip rings. The timer controls the sequence of current to each of the de-icers. The heating sequence optimally loosens the ice so that centrifugal force can carry it away. The same portion of each blade is heated at the same time to prevent an out-of-balance condition.

The ammeter is used to monitor the system's operation. When each heater element is taking an equal amount of current, the pilot knows that the propellers are de-icing evenly. The advantage of the electric heat system is its minimal weight, but in some cases, it can place a high electrical load on the generator or alternator.

ELECTRO-EXPULSIVE SEPARATION SYSTEM

The Electro-Expulsive Separation System was developed by NASA and licensed to Ice

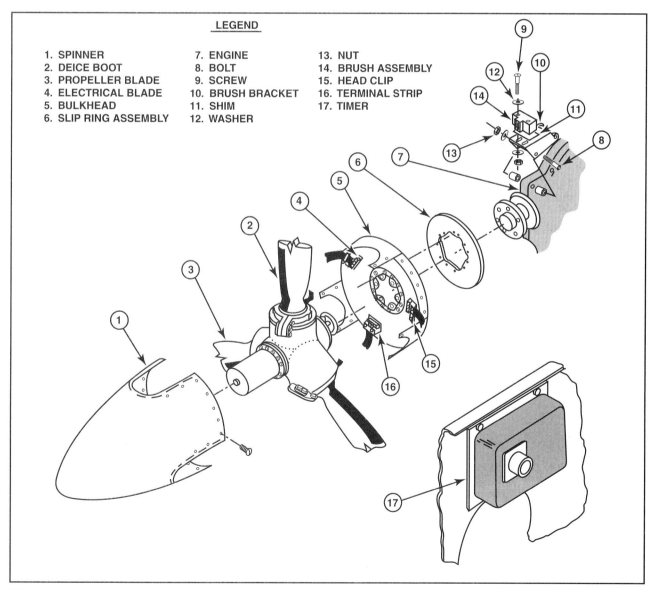

Figure 13-14. An electric de-icing system is used on many propellers.

Management Systems, Inc. This relatively lightweight system provides ice protection at power levels far below those required by bleed air and electro-thermal systems. It also offers a reliable alternative to pneumatic or electrical de-icing boots on wing leading edges, thus reducing airfoil drag and surface erosion.

In a typical installation, the electro-expulsive de-icing system employs a de-icing control unit linked to two energy storage banks, one for each wing. Actuators are mounted directly beneath a metal leading edge. Energy storage banks deliver high-current electrical pulses to the actuators in timed sequences. This generates opposing electro-magnetic fields, which cause the actuators to change shape rapidly. An erosion shield on the wing's leading edge flexes and vibrates at a high frequency. Rapid motion results in acceleration-based debonding of accumulated ice on the erosion shield. The system essentially pulverizes ice into small particles and removes layers as thin as frost or as thick as an inch of glaze ice. Though the system is not yet used on many aircraft, it promises to be useful as the particles of ice are small enough not to damage the airframe or engines.

GROUND DE-ICING OF AIRCRAFT

Aircraft ground de-icing/anti-icing plays a vital role in cold weather procedures to insure that an aircraft is free of ice, frost, and snow contamination before takeoff. Procedures range from a low-tech broom to sophisticated chemicals. Airlines most commonly use a truck-mounted mobile de-icer/anti-icer. These units generally consist of one or more fluid tanks, a heater to bring the fluid to the desired application temperature, an aerial device (boom and basket) to reach remote areas, and a fluid-dispensing system (including pumps, piping, and a spray nozzle). The dispensing system is capable of supplying fluid at various pressures and flow rates with an adjustable spray pattern at the nozzle.

ICE AND SNOW REMOVAL

There are several different types of freezing point depressants (FPDs), but some common aircraft FPDs are Type I (unthickened) fluids and Type II and IV (thickened) fluids. Type I fluids have a relatively low viscosity except at very cold temperatures. Viscosity is a measure of a fluid's ability to flow freely. For example, water has low viscosity, and honey has relatively high viscosity. Because the rate at which fluid flows off the wing depends on the fluid viscosity, a high viscosity fluid is likely to have larger aerodynamic effects than a low viscosity fluid. The viscosity of Type I fluids depends only on temperature. For the fluids to be effective, handle them properly at every step in the de-icing/anti-icing process, including transportation and storage. Follow FPD fluid manufacturer's guidelines for the entire process.

Holdover time, which is the length of time that the surface will be protected from ice, frost, and snow, is relatively short for Type I fluids. Because taxi times are often much longer than the holdover times that Type I fluids provide, fluid manufacturers developed Type II or IV (thickened) de-icing/anti-icing fluids in cooperation with the airlines. Type II and IV fluids have significantly longer holdover times than Type I fluids. Systems that mix various FPDs with compressed air are now being used in the aviation industry as well. These systems use fewer chemicals, thus reducing recovery systems costs in preventing chemicals from reaching water supplies.

The International Standards Organization (ISO), Society of Automotive Engineers (SAE), and Association of European Airlines (AEA) publish holdover times for each fluid. They are based on temperature, fluid mixture, and general weather conditions. Holdover times are only guidelines and other variables that reduce fluid effectiveness must be taken into account. High wings or jet blast may cause the fluid to flow off. Wet snow has a higher moisture content than dry snow, so it dilutes the fluid faster. The heavier the precipitation, the shorter the holdover time. Airplane skin temperatures that are significantly lower than the outside air temperature can also decrease holdover time. De-ice and anti-ice airplanes as close to the departure time as possible, especially when bad weather conditions cause shorter holdover times.

In general, there are two methods for de-icing/anti-icing an airplane with a mobile unit: a one-step and a two-step process. The former consists of applying heated fluid onto the airplane to remove accumulated ice, snow, or frost and prevent their subsequent buildup. The primary advantage of this method is that it is quick and uncomplicated, both procedurally and in terms of equipment requirements. However, in conditions where large deposits of ice and snow must be flushed off the airplane, the total fluid usage will be greater than for a two-step process, which uses a more dilute fluid. [Figure 13-15]

The two-step process consists of separate de-icing and anti-icing steps. In the de-icing step, a diluted fluid, usually heated, is applied to the airplane surfaces to remove accumulated ice, snow, or frost. The dilution must protect from refreezing long enough for the second step (anti-icing) to be completed.

Figure 13-15. Spraying an airplane with a mixture of isopropyl alcohol and ethylene glycol removes frost and ice from the surface and prevents its refreezing for a period of time (holdover time).

During the anti-icing step, a more concentrated fluid is applied to the uncontaminated surfaces. The concentrated solution is either 100% or diluted appropriately, depending upon weather conditions and applied cold. Type I or Type II/IV fluids can be used for both steps, or Type I for the first step and Type II/IV for the second. This choice depends upon weather conditions, required holdover time, availability of fluids at a particular station, and equipment capability.

Before spraying fluid, configure the airplane for this procedure. If there is a heavy accumulation of contamination, remove as much as possible with a rope or a squeegee before applying the fluid. This reduces the amount of fluid required to get the surface clean. Make sure the flaps are free from snow or ice obstructions before moving them. When the flaps are clean, retract them to keep snow, ice, or slush from being washed off the wing surface into the flap mechanisms. Turn off the air-conditioning packs to keep fluid fumes from getting into the passenger cabin. Shut down the engines. If they must continue running, close the bleed valves to prevent fluid in the engine inlet from entering the bleed system. If the APU is operating, close the bleed valve. Several operators have reported instances of smoke in the cabin and sometimes activation of the smoke detectors attributed to ingestion of de-icing fluids into the engine or APU inlets.

Each airplane has a specific configuration for de-icing/anti-icing. For example, the stabilizer setting varies with airplane models. Check the operations manual for model-specific procedures. Specific de-icing/anti-icing procedures are determined by a combination of common sense and airplane considerations. Maintenance manuals for each type of airplane provide specific procedures. General precautions include:

Do not spray de-icing/anti-icing fluid directly at or into pitot inlets, TAT probes, or static ports.

Do not spray heated de-icing/anti-icing fluid or water directly onto cold windows.

Do not spray de-icing/anti-icing fluid directly into engine, APU inlets, air scoops, vents or drains. If an engine compressor will not turn because of ice, direct hot air into the engine.

Be sure that ice and/or snow is not forced into areas around flight controls during ice and snow removal.

Remove all ice and snow from doors and door operating mechanisms before closing any door.

Do not use hard or sharp tools to scrape or chip ice from an airplane's surface.

Open cargo doors only when necessary and clear cargo containers of ice or snow prior to loading. Apply de-icing/anti-icing fluid on pressure relief doors, lower door sills, and bottom edges of doors prior to closing them for flight.

FROST REMOVAL

When possible, remove frost from the wing and tail surfaces by brushing it off with a long handled T-broom. Better yet, prevent ice from forming on the surfaces by covering them with nylon or canvas when the airplane is secured for the night. Spray the surfaces with a de-icing solution of ethylene glycol and isopropyl alcohol just before flight to effectively remove all traces of frost.

SNOW REMOVAL

Remove wet snow with a brush or squeegee. Be careful not to push the snow into the gaps around control surfaces and doors. If the snow is sticking to the aircraft, removal by chemical means is necessary.

SECTION B

RAIN CONTROL SYSTEMS

Rain on the windscreen can obstruct the pilot's visibility during the landing and taxi phases of flight. Most small, general aviation aircraft are not equipped with any sort of rain-removal equipment. Airplanes typically do not have windshield wipers because windscreens are usually constructed of soft, transparent acrylic plastic, which is easily damaged or scratched. Rain normally beads up on the windscreen while the propeller slipstream or relative wind blows it off.

If an acrylic plastic windscreen is rough and scratched to the point that rain spreads out, waxing the windscreen will allow the rain to bead up and blow off. Larger and faster aircraft are equipped with elaborate rain control systems such as mechanical windshield wipers, chemical rain repellant, and pneumatic rain-removal systems.

WINDSHIELD WIPER SYSTEMS

Windshield wipers for aircraft are similar to those on automobiles except they must be able to withstand the air loads that are caused by high speeds of operation. A DC motor that drives a converter usually operates electrical windshield wipers. Converters change the rotary output of the motor into the reciprocating motion needed for the wiper blades. When the windshield wiper switch is turned OFF, the control circuit is open. However, the motor continues to run until the blades are driven to the PARK position. The motor then stops, but the control circuit is armed so the motor will start when the windshield wiper switch is turned on. Some installations have a separate position on the speed selector switch that allows the pilot to position the wiper blades in the PARK position before putting the switch in the OFF position. Common problems encountered with the use of windshield wipers include insufficient wiper pressure caused by aerodynamic forces and wiper systems that fail to oscillate fast enough to keep the windshield free of moisture. [Figure 13-16]

Some aircraft are equipped with a system for washing dirt and bugs from the windscreen. On the Lockheed

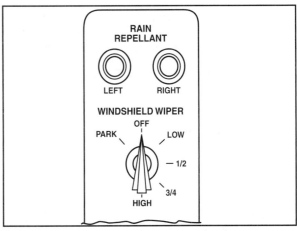

Figure 13-16. The rain control panel of a large transport aircraft controls the wiper speed as well as the rain repellant system.

L-1011, four spray nozzles supply windshield washer fluid from a reservoir/pump assembly, two nozzles for each windshield. A latching-type push-button switch on the captain's wiper control panel controls the system. A "PUMP ON" light adjacent to the switch illuminates when the pump is powered. The motor/pump assembly, submersed in washer fluid, consists of a 3-phase AC motor and centrifugal pump. The pump provides a fluid flow rate of 10 GPH at 30-35 PSIG. The flow rate to each pair of spray nozzles can be adjusted with a manually operated flow control valve in the left or right side console. Each spray nozzle head has two outlets that direct fluid to each side of the associated windshield. [Figure 13-17]

Some aircraft use hydraulic windshield wipers that derive pressure from the main hydraulic power system. Hydraulic fluid flows under pressure into the control unit, which periodically reverses the direction of the flow of fluid to the actuators. Inside the actuators are pistons that move a rack and pinion gear system. As the pistons move in one direction, the wiper will move. When the flow is reversed, the piston and the wiper blades move in the opposite direction. When the control valve is turned OFF, the blades are driven to, and held in the PARK position. Speed control is accomplished by varying the flow rate through a variable orifice in the fluid line. [Figure 13-18]

Airframe Ice and Rain Control 13-17

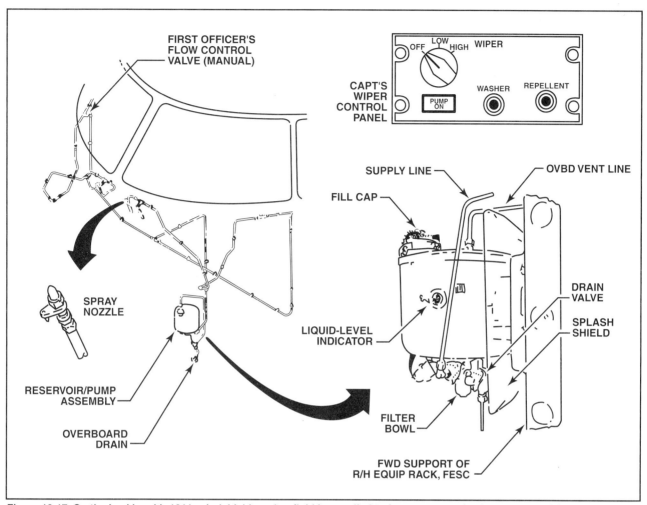

Figure 13-17. On the Lockheed L-1011, windshield washer fluid is supplied to four spray nozzles from a reservoir/pump assembly.

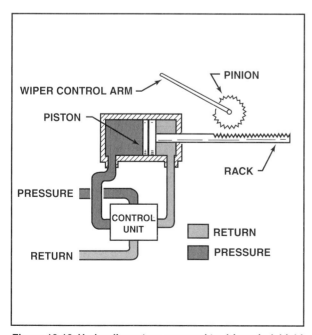

Figure 13-18. Hydraulic motors are used to drive windshield wipers on some aircraft.

Never operate windshield wipers on a dry windshield. Keep the blades clean and free of any contaminants that could scratch the windshield. If the windshield wiper should ever have to be operated for maintenance or adjustment, flood the windshield with ample quantities of fresh, clean water and keep it wet while the wiper blades are moving.

CHEMICAL RAIN REPELLANT

Many jet transport aircraft utilize a liquid chemical rain repellent that is sprayed on the windshield to prevent water from reaching the surface of the glass. The water forms into beads and the wind carries it away, leaving the glass free of water.

Rain repellent is a syrupy liquid contained in pressurized cans connected to the rain repellent system. If an aircraft is flying in rain too heavy for the windshield wipers, the pilot can depress the rain repellent button. A single timed application of the liquid will spray out onto the windshield, and the wipers will spread it out evenly over the surface.

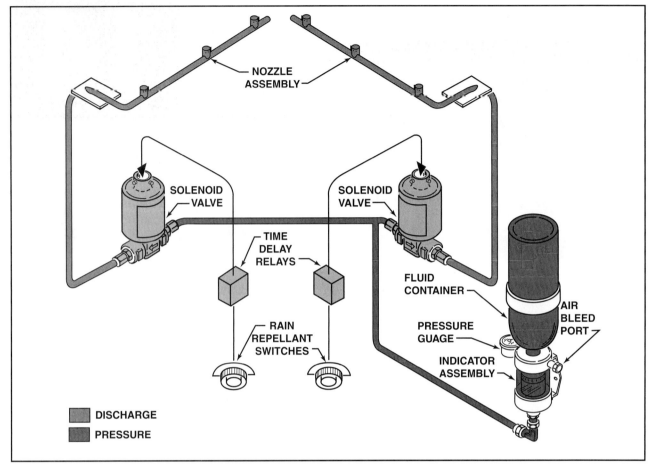

Figure 13-19. Chemical rain repellant systems are used primarily on large transport aircraft.

Repellant liquid should never be sprayed onto the windshield unless the rain is sufficiently heavy. Too much repellent can smear the windshield and be difficult to see through. It is difficult to remove if sprayed onto a dry windshield.

The operating system consists of two pressurized containers of repellent and two DC solenoid valves that, once actuated, are held open by a time-delay relay. When the rain repellent push-button switch is depressed, fluid flows for the required period of time and then the valve closes until the push button is pressed again. Rain intensity determines the number of times to depress the button. [Figure 13-19]

PNEUMATIC RAIN REMOVAL SYSTEMS

In pneumatic rain removal systems, high-pressure compressed air is ducted from the engine bleed air system into a plenum chamber. It is then directed up against the outside of the windshield as a high-velocity sheet of air. This air blast effectively prevents the rain from reaching the windshield surface. [Figure 13-20]

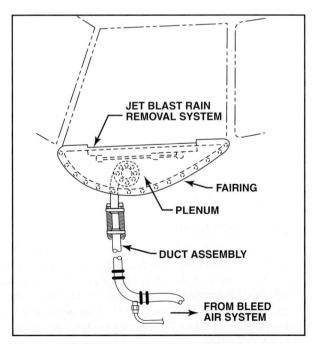

Figure 13-20. Pneumatic rain removal systems use bleed air from the turbine engines to blow high velocity air across the windscreen, preventing rain from obscuring the pilots' vision.

CHAPTER 14

CABIN ATMOSPHERE CONTROL

INTRODUCTION

The crew and passengers of modern, high-performance aircraft are physically unable to survive the extreme environment in which these airplanes fly without some sort of conditioning of the air within the cabin and cockpit. Primarily because of the various altitudes at which an aircraft operates, the cabin atmosphere must be controlled to increase the comfort of the occupants or even to sustain their lives. This chapter will discuss the physiology of the human body that determines the atmospheric conditions required for life, how oxygen and cabin altitude are controlled to provide a livable atmosphere for the aircraft occupants, and how the comfort needs of the passengers and crew are met.

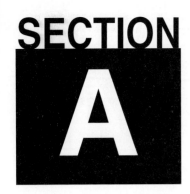

SECTION A

FLIGHT PHYSIOLOGY

In order to understand the reasons for controlling the cabin atmosphere or environment, it is necessary to understand both the characteristics of the atmosphere and the physiological needs of the persons flying within that atmosphere. Each type of aircraft will have specific requirements according to the altitudes and speeds at which the aircraft is flown.

THE ATMOSPHERE

The atmosphere envelops the earth and extends upward for more than 20 miles, but because air has mass and is compressible, the gravity of the earth pulls on it and causes the air at the lower levels to be more dense than the air above it. This accounts for the fact that more than one-half of the mass of the air surrounding the earth is below about 18,000 feet.

The atmosphere is a physical mixture of gases. Nitrogen makes up approximately 78% of the air, and oxygen makes up 21% of the total mixture. The remainder is composed of water vapor, carbon dioxide and inert gases. Oxygen is extremely important for both animal and plant life. It is so important for animals that if they are deprived of oxygen for even a few seconds, permanent damage to the brain or even death may result. Water vapor and carbon dioxide are also extremely important compounds. The other gases in the air, such as argon, neon, and krypton are relatively unimportant elements physiologically.

The density of air refers to the number of air molecules within a given volume of the atmosphere. As air pressure decreases, the density of the air also decreases. Conversely, as temperature increases the density of the air decreases. This change in air density has a tremendous effect on the operations of high altitude aircraft as well as physiological effects on humans. [Figure 14-1]

Turbine engine-powered aircraft are efficient at high altitudes, but the human body is unable to exist in this cold and oxygen-deficient air, so some provision must be made to provide an artificial environment to sustain life.

Standard conditions have been established for all of the important parameters of the earth's atmosphere. The pressure exerted by the blanket of air is considered to be 29.92 inches, or 1013.2 hectoPascals (millibars), which are the same as 14.69 pounds per square inch at sea level, and decreases with altitude as seen in figure 14-1. The standard temperature of the air at sea level is 15° Celsius, or 59° Fahrenheit. The temperature also decreases with altitude, as illustrated in figure 14-1. Above 36,000 feet, the temperature of the air stabilizes, remaining at –55° C (–69.7° F).

HUMAN RESPIRATION AND CIRCULATION

The human body is made up of living cells that must be continually supplied with food and oxygen and must have their waste carried away and removed from the body. Blood, circulated through the body by the heart, carries food and oxygen to the cells and carries away waste products.

When people inhale, or take in air, the lungs expand and the atmospheric pressure forces air in to fill them. This air fills millions of tiny air sacs called alveoli, and the oxygen in the air diffuses through the extremely thin membrane walls of these sacs into blood vessels called arteries. Nitrogen is not able to pass through these walls. The blood circulates through the body in the arteries and then into extremely thin capillaries to the cells, where the oxygen is used to convert the food in the blood into chemicals that are usable by the cells. The waste product, carbon dioxide, is then picked up by the blood and carried back into the lungs through blood vessels called veins. The carbon dioxide is able to diffuse through the

Cabin Atmosphere Control

FEET	IN. OF HG.	MM OF HG.	PSI	C°	F°
0	29.92	760.0	14.69	15.0	59.0
2,000	27.82	706.7	13.66	11.0	51.9
4,000	25.84	656.3	12.69	7.1	44.7
6,000	23.98	609.1	11.77	3.1	37.6
8,000	22.23	564.6	10.91	−0.8	30.5
10,000	20.58	522.7	10.10	−4.8	23.4
12,000	19.03	483.4	9.34	−8.8	16.2
14,000	17.58	446.5	8.63	−12.7	9.1
16,000	16.22	412.0	7.96	−16.7	1.9
18,000	14.95	379.7	7.34	−20.7	−5.1
20,000	13.76	349.5	6.76	−24.6	−12.3
22,000	12.65	321.3	6.21	−28.6	19.4
24,000	11.61	294.9	5.70	−32.5	−26.5
26,000	10.64	270.3	5.22	−36.5	−33.6
28,000	9.74	237.4	4.78	−40.5	−40.7
30,000	8.90	226.1	4.37	−44.4	−47.8
32,000	8.12	206.3	3.99	−48.4	−54.9
34,000	7.40	188.0	3.63	−52.4	−62.0
36,000	6.73	171.0	3.30	−55.0	−69.7
38,000	6.12	155.5	3.00	−55.0	−69.7
40,000	5.56	141.2	2.73	−55.0	−69.7
42,000	5.05	128.3	2.48	−55.0	−69.7
44,000	4.59	116.6	2.25	−55.0	−69.7
46,000	4.17	105.9	2.05	−55.0	−69.7
48,000	3.79	96.3	1.86	−55.0	−69.7
50,000	3.44	87.4	1.70	−55.0	−69.7
55,000	2.71	68.8	1.33		
60,000	2.14	54.4	1.05		
64,000	1.76	44.7	.86		
70,000	1.32	33.5	PSF 113.2	TEMPERATURE REMAINS CONSTANT	
74,000	1.09	27.7	77.3		
80,000	.82	20.9	58.1		
84,000	.68	17.3	47.9		
90,000	.51	13.0	35.9		
94,000	.43	10.9	29.7		
100,000	.33	8.0	22.3		

Figure 14-1. This chart illustrates that as altitude increases from sea level, the pressure decreases. It also shows that to a point, temperature decreases before finally leveling off.

membrane walls into the alveoli, where it is expelled during exhalation. [Figure 14-2]

There are two important considerations in providing sufficient oxygen for the body. There must be enough oxygen in the air to supply the body with the amount needed, and it must have sufficient pressure to enter the blood by passing through the membrane walls of the alveoli in the lungs.

Oxygen makes up approximately 21% of the mass of the air, and so 21% of the pressure of the air is caused by the oxygen. This percentage remains almost constant as the altitude changes, and is called the partial pressure of the oxygen. It is the partial pressure of the oxygen in the lungs that forces it through the alveoli walls and into the blood. At higher altitudes there is so little total pressure that there is not enough partial pressure of the oxygen to force it into the blood. This lack of oxygen in the blood is called hypoxia.

HYPOXIA

Any time the body is deprived of the required amount of oxygen, it will develop hypoxia. As hypoxia becomes more severe, a person's time of useful consciousness decreases. Time of useful consciousness is defined as the time a person has to take corrective action before becoming so severely impaired that they cannot help themselves. One of the worst things about hypoxia is the subtle way it attacks. When the brain is deprived of the needed oxygen, the first thing people lose is their judgment. The effect is similar to intoxication; people are unable to recognize how badly their performance and judgment are impaired. Fortunately, hypoxia affects every individual the same way each time it is encountered. If a person can experience hypoxia symptoms in an altitude chamber under controlled conditions, they are more likely to recognize the symptoms during subsequent encounters.

Two of the more common first indications of hypoxia occur at about ten thousand feet altitude. These are an increased breathing rate and a headache. Some other signs of hypoxia are light-headedness, dizziness with a tingling in the fingers, vision impairment, and sleepiness. Coordination and judgment will also be impaired, but normally this is difficult to recognize. Because it is difficult to recognize hypoxia in its early stages, many pressurized aircraft have alarm systems to warn of a loss of pressurization.

CARBON MONOXIDE POISONING

Carbon monoxide is the product of incomplete combustion of fuels which contain carbon and is found in varying amounts in the smoke and fumes from

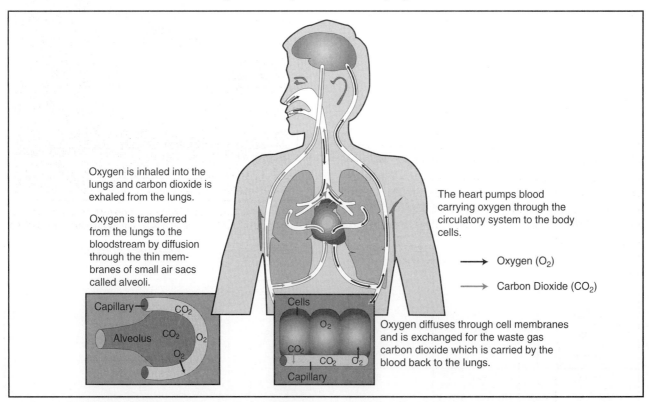

Figure 14-2. The cardiovascular system is made up of the heart, lungs, arteries, and veins. This system transports food and oxygen to the cells of the body and transports waste in the form of carbon dioxide from the cells back out of the body.

burning aviation fuel and lubricants. Carbon monoxide is colorless, odorless and tasteless, but since it is normally combined with other gases in the engine exhaust, you can expect it to be present when exhaust gases are detected.

When carbon monoxide is taken into the lungs, it combines with the hemoglobin in the blood. It is the hemoglobin that carries oxygen from the lungs to the various organs of the body. Since hemoglobin has a far greater attraction for carbon monoxide than it has for oxygen, it will load up with carbon monoxide until it cannot carry the much-needed oxygen. This results in oxygen starvation, and when the brain is deprived of oxygen the ability to reason and make decisions is greatly impaired. Exposure to even a small amount of carbon monoxide over an extended period of time will reduce the ability to operate the aircraft safely. The effect of carbon monoxide is cumulative, so exposure to a small concentration over a long period of time is just as bad as exposure to a heavy concentration for a short time.

The decrease in pressure as altitude increases makes it more difficult to get the proper amount of oxygen. If there is carbon monoxide in the cabin, or if a person is smoking tobacco while flying, it will intensify the problem and even further deprive the brain of the oxygen it needs.

Most small single-engine airplanes are heated with exhaust-type heaters in which the cabin ventilating air passes between a sheet metal shroud and the engine exhaust pipes or muffler. If a crack or even a pinhole size leak should exist in any of the exhaust components, carbon monoxide can enter the cabin. The possibility of this type of poisoning is most likely in the winter months when heat is most needed and when the windows and vents are usually closed to keep out cold air. Combustion heaters that burn fuel from the aircraft tanks to produce heat can also be a source of carbon monoxide. This type of heater is found on many small and medium-sized twin engine general aviation aircraft as well as on older airliners.

Early symptoms of carbon monoxide poisoning are similar to those of other forms of oxygen deprivation; sluggishness, a feeling of being too warm, and a tight feeling across the forehead. These early symptoms may then be followed by a headache and a throbbing in the temples and ringing in the ears. Finally, there may be severe headaches, dizziness, dimming of the vision and if something is not done soon, this can continue until unconsciousness and death.

If carbon monoxide poisoning is suspected, the heater should be shut off and all possible vents opened. If the aircraft is equipped with oxygen, 100% oxygen should be breathed until the symptoms disappear, or until landing. Carbon monoxide detectors are available that can be installed on the instrument panel. These are simply small containers of a chemical that changes color, generally to a darker color, when carbon monoxide is present. As an example, light yellow ones will turn dark green and white ones will turn dark brown or black. If there is any indication of carbon monoxide in the cabin, every part of the exhaust system should be checked to find and repair the leak before the aircraft is returned to service. [Figure 14-3]

Figure 14-3. One type of carbon monoxide detector consists of a tablet that changes color when exposed to carbon monoxide.

SECTION B

OXYGEN AND PRESSURIZATION SYSTEMS

As an aircraft climbs from sea level to increasingly high altitudes, the crew and passengers move further and further from an ideal physiological condition. In order to compensate for an atmosphere that becomes thinner as altitude increases, two different approaches have been developed. One of these is to provide pure oxygen to supplement the ever-decreasing amount of oxygen available in the atmosphere. The other is to pressurize the aircraft to create an atmosphere that is similar to that experienced naturally at lower altitudes. For aircraft that fly at extremely high altitude, a combination of pressurization and supplementary oxygen for emergencies is required.

OXYGEN SYSTEMS

At higher altitudes (generally above 10,000 feet) the air is thin enough to require supplemental oxygen for humans to function normally. Modern aircraft with the capability to fly at high altitudes usually have oxygen systems installed for the use of crew and/or passengers.

CHARACTERISTICS OF OXYGEN

Oxygen is colorless, odorless and tasteless, and it is extremely active chemically. It will combine with almost all other elements and with many compounds. When any fuel burns, it unites with oxygen to produce heat, and in the human body, the tissues are continually being oxidized which causes the heat produced by the body. This is the reason an ample supply of oxygen must be available at all times to support life.

Oxygen is produced commercially by liquefying air, and then allowing nitrogen to boil off, leaving relatively pure oxygen. Gaseous oxygen may also be produced by the electrolysis of water. When electrical current is passed through water (H_2O), it will break down into its two elements, hydrogen and oxygen.

Oxygen will not burn, but it does support combustion so well that special care must be taken when handling. It should not be used anywhere there is any fire, hot material or petroleum products. If pure oxygen is allowed to come in contact with oil, grease or any other petroleum product, it will combine violently and generate enough heat to ignite the material.

Commercial oxygen is used in great quantities for welding and cutting and for medical use in hospitals and ambulances. Aviator's breathing oxygen is similar to that used for commercial purposes, except that it is additionally processed to remove almost all of the water. Water in aviation oxygen could freeze in the valves and orifices and stop the flow of oxygen when an aircraft is flying in cold conditions found at high altitude. Because of the additional purity required, aircraft oxygen systems must never be serviced with any oxygen that does not meet the specifications for aviator's breathing oxygen. This is usually military specification MIL-O-27210. These specifications require the oxygen to have no more than two milliliters of water per liter of gas.

SOURCES OF SUPPLEMENTAL OXYGEN

Aircraft oxygen systems employ several different sources of breathing oxygen. Among the more common ones are gaseous oxygen stored in steel cylinders, liquid oxygen stored in specially constructed containers called Dewars, and oxygen generated by certain chemicals that give off oxygen when heated. A Dewar, sometimes called a Dewar flask, is a special type of thermos bottle designed to hold extremely cold liquids. Recently, a system using microscopic filters to separate oxygen from other gases in the air has been developed for medical uses, and is being investigated for use in aircraft.

GASEOUS OXYGEN

Most of the aircraft in the general aviation fleet use gaseous oxygen stored in steel cylinders under a pressure of between 1,800 and 2,400 psi. The main reason for using gaseous oxygen is its ease of handling and the fact that it is available at most of the airports used by these aircraft. It does have all the disadvantages of dealing with high-pressure gases, and there is a weight penalty because of the heavy storage cylinders. [Figure 14-4]

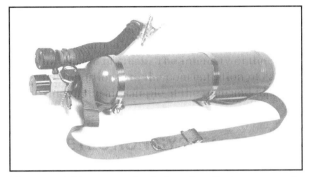

Figure 14-4. Most general-aviation aircraft store oxygen in steel, high-pressure cylinders.

LIQUID OXYGEN

Most military aircraft now carry their oxygen in a liquid state. Liquid oxygen is a pale blue, transparent liquid that will remain in its liquid state as long as it is stored at a temperature of below −181°F. This is done in aircraft installations by keeping it in a Dewar flask that resembles a double-wall sphere having a vacuum between the walls. The vacuum prevents heat transferring into the inner container.

Liquid oxygen installations are extremely economical of space and weight and there is no high pressure involved in the system. They do have the disadvantage, however, of the dangers involved in handling the liquid at its extremely low temperature, and even when the oxygen system is not used, it requires periodic replenishing because of losses from the venting system. [Figure 14-5]

Figure 14-5. Military aircraft usually use liquid oxygen, stored in special insulated containers called Dewars.

CHEMICAL, OR SOLID, OXYGEN

A convenient method of carrying oxygen for emergency uses and for aircraft that require it only occasionally is the solid oxygen candle. Many large transport aircraft use solid oxygen generators as a supplemental source of oxygen to be used in the event of cabin depressurization.

Essentially, a solid oxygen generator consists of a shaped block of a chemical such as sodium chlorate encased in a protective steel case. When ignited, large quantities of gaseous oxygen are released as a combustion by-product. They are ignited either electrically or by a mechanical igniter. Once they start burning, they cannot be extinguished and will continue to burn until they are exhausted. Solid oxygen candles have an almost unlimited shelf life and do not require any special storage conditions. There are specific procedures required for shipping these generators and they may not be shipped as cargo aboard passenger carrying aircraft. They can be shipped aboard cargo only aircraft and must be properly packaged, made safe from inadvertent activation, and identified properly for shipment. They are safe to use and store because no high pressure is involved and the oxygen presents no fire hazard. They are relatively inexpensive and lightweight. On the negative side, they cannot be tested without actually being used, and there is enough heat generated when they are used that they must be installed so that the heat can be dissipated without any damage to the aircraft structure. [Figure 14-6]

Figure 14-6. Solid oxygen generators, called candles, are used in many large aircraft to provide supplemental oxygen for the passengers in case of depressurization. They are also found in some smaller business aircraft.

MECHANICALLY-SEPARATED OXYGEN

A new procedure for producing oxygen is its extraction from the air by a mechanical separation process. Air is drawn through a patented material called a molecular sieve. As it passes through, the nitrogen and other gases are trapped in the sieve and only the oxygen passes through. Part of the oxygen is breathed, and the rest is used to purge the nitrogen from the sieve and prepare it for another cycle of filtering. This method of producing oxygen is currently being used in some medical facilities and military aircraft. It appears to have the possibility of replacing all other types of oxygen because of the economy of weight and space, and the fact that the aircraft is no longer dependent upon ground facilities for oxygen supply replenishment.

OXYGEN SYSTEMS AND COMPONENTS

The aviation maintenance technician will encounter oxygen systems during the course of servicing and repairing aircraft. Actual servicing or repair of the oxygen system itself must be accomplished in accordance with the manufacturer's instructions, but a general knowledge of gaseous, liquid, and chemical oxygen systems and how they operate will enable the technician to better prepare the aircraft for flight.

GASEOUS OXYGEN SYSTEMS

Gaseous oxygen systems consist of the tanks the oxygen is stored in, regulators to reduce the pressure from the high pressure in the tanks to the relatively low pressure required for breathing, plumbing to connect the system components, and masks to deliver the oxygen to the crewmember or passenger.

Storage Cylinders

Most military aircraft at one time used a low-pressure oxygen system in which the gaseous oxygen was stored under a pressure of approximately 450 psi in large yellow-painted low pressure steel cylinders. These cylinders were so large for the amount of oxygen they carried that they never became popular in civilian aircraft, and even the military has stopped using these systems.

Today, almost all gaseous oxygen is stored in green painted high-pressure steel cylinders under a pressure of between 1,800 and 2,400 psi. All cylinders approved for installation in an aircraft must be approved by the Department of Transportation (DOT) and are usually either the ICC/DOT 3AA 1800 or the ICC/DOT 3HT 1850 type. Aluminum bottles are also available, but are much less common. Newer, light-weight "composite" bottles that comply with DOT-E-8162 are becoming more common. These bottles are made of lighter, thinner metals combined with a wrapping of composite material.

Cylinders must be hydrostatically tested to 5/3 of their working pressure, which means that the 3AA cylinders are tested with water pressure of 3,000 psi every five years and stamped with the date of the test. 3HT cylinders must be tested with a water pressure of 3,083 psi every three years, and these cylinders must be taken out of service after 24 years, or after they have been filled 4,380 times, whichever comes first. E-8162 cylinders are tested to the same standards as the 3HT cylinders, but must be taken out of service after 15 years or 10,000 filling cycles, whichever occurs first.

All oxygen cylinders must be stamped near the filler neck with the approval number, the date of manufacture, and the dates of all of the hydrostatic tests. It is extremely important before servicing any oxygen system that all cylinders are proper for the installation and that they have been inspected within the appropriate time period.

Oxygen cylinders may be mounted permanently in the aircraft and connected to an installed oxygen plumbing system. For light aircraft where oxygen is needed only occasionally, they may be carried as a part of a portable oxygen system. The cylinders for either type of system must meet the same requirements, and should be painted green and identified with the words AVIATOR'S BREATHING OXYGEN written in white letters on the cylinder. Many high-pressure oxygen systems use pressure-reducing valves between the supply cylinders and the flight deck or cabin equipment. These valves reduce the pressure down to 300-400 PSI. Most systems incorporate a pressure relief valve that prevents high-pressure oxygen from entering the system if the pressure-reducing valve should fail.

On a hot day, the temperature inside a parked aircraft can cause the pressure in an oxygen cylinder to rise to dangerous levels. Permanently mounted gaseous oxygen systems, especially in large aircraft, normally have some type of thermal relief system to vent oxygen to the atmosphere if the cylinder pressure becomes too high. Venting systems may be temperature or pressure activated. To alert the crew that a thermal discharge has occurred, many systems use a "blow-out" disk as a thermal discharge indicator. A flush-type fitting containing a green plastic disk about 3/4 inch in diameter is mounted on the outside of the aircraft near the location of the oxygen bottles. If a thermal discharge occurs, the disk blows out of the fitting, and leaves the vent port visible. If the disk is found missing, there is no oxygen in the system and the aircraft must not be flown in conditions where supplemental oxygen might be required.

A thermal discharge requires maintenance on the oxygen system. The discharge mechanism must be reset or replaced, the indicator disk replaced and the system serviced with oxygen to the correct pressure. Consult the maintenance manual for the particular aircraft to determine the proper procedures.

Regulators

There are two basic types of regulators in use, and each type has variations. Low-demand systems, such as are used in smaller piston-engine powered general aviation aircraft, generally use a continuous flow regulator. This type of regulator allows oxygen to flow from the storage cylinder regardless of whether the user is inhaling or exhaling. Continuous flow systems do not use oxygen economically, but their simplicity and low cost make them desirable when the demands are low. The emergency oxygen systems that drop masks to the passengers of large jet transport aircraft in the event of cabin depressurization are of the continuous flow type.

Continuous Flow Regulators are of either the manual or automatic type. Both of these are inefficient in that they do not meter the oxygen flow according to the individual's needs.

Manual Continuous Flow Regulators typically consist of two gauges and an adjustment knob. One typical regulator has a gauge on the right that shows the pressure of the oxygen in the system and indicates indirectly the amount of oxygen available. The other gauge is a flow indicator and is adjusted by the knob in the lower center of the regulator. The user adjusts the knob so that the flow indicator needle matches the altitude being flown. The regulator meters the correct amount of oxygen for the selected altitude. If the flight altitude changes, the pilot must remember to readjust the flow rate. [Figure 14-7]

Figure 14-7. Manual continuous flow regulators must be reset as altitude changes.

Automatic Continuous Flow Regulators have a barometric control valve that automatically adjusts the oxygen flow to correspond with the altitude. The flight crew need only open the valve on the front of the regulator, and the correct amount of oxygen will be metered into the system for the altitude being flown. [Figure 14-8]

Figure 14-8. Automatic continuous flow regulators adjust oxygen flow automatically as altitude changes.

Oxygen is usually supplied to the flight crew of an aircraft by an efficient system that uses one of several demand-type regulators. Demand regulators allow a flow of oxygen only when the user is inhaling. This type regulator is much more efficient than the continuous flow type. [Figure 14-9]

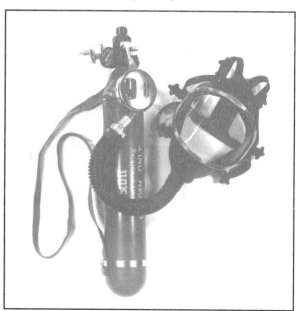

Figure 14-9. This demand-type regulator is fitted to a portable oxygen bottle and a full-face type mask. This type of system is often used aboard cargo aircraft as a smoke combat unit to allow a crewmember to locate and extinguish a cargo fire.

Diluter Demand Regulators are used by the flight crews on most commercial jet aircraft. When the supply lever is turned on, oxygen can flow from the supply into the regulator. There is a pressure reducer at the inlet of the regulator that decreases the pressure to a value that is usable by the regulator. The demand valve shuts off all flow of oxygen to the mask until the wearer inhales and decreases the pressure inside the regulator. This decreased pressure moves the demand diaphragm and opens the demand valve so oxygen can flow through the regulator to the mask. [Figure 14-10]

A diluter demand regulator dilutes the oxygen supplied to the mask with air from the cabin. This air enters the regulator through the inlet air valve and passes around the air-metering valve. At low altitude, the air inlet passage is open and the passage to the oxygen demand valve is restricted so the user gets mostly air from the cabin. As the aircraft goes up in altitude, the barometric control bellows expands and opens the oxygen passage while closing off the air passage. At an altitude of around 34,000 feet, the air passage is completely closed off, and every time the user inhales, pure oxygen is metered to the mask.

If there is ever smoke in the cabin, or if for any reason the user wants pure oxygen, the oxygen selector on the face of the regulator can be moved from the NORMAL position to the 100% position. This closes the outside air passage and opens a supplemental oxygen valve inside the regulator so pure oxygen can flow to the mask.

An additional safety feature is incorporated that bypasses the regulator. When the emergency lever is placed in the EMERGENCY position, the demand valve is held open and oxygen flows continuously from the supply system to the mask as long as the supply lever is in the ON position.

When a person breathes normally, the lungs expand and atmospheric pressure forces air into them. But at altitudes above 40,000 feet not enough oxygen can get into the lungs even with the regulator on 100%. Operation of unpressurized aircraft at and above 40,000 feet requires the use of pressure demand regulators. These regulators have provisions to supply 100% oxygen to the mask at higher than ambient pressure, thus forcing oxygen into the user's lungs.

Pressure Demand Regulators operate in much the same way as diluter demand regulators except at extremely high altitudes, where the oxygen is forced into the mask under a positive pressure. Breathing at this high altitude requires a different technique

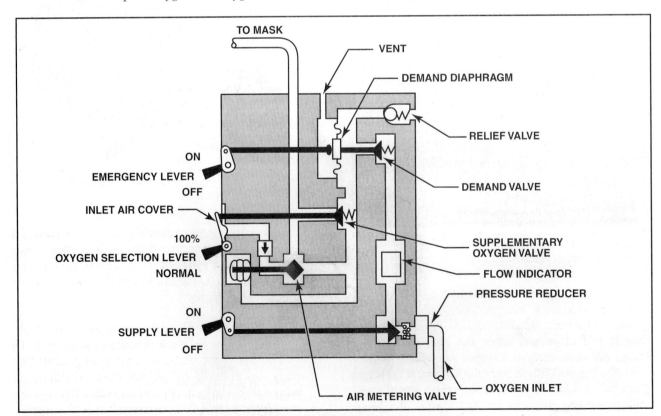

Figure 14-10. The flight crews of most commercial aircraft use diluter demand oxygen systems.

from that required in breathing normally. The oxygen flows into the lungs without effort on the part of the user, but muscular effort is needed to force the used air out of the lungs. This is exactly the opposite of normal breathing. [Figure 14-11]

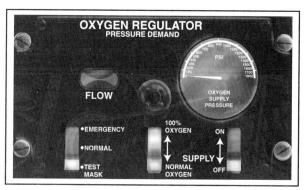

Figure 14-11. This pressure demand regulator supplies oxygen under pressure for flights above 40,000 feet.

Masks

Masks are used to deliver the oxygen to the user. These are either of the continuous flow or demand type.

Continuous flow masks are usually the rebreather type and vary from a simple bag-type disposable mask used with some of the portable systems to the rubber bag-type mask used for some of the flight crew systems. [Figure 14-12]

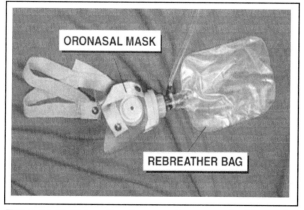

Figure 14-12. Rebreather type masks are used with continuous flow oxygen systems.

Oxygen enters a rebreather mask at the bottom of the bag, and the mask fits the face of the user very loosely so air can escape around it. If the rebreather bag is full of oxygen when the user inhales, the lungs fill with oxygen. Oxygen continues to flow into the bag and fills it from the bottom at the same time the user exhales used air into the bag at the top. When the bag fills, the air that was in the lungs longest will spill out of the bag into the outside air,

and when the user inhales, the first air to enter the lungs is that which was first exhaled and still has some oxygen in it. This air is mixed with pure oxygen, and so the wearer always breathes oxygen rich air with this type of mask. More elaborate rebreather-type masks have a close-fitting cup over the nose and mouth with a built-in check valve that allows the air to escape, but prevents the user from breathing air from the cabin.

The oxygen masks that automatically drop from the overhead compartment of a jet transport aircraft in the event of cabin depressurization are of the rebreather type. The plastic cup that fits over the mouth and nose has a check valve in it, and the plastic bag attached to the cup is the rebreather bag.

With **demand-type masks** the regulator is set up to meter the proper amount of oxygen to the user, so outside air would upset the required ratio of air to oxygen. Demand-type masks must fit tightly to the face so no outside air can enter. [Figure 14-13]

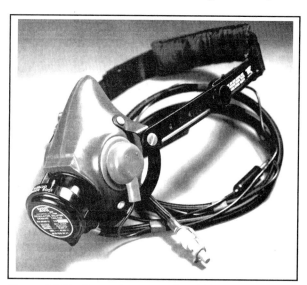

Figure 14-13. Demand-type masks deliver oxygen only when the wearer inhales.

A full-face mask is available for use in case the cockpit should ever be filled with smoke. These masks cover the eyes as well as the mouth and nose, and the positive pressure inside the mask prevents any smoke entering.

Most of the rigid plumbing lines that carry high-pressure oxygen are made of stainless steel, with the end fittings silver soldered to the tubing. Lines that carry low-pressure oxygen are made of aluminum alloy and are terminated with the same type fittings used for any other fluid-carrying line in the aircraft. The fittings may be of either the flared or flareless type. It is essential in any form of aircraft maintenance that only approved components be used. This is especially true of oxygen system components. Only valves carrying the correct part number should be used to replace any valve in an oxygen system.

Many of the valves used in oxygen systems are of the slow-opening type to prevent a rapid in-rush of oxygen that could cause excessive heat and become a fire hazard. Other valves have restrictors in them to limit the flow rate through a fully open valve.

Typical Installed Gaseous Oxygen Systems

If an aircraft has an installed oxygen system, it will be one of three types: the continuous flow type, the diluter demand type or the pressure demand type.

Most single engine aircraft utilize a continuous flow oxygen system. The external filler valve is installed in a convenient location and is usually covered with an inspection door. It has an orifice that limits the filling rate and is protected with a cap to prevent contamination when the charging line is not connected. The DOT approved storage cylinder is installed in the aircraft in a location that is most appropriate for weight and balance considerations. The shutoff valve on the cylinder is of the slow-opening type and requires several turns of the knob to open or close it. This prevents rapid changes in the flow rate that could place excessive strain on the system or could generate too much heat. Some installations use a pressure-reducing valve on the cylinder. When a reducer is used, the pressure gauge must be mounted on the cylinder side of the reducer to determine the amount of oxygen in the cylinder. [Figure 14-14]

The pressure gauge is used as an indication of the amount of oxygen in the cylinder. This is not, of course, a direct indication of quantity, but within the limitations seen when discussing system servicing, it can be used to indicate the amount of oxygen on board.

The pressure regulator reduces the pressure in the cylinder to a pressure that is usable by the masks. This regulator may be either a manual or an automatic type. There must be provision, one way or another, to vary the amount of pressure supplied to the masks as the altitude changes.

The mask couplings are fitted with restricting orifices to meter the amount of oxygen needed at each mask. In figure 14-14 the pilot's coupling has an orifice considerably larger than that provided for the passengers. The reason is that the pilot and other flight crewmembers require more oxygen since they are more active, and their alertness is of more vital importance than that of the passengers.

Some installations incorporate a therapeutic mask adapter. This is used for any passenger that has a health problem that would require additional oxygen. The flow rate through a therapeutic adapter is approximately three times that through a normal passenger mask adapter.

Each tube to the mask has a flow indicator built into it. This is simply a colored indicator that is visible when no oxygen is flowing. When oxygen flows, it pushes the indicator out of sight.

Pressurized aircraft do not normally have oxygen available for passengers all of the time, but FAR Part 91 requires that under certain flight conditions, the pilot operating the controls wear and use an oxygen mask. Because of this requirement, most executive aircraft that operate at high altitude are equipped with diluter demand or pressure demand oxygen regulators for the flight crew and a continuous flow system for the occupants of the cabin. Aircraft operating at altitudes above 40,000 feet will usually have pressure demand systems for the crew and passengers. [Figure 14-15]

The masks for the flight crew normally feature a quick-donning system. The mask is connected to a harness system that fits over the head. This system is designed so that the mask can be put on with one hand and be firmly in place, delivering oxygen, within a few seconds.

LIQUID OXYGEN SYSTEMS

Civilian aircraft do not generally use liquid oxygen (LOX) systems because of the difficulty in handling this form of oxygen, and because it is not readily available to the fixed-base operators who service general aviation aircraft. The military, on the other hand, uses liquid oxygen almost exclusively because of the space and weight savings it makes possible. One liter of liquid oxygen will produce approximately 860 liters of gaseous oxygen at the pressure required for breathing.

The regulators and masks are the same as those used for gaseous oxygen systems, the difference in the systems being in the supply. Liquid oxygen is held in a spherical container and in normal operation the buildup and vent valve is back-seated so some of the LOX can flow into the buildup coil where it absorbs enough heat to evaporate and pressurize the system to the amount allowed by the container pressure

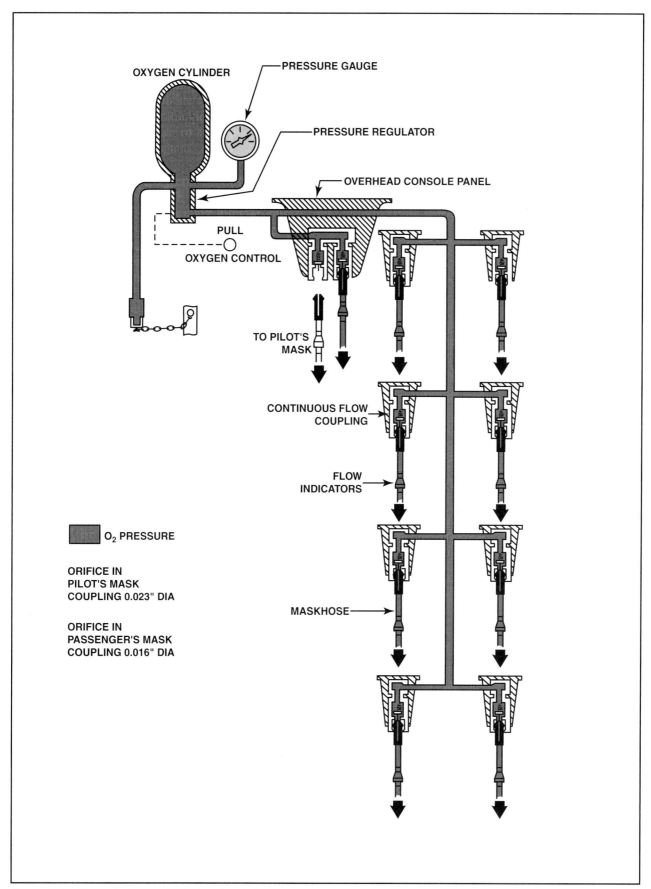

Figure 14-14. The typical general aviation aircraft has an installed system similar to this one.

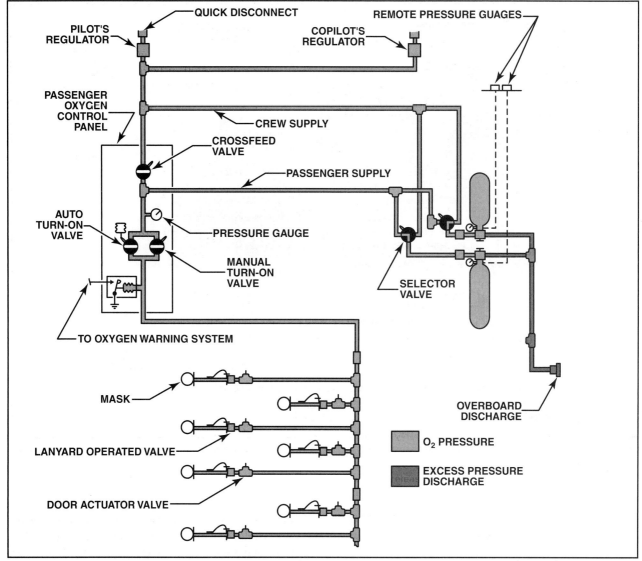

Figure 14-15. Aircraft that do not require oxygen to be constantly available to passengers will have a diluter or pressure demand regulator for the flight crew.

regulator, normally about 70 psi. This gaseous oxygen maintains a relatively constant pressure in the container and supplies the oxygen to the regulator. [Figure 14-16]

When the supply valve on the regulator is turned on, LOX flows from the container into the supply evaporator coil where it absorbs heat and turns into gaseous oxygen. If, for any reason, excessive pressure should build up in the system, it will vent overboard through one of the relief valves.

CHEMICAL OXYGEN SYSTEMS

Another source of oxygen is the chemical system. This system uses chemical oxygen generators also called "oxygen candles" to produce breathing oxygen. The size and simplicity of the units, and minimal maintenance requirements make them ideal for many applications. The chemical oxygen generator requires approximately one-third the space for equivalent amounts of oxygen as a bottled system. The canisters are inert below 400°F, even under severe impact. Oxygen candles contain sodium chlorate mixed with appropriate binders and a fuel formed into a block. When the candle is activated, it releases oxygen. The shape and composition of the candle determines the oxygen flow rate. As the sodium chlorate decomposes, it produces oxygen by a chemical action. [Figure 14-17]

An igniter, actuated either electrically or by a spring, starts the candle burning. The core of the candle is insulated to retain the heat needed for the chemical action and to prevent the housing from getting too hot. Filters are located at the outlet to prevent any contaminants entering the system.

Cabin Atmosphere Control

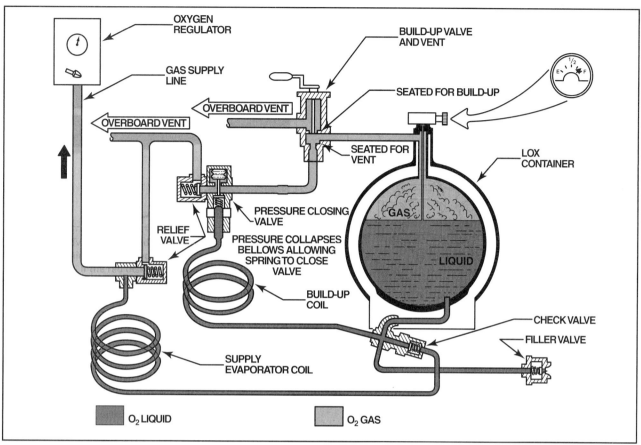

Figure 14-16. Liquid oxygen systems require specialized plumbing to handle the conversion from liquid to gas and the venting of excess pressures.

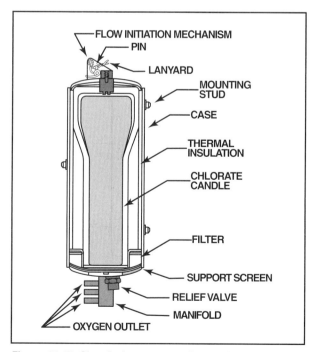

Figure 14-17. Chemical oxygen candles produce oxygen by heating sodium chlorate. The sodium chlorate is converted to salt and oxygen.

The long shelf life of unused chemical oxygen generators makes them an ideal source of oxygen for occasional flights where oxygen is needed, and for the emergency oxygen supply for pressurized aircraft where oxygen is required only as a standby in case cabin pressurization is lost.

The emergency oxygen systems for pressurized aircraft have the oxygen generators mounted in either the overhead rack, in seat backs, or in bulkhead panels. The masks are located with these generators and are enclosed, hidden from view by a door that may be opened electrically by one of the flight crew members or automatically by an aneroid valve in the event of cabin depressurization. When the door opens, the mask drops out where it is easily accessible to the user. Attached to the mask is a lanyard that, when pulled, releases the lock pin from the flow initiation mechanism, so the striker can hit the igniter and start the candle burning. Once a chemical oxygen candle is ignited, it cannot be shut off. It must burn until it is exhausted, and the enclosure must not be closed until the cycle has completed. [Figure 14-18]

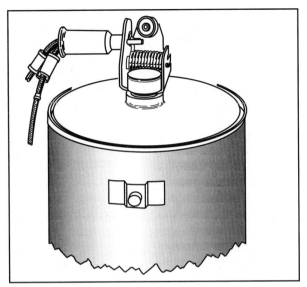

Figure 14-18. Pulling a lanyard on some chemical oxygen candles removes the safety pin to allow the spring to actuate the igniter.

OXYGEN SYSTEM SERVICING

Care and attention to detail is the mark of professional aviation maintenance, and nowhere is this characteristic more important than when servicing aircraft oxygen systems. Compressed gaseous oxygen demands special attention because of both its high storage pressure and its extremely active chemical nature.

When possible, all oxygen servicing should be done outdoors, or at least in a well-ventilated area of the hangar. Removable or portable supply cylinders should be removed from the aircraft for servicing. When oxygen servicing is performed in the aircraft, suspend all electrical work. In all cases the manufacturer's service information must be used while performing service, maintenance or inspection of aircraft oxygen systems.

SERVICING GASEOUS OXYGEN SYSTEMS

Some generic procedures are listed here as an orientation to the oxygen system servicing.

Leak Testing Gaseous Oxygen Systems

Searches for leaks are made using a special leak detector. This material is a form of non-oily soap solution. This solution is spread over every fitting and at every place a leak could possibly occur, and the presence of bubbles will indicate a leak. If a leak is found, the pressure is released from the system, and the fittings checked for proper torque. Flareless fittings can leak from both under and overtightening. If the fitting is properly torqued and still leaks, remove the fitting and examine all of the sealing surfaces for indications of damage. It may be necessary to replace the fitting and reflare the tube or install a new flareless fitting.

Draining the Oxygen System

Draining of the oxygen system should normally be done after the high-pressure bottle has been removed or isolated from the system. Either outdoors or in a well-ventilated hangar, the system's pressure should be bled off by opening the appropriate fitting. Normally a system will require purging after the system has been drained. All the safety precautions mentioned later in this chapter should be followed during any oxygen draining procedure.

Filling an Oxygen System

Fixed base operators who do a considerable amount of oxygen servicing will usually have an oxygen servicing cart. Such carts usually consist of six large cylinders, each holding approximately 250 cubic feet of aviator's breathing oxygen. A seventh cylinder, facing the opposite direction and filled with compressed nitrogen, is normally carried to charge hydraulic accumulators and landing gear struts. Fittings on the nitrogen cylinders are different from those on the oxygen cylinders to minimize the possibility of using nitrogen to fill the oxygen system, or of servicing the other systems with oxygen. [Figure 14-19]

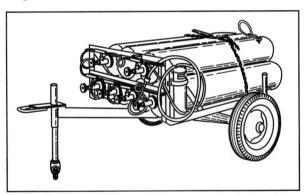

Figure 14-19. Oxygen service carts consist of a battery of O_2 bottles hooked to a common service manifold. Sometimes a nitrogen bottle is on the same cart, but with different hose fittings to prevent inadvertently interchanging the two systems.

Each oxygen cylinder has its own individual shutoff valve, and all of the cylinders are connected into a common service manifold that has a pressure gauge. A flexible line with the appropriate fittings connects the charging manifold to the aircraft filler valve.

Various manufacturers of oxygen equipment use different types of connections between the supply and the aircraft, and a well-equipped service cart should have the proper adapters. These adapters must be kept clean and protected from damage. Leakage during the filling operation is not only costly, but it is hazardous as well. Before filling any aircraft oxygen system, all of the cylinders being refilled must be checked to ensure that they are of the approved type, and have been hydrostatically tested within the required time interval.

No oxygen system should be allowed to become completely empty. When there is no pressure inside the cylinder, air can enter, and most air contains water vapor. When the water vapor mixes with the oxygen the mixture expands as it is released through the small orifices in the system. This expansion lowers the temperature and the water is likely to freeze and shut off the flow of oxygen to the masks. Water in a cylinder can also cause it to rust on the inside and weaken it so it could fail with catastrophic results. A system is considered to be empty when the pressure gets down to 50 to 100 psi. If the system is ever allowed to get completely empty, the valve should be removed and the cylinder cleaned and inspected by an FAA-approved repair station.

When an aircraft's oxygen system is being filled from a large supply cart, the cylinder having the lowest pressure should be used first. (The pressure in each tank should have been recorded on the container with chalk or in a record kept with the cart.) The valve on the cylinder should be opened slightly to allow some oxygen to purge all of the moisture, dirt and air from the line; then the line should be connected to the aircraft filler valve and the valve on the cylinder opened slowly. Most filler valves have restrictors that prevent an excessively high flow rate into the cylinder. When the pressure in the aircraft system and that in the cylinder with the lowest pressure stabilizes and there is no more flow, this new pressure should be recorded and the cylinder valve closed. The valve on the cylinder having the next lowest pressure should be opened slowly and oxygen allowed to flow into the system until it again stabilizes. Continue this procedure until the aircraft system has been brought up to the required pressure. [Figure 14-20]

The ambient temperature determines the pressure that should be put into the oxygen system, and a chart should be used to determine the pressure needed. For example, if the ambient temperature is 90° F and a stabilized pressure in the system of 1,800 psi is desired, the oxygen should be allowed to flow until a pressure of 2,000 psi is indicated on the system pressure gauge. When the oxygen in the system drops to the standard temperature of 70°F, the pressure will stabilize at 1,800 psi. If the ambient temperature is low, the filling of the system must be stopped at a lower pressure, because the

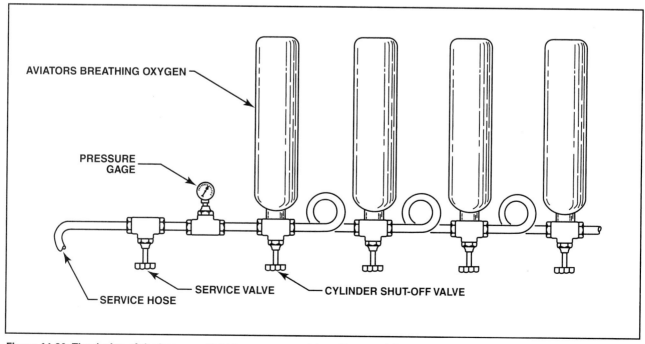

Figure 14-20. The design of the hose manifold for oxygen servicing allows each bottle to be connected in turn to the receiving system. Over time, this setup allows each cylinder to be expended down to its allowable limit. Normally a small amount of pressure is kept in each tank, even when "empty."

oxygen will expand and the pressure will rise when it warms up to its normal temperature. [Figure 14-21]

AMBIENT TEMP. DEGREES FAHRENHEIT	FILLING PRESSURE	
	FOR 1800 PSI AT 70°	FOR 1850 PSI AT 70°
0	1600	1650
10	1650	1700
20	1675	1725
30	1725	1775
40	1775	1825
50	1825	1875
60	1875	1925
70	1925	1975
80	1950	2000
90	2000	2050
100	1050	2100
110	2100	2150
120	2150	2200
130	2200	2250

Figure 14-21. Charts such as this one are used to determine the proper filling pressure for various ambient temperatures.

Purging A Gaseous Oxygen System

If the oxygen system has been opened for servicing, it should be purged of any air that may be in the lines. To purge a continuous flow system, oxygen masks are plugged into each of the outlets and the oxygen supply valve turned on. Oxygen should be allowed to flow through the system for about ten minutes. Diluter demand and pressure demand systems may be purged by placing the regulators in the EMERGENCY position and allowing the oxygen to flow for about ten minutes. After the system has been thoroughly purged, the cylinders should be filled to the required pressure.

FILLING A LIQUID OXYGEN SYSTEM

Service carts for liquid oxygen normally carry the LOX in 25- or 100-liter containers. Servicing systems from these carts is similar to that described in the previous section on gaseous oxygen systems. Protective clothing and eye protection must be worn since liquid oxygen has such a low boiling point that it would be sure to cause serious frostbite if spilled on the skin. Any empty LOX system or one that hasn't been in use for some time should be purged for a few hours with heated dry air, or nitrogen.

The service cart should be attached to the aircraft system and, after placing the buildup and vent valve in the vent position, the valve opened on the service cart. As the LOX flows from the service cart into the warm converter, it vaporizes rapidly and cools the entire system. Considerable gaseous oxygen is released during the filling procedure, and it vents to the outside air through the buildup and vent valve. This venting of the gaseous oxygen will continue until liquid oxygen starts to flow out of the vent valve. A steady stream of liquid indicates that the system is full.

The system should vent freely as it is being filled and frost should form only on the outlet and the hoses. If any frost forms on the supply container, it could be an indication of an internal leak, and since the pressure can build up extremely high, any trace of a leak demands that the equipment be shut down immediately and the cause of the frosting determined.

When the liquid oxygen cart is attached to the aircraft system, the valve should be fully opened, then closed slightly. If it is not, it is possible that the oxygen flowing through the valve could cause the valve to freeze in the open position and be difficult or impossible to close.

There are two ways LOX converters are serviced. Some are permanently installed in the aircraft and are serviced from an outside filler valve. The buildup and vent valve is placed in the vent position, the service cart is attached to the filler valve, and liquid oxygen is forced into the system until liquid runs out of the vent line. When the system is full, the buildup and vent valve is returned to the buildup position to build up pressure in the converter. Other installations have quick-disconnect mounts for the converters so the empty converter can be removed from the aircraft and replaced with a full one. Exchanging converters allows oxygen servicing to be done much more quickly and safely than can be done by filling the converter in the aircraft.

INSPECTING THE MASKS AND HOSES

Disposable masks such as those used with many of the portable systems should be replaced with new masks after each use, but the permanent masks used by crew members are normally retained by each individual crewmember. These masks are fitted to the face to minimize leakage and are usually treated as personal flight gear. They should be occasionally cleaned by washing them with a cloth wet with a lukewarm detergent solution and then allowing them to dry at room temperature. The face portion of the mask may be disinfected with a mild antiseptic.

The quick-donning masks for use by airliner flight crews are part of the aircraft and not crew personal equipment. Most airlines require each crewmember to don and test the mask as part of the required preflight inspection. Alcohol swabs in small sealed packets are provided to sterilize the mask before the crewmember dons the mask.

The masks and hoses should be checked for leaks, holes or rips, and replaced rather than repaired. When storing the mask in the airplane, it should be protected from dust and dampness, and especially from any type of grease or oil.

REPLACING TUBING, VALVES AND FITTINGS

It is extremely important when installing any oxygen line in an aircraft that no petroleum product is used as a thread lubricant, and that the lines are thoroughly cleaned of any trace of oil that was used in the flaring or presetting operation. Trichlorethylene or some similar solvent may be used to clean the tubing and fittings. After they are thoroughly clean, they should be dried either with heat or by blowing them with dry air or dry nitrogen.

Tapered pipe threads must never be lubricated with a thread lubricant that contains any form of petroleum. Oxygen-compatible thread lubricant that meets specification MIL-G-27617 may be used, or the male threads may be wrapped with Teflon tape and the fittings screwed together.

Before any tubing or fitting is replaced in an oxygen system, the part must be thoroughly cleaned and inspected. The part should be checked for evidence of corrosion or damage, and degreased with a vapor degreaser or ultrasonic cleaner. The new line should be flushed with stabilized trichlorethylene, acetone, or some similar solvent, and dried thoroughly with dry air or nitrogen. If neither dry air nor nitrogen are available, the part may be dried by baking it at a temperature of about 250° F until it is completely dry. When the parts are dry, close them with properly fitting protective caps or plugs, but never use tape in any form to seal the lines or fittings, as small particles of the tape are likely to remain when it is removed.

PREVENTION OF OXYGEN FIRES OR EXPLOSIONS

Safety precautions for oxygen servicing are similar to those required for fueling or defueling an aircraft. The airplane and service cart should be electrically grounded and all vehicles should be kept a safe distance away. There should be no smoking, open flame or items which may cause sparks within 50 feet or more depending upon the ventilation of the area during servicing operations. Since the clothing of a person involved in servicing an oxygen system is likely to be permeated with oxygen, smoking should be avoided for ten to fifteen minutes after completing the oxygen servicing.

The most important consideration when servicing any type of oxygen system is the necessity for absolute cleanliness. The oxygen should be stored in a well ventilated part of the hangar away from any grease or oil, and all high pressure cylinders not mounted on a service cart should be stored upright, out of contact with the ground and away from ice, snow or direct rays of the sun.

Protective caps must always be in place to prevent possible damage to the shutoff valve. The storage area for oxygen should be at least 50 feet away from any combustible material or separated from such material by a fire resistant partition. When setting up an oxygen storage area, you should be sure that it meets all insurance company and Federal/State Occupational Safety and Health Act (OSHA) requirements.

Because of the extreme incompatibility of oxygen and any form of petroleum products, it is a good idea to dedicate all necessary tools to be used exclusively with oxygen equipment. Any dirt, grease or oil that may be on the tools or on any of the hoses, adapters, cleaning rags, or even on clothing is a possible source of fire.

PRESSURIZATION SYSTEMS

The air that forms our atmosphere allows people to live and breathe easily at low altitudes, but flight is most efficient at high altitudes where the air is thin and the aerodynamic drag is low. In order for humans to fly at these altitudes, the aircraft must be pressurized and heated so that it is comfortable for the aircraft occupants.

PRESSURIZATION PROBLEMS

Turbine engines operate effectively at these high altitudes, but piston engines (as well as human occupants of an aircraft) require a supply of additional oxygen. Superchargers compress the air before it enters the cylinders of a reciprocating engine, and the occupants can be furnished supplemental oxygen to maintain life at these high, but aerodynamically efficient, altitudes.

Oxygen was used by some flight crews as early as World War I, and its inconvenience was tolerated by the crew as a necessary part of the flight. By the middle 1930s, airplanes and engines had been built that could carry passengers to altitudes where supplemental oxygen was needed, but the inconvenience of requiring passengers to wear oxygen masks proved to be a real deterrent to high altitude passenger flights.

In 1934 and 1935, the American aviator Wiley Post made a series of flights in his Lockheed Vega, the Winnie Mae, to altitudes near 50,000 feet. When flying at these altitudes, Post wore a rubberized pressure suit that resembled the suit worn by a deep-sea diver. As a result of his experiments, Post felt that flights at altitudes up to 30,000 feet would be practical if some method were possible to enable the passengers to breathe. Pressurization was the answer, but the technology of aircraft structures in the 1930s did not allow for pressurization of the aircraft itself.

During World War II, the need for bombers to operate at extremely high altitudes for long flights caused some of the manufacturers, notably the Boeing Airplane Company, to develop a pressurized structure — or pressure vessel, as it is called. This allows the occupants of the aircraft to operate in a cabin that is artificially held at an altitude far below the flight altitude of the airplane. This means the pressure inside the aircraft's cabin is much higher than the ambient pressure (outside pressure) when the aircraft is at high altitude.

Many of the piston engine transport aircraft of the 1950s had pressurized cabins and were able to carry passengers in comfort over the top of most bad weather. This type of aircraft made flying truly practical as a means of mass transportation.

Piston engines were limited to relatively low altitudes and did not require a high cabin pressure, so no great structural problems showed up with pressurized piston engine airplanes. But, when the jet transport airplane began to fly in the early 1950s, the large pressure differential required for the altitudes they flew created metal fatigue. Metal fatigue caused by the repeated pressurization and depressurization cycles caused several disastrous accidents.

Today, aircraft structural design has advanced to the point that pressurized aircraft are able to safely and comfortably carry large loads of passengers at efficiently high altitudes for long distance flights. Flight crews must be aware of the structural limitations of the aircraft and not exceed the maximum allowable differential between the pressure inside the structure and the pressure on the outside. The amount of air pumped into the cabin is normally in excess of that needed, and cabin pressure is controlled by varying the amount of air leaving the cabin through outflow valves controlled by the cabin pressure controller.

Most cabin pressurization systems have two modes of operation; the isobaric mode in which the cabin is maintained at a constant altitude (iso means same, and baric means pressure), and the constant pressure differential mode. In the isobaric mode, the pressure regulator controls the outflow valve as the aircraft goes up in altitude to maintain the same pressure in the cabin. When the pressure differential between that inside the cabin and that outside reaches the maximum structural pressure limitation, the pressure controller shifts to the constant differential mode and maintains a constant pressure differential. As the flight altitude increases, so does the cabin altitude, always maintaining the same differential pressure between the inside and the outside.

SOURCES OF PRESSURIZING AIR

The pressurization of modern aircraft is achieved by directing air into the cabin from either the compressor section of a jet engine, from a turbosupercharger, or from an auxiliary compressor.

RECIPROCATING ENGINE AIRCRAFT

When pressurization was first used, it was for large aircraft such as the Lockheed Constellation and the Douglas DC-6. These large cabins required great volumes of compressed air, and this was provided by a positive displacement Roots-type compressor or by a variable displacement centrifugal compressor driven by one of the engines. Pressurization air for smaller piston-engine aircraft is provided by bleed air from the engine turbochargers. [Figure 14-22]

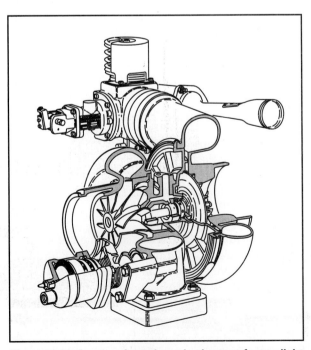

Figure 14-22. Pressure from the turbocharger of some light aircraft is used to provide cabin pressurization.

Cabin Atmosphere Control

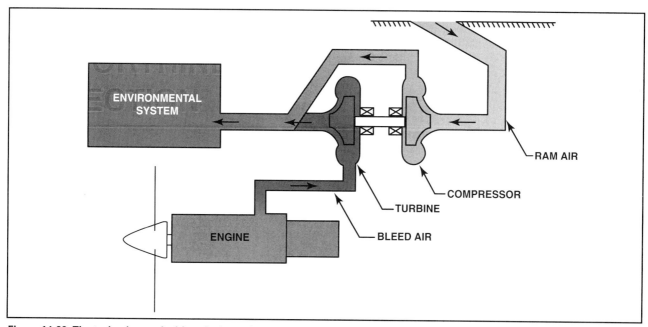

Figure 14-23. The turbocharger is driven by hot exhaust gases. The compressor portion of this system is turned by a shaft attached directly to the turbine. The turbine is driven by bleed air from the turbine engine. The ram air is compressed by the compressor and then blended with bleed air to the correct pressure and temperature.

TURBINE ENGINE AIRCRAFT

The compressor in a turbine engine is a good source of air to pressurize the cabin, and since this air is quite hot it is used to provide heat as well as pressurization. Engine power is required to compress this air, and this power is subtracted from that available to power the aircraft.

Compressor bleed air may be used directly, or it may be used to drive a turbocompressor. Outside air is taken in and compressed, and then, before it enters the cabin, it is mixed with the engine compressor bleed air that has been used to drive the turbocompressor. [Figure 14-23]

A jet pump flow multiplier can provide cabin pressurization air without the complexity of the turbocompressor. Compressor bleed air flows through the nozzle of a jet pump at high velocity and produces a low pressure that draws air in from the outside of the aircraft. The bleed air and the outside air mix and flow into the cabin to provide the air needed for pressurization. [Figure 14-24]

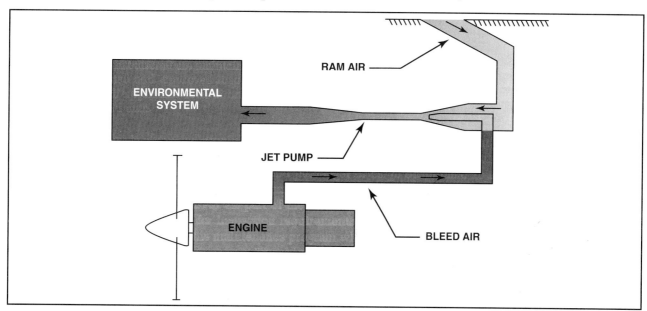

Figure 14-24. The jet-pump type pressurization uses aerodynamic principals to eliminate most moving parts.

Air Cycle Machines are used by many modern turbine-engine aircraft to provide both pressurization and temperature control. Air Cycle Machines use a clever application of the laws of physics to cool hot engine bleed air. Boeing calls these systems "packs," an acronym for pneumatic air conditioning kit. The theory and operation of air cycle machines will be discussed in detail in Section C of this Chapter. Three identical packs are installed on each 747 and any one can supply all the conditioned air needed for pressurization, temperature control and ventilation.

CONTROL OF CABIN PRESSURE

It would be impractical to build the pressure vessel of an aircraft that is airtight, so pressurization is accomplished by flowing more air into the cabin than is needed and allowing the excess air to leak out. There are two types of leakage in an aircraft pressure vessel; controlled and uncontrolled. The uncontrolled leakage is the air that escapes around door and window seals, control cables and other openings in the sealed portion of the structure. Controlled leakage flows through the outflow valve and the safety valve. This controlled leakage is far greater than the uncontrolled, and it determines the amount of pressure in the cabin. Pressurization control systems can be of the pneumatic or electronic type, with the electronic type incorporating electrically controlled outflow valves.

PRESSURIZATION COCKPIT CONTROLS

Most pressurization systems have cabin altitude, cabin rate-of-climb, and pressure differential indicators. The cabin altitude gauge measures the actual cabin altitude. The cabin altitude is almost always much below that of the aircraft, except when the aircraft is on the ground. An example would be an aircraft cruising at 40,000 feet would normally have a cabin altitude of about 8,000 feet. The cabin rate-of-climb indicator allows the pilot or flight engineer to adjust the rate the cabin altitude is climbing or descending to levels that are comfortable for the passengers. Normal climb rate is 500 feet per minute and normal descending rate is 300 feet per minute. The cabin rate-of-climb can be automatic or manual according to the type of aircraft. The differential pressure gauge reads the current difference in pressure between the aircraft's cabin interior and the outside air. The modes of operation of the pressurization system are generally automatic and manual control. In the manual control mode, the pilots can control the outflow valves directly through switches and indicators that are used to position the outflow valves if the automatic mode fails. If the cabin altitude exceeds 10,000 feet, on most aircraft, an alarm (intermittent horn) will sound, alerting the flight crew to take action. [Figure 14-25]

CABIN AIR PRESSURE REGULATOR AND OUTFLOW VALVE OPERATION

Cabin pressure regulators and outflow valves may be pneumatically or electrically operated. Modern systems are almost entirely electronically controlled. The outflow valve is controlled by the cabin pressure regulator and can be closed, open or modulated. This means that it is working at a position somewhere between the two extremes to maintain the pressure called for by the controller. The cabin pressure regulator contains an altitude selector and a rate controller.

Pneumatic Regulator and Outflow Valve Operation

Pneumatic regulators use variations in air pressure to activate the outflow and safety valves. The outflow valve and the safety valve are normally located in the pressure bulkhead at the rear of the aircraft cabin. The safety valve is normally closed (except on the ground) and is used primarily as a backup in case of a malfunction of the outflow valve. [Figure 14-26]

When the aircraft is on the ground and prepared for flight, the cabin is closed and the safety valve is held off its seat by vacuum acting on the diaphragm. The dump solenoid in the vacuum line is held open because the circuit through the landing gear safety switch is completed when the weight of the aircraft is on the landing gear. As soon as the aircraft takes off, the safety switch circuit opens and the dump solenoid shuts off the vacuum line to the safety valve, which allows the valve to close. If for any reason the pressure in the cabin should exceed a set limit, the safety valve will open fully. This will prevent cabin over-pressurization that could cause the structure of the aircraft to fail.

The outflow valve is closed until it receives a signal from the controller, and as soon as the safety valve closes, the cabin begins pressurize at the rate allowed by the rate controller. This increase in pressure is sensed by the controller. When the cabin reaches the selected altitude, the diaphragm in the controller moves back and vacuum is sent into the outflow valve to open it and allow some of the pressurizing air to escape from the cabin. This modulation of the outflow valve will maintain the cabin pressure at the altitude selected. As the flight altitude increases, the outside pressure decreases. When ambient pressure becomes low enough that the cabin differential pressure nears the structural limit, the upper diaphragm in the outflow valve

Cabin Atmosphere Control

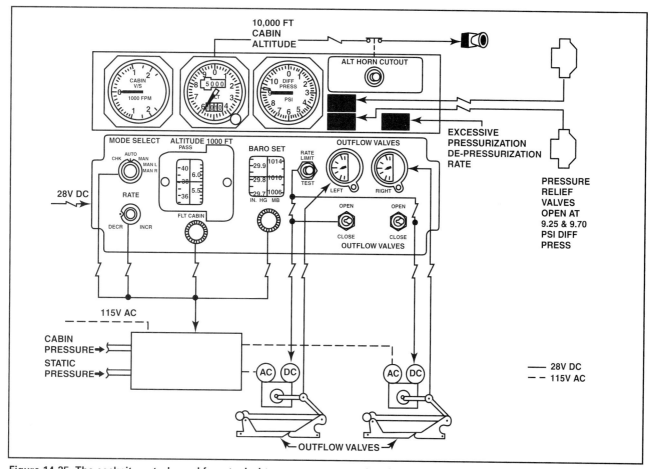

Figure 14-25. The cockpit control panel for a typical transport-category aircraft pressurization system displays information on the cabin vertical speed, cabin altitude and differential pressure and provides controls for selecting automatic or manual mode, setting the desired cabin altitude and the reference barometric pressure. A means of manually controlling the outflow valve position and system warning indications are also provided.

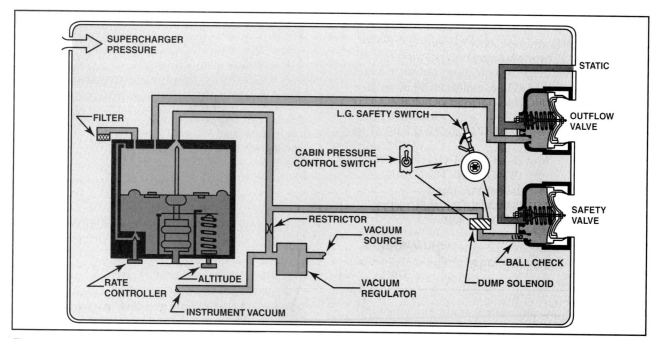

Figure 14-26. The cabin pressure is set at the control panel in the cockpit and controlled by the outflow valve. The safety valve is similar to the outflow valve and functions as a backup for the outflow valve, and to dump pressurization when the wheels are on the ground.

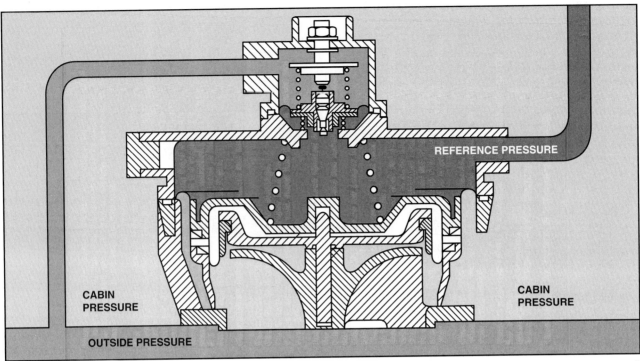

Figure 14-27. The outflow valve maintains a set altitude until the pressure differential with outside air approaches the structural limit of the aircraft. It then maintains a differential with the outside pressure.

moves up until the adjusting screw depresses the valve and releases some of the reference pressure to the outside air. This decrease in pressure allows the outflow valve to open so it can maintain the cabin pressure at a constant amount above the outside air pressure. [Figure 14-27]

Electronic Regulator and Outflow Valve Operation.
Electronic regulators and electrically actuated outflow valves perform the same function as pneumatic systems, only the power source is different. Electrical signals are sent to the cabin pressure controller from the cockpit control panel to set the mode of operation, the desired cabin altitude and either standard or local barometric pressure. In automatic mode, the cabin pressure controller sends signals to the AC motors, which modulate as required to maintain the selected cabin altitude. In manual mode, the controller uses the DC motors to operate the outflow valves. Interlocks prevent both motors from operating at the same time. All pressurized aircraft require some form of a negative pressure-relief-valve. This valve opens when outside air pressure is greater than cabin pressure. The negative pressure-relief-valve prevents accidentally obtaining altitude, which is higher than the aircraft altitude. This possibility would exist during descent. The outflow valves automatically drive to the full-open position whenever the aircraft weight is on the wheels. Pneumatically operated pressure relief valves open automatically if the cabin differential pressure becomes too great.

These valves are completely independent of the rest of the pressurization system. [Figure 14-28]

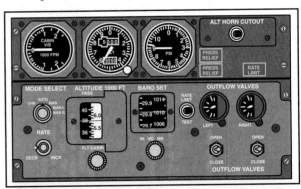

Figure 14-28. The pressurization control system regulates and maintains cabin pressure, and the rate of cabin pressure change, as a function of settings on the control panel. This is accomplished by regulating the flow of air vented from the cabin through motor driven outflow valves.

AIR DISTRIBUTION

The air distribution system on most aircraft mixes cold air from the air-conditioning packages (packs) and hot engine bleed air in the conditioned air manifold according to the temperature called for by the flight crew. This pressurized air passes through a combination check valve/shutoff valve on its way to the delivery air ducts. This check valve prevents the air pressure from being lost through an inoperative compressor. The pressurized air is then distributed

Cabin Atmosphere Control

to side wall or overhead vents in the cabin. The cabin air is then drawn back into the conditioned air manifold by recirculating fans, mixed with new incoming air, then redistributed to the aircraft cabin. Each passenger can turn the conditioned air "on" or "off" by adjusting the air outlet control on the gasper fan located in the overhead panel above each seat. [Figure 14-29]

CABIN PRESSURIZATION TROUBLESHOOTING

If a malfunction occurs in the pressurization system, the aircraft manufacturer's service manual

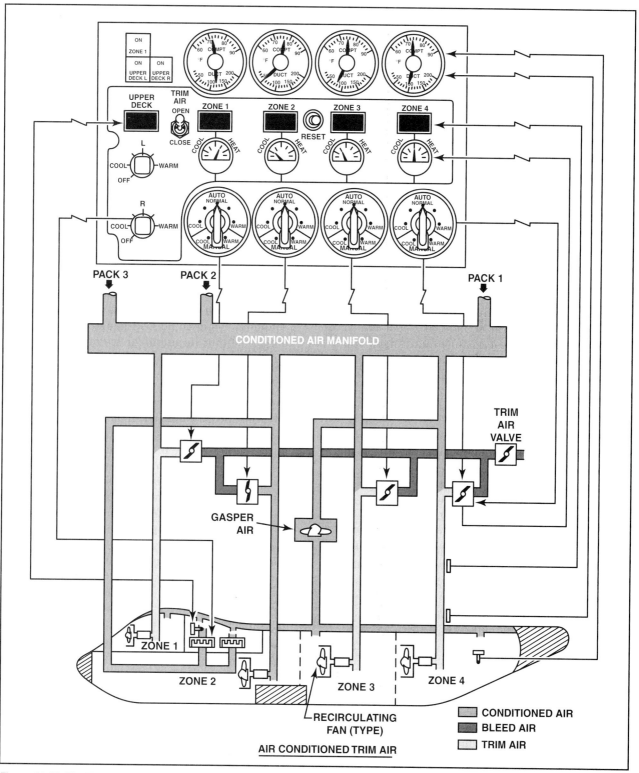

Figure 14-29. The Boeing 747 air distribution system is typical of systems found on large aircraft.

should always be used to troubleshoot and repair the system. Fault isolation systems and troubleshooting charts can be very helpful in isolating the defective system components. [Figure 14-30]

| | PRESSURIZATION MALFUNCTIONS | |
MALFUNCTION	CONDITION	POSSIBLE CAUSE
1. Pack trip	Generally an air-conditioning package overheat	Pack temperatures too high
2. Bleed trip	Bleed air from engine shut off	Overheat
3. Supply duct overheat	Supply duct overheated	Mixing valve or pack failure
4. Automatic cabin pressure control failure	Manual operation	Control failure
5. Rapid depressurization	Too much air exiting cabin	Major hole in aircraft (door or window blowout)
6. Explosive depressurization	All air exiting aircraft cabin	Structure failure
7. Single pack operation	Low pressurization capabilities	Inop pack or packs

Figure 14-30. Most aircraft service manuals have troubleshooting charts to assist the technician in locating problems within the cabin pressurization system.

SECTION C

CABIN CLIMATE CONTROL SYSTEMS

Aircraft fly in a wide variety of climatic conditions. Flights might begin on the ramp at 95° Fahrenheit (35° Celsius) and then climb to cruise at a temperature of −40° Fahrenheit (−40 degrees Celsius). Climate control systems then must be able to provide comfortable cabin temperatures, regardless of the outside air temperature. The quality of the air supply is also important: it must be free of contaminants, fumes, odors or other factors that might affect the health or comfort of the passengers or crew.

VENTILATION SYSTEMS

Most small general aviation aircraft have relatively simple systems to supply unconditioned ambient air to the cabin, primarily for cooling. The system may consist simply of a window that can be opened in flight or by any of several types of air vents that deliver ram air to the occupants. Occasionally, the system may include a fan to assist in moving air when the aircraft is on the ground.

Business jets and airliners generally have a system that supplies cool, conditioned air to individual air vents at each seat. The air vent system (sometimes called the gasper system) consists of a gasper fan, ducts and the overhead ventilating air outlets above the passenger seats. Cooling air is blown over the passengers, which is refreshing, but only when the passenger opens the air outlet for that seat.

HEATING SYSTEMS

EXHAUST SHROUD HEATERS

The most common type of heater for small single-engine aircraft is the exhaust-shroud heater. A sheet-metal shroud is installed around the muffler in the engine exhaust system. Cold air is taken into this shroud and heat that would otherwise be expelled out the exhaust is transferred to the ambient air. This air is then routed into the cabin through a heater valve in the firewall. When the heater is not on, this air is directed overboard. This type of heater is quite economical for small aircraft, as it utilizes heat energy that would otherwise be wasted.

One of the problems with this type of heater is the possibility of carbon monoxide poisoning if there should be a leak in the exhaust system. For this reason, it is very important that the shrouds be removed and the exhaust pipes and mufflers carefully inspected on the schedule recommended by the aircraft manufacturer. Some leaks may be present but not large enough to show up clearly when the metal is cold, so these components should be tested with air pressure. It is possible to test some of them on the aircraft by connecting the output of a vacuum cleaner to the exhaust stack and covering the muffler with a soapy water solution and watching for bubbles. Some aircraft have Airworthiness Directives that require the mufflers to be removed, submerged in water, and pressurized with air to search for leaks.

The surface area of the muffler determines the amount of heat that is transferred to the air from the muffler. Some manufacturers have increased this area by using welded-on studs. This type of muffler is more efficient but it must be checked with special care as it is possible for minute cracks to start where the studs are welded onto the muffler. [Figure 14-31]

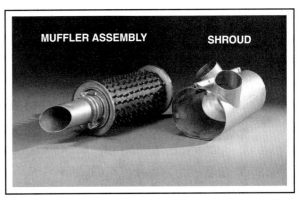

Figure 14-31. Some exhaust shroud heaters utilize welded-on studs to increase the effective surface area for heat transfer.

ELECTRIC HEATING SYSTEMS

Electric heating on aircraft is generally a supplemental heating source. The heaters use heating elements that create heat through electrical resistance. Some aircraft use this type of heat when the aircraft is on the ground and the engines are not running. A fan blows air over the heating coils to heat and circulate the air back into the cabin. Safety devices are installed in these systems to prevent them from overheating if the ventilating fan should become inoperative.

COMBUSTION HEATERS

Exhaust shroud heaters are used for small single-engine aircraft, and compressor bleed air heating is primarily used on large turbine-powered aircraft. Light and medium twin-engine aircraft are often heated with combustion heaters. [Figure 14-32]

Combustion heaters consist of two stainless steel cylinders, one inside the other. Air from outside the aircraft is directed into the inner cylinder, and aviation gasoline drawn from the fuel tank is sprayed over a continually sparking igniter plug. The combustion gases are exhausted overboard. Ventilating air flows through the outer cylinder around the combustion chamber, picks up the heat, and is distributed throughout the cabin.

The hot air ducts are normally located where they will blow warm air over the passengers' feet and the lower parts of their bodies. This type of heater has a number of safety features that prevent it creating a fire hazard in the event of a malfunction. [Figure 14-33]

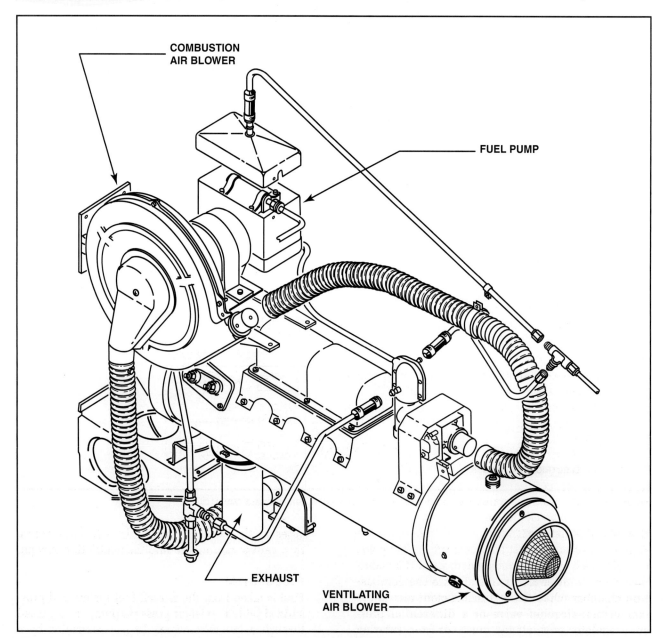

Figure 14-32. Combustion heaters that utilize the same fuel as the engines are installed in many twin-engine aircraft.

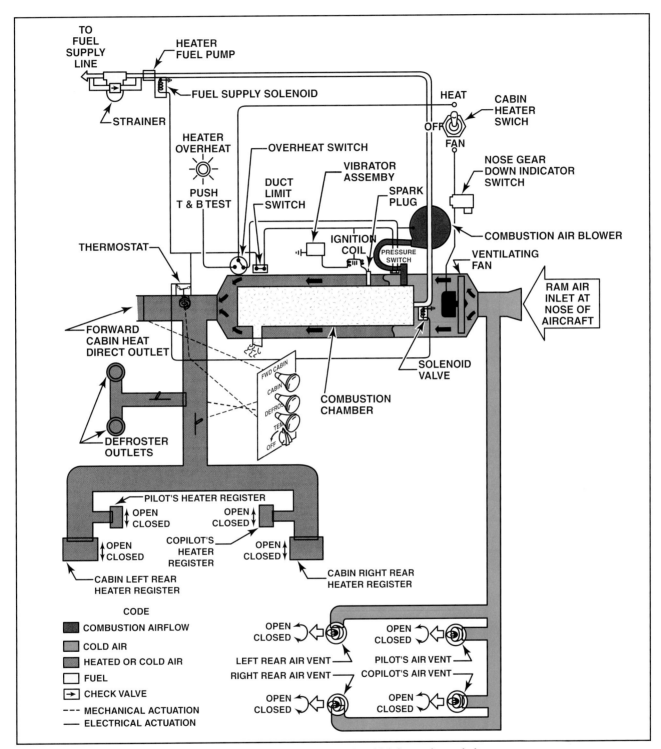

Figure 14-33. The combustion heater uses engine fuel to heat ram air, which heats the cockpit.

COMBUSTION AIR SYSTEM

A scoop on the outside of the aircraft picks up the air that used in the combustion process. The combustion air blower forces this air into the combustion chamber when there is insufficient ram air. A combustion-air-relief valve or a differential pressure regulator prevents too much air from entering the heaters as air pressure increases. The exhaust gases are then vented overboard at a location where they cannot recirculate into the ventilation system.

FUEL SYSTEM

Fuel is taken from the aircraft fuel system and pressurized with a constant pressure pump, and passed through a fuel filter. Fuel flow is controlled by a solenoid valve that may be turned off by the overheat switch, the limit switch, or by the pressure switch. There is a second solenoid valve in the fuel line that is controlled by the cabin thermostat. It shuts off the fuel at a point just before it enters the combustion chamber.

VENTILATION AIR SYSTEM
Ram air enters the heater from outside the aircraft, and flows over the outside of the combustion chamber, where it picks up heat and carries it inside the aircraft. There is a ventilating fan in the heater that operates when the aircraft is on the ground. When the aircraft becomes airborne, a switch on the landing gear shuts off the ventilating fan and all airflow is provided by ram air. The ventilating air pressure is slightly higher than the pressure of the combustion air, so in the event of a crack in the combustion chamber, ventilating air will flow into the combustion chamber rather than allowing the combustion air that contains carbon monoxide to mix with the ventilating air.

CONTROLS
The only action required to start the combustion heater is to turn the cabin heater switch ON and adjust the cabin thermostat to the desired temperature. When the cabin heater switch is turned on, the fuel pump starts, as well as the blowers for ventilation air and combustion air. As soon as the combustion air blower moves the required amount of air, it trips a pressure switch that starts the ignition coil supplying current to the igniter plug. The fuel supply solenoid valve is opened and fuel can get to the heater. When the thermostat calls for heat, the second fuel solenoid valve opens and fuel sprays into the combustion chamber and burns. As soon as the temperature reaches the value for which the thermostat is set, the contacts inside the thermostat open and de-energize the fuel solenoid valve, shutting off the fuel to the heater, and the fire goes out. The ventilating air cools the combustion chamber, and the cool air causes the thermostat to call for more heat. The cycle then repeats itself.

SAFETY FEATURES
The duct limit switch is in the circuit to the main fuel solenoid, and will shut off the fuel to the heater if for any reason there is not enough air flow to carry the heat out of the duct, or if the duct temperature reaches the preset maximum value.

The overheat switch is the final switch in the system. It is set considerably higher than the duct limit switch, but below a temperature that could cause a fire hazard. If the temperature put out by the heater reaches the limit allowed by this switch, the switch will close the fuel supply solenoid valve and will also shut off the combustion air flow and the ignition. A warning light will illuminate, alerting the pilot that the heater has been shut down because of an overheat condition. This switch, unlike the others, cannot be reset in flight, but can only be reset on the ground at the heater itself.

MAINTENANCE AND INSPECTION
Combustion heaters are relatively trouble-free, but they should be carefully inspected in accordance with the recommendations of the aircraft manufacturer and should be overhauled according to the schedule established by the heater manufacturer. The fuel filter should be cleaned regularly and the spark plug should be cleaned and gapped at the recommended interval. The entire system should also be checked for any indication of fuel or exhaust leakage.

COMPRESSOR BLEED AIR HEATERS
Turbine engines have a large amount of hot air in their compressors that is available for heating the cabin. The hot bleed air is mixed with cold ambient air to provide air of the proper temperature to the cabin. This form of heating is usually combined with an air-cycle air-conditioning system. The air-conditioning system of a large jet transport aircraft provides a means to cool or heat the pressurizing air as required.

AIRCRAFT AIR CONDITIONING SYSTEMS
Air conditioning is more than just the cooling of air. A complete air-conditioning system for an aircraft should control both the temperature and humidity of the air, heating or cooling it as is necessary. It should provide adequate movement of the air for ventilation, and there should be provision for the removal of cabin odors.

AIR-CYCLE AIR CONDITIONING
In a jet transport aircraft, hot compressor bleed air is taken from the engine compressors. An air-cycle machine (ACM) applies several basic laws of physics to cool this bleed air and then mix it with hot bleed air to provide air at the desired temperature for ventilation and pressurization. The air-cycle machine and its associated components are often referred to as a "pack." [Figure 14-34]

SHUTOFF VALVE
The air-conditioning shutoff valve, often called the pack valve, is used to control the flow of air into the system. It can either shut off the air flow or modulate the flow of air to provide that which is needed to operate the air-conditioning package.

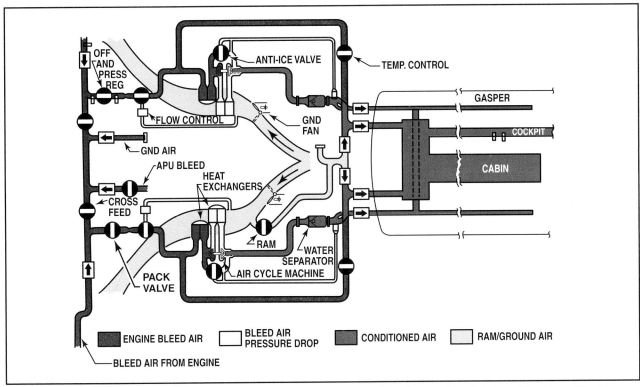

Figure 14-34. The air cycle system utilizes bleed air from the turbine engine(s) to heat and cool air for cabin air conditioning.

PRIMARY HEAT EXCHANGER

The primary heat exchanger is a radiator through which cold ram air passes to cool the hot bleed air from the engines. As the cold ram air passes over the radiator's fin-like tubes, bleed air passing through the tubes is cooled. The flow of ram air through the heat exchangers is controlled by moveable inlet and exit doors, which modulate in flight to provide the required cooling. On many aircraft, the heat exchangers are sized to provide most, if not all, of the necessary cooling in flight. On the ground there is not enough air passing through the cooling doors, so fans called pack fans provide adequate airflow to cool the heat exchangers.

AIR CYCLE MACHINE BYPASS VALVE

When cooling requirements are low, some or all of the hot bleed air from the engines can be bypassed around the ACM (the compressor and turbine) if warm air is needed in the cabin. There would be no purpose in cooling all the air if warm air is called for by the temperature controls. This outlet air from the primary heat exchanger may be routed directly to the inlet side of the secondary heat exchanger in some systems to provide additional cooling.

SECONDARY HEAT EXCHANGER

As cooling requirements increase, air exiting the primary heat exchanger is routed to the compressor side of the ACM. The compressor raises both the pressure and temperature of the air passing through it. The warmer, high pressure air is then directed to the secondary heat exchanger. This heat exchanger provides an additional stage for cooling the hot engine bleed air after it has passed through the primary heat exchanger and the compressor of the ACM. It operates in the same manner as the primary heat exchanger.

REFRIGERATION BYPASS VALVE

Some systems use a refrigeration bypass valve to keep the temperature of the air exiting the ACM from becoming too cold. Generally this air is kept at about 35° F (2° C) by passing warm bleed air around the ACM and mixing it with the output air of the ACM. The primary purpose of this valve is to prevent water from freezing in the water separator.

REFRIGERATION TURBINE UNIT

Pressure and temperature, are interchangeable forms of energy. A turbine engine extracts energy from the burning fuel to turn the compressor, and this energy raises both the pressure and the temperature of the engine inlet air. Compressed air with this energy in it is taken from the engine and passed through the primary heat exchanger, where some of the heat is transferred to ram air passing around the tubes in the radiator-like cooler. The high-pressure air, somewhat cooled, is then ducted into the air cycle machine where most of the remainder of its energy is extracted by the air cycle machine. It consists of a centrifugal air compressor and an expansion turbine that drives the compressor. When the compressor bleed air passes through the primary heat exchanger, it loses some of its heat but almost none of its pressure. This air then enters the compressor of the air cycle machine, and its pressure is further increased. With the increase in pressure, there is some increase in its temperature, but this is removed by the secondary heat exchanger. Now the somewhat cooled high-pressure air flows into the expansion turbine where a large percentage of its remaining energy is used to drive the compressor. As this air expands across the turbine, there is a large decrease in pressure. The decrease in pressure, coupled with the energy extracted to drive the compressor, results in a very large decrease in temperature. There are two forms of cooling used in this system. Some is done by transferring heat to the ram air, but most of the heat is removed by expansion and converting it into work to drive the compressor. This type of cooling system is called a bootstrap system. [Figure 14-35]

WATER SEPARATORS

The rapid cooling of the air in the turbine causes moisture to condense in the form of a fog, and when this foggy air passes through the water separator, the tiny droplets of water coalesce in a fiberglass sock and form large drops of water. The louvers over which the sock fits are shaped to impart a swirling motion to the air, and the drops of water are slung to the sides of the container by centrifugal force, where they are carried overboard through the drain valve.

This water is kept from freezing by mixing the air in the separator with warm air. A temperature sensor in the outlet of the water separator regulates a temperature control valve in a bypass line around the air cycle machine. If the temperature of the air at the outlet of the water separator ever drops below 38° F, the control valve opens so warm air can mix with that in the water separator. This precludes cabin airflow blockage and possible damage to the separator.

RAM AIR DOOR

Some aircraft are equipped with a ram air door to allow cool outside air to ventilate the cabin with fresh air during unpressurized flight. It is generally fully

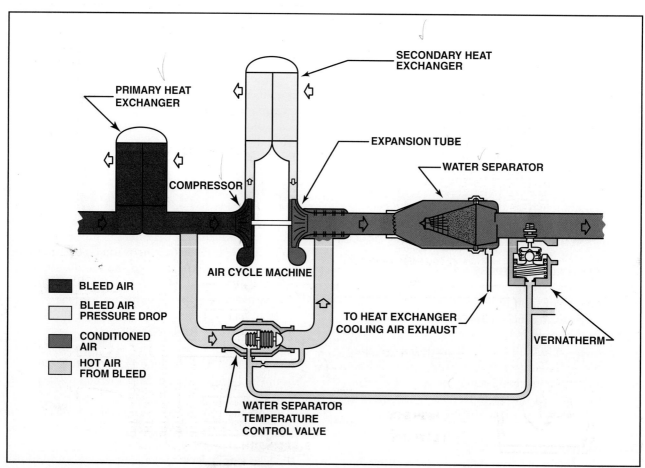

Figure 14-35. The air-cycle air conditioner utilizes bleed air to heat and cool the cabin.

open or closed and is seldom used on pressurized aircraft except in emergencies. An electric heater may be provided to warm this ram air as necessary.

CABIN TEMPERATURE PICKUP UNIT
Normally, temperature sensors are located in each passenger zone in the aircraft's cabin for the purpose of controlling the zone temperature. The cabin zone controller uses the sensed difference between the temperature demand signal from the selector and the actual supply temperature to position the associated air mix valve.

ZONE TEMPERATURE CONTROLLER
Some aircraft, such as the Boeing 747, use a slightly different method to provide conditioned air at the proper temperature to each cabin zone. In this aircraft, the zone temperature controllers have two modes of operation, automatic and manual, and send signals to each pack controller. If all the zone temperature controllers are in auto, the zone calling for the coldest temperature sets the output temperature for all the air-conditioning packs. The output air from each pack enters the conditioned air manifold. For each zone requiring a warmer temperature, the controller for that zone adjusts the position of a trim air valve to mix warm bleed air with the cold air from the conditioned air manifold. If all the zone temperature controllers are in manual mode, the pack controllers will set the ACM outlet temperature to 35° F (2° C). If any one zone controller is in auto with the remainder in manual, the controller in auto determines the pack outlet temperature.

VAPOR-CYCLE AIR CONDITIONING
Temperature is a measure of the effect of heat on a body or material, and is a convenient way of expressing this physical phenomenon numerically. While there is a relationship between heat and temperature, heat can be added to or removed from a refrigerant without changing its temperature. The heat put into a material as it changes its state without changing its temperature is called latent heat, and this heat will be returned when the material reverts to its original state. This process acts in a continuous cycle. A refrigerant changes state from a liquid into a vapor, and in doing so, it absorbs heat from the cabin. This heat is taken outside of the aircraft and is given off to the outside air as the refrigerant returns to a liquid state. [Figure 14-36]

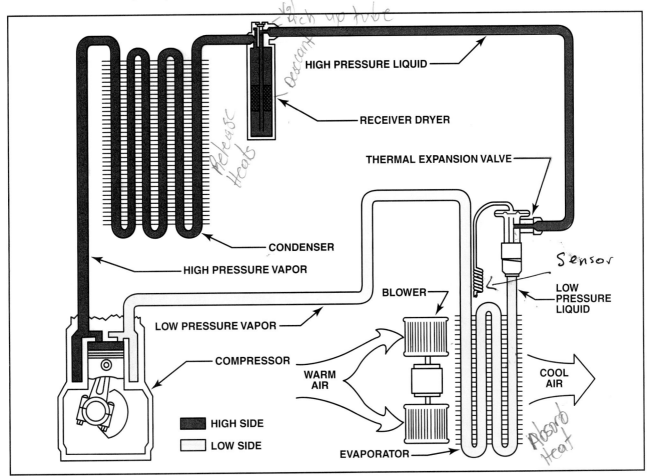

Figure 14-36. The vapor-cycle air conditioning system is the same basic system as used in modern automobiles.

TRANSFER OF HEAT

Heat is a form of energy, and can neither be created nor destroyed. It can, however, be transformed or moved from one place or material to another. This energy continues to exist regardless of its form or location. Heat will flow from an object having a certain level of energy into an object having a lower level. Any material that allows this transfer easily is said to be a conductor of heat, while any material that impedes the transfer is called an insulator.

The refrigerant used in an aircraft air-conditioning system is a liquid under certain conditions. When it is surrounded by air having a higher level of heat energy, heat will pass from the air into the liquid. As the liquid absorbs the heat, it changes state and becomes a gas. The air that gave up its heat to the refrigerant is cooled in the process.

The system is divided into two sides, one that accepts the heat and the other that disposes of it. The side that accepts the heat is called the low side, because here the refrigerant has a low temperature and is under a low pressure. The heat is given up on the high side, where the refrigerant is under high pressure and has a high temperature. Notice in figure 14-36, that the system is divided at the compressor where the refrigerant vapor is compressed, increasing both its pressure and temperature, and at the expansion valve where both pressure and temperature drop.

The refrigeration cycle starts at the receiver-dryer which acts as a reservoir to store any of the liquid refrigerant that is not passing through the system at any given time. If any refrigerant is lost from the system, it is replaced from that in the receiver-dryer. A desiccant agent is used in the receiver-dryer to trap and hold any moisture that could possibly be in the system. This is necessary since a tiny droplet of water in the refrigerant is all that is needed to freeze in the orifice of the expansion valve, completely stopping operation of the system.

Liquid refrigerant leaves the receiver-dryer and flows under pressure to the expansion valve where it sprays out through a tiny metering orifice into the coils of the evaporator. The refrigerant is still a liquid, but it is in the form of tiny droplets, affording the maximum amount of surface area so the maximum amount of heat can be absorbed.

The evaporator is the unit in an air-conditioning system that produces the cold air. Warm air is blown through the thin metal fins that fit over the evaporator coils. This heat is absorbed by the refrigerant, and when the air emerges from the evaporator, it is cool. When heat is absorbed by the refrigerant, it changes from a liquid into a gas without increasing its temperature. The heat remains in the refrigerant in the form of latent heat.

The refrigerant vapor that has the heat from the cabin is taken into the compressor, where additional energy is added to it to increase both its pressure and temperature. It leaves the compressor as a hot, high-pressure vapor. The heat trapped in the refrigerant vapors in the condenser escapes into the walls of the coil and then into the fins that are pressed onto these coils. Relatively cool air from outside the aircraft flows through these fins and picks up the heat that is given up by the refrigerant. When it loses its heat energy, the refrigerant vapor condenses back into a liquid and then flows into the receiver-dryer where it is held until it passes through the system for another cycle.

REFRIGERANT

Almost any volatile liquid can be used as a refrigerant, but for maximum effectiveness, it must have a very low vapor pressure and therefore a low boiling point. The vapor pressure of a liquid is the pressure that will exist above a liquid in an enclosed container at any given temperature. For example, a particular liquid refrigerant in an open container boils vigorously as the liquid turns into a gas at a temperature of 70° F. If the container is closed, the liquid will continue to change into a vapor and the pressure of the vapor will increase. When the pressure reaches 70.1 psi, no more vapor can be released from the liquid. The vapor pressure of this particular material is then said to be 70.1 psi at 70° F.

Many different materials have been used as refrigerants in commercial systems, but for aircraft air-conditioning systems, dichlorodifluoromethane is almost universally used. It is a stable compound at both high and low temperatures and does not react with any of the materials in an air-conditioning system. It will not attack the rubber used for hoses and seals, and is colorless and practically odorless. Rather than calling this refrigerant by its long chemical name, it is just referred to as Refrigerant-12, or, even more simply as R-12. It may also be known by one of its many trade names such as Freon-12®, Genetron-12®, Isotron-12®, Ucon-12®, or by some other proprietary name. The important thing to remember is the number. Any of these trade names associated with another number is a different product. Freon-22®, for example is similar to Freon-12, except that its vapor pressure is different. It is the refrigerant commonly used in commercial refrigerators and freezers. When servicing an aircraft

Cabin Atmosphere Control

air-conditioning system, it is extremely important to use only the refrigerant specified in the aircraft manufacturer's service manual. R-12 is the refrigerant discussed in this section, but the procedures used for other refrigerants are basically the same.

One of the characteristics of R-12 that makes it desirable for aircraft air-conditioning systems is its temperature-vapor pressure relationship. In the temperature range between 20° and 80° F, the range where most air conditioning occurs, there is an approximate relationship of one psi of vapor pressure for each degree of Fahrenheit temperature. While this relationship is not exact, it is close enough to make servicing relatively easy. If the low-side pressure is 28 psi, the temperature of the refrigerant in the evaporator coils is about 30° F. This is the temperature of the refrigerant and not of the air passing through the evaporator. That will be somewhat higher (34 or 35° F.) This temperature will give the most effective cooling, since the evaporator coils will be cold enough to cool the air, but not cold enough to cause ice to form. [Figure 14-37]

Refrigerant-12 boils at normal sea level pressure at 21.6° F, and if a drop of liquid R-12 contacts skin, it will cause frostbite. Even a tiny drop of liquid R-12 in the eye is hazardous. If liquid refrigerant gets in the eye, the eye should be flooded with cool water, treated with mineral oil or petroleum jelly and a physician seen immediately. It is extremely important to wear eye and skin protection any time air conditioning systems are being serviced. R-12 is not normally toxic. However, when R-12 is burned its characteristics change drastically, becoming deadly phosgene gas.

REFRIGERATION OIL

Since the air-conditioning system is completely sealed, the oil used to lubricate the compressor seals and expansion valve must be incorporated within a sealed a system. The oil is a special, highly refined mineral oil, free from such impurities as water,

TEMP °F	PRESSURE PSI	TEMP °F	PRESSURE PSI	TEMP °F	PRESSURE PSI	TEMP °F	PRESSURE PSI	TEMP °F	PRESSURE PSI
0	9.0	35	32.5	60	57.7	85	917	110	136.0
2	10.1	36	33.4	61	58.9	86	93.2	111	138.0
4	11.2	37	34.3	62	60.0	87	94.8	112	140.1
6	12.3	38	35.1	63	61.3	88	96.4	113	142.1
8	13.4	39	36.0	64	62.5	89	98.0	114	144.2
10	14.6	40	36.9	65	63.7	90	99.6	115	146.3
12	15.8	41	37.9	66	64.9	91	101.3	116	148.4
14	17.1	42	38.8	67	66.2	92	103.0	117	151.2
16	18.3	43	39.7	68	67.5	93	104.6	118	152.7
18	19.7	44	40.7	69	68.8	94	106.3	119	154.9
20	21.0	45	41.7	70	70.1	95	108.1	120	157.1
21	21.7	46	42.6	71	71.4	96	109.8	121	159.3
22	22.4	47	43.6	72	72.8	97	111.5	122	161.5
23	23.1	48	44.6	73	74.2	98	113.3	123	163.8
24	23.8	49	45.6	74	75.5	99	115.1	124	166.1
25	24.6	50	46.6	75	76.9	100	116.9	125	168.4
26	25.3	51	47.8	76	78.3	101	118.8	126	170.7
27	26.1	52	48.7	77	79.2	102	120.6	127	173.1
28	26.8	53	49.8	78	81.8	103	122.4	128	175.4
29	27.6	54	50.9	79	82.5	104	124.3	129	177.8
30	28.4	55	52.0	80	84.0	105	126.2	130	182.2
31	29.2	56	53.1	81	85.5	106	1281	131	182.6
32	30.0	57	55.4	82	87.0	107	130.0	132	185.1
33	30.9	58	56.6	83	88.5	108	132.1	133	187.6
34	31.7	59	57.1	84	90.1	109	135.1	134	190.1

Figure 14-37. Each refrigerant has its own temperature-vapor pressure chart. This chart allows the technician to estimate the temperature at the evaporator coils by measuring low side pressure.

sulfur or wax. The identification number of the oil refers to its viscosity. The lower the number, the less viscous the oil. It is very important to use the oil specified in the aircraft manufacturer's service manual when servicing the system. Whenever opening the system, you must purge all of the refrigerant. However, when purging the system, do not open the valves to the point at which the refrigerant can escape fast enough to blow out the oil with the vapor. To reduce the chance of contamination, tightly close the oil container when it is not in use. Never pour refrigerant oil from one container into another, and, discard oil removed from the system. Always service refrigerant systems with new oil.

RECEIVER-DRYER

The receiver-dryer is the reservoir for the system and is located in the high side between the condenser and the expansion valve. Liquid refrigerant enters from the condenser and is filtered and passed through a desiccant such as silica-gel to absorb any moisture that might be in the system. A sight glass is normally installed in the outlet tube to indicate the amount of charge in the system. Bubbles can be seen in the glass when the charge is low. A pickup tube extends from the top of the receiver-dryer to near the bottom where the liquid refrigerant is picked up. A filter is installed either on the end of the pickup tube or between the tube and the desiccant to prevent any particles getting into the expansion valve. It is of extreme importance that all moisture be removed from the system, as a single drop can freeze in the expansion valve and stop the entire air conditioning process. Water will also react with the refrigerant to form hydrochloric acid that is highly corrosive to the metal in the system. [Figure 14-38]

THERMAL EXPANSION VALVE

The thermal expansion valve is the control device which meters the correct amount of refrigerant into the evaporator. The refrigerant should evaporate completely by the time it reaches the end of the coils. The heat load in the aircraft cabin controls the opening, or orifice, in the valve. There are two types of thermal expansion valves, the internally equalized valve, and the externally equalized valve.

The internally equalized thermal expansion valve is controlled by the amount of heat in the evaporator. A capillary tube to the evaporator connects the diaphragm chamber of the valve. The end of the capillary is coiled into a bulb and is held tightly against the discharge tube of the evaporator. Coiling this tube allows a greater area to be held in intimate contact with the tube, allowing for a more accurate temperature measurement. If the liquid

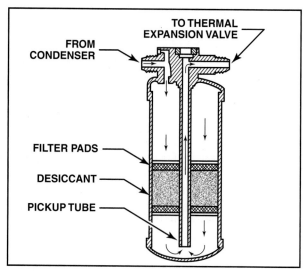

Figure 14-38. The receiver-dryer removes moisture from the system. If moisture remains in the system, the low temperatures will cause it to freeze, clog the small orifices within the system, and cause the system to stop working.

refrigerant completely evaporates before it reaches the end of the evaporator, it will continue to absorb heat and become superheated. It is still very cold to touch, but it is considerably warmer than it would be if it had not absorbed this additional heat. The expansion valve is adjusted to a given amount of superheat. When the pressure of the refrigerant vapor reaches this value, the diaphragm pushes down against the superheat spring and opens the valve, allowing more refrigerant to enter the evaporator. A balance between the vapor pressure on the diaphragm and the superheat spring controls the amount of refrigerant flow. These valves are calibrated at the factory and cannot normally be adjusted in the field. If there is a lot of heat in the cabin, the liquid refrigerant will evaporate quickly, and more superheat will be added to the vapor, so the valve will open and allow more refrigerant to flow into the evaporator. When the heat load is low, the liquid will use most of the evaporator length to evaporate. Little superheat will be added, and a smaller amount of refrigerant will be metered into the coils. [Figure 14-39]

The externally equalized expansion valve equalizes temperature against high-side temperature. There is a noticeable pressure drop across large evaporators because of the opposition to the flow of refrigerant. An externally equalized expansion valve has an additional port to adjust for this loss of pressure. This increased flow provided by the pressure equalization will maintain a constant pressure across the evaporator. The temperature sensing function of the valve is therefore able to meter the refrigerant as a function of the actual heat load in the cabin. [Figure 14-40]

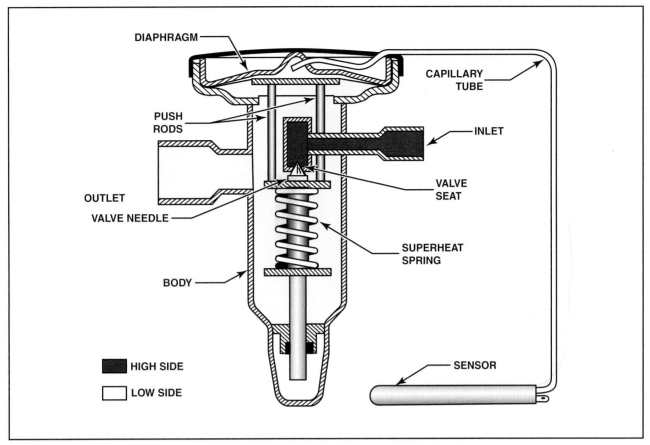

Figure 14-39. The internally equalized thermal expansion valve adjusts the amount of refrigerant so it finishes turning to a gas as it leaves the evaporator coils.

EVAPORATOR

The evaporator is the actual cooling unit in a vapor-cycle air-conditioning system. An evaporator consists of one or more circuits of copper tubing arranged in parallel between the expansion valve and the compressor. These tubes are silver-soldered into a compact unit, with thin aluminum fins pressed onto their surface. The evaporator is usually mounted in a housing with a blower. The blower forces cabin air over the evaporator coils. The refrigerant absorbs heat from the cabin air, thereby cooling it before it returns to the cabin. A drip pan is mounted below the evaporator to catch water that condenses out of the air as it cools. The capillary of the thermostat is placed between the fins of the evaporator core to sense the temperature of the coil, and it is this temperature that controls the cycling of the system. [Figure 14-41]

COMPRESSOR

The compressor circulates the refrigerant through the system. Refrigerant leaves the evaporator as a low-pressure, low-temperature vapor and enters the compressor. The compressor provides the energy necessary to operate the system. The gas leaving the compressor is at a high temperature and pressure. Aircraft air-conditioning systems usually use reciprocating-type compressors, which have reed valves and a lubricating system that uses crankcase pressure to force oil into its vital parts. [Figure 14-42]

On small aircraft, these compressors are usually belt driven by the engine, very similar to the arrangement used in an automobile. The compressors in systems used on larger aircraft are driven by electric or hydraulic motors, or by compressor bleed air powered turbines. Engine-driven compressors are single speed pumps whose output is controlled by a magnetically actuated clutch in the compressor drive pulley. When no cooling is needed, the clutch is de-energized and the compressor does not operate. When the air conditioner is turned on, and the thermostat calls for cooling, the magnetic clutch is energized, causing the drive pulley to turn the compressor and pump refrigerant through the system. [Figure 14-43]

Electric motor-driven compressors are controlled by a thermostat that turns the compressor motor on and off as required. Hydraulic motors are turned off and

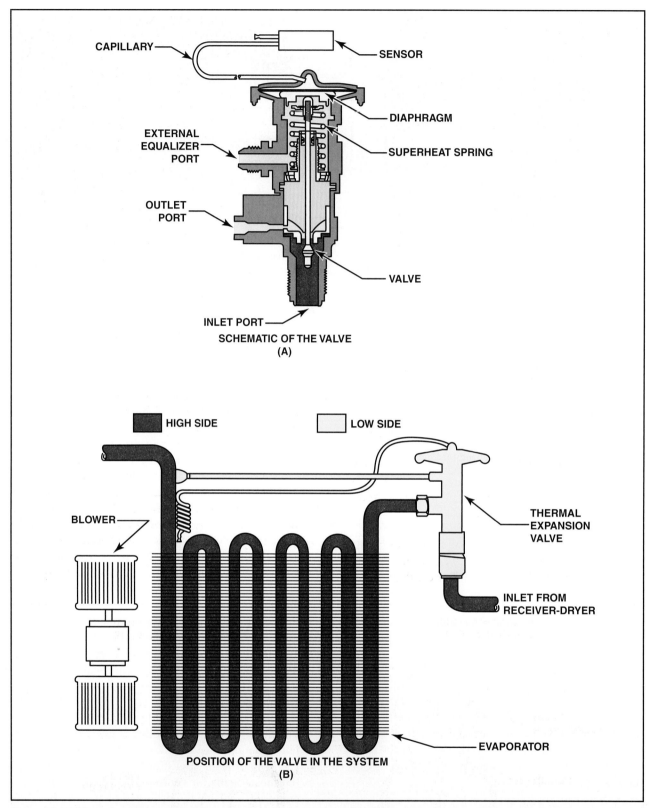

Figure 14-40. The externally equalized thermal expansion valve is used on large evaporators to compensate for pressure loss due to length of the evaporator coils.

on by solenoid valves controlled by the thermostat. When the valve is opened, hydraulic fluid is directed under pressure to the motor. When the motor is not being driven, the output of the engine-driven hydraulic pump is returned to the reservoir. In all of these systems, the cabin blower operates

Figure 14-41. The evaporator removes heat from the cabin air and transfers it to the refrigerant flowing through the evaporator coils.

Figure 14-43. A magnetically controlled clutch turns the compressor on and off as required to cool the cabin. This is similar to the system used on most modern cars.

continually, forcing the cabin air over the evaporator so heat from cabin air can be transferred into the refrigerant.

CONDENSER

The condenser is the radiator-like component that receives the hot, high-pressure vapors from the compressor and transfers the heat from the refrigerant vapors to the cooler air flowing over the condenser coils. When heat is removed from the vapor, the refrigerant returns to a liquid state.

The condenser is made of copper tubing with aluminum fins pressed onto it, formed into a set of coils, and mounted in a housing. The condenser and the evaporator are similar in both construction and appearance, differing primarily in strength. Since the condenser is in the high side of the system, it must be capable of withstanding the high pressure found there. Condensers normally operate at a pressure of about 300 psi and have a burst pressure in excess of 1,500 psi.

In some of the smaller airplanes the condenser is mounted under the fuselage where it can be extended down into the air stream when the system is operating, and retracted into the fuselage when the system is off. An interlock switch on the throttle retracts the condenser and de-energizes the compressor clutch when the throttle is opened for full power, to prevent the compressor loading the

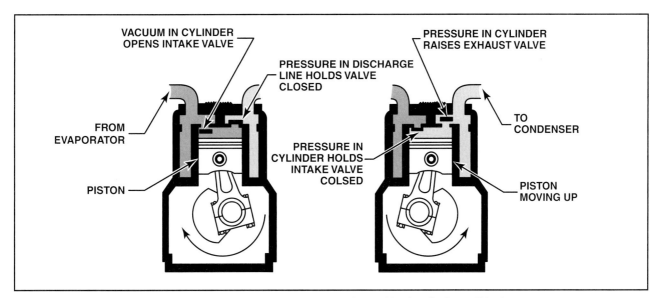

Figure 14-42. The reciprocating, piston-type, compressor is commonly used in aircraft air conditioning systems.

engine and the condenser causing drag when the airplane needs maximum performance such as for takeoff.

In larger aircraft the condenser is mounted in an air duct where cooling air can be drawn in from the outside and blown over the coils. In flight, ram air usually provides sufficient airflow over the condenser for proper operation. For ground operation, a fan must be used to supply the necessary cooling airflow.

Many vapor-cycle-cooling systems incorporate a sub-cooler that cools the liquid before it enters the compressor in addition to cooling the vapor after the condenser. Cooling the refrigerant in a sub-cooler prevents premature vaporization or flashoff.

SERVICE VALVES

The refrigeration system is sealed so that there is no opening to the atmosphere, but there must be some provision made to service it with refrigerant. Service valves provide access to the system, and there are two types of valves commonly found in aircraft air-conditioning systems: Schrader valves and compressor isolation service valves.

Schrader valves are often used when it is not convenient to service an aircraft system at the compressor because of the proximity of the propeller. The valves are mounted on either side of the evaporator or in some other part of the system where they can be reached for servicing. One of the valves is in the high side of the system, and the other is in the low side.

Schrader valves have a core similar to that used in a tire valve, and have only two positions, seated and open. To enter a system using Schrader valves, the service hose is screwed onto the valve, and a pin inside the hose will depress the valve core stem. When the hose is removed, the valve seats. A protective cap must be in place any time a hose is not attached to the valve to keep out contaminants. [Figure 14-44]

Compressor isolation service valves are normally mounted on the compressor itself, and in addition to allowing entry into the system for the service hoses, this valve can also be used to isolate the compressor from the system for servicing without losing the refrigerant charge. [Figure 14-45]

This valve has three positions. The valve is back-seated for normal system operation, and when back-seated, the service port is closed, and the passage from the system line to the compressor is open. To open the system for servicing, the service hose is

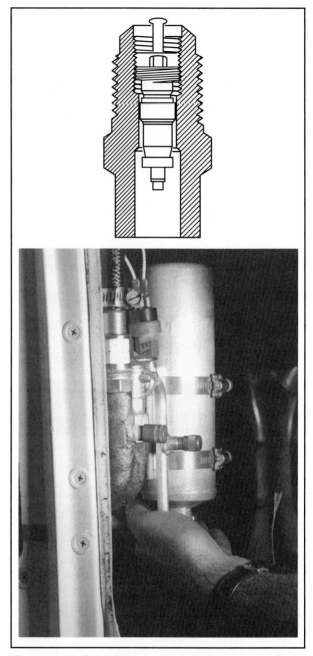

Figure 14-44. Schrader-type service valves are similar to those found on automobile tires.

screwed onto the service port and the valve turned three or four turns. In this intermediate position, the system can function normally, and the service line has access to the system for measuring the pressure or for adding or removing refrigerant. When the valve is front seated, turned all the way clockwise, the line to the system is closed and the compressor is isolated. The service valves must both be front-seated when checking the refrigerant oil in the compressor or for any other type of compressor servicing. When the system is closed for normal operation, a protective cap should be screwed onto the

Cabin Atmosphere Control

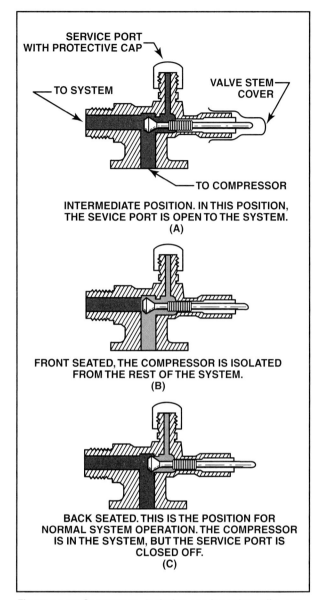

Figure 14-45. Compressor isolation service valves allow servicing of the system with the compressor isolated.

service port and a plastic or metal cap installed over the squared drive end of the valve stem.

SERVICE EQUIPMENT

Specialized equipment is required to perform servicing on air conditioning systems. Manifold sets that allow selective measuring of pressures as well as selective pressurizing and depressurizing of the two sides of the system have been in use for some time. In recent years, environmental concerns have dictated that refrigerants not be released into the atmosphere. Recycling/recovery equipment has been combined with the vacuum pump that has previously been used to depressurize the system. These units allow the reuse of the refrigerant that formerly was released to the atmosphere.

MANIFOLD SET

The manifold set consists of three fittings to which the service hoses are attached, two hand valves with O-ring seals, and two gauges, one for measuring the pressure in the low side of the system and one for the pressure in the high side. [Figure 14-46]

Figure 14-46. A manifold set is required to do most servicing of air conditioning systems.

The low-side gauge is a compound gauge, meaning that it will read pressure on either side of atmospheric pressure. Its range is from 30 inches of mercury (approximately 60 psi) gauge pressure below that of the atmosphere, to about 30 inches of mercury (approximately 60 psi) gauge pressure above atmospheric. **The high-side gauge** is a high-pressure gauge that has a range of from zero up to around 600 psi, gauge pressure.

The manifold connects the gauges, the valves, and the charging hoses. The low-side gauge is connected to the manifold directly at the low-side fitting. The high-side gauge is likewise connected directly to the high-side fitting. The center fitting of the manifold can be isolated from both of the gauges and from the high- and low-side service fittings by the hand valves.

When these valves are turned fully clockwise, the center fitting is isolated. When the low-side valve is opened by turning it counterclockwise, the center fitting is opened to the low-side gauge and the low-side service line. The same is true for the high side when the high-side valve is opened.

The charging hoses are attached to the fittings of the manifold set for servicing the system. The high side fitting may be located either at the compressor discharge, the receiver-dryer, or on the inlet side of the thermal expansion valve. The low-side service valve may be located at the compressor inlet, or at the discharge side of the expansion valve. The center hose attaches to the recovery/recycling/vacuum unit for evacuating the system, or to the refrigerant supply for charging the system. Charging hoses used with Schrader valves must have a pin to depress the valve, and these hoses are normally color-coded to quickly identify them. The high-side hose is red, the low-side hose is blue, and the center hose is usually yellow. When not using the manifold set, the hoses should be capped to prevent moisture from contaminating the valves. The caps are sometimes an integral part of the charging hoses. [Figure 14-47]

Figure 14-47. When the manifold set is not in use, the charging hoses should be protected by screwing their end fittings into plugs provided with the set.

REFRIGERANT SOURCE

The refrigerant used in aircraft air-conditioning systems has, until recently, been Refrigerant-12 (R-12). This material can be purchased in handy one-pound or two-and-a-half pound cans, ten- or twelve-pound disposable cylinders, or in larger returnable cylinders. The exact amount of refrigerant put into a system is determined by its weight rather than by its volume. In the late 1990s, concern for the ozone layer caused a shift away from use of R-12 refrigerant. New refrigerants do not have the same damaging chlorofluorocarbons (CFCs) as R-12. In most cases, new refrigerants are not compatible with systems designed to use R-12 and components of the system must be replaced with components compatible with the new refrigerants. Procedures for servicing systems using the new refrigerants are similar to those for R-12, but service manuals should always be consulted for exact procedures.

The smaller cans of refrigerant are opened with a special valve tap that is screwed onto the can to attach the manifold set. When the valve is attached to the can, the seal is pierced and the refrigerant can flow into the manifold set. Larger cylinders have a built-in shutoff valve to which the service hoses attach directly. A charging stand is the preferred way of handling the refrigerant since it provides all of the needed equipment and tools for servicing aircraft air-conditioning systems. [Figure 14-48]

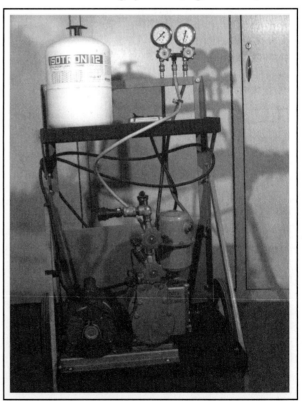

Figure 14-48. A charging stand provides all of the needed equipment and tools for servicing aircraft air-conditioning systems.

R-12 is normally put into the system in its vapor form after the system has been evacuated and is still under vacuum. Vapor is put into the system on the low side by holding the container upright. Heat will hasten the discharge of the refrigerant vapor, but care must be taken when you use heat. Use only water heated to about 120° F. NEVER USE DIRECT FLAME OR AN ELECTRIC HEATER. When the container is inverted, liquid will flow out. If allowed by the service manual, liquid should go into the high side, where it can go directly into the receiver-dryer. (NOTE: In some systems having the service valves quite a long way from the compressor, it is permissible to put liquid into the low side when the low-side pressure is low enough and the outside air temperature is high enough. Be sure to use only the procedures recommended in the aircraft service manual.) Special precautions should be observed to be sure that only the correct refrigerant is used. Refrigerant-22 is similar to R-12, but its pressure is higher for the same temperature, and there is a danger of damaging the system or causing leaks due to excessive pressure.

VACUUM PUMP

Just a few drops of water is all that is necessary to completely block an air-conditioning system. If this water freezes in the thermal expansion valve, the vapor cycle ceases. To eliminate any water from the system, the system must be evacuated. In this procedure, a vacuum pump (a part of the recovery/recycling/vacuum unit) is attached to the manifold set and all of the air, refrigerant and water vapor is pumped out of the system. As the vacuum pump reduces the system pressure below atmospheric pressure, the boiling point of water decreases and the resulting vapor will be drawn from the system. The vacuum pump used for the evacuation must produce an extremely low pressure, but the flow is of little importance. A typical pump used for evacuating air-conditioning systems pumps about 0.8 cubic foot of air per minute and will evacuate a system to about minus 29 inches of mercury, that is, below standard sea-level pressure. At this low pressure, water boils at temperatures as low as 45° F, becomes vapor, and is evacuated. [Figure 14-49]

LEAK DETECTOR

The continued operation of an air-conditioning system depends upon the system maintaining its charge of refrigerant, and all of the charge can be lost to even a tiny leak. Naturally a small leak of a colorless, odorless gas is difficult to find, and without the aid of a leak detector, it would be almost impossible.

Of the several types of leak detectors available, the most simple is a soap solution. A relatively thick solution of soap chips and water is applied with a

Figure 14-49. A vacuum pump is attached to the center hose of the manifold set to evacuate the air-conditioning system in the airplane.

paintbrush to any part of the system where a leak is suspected. Bubbles will indicate the presence of a leak.

A common type of leak detector used in automotive air conditioning and commercial refrigeration service is a propane-burner-type detector. The torch type leak detector is definitely NOT RECOMMENDED for use with an aircraft air-conditioning system, because of the danger of an open flame around the aircraft. The most acceptable type of leak detector for aircraft air-conditioning servicing is an electronic oscillator that produces an audible tone. The presence of R-12 will cause the frequency to increase to a high-pitched squeal. This type of detector is recommended because it is both safe and sensitive. A good electronic leak detector can detect leaks as small as one-half ounce per year. [Figure 14-50]

Figure 14-50. An electronic-type leak detector is being used to search for a leak around the service valves on the compressor.

SYSTEM SERVICING

Understanding the operation of an aircraft air-conditioning system and the function of each component makes servicing the system less difficult. There is not a great deal involved in maintaining these systems. An aircraft maintenance technician will normally only inspect the system and replace/repair portions of the air delivery systems. A certified refrigeration technician will normally do any service and repair to the vapor-cycle components themselves.

TESTS AND INSPECTION

A visual inspection of the aircraft air conditioning system will reveal most defects. All of the units in the system should be checked for indications of looseness, misalignment, and any indications of leakage. Since the refrigerant oil is dispersed throughout the system, it is quite possible for a leak to be indicated by oil seeping out at the point of the leakage. All of the air ducts should be inspected for indication of obstructions or deformation, and the blower motor should spin freely without any binding or excessive noise. The evaporator fins should be clean and free from dust, lint, or any other obstruction, and any fins that are bent over enough to obstruct airflow should be straightened with a fin comb. Distorted fins on the evaporator will block the air flow and prevent heat being absorbed by the refrigerant. Too much blockage can cause the evaporator to ice up. [Figure 14-51]

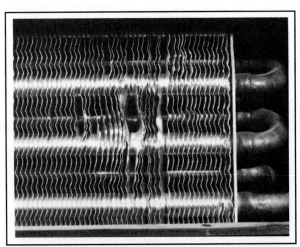

Figure 14-51. Distorted fins on the evaporator can reduce air flow and can possibly cause the evaporator to ice up.

The condenser should be checked for obstructions and security of mounting. If it is a retractable condenser, the mechanism that extends and retracts it should be checked, and it should come up streamlined with the structure when the system is turned off. On this type of installation, the condenser retracts and the compressor clutch de-energizes when the throttle is fully opened. The microswitch on the throttle should be checked for the proper adjustment and positive operation.

The compressor mounting brackets should all be checked, since the compressor is subject to extremely hard service. If the compressor is belt driven, the belt should be checked for tension and condition. A belt tension gauge should be used if available; otherwise the belt should be adjusted until there is about a half-inch deflection between pulleys when the pressure specified in the manual is put on the belt.

The entire run of hose from the compressor and condenser into the cabin should be checked for chafing or interference with the structure or any of the components. Grommets should be installed anywhere chafing could occur.

A leak test is used to detect a loss of refrigerant. Lack of refrigerant is one of the most common causes of failure to cool. The sight glass on the receiver-dryer should be inspected while the system is in operation. If bubbles are visible in the sight glass, there is not enough refrigerant in the system.

A complete absence of cooling with no bubbles in the sight glass could mean that there is no refrigerant in the system. In order to find the leak that caused the loss of refrigerant, the system must be at least partially charged.

The manifold set should be connected into the system with both the high- and the low-side valves closed. There should be at least 50 psi refrigeration pressure in the system. If there is not enough pressure for the test, refrigerant must be added. The high-side valve is opened and the proper type of refrigerant is allowed to flow into the system until the low-side gauge indicates about 50 psi; then the high side valve is closed.

The entire system is then checked with a leak detector. The probe should be held under every fitting where a leak could be present, especially at any point in the system where there is an indication of oil seepage. It is possible for there to be a very small leak at the front end of the compressor through the front seal, and since this seal is lubricated with refrigeration oil that is full of refrigerant, it may show up as a leak. To prevent this false indication, wash the oil out of the seal cavity with some solvent such as Xylene. A leakage of about one ounce of refrigerant per year is normally permissible through these seals, and this small leak should not be cause for worry.

One source of leakage which can cause a refrigerant loss without being found by a leak detector is the loss through the flexible hoses used in the system. Even though this type of hose is in good condition, it can allow several ounces of refrigerant to seep out each year through its pores. Since this leakage is spread throughout over the length of the hose, it is difficult to detect.

If a leak is found, the system should be evacuated and the leak repaired. The compressor oil should be checked whenever the system is evacuated.

A performance test is used to determine how well the system is functioning. The manifold set should be connected into the system with both valves closed. Run the engine at approximately 1,250 rpm, with the air-conditioning controls set for maximum cooling. Place a thermometer into the evaporator as near the coil as possible, and then turn on the blower to low or medium speed. [Figure 14-52]

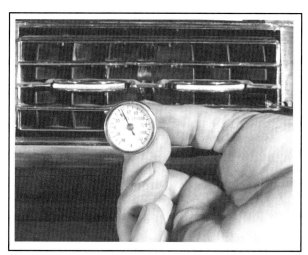

Figure 14-52. Part of the performance test of the air-conditioning system is to measure the temperature of the output air.

After the system has operated for a few minutes, the low-side gauge should read between 20 and 30 psig, and the high-side gauge should read in the range of 225 to 300 psig. The evaporator temperature should be somewhere around 40° to 50° Fahrenheit.

Touch can also used to determine that the system is operating normally. If the system is operating correctly, there should be no appreciable temperature difference between the inlet and the outlet side of the receiver-dryer; both sides should be warm to the touch. All of the lines and components in the high side of the system should be warm, and all of the lines and components in the low side of the system should be cool.

If the atmospheric conditions are especially humid, the amount of cooling will be reduced because of the water that condenses on the evaporator. When water changes from a vapor into a liquid, it gives off heat which goes into the refrigerant and decreases the amount of heat the refrigerant can absorb from the air in the cabin.

PURGING THE SYSTEM

Any time the system is to be opened, all of the refrigerant must be purged. To accomplish this, the manifold set is connected into the system with both valves closed, and the center hose attached to the recycling/recovery equipment. Environmental Protection Agency (EPA) regulations require that all CFC-12 refrigerants and substitute refrigerants be reclaimed when servicing air conditioning equipment. When the system is empty, it may be opened. Any time a system is opened, all of the lines should be capped to prevent the entry of water vapor, dirt or foreign matter.

CHECKING COMPRESSOR OIL

In order to check the oil, the system should be operated for at least fifteen minutes, then completely evacuated. When there is no pressure in the system, remove the oil filler plug from the compressor and use a special oil dipstick made according to drawings furnished by the airframe manufacturer. A range of oil level is indicated in the compressor service manual, and it should not be allowed to go below the minimum level, nor should it be filled above the maximum. [Figure 14-53]

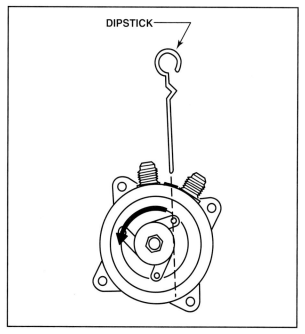

Figure 14-53. The correct amount of compressor oil is essential for proper functioning of the compressor.

Only oil recommended by the compressor manufacturer should be used, and should be kept tightly capped at all times it is not being used. After the proper amount of oil has been added to the system, the filler plug should be replaced and the system charged.

SYSTEM EVACUATION

Any time an air-conditioning system has been opened, it must be evacuated before it is recharged. Evacuating the system simply means "pumping the system down" by attaching a recovery/recycling/vacuum unit to the system and lowering the pressure so any water in the system will turn into a vapor and be drawn out.

Water boils at 212° F at the standard sea level pressure of 29.92 inches of mercury absolute (zero inches of mercury, gauge pressure). If the pressure is lowered 27.99 inches of mercury, gauge pressure (1.93″ ABS) the water will boil at 100° F. At 0.52 inches of mercury, gauge pressure (ABS), it will boil at 60° F. And, at 0.04 inches of mercury, gauge pressure(ABS), it will boil at 0° F. [Figure 14-54]

With the manifold set connected in the system, a recovery/recycling/vacuum unit is attached to the center hose. The low-side manifold valve should be opened and the gauge should indicate a vacuum. After pumping for about five minutes, the high-side gauge should indicate somewhat below zero, but the high range of this gauge will prevent any readable indication. After about fifteen minutes, the system should be down to around 25 inches of mercury gauge pressure, but the pump should be run for at least thirty minutes, and longer if possible. After closing both manifold valves, the recovery/recycle/vacuum unit should be removed, and the protective caps replaced on the pump fittings.

CHARGING THE SYSTEM

With the system still under vacuum from the evacuation process, both valves should be closed on the manifold set and the refrigerant source connected to the center hose. The high-side valve is then opened and the low-side gauge observed. As refrigerant flows into the system, the low-side gauge should come out of a vacuum, indicating that the system is

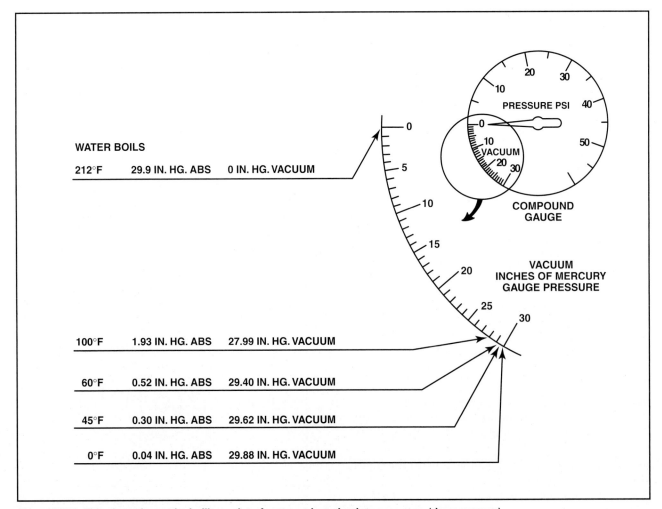

Figure 14-54. This chart shows the boiling point of water at low absolute pressures (deep vacuum).

clear of any blockage and is taking the charge of refrigerant. [Figure 14-55]

Both manifold valves are then closed and the engine started and run at about 1,250 rpm. The air-conditioning controls should be set for full cooling. With the R-12 container upright so vapor will come out, the low-side valve is opened to allow the vapor to enter the system. When the low-side pressure is down to below 40 psig, the can may be inverted and liquid allowed to enter the system. At this pressure, the liquid will turn into a vapor before it enters the compressor and will do no damage. NOTE: Do not invert the can if the outside air temperature is below 80° F. All of the R-12 may not be vaporized by this cool air.

The system should be charged with as many pounds of refrigerant as called for by the system specifications. A full charge will be indicated by the absence of bubbles in the sight glass in the receiver-dryer. Usually an additional quarter- or half-pound of refrigerant is added after the bubbles stop. When the charge is completed, the manifold valve is closed and performance tests performed.

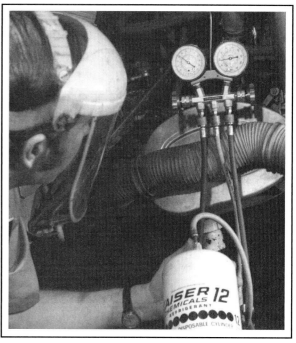

Figure 14-55. This air-conditioning system is being charged from a can of R-12.

CHAPTER 15

AIRCRAFT FUEL SYSTEMS

INTRODUCTION

Aircraft fuel systems vary in complexity from the extremely simple systems found in small, single-engine airplanes to the complex systems in large jet transports. Regardless of the type of aircraft, all fuel systems share many of the same common components. Every system has one or more fuel tanks, tubing to carry the fuel from the tank(s) to the engine(s), valves to control the flow of fuel, provisions for trapping water and contaminants, and a method for indicating the fuel quantity. Although fuel systems in modern aircraft are relatively simple, the safety and reliability of these systems is dependent upon proper inspection and maintenance.

All powered aircraft, whether rotary or fixed wing, depend upon a continuous, uninterrupted flow of uncontaminated fuel under all operating conditions. The weight of the fuel constitutes a large percentage of the aircraft's total weight. This may range from about 10% of the gross weight of small personal airplanes, to more than 40% for jet aircraft used on long overseas flights.

The weight of the fuel requires that the structure be strong enough to carry it in all flight conditions. The aircraft designer locates the fuel tanks so that the decreasing weight from fuel consumption will not cause balance problems. To reduce stresses on the airframe and improve structural life, many jet transports have fuel management procedures that specify how the fuel is to be used from the various tanks. For example, a Boeing 747 will first use the fuel in the center wing tank, followed by fuel in the inboard tanks until their quantities are equal to the outboard tanks.

Improper management of the fuel system has caused more aircraft accidents than failures of any other single system. Engine failure will occur if all of the fuel in the tanks has been burned, but engines will also stop if an empty tank is selected, even though there is fuel in the other tanks.

Contamination in the fuel may clog strainers or filters and shut off the flow of fuel to the engines. Contamination may take many forms, including solid particles, water, ice and bacterial growth. Water that condenses in partially filled tanks will stop the engine when it flows into the metering system. Water in turbine-powered aircraft is a special problem, as the more viscous jet fuel will hold water entrained in such tiny particles that it does not easily settle out. When the fuel temperature drops at high altitude, the water may form ice crystals that can freeze on the fuel filters and shut off the flow of fuel. Many jet engines have fuel heaters to prevent ice formation on the fuel filters and fuel metering system components.

The type or grade of aircraft fuel must be carefully matched to the engine. It is the responsibility of the pilot in command to verify before a flight is started that the aircraft is adequately supplied with the proper fuel. The person doing the refueling can assist by being vigilant for problems with fuel quality and type. A significant problem is the introduction of jet fuel, which is designed for turbine engines, into the fuel tanks of turbocharged piston engine aircraft. These engines are designed to operate on high-octane gasoline. At high power settings, jet fuel contamination causes severe detonation, which can lead to catastrophic engine failure, usually on takeoff. The potential for a disastrous accident is obvious.

SECTION A

AVIATION FUELS AND FUEL SYSTEM REQUIREMENTS

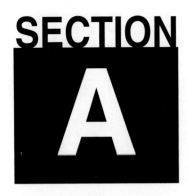

Aviation fuel is a liquid containing chemical energy that, through combustion, is released as heat energy and then converted to mechanical energy by the engine. Gasoline and kerosene are the two most widely used aviation fuels.

CHARACTERISTICS OF AVIATION FUELS

Weight is always a primary consideration in aircraft operation. Every extra pound (kilogram) used in the airframe and powerplant subtracts one pound (kilogram) from the aircraft's useful load. For this reason, aviation fuels must have the highest possible energy, or heat value per pound. Typical 100LL aviation gasoline (avgas) has 18,720 British Thermal Units (BTUs) per pound. Jet A turbine fuel has about 18,401 BTUs per pound. However, Jet A weighs 6.7 pounds per gallon while a gallon of 100LL weighs 6 pounds. In other words, jet fuel is denser than avgas, and as a result, Jet A supplies 123,287 BTUs per gallon whereas 100LL supplies 112,320 BTUs per gallon. The density of jet fuel varies more widely than the density of avgas with variations in temperature. Jet fuel density becomes an important factor when fueling large jet transport aircraft.

The dynamics of the internal combustion cycle demand certain properties from gasolines. Aircraft engines compound these demands because they must operate under a wide range of atmospheric conditions. One of the most critical characteristics of aviation gasoline is its volatility, which is a measure of a fuel's ability to change from a liquid into a vapor. Volatility is usually expressed in terms of Reid vapor pressure. Vapor pressure represents the required pressure above the liquid to prevent vapors from escaping at a given temperature. The vapor pressure of 100LL aviation gasoline is approximately seven pounds per square inch at 100° F. Jet A, on the other hand, has a vapor pressure of less than 0.1 psi at 100° F, and Jet B has a vapor pressure of between two and three pounds per square inch at 100° F. Automotive gasoline, by comparison, can produce vapor pressures as high as 14 psi at 100°F.

A fuel's volatility is critical to its performance in an aircraft engine. For example, in a piston engine, the fuel must vaporize readily in the carburetor to burn evenly in the cylinder. Fuel that is only partially atomized leads to hard starting and rough running. On the other hand, fuel that vaporizes too readily can evaporate in the fuel lines and lead to vapor lock. Furthermore, in an aircraft carburetor, an excessively volatile fuel causes extreme cooling within the carburetor body when the fuel evaporates. This increases the chances for the formation of carburetor ice, which can cause a rough running engine or a complete loss of engine power. Fuel injection systems reduce the problems associated with partial atomization and icing caused by fuel cooling but are also susceptible to vapor lock. Therefore, the ideal aviation fuel has a high volatility that is not excessive to the point of causing vapor lock.

RECIPROCATING ENGINE FUEL

Aviation fuels are distilled from crude oil by fractional distillation. Each different product to be extracted from the crude oil has a distinct boiling temperature. Each product is boiled off or separated from the crude oil as it is heated to increasingly higher temperatures. Gasoline boils at a relatively low temperature and is taken off first; then the heavier fractions are boiled off to become turbine engine fuel, diesel fuel, and furnace oil.

Aviation gasoline consists almost entirely of hydrogen and carbon compounds. Some impurities in the form of sulfur and dissolved water will be present. This water cannot be avoided since the gasoline is exposed to moisture in the atmosphere. A small amount of sulfur, always present in crude oil, is left in during manufacture.

The characteristics and properties of aviation gasoline govern the following: the selection of the proper fuel; the installation of hoses, gaskets, and seals; and the ability to operate the engine reliably, efficiently, and without damage. Though the aviation maintenance technician does not normally choose the system components, the technician will benefit from a thorough understanding of the various factors that go into such selection.

VOLATILITY

Volatility is a measure of a liquid's tendency to vaporize under given conditions. Gasoline is a complex blend of volatile hydrocarbon compounds that have a wide range of boiling points and vapor pressures. It is blended in such a way that a straight chain of boiling points is obtained. This is necessary to obtain the required starting, acceleration, power, and fuel mixture characteristics for the engine.

VAPOR LOCK

If the fuel does not vaporize readily enough, it can result in hard starting, slow warm-up, poor acceleration, uneven fuel distribution to the cylinders, and excessive crankcase dilution. If the gasoline vaporizes too readily, fuel lines may become filled with vapor and deliver a reduced supply of gasoline to the engine. In severe cases, this may result in engine stoppage. This phenomenon is referred to as vapor locking. The Reid vapor pressure test gives a measure of a gasoline's tendency to vapor lock. In this test, a sample of the fuel is sealed in a "bomb" equipped with a pressure gauge. The apparatus is then immersed in a constant-temperature bath and the indicated pressure is noted. The higher the corrected vapor pressure of the sample under test, the more susceptible it is to vapor locking. Aviation gasolines are limited to a maximum of 7 psi to minimize the tendency to vapor lock at high altitudes. [Figure 15-1]

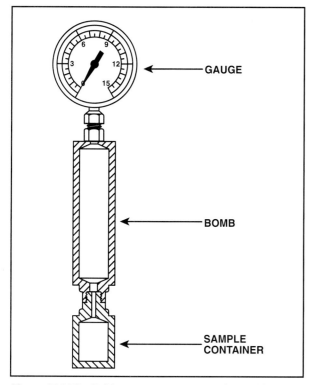

Figure 15-1. The Reid vapor pressure tester is used for measuring fuel samples. Vapor pressure is a major factor in the susceptibility of a fuel to vapor lock.

CARBURETOR ICING

Carburetor icing is also related to volatility. When fuel changes from liquid to vapor, it extracts heat from its surroundings. The more volatile the fuel, the more rapid the heat extraction. As gasoline vaporizes leaving the discharge nozzle of a float-type carburetor, it can freeze the water vapor in the incoming air. This moisture may freeze on the walls of the induction system, the venturi throat, or the throttle valve. This type of ice formation restricts the fuel and air passages of the carburetor. It can cause loss of power, and if not eliminated, eventual engine stoppage. This icing condition is most severe in temperatures ranging from 30° to 40° F (-1° C to +4° C) outside air temperature, but may occur at much higher temperatures.

AROMATIC FUELS

Some fuels may contain considerable quantities of aromatic hydrocarbons, which are added to increase the rich mixture performance rating of the fuel. Such fuels, known as aromatic fuels, can swell some types of hoses and other rubber parts of the fuel system. For this reason, aromatic-resistant hoses and rubber parts have been developed for use with aromatic fuels. The use of aromatic fuels is associated with the high-horsepower, reciprocating engines used on military and large transport-category aircraft. These aircraft are disappearing from the active fleet, and this type of fuel is no longer available.

DETONATION

Reciprocating engine aircraft require high-quality aviation gasolines to ensure reliable operation. These fuels are specially formulated to possess certain characteristics that allow them to function reliably in aircraft. To understand the different numbers used to designate fuel grades, the aircraft technician must first be familiar with detonation in reciprocating engines.

When a fuel-air charge enters the cylinder of a piston engine, it is ignited by the spark plugs. Ideally, the fuel burns at a rapid but uniform rate. The expanding gases then push the piston downward, turning the crankshaft and creating power.

Detonation is the explosive, uncontrolled burning of the fuel-air charge. It occurs when the fuel burns unevenly or explosively because of excessive temperature or pressure in the cylinder. Rather than smoothly pushing the piston down, detonation slams against the cylinder walls and the piston. The pressure wave hits the piston like a hammer, often damaging the piston, connecting rods, and bearings.

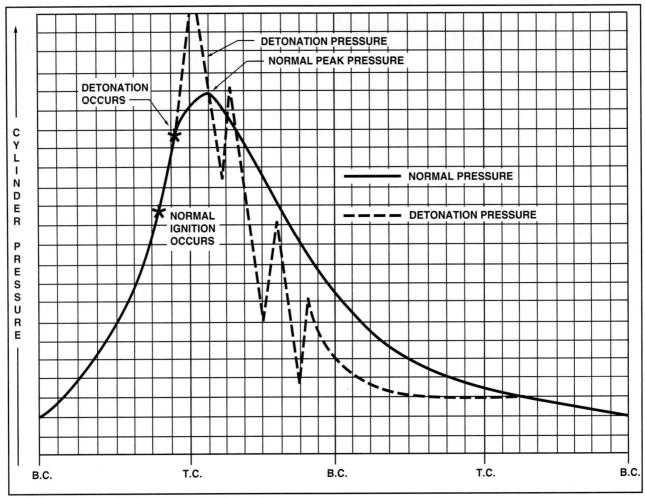

Figure 15-2. This chart illustrates the pressure created in a cylinder as it passes through its various strokes. As you can see, when normal combustion occurs, cylinder pressure builds and dissipates evenly. However, when detonation occurs, cylinder pressure fluctuates dramatically.

This is often heard as a knock in the engine. Detonation also causes high cylinder head temperatures, and if allowed to continue, can melt engine components. [Figure 15-2]

Detonation can happen any time that an engine overheats. It can also occur if an improper fuel grade is used. The potential for engine overheating is greatest under the following conditions:

Use of fuel grade lower than recommended

Takeoff with an engine that is above or very near the maximum allowable temperature

Operation at high rpm and low airspeed

Extended operations above 75 percent power with an extremely lean mixture

PREIGNITION

Combustion is precisely timed in a properly functioning ignition system. In contrast, preignition is when the fuel/air mixture ignites too soon. Preignition is caused by hot spots in the cylinder. A hot spot may be caused by a small carbon deposit, a cracked ceramic spark plug insulator, or almost any damage within the combustion chamber. When preignition exists, an engine may continue to operate even though the ignition has been turned off. In extreme cases, preignition can cause serious damage to the engine in a short period of time.

Preignition and detonation often occur simultaneously, and one may cause the other. Inside the aircraft, it will be difficult to distinguish between the two since both are likely to cause engine roughness and high engine temperatures.

OCTANE AND PERFORMANCE NUMBERS

Aviation gasoline is formulated to burn smoothly without detonating, or knocking, and fuels are numerically graded according to their ability to resist detonation. The higher the number, the more resistant the fuel is to detonation. The most common grading system is octane rating. The octane number assigned to a fuel compares the anti-knock properties of that fuel to a mixture of iso-octane and normal heptane. For example, grade 80 fuel has the same anti-knock properties as a mixture of 80 percent iso-octane and 20 percent heptane. 100 Octane fuel has the same anti-knock properties as pure iso-octane gasoline. Engines having high compression ratios and/or high horsepower output require higher-octane fuel.

Some fuels have two performance numbers, such as 100/130. The first number is the lean-mixture rating, whereas the second number represents the rich-mixture rating. To avoid confusion and to minimize errors in handling different grades of aviation gasolines, it has become common practice to designate the different grades by the lean mixture performance numbers only. Therefore, aviation gasolines are identified as Avgas 80, 100, and 100LL. Although 100LL performs the same as grade 100 fuel, the "LL" indicates it has a low lead content.

Another way petroleum companies help prevent detonation is to mix tetraethyl lead into aviation fuels. However, it has the drawback of forming corrosive compounds in the combustion chamber. For this reason, additional chemicals such as ethylene bromide are added to the fuel. These bromides actively combine with lead oxides produced by the tetraethyl lead allowing the oxides to be discharged from the cylinder during engine operation.

PURITY

Aviation fuels must be free from impurities that would interfere with the operation of the engine or the components in the fuel and induction system. Even though many precautions are observed in storing and handling gasoline, it is not uncommon to find a small amount of water and sediment in an aircraft fuel system. A small amount of such contamination is usually retained in the strainers of the fuel system. Generally, this is not considered dangerous if the strainers are drained and cleaned at frequent intervals. However, the water can present a serious problem because it settles to the bottom of the fuel tank and can then be circulated through the fuel system. A small quantity of water flowing with the gasoline through the carburetor jets will not be especially harmful. An excessive amount of water will displace the fuel passing through the jets, causing loss of power and possible engine stoppage.

Under certain conditions of temperature and humidity, condensation of moisture (from the air) occurs on the inner surfaces of the fuel tanks. Since this occurs on the portion of the tank above the fuel level, it is obvious that servicing an airplane immediately after flight will do much to minimize this hazard.

FUEL INDENTIFICATION

In the past, there were four grades of aviation gasoline, each identified by color. The only reason for mentioning the old ratings is because manuals on older airplanes may still contain references to these colors. The old color identifiers were:

    80/87 — Red

    91/96 — Blue

    100/130 — Green

    115/145 — Purple

The color code for the aviation gasoline currently available is as follows:

    80 — Red

    100 — Green

    100LL — Blue

A change in color of an aviation gasoline usually indicates contamination with another product or a loss of fuel quality. A color change can also be caused by a chemical reaction that has weakened the dye component. This color change itself may not affect the quality of the fuel, but if one has occurred, determine the cause before releasing the aircraft for flight.

The most positive methods of identifying the type and grade of fuel include the following:

1. Marking of the Hose. A color band not less than one foot wide is painted adjacent to the fitting on each end of the hose used to dispense fuel. The bands completely encircle the hose and the name and grade of the product is stenciled longitudinally in one-inch letters over the color band.

2. Fuel trucks and hydrant carts are marked with large fuel identification decals on each side of the tank or body and have a small decal on the dashboard. These decals utilize the same color code. The fixed ring around both the dome covers and hydrant box lids are also painted in accordance with the color code. In short, all parts of the fueling facility and equipment are identified and keyed into the same marking and color code. The

delivery pipes of truck fill stands are banded with colors corresponding to those on the dispensing hose. [Figure 15-3]

3. In addition to coloring fuels, a marking and coding system has been adopted to identify the various airport fuel handling facilities and equipment, according to the kind and grade of fuel they contain. For example, all aviation gasolines are identified by name, using white letters on a red background. In contrast, turbine fuels are identified by white letters on a black background.

TURBINE ENGINE FUELS

Aircraft gas turbine engines are designed to operate on a distillate fuel, commonly called jet fuel. Jet fuels are also composed of hydrocarbons with a little more carbon and a higher sulfur content than gasoline. Inhibitors may be added to reduce corrosion, oxidation, and the growth of microbes or bacteria. Anti-icing additives are also added. Turbine engines can operate for limited periods on aviation gasoline. However, prolonged use of leaded avgas forms tetraethyl lead deposits on turbine blades and decreases engine efficiency. Turbine engine manufacturers specify the conditions under which gasoline can be used in their engines, and these instructions should be strictly followed. Reciprocating engines will not operate on turbine fuel. Jet fuel should never be put into a piston engine aircraft.

VOLATILITY

One of the most important characteristics of jet fuel is its volatility. It must, of necessity, be a compromise between several opposing factors. A highly volatile fuel is desirable to aid starting in cold weather and to make aerial restarts easier and surer. Low volatility is desirable to reduce the possibility of vapor lock and to reduce fuel loss to evaporation.

At normal temperatures, gasoline in a closed container or tank can give off so much vapor that the

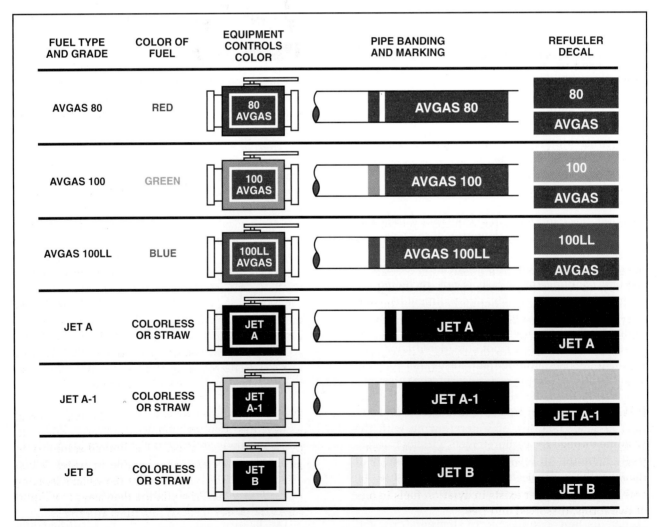

Figure 15-3. Labeling and color-coding of fuel carriers, hoses, and equipment helps to prevent filling the aircraft with the wrong fuel.

fuel/air mixture may be too rich to burn. Under the same conditions, the vapor given off by Jet B fuel can be in the flammable or explosive range. Jet A fuel has such a low volatility that at normal temperatures it gives off very little vapor and does not form flammable or explosive fuel/air mixtures. [Figure 15-4]

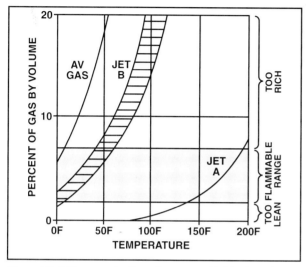

Figure 15-4. The vaporization of avgas and jet fuels varies as the temperature changes. This chart shows that at normal temperatures, avgas is too rich to burn, Jet A is too lean to burn, and Jet B is very flammable, at least in the low-normal temperature range.

FUEL TYPES

Because jet fuels are not dyed, there is no color identification for them. They range from colorless to straw-colored (amber), depending on age or the crude petroleum source.

There are currently two types of turbine fuel in use: JET A and JET A-1, which are kerosene types, and JET B, which is a blend of gasoline and kerosene fractions. Jet A-1 specifies a freeze point of -52.6° F (-47° C). Jet A specifies a freeze point of -40° F (-40° C). JP-4, similar to Jet B, is normally used by the military, particularly the Air Force. This fuel has an allowable freeze point of -50° C (-58° F). Jet fuel designations, unlike those for avgas, are merely numbers that label a particular fuel and do not describe any performance characteristics.

PROBLEMS WITH WATER IN TURBINE FUEL

Water has always been one of the major contamination problems with aviation fuel. It condenses out of the air in storage tanks, fuel trucks, and even in aircraft fuel tanks. Water exists in aviation fuels in one of two forms, dissolved and free.

Water dissolves in aviation fuels in varying amounts depending upon the fuel composition and temperature. This can be likened to humidity in the air. Undissolved, excess water is called "free water." Lowering the fuel temperature causes dissolved water to precipitate out as free water, somewhat similar to the way fog is created. Typically, dissolved water does not pose a problem to aircraft and cannot be removed by practical means.

Free water can appear as water slugs or as entrained water. A water slug is a relatively large amount of water appearing in one body or layer. A water slug can be as little as a pint or as much as several hundred gallons. Entrained water is suspended in tiny droplets. Individual droplets may or may not be visible to the naked eye, but they can give the fuel a cloudy or hazy appearance, depending upon their size and number. Entrained water usually results when a water slug and fuel are violently agitated, as when they pass through a pump. Lowering the temperature of a fuel saturated with dissolved water. Because of its high viscosity, entrained water is often visible in turbine fuel as a haze. Entrained water usually settles out with time.

Most aircraft engines can tolerate dissolved water. However, large slugs of free water can cause engine failure, and ice from slugs and entrained water can severely restrict fuel flow by plugging aircraft fuel filters and other mechanisms.

Jet engine fuel control mechanisms contain many small parts that are susceptible even to small accumulations of ice. Fuel heaters protect fuel systems that are subject to ice crystals. These devices can satisfactorily deal with dissolved and even entrained water; however, there is little margin for handling large amounts of free water. Some fuel filters are equipped with a differential pressure sensor across the filter element. This sensor will illuminate a warning light on the instrument panel if the filter ices up and the pressure drop across the element rises to the preset value. To further minimize the ice problem, most jet fuel is treated with an anti-icing additive that mixes with the water in the fuel and lowers its freezing point so it will remain in its liquid state.

One commonly used anti-icing additive is Prist, manufactured by PPG Industries. Prist is added to jet fuel during refueling. It has limited solubility in jet fuel but is completely soluble in water. When dissolved in water, Prist lowers the water's freezing point. The water/Prist mixture then stays in a liquid state and passes through fuel lines and filters.

High-flying jet aircraft are often equipped with a

temperature sensor in one of the outboard fuel tanks. The Aircraft Flight Manual may contain a restriction for maintaining the fuel temperature at a specified value above the fuel freezing point. If the fuel temperature becomes too cold, the aircraft may have to descend to warmer air to avoid problems with ice formation in the fuel.

Water can enter an airport fuel system through leaks in the seals of equipment, or brought in when fuel is delivered. The best means of minimizing the amount of water entering a system is through inspection and maintenance of equipment, and by making certain that only dry fuel is received.

Water can be detected in many ways. To find free water lying in the bottom of underground storage tanks, apply a water-finding paste to the end of a gauge stick and place it in the tank. Allow at least 30 seconds for the paste to react since other contaminants can slow its reaction time. For above ground tanks and equipment, draw a sample into a glass container and simply look for free water. A small amount of liquid vegetable dye (cake coloring) is helpful to highlight free water in a sample. It mixes with and colors the water but is insoluble in fuel. Water is removed from fuel by providing adequate filtration or separation equipment will remove water from the fuel.

The other problem with water in turbine engine fuel is that it may serve as a home for microscopic-sized animal and plant life. Microbial growth, or contamination with bacteria, or "bugs," has become a critical problem in some turbine fuel systems and some aircraft. Because microbes thrive in water, a simple and effective method to prevent or retard their growth is to eliminate the water.

Sometimes microbial growth occurs despite efforts to eliminate water from the fuel tanks. Microbiocides are introduced into fuel storage tanks to combat microbial growth. It should be introduced into the fuel when the tank is about half filled to ensure faster and more complete dispersion. Normally the microbiocide is introduced initially at a high concentration to kill the growth. Once the initial treatment is completed, the concentration is cut in half for long-term maintenance of fungus-free fuel.

FUEL CONTAMINATION

Contaminants can include either soluble or insoluble materials or both. The more common forms of aviation fuel contamination include solids, water, surfactants, and microorganisms. Fuel can be contaminated by mixing with other grades or types of fuels, by picking up compounds from concentrations in rust and sludge deposits, by additives, or by any of a number of other soluble materials.

The greatest single danger to aircraft safety from contaminated fuels cannot be attributed to solids, exotic microorganisms, surfactants, or even water. It is contamination resulting from human error. Any human error that fills an aircraft with the wrong grade or type of fuel or mixes different types of fuel is cause for serious concern. An accident may be the end result. The possibility of human error can never be eliminated, but it can be minimized through careful design of fueling facilities, good operating procedures, and adequate training.

BASIC FUEL SYSTEM REQUIREMENTS

Requirements for fuel system design are specified in detail in the parts of the Federal Aviation Regulations under which the aircraft was built. The vast majority of airplanes in the general aviation fleet are built under FAR Part 23, (*Airworthiness Standards: Normal, Utility, and Acrobatic Category Airplanes*). Awareness of the basic fuel system requirements for these airplanes will help the aircraft maintenance technician better understand the function of an aircraft fuel system.

1. No pump can draw fuel from more than one tank at a time, and provisions must be made to prevent air from being drawn into the fuel supply line. (23.951)

2. Turbine-powered aircraft must be capable of sustained operation when there is at least 0.75 cc of free water per gallon of fuel, and the fuel is cooled to its most critical condition for icing. The system must incorporate provisions to prevent the water that precipitates out of the fuel from freezing on the filters and stopping fuel flow to the engine. (23.951)

3. Each fuel system of a multi-engine aircraft must be arranged in such a way that the failure of any one component (except the fuel tank) will not cause more than one engine to lose power. (23.953)

4. If multi-engine aircraft feed more than one engine from a single tank or assembly of interconnected tanks, each engine must have an independent tank outlet with a fuel shutoff valve at the tank. (23.953)

5. Tanks used in multi-engine fuel systems must have two vents arranged so that they are not likely to both become plugged at the same time. (23.953)

Aircraft Fuel Systems

6. All filler caps must be designed so that they are not likely to be installed incorrectly or be lost in-flight. (23.953)

7. The fuel systems must be designed to prevent the ignition of fuel vapors by lightning. (23.954)

8. A gravity feed system must be able to flow 150% of the takeoff fuel flow when the tank contains the minimum fuel allowable, and when the airplane is positioned in the attitude that is most critical for fuel flow. (23.955)

9. A pump-feed fuel system must be able to flow 125% of the takeoff fuel flow required for a reciprocating engine. (23.955)

10. If the aircraft is equipped with a selector valve that allows the engine to operate from more than one fuel tank, the system must not cause a loss of power for more than ten seconds for a single-engine or twenty seconds for a multi-engine airplane when switching from a dry tank. (23.955)

11. Turbine-powered aircraft must have a fuel system that will supply 100% of the fuel required for its operation in all flight attitudes, and the flow must not be interrupted, as the fuel system automatically cycles through all of the tanks or fuel cells in the system. (23.955)

12. If a gravity-feed system has interconnected tank outlets, it should not be possible for fuel feeding from one tank to flow into another tank and cause it to overflow. (23.957)

13. The amount of unusable fuel in an aircraft must be determined and this must be made known to the pilot. Unusable fuel is the amount of fuel in a tank when the first evidence of malfunction occurs. The aircraft must be in the attitude that is most adverse for fuel flow. (23.959)

14. The fuel system must be so designed that it is free from vapor lock when the fuel is at a temperature of 110° F under the most critical operating conditions. (23.961)

15. Each fuel tank compartment must be adequately vented and drained so no explosive vapors or liquid can accumulate. (23.967)

16. No fuel tank can be on the engine side of the firewall, and it must be at least one-half inch away from the firewall. (23.967)

17. No fuel tank can be installed inside a personnel compartment of a multi-engine aircraft. (23.967)

18. Each fuel tank must have a 2% expansion space that cannot be filled with fuel, and it must also have a drainable sump where water and contaminants will normally accumulate when the aircraft is in its normal ground attitude. (23.969 and 23.971)

19. Provisions must be made to prevent fuel spilled during filling of the tank from entering the aircraft structure. (23.973)

20. The filler opening of an aircraft fuel tank must be marked with the word "AVGAS" and the minimum grade of fuel for aircraft with reciprocating engines. For turbine-powered aircraft, the tank must be marked with the word "JET FUEL" and the permissible fuel designation. If the filler opening is for pressure fueling, the maximum permissible fueling and defueling pressure must be specified. (23.1557)

21. If more than one fuel tank has interconnected outlets, the airspace above the fuel must also be interconnected. (23.975)

22. If the carburetor or fuel injection system has a vapor elimination system that returns fuel to one of the tanks, the returned fuel must go to the tank that is required to be used first. (23.975)

23. All fuel tanks are required to have a strainer at the fuel tank outlet or at the booster pump. For a reciprocating engine, the strainer should have an 8- to 16-mesh element, and for turbine engines, the strainer should prevent the passage of any object that could restrict the flow or damage any of the fuel system components. (23.977)

24. For engines requiring fuel pumps, there must be one engine driven fuel pump for each engine. (23.991)

25. There must be at least one drain that will allow safe drainage of the entire fuel system when the airplane is in its normal ground attitude. (23.999)

26. If the design landing weight of the aircraft is less than that permitted for takeoff, there must be provisions in the fuel system for jettisoning fuel to bring the maximum weight down to the design landing weight. (23.1001)

27. The fuel-jettisoning valve must be designed to allow personnel to close the valve during any part of the jettisoning operation. (23.1001)

SECTION B

FUEL SYSTEM OPERATION

The aircraft fuel system stores fuel and delivers the proper amount of clean fuel at the right pressure to meet the demands of the engine. The fuel system must be designed to provide positive and reliable fuel flow through all phases of flight. This must include changes in altitude, violent maneuvers, and sudden acceleration and deceleration. Furthermore, the system must be reasonably free from any tendency to vapor lock. Indicators such as tank quantity gauges, fuel pressure gauges, and warning signals provide continuous monitoring of how the system is functioning.

SMALL SINGLE-ENGINE AIRCRAFT FUEL SYSTEMS

Single-engine aircraft may utilize any of several types of fuel systems, depending upon whether a carburetor or fuel injection system is used, and whether the aircraft is a high-wing or low-wing design.

GRAVITY-FEED SYSTEMS

The most simple aircraft fuel system is found on small, high-wing single-engine training-type airplanes. These systems normally use two fuel tanks, one in either wing. The two tank outlets are connected to the selector valve. Fuel can be drawn from either tank individually, or both tanks can feed the engine at the same time. A fourth position on the selector valve turns off all fuel to the engine. Since both tanks can feed the engine at the same time, the space above the fuel in both tanks must be interconnected and vented outside of the airplane. The vent line normally terminates on the underside of the wing where the possibility of fuel siphoning is minimized. [Figure 15-5]

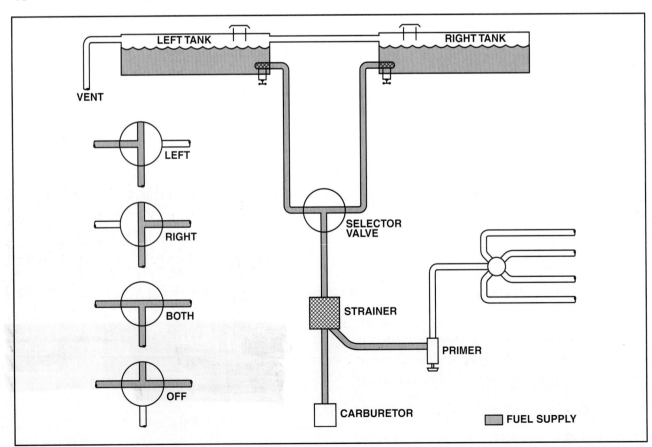

Figure 15-5. A typical small, single-engine high-wing airplane uses a gravity-feed fuel system to deliver fuel to the engine.

Aircraft Fuel Systems

After the fuel leaves the selector valve, it passes through the main strainer and on to the carburetor inlet. Fuel for the primer is taken from the main strainer.

PUMP-FEED SYSTEMS

Low-wing airplanes cannot use gravity to feed fuel to the carburetor. An engine-driven and/or electric pump must be used to provide adequate fuel pressure. The selector valve in these systems can normally select either tank individually, or shut off all flow to the engine. The selector valve does NOT have a "Both" position, because the pump would pull air from an empty tank rather than fuel from a full tank. After leaving the fuel selector valve, the fuel flows through the main strainer and into the electric fuel pump. The engine-driven pump is in parallel with the electric pump, so the fuel can be moved by either. There is no need for a bypass feature to allow one pump to force fuel through the other. To assure that both pumps are functioning, note the fuel pressure produced by the electric pump before starting the engine, and then, with the engine running, turn the electric pump off and note the pressure that is produced by the engine-driven pump. [Figure 15-6]

The electric pump is used to supply fuel pressure for starting the engine and as a backup in case the engine-driven pump should fail. It also assures fuel flow when switching from one tank to the other.

HIGH-WING AIRPLANE USING A FUEL INJECTION SYSTEM

The Teledyne-Continental system returns part of the fuel from the engine-driven fuel pump back to the fuel tank. This fuel contains any vapors that could block the system, and by purging all of these vapors from the pump and returning them to the tank, they cannot cause any problems in the engine. Fuel flows by gravity from the wing tanks through two feed lines, one each at the front and rear of the inboard end of each tank, into two small accumulator (reservoir) tanks, and from the bottom of these tanks to the selector valve. [Figure 15-7]

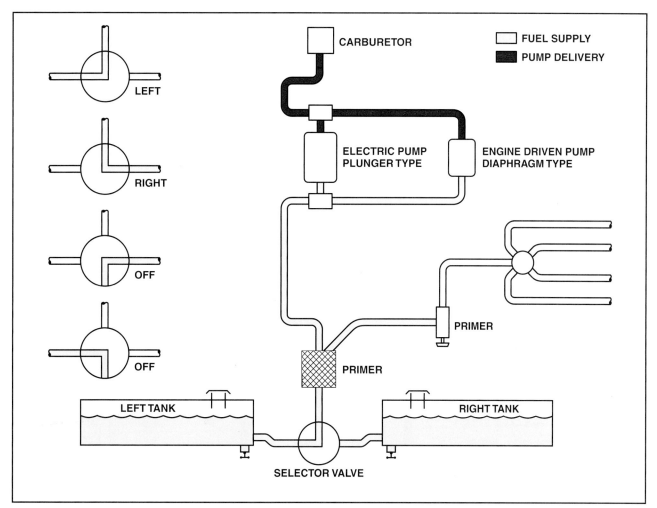

Figure 15-6. A typical small, single-engine, low-wing airplane uses a pump-feed fuel system to deliver fuel to the engine.

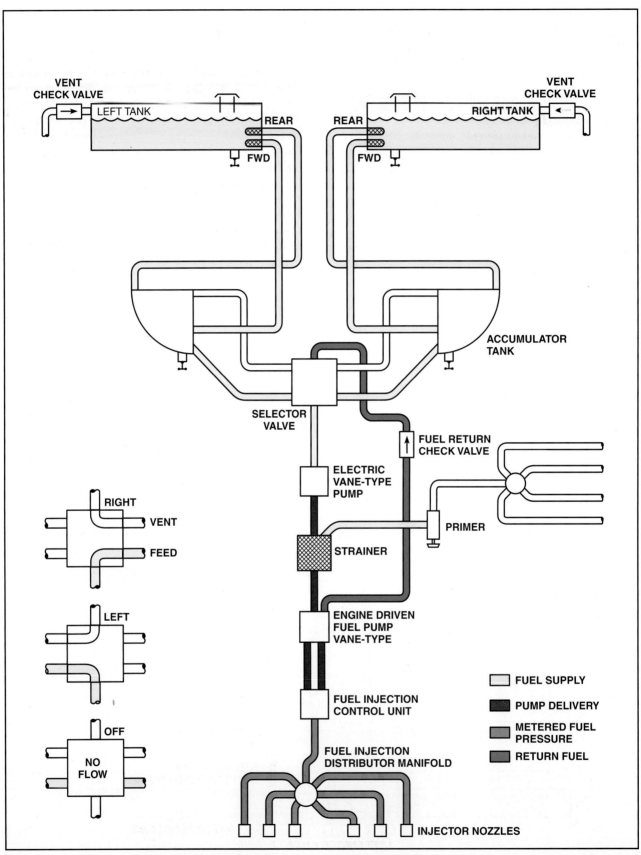

Figure 15-7. A Teledyne-Continental fuel injection system used on some high-performance single-engine airplanes. This system uses a combination of gravity feed, an electric vane-type pump, and an engine-driven fuel pump to deliver fuel to the engine. Excess fuel is returned to the selector valve in this system.

The selector valve directs fuel from the desired accumulator tank to the engine, and at the same time directs the fuel vapor from the engine-driven pump back to the selected reservoir tank. This vapor then returns to the wing tank that supplies the reservoir tank.

The electric auxiliary fuel pump picks up the fuel at the discharge of the selector valve and forces it through the strainer to the inlet of the engine-driven fuel pump. From the engine-driven fuel pump, the fuel flows to the fuel-air control unit where the fuel that is needed for engine operation goes to the cylinders, and all of the excess fuel returns to the inlet side of the pump. Some of this fuel has vapor in it and is returned to the selector valve through the fuel-return check valve.

SMALL MULTI-ENGINE AIRCRAFT FUEL SYSTEMS

The RSA fuel injection system does not return fuel to the tank like the Teledyne-Continental system just discussed. This system is used on both single and multi-engine aircraft.

When used on a multi-engine aircraft, each wing has two fuel tanks connected together, which serve as a single tank, and the selector valves allow any engine to operate from the tanks in either wing. The term "cross-feed" indicates that an engine is drawing fuel from the opposite wing. Fuel flows from the selector valve to the fuel filter, then to the electric fuel pump, on to the engine-driven pump, into the fuel injection system, and to the cylinders. [Figure 15-8]

Instrumentation for this system consists of fuel quantity, fuel pressure, and fuel flow gauges. The fuel quantity gauges show the total amount of fuel in the two tanks in each wing. The fuel pressure gauges show the pressure produced by each fuel pump. This pressure is measured at the inlet of the fuel metering unit. The fuel flow indicator is a differential pressure gauge that reads the pressure drop across the fuel injector nozzles and is calibrated in either gallons per hour or in pounds per hour of fuel burned.

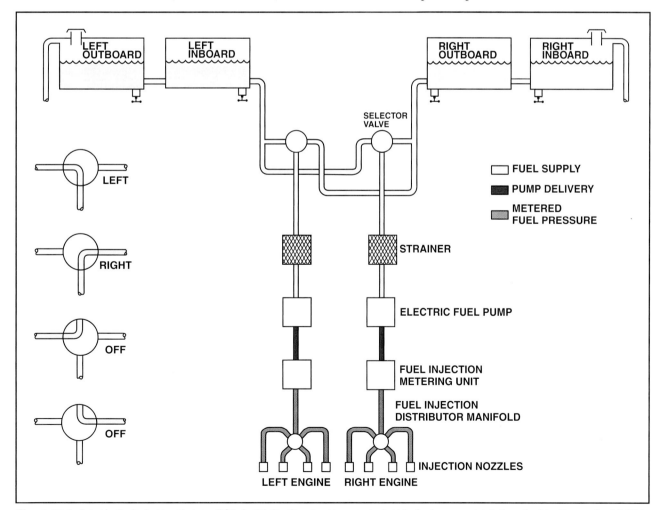

Figure 15-8. A typical airplane using an RSA fuel injection system uses electric fuel pumps to deliver fuel to the engine(s). This system does not return fuel to the tanks as in the Teledyne-Continental fuel injection system.

LARGE RECIPROCATING-ENGINE AIRCRAFT FUEL SYSTEMS

Transport-category aircraft powered by reciprocating engines are rapidly disappearing from the active fleet. One exception seems to be the venerable Douglas DC-3. This aircraft has seen a working life of more than 60 years and is still being used for passenger and cargo applications. The fuel system installed on the DC-3 is typical for aircraft using large radial-type engines. [Figure 15-9]

Two main tanks and two auxiliary tanks mounted in the center wing section of the airplane supply the fuel. The capacity of each main tank is 202 gallons, and the auxiliary tanks hold 200 gallons each. Provisions are made for the installation of 2 to 8 long-range tanks, each holding 100 gallons. This makes it possible to carry a fuel load as large as 1604 gallons in 12 tanks.

Fuel quantity is measured by a liquidometer system, which consists of a float assembly and a liquidometer unit in each tank. These are connected electrically to the fuel gauge on the right instrument panel in the pilots' compartment. There are two tank selector valves, operated by dial and handle controls in the pilots' compartment. Ordinarily the left-hand engine draws fuel from the left tanks, and the right engine draws fuel from the right tanks, but by using the selector valves, any tank may supply fuel to either engine.

Two hand-operated wobble pumps are used to raise the fuel pressure when starting the engines, or before the engine-driven pumps are in operation. Fuel flows from the wobble pumps through lines to the strainers located in each nacelle, through the engine-driven pumps, and from there, under pressure, into the carburetors. A cross-feed line is connected on the pressure side of each engine-driven pump, and the two cross-feed valves in this line are operated by a single control in the pilots' compartment. The cross-feed system enables both engines to receive fuel from one engine-driven pump in case either pump fails.

On later model airplanes, two electric booster pumps replace the wobble pumps. A fuel strainer is in the center wing near each selector valve. The fuel, therefore, flows from the selector valves, through the strainers, through the booster pumps, through the engine-driven fuel pumps into the carburetor. There is no crossfeed system on airplanes equipped with electric booster pumps. The booster pumps will furnish ample pressure and supply for operation of the airplane in case either engine-driven pump fails.

A vapor overflow line connects from the top chamber of the carburetor to the main tanks, and a fuel line from the back of each carburetor operates the fuel pressure gauge in the pilots' compartment. This pressure gauge normally shows from 14 to 16 pounds of pressure. On some airplanes, a pressure-warning switch is installed in the fuel pressure gauge line. When the fuel pressure drops below 12 pounds, the switch illuminates a warning light on the instrument panel.

A restricted fitting on the fuel-pressure gauge line connects to the oil-dilution solenoid. This unit releases fuel into the engine oil system and the propeller feathering oil to aid in cold weather starting. Another solenoid valve in the fuel-pressure gauge line releases fuel into the eight upper cylinders of the engines for priming.

Vent lines from each tank vent overboard, and a vapor line connects each main tank with its corresponding auxiliary tank.

JET TRANSPORT AIRCRAFT FUEL SYSTEMS

A large jet transport aircraft such as the Boeing 727 has a relatively simple fuel system that supplies its three engines from three fuel tanks. Tanks No. 1 and No. 3 are integral tanks, that is, part of the wing is sealed off and fuel is carried in the wing structure itself. Each of these tanks holds about 12,000 pounds of fuel. A fuselage tank, consisting of either two or three bladder-type fuel cells, holds another 24,000 pounds of fuel. [Figure 15-10]

Each of the wing tanks has two 115-volt AC electric boost pumps, and the fuselage tank has four of such pumps. Each of the three engines may be fed directly from one of the three fuel tanks, or all of the tanks and engines may be opened from the cross-feed manifold.

Perform pressure fueling by connecting the fuel supply to a single-point-fueling receptacle located under the leading edge of the right wing. Fuel flows from this receptacle, through the fueling and dump manifold, and into all three tanks through the appropriate fueling valves. When the tanks are completely filled, pressure shutoff valves sense the amount of fuel and shut off the fueling valve, which prevents the tank from being overfilled or damaged. If a partial fuel load is required, the person fueling the aircraft can monitor a set of fuel quantity gauges at the fueling

Aircraft Fuel Systems

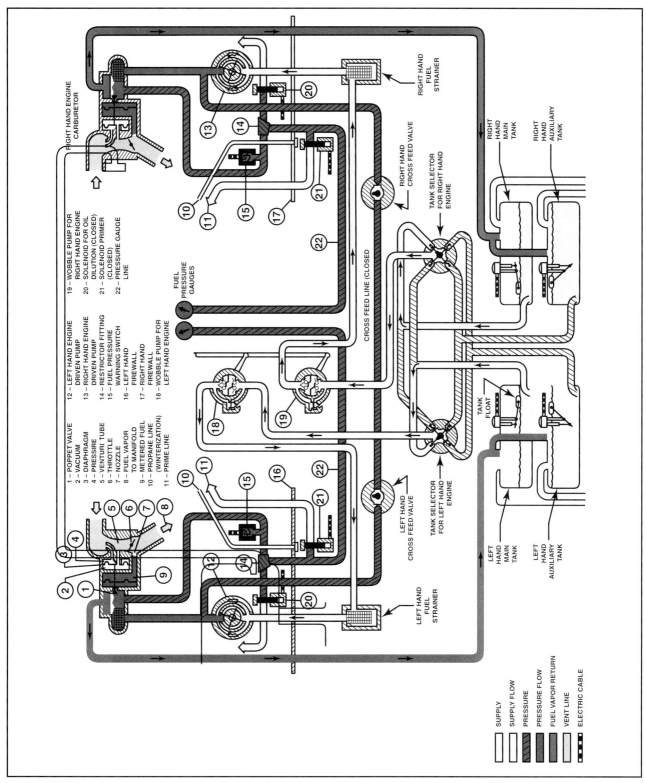

Figure 15-9. Older aircraft, such as the DC-3 shown here, had a complex, manual-fuel-management system.

station and shut off the fuel flow to any tank when the desired level is reached.

Defuel the airplane tanks by connecting the fuel receiving truck to the manual defueling valve, close the engine shutoff valves, and open the cross-feed valve from the tank to be emptied. Either pump out the fuel from the tank with the boost pumps, or pull it from the tank by suction from the receiving truck. If it is pulled out by

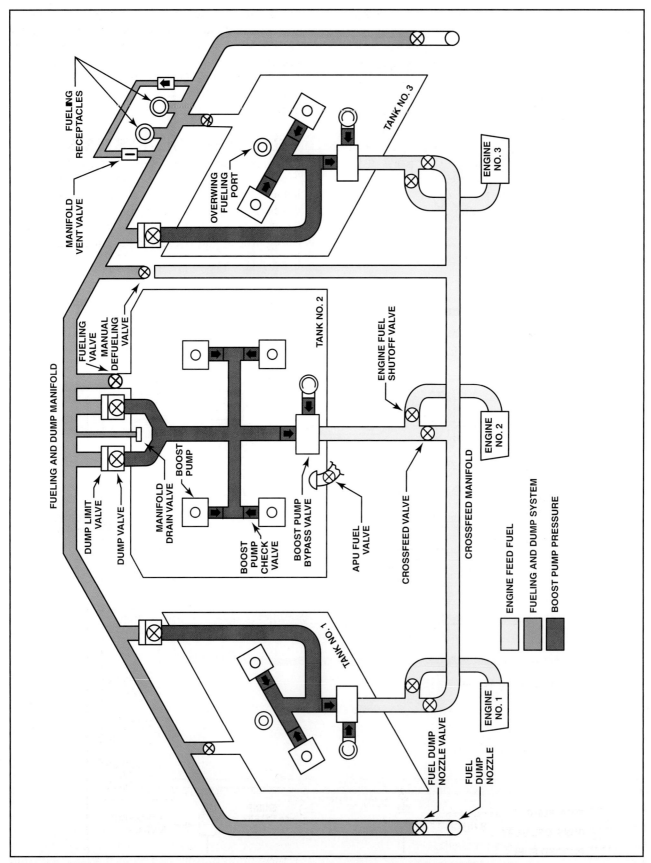

Figure 15-10. Modern jet aircraft have relatively simple systems. This system for the Boeing 727 normally feeds the three engines from individual systems. Any of the tanks can be used to supply the other engines, and a combination fueling and dump manifold is used to both fill and empty the tanks.

suction, it leaves the tank through the boost pump bypass valve.

Fuel may be dumped in-flight by opening the specific fuel dump valve and then opening the fuel dump nozzle valve in the wing tip through which the fuel is to leave the airplane. Fuel can be dumped from individual wing tips or from both tips at the same time. In-flight fuel jettison systems are divided into two separate systems, one for each wing. Dumping fuel from individual wings allows the pilot to control fuel balance. Fuel is dumped from locations that allow it to remain clear of any part of the aircraft.

There is a fuel dump limit valve in each of the three systems that will shut off the flow if the pressure drops below what is needed to supply the engine with adequate fuel. It will also shut off the dump valve when the level in the tank gets down to the preset dump shutoff level. This system is capable of dumping about 2,300 pounds of fuel per minute when all of the dump valves are open and all of the boost pumps are operating.

HELICOPTER FUEL SYSTEMS

A typical light turbine-powered helicopter system incorporates a single, bladder-type fuel cell, located below and aft of the rear passenger seat. Installed in the fuel cell are two submersible, centrifugal-type boost pumps, upper and lower fuel-quantity indicating probes, and a solenoid-operated sump drain. [Figure 15-11]

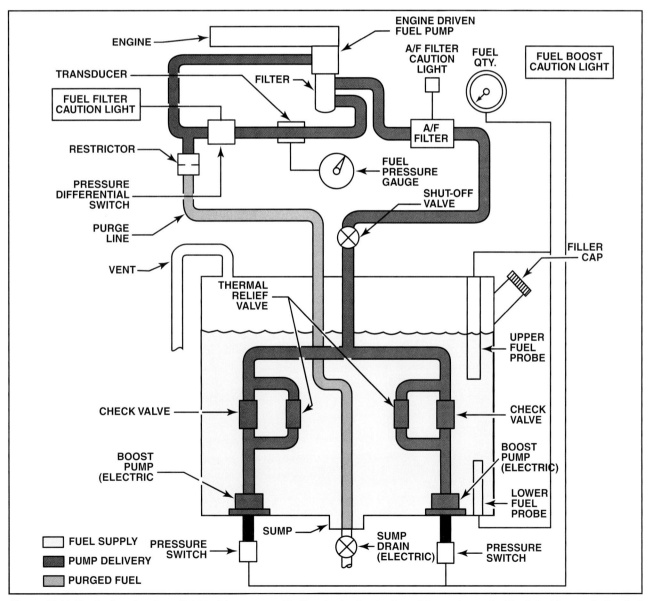

Figure 15-11. The fuel system for a light turbine-powered helicopter is simple since it has only one fuel tank that is usually located on the center of gravity of the helicopter.

The boost pumps are connected so that their outlet ports join to form a single line to the engine. Either pump is capable of supplying sufficient fuel to operate the engine. Check valves are installed at the outlet of each pump, and a pressure switch located in the outlet port of each pump will illuminate the FUEL BOOST CAUTION LIGHT in case of a pump failure. An electrically operated shut-off valve is installed in the fuel line running from the tank to the engine. A fuel selector valve is not necessary because only one tank is used in this system.

Fuel is filtered twice before entering the engine, and each filter is equipped with a warning light to indicate clogging. Additional provisions are made in the system for a fuel pressure gauge, vent system, and a fuel quantity indication.

AIRCRAFT FUEL SYSTEM COMPONENTS

The basic components of a fuel system include tanks, lines, valves, pumps, filtering units, gauges, warning systems, and for piston-engine aircraft, a primer. Some systems will include central refueling provisions, fuel dump valves, and a means for transferring fuel. In order to clarify the operating principles of complex aircraft fuel systems, the various units are discussed in the following paragraphs.

TANKS

The location, size, shape, and construction of fuel tanks vary with the type and intended use of the aircraft.

Fuel tanks are manufactured from materials that will not react chemically with any aviation fuel and have a number of common features. Usually a sump and drain are provided at the lowest point in the tank, and the top of each tank is vented to the atmosphere. All except the smallest of tanks are fitted with **baffles** to resist fuel surging caused by changes in the attitude of the aircraft. Many fuel tanks incorporate **flapper valves** to prevent fuel from flowing away from the boost pump or tank outlet when the aircraft is in a high "G" maneuver. In this capacity, the flapper valves serve as check valves. An expansion space is provided in fuel tanks to allow for an increase in fuel volume due to increases in its temperature.

Some fuel tanks are equipped with dump valves that make it possible to jettison fuel during flight in order to reduce the weight of the aircraft to its specified landing weight. In aircraft equipped with dump valves, the operating control is located within reach of the pilot, copilot, or flight engineer. Dump valves are designed and installed to afford safe, rapid discharge of fuel.

WELDED OR RIVETED FUEL TANKS

Most older aircraft use welded or riveted gasoline tanks to hold their fuel, but because of the limitations of weight and space, these tanks have been replaced almost totally by either integral or bladder-type tanks.

The smaller fuel tanks are made of thin sheet steel coated with an alloy of lead and tin. This material is called terneplate. Terneplate sheets are formed into the shapes needed to construct the tank, and all of the seams are folded in the best tradition of commercial sheet metal practice. Solder is sweated into the seams. This provides a good leak-proof joint, and the tanks are relatively low cost. The weight of a terneplate tank is more than that of an aluminum alloy tank, but for the type of airplane in which it is installed, the low cost advantage overcomes its weight disadvantage. Most terneplate tanks are of such small capacity that they seldom require baffles.

The larger fuel tanks of older aircraft are generally made of either 3003 or 5052 aluminum alloy. Both of these metals are relatively lightweight and easily welded. The parts of the tank are stamped out of sheet metal and formed to the required shape, and the tank is often riveted together with soft aluminum rivets to hold its parts in position. All of the seams are torch-welded to provide a fuel-tight seal.

Many tanks are large enough to require baffles to prevent the fuel from sloshing around in flight and either damaging the tank or causing balance problems.

All rigid fuel tanks must be supported in the aircraft structure with hold-down straps that will prevent the tank from shifting during any maneuver. All of the tank mounts must be padded with some type of material, usually felt, to prevent the tank from chafing against the structure.

Some modern small airplanes use fuel tanks that actually form part of the leading edge of the wing. Sealant is usually applied before these tanks are assembled. Others are welded together by electric resistance welding. Either of these tanks sometimes leak, and the manufacturer has approved a sloshing procedure to seal them. A sloshing compound is a liquid sealant that is poured into the tank, which flows over the entire inside surface and into the seams and crevices. The bulk of the compound is then poured out, and the sealant that remains inside the tank is allowed to cure. [Figure 15-12]

Figure 15-12. In this built-up fuel tank that forms the leading edge of a wing, sealant is placed in all of the seams before the tank is assembled.

INTEGRAL FUEL TANKS

Rigid tanks require a large open space in the aircraft structure for their installation, and very few aircraft have this amount of space that is not crossed with structural members. However, most wings have large empty spaces, and with the availability of new, space-age sealants, it has become standard practice for many aircraft manufacturers to seal off a portion of the wing to form a fuel tank. This type of tank has the advantage of using a maximum amount of space for the fuel and having a minimum amount of weight. A typical light-aircraft integral fuel tank occupies the leading edge portion of the wing from the front spar forward, and it is sealed at both ends and all along the spar with a two-part sealant. All of the rivets and nutplates are sealed, as well as around all of the inspection openings. The sealant is spread along each seam individually rather than sloshing the entire tank. [Figure 15-13]

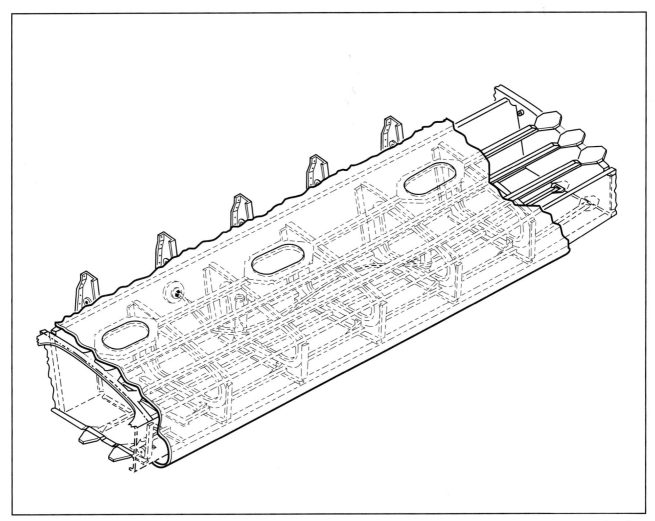

Figure 15-13. This drawing of an integral fuel tank illustrates a tank that makes up the bulk of the leading edge structure of a wing. All seams are sealed during assembly.

Some airplanes have the leading edge of the wing made of formed honeycomb, with facings of sheet aluminum or fiberglass on both the inside and outside. This makes an excellent fuel tank with minimal sealing. [Figure 15-14]

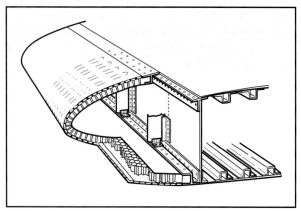

Figure 15-14. Leading edges constructed of aluminum honeycomb can be easily sealed to form a fuel tank.

BLADDER TANKS

The bladder tank is an excellent substitute for a welded fuel tank. Bladder tanks have been successfully used for both small and large aircraft. Prepare the fuel bay by covering all sharp edges of the metal structure with a chafe-resisting tape and install a bladder made of thin fabric, which is impregnated with neoprene or some similar material that is impervious to fuel. [Figure 15-15]

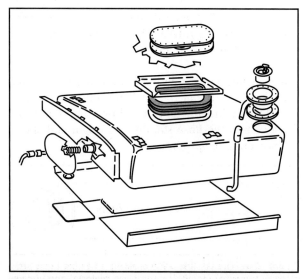

Figure 15-15. Bladder tanks are made of neoprene impregnated cloth that is snapped or laced into the fuel cell cavity in the wing of the airplane.

Put the bladder into the cavity prepared for it by folding and inserting it through an inspection opening. Snap or clip it in place, or in some instances, lace it to the structure. Secure an opening in the bladder to the inspection opening and cover it with an inspection plate.

There are a few considerations that must be observed with aircraft bladder tanks. On each inspection, be sure that the bladder has not pulled away from any of its attachment points. If it has pulled away, the amount of fuel the tank can hold will be decreased and the fuel quantity gauge will be inaccurate. Inspect bladder tanks for wrinkles that can trap water and prevent it from reaching the sump for removal. Also, never allow these tanks to stand empty for any extended period. If it is ever necessary to leave the tank empty for an extended period, wipe the inside of the bladder carefully with an oily rag, leaving a film of engine oil on its inside surface.

FUEL TANK FILLER CAPS

The filler cap on a fuel tank is perhaps one of the least noticed, but most important components on an aircraft. Take care when installing a fuel tank cap, and carefully examine it on each routine maintenance inspection.

Almost all fuel tank caps are located on the upper surface of the wing, and it is possible for fuel to be siphoned from the tank if the cap is leaking or improperly installed. Some fuel tank caps are vented, and it is important that the vent hole be clear. Some caps have a gooseneck tube on the vent that sticks up above the tank cap, and it is extremely important that these tubes point forward to provide a slight positive pressure inside the tank while in flight.

There are numerous types of fuel tank caps used on modern aircraft, and only the tank cap approved for that particular aircraft should be used. The cap is actually part of the fuel tank filler adapter assembly, and the replacement of one type of adapter with another usually constitutes a major alteration, requiring approval of the aircraft manufacturer and/or the FAA. Complete the appropriate paperwork after making the alteration. [Figure 15-16]

Locking fuel tank caps are popular on aircraft since vandalism has become so rampant. Foreign material put into a fuel tank can cause an expensive servicing problem, easily prevented with locking fuel tank caps.

Lightning-safe fuel tank caps are often installed on aircraft that fly in all types of weather. These caps have no metal exposed on the inside of the tank and

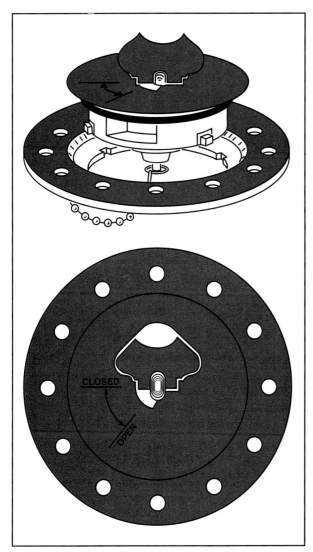

Figure 15-16. A typical fuel filler cap is part of a unit. If any part is damaged or lost, replace it with the proper part for that assembly.

will not conduct the lightning charge to the fuel. Even the lanyard that prevents the tank cap from being misplaced is made of a strong, nonconductive plastic material.

Non-siphoning fuel tank cap adapters have a small spring-loaded flapper inside the adapter that is pushed open by the fuel nozzle. The flapper closes when the nozzle is withdrawn from the tank. No fuel can siphon out of the tank even if the cap is left off the adapter.

FUEL LINES AND FITTINGS

The plumbing in aircraft fuel systems must be constructed of the highest quality material, and all of the workmanship must conform to approved aircraft practices. The metal tubing is usually made of aluminum alloy, and the flexible hose is made of synthetic rubber or Teflon. The engine's fuel flow requirements govern the diameter of the tubing.

Most of the rigid fuel lines used in an aircraft are made of 5052 aluminum alloy, but in some aircraft, the lines that pass through the wheel wells and some of the lines in the engine compartments are made of stainless steel as insurance against damage from either abrasion or heat. The fittings used on the lines may be of either the AN or MS flare type or a flareless type, depending upon the system installed by the manufacturer.

Replacement of a fuel line is normally done by installing a new line furnished by the aircraft manufacturer. If it is ever necessary to fabricate a line in the shop, use only the correct material for the line, and do not use substitute fittings without specific approval of either the manufacturer or the FAA.

Both the flare-type and the flareless fittings provide a good leak-proof connection if they are properly installed, and they will not usually develop a leak unless subjected to abuse or mistreatment. If a leaky fitting is found, remove the pressure from the system and re-tighten the fitting to proper torque specifications. If it is sufficiently tight, check the sealing surface for any indication of damage. If there are scratches or damage, replace the fitting. Make no attempt to stop a leak by overtightening a fitting. This is especially true of the flareless fittings, as they are highly susceptible to damage caused by excess torque. Tighten a flareless fitting finger-tight and then turn the fitting with a wrench only one-sixth to, at the most, one-third of a turn.

When installing flexible hoses in a fuel system, be sure that they do not twist when tightening the fitting. The lay line (often a line of printing) that runs the length of the hose should be straight and show no indication of spiraling. [Figure 15-17]

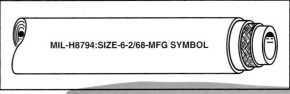

Figure 15-17. The lay line printed along a flexible hose provides a means of identifying the hose material and shows whether it has been twisted during installation.

Many of the fuel lines in an engine compartment are encased in a fire sleeve. If the aircraft requires this protection, install the proper type of fire sleeve in the manner specified by the aircraft manufacturer.

Route the fuel lines in accordance with the manufacturer's recommendations. In case of an alteration requiring new or re-routed fuel lines, there are certain basic requirements for routing fuel lines in an aircraft:

1. If it is impossible to physically separate fuel lines from electrical wire bundles, locate the fuel lines below the wiring and clamp the wire bundle securely to the airframe structure. It is never permissible to clamp a wire bundle to a fuel line.

2. Support all fuel lines so there will be no strain on the fittings, and never pull a line into place by the fitting.

3. There must always be at least one bend in rigid tubing between fittings. This allows for slight misalignment of the ends, for vibration, and for expansion/contraction caused by temperature changes.

4. Electrically bond all metal fuel lines at each point that they are attached to the structure. Do this by using bonded cushion clamps to hold the tubing. [Figure 15-18]

5. Protect all fuel lines from being used as a handhold.

6. To protect fuel lines from being stepped on or damaged by baggage or cargo, route them along the sides or top of compartments where this type of damage could occur.

FUEL VALVES

Selector valves are installed in the fuel system to provide a means for shutting off the fuel flow, for tank and engine selection, for crossfeed, and for fuel transfer. The size and number of ports (openings) vary with the installation. Valves may be hand-operated, motor-operated, or solenoid-operated. Valves must accommodate the full flow capacity of the fuel line, must not leak, and must operate freely. A manually operated valve must have a definite "feel" or "click" when it is in the correct position.

HAND OPERATED VALVES

Hand operated valves may be found on small and medium-sized aircraft, and will likely be either cone-type or poppet-type selector valves.

All fuel systems provide for shutting off the flow of fuel from the tanks to the engine, and most of the smaller aircraft use hand operated valves. The simplest valve is the **cone-type**, in which a cone, usually made of brass, fits into a conical recess in the valve body. The cone is drilled so it will allow flow from the inlet of the valve to any one of the outlets that is selected. A detent plate is installed on the shaft that is used to turn the cone, and a spring-loaded pin slips into the detent when the hole in the cone is accurately aligned with the holes in the valve body. This allows the pilot to tell by feel when the valve is in any given position. [Figure 15-19]

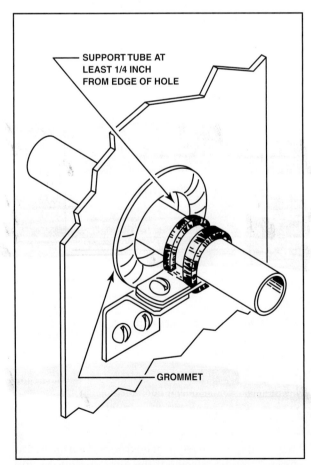

Figure 15-18. Support the fuel lines in an aircraft with bonded-type cushion clamps. Protect the edges of any hole that the tube passes through with a rubber or nylon grommet.

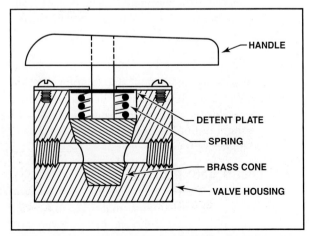

Figure 15-19. A cone-type fuel selector valve is used on some aircraft.

One problem with cone-type valves is that they can become difficult to turn. This can prevent the detent from providing a positive feel when the valve is properly centered in a selected position. In several instances, this condition has led to engine damage by preventing adequate fuel flow. The poppet-type valve overcomes this problem by using a camshaft, operated by the selector valve handle, to open the correct poppets and control the flow of fuel through the valve. The positive shutoff of fuel is provided by the spring on the valve, and it is easy for the pilot to tell by feel when the valve is centered in a selected position. [Figure 15-20]

MOTOR-OPERATED VALVES

Larger aircraft must use remotely operated valves in the fuel system. There are two basic types of remotely operated valves in popular use today: those driven by an electric motor, and those operated by a solenoid.

There are two types of motor-operated valves. In one of them, the motor drives a drum that has holes cut in it. This is so fuel can flow through the drum when it is in one position, and is shut off when it is rotated ninety degrees.

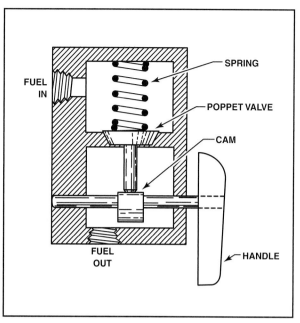

Figure 15-20. Poppet-type valves use a cam to open the poppet valves. This type of valve is much less likely to become difficult to turn than the cone-type valve.

The other valve uses a motor-driven sliding gate. The gate is drawn to let fuel through and shut to stop it. [Figure 15-21]

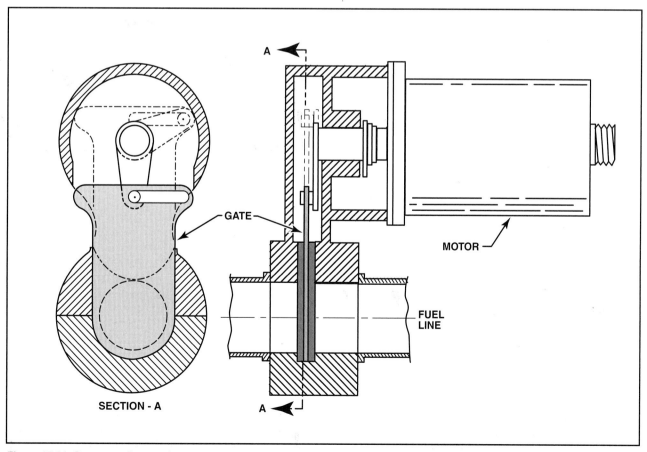

Figure 15-21. One type of remotely operated valve used in large aircraft is the motor-driven, gate-type fuel-shutoff valve.

SOLENOID-OPERATED VALVES

A solenoid valve has the advantage over a motor-driven valve of being much quicker to open or close. When electrical current momentarily flows through the opening solenoid coil, it exerts a magnetic pull on the valve stem to open the valve. When the stem rises high enough, the spring-loaded locking plunger of the closing solenoid is forced into the notch in the valve stem. This holds the valve locked open until current is momentarily directed into the closing solenoid coil. The magnetic pull of this coil pulls the locking plunger out of the notch in the valve stem, and the spring closes the valve and shuts off the flow of fuel. [Figure 15-22]

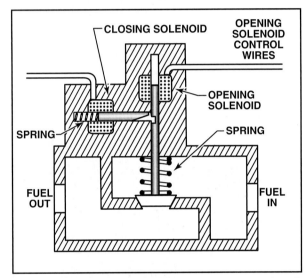

Figure 15-22. A solenoid operated poppet-type valve. A plunger holds open the valve until a closing solenoid pulls the plunger, allowing the valve to close.

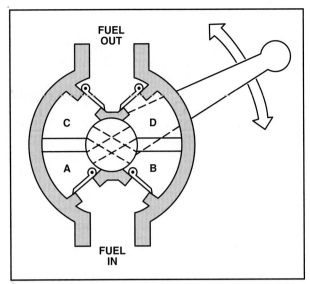

Figure 15-23. Larger aircraft that have the capability of transferring fuel from one part of the system to another are often equipped with a manual fuel-transfer pump called a "wobble pump."

FUEL PUMPS

The purpose of an engine-driven fuel pump is to deliver a properly pressured, continuous supply of fuel during engine operation. Auxiliary fuel pumps may be installed in the system to aid in engine starting and to assure a positive pressure to the inlet of the engine-driven fuel pump.

HAND-OPERATED PUMPS

Hand-operated fuel pumps are often called "wobble pumps." The name comes from the method of operation of one of the early types of hand fuel pumps. These pumps are used for backing up an engine-driven pump and for transferring fuel from one tank into another. [Figure 15-23]

When the handle is moved up and down, the vane inside the pump rocks back and forth. When the handle is pulled down, the left side of the vane moves up, pulling fuel into chambers A and D through the flapper-type check valve and the drilled passage between the chambers. Fuel in chamber B is forced into chamber C through the passage drilled through the center of the vane, and out the pump discharge line through the check valve. When the handle is moved up, the vane moves in the opposite direction, pulling fuel into chambers B and C. The fuel in chamber A is forced out of the pump through chamber D. This is a double-acting pump since it moves fuel on each stroke of the handle.

CENTRIFUGAL BOOST PUMP

By far the most popular type of auxiliary fuel pump in use in modern aircraft is the centrifugal boost pump. These pumps are installed on either the inside or the outside of the fuel tank. [Figure 15-24]

An electric motor drives a centrifugal pump, and it uses a small impeller to sling fuel out into the discharge line. These pumps are not of the constant displacement type, so restricting their outlet does not affect them. Many are two-speed types that use an electrical resistor in series with the motor to vary its speed.

Some centrifugal boost pumps have a small agitator on the pump shaft that stirs up the fuel being drawn into the impeller. Any of the tiny vapor bubbles that form in the fuel are forced to coalesce into larger bubbles and rise to the top of the tank rather than enter the fuel line.

This boost pump is used in its low-speed position for starting the engine and for minor vapor purging.

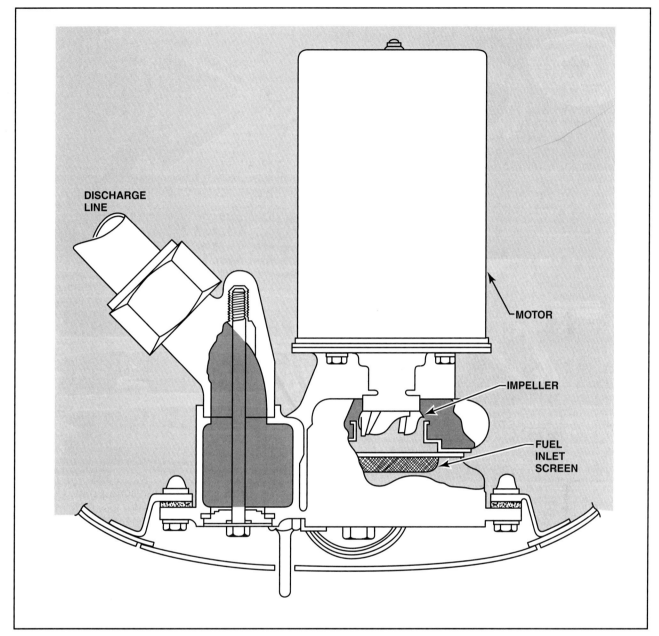

Figure 15-24. Centrifugal boost pumps are often submerged inside a fuel tank.

In its high-speed position, it is used as a backup for the engine-driven pump during takeoff and high power engine operation. It is also used in its high-speed position for major purging of fuel vapors. Some installations are quite critical with regard to the pressure delivered by the boost pump to the engine, and these systems have resistors in the boost pump circuit that are controlled by a precision switch on the throttle. When the throttle is opened and the boost pump switch is on, the pump operates at its high speed, but when the throttle is retarded, the pump speed will automatically decrease. This lowers the output pressure enough so that the boost pump will not flood the engine.

FUEL EJECTORS

To assure that there will always be an adequate supply of fuel available, boost pumps are sometimes located in a fuel collector can. This is an area of the fuel tank that has been partitioned off and equipped with a flapper-type valve to allow fuel to flow into the collector from the tank. A fuel ejector system uses the venturi principle to supply additional fuel to the collector can, regardless of aircraft attitude.

The submerged motor-driven boost pumps supply fuel from each tank to their respective engines. During operation of the boost pumps, a portion of their output is routed to the fuel ejectors. The flow of fuel through a venturi supplies the low pressure needed to draw additional fuel from the ejector location. Fuel is then routed to the fuel collector cans.

PULSATING ELECTRIC PUMPS

Cost is always a critical concern when producing smaller personal airplanes for the general aviation fleet. Rather than using the more expensive centrifugal boost pumps, many of the smaller low-wing airplanes use a pulsating pump that is similar in operation to the electric pumps used in automobiles and trucks for many years.

This simple type of pump consists of a solenoid coil installed around a brass tube that connects the two fuel chambers. In the core of the coil, a steel plunger rides up and down inside the brass tube. A calibrated spring forces the plunger upward, and the solenoid's magnetism pulls it down. One check valve is installed inside the plunger and another is in the extension of the brass tube in the fuel inlet chamber. [Figure 15-25]

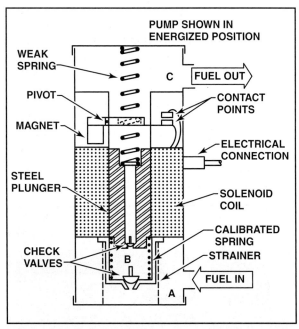

Figure 15-25. The plunger-type auxiliary fuel pump uses a solenoid and spring to move a plunger up and down within the barrel of the pump.

When the pump is not turned on, the calibrated spring forces the plunger up in the brass tube where it attracts the magnet through the tube, pulling the points closed. When the pump is turned on, current flows through the contact points and energizes the solenoid coil. This pulls the plunger down into the coil. The fuel in chamber B passes up through the check valve into the plunger. When the plunger is centered in the coil, it no longer has any effect on the magnet attached to the contact points, springing

these points open and stopping the current flow into the coil. The calibrated spring forces the plunger up and the fuel out through chamber C and the discharge line to the engine. As the plunger moves up, fuel flows from the inlet through chamber A and the lower check valve into chamber B, and the cycle is ready to start over. This type of pump will pulsate rapidly when the engine is accepting all of the fuel it pumps. But when the needle valve on the carburetor is closed, the pressure will build up in the line between the carburetor and the pump, and the pump will pulse slowly.

A plunger type pump is normally installed in parallel with a diaphragm-type engine-driven pump so that either or both pumps can supply fuel pressure to the engine.

VANE-TYPE FUEL PUMPS

An accessory drive on the engine or an electric motor drives the rotor in a vane-type fuel pump. Four steel vanes slide back and forth in slots cut in the rotor. A hard steel pin floats in the hollow center of the rotor, holding the vanes against the wall of the pump cavity in which the rotor fits eccentrically. As the rotor turns, fuel is drawn into the pump through the inlet and into the space between the vanes. It is then forced out of the pump on the discharge side. This is a constant displacement type of pump, and every time the rotor makes one revolution, it moves a given amount of fuel. Fuel systems are not designed to accept all of the fuel that this type of pump can move, so provision must be made to relieve the discharge pressure back to the pump inlet. This prevents the output pressure from building up beyond the desired value when the pump discharge is restricted. [Figure 15-26]

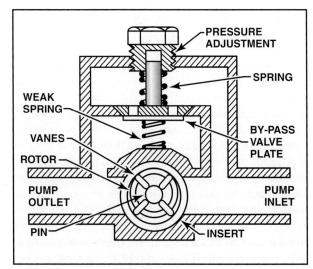

Figure 15-26. The vane-type fuel pump is a constant displacement pump.

Aircraft Fuel Systems

Vane-type pumps have a relief valve that opens when the discharge pressure reaches a set value. A screw adjusts the tension on the relief valve spring to control the discharge pressure setting. When the valve lifts off its seat, the fuel flows back to the inlet side of the vane assembly, and the discharge pressure is maintained at the value for which the relief valve is set. [Figure 15-27]

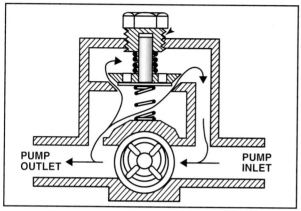

Figure 15-27. Vane-type fuel pumps utilize a relief valve to maintain the proper system pressure.

These pumps, when installed on the engine, are in series with the boost pump. They must be capable of bypassing all of the fuel when the engine is being started and in case the vane-type pump should ever fail. The bypass valve is usually a spring-loaded disk on the lower face of the relief valve. If the pressure at the pump inlet is ever greater than its outlet pressure, the fuel will force the disk away from the relief valve and fuel will flow through this opening to the engine. The bypass valve spring is so weak that there is negligible opposition to the flow of fuel. There are other features of vane-type pumps that are of no concern when they are used as boost pumps, but will be discussed when they are used as engine-driven pumps. [Figure 15-28]

FILTERS

It is essential that the fuel supplied to an aircraft engine be free from contamination. Because of this, every aircraft fuel system requires a series of strainers and filters. In addition to strainers, there must be a provision for draining a sample of fuel from all of the tanks and from the main strainer. This is to examine the fuel for the presence of any water or solid particles that could have condensed in the tank or been introduced during fueling.

Almost all fuel tanks used in smaller aircraft have a rather coarse mesh finger strainer at the tank outlet. This strainer increases the area of the discharge port of the tank and helps prevent contaminants from

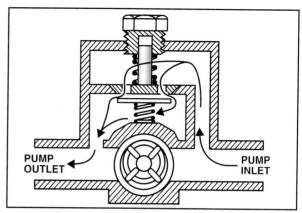

Figure 15-28. If fuel inlet pressure is greater than pump pressure, the bypass valve opens to allow fuel to flow around the pump vanes.

shutting off the flow of fuel. If a boost pump is installed in the tank, the screen is around the inlet to the pump. Here it serves the same function: to prevent the fuel flow from being stopped by anything less than an excessive amount of contamination.

All fuel tanks should have a drain valve (sump) or drain plug at their lowest point. Most operators take a sample of fuel before the first flight of the day, after each refueling, and in some instances, before each flight. Care should be taken to wait an appropriate amount of time before testing for contamination since it takes time for water to settle out of fuel, depending on how much agitation had occurred, and the thickness of the fuel. Any water or solid contaminants in the fuel settle around this valve where they can be drained out. [Figure 15-29]

Figure 15-29. Fuel tank sump drains are used to rid the system of water and other contaminants that have settled to the sump area of the system.

Many smaller fuel strainers use a filter element, consisting of a simple disk of relatively fine mesh screen wire at the top of the strainer bowl. Fuel from the tank enters the bowl through the center of the screen, and to get to the carburetor, it must flow upward through the screen. Water and solid contaminants cannot pass through the screen, and so they collect in the bowl. Most of these bowls are equipped with a quick-drain valve to drain out a sample of the fuel to check it for water or solid particles.

Many larger fuel strainers use a cylindrical screen wrapped around a coarse mesh screen that gives the strainer its shape and physical strength. Inside the cylinder is a cone, also made of screen wire, providing additional surface area for the strainer. Fuel flows into the strainer around the outside of the screen and up through the inside of the cone. Any water or contaminants will collect in the bottom of the strainer housing where they can be drained out on a routine maintenance inspection or during a preflight inspection. The main fuel strainer is normally located at the lowest point in the fuel system. [Figure 15-30]

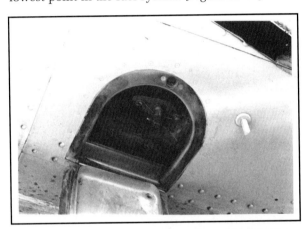

Figure 15-30. The main fuel strainer of an aircraft fuel system also employs a drain for ridding the system of contaminants.

FUEL HEATERS AND ICE PREVENTION SYSTEMS

Turbine-powered aircraft that operate at high altitudes and low temperatures for extended periods of time have a problem with water condensing out of the fuel and freezing. Ice crystals may collect on the fuel filters and shut off fuel flow to the engine. To prevent this, these aircraft have a fuel temperature gauge to inform the flight engineer when there is a danger of ice formation. [Figure 15-31]

Fuel filters have a pressure switch connected across the filter element that will close if the element clogs with enough ice to obstruct the fuel flow. Illumination of the fuel icing light informs the flight engineer that one of the fuel filters is clogging. The engineer can then open the fuel heat valve and compressor bleed air will flow through the fuel/air heat exchanger. This will raise the temperature of the

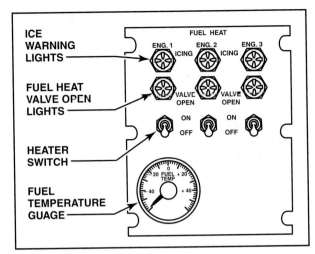

Figure 15-31. Some jet transports are equipped with a fuel heat control panel that provides indicator lights and controls to monitor and eliminate potential fuel-icing conditions.

fuel enough to melt any accumulated ice. When the ice has melted, restoring full flow, the fuel ice light will go out. If the filter is clogged with dirt or unmelted ice, a bypass valve will open and the fuel will bypass the filtering element. Fuel heat may also be automatic and the system may use a fuel/engine-oil heat exchanger, or turbine engine bleed air.

FUEL SYSTEM INDICATORS

Operating principles of the various instruments used in fuel systems were covered in Chapter 11. The various instruments and their purposes in a fuel system are listed here.

FUEL QUANTITY INDICATING SYSTEMS

This is one of the required instruments for all powered aircraft. It may be as simple as a cork float riding on top of the fuel in the tank, projecting a wire out through the tank filler cap. There is no requirement for these simple systems to be calibrated in discreet amounts, and they show only the relative amount of fuel in the tank. Somewhat more elaborate indicators have the float driving a pointer that shows whether the fuel level in the tank is 1/4, 1/2, 3/4, or Full. [Figure 15-32]

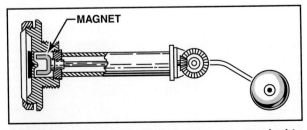

Figure 15-32. All aircraft with fuel systems are required to have a fuel indicating system. Many light aircraft have a simple float and gear assembly to read out relative quantities of fuel.

Electrical ratiometer-type fuel quantity gauges are used in many reciprocating engine aircraft. These gauges show the fuel level in the tanks by converting the position of the float into resistance in a fuel tank transmitter unit. The indicator is calibrated in gallons of fuel.

A capacitance-type fuel quantity system is most frequently used when it is necessary to know the mass of the fuel in the tank, rather than just its level, as is needed for turbine-engine aircraft.

Concentric metal tubes serve as capacitors and extend across the fuel tanks from top to bottom. Either air or fuel can act as a dielectric, but values, or *dielectric constants* of the two are significantly different. As the quantity of fuel in the tank changes, the portion of the probe immersed in fuel or exposed to air varies. The difference in dielectric values changes the electrical capacitance of the probes, and a capacitance bridge measures this quantity. The density of the fuel also affects its dielectric constant and thus the probe's capacitance. Because of this, the system can be calibrated to indicate the amount of fuel remaining in the aircraft in pounds or kilograms. [Figure 15-33]

Most large jet transport aircraft have a means for personnel to manually gauge the amount of fuel in a tank. One method uses a magnetically locked fuel measuring stick. To check the fuel quantity using one of these devices, turn the latching cam on the bottom surface of the wing tank a quarter of a turn, and pull down the measuring stick until the magnet on the measuring stick meets the magnet in the float. You can feel when the magnets grip together and can read the number on the measuring stick that relates to the fuel level in the tank. [Figure 15-34]

Another type of measuring stick shows the fuel level by fuel dripping from the hollow measuring stick when it is pulled down to the top of the fuel in the tank. Still another type uses a measuring stick made of transparent acrylic that has a wedge-shaped top. The technician can look up through the transparent rod as it is lowered, and when the end of the rod appears as a sharply defined line, it is at the top of the fuel in the tank. Again, the amount the rod sticks out from the bottom of the tank indicates the quantity of fuel.

FLOWMETERS

Small reciprocating-engine aircraft using carburetors seldom have fuel flowmeters. The pilot

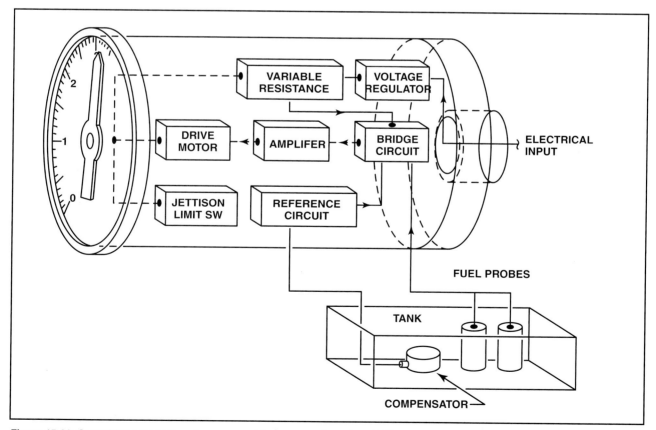

Figure 15-33. Capacitive-type fuel probes are used in many modern aircraft. The capacitance of each individual probe is fed into a central unit that can indicate the amount of fuel in each tank or the total fuel in the system.

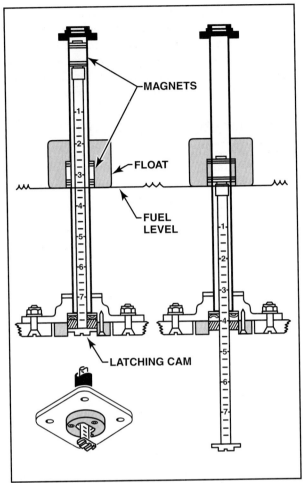

Figure 15-34. One type of on-the-ground fuel indicating system utilizes magnets that attract to each other when the level of the measuring stick meets the level of the fuel tank float. The portion of the stick pulled out of the bottom of the wing shows the fuel quantity.

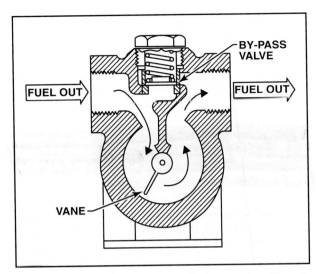

Figure 15-35. One kind of flowmeter is the vane type. The relative flow of the fuel moves the vane, which is attached to an Autosyn transmitter connected to the gauge in the cockpit.

assumes a flow rate based on the engine RPM and manifold pressure, and checks it against the amount of fuel used in a given period of time as determined from performance charts.

Larger reciprocating engines use a fuel flowmeter between the fuel pump and the carburetor. The fuel flowing to the carburetor moves a spring-loaded vane. The greater the flow, the further the vane will move. The movement of the vane is transmitted to the indicator, which may be calibrated in gallons or pounds per hour. This is only an approximation since it assumes that the fuel is at a standard temperature and has a standard density. [Figure 15-35]

Turbine-powered aircraft are concerned with the mass of the fuel flowing into the engine rather than just its volume. Flowmeters for these engines actually compensate for the density of the fuel. The fuel flow indicators will then accurately reflect the number of pounds or kilograms flowing per hour to each engine.

Reciprocating engines that are equipped with fuel injection systems have a flowmeter indicator that is actually a fuel pressure gauge. For normally aspirated engines, this is a bourdon tube instrument that measures the pressure drop across the fuel injector nozzles. The greater the flow, the greater the pressure drop. Turbocharged engines use a differential pressure gauge to measure the flow. They measure pressure at the distributor, or manifold valve, and compare it with the upper deck air pressure (the air pressure as it enters the fuel metering system). One major problem with this type of flow indicator is that a clogged injector nozzle will decrease the fuel flow, but the pressure drop across the nozzle will increase, appearing to the pilot as an increased fuel flow.

The latest development in fuel flow instruments is the digital-type system that uses a small turbine wheel in the fuel line to the fuel control unit. As fuel flows through this line, it spins the turbine and a digital circuit reads the number of revolutions in a specified period of time and converts this into a fuel flow rate. This flow rate may be electronically compensated to correct minor discrepancies in actual fuel flow amounts.

When electronic fuel flowmeters are connected to other electronic equipment in the aircraft, they can be made to present a running total of the fuel on board, and to predict the amount of time the fuel will last at the present rate of consumption. When this equipment is linked to the distance measuring equipment (DME), it can even show the range of the aircraft at the present power setting. [Figure 15-36]

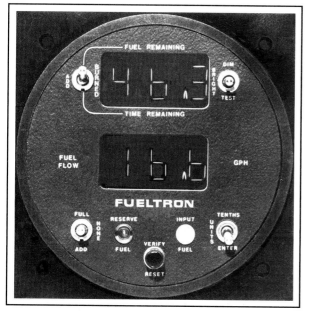

Figure 15-36. Electronic flowmeters are often connected to the computers in the aircraft. They can then indicate time and even distance until fuel exhaustion.

FUEL TEMPERATURE GAUGES

High flying jet aircraft are equipped with a ratiometer-type fuel temperature measuring system. This unit measures the temperature of the fuel in the tanks and displays it on the flight engineer's instrument panel. The flight engineer monitors the fuel temperature and uses the fuel heaters as necessary.

FUEL PRESSURE GAUGES

It is necessary to know that a pump fed fuel system is delivering the proper amount of fuel to the fuel metering system. To provide this information, simple aircraft generally use a bourdon tube pressure gauge connected to the inlet of the fuel metering system to measure the pressure. The pressure read here before starting the engine shows the output of the boost pump, and when the engine is running and the boost pump is turned off, the gauge shows the pressure that the engine-driven pump is producing.

Large reciprocating engines equipped with pressure carburetors are not concerned with the actual pressure produced by the pump but with the difference between the inlet fuel pressure and the inlet air pressure. This pressure is measured at the carburetor inlet, but instead of using a simple bourdon tube indicator, it uses a differential bellows-type instrument.

VALVE-IN-TRANSIT INDICATOR LIGHTS

On large multi-engine aircraft, each of the fuel crossfeed and line valves may be provided with a valve-in-transit indicator light. This light is on only during the time the valve is in motion and is off when movement is complete.

JET TRANSPORT AIRCRAFT FUEL SYSTEMS

A large jet transport aircraft such as a Boeing 747 has a relatively simple fuel system that supplies fuel to its four engines from five main tanks and two reserve tanks. All of the tanks are integral, that is, part of the wing is sealed off and the fuel is carried in the wing structure itself. Each reserve tank holds about 500 gallons, the outboard main tanks hold about 4,300 gallons, the inboard mains hold about 12,500 gallons, and the center wing tank holds about 17,000 gallons. At each wingtip is a surge and vent tank that provides venting for all the fuel tanks and contains a float switch to shut off refueling if a main tank is overfilled. A fully fueled airplane contains nearly 347,000 pounds (157,000 kilograms) of fuel.

Each main wing tank has two 115-volt AC electric boost pumps that provide fuel to the engines at 17 psi. The inboard main tanks also have two 115-volt AC jettison pumps with an output pressure of 37 psi. The center wing tank has two 115-volt AC override/jettison pumps with an output pressure of 37 psi. The center wing tank also has a scavenge pump that transfers residual fuel to the #2 main tank at a low rate. Fuel for an Auxiliary Power Unit (APU) is normally provided by the #2 main tank aft boost pump whenever AC power is available. The #2 main tank also has a 28-volt DC electric pump to supply fuel to the APU when AC power is turned off. All pumps have check valves in the output side to prevent reverse-flow back through the pump into the tank. A suction-feed port is provided to allow an engine-driven pump to draw fuel from its respective tank if both boost pumps fail. Each engine has a motor-actuated fuel shut-off valve to cut off fuel flow to the engine during an emergency or at engine shutdown.

Fueling is accomplished by connecting the fuel supply to single-point fueling receptacles located on the underside of each wing between the inboard and outboard engines. Single-point-fueling reduces fueling time, eliminates damage to aircraft skin, and reduces chances of fuel contamination. There are two hose connections on each side, but the refueling system controls are located only on the left side. Fuel flows from the receptacles into the same manifold used to jettison fuel, and from there through electrically actuated refueling valves into each fuel tank. When the tanks are completely filled, a sensing system closes the refueling valve(s) to prevent spillage or damage to the tank. If a partial fuel load is desired, the fueling personnel can monitor a set of fuel quantity gauges at the refueling station and shut off the fuel flow to each tank as it reaches the desired quantity. On newer airplanes equipped with digital fuel-quantity-indicating systems, a pre-selected total fuel quantity can be set at the refueling station and the system will automatically shut off fuel flow to each tank when it reaches the proper quantity.

The airplane may be defueled by connecting a fuel receiving truck to the manual defueling valves at either fueling station. Open the manual defueling valves and either pump the fuel out of the tanks using the boost or jettison pumps, or use suction from the receiving truck.

The crew may jettison fuel while in flight by actuating one or both jettison pumps in the inboard main tank and then opening the jettison nozzle valve in either or both wingtips. Fuel jettison from the center wing tank is accomplished by turning on the override/jettison pumps and opening the center wing tank jettison valves. Jettison fuel from the reserve tanks and outboard main tanks is accomplished by opening the reserve transfer valves and the #1 and #4 jettison transfer valves. Gravity drains the fuel into the respective inboard main tank, and from there, it is pumped overboard. A standpipe in the inboard main tanks prevents jettison of the last 25,000 pounds (11,250 kilograms) of fuel. With all six jettison pumps operating, fuel is dumped at a rate of approximately 5,000 pounds (2,300 kilograms) per minute.

The fuel system has provisions for heating the fuel before it enters the fuel filter and engine fuel control if its temperature is low enough for there to be danger of ice forming. Operation varies depending on the type of engine installed. Aircraft powered by General Electric (GE engines have automatic fuel heat, Pratt & Whitney powered airplanes have fuel heaters controlled by the Flight Engineer.

Expect system variations between different models of the 747, and between individual airplanes of the same model, as upgrades and improvements are routinely made during the service life of the type.

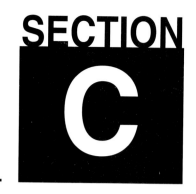

SECTION C

FUEL SYSTEM REPAIR, TESTING, AND SERVICING

The potential disaster of fuel leaks, fires, and explosions make it imperative to follow precise and exact procedures while working on or servicing fuel systems. In many cases, the manufacturer of the aircraft or fueling equipment provides instructions. Follow these at all times to ensure a safe working and operating environment. This section includes generic procedures for use when no specific guidance is available.

FUEL TANK REPAIR AND TESTING

Deal with any indication of leaking fuel immediately. Quickly and accurately troubleshoot and repair the fuel system to eliminate all possible faults that could result in fuel starvation or fire.

TROUBLESHOOTING THE FUEL SYSTEM

To become proficient at the art of troubleshooting, become familiar with the complete fuel system. Do this by reviewing the schematics of various portions of the system and the nomenclature of the units. Study aircraft and engine maintenance manuals to better understand any particular function within the system.

LOCATION OF LEAKS AND DEFECTS

Determining the location of leaks and defects within the fuel system's internal portions is usually a matter of observing the pressure gauge and operating the selector valves. Visualizing the path of fuel flow from the tank to the fuel-metering device, noting the location of the pump(s), selector valves, emergency shut-off valves, etc. can aid in troubleshooting the internal fuel system. The fuel pump must be operating in order to pressurize the system.

Locating leaks or defects in the fuel system's external portions involves very little time in comparison. Usually, stains or newly developed wet spots, as well as the presence of fuel odor, indicate fuel leaks. Carefully examine the plumbing, clamps, gaskets, supports, etc. at each inspection period. Any defect or leak in the internal or external fuel system is a potential hazard.

REPLACEMENT OF GASKETS, SEALS, AND PACKINGS

To prevent fuel leakage, it is of utmost importance to properly install all gaskets, seals, and packings. When replacing units of the fuel system, check each part for cleanliness, ensure that all of the old gasket material is removed, and ensure that none of the old seal remains in the groove seat. Replace old gaskets and seals with new ones and check the new gaskets and seals for cleanliness and integrity. The replacement must be the right part for the job, and its cure date must be current. Many gaskets and seals have a finite shelf life and must not be installed if they have exceeded that time. Mating surfaces should be perfectly flat so that the gasket can do the job for which it was designed. Evenly tighten or torque screws, nuts, and bolts that hold units together to prevent leakage past the gasket or seal.

TANK REPAIRS

There are three basic types of fuel cells used in aircraft: welded sheet metal tanks, integral, and bladder-type. Any fuel system that will not contain fuel is unairworthy. Inspecting the fuel bays or aircraft structure for evidence of fuel leaks is an important part of the preflight inspection.

WELDED FUEL TANKS

If a welded tank cracks or is damaged in such a way that it leaks, drain and remove the tank from the aircraft and weld the leak to repair it. Before welding any fuel tank, thoroughly purge it of all explosive fumes by steam cleaning, or some other means to remove all of the danger. To steam a fuel tank, pass live steam through it from the bottom and let it flow out the top. Continue steaming for at least one hour.

It may be convenient to have the damaged tank purged by a facility that specializes in repairing automobile or truck fuel tanks. These facilities usually have large vats of chemical compounds that will make the fuel vapors inert and the tank safe for welding. If following this procedure, remove all traces of the chemical from the tank with ample quantities of hot water and the neutralizing agent recommended for the chemical.

When welding tanks for repair, use flux to remove corrosion in the weld area to make it stronger. Remove flux with hot water and neutralize it with the manufacturer's recommended solution.

After completing the repair, test the tank to be sure that there are no more leaks. Welded fuel tanks are normally tested with 3.5-psi of air pressure inside the tank. Measure the pressure with an accurate gauge connected to the tank and maintain it with a good pressure regulator. As a safety feature, cover the filler opening by hand to be able to immediately release the pressure if there is any possibility of tank damage. Mount some of the larger tanks in a frame in the same way they are supported in the aircraft. This preserves the seams and baffles from any distortion the air pressure may cause. Search for leaks with a soap and water solution, and check carefully around the repaired area and along every weld in the tank.

SOLDERED TANKS

Terneplate tanks are found in some of the older small airplanes. When these tanks leak, remove them from the aircraft and make the fumes inert with either live steam, or some other equally effective method. Examine the tank carefully. Terneplate is sheet steel with a thin coating of lead and tin, and if any of the coating scratches off, rust formation is possible. If the aircraft has not been flown for some time and the tank has been left only partially filled, it is possible that water has accumulated and formed rust in the bottom of the tank. Repair terneplate tanks by sweat-soldering a patch of terneplate over the leaking area. Carefully examine terneplate tanks before repairing them because a severely rusted tank may be replaced more economically than repaired if a replacement tank is available.

BLADDER TANKS

Many modern aircraft use thin, synthetic, rubber-impregnated cloth bladders to hold fuel. When fuel is put in a bladder, some plasticizers in the synthetic rubber leach out into the fuel. If the tank is allowed to stand empty or even partially empty for more than a week, the bladder material will become brittle and may develop cracks that will shorten its service life. Any time a tank of this type is required to remain empty, wipe its inside surface with clean, lightweight engine oil. All tank bladders have a finite life and will eventually develop leaks. When a tank begins to seep fuel, remove it from the aircraft and either recondition it at an FAA approved repair station, or replace it with a new bladder. Remove a damaged bladder and repair it with a repair kit that is specially approved for this type of repair. Follow the instructions furnished with the kit in detail.

INTEGRAL FUEL TANKS

Integral tanks, or "wet wings," have become possible because sealants have been developed that are effective in sealing around all of the seams. It is possible for some of the sealing material to pull away from a seam and allow fuel to leak. Because of this possibility and the fact that not all leaks are the same, most manufacturers have agreed upon a classification of leaks. The amount of leakage in tanks determines what action to take. Any type of leak that allows fuel vapors to accumulate must be repaired before the aircraft is allowed to fly. Although many leaks are found on the surface of the aircraft, or on the ground beneath the aircraft, it is possible for an internal component such as a valve to leak without being externally apparent. These leaks are potentially as hazardous as external fuel leaks.

If a leak is discovered that does not constitute a hazard to flight, the aircraft may continue to fly until it is taken out of service for other maintenance and then be repaired. However, the leak must be watched to be sure it does not increase in size.

When repairing an integral tank, drain and purge it with either argon or carbon dioxide until a vapor detector shows that the tank is free of explosive vapors and safe to repair. Both of these gases are heavier than air and will remain in the tank during the repair.

Remove old sealant from around the leak with a chisel-shaped piece of hard plastic to remove the bulk of the sealant, and then aluminum wool to clean away all remaining traces. Do NOT use steel wool or sandpaper to remove the old sealant. Particles from either would remain and cause intergranular corrosion in the future. After removing all of the old sealant, vacuum out the debris and scrub the area with a cloth dampened with some approved solvent such as acetone. When the area is completely clean and dry, mix the new sealant according to the directions, and apply it in the manner specified by the manufacturer.

Check the repaired tank for leaks by applying the air pressure recommended by the airframe manufacturer. This pressure will be in the range of one-half pound per square inch for an integral tank. Apply pressure with a good regulator and monitor it with a water manometer.

FUEL LEAK CLASSIFICATION

The size of the surface area that a fuel leak moistens in a 30-minute period is used as the classification standard. Wipe the leak area completely dry with clean cotton cloths. Compressed air may also be used to dry the leak areas that are difficult to wipe. Always wear goggles when using compressed air to dry the leak areas. Then dust the leak area with dyed red talcum powder. The talcum powder turns bright red as fuel wets it, making the wet area easier to see.

At the end of 30 minutes, each leak is classified into one of four classes: slow seep, seep, heavy seep, or running leak. A slow seep wets an area around the leak source not over 3/4 of an inch in diameter. A seep wets an area from 3/4 inches to 1 1/2 inches in diameter. A heavy seep wets an area from 1 1/2 inches to 3 inches in diameter. Fuel does not run, flow, or drip at the end of a 30-minute period for any of these three leak classifications. [Figure 15-37]

The last classification, a running leak, is the most severe and the most dangerous. It may drip from the aircraft surfaces, or it may even run down the inspector's finger when the wet area is touched. The aircraft is unsafe for flight and must be grounded for repair. When possible, remove the fuel from the leaking tank after marking the leak location. If it is impossible to defuel the tank immediately, isolate the aircraft in an approved area. Place appropriate warning signs around the aircraft until qualified personnel can defuel the leaking tank.

FIRE SAFETY

The first and most difficult step in achieving fire safety is to correct the misconceptions about turbine fuel "safety." At the time these fuels were first introduced, many people said, "fire problems in aircraft are over, turbine fuel is completely safe." This is obviously nonsense but it has been persistent nonsense. Flight line personnel have agreed that aviation gasoline will burn, and therefore they have exercised reasonable care and caution in handling it. However, it has been difficult to convince many people that turbine fuels are dangerous under some circumstances.

Turbine fuel characteristics do vary from those of gasoline. For example, kerosene has a slow flame propagation and burning rate, which makes it less hazardous in the event of a spill or a ground accident. However, it does ignite readily when vaporized or misted, as when sprayed through a small leak in a service hose. Consequently, follow the fire safety precautions in handling turbine fuels with the same care observed in handling gasoline.

FIRE HAZARDS

Any facility that stores or handles fuel represents a major fire hazard. This also holds true for facilities that store or handle aviation fuels. Therefore, all

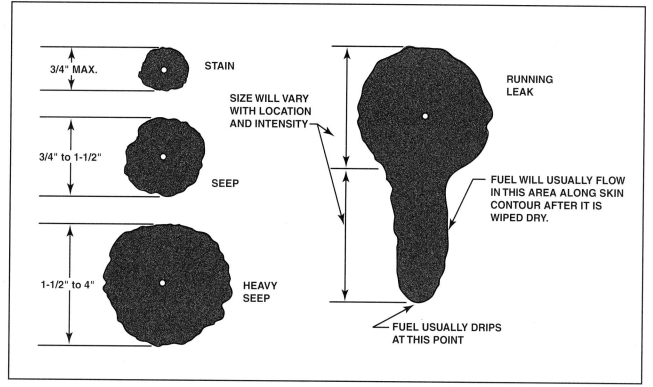

Figure 15-37. Fuel leaks are classified according to the surface area wetted during a 30-minute period.

personnel should be aware of the danger and be trained on how to handle fuel. Aviation fuels are both highly flammable and volatile. Take special care when transferring them into or out of an aircraft. Have the proper type of fire extinguisher available at the aircraft. It must be properly serviced and unused, since it was last serviced.

VOLATILITY

When an aircraft is fueled, vapors rise from the tank. The more volatile the fuel (the higher its vapor pressure) and the higher the outside temperature, the more vapors are released, requiring more caution when fueling.

Because of the flammable nature of fuel vapors, do not fuel or defuel in a hangar or an enclosed area. Furthermore, if fuel spills, wipe it up or wash it away with water as soon as possible. Never sweep away spilled fuel with a dry broom, as the static electricity generated by the broom can ignite fuel vapors.

Store aviation fuel in approved containers only. Keep these containers closed and stored in a cool and isolated area that has been approved for fuel storage.

STATIC ELECTRICITY

All aviation fuels burn under conditions where there is sufficient oxygen and a source of ignition. Sufficient air and fuel vapors to support combustion are normally present during any fuel-handling operation. Therefore, it is vital that all ignition sources be eliminated in the vicinity of any fuel-handling operation. Obvious sources of ignition include matches, cigarette lighters, smoking, open flames, even backfires from malfunctioning vehicles. However, one source of ignition that may not be so obvious is the sparks created by static electricity.

Static electrical charges are generated in various degrees whenever one body passes through or against another. For example, an aircraft in flight through the air, a fuel truck driving on a roadway, the rapid flow of fuel through a pipe or filter, and even the splashing of fuel into a fuel truck or aircraft during fueling operations can all generate static electricity.

To minimize this hazard, it is necessary to eliminate static electrical charges before they can build up to create a spark. To do this, bond and ground all fueling system components together with static wires and allow sufficient time for the charge to dissipate before performing any act that could draw a spark.

Contrary to popular belief, bleeding off an electrical charge from a body of fuel is not always an instantaneous act. In fact, it can take several seconds to bleed off all static charges from some fuels. Because of this, it is essential to carry out the following procedures when bleeding off static charges.

When handling aviation fuels:

1. Connect a grounding cable (static wire) from the fuel truck or hydrant cart to ground. Furthermore, when loading a fuel truck, connect the static wire from the loading rack to the fuel truck before opening the dome cover.

2. Connect a static wire from the fuel truck, hydrant cart, pit or cabinet to the aircraft. [Figure 15-38]

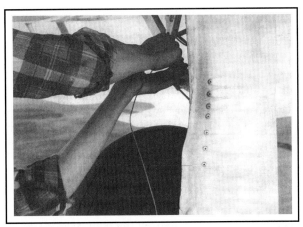

Figure 15-38. Before opening the fuel tanks or connecting the fuel hose, connect the aircraft and the fuel truck together with an electrical bonding cable.

3. When conducting overwing fueling, connect the fuel-nozzle static wire to the aircraft before opening the tank cover. If not equipped with a static wire, ground the nozzle against the side of the fuel opening before dispensing fuel. Underwing nozzles need not be bonded to the aircraft.

4. In general, the dispensing unit should be grounded first, and should ultimately be bonded to the receiving unit. Never open the dome or tank covers during a fuel transfer unless all grounds and bonds are in place.

When handling turbine fuels:

1. Minimize splashing during the loading of a fuel truck by placing the end of the loading spout at, or as near as possible to the compartment bottom.

2. Do not suspend or lower any metal or conductive objects such as gauge tapes, sample containers, or

thermometers into a tank or fuel truck while filling it. Give any possible static charge a few minutes to bleed off after filling before using these devices.

3. When filling large storage tanks, minimize splashing with a slow initial flow rate until the end of the tank inlet line is covered with at least two feet of fuel.

FUEL SYSTEM SERVICING

Everyone involved with handling and dispensing aviation fuels should realize that the safety of an aircraft may depend upon their skill, knowledge, and ability to deliver the correct grade of clean dry fuel into it. This is one of the prime factors contributing to flight safety.

Fuels, fueling methods, and equipment are continually being developed and improved to meet the ever-increasing demands of modern aircraft and the aviation industry. However, one thing never changes — the vital importance of supplying the correct grade of uncontaminated fuel to the aircraft. The possibility of human error can never be eliminated, but it can be minimized through careful design of fueling facilities, good operating procedure, and adequate training of personnel.

CHECKING FOR FUEL SYSTEM CONTAMINANTS

Draining a fuel sample from the main strainers of an aircraft has previously been considered an acceptable method for assuring that the fuel in the system is clean. In most cases, this practice is no longer considered adequate. The introduction of turbine-powered aircraft has made the need for fuel cleanliness much more important, and at the same time more difficult to maintain.

The combustion process in the jet engine must be carefully controlled. This requires complicated, precision fuel-control systems that are much more sensitive to fuel quality, and cleanliness in particular, than those for piston engines. Besides this, the fuel quantity passing through these systems for each hour of flight is considerably greater than with piston engines; hence, any slight contamination in the fuel accumulates at a much faster rate.

Along with the introduction of the more critical jet engine has come the utilization of a fuel that is harder to keep free of contamination. For example, a particle of dirt or rust, or a drop of water settles out of aviation gasoline four times faster than it does in turbine fuels. To better understand what is required to maintain fuel quality, it is first necessary to have a basic knowledge of aviation fuels, the common forms of fuel contaminants, how they get into the fuel, and how they can be detected and removed. In all cases, it is necessary to wait a period of time for the contaminant to drop out of suspension. This depends on factors such as type of fuel, temperature, and how much the fuel has been agitated.

TYPES OF CONTAMINANTS

The most common forms of aviation fuel contaminants are solids, water, surfactants, microorganisms, and the intermixing of grades or types of fuel. Surfactants and microorganisms, virtually unknown contaminants in aviation gasolines, have become critical with the advent of turbine fuels.

Solid contaminants are those which do not dissolve in fuel. The most common contaminants are iron rust and scale, sand, and dirt. However, other debris such as metal particles, dust, lint, particles of filter media, rubber, valve lubricants, and even bacterial sludge can be encountered. Solid contaminants are typically introduced inadvertently into fuel at every stage of its movement from the refinery to the aircraft.

The maximum amount of solids that an aircraft can tolerate depends on the type of aircraft and fuel system, and the number and size of the solid contaminants. Particles as small as 1/20 the diameter of a human hair can damage close tolerance mechanisms in modern turbine engines.

The best method of controlling solids is to limit their introduction into the fuel. Obviously, do not use rusty lines, tanks, and containers. Furthermore, keep covers and caps tightly closed until ready to begin pumping fuel. Take care to keep wiping rag lint, wind-blown sand, dirt, and dust from entering the system during filling or fueling operations. Clean fueling nozzles and loading spouts before use, and replace dust caps and other protective devices after using them. Furthermore, regularly inspect and maintain filters in accordance with operating specifications.

"**Surfactants**" is a contraction of the words SURFace ACTive AgeNTS. Surfactants consist of soap or detergent-like materials that occur naturally in fuel, or can be introduced during refining or handling. Surfactants are usually more soluble in water than in fuel and reduce the surface tension between water and fuel. This stabilizes suspended water droplets and contaminants in the fuel. They are attracted to the elements of filter/separators, which can make them ineffective. Surfactants, in large concentrated quantities, usually appear as a tan to dark brown liquid with a sudsy-like consistency.

Surfactants alone do not constitute a great threat to aircraft. However, because of their ability to suspend water and dirt in the fuel and inhibit filter action, they allow these contaminants to get into an aircraft's fuel system. Surfactants have become one of the major contaminants in aviation turbine fuels and can cause fuel gauge problems. There is no established maximum limit on the level of surfactants that can be safely contained in a fuel, and there are no simple tests for determining their concentration in fuel. Some common danger signals of a surface contaminated facility are:

1. Excess quantities of dirt and/or free water going through the system.

2. Discovery of sudsy-like liquid in tank and filter/separator sumps.

3. Malfunctioning of filter/separators.

4. Slow effective settling rates in storage tanks.

Microorganisms have become a critical problem in some turbine fuel systems. There are over 100 different varieties of microorganisms that can live in the free water that accumulates in sumps and on the bottom of storage and aircraft tanks. Many microorganisms are airborne, and therefore, fuel is constantly exposed to this type of contaminant.

The principle effects of microorganisms are:

1. Formulation of a sludge or slime that can foul filter/separators and fueling mechanisms.

2. Emulsification of the fuel.

3. Creation of corrosive compounds and offensive odors.

Severe corrosion of aircraft fuel tanks has been attributed to microorganisms, causing considerable expense in removing these growths and repairing their damage. The actual determination of microbial content, or number of colonies, is reserved for the laboratory. Remove any evidence of black sludge or slime, or even a vegetative-like mat growth. Growths also appear as dark brown spots on some filter/separator element socks. Replace the socks whenever this condition is discovered.

Because microbes thrive in water, a simple and effective method to prevent or retard their growth is to eliminate the water. Introducing a fuel additive during the fueling process is a common way of doing this. [Figure 15-39]

Figure 15-39. A biocidal agent is added to turbine fuel to prevent microbial growths that cause corrosion in the fuel tanks.

DETECTION OF CONTAMINANTS

Because solid contaminants generally appear in relatively small numbers and sizes relative to the fuel volume, their detection can be difficult. Aviation gasoline is generally considered "clean" if a one-quart sample is clear of any sediment when viewed in a clean and dry glass container. It may be helpful to swirl the container to create a vortex. The solid contaminants, if present, will tend to collect at the bottom beneath the vortex.

Turbine fuels must be several orders of magnitude cleaner than aviation gasoline. While the above visual test is adequate for operational checks, it is occasionally necessary to check the operation efficiency and cleanliness level of a turbine fuel system with a more accurate tool. The aviation industry has adopted the Millipore test for this purpose.

The Millipore is a filter-type test capable of detecting microscopic solid contaminants down to .8 of a micron in size, which is approximately 1/120 the diameter of a human hair. An evaluation guide is provided, containing instructions for conducting these tests, along with the means for evaluating the results.

The "white bucket" test is particularly helpful in detecting surfactant concentrations in turbine fuel. All that is required is a clean white porcelain bucket and water that has been in contact with the fuel in tank bottoms, filter/separators, or other points where surfactants are likely to accumulate. Surfactants, if present, will appear as a brown, sudsy water layer on the bottom of the bucket or at the fuel-water interface.

CONTAMINATION CONTROL

Miscellaneous contaminants can include either soluble or insoluble materials or both. Fuel can be contaminated by mixing it with other grades or types of fuels, by picking up compounds from concentrations in rust and sludge deposits, by additives, or by a number of soluble materials.

The greatest single danger to aircraft safety from contaminated fuels cannot be attributed to solids, exotic microorganisms, surfactants, or even water. It is contamination resulting from human error. It is placing the wrong grade or type of fuel into an aircraft, mixing grades, or any other type of human error that allows placement of off-specification fuels aboard the aircraft.

Do not put aboard an aircraft any fuel that is suspected to be off-specification because of contaminants or mixing with other fuels. If in doubt, utilize laboratory and other tests to definitely establish whether the fuel is usable for aviation purposes.

FUELING PROCEDURES

The fueling process begins with the delivery of fuel to the airport fueling facility, usually by tank truck. Quality control begins by checking the bill of lading for the proper amount and grade of fuel.

Fuel testing should begin with the tank truck. The personnel receiving the fuel delivery must determine that the proper type of fuel is in the truck, as well as take samples and check for visible contamination. Once all checks are completed, connect the truck to the correct unloading point and proceed.

Allow turbine fuel to settle for a minimum of two hours after any disturbance before pumping it into an aircraft. Aviation gasolines do not need time to settle before being withdrawn for use. However, make no withdrawals from a tank while it is receiving fuel from a transport truck.

FROM A FUEL TRUCK

Aircraft can have fuel pumped directly into their tanks from over-the-wing tank openings, or from a single point source under the wing. Typically, over-wing fueling is done with a fuel truck, whereas underwing fueling is done from a pit through single-point fueling.

Before driving a fuel truck to an aircraft, drain the sumps and check the fuel to be sure it is bright and clear. Furthermore, be certain that fire extinguishers are in place and fully charged. The fuel truck should approach the aircraft parallel to the wings and stop in front of it. Engage the parking brake, chock the wheels, and connect the static bonding wire between the truck and the aircraft.

Prior to removing the aircraft's fuel tank cap, verify the proper grade of fuel. Do this by reading the placard near the filler cap and compare it with the grade being delivered. [Figure 15-40]

Figure 15-40. An aircraft fuel tank must be clearly marked with the proper grade of fuel required.

Place a mat over the wing so the fuel hose can not scratch the finish. Connect the static bonding wire between the nozzle and the aircraft and remove the fuel tank cap. Do not contact the bottom of the tank with the end of the nozzle when inserting it into the tank because it could dent the thin metal. If the tank is a fuel cell, contact with the nozzle could puncture it and cause a serious leak.

Misfueling is a constant danger that can result in complete engine failure. To help prevent misfueling accidents, the nozzles used to pump turbine fuel are larger than the nozzles used to pump aviation gasoline. Furthermore, FAR 23.973 specifies that all general aviation aircraft utilizing aviation gasoline have restricted fuel tank openings that will not allow the nozzle used to pump Jet A to fit in the tank opening. While it is possible for a jet or turbine engine to run on gasoline, a piston engine will not run on Jet A.

UNDERGROUND STORAGE

Most large airports that service transport category aircraft have underground storage tanks and buried fuel lines. This arrangement allows aircraft to be fueled without having to transport it in trucks. Since most aircraft that are fueled from this type of system use under-wing fueling, the method is discussed here.

After driving a service truck that has filters, water separators, and a pump to the aircraft, connect its inlet hose to the underground hydrant valve. Attach the discharge hose or hoses from the servicer to the fueling ports on the aircraft. With a qualified maintenance person in the aircraft monitoring the fuel controls, open the valves and start the pumps. The person monitoring the fuel controls can determine the sequence in which the tanks are filled and can shut off the fuel when the correct load has been taken on board.

Some large corporate aircraft also have single point refueling systems. However, in most cases, control of the fueling sequence comes from an outside control panel located under an access cover. A service technician must be checked out on these systems before operating them. Should there be any questions about the operation, the technician should ask for assistance from the pilot-in-command of the aircraft. [Figure 15-41]

Figure 15-41. Fuel can be delivered simply and quickly through the underwing fueling ports.

The aircraft mechanic may be called upon to fuel or defuel aircraft or to assist in training ground service personnel. The steps outlined below represent general procedures to carry out when fueling any aircraft (specific instructions from the pilot may at times supersede these instructions):

1. Verify the grade and quantity of fuel required.

2. Confirm that the fueler or system contains the correct grade and quantity required.

3. Check the fueler tank sumps for water before fueling. Drain if necessary.

4. Approach the aircraft carefully. Try to position the fueler so that it can be quickly driven or pulled away in case of emergency. Avoid backing up to the aircraft; if this is necessary, position a guide near the rear of the fueler. Set the brake and chock the tires.

5. Bond and ground the aircraft and equipment in the proper sequence: fueler to ground, then fueler to aircraft. Before opening the overwing fuel-filler cap, connect the nozzle ground to the aircraft. Keep a constant contact between an overwing nozzle and the filler neck spout while filling. It is not required to ground underwing nozzles. After fueling, reverse the steps above.

6. Do not drag hoses across deicer boots or wing edges. Always place drop-deck ladders so that the pads rest squarely on the leading edge of the wings. When on the wings, walk only where designated. Clean off all greasy marks and dirt before leaving the wing. [Figure 15-42]

Figure 15-42. Filling an aircraft fuel tank at the filler neck is called over-the-wing fueling.

7. Never prop open or leave nozzles unattended while fueling aircraft. Never drop or drag nozzles across the pavement. Immediately replace nozzle dust caps after fueling.

8. It can be very dangerous to leave a filler cap off an aircraft fuel tank. Never open a cap until you are ready to fuel that specific tank, then lock it and close the flap immediately after fueling. Before leaving the wing, check each filler cap for security. Notify maintenance or the pilot if the filler cap or flap is working improperly.

9. Check the fueler or other fueling equipment and filter/separator sumps for water after fueling is complete. If more than a trace of water or other contaminants are found, notify the pilot and make arrangements to sample the fuel in the aircraft tanks.

10. Never pull away from an aircraft without first checking to make sure that there is no one left on the top deck and that all hoses and ground wires are properly stowed.

DEFUELING

It is sometimes necessary to remove fuel from an aircraft, either for maintenance reasons or because of a change in flight plans after the aircraft was serviced. Defueling is accomplished in much the same manner as fueling, using many of the same safety precautions. On swept-wing aircraft, defuel outboard wing tanks first and fuel them last since these tanks are usually aft of the center of gravity.

Never defuel an aircraft inside a hangar or in any area with inadequate ventilation. Take all proper precautions with regard to neutralizing any static electricity that builds up when the fuel flows through the lines.

If only a small quantity of fuel is off-loaded and there is no reason to suspect contamination, take the fuel back to stock. On the other hand, the quality of the off-loaded fuel could be suspect if it was removed because of an engine problem. If removing a large quantity of fuel, it could become suspect by drawing fuel from the bottom of the tank. Segregate this fuel, preferably in a fuel truck, and quarantine it until assured of its quality. Never return fuel suspected of contamination to storage or place it aboard another aircraft. If returning acceptable fuel to storage, be certain to put it back into a tank containing the same grade of fuel.

If defueling an aircraft into drums, use clean drums and replace and tighten the bungs immediately after filling. Some companies, and some aircraft operations manuals, do not allow the reuse of drum stored fuel in an aircraft. Frequently, this fuel is used in ramp vehicles, space heaters, and GPUs.

REVIEW OF SAFETY PROCEDURES

These safety procedures are suggested to minimize the potential for accidents during fueling or defueling of an aircraft. Specific guidance for particular aircraft and fueling/defueling equipment may require further procedures. Manufacturer's specifications should always take precedence.

1. Be sure that only the correct grade of fuel is put into an aircraft. Remember that aviation gasoline comes in various grades, and the wrong grade can cause severe damage to the engine. Turbine fuel in a reciprocating engine can cause severe detonation and engine failure, and improper use of aviation gasoline in a turbine engine can also be harmful.

2. Be sure to properly bond the fuel truck or servicer to the aircraft, and bond the fuel nozzle to the structure before taking the cover from the fuel tank.

3. Wipe up spilled fuel or flood it with water. Do not sweep spilled fuel with a dry broom.

4. Be sure that there are no open fires near the fueling or defueling operations.

5. Be sure that fire extinguishers suitable for a Class B fire are available. Either CO_2 or dry powder units are generally used.

6. Protect the aircraft structure from damage by the fuel hose and nozzle.

7. Be sure that the radio or radar are not used during fueling or defueling, and that no electrical equipment is turned on or off, except for the equipment needed for the fueling operation.

8. When defueling, be sure that the fuel is not contaminated if it is to be used again.

9. Be sure that the filters in the tank truck or servicer remove all traces of water and contamination and that the fuel pumped into an aircraft is bright and clear.

10. If a biocidal additive is required, mix it with the fuel in the proper concentration.

11. If fueling the aircraft in the rain, cover the tank opening to exclude water from the tank.

12. Place dust covers and caps over the end of the fuel nozzles and any unused open fuel lines.

13. Park the tank truck parallel to the wing of the aircraft, set the parking brake, and chock the wheels so the truck cannot roll into the aircraft.

14. When conducting underwing pressure fueling, adhere to the airplane manufacturer's specified pressure and delivery rate.

15. If any fuel is spilled onto skin, wash it off with soap and water as soon as possible. Do not wear fuel soaked clothing.

CHAPTER 16

FIRE PROTECTION SYSTEMS

INTRODUCTION
Since fire is one of the most dangerous threats to the safe operation of an aircraft, manufacturers and operators install a variety of overheat, fire detection, smoke detection, and extinguishing devices. Although the majority of aircraft fire-protection systems are installed around the powerplant section, it is typically the responsibility of an airframe technician to maintain all fire-protection systems regardless of where they are installed. To maintain the highest level of reliability from these systems, a technician must be familiar with the basic operating principles, troubleshooting, and repair of the various types of fire protection devices used on modern aircraft.

SECTION A

FIRE DETECTION

On early aircraft, the task of detecting smoke and fire was reasonably easy because the pilot could see most areas of the aircraft from the cockpit. However, as larger and more complex aircraft were built, it became nearly impossible for the crew to observe all parts of an aircraft, and smoke and fire were often not detected until the hazard was beyond control. To resolve this problem, modern aircraft have overheat and fire-detection systems installed to provide an early warning of hazards so the crew can take appropriate actions to reduce or eliminate them.

Overheat and fire-detection systems are designed with components developed for specific tasks; so, compared to other aircraft systems, maintenance requirements for fire detection components are somewhat specialized. To be able to keep these systems operating properly, a technician must understand the basic operating principles and maintenance practices used by various fire-detection system manufacturers.

PRINCIPLES OF FIRE-DETECTION SYSTEMS

For a fire to occur, three conditions must be met. There must be fuel, oxygen, and enough heat to raise the temperature of the fuel to its ignition or kindling point. If any of these elements is missing or removed, fire will not be sustained. [Figure 16-1]

Chemically, fire is a reaction between oxygen and fuel. This reaction reduces fuel to its basic chemical elements and in the process produces tremendous amounts of heat. Paper, for example, is an organic material composed primarily of carbon and hydrogen. When the paper is heated to its kindling temperature in the presence of air, the carbon and hydrogen will unite with oxygen to form carbon dioxide (CO_2) and water (H_2O). Other elements in the paper, and the products of incomplete combustion, show up as ash and black carbon to form smoke.

In the case of smoke and fire hazards aboard aircraft, the emission of smoke or the presence of flames and heat makes it reasonably easy for a person to physically detect a fire or overheat condition. The smoke

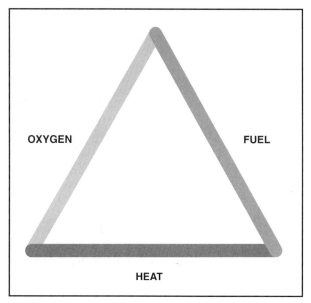

Figure 16-1. A fire triangle illustrates that a fire requires fuel, oxygen, and enough heat to cause the fuel and oxygen to ignite. If any of these elements is missing, a fire will not ignite or continue to burn.

produced by combustion produces strong odors and is readily visible in most circumstances, so the crew of an aircraft can physically detect a fire hazard in its early stages, provided they are in the same compartment or area of the aircraft where the fire occurs. However, many aircraft areas are inaccessible to the crew, and, because of the design of the aircraft, airflow around and through various compartments may prevent the hazard from being detected until it is too late to remedy the problem.

To provide a more thorough means of monitoring remote locations of an aircraft for smoke or fire, detection systems are mounted in areas the crew does not have access to in flight. Some examples of areas where these systems may be installed include engine nacelles, baggage compartments, electrical or electronic equipment bays and passenger lavatories. Depending on the types of combustible materials that may smolder or ignite, the systems are designed to activate by various means to provide the most accurate indication of an actual hazard. These systems monitor areas, commonly called fire zones, for heat, flames, the rate of temperature rise, or the presence of smoke.

CLASSES OF FIRES

To understand how and why different types of fire-detection systems are better suited for certain applications, you need to be familiar with the classifications of fire as identified by the National Fire Protection Association. These fires are identified in conjunction with the types of materials consumed by a fire and are assigned different letter classifications as follows:

A Class A fire is one in which solid combustible material burns, such as wood, paper, or cloth. Control cabins or passenger compartments are examples of locations where Class A fires are likely to occur. Since the interiors of the passenger compartment and of the cockpit are readily accessible to the crew, fire detection in these areas is generally accomplished by visual surveillance. On the other hand, such fires can also occur in baggage compartments, where crew access is limited or even impossible during flight. In these areas, monitoring is primarily accomplished with electrically powered smoke- or flame-detector systems.

Class B fires are composed of combustible liquids such as gasoline, oil, jet fuel, and many of the paint thinners and solvents used in aviation maintenance. On an aircraft, these classes of fires typically occur in engine compartments or nacelles, and in compartments that house an auxiliary power unit (APU). Since operating temperatures within these areas can be extreme, overheat detection systems, which sense the rate of temperature rise, are often used to monitor the zone for the presence of fire or overheat conditions. With these types of monitoring devices, false alarms are less likely than with other types of detection systems.

Class C fires are those that involve energized electrical equipment. These fires require special care because of the dangers from the electricity, in addition to those from the fire itself. Such fires are generally confined to electrical and electronic equipment bays and to areas behind electrical control panels. Since the initial stages of electrical equipment fires are usually preceded by large amounts of smoke, these areas of an aircraft are generally monitored by smoke-detection systems.

Class D fires involve burning metals such as magnesium, and are difficult to extinguish. Using the wrong type of extinguishing agent with these fires may not only be ineffective, but may even cause the fire to spread. Although these types of fires are not common in aircraft during flight, they can occur in maintenance shops, where metal shavings may ignite when exposed to intense heat such as from a welding torch or high-voltage source.

FIRE ZONES

Various compartments in an aircraft are classified into fire zones based on the amount and characteristics of airflow through them. The airflow through a compartment determines the effectiveness of fire-detection systems, as well as the effectiveness of suppressant materials used to extinguish a fire. Fire zones are primarily classified by the amount of oxygen that is available for combustion and are identified as A, B, C, D, or X zones.

Class A zones have large quantities of air flowing past regular arrangements of similarly shaped obstructions. The power section of a reciprocating engine is a common example of this zone. For these areas, a fire-extinguishing system is usually installed, but may not prove adequate since the suppressant may be carried out into the air-stream before extinguishing the fire.

Class B zones have large quantities of air flowing past aerodynamically clean obstructions. Heat-exchanger ducts and exhaust manifold shrouds are usually of this type, as are zones where the inside of the cowling or other enclosure is smooth, free of pockets, and adequately drained so that leaking flammables cannot puddle. For example, turbine engine compartments are in this zone class, if the engine surfaces are aerodynamically clean and a fireproof liner is installed to produce a smooth enclosure surface over any adjacent airframe structure. Class B zones are usually protected by temperature sensing elements or flame and smoke detection systems as well as extinguishing equipment, to provide a means of controlling a fire if one should occur.

Class C zones have relatively low airflow through them. An auxiliary power unit (APU) compartment is a common example of this type of zone. These may be protected by a fire-detection and extinguishing system, or the compartment may have provisions for isolating flammable materials such as fuel, oil, and hydraulic fluids.

Class D zones have very little or no airflow. These include wing compartments and wheel wells, where little ventilation is provided. Due to the lack of airflow, fire-extinguishing systems are usually not necessary since the fire will self-extinguish as it consumes the atmosphere. However, fire-detection systems are often installed in Class D zones to warn the crew that damage may have occurred to airframe components, so that corrective actions may be taken. For example, a fire in a wheel well should self-extinguish due to lack of air, but the wheels and tires may be damaged. A fire-detection system will warn the flight crew, so that special precautions may be taken during the landing to preclude further hazards.

Class X zones have large quantities of air flowing through them and are of unusual construction, making fire detection and uniform distribution of an extinguishing agent very difficult. Zones containing deeply recessed spaces and pockets between large structural formers are of this type. Fires in Class X zones will need twice the amount of extinguishing agent normally used in a Class A zone.

REQUIREMENTS FOR OVERHEAT AND FIRE-DETECTION SYSTEMS

Modern detection systems have been proven to be highly reliable when properly maintained. These systems consist of electrical or electronic sensors that are installed in remote locations. The sensors warn the operator of impending hazards by sounding an audible alarm and illuminating a warning light that indicates the location of the hazard. Before these systems are approved by the FAA for installation in an aircraft, the manufacturer must prove that the fire-detection system design meets the following criteria:

1. The system must be constructed and installed in a manner that prevents false warnings under all flight and ground operating conditions.

2. There must be a rapid indication of a fire and an accurate indication of the fire's location.

3. The system must have an accurate indication that a fire has been extinguished.

4. The system must automatically reset once a fire is extinguished, to provide an immediate indication if the fire re-ignites.

5. When there is a fire, there must be a continuous indication for its duration.

6. The detection system must have a means for electrically testing the integrity of the detection-system circuitry from the cockpit.

7. The detector or sensing units must be able to resist exposure to oil, water, vibration, extreme temperatures, and maintenance handling. The units should also be lightweight and easily adaptable to any mounting position and must also operate directly from the aircraft power system, without inverters. In addition, when the detectors are not sensing a hazard, there should be minimal requirements for electricity to power the system.

8. Each detection system must actuate a cockpit light indicating the location of the fire, as well as an audible alarm.

9. In the case of multi-engine aircraft, the detection system must consist of a separate sensing circuit for each engine.

There are a number of overheat and fire-detection systems that satisfy these requirements, and a single aircraft may utilize more than one type.

FIRE-DETECTION/OVERHEAT SYSTEMS

Engine fire-detection systems generally fall into two categories: spot-detection type systems and continuous-loop type systems. With a **spot-detection type system**, individual fire detectors, or switches, are used to detect a fire. Such detectors must be placed in locations where a fire is likely to occur, because with this type of system a fire warning sounds only when a fire exists in the same location as the detector. The **continuous-loop type system** works on the same basic principle as the spot-type fire detectors, except that a single switch in the form of a long inconel tube is used instead of several individual switches. The small-diameter inconel tube is run completely around an engine nacelle or an area that surrounds an auxiliary power unit, thus allowing more complete coverage than spot-type detection systems.

The most common types of fire detection systems found in modern aircraft include Fenwal, the Kidde, the Lindberg, the Systron-Donner, and the flame-detector system.

FENWAL SYSTEMS

Fenwal produces a thermoswitch fire-detection system, a thermocouple fire-detection system, and a continuous-loop fire-detection system.

THERMOSWITCH DETECTOR

A thermoswitch fire detection system is a spot-type detection system that uses a number of thermally activated switches. Each switch, or sensor, consists of a **bimetallic thermoswitch** that closes when heated to a predetermined temperature. [Figure 16-2]

Fire Protection Systems

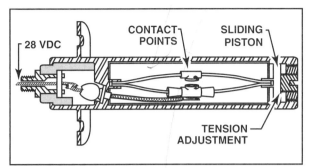

Figure 16-2. With a thermoswitch detector, the actual switch is mounted inside a stainless steel housing. If a fire starts, the switch housing heats up and elongates, causing the contact points to close. To adjust a thermoswitch, the housing must be heated to a specified temperature and then a tension adjustment is turned in or out until the contacts just close. In most cases, this adjustment is set by the detector manufacturer and is not adjusted in the field.

There are two basic types of thermoswitch systems, the single loop and the double loop.

Single-Loop System

With a Fenwal **single-loop** system, all of the thermoswitches are wired in parallel with each other, and the entire group of switches is connected in series with an indicator light. In this arrangement, once a thermoswitch closes, the circuit is completed and power flows to the warning light. [Figure 16-3]

To provide for circuit testing, a test switch is installed in the cockpit. Once the test switch is depressed, power flows to a relay that provides a ground to the warning light, simulating a closed thermoswitch. Once grounded, the warning light illuminates only if there is no break in the warning circuit. In addition to the test feature, most fire-detection circuits include a dimming relay for night operations that, when activated, alters the warning circuit by increasing resistance. The increased resistance reduces the amount of current flowing to the light. In most airplanes, several circuits are wired through the dimming relay so all the warning lights may be dimmed at the same time.

Double-Loop System

In a **double-loop** system, all of the detectors are connected in parallel between two complete loops of wiring. The system is wired so that one leg of the circuit supplies current to the detectors while the other leg serves as a path to ground. With this double-loop arrangement the detection circuit can withstand one fault, either an open or short circuit, without causing a false fire warning. For example, if the ground loop should develop a short, a false fire warning will not occur, because the loop is already grounded. On the other hand, if the powered loop shorts, the rapid increase in current flow would trip a relay that causes the powered loop to become the

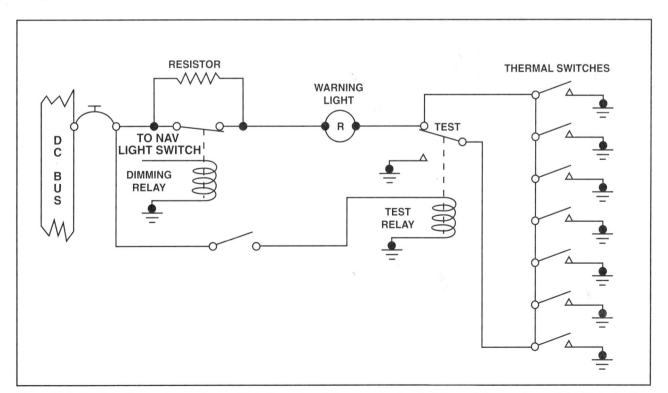

Figure 16-3. Fire detection systems using multiple thermal switches are wired so that the switches are in parallel with each other and the entire group of switches is in series with the indicator light. When one switch closes, a ground is provided for the circuit and the warning light illuminates.

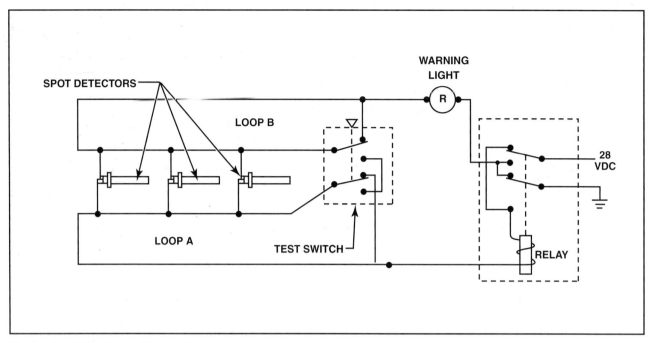

Figure 16-4. With the double-loop thermoswitch system shown here, wire loop A is positive while wire loop B is negative. However, if an open or short develops in wire loop A, the sudden rush of current will activate a relay that causes the positive loop to become negative and the negative loop to become positive.

ground and the grounded loop to become powered. [Figure 16-4]

THERMOCOUPLE DETECTOR

A thermocouple-type, Edison fire-detector system is similar to a thermoswitch system in that they are both spot-type detection systems. However, a thermocouple detector initiates a fire warning when the temperature of the surrounding air rises too rapidly (warms too fast), rather than responding to a preset temperature as does the thermoswitch detector.

A thermocouple consists of a loop of two dissimilar metal wires such as chromel and constantan that are joined at each end to form two junctions. When a temperature difference exists between the two junctions, electrical current flows and a warning light is activated. In a typical thermocouple system, one or more thermocouples, called **active thermocouples** are placed in fire zones around an engine while a separate thermocouple, called the **reference thermocouple**, is placed in a dead-air space between two insulated blocks. Under normal operations, the temperature of the air surrounding the reference thermocouple and the active thermocouples are relatively even, and no current is produced to activate a warning light. However, when a fire occurs, the air temperature around the active thermocouples rises much faster than the air temperature around the reference thermocouple. The difference in temperature produces a current in the thermocouple circuit and activates a warning light and horn. [Figure 16-5]

In most thermocouple systems, the sensitive relay, slave relay, and a thermal test unit are contained in a **relay box**. A typical relay box can contain from one to eight identical circuits, depending on the number of potential fire zones. The thermocouples control the operation of the relays, while the relays control the warning lights. The test circuit includes a special **test thermocouple** that is wired into the detector circuit and a small electric heater. The test thermocouple and heater are mounted inside the relay housing and, when the test switch in the cockpit is closed, current flows through the heater, which heats the test thermocouple. The temperature difference between the test thermocouple and the reference thermocouple produces a current flow that closes the sensitive relay and slave relay so the warning light can illuminate. Approximately 4 milliamperes of current is all that is needed to close the sensitive relay and activate the alarm.

The total number of thermocouples used in a particular detector circuit depends on the size of the fire zone and the total circuit resistance. Typically, circuit resistance is less than five ohms. In addition, most thermocouple circuits contain a resistor connected across the slave relay terminals. This resistor absorbs the coil's self-induced voltage when current

Fire Protection Systems

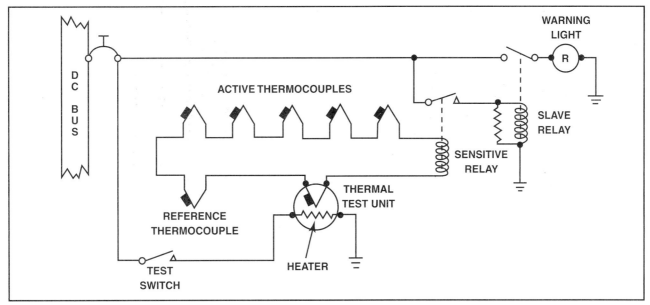

Figure 16-5. In a thermocouple fire detection circuit, the wiring system is typically divided into a detector circuit, an alarm circuit, and a test circuit. When a temperature difference exists between an active thermocouple and the reference thermocouple, current flows through a sensitive relay coil. When the sensitive relay closes it trips the slave relay, which, in turn, allows current to flow to the warning light.

ceases to flow through the coil and the magnetic field collapses. If this self-induced voltage were not absorbed, arcing would occur across the sensitive relay contacts, causing them to burn or weld.

CONTINUOUS-LOOP DETECTOR

In addition to a thermoswitch detection system, Fenwal also produces a continuous-loop type system that consists of a single fire or **overheat-sensing element** that varies in length, depending on the size of the fire zone. A typical sensing element can be anywhere from 1 foot to 15 feet long. As mentioned earlier, the sensing element used in a continuous-loop fire detection system consists of a flexible, small-diameter inconel tube. [Figure 16-6]

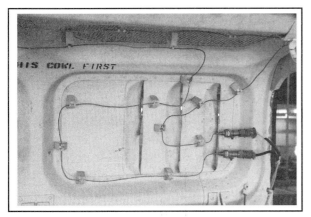

Figure 16-6. Fenwal continuous-loop fire detection elements sense a large area for fire and overheat conditions. In this picture, a continuous-loop detector can be seen running through an area inside an engine cowling.

In the Fenwal system, the metal inconel tube uses a single wire electrode made with pure nickel. The pure-nickel electrode is surrounded by ceramic beads to prevent the electrode and conductor from touching each other. The beads in this system are wetted with a **eutectic salt**, which has an electrical resistance that varies with temperature. [Figure 16-7]

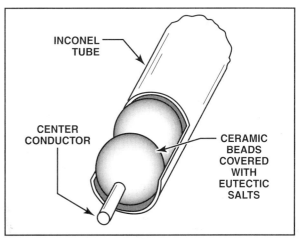

Figure 16-7. A Fenwal continuous-loop sensing element consists of a sealed inconel tube containing a single center conductor and ceramic beads wetted with a eutectic salt.

The center conductor protrudes out each end of the inconel tube where an electric terminal is affixed to the electrode. Current is then applied to the conductor while the outer tube is grounded to the aircraft structure. At normal temperatures, the eutectic salt core material prevents electrical current from flowing between the center conductor and the tube. However, when a fire or overheat condition occurs, the core resistance drops and current flows between the center conductor and ground, energizing the alarm system.

The Fenwal system uses a magnetic amplifier control unit. This unit is a non-averaging controller that supplies power to the sensing element and sounds an alarm when the circuit to ground is completed through the inconel tube. [Figure 16-8]

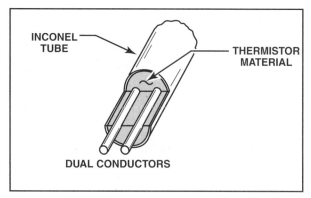

Figure 16-9. A Kidde sensing element consists of a sealed inconel tube containing two conductors embedded in a thermistor material.

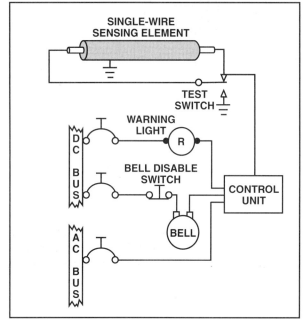

Figure 16-8. With a Fenwal continuous-loop fire detection system, AC voltage is applied to the sensing element through the control unit. Once the air surrounding the sensing element reaches a predetermined temperature, the resistance of the eutectic salt within the element decreases enough to allow current to flow to ground. The control unit then senses the flow of AC current and closes a relay which grounds the warning circuit and illuminates the warning light.

KIDDE SYSTEM

The Kidde system is also a continuous-loop type system consisting of a single overheat-sensing element that varies in length. The sensing element consists of a rigid, preshaped inconel tube with two internal wire conductors. The conductors are embedded in a **thermistor**, or **thermal resistor material**, to prevent the two electrodes from touching each other and the exterior casing. Like the eutectic salt used in the Fenwal system, the thermistor material has an electrical resistance that decreases as the temperature increases. [Figure 16-9]

One of the wires is electrically grounded to the outer tube at each end and acts as an internal ground, while the second wire is a positive lead. When a fire or overheat occurs, the resistance of the thermistor material drops, allowing current to flow between the two wires to activate an alarm.

Each conductor is connected to an **electronic control unit** mounted on separate circuit cards. In addition to constantly measuring the total resistance of the full sensing loop, the dual control unit provides for redundancy even if one side fails. In fact, both the Fenwal and Kidde systems will detect a fire when one sensing element is inoperative, even though the press-to-test circuit does not function, indicating that there is a fault in the system.

LINDBERG SYSTEM

The Lindberg fire detection system is a **pneumatic continuous-loop type system** consisting of a stainless steel tube filled with an inert gas and a discrete material that is capable of absorbing a portion of the gas. The amount of gas the material can absorb varies with temperature. One end of the tube is connected to a pneumatic pressure switch called a **responder**, which consists of a diaphragm and a set of contacts. [Figure 16-10]

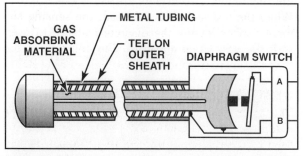

Figure 16-10. The sensing element used with a Lindberg continuous-loop system consists of a stainless steel tube that is filled with an inert gas and a gas absorbing material. One end of the tube is sealed while the other end is connected to a diaphragm switch.

Fire Protection Systems

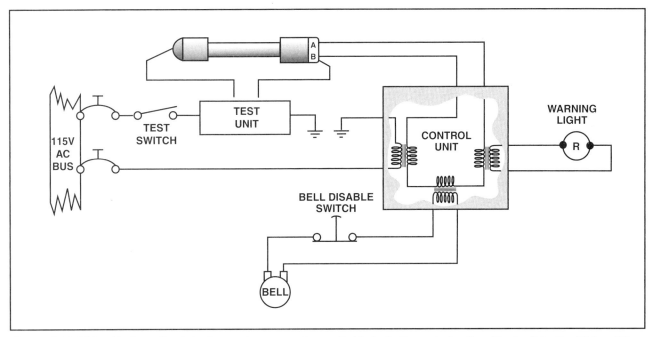

Figure 16-11. With a Lindberg fire-detection system, power is supplied to both the control unit and test unit by the AC bus. When a fire or overheat condition exists, the diaphragm switch closes, completing the circuit for both the warning light and the bell.

When the temperature surrounding the sensing element rises because of a fire or overheat condition, the discrete material within the tube also heats up and releases the absorbed gas. As the gas is released, the gas pressure within the tube increases and mechanically actuates the diaphragm switch in the responder unit. Once the diaphragm switch closes, the warning light illuminates and the alarm bell sounds. Because the Lindberg system works on the principle of gas pressure, it is sometimes referred to as a pneumatic system. [Figure 16-11]

To test a Lindberg system, low-voltage alternating current is sent through the element's outer casing. This current heats the casing until the discrete material releases enough gas to close the contacts in the diaphragm switch and initiate a fire warning. When the test switch is released, the sensing element cools allowing the discrete material to reabsorb the gas. Once absorbed, the contacts in the diaphragm switch open and the fire warning stops.

SYSTRON-DONNER SYSTEM

The Systron-Donner system is another pneumatic continuous-loop system that utilizes a gas filled tube with a titanium wire running through its center as a sensing element. The tube itself is made of stainless steel and is filled with helium gas. The titanium wire, on the other hand, acts as a gas-absorption material that contains a quantity of hydrogen. For protection, the wire is either wrapped with an inert metal tape or inserted in an inert metal tube. One end of the sensor tube is connected to a responder assembly containing a diaphragm switch that provides a warning for both an overheat condition and a fire.

Like the Lindberg system, the Systron-Donner system's principle of operation is based on the gas law: if the volume of a gas is held constant and the temperature increases, gas pressure also increases. The helium gas surrounding the titanium wire provides the systems **averaging** or **overheat** function. At normal temperatures, the helium pressure in the tube exerts an insufficient amount of force to close the overheat switch. However, when the average temperature along the length of the tube reaches an overheat level, the gas pressure increases enough to close the diaphragm switch, which activates the alarm. Once the source of the overheat condition is removed, the helium gas pressure drops and the diaphragm switch opens.

The system's fire detection, or **discrete**, function is provided by the gas-charged titanium wire. When exposed to a localized high temperature, such as a fire or turbine engine compressor bleed air leak, the titanium wire releases hydrogen gas. This increases the sensor's total gas pressure, which closes the diaphragm switch and trips the fire alarm. A typical Systron-Donner system sensor activates a fire alarm when exposed to a 2,000°F flame for five seconds.

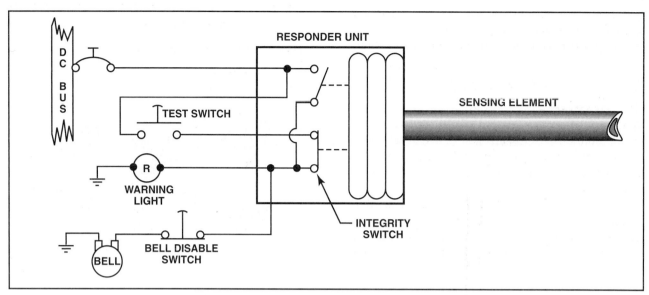

Figure 16-12. The Systron-Donner fire detection and overheat system consists of a helium-filled sensor tube surrounding a hydrogen-charged core. With this system, excessive temperatures increase the gas pressure which forces a diaphragm switch closed. Once closed, power flows to the warning light and bell.

After a fire is extinguished, the sensor core material reabsorbs the hydrogen gas and the responder automatically resets the system. [Figure 16-12]

To check system integrity, the responder unit of a Systron-Donner system contains an **integrity switch** that is held closed by the normal gas pressure exerted by the helium. When the integrity switch is closed, depressing the test switch results in a fire warning. However, if the sensing element should become cut or severely chafed, the helium gas will escape and the integrity switch remains open. In this situation, depressing the test switch provides a "no test" indication.

Systron-Donner sensor elements are quite durable and can be flattened, twisted, kinked, and dented without losing their overheat and fire detection abilities. A typical sensing system consists of two separate sensing loops for redundancy. Both loops are required to sense a fire or overheat before an alarm will sound. However, if one loop fails, the system logic will isolate the defective loop and reconfigure to a single loop operation using the good loop. [Figure 16-13]

FLAME DETECTORS

Another type of fire detection system that is used on an aircraft is a flame detector system. Most flame detectors consist of a photoelectric sensor that measures the amount of visible light or infrared radiation in an enclosed area. The sensor is placed so it can see the surrounding area, and anytime there is an increase in the amount of light that strikes the

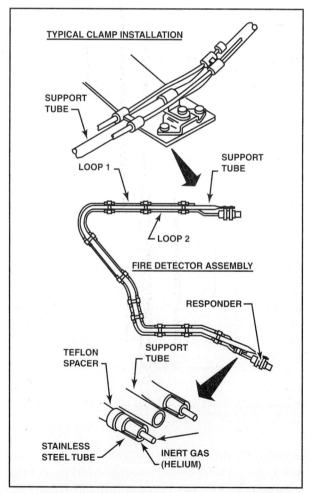

Figure 16-13. A typical installation of a Systron-Donner system consists of two independent loops attached to a support tube. The support tube establishes the routing of the detector element and provides attach points to the airplane.

cell, an electrical current is produced. Once enough current is produced and channeled through an amplifier, a fire warning light and bell are activated.

SMOKE AND TOXIC GAS DETECTION SYSTEMS

The smoke-detection system of the aircraft monitors certain areas of the aircraft for the presence of smoke, which can be an indication of an impending fire condition. These may include, but are not limited to, cargo and baggage compartments, and the lavatories of transport category aircraft. A smoke-detection system is used where the type of fire anticipated is expected to generate a substantial amount of smoke before temperature changes are sufficient to actuate an overheat-detection system.

The presence of carbon monoxide gas (CO) or nitrous oxides are dangerous to flight crews and passengers, and may indicate a fire condition. Detection of the presence of either or both of these gases could be the earliest warning of a dangerous situation.

SMOKE DETECTORS

To be reliable, smoke detectors must be maintained so that smoke in a compartment will be indicated as soon as it begins to accumulate. In order for the detector to operate properly, smoke detector louvers, vents, and ducts must not be obstructed. Smoke detection instruments are classified by method of detection, and, in some cases, an aircraft will have different types of detectors installed in various locations.

LIGHT REFRACTION TYPE

This type of detector consists of a photoelectric cell, a beacon lamp, and a light trap, all mounted on a labyrinth. Air samples are drawn through the detector unit, usually by a small circulating fan. When smoke particles are present, they refract light into the photoelectric cell. An accumulation of 10% smoke in the air causes the photoelectric cell to conduct current. When activated by smoke, the detector supplies a signal to a smoke detector amplifier, which activates a warning light and aural warning in the cockpit. [Figure 16-14]

A test switch permits checking the operation of the smoke detector. Closing the switch connects 28 VDC electricity to the test relay. When the test relay energizes, voltage is applied through the beacon lamp and test lamp in series to ground. A fire indication will be observed only if the beacon and test lamp,

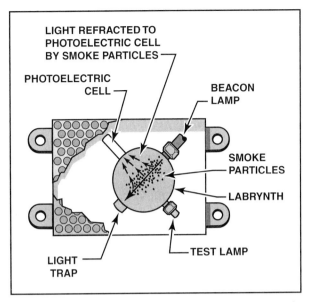

Figure 16-14. Smoke particles, drawn into the photoelectric smoke detector, refract light into a photocell, setting off an alarm.

the photoelectric cell, the smoke detector amplifiers, and associated circuits are all operable. [Figure 16-15]

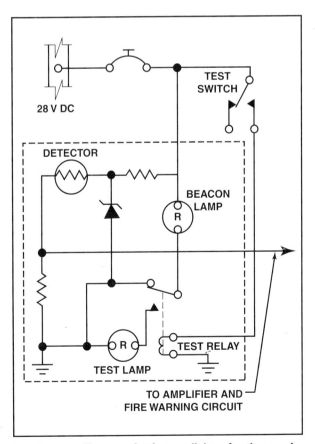

Figure 16-15. The test circuit on a light refraction smoke detector simultaneously checks the beacon light, the photocell and the associated circuits.

With some light-refraction smoke detectors, the detector can be functionally tested with a flashlight equipped with a red-colored lens. Directing the light beam into the detector simulates the light condition that would be produced with smoke. However, when conducting a test in this manner, ambient light must be shielded from entering the detector for the test to be effective.

IONIZATION TYPE

Ionization-type smoke detectors use a small amount of radioactive material to ionize some of the oxygen and nitrogen molecules in the air sample drawn into the detector cell. These ions permit a small electrical current to flow through the detector chamber test circuit.

If smoke is present in the air sample being drawn through the detector, small particles of the smoke will attach themselves to the oxygen and nitrogen ions, reducing the electrical current flow in the test circuit. If the current flow falls below a preset value, the alarm circuit will activate visual and aural cockpit alarms. [Figure 16-16]

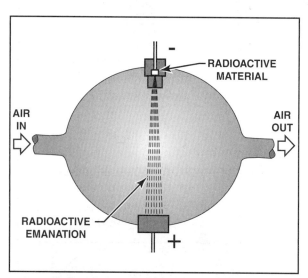

Figure 16-16. The ionization-type detector conducts electricity until smoke particles in the detection system cause a decrease in the amount of current flow.

SOLID-STATE TYPE

Solid-state smoke or toxic gas warning systems operate by comparing signals from two detecting elements, one located in the area being monitored, the other exposed to outside air.

These detecting elements consist of a heating coil encased in a coating of semiconductor material. Carbon monoxide or nitrous oxides, if present, will be absorbed into this coating and change the electrical current-carrying capability of the detector. These elements are connected into a type of bridge circuit so that when both elements are conducting evenly the bridge will be balanced, and no warning signal will be present. If the element in the area being monitored is subjected to CO gas or nitrous oxides, an unbalanced condition will be created across the bridge and the warning circuit will illuminate the cockpit warning lamp.

CARBON MONOXIDE DETECTORS

CO detectors are used to sense the presence of deadly carbon monoxide gas, and are primarily found in aircraft cabins or cockpits. CO is a colorless, odorless, tasteless, non-irritating gas that is a byproduct of incomplete combustion, and is found in varying degrees in smoke and fumes from burning substances. Exposure to even small amounts of the gas is dangerous. A concentration of 0.02% (2 parts in 10,000) may produce headache, mental dullness, and some degree of physical impairment within a few hours. Higher doses or prolonged exposure may cause death.

Probably the simplest and least expensive CO indicator is a button, worn as a badge or installed on the instrument panel or cockpit wall. The button contains a tablet that changes from a normal tan color to progressively darker shades of gray-to-black when exposed to CO gas. The color transition time is relative to the concentration of CO. At a concentration of 50 ppm (0.005%), the first discoloration will be apparent within 15 to 30 minutes, while a concentration of 100 ppm (0.01%) will change the color of the tablet in 2 to 5 minutes, and to dark gray or black in 15 to 20 minutes. The buttons are effective, but must be replaced at the manufacturer's recommended intervals to keep them at the highest level of performance.

Other types of CO detectors are installed to maintain a constant sampling of the cabin and cockpit air when the aircraft is in operation. Such detectors are especially useful in reciprocating-engine aircraft that use either internal combustion heaters or shrouded exhaust manifold systems for cabin heat. Such CO-detection systems electronically sample the cabin air, and sound an aural warning if CO is present in hazardous amounts. CO can be discharged into the cabin if the heater leaks from the combustion side of the system into the ventilating air stream.

Occasionally a manufacturer may require that an area of an aircraft be checked for the presence of CO after a repair. To perform this testing, there are several types of portable CO detectors, commonly called sniffers, that are available for use. One type has a replaceable indicator tube that contains a yellow silica gel. During operation, a sample of air is drawn through the detector tube. When the air sample contains carbon monoxide, the yellow silica gel turns to a shade of green. The intensity of the green color is proportional to the concentration of carbon monoxide in the air sample at the time and location of the tests.

FIRE-DETECTION SYSTEM INSPECTION AND TESTING

Although the airframe structure and engine cowl provide some protection for the sensing elements of fire-detection systems, damage can still result from vibration and handling during removal and installation. This, combined with the relatively small size of sensing elements, dictates the need for a regular inspection program. The following procedures are provided as examples of some general inspection practices that should be periodically accomplished on a typical fire-detection system. However, these procedures should not be used in lieu of the manufacturer's approved maintenance directives or applicable instructions.

SPOT-TYPE AND THERMOCOUPLE MAINTENANCE

Spot-type and thermocouple detection systems are relatively simple to inspect and maintain. The individual sensing units should be inspected for security of attachment, dented or distorted housings, and electrical wire connections. However, when it is necessary to splice electrical wire between sensing elements, care should be exercised to only use the materials and splicing techniques that are authorized by the detection-system manufacturer. In some installations, wire splices may cause a change in the electrical resistance of the sensing circuit, causing the system to malfunction.

Thermocouple detector mounting brackets should be repaired or replaced when cracked, corroded, or damaged. When replacing a thermocouple detector, note which wire is connected to the plus (+) terminal of the defective unit and connect the replacement detector in the same way.

After the components of a fire-detection system have been inspected, the system must be tested. To test a typical fire-detection system, power is turned on in the cockpit and the fire detection test switch is placed in the "TEST" position. Once this is done, the red warning light should illuminate within the time period established for the system. On some aircraft, an audible alarm will also sound.

For some spot-type and thermocouple detection systems, as well as continuous-loop systems, a **Jetcal Analyzer** unit may be used to physically test a sensing element. A Jetcal Analyzer consists of a heating element that is used to apply a known heat value to a fire-detector element. The heat value displays on the potentiometer of the Jetcal control panel. When the alarm temperature is reached, the cockpit warning light will illuminate. If the light illuminates before the prescribed temperature setting, the entire detector circuit should be inspected for dented sensing elements, kinked wires and sensing tubes, or other damage that could affect the electrical resistance of the circuit. [Figure 16-17]

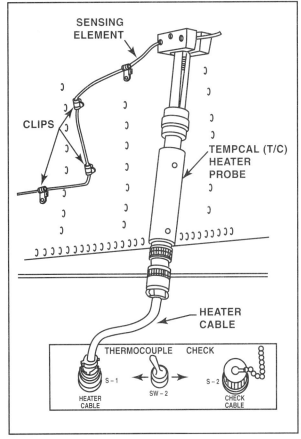

Figure 16-17. A Jetcal Analyzer can be used to heat a fire detector element to test the fire warning system.

CONTINUOUS-LOOP MAINTENANCE PRACTICES

One of the first items that must be periodically checked on continuous-loop detection systems is the routing and security of the detector elements. Long, unsupported sections can vibrate excessively and cause damage to the element. Common locations of cracked or broken elements are near inspection plates, cowl panels, engine components, or cowl supports.

The distance between clamps on straight runs is usually between 8 and 10 inches and is specified by each manufacturer. To ensure adequate support when a sensing element ends at a connector, a support clamp should be located about four to six inches from the connector fitting. On elements that are routed around certain components, a straight run of one inch is typically maintained from all connectors before a bend is started. The optimum bend radius for most continuous-loop type sensing elements is three inches. [Figure 16-18]

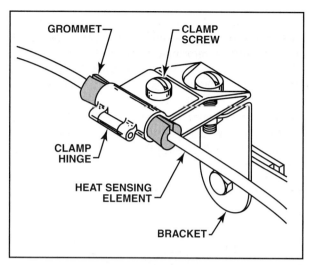

Figure 16-19. Grommets should be installed on the sensing element so both ends are centered on its clamp. The split end of the grommet should face the outside of the nearest bend. Clamps and grommets should fit the element snugly.

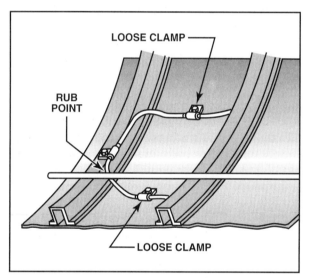

Figure 16-18. A loose clamp can result in interference between a cowl brace and a sensing element. This interference can cause the sensing element to wear, which could create a short.

The clamps used to support most continuous-loop sensing elements consists of a small hinged piece of aluminum that is bolted or screwed to the aircraft structure. To help absorb engine vibration, most support clamps use a rubber grommet, wrapped around the sensing element. Rubber grommets often become softened from exposure to oils and hydraulic fluid, or hardened from excessive heat. Such grommets should be inspected on a regular basis and replaced as necessary. [Figure 16-19]

A continuous-loop sensing element should be checked for dents, kinks, or crushed areas. Each manufacturer establishes the limits for acceptable dents or kinks as well as the minimum acceptable diameter for a sensing element. It is important to note that if a dent or kink exists that is within the manufacturer's limits, no attempt should be made to

straighten it. By attempting to unnecessarily straighten a sensing element, stresses may be set up that could cause the tubing to fail. [Figure 16-20]

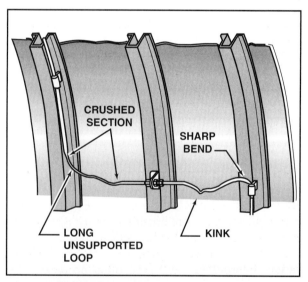

Figure 16-20. Fire-sensing elements are located in exposed areas and, therefore, are subject to impact and abrasion. When inspecting fire detection elements, be alert for sharp bends, kinks, and crushed sections.

If shielded flexible leads are used on the ends of the sensing element, they should be inspected for fraying. The braided sheath is made up of many fine metal strands, woven into a protective covering and surrounding the inner insulated wire. Continuous bending or rough treatment can break the wire strands, especially those near the connectors, and cause a short circuit.

Nuts at the end of a sensing element should be inspected for tightness and proper safetying. Loose nuts should be retorqued to the value specified by the manufacturer. Some connection joints require the use of copper crush gaskets. If this type of gasket is present on a joint, it should be replaced anytime the connection is separated. Additional items to look for include pieces of safety wire or other metal particles that could short the sensing element. [Figure 16-21]

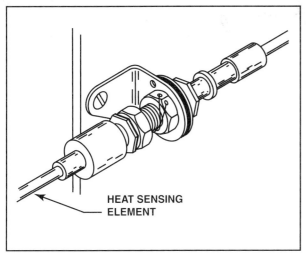

Figure 16-21. When inspecting an electrical connector joint such as this one, verify that the retaining nut is properly torqued and the safety wire is secure.

TROUBLESHOOTING

Intermittent alarms or false alarms are probably the most common problems associated with a fire-detection system. Most intermittent alarms are caused by an intermittent short circuit in the detector system wiring. Electrical shorts are often caused by a loose wire that occasionally touches a nearby terminal, a frayed wire brushing against a structure, or a sensing element that has rubbed against a structural member long enough to wear through the insulation. Intermittent faults can often be located by applying power to the system and moving wires to recreate the short.

False alarms can typically be located by disconnecting the engine sensing loop from the aircraft wiring. If the false alarm continues, a short exists between the loop connections and the control unit. If the false alarm ceases when the engine sensing loop is disconnected, the fault is in the disconnected sensing loop. The loop should be examined to verify that no portion of the sensing element is touching the hot engine. If there is no contact, the shorted section can be located by isolating and disconnecting elements consecutively around the entire loop. Kinks and sharp bends in the sensing element can cause an internal wire to short intermittently to the outer tubing. The fault can be located by checking the sensing element with a megohm meter, or megger, while tapping the element in the suspected area to produce the short.

Moisture in the detection system seldom causes a false fire alarm. However, if moisture does cause an alarm, the warning will persist until the contamination is removed or boils away and the resistance of the loop returns to its normal value.

Another problem that could be encountered is the failure to obtain an alarm signal when the test switch is actuated. Such failure could be caused by a defective test switch or control unit, the lack of electrical power, an inoperative indicator light, or an opening in the sensing element or connecting wiring. Kidde and Fenwal continuous-loop detectors will not test if a sensing element is shorted or broken; however, they will provide a fire warning if a real fire exists. When the test switch fails to provide an alarm, the continuity of a two-wire sensing loop can be determined by opening the loop and measuring the resistance of each wire. In a single-wire continuous-loop system, the center conductor should be grounded.

SECTION B

FIRE-EXTINGUISHING SYSTEMS

Hand-held fire extinguishers and extinguishing systems are installed in many aircraft to provide the flight crew and maintenance personnel with the ability to fight fires while the aircraft is operating on the ground or in flight. Portable extinguishers are commonly installed in the cockpit and passenger cabin of many aircraft. More elaborate extinguishing systems are installed in transport category and corporate airplanes to extinguish fires in the engine, auxiliary power unit, baggage, and electronic equipment compartments. In addition, many transport category airplanes have fire-extinguishing systems located in trash receptacles to protect against fires that may occur in the lavatories of passenger-carrying aircraft.

FIRE-EXTINGUISHING AGENTS

As previously mentioned, the three elements that are needed to support combustion are a combustible fuel, oxygen, and heat. If any one of these elements is removed, a fire will not burn. The portable and fixed fire-extinguisher systems used in most aircraft are designed to displace the oxygen with an inert agent that does not support combustion. The most common types of aircraft extinguishing agents that are used include carbon dioxide and halogenated hydrocarbons.

CARBON DIOXIDE

Carbon dioxide (CO_2) is a colorless, odorless gas that is about one and one-half times heavier than air. To be used as an extinguishing agent, carbon dioxide must be compressed and cooled until it becomes a liquid that can be stored in steel cylinders. When released into the atmosphere, carbon dioxide expands and changes to a gas that cools to a temperature of about $-110°F$. Because of the cooling effect, the water vapor in the air immediately condenses to form "snow," which causes the CO_2 to appear to settle over the flames and smother them. However, the fire is actually extinguished by the CO_2 displacing the oxygen in the atmosphere, interrupting the chemical reaction between the fuel and the oxygen. Once the "snow" warms, it evaporates, leaving almost no residue.

Carbon dioxide is effective on both Class B and Class C fires. A carbon dioxide hand held fire extinguisher can be used on an electrical fire, provided the discharge horn is constructed of a nonmetallic material. A metallic horn would tend to transfer an electrical charge back to the fire extinguisher and to ground through the person holding the extinguisher. In addition, since carbon dioxide leaves almost no residue, it is well suited for engine intake and carburetor fires. Furthermore, carbon dioxide is nontoxic and does not promote corrosion. However, if used improperly, carbon dioxide will dissipate oxygen uptake in the lungs, which can cause physiological problems such as mental confusion and suffocation. Because of its variation in vapor pressure with temperature, it is necessary to store CO_2 in stronger containers than required for most other extinguishing agents.

HALOGENATED HYDROCARBONS

A **halogen** element is one of the group that consists of chlorine, fluorine, bromine, or iodine. Some hydrocarbons combine with halogens to produce very effective fire-extinguishing agents that work by excluding oxygen from the fire source and by chemically interfering with the combustion process. Halogenated hydrocarbon fire-extinguishing agents are most effective on Class B and C fires but can be used on Class A and D fires as well. However, their effectiveness on Class A and D fires is somewhat limited.

Halogenated hydrocarbons are numbered according to their chemical formulas with five-digit Halon numbers, which identify the chemical makeup of the agent. The first digit represents the number of carbon atoms in the compound molecule; the second digit, the number of fluorine atoms; the third digit, the number of chlorine atoms; the fourth digit, the number of bromine atoms; and the fifth digit, the number of iodine atoms, if any. If there is no iodine present the fifth digit does not appear. For example, bromotrifluoromethane CF_3Br is referred to as Halon 1301, or sometimes by the trade name **Freon 13**™.

Halon 1301 is extremely effective for extinguishing fires in engine compartments of both piston and turbine powered aircraft and is also considered to be one of the best extinguishing agents for aircraft interior fires. In engine compartment installations, the Halon 1301 container is pressurized by compressed nitrogen and is discharged through spray nozzles. Halon 1301 is also widely used as the agent for portable fire extinguishers. [Figure 16-22]

Figure 16-22. Halogenated hydrocarbon fire-extinguishing agents provide effective fire suppression in aircraft.

A number of halogenated hydrocarbon agents have been used in the past but are no longer in production. The reason for this is that some early Halon extinguishing agents produced toxic or corrosive gases when exposed to fire. For example, carbon tetrachloride (Halon 104) was the first generally accepted Halon extinguishing agent and was very popular for electrical hazards. However, when exposed to heat, its vapors formed a deadly phosgene gas, which is a form of nerve gas.

Another once-popular agent was methyl bromide (Halon 1001). However, methyl bromide is toxic to personnel and corrosive to aluminum alloys, magnesium, and zinc. Of all the halogenated hydrocarbon extinguishing agents, Halon 1301 is the safest to use from the standpoint of toxicity and corrosion hazards. In small dosage amounts, the gas has a low toxicity, but has similar effects of depriving oxygen from the lungs.

Because of changing regulations and developing environmental impact data, you should keep abreast of current developments pertaining to the use of halogenated hydrocarbons as fire-extinguishing agents. For example, several studies suggest that chloroflourocarbons (CFCs), such as Halon, damage the ozone layer in the stratosphere, allowing higher levels of ultraviolet radiation to reach the earth. To reduce damage to the ozone layer, the Environmental Protection Agency banned the production of CFCs after December 31, 1995. However, existing stocks of CFCs are still allowed to be used after this date. Several alternatives to CFCs have recently been developed and will most likely find applications as aviation fire-extinguishing agents. For example, DuPont FE-25™ has proven to be an acceptable substitute for Halon 1301 as an extinguishing agent and has no harmful affect on the earth's ozone layer. Other replacement extinguishing agents being researched include water mist sprays, which have been proven to be effective in combating many A, B, and C class fires.

As an aviation maintenance technician, it is important to be aware of EPA and FAA regulations governing the use and disposal of CFCs. Improper handling or disposal of halogenated hydrocarbons can lead to civil and criminal penalties.

PORTABLE FIRE EXTINGUISHERS

Portable, or hand-held, fire extinguishers are installed in many aircraft inside the cockpit and passenger compartments where they can be readily accessed in the event of a fire. For installations on commercial passenger-carrying aircraft, the number and location of extinguishers may be mandated by FAR requirements, while the owners of smaller general aviation airplanes are given the option to have portable extinguishers installed.

PORTABLE FIRE EXTINGUISHER INSTALLATIONS

The FAA accepts the installation of most portable fire extinguishers that are approved by certifying organizations such as Underwriters' Laboratories, Inc., Factory Mutual Research Corp., or by the U.S. Coast Guard under title 46 of the CFR for use in aircraft. In most cases, these extinguishers are mounted in brackets supplied by the extinguisher manufacturer and can tolerate the inertia forces that may be encountered during flight or due to an accident. When evaluating the installation of a new or previously installed portable extinguisher, consideration should be given to the following items:

1. Portable extinguishers should be mounted as near as possible to the hazardous areas they are intended to protect. If no obvious hazard areas exist, the extinguisher should be mounted near the passenger entrance door or in a flight-attendant station, if one is provided.

2. When two or more extinguishers are installed, they should be located with one at each end of the passenger compartment and spaced uniformly throughout the remainder of the cabin.

3. The extinguisher should be positioned in a location that makes it readily visible and accessible.

When this is not possible, a placard may be installed with letters at least 3/8-inch high indicating the location of the extinguisher.

4. The extinguisher manufacturer's mounting bracket should be used only after determining that it is capable of sustaining the inertia force requirements designated in FAR Parts 23.561, 25.561, 27.561, or 29.561 with regard to the type of aircraft. Meeting these requirements helps to ensure that the extinguisher will not become dislodged while in flight or during a hard landing or accident, which could cause severe injuries to the occupants if struck by the extinguisher. When evaluating the mounting bracket installation, verify that it does not obstruct or damage the aircraft structure. For example, make sure the mounting hardware does not penetrate into electric cables, control cables or fluid carrying hoses. Also, check movable items such as flight controls and seat travels to verify that the extinguisher does not hinder full movement.

5. Verify that all maintenance documentation detailing the installation is complete. Required record entries include amended empty weight and empty weight C.G. data, as well as the equipment list and permanent maintenance records. If the extinguisher is installed in an aircraft that does not have type certificate approval for the installation, also verify that there is an appropriate FAA Form 337, detailing the major alteration.

PORTABLE EXTINGUISHER MAINTENANCE

Most portable fire extinguishers are vendor supplied components and may not be covered in the aircraft manufacturer's maintenance instructions. In these situations, the best resource for servicing and maintenance information may be directly from the extinguisher manufacturer. However, for extinguishers installed in aircraft that are used in commercial operations, the maintenance requirements for portable extinguishers are often detailed in the carrier's individual operating specifications.

In some situations, the extinguisher manufacturer may provide basic servicing and maintenance information on the identification label. This information should be reviewed during any inspection. Items that are typically checked include weighing the container to determine the quantity of extinguishing agent, and checking a pressure gauge to determine the propellant charge. The information on the label may also indicate any time or life limits on the serviceability of the extinguisher or the requirements for hydrostatic testing, if required. Any servicing or maintenance performed on the extinguisher must be recorded in the aircraft's permanent maintenance records and is often also indicated directly on the extinguisher container. [Figure 16-23]

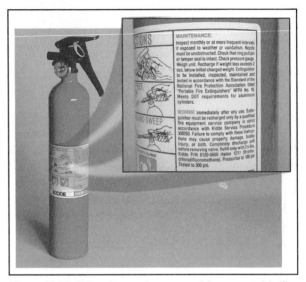

Figure 16-23. When inspecting or servicing a portable fire extinguisher, always review the information provided on the identification label. Also, verify that the extinguishing agent is appropriate for the type of fire that is likely to occur in the fire zone where the extinguisher is installed.

Additional items to check include releasing the extinguisher from its mounting bracket to determine its ease of removal and checking that the activation-trigger safety pin is properly installed. If there is any doubt as to the integrity of the extinguisher's condition, it should be replaced or sent to a certified repair station that is authorized to perform full servicing and maintenance on portable fire extinguishers.

FIXED FIRE-EXTINGUISHING SYSTEMS

In an aircraft, it is important that the type of fire-extinguishing system be appropriate for the class of fire that is likely to occur. There are two basic categories of fixed fire-extinguishing systems: **conventional systems**, and **high-rate-of-discharge (HRD) systems**. Both systems utilize one or more containers of extinguishing agent and a distribution system that releases the extinguishing agent through perforated tubing or discharge nozzles. As a general rule, the type of system installed can be identified by the type of extinguishing agent used. For example, conventional systems usually employ carbon dioxide as the extinguishing agent while HRD systems typically utilize halogenated hydrocarbons.

CONVENTIONAL SYSTEMS

The fire-extinguishing installations used in most older aircraft are referred to as conventional systems. Many of these systems are still used in some aircraft, and are satisfactory for their intended use. A conventional fire-extinguisher system consists of a cylinder that stores carbon dioxide under pressure and a remotely controlled valve assembly that distributes the extinguishing agent.

Carbon dioxide cylinders come in various sizes, are made of stainless steel, and are typically wrapped with steel wire to make them shatterproof. In addition, the normal gas storage pressure ranges from 700 to 1,000 psi. Since the freezing point of carbon dioxide is so low, a storage cylinder does not have to be protected against cold weather. However, cylinders can discharge prematurely in hot climates. To prevent this, manufacturers sometimes charge a cylinder with about 200 psi of dry nitrogen before they fill the cylinder with carbon dioxide. When treated in this manner, most CO_2 cylinders are protected against premature discharge up to 160°F. The nitrogen also provides additional pressure during normal release of the agent.

Carbon dioxide cylinders are equipped internally with one of three types of **siphon tubes**. The cylinders used in aircraft typically utilize either a straight-rigid, or a short-flexible siphon tube. The type of siphon tube installed in the cylinder is determined by the cylinder's mounting position. [Figure 16-24]

The CO_2 within a cylinder is distributed through tubing from the CO_2 cylinder valve to the control valve assembly in the cockpit. Once past the control valve, the CO_2 proceeds to the fire zone via solid tubing installed in the fuselage or wing. Inside the fire zone, the tubing is perforated so the carbon dioxide can be discharged. [Figure 16-25]

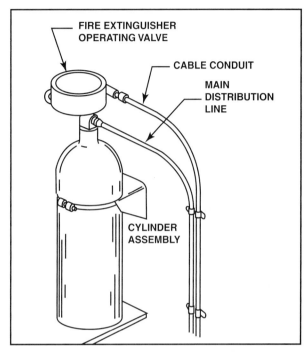

Figure 16-25. In a fire-extinguishing system that utilizes carbon dioxide as an extinguishing agent, a sturdy cylinder assembly is mounted to the airframe and connected to a distribution line. In addition, an operating valve that is controlled from the cockpit is installed to hold the carbon dioxide in the cylinder until it's needed.

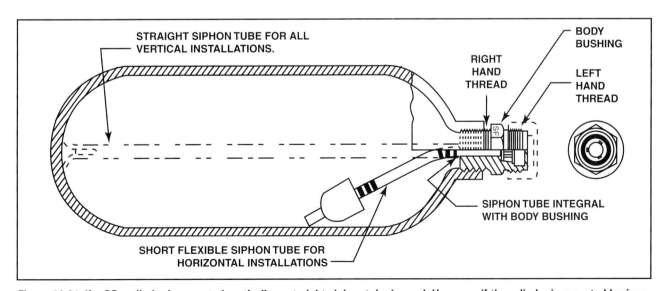

Figure 16-24. If a CO_2 cylinder is mounted vertically, a straight-siphon tube is used. However, if the cylinder is mounted horizontally, a short-flexible siphon tube must be used. The type of siphon tube installed is typically indicated by a stamped code on the body bushing. If an "SF" appears, a short-flexible siphon is installed. If an "S" appears, a straight siphon tube is installed. Other manufacturers stamp or stencil the type of siphon used on the cylinder body.

To operate a conventional fire extinguisher system used to protect an engine compartment, a selector valve in the cockpit must be manually set for the engine that is on fire. Once this is done, a T-shaped control handle located next to the selector valve is pulled upward to actuate the release lever in the CO_2 cylinder valve. Once released, the compressed carbon dioxide flows in one rapid burst to the outlets in the distribution line of the affected engine compartment. Contact with the air converts the liquid CO_2 into a visible gas, which extinguishes the flames by displacing oxygen.

Some CO_2 systems designed to protect engine fire zones have multiple bottles, which gives the system the capability of delivering extinguishing agent twice to any of the engine compartments. Each bank of CO_2 bottles is equipped with a red **thermo-discharge indicator disk** and a yellow **system-discharge indicator disk**. The red thermo-discharge disc is set to rupture and discharge the carbon dioxide overboard if the cylinder pressure becomes excessively high (about 2,650 psi). On the other hand, the yellow system-discharge disk ruptures whenever a bank of bottles has been emptied by a normal discharge. These disks are mounted so that they are visible on the outside of the fuselage. This way, during a preflight inspection, the flight crew can identify the condition of the system.

HIGH-RATE DISCHARGE SYSTEMS

High-rate-of-discharge (HRD) is the term applied to the fire-extinguishing systems found in most modern turbine engine aircraft. A typical HRD system consists of a container to hold the extinguishing agent, at least one bonnet assembly, and a series of high-pressure feed lines.

The containers used in an HRD system are typically made of steel and spherically shaped. There are four sizes commonly in use today, ranging from 224 cubic inches to 945 cubic inches. The smaller containers generally have two openings, one for the **bonnet assembly** or **operating head**, and the other for a **fusible safety plug**. The larger containers are usually equipped with two bonnet assemblies.

Each container is partially filled with an extinguishing agent, such as Halon 1301, and sealed with a **frangible disk**. Once sealed, the container is pressurized with dry nitrogen. A container pressure gauge is provided so you can quickly reference the container pressure. The bonnet assembly contains an electrically ignited discharge cartridge, or **squib**, which fires a projectile into the frangible disk. Once the disk breaks, the pressurized nitrogen forces the extinguishing agent out of the sphere. A strainer is installed in the bonnet assembly to prevent the broken disk fragments from getting into the distribution lines. [Figure 16-26]

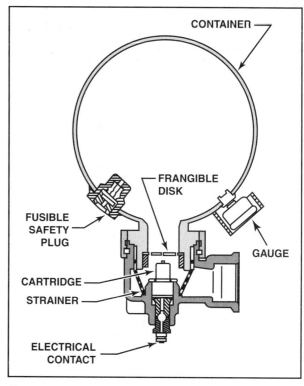

Figure 16-26. In a typical HRD container, the extinguishing agent is released by an electrically actuated explosive that ruptures a frangible disk. Once broken, the disk fragments collect in a strainer while the extinguishing agent is directed to the engine nacelle.

As a safety feature, each extinguishing container is equipped with a thermal fuse that melts and releases the extinguishing agent if the bottle is subjected to high temperatures. If a bottle is emptied in this way, the extinguishing agent will blow out a red indicator disk as it vents to the atmosphere. On the other hand, if the bottle is discharged normally, a yellow indicator disk blows out. Like a conventional system, the indicator disks are visible from the outside of the fuselage for easy reference. [Figure 16-27]

When installed on a multi-engine aircraft, the fire-extinguishing-agent containers are typically equipped with two firing bonnets. The two discharge ports allow one container to serve both engines. [Figure 16-28]

On large, multi-engine aircraft, two extinguishing-agent containers are generally installed, each with two firing bonnets. This allows twin-engine aircraft

Fire Protection Systems

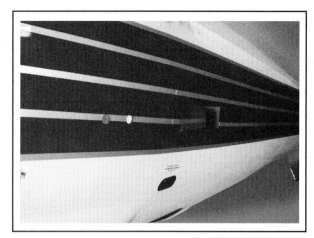

Figure 16-27. Two colored indicator disks are visible on the exterior of an aircraft equipped with CO_2 or HRD extinguisher system bottles. If the red disk is missing, it indicates that the fire bottles have discharged because the bottle pressure exceeded limits due to thermal heating. If the yellow disk is missing, it indicates that the bottles were discharged through activation of the system from the cockpit controls.

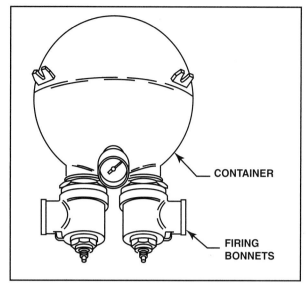

Figure 16-28. A typical extinguishing-agent container on a multi-engine aircraft has two firing bonnets.

to have a dedicated container for each engine. In addition, the two discharge ports on each bottle provide a means of discharging both containers into one engine compartment. [Figure 16-29]

INSPECTION AND SERVICING

Regular maintenance of fire-extinguishing systems includes inspecting and servicing the fire-extinguisher bottles, removing and re-installing discharge cartridges, testing the discharge tubing for leaks, and testing electrical wiring for continuity. The following discussion looks at some of these common maintenance procedures to provide an understanding of the operations involved. However, as an aviation maintenance technician, you must understand that fire-extinguishing-system maintenance procedures vary substantially, depending on

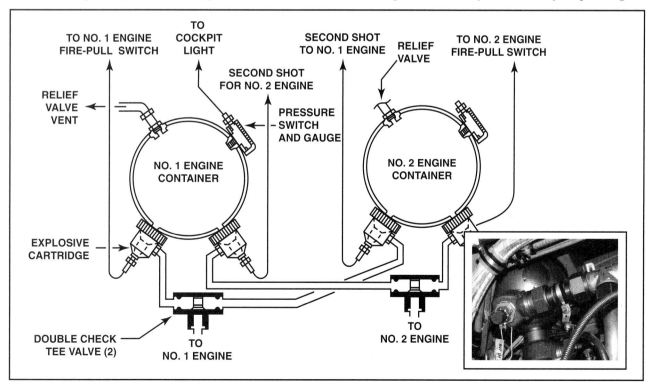

Figure 16-29. A typical high-rate-of-discharge extinguishing system installed on a twin-engine, turbine-powered aircraft utilizes two agent containers, each with two discharge ports. This permits two applications of extinguishing agent to any one engine.

the design and construction of the particular unit being serviced. Therefore, the detailed procedures outlined by the airframe or system manufacturer should always be followed when performing maintenance.

CONTAINER PRESSURE CHECK

A pressure check of fire-extinguisher containers is made periodically to determine that the pressure is between the minimum and maximum limits prescribed by the manufacturer. Aircraft service manuals contain pressure/temperature curves or charts that provide the permissible gauge readings corrected for temperature. If the pressure does not fall within the appropriate limits, the container must be removed and replaced with a properly charged container. [Figure 16-30]

Once it has been determined that a bottle is properly charged, check to make certain that the glass on the pressure gauge is not broken. In addition, verify that the bottle is securely mounted to the airframe.

The only way to determine if the appropriate amount of extinguishing agent is in a given container is to weigh the container. Therefore, most fire-extinguishing containers require re-weighing at frequent intervals. In addition to the weight check, fire-extinguisher containers must be hydrostatically tested at five-year intervals.

DISCHARGE CARTRIDGES

The discharge cartridges used with HRD containers are life-limited and the service life is calculated from the manufacturer's date stamped on the cartridge. The manufacturer's service life is usually expressed in terms of hours and is valid as long as the cartridge has not exceeded a predetermined temperature limit. Many cartridges are available with a service life of up to 5,000 hours. To determine a cartridge's service life, it is necessary to remove the electrical leads and discharge hose from the bonnet assembly. Once this is done, the bonnet assembly can be removed from the extinguisher container so the date stamped on the cartridge can be seen.

Most new extinguisher containers are supplied with their cartridge and bonnet assembly disassembled. Therefore, care must be taken in assembling or replacing cartridges and bonnet assemblies. Before installation on an aircraft, the cartridge must be

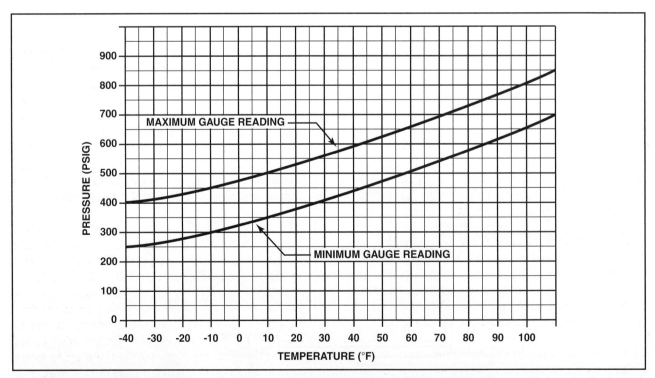

Figure 16-30. This pressure/temperature chart allows you to determine if a specific fire-extinguishing bottle is properly charged. As an example, assume the ambient temperature is 70°F and the fire-extinguishing container needs to be checked to see if it is properly charged. To do this, find 70 degrees at the bottom of the chart and follow the line up vertically until it intersects the minimum gauge-reading curve. From here, move left horizontally to find a minimum pressure of about 540 psig. Next, go back to the 70° line and follow it up vertically until it intersects the maximum gauge-reading curve. From this point, follow the horizontal line to the left to determine a maximum pressure of approximately 690 psig. The container is properly charged as long as the pressure gauge on the container indicates between 540 psig. and 690 psig.

Fire Protection Systems

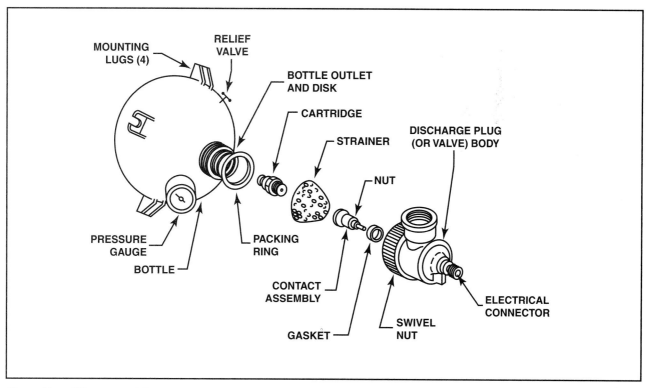

Figure 16-31. When assembling a discharge cartridge into a bonnet assembly, it is best to use an exploded view drawing like the one above. Once assembled, the entire bonnet assembly is attached to the container by means of a swivel nut that tightens against a packing ring gasket.

properly assembled into the bonnet and the entire assembly connected to the container. [Figure 16-31]

If a discharge cartridge is removed from a bonnet assembly, it should not be used in another bonnet assembly. In addition, since discharge cartridges are fired electrically, they should be properly grounded or shorted to prevent accidental firing. Wrapping a piece of safety wire between the two electrical terminals of the discharge cartridge is sometimes done to keep both terminals electrically neutral.

727 FIRE-PROTECTION SYSTEM

The following discussion is intended to provide an overview of a typical fire-extinguishing system installed on a transport-category aircraft. The fire-protection system used on a Boeing 727 is typical of those found on several aircraft in service today.

In the Boeing 727 powerplant fire-extinguishing system, all three powerplant areas are protected by two high-rate-of-discharge bottles. Each of the two-agent bottles has a gauge to indicate its pressure. An electrical pressure switch is mounted on each bottle to activate a bottle discharge light on the instrument panel when the pressure on the agent bottle is below limits.

Once the extinguishing agent leaves a bottle, it proceeds to a two-way shuttle valve that channels the extinguishing agent into the distribution system. Once in the distribution system, the extinguishing agent passes through the appropriate engine selector valve to a series of discharge nozzles within the engine compartment. If the fire is not extinguished after discharging one bottle, the second bottle can be discharged and the extinguishing agent routed to the same engine. [Figure 16-32 on page 16-24]

The controls for the 727 fire-protection system consist of three engine fire-warning lights, one wheel-well fire-warning light, a bottle transfer switch, a fire-bell cutout switch, a fire-detection-system test switch, and a detector-inoperative test switch. The fire warning lights are part of the fire-detection system and illuminate whenever one of the fire detectors detects a fire. On the other hand, the bottle transfer switch allows the pilot to select which bottle of extinguishing agent is discharged. The fire-bell disable switch silences the fire bell after it has been activated by a fire indication. The fire-detection-system test switch checks the continuity of the detectors and operation of the warning system. The detector-inoperative test switch tests the circuits that activate the "Detector Inop"

16-24   **Fire ProtectonSystems**

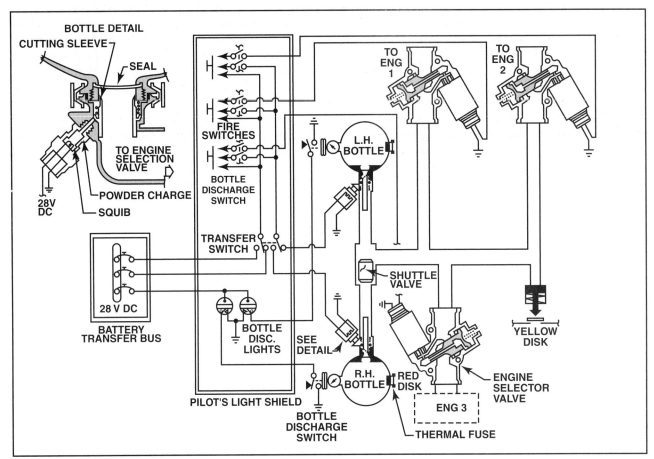

Figure 16-32. The Boeing 727 aircraft utilizes two fire bottles and three selector valves to provide fire suppression to all three engines. With this arrangement, the cockpit crew can discharge both bottles to a single engine.

lights and, if the systems are functioning properly, momentarily illuminate the "Detector Inop" lights. [Figure 16-33]

When a fire is sensed, a red warning light inside the engine-fire switch illuminates and the fire bell rings. When the warning light comes on, the pilot

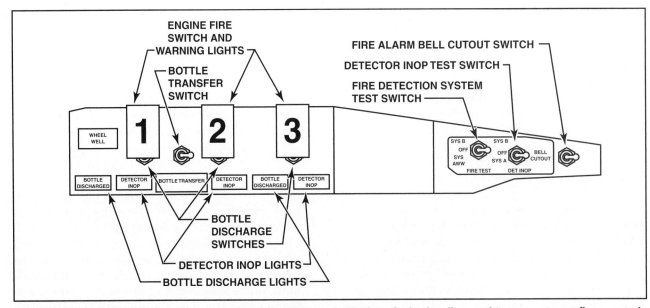

Figure 16-33. A typical Boeing 727 fire control panel provides an indication of wheel-well or engine-compartment fires, controls fire bottle discharge, and permits testing of the fire detector system.

pulls the appropriate engine-fire handle. This arms the fire-extinguisher-bottle discharge switch, disconnects the generator field relay, stops the flow of fuel and hydraulic fluid to the engine, and shuts off the engine bleed air. It also deactivates the engine-driven hydraulic pump low-pressure lights and uncovers the bottle discharge switch. If the pilot determines that a fire actually exists in the engine compartment, the extinguishing agent is released by depressing and holding the bottle discharge switch. Once the discharge switch is depressed, electrical current causes the discharge cartridge to explode and shatter the frangible disk. With the frangible disk broken, the extinguishing agent is released into the appropriate engine compartment. Once the extinguishing agent is discharged, the fire warning light should go out within thirty seconds. If the warning light does not go out, the pilot can move the bottle transfer switch to its opposite position to select the second bottle of extinguishing agent, and again push the bottle discharge switch. Once the fire-extinguisher bottle has been discharged, or when its pressure is low, the appropriate bottle discharge light illuminates.

AIRCRAFT AIRWORTHINESS INSPECTION

CHAPTER 17

INTRODUCTION

In order to ensure that aircraft are maintained to the highest standard of airworthiness, they are managed and inspected under FAA-mandated and -approved inspection programs. Inspection programs must ensure the aircraft is airworthy and conforms to all applicable FAA aircraft specifications, type certificate data sheets, airworthiness directives, and other FAA approved data.

Inspection planning is organized around an aircraft's age, utilization, environmental conditions, and the type of operation. Examples include changes in temperature, frequency of landings and takeoffs, operation in areas of high industrial or environmental pollutants, and passenger or cargo operations. To assure proper maintenance, each inspection interval must be stated in terms of flight hours, calendar times, and cycles (the number of takeoffs and landings the aircraft makes). As part of the aircraft's certification process, the aircraft manufacturer and the FAA agree on the frequency for inspection requirements on the aircraft as well as functional checks of each system. This forms the basis for the maintenance program when the aircraft is in service. Every system on the aircraft has its own inspection requirements. Typically, major system-inspection requirements are synchronized to minimize aircraft downtime and to eliminate a duplication of effort. However, it is common to have completely separate inspection cycles for the primary aircraft structure and its engines.

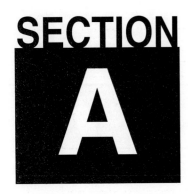

SECTION A

REQUIRED AIRWORTHINESS INSPECTIONS

On a base level, **"inspect"** means to examine by sight and touch. When performing inspections, the inspector measures and checks conditions against established guidelines. An inspector must be able to recognize defects and be aware of failure modes. Aircraft inspections include manual tasks such as initiating the inspection, accessing the aircraft, and responding to problems. In addition, cognitive tasks, such as search and decision making skills, are also used in the inspection process. An inspector should be able to identify and determine the acceptable degree of deterioration or defects permitted by the manufacturer's manuals or other approved data.

Initiating the inspection can begin by reviewing a maintenance checklist or work card, and understanding the area or item to be inspected. Maintenance checklists for small aircraft (under 12,500 lbs. gross takeoff weight) must conform to FAR Part 43, Appendix D. Most aircraft manufacturers provide inspection checklists regarding the specific aircraft they produce. Small aircraft manufacturers' inspection schedules meet the minimum requirements of Appendix D and contain many details covering specific items of equipment installed on a particular aircraft. In addition, they often include references to service bulletins and service letters, which might otherwise be overlooked. As long as they meet the minimum requirements of Part 43 Appendix D, approved inspection checklists may also be customized and made more extensive to meet the needs of an individual owner/operator. Large and turbine powered aircraft are inspected under more encompassing inspection programs tailored to their specific type of aircraft and operating conditions.

Aircraft are subject to many required inspections. These range from the basic pre-flight inspection, a daily walk-around inspection, to extensive heavy maintenance checks, which involve significant disassembly and detailed inspection of the aircraft.

PRE-FLIGHT INSPECTIONS

An FAA approved **Minimum Equipment List (MEL)** includes equipment that, if inoperative, may either ground the aircraft or allow it to be flown with flight restrictions deferring maintenance for specific periods of time. An aircraft's MEL is specific to its precise configuration and serial number. When a MEL item is discovered inoperative, it is reported by making an entry in the aircraft's maintenance record. The inoperative equipment is either repaired or deferred according to the MEL instructions prior to further flight. After repair, record an airworthiness release or aircraft maintenance entry to remove the flight restrictions. [Figure 17-1]

During a pre-flight inspection, all of the aforementioned items are verified by the pilot along with performing a visual walk-around inspection. The walk-around entails referencing a pre-flight checklist and looking for obvious problems such as nicks and cracks on the propeller, missing hardware, properly inflated tires, and flight control damage. Although pre-flight checklists are primarily designed for the flight crew, an aircraft technician should also perform these checks before operating an aircraft. [Figure 17-2]

FAR PART 91 REQUIRED INSPECTIONS

FAR Part 91 contains the General Operating and Flight Rules of aircraft and specifies the inspections required to determine the airworthiness of an aircraft. Subpart E of Part 91 deals with and describes the approved inspection programs for aircraft operations.

Small aircraft are governed by subpart E and must have a complete annual inspection every 12 calendar months. If the aircraft is operated for compensation or hire, it must have a "100-hour" inspection of the same scope as an annual inspection performed every one hundred hours of operation. Large and turbine powered, multi-engine aircraft require more specific detailed inspections that are tailored to their particular flight operations.

ANNUAL INSPECTION

The most common type of inspection required for small general aviation aircraft is the annual inspec-

Aircraft Airworthiness Inspection   17-3

ROCKWELL COMMANDER 500A REGISTRATION NO. 00XYZ SERIAL NO. 500A3848Q			21-1 **REVISION 2** 4/20/00
1. **SYSTEM & SEQUENCE NUMBERS**	\multicolumn{3}{l}{**2. NUMBER INSTALLED**}
	\multicolumn{3}{l}{**3. NUMBER REQUIRED FOR DISPATCH**}
	\multicolumn{3}{l}{**4. REMARKS OR EXCEPTION**}
21. AIR CONDITIONING			

1. COMBUSTION HEATER C	1	0	• (M) MAY BE INOPERATIVE PROVIDED:
			  a. MAINTENANCE PULLS AND CAPS JANITORIAL HEATER CIRCUIT BREAKER.
			
			MAINTENANCE:
			
			A certificated mechanic shall perform an inspection of the combustion heater. Remove nose section top access panel.
			  a. Inspect the heater for general security, damage and fuel leaks. Inspect for damage to any of the associated systems adjustment to the heater.
			  b. Replace access cover.
			  c. Inspect heater fuel pump located inside the nose wheel well area mounted against the top of the wheel well. Check for security, damage and fuel leaks to heater fuel pump.
			  d. Inspect fuel cycling solenoid valve and fuel safety solenoid valve located in the wheel well area against the lower bulkhead area looking aft. Check for security, damage and fuel leaks to any associated fuel lines in this area.
			
			Enter a statement of work performed in the aircraft flight log and install placard.

2. HEATER VENTILATION FAN C	1	0	• MAY BE INOPERATIVE PROVIDED:
			  a. COMBUSTION HEAER IS NOT UTILIZED ON THE GROUND.
			  b. HEATER IS TURNED OFF PRIOR TO LANDING.
			  c. WINDSHIELD DEFOGGING IS NOT REQUIRED ON THE GROUND.
			             FINAL APPROVAL
			   FEDERAL AVIATION AMINISTRATION
			             NM- FSDC -
			SIGNATURE . . . .
			
			DATE . . . . . . SEP 5 2000 . . . .

MINIMUM EQUIPMENT LIST	\multicolumn{2}{c}{13-12}	04/20/00 REVISION 3

Figure 17-1. A Minimum Equipment List (MEL) includes items of equipment related to the aircraft's airworthiness. It does not contain items such as wings, flaps, and rudders, which are obviously required. MELs also list equipment that may be deferred with flight limitations.

EXTERIOR INSPECTION

Note

Visually check aircraft for general condition during walk-around inspection. In cold weather, remove even small accumulations of frost, ice or snow from wing, tail and control surfaces. Also, make sure that control surfaces contain no internal accumulations of ice or debris. If night flight is planned, check operation of all lights, and make sure a flashlight is available.

① a. Remove control wheel lock.
  b. Check ignition switch off.
  c. Turn on master switch and check fuel quantity indicators; then turn off master switch.
  d. Check that fuel selector valve handle is on fuller tank.

② a. Remove rudder gust lock, if installed.
  b. Disconnect tail tie-down.
  c. Check control surfaces for freedom of movement and security.
  d. Check cargo doors securely latched and locked (right side only). If cargo load will not permit access to the front cargo door inside handle, lock the door from the outside by means of the T-handle stored in the map compartment.

IMPORTANT

The cargo doors must be fully closed and latched before operating the electric wing flaps. A switch in the upper door sill of the front cargo door interrupts the wing flap electrical circuit when the front door is opened or removed, thus preventing the flaps being lowered with possible damage to the cargo door or wing flaps when the cargo door is open. If operating with the cargo doors removed and the optional spoiler kit installed, check that the wing flap interrupt switch cover plate is installed so that the wing flaps can be lowered in flight.

③ a. Check aileron for freedom of movement and security.

④ a. Disconnect wing tie-down.
  b. Check fuel tank vent opening for stoppage.
  c. Check main wheel tire for proper inflation.
  d. Visually check fuel quantity; then check fuel filler cap secure and vent unobstructed.

⑤ a. Inspect flight instrument static source opening on side of fuselage for stoppage (both sides).
  b. Check propeller and spinner for nicks and security, and propeller for oil leaks.
  c. Check nose wheel strut and tire for proper inflation.
  d. Disconnect nose tie-down.
  e. Check oil level. Do not operate with less than nine quarts. Fill to twelve quarts for extended flight.
  f. Before first flight of day and after each refueling, pull out strainer drain knob for about four seconds, to clear fuel strainer of possible water and sediment. Check strainer drain closed. If water is observed, there is a possibility that the wing tank sumps contain water. Thus, the wing tank sump drain plugs and fuel reservoir drain plugs should be removed to check for the presence of water.

⑥ a. Check main wheel tire for proper inflation.
  b. Visually check fuel quantity; then check fuel filler cap secure and vent unobstructed.

⑦ a. Remove pitot tube cover, if installed, and check pitot tube opening for stoppage.
  b. Disconnect wing tie-down.
  c. Check fuel tank vent opening for stoppage.

⑧ a. Check aileron for freedom of movement and security.

Figure 1-1.

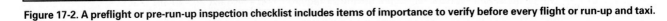

Figure 17-2. A preflight or pre-run-up inspection checklist includes items of importance to verify before every flight or run-up and taxi.

tion. Within every 12 calendar months, the aircraft must have a complete inspection performed to determine if the aircraft meets all the requirements for its certification. A calendar month is one that ends at midnight of the last day of the month. For example, if the inspection was completed on January 14, it will remain valid until midnight January 31, the following year. An aircraft may not be over flown beyond the annual due date unless a special flight permit is obtained authorizing the aircraft to be flown to an inspection facility.

The FAA specifies the details of both an annual and a 100-hour inspection in Appendix D of 14 CFR Part 43. Appendix D includes a list of items entitled, "Scope and detail of items (as applicable to the particular aircraft) to be included in annual and 100-hour Inspections." This list is not all-inclusive to each aircraft manufactured, but typical of the scope of inspection the FAA requires. The manufacturer of the aircraft provides a detailed inspection checklist, which meets the minimum requirements of Appendix D, in the service manual for each aircraft it produces.

Figure 17-3 represents a portion of a typical manufacturer's inspection checklist. The checklist shows the recommended time intervals of items inspected under a progressive inspection program, a complete inspection, or annual, including all 50, 100 and 200-hour items in addition to any special inspection items.

Annual inspections must be performed by an A&P technician holding an Inspection Authorization (IA) or an inspector authorized by a certified repair station with an airframe rating. If the aircraft passes the inspection, the inspectors must write up the inspection results in the maintenance records, and approve the aircraft for return to service. If for any reason the aircraft does not meet all of the airworthiness requirements, the inspector must provide a list of discrepancies and unairworthy items to the aircraft owner. The inspector may not delegate any inspection responsibility to another A&P or repairman, nor may the inspector merely supervise the inspection.

However, as long as the discrepancy found does not require a major repair, any certified A&P technician may correct each discrepancy the inspector listed, and then approve the aircraft for return to service. The due date of the next annual inspection is then based on the date of the original inspection and not on the date the discrepancies were corrected. For example, if an aircraft's annual was completed on March 20, but a discrepancy repair was not completed until April 15, the next annual is still due March 30 the following year.

If the aircraft does not pass the annual inspection, it may not be flown until the unairworthy condition is corrected. However, if the owner wants to fly the aircraft to a different repair location, a special flight permit may be obtained to ferry the aircraft to that alternate repair location.

100-HOUR INSPECTION

If the aircraft is operated for compensation or hire, it must be given a complete inspection of the same scope and detail as the annual inspection every 100 hours of operation unless it is maintained under an FAA-approved, alternative inspection program such as a progressive inspection program. In the case of a 100-hour inspection, the time limitation may be exceeded by no more than 10 hours of flight operation while enroute to an inspection facility. However, the excess time used to reach the inspection location must be included in computing the next 100 hours of time in service. For example, if a 100-hour inspection was due at 1000 hours and the pilot over-flew the aircraft to 1008 hours to reach an inspection facility, the next 100-hour inspection is still due at 1100 hours of operation.

The difference between a 100-hour and an annual inspection is that a certified A&P technician may conduct the 100-hour inspection and approve the aircraft for return to service. The A&P technician who inspected the aircraft must make the proper entries in the aircraft's maintenance records and approve the aircraft for return to service before the 100-hour inspection is considered complete.

Like the inspector performing an annual, the A&P inspecting the aircraft may not merely supervise the inspection process. The maintenance technician performing the 100-hour inspection is responsible for approving the aircraft for return to service. In other words, the A&P signing off the aircraft must be the one who actually performed the inspection. However, the inspector may utilize other A&Ps or repairmen in the preparation for the inspection such as removing inspection panels, cowlings, and fairings. In addition, any certified A&P technician may repair and sign off any discrepancies found by the inspector as long as they are not major repairs or major alterations.

100-hour inspections may be signed off as annual inspections if an A&P mechanic holding an inspection authorization (IA) completed the inspections. In a sense, the aircraft could have several annuals performed in one calendar year at each 100 hours of operation. However, a 100-hour inspection may not take the place of an annual inspection. If an aircraft is operated under the requirements of an annual

	EACH 50 HOURS	EACH 100 HOURS	EACH 200 HOURS	SPECIAL INSPECTION ITEM
AIRFRAME				
1. Aircraft Exterior	•		•	
2. Aircraft Structure			•	
3. Windows, windshield, doors and seals	•		•	
4. Seat stops, seat rails, upholstery, structure and mounting			•	
5. Seat belts and shoulder harnesses	•		•	
6. Control column bearings, sprockets, pulleys, cables, chains and turnbuckles			•	
7. Control lock, control wheel and control column mechanism			•	
8. Instruments and markings	•		•	
9. Gyros central air filter			•	13
10. Magnetic compass compensation				5
11. Instrument wiring and plumbing			•	
12. Instrument panel, shock mounts, ground straps, decals and labeling			•	
13. Defrosting, heating and ventilating systems and controls	•		•	
14. Cabin upholstery, trim sun visors and ash trays			•	
15. Area beneath floor, lines, hose, wires and control cables			•	
16. Lights, switches, circuit breakers, fuses and spare fuses	•		•	
17. Exterior lights	•			
18. Pitot and static systems			•	
19. Stall warning unit and pitot heater			•	
20. Radios, radio controls, avionics and flight instruments	•		•	
21. Antennas and cables			•	
22. Battery, battery box and battery cables	•			
23. Battery electrolyte				14
24. Emergency locator transmitter		•		15
25. Oxygen system			•	
26. Oxygen supply, masks and hose	•			16
27. Deice system plumbing			•	
28. Deice system components			•	
29. Deice system boots			•	
CONTROL SYSTEMS				
In addition to the items listed below, always check for correct direction of movement, correct travel and correct cable tension.				
1. Cables, terminals, pulleys, pulley brackets, cable guards, turnbuckles and fairleads			•	
2. Chains, terminals, sprockets and chain guards			•	
3. Trim control wheels, indicators, actuator and bungee	•			
4. Travel stops			•	
5. Decals and labeling			•	
6. Flap control switch, flap rollers and flap position indicator	•			
7. Flap motor, transmission, limit switches, structure, linkage, belt cranks, etc.			•	
8. Flap actuator jackscrew threads				17
9. Elevators, trim tab, hinges and push-pull tab	•			
10. Elevator trim tab actuator lubrication and tab free-play inspection				18
11. Rudder pedal assemblies and linkage			•	
12. External skins of control surfaces and tabs	•			
13. Ailerons, hinges, and control rods	•			
14. Internal structure of control surfaces			•	
15. Balance weight adjustment			•	

Figure 17-3. (1 Of 2)

SPECIAL INSPECTION ITEMS

1. First 25 hours, refill with straight mineral oil (MIL-L-6082) and use until a total of 50 hours have accumulated or oil consumption has stabilized; then change to ashless dispersant oil. Change filter element each 50 hours, or every six months.
2. Clean filter, replace as required.
3. Replace hoses at engine overhaul or after 5 years, whichever comes first.
4. General inspection every 50 hours.
5. Each 1000 hours, or to coincide with engine overhaul.
6. Each 100 hours for general condition, lubrication and freedom of movement. These controls are not repairable. Replace every 1500 hours or sooner if required.
7. Each 500 hours.
8. Internal timing and magneto-to-engine timing limits are described in the engine service manual.
9. Remove insulation blanket or heat shields and inspect for burned area, bulges or cracks. Remove tailpipe and ducting; inspect turbine for coking, carbonization, oil deposits and impeller for damage.
10. First 100 hours and each 500 hours thereafter. More often if operated under prevailing wet or dusty conditions.
11. If leakage is evident, refer to Governor Service Manual.
12. At first 50 hours, first 100 hours, and thereafter each 500 hours or one year, whichever comes first
13. Replace each 500 hours.
14. Check electrolyte level and clean battery compartment each 50 hours or each 30 days.
15. Refer to manufacturer's manual.
16. Inspect masks, hose and fittings for condition, routing and support.
17. Refer to maintenance manual.
18. Lubrication of the actuator is required each 1000 hours or three years.
19. Each five years replace all rubber packings, back-ups and hydraulic hoses in both the retraction and brake systems. Overhaul all retraction and brake system components.
20. Replace check valves in turbocharger oil lines each 1000 hours.
21. Check alternator belt tension.

Figure 17-3. (2 Of 2) An excerpt of a typical manufacturer's inspection checklist utilized during annual inspections that outlines the required inspection items. This inspection checklist is multi-functional. It outlines 50-hour, 100-hour, 200-hour, and annual inspection intervals.

inspection, it must be inspected by an A&P who holds an IA rating, or certified repair station inspector and be signed off as an annual inspection only.

PROGRESSIVE INSPECTION

At times, aircraft operators may feel that it is not economical to keep the airplane out of commission long enough to perform a complete annual inspection at one time. In which case, the owner may elect to use a progressive inspection schedule. A progressive inspection is exactly the same in scope and detail as the annual inspection but allows the workload to be divided into smaller portions and performed in shorter time periods. For example, the engine may be inspected at one time, the airframe inspection may be conducted at another time, and components such as the landing gear at another. Progressive inspection schedules must ensure that the aircraft will be airworthy at all times and conform to all applicable FAA aircraft specifications, type certificate data sheets, airworthiness directives, and other data such as the manufacturer's service bulletins and service letters.

The manufacturer provides guidelines to help an operator select an appropriate inspection program for their specific operation. For example, if an aircraft is flown more than 200 hours per calendar year, a progressive inspection program is most likely recommended to reduce aircraft downtime and overall maintenance costs.

Again referring to Figure 17-3, this aircraft inspection chart outlines a typical schedule used in a progressive inspection program. As shown in the chart, there are items inspected at 50, 100, and 200 hours, in addition to special inspection items that require servicing or inspection at intervals other than 50, 100 or 200 hours. The inspection intervals are separated in such a way to result in a complete aircraft inspection every 200 flight hours. This particular inspection program would not be recommended or practical unless the aircraft is flown more than 200 hours per year.

Before a progressive inspection schedule may be implemented, the FAA must approve the inspection program. The owner must submit a written request outlining their intended progressive inspection guidelines to the local FAA Flight Standards District Office (FSDO) for approval. After approval, and before the progressive inspection program may begin, the aircraft must undergo a complete annual inspection. After the initial complete inspection, routine and detailed inspections must be conducted as prescribed in the progressive inspection schedule. Routine inspections consist of visual and operational checks of the aircraft, engines, appliances, components and systems normally without disassembly. Detailed inspections consist of thorough checks of the aircraft, engines, appliances, components and systems including necessary disassembly. The overhaul of a component, engine, or system is considered a detailed inspection.

A progressive inspection program requires that a current and FAA-approved inspection procedure manual for the particular airplane be available to the pilot and maintenance technician. The manual explains the progressive inspection and outlines the required inspection intervals. All items in the inspection schedule must be completed within the 12 calendar months that are allowed for an annual inspection. The progressive inspection differs from the annual or 100-hour inspection in that a certified mechanic holding an inspection authorization, a certified repair station, or the aircraft manufacturer may supervise or conduct the inspection.

If the progressive inspection is discontinued, the owner or operator must immediately notify, in writing, the local FAA Flight Standards District Office (FSDO) of the discontinuance. In addition, the first complete inspection is due within 12 calendar months or, in the case of commercial operations, 100 hours of operation from the last complete inspection that was performed under the progressive inspection schedule.

LARGE and TURBINE POWERED MULTI-ENGINE AIRCRAFT

Large (over 12,500 lbs. gross takeoff weight) and multi-engine turbine aircraft operating under FAR Part 91, require inspection programs tailored to the specific aircraft and its unique operating conditions. These unique conditions would include scenarios such as high flying times, aircraft operated in extremely humid environments, or in extremely cold or wet climates. Because of the size and complexity of most turbine-powered aircraft, the FAA requires a more detailed and encompassing inspection program to meet the needs of these aircraft and flying conditions. Although they may be operated under Part 91, large and turbine-powered aircraft are often inspected under programs normally utilized by air carrier or air taxi operations.

The registered owner or operator of a large or turbine-powered aircraft operating under Part 91 must select, identify in the aircraft maintenance records, and use one of the following inspection programs: a continuous airworthiness inspection program, an approved aircraft inspection program (AAIP), the manufacturer's current recommended inspection program, or any other inspection program developed by the owner/operator and approved by the FAA. The exception is in the case of turbine-powered rotorcraft operations, in which case, the owner/operator may choose to use the inspection provisions set out for small aircraft: annual, 100-hour, or progressive inspection programs. After selection, the operator must submit an inspection schedule, along with instructions and procedures regarding the performance of the inspections, including all tests and checks, to the local FAA FSDO for approval.

A **continuous airworthiness inspection program** is designed for commercial operators of large aircraft operating under FAR Part 121, 127, or 135. It is one element of an overall continuous airworthiness maintenance program (CAMP) currently utilized by an air carrier that is operating that particular make and model aircraft. [Figure 17-4]

Figure 17-4. Large turbine powered corporate jet owners may elect to use a continuous airworthiness inspection program because of the complexity of the aircraft and its systems.

A continuous airworthiness inspection program might be chosen under Part 91 operations when an air carrier purchases or leases an aircraft operating under another air carrier's 121 certificate. For example, Airline B purchases an aircraft from Airline A. The aircraft must be operated under an inspection program during the transition from Airline A to Airline B. Instead of creating an entirely new inspection program tailored to the specific aircraft during this transition period, Airline B may choose to keep the aircraft on its current continuous airworthiness inspection program until it is placed on the new owner's Part 121 operating certificate.

An **approved aircraft inspection program (AAIP)** may be chosen by on-demand operators who operate under Part 135. If the FAA determines that annual, 100 hour, or progressive inspections are not adequate to meet Part 135 operations, they may require or allow the implementation of an AAIP for any make and model aircraft the operator exclusively uses. The AAIP is similar to the CAMP utilized by most Part 121 air carriers. This program encompasses maintenance and inspection into an overall continuous maintenance program. [Figure 17-5]

Figure 17-5. Turbo-prop aircraft typical of the type operated by air-taxi operators. Each aircraft operated by air-taxi operators may be maintained under an AAIP designed specifically to that particular aircraft by registration number.

A **complete manufacturer's recommended inspection program** consists of the inspection program supplied by the airframe manufacturer and supplemented by the inspection programs provided by the manufacturers of the engines, propellers, appliances, survival equipment, and emergency equipment installed on the aircraft. A manufacturer's inspection program is used more frequently when an aircraft is factory new. If an aircraft has several modifications, updated systems, or custom avionics not installed at the factory, the manufacturer's inspection program alone may not be adequate in the overall inspection of the aircraft and all of its installed equipment and components. In this case, another method of inspection must be chosen.

The owner of an aircraft may choose to develop their own inspection program. The recommended manufacturer's inspection program is generally used as the basis of an **owner developed inspection plan**. However, deviation from the manufacturer's inspection program must be supported and approved by the FAA. The customized plan must include the inspection methods, techniques, practices, and standards necessary for the proper completion of the program. Most owner developed inspection programs include inspection and repair requirements only, and do not require continual maintenance performed to their aircraft.

CONFORMITY INSPECTIONS

Aircraft are manufactured to FAA approved specifications. Alterations made to the original design specifications of the aircraft require approval in the form of a sign-off from a certificated maintenance technician or, in the case of a major repair or alteration, approval from the FAA on form 337. The absence of approval for any alteration renders the aircraft unairworthy. A **conformity inspection** is an essential element of all aircraft inspection programs and performed to determine whether the aircraft conforms to or matches its approved specifications.

A conformity inspection is essentially a visual inspection that compares the approved aircraft specifications with the actual aircraft and associated engine and components. A list is compiled outlining the information gathered from the type certificate data sheets (TCD), applicable supplemental type certificate data sheets (STC), major repair & alteration information (FAA Form 337), aircraft equipment list, airworthiness directive compliance record, etc. The list includes model numbers, part numbers, serial numbers, installation dates, overhaul times, and any other pertinent information obtained in the above reference documents. The mechanic performs a visual inspection and compares the aircraft with the compiled list of information making note of any deviation from the aircraft specifications. [Figure 17-6]

A conformity inspection is not specifically required by name, but it is inherently required at every inspection interval due to the nature of the inspection; to determine whether the aircraft conforms to its certification specifications. However, a conformity inspection is specifically required when an aircraft is exported to or imported from another country with the intention of becoming registered in that respective country. Further, a conformity inspection is highly recommended when performing a pre-purchase inspection for a prospective aircraft buyer.

CONFORMITY INSPECTION CHECKLIST

NAME AND ADDRESS OF OPERATOR: CERTIFICATE NO.

_____   _____
_____
_____

AIRCRAFT TYPE: _____
REGISTRATION NO. _____ SERIAL NO. _____
AIRCRAFT TOTAL TIME: _____ TACH: _____ HOBBS: _____

GENERAL INSPECTION INFORMATION:

Last Annual Inspection:
Date: _____ A/C TT: _____ Mech/Fac _____

Last 100 Hour Inspection:
Date: _____ A/C TT: _____ Mech/Fac _____

Transponder/Encoder Last Test Date _____ Altimeter/Static Last Test Date _____

ELT Battery Due Date _____ Life Jackets Last Inspected _____

Fire Extinguisher(s) Last Inspection _____

ITEMS REQUIRED IN/ON THE AIRCRAFT AT ALL TIMES DURING OPERATION:

_____ Airworthiness certificate   _____ Current aircraft registration
_____ Aircraft flight manual   _____ Flight-deck checklist
_____ Compass correction card   _____ Passenger briefing cards
_____ Discrepancy log   _____ External Data Plate
_____ MEL (if applicable)   _____ Cargo & baggage restraints
_____ Seat Belts (Metal/Metal)(TSO)   _____ Crew shoulder harness (TSO)
_____ Current equipment list   _____ Current weight & balance & equipment
_____ Placards and markings required by the TCD, STC, flight manual, and airworthiness directives.

REVIEW THE AIRCRAFT RECORDS FOR THE FOLLOWING INFORMATION:

AIRFRAME:

MAKE: _____ MODEL: _____ S/N: _____
Type Certificate No. _____ Maintenance Doc & Rev. no. _____
STC's installed: _____
_____
_____

Applicable airworthiness directives _____
_____
_____

Page 1 of 2

Figure 17-6. (1 of 2) Typical conformity inspection checklist. A mechanic visually inspects the aircraft then documents the actual aircraft and equipment information on a conformity checklist. The checklist is then compared to the aircraft's specifications to determine airworthiness compliance.

ENGINE:
MAKE: _____ MODEL: _____ S/N: _____
TBO hours/years: _____ Maintenance Doc & Rev. no. _____
Engine TT: _____ TSO _____ Date of last Overhaul: _____

STC's installed: _____
_____
_____

Applicable airworthiness directives _____
_____
_____

PROPELLER:
MAKE: _____ MODEL: _____ S/N: _____
TBO hours/years: _____ Maintenance Doc & Rev. no. _____
Prop TT: _____ TSO _____ Date of last Overhaul: _____

STC's installed: _____
_____
_____

Applicable airworthiness directives _____
_____
_____

GOVERNOR:
MAKE: _____ MODEL: _____ S/N: _____
TBO hours/years: _____ Maintenance Doc & Rev. no. _____
Gov. TT: _____ TSO _____ Date of last Overhaul: _____

STC's installed: _____
_____
_____

Applicable airworthiness directives _____
_____
_____

MAGNETOS:
LH MAKE: _____ MODEL: _____ S/N: _____
RH MAKE: _____ MODEL: _____ S/N: _____
TBO hours/years: left _____ right _____ Maintenance Doc & Rev. no. _____
Mageto TSO left _____ right _____ Date of last Overhaul: left _____ right _____

STC's installed: _____
_____
_____

Applicable airworthiness directives _____
_____
_____

Page 2 of 2

Figure 17-6. (2 0f 2)

Although the conformity inspection is an important part of the overall inspection process, it is one of the most common inspections overlooked or not entirely carried out. For example, an IA performing an annual inspection is responsible for determinig the airworthiness of the aircraft. Many times, inspectors fail to visually verify the equipment installed on the aircraft with the equipment list. In doing so, the IA may overlook a piece of equipment installed on the aircraft but not documented in the maintenance records, which could render the aircraft unairworthy. The verification of the presence of equipment installed in the aircraft, but not verifying that the installation was properly performed may also render the aircraft technically unairworthy. The inspector must not only verify the physical presence of items but also confirm whether the installation of the equipment was properly performed, especially if the installation was done without proper documentation.

A skilled and effective inspector meticulously verifies the installation of equipment list items. Not only verifying that they are physically in the aircraft, but also that they were properly installed and, in the case of a major repair or alteration, that a form 337 was created and approved by the FAA.

AIR CARRIER & AIR CHARTER OPERATIONS

Aircraft operators regulated under FAR Part 121 or 135 must maintain their aircraft under comprehensive maintenance and inspection programs. One of the differences between Part 91 operations and Air Carrier operations is that Part 121 operators must continually maintain and inspect their aircraft.

Ongoing maintenance is not required on aircraft operated under Part 91. The operating rules of Part 91 only require an owner to correct discrepancies found during inspection intervals. Air carriers, on the other hand, must perform aircraft maintenance and inspection on a continual basis.

Air charter operations regulated under Part 135 offer another unique operating environment. Depending on the type of operation, and the size and complexity of aircraft operated, a range of inspection rules apply. Part 135 operators may choose from several different inspection programs depending on the number of seats and complexity of the aircraft.

PART 121 AIR CARRIER INSPECTIONS

Air carriers operating under Part 121 must maintain their aircraft under a Continuous Airworthiness Maintenance Program (CAMP). A **continuous airworthiness inspection program** is one element of an overall CAMP. The basic requirements of a CAMP include inspection, scheduled and unscheduled maintenance, overhaul and repair, structural inspection, required inspection items (RII), and a reliability program. Specific instructions, standards, and operations specifications for each element of the continuous airworthiness maintenance program must be included in the air carrier's maintenance manual for the specific aircraft for which it is applicable. A CAMP is a fleet program and encom-

MAINTENANCE CHECK SCHEDULE		
CHECK	SCOPE	INTERVAL
Service check	Log book and maintenance forms review (for example: time control items). Exterior visual checks and routine aircraft servicing such as hydraulic fluids, engine oil, & general lubrication. Operational checks.	Required no more than 48 elapsed calendar hours from the last Service Check, A-1, A-2, A-3, A-4, or C check.
A Check: A-1 check A-2 check A-3 check A-4 check	Log book and maintenance forms review. Exterior visual check, routine and specific inspections, and routine aircraft servicing. Replacing time-limited items. Operational checks.	Required no more than 125 flight hours from the last equalized A and/or C check.
C Check	Includes "A: check items in addition to detailed inspections of aircraft, engines, components, and appliances.	Required no more than 3600 flight hours from the last C check.
D Check: D-1 check D-2 check D-3 check D-4 check	Includes "C" check items in addition to extensive dissassemby and opening up of the aircraft, and weight & balance. Flight test after operational checks.	Required to be performed at no more than 9000 flight hours or 3 calendar years, whichever occurs first from the last phase D check.

Figure 17-7. Typical air carrier maintenance "letter check" schedule outlining the scope and time intervals of required inspections for a specific type of aircraft. The maintenance schedule outline is used in conjunction with the specific work cards to maintain the airworthiness of the aircraft and all installed equipment.

passes the entire group of aircraft versus inspection programs regarding individual aircraft such as an AAIP, which is utilized under Part 135 air charter operations.

Like a progressive inspection program, the FAA must approve a continuous inspection program. This inspection program is extremely comprehensive, specific to the operator's aircraft, and requires complex maintenance facilities and large numbers of technical personnel. A continuous airworthiness inspection program is a program of FAA-approved inspection schedules which allow aircraft to be continually maintained in a condition of airworthiness without being taken out of service for long periods of time. This program keeps aircraft downtime to a minimum due to segmented maintenance or inspections intervals, thereby keeping the aircraft in service in a more efficient and convenient manner.

The continuous inspection program for a large air carrier may, as an example, consist of "letter check" inspection schedules. An example of a typical letter check inspection schedule is outlined in Figure 17-7. Letter checks are normally scheduled prior to due times or cycles. Over-flying due times or cycles of any required inspection is a direct violation of FAA regulations and may include large monetary fines. [Figure 17-7].

It is difficult to provide an overall description of a general air carrier inspection program because each air carrier's CAMP is designed specifically to its aircraft and type of operating conditions. Hence, every air carrier operating in the U.S. utilizes a different CAMP designed specifically for its individual needs and specific flight operations.

There are many different methods of inspection scheduling, inspection frequency, and terminology used throughout the airline industry. For example, one airline may refer to cursory line maintenance as a "daily" check, while another may refer to the same type of line check as a "service" check. The scope of these types of inspections is also designed explicitly for the particular aircraft. What is included in a daily check for one specific type of aircraft may not be comprehensive enough for another. Again, figure 17-7 illustrates a letter check schedule including phase inspections within the "A" and "D" checks regarding a specific type of aircraft.

In this schedule arrangement, service checks are based on calendar hours while all other letter checks are based on flight hours. The completion of an "A" check eliminates the need for a service check due at the identical time interval. In other words, if a more detailed inspection is performed, it may zero out the less-encompassing inspection due time. A service check is due 48 calendar hours from the completion period of a "service", "A", "C", or "D" check. The next "A" check phase is due 125 flight hours from a completed "A", "C", or "D" check. This inspection schedule shows a series of "A" checks between each "C" check. There are twelve sets of "A" checks (A-1, A-2, A-3, A-4) between each complete heavy "C" check. "C" checks are due every 3600 flight hours and two comprehensive "C" checks are due between every heavy "D" check.

Each level of inspection must be clearly defined in the operator's continuous airworthiness inspection program. For example, a specific area of the aircraft may require only a visual inspection during pre-flight, "service checks", and "A" checks but may require a detailed inspection in the same area for a heavy "C" or "D" check. In most letter check maintenance schedules, the inspection and maintenance become more detailed and build upon the prior letter check performed.

Work cards act as control documents in the continuous inspection process. Job cards are issued for all aspects of CAMP inspections and are used to organize inspection instructions and account for the specific steps involved. Depending on the scope of inspection, several work or job cards are referenced. Each work card outlines one specific area of the inspection. Figure 17-8 is an example of a work card used during a heavy "C" check regarding an air carrier aircraft. The work card provides an outline of a specific area of the aircraft inspection. Recurring airworthiness directives and manufacturer's service bulletins are usually incorporated on work cards also. The work card provides accountability columns where the inspector or maintenance technician signs off each step as it is inspected or serviced. In addition, specific instructions, including reference figures, may be included with each work card. The completed work card becomes part of the aircraft's maintenance record. [Figure 17-8]

FAR Part 121 outlines the specific approval for return-to-service requirements for air carrier operations.

FLY HIGH AIRLINES			
B737-200 C - Check			
INSPECT LEFT ELEVATOR/TAB STRUCTURE AND HINGE FITTINGS		CARD NUMBER	6-4008
A/C NUMBER	STATION	DATE	

		INSPECT LEFT ELEVATOR/TAB STRUCTURE AND HINGE FITTINGS	
	M	1.	INSPECT LEFT ELEVATOR UPPER AND LOWER SKIN (service bulletin AOT-53-02)
	M	2.	INSPECT LEFT ELEVATOR INTERNAL STRUCTURE
		3.	INSPECT THE FOLLOWING L/H ELEVATOR TAB HINGES
I			a. Inspect L/H elevator tab hinge no. 1
I			b. Inspect L/H elevator tab hinge no. 2
		4.	INSPECT THE FOLLOWING L/H ELEVATOR AND HORIZONTAL STABILZER HINGE FITTINGS AND BEARINGS
I			a. Inspect L/H elevator No. 1 hinge bearing and bolt (AD 97-08-22)
I			b. Inspect L/H elevator No. 2 hinge bearing and bolt (AD 97-08-22)
		5.	INSPECT THE FOLLOWING L/H ELEVATOR AND HORIZONTAL STABILZER HINGE FITTINS, BEARINGS AND PLATE ASSEMBLIES
I			a. Inspect No. 3 hinge bearing plate assembly
I			b. Inspect No. 4 hinge bearing plate assembly

INSPECT LEFT ELEVATOR/TAB STRUCTURE AND HINGE FITTINGS

APPLICABLE FIGURES: FIG.1

1. **INSPECT LEFT ELEVATOR UPPER AND LOWER SKIN FOR DELAMINATION, CRACKS AND SIGNS OF BONDED SKIN SEPARATION.**

2. **INSPECT LEFT ELEVATOR INTERNAL STRUCTURE FOR CONDITION INCLUDING:**
   a. Check internal spars, webs, ribs and stiffeners.
   b. Check condition of structure at front spar hinge attachment to elevator.
   c. Check tab lock mechanism for condition.

3. **INSPECT THE FOLLOWING L/H ELEVATOR TAB HINGES FOR GENERAL CONDITION AND OBVIOUS DAMAGE.**
   a. Inspect L/H elevator tab hinge No. 1.
   b. Inspect L/H elevator tab hinge No. 2.

4. **INSPECT THE FOLLOWING L/H ELEVATOR AND HORIZONTAL STABILZER HINGE FITTINGS AND BEARINGS FOR GENERAL CONDITION AND OBVIOUS DAMAGE. (Refer to figure 1)**
   a. Inspect L/H elevator No. 1 hinge bearing and bolt.
   b. Inspect L/H elevator No. 2 hinge bearing and bolt.

5. **INSPECT THE FOLLOWING L/H ELEVATOR AND HORIZONTAL STABILER HINGE FITTINGS, BEARINGS AND PLATE ASSEMBLIES FOR GENERAL CONDITION AND OBVIOUS DAMAGE.**
   a. Inspect No. 3 hinge bearing plate assembly.
   b. Inspect No. 4 hinge bearing plate assembly.

Figure 17-8. (1 of 2) Work/job card which references the "Left elevator/tab structure and hinge fitting" inspection required at a heavy "C" check. The work card includes the specific inspection steps along with supporting documentation helpful in the completion of the inspection.

Aircraft Airworthiness Inspection 17-15

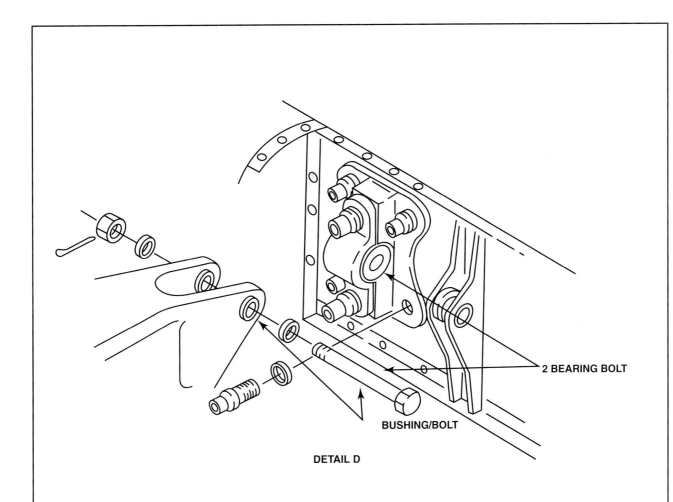

DETAIL D

INDEX NO.	PART NAME	DIM.	DESIGN LIMITS DIAMETER		WEAR LIMITS		REPLACE WORN PART	REPAIR WORN PART	REPAIR INSTR.
			MIN	MAX	MAX WEAR CM.	MAX DIAM CLEAR-ANCE			
1	BEARING	ID	0.3120	0.3125	0.3170	0.0050	X		
	BOLT	OD	0.3110	0.3120	0.3070		X		
2	BEARING	ID	0.3745	0.3750	0.3825	0.0050	X		
	BOLT	OD	0.3735	0.3745	0.3665		X		
3	BUSHING	ID	0.3120	0.3130	0.3170	0.0050	X		
	BOLT	OD	0.3110	0.3120	0.3070		X		
4	BUSHING	ID	0.3745	0.3735	0.3825	0.0050	X		
	BOLT	OD	0.3735	0.3745	0.3675		X		

FIG 1

Figure 17-8. (2 of 2)

PART 135 AIR CHARTER INSPECTIONS

Part 135 on-demand air charter operators have several different options regarding the type of inspection programs with which they must comply. Air charter companies that operate aircraft with less than 9 seats may choose to inspect these aircraft under FAR Part 91 and Part 43 rules, 100-hour or progressive inspection programs. In other words, they are not required to perform continual maintenance on their aircraft, only inspection and discrepancy repair. Air charter operators that operate aircraft with 10 or more seats are required to implement a more-encompassing continual maintenance and inspection program. They may choose to implement a Continuous Airworthiness Maintenance Program (CAMP), an Approved Aircraft Inspection Program (AAIP), a current manufacturer's inspection program, or an operator developed inspection and maintenance program approved by the FAA.

An approved aircraft inspection program (AAIP) is the inspection program most often implemented by FAR Part 135 operators. It is similar to a continuous airworthiness maintenance program used by Part 121 air carriers. However, AAIPs are not fleet inspection programs and do not require continual maintenance. They require continual inspection and are set up for the individual aircraft by registration number and serial number. Air charter operations may have several different AAIPs for different aircraft operated.

For example, an air charter operation that operates an aircraft with 9 or fewer seats may inspect that particular aircraft under 100-hour or progressive inspection intervals. The same operation may also operate several larger, complex aircraft and inspect them under separate AAIPs. It is possible for an air charter operator to use a different inspection program for each of its aircraft, progressive for one, AAIP for another, etc. [Figure 17-9]

Manufacturers' inspection programs are more specific than the 100-hour or annual inspections but lack the ease and control provided by the approved aircraft inspection program. An AAIP allows the operator to choose their own maintenance and inspection schedules. An AAIP is not considered better than a manufacturer's program, however, an AAIP provides the FAA inspector with more control of the program's content. It requires the operator to validate its programs and revisions to the inspector which manufacturer's programs do not require. This

Figure 17-9. Air medical operators may operate several different types of airplanes and helicopters and inspect each under separate inspection programs. AAIPs are not fleet programs; they are inspection programs designed for individual aircraft. A charter company that owns and operates five different aircraft could conceivably operate them under five different AAIPs; each specific to an individual aircraft.

is not to say that a manufacturer's program cannot be used, but it must be identified as an AAIP and approved for a particular operator as that operator's program, not the manufacturer's.

When establishing an approved aircraft inspection program (AAIP), it should include avionics, instrument systems, and appliances. These types of systems are not always installed by the aircraft manufacturer and may not be included in their recommended inspection program. The AAIP must include instructions and procedures for all installed systems.

Approved aircraft inspection programs are similar to continuous airworthiness inspection programs in that they both differ tremendously from operator to operator and aircraft to aircraft. An example of an AAIP might contain a daily service check, a 50-hour Preventative Maintenance Inspection (PMI), a series of 5 separate phase inspections conducted 150 hours apart, a 2500-hour major airframe inspection, and additional maintenance items that include stand-alone inspections. [Figure 17-10] [Figure 17-11]

SPECIAL INSPECTIONS

Special inspections are scheduled inspections with prescribed intervals other than the normally established inspection intervals set out by the manufacturer. Special inspections may be scheduled by flight hours, calendar time, or aircraft cycles. For instance, in the case of a progressive inspection schedule for a small Cessna, special inspections occur at intervals other than 50, 100, or 200 hours.

INSPECTION SCHEDULE OUTLINE

A/C time flight hrs	PHASE 1	PHASE 2	PHASE 3	PHASE 4	Type of Inspection
200	X				Nose landing gear area, nose gear, pilot's compartment, cabin section, rear fuselage & empennage, wings, main gear area, engines, landing gear retraction, operational inspection, post inspection.
400		X			Nose section, nose avionics compartment, nose landing gear area, nose gear, pilot's compartment, cabin section, rear fuselage & empennage, wings, main landing gear area, engines, landing gear retraction, operational inspection, post inspection.
600			X		Nose landing gear area, nose gear, pilot's compartment, cabin section, rear fuselage & empennage, wings, main gear area, engines, landing gear retraction, operational inspection, post inspection
800				X	Nose section, nose avionics compartment, nose landing gear area, nose gear, pilot's compartment, cabin section, rear fuselage & empennage, wings, main landing gear area, engines, landing gear retraction, operational inspection, post inspection.

After "phase 4" inspection is completed, repeat inspection sequence. The complete program must be accomplished at least one time every 24 calender months. Any part of the inspection not completed is due immediately. Completion of phases 1-4 is considered a "complete inspection."

Figure 17-10. An example of a typical AAIP phase inspection schedule outline.

Special inspection items are usually explained in the notes section of the service manual inspection chapter.

Examples of special inspection items may include oil change information after an engine overhaul, the inspection and replacement of hoses at engine overhaul, and magnetic compass compensation every 1000 hours. Additionally, inspection and replacement of the rubber packings on each brake at 5-year intervals, and inspection and lubrication of the elevator trim tab actuator at 500-hour intervals may also constitute special inspection items. Each manufacturer outlines special inspection items specific to each model of aircraft.

Altimeter and static system inspections and certifications are considered special inspections. Every aircraft operated under Instrument Flight Rules must have its altimeters and static systems inspected and certified for integrity and accuracy every 24 calendar months as required by FAR Part 91.411. The scope of the altimeter and static system certification is outlined in FAR Part 43, Appendix E. The altimeter is checked for operation and accuracy up to the highest altitude it is used, usually the aircraft's service ceiling, and a record made of this inspection and certification in the aircraft maintenance records.

The altimeter certification may be conducted by the manufacturer of the aircraft, or by a certificated repair station (CRS) holding an appropriate rating that authorizes this particular inspection. However, a certified airframe technician may perform the static pressure system leakage tests and integrity inspection but cannot perform the certification.

ATC transponder inspections are also considered special inspections. The radar beacon transponder that is required for aircraft operating in most areas of controlled airspace must be inspected each 24 calendar months by any of the following: a certificated repair station approved for this inspection, a holder of a continuous-airworthiness maintenance program, or the manufacturer of the aircraft on which the transponder is installed. This test is required by FAR Part 91.413 and described in FAR Part 43, Appendix F.

The **emergency locator transmitter (ELT) inspection** is also considered a special inspection. FAR Part 91.207 requires the ELT inspection every 12 months. The inspection entails checking for proper

SCHEDULED INSPECTION PROGRAM

Owner _____, LH Eng S/N _____,
W/O number _____,   Total time _____ Total cycles _____,
Date in _____ Date out _____, TSOH _____ Cycles SOH _____,
Serial no. _____ Reg. No. _____, TSHSI _____ Cycles SHSI _____,
Last Inspection _____ phase   RH Eng S/N _____,
Last Inspection date _____,   Total time _____ Total cycles _____,
Last Inspection hours _____,   TSOH _____ Cycles SOH _____,
Hourmeter _____ Total time _____, TSHSI _____ Cycles SHSI _____,
Total cycles _____,   LH Prop S/N _____,
Researched by: _____,   LH Prop total time _____ TSOH _____,
                                         RH Prop S/N _____,
                                         RH Prop total time _____ TSOH _____,
                                         RH Prop total time

PHASE 1 INSPECTION

A. NOSE SECTION	ATA ref.	Mech	Insp
1. Combustion heater			
a. Check the gap and condition of the heater spark plug	21-40-00		
b. Check fuel plumbing, pump and regulator for leakage, damage, and security of attachment	21-40-00		
c. Clean and inspect the system fuel filter at the inlet port of the fuel control valve	21-40-00		
B. NOSE AVIONICS COMPARTMENT			
NOTE: There are no inspections required in this section during this phase			
C. NOSE LANDING GEAR AREA			
1. Electrical wiring and equipment- inspect all exposed wiring & equip for chafing & damage	AC 43.13		
D. NOSE GEAR			
1. Wheel			
a. Inspect wheel for wear, damage, and corrosion	32-40-00		
b. Inspect wheel bearings and races for wear, pitting, cracks, discoloration, rust, or damage	32-40-00		
2. Tire			
a. Inspect for wear and deterioration	12-20-00		
b. Check for correct inflation	12-20-00		
3. Shimmy damper - Inspect for leaks, security, and attachment			
4. Nose gear brace stop lugs - Inspect for cracks, damage or deterioration			
5. Nose gear steering stop - Inspect steering stop for damage or distortion			
6. Landing & taxi lights - Inspect for broken lens or bulbs	33-40-00		
7. Steering linkage- Inspect nose gear steering mechanism & attaching hardware for wear	32-50-00		
8. Nose landing gear strut - Check strut for leakage and correct extension	32-20-00		

Phase 1 Inspection (page 1 of 5)

Figure 17-11. An example of an AAIP phase 1 inspection job card and control document.

installation, battery corrosion, operation of the controls and crash sensor, and the ELT signal. Check the ELT battery's expiration date and record the expiration date for replacing or recharging the battery in the maintenance record. The expiration date must also be legibly marked on the outside of the ELT.

CONDITIONAL INSPECTIONS

A conditional inspection is an unscheduled inspection conducted as a result of a specific over-limit, or abnormal event. Examples of events requiring special inspections include:

- Hard landings
- Overstress conditions
- Flight into severe turbulence
- Flight into volcanic ash
- Overtemp conditions
- Overweight landings
- Exceeding placarded speed of flaps and landing gear
- Bird strike
- Lightning strike
- Foreign object damage (FOD)

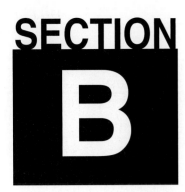

SECTION B
INSPECTION GUIDELINES AND PROCEDURES

The inspection of an aircraft to determine its airworthiness requires a great amount of skill and judgment. For the most part, the items to be inspected are listed in an inspection checklist. However, how well an inspector evaluates an item's airworthiness is up to the judgement and skill of the individual. These factors combine to require the inspector to develop a system or procedure for effectively inspecting an aircraft.

It is imperative that inspectors set up a set of standards in order to determine an item's airworthiness. These standards must be high enough to guarantee the airworthiness of the aircraft, but not so high to cause needless expense to the owner. The inspector must also withstand pressures applied by others to lower those standards by representing items as being airworthy when they are not. Once these standards are compromised, it is very difficult to restore the integrity of an aircraft inspector.

INSPECTION FUNDAMENTALS

Aircraft inspectors should be familiar with the visible, measurable or otherwise detectable effects of wear and tear on an aircraft. An effective inspector is able to recognize and determine the cause of the wear and tear that is found during inspection, which makes the subsequent repair straightforward. The five most common sources of wear and tear are weather, friction, stress overloads, heat, and vibration.

The damaging effects of weather can vary widely and range from surface corrosion, oxidation, wood rot, wood decay, fabric decay, fabric brittleness, fabric mildew and cracks, and interior damage and exterior paint oxidation due to ultra-violet rays. In addition, physical damage due to weather can range from lightning damage, hail dents, wind damage to control stops and control rigging, to surface damage due to sand and dirt erosion. Atmospheric moisture content is another consideration when inspecting an aircraft. The amount of water and salt the air holds may directly influence the potential corrosion found on the aircraft, especially aircraft based near large bodies of water and oceans. For further information regarding the identification and treatment of corrosion, see chapter 12 of the *A&P Technician General Textbook*.

Friction damage manifests in many different forms such as abrasions, burnishing, chafing, cuts, dents, elongation, erosion, galling, gouging, scratches, scoring, and tearing. In the context of this section, friction is the rubbing of one object against another that causes a destructive result. [Figure 17-12]

- **Abrasion** is caused by a rough substance between two moving surfaces.

- **Burnishing** is the polishing of a surface by the sliding contact with another, smoother, harder, metallic surface. Bearings have a tendency to burnish and should be checked and lubricated regularly.

- **Chafing** is the wear between two parts rubbing, sliding, or bumping into each other that are not normally in contact.

- **Elongation** is the oval-shaped wear of a bearing surface around bolts, hinge pins, clevis pins, etc.

- **Erosion** is the loss of metal from the surface by the mechanical action of materials such as dirt, sand, or water. Propellers, leading edges of the wings and empennage, wheel fairing, landing gear, and cowlings are susceptible to erosion damage.

- **Galling** is the breakdown or buildup of the metal surface due to excessive friction between two parts in motion. Particles of the softer metal are torn loose and welded to the harder metal surface.

Overloading the aircraft may result in the failure or deformation of the structure, either slightly or prominently, but usually produces visible damage. The types of stress overloads that an inspector must

Aircraft Airworthiness Inspection

This nose strut shows signs of abrasion due to a lack of lubrication on the strut surface. The protective plating has also been rubbed away at the base exposing the metal underneath. The unprotected portion of the strut also shows signs of oxidation corrosion. Cleaning and lubricating the strut surface extends the life and appearance of the strut

Wheel bearings have a tendency to burnish with a lack of lubrication. The bearing race in this example shows signs of burnishing. Detailed inspection and lubrication of the bearing assembly will extend the life of the bearings.

This example illustrates chafing caused by the control cables rubbing the ducting found under the floor panels of a Beech King Air.

Elongation is a defect that needs to be checked at attach points on the aircraft. The attachment plate of this hydraulic actuator shows signs of elongation of both bolt holes. The continuation of the elongation will eventually fatigue the metal to the point of failure if not detected.

This propeller shows signs of erosion on the leading edge due to sand, dirt, and foreign objects wearing away the surface metal

Figure 17-12. Examples of friction damage.

become familiar with are tension, compression, torsion, shear, and bending overloads. [Figure 17-13]

- **Tension** overloads usually occur after hard landings, taxiing on rough fields, or flight in turbulent air. Failure is indicated by signs of the pulling away of fittings from the fuselage, failure of welded areas, wrinkling of metal skin, and deformed or cracked fittings.

- **Compression** overloads may manifest as bulges in the metal skin, breaks in paint, and bows or bends in the long members such as wing struts. Wood compression may be detected by a slight ridge across the face of the member at right angles to the grain.

- **Torsion** or twisting overloads will turn one end of a part around its longitudinal axis

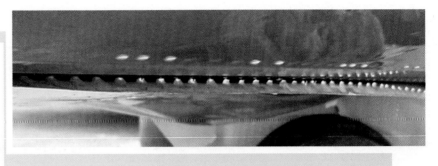

Tension or stretching damage may be exhibited by the pulling away of the skin from the structure of an aircraft. In this example, the lower wing skin of a damaged Beech Bonanza has been pulled away from the riveted seam exposing the interior wing area.

The wing tip of this aircraft is bent in an upward direction illustrating bending stress overload. The inboard portion of the wing was held in place while a bending force was applied to the wing tip.

The firewall of this small aircraft was compressed in a hard landing. The firewall is constructed of stainless steel requiring a large amount of compression stress overload

Figure 17-13. Examples of stress-overload damage.

while the other end is held fixed or turned in the opposite direction. Wheels caught in frozen ruts during landing may twist the landing gear and cause torsion damage. Careful inspection of the landing-gear torque links should be made after landing on rough or rutted fields. Severe air loads imposed upon the aircraft during flight through turbulent air may twist the control surfaces. Improper rigging of the wing and tail control surfaces may also cause torsion overloads by producing a positive load on one side of the surface at all times.

- **Shear** overloads result from forces that are applied to an object in an opposite but parallel direction. When a shear overload is applied, the part having the least resistance to the force will fail first. Because bolts, rivets, and clevis pins are used in areas subject to shear forces, they should be inspected for shear failure. Bent, torn, or deformed bolts, rivets, or clevis pins are good indications of shear damage.

- **Bending** overloads cause rigid members to curve or bow away from a straight line. Hard landings, abnormal flight loads, and improper ground handling may cause bending damage. Wood or metal skin may show signs of wrinkling, cracking, or distortion. On fabric covered airplanes, a bent member may be detected by looseness or wrinkling of the fabric.

The primary source of heat damage affecting the aircraft is the powerplant. Inspectors must be familiar with direct and indirect heat sources that cause damage. Direct heat damage is normally caused by leaking exhaust gases, and, in the case of severe leaks, may allow flames to escape resulting in devastating consequences. Indirect heat damage may result from excessive engine compartment heat indicated by high oil and cylinder head temperatures, blistering paint on the engine cowling, and odors of burned oil or rubber during or after engine run-up.

Improperly installed or leaking engine baffles, misaligned cowlings, improper carburetor-heat control rigging, improper cowl-flap door rigging, and dirty air coolers may cause indirect heat damage. In addition, the use of an improper grade of oil, and oil leakage, may also cause indirect heat damage to the aircraft and engine. [Figure 17-14]

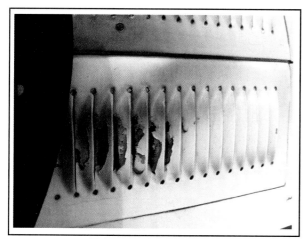

Figure 17-14. The bubbling of the paint in addition to the exhaust trail exiting the engine cowling vent illustrates indirect heat damage.

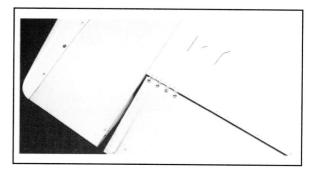

Figure 17-15. The cracks on the wing skin of this Piper Seneca were caused by the excessive play in the aileron hinge. During flight, vibration or "flutter" of the ailerons occurred which stressed the aircraft structure and caused stress cracks to manifest on the upper wing surface.

Vibration causes many malfunctions and defects throughout the life of the aircraft. Vibration affects loose or improperly installed parts and accelerates wear to the point of failure in some cases.

Low frequency vibration can be felt or noticed by the pilot or mechanic. It is usually caused by a malfunctioning powerplant, propeller, worn engine-mount pads, loose aircraft structure joints, or improper rigging. Noticeable vibration causes abnormal wear between moving parts. Excessive clearances and poor installation are also factors affecting the level of vibration damage and should be considered when inspecting the aircraft.

For example, control surface and trim tab "free-play" limits may be extreme due to excessively worn hinges and actuator damage. Excessive free-play causes the control surface to vibrate or "flutter" in flight. The vibration then transfers through the airframe structure and causes fractures and fatigue to appear in locations remote from the source. [Figure 17-15]

INSPECTION GUIDELINES

In addition to the aforementioned wear and tear effects, following is a brief outline of some of the most common deficiencies to look for in an aircraft inspection. An inspector must be familiar with each of these areas in order to perform effective and high-quality inspections.

- **Movable Parts**: proper lubrication, security of attachment, binding, excessive wear, proper safety wiring, proper operation and adjustment, proper installation, correct travel, cracked fittings, security of hinges, defective bearings, cleanliness, corrosion, deformation, and sealing and tension.

- **Fluid lines and hoses**: proper hose or rigid tubing material, proper fittings, correct fitting torque, leaks, tears, cracks, dents, kinks, chafing, proper bend radius, security, corrosion, deterioration, obstructions and foreign matter, and proper installation.

- **Wiring**: proper type and gauge, security, chafing, burning, defective insulation, loose or broken terminals, heat deterioration, corroded terminals, and proper installation.

- **Bolts**: Correct torque, elongation of bearing surfaces, deformation, shear damage, tension damage, proper installation, proper size and type, and corrosion.

- **Filters, screens, and fluids**: cleanliness, contamination, replacement times, proper types, and proper installation.

- **Powerplant Run-up**: Engine temperatures and pressures, static RPM, magneto drop, engine response to changes of power, unusual engine noises, ignition switch operation, fuel shut-off/selector valves, idling speed and mixture settings, suction gauge, fuel flow indicator operation, engine mount security, mount bolt torque, spark plug security, ignition harness security, oil leaks, exhaust leaks, muffler cracks and wear, security of all engine accessories, engine case cracks, oil breather obstructions, firewall condition, and proper operation of mechanical controls.

- **Propellers**: nicks, dents, cracks, cleanliness, lubrication, gouges, proper blade angles, blade tracking, proper dimensions, governor leaks and operation, and control linkages for proper tension and installation. Nicks on the leading edge of the blade are an important item to inspect for; they produce stress concentrations that need to be removed immediately upon discovery in order to prevent the blade from separating at the nick.

INSPECTION PROCEDURES

The inspection of aircraft requires a great deal of organization and planning. Effective inspections must be performed in a logical and orderly sequence to ensure that no inspection item is overlooked or forgotten. The accepted method of performing an inspection that is used by the aircraft maintenance industry also includes the service and repair activities that are necessary to approve the aircraft for return to service.

The inspection of an aircraft is divided into five basic phases: pre-inspection, examination, service and repair, functional check, and the return to service phase.

PRE-INSPECTION PHASE

The pre-inspection phase is very important and serves to organize the paperwork, records, tools, and equipment needed for the inspection. This phase usually includes: work order completion, compilation of the aircraft specifications, review of maintenance records, airworthiness directive research, manufacturer service bulletin and letter research, airworthiness alert research, producing the inspection checklist, and aircraft preparation.

The pre-inspection phase begins with the completion of the work order which outlines and authorizes the performance of the services. The maintenance records, airworthiness directives, service bulletins, and any other relevant service information are researched and, if applicable, added to the inspection checklist. The aircraft is cleaned, and the engine is usually run-up to check engine parameters and to set a base line for the post-inspection run-up. Removal of inspection panels, engine cowling, and interior, if required, are done during the pre-inspection phase. In addition, tools and equipment are made ready, and any known parts that are needed are ordered. [Figure 17-16]

WORK ORDER

The work order is the agreement between the shop or mechanic and the owner of the aircraft concerning the work to be performed. It describes the work

Figure 17-16. During the pre-inspection phase, the aircraft is prepared for the inspection by removing all applicable inspection panels and completely uncowling the engine compartment. It is important to have access to as much of the aircraft, systems, and components as possible for a complete inspection.

requested and serves as a record of parts, supplies, and labor expended on the aircraft. While interviewing the owner, describe the work requested and any discrepancies that the owner wants repaired. The owner then signs the work order before work begins on the aircraft. [Figure 17-17]

Figure 17-17. Preparing the work-order with the customer is an important step in the pre-inspection phase of any inspection.

Clearly explain to the customer that additional charges may apply regarding maintenance performed to correct any discrepancy found during the inspection. It should be noted that the work order normally only estimates the total cost of the inspection and any subsequent maintenance repair. It is impossible to determine the labor and parts expense of unknown discrepancies. Certain shops charge a flat rate for the inspection and charge separately for parts and labor regarding any maintenance done to the aircraft. Others may charge on an hourly basis along with any expenses for parts and supplies that are incidental to the inspection and maintenance.

At times, discrepancies are detected upon inspection. It is wise to provide the owner the opportunity to choose to fix the discrepancies or not. If the owner chooses to repair any discrepancy that is found during the inspection, revise the work order with reference to the needed repairs. Have the owner sign the revision order before beginning the repairs.

MAINTENANCE RECORDS AND AIRCRAFT SPECIFICATION REVIEW

The maintenance record and aircraft specifications review is a very important part of any inspection and takes place before the aircraft is physically examined. Maintenance records can reveal quite a bit about the care and maintenance of an aircraft. The maintenance history of the aircraft is carefully examined to determine repetitive maintenance problems, airworthiness directive compliance, any major repairs and alterations done to the aircraft,

and, on a base level, whether the aircraft has had maintenance performed in a consistent manner. Maintenance records are researched to determine information such as the type of oil in use, ELT battery expiration and operational test date, altimeter and transponder test due dates, when the spark plugs were last changed, age of the battery, when vacuum system filters were last changed, life-limited parts status, aircraft total time, major repair and alteration information, and engine time since overhaul (TSO). [Figure 17-18]

Figure 17-18. Thorough maintenance information research is key to an effective inspection. Without complete and correct aircraft information, important items may be overlooked during an inspection.

All aircraft must conform to their certification requirements. Therefore, the research and compilation of the aircraft specifications is essential to a proper conformity inspection. A conformity inspection entails a visual inspection of the entire aircraft, engine, propeller, avionics, and appliances using information gathered from the TCD, STCs, aircraft equipment list, and applicable airworthiness directives. A thorough inspection starts with the research of the aircraft specifications and maintenance information.

In addition to the records review, the inspection checklist must be obtained that is specific to the aircraft make and model. When performing annual or 100-hour inspections, the use of a checklist is required by FARs. The technician may design a checklist that is specific to the aircraft being inspected, or use a checklist provided by the manufacturer of the aircraft, engine, propeller, and installed components as long as it meets the minimum requirements outlined in 14 CFR part 43 Appendix D.

SERVICE BULLETINS AND LETTERS

A thorough inspection includes the research and documentation of applicable service bulletins and service letters. During the records review, manufacturer's service information is researched to verify any possible changes that were made to improve the service life or efficiency of the aircraft, engine, propellers, or appliances. The manufacturer of the aircraft publishes service bulletins and letters to inform the owner of any design changes, malfunctions, or servicing requirements. They may require an inspection or repair to correct an unsatisfactory condition. The FAA does not require the compliance of service letters or bulletins. However, the owner should be encouraged to comply. Many times, service bulletin information is a precursor to a mandatory airworthiness directive (AD). Additionally, airworthiness directives may reference service bulletin information for specific instructions regarding inspection and/or repair when complying with the AD.

Each manufacturer has a different service bulletin numbering system. Most will include the year in the service bulletin reference number but some do not. To perform a service bulletin search using microfiche, review the index of service bulletins that apply to the type and model of the aircraft, engine, propeller, and appliances and compile a list of applicable service bulletins. In addition, updated service bulletins that are received after the publish date of the microfiche are referenced in the "service bulletins received after cutoff" section of the master reference fiche and are usually referenced by service bulletin number. [Figure 17-19]

Figure 17-19. Inspections begin with service information research. This mechanic is researching service bulletin information using microfiche. Although microfiche is a valid and accurate way to research and compile current maintenance information, it is being replaced with computerized search programs that do not require extensive microfiche libraries. One CD-ROM takes the place of hundreds of microfiche making it easier and more efficient to update and research information.

If the maintenance facility utilizes a computerized search program, a search may be made that is specific to the make and model of the aircraft, engine, propellers, and appliances. Most computerized search programs allow the technician to enter specific search criteria, such as make and model, which makes this method more efficient and less time consuming than microfiche. The program searches by the criteria entered into a search field and produces a list of applicable service bulletins. Again, most computerized service bulletin subscriptions will reference updated information that is received after the publish date in a specific section such as "New service bulletins." Furthermore, a number of computerized maintenance-information-services offer real time search capabilities over the Internet offering daily updated service bulletin information.

Once the service bulletin list is compiled, confirm the applicability by comparing the serial number of the aircraft or equipment against the relevant serial numbers in the "effectivity" or "models affected" section of the service bulletin. A list of bulletins that apply by serial number is then compared to the service bulletins complied with and referenced in the maintenance records. Those that have not been complied with would then be due. Because the FAA does not require mandatory compliance with service bulletins, they should be discussed with the aircraft owner to determine if compliance is desired.

Compliance with service bulletin information at regular inspection intervals may save time and money in the long run. Many times, service bulletins bring attention to malfunctions and design changes that eventually become important enough to warrant publishing of an airworthiness directive (AD). If the service bulletin is complied with during a regularly scheduled inspection, it may eliminate the need to perform the inspection and/or repair again to comply with a subsequent AD. The "compliance" section of the AD will clarify whether the accomplishment of the service bulletin satisfies the AD compliance.

AIRWORTHINESS DIRECTIVES

Airworthiness directives (AD) are issued by the FAA to correct unsafe conditions that affect the safety of an aircraft. ADs are mandatory and require compliance. Thus, it is imperative to comply with all ADs that apply to the aircraft. At the beginning of every inspection, research and compile a listing of all airworthiness directives that are applicable to the aircraft, its engine, propeller and any installed component.

In the case of airworthiness directives, they may be researched in the same manner as service bulletins: manually through microfiche or computer search programs. In addition, several maintenance information companies provide detailed searches that are applicable to the particular model of airplane and its installed equipment for nominal fees, thus eliminating the need for inefficient research time.

Aircraft owners are required to maintain the current airworthiness directive (AD) status of their aircraft and all installed equipment. Included in the AD status is the method of compliance, AD number and revision date, whether the ADs are recurring or one-time only, and finally, the time and date when the next action is required. To improve the ability to track AD compliance, most aircraft records include a separate airworthiness directive compliance record, which keeps a cumulative record of the current AD status for a particular aircraft. Instead of looking through logbooks page by page, the AD compliance record makes researching AD information much easier by compiling AD compliance in one convenient location.

In addition to compiling the applicable AD information, the technician must be able to interpret the applicability and compliance sections in the body of the AD. Every AD applies to each aircraft or component as identified in the applicability statement regardless of the classification or category. The serial number range or series of aircraft or component that is listed in the applicability statement determines whether the AD is valid for that particular aircraft or component. When there is no serial number range specified, the AD applies to all serial numbers. [Figure 17-20]

INITIAL RUN-UP

After completing the pre-inspection paperwork and maintenance records review, perform an engine run-up to provide a baseline of engine parameters to compare to the post-inspection run-up indications. A pre-inspection run-up also warms the engine and provides proper lubrication. Perform an engine run-up to determine whether the engine develops proper static rpm and manifold pressure, if applicable, and to check pressures and temperatures to be sure that they are within proper operating ranges. Check the magnetos, carburetor heat, and propellers for the proper operation, and test the generator or alternator for proper output.

During the run-up, check the operation of electrical flaps for symmetrical movement and smooth operation through the entire range of travel. Also, verify that the flap indicator agrees with the actual flap position. Check flight control movement and travel, making note of any roughness or malfunction. Verify that the ailerons move in the proper direction with alternating control inputs; rotating the yoke to the right moves the right aileron up and left aileron down and vice versa when rotating the yoke to the left.

Also, while the electrical power is available, check the radios for proper operation, listening for any noise that may be caused by the interference of the engine or any aircraft system. Check the magnetic compass reading for any deviation caused by electrical interference while the electrical systems are operating. Make sure the compass correction card, if required, is placarded.

Set the altimeters to the current barometric pressure and compare the altimeter indication with the actual field elevation where the aircraft is located making note of any discrepancy. In addition, while operating at high RPM, check the instrument pressure or vacuum for proper operating range indications.

Check the operation of the fuel selector valve by selecting each fuel tank to verify consistent engine function when drawing fuel from individual tanks. Make note of any changes in engine RPM, and fuel flow or pressure fluctuations. Check the fuel pressure produced by the engine-driven pump and, after shutdown, by the electric boost pumps. After engine shutdown, listen to the gyro instruments as they run down to detect any bearing roughness.

After the engine is shutdown, uncowl the engine and look for any loose or disconnected lines, oil and fuel leaks, or any other irregularity. Finally, once the run-up is concluded, completely wash down the engine to remove all oil and dirt that might hinder a complete inspection.

EXAMINATION PHASE

The primary purpose of the examination phase is to physically evaluate the airworthiness of the aircraft and its components. All of the subsequent activities of the inspection are dependent upon, and in support of, the examination phase of the inspection.

The examination phase is the actual inspection of the aircraft. It starts with a conformity inspection, which compares the actual aircraft with its certification specifications. It then proceeds to looking at, feeling, checking, measuring, operating, moving, testing, and whatever else is needed to determine

99-16-06 - Failure of the wing attach fittings
The New Piper Aircraft, Inc. **Category** – Airframe **Effective date** – 09/24/1999 **Recurring** – No **Supersedes** – N/A **Superseded by** – N/A
Amendment 39-11241; Docket No. 99-CE-01-AD **Applicability:** Model PA-46-350P airplanes, serial number 4622191 through 4622200 and 4636001 through 4636175, certificated in any category. **Note 1:** The affected serial numbers refer to airplanes that have been delivered since January 1995 and could have insufficientstrength wing attach fittings installed. Airplanes manufactured after serial number 4636175 have this problem corrected prior to delivery. **Note 2:** This AD applies to each airplane identified in the preceding applicability provision, regardless of whether it has been modified, altered, or repaired in the area subject to the requirements of this AD. For airplanes that have been modified, altered, or repaired so that the performance of the requirements of this AD is affected, the owner/operator must request approval for an alternative method of compliance in accordance with paragraph (c) of this AD. The request should include an assessment of the effect of the modification, alteration, or repair on the unsafe condition addressed by this AD; and, if the unsafe condition has not been eliminated, the request should include specific proposed actions to address it. **Compliance:** Required within the next 100 hours time-in-service (TIS) after the effective date of this AD, unless already accomplished. To prevent the potential for failure of the wing attach fittings caused by the utilization of substandard material, which could result in the wing separating from the airplane with consequent loss of control of the airplane, accomplish the following: (a) Install reinforcement plates to the wing forward and aft attach fittings by incorporating the Wing to Fuselage Reinforcement Installation Kit, Piper part number 766-656. Accomplishment of the installation is required in accordance with the instructions to the above referenced kit, as referenced in Piper Service Bulletin No. 1027, dated November 19, 1998. (b) Special flight permits may be issued in accordance with Secs. 21.197 and 21.199 of the Federal Aviation Regulations (14 CFR 21.197 and 21.199) to operate the airplane to a location where the requirements of this AD can be accomplished. (c) An alternate method of compliance or adjustment of the compliance time that provides an equivalent level of safety may be approved the Manager FAA, Atlanta Aircraft Certification Office (ACO). The request shall be forwarded through an appropriate FAA Maintenance Inspector, who may add comments and then send it to the Manager, Atlanta ACO. **Note 3:** Information concerning the existence of approved alternative methods of compliance with this AD, if any, may be obtainedfrom the Atlanta ACO. (d) The installation required by this AD shall be done in accordance with the instructions to the Wing to Fuselage Reinforcement Installation Kit, Piper part number 766-656, dated November 6, 1998, as referenced in Piper Service Bulletin No. 1027, dated November 19, 1998. This incorporation by reference was approved by the Director of the Federal Register in accordance with 5 U.S.C. 552(a) and 1 CFR part 51. Copies may be obtained from the New Piper Aircraft, Inc. Customer Services. Copies may be inspected at the FAA, Central Region, Office of the Regional Counsel or at the Office of the Federal Registry in Washington DC. (e) This amendment becomes effective on September 24, 1999.

Figure 17-20. Example of an airworthiness directive regarding a Piper PA-46. ADs are set up in the same format: the heading showing the AD number, revision date and subject, the "applicability statement" that distinguishes the aircraft or component applicability, and the "compliance statement" that specifies the time and procedural requirements for AD compliance.

the condition of the aircraft and its components. A checklist is followed with a planned sequence or order in which items of the aircraft are inspected. Note the needed service and discrepancies that are discovered during the examination phase on a discrepancy list. The discrepancy list is used for follow-up repair either during the inspection or by another certificated technician after the completion of the inspection. [Figure 17-21]

One of the most important considerations for an efficient inspection is that it must be systematic. Using the checklist, inspect one complete system before going to the next. For example, check the complete aileron system from the control wheel to each aileron and back to the wheel. Then check the complete elevator system. Jumping from one part of a system or component to another leaves room for mistakes and the possibility of overlooking problem areas.

SERVICE AND REPAIR PHASE

The service and repair phase of the inspection includes the necessary maintenance that is required to approve the aircraft for return to service and to preserve its airworthy condition. Servicing consists of tasks such as lubricating wheel bearings and

Aircraft Airworthiness Inspection

A conformity inspection requires the thorough inspection of the aircraft's specifications.

A differential compression test is used to check cylinder condition. Perform the compression check while the engine is still hot to gain accurate compression indications. The increased clearances of the pistons, rings, and valves of a cold engine, in addition to the lack of lubrication, may result in air compression leakage and inaccurate indications.

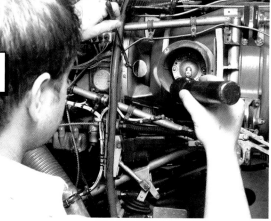

Inspection mirrors and flashlights are essential tools used during the inspection process.

Figure 17-21. Examples of areas checked during typical inspection intervals.

A de-ice boot inspection includes checking for items such as nicks, cuts, delamination, cracking, and holes along the boot surface.

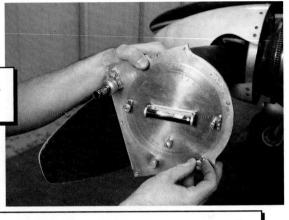

An improper blade angle may be indicated by poor engine performance. The propeller blade angle is checked while it is on the engine.

Spark plugs are checked for proper electrode gap and carbon deposits.

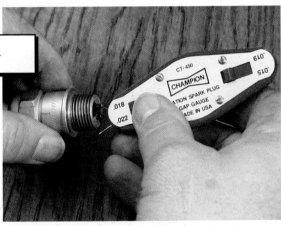

The nose strut, torque links, and shimmy dampener are checked for discrepancies such as free play in the system, abrasion damage due to lack of lubrication, and corrosion.

Erosion, nicks, dents, and scratches are commonly found on propellers. If left unchecked, these deficiencies can lead to stress concentrations and fatigue failure.

Aircraft Airworthiness Inspection

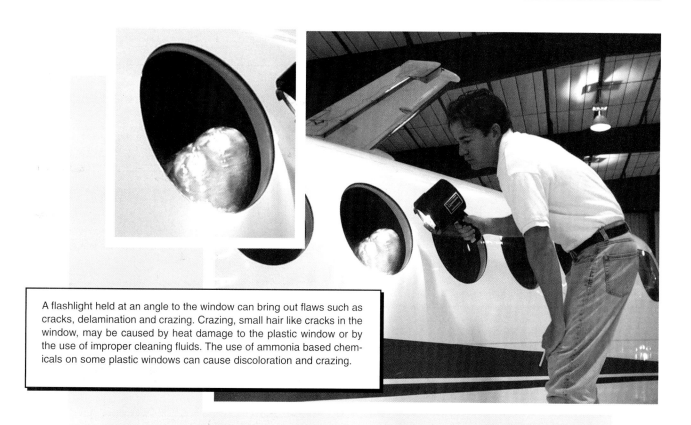

A flashlight held at an angle to the window can bring out flaws such as cracks, delamination and crazing. Crazing, small hair like cracks in the window, may be caused by heat damage to the plastic window or by the use of improper cleaning fluids. The use of ammonia based chemicals on some plastic windows can cause discoloration and crazing.

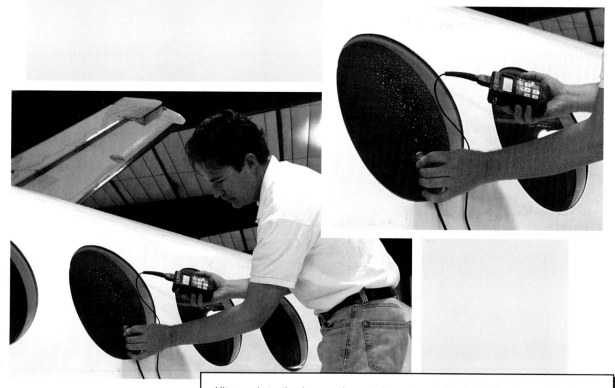

Ultra-sonic testing is a good way to inspect a window for hidden flaws, delamination and to check window thickness. The use of a coupling gel is essential for proper indications.

This King Air 200 is going through a phase 3 inspection as part of a pre-purchase examination. The interior is completely removed to check control cables, wiring, and ducting for deficiencies.

A tensiometer is used to check the tension of control cables. Specific control cable tensions are required and outlined in the aircraft's service manual.

Inspecting behind the instrument panel can be complicated but it is an important part of the inspection. Wiring harnesses, electrical connections, fittings, and component integrity can be checked from behind the instrument panel.

During an inspection, the hydraulic accumulator, reservoir, and lines are checked for integrity, leakage, and accumulator pre-charge. A gauge indicates the pre-charge pressure of the accumulator when the hydraulic pumps are not operating. With the pumps operating, the gauge indicates system pressure.

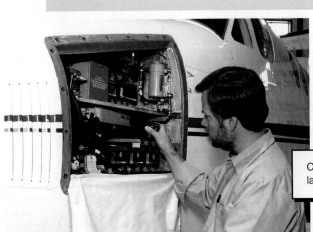

Check the mounting, security, and attachment of avionics installations.

moving parts, replacing and cleaning filters and screens, adding fluids, servicing the battery, and cleaning the aircraft.

Although discrepancy repair is not part of the inspection itself, it is closely related and usually done concurrently with the inspection. The repair phase may include replacement, repair, and overhaul of the aircraft components and systems that are found to be deficient or unairworthy. Additionally, modifications to the aircraft that require a Supplement type certificate (STC), are often done in conjunction with an annual or 100 hour inspection. Modifications that require an STC are considered major alterations, therefore, must be returned to service by an A&P mechanic with an Inspection Authorization.

FUNCTIONAL CHECK PHASE

After the inspection is accomplished and all needed maintenance is completed, the maintenance technician conducts functional or operational checks on the aircraft and systems. When performing an annual or 100-hour inspection, FAR part 43.15 requires a functional check on the aircraft engines. Therefore, perform a post-inspection engine run-up to determine whether the power output (static and idle rpm), magnetos, fuel and oil pressures, and cylinder and oil temperatures meet the manufacturer's specifications. The engine functional check phase also allows the technician to check for fuel leaks, oil leaks, and any other irregularity that may indicate something left open or loose during the inspection.

Additional functional checks are recommended to ensure that the installed systems or subsequent discrepancy repairs are airworthy according to the manufacturer's specifications.

After all maintenance is completed, a good wash of the aircraft to remove any trace of oil or grease left on it from the inspection is recommended. Include the windshield and all of the windows remembering to only use the proper cleaning fluids. Using the wrong type of chemical can damage or destroy the aircraft finishes and windows. Carefully clean the inside of the aircraft to remove any fingerprints or grease marks left during the inspection. Vacuum the carpet, straighten the seat belts, and make the inside of the cabin appears neat and organized.

An aircraft owner may not understand the intricacies of an inspection, but are sure to notice any grease spots or smudge marks left behind. They may feel that a person who is careless enough to leave a disorganized and unclean airplane may have been equally careless in the inspection.

RETURN-TO-SERVICE PHASE

After the inspection is accomplished, you must complete the paperwork before the aircraft is approved for return to service. In the case of a 100-hour inspection, the work order is completed, the AD compliance record is filled out, and inspection entries are recorded in the maintenance logs.

Complete the work order to detail the inspection and all of the work and servicing that was performed. In most cases, the work order is very detailed and may be recognized as part of the aircraft records. Tally up all labor charges, cost of parts and supplies, and any special charges such as outside labor and telephone calls related to the job. Complete and systematic maintenance documentation not only protects the maintenance technician if a question ever arises concerning work that was or was not done, but it is assumed that good records normally accompany good work.

Before an aircraft can be legally flown, entries must be made in the maintenance records and signed off by the appropriately rated maintenance technician. The inspection entry and sign-off constitutes "approving the aircraft for return to service." FARs require a separate 100-hour inspection entry for each log if the owner maintains separate logbooks for the airframe, engine, and propellers. In the case of an annual inspection, an entry is only required in the airframe log. However, most inspectors enter an annual inspection entry in all logbooks, thus making maintenance record research more efficient and easy. Again, the inspection is not complete and the aircraft is not approved for return to service without the proper logbook sign-offs. Refer to the "Aircraft Maintenance Records" section of this chapter for a detailed explanation of entry requirements. [Figure 17-22] [Figure 17-23]

Figure 17-22. The inspection is not complete without the proper maintenance entry. Make sure the entry is legible and that it details the inspection performed according to the FAR requirements specific to the inspection entries.

AVIATION SERVICE COMPANY
285 Airport Rd.
Somewhere, USA
(555) 555-5555

Work Order No. 13584

I hereby authorize the following repair work along with the necessary materials and hereby grant you and your employees' permission to operate and fly the aircraft herein described for testing and or inspection. An express mechanics lien is hereby acknowledged on this aircraft or articles left in case of fire, theft, or any other cause beyond your control. If it becomes necessary for you to employee a collection agency and or an attorney to collect this account, I the undersigned agree to pay all court costs plus a reasonable attorney's fee.

X _Jennifer Brown_

Quan.	MATERIALS DESCRIPTION	PART NUMBER	UNIT $	TOTAL
1	oil filter	CH48108	11.90	11.90
1	prop shaft seal	2846A	6.50	6.50
1	6.00 x 6, 4 ply tire	4891BTY	75.00	75.00
1	strobe light	TJ345	8.75	8.75
1	ELT Battery	AT874	87.50	87.50
			TOTAL PARTS	189.65

OUTSIDE WORK

Service Description	Hrs.	Charge	$ amount
			14.24
		TOTAL OUTSIDE WORK	14.24

OIL, FUEL, & SOLVENT

Qty		
	Gallons FUEL ___ octane	
8	Quarts OIL _100_ weight	
0	Solvent	

REMARKS AND RECOMMENDATIONS FOR REPAIRS:
Repair all discrepancies found during 100-hr insp. reference attached discrepancy list. _Jennifer Brown_

DATE _1/16/00_

NAME _Jennifer Brown_
ADDRESS _121st Ave, Somewhere, USA_
CITY, STATE, ZIP
MAKE & MODEL _Cessna 172R_   REGISTRATION NUMBER _N00XYZ_   TOTAL TIME _1140.8_
SERIAL NUMBER _17280044_   LABOR RATE _60.00/hr_   TAX RATE _4.8%_
RECEIVED ___ Am _1:12_ PM   PHONE WHEN READY YES ☒ NO ☐   # _555-6666_   WRITTEN BY: _DT_

DESCRIPTION
CHANGE OIL ☒
FILTER ☒
INSPECTIONS:
25-50-75 ☐
Annual ☐
100 Hour ☒
Progressive ☐
Periodic ☐
WASH ☐
BRIGHTWORK ☐

SERVICE DESCRIPTION

	HRS.	SERVICE
Perform 100-hour inspection	15	900.00
Replaced propeller shaft seal	3	180.00
Blended # propeller blade	1.5	90.00
Created + installed fuel placard	1	60.00
Replaced left main tire	2	120.00
Replaced left strobe light	1	60.00
Replaced ELT battery	.5	30.00
Total Hrs. & Svc.	24	1440.00

ACCEPTANCE OF REPAIRED PLANE: At time of delivery of said aircraft the undersigned has read and examined all notations in the "remarks and recommendations section" and accepts the same as modifications of the repair and service work originally ordered as set forth herein above, hereby ratifying the original repair service except as modified.

Owners signature: _Jennifer Brown_

SALES SUMMARY

Total Service	1440.00
Total Parts	189.65
Outside Work	14.24
Gas, Oil, solvent	
Tax	78.91
TOTAL	1722.80

Figure 17-23. (1 of 9) Example of the completed paperwork regarding a 100-hour inspection on a Cessna 172 including the work order, inspection checklist, discrepancy list, AD compliance record, and inspection and repair logbook entries.

100-hour Inspection Checklist
1996 Cessna 172R

DATE: 1/20/00   REGISTRATION NO.: N00XYZ   TOTAL TIME: 1140.8
MAKE: Cessna   MODEL: 172R   S/N: 17280044
ENGINE MAKE, MODEL, S/N: Lycoming IO-360-L2A   TSMO: 0
PROPELLER MAKE, MODEL, S/N: McCauley IC235/LFA757   TSO: 0
DATE PREVIOUS 100-HOUR: 11/2/99   TIME SINCE INSPECTION: 98.4

OWNER: Ms. Jennifer Brown
ADDRESS: 12 1st Ave, Somewhere USA

INSPECTION PERFORMED BY: Beth Collins
CERTIFICATE NO.: A+P 123456789
ADDRESS: 285 Airport Rd, Somewhere, USA

Instructions: Initial "inspected" column after inspecting item. Check discrepancy column if discrepancies found. Document measurements, observations, and comments including corrective actions on discrepancy worksheet.

Insp	Des	Item
BC		Airworthiness directive list
BC		Service bulletin & letter information
BC		Equipment list
BC		Maintenance logs: airframe, engine, propeller
BC		Form 337s – no major repair or alterations noted
BC		Weight & balance
BC		Approved flight manual
BC		Aircraft registration
BC		Airworthiness certificate (does it match model and log book entry?)
Pre inspection engine run-up		
BC		Engine temperatures and pressures
BC		Static RPM
BC		Magneto drop
BC		Engine response to changes in power
BC		Unusual engine noises
BC		Fuel selector valve: operate engine on each tank position and OFF position long enough to ensure proper selector valve function.
BC		Idling speed and mixture; proper idle cut-off
BC		Alternator and ammeter operation
BC		Suction gage:
BC		Fuel flow indicator
BC	X	After shut-down, check for fuel and oil leaks
BC		Clean engine compartment

Cessna 172R 100-hour inspection checklist   page 1 of 4

Insp	Des	Item
		Propeller
BC		Spinner and spinner bulkhead
BC	X	Blades
BC		Hub
BC		Bolts and nuts
		Engine compartment
		Check for evidence of oil and fuel leaks, then clean entire engine compartment prior to inspection.
BC		Engine oil, screen, filler cap, dipstick, drain plug and external filter element
BC		Oil cooler
BC		Induction air filter: clean, inspect, replace if needed.
BC		Induction air box, air valves, doors, and controls
BC		Cold and hot air hoses
BC		Engine baffles
BC		Cylinders, rockers box covers, and push rod housings
BC		Crankcase, oil sump, accessory section, and front crankshaft seal
BC		All lines and hoses
BC		Intake and exhaust systems
BC		**AD 97-12-06: Gascolator, tailpipe, and cowling area**
BC		Ignition harness
BC		Spark plugs
BC		Compression check: Cyl 1: 78/80 Cyl 4 80/80 Cyl 2: 79/80 Cyl 5 79/80 Cyl 3 79/80 Cyl 6 79/80
BC		Crankcase and vacuum system breather lines
BC		Electrical wiring
BC		Vacuum pump and relief valve
BC		Vacuum relief valve filter
BC		Engine controls and linkage
BC		Engine shock mount pads, mount structure, and ground straps
BC		Cabin heater valves, doors, and controls
BC		Starter, solenoid and electrical connections
BC		Starter brushes, brush leads, and commutator
BC		Alternator, and electrical connections
BC		Alternator brushes, brush leads, and commutator or slip ring
BC		Voltage regulator mounting and electrical leads
BC		Magnetos (external inspection) and electrical connections
BC		Magneto timing
BC		Injection system
BC		Firewall
BC		Engine cowling
		Fuel System
BC		Fuel strainer, drain valve, and control
BC		Fuel strainer screen and bowl

Cessna 172R 100-hour inspection checklist

Aircraft Airworthiness Inspection

Insp	Des	Item
BL	X	Fuel tanks, fuel lines, sump drains, filler caps, and placards
BC		Drain fuel and check tank interior, attachment, and outlet screens
BC		Fuel vents, vent valves, & vent line drain
BC		Fuel selector valve and placards
BC		Fuel valve drain plug
BC		Engine primer

Landing Gear

Insp	Des	Item
BC		Brake fluid, lines and hoses, linings, disc, brake assemblies, and master cylinders
BC	X	Main gear wheels, wheel bearings, step and spring strut, tires, and fairings
BC		Main and nose gear wheel bearing lubrication: clean, repack, & lubricate
BC		Steering arm lubrication
BC		Torque link lubrication
BC		Nose gear strut servicing
BC		Nose gear shimmy dampener servicing
BC		Nose gear wheels, wheel bearings, strut, steering system, shimmy dampener, tire, fairing, and torque links
BC		Tires
BC		Parking brake and toe brake operational check

Airframe

Insp	Des	Item
BC		Aircraft exterior
BC		Aircraft structure
BC		Windows, windshield, and doors
BC		Seats, stops, seat rails, upholstery, structure, and seat mounting
BC		Safety belts and attaching brackets
BC		Control "U" bearings, sprockets, pulleys, cables, chains, and turnbuckles
BC		Control lock, control wheel, and control "U" mechanism
BC		Instruments and markings
BC		Gryos central air filter: plug vacuum line when removing filter
BC		Magnetic compass compensation
BC		Instrument wiring, and plumbing
BC		Instrument panel, shockmounts, ground straps, cover, and decals and labeling
BC		Defrosting, heating, and ventilating systems, and controls
BC		Cabin upholstery, trim, sun visors, and ash-trays
BC		Area beneath floor, lines, hoses, wires, and control cables
BC		Lights, switches, circuit breakers, fuses, and spare fuses
BL	X	Exterior lights
BC		Pitot and static systems
BC		Stall warning system
BC		Radios and radio controls
BC		Radio antennas
BC		Avionics and flight instruments
BC		Antennas and cables
BC		Battery, battery box, and battery cables

Cessna 172R 100-hour inspection checklist

Figure 17-23. (4 of 9)

Insp	Des	Item
BC		Battery electrolyte level: only use distilled water to maintain electrolyte level
BC	X	Emergency locator transmitter (ELT): attachment, test, & expiration date – replace if expired

Control System

In addition to the items listed below, always check for correct direction of movement, correct travel, and correct cable tension.

Insp	Des	Item
BC		Cables, terminals, pulleys, pulley brackets, cable guards, turnbuckles, and fairleads
BC		Chains, terminals, sprockets, and chain guards
BC		Trim control wheels, indicators, actuator, and bungee
BC		Travel stops
BC		All decals and labeling
BC		Flap control switch, flap rollers and tracks, flap position pointer and linkage, and flap electric motor and transmission
BC		Flap actuator jack screw threads
BC		Elevator and trim tab hinges, tips and control rods
BC		Elevator trim tab actuator lubrication and tab free-play inspection
BC		Rudder pedal assemblies and linkage
BC		Skin and structure of control surfaces and trim tabs
BC		Balance weight attachment

Post inspection engine run-up

Insp	Des	Item
BC		Engine temperatures and pressures
BC		Static RPM
BC		Magneto drop
BC		Engine response to changes in power
BC		Unusual engine noises
BC		Fuel selector valve: operate engine on each tank position and OFF position long enough to ensure proper selector valve function.
BC		Idling speed and mixture; proper idle cut-off
BC		Alternator and ammeter operation
BC		Suction gage:
BC		Fuel flow indicator
BC		After shutdown, check for fuel and oil leaks.

Clean up

Insp	Des	Item
BC		Reinstall all inspection panels and cowlings
BC		Wash exterior
BC		Clean windows and windshield
BC		Clean and vacuum interior
BC		Straighten seat belts

Final paperwork

Insp	Des	Item
BC		Update AD compliance list
BC		Complete work-order: regarding parts, supplies, and final labor figures
BC		Produce log book entries: airframe, engine, and propeller logs
BC		Produce list of discrepancies, if applicable

Cessna 172R 100-hour inspection checklist

Figure 17-23. (5 of 9)

Aircraft Airworthiness Inspection

100-hour inspection – 172R

#	Discrepancy	Action Taken		
1	Propeller shaft seal leaking	Replaced propeller shaft seal P/N 2846A	Removed S/N 3484-12-A Installed S/N QB146-8	MECH BT	INSP	DATE 1/18/00
2	#2 propeller blade S/N B3841L 1/4 inch nick found on leading edge 6 inches from base	Check McCauley specifications. Blended #2 propeller to McCauley specs.	Removed S/N Installed S/N	MECH BT	INSP	DATE 1/18/00
3	Right hand fuel tank placard missing.	Replaced fuel placard on right hand tank.	Removed S/N Installed S/N	MECH BT	INSP	DATE 1/18/00
4	Left main tire worn with threads showing on outboard sidewall.	Replaced left main tire	Removed S/N 4279U00601 Installed S/N 8932762	MECH BT	INSP	DATE 1/19/00
5	Left strobe light inop.	Replaced left strobe light.	Removed S/N Installed S/N	MECH BT	INSP	DATE 1/19/00
6	ELT battery expired	Replaced ELT battery	Removed S/N B89-A-123 Installed S/N 3924-Q587	MECH BT	INSP	DATE 1/19/00

Figure 17-23. (6 of 9)

AIRWORTHINESS DIRECTIVE COMPLIANCE RECORD

Aircraft Make **Cessna**   Model **172R**   S/N **17280044**   Registration No. **N00XYZ**

AD #	Effectivity Date	Subject	Date & hours @ compliance	Method of Compliance	One time	Recurring	Next comp. @ hrs/date	Authorized signature, cert. type, & number
96-09-10	7/15/96	Oil pumps	1/15/97, 128.7	AD 96-09-10 not applicable to IO-360-L2A engines. No further action required.	X			Bob Collins A+P 123456789
96-23-03	12/17/96	High pressure fuel pump	1/15/97, 128.7	Previously complied with in accordance with paragraph 2 and service bulletin no. 525A. Mfg. date code does not meet applicable codes listed in AD. No further action required.	X			Bob Collins A+P 123456789
97-01-03	1/21/97	Piston pins	2/2/97, 140.2	AD 97-01-03 not applicable to N00XYZ due to serial number range and model affectivity. No further action required.	X			Bob Collins A+P 123456789
97-12-06	7/15/97	Gascolator, tailpipe, and cowling area	1/20/00, 1140.8	Complied with by modifying cowling in accordance with paragraph (b) and service bulletin no. SB97-2861. No further recurring action required.	X			Bob Collins A+P 123456789
97-15-11	8/12/97	Piston pins	1/20/00, 1140.8	AD 97-15-11 not applicable to IO-360-L2A engines. No further action required.	X			Bob Collins A+P 123456789
98-01-01	2/2/98	Static air source valve	1/5/99, 690.5	Previously complied with in accordance with paragraph (b)(2)(i). No further action required.	X			Bob Collins A+P 123456789
98-13-41	8/18/98	Aileron control cable	7/20/98, 503.2	Previously complied with in accordance with paragraph (2) and service bulletin no. 98-27-02. No further action required.	X			Bob Collins A+P 123456789
98-17-11	8/28/98	Crankshafts	10/28/98, 593.1	AD 98-17-11 not applicable to IO-360-L2A engines. No further action required.	X			Bob Collins A+P 123456789
98-18-12	10/19/98	Torque check inspection	2/20/99, 768.5	Previously complied with in accordance with paragraph (3)(ii). No further recurring action required.	X			Bob Collins A+P 123456789
98-25-03	12/18/98	Prevent loss of aileron control caused by a damaged or frayed control cable.	1/5/99, 690.5	98-25-03 not applicable to N00XYZ due to serial number range and model affectivity. No further action required	X			Bob Collins A+P 123456789

Figure 17-23. (7 of 9)

Airframe maintenance log entry

January 18, 2000   Total time: 1140.8 hours

Installed fuel placard next to right hand fuel filler opening in accordance with Cessna 172R service manual. _[signature]_ A&P. No. 987654321
Brian Thomas

Propeller maintenance log entry

January 18, 2000   Total time: 1140.8 hours; Propeller TSO: 0 hours

Removed propeller-shaft seal and replaced with seal P/N 12D4901 in accordance Cessna service manual section 11-4 through 11-5. Blended 1/16th inch nick located on the leading edge 6 inches from the base of the #2 blade in accordance with McCauley overhaul manual section 6-11. Post repair dimensional inspection performed in accordance with McCauley overhaul manual section 6-13 and found to be within specifications. _[signature]_ A&P. No. 987654321
Brian Thomas

Airframe maintenance log entry

January 19, 2000   Total time: 1140.8 hours

Removed emergency locator transmitter battery S/N B89-A-123 and replaced with emergency locator transmitter battery S/N 3924-Q587. ELT operational check acceptable. New ELT battery expires January 20, 2002. Removed left main tire (S/N 4279N00601). Balanced and installed left main tire (Michelin 6.00 x 6, 4 ply tire, S/N 8932TG22). Replaced left strobe light with P/N T2345; operational check good. All work performed in accordance with a Cessna 172R service manual.
[signature] A&P. No. 987654321
Brian Thomas

Airframe maintenance log entry

January 20, 2000   Total time: 1140.8 hours

Complied with AD 97-12-06 by modifying engine cowling in accordance with paragraph (b) of AD 97-12-06 and Cessna service bulletin no. SB97-2861. ---------------------------END--------------------
[signature] A&P. No. 987654321
Brian Thomas

Figure 17-23. (8 of 9)

Airframe 100-hr log entry

January 20, 2000   Aircraft Total time: 1140.8 hours

Performed 100-hour inspection In accordance with FAR part 43 appendix D and Cessna 172 service manual, section 2-6 through 2-12. Airworthiness Directive compliance may be found in aircraft records. I certify this aircraft has been inspected in accordance with a 100-hour inspection and was determined to be in airworthy condition. *Beth Collins* A&P. No. 123456789
Beth Collins

Engine 100-hr log entry

January 20, 2000   Aircraft Total time: 1140.8 hours, Engine TSMO: 0 hours

Performed 100-hour inspection in accordance with FAR part 43 appendix D and Cessna 172 service manual, section 2-6 through 2-12. Airworthiness Directive compliance may be found in aircraft records. Compression test results: #1-78/80, #2-79/80, #3- 79/80, #4- 80/80, #5- 79/80, #6- 79/80. I certify this engine has been inspected in accordance with a 100-hour inspection and was determined to be in airworthy condition. *Beth Collins* A&P. No. 123456789
Beth Collins

Propeller 100-hr log entry

January 20, 2000   Aircraft Total time: 1140.8 hours, Propeller TSO: 0 hours

Performed 100-hour inspection in accordance with FAR part 43 appendix D and Cessna 172 service manual, section 2-6 through 2-12. Airworthiness Directive compliance may be found in aircraft records. I certify this propeller has been inspected in accordance with a 100-hour inspection and was determined to be in airworthy condition. *Beth Collins* A&P. No. 123456789
Beth Collins

Figure 17-23. (9 of 9)

SECTION C

AIRCRAFT MAINTENANCE RECORDS

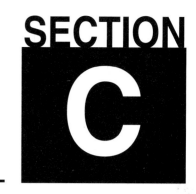

INTRODUCTION

Aircraft maintenance records provide evidence that the aircraft conforms to its airworthiness requirements, therefore, incomplete or missing records may render the aircraft unairworthy. Aviation maintenance technicians are required to record maintenance entries and aircraft owners are required to maintain them.

To keep the maintenance history of the aircraft, engines, propeller, components, and appliances clear and easy to research, maintenance record entries and inspection entries should be separated. Maintenance and inspection records document different events altogether. Individual FARs outline the requirements of maintenance and inspection record entries; Part 43.9 outlines maintenance entry requirements and Part 43.11 outlines inspection entry content. According to FAR Part 43.9, inspection events are specifically excluded from the required maintenance record entries; again, reinforcing the idea that maintenance events and inspection events need separate maintenance log entries.

MAINTENANCE RECORD FORM AND CONTENT

Except for Air Carrier and some Air Charter operators, technicians who maintain, perform preventive maintenance, rebuild, or alter an aircraft, airframe, aircraft engine, propeller, appliance, or component are required to make an entry in the maintenance record containing the following:

- **A description of the work performed or reference to FAA acceptable data.** The description should describe the work performed so that a person unfamiliar with the work may understand what was done, along with the methods and procedures used in performing it. When the work becomes extensive, it could result in a very large record. To prevent this, the rule permits reference to technical data that is acceptable to the FAA in lieu of making the detailed entry. Manufacturer's manuals, service bulletins, service letters, work orders, airworthiness directives, advisory circulars, and other acceptable data that accurately describe what was done may be referenced.

- **The completion date of the work performed.** Normally, this is the date the work was completed. However, the dates may differ when work is accomplished by one person and approved for return to service by another.

- **The name of the person performing the work if it is someone other than the person approving the return to service.**

- **The signature, certificate number, and type of certificate held by the person approving the work for return to service.** Two signatures may appear in the case of one person performing the work and another returning the aircraft to service, however, a single entry is acceptable.

As discussed earlier, the FARs require the maintenance technician to produce maintenance records that contain specific information. The owner, however, is responsible for maintaining records that contain additional information. According to FAR Part 91.417, owners must maintain the following information:

- **The total time-in-service of the airframe, engines, propellers, and each rotor.** Time in service, with respect to maintenance time records, is defined as the time from the moment an aircraft leaves the surface of the earth until it touches down at the next point of landing. Part 43.9 does not require time in service to be part of maintenance record entries. However, Part 43.11 requires time-in-service to be recorded in the inspection entries under Part 91 and Part 125.

- **The status of life-limited parts of each airframe, engine, propeller, rotor, and appliance.** If the total time of the aircraft and the time-in-service of a life-limited part are both recorded in a maintenance entry, then the normal record of time-in-service automatically meets this requirement.

- **The time since the last overhaul of all items installed on the aircraft which are required to be overhauled on a specified time basis.**

Again, if the total time of the aircraft and the time since the last overhaul are both recorded in a maintenance entry, then the normal record of time-in-service automatically meets this requirement.

- The current inspection status of the aircraft, including the time since the last inspection that was required by the inspection program under which the aircraft and its appliances are maintained.
- The status of applicable airworthiness directives (AD) including the method of compliance, the AD number, revision date, whether or not the AD involves recurring action, and, if applicable, the time and date when the next action is required.
- Copies of FAA Form 337 for each major alteration to the airframe and currently installed engines, rotors, propellers, and appliances.

The list of information that the owner must maintain varies from the list of information that the maintenance technician must record. Although the technician is not required to record the above listed information, thorough technicians include it in the maintenance logbook entries. Figure 17-24 and figure 17-25 illustrate typical maintenance record entries. [Figure 17-24] [Figure 17-25]

July 7, 2001   2345 TT; 2287 Tach time
Removed vacuum pump S/N AP1234 and replaced with vacuum pump P/N 4563, S/N 234-53. All work performed in accordance with Cessna service manual, page 6-3.
Operational check ok. ---------END---------------------
                       A&P 2345678
Joe S. Brown

Figure 17-24. Sample entry—maintenance record entry regarding the replacement of a vacuum pump and entered in the airframe log.

January 18, 2000   Total time 1245.7
Removed emergency locator transmitter (ELT) battery S/N 234-345Q and replaced with emergency locator transmitter battery P/N TL342, S/N 34AQ456 in accordance with Piper service manual page C23-24. Functional check good. ELT replacement due on January 18, 2002.   A&P no. 23456766
                John D. Brown

Figure 17-25. Sample entry—maintenance record entry regarding the ELT battery replacement recorded in the airframe log.

Aircraft owners are not required to keep separate logbooks for the airframe, engines, propellers, or appliances; however, most do. Most owners, who operate under Part 91 rules, maintain airframe, engine, and propeller logs. This practice helps in the research and tracking of the aircraft history, time-limited items, inspection times, airworthiness directive compliance, etc. The maintenance technician must know where to record specific types of maintenance and inspection information. For example, an engine oil and filter change would be recorded as a maintenance entry in the engine logbook. However, the repair of the exhaust system would be recorded as a maintenance entry in the airframe logbook.

INSPECTION RECORD FORM AND CONTENT

Before any inspection is considered complete, the inspection record entry must be recorded in the aircraft's maintenance records. The inspection record requirements of FAR Part 43.11 apply to the annual, 100-hour, and progressive inspections under Part 91. FAR Part 43.11 also applies to inspection programs under Part 125, approved aircraft inspection programs (AAIP) under Part 135 and the 100-hour and annual inspections under Part 135.411. Inspections performed on transport category aircraft require record entries outlined in FAR Part 121.709.

According to FAR Part 43.11, the person approving or disapproving for return-to-service the aircraft, or any item after any inspection, is required to make an entry in the maintenance record containing the following information:

- The type of inspection and a brief description of the extent of the inspection.
- The date the inspection was completed.
- The aircraft total time-in-service.
- Certification statement.
- The signature, the certificate number, and the type of certificate held by the person approving or disapproving the aircraft for return to service.
- If an inspection is conducted under an program that is allowed in Part 91, 123, 125, or 135, such as Progressive or Approved Aircraft Inspection Programs (AAIP), then the logbook entry must identify the inspection program, the part of the inspection program that was accomplished, and also contain a statement that the inspection was performed in accordance with the procedures for that particular program.

Aircraft Airworthiness Inspection

In addition, if the person performing any inspection that is required by Part 91, 125, or 135 should find the aircraft unairworthy, the inspector must provide the owner with a signed and dated list of discrepancies. When a discrepancy list is provided to an owner, it basically means that, with the exception of the listed discrepancies, the aircraft inspected is airworthy. When an inspection is terminated before it is completed, the maintenance record must clearly indicate that the inspection was discontinued. Although it is no longer required to forward a copy of the discrepancy list to the local flight standards district office (FSDO), it becomes part of the maintenance record and the owner is responsible for maintaining it accordingly. The inspection entry must reference the discrepancy list if one is provided to the owner.

Many times, discrepancies that are found during inspections are repaired and signed off as the inspection progresses. In this circumstance, a list of discrepancies is not needed and the inspection may be signed off as being airworthy. In the case of a 100-hour inspection, while the certified mechanic inspects the aircraft, another certified mechanic may repair and sign off any discrepancies prior to the completion of the inspection. In this case, maintenance record entries are produced regarding the repairs and separate entries are produced documenting the inspection. Keeping the maintenance entries and inspection entries separate helps keep the aircraft logs clear and easy to follow.

If the owner maintains separate records for the airframe, engines, and propellers, the entry for the 100-hour inspection is entered in each, while the annual inspection is only required to be entered into the airframe record.

ANNUAL INSPECTION ENTRIES

An annual inspection may be signed off in the maintenance records as airworthy or unairworthy depending on the condition of the aircraft. Whether or not the aircraft owner keeps separate logs, FARs stipulate the annual inspection need only be recorded in the airframe logbook. However, it is good practice to enter an annual inspection record in all maintenance logbooks making maintenance information easier to research and compile. Figure 17-26 illustrates a typical annual inspection entry regarding an airworthy aircraft. [Figure 17-26]

Figure 17-27 represents an unairworthy annual inspection entry. In addition, a list of discrepancies outlining the unairworthy items found during the inspection must be provided to the aircraft owner and referenced in the inspection record. [Figure 17-27] [Figure 17-28]

September 27, 2000   Total time: 2780 hours

Performed an annual inspection in accordance with FAR part 43 appendix D and manufacturer's maintenance manual, section 2-1 through 2-11. Airworthiness Directive compliance may be found in aircraft records. I certify that this aircraft has been inspected in accordance with an annual inspection and was determined to be in airworthy condition.

Joe Smith I.A. No. 13453234
Joe L. Smith

Figure 17-26. Sample entry—airworthy annual inspection.

May 12, 2000   Total time: 3245 hours

Performed an annual inspection in accordance with FAR part 43 appendix D and Cessna 172G service manual, section 2-1 through 2-11. Airworthiness Directive compliance may be found in aircraft records. I certify that this aircraft has been inspected in accordance with an annual inspection and a list of discrepancies and unairworhty items dated May 12, 2000 has been provided to the aircraft owner.

Joe Smith I.A. No. 13453234
Joe L. Smith

Figure 17-27. Sample entry—unairworthy annual inspection entry.

May 12, 2000

Ms. Rhonda Jones
1234 W. 1st St.
Denver, CO 23456

Re: Unairworthy items found during annual inspection

Dear Ms. Jones,

This letter is to certify that on April 10, 2000, I completed a 100-hour inspection on your Cessna 172F, N123BC, and found it to be in an unairworthy condition for the following reasons:
  1) Compression in cylinder #4 read 40/80, which is below the manufacturer's recommended limits.
  2) A 1/8th inch nick was found on the leading edge of the propeller.
Your aircraft will be considered airworthy when the above listed discrepancies are corrected and approved for return to service by a person authorized in FAR part 43.

Thank you,
Joe Smith
Joe L. Smith, A&P no. 13453234
Aviation Services Inc.
1234 2nd Ave.
Denver, CO 23456

Figure 17-28. Typical letter to aircraft owner itemizing the discrepancies found during an annual inspection.

100-HOUR INSPECTION ENTRIES

A 100-hour inspection may also be signed off in the maintenance records as airworthy or unairworthy. The required inspection entry items are listed in FAR Part 43.11. If the aircraft owner maintains separate logs, the 100-hour inspection must be recorded in each applicable maintenance log unlike the annual, which is only required to be recorded in the airframe logbook. Figure 17-29 and Figure 17-30 illustrate airworthy and unairworthy 100-hour inspection entries. [Figure 17-29] [Figure 17-30]

```
June 30, 2000          Total time: 1459 hours
Performed 100-hour inspection in accordance
with FAR part 43 appendix D and
manufacturer's maintenance manual, section B1
through B10. Airworthiness Directive compliance
may be found in aircraft records. I certify that
this aircraft has been inspected in accordance
with a 100-hour inspection and was determined
to be in airworthy condition.
        Linda Brown       A&P. No. 1347890
Linda P. Brown
```

Figure 17-29. Sample entry—airworthy 100-hr inspection record entry.

```
April 10, 2000         Total time: 1002 hours
Performed 100-hour inspection in accordance with
FAR part 43 appendix D and manufacturer's
maintenance manual, section B1 through B10.
Airworthiness Directive compliance may be found
in aircraft records. I certify that this aircraft has
been inspected in accordance with a 100-hour
inspection and a list of discrepancies and
unairworthy items dated April 10, 2000 has been
provided for the aircraft owner.
        Linda Brown       A&P. No. 1347890
Linda P. Brown
```

Figure 17-30. Sample entry—unairworthy 100-hr inspection record entry.

PROGRESSIVE INSPECTION & APPROVED AIRCRAFT INSPECTION PROGRAM (AAIP) ENTRIES

FAR Part 43.11(a)(7), which refers to inspection programs such as AAIPs and progressive inspections, now requires a more specific statement than previously required. The entry must identify the inspection program used, identify the portion or segment of the inspection program accomplished, and contain a statement that the inspection was performed in accordance with instructions and procedures for that particular program. Samples of a progressive inspection entry and an AAIP inspection entry follow. [Figure 17-31] [Figure 17-32]

```
January 5, 2000         Total time: 540
Performed progressive inspection in accordance
with FAR part 43 appendix D and Cessna 421G
service manual page 2-1 through 2-10.
Airworthiness directive compliance may be
found in aircraft records. I certify that in
accordance with a progressive inspection
program, a routine inspection of the left wing
and a detailed inspection of the right hand
engine were performed and the aircraft is
approved for return to service.
        John B. Boone      A&P no. 1589432
John B. Boone
```

Figure 17-31. Sample entry—Progressive inspection entry.

```
October 6, 2000         Total time: 4567 hours
Performed a phase 3 inspection in accordance
with the StarJet Charter approved inspection
manual, section 3-1 to 3-16. Airworthiness
Directive compliance may be found in aircraft
records. I certify that this aircraft has been
inspected in accordance with an Approved
Aircraft Inspection Program and was determined
to be in airworthy condition.
        James Blue         A&P. No. 1347890
James T. Blue
```

Figure 17-32. Sample entry—Approved Aircraft Inspection Program (AAIP) entry.

AIRWORTHINESS DIRECTIVE COMPLIANCE ENTRIES

Although it is the owner's primary responsibility to maintain their aircraft in an airworthy condition, including airworthiness directive compliance, maintenance professionals may also have direct responsibility for AD compliance. When 100-hour, annual, or progressive inspections are performed on an aircraft, the technician performing the inspection is required to determine that all applicable airworthiness requirements are met, including the compliance of any applicable airworthiness directives.

When airworthiness directives are accomplished, maintenance personnel are required to include the completion date, name of the person complying with the AD, signature, certificate number, and kind of certificate held by the person approving the work, and the current status of the applicable "AD" in the maintenance record entry. The owner is required by FAR Part 91.417 to maintain AD compliance information including the current status of the AD along with the method of compliance, the AD number, and revision date and, if the AD is recurring, the time and date when the next action is required.

Aircraft Airworthiness Inspection

The recording of the airworthiness directive compliance may either be recorded in the maintenance logbook and/or kept as a separate listing in the maintenance records in the form of a running AD log.

> June 30, 1987   Aircraft total time 2345.5
>
> Performed visual inspection of the flap sector upper mounting brackets in compliance with AD 99-22-05 paragraph A and Gates Learjet airplane modification kit 55-86-2. No cracks found upon inspection. Replacement of flap upper mounting brackets due at 2398.5TT in accordance with paragraph (A)(2).
>
> *Jane Brightz* A&P 349024434
> Jane P. Brightman

Figure 17-33. Sample entry—Airworthiness directive (AD) compliance logbook entry regarding an AD that required an inspection and subsequent repair within 100 flight hours of the effective date of the AD and recorded in the airframe logbook.

Figure 17-33 illustrates an AD listing format that keeps track of all ADs complied with on a specific aircraft. [Figure 17-33] [Figure 17-34] [Figure 17-35]

> May 2, 2000   1449.2 Total time; 151.3 TSMO
>
> Performed visual inspection of the oil adapter locking nut installation in accordance with AD 96-12-22 paragraph (2)(a)(1). Correct torque noted and no oil leakage found upon inspection. Next inspection due at 1539.2 hours or when the engine oil filter is removed, whichever occurs first.
>
> *Porter B* I.A. 123456789
> Porter Berryman

Figure 17-34. Sample entry—AD compliance maintenance entry regarding a recurring AD entered in the engine logbook.

| AIRWORTHINESS DIRECTIVE RECORD ||||||||||
| AIRCRAFT MAKE: Cessna | | MODEL: 4146 | | | SERIAL NUMBER: 1895A | | | N NUMBER: 1386V | |
AD NUMBER	DATE OF REVISION	SUBJECT	DATE OF COMPLIANCE	TOTAL TIME AT COMPLIANCE	METHOD OF COMPLIANCE	ONE TIME	RECURRING	NEXT DUE DATE/TIME	NAME SIGNATURE NUMBER
76-03-05	2-17-90	Bracket Installation	6-30-92	826 HRS.	Install bracket i/a/w Para. b	X			*John M. Smith* John Smith A&P 00110000
76-05-04	7-12-90	Stab. Attach. Inspection	6-26-92	1248 HRS.	Inspected Stab. Attach i/a/w Para a & b		X	2248 HRS.	*Robert B. Johnson* Robert Johnson A&P 110011000

Figure 17-35. Sample airworthiness directive (AD) record listing.

INDEX

A

ABSOLUTE ALTITUDE 11-7
AC GENERATORS 7-2
ACCELOROMETER 11-15
    G-meter 11-15
ACCUMULATOR 8-12, 8-36
    preload pressure 8-36
    types of 8-36
ACTUATORS 8-39
    linear 8-39
    rotary 8-41
AILERONS 1-12
AIR CONDITIONING SYSTEMS 14-30
    air-cycle 14-30
    charging 14-46
    distribution 14-24
    evaporator 14-37
    purging 14-45
    receiver-dryer 14-36
    refrigerant 14-34
    service equipment 14-41
    servicing 14-41, 14-44
    system evacuation 14-46
    troubleshooting 14-25
    vapor-cycle 14-33
AIRFOIL 1-4
    downwash 1-6
AIRLOC FASTENER 2-50
AIRSPEED 11-9
    indicators 11-9
    types of 11-9
ALLOYS 2-6
    aluminum 2-6
ALTERNATING CURRENT (AC) 7-44
    automated AC power system 7-51
ALTERNATOR 7-22
    AC alternators 7-22
    circuits 7-38
    components of 7-19
    control unit (ACU) 7-21
    DC alternators 7-19
    drive unit 7-50
    frequency 7-25
    ratings 7-25
    rectifiers 7-20
    revolving-armature type 7-23
    revolving-field type 7-23
    service and maintenance 7-21
    types 7-19, 7-23
ALTIMETER 11-6, 12-26
    encoding 11-8
    radar 12-26
    radio 11-8
    servo 11-8
    tests 11-8
ALUMINUM HYDROXIDE 2-14
AMERICAN WIRE GAGE (AWG) 7-54
ANEROID BELLOWS 11-4
ANGLE OF INCIDENCE 1-39
ANGLE-OF-ATTACK INDICATOR 10-10
ANODIC 2-13
ANTENNAS 12-7
    ADF 12-60
    DME/transponder 12-31
    ELT 12-31
    glideslope 12-60
    groundplane 12-57
    Hertz 12-7
    HF comm 12-31
    installation 12-56
    interference 12-58
    loop 12-8
    Marconi 12-7
    marker-beacon 12-30
    polarization 12-8
    radiotelephone 12-32
    satellite communications 12-31
    TCAS 12-31
    VHF comm 12-31, 12-59
    VOR 12-28, 12-59
    wire routing 12-7
ANTI-ICING 13-3
ANTI-SKID BRAKE CONTROL SYSTEMS
    9-38, 10-2
    components 10-3
    control unit 10-5
    maintenance 10-7
    testing 10-6
    warning lights 10-6
ANTI-TORQUE ROTOR 1-53
ARMATURE REACTION 7-10
ARTIFICIAL FEEL 1-35
AUTOMATIC FLIGHT CONTROL SYSTEM
    11-20
AUTOMATIC TENSION ADJUSTERS 1-46
AUTOPILOTS 12-38
    flight control panel 12-44
    flight director 12-45
    maintenance 12-46
    operation 12-35
    servo 12-38
AUTOSYN® SYSTEMS 10-11
AUXILIARY POWER UNIT (APU) 7-48
AVIATION SNIPS 2-20
AVIONICS 7-40, 12-2
    amplifiers 12-6
    audio control panels 12-13
    bonding and shielding 12-50
    carrier wave 12-3
    cockpit voice recorders and flight data
        recorders 12-25
    communications radios 12-9
    electromagnetic waves 12-2
    filters 12-7
    frequency 12-2
    ground, sky, and space waves 12-4
    installation 12-48, 12-55
    maintenance 12-48
    modulation 12-3
    oscillator 12-6
    power circuit 7-40
    radio waves 12-2
    receivers 12-8
    speakers and microphones 12-9
    static dischargers 12-53
    Stormscope 12-27
    TCAS 12-28
    transmitters 12-6
    transponders 12-21
    tuning circuits 12-8
AXES 1-20
    lateral 1-20
    longitudinal 1-20
    vertical 1-20

B

BACKUP RINGS 8-43
BALANCE WEIGHTS 9-10
BAND SAW 2-23
BATTERY 7-26
    capacity 7-27
    circuits 7-35
    deep-cycling 7-33
    installlation 7-30
    lead-acid 7-26
    nickel-cadmium 7-30
    ratings 7-27
    servicing and charging 7-28
    troubleshooting 7-34
BEARING LUBRICATION 9-9
BEARING STRENGTH 2-5
BEND ALLOWANCE 2-73
BEND RADIUS 2-70
BEND TANGENT LINES 2-71
BIAS CUT 5-6, 5-8

BIMETALLIC THERMOSWITCH 16-4
BIPLANE COMPONENTS 1-49
BLEED AIR HEATERS 14-30
BONDING AND SHIELDING 12-50
BOOST PUMP 15-24
BOUNDARY LAYER 1-13
BOURDON TUBE 11-4
BOX BRAKE 2-30
BRAKES 9-18
    actuators 9-25
    aircraft 9-18
    anti-skid 9-38
    bleeding 9-31
    boosted 9-27
    carbon 9-20
    construction 9-20
    damage 9-37
    deboosters 9-28
    expander tube 9-24
    inspection 9-30
    linings 9-18, 9-36
    malfunctions 9-37
    master cylinders 9-25
    multiple-disc 9-19
    power 9-27
    seals 9-35
    segmented rotor-disc 9-20
    servicing 9-30
    single-disc 9-18
BUCKING BARS 2-64
BUILT IN TEST EQUIPMENT (BITE) 7-51
BYPASS VALVE 14-31

C

CABLE 1-41, 7-64, 12-55
    coaxial 7-64, 12-55
    construction 1-41
    control 1-41
    inspection 1-43
    installation 1-44
    sleeves 1-42
    tension 1-45
    terminals 1-43
CAMLOCK 2-50
CANARD 1-33
CANTILEVER 1-9, 1-39
CAPACITANCE FUEL QUANTITY
  SYSTEMS 11-34
CARBON MONOXIDE DETECTORS 16-12
CARBON-PILE 7-14
CARBURIZING 4-15
CELLULOSE ACETATE BUTYRATE 5-3
CENTER OF LIFT 1-6
CENTER SECTION 1-49
CHARGING 14-46
CHIP CHASERS 2-35
CIRCUIT BREAKERS 7-70, 12-49
CIRCULAR MIL 7-54
CLECO FASTENERS 2-33

COANDA EFFECT 1-66
COAXIAL CABLE 7-64
COLD WORKING 2-10
COMBINATION SQUARE 2-16
COMBUSTION HEATING 14-28
COMING-IN SPEED 7-12
COMMUTATOR 7-4
COMPASS SWING 11-25, 11-48
COMPOSITES 3-39
    acoustic emission testing 3-41
    adhesives 3-28
    aluminum alloy-faced honeycomb 3-61
    bias 3-25
    compression molding 3-37
    core materials 3-30
    curing 3-54
    cutting fabrics 3-42
    damage assessment 3-45
    delamination repair 3-59
    drilling 3-42
    dye penetrant inspection 3-41
    electrical bonding 3-39
    elements of 3-22
    fabric 3-24, 3-52
    fiber orientation 3-22
    fiberglass lay-up 3-39
    fibers/thixotropic agents 3-30
    filament winding 3-38
    finishes/paint 3-39
    Honeycomb core repair 3-61
    hydraulic press cutting 3-45
    impregnating raw fabric 3-52
    inspection 3-39
    laminate repair 3-59
    laser cutting 3-45
    machining 3-41
    maintenance entries 3-62
    manufacturing processes 3-37
    materials preparation 3-47
    matrix systems 3-26
    pot life 3-27
    pre-impregnated materials 3-28
    pressure application methods 3-53
    repair 3-45
    resins 3-27
    safety 3-35
    sanding 3-43
    sandwich construction 3-32
    sandwich structure repairs 3-61
    selvage edge 3-25
    shelf life 3-35
    structures 3-22
    surface preparation 3-47
    tap test 3-40
    vacuum bagging 3-38, 3-56
    warp 3-24
    warp compass 3-52
    water removal 3-51
    weft/fill 3-25
    wet lay-up 3-38
COMPOUND CURVES 2-78

COMPRESSION RIVETING 2-61
COMPRESSOR 14-37
CONDENSER 14-39
CONING 1-55
CONSTANT-SPEED DRIVE UNIT (CSD) 7-25
CONTINUOUS-LOOP DETECTOR 16-7
CONTROL LOCKS 1-11
CONTROL SURFACE 1-10
CONTROL SYSTEMS 1-24
CONTROL YOKE 1-26
COPPER LOSSES 7-84
CORIOLIS EFFECT 1-56
CORNICE BRAKE 2-29
CORROSION 2-13, 9-10
    galvanic 2-13
    prevention 2-13
    wheel 9-10
COUNTERSINKING 2-58
CRIMPING 7-59
CROWN FLUSH RIVET 2-37
CUMULATIVE-COMPOUNDED 7-10
CURRENT LIMITER 7-13, 7-70
CUTTING TOOLS 2-19
    sheetmetal 2-19
CYLINDERS 14-8
    high-pressure 14-8
    low pressure 14-8

D

DC GENERATORS 7-2
DE-ICING SYSTEMS 13-8
    de-icer boot 13-8, 13-11
    ground de-icing 13-14
DEBOOSTERS 9-28
DEBURRING 2-57
    tools 2-21
DECALAGE 1-49
DEGREE OF COMPOUNDING 7-10
DETECTION SYSTEMS 16-4
DETONATION 15-3
DEVIATION 11-24
DIFFERENTIALLY-COMPOUNDED 7-10
DIHEDRAL ANGLE 1-39
DIMPLING 2-37, 2-59
DIRECTION-INDICATING INSTRUMENTS
  11-21
DISC SANDER 2-24
DIVERTER 7-10
DIVIDERS 2-17
DOPE 5-2, 6-2
DOUBLE-LOOP SYSTEM 16-5
DRILLS 2-24
    jigs 2-25
    motor 2-24
    press 2-26
    stops 2-26
    twist drills 2-27
DROOP 1-54
DROP HAMMER 2-31

Index

DRUM-TYPE ARMATURE 7-6
DRY-BULB TEMPERATURE 6-9
DZUS FASTENERS 2-48

E

EDDY CURRENT LOSSES 7-84
ELASTOMERIC BEARINGS 1-54
ELECTRIC HEATING 14-27
ELECTRICAL SYSTEM 7-1
    circuit breakers 7-70
    circuits 7-35
    losses 7-84
    relays and solenoids 7-68
    switches 7-66
ELECTRO STATIC DISCHARGE SENSITIVE (ESDS) 7-52
ELECTROCHEMICAL ACTION 2-13
ELECTRONIC ATTITUDE DIRECTOR INDICATOR (EADI) 11-37
ELECTRONIC FLIGHT INSTRUMENTS SYSTEMS (EFIS) 11-36
ELEVONS 1-12
EMERGENCY EXTENSION SYSTEMS 9-16
EMERGENCY LOCATOR TRANSMITTERS (ELT) 12-25
EMPENNAGE 1-2, 1-13
ENGINE INDICATION AND CREW ALERTING SYSTEM (EICAS) 10-13
ENGINE INDICATOR AND CREW ALERTING SYSTEM (EICAS) 11-41
ENGINE MOUNTS 1-19
ENGINE PRESSURE RATIO INDICATOR (EPR) 11-6
EUTECTIC SALT 16-7
EVAPORATOR 14-37
EXTERNAL POWER CIRCUIT 7-38

F

FABRIC COVERING 5-1
    approval criteria 5-3
    bias 5-6
    fabric grades 5-6
    fabric materials 5-6
    finishing 5-8
    inspection 5-14, 5-26
    repair 5-26
    selvage edge 5-6, 5-24
    sewed-in patch 5-28
    supplemental type certificates 5-4
    synthetic 5-7
    warp 5-6
    weft 5-6
FAIRLEADS 1-44
FARADAY CAGE EFFECT 6-17
FASTENERS 2-33
    cleco 2-33
    dzus 2-48
    structural 2-35
    wing nut 2-34
FENWAL SYSTEMS 16-4
FIELD COILS 7-6
FIELD SATURATION 7-9
FILES 2-20
FILLER 6-14
FILLER RODS 4-25
FINISHES 6-7
    enamel 6-9, 6-13
    polyurethane 6-4
    primers 6-7
    seam paste 6-13
    thinners and reducers 6-13
    tube oil 6-13
    wing walk compound 6-11
    wrinkle 6-11
    zinc chromate 6-7
FINISHING EQUIPMENT 6-15
FIRE SYSTEMS 16-1
    classes of 16-3
    detection 16-2
    fire detection/overheat systems 16-4
    fire-detection system inspection 16-13
    fire-extinguishing agents 16-16
    flame detectors 16-10
    high rate discharge systems 16-20
    inspection 16-13, 16-21
    Linberg system 16-8
    maintenance 16-13
    overheat/detection systems 16-4
    servicing 16-21
    smoke detectors 16-11
    troubleshooting 16-15
    zones 16-3
FLAPERONS 1-12
FLAPS 1-30
FLASH POINT 8-7
FLASHING THE FIELD 7-15
FLASHOFF 14-40
FLAT-COMPOUND 7-10
FLEXATIVE, PRIMER 6-4
FLEXTURES 1-54
FLIGHT CONTROLS 1-35
    artificial feel 1-35
FLIGHT DATA RECORDER (FDR) 12-25
FLIGHT DIRECTOR 12-45
FLIGHT MANAGEMENT SYSTEM (FMS) 12-39
FORMING BLOCK 2-79
FRANGIBLE DISK 16-20
FUEL SYSTEMS 11-35, 15-1
    avgas 15-2
    baffles 15-18
    contamination 15-37
    detonation 15-3
    dumping 15-17
    filters 15-27
    fittings 15-21
    flapper valves 15-18
    flowmeters 15-29
    fueling procedures 15-39
    gravity feed 15-10
    heaters 15-28
    identification 15-5
    injection system flowmeter 11-35
    instruments 11-35
    Jet A 15-7
    leak classification 15-35
    lines 15-21
    octane 15-5
    pumps 15-24
    repair 15-33
    safety 15-35
    single-point-fueling 15-14, 15-31
    sump 15-27
    systems 15-1
    tanks 15-18, 15-34
FUNGICIDAL PASTE 5-10
FUSELAGE 1-2
FUSES 7-70

G

GAS CYLINDERS 4-26
GENERATORS 7-2
    armature 7-6, 7-10
    brushes 7-7
    circuits 7-36
    commutator 7-6
    construction 7-4
    field frame 7-5
    inspection and repair 7-16
    instrumentation 7-51
    integrated drive generator (IDG) 7-25
    KVARS 7-50
    left-hand rule 7-2
    service and maintenance 7-14
    starter 7-10
    testing 7-18
    theory of operation 7-2
    types 7-8
    voltage regulation 7-12
GLUES, TYPES 3-6
GRAB TEST 5-13
GRAIN PATTERN 2-69
GREIGE 5-23
GROUNDPLANE 12-57
GROWLER 7-16
GUSSETS 1-8
GYRO, ATTITUDE 11-18
GYROPLANE 1-52
GYROSCOPIC INSTRUMENTS 11-16

H

HANGAR RASH 5-26
HEAT TREATMENT 2-8
    aging 2-8, 2-9
    annealing 2-9
    precipitation heat treatment 2-9

quenching 2-9
Solution Heat Treatment 2-8
HEATERS 14-27
HELICOPTERS 1-52
    autorotation 1-61
    Coanda effect 1-66
    direct rotor head tilt 1-62
    dual-rotor 1-53
    hovering flight 1-57
    lift 1-55
    NOTAR®/no tail rotor system 1-66
    power systems 1-70
    retreating blade stall 1-60
    single-rotor 1-52
    stabilizer systems 1-66
    swash plate control system 1-62
    throttle control 1-63
    torque 1-57
    translating tendency 1-57
    translational lift 1-60
    transmissions 1-71
    turboshaft engines 1-70
    vertical ascent and descent 1-58
    vibration 1-68
HELIX 1-52
HI-LITE FASTENERS 2-46
HI-LOK FASTENERS 2-45
HOLE FINDERS 2-35
HONEYCOMB, ALUMINUM ALLOY-FACED 2-13
HYDRAULIC FUSES 8-33
HYDROPRESS 2-31
HYDRULIC POWER SYSTEMS 8-11
    accumulator 8-12, 8-36
    actuator 8-39
    components of 8-16
    filter 8-19
    fluid 8-6
    large-aircraft 8-46
    open-center 8-15
    power pack 8-15
    pressure regulators 8-35
    pumps 8-22
    seals 8-42
    unloading valve 8-12
HYPOXIA 14-4
HYSTERESIS LOSSES 7-84

I

ICE DETECTION SYSTEMS 13-2
IN-SLIPSTREAM SPACING 5-20
INCIDENCE ANGLE 1-39
INDICATED ALTITUDE 11-7
INDICATING SYSTEMS 10-9, 11-33
    angle-of-attack 10-10
    engine Indication and crew alerting (EICAS) 10-13
    fuel pressure 11-35
    fuel quantity 11-33

    remote position 10-10
INDICATOR DISK 16-20
INSPECTION 9-8, 14-16, 17-1
    AD (airworthiness directive) compliance 17-26
    altimeter and static systems 17-17
    ATC transponder 17-17
    bearing 9-8
    brake 9-30
    examination phase 17-27
    friction damage 17-20
    fundamentals 17-20
    fusible plug 9-10
    guidelines 17-23
    minimum equipment list (MEL) 17-2
    overload damage 17-20
    pre-flight 17-2
    pre-inspection phase 17-24
    procedures 17-24
    records review 17-25
    required 17-2
    return-to-service phase 17-33
    routine 17-8
    service and repair phase 17-28
    service bulletin search 17-26
    special 17-16
    tire 9-43
    vibration damage 17-23
    visual 14-44
    weather damage 17-20
    wheel 9-7
    work/job cards 17-13
INSPECTION PROGRAMS 17-1
    100-hour 17-5
    air carrier 17-12
    annual 17-2
    approved aircraft inspection program (AAIP) 17-9, 17-16
    complete 17-5
    conditional 17-19
    conformity 17-9
    continuous airworthiness inspection program 17-8, 17-12
    letter check 17-13
    progressive 17-7
INSTRUMENTS 11-1
    accelerometers 11-15
    air temperature 11-42
    altimeter 11-6
    aneroid bellows 11-4
    Bourdon tube 11-4
    differential bellows 11-5
    direction-indicating 11-21
    electronic 11-35
    filters 11-29
    gyroscopic 11-16
    installation 11-43
    magnetic compass 11-22
    manifold pressure gauge 11-6
    pneumatic 11-26
    pressure-measuring 11-2

    range markings 11-45
    servicing 11-30
    temperature-measuring 11-12
    thermocouple 11-15
    vertical speed indicator 11-11
INTEGRATED DRIVE GENERATOR (IDG) 7-25
INTER-RIB BRACING 5-15
INTERPOLES 7-11
INVERTERS 7-44
IRREVERSIBLE CONTROL SYSTEM 1-35

J

JETCAL ANALYZER 16-13
JO-BOLTS 2-47
JOGGLING 2-82

K

KETTS SAW 2-19
KIDDE SYSTEM 16-8

L

LANDING GEAR 7-40, 9-1, 10-12
    alignment 9-12
    camber 9-13
    circuit 7-40
    configuration warning system 10-12
    ground locks 9-17
    indicators 9-17
    nosewheel centering 9-17
    retraction check 9-17
    retraction systems 9-14
    rigging 9-17
    safety switch 9-16
    servicing 9-5
    shock struts 9-4
    toe-in/toe-out 9-12
    torque links 9-12
    types 9-2
LAP WINDING 7-7
LAYOUT TOOLS 2-16
LEAK DETECTOR 14-43
LEFT-HAND RULE 7-2
LIFT 1-6, 1-55
LIGHTENING HOLES 2-81
LIGHTS 7-71
    anticollision 7-73
    exterior 7-71
    fluorescent 7-75
    incandescent 7-74
    interior 7-74
    maintenance and inspection 7-76
    position 7-72
    xenon 7-72
LINDBERG SYSTEM 16-8

LINE REPLACEABLE UNITS (LRU) 7-51
LOCK BOLTS 2-44
LONG-SHUNT CONNECTION 7-10

M

MACHMETER 11-11
MAGNESIUM ALLOYS 2-11
MAGNESYN® SYSTEMS 10-12
MAGNETIC COMPASS 11-22
MAGNETOMOTIVE FORCE 7-9
MAINTENANCE 17-12
    logbook entries 17-33
    work/job cards 17-13
MANIFOLD PRESSURE GUAGE 11-6
MANIFOLD SET 14-41
MASKING 6-25
MAULE TESTER 5-12
MECHANICAL LOSSES 7-84
MEDIA BLASTING 6-7
MERCERIZED 5-6
METAL; ALLOYS 2-6
MICRONIC FILTERS 8-20
MICROSHAVER 2-37
MINIMUM EQUIPMENT LIST (MEL) 17-2
MOLD LINE 2-70
MOLD POINT 2-71
MOLTEN PUDDLE 4-27
MONOCOQUE 1-3, 1-15, 2-2
MOTOR SLIP 7-86
MOTORING 7-13
MOTORS 7-85
    AC motors 7-85
    construction 7-79, 7-86
    DC motors 7-77, 7-85
    reversing direction 7-81
    right-hand rule 7-78
    theory 7-77
    torque 7-78
    types 7-82, 7-85

N

NACA RIVETING 2-68
NAP 5-22
NAVIGATIONAL SYSTEMS 12-13
    area navigation 12-17
    automatic direction finder (ADF) 12-13
    distance measuring equipment (DME) 12-16
    glideslope 12-24
    global positioning system (GPS) 12-20
    inertial navigation system (INS) 12-19
    instrument landing system (ILS) 12-23
    long range navigation (LORAN) 12-19
    marker beacons 12-24
    primary radar 12-21
    radar altimeter 12-26
    RNAV 12-17
    secondary radar 12-21
    transponders 12-21
    VOR navigation 12-14
NEUTRAL AXIS 2-70
NEUTRAL FLAME 4-28
NIBBLERS 2-20
NICOPRESS SLEEVES 1-42
NONTAUTENING DOPE 5-24
NOSE WHEEL STEERING SYSTEM 9-11

O

O-RING 8-43
OCCUPATIONAL SAFETY AND HEALTH ADMINISTRATION (OSHA) 6-15
OLYMPIC-LOK FASTENERS 2-41
ORLON 5-23
OVER-COMPOUNDED 7-10
OVERHEAT SYSTEMS 16-4
OXIDE FILM 2-14
OXYGEN SYSTEMS 14-6
    candles 14-14
    characteristics 14-6
    chemical (solid) 14-7
    inspection 14-18
    liquid 14-7, 14-12
    masks 14-11
    regulators 14-9
    servicing 14-16

P

PAINTING PROCESSES 6-5
    equipment 6-15
    finish, application 6-21
    finishing materials 6-7
    masking 6-25
    problems 6-24
    safety 6-15, 6-29
PAR-AL-KETONE 1-44
PARALLELING CIRCUIT 7-45
PASCAL'S LAW 8-2
PENDULOUS VANES 11-19
PHYSIOLOGY 14-2
PIG-TAIL 7-7
PITOT-STATIC SYSTEM 11-30
    static system testing 11-32
    tester 11-49
PLANE OF COMMUTATION 7-11
PLASTIC 3-63
    cementing 3-66
    drilling 3-65
    forming 3-63
    types 3-63
PLASTICIZERS 5-10
PNEUMATIC SYSTEMS 8-49, 11-26, 16-8
    backup system 8-55
    components 8-51, 8-56
    continuous-loop system 16-8
    filters 8-54
    high pressure systems 8-49
    instruments 11-26
    low-pressure systems 8-50
    maintenance 8-58
    medium-pressure systems 8-50
    moisture separator 8-54
    pumps 8-51
    vacuum pump systems 11-26
    valves 8-53
POLE SHOES 7-6
POLYPHASE 7-23
POLYURETHANE 6-9
POT LIFE 3-27, 6-11
PRESSURE 1-44, 8-36, 11-2
    absolute 11-3
    altitude 11-7
    atmospheric 11-4
    differential 11-4
    gauge 11-4
    reducer 8-36
    regulator 8-35
    relief valves 8-34
    seals 1-44
PRESSURIZATION SYSTEMS 14-19
    outflow valve 14-22
PROOF LOAD TEST 1-43
PULL TEST 5-13
PULLEYS 1-44
PUMPS, HYDRAULIC 8-23
PUNCH TESTER 5-12
PUNCHES, TYPES 2-17
PURGING 14-45
PUSH-PULL RODS 1-47
PYROLYTIC METHODS 6-5

R

R-12 14-35
RADAR, AIRBORNE WEATHER 12-26
RAIN CONTROL SYSTEMS 13-16
    chemical rain repellant 13-17
    pneumatic rain removal systems 13-18
RAM AIR DOOR 14-32
RATE-OF-CLIMB INDICATOR 11-11
RECEIVER-DRYER 14-36
RECIPROCATING SAWS 2-19
RECORDS 17-43
    work order 17-24
REFRIGERANT 14-34
REGISTRATION NUMBERS 6-26
    decals 6-29
    pressure-sensitive vinyl 6-27
    Regulators 6-15
REGULATOR 8-35
    pressure 8-35
REGULATORS 14-9
    continuous flow 14-9
    demand 14-9
    diluter demand 14-10
    pressure demand 14-10
REJUVENATOR 5-11

RELAYS 7-68
REMOTE INDICATING COMPASS 11-25
REMOTE POSITION INDICATING SYSTEMS 10-10
    Autosyn® 10-11
    Magnesyn® 10-11
RESPIRATORS 6-18
RETARDER 5-10
RETRACTION SYSTEMS 9-14
    large aircraft 9-16
    small aircraft 9-14
REVERSE-CURRENT CUTOUT 7-13
REVOLVING-ARMATURE TYPE ALTERNATOR 7-23
REVOLVING-FIELD TYPE ALTERNATOR 7-23
RIGGING SPECIFICATIONS 1-37
RIGHT-HAND RULE 7-78
RIPPLE 7-4
RISER 7-7
RIVETS 2-36
    alloys 2-38
    blind 2-39
    cutters 2-53
    edge distance 2-53
    friction-lock rivets 2-40
    guage 2-54
    gun 2-62
    Hi-Shear 2-43
    icebox 2-39
    installation 2-51, 2-60, 2-65
    joints 2-5
    layout patterns 2-53
    mechanical-lock rivets 2-40
    pitch 2-54
    pop rivets 2-39
    removal 2-67
    sets 2-63
    single-shear strength 2-51
    solid shank 2-36
    team riveting 2-68
RIVNUTS 2-48
ROD-END BEARINGS 1-48
ROLL FORMER 2-31
ROTOR BLADES, CONING 1-55
ROTOR SYSTEMS 1-53
    antitorque 1-57
    elastomeric bearings 1-54
    flextures 1-54
    fully articulated 1-53
    gyroscopic precession 1-55
    rigid 1-54
    semirigid 1-53
    types 1-53
ROTOR, ALTERNATOR 7-19
RUDDER PEDALS 1-28
RUDDERVATORS 1-14

S

SALIENT POLES 7-6
SAYBOLT UNIVERSAL VISCOSIMETER 8-6
SCALES 2-16
SCRIBES 2-17
SCROLL SHEARS 2-24
SEALS 8-42
    installation 8-44
    materials 8-44
SERVICE VALVES 14-40
SERVO 11-8, 12-38
SETBACK 2-71
SEYBOTH 5-12
SHEAR SHAFT 8-23
SHEET METAL 2-69
    bend radius 2-70
    clad 2-8
    corrosion prevention 2-13
    fabrication 2-51
    layout and forming 2-69
    materials 2-6
    tools 2-16
SHEET METAL REPAIRS 2-83
    corrugated skin 2-92
    flush patch 2-90
    former and bulkhead 2-94
    inspection/assessment 2-83
    leading edge 2-95
    pressurized structure 2-96
    spar 2-95
    stringers 2-91
    trailing edge 2-92
SHIMMY DAMPERS 9-12
SHORT-SHUNT CONNECTION 7-10
SINGLE-LOOP SYSTEM 16-5
SINGLE-PHASE ALTERNATOR 7-23
SLIP RING 7-19
SLIPPAGE MARK 9-51, 11-45
SMOKE DETECTORS 16-11
    ionization type 16-12
    light refraction type 16-11
    solid-state type 16-12
SOLENOIDS 7-68
SOLVENT 3-66, 6-14
SPARS 1-6
SPLIT FIELD MOTOR 7-81
SPOILERS 1-12
SPRAY GUNS 6-17
    air cap 6-17
    fluid tip 6-17
    pressure-feed 6-17
    suction-cup-type 6-2
    suction-feed 6-17
SPRAY PAINTING 6-16
    airless spraying 6-18
    electrostatic systems 6-16
    spray pattern 6-21
SQUARE MIL 7-54
SQUARING SHEAR 2-22
STABILATOR 1-14
STABILITY 1-21
STAGGER 1-49
STAINLESS STEEL 2-12
STALL WARNING INDICATORS 10-9
    electric 10-9
    non-electric 10-9
STALLS 1-13
STARTER CIRCUIT 7-39
STATIC DISCHARGERS 12-53
STATIC PORT 11-30
STOP-DRILL 2-6
STRAIN-HARDENING 2-10
STREAMLINED WIRES 1-40
STRESSES 2-3
    bending 2-4
    compression 2-3
    shear 2-4
    tension 2-3
    torsion 2-4
STRETCH PRESS 2-31
STRUCTURAL COMPONENTS 1-4
    formers 1-2
    ribs 1-6
    rip-stop doublers 1-4
    spar 1-6
    stringers 1-2
STRUCTURAL DESIGNS 1-2
    airfoil 1-4
    empennage 1-2
    fuselage 1-2
    laminated 1-3
    monocoque 1-3
    semi-monocoque 1-3
    stressed-skin structure 1-3
    truss structure 1-2
    types 1-4
STRUTS 1-50
SUBSTRATE 6-6
SUCTION RELIEF VALVES 11-28
SURFACE PATCH 3-18
SWAGED TERMINALS 1-43
SWITCH METHOD 7-82
SWITCHES 7-66, 11-6, 12-49
    derating factor 7-66
    installation 7-67
    precision 7-68
    pressure 11-6
    rotary 7-68
    toggle 7-67
SYNCHROSCOPES 11-16
SYSTRON-DONNER SYSTEM 16-9

T

TACHOMETERS 11-16
TAPER-LOK FASTENER 2-46
TAXIING 9-52
TEMPERED GLASS 13-6
TENSIOMETER 1-46
THERMAL EXPANSION VALVE 14-36
THERMISTOR 16-8
THERMOCOUPLE 11-15, 16-6
    detector 16-6

Index

THINNERS AND REDUCERS 6-13
THREE-PHASE 7-23
THREE-UNIT REGULATOR 7-13
THROATLESS SHEARS 2-22
TIN SNIP 2-20
TINNING 7-62
TIRES 9-39
    balancing 9-51
    bead 9-40
    inflation 9-43
    inspection 9-43, 9-46
    mounting 9-49
    ply rating 9-39
    repair 9-47
    slippage mark 9-51
    storage 9-47
    tread 9-41
    tubes 9-39, 9-48
    types 9-39
TITANIUM 2-11
TOE IN/TOE-OUT 9-12
TOOLS 2-16
    forming 2-29
    sheet metal 2-16
    shop 2-22
    special assembly 2-33
TORQUE 1-57
TORQUE LINKS 9-12
TORQUE TUBES 1-48
TOXIC CHEMICALS 6-29
TRAFFIC ALERT AND COLLISION AVOIDANCE SYSTEM (TCAS) 12-28
TRANSPONDER 12-21
TRANSVERSE PITCH 2-54
TRAVEL ADJUSTMENT 1-45
TRIM CONTROLS 1-28
TRUE AIRSPEED 11-9
TUNGSTEN INERT GAS WELDING (TIG) 4-4
TURBOSHAFT ENGINES 1-70
TURN INDICATORS 11-20
TWO-PHASE ALTERNATORS 7-23

U

UNDER-COMPOUND GENERATOR 7-10
UNDERCUT 7-17

V

V-BLOCK 2-79
VACUUM PUMPS 11-27, 14-43
VALVES 8-38, 9-28, 10-4, 11-28, 14-31
    air 8-38
    bypass 14-31
    check 8-29, 8-53
    closed-center 8-28
    control 8-53, 10-4
    flapper 10-4
    flow control 8-28
    open-center 8-28
    outflow valves 14-22
    pack (shutoff) 14-30
    power brake control 9-28
    priority 8-31
    pump-control 8-12
    quick-disconnect 8-32
    relief 8-34, 8-53, 11-28
    selector 8-28
    sequence 8-31
    shuttle 8-55
    unloading 8-12
    wing flap overload 8-47
VAPOR LOCK 15-3
VARIABLE SPEED MOTOR 7-80
VARIATION, MAGNETIC 11-22
VENTURI SYSTEMS 11-26
VERTICAL SPEED INDICATOR (VSI) 11-11
VIBRATING-TYPE VOLTAGE REGULATOR 7-12
VISCOSITY 6-19, 8-6, 13-14
    viscosity cup 6-19
VOLATILITY 15-3
VOLTAGE REGULATORS 7-45
    carbon pile 7-45
    paralleling 7-45
    vibrator-type 7-45
VORTEX GENERATORS 1-13

W

WARNING SYSTEMS 10-1, 12-26
    engine indication and crew alerting system (EICAS) 10-13
    ground proximity warning system (GPWS) 10-13, 12-26
    landing gear configuration warning 10-12
    mach/airspeed warning systems 10-13
    takeoff configuration warning system 10-12
WARP 5-6
WASH-IN/WASH-OUT 1-50
WATER SEPARATORS 14-32
WATER TRAP 6-15
WEFT 5-6
WELDING 4-1
    aluminum 4-15
    brazing 4-11
    cutting 4-28
    gas metal arc welding (GMAW) 4-4
    goggles 4-25
    oxyacetylene 4-3
    shielded metal arc welding (SMAW) 4-3
    soldering 4-12
    spot 4-6
    torches 4-24
    tube splicing with reinforcement 4-18
    tungsten inert gas (TIG) 4-4
WHEELS 9-6
    alignment 9-12
    balance weights 9-10
    bead seat 9-9
    bearings 9-8
    construction 9-6
    corrosion 9-9
    disassembly 9-7
    fusible plug 9-10
    inspection 9-7
    lubrication 9-9
WINDSCREEN 13-16
WINDSHIELD 3-68
    installation 3-68
WINDSHIELD WIPER SYSTEMS 13-16
WING 1-39, 5-16
    alignment 1-39
    cantilever 1-9
    covering 5-16
    fence 1-33
    strut-braced 1-39
WING TIP VORTICES 1-33
WINGLETS 1-33
WIRE-BRACED TAIL SURFACES 1-40
WIRING 7-53
    bonding 7-63
    clamping 7-58
    identification 7-54
    installation 7-56
    markings 7-54
    routing 7-58
    shielding 7-59
    size 7-54
    splicing repairs 7-62
    stripping 7-60
    terminals 7-59
    types 7-53
WOOD 3-1
    aircraft structures 3-2
    approved substitutions 3-2
    assessment 3-3
    compression failure 3-4, 3-11
    defects 3-4
    inspection 3-9
    laminated 3-3
    plywood 3-2
    sitka spruce 3-2
    types of deterioration 3-9
WOOD BONDING PROCESS 3-7
    applying the adhesive 3-7
    bonding time-periods 3-7
    clamping pressure 3-8
    closed-assembly time 3-7
    methods of applying pressure 3-8
    nailing strips 3-8
    open-assembly time 3-7
    pot life 3-7
    pressing time 3-8
    wood preparation 3-7
WOOD STRUCTURE REPAIRS 3-11
    cap strip 3-16

plug patch 3-18
plywood skin 3-17
scarf repair 3-12
scarfed patch 3-18
spar longitudinal cracking 3-12
splayed patch 3-18
wing rib 3-15

WORK HARDENING 2-10

Z

ZIRCONIUM TUNGSTEN 4-6